The *First* Sourcebook on Nordic Research in Mathematics Education

Norway, Sweden, Iceland, Denmark, and contributions from Finland

Monograph 10

in *The Montana Mathematics Enthusiast: Monograph Series in Mathematics Education*

Bharath Sriraman, *Series Editor*

THE MONTANA MATHEMATICS ENTHUSIAST
MONOGRAPH SERIES IN MATHEMATICS EDUCATION

Series Editor
Bharath Sriraman, The University of Montana

International Perspectives on Social Justice in Mathematics Education
Edited by Bharath Sriraman

Mathematics Education and the Legacy of Zoltan Paul Dienes
Edited by Bharath Sriraman

Beliefs and Mathematics: Festschrift in Honor of Guenter Toerner's 60th Birthday
Edited by Bharath Sriraman

Creativity, Giftedness, and Talent Development in Mathematics
Edited by Bharath Sriraman

Interdisciplinary Educational Research in Mathematics and its Connections to the Arts and Sciences
Edited by Bharath Sriraman, Claus Michelsen, Astrid Beckmann, Viktor Freiman

Critical Issues in Mathematics Education
Edited by Paul Ernest, Brian Greer, Bharath Sriraman

Interdisciplinarity, Creativity, and Learning: Mathematics with Literature, Paradoxes, History, Technology, and Modeling
Edited by Bharath Sriraman, Viktor Freiman, Nicole Lirette-Pitre

Relatively and Philosophically E[a]rnest: Festschrift in Honor of Paul Ernest's 65th Birthday
Edited by Bharath Sriraman, Simon Goodchild

The Role of Mathematics Discourse in Producing Leaders of Discourse
By Libby Knott

The *First* Sourcebook on Nordic Research in Mathematics Education

Norway, Sweden, Iceland, Denmark, and contributions from Finland

EDITED BY

Bharath Sriraman

Christer Bergsten

Simon Goodchild

Guðbjörg Pálsdóttir

Bettina Dahl

Lenni Haapasalo

IAP

Information Age Publishing • Charlotte, NC • infoagepub.com

Library of Congress Cataloging-in-Publication Data

The first sourcebook on Nordic research in mathematics education : Norway, Sweden, Iceland, Denmark, and contributions from Finland / edited by Bharath Sriraman ... [et al.].
p. cm. – (The Montana mathematics enthusiast)
Includes bibliographical references.
ISBN 978-1-61735-098-6 (pbk.) – ISBN 978-1-61735-099-3 (hardcover) – ISBN 978-1-61735-100-6 (e-book)
1. Mathematics–Study and teaching–Research–Scandinavia. I. Sriraman, Bharath.
QA14.S34F57 2010
510.71'048–dc22

2010021956

Bharath Sriraman, Ph.D.
Editor, *The Montana Mathematics Enthusiast*
The University of Montana
Missoula, MT 59812
Email: sriramanb@mso.umt.edu
(406) 243-6714

Printed and bound by CPI Group (UK) Ltd, Croydon, CR0 4YY

To my children Sarah, Jacob and Miriam & the North—
always true, inspiring and guiding…

CONTENTS

Section V: Contributions from Finland

FOREWORD

Bharath Sriraman

THE UNIVERSITY OF MONTANA

It is difficult to believe that a task that was once perceived as impossible by many skeptics has indeed been accomplished. The idea of a single comprehensive Sourcebook to showcase the mathematics education research done in the Nordic countries came to me a few years ago, and upon broaching this suggestion to those in the hierarchical orthodoxy that characterizes any field in various parts of the world during my travels in the last 7 years, I was told that it was too ambitious a project to orchestrate. Yet this book has come to Be!

My connection to the far North first began as a young sailor and continued as an undergraduate student in Alaska, a traveler in Yakutsk, Russia and the Canadian North, and has carried on in my professional life with numerous colleagues in Iceland, Norway, Denmark and Sweden, who have been like a wonderful extended family. Working professionally in North America for most of my career led me to the realization that a considerable amount of valuable research from "elsewhere" was often overlooked by researchers in the field taking a rather myopic view of the literature. Taking inspiration from Marcia Ascher's influential book *Mathematics from Elsewhere* and from Claudia Zaslavsky's classic *Africa Counts*, which are books that changed mainstream mathematicians perceptions of the heritage of mathematics, I thought it was of value to the mathematics education community *outside the Nordic world*, to pay attention to the considerable efforts of our colleagues in Norway, Sweden, Iceland, Denmark and Finland, to advance our field. Another reason for putting together a book such as this was to create a visible platform for researchers within the Nordic world to have an international audience become familiar with their work, since a large portion of the extant Nordic research, with some notable exceptions, has appeared within reports and journals that are not accessed by those "elsewhere."

The book is organized in 5 sections, namely Norway, Sweden, Iceland, Denmark and Finland. Each of the sections was compiled and edited by a researcher situated within the local national context. For the Norwegian and Swedish sections, I initially played a coordinating/organizational role after which I handed over the reins to the two section editors Simon Goodchild and Christer Bergsten respectively. The Norwegian section, compiled by Simon Goodchild, is the largest section of the book. Simon provided extensive feedback to the numerous authors in Norway, which occurred in numerous iterations to raise the quality of the presentation in the chapters. I received several e-mails from Norwegian authors who indicated that Simon's guidance, skill and patience be recognized. The 17 chapters of the Norwegian section are the result of his dedicated effort for which they are very thankful. Christer Bergsten was entrusted with the Swedish section of the book that again required juggling a large cast of authors with varied backgrounds, and in his inimitable style Christer brought his section consisting of 13 chapters (including his introductory chapter) to completion with diligence and care. I thank both Simon and Christer for the two larger sections of this book.

The Icelandic section was put together by Gudbjorg Palsdottir and myself with advice from Gudny Gunnarsdottir. Mathematics education research in Iceland is in its infancy in comparison to the other Nordic countries, however Iceland's infancy confers the advantage of learning from mistakes made elsewhere and being able to glimpse at developments on either side of the Atlantic [*alas their banking system did not do so!*]. Indeed, Iceland is located on the mid-Atlantic rift and as a result is geographically located both on the continents of North America and Europe. The 9 chapters in the Icelandic section reveal efforts aimed at mathematics teacher, researcher and curricular development. I feel privileged to have been privy to this for the last many years and thank all the authors for their willingness to write about their efforts in mathematics education. The Danish section of the book was organized by Claus Michelsen and myself, but with the growing pressures of time and other factors, Claus handed over the section to Bettina Dahl, who then went on to produce a very impressive section of the book consisting of 10 chapters. As I write this introduction to the book in Tromsø, Norway—a conference to celebrate the work of Ole Skovsmose on his retirement took place in Aalborg, Denmark. Ole's work on critical mathematics education is known outside the Nordic world and has inspired many researchers around the world including myself to carry on his bold vision for mathematics education. Finally, the Finnish section of the book contains four contributions compiled by Lenni Haapasalo, who is to be thanked for his effort.

This book taken as a whole serves as a valuable addition and complement to the present mathematics education library of Information Age Publishing which also published the Second Handbook of Research on Mathematics Teaching and Learning in 2007. I wish to thank the large cast of authors of this book for their patience and willingness to stick with this ambitious project and see it to its completion. In particular I would like to thank the following colleagues in the Nordic world for their camaraderie and support, namely Gudbjorg Palsdottir, Gudny Gunnarsdottir, Olof Steinthorsdottir in Iceland; Simon Goodchild, Barbro Grevholm, Anne-Birgitte Fyhn, Raymond Bjuland, Reidar Mosvold in Norway; Claus Michelsen, Bettina Dahl in Denmark; and Christer Bergsten in Sweden, among others. Last but not least the support and faith of George Johnson, the founder of Information Age Publishing, in this book project is much appreciated. The Montana Mathematics Enthusiast Monograph Series in Mathematics Education is pleased to present this massive compilation of 53 chapters focused on Nordic mathematics education research to the reader. I hope this book is of use to many generations of mathematics education researchers inside and outside the Nordic world.

Tromsø, Norway
January 17, 2010

ACKNOWLEDGMENT

Work on this book was made possible with support from The University of Montana Faculty Exchange Program for research travels in Norway and Denmark as well as the Sabbatical Leave in 2009–2010, in addition to mini-sabbatical support granted in 2007.

SECTION I

Norwegian Research in Mathematics Education

Simon Goodchild, *Section Editor*

Dedicated to the memory of
Stieg Mellin-Olsen (1939–1995)
and
Gard Brekke (1943–2009)
Luminaries in the Norwegian community of mathematics educators

1

NORWEGIAN RESEARCH IN MATHEMATICS EDUCATION

Introduction to the Norwegian Section of the Sourcebook of Nordic Research in Mathematics Education

Simon Goodchild

UNIVERSITY OF AGDER

INTRODUCTION

The chapters in this Norwegian section of the Sourcebook are intended to offer the reader a representative view of mathematics education research in Norway. The chapters are organized into three parts: The first part sets out the context of mathematics education and research in the field in Norway, and the second describes some of the larger research projects that have been pursued in recent years. The third part comprises chapters written by researchers who have completed their doctoral studies within the last five years. This introduction to the Norwegian part begins with a brief outline of the national context, which is intended to help the reader understand some of the challenges and opportunities that face Norwegian researchers in establishing and maintaining their professional network. Following this, each chapter is introduced.

GEOGRAPHICAL AND STRUCTURAL CONTEXT

To understand mathematics education research in Norway, it may help to have a sense of the national context. The population of Norway in 2010 is about 4.9 million.[1] There are seven universities (Tromsø, Norwegian University of Science and Technology in Trondheim, Bergen, Oslo, University of Life Sciences in Oslo, Stavanger, and Agder), and 25[2] university colleges, all with teacher education programs. In 2007, there were 7,700 first-year admissions to teacher education programs for the whole of Norway (DKK, 2009); this figure includes all the aforementioned institutions, with the largest intakes at Oslo University College (852) and Bergen University College (525). Clearly only a fraction of these students will be specializing in mathematics, although many of the individual programs include a compulsory mathematics component. One further fact is relevant: The straight line distance between the University of Agder in the south of the country and the University of Tromsø in the north (the most northerly university in the world) is nearly 1,400 km. Given the distance, the mountainous landscape, and broken coastline, flying is the only practical means of travelling between the major centers. The Norwegian mathematics education community is, therefore, relatively small, fragmented, and widely dispersed.

The First Sourcebook on Nordic Research in Mathematics Education, pages 3–9

A DEVELOPING COMMUNITY OF RESEARCHERS

The last two decades have seen significant growth in the mathematics education community in Norway; this is especially evident from the Norwegian contributions to the Nordic regional conferences in mathematics education, which are held on a three-yearly cycle. The conference, NORMA, aims to serve, particularly, the Nordic (Iceland, Finland, Norway, Sweden and Denmark) and Baltic (Estonia, Lithuania and Latvia) mathematics education research communities. In the introduction to the proceedings of the most recent conference, held in 2008 in Copenhagen, Carl Winsløw (2009) notes that the first NORMA conference held in 1994 included only one research report from Norway, out of a total 33 presented in the conference.The corresponding figures for 2008 were 21 Norwegian papers out of a total 62. Winsløw remarks, "The strong appearance of Norway is impressive and reflects a strong development of mathematics education research in this country over the past decade" (2009, p. 3). Winsløw's remarks are made in the context of the number of contributions of the other major national groups (Sweden, 12 papers, national population approximately 9 million; Denmark, 7 papers, national population approximately 5.5 million; and Finland, 8 papers, national population approximately 5.3 million).

The impressive appearance of Norwegian mathematics education research that Winsløw recognizes in the context of the Nordic countries may also be recognized by those who participate regularly in the major international conferences (PME[3] and CERME[4]); and it is noted that Bergen University College was host to the PME conference in 2004. However, developments in Norwegian mathematics education research are yet to be reflected by publication in the top-rank international journals. For example, in a recent editorial in *The Journal for Research in Mathematics Education*, Heid (2009) notes that only one manuscript was submitted from Norway over an eighteen-month period. Similarly, only three papers with Norwegian authors were published in *Educational Studies in Mathematics* in the last six years.

THE SIGNIFICANT ROLE OF RESEARCHER EDUCATION

The preceding paragraphs reflect the national milieu in mathematics education as relatively young, growing steadily, and aspiring to take its place on the international stage. The chapters from Norwegian researchers in this volume reflect this milieu. Many of the chapters are based upon research undertaken as part of doctoral studies (Carlsen, Fyhn, Kleve, Mosvold, Streitlien, Torkildsen, Wæge) that have been completed within the last five years. Also, several of the chapters—in particular, Breiteig and Grevholm; Bjuland, Cestari and Borgersen; Breiteig and Goodchild—demonstrate how the research carried out by students in masters' programs has contributed to the development of the milieu. A major contribution to the development of the mathematics education community in Norway, which is not reflected adequately in the chapters in this Norwegian section, is that which has been made by the Nordic Graduate School in Mathematics Education (NoGSME), which received funding from the Nordic Academy of Advanced Studies (2004–2008).

The Nordic Graduate School in Mathematics Education[5] was led by Barbro Grevholm and based at the University of Agder, and it has had a major impact in several important respects. Eleven seminars for supervisors were held in different locations in the Nordic countries; these offered an opportunity for developing the network of researchers in the field in addition to their explicit purpose of developing supervisor competencies. Summer schools (and one winter school in 2008 to avoid clashing with ICME11, which took place that summer) were held principally for doctoral students. The summer schools invited first-rank international scholars to lecture and provide additional supervision for the students. Again, this provided an opportunity for networking as well as ensuring that the quality of the research and dissertations of Nordic students compared favorably with international standards. The Graduate School also supported doctoral students' mobility, to enable them to make extended visits to international centers of excellence in their field of inquiry, and smaller travel grants that enabled students to attend courses at other universities. It is difficult to provide a measure of the impact of the Graduate School,[6] or assess its value to the mathematics education research community. However, it is fair to say that the impact has been considerable and it has been highly valued both in terms of the network made possible and the personal competencies developed. As noted in the opening paragraph, given the relatively small, fragmented, and dispersed nature of the mathematics education research community in Norway, the opportunities for networking and creating a critical mass of researchers has been important. The period of funding for NoGSME is now ended, and a new organization has been established to sustain the network and promote the activities formerly undertaken by NoGSME—this is the Nordic society for Research in Mathematics Education (NoRME).

ORGANIZATION OF THE NORWEGIAN CHAPTERS

As indicated above, the chapters in this Norwegian section of the Sourcebook are collected together into three parts. Part One comprises three chapters that provide a historical and present account of the mathematics education milieu in the country. Part Two comprises five chapters that represent collaborative effort, most based on research that has been funded, at least in part, by The Research Council of Norway. Part Three comprises seven chapters that are based on doctoral research; these chapters, which report small-scale case studies, are of special value in this present volume. The case study research offers insights into teaching and learning mathematics in school in Norway that are made possible through in-depth qualitative studies. The chapters also introduce some of the up-coming generation of Norwegian researchers in the field, revealing some of the scope of their interests and to a large extent offering an insight into the future of mathematics education research in Norway.

The Norwegian section concludes with a tentative glance into an uncertain future. Here, Goodchild and Grønmo take stock of the present situation as presented by the authors of this Norwegian section, and combine this with the signals regarding teaching and learning and educational research that are coming from Parliament, The Research Council and other places.

Part One

Part One opens with an overview of the development of mathematics education in Norway. In this chapter Trygve Breiteig and Simon Goodchild (Chapter 2) chart the major moments in the development of the school curriculum from the 12th century establishment of Cathedral Schools. Research specifically in mathematics education has been carried out since the 1920s, and this, it appears, has been driven by recurrent curriculum revisions and persistent concerns about improving pupils' performance in the subject. The chapter also provides an account of the development of the community of mathematics education researchers since the early 1990s.

One of the founding fathers of mathematics education research in Norway was Stieg Mellin-Olsen. The impact of his work is evident in several of the chapters in this Norwegian section. Mellin-Olsen opened up a number of important lines of research (affect and motivation, language, discourse and semiotics, socio-cultural theory, political dimensions, and ethno-mathematics). Mellin-Olsen was certainly making an impact in the international community during his lifetime, and even now, some 15 years after his death, the Norwegian community still recognizes his forward-thinking contribution to the field. It would be difficult to understand the development of Norwegian research in mathematics education without knowing something of his life, work, and contribution. Christoph Kirfel and Nora Lindén provide this necessary account in Chapter 3. Nora Lindén was Mellin-Olsen's wife and shared in much of his work, professional interests, and international collaboration. Christoph Kirfel was a close colleague of Mellin-Olsen (as also in this volume were Marit Johnsen-Høines and Ole Einar Torkildsen).

The next chapter in this opening part, written by Liv Sissel Grønmo (Chapter 4) offers a commentary on the performance of Norwegian pupils in the international comparative studies, TIMSS and PISA. These international studies present the performance of Norwegian pupils as below international averages, and well below the other countries with which it is natural to compare Norway. The concern to improve the performance of Norwegian pupils in mathematics has been one factor driving research in the field. There are characteristics of Norwegian education that are deeply rooted in the social and political culture of the country, such as inclusive education (there are very, very few special schools). Further, the culture and legal framework of equal rights means that children may not be organized within or across schools on the basis of any criteria, apart from age. While most pupils attend publicly financed schools controlled by local authorities, there are some schools with a religious foundation, some international schools, about 30 Steiner Schools and 60 Montessori schools and kindergartens; thus there has not developed a culture of elite schools serving the wealthier sectors of society. For an English mathematics education researcher working in Norway, it is refreshing to be part of a system where social and cultural ideals of equal rights and equal opportunities are not sacrificed to simplistic and rather instrumental approaches that address only one narrow measure of pupils' educational development.

Part Two

Breiteig and Goodchild in Chapter 2 explain the circumstances of the development of a large mathematics education research group at Agder University College, later to become the University of Agder (UiA). This group has been at the front line in many of the developments in the field within Norway, notwithstanding the important contributions made by other active research groups in Oslo, Bergen, and Notodden (Telemark). Much of the masters' research in mathematics education has been supervised by UiA researchers, and the doctoral program there has initiated a number of studies that have made a significant contribution towards the growth of the field in Norway, recognized by Winsløw as noted above. The large research group established at UiA has also been successful in drawing in a substantial amount of research funding, especially over the last five years, much of which has been used to pursue developmental research in mathematics teaching. Four of the chapters in this section emerge from the group at Agder (Jaworski, Chapter 5; Fuglestad, Chapter 6; Breiteig & Grevholm, Chapter 8; and Borgersen, Bjuland, and Cestari, Chapter 9). Barbara Jaworski was Professor of Mathematics Education at the University of Agder from 2003 to 2009, the latter two years in an adjunct (20%) position. Jaworski initiated and led two major projects based on developing collaborative communities of inquiry, including the university research group and teachers in local schools. In Chapter 5, Jaworski describes these projects, their theoretical roots in inquiry and community, and their developmental research methodology. The chapter also includes some examples of mathematical tasks. Jaworski concludes the chapter by taking a critical look at how the projects have developed and what has been achieved.

Anne Berit Fuglestad (Chapter 6) reports another project, ICT and Mathematics Learning (ICTML), which was hosted at the University of Agder in partnership with the LCM project described by Jaworski (Chapter 5). As with LCM, the ICTML project was rooted in socio-cultural theory and fundamental principles of community and inquiry as agents of mathematics teaching development. Fuglestad also makes use of the "instrumental approach" developed especially by French researchers (e.g., Vérillon, & Rabardel, 1995). The essential difference between the projects was that ICTML was specifically concerned with the development of the use of digital technologies in the teaching of mathematics. Fuglestad sets out by briefly reviewing the level of adoption and use of digital technologies in Norwegian schools and exposes this to be rather disappointing given the national emphasis on using ICT, the investment in equipment, and the period through which the (slow) development has taken place. Fuglestad examines evidence from the teacher teams in three schools that participated in ICTML, and notes the case of a fourth school. The small teams of teachers responded in different ways to the project's initiatives. This provides Fuglestad with the opportunity to identify and discuss reasons for the slow pace of development of the use of computers in the teaching and learning of mathematics.

Marit Johnsen-Høines (Chapter 7) describes the Learning Conversation in Mathematics Practice (LCMP) project, which is led from Bergen University College. In the chapter, Johnsen-Høines describes a multilayered approach in which the participants include researchers and teacher educators at Bergen University College, student-teachers, school teachers, and pupils. Participants at all levels contribute to the data—conversations about mathematics and mathematical activity—and towards providing partial analysis of the data; thus all are included in an inquiry community. One special dimension in the project is that the pupils, as part of another project being led by the local education authority, spend time in local businesses and industry to experience mathematics used in the work place. The research is then concerned to explore and expose the characteristics of conversations that can be described as "learning conversations in mathematics."

Trygve Breiteig and Barbro Grevholm (Chapter 8) report several studies that have been conducted on the basis of a longitudinal test that was constructed as part of the LCM project described by Jaworski (Chapter 5) and Fuglestad (Chapter 6). The tests were constructed to expose features of pupils' mathematical performance at grades 4, 7, 9, and 11. Pupils in the project schools took the tests in the autumn (near the start) and in the late spring (towards the end) of their time in the chosen grades. Breiteig and Grevholm describe how the tests were used as a developmental tool in many different ways. These included their use as originally intended to explore pupils' performance and thus provide valuable information for developing teaching and learning. They were also used as a resource for training in inquiry and research and used in that context with initial teacher education students as well as masters and doctoral students.

The chapter by Hans Erik Borgersen, Maria Luiza Cestari, and Raymond Bjuland (Chapter 9) is rather different from the other four chapters in this part, as it describes the way one line of research has developed over several decades. They are able to identify three distinct phases of activity: research into small group learning and a focus on open-ended problem solving in geometry is the first phase; the development of a dialogical approach as a methodological perspective marks the second phase; and the third and current phase focuses on a multimodal approach. The authors of this chapter outline the main theoretical developments in each of these phases and show how the research activity of masters and doctoral students has contributed to development and advances in knowledge.

Part Three

In their historical overview of the development of mathematics education in Norway, Breiteig and Goodchild draw attention to the frequent changes to the Norwegian national curriculum for schools and consequent changes to the mathematics syllabi that have taken place over the past 50 years. The chapter also reveals how officially sponsored, large-scale evaluations have often led to the critique of a curriculum that has not been properly implemented. The chapters in this part by Bodil Kleve and Reidar Mosvold, based on small-scale case study research, expose some of the significant issues that interfere with the implementation of a new curriculum.

Bodil Kleve (Chapter 10) reports a case study of three teachers and their responses to the curriculum reform that took place in 1997. The research was undertaken as part of Kleve's doctoral education; it is a qualitative study based upon classroom observation, focus groups, and individual interviews. Between the time that Kleve started her research and when she reported its findings, the decision was made to introduce a completely new curriculum, with a different theoretical foundation. The new curriculum was in place before Kleve's study was published. This offers an insight into the challenges faced by curriculum developers who are not given the opportunity to reflect properly on the implementation of one curriculum before introducing another. Although only three teachers are included in Kleve's case study, her chapter offers some interesting insights into the practical reality of teaching mathematics in Norwegian lower secondary schools and the differences between the intended and implemented curriculums. Kleve reveals how teachers' knowledge, attitudes, and professional competencies interfere with the implementation of the curriculum as intended. Her research provides a strong argument for the type of mathematics teaching developmental research described by Jaworski in this volume.

Reidar Mosvold (Chapter 11) also describes case study research that focuses on three lower secondary school teachers' implementation of the 1997 curriculum reform, but he chooses to focus on one specific element within the curriculum—mathematics in everyday life. Mosvold's account of how this element of the curriculum has been interpreted in successive revisions, and then omitted completely from the latest curriculum reform of 2006, demonstrates an aspect of the nature of changes with which teachers are expected to cope. Mosvold focuses on teachers' beliefs and practices and, in particular, the use of project work and textbooks in their implementation of this part of the curriculum. He uses classroom observation, interviews, and questionnaires to address two principal questions: What are the teachers' beliefs about connecting school mathematics and everyday life? What ideas are carried out in their teaching practice? Mosvold's account of lower secondary mathematics classes appears consistent with that set out by Kleve in her chapter; Mosvold, however, notes a further significant feature, that of teachers' reliance upon a textbook to interpret the curriculum. He also observes that teachers struggle to implement the curriculum as intended, and this raises a question about suitability for purpose. As noted earlier in this introductory chapter, there are deep-seated political and cultural conditions regarding the structure and governance of schools—inclusiveness and equal opportunities were mentioned, and one can also add local control. Mosvold's analysis raises questions about the extent to which the content and structure of the curriculum are in harmony with the conditions of schooling and teacher education.

Ole Einar Torkildsen (Chapter 12) describes his research into pupils' responses to open-ended investigational tasks set in the Norwegian journal for mathematics teachers, *Tangenten*. The tasks were set as a form of competition in which groups or classes of pupils were expected to collaborate with minimum support from their teachers. In each case, pupils at all grade levels were confronted with the same investigation. Torkildsen describes the nature of the tasks considered suitable to offer appropriate mathematical challenges across the grades levels. He then goes on to explain how he "decoded" pupils' solutions into a recognizable mathematical form. This entails what he describes as a form of "mathematical archaeology," a term he attributes to Skovsmose (1994); it is a process of digging into the pupils' solution to expose the mathematical structure. Torkildsen's claim that his study reveals pupils using "mathematical structures not usually taught in schools" raises many questions about the suitability of content and the nature of challenges within the school curriculum.

Anne Fyhn (Chapter 13) writes about how she uses pupils' experience gained in climbing activity to develop their understanding of geometry. Her chapter reveals another form of mathematical archaeology that contrasts with that described by Torkildsen. Fyhn's approach is to dig into the cultural activities of pupils to expose the mathematical structures embedded in them, and then use these as a context to teach the mathematical concepts. Fyhn relates her research conceptually to the notions of embodied cognition and the Dutch tradition of realistic mathematics education. The research described in this chapter is a design study with several phases, commencing with a case study of one 12-year-old girl with whom Fyhn talks following a climbing trip. This leads into the design of mathematics lessons based on pupils' experiences on a climbing wall in a sports hall. Much of this work has been published internationally, and Fyhn uses this chapter to report a new phase in her design research, which is to reflect on and critique the previous research that has been reported elsewhere.

Anne Fyhn is based at the University of Tromsø, and thus she must have the distinction of being one of (if not *the*) world's most northerly mathematics educators. She has explored other situations in which mathematics can be exposed, such as skiing, and she is interested to pursue research into the cultural practices (music, crafts, and design) of the Sami people who are indigenous in this remote northern part of Norway and neighboring Finland and Russia. This interest in ethnomathematics is reassuring evidence that this line of research, opened up in Norway by Stieg Mellin-Olsen, remains alive and active in the country.

Åse Streitlien (Chapter 14) takes a socio-cultural perspective as she reports her research into participation patterns in primary classrooms. Her chapter is based upon episodes of dialogue between teacher and pupils, six to seven years of age, in which she attempts to make sense of the dynamics of interaction and how the pupils engage in creating mathematical meaning. Streitlien observes that these young pupils are being encultured into the practice of mathematics learning and teaching in school, and notes that the way the dialogues are managed may have a profound impact on their understanding of mathematics and the nature of classroom practice. The episodes that Steitlien chooses to illustrate her chapter provide a compelling picture of the complexity of the classroom situation in which the teacher has to manage social, emotional and cognitive (mathematical) factors.

Martin Carlsen's research (Chapter 15), like Streitlien's, is a case study of pupils working in a mathematics classroom. He is also working within a sociocultural framework and uses pupil dialogue as his main source of data. However, where the participants in Streitlien's research are grade 1 primary children, Carlsen's focus is on grade 13 students—that is, in their final year of upper secondary school. The pupils[7] in Carlsen's research are organized by their teacher to work in groups, and a problem-oriented approach to the curriculum is adopted. Carlsen adopts an ethnographic style approach to data collection, and his data comprises recordings of the naturally occurring conversation between pupils during the group activity. His analysis of the resulting dialogues reveals the pupils' appropriation of the language and concepts of scalar product and geometric series.

Kjersti Wæge's chapter (16) reports her research focused on pupils' motivational goals in the context of a teaching approach characterized by "inquiry" into mathematics. The research was set in an upper secondary school and concerned pupils aged 16, in the year following their transfer from lower secondary. Wæge's study addressed the question, "How can pupils' motivation for learning mathematics develop when they experience an inquiry mathematics teaching approach?" Qualitative methods were used that included instructional design and data arising from observation, interviews, and conversations with pupils and teacher to seek evidence of pupils' motivational goals. The chapter additionally offers a review of two other important Norwegian research contributions on pupils' motivational goals: Mellin-Olsen's theorization of pupils' rationales for learning, and a series of studies reported by Skaalvik and colleagues that were carried out within the tradition of positivistic cognitive psychology. Wæge observes that as the small group of seven pupils in her study revealed evidence of change in their motivational goals as they experienced a new, for them, inquiry approach to learning mathematics, the pupils focused more on developing relational understanding. Wæge also exposes a relationship between pupils' need for competence and their need for autonomy in mathematical learning activity. An important contribution of Wæge's chapter is her effort to synthesize concepts and categories of motivational goals that have been developed within contrasting paradigms using radically different methodologies.

Conclusion

In the final chapter of this Norwegian section, Simon Goodchild and Liv Sissel Grønmo (Chapter 17) attempt to see what the future holds for mathematics education research in Norway. The chapter reflects on the present position, which is well represented by the chapters in this book. It then takes a look at the signals coming from government sources that set out priorities and directions for national development. There are encouraging signs that mathematics education will continue to develop, especially in terms of doctoral education and raising the level of teachers' knowledge and competencies. At the same time, there are signs that research may be redirected and that the nature of studies supported and encouraged by the

Research Council will be quite different from those reported in this chapter.

It is hoped that the chapters provide the representative view of mathematics education research in Norway at the end of the first decade of the 21st Century. The contributors represent most of the major centers of research in the field, and attempts have been made to ensure that they refer to other related Norwegian research. The chapters provide insights into school classrooms through primary, lower secondary, and upper-secondary. In some cases these might be recognized as "regular classrooms," such as in the chapters by Streitlien, Kleve, and Mosvold. In other cases, the chapters describe classroom activity in which there is an effort to reform or develop mathematical teaching, such as Jaworski, Fuglestad, Bjuland and colleagues, Carlsen, and Wæge. The English editor of this Norwegian part is grateful to all the contributors who have provided him with a wonderful learning experience providing fresh insights into Norwegian mathematics education and research in the field. The hope is that all readers will enjoy the same rich experience.

NOTES

1. *Source:* Statistics Norway: http://www.ssb.no/english/subjects/02/ retrieved May 27, 2010.
2. By the time of publication this will be 24 as Tromsø University College amalgamates with the University of Tromsø.
3. The International Group for the Psychology of Mathematics Education
4. Congress of the European Society for Research in Mathematics Education.
5. See http://www.uia.no/no/portaler/om_universitetet/teknologi_og_realfag/phd/matdrg/forskerskolen
6. In response to an early draft of this introductory chapter, Anne Fyhn, one of the contributors to this Norwegian section of the Sourcebook wrote: "I do not think I would have completed my thesis on time if not for the graduate school because I needed a milieu where I could discuss my work; the summer school was almost crucial for my development as a doctoral student."
7. For many readers it will be strange to refer to these 19-year-old learners as "pupils," but this is the custom in Norway, where *pupils* refers to learners in school and students in post-school education. We have followed this custom in the Norwegian chapters to reflect the national scene and because it helps to distinguish between pupils in school, student-teachers in teacher education programs, and experienced teachers.

REFERENCES

DKK (2009). *Læreren: Rollen og utdanningen* [The teacher: The role and the education]. St. meld. Nr. 11 [Parliament white paper 11]. Oslo, Norway: Det Kongelige Kunnskapsdepatement. Retrieved July 13, 2009, from http://www.regjeringen.no/en/dep/kd/Documents/Bills-and-white-papers/Reports-to-the-Storting/2008-2009/stmeld-nr-11-2008-2009-.html?id=544920

Heid, M. K. (2009). Editorial: Mathematics education research as an international enterprise. *Journal for Research in Mathematics Education 40*(3), 214–215.

Skovsmose, O. (1994). *Towards a philosophy of critical mathematics education*. Dordrecht, Holland: Kluwer Academic.

Vérillon, P. & Rabardel, P. (1995). Cognition and Artefacts: a contribution to the study of thought in relation to instrumented activity. *European Journal of Psychology for Education, 10*, 77–101.

Winsløw, C. (2009). Nordic research in mathematics education: from NORMA08 to the future. In C. Winsløw (Ed.), *Nordic research in mathematics education: Proceedings from NORMA08 in Copenhagen, April 21–April 25, 2008* (pp. 1–4). Rotterdam, Holland: Sense Publishers.

2

THE DEVELOPMENT OF MATHEMATICS EDUCATION AS A RESEARCH FIELD IN NORWAY

An Insider's Personal Reflections[1]

Trygve Breiteig and Simon Goodchild

UNIVERSITY OF AGDER

Abstract: This chapter describes the development of mathematics education as a research field in Norway. The development is explained in terms of changes in the school curriculum, the political and economic development of Norway, local institutional developments, and the individual contributions of influential Norwegian researchers. The introduction of modern mathematics to Norway and the perceived impact this had on mathematics teaching, learning, and pupils' performance is identified as a major moment in the historical development of mathematics education as a research field. Concurrent and related influences from without Norway are also noted. Other significant moments in the development of the research field are described; these relate to concerns about pupils' performance in the subject and gender balance. The chapter includes brief summaries of the work of some influential researchers in the field: Bernhof Ribsskog, John Hofseth, and Stieg Mellin-Olsen, in addition to a list of (most of the) doctoral dissertations in mathematics education, completed at Norwegian universities up to mid-2010.

Keywords: Mathematics education research, Mathematics curriculum, Norway, Norwegian research

INTRODUCTION

The purpose of this chapter is to provide an overview of events and activities that have contributed to the development of mathematics education as a research field within Norway. A number of events took place during and around 1976 that mark this as a time of significant transition in the development of mathematics education in Norway. It seems to us that the nature of the field of study, although developing over decades, was so transformed that we can reasonably describe two major periods of mathematics education research: one leading up to 1976, and the other in the years that have followed. In this chapter we begin by describing the events around 1976; following this we describe the major developments in the field in the four decades before, and then in the three decades since.

THE EMERGENCE OF MATHEMATICS EDUCATION AS A RESEARCH FIELD IN NORWAY

In May 1976, Ernst S. Selmer, professor of mathematics and academic teacher at University of Bergen, organized a seminar at Utstein Kloster, a conference center housed in buildings preserved from a monastery

The First Sourcebook on Nordic Research in Mathematics Education, pages 11–33

of the middle-age period situated between Stavanger and Bergen. The seminar was convened to provide an opportunity to discuss the nature of mathematics in the school curriculum and teacher education, and the direction of and responsibility for mathematics teacher education. The main concern to be addressed was closely related to curriculum developments in "modern mathematics" (the "New Math"). The seminar brought together mathematics teachers, mathematics teacher educators, and university mathematicians. Feelings were running high, as had been observed for half a decade in mathematics education communities. Emotions were so strong that people on opposite sides of the dominant issue had difficulty even to sit at the same table. Was this, in the land that awards the Nobel Peace Prize, a Norwegian "math war"?

A rift between modernizers and traditionalists had been emerging over several years; it seemed that those supporting modern mathematics were gaining ground, which can be seen in actions from public school authorities from the late 1960s. In 1971, an interim revision of the national curriculum had been introduced in which two alternative mathematics curricula were included. Alternative 1 was based on a traditional mathematics curriculum. Alternative 2, influenced by developments in the U.S. and Western Europe, was based on modern mathematics, with a foundation in set theory, algebraic structure, and unifying concepts. The idea behind the 1971 curriculum revision was that eventually modern mathematics would replace the traditional mathematics curriculum entirely. However, this could not be achieved merely by edict or the issue of a new plan because it would take several years to ensure all teachers were prepared to teach the new syllabus. Hence a substantial program for in-service mathematics teacher education was launched. Not all agreed that modern mathematics was the route to take. Criticism of "modern mathematics in school" had been around for many years internationally, in particular initiated by Morris Kline in 1961 (Valås, 1977), where opponents accused the reform movement of replacing the fertile and rich essence of mathematics with sterile, peripheral, and pedantic detail. This sharp and relentless debate spread to Norway shortly after the implementation of the curriculum of 1971 (Gjone, 1985, 2006). Even the Norwegian Parliament discussed the issue and decided to use more time and ask for a deeper report, which came in 1973. In this unsettled situation, it was proposed to make mathematics optional in the final grade of elementary school. This radical and, for many educators, absurd proposal naturally caused substantial protests and discussion.

In this context some research, detailed expositions, and developmental work were taking place, mostly initiated by the Research Council for Schooling (Forsøksrådet for skoleverket 1954–1984). We categorize some of the projects as decision-oriented inquiry in mathematics education. Especially we acknowledge the substantial Individualized Mathematics Education project (IMU—Individualisert matematikkundervisning) from 1970–1975. The purpose was to try out a model to address the challenges of mixed ability classes and diversity in secondary schools, by introducing individual work units at different levels, with the choice of levels being guided by diagnostic tests. Consequently, an extensive system of material had to be developed.

The seminar at Utstein Kloster in 1976 was important because it brought mathematics educators together in a context where they could begin to discuss the issues at the heart of the prevailing tensions face to face and not just criticize opinions within their isolated partisan groups. It would be misleading to suggest that "modern mathematics" was the only divisive issue at the time. There was also tension arising from concerns about where the education of secondary mathematics teachers should take place—in the teachers' colleges or the universities. Criticism was voiced from universities, in particular, when Oslo and Hamar teachers' colleges introduced a mathematics extension course, which was held to be the preserve of the universities. The national curriculum had extended compulsory education from seven to nine years in 1969 and created a separation between primary schools and secondary schools, the former covering the first six years of children's education from ages 7 to 13, followed by three years of secondary school.[2] The secondary mathematics teachers required an extended study of mathematics. The teachers union's Committee for Teacher Education in Mathematics, which also strongly supported the introduction of modern mathematics, was concerned that the teachers' colleges would not teach mathematics to the required standard.

The Utstein seminar was successful in bringing opposing parties together and pushing forward the process through which, eventually, a consensus was achieved (modern mathematics emphasizing set theory and structure was toned down). The success may, in part, be accounted for by the relatively small size of the Norwegian mathematics education community, proportional to the national population, and thus it was possible to bring together a group that was truly representative of the country's mathematics educators. It may be, and this is only our conjecture, that the seminar marks the birth of mathematics education as a recognizable and acknowledged field of scientific inquiry in Norway following several decades of gestation.

The seminar was not the only event around this time that marked a coming together of educators and mathematicians, which was to form the embryo of the mathematics education research community in Norway. New positions and research fellowships devoted to mathematics education were established at some Norwegian universities around this time. Also, the Third International Congress on Mathematics Education (ICME3) held in Karlsruhe made an important contribution.[3] It was here that the PME study group was established (later the International Group for the Psychology of Mathematics Education—which in 2009 held its 33rd annual conference). Peter Hilton's lecture at ICME3 on "spread of false dichotomies" focused participants on many of the issues that needed attention and urged for reconciliation of pure and applied mathematics, of modern and traditional school mathematics, of skills and understanding in mathematics. In addition Michael Atiyah and Hans Freudenthal recommended participants to concentrate on a fruitful and rich core ideas of mathematics. Heinrich Bauersfeld expressed the view on mathematics education as a special academic field "since it is difficult to get a discipline to adapt its research strategies for use in other disciplines, mathematics education will have to develop their own approaches" (Athen & Kunle, 1976, p. 240)

The Norwegian Council of Mathematics (Norsk matematikkråd, NMR) was interested in the debate within the international mathematics education community. In March 1977, the Board of NMR commissioned a group in Kristiansand to elaborate the education to be included in a study program for mathematics teacher education. The group considered issues such as how education themes should relate to the mathematics courses given in teacher education, which institutions should be responsible for different components of the programs, and the forms of cooperation that might fruitfully be established between universities and colleges of education. The report (Bekken, Borgersen, & Breiteig, 1978) advocated a closer relation between subject knowledge and education theory, as well as the inclusion of a meta-perspective on mathematics, including historical and cultural aspects of the subject, how mathematics can serve to address real problems by modelling and applications, and what constitutes the nature and methodology of mathematics. Thus mathematical competency for teachers was presented as being more than knowledge *in* mathematics; it should also include insight and knowledge *about* mathematics, its development, culture, place in science and society and its progress, as well as its didactics.

Prior to the events around 1976 it seems that the development of the compulsory school curriculum had been influenced by the empirical research of psychologists and pedagogues, and post-compulsory education by mathematicians with limited background knowledge of psychology and learning theories. Independently, we believe, both groups were impoverished because of the limited boundaries of their disciplines. Some of the empirical studies of the former group might today be considered to be mathematics education research, and there is much in them that deserves such a label. However, we believe it is their coming together in the mid 1970s that marks the emergence of the field of mathematics education in Norway. It is interesting to note that the catalyst for bringing the groups together was not differences of opinion about approaches to teaching and learning mathematics; rather it was consideration of the nature of mathematics to be taught and learned. We now turn to the two periods that we believe can be distinguished as before and after the events described above. To place these properly in their Norwegian context, we briefly outline the development of education in Norway.

A BRIEF OUTLINE OF (MATHEMATICS) EDUCATION IN NORWAY

To understand the development of education, and more recently mathematics education in Norway, it may be helpful to have a sense of the recent history and economic development of the country. Up to the beginning of the nineteenth century, Denmark was the dominant power in the region and Norway was "ruled" by the Danish monarch and government. Following the defeat of the Napoleonic army in 1814, Denmark was forced to cede sovereignty of Norway to a union of Norway and Sweden in which the center of power resided in Sweden. Sweden's monarch became Norway's head of state, and foreign policy was controlled by the Swedish government. However, Norway was able to establish its own constitution and parliament, and a movement for establishing a Norwegian identity began. Value was placed on national culture as expressed through interest in history and folklore, art and architecture, folk music, and literature. Especially significant was the development of a Norwegian written language. Today there are two distinct official forms of the (written) Norwegian language: Nynorsk and bokmål. Nynorsk (New Norwegian) is based on the many dialects found in Norway. Bokmål appears closer to the Danish language, which had been the official language in Norway up to 1814. Spoken Norwegian is categorized in 17 ma-

TABLE 2.1 Chronology of Notable Events in the Development of School Education in Norway Focused on Mathematics[a]

12th Century	Cathedral schools were established for the education of priests in Trondheim, Oslo, Bergen, and Hamar.
1537	Cathedral schools turned into "Latin schools," which were to be set up in every market town.
1739	First education act in Norway—introduces a curriculum based on reading to support a Christian education, compulsory for all children.
1802–1829	Niels Henrik Abel (possibly Norway's greatest mathematician). In 1815 Abel enrolled at the Christiania (Oslo) Cathedral School.
1814	The alliance of Denmark (and Norway) with the defeated Napoleon leads to the separation of Norway from Denmark. Norway is joined in an amicable union with Sweden.
1827	Writing, singing and mathematics (calculations) are introduced to the folkeskole (elementary school) curriculum
1889	Introduction of compulsory seven years (ages 7 to 14) elementary education.
1905	Norway achieves complete independence from Sweden.
1922	The first national curriculum for elementary schools in rural areas (landsfolkeskolen) is introduced.
1925	A national curriculum for elementary schools in urban areas (byfolkeskolen) is introduced[b]
1939	A revised national curriculum is introduced, informed by research in experimental psychology and influenced especially by the German notions of *Bildung* and *Arbeitsschule* and the term "Learning by doing."
1940–1945	Second World War, Norway is occupied by the German army.
1945–1960	The post-war period was dominated by reconstruction following damage and destruction caused by the retreating German army, building houses, and rebuilding infrastructure and services. There was also a struggle to address food shortages.
1960	"Trial" plan for nine years schooling (Forsøksplan for niårig skole) was presented (eventually required by the national department of education in 1969). Compulsory school extended from 7 to 9 years. Discussion about whether it should be 6 years primary + 3 years secondary or 7 + 2, and what the organizational structure should be. First schooling was organized according to potential employment; later a basic common curriculum was introduced but with different depths of specialization in Mathematics, Norwegian, English, and German, finally as pace differentiated mixed ability teaching ("linjedelt" versus "plandelt" or "tempodifferensiert, sammenholdt"). The curriculum subject term "calculation" (regning) was changed to "mathematics" in the 1960 plan.
1963–1967	Period of activity of the Nordic Committee for Modernizing Mathematics (NKMM)[c]
1971	A preliminary revised national curriculum is implemented. (Mønsterplan for grunnskolen). Mathematics is provisionally presented as two alternatives:
	Alternative 1 Traditional mathematics (calculation and mathematics / regning og matematikk)
	Alternative 2 Modern mathematics influenced by the international modern mathematics movement, built on reform work in the Nordic countries since the early sixties.
1974	Revised National curriculum for compulsory school grades 1–9. (Mønsterplan for grunnskolen, M74). Upper secondary school, grade 10–12, reorganized—a common first year where mathematics is included, then streaming into three programs, one of which is science and mathematics (naturfaglig linje).
1981	A publishing company, *Caspar*[d], is established by Stieg Mellin-Olsen and Marit Johnsen-Høines with the aim of stimulating mathematics education as an academic and professional field. (www.caspar.no).
1987	A new national curriculum for compulsory school grades 1–9 is introduced (M87 Mønsterplan for grunnskolen).
1989	Mathematics is included as compulsory in general teacher education built on subject knowledge developed in upper secondary school.
1990	*Tangenten*, a professional journal for mathematics teachers at all levels of school, is established by the Caspar publishing company. (See Torkildsen, this volume, Chapter 12, & www.caspar.no/tangenten.php).
1991	Stieg Mellin-Olsen appointed Professor of Practical Pedagogy at the University of Bergen.
1993	Gunnar Gjone appointed Professor of Mathematics Education at University of Oslo. This is the first professorship in mathematics education in Norway and thus demonstrates growing recognition of the emerging field of research.
1994	A new curriculum for upper secondary school gives right for all by law to receive a three years upper secondary education, aiming at obtaining competency for tertiary studies or a vocational education.
1997	Curriculum for Compulsory School (L97 Læreplan for grunnskolen). Compulsory school extends to 10 years by requiring the enrollment of six-year-old children.
1997	National Association for Mathematics in School, a Norwegian association of mathematics teachers and teacher educators, is founded by participants at the annual *matematikklærerkonferanse* (Mathematics Teachers Conference).[e]

TABLE 2.1 Chronology of Notable Events in the Development of School Education in Norway Focused on Mathematics[a] (continued)

2002	*Nasjonalt Senter for Matematikk i Opplæringen* (NSMO, National Centre for Mathematics Teaching) is established by the Norwegian Department of Education—based in Trondheim. (http://www.matematikksenteret.no/)
2006	Knowledge Promotion (KL06 Kunnskapsløftet), a common curriculum framework for primary and all secondary education, grade 1–13, is organized around the elaborated notion of "competency."

[a] The details in this table are drawn from a variety of sources and the personal knowledge of the authors. One important source that deserves recognition is Gjone (2006).

[b] There was a notable distinction between the curricula for rural and urban schools, which existed from 1922/1925 into the 1960s: In the rural schools children would attend alternate days, just three days a week. On the days when they were not at school, children would be able to help on the family farm or the family fishing business and engage in school homework set by their teacher.

[c] The Nordic Committee also stimulated an individualised mathematics project in Sweden at the same time as the above mentioned IMU project in Norway, see Larsson (1973).

[d] Named in memory of and to honor the Norwegian Mathematician Caspar Wessel (1745–1818).

[e] *Landslaget for matematikk i skolen* (LAMIS), www.lamis.no

jor, geographically located dialects, which can sound quite different from either of the written varieties of the language. The union with Sweden was dissolved amicably in 1905 when Norway became an autonomous parliamentary democracy with a constitutional monarch as head of state.[4] Following the dissolution of the union with Sweden, the quest to establish Norway's individual identity and culture continued.

As a young independent nation, Norway faced many challenges in building and strengthening the economy and national infrastructure, building the defense system, and facilitating the transition from an agrarian to an industrial society. The early decades of Norway's independent development were marked by periods of unemployment, food shortages, and a relatively high rate of emigration leading to the loss of human resources. It is only in the last three decades that Norway has developed into one of the world's richest countries as a result of the exploitation of North Sea oil reserves. Prior to this time Norway's economy has been dependent mostly on the fishing industry and subsistence agriculture—rather demanding occupations given the harsh northern climate. There was also shipping and some industry, which over the last 100 years has benefited from an abundance of hydro-generated electricity.

Table 2.1 provides a chronology of major events in the development of school education in Norway together with a few other significant events. The following sections provide some details of the development of mathematics education in schools over the decades preceding the mid-1970s events.

Mathematics Education in Norway 1935–1975

The revised national curriculum for schools implemented in 1939 was based on seven years of compulsory (elementary) education with children starting school at age 7. Children who wanted to continue with their education after age 14 had the chance to attend *realskole* and follow a three-year program, or possibly after two years at *realskole* to transfer to upper secondary school *gymnas*, which was the normal route into a university course. Up to the Second World War, general teacher education was provided principally by a four-year program in teachers' colleges (*lærerskoler*), which normally enrolled students at the end of the *realskole* program at the age of 17+. There was a substantial element of mathematics in this program. However, from the 1940s primary teacher education was increasingly dominated by a two-year program designed to build on upper secondary school education (*gymnas*) at the age of 19+. In 1973 this was extended to three years, and in 1992 to a four-year program. Mathematics was absent in this primary teacher education program until as recently as 1990. Alternatively, there was a specialized teacher education for those candidates completing a university degree; they could train to teach by following a six-month program referred to as "pedagogic seminar." This structure is presented in Figure 2.1.

Bernhof Ribsskog

Among the pedagogues that strongly influenced the development of education in this period is Bernhof Ribsskog. We believe his work deserves to be singled out for attention because of the impact it had at the time and its sustained influence over many years.

Bernhof Ribsskog (1883–1963) was educated as a teacher; he achieved a university degree in sciences with astronomy as his main subject. He went on to study experimental psychology with professor Otto Klemm in Leipzig, where he used scientific methods (or, "exact methods") to investigate psychological

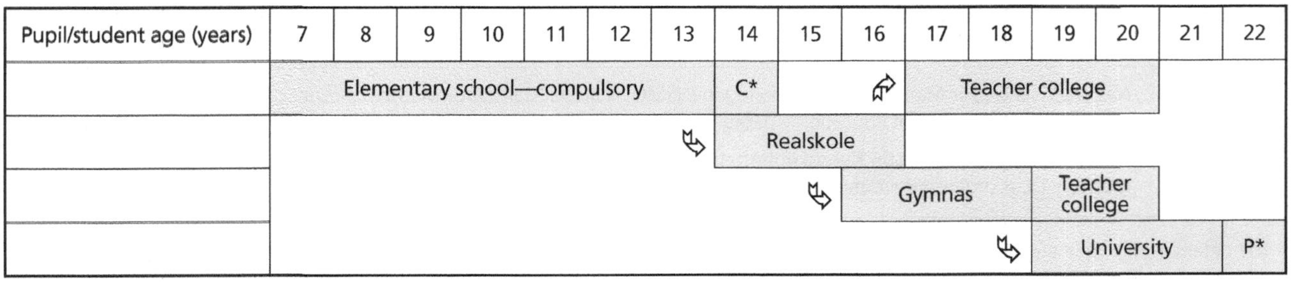

C* = Framhaldsskole, one year optional continuing school.
P* = Pedagogic seminar, a post degree programme, six months (from 1992, one year) of teacher education for university graduates.

Figure 2.1 The structure of teacher education in Norway 1922–1973.

phenomena. Later, in the 1930s, Ribsskog became heavily involved in initiatives to strengthen educational research in Norway. He drew attention to many key issues in Norwegian school, such as curriculum, evaluation, and teaching aids, and the activities he initiated were, to an increasing degree, built on research. He argued that reliable progress in schooling will only happen if informed by empirical research, by which he meant different sorts of experiments that were closely investigated. Ribsskog asserted that teachers should have a scientific base for their profession. Just as a medical doctor builds her competency and practice on a firm scientific foundation and is professionally respected because of this, so should be the case for a teacher and a pedagogue.

Two of Ribsskog's own substantial works: *Achievement Tests in Calculation* (*Standpunktprøver i regning*) (Ribsskog, 1936); and *Mathematics* (*Regning*) (Ribsskog, 1935) are of special interest for development of mathematics education.

Ribsskog and Aall (1936) laid foundations for the national curriculum of 1939. Indeed, a main goal of Ribsskog's research at this time was to study issues that should inform the revised national curriculum, which was to be implemented in 1939. Earlier, with Wiborg, Ribsskog had published results of a large-scale investigation in school mathematics conducted in several urban schools in Norway. The study enabled them to present an overview of the kinds of mistakes made by pupils in classroom tasks and pupils' use of symbols (Ribsskog & Wiborg, 1932). The investigation also focused on other subjects including history, religion, and geography and explored the degree to which pupils attained the goals set in the curriculum. The study revealed very low levels of achievement.

From the early 1930s, Fritz Wigforss in Sweden worked on standardizing different achievement tests (*standpunktprøver*) in mathematics for basic school (Wigforss, 1930, 1939). One of these, a test on problem solving in mathematics, calculations in "applied" tasks, was adapted for the Norwegian context by Ribsskog and used in June 1935 with nearly 16,000 pupils from grade 2 to 7 in Oslo schools. Each test comprised 20 tasks, of varying complexity from the very easy to the very difficult for every grade. Examples of the type of question (our translation):

Example 1. Petter has 1 crown and 10 pennies. He buys 10 (pieces of) 6-penny cakes. With the money left over he buys 2-penny cakes. How many does he get?[5]

Example 2. What is the biggest number that gives a remainder 2 when we divide 212 by the number and 296 by the number?

The examples above indicate how the nature of the challenge varied; in the first task, 52% of pupils were successful. The second example is of a more demanding type of task as it requires deeper conceptual understanding and working with both calculations and the properties of number; these were solved by very few pupils. Overall, the facility of different tasks varied from 1% of pupils getting a task correct to 98%.

One intention was to offer an insight into differences in attainment between boys and girls and between classes, the effect of social conditions and ability of individual pupils. A main aim was to provide teachers with a tool that would support them in their effort to adjust teaching to pupils' abilities. The study resulted in standards that could be achieved by an "average pupil" for each grade. Different kinds of mistakes in the conceptual bases for choice and use of operations in practical contexts, and use of symbols, were revealed. The study thus gave information about expectations of pupils' competencies in context-based calculations. Based upon this material, Ribsskog was able to elaborate norms for pupils and classes in each grade. Results from these tests revealed that at all levels, boys on average performed better than girls, and the difference increased with age. Further, a large spread in attainment was documented, revealing, for example, that the best pupils in grade 2 outperformed the weakest pupils in grade 7.[6] The performance in well-provided socioeconomic areas was much better that in poorly provided areas. Ribsskog concluded that in arithmetic it is meaningless to have the same expectations of all pupils of the same class or the same grade.

As noted above, Ribsskog (1935) also published a handbook for teachers on working methods in school mathematics. In this he made connections between international research and development with practical work in classrooms. The scope of the handbook is extensive, comprising 474 pages and 86 references, many of which are from German and some English and Swedish sources. We find it remarkable not only that this book was produced at all but also for its content, which could clearly be of value for teachers today. We reflect upon why such constructive and well-founded work is now hardly visible and largely forgotten. Possibly it is another victim of the turbulent and difficult period through which Norway and most of Western Europe passed in the 1940s. We wonder if it might also reflect an underlying issue relating to the continuity of educational research in which the aim and rewards appear to be focused on novelty rather than consolidation.

The ideas of the German "Arbeitsschule," from Georg Kerschensteiner, and those of John Dewey and the reform pedagogy were implemented in the curriculum of 1939. Also influential was the French tradition from Célestin Freinet and Ovide Decroly, which emphasized cooperative learning, enquiry-based learning, the natural method (inductive approach), and the children's interest (Mosvold, 2006). Mosvold (2006) claims that the national curriculum of 1939 is still viewed by many as the best national curriculum Norway ever had. The plan for each subject was supported by research and followed by teachers' guides with further elaboration of the main ideas, strategies, and teaching methods supported by research reports. The curriculum developers conducted their own research, but also considered results from international research (Mosvold, 2006). Thus this research had common goals with the developmental research that characterizes a major body of mathematics education research carried out in Norway in the first decade of the 21st century (see for example Jaworski, this volume, Chapter 5).

1940s–1950s: John Hofseth

John Hofseth (1950) refers to psychological research on pupils' language and vocabulary and their understanding of terms used in the different school subjects. Hofseth recognized that a substantial reason for difficulties experienced by pupils is that they cannot read with understanding the language used to present the topic. For example, in algebra, much of the content is new to the pupils, and familiar words may have a different or specialized meaning when used in the context of algebra. The teacher thus has a substantial task to teach pupils to understand the specialized meaning of the terms and symbols used.

Hofseth refers to analyses of technical vocabulary in basic school mathematics, grouping technical vocabulary into symbols and terms for:

1. calculations—such as add, sum, multiply, multiplier, product, difference, divisor, quotient, remainder; or terms connected to numbers—such as fraction, denominator, numerator, common denominator, lowest terms, invert, cancel, decimal number, cipher, decimal fraction
2. measuring and units
3. applications of mathematics—such as measure, width, depth, distance, area, capacity, average, cost, freight, order, sale, percentage, interest, interest rate
4. vocabulary of plane and three-dimensional geometry.

He asserted that when using the technical terms in mathematical education, the teacher should not regard them as "isolated words, but connected to elementary ideas expressed in a language that pupils may use in their thinking, and necessary to be able to think efficiently" (Hofseth, 1950, p. 32, our translation). A list of terms for basic school was produced, specified for the different grades, and offered both for research purposes and inclusion in the Norwegian curriculum. To produce such a list requires knowledge about the central ideas of mathematics, as well as an appreciation of language, and application of the principle that only didactically necessary and purposeful terms should be introduced, Hofseth claims (1950).

The national curriculum of 1939 (Normalplanen) contains a list of 41 terms, foreign words (fremmedord), which pupils needed in their school mathematics. Hofseth studied pupils' recall, understanding, and knowledge of vocabulary for their mathematics. In his study he included 29 of these foreign terms from the curriculum. The data were collected in June 1943, and this was the first survey of its kind into pupils' use of language in both primary and secondary schools in Norway. He used questionnaires with open questions. The tests were structured with two categories, A-tasks and B-tasks. As a means of establishing a measure of validity of the test, pupils were required to first complete the A-tasks and then the B-tasks without referring back to the A-tasks. The test was carried out with 541 pupils in Oslo equally distributed between grades 4, 5, 6 and 7; boys and girls (in separate classes); and two socio-economic groups: geographically defined by schools located in Oslo west (more affluent middle classes) and Oslo east (less affluent manual working classes).

Examples of A-tasks (authors' translation):

24. In the calculation task[7] 126 : 14 = 9 we call 126 ________________
25. 14 we call ________________
26. and the result, 9, we call ________________
27. We call this type of calculation sharing. But you know another name that means the same? What is that? ________________

Examples of B-tasks (authors' translation):

9. Below some numbers are listed. Underline those which are decimal numbers:

   $6\frac{1}{3}$ 8,45   206   0,34   2160 $\frac{2}{5}$

52. Underline the number that is the least common multiple of 3 and 4.

   16   24   12   9   20   18

His report includes a list ranking terms from the easiest to the hardest, from the most familiar to the least. The terms common to all grades 4–7 were studied separately. Here he exposed that terms that cause difficulties for the whole group also create problems for the grade 7 subgroup, thus indicating that the difficulties might be imbedded in the concept itself more than in the teaching. The research revealed that the operation verbs are easier to recognize (divide, multiply, add, subtract—in this order) than the names of the operations (division, multiplication, addition, subtraction), and most difficult to recognize are the terms of the outcomes from the operations (sum, product, quotient, difference, minuend—in decreasing order of solution frequency). The data revealed that the pattern of difficulty was consistent through the grades.

When taking the data from grade 7 pupils separately, Hofseth observed that rates of successful solution ranged widely. The frequency of success varies from divide (91,3% correct) and multiply (91% correct)—to the concepts difference (18.8% correct), prime factor (17.8% correct), and minuend (14.1% correct). Thus, less than one out of five pupils leaving school at the time could use with understanding the terms "difference" and "prime factor." Girls scored higher on average than boys for grades 4, 5, and 7. While the boys had a linear learning curve through grades 4 to 7, the girls had a decreasing trend for grade 6, especially for those in schools in the low socioeconomic area. However, their performance rose substantially in grade 7 where girls again outperformed boys.

1960s: Influence of "Modern Mathematics" on Mathematics Education in Norway

The 1939 curriculum remained in place for over 30 years. In the Norwegian context, this now appears to be a remarkably stable period given that since 1970 there have been five major revisions to the school curriculum. We want to suggest, tentatively, that the stability reflects the quality of the curriculum that was based on the serious research of Ribsskog and others. However, we must also note the stagnation that occurred during and following the Second World War, first while Norway was occupied and then during a substantial period of reconstruction following the liberation. A further factor, to which we have also drawn attention, is that until the discovery of North Sea oil reserves, Norway was a rather poor country and perhaps "tinkering" with the school curriculum is a preoccupation of politicians in countries that have the resources to support change—such as the introduction of new textbooks and teachers' in-service professional development. The stability of the curriculum, however, began to be disturbed during the 1960s.

One source of disturbance was the modern mathematics movement that originated in the United States. "Modern mathematics" had emerged in the early 1950s in a project based at the University of Illinois led by Professor Max Beberman. However, it did not gain momentum until the launch of Sputnik by the Soviet Union stimulated serious and urgent reflection on the quality of mathematics and science education in the United States. The developments in the U.S. spread to Europe, notably through a seminar in 1959 organized by the Organization for European Economic Cooperation (later to become the Organization for Economic Cooperation and Development); this has since become known as the 'Royaumont seminar' (OEEC, 1961). Norway was represented at the seminar by Kay Piene, Rector of the Oslo Pedagogic Seminar, responsible for teacher education of university graduates; Ingebrigt Johansson, Professor of Mathematics at the University of Oslo; and Ingvald Gjelsvik, Rector of Gudbrandsdal upper secondary school. This representation is interesting because it reveals where influence in mathematics education might have been located at the time; the group was certainly not representative of the wider community who were engaged in mathematics education.

The Royaumont seminar in 1959 influenced the development of school mathematics in Norway: It provided the momentum for the Nordic committee for modernizing the mathematics education (NKMM, Den nordiske komiteen for modernisering av matematikkundervisningen), which included two of the Norwegian participants at Royaumont.[8] The committee's work resulted in preliminary curricula including new textbooks intended for practical investigation. Pedagogues, school leaders, and mathematics professors participated in this work. The extensive work is reviewed in a final report (NU, 1967), which included an extensive discussion of the goals for school mathematics written by Professor Bent Christiansen from Denmark, valuable documentation of curricula, and an extensive bibliography of journals and books.

A desire for change was present; perhaps the remarkable stability of the mathematics content in the compulsory school curriculum was one motivation for change. There was an expression of a general dissatisfaction with mathematics in the school curriculum, which many regarded as outdated. It was recognized that mathematics had undergone considerable development; new areas, with new applications, newer work on its history and philosophy had accumulated, and it was felt that these should influence and vitalize the subject in school. There also appeared an increasing distance between school mathematics and university mathematics. Thus there was a perceived need to bring mathematics in school up to date.

The curriculum reform movement in mathematics and the influence by the "modern mathematics" thinking or ideology and its implementation in reform curricula and texts have been analyzed in separate studies by Harald Valås and Gunnar Gjone. They both documented analysis on the modernizing wave in American Reform Pedagogy from the 1960s, including the SMSG (School Mathematics Study Group), and the aforementioned NKMM. Gjone has since engaged in further studies of the development of the school mathematics curriculum (Gjone, 1983, 1985, 1992). Valås studied SMSG in USA from its beginning in the late fifties, because this was the biggest and most extensive reform project in school mathematics and one of the first attempts to build school curriculum content and exposition on mathematics as a scientific discipline (Valås, 1977). There are similarities between curriculum developments at this time in Norway with the earlier developments led by the SMSG in the U.S. Hence a brief look at the SMSG is relevant here. In the SMSG work chaired by Edward G Begle from Yale University, mathematicians took a substantial role, and they based their work on the fact that to influence the content of school mathematics most efficiently, one should produce good and comprehensive textbooks, employing a large group of authors. Trial texts should be first used with pilot classes and then revised for the published versions. The texts should finally have a form that connects to existing curricula so that teachers, supported by in-service training, could use these texts in their teaching. Material for teachers' in-service training was also produced. The SMSG started their reform implementation at the top with respect to age (grade 7 and 8, and then 9 to 12) and abilities ("college capable"); however, the materials were extended to elementary schools as early as 1960. In Norway a similar approach was adopted by the aforementioned Nordic committee, NKMM, with trial texts in mathematics (Forsøkstekster i matematikk) developed for lower and upper secondary school.

The principles that influenced the work of NKMM included

- integration of arithmetic and algebra
- exploration of unifying concepts to reduce the total number of basic ideas
- standardization of language, which should be used precisely
- promotion of understanding rather that mere mastery of skills
- motivation and retention improved through discovery approaches (Valås, 1977, pp. 10–12).

Key ideas pursued in the work were "basic concepts," "structure," and "deeper understanding." For example, to achieve deeper understanding of the number system, pupils might work on calculations with numbers in bases other than ten, and on tasks that would highlight the structure of the number system. Also, the structure of the whole numbers might be used to develop the set of rational numbers. There was less attention given to the arithmetical operations than on the field structure for the number system and connections of basic operations. Axioms for field and order relations were emphasized in order to clarify and organize the mathematical content, and to support the understanding that mathematics is logically based. Concepts from set theory would unify different mathematical areas such as algebra and geometry and serve as a pedagogic tool to understand the logical structure. The concept of function was another unifying concept.

In retrospect, it appears that the influence of NKMM was limited, and modern mathematics did not gain a strong foothold in Norway. Possibly this was due to the combined impact of the public discussion from 1972 onwards, heavy criticism from university scholars, political treatment of the dilemma, and the official report (NOU, 1973) that resulted in a revised curriculum in 1974. Modern mathematics was also considered as an academic movement and thus contradictory of the idea of *Arbeitsschule* in the preceding curriculum. Also influential, we believe, was the coming together of groups holding contrary opinions at the Utstein seminar, which we described at the start of this chapter. There were other influences present at the time—for example, the School Mathematics Project and Mathematics for the Majority project in England—emphasizing experience, investigations, and inductive thinking and bridging subject boundaries. Also having an impact was the work being developed at the IOWO[9] (later Freudenthal Institute) in Holland, leading towards the "realistic mathematics education" and other perspectives that emphasized mathematics as a process, emphasizing investigations, inquiry, reasoning, and proving.

We need to give a brief view of other influences on mathematics and the school curriculum that had emerged during the 1960s. There was pressure for increasing levels of democracy in public life and a resistance against centralization. For schools this was written into national regulations in 1969, and each school was required to have a parents' council, a pupils' council, and class councils. An early example of mathematics education research that stands in contrast to the modern mathematics movement appears finely tuned with these local pressures. In this, Stieg Mellin-Olsen focused on mathematics for the "slow learners." (Mellin-Olsen, 1972). See more on Mellin-Olsen's work in Chapter 3 (this volume).

It was also around this time that Stieg Mellin-Olsen began to make an impact on mathematics education in Norway and further afield. He was a competent mathematician who early in the 1970s had started to engage in theoretical analyses of mathematics education based on empirical data. In contrast to the SMSG and the Norwegian trial texts, he argued for the democratic right of access to mathematics including the "less academic" pupils, and those holding alternative views of knowledge from different social classes. Right of access had consequences, he argued, for the content and working style of mathematics.[10] Mellin-Olsen was active and influential in mathematics education research from the early 1970s through to the mid 1990s. He made original contributions to the field of mathematics education, for example, by elaborating the idea of instrumentalism in education, and we note that Richard Skemp gives Mellin-Olsen the credit for distinguishing between relational and instrumental understanding (Skemp, 1976). Mellin-Olsen exposed differences in students' rationales for their learning, which he theorized in ground-breaking manner by synthesizing Soviet activity theory and the work of the anthropologist Gregory Bateson and social psychologist George Herbert Mead (see Mellin-Olsen, 1987, also Wæge, this volume, Chapter 16). The contribution of Stieg Mellin-Olsen to the development of mathematics education in Norway is, we believe, of huge significance. Sadly, he was a victim of cancer and died in 1995, but his influence in Norway continues; his work remains fresh and reveals remarkable foresight and originality. Given his position in the Norwegian context, Chapter 3, this volume by Kirfel and Linden, is devoted to Mellin-Olsen's contribution to mathematics education and thus we limit the discussion of his work here.

Mathematics Education Research After 1976

Besides Mellin-Olsen's pioneering work, other activity in mathematics education research was undertaken, of which two significant projects are described here.

The Calculation Skill Investigation (Østfold og Telemark-undersøkelsen)

Hammervoll and Melbye in 1980–1981 conducted research on pupils' performance in the basic algorithms for addition, subtraction, multiplication, and division based on a diagnostic instrument developed in the mid-1970s as part of PUMP[11]-project in Sweden (for a brief description, see Johansson, 1993). The instrument was a two-dimensional matrix of tasks for each operation based upon an analysis of the subtasks within each operation algorithm, and evidence from empirical studies in pupils' calculation skills and theories of mental processes. The matrices were organized for each operation (addition, subtraction, multiplication and division) from top to bottom (by number of sub-operations in the task) and left to right according to the complexity of calculation (such as in a written algorithm the need to "carry a tens digit to the adjacent left hand column"); see Figure 2.2. Thus each operation algorithm was organized in steps, and tasks were given to test pupils' competence at each step. The pupils tested were enrolled in a random sample of schools in the Østfold county in south-east Norway (Hammervold & Melbye, 1980). Roughly one thousand pupils in each of grades 4, 5, and 6 took the test. Traditionally these are the grades where calculation skills are emphasized. The division algorithm was only tested at grades 5 and 6, with grade 5 pupils limited to one-digit divisors. Four questions were set for each part type of task (e.g., addition of two three-digit numbers, which required carrying a tens digit three

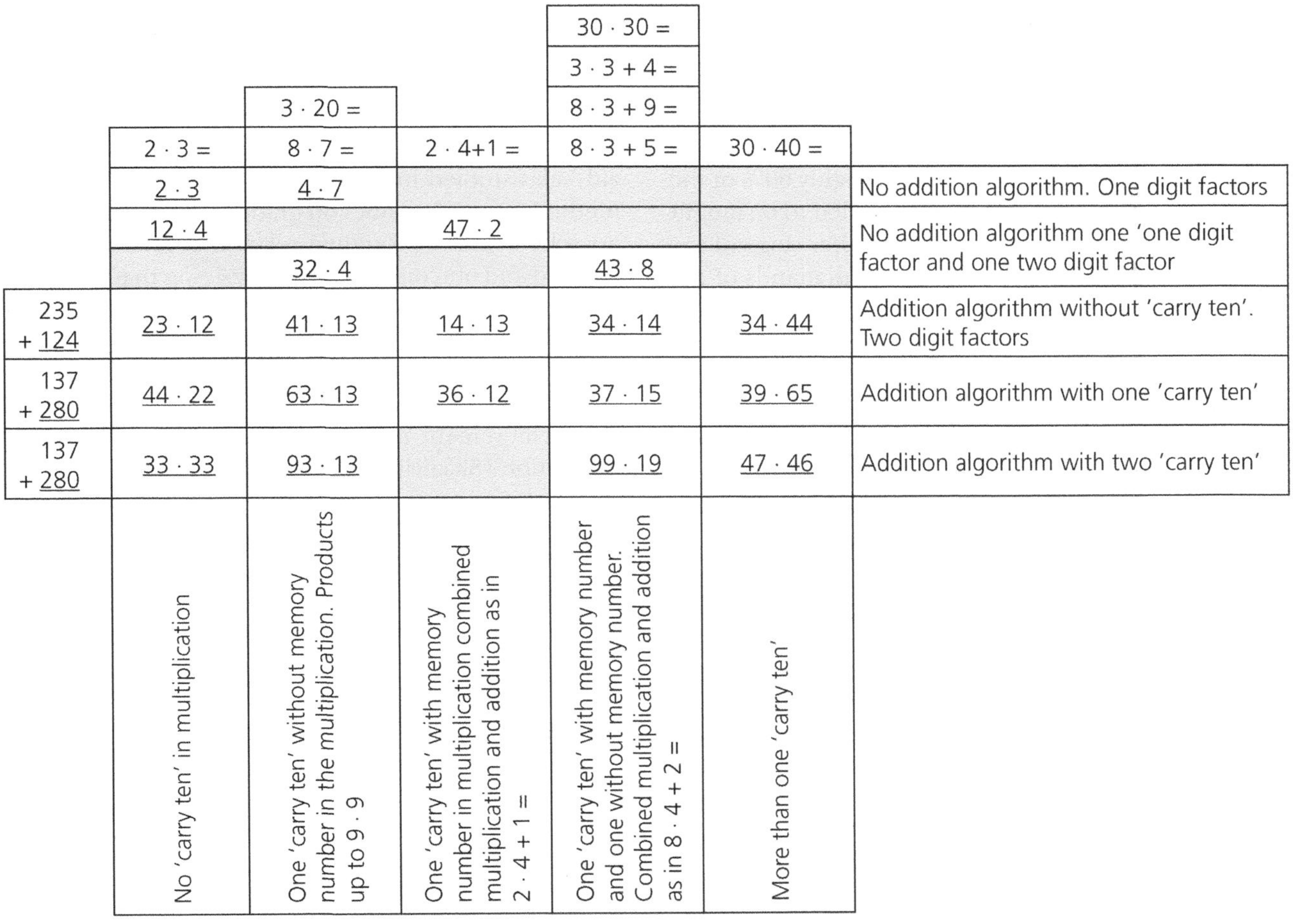

Figure 2.2 Structure of tasks in the multiplication matrix adapted from Melbye and Hammervoll (1990, p. 128).

times. Four calculations were presented: 554 + 947, 593 + 678, 384 + 828, 666 + 585[12]), and a pupil needed to get all four questions correct if she were to be counted as competent in that type.

The researchers claimed that the structure of the tests enabled them to identify where pupils' problems in using the standard algorithms might arise. The results suggested that relatively many pupils were unable to reach the level that they could rely safely upon their calculation result. For example 88, 77, and 66% of the grades 4, 5, and 6 pupils, respectively, were found *uncertain* on multiplication of the type 36×84 (requiring tens to be carried: twice in the multiplication stage and once in the addition). The disappointing performance of pupils was the cause of much discussion in the media, with teachers, and among politicians. The research was cited in debates about schools and the curriculum and many, especially parents, laid the blame for pupils' poor performance on modern mathematics and reduced emphasis on skill and practice. The Østfold investigation was, however, criticized for its focus on skills, on isolated context-free calculations, which could be performed easily using technological aids, and for ignoring other crucial aspects such as building concepts and relations between concepts, developing strategic skills for problem solving, and tackling everyday life situations.

Hammervoll and Melbye also carried out a comparison study between Norway, Sweden, and Holland and found that pupils from Holland scored higher. Further, they found that pupils performed better in Norway than Sweden in addition and subtraction; however, the opposite was observed in multiplication and division. The reason for Dutch pupils' superior performance could be due to the time allocated to mathematics in school, or the textbooks used, but the researchers hypothesized that it might be due to the resistance to the introduction of "new mathematics" in Holland, which had been articulated especially by Hans Freudenthal. The results were thus used by many teachers and parents to argue for "back to basics" and to systematize the syllabus by a small step method, where content is broken down into small pieces in a behavioristic approach.

We insert an ironic note here by referring to research by Jostein Våge (1979) studying the official examinations that grade 9, lower secondary, pupils were set during the period 1976–1979—that is, the examination at the end of compulsory school. His analysis of the examination revealed heavy emphasis on basic calculations. He matched the questions to the syllabus for different grades and found roughly 60% of the tasks related to content that was intended to be taught in the primary school grades. In a further stage of the analysis he examined the intellectual demands of the tasks against Bloom's five-level model. Here he found that roughly 20% of the questions demanded basic recall; over half tested mechanical calculation skills, and the remainder understanding. None of the questions demanded problem solving or creative reasoning. The concepts of modern mathematics that had been introduced to the syllabus were not being tested. Våge suggested that the absence of questions testing modern mathematics was probably due to some recognition that many schools had not adapted to teaching the new curriculum topics introduced around 1970. He also expressed his belief that as long as the "modern" topics were not being tested, there would be little motivation for schools to adapt to the new curriculum. Våge was articulating the widely held belief that the content of official tests has a stronger influence on what is taught than the content of the syllabus. It seems to us, then, that the tests used by Hammervoll and Melbye possibly reflected quite well what was happening in schools and yet the results were used to criticize a syllabus that was not being implemented! Våge also examined 452 pupils' performance in the examination and revealed, in addition, that a majority of the questions set were not good at discriminating between students, being either too easy or too difficult.

Ten years later, in 1990, Hammervoll and Melbye repeated their study, this time with a smaller sample of pupils taken from Østfold schools in the southeast of Norway (165 in grade 4, 222 in grade 5, and 217 in grade 6) and a sample of pupils from schools in Nordland, a county in the north of Norway (186 in grade 4 and 163 in grade 6). Improved pupil performance, especially in grades 5 and 6, led the researchers to assert that "more systematic and well-founded teaching of the four basic operations had led to the improvement" (Melbye & Hammervoll, 1990, p. 111, authors' translation). They also claimed "the results (of the 1990 study) were encouraging and gave clear signals that school mathematics in Østfold were moving in the right direction" (1990, p. 112, authors' translation). The repeat study was considered important because in the meantime a new curriculum had been introduced (M87 Mønsterplan for grunnskolen—compulsory school, grades 1–9). The results of the 1990 study were used to inform an evaluation of the revised curriculum.

Somewhat complementary to Hammervoll and Melbye's study, other research on teaching methods was carried out in intitutes of pedagogy. For instance, Harald Valås studied inductive versus traditional teaching methods in mathematics on grade 7–9 in lower secondary school, and he sought a relation between teaching method and outcome. His hypothesis was that inductive teaching was to be preferred above traditional teaching approaches—with respect to insight and understanding obtained, knowledge and skill, retention, and motivation. Valås took this issue even further in a study of how achievement may be related to self-concept and motivation (Skaalvik & Valås, 1999).

Cognition and Awareness: The KIM Project

In 1993, a project was commissioned by the Norwegian ministry of education with aims, among others, to develop diagnostic tests and teacher development materials. The project Quality in Mathematics Teaching (*Kvalitet i matematikkundervisninga*, KIM) has continued as a collaboration between Telemarksforsking[13]–Notodden and Senter for lærerutdanning og skoletjeneste,[14] University of Oslo. It is founded on constructivist learning theory; the first project leader was Gard Brekke, who gained his PhD in 1991 following work at the Shell Centre, University of Nottingham, UK (Brekke, 1991), with Alan Bell and Malcolm Swan.

Brekke and his colleagues produced a set of materials that were tested in a pilot study reported in Breiteig and Brekke (1993). As the intention of the KIM project implies, it is based upon the notion of conceptions that pupils develop during their learning. The materials produced within the project are intended to help teachers expose pupils' misconceptions through diagnostic tests and teaching, and then provide pupils with learning experiences that enable them to become aware of and challenge their own misconceptions. The materials produced synthesized and applied results, reported by the UK Assessment of Performance Unit (1982) and the work pioneered at the Shell Centre by Alan Bell and colleagues, and they were informed by Gérard Vergnaud's articulation of conceptual fields. This included, for example, the project Diagnostic Teaching and the Acquisition of General Strategies in Mathematics (Bell & Shiu, 1981), which demonstrated that pupils may obtain correct answers to common mathematical tasks, even with a misconception that may hinder their future learning and understanding. Correct answers then might provide feedback that strengthens the misconception. It is thus valuable in the learning process to explicitly identify common misconceptions and have tools to expose each of them in order to overcome them and build robust knowledge that will support future learning. The KIM project resulted in a large volume of material including diagnostic tests and teachers' guides with components including numbers, calculations, algebra, functions, graphs, and geometry. Diagnostic test items for pupils in grades 4, 6, and 8 were subject to national standardizing based on a random sample, balanced for region and school size, including roughly 100 classes in each grade tested. In total around 2,000 pupils at each level took the test (the numbers varied slightly across the different content domains—functions, number, etc.), and the analysis was based on about 500 pupils in each grade randomly chosen from the circa 2,000 that participated (see Brekke, 1995). The latest phase of the continuing project is to make this material available on the internet.[15]

Development of National Curricula

The next major revision to the Norwegian national curriculum occurred in 1997 (L97). This time the revision was inspired by constructivist learning theory and informed by the KIM project and other constructivist sources. Emphasis was laid on five principles: mathematics as discovery activity (investigation, experimenting, playing and games); mathematics as a tool (experience from everyday life, practical situations, realistic problems); mathematics as a creative field (concept forming, aesthetic features, creativity and imagination); mathematics as a reflective and metacognitive activity (discussing, informing, articulating, sharing, thinking); and mathematics in community and culture (the role of mathematics in culture, science, and the history of mathematics). The introduction was accompanied by a program of in-service professional development (in the form of written guidance and short courses) that was aimed to enable teachers to interpret and implement the curriculum as intended. The 1997 mathematics curriculum L97 was evaluated, as part of a program to evaluate the whole of the 1997 reform (R97), by a team led by Gard Brekke (Alseth, Breiteig & Brekke, 2003). The evaluation was multifaceted and included separate studies of the mathematical content intended, the text books being used in school, the in-service professional development that had been offered, classroom observations, teacher interviews, and pupils' competencies at grades 4, 7 and 9, with a comparison of results of grade 9 pupils who took the same test before the new plan was implemented.

The evaluation revealed that school textbook authors had made a serious attempt to respond to the changes in the revised curriculum, and the five principles upon which the curriculum was based were reflected within the school texts. Nevertheless, the textbooks initiated pupils' activity through short closed tasks that required no more than pencil-and-paper activity. Classroom observation revealed, for the most part, that little had changed with regards to teaching practices, which were characterized by teacher introductions to topics and explanation of specific tasks and pupils working individually on exercises. This accords with Bodil Kleve's case study of three teachers, which is reported in this volume Chapter 10.

Inevitably, a central concern of the evaluation was to expose any impact on pupils' performance in mathematics that might be ascribed to the new curriculum. We take a short detour to describe the background for the study of impact because it will give a view of significant aspects of mathematics education research in Norway around this time. The curriculum had not been introduced as an experimental treatment with planned pre-tests and post-tests. The researchers in the evaluation had to work with the pre-revision data that were available despite the fact that the data arose from sources that were not operationalized to match the intentions and principles of new curriculum. For the evaluation, pupils were tested in grades 4, 7, and 9. For grades 4 and 7, a test was constructed from questions that had been used in the TIMSS study (Brekke, Kobberstad, Lie, & Turmo, 1998) and the investigation of pupils' performance that had been carried out as part of the KIM project (Brekke, 1995) described above. The study included over 1,000 pupils from each of grades 4 and 7 drawn from schools in six counties. Comparison was made with pupils' performance before the curriculum reform based on data from responses to questions as they had been given in the KIM study and Norwegian pupils included in the TIMSS study.

The comparison for pupils in grade 9 was made using a selection of tasks from the Kassel-Exeter study, which was an international longitudinal study of pupils' development that had been carried out in Norway in 1994 and 1995 (Hinna, 1996). The Kassel-Exeter study originated, as the name suggests, as a joint research project between Germany and the UK and expanded to include more countries (Burghes, 1999).[16] The idea was to monitor the progress in mathematics of pupils who were starting the school year corresponding to the age 14+ over a two-year period (1993–1994). Norway joined the project one year later and thus pupils were only monitored over one year. The Norwegian part of the study is covered in the *hovedoppgave i matematikk* of Kristin Hinna (1996).[17] Hinna examined the progress of just under 300 pupils between autumn 1994 shortly after beginning grade 8, and the autumn of 1995. The tests used covered number, algebra, shape and space, and handling data. Hinna was able to identify pupils' progress in mathematics for 90.4% of the sample. She explored the effect of different factors that might contribute to progress such as the textbook, teachers' qualifications, teachers' experience, and pupils' interaction within the classroom. Only the textbook was a significant predictor in all the tests. In comparison with the 13 other countries included in the study, pupils in grade 8 in Norway were at about the median rank at the start of grade 8 and made similar progress to pupils in the other countries.[18]

We note that the tests used to evaluate the impact of the curriculum reform of 1997 with respect to pupils' outcome were based upon tests that were not specifically designed for the purpose. Pupils in grade 4, 7, and 9 in 2001 were compared with those of the same age from 1995. Pupils in the higher grades had similar lengths of school experience, whereas in the lowest grade the post-reform pupils had received an extra year of schooling because the reform extended the compulsory schooling by one year, with children starting a year earlier. It is perhaps unsurprising that the grade 4 pupils performed better in the tests than the similar pupils before the reform. However, the reverse was the case for the older pupils. At grade 7 there was a general decline in performance in what the researchers described as procedural knowledge. However, there was no appreciable decline in conceptual knowledge, and evidence of improvement in areas that had greater emphasis in the new curriculum such as the dynamic elements of geometry, and probability and data handling. Pupils at grade 9 also showed a marked decline in performance except in the topic of probability and data handling. The authors of the review note, however, that at the time of the evaluation the grade 9 pupils had started school several years before the reform took place.

The TIMMS study of 1995 had reported disappointing performance of Norwegian pupils. In 2003, the PISA study of OECD countries also revealed that Norwegian pupils' performance was below average, especially disappointing in relation to those countries with which it is natural to draw comparisons. These studies are reported in more detail in the chapter by Liv Sissel Grønmo (this volume, Chapter 4). Pressure was on the Norwegian government to do something, and a new curriculum was introduced in 2006, this time based on the notion of "competencies" influenced by Mogens Niss and colleagues' work in the Danish KOM project (Niss, 2004; Niss & Jensen, 2002). We note here that the evaluation of the 1997 revision of the curriculum, and Kleve's research, had not demonstrated that the intended curriculum was ineffective—rather that it was not being implemented as intended. Further, the decline in pupils' performance in the tests could not be used as evidence of a negative impact of the curriculum because the tests were not operationalized and validated in the first instance as measures to demonstrate impact.

The foregoing account reflects an interpretation of events that suggests the recent evolution of mathematical education in Norwegian schools has occurred in reaction to outcomes of studies of pupils' performance. Further, that the tests have assessed pupils' performance in the context of an implemented curriculum rather than what has been intended. Nevertheless, pupils' disappointing performance has been blamed on the intended curriculum, thus leading to fresh curriculum revisions. Recent international studies of pupils performance (PISA 2003, PISA 2006) repeat the picture of Norwegian pupils' disappointing performance in mathematics, on average below the performance of pupils in those countries it is most natural to make comparisons with (such as in north-west Europe and especially the other Nordic countries). In terms of impact on national policy and development, the studies based on national and international assessment are an important part of the recent history and development of mathematics education research in Norway. For more on this theme see Grønmo, Chapter 4, this volume. Since the early 1990s, mathematics education research has also developed as it has become recognized and gradually more visible as a field of academic endeavor in Norwegian universities and university colleges; it is to this development that we now turn.

THE DEVELOPMENT OF MATHEMATICS EDUCATION AS A RESEARCH FIELD IN HIGHER EDUCATION

Gunnar Gjone was appointed the first professor of Mathematics Education in Norway by the University of Oslo in 1993. Gjone had ten years earlier completed his PhD degree at the University of Oslo based on an analysis of documentary evidence in a study of the development of the modern mathematics in school (Gjone 1983, 1985). Two years before Gjone's appointment as professor, Stieg Mellin-Olsen had been appointed professor of practical pedagogy at the University of Bergen; in spirit, if not in title, Mellin-Olsen might then be considered the first professor in the field of mathematics education in Norway.

Stieg Mellin-Olsen

We noted earlier that Mellin-Olsen's "remarkable foresight and originality" led him, based on experiences in mathematics learning, to investigate traditional learning theories inspired by Piaget. He found that it could not fully explain learning in the classroom and the development of pupils' instrumentalism. Thus he analysed social and political aspects based on theories from social anthropology and sociology. Mellin-Olsen's iconoclastic work was subsequently published as *The Politics of Mathematics Education* (1987), which in the foreword he describes as "the result of a twenty-year long search to find out why so many intelligent pupils do not learn mathematics whereas, at the same time, it is easy to discover mathematics in their out-of-school activities" (p. xiii).

Gjone accounts for the appointments of professors in mathematics education and related fields at this time as an outcome of reorganization of the "pedagogic seminar"—that is, the program followed by university graduates who wanted teacher education. Gjone explains that the pedagogic seminar had been founded by the University of Oslo in 1907, but there had been some dispute about where this teacher education was best placed, in the schools or the university. In 1923 the program had been taken out of the university control but then reinstated in 1988. The establishment of professorial positions in didactics and pedagogy followed this development. Gjone gives a more detailed account (in Norwegian) in his paper celebrating the 25th anniversary of the publishing company Caspar founded by Stieg Mellin-Olsen and Marit Johnsen-Høines, (Gjone, 2006). Gjone also points to other concurrent developments in teacher education at the colleges of education, which created a need for growth. In 1989, by a decree of Parliament (Stortinget), mathematics became a compulsory part of general teacher education to be extended about ten years later. This led to a substantial demand for people qualified to teach these mathematics courses within teacher education. The new circumstances gave Norwegian mathematics educators the incentive to establish links between research, development in international settings, and practical teacher education (Gjone, 2006).

Before proceeding further we want to acknowledge doctoral programs such as *realfagdidaktikk* (which encompasses the natural sciences and mathematics education) at the University of Oslo, and others in pedagogy and education at the University of Tromsø, the Norwegian University of Science and Technology (NTNU) in Trondheim, the University of Bergen, and the University of Stavanger. Other chapters in this volume reflect the research in mathematics education that is pursued in these places (e.g., Fyhn, Chapter 13; Grønmo, Chapter 4; Johnsen-Høines, Chapter 7; Streitlien, Chapter 14, Torkildsen, Chapter 12; and Wæge, Chapter 16). However, at the time of writing there is just one doctoral program with the specific title "mathematics education" in Norway, and this has evolved as an outcome of one institution's effort to establish a research group in the field. An account of the recent history of mathematics education research in Norway would not be complete without writing something about this development.

In May 1992 the "*Sånn Ja!*"—"Such, yes!"—conference was held at Kristiansand in the south of Norway[19] on mathematics education and women in mathematics, organized by collaborative effort of The Research Council of Norway, Agder Regional University College, or simply Agder College (Agder distrikthøgskole[20]), and Teachers' University College, or Kristiansand College of Education (Kristiansand lærerhøgskole). This conference was important both because of the agenda and those who participated, including senior personnel from the universities, research institutions, teacher education colleges, and the ministry of education. International guests were also in attendance. Anne-Lise Hilmen, Assistant Director of the Research Council of Norway, opened the conference with the following words:

> Ladies and gentlemen,
>
> Something must be done! This is the reason why responsible bodies and institutions have taken the initiative to arrange this conference.
>
> Something must be done about the shortage of women in mathematics and something must be done about the content and quality of the teaching of mathematics at all levels of our educational system. (Sånn Ja!, 1993, p. 7)

The conference included addresses by professors Mogens Niss (Roskilde University, Denmark) on "Mathematics Education as an Academic Field," which stressed the contact with the international research field; Gila Hanna (The Ontario Institute for Studies in Education) on "Retaining Women in Mathematics and Science," which emphasized gender issues; and Alan H. Schoenfeld (University of California, Berkeley) on "Aspects of Mathematical Thinking and Problem Solving," which focused on the process aspects of mathematics. Schoenfeld also gave an address entitled "Thoughts About a Master's Degree in Mathematics Education."

The "Sånn Ja!" conference was held at a time of national reorganization of institutions of higher education. In Kristiansand, the Agder College and the Kristiansand College of Education, together with four other colleges, were amalgamated to form Agder University College (AUC). One outcome was the permission given to AUC by the Ministry of Education to establish a *cand scient* program in mathematics education (*hovedfag* a higher academic degree—a postgraduate program considered more demanding than a present-day master's). Up to 1992, this qualification was the preserve of the established universities (Oslo, NTNU in Trondheim, Bergen, Tromsø). It was decided to establish a *hovedfag* in mathematics education at AUC, which was launched in 1994. This was the first regular program of its kind in Norway, and in itself it initiated a significant body of research in the field of mathematics education. The *hovedfag* in mathematics education was met with considerable and accumulated interest among teachers and mathematics teacher educators who had been looking for the opportunity for further study and research in mathematics education. The first eight candidates were enrolled in 1994 and completed their study in 1996. Among these was Kristin Hinna, whose research in connection to the Kassel-Exeter project was mentioned above. Again we want to stress that we have no wish to devalue earlier candidates with extended master's theses (*hovedfag*) in mathematics education, or in pedagogy with mathematics education direction, that were pursued within the established universities. Examples are Egil Frøyland's historical analysis of mathematics in school emphasizing development in the 19th century (Oslo, 1965); Liv Sissel Grønmo's comparative studies of pupils' achievement (Oslo, 1992), which furthers, for instance, that of Jan Einar Nordgreen (Oslo, 1982)—"Invitation to Creative Mathematics"; Geir Botten (NTNU, 1992) with "Mathematics: A Subject to Acknowledge: In School, in University and in the Community" (*Matematikk—et fag å regne med: både i skolen, på universitetet og i samfunnet*); and more recently Rune Herheim (Bergen, 2004) on "Class Teaching: Dialogical, Sociocognitive Learning" (*Klasseundervising. Dialogisk, sosiokognitiv læring*), and Hege Marie Mandt on "Mathematics in General Teacher Education" (Mandt, 2006).

Although important for development of mathematics education at the universities, we see these as isolated cases. At Agder University College, the program was run regularly starting in 1992. It was a bold decision on the part of the Norwegian Education Department to allow a university college to introduce an extended master's degree (*hovedfag*) program in a "university subject." Lines of cooperation with universities and other resource agents were set up, relations between mathematics and mathematics education discussed, and an intention to utilize different competencies, not to compete but to complement each other, was emphasized. Stringent quality controls were set up, and the program proposal, and later its implementation, was externally evaluated regularly to determine its quality and effectiveness and to ensure that no dilution of standards occurred. The program continues, at the time of writing in its fifteenth year, although it has now been transformed into a more internationally recognizable Master of Mathematics Education. By 2009, more than 80 candidates have completed this degree.[21] The program remains popular, especially among lower and upper secondary school teachers and future teacher educators, and it is now presented in English, thus opening it to wider, international participation. It also provides a foundation for the doctoral program.

A subsequent decision by the Norwegian government introduced the possibility for the university colleges to gain full university status. As well as bringing the prestige (and marketing power) of the university title, this would also allow a greater degree of autonomy. A number of criteria were set up that the institutions had to meet, subject to rigorous external evaluation. One of the criteria was that the institution should have graduates from at least four mature doctoral programs. There was a great desire for Agder University College to achieve the university title, within the institution and locally as the community at large wanted its own university. The mathematics education group at Agder (MERGA) were well placed to make a successful proposal to establish one of the first doctoral programs at the institution. A doctoral program in mathematics education was established in 2002. To support the new program, three professors in mathematics education were appointed: Anna Kristjánsdòttir (from Iceland), Barbro Grevholm (from Sweden), and Barbara Jaworski (from England). Thus a substantial research group in the field was formed together with established colleagues including Maria Luiza Cestari who was promoted to professor at the same time as the new appointments (originally from Brazil, she had worked within the institution from the early 1990s and gained her PhD from the University of Oslo in 1997), Trygve Breiteig (*høgskoledosent*—translated into English: professor), Anne Berit Fuglestad (also *høgskoledosent*/professor, who gained her PhD from the University of Nottingham), and Hans Erik Borgersen (Associate Professor in Mathematics). Additionally, an important member of the group was Reinhard Siegmund-Schultze (Professor of History of Mathematics, from Germany); he filled this position in 2000 following the retirement of Professor Bob Burn (from England) who joined the group first in 1995 then in 1997 to 2000 and was appointed professor in 1998.[22]

The names above do not include all those who worked hard to establish the doctoral program; they became the core contributors to the program once it started. It is interesting to note that the international dimension of the group extended in 2007 to include Professor Cyril Julie from South Africa. Also we draw attention to the proportion of women holding senior research positions within the group. The words of Anne-Lise Hilmen as she opened the "Sånn ja!" conference were being heeded.

Agder University College became the University of Agder (UiA) in September 2007, the seventh university in Norway after Oslo, Bergen, Tromsø, the Norwegian University of Science and Technology (NTNU) in Trondheim, Norwegian University of Life Sciences (UMB) in Ås, and Stavanger. We note with pleasure the contribution of mathematics education to this widely welcomed development. The doctoral program in mathematics education at UiA is the only program specifically *named* mathematics education in Norway, although we respectfully acknowledge the doctoral program in *realfagdidaktikk* (natural sciences & mathematics education) that is offered at the University of Oslo. The mathematics education group at Agder (MERGA) is the largest research group in the field at any of the Nordic universities, and the doctoral courses attract students from all of the Nordic countries and further afield.

In 2003, the mathematics education group at Agder made a successful application to the Research Council of Norway (RCN) for a substantial grant (approximately US$3 million) from the KUL program (Knowledge, education and learning—*Kunnskap, utdanning og læring*). This was to support a mathematics teaching developmental research project, Learning Communities in Mathematics. The project involved collaboration between members of the MERGA group and teachers in eight schools (reported in this volume by Jaworski, Chapter 5). Within the same KUL program, a parallel successful application received funding (approximately US$1 million) for the project ICT and Mathematics Learning (ICTML), reported by Fuglestad in this volume (Chapter 6). These projects had a funded life of four years (2004–2007). In 2006 the group made another successful application to RCN for a grant (approximately US$1.5 million) within the PFoU program (practice-directed research and development—*Praksisrettet forskning, og utvikling*). This supported a new mathematics teaching development project called Teaching Better Mathematics (TBM). The TBM project is both an extension and development of the preceding LCM and ICTML projects; it extends the period of working with teachers in schools and draws in new schools and kindergartens. TBM develops the activity through the establishment of a consortium of mathematics education at UiA, and university colleges in Bodø, Trondheim, Bergen, and Oslo. The research groups in each place established independent research projects with a common principle of *inquiry* in mathematics, teaching mathematics, and development in teaching mathematics. The group at UiA and local schools gained additional support (approximately US$1 million) from a local fund that had been established to develop the research and development within the university. This enabled the establishment of a project to partner with TBM called Learning Better Mathematics (LBM), which specifically supported the work of teachers in schools and members of the MERGA group to work with the teachers. To complete the picture, a further successful grant application (approximately US$1 million) was made to RCN within the Strategic University College (*Strategisk høgskole*) program, which was intended to develop research capacities in the university colleges. Within Agder University College this grant was used in a research project concerned with the development of mathematics teaching in primary schools as a part of whole school development.

Significant national and international events and programs have also been centered in Bergen, partly through the active engagement of Stieg Mellin-Olsen and partly by those who try to sustain the work following his untimely death. In the mid-1990s, Mellin-Olsen established an enduring link with the University of the Western Cape (Slammert, 1996), which more recently has resulted in a formal collaboration, GRASSMATE (Graduate Studies in Science, Mathematics and Technology Education). The project, funded by the NORAD (Norwegian Agency for Development Cooperation) through the Norwegian Committee for Development Research and Education (NUFU) for the period 2002–2006, was set up with the aim of competence building in sub-Saharan Africa and also in Norway. For an overview, see Mikalsen & Julie (2008).[23]

Stieg Mellin-Olsen must be given much of the credit for the attention to national and international social and political issues within mathematics education research. Shortly after his death, the Third International Conference on the Political Dimensions of Mathematics Education was held in Bergen (the first had been held in London, the second in Johannesburg). The fact that Bergen was chosen as the third venue for the conference was due in no small part to it being Mellin-Olsen's home city. Mellin-Olsen's close colleague, Marit Johnsen-Høines, has worked hard to sustain the engagement in mathematics education in the Bergen academic community. An outcome of this was the hosting, by Bergen University College, of the 28th Annual Conference of the International Group for the Psychology of Mathematics Education (PME) in 2004.

In September 2008, Marit Johnsen-Høines was appointed Professor of Mathematics Education at Bergen University College. The research activity of the group she leads is mainly from socio-cultural, socio-political, and dialogical perspectives and is applied to action research and developmental research projects that focus on teaching and learning in kindergartens, schools, and teacher education. In 2007 the group was successful in its application to RCN for a grant (approximately US$1 million) for the research project Learning Conversation in Mathematics Practice (LCMP). This project is based on collaboration between researchers, teachers, student teachers, and mentors; see Chapter 7 by Johnsen-Høines, this volume.[24] LCMP is a separately funded part of the TBM consortium led by UiA. Also, Ole Einar Torkildsen was appointed professor in mathematics education at Volda University College in 2008. His area of work is mentioned later and in Chapter 12 this volume.

At the University of Oslo, mathematics education research has recently focused mainly on participation in the large international studies of pupils' performance, TIMSS, PISA, and TEDS-M. Chapter 4, by Liv Sissel Grønmo, provides an account of this activity, which uses quantitative methods to create knowledge within mathematics education. The aim of equal rights and opportunities for the same education for all inhabitants is strong and explicit in Norwegian educational policy. Connected to the issue of equal opportunity of education, the Department of Special Needs Education at the University of Oslo has developed responses to teaching problems to pupils with disabilities in school. Special needs education has a multidisciplinary approach to the learning, teaching, and rehabilitation of children or young people with disabilities. Snorre A. Ostad has carried out extensive research on mathematics learning for pupils having problems with mathematics. His research ranges from cognitive studies of children at risk from a learning disability, to basic research on cognitive systems that may contribute to individual differences in mathematics competencies. His research includes mathematically "normal functioning pupils" (for research purposes he called them MN pupils), children with difficulties in mathematics (MD pupils), and also blind children and deaf children. His work includes comparative studies between the two groups of MN and MD pupils. He has studied metacognitive aspects, such as developmental characteristics of subvocal speech and choice of solution strategies on mathematical tasks. Ostad engaged with the problem about MD pupils (dys-mathematicians); are they "slow learners" or "different learners"? Do they follow a developmental delay or a different developmental model? He found evidence for the latter in most cases.

Research on pupils with special needs in mathematics has been carried on in Norway in a Nordic cooperation and with a support network. This is documented by the Nordic research seminars, held every second year since 2001. The 2007 conference had the substantial title "Different learners—different Math?," which indicates a burning issue.[25] There is a line from the first conference with the title: "*En matematikk for alle i en skole for alle*" ("A mathematics for all in a school for all.")

The first years of the twenty-first century have been marked by the award of several doctoral degrees (Dr. Philos., Dr Polit., or the newer Ph.D.) in the field of mathematics education. The following list of completed theses provides an insight into the breadth of research that has been pursued.

Marit Johnsen-Høines (2002) *Fleksible språkrom. Matematikklæring som tekstutvikling* [Flexible language spaces. Mathematics learning as text development]. University of Bergen

Raymond Bjuland (2002) *Problem solving in geometry. Reasoning processes of student teachers working in small groups: A dialogical approach.* University of Bergen

Reidar Mosvold (2006) *Mathematics in everyday life. A study of beliefs and actions.* University of Bergen

Sharada Gade (2006) *The micro-culture of a mathematics classroom: Artefacts and activity in meaning making and problem solving.* Agder University College

Ole Einar Torkildsen (2006) *Mathematical archaeology on pupil's mathematical texts. Unearthing of mathematical structures.* University of Oslo

Åse Streitlien (2006) *Rom for deltakelse. En studie av interaksjon og kommunikasjon i matematikkundervisningen [Space for participation. A study of interaction and communication in mathematics teaching].* University of Oslo

Bodil Kleve (2007) *Mathematics teachers' interpretation of the curriculum reform, L97, in Norway.* Agder University College

Elin Reikerås (2007) *Aspects of arithmetical performance related to reading performance: A comparison of children with different levels of achievement in mathematics and reading at different age levels.* University of Stavanger

Kjersti Wæge (2007) *Elevenes motivasjon for å lære matematikk og undersøkende matematikkundervisning [Pupils' motivation to learn mathematics and investigating mathematics teaching].* NTNU, Trondheim

Anne Fyhn (2007) *Vinkler som redskap for å begripe rommet: Hvordan undervisning om vinkler kan bygge på elevenes erfaringer fra fysisk aktivitet og bevegelse [Angles as tools to comprehend space: How teaching about angles can build on pupils experiences from physical activity and movement].* University of Tromsø

Martin Carlsen (2008) *Appropriating mathematical tools through problem solving in collaborative smallgroup settings.* University of Agder

Tone Bulien (2008) *Matematikkopplevelser i lærerutdanningen: En fenomenologisk orientert narrativ analyse av studenttekster [Mathematical experiences in teacher education: a phenomenologically oriented narrative analysis of student texts].* University of Tromsø

Ingvald Erfjord (2008) *Teachers' implementation and orchestration of Cabri-use in mathematics teaching.* University of Agder.

Unni Wathne (2009) *Barns tilnærming til analogiske og kombinatoriske resonnement : en longitudinell studie av samspill i smågrupper [Children's approach to analogical and combinatoric reasoning: a longitudinal study of interaction in small groups].* University of Agder.

Claire Vaugelade Berg (2009) *Developing algebraic thinking in a community of inquiry: collaboration between three teachers and a didactician.* University of Agder

Mary Billington (2010) *Processes of instrumental genesis for teachers of mathematics.* University of Agder

Per Sigurd Hundeland (2010) *Matematikklærerens kompetanse [Mathematics teachers' competence].* University of Agder

We are aware that the production of such a list is beset by problems. We are concerned that we may have inadvertently missed scholars from the list; if this is the case, we offer our profound apologies. Second, we are aware that a list such as this is quickly out of date. Despite these issues, we believe it is worth including it because it provides readers with a sense of the range of topics in doctoral research in the field of mathematics education. It also demonstrates the expansion of mathematics education doctoral research that is being pursued within all the universities in Norway. Many of the people in this list have contributed chapters to this volume, but we think a further purpose in producing such a list is that some of these researchers will form the nucleus of the next generation of mathematics education researchers.

SUMMARY: SOME PERSONAL REFLECTIONS

In this chapter we have tried to set out the development of mathematics education research in Norway within the wider context of developments within schools, colleges, and universities. In this summary we want to try to offer readers some characterization of the nature or culture of the Norwegian mathematics education community and how this is evident through the research that is pursued. We want to note that the first author of this chapter has been part of this development. His school years were within the context of the 1939 curriculum, in rural area schools (*landsfolkeskolen*), where attendance was just three days each week, and in the other days he helped on the family farm (and did his homework!). He then trained as a teacher and qualified in the early 1960s. He was thus part of the community that was influenced by the modern mathematics movement and the traditionalist reaction and contributed to the Utstein kloster seminar that we described at the beginning of this chapter. Later, he was a mathematics teacher educator participating in the development of mathematics education; he was part of the committee that designed the constructivist-based school curriculum of 1997, and then a member of the team asked to evaluate that curriculum. He experienced first-hand the critique of a curriculum based on outcomes of an implementation that did not match the intention. In many respects, this account of the development of mathematics education has also been an account of the professional experience of Trygve Breiteig and his peers, such as Gard Brekke, Gunnar Gjone, Ole Einar Torkildsen and others whose names have not appeared in this chapter. Therefore we make it explicit that our views are not from a remote and objective standpoint. It might be asked what value it serves for us to attempt the characterization we promise. Perhaps, later, someone will consider it of interest to write an outsider's account of the development of mathematics education research in Norway; we believe there would be value in such an exercise. However, the account that we offer here reflects how it has *felt* to be part of this development and the values that have developed and now exist within our professional milieu. The feelings are most easily expressed by identifying a number of trends in the development.

First we can perceive a trend to *build communities* in learning and in research. Cooperative learning in problem solving has been the focus of several researchers, for example, by Stieg Mellin-Olsen, Hans Erik Borgersen (e.g., Borgersen, 1994, 2004) and Henrik Dahl (1994). Mellin-Olsen invited classes of school pupils to submit their common solution to open-ended tasks to the journal *Tangenten* (cf. page 15). The purpose was to challenge class communities to solve the given tasks rather than individuals. Ole Einar Torkildsen later analyzed the solutions of all contributing classes, looking for seeds of deep mathematical structures in the pupils' work (see Torkildsen, this volume Chapter 12). With a sharp mathematical eye, he uncovered evidence of interesting mathematical structures. The LCM projects funded by the Research Council of Norway (RCN) at the University of Agder (cf. page 27 and Jaworski, this volume, Chapter 5) are premised on the value of communities as the basis for learning and development. In this mutual respect between teachers and didacticians, their collaborations are based on the dynamics of learning communities with the intention of illuminating the complex problems of teaching and learning mathematics. Interest in collaborative learning is also clearly evident in the doctoral work of Maria Luiza Cestari (Cestari, 1997), Bjuland (cf. page 28), and Johnsen-Høines (cf. page 28), in which aspects of dynamics in a learning group are explored, and where ideas of communication and the "space" of language are refined, grounded theoretically, and explored (see Borgersen, Cestari, & Bjuland, this volume, Chapter 9). We can also point to the GRASSMATE project (cf. page 27) as an example of the concern for "community." In this project two quite distant cultures meet in a cooperative effort with the goal of raising research capacity in mathematics and science education. The project emphasizes the social constructive foundation of knowledge building and the dynamics of cooperation.

Especially since the Utstein kloster seminar in the mid 1970s, close collaboration and shared values between university mathematicians and mathematics educators have been evident. This leads us to perceive a trend to include a strong *mathematical perspective* in research and developmental work in mathematics education. To some degree this perspective turns the question from "How can theories of education help us to communicate mathematics?" into "How can mathematics help us to educate?" In this respect, mathematics and the history of mathematics may contribute as sources of rich ideas for development and education. This trend may be inspired by works of notable thinkers and expositors such as Felix Klein, George Polya, and Hans Freudenthal. One issue here relates to the possibility that reinvention of parts of mathematics opens opportunities for personal experience of the mathematical knowledge.

Mathematics in teacher education has played an increasingly important role for the development of mathematics education in Norway. In the years 1977–1980 the process of creating a new national curriculum for teacher education in mathematics took place. Thus seminars were arranged in 1977, with international input from the UK and the Netherlands (Hilary Shuard from Cambridge and Fred Goffree from Utrecht). We think that the *process aspect of mathematics* was focused and elaborated in these seminars. "Process"—including investigation, looking for pattern, reinvention, inquiry, mathematization—was enlightened, and started to come significantly into mathematics educational thinking in Norway. From this time this aspect of *mathematics as an activity* has been developed further and can be traced in later work by for example by Marit Johnsen-Høines and Ole Einar Torkildsen and in the KUL projects at the University of Agder.

Thus Johnsen-Høines, in her doctoral work (cf. page 28), explores the development of flexible "spaces" of language using a mathematical question as the point of departure. The question is crucial and should be mentioned here. It was a deep issue about exponential growth: the number of members in a cycle club, where each member regularly recruits more members. The problem posed was interpreted differently by two individuals, and the tension rising from different interpretations might create new knowledge about student teachers' thinking. Also Ole Einar Torkildsen, in his PhD study, takes ideas based on Pascal's (arithmetic) triangle to develop activities for small children resulting in their spontaneous physical exercise: The children arrange themselves in a pattern and each child connects to two others by a scarf in each hand. A good background in mathematics is necessary to see the potential in the children's spontaneous work and in their ways of communication their findings. Yet another example is "the vice" for developing ideas of limit, discussed by Bob Burn (Burn, 2005). Ideas from the history of mathematics, for example from calculus, are developed to inspire educational endeavor. Ideas representing milestones in development of mathematics may be extracted, milestones such as, for example, Jakob Bernoulli's *Ars conjectandi* in algebra from 1713 or Jaques P.M. Binet's formula for fibonacci numbers from 1843. Educators may look on historical issues from a more recent perspective, as did, for example, Roland J.K. Stowasser and Trygve Breiteig (Breiteig & Stowasser,1979; Stowasser & Breiteig, 1984). Such elaboration may condense a long stretch of historical development. A key issue here concerns the sources and treasures that can be found in mathematics and its history to support education in mathematics.

At the beginning of 2008 there were nine professors of mathematics education in Norway, and only three of these were Norwegian. During the course of 2008, another three professors were appointed, one of these from outside Norway. These figures do not include the professors of mathematics who make valuable contributions to mathematics education. The statistics reveal that the development of mathematics education research in Norway has been supported by the employment of foreigners; it is hoped that the most recent appointments indicate the development of the Norwegian community and a balance between Norwegian and foreign researchers. Nevertheless, we can see a clear development of a truly international community, made possible in part by the general proficiency of Norwegians in English, and the common ground of Scandinavian languages that enables effective communication. Many doctoral students choose to write their dissertation in English for a number of reasons beyond a student's personal preference. We name two: First, since the main body of literature in the field is in English, it saves the necessity of translating complex language and theoretical ideas into Norwegian. Second, by writing in English, authors are immediately making their work accessible internationally and developing skills useful for presentations in international conferences. It is, moreover, possible for non-Norwegian speakers to carry out field work in Norwegian schools as many, especially upper secondary schools, will have classes (of Norwegian students) who have opted to be taught in English. One such example of a non-Norwegian speaker researching in a Norwegian classroom is Sharada Gade, from India, (cf. page 28) who has the distinction of being awarded the first Mathematics Education PhD at the University of Agder. The international network of mathematics education researchers is very important to a relatively small community as exists in Norway.

NOTES

1. The "insider" is Trygve Breiteig, who, in 2008, retired from paid employment (not from mathematics education) after over 40 years of participation in the Norwegian Mathematics Education Community.
2. It seems that the transition to nine years compulsory education was not achieved as a sharp break from one pattern to another. Some local authorities had anticipated the changes and were prepared; others required time to adjust buildings and accommodation to the new requirements.
3. Athen, H. & Kunle, H. (Eds.). (1976). *Proceedings of the third International Congress on Mathematical Education.* Karlsruhe: Organising committee of the 3rd ICME.
4. There was no "obvious" candidate for the throne and thus a Danish prince was invited to become King Haakon VII of Norway.
5. Petter har 1 kr 10 øre. Han kjøper 10 stykker 6-øres kaker. For de pengene som blir til overs, skal han ha 2-øres kaker. Hvor mange får han?
6. We note that nearly fifty years later, the Committee of Inquiry, set up to investigate mathematics teaching in schools in the UK, made a similar observation (D.E.S., 1982, para. 342).
7. In Norway, the colon (:) is the symbol used to denote division.
8. The Norwegian members of the committee were: School Principal Torgeir Bue, Kristiansand, after September 1963 replaced by School Principal Karsten Kjelberg, Oslo; Professor Ingebrigt Johansson, Oslo; Rector Kay Piene, Pedagogisk seminar, Oslo; Rector Henrik Halvorsen, Notodden, after June 1965 replaced by lecturer Ragnar Solvang, Oslo.
9. Instituut voor Ontwikkeling van het Wiskunde, Utrecht.
10. For more on this issue and Mellin-Olsen's influence see Torkildsen, this volume, Chapter 12
11. Processanalys av Undervisning i Matematik/Psykolingvistik
12. The tasks were presented vertically with units, tens, and hundreds aligned respectively.
13. Telemark Research www.tfn.no/article/frontpage/2/
14. Centre for Teacher Education and School Service (later Department of Teacher Education and School Development—Institutt for lærerutdanning og skoleutvikling, ILS), www.ils.uio.no/english/index.html
15. Further information available at http://www.tfn.no/en/mainmenu/1219844051/1219913611 (retrieved August 5, 2009)
16. More details covering the international breadth of the study can be found at http://www.cimt.plymouth.ac.uk/projects/kassel/default.htm [retrieved July 18, 2008]
17. Hovedoppgave, the independent research element of the Nowegian *hovedfag*—this program has been replaced by master study. The two-year hovedfag is generally considered to be more demanding than master study. We use the term "extended master."
18. Visual interpretation of graphs at http://www.cimt.plymouth.ac.uk/projects/kassel/monograph/year_results.pdf [retrieved July 18, 2008]
19. The Norwegian title "sånn ja!" is a play on words, which when uttered sound very much like the name Sonja. The conference title was intended to honour the work of Sonja Kovalevskij (1850–1891), who in 1883 was appointed the world's first female professor of mathematics at the University of Stockholm.
20. The two administrative counties in the south of Norway are East and West Agder. Kristiansand is the principal town in West Agder. Before the 'Sånn ja!' conference in 1992 there had been growing contact and mutual support

between the mathematics tutors in Agder distriktshøgskole and Kristiansand lærerhøgskole.

21. For a classification and abstract of each thesis, see http://www.uia.no/no/portaler/om_universitetet/teknologi_og_realfag/matematiske_fag/mastmat5/masteroppgaver
22. Other significant contributors to developing the program were Professors of Mathematics Leiv Storesletten and Åsvald Lima and Associate Professor of Mathematics Otto B. Bekken.
23. GRASSMATE website at http://www.uib.no/people/pprsk/GRASSMATE/
24. cf. http://www.hib.no/fou/limp/index.html
25. For more information about this, see http://www.vasa.abo.fi/mathconf/

ACKNOWLEDGEMENT

The authors wish to express their sincere gratitude to Professor Gunnar Gjone, University of Oslo, for his advice in the preparation of this chapter.

REFERENCES

Alseth, B., Breiteig, T., & Brekke, G. (2003). *Endringer og utvikling ved R97 som bakgrunn for videre planlegging og justering: Matematikkfaget som kasus* [Changes and development with R97 as background for further planning and revision: The case of Mathematics]. Notodden, Norway: Telemarksforsking.

Assessment of Performance Unit. (1982). *Mathematical development. A review of monitoring in mathematics 1978 to 1982.* London: HMSO.

Athen, H. & Kunle, H. (Eds.). (1976). *Proceedings of the third International Congress on Mathematical Education.* Karlsruhe: Organising committee of the 3rd ICME.

Bekken, O.B., Borgersen, H.E., & Breiteig, T. (1978). Matematikk som undervisningsfag. Rapport fra Norsk Matematikkråds arbeidsgruppe i fagmetodikk. Fagseksjon for matematikk. [Mathematics as a teaching subject. Report from the Norwegian Mathematics Council working group in subject method. Mathematics section]. *Skrifter, 1978,* 2.

Bell, A., & Shiu, C. (1981). *Diagnostic teaching in mathematics: Part 1 of SSRC research project "Diagnostic teaching and the acquisition of general strategies in mathematics."* Nottingham, UK: Shell Centre, University of Nottingham.

Borgersen, H. E. (1994). Open ended problem solving in geometry. *Nordisk Matematikkdidaktikk, 2,* 6–35.

Borgersen, H. E. (2004). Open ended problem solving in geometry revisited. *NOMAD—Nordisk Matematikkdidaktikk, 9*(3), 35–65.

Breiteig, T., & Brekke, G. (1993). *Ansvar for egen læring i matematikk* [Responsible for own learning in mathematics]. *Report in Program for utdanningsforskning, Norges forskningsråd.* Oslo, Norway: Norges forskningsråd.

Breiteig, T., & Stowasser, R.J.K. (1979). *Problemorientert matematikkundervisning—med eksempler fra rekursjon.* [Problem-oriented mathematics teaching—with examples from recursion] *Normat, 27*(2), 56–67.

Brekke, G. (1991). *Multiplicative structures at age seven to eleven. Studies of children's conceptual development and diagnostic teaching experiments.* Unpublished doctoral dissertation, Shell Centre for Mathematical Education, University of Nottingham, UK.

Brekke, G. (1995). *Veiledning til tall og tallregning. Kartlegging av matematikkforståelse.* [Guidance on number and calculation. Charting mathematics understanding]. Oslo, Norway: Nasjonalt læremiddelsenter.

Brekke, G. Kobberstad, T., Lie, S., & Turmo, A. (1998). *Hva i all verden kan elevene i matematikk?* [What in all the world can pupils (do) in mathematics?]. Oslo, Norway: Universitetsforlaget.

Burghes, D. (1999). The Kassel Project: An international longitudinal comparative project in secondary mathematics. In B. Jaworski & D. Phillips (Eds.), *Comparing standards internationally* (pp. 135–153). Oxford: Symposium Books.

Burn, B. (2005). The Vice: Some historically inspired and proof-generated steps to limits of sequences. *Educational Studies in Mathematics 60,* 269–295.

Cestari, M. L. (1997). *Communication in mathematics classrooms: A dialogical approach.* Unpublished doctoral dissertation, University of Oslo.

Dahl, H. (1994). *Teaching small groups: Experiences at Agder College 1973–1993.* Kristiansand, Norway: Agder distriktshøgskole.

D.E.S. (1982). *Mathematics counts: Report of the Committee of Inquiry into the Teaching of Mathematics in Schools under the chairmanship of Dr. W. H. Cockcroft.* London: HMSO.

Gjone, G. (1983). *"Moderne matematikk" i skolen: Internasjonale reformbestrebelser og nasjonalt læreplanarbeid* ["Modern mathematics" in school: International reform efforts and national curriculum work]. Unpublished doctoral dissertation, University of Oslo.

Gjone, G. (1985). *"Moderne matematikk" i skolen: Internasjonale reformbestrebelser og nasjonalt læreplanarbeid.* ["Modern mathematics" in school: International reform efforts and national curriculum work]. Oslo, Norway: Universitetsforlaget.

Gjone, G. (1992). Norsk matematikkundervisning på vei mot år 2000. [Norwegian mathematics teaching towards the year 2000] In G. Emanuelson, B. Johansson & T. Lingefjärd (eds.), *Matematikämnet i internationell belysning* [Mathematics in an international context] (pp. 55–88). Institutionen för ämnesdidaktik. Rapport nr 1992:01 Gothenburg: University of Gothenburg.

Gjone, G. (2006). Den matematikkdidaktiske forskningen i Norge: Nordisk utvikling og internasjonal påvirkning [Mathematics education research in Norway: Norwegian development and international influence]. In Caspar forlag (Ed.), *Stifinneren* [The Pathfinder, published to mark the 25th anniversary of Caspar forlag] (pp. 17–31). Bergen, Norway: Caspar forlag.

Hammervold, T. & Melbye, P.E. (1980). *Regneferdighetsundersøkelse i Østfold: En registrering av baseferdighetene i matematikk i grunnskolens 4., 5. og 6. klassetrinn.* [Calculation skills investigation in Østfold: A record of basic skills in mathematics in primary school grades 4, 5 and 6]. Halden, Norway: Skrifter 5, Halden lærerhøgskole.

Hinna, K. (1996). *Elevers kunnskaper og fremgang i matematikk. En toårig undersøkelse av 300 norske ungdomskoleelever.* [Pupils' knowledge and progress in mathematics. A two-year investigation of 300 Norwegian secondary school pupils]. Hovedoppgave i matematikkdidaktikk. Kristiansand, Norway: Høgskolen i Agder.

Hofseth, J. (1950). *Barns forståelse av regneuttrykk.* [Children's understanding of calculation expressions]. Oslo: Cappelen.

Johansson, B. (1993). Diagnostic assessment in arithmetic. In M. Niss (Ed.), *Investigations into assessment in mathematics education: An ICMI study* (pp. 169–184). Dordrecht, Holland: Kluwer.

Larsson, I. (1973). Individualised mathematics teaching: Results from the IMU project in Sweden. Lund, Sweden: CWK Gleerup. Retrieved July 01, 2009, from http://www.eric.ed.gov/ERICDocs/data/ericdocs2sql/content_storage_01/0000019b/80/38/bc/4b.pdf.

Mandt, H. M. (2006). En kvalitativ tilnærming til Matematikk 1-kurset i allmennlærerutdanningen høsten 2005 [A qualitative approach to the (first) course in mathematics in general teacher education autumn 2005]. Masteroppgave i matematikkdidaktikk [Masters dissertation in mathematics education]. Universitetet i Bergen, Det matematisk-naturvitenskapelige fakultet. Retrieved August 5, 2009, from https://bora.uib.no/bitstream/1956/1867/1/Masteroppgave-Mandt.pdf

Melbye, P.E., & Hammervoll, T. (1990). *Regneferdighetsundersøkelse i Østfold–10 år etter.* [Calculation skills investigation in Østfold – 10 years on]. Halden, Norway: Skrifter 4, Halden lærerhøgskole.

Mellin-Olsen, S. (1972). *Mathematics for the "slow learners."* Bergen, Norway: University of Bergen.

Mellin-Olsen, S. (1987). *The politics of mathematics education.* Dordrecht, Holland: Reidel.

Mikalsen, Ø. & Julie, C. (2008). *GRASSMATE: The story of a competence building programme in Southern Africa.* Bergen, Norway: University of Bergen.

Mosvold, R. (2006). *Mathematics in everyday life: A study of Norwegian teachers' beliefs and actions concerning the connection with mathematics and everyday life.* Unpublished doctoral dissertation, University of Bergen, Norway.

Niss, M. (2004). The Danish "KOM" project and possible consequences for teacher education. In R. Strässer, G. Brandell, B. Grevholm, & O. Helenius (Eds.), *Educating for the future: Proceedings of an international symposium on mathematics teacher education* (pp. 179–190). Göteborg, Sweden: NCM, University of Gothenburg.

Niss, M. & Jensen, T.H. (Eds.). (2002). *Kompetanser og matematiklæring: Ideer og inspirasjon til udvikling af matematikundervisning i Danmark.* [Competence and learning mathematics: Ideas and inspiration for the development of mathematics teaching]. Copenhagen, Denmark: Undervisningsministeriet.

Norsk lektorlag og Norsk lærerlags felles fagkomite for matematikk. (1970). Innstilling. [Report]. *Vår skole 56*(4), 146–166.

NOU. (1973). *Matematikk i grunnskolen* [Mathematics in basic school]. Norges offentlige utredninger, *1973*, 33 [Norway's Public Report *1973*, 33].

NU. (1967). *Nordisk skolmatematik.* [Nordic school mathematics]. Stockholm: Nordisk udredningsserie.

OEEC. (1961). *New thinking in school mathematics.* Paris: Organization for European Economic Co-operation (Now OECD).

Ribsskog, B. (1935). *Regning. In the series: Arbeidsmåten i folkeskolen. Håndbok for lærere.* [Calculation. In the series: The work style in the public school] Oslo, Norway: Gyldendal norsk forlag.

Ribsskog, B. (1936). *Standpunktprøver i regning.* [Achievement tests in calculation]. Oslo: Gyldendal norsk forlag.

Ribsskog, B. & Aall, A. (1936). *Undervisningsplanene i folkeskolen. Et pedagogisk-psykologisk forarbeide til de nye planer for orienteringsfagene.* [Curriculum in basic schools. A pedagogical-psychological base for the new plans for social studies and natural science]. Oslo: Gyldendal norsk forlag.

Ribsskog, B. & Wiborg, N. (Eds.). (1932). *Arbeidsmåten i folkeskolen: Tegning og Skrivning. Regning.* [Approaches in basic school: Drawing, writing and calculation]. Oslo: Gyldendal norsk forlag.

Skaalvik, E.M. & Valås, H. (1999). Relations among achievement, self-concept, and motivation in mathematics and language arts: A longitudinal study. *Journal of Experimental Education, 67*, 135–149.

Skemp, R. R. (1976). Relational understanding and instrumental understanding. *Mathematics Teaching, 77*, 20–26.

Slammert, L. (1996). Tribute to Stieg Mellin-Olsen (1939-1995). In T. Kjærgård., A. Kvamme, & N. Lindén (Eds.), *Numeracy, race, gender and class: Proceedings of The Third International Conference on the Political Dimensions of Mathematics Education – PDME III* (pp. 9–12). Bergen, Norway: Caspar.

Stowasser, R.J.K. & Breiteig, T. (1984). An idea from Jakob Bernoulli for the teaching of Algebra: A challenge for the interested pupil. *For the Learning of Mathematics, 4*(3), 30–38.

Sånn, ja! (1993). *Rapport fra en konferanse om matematikk-didatikk og kvinner i matematiske fag, Kristiansand 8.-10. mai 1992* [Such, yes! Report from a conference on mathematics education and women in mathematics, Kristiansand 8–10 May, 1992]. Oslo: Norges forskningsråd, avd. NAVF, Sekretariatet for kvinneforskning.

Valås, H. (1977). *Struktur og enhet i skolematematikken for junior high school. Noen sider ved det praktisk-pedagogiske utviklingsarbeid i skolematematikken de siste to decennier belyst ved tre reformforslag. Rapport nr 1 fra forskningsprosjektet Faglig kunnskap og individuell og sosial relevans, NAVF B75.10-027.* [Structure and unity in school mathematics for junior high school. Some pages with practical-pedagogical development work in school mathematics the last decade revealed with three reforms. Report no.1 from the research project Subject knowledge and individual and social relevance]. Oslo: NAVF/NFR.

Våge, J. (1979). En undersøkelse og vurdering av oppgaver og besvarelser fra grunnskolens avgangseksamen i matematikk 1977. [An investigation and evaluation of tasks and responses from the compulsory schools' leaving examination in mathematics 1977]. *Skoleforum, 5*, FF14–FF18.

Wigforss, F. (1930). *Inträdesprovet i räkning till den fyraåriga realskolan.* [The entrance exam in calculation for the four year middle school]. *Skola och Samhälle, 11*(3–4), 174–181.

Wigforss, F. (1939). *Inträdesprov eller ej?* [Should there be an entrance exam or not?]. *Skola och Samhälle, 20*(3), 90–107.

3

THE CONTRIBUTION OF STIEG MELLIN-OLSEN TO MATHEMATICS EDUCATION: AN INTERNATIONAL LUMINARY

A Will to Explore the Field, an Ability to Do It

Christop Kirfel
UNIVERSITY OF BERGEN

Nora Lindén
BERGEN UNIVERSITY COLLEGE

Abstract: Stieg Mellin-Olsen lived and worked in Bergen, Norway most of his life. His academic work included teaching, research, and publication in the field of mathematics education in school, at Bergen College of Education and Bergen University, where he held a post as professor of education starting in 1990. He explored activity theory as a theoretical means to understand the field of teaching and learning mathematics. Mellin-Olsen believed in the significance of activity and could see little value in a teaching practice where everything was told by the teacher and nothing discovered by the learners themselves. His work on pupils' rationales for learning was an example of his belief in pupils' own resources.

Keywords: Activity theory, Ethnomathematics, Mathematics as language, Mathematics as a tool, Rationales for learning

STIEG THEODOR MELLIN-OLSEN 1939–1995: INTRODUCTION

When Stieg Mellin-Olsen was preparing one of his most important books, *The Politics of Mathematics Education* (1987), the political situation in the world was characterized by a number of ongoing struggles. In the streets of Europe there were battles and protests, in South Africa the liberation process was going on with its brutal scenes, the cold war held large parts of the world in its grip, and through television and newspapers all of this was continuously in our consciousness. Many developing countries were still suffering from the consequences of colonization. The struggle for civil rights and resistance against oppression and hegemony were—not unlike today—crucial ingredients of the ongoing political debate.

Mellin-Olsen worked as a mathematics teacher at different levels and observed first-hand how many young people did not manage to learn mathematics in school, in spite of being able to cope with the mathematics embedded in the activities they met in their own life situations outside school. He recognized that the major concepts that can be used to categorize political

The First Sourcebook on Nordic Research in Mathematics Education, pages 35–47

situations also fit the domain of schooling. Hence, notions such as oppression, resistance, or civil rights can be used to understand better the situations many pupils experience in school. Moreover, they could be helpful in explaining the difficulties pupils experience and show a way out of their constrained situations. For Mellin-Olsen, good mathematics education included an element of liberation and thereby had a political dimension.

In this chapter, the authors present a short outline of Stieg Mellin-Olsen's experiences as a teacher in secondary school in Bergen in the 1960s, followed by an exposition of his activity in higher education at the University of Bergen and the Teacher's College in Bergen. His theoretical interest and academic activity in the field of mathematics education are major issues throughout the chapter. The authors' intention is to describe how Mellin-Olsen's ideas of mathematics education and teaching led him into research in which the interests and development of pupils and students were guiding principles.

A Short Biography

Stieg Mellin-Olsen was born in Bergen, Norway, April 30, 1939. He spent his first school years in a small school in a suburb of the city before his family moved closer to the city center, where he finished upper secondary school. Following compulsory military service, he studied mathematics and geophysics at the University of Bergen, after which he went to Oslo to study for a master's degree in mathematics (see Table 3.1). In 1965 he returned to Bergen. He had a genuine desire to teach, and his prevailing aim, he often remarked, was to let pupils and students *discover* the beauty of mathematics for themselves and to *explore* the importance, relevance, and usefulness of mathematics in their lives. He perceived the mathematics taught in school as traditional with respect to content and teaching methods, notwithstanding a number of experimental curricula at the time (see Breiteig and Goodchild, this volume, Chapter 2). In 1978, with Ragnar Solvang, Mellin-Olsen wrote *Matematikk fagmetodikk*, (Mathematics Education, subject didactics) in which they describe the mathematics curriculum for schools in Norway from 1960 as "Totally traditional."

TABLE 3.1 Important Mileposts in the Academic Life of Stieg Mellin-Olsen

1964	Master's degree in pure mathematics
1965–1975	Teacher posts at different elementary and secondary schools in the Bergen area
1967	Part time lecturer at the University of Bergen
1973	University lecturer at University of Bergen (full time)
1977	Associate professor at Bergen College of Education
1991–1995	Professor at the Institute of Practical Education (University of Bergen)

STARTING RESEARCH

Mellin-Olsen began to engage in mathematics education research while he was teaching in school. Together with Rolf Rasmussen he embarked on a study related to "the views of parents about the schooling experiences of underachieving pupils." The report (Mellin-Olsen & Rasmussen, 1975) was titled "Skolens Vold" (The Violence of School). The title is significant in that it does not pathologize the pupils or their parents, but draws attention to the structure of schooling as contributing to the unsatisfactory experiences some learners have with schooling. A major finding of the study was that schooling was alienating itself from pupils and parents in a way that left parents feeling that their support given to their children did not satisfy the demands of the school and was not wanted or appreciated by the school authorities. The study brought to attention the importance of listening to the voices of pupils and their parents when seeking to understand pupils' experiences of schooling. This influenced Mellin-Olsen's search for explanation and understanding of pupils' difficulties in learning in school throughout the remainder of his academic life.

Mellin-Olsen believed in the significance of activity and could see little of value in a teaching practice where everything was told by the teacher and nothing discovered by the learners themselves. When he talked about mathematics it was by emphasizing knowledge as a tool, not an aim, even in abstract mathematics itself. In this respect he saw great value in the work of the Russian scholars A. Leont'ev and L. Vygotsky, which was beginning to make an impact on thinking in the west following the translation into English of Vygotsky's *Thought and Language* in 1962, and *Mind in Society* in 1978. Studying activity theory as a theoretical approach gave Mellin-Olsen more confidence in his search for support for his own ideas. Mathematics education was at that time a young field of inquiry, and therefore it was natural for Mellin-Olsen to look for support from other academic disciplines that were not immediately connected with mathematics teaching and learning.

One of Stieg Mellin-Olsen's main contributions to the field of mathematics education is the further development of activity theory. True enough, there are a number of theories that put human activity into the center of the didactic debate today. Mellin-Olsen started exploring activity theory, but he pursued his inquiry critically, which led to his own interpretation and contribution:

"Activity will refer to actions emerging from the individual's own motivation." (Mellin-Olsen, 1987, p. 30).

The novelty in Mellin-Olsen's interpretation of activity is its recognition of the importance of activity in the life of the individual and society, and hence of the learning process in its own right. In activity theory, as it first was developed by Leont'ev; "activity" describes the total life of an individual within a society. Activity explains the individuals' motivation for engaging in an enterprise. The individual may be excluded from certain activities, and she or he may fight against the structures that cause the exclusion. Sometimes an individual does not manage to learn the content in a school subject like mathematics. According to Mellin-Olsen, this might be due to the individual's lack of ability to see the relevance of the content for her or his own activities. Therefore, an outcome from an earlier learning experience may be decisive for how relevant the learner finds new content in a lesson.

The relationship between knowledge acquired inside or outside school has to be interpreted dialectically. Mellin-Olsen claimed that experiences with earlier learning episodes could, over time, be crucial for how the pupil will meet a new learning episode. In the same way, the new theme will be important for the pupil's evaluation of the whole experience. Therefore, the teacher's task is to create situations where pupils experience the relevance of the subject matter for their own lives and possibly perceive how this content may change their lives. In his teaching and research, Mellin-Olsen saw this close relationship between knowledge already acquired and new learning as decisive for creating a productive classroom environment. Mellin-Olsen describes a relevant dilemma in this respect in the project described below.

THE FLESLAND PROJECT: A MATTER OF LEARNING IN SCHOOL OR OUTSIDE SCHOOL

From the inner city school, Mellin-Olsen moved to a position in a secondary school in a rural area just outside Bergen in 1967, and at the same time was also appointed to a part-time position as a lecturer in mathematics education at the University of Bergen. In his new positions he was able to combine school teaching and research. Coincidentally, these positions gave him the opportunity to use a current and highly pertinent local episode to develop his ideas based on activity theory within his own teaching.

Some of the farmland in the area adjacent to the school in which he was working was to be expropriated to develop a new airport for Bergen. The pupils he taught were to suffer a reduction of their own playing space, and there would be more traffic as well as a lot of noise from planes. Their parents, many of whom were farmers in the area, were to lose land and consequently have reduced incomes. Mellin-Olsen saw the plans as a useful theme for teaching. The issue was relevant for people living in the area. The ongoing building of the airport could be seen and heard by everyone, and Mellin-Olsen noticed that the pupils´ experiences were discussed in breaks and out-of-classroom time. Over the course of two months in 1975, the Flesland project was taken as point of departure in mathematics and natural science classes.

Pupils worked with measurement, speed, and different algebraic and geometric tools, and some of the teaching was conducted outside the classroom to link the mathematics to what was happening in "real life," outside of school. After the teaching period was completed, Mellin-Olsen saw that very little engagement and visible learning of mathematics among the pupils had resulted, and he sought an explanation. He later said that he expected success from this project, but met failure.

One reason could be, according to Mellin-Olsen, that the issues were presented as adults' problem statements. The pupils did not see school as part of their environment. They could not engage in the discussion about the effect the airport would have on them and relate it to school mathematics. Another possible reason, Mellin-Olsen claims, was that the issue was not introduced as a class topic by the pupils themselves. To them the development of the airport was an issue discussed in their free time. School was just academic and separated from the other events going on in their lives. School was a way of getting good marks for a good future job, not a tool for critical analysis of the environment (Mellin-Olsen, 1975b).

After nearly 10 years of teaching in secondary school, in 1973 Mellin-Olsen accepted a position as a full-time assistant lecturer at the University of Bergen, where he was responsible for teaching students how to teach mathematics following completion of their bachelor's or master's degree. This was a finishing course before starting teaching in school. In this position he combined his deep knowledge of mathematics with experiences and knowledge derived from his years of teaching pupils in school as well as his research in the field of mathematics education.

RATIONALES FOR LEARNING

Arising from his observations during the Flesland-project, Mellin-Olsen (1987) identified two kinds of rationales that pupils might hold, which are quite decisive for their learning:

> Two major rationales for learning can be identified as drives for school learning.
>
> It is the rationale which is related to school's influence on the future of the pupil, by the formal qualifications it can contribute. This role as an *instrument* for the pupil will provide the pupil with an *instrumental rationale* (I-rationale). In its purest form the I-rationale will tell the pupil that she/he has to learn, because it will pay out in terms of marks, exams, certificates and later a successful job.
>
> On the other hand there is a rationale which relates to knowledge as such, saying that this knowledge has a value besides its importance for the external examination. It is a rationale which says that knowledge has an importance beyond its status as school knowledge. I call this rationale the S-rationale. (Mellin-Olsen, 1987, pp. 157–158)

Mellin-Olsen saw Mead's theory as useful in order to describe the social-psychological aspect of activity theory, especially when it comes to rationales for learning. Mead (1934) claims that the individual uses society and the groups around herself/himself as a mirror in his own development in such a way that the information the individual has about herself/himself stems from the others' reactions. This feedback that the individual receives upon her or his own conduct is decisive for her or his self-image and development. Mead calls people or organizations whom the individual meets "the Generalized Others" (GO). School is one of these GOs. The fact that our social surroundings have some influence upon us might be considered to be common sense, but to claim that the reactions we receive upon our own behavior from persons and groups around us are to constitute our own self-esteem and the image we have of our own person is much stronger. The behavior of an individual is not random action, but always built upon some kind of logic or rationale. The construction of such a rationale is strongly dependent on the impact of the different GOs to which the individual relates. This relates strongly to the I- and S-rationales outlined above. Having these considerations by Mead in mind, we continue the above quotation from Stieg Mellin-Olsen (1987, pp. 157–158).

> The S stands for "social," indicating that such an evaluation of knowledge is made by the individual as a social subject, that is, by the individual through a reference to his GOs, which goes beyond the I-rationale. Obviously it is not the case that either the I- or the S-rationale will work as a drive for learning. Usually they will work together, and the pupil's rationale for learning may be seen as a resultant of the I and S: knowledge is regarded important by the pupil for external examinations and for itself. (Mellin-Olsen, 1987, pp. 157–158)

Understanding which rationale is dominant in a specific pupil may help the teacher when the pupil does not learn. The teacher has to find the pupils' rationale for learning and build the teaching situation on that foundation. Mellin-Olsen looked upon the pupils' rationales as dynamic. An instrumental rationale in one field may be transformed into an S-rationale when suddenly the pupil's eyes are opened for the significance the subject has for the pupil's own activities. It might be challenging for the teacher when a pupil reveals different rationales for his learning at the same time. On the other hand, the different rationales refer to both the individual and to the surrounding society. They show how an individual can relate to society and possibly engage in problems in her or his own surroundings that need to be challenged. In that sense, the two rationales also have a political dimension. Mellin-Olsen later presented a wider understanding of the theory of rationales.

Mellin-Olsen warned against instrumental knowledge developed through practice without an engagement and rationale for learning. Pupils memorizing mathematical rules without seeing the learning as important in their lives he called instrumental learning. He saw instrumental learning as a reason for pupils not being able to learn mathematics. This engagement in researching the issue of teaching mathematics led him to cooperation with the English scholar Richard R. Skemp in Manchester.[1]

In 1973, Mellin-Olsen and Skemp carried out an investigation of instrumentalism in a Norwegian secondary school and an English grammar school. They found that instrumental learning was more common among the pupils tested in Norway than in England. The tests contained 20 mathematical problems where the pupils had to choose what explanation was convincing for them. Some explanations were instrumental; others, it appeared, emerged from an S-rationale. During this research Mellin-Olsen identified the two rationales, mentioned above, seen as drives for school learning.

From the study Mellin-Olsen and Skemp concluded that there were certain important factors that might contribute towards the reduction of instrumentalism in pupils' learning. Some of these factors are

- Constructive teaching approach
- Syllabus linked to the pupils' own culture of knowledge

- Certain degree of freedom for the teachers in their choice of teaching methods (Mellin-Olsen, 1976a)

The theory of rationales for learning has had an impact on later scholars in the field of mathematics education. One example is the work of Simon Goodchild, who set out to expose the incidences of I- and S-rationales in an English secondary (year 10) class. His work, first described in his PhD thesis (Goodchild, 1997), also presents a third rationale, which he labels a P-rationale. When a student accepts to engage in the classroom activity presented for the class according to an understanding of her or his role in the class, she or he possesses a P-rationale. According to Goodchild, a P-rationale is not a rationale for learning. It is a rationale for engaging in activity because that is what is expected in the classroom (Goodchild, 2001), Kjersti Wæge makes strong references to the rationales mentioned above in her PhD thesis (see also Wæge, this volume, Chapter 16).

An Early Norwegian Critique of Piaget's Constructivism

In 1977, Mellin-Olsen wrote the book *Læring som Sosial Prosess* (Learning as a social process). In this he discussed reasons why clever pupils often lack interest in the issues to be learned in school. Here he applies Bateson's theory as a tool for understanding the reason pupils fail to learn, in addition to using Piagetian theory as an introduction to a wider theory of knowledge. Piaget connects the process of learning with the notion of logical operational thinking. The individual constructs her or his knowledge through operations. By trying out, moving things around, putting bits together, the child learns about things and about her/himself. Knowledge is constructed during the process in which the individual is actively exploring.

Mellin-Olsen perceived a number of deficits in Piaget's exclusive focus on practical actions (operations). The psychological component, the meaning, the motivation, and the aim and relevance are neglected. The main emphasis of Piaget (or maybe the Piagetian tradition) is on cognition and cognitive development. Later research criticizes Piaget on this point and puts his findings into an extended perspective when taking the dimension of meaning into account. Pupils do not act and think only *in* the situation they are put in, they also act and think *about* the situation, by means of knowledge about the situation. According to Mellin-Olsen, Piaget does not put sufficient emphasis on this sociological aspect in his theory of learning and on epistemology. Mellin-Olsen was one of the early Norwegian critics of the Piagetian approach, and he often felt he stood alone with these critical ideas (Mellin-Olsen, 1977).

In *The Politics of Mathematics Education* (1987) Mellin-Olsen wrote the following critique on Piaget's epistemology:

> I miss the pupil's evaluation of the kind of learning situation she is confronted with. I miss recognition of the fact that she is in a position to reject the kind of activity she is invited to participate in. I miss the relationship between her own judgement of the educational situation she is part of, how this affects her learning behaviour, and how all this relates to the design of the educational situation.
>
> Such a relationship is dialectic between two levels: the judgement of the particular learning task she is confronted with and her meta-knowledge about the subject, telling her that the tasks she is usually confronted with prove to be valuable to her (or not). These two levels influence each other dialectically. (Mellin-Olsen, 1987, p. 20)

LANGUAGE AS TOOL

Another important contribution of Vygotsky that Stieg Mellin-Olsen investigated in his research is the concept of language as a tool. Vygotsky considers language as a tool to solve problems, in addition to the communicative role language apparently has. In the same manner, other sets of signs, such as mathematical symbols, can be interpreted as tools. The important thing is that the child discovers this function of language and mathematics; that mathematics is a tool that may be used in order to follow one's own objectives, and that may support one in one's personal activity. This discovery may only occur when the pupil himself sees the functionality of the tool in connection with his own motivation. Traditional mathematics education has focused on providing pupils with a continuous chain of new tools during their years at school. The pupil puts these tools on her intellectual "shelf" as decoration. Nevertheless, an important aspect is usually forgotten in this respect. If the pupil does not discover the functionality, the usefulness of the tool, the whole remains an amusing collection of knick-knacks. This point of view has a number of consequences for teaching, especially about the way a teacher needs to serve pupils who do not cope with mathematics. According to Vygotsky, the explanation for a pupil's failure to cope is that the pupil in question does not manage to connect her or his own activities to the educational situation. The teacher's task is therefore to build the connection between the pupil's wish for knowledge and the teacher's knowledge as it is presented in her or his teaching.

In developing these ideas of Vygotsky, Mellin-Olsen uses Bateson's theory of double bind (Bateson, 1972). In education, the double bind occurs when the pupils are not able to distinguish between the teacher's expectations of them and the pupils' possibilities of developing the knowledge that the teacher wants. This is a contradiction between the "private" understanding of the pupil, and the "social" demands experienced by the teacher. A possible solution to this tension could be, according to Mellin-Olsen (1987), to help the pupil develop a meta-knowledge about the situation, such that she or he can see, and accept, that her or his functional language exists in relation to the standard language of the teacher.

Communication and the learning situation have to be taken into account, not only the mathematical content in the learning situation. This became a vital part of Mellin-Olsen's work. He asserts that language can lead and sometimes determine human activity. He saw that a pupil in her language development moves from the audible voice with which the pupil uses to solve problems to the inner voice, an internalization of the outer voice: thought. Vygotsky calls the communicative activity in a group of people interpersonal activity. This is the point of departure for what the pupil internalizes and what develops to the intrapersonal activity: thinking. Internalization is the inner reconstruction of the outer behavior.

The major tool in this process is language. In the same manner, "meaning" is transferred. In the beginning meaning is included in the activity of the group. There it exists in some objective manner. Then it is internalized by the pupil through activity and interpreted by language and transformed into a subjective meaning. We can think of a mathematical activity where the pupils use scissors in order to produce some shapes. The common meaning is the product to be cut out, using the scissors. The knowledge and the meaning in the activity are communicated by language and the activity itself. In the end both of them are internalized in the child's mind: The child has acquired new knowledge about geometrical shapes (Mellin-Olsen, 1987).

MATHEMATICS AND LANGUAGE

When a teacher starts teaching a new subject, she or he often experiences that the pupils are in possession of considerable knowledge about the subject, but that they have difficulties in expressing this knowledge in standard terms and they may not be able to use it. The pupils' knowledge may be "coded" in a different way from what the world of school expects and therefore it is not acknowledged and appreciated. It shows that language plays a bigger role than many have assumed, and that quite a big part of the mathematical learning process is a piece of translation work.

Finger counting is quite common among small children, and when a child develops her or his own written number system, the symbols are often similar to fingers, coins, or stones that might be the starting point. The crucial thing in this process—according to Mellin-Olsen—is not the content of a specific symbol, but the understanding that it is possible to recall, remember, and communicate by the symbol. The pupil can invent the symbols herself and discover that the symbols together form a system. The pupils' drawings may therefore become a vital tool in the mathematics lessons. The drawings may become dynamic—somebody can enter the picture, and somebody can leave it again. This gives the possibility of language training. In that manner one can create situations where mathematical concepts of addition and subtraction are necessary and useful before one starts with the written algorithm. This kind of work with mathematical concepts is necessary in order to be able to learn the algorithm later. It was on these ideas that Mellin-Olsen inspired his students to develop and try out methods for teaching mathematics in primary school.

ETHNOMATHEMATICS

Many young people at the age of 12–14 have difficulties with mathematics in school. According to a Piagetian approach, one would say that formal operational thinking is not yet developed in these pupils. At the same time, we see that the same pupils manage similar tasks when occurring in their own lived world, according to Stieg Mellin-Olsen. Strategic thinking, games, gambling, building activities, stability questions, buying, and selling may be some examples here of pupils' tasks from their own lived world.

Stieg Mellin-Olsen argues that an explanation following a Piagetian tradition is not satisfactory. Piaget himself was aware of this and he mentions the possibility that culture may play a more decisive role. Alan Bishop's research among the Papua New Guinean people (Bishop, 1979) supports these ideas. In Papua New Guinea, the problem with those pupils who did not cope with school was not necessarily due to lack of knowledge but to big differences in culture and language. There is a large distance between school culture and the culture in which the pupils live. The question is how to build bridges between these two worlds. Here, ethnomathematics has a contribution to make. Ethnomathematics projects investigate and set out to discern the mathematics embedded in a culture, the artefacts and techniques that are common in this culture, and make them an issue in education at school. In this way one can show the mathematics in these objects to the pupils and develop mathematical notions and concepts on that foundation. This claims knowledge about the culture, and the teacher has to dive into the young people's culture in order to find fitting objects and scenes where the young people's world gives them the opportunity of mathematical analysis.

Culturally related objects or techniques often involve special activities, and perhaps some crafting activity (weaving, plaiting, decorating patterns, games, origami) becomes the starting point of the analysis. Hence the mathematical concept does not occur alone as an abstraction but is woven into a cultural context that creates meaning—and thereby it is hoped—motivation for the pupils. The pupil sees that the knowledge he already possesses is recognized and appreciated—a first step in a direction of liberating knowledge.

In the 1960s and 1970s, the first ethnomathematical projects were published. At that time studies from developing countries were predominant. But soon researchers realised that the idea could be transferred to the western world. Stieg Mellin-Olsen and his students carried out research on ethnomathematical phenomena in the Norwegian folk culture and showed how deeply rooted and known patterns and techniques (Figure 3.1), such as knitting, embroidery, and carving, could give rise to meaningful mathematics education (Johnsen-Høines & Mellin-Olsen, 1985).

An ethnomathematical approach to education does not only comprise a new selection of themes and objects for the mathematics lessons but also a new role for the teacher as a kind of anthropologist who is eager to study the pupils' world and who seeks out activities that might build a bridge between the world of school, where the subjects are in focus, and the real-life world, where meaning is obvious. The teacher becomes a social anthropologist. Education becomes bridging and translation. (For example see Fyhn, this volume, Chapter 13).

Figure 3.1 Knitted mitten. Eight leaves roses.

Mellin-Olsen worked for many years together with Paulus Gerdes from Mozambique, investigating cultural artefacts as didactic tools in mathematics. They both developed and used ethnomathematics in exactly that sense in Mozambique. Gerdes studied a number of games and craft techniques in Southern Africa, uncovered their mathematical aspects, and incorporated them into the curriculum. His numerous books on ethnomathematics are widely distributed. As early as the 1980s, when the idea of ethnomathematics still was not fully recognized, Gerdes and Mellin-Olsen met and discussed this topic. There was a cordial contact between Mellin-Olsen and Gerdes. They arranged conferences together and were involved in the same program for PhD students from Southern Africa, which Mellin-Olsen led until his death.

MATHEMATICS AND PSYCHOLOGY

Mellin Olsen also shows how Bateson's theory (1972) about meta-learning may shed light on some aspects of pupils' activity during the learning process. Meta-learning is learning about learning, containing the context of the learning situation among other aspects. Knowledge about the learning situation is often neglected when we teach mathematics, according to Stieg Mellin-Olsen. For example, it is practicable to extend the notion of addition and subtraction from integers to fractions with common denominator. But a similar extension of the conception of multiplication will represent a serious shift of context that the teacher often forgets to work on consciously. The pupil may rather quickly give up. There are other forms of context learning, too. Mellin-Olsen describes the following situation: If a pupil does not perform well in mathematics and this situation continues over time and she or he is confronted with the bad results, the context of mathematics education as an arena of defeat becomes a part of the pupil's meta-learning. If such a pupil is to receive support, it is crucial to take this meta-knowledge as a point of departure and arrange the educational situation on new premises. This may happen outside the classroom, without textbooks, such that the pupil's meta-knowledge about herself as a learner and about mathematics education does not destroy her learning potential in the new situation.

DOUBLE BIND

Mellin Olsen continued his theoretical investigation in engaging with Bateson's notion of double bind (1987). It is interesting to reflect on the fact that it is in the same year that Engeström's book *Learning by Expanding* (1987) was published, where Engeström refers to the double bind in the context of activity theory. Both *Learning by Expanding* and *the Politics of Mathematics Education* have had an impact on later research in the field of understanding learning. Contradictions mediated in the same communication process and which do not become conscious for the recipient may create a double bind. This may—as one can see in psychoanalysis—create distortions and misbehavior. Mellin-Olsen (1991a) adapted this concept of double bind to the educational situation, especially mathematics education. The contradictory messages pupils receive about the relevance and importance of mathematics may create such an unconscious double bind, which may lead to bad results and a blockage of the learning process. One of these double communications may arise from the focus on the examinations in the subject on the one hand, and use of more communicative learning forms on the other hand. He also refers to Brousseau (1984), who takes this concept even further and develops the notion of the *didactical contract,* which both the pupil and the teacher are woven into without being able to negotiate it. The contract is given in advance by the school and the structures around. If genuine learning is to occur—according to Brousseau—the contract has to be broken. Mellin-Olsen claims that not only the contract in the didactical situation in the classroom has to be broken, but also the contract on a higher level concerning the subject content and organization of school in society.

MATHEMATICS AND POWER

Mellin-Olsen was always interested in the relationship between power and learning. He built his own ideas on the principles expounded in Foucault's (1985) account of discourse in mathematics education. A discourse is a structure or network of meanings. The discourse mediates both a message and inclusion into a system of values belonging to a social group. Mathematics education has got its own discourse, which again is part of the larger discourse of schooling and education. The language of mathematics teachers and their thoughts about their own teaching was an important area of research for Stieg Mellin-Olsen. He interviewed a number of mathematics teachers and tried to reconstruct vital elements in their discourse (Mellin-Olsen, 1991a,b). One of these elements is the use of the metaphor of travelling, of driving, of speed and velocity. Education is perceived as a journey, a trip, a tour.

Distance (number of pages in the textbook) and time become crucial variables and the decisive vehicle. The exam is the final aim for the trip. This fact influences a great number of other aspects of learning, as for example differentiation, assessment, and the teachers' image of the pupils. Several of the teachers in the study expressed feelings of being trapped in this language cage, and some of them had ideas of an alternative discourse such as project work or open problems, which they were never given the chance to try out.

It is not easy to break free from a discourse. It becomes a part of us. At the same time, the discourse has a great power and influence over those who live within it, and it raises barriers against renewal and change. Therefore it was important for Stieg Mellin-Olsen to uncover the discourse of mathematics education, show the weaknesses and limitations, and, at the same time, suggest an alternative discourse that might lead to a wider understanding of the subject and the activities belonging to it. One alternative that Mellin-Olsen described is to take pupils' interest and experience as a point of departure for the teaching of mathematics. In that way the traditional discourse of mediating knowledge from teacher to pupil might be broken (Mellin-Olsen, 1991a,b).

MATHEMATICS AND DIALOGUE

Mellin-Olsen emphasized the use of dialogue as a tool to develop and handle various forms of knowledge. His contribution at the Political Dimensions of Mathematics Education conference in Johannesburg in 1993 focused on dialogue as a teaching tool, and a tool for understanding the activity in the classroom (Mellin-Olsen, 1993a). Two years earlier he presented his research on teachers' thinking in the book about how teachers think about mathematics teaching (Mellin-Olsen, 1991b), where he applied interview as a research method. Mellin-Olsen had almost finished a book about qualitative research methods when he died. The book was published post mortem in 1995.

MATHEMATICS AND POLITICS

Mellin-Olsen was deeply inspired by Paulo Freire, who worked on developing a pedagogy of liberation in

Brazil, in which, at the time, some of the world's poorest and most oppressed peoples lived, where one of the major problems was illiteracy. Freire (1975) developed his theory about *conscientisation*. He developed his theoretical framework in a country where most of the population suffered heavy oppression. Freire distinguished between man's *integration* and *adaptation* to society and relates these two processes to respectively a *subject* or an *object* in society. To the extent that man is not prevented from acting in his world, from choosing and making decisions about his actions, he is in a position where he can both adapt to society and transform this reality. He is then integrated into society. If not, he is subjected to the choices of others: his decisions do not belong to himself but result from external prescriptions. In this case he is only adapted to society, being objectified in it (Mellin-Olsen, 1972, 1975).

In his work Mellin-Olsen adapts the ideas of Freire that education of illiterate people is not only about reading and writing but also about liberation, human dignity, participation in society, and the possibility to change it. Mellin-Olsen then transfers parts of Freire's work to mathematics, a field he looks at as a crucial tool for participation in and change of society as well. At the same time he criticizes Freire for not going far enough. Conscientisation, critical awareness and knowing about oppression and liberating forces are not enough. The liberation process itself, the action, belongs to the learning process. We easily see this wish in Stieg Mellin-Olsen's school projects. The Flesland-project (see above) may serve as a typical example that enlightens this aspect of the thinking of Mellin-Olsen:

> Both Giroux and Freire ideologically regard Activity as a necessary and desirable step following a critical awareness of society. In theory they do not see this as a task for the school to initiate. One problem of such a position for the teacher in the classroom is a problem of *faith*: 'See Teach, all this talk, it is certainly right, but what then?' It is hard to see how pupils who may be in a difficult life situation can have faith in their future curriculum if their meta concept of it tells them that school helps them to understand what is wrong, but does not stand up for them in a common attempt to challenge the difficulties. A call for a critical awareness excluding action may be interpreted as just another discussion, which will have been heard too many times before. There is a profound metalearning in young people who are fed up with those who can talk about society but do no more than that. (Mellin-Olsen, 1987, p. 203)

> My paradigm for building a mathematical education including these pupils will be to start with their resistance, taking advantage of it, rather than disregarding it or moralising. It is in this way that I *politicise* mathematics education. (Mellin-Olsen, 1987, p. 190)

BUILDING NETWORKS

Coming from a small research community, Stieg Mellin-Olsen saw the necessity of collaboration with colleagues nationally as well as internationally. Through his inspiring and engaging attitude he succeeded to support teachers and colleagues he met through his work in a way that made them successful in their own projects.

In 1977 Mellin-Olsen left the university for a post as associate professor at Bergen College of Education. At that time, mathematics was taught very traditionally with respect to both content and method, with courses only preparing the teacher students for whole class teaching. Mellin-Olsen found this very unsatisfactory. He met regularly with some students and school teachers interested in developing group work as an approach to teaching and learning mathematics.

He also taught additional mathematics, not included in the curriculum, which he himself thought important. The group of students worked towards a master's degree in mathematics education. Mellin-Olsen worked as their unofficial supervisor and mentor. In this activity he supported the development of the next generation of teachers, in seeing mathematics and education as important features of the same issue. The work in the mathematics group led to a wider network of students and scholars in Norway and beyond. While working with this group, Stieg Mellin-Olsen developed dialogue as a research method (dialogue as a tool to handle various forms of knowledge). He used qualitative research methods in mathematics education quite early, and this led to some interesting publications in the field mentioned below.

In addition to smaller networks, in which he participated for shorter periods, Mellin-Olsen worked throughout the years in five networks in particular. These networks were:

- BACOMET, a group of European scholars that focused discussion on their research in mathematics education.[2] The BACOMET seminars were an important base for the development of ideas and research in the field. The cooperation with A. G. Howson from England meant a lot to Mellin-Olsen's own research (Howson & Mellin-Olsen, 1986).

  From the cooperation within the BACOMET group another book—*Mathematical knowledge:*

Its growth through teaching—was published (Bishop, Mellin-Olsen, & Van Dormolen, 1991). Mellin-Olsen published *Hvordan Tenker Lærere om Matematikkundervisning* (How do Teachers Think About Mathematics Teaching) in 1991 and *Samtalen som Forskningsmetode* (Conversation as a Research Tool) in 1995.

- The Democracy Symposium, where the participants came from Denmark, Finland, Iceland, Norway and Sweden. Here the cooperation with O. Skovsmose was an important stimulation for Mellin-Olsen (Nissen & Bjørneboe, 1990).
- The PhD program in the Southern Africa. This project was based at the University of the Western Cape, the University of Bergen, and Bergen College of Education under the leadership of Cyril Julie, Lionel Slammert, Paulus Gerdes, and Stieg Mellin-Olsen. From Bergen College of Education, Vetle Rohde, Christoph Kirfel, and Nora Lindén participated in the program.
- PME (International Group for the Psychology of Mathematics Education) came into existence in 1976. Mellin-Olsen participated in these conferences on a regular basis, and shared his research interests with colleagues from the whole world, as well as being stimulated by new ideas about mathematics education from abroad.
- PDME (Political Dimension of Mathematics Education). The Network was initiated at the ICME6 conference in Budapest in 1988. Stieg Mellin-Olsen strongly contributed to the establishment of this network. Three PDME conferences were held: the first in London in 1990, then in Johannesburg in 1992, and the last in Bergen in 1995, where Mellin-Olsen led the preparations and was supposed to be the host of the conference. He died six months before the conference took place.

In the PME and PDME networks, the cooperation with colleagues from the international community of mathematics education was of great importance to Mellin-Olsen. The work together with Cyril Julie and Paulus Gerdes from the Southern Africa; Munir Fasheh from Palestine; Alan Bishop, Marilyn Nickson, Celia Hoyles, and Leone Burton from UK; and Scandinavian colleagues gave him encouragement and stimulation to continue with his writing. In 1994 he started writing a book about women in the international field of mathematics education. He interviewed ten female scholars in the field, and the book presents a good picture of the development at that time of a female view of mathematics education. He called the book *Women's Talk,* indicating that the voices of female mathematics educators are important (Mellin-Olsen, 1995a).

Mellin-Olsen was not only a brilliant author. He was also genuinely fascinated with the process of editing and publishing. In 1981 he, together with his colleague Marit Johnsen Høines, founded a publishing house, *Caspar forlag.* They chose the name Caspar in honour of the first Norwegian mathematician and surveyor, Caspar Wessel, who in 1797 described the mathematics of representing complex numbers, some time before Gauss.

Caspar forlag soon became a viable publishing house, issuing books in mathematics education as well as general education. In 1990, the first issue of the journal *Tangenten,* a journal for mathematics teachers on all levels at school, primary and secondary, was published and sent out to all schools in Norway.

Tangenten continues to have a healthy circulation at the time of writing. Mellin-Olsen wrote many articles in the first issues where he presented the work of a number of foreign researchers in mathematics education for Norwegian teachers. He encouraged many of his colleagues to write in *Tangenten* about alternative ideas within mathematics education (see Torkildsen, this volume, Chapter 12). In this way Mellin-Olsen hoped to raise the quality of mathematics education in Norway.

In 1991 Mellin-Olsen moved back to the University of Bergen, as Professor in Practical Education at the Institute of Practical Education (IPP) where he became dean of the institute, a position he occupied until his death in 1995. The institute of Practical Education at the University of Bergen had as its main task to provide students with a formal didactical competency, after finishing their Bachelor, or Master's studies. Here he worked closely together with colleagues representing a variety of subjects. His interest in education as such, not only mathematics education, was an important driving force for him during this period of time.

Through presentations at conferences, articles in journals as well as his lecturing, Mellin-Olsen's concern for pupils who struggled with learning mathematics remained an important issue in his work (Mellin-Olsen, 1975, 1976a, 1978, 1997). In this respect, Mellin-Olsen extended his academic writings to include books for young readers. His meetings with pupils, or young adults, in school and the streets of Bergen made him wish to engage them in reading, in an attempt to give them wider perspectives on adolescence. In the three books, *Skillelinjer* (Dividing Lines, 1978), *Guttene i Obsen* (The Boys in Observation Class, 1980) and *Sammys festning* (Sammy's Fortress, 1984) Mellin-Olsen combined his knowledge of learning with his knowledge about being young in the city of Bergen.

THE SIGNIFICANCE OF STIEG MELLIN-OLSEN'S IDEAS

Stieg Mellin-Olsen challenged traditional ideas in the field of mathematics education. His new ideas were developed through the years in Norwegian schools, research, and cooperation with colleagues abroad. His research and teaching as well as his writings introduced new ideas and new thinking to the Norwegian academic community. He challenged the academic language by introducing ideas from Marx, Vygotsky, and Freud into his academic writing about mathematics education. He wished to be critical of given rules that "authorities" claimed were the only approach to teaching. He brought new ideas into the field we today call adapted teaching.

He always wanted to be on the side of people who struggled for a better life, supporting pupils and students in their search for meaningful learning. His teaching and research engagement ranged from pupils in primary and secondary school with special needs, to gifted secondary pupils, and further to students and coming teachers who searched for alternative ideas of teaching and learning.

New knowledge must continuously be developed; knowledge obliges, was his claim. Mellin-Olsen had a challenge and advice for us all, when he encouraged us to *dare to use our understanding*. This is a strong invitation and must be seen as his wish for a better mathematics education developed by young people's own understanding of life.

NOTES

1. Readers familiar with Skemp's seminar article "Relational Understanding and Instrumental Understanding" (Skemp, 1976) will be aware that Skemp credits Mellin-Olsen with these terms.
2. The BACOMET (Basic Components of Mathematics Education for Teachers) Project was established in 1979. A number of papers written by the group are collected in the book *Perspectives on Mathematics Education* edited by B. Christiansen, A. G. Howson, and M.Otte (1986).

BIBLIOGRAPHY OF MELLIN-OLSEN'S MAJOR WORKS

Mellin-Olsen, S. (1964). *Metrisering av topologiske rom* [Metrisation of topological space. Masters thesis]. Hovedoppgave. Oslo, Norway: Matematisk institutt, Universitetet i Oslo.

Mellin-Olsen, S. (1966a). Strukturelle undervisningsmetoder i matematikken [Structural teaching methods in mathematics]. *Norsk Pedagogisk Tidskrift, Oslo: Universitetsforlaget 50*(9), 372–383.

Mellin-Olsen, S.(1966b). *Rapport fra skoleforskning ved Sosiologisk Institutt* [Report from school research at the Institute of Sociology] Universitetet i Bergen, nr.1.1966.

Mellin-Olsen, S. (1970a). *Undervisningsprosjekter i matematikk* [Teaching projects in mathematics]. Oslo: Universitetsforlaget.

Mellin-Olsen, S. (1970b). *Matematikk i ungdomsskolen. Fagmetodikk* [Mathematics in secondary school. Subject didactics]. Oslo: Olaf Norlis forlag.

Mellin-Olsen, S. (1971a). Piaget om kunnskap og læring [Piaget about knowledge and learning]. Oslo: *Den Høgre Skolen Tidskrift for Norsk Lektorlag70*(9), 457–463.

Mellin-Olsen, S. (1971b). *Strukturert matematikk* [Structured mathematics]. Bergen, Norway: J.W. Eide forlag.

Mellin-Olsen, S. (1971c). *Cappelens regneverk for 9. skoleår* [Cappelen's mathematics textbook for year 9]. Oslo: Cappelens forlag.

Mellin-Olsen, S.(1972). *Mathematics for "slow" learners.* Det Pedagogiske Seminar, Universitetet i Bergen.

Mellin-Olsen, S. (1974). *Matematikk, produksjonsforhold og undervisning. Et marxistisk perspektiv på moderne matematikk* [Mathematics, production and teaching. A Marxist perspective on modern mathematics]. Oslo: Pedagogen, Norsk Lærerskolelag nr.2/1974.

Mellin-Olsen, S. (1975a). *Instrumentalisme som pedagogisk begrep* [Instrumentalism as an educational concept] Det Pedagogiske seminar, Universitetet i Bergen.

Mellin-Olsen S. (1975b). *Dialektisk pedagogikk- idealistisk pedagogikk?* [Dialectical education-idealistic education?]. Oslo: Praxis Tidskrift for skole og samfunn nr.6, 4 årgang.

Mellin-Olsen, S., & Rasmussen, R. (1975). *Skolens Vold* [The violence of the school] Oslo: Pax forlag.

Mellin-Olsen, S. (1976a). *Eleven, matematikken og samfunnet* [The pupil, the Mathematics and Society]. Oslo: NKI forlaget.

Mellin-Olsen, S. (1976b). Enhetsskolens motsetninger [The antagonism of the comprehensive school]. In *Nordisk forum* nr.9/1976. Nordisk Sommeruniversitet, Roskilde.

Mellin-Olsen, S. (1977). *Læring som sosial prosess* [Learning as a social process]. Oslo: Gyldendals fakkelbøker.

Mellin-Olsen, S., & Solvang, R. (1978a). *Temahefter i matematikk fagmetodikk* [Theme booklets in mathematics subject didactics. Special needs in Mathematics] Oslo: NKI forlaget.

Solvang, R. & Mellin-Olsen, S. (1978b). *Matematikk fagmetodikk* [Mathematics subject didactics]. Oslo: NKI forlaget.

Mellin-Olsen, S. (1979a). *Rapport fra fagdidaktisk verksted om Lærebøker i matematikk* [Report from a didactics seminar about textbooks in mathematics]. Bergen: lærerhøgskole.

Mellin-Olsen, S. (1979b). Innholdet i enhetsskolen- innhold for arbeiderklassen? [The content of the comprehensive school, the content for the working class?] Oslo: Universitetsforlaget *Norsk pedagogisk tidskrift* nr.2/1979.

Mellin-Olsen, S. (1981). Didaktikk for integrering—arbeid teori i grunnskole [Didactics for integration of practice—theory in primary/secondary school]. Bergen, Norway: Caspar forlag.

Mellin-Olsen, S.(1982). Om å forankre fagdidaktikken teoretisk [To anchor subject didactics theoretically]. *Norsk pedagogisk tidskrift*

Mellin- Olsen, S. (1983). Vi vil ha et rusfritt miljø. Undervisningslære. [We want an environment free from intoxication. Didactics]. Oslo, Norway: Universitetsforlaget.

Bjorøy, D.O. & Mellin-Olsen, S. (1985a). *Strukturert problemløsning* [Structured problem solving]. Bergen, Norway: Caspar forlag.

Johnsen-Høines, M. J. & Mellin-Olsen, S. (Eds.). (1985b). M*athematics and culture*, Bergen, Norway: Caspar forlag.

Howson, A. G. & Mellin-Olsen S. (1986). Social norms and external evaluation. In B. Christiansen, A. G. Howson, & M. Otte (Eds.), *Perspectives on mathematics education* (pp. 1–48). Dordrecht, Holland: D. Reidel Publ. Co.

Mellin-Olsen, S. (1987). *The politics of mathematics education,* Dordrect, Holland: Kluwer Academic Publishers.

Mellin-Olsen, S. (1988). *The double bind as a didactical trap,* Paper for Bacomet 2. Bergen College of Education.

Mellin-Olsen, S. (1989a). *Kunnskapsformidling. Virksomhetsteoretiske perspektiver,* [Mediation of Knowledge. Perspectives of Activity Theory]. Bergen, Norway: Caspar forlag.

Mellin-Olsen, S. (Ed.) (1989b). *Om kunnskap. Fagdidaktiske perspektiver* [About knowledge. Subject didactics perspectives]. Bergen Lærerhøgskole.

Mellin-Olsen, S. (1991a). Liberation of Knowledge. In *Proceedings from PDME 1,* London Dept. of Mathematics, Statistics & computing, Institute of Education, University of London.

Mellin-Olsen, S.(1991b). *Hvordan tenker lærere om matematikkundervisning?* [How do teachers think about mathematics teaching?]. Bergen Lærerhøgskole.

Mellin-Olsen, S. (1993a). A Constructivist's Approach to Teacher Education: Views and Considerations. In *Proceedings from PDME 2,* Johannesburg, South Africa: Maskew Miller and Longman.

Mellin-Olsen, S. (1993b). Dialogue as a tool to handle various forms of knowledge. In Proceedings from PDME II Johannesburg 2–5. April 1993 (pp. 3–15).

Mellin- Olsen, S.(1993c). Semesteroppgaver, hovedfag og dr. grad. Hvordan sikre nødvendig kvalitet? [Term papers, Master Degree and PHD. How to secure quality?]. *Tangenten, 4*(3/4), 6–15.

Mellin-Olsen, S. (1995a). *Mathematics education: Women's Talk,* Bergen, Norway: Caspar forlag.

Mellin-Olsen, S. (1995b). *Samtalen som forskningsmetode* [Conversation as a research tool]. Bergen, Norway: Caspar forlag.

Mellin-Olsen, S. & Lindén, N. (Eds.) (1997). *Perspektiver på matematikkvansker.* [Perspectives on special needs in mathematics] Bergen, Norway: Caspar forlag.

Fiction

Mellin-Olsen, S. (1978). *Skillelinjer* [Dividing Lines]. Oslo: Gyldendal Norsk forlag.

Mellin-Olsen, S. (1980). *Guttene i Obsen* [The Boys in the Observation Class]. Oslo: Gyldendal Norsk forlag.

Mellin-Olsen, S. (1984). *Sammys festning* [Sammy's Fortress]. Oslo: Tiden Norsk forlag.

REFERENCES

Bateson, G. (1972). *Steps to ecology of mind.* New York: Paladin Books.

Bishop, A. J. (1979). Visualising and mathematics in a pre-technological culture. *Educational Studies in Mathematics 8,* 118–135.

Bishop, A., Mellin-Olsen, S., & Van Dormolen, J. (Eds.). (1991). *Mathematical knowledge: Its growth through teaching.* Dordrecht, Holland: Kluwer Academic Publishers.

Brousseau, G. (1984). The Crucial Role of the Didactical Contract in the Analysis and Construction of Situations in Teaching and Learning Mathematics, H.G. Steinar (ed.), Theory of Mathematics Education, Occasional Paper 54, IDM, Bielefeld.

Engeström, Y. (1987). *Learning by expanding: An activity-theoretical approach to developmental research.* Helsinki: Orienta-Konsultit.

Foucault, M. (1985). *The archaeology of knowledge.* London: Tavistock Publications.

Freire, P. (1975). *Education for critical consciousness.* London: Continuum International Publishing.

Goodchild, S. (1997). *An exploratory study of year ten students' goals in the mathematics classroom.* Unpublished doctoral thesis, University of Exeter.

Goodchild, S. (2001). *Students' goals: A case study of activity in a mathematics classroom.* Bergen, Norway: Caspar.

Howson, A. G. & Mellin-Olsen S. (1986). Social norms and external evaluation. In B. Christiansen, A. G. Howson, & M. Otte (Eds.), *Perspectives on mathematics education* (pp. 1–48). Dordrecht, Holland: D. Reidel Publ. Co.

Johnsen-Høines, M. J. & Mellin-Olsen, S. (Eds.). (1985). M*athematics and culture*, Bergen, Norway: Caspar forlag.

Mead, G. H. (1934). *Mind, self & society,* Chicago: The University of Chicago Press.

Mellin-Olsen, S.(1972). *Mathematics for "slow" learners.* Det Pedagogiske Seminar, Universitetet i Bergen.

Mellin-Olsen S. (1975b). *Dialektisk pedagogikk- idealistisk pedagogikk?* [Dialectical education-idealistic education?]. Oslo: Praxis Tidskrift for skole og samfunn nr.6, 4 årgang.

Mellin-Olsen, S., & Rasmussen, R. (1975). *Skolens Vold* [The violence of the school] Oslo: Pax forlag.

Mellin-Olsen, S. (1976a). *Eleven, matematikken og samfunnet* [The pupil, the Mathematics and Society]. Oslo: NKI forlaget.

Mellin-Olsen, S. (1977). *Læring som sosial prosess* [Learning as a social process]. Oslo: Gyldendals fakkelbøker.

Mellin-Olsen, S. (1987). *The politics of mathematics education,* Dordrect, Holland: Kluwer Academic Publishers.

Mellin-Olsen, S. (1991a). Liberation of knowledge. In *Proceedings from PDME 1,* London Dept. of Mathematics, Statistics & computing, Institute of Education, University of London.

Mellin-Olsen, S.(1991b). *Hvordan tenker lærere om matematikkundervisning?* [How do teachers think about mathematics teaching?]. Bergen: Lærerhøgskole.

Mellin-Olsen, S. (1993a). A Constructivist's Approach to Teacher Education: Views and Considerations. In *Proceedings from*

PDME 2, Johannesburg, South Africa: Maskew Miller and Longman.

Mellin-Olsen, S. (1995a). *Mathematics education: Women's talk,* Bergen, Norway: Caspar forlag.

Nissen, G., & Bjørneboe, J. (Eds.) (1990). *Matematikkundervisninge og demokrati* [Mathematics education and democracy] IMFUFA Roskilde Universitetscenter

Skemp, R. R. (1976). Relational understanding and instrumental understanding. *Mathematics Teaching, 77,* 20–26.

Wæge, K. (2007). *Elevenes motivasjon for å lære matematikk og undersøkende matematikkundervisning [Pupils' motivation to learn mathematics and investigative mathematics teaching].* Trondheim: Norwegian University of Science and Technology.

4

NORWAY

Low Achievement in Mathematics in Compulsory[1] School as Evidenced by TIMSS and PISA

Liv Sissel Grønmo

ILS, UNIVERSITY OF OSLO

Abstract: International comparative studies such as TIMSS and PISA have attracted an influential audience of politicians, policy makers, curriculum developers, and researchers in mathematics education, in Norway as well as in many other countries. Comparative studies are carried out primarily in order to offer insight into pupils' achievement that will support reflection on a country's own system and practice. The results of analyses of data from international comparative studies have to be interpreted by each country according to its specific national needs and goals in education. This overview of international comparative research in Norway takes a Norwegian perspective, which includes an emphasis on similarities and differences between the Nordic countries. Special attention is given to the relation between pure and applied mathematics, and how it may relate to achievement in mathematics.

Keywords: Mathematics, Nordic Profile, Pure Mathematics, Applied Mathematics, TIMSS, PISA

INTRODUCTION

This chapter offers an overview of several approaches to research in mathematics education in Norway, mainly related to two large-scale international studies: TIMSS (an IEA study: *Trends in International Mathematics and Science Study*, Grades 4 and 8) and PISA (an OECD study: *Programme for International Student Assessment*, Grade 10). Principal results from these studies are discussed and related to other research in mathematics education. Attention is drawn to research based on first and secondary analyses of data, including research where data are re-categorized for secondary analysis. Some research is included that uses qualitative or mixed methods to contribute to a deeper understanding of the results.

A number of research questions such as "Does it make sense to talk about different profiles in school mathematics in different countries or groups of countries?" and "Are these profiles stable over time and across studies?" are addressed by referring to a number of analyses and articles. Special attention is given to understanding the characteristics of what may be called a Nordic tradition in mathematics education, and this is discussed in relation to traditions in other groups of countries, such as English-speaking, East European, and East Asian countries. This also includes research on more theoretical aspects of mathematics education, such as the relationship between pure and applied mathematics, showing how international comparative studies have contributed to discussions about such issues. In this context, analysis of the differences between TIMSS and PISA is important. Attention is drawn to the comple-

The First Sourcebook on Nordic Research in Mathematics Education, pages 49–69

mentary aspects of TIMSS and PISA, and how these complementarities can be used to get better insight into mathematics education. Re-categorizing data is undertaken to address questions that are particularly interesting from the perspective of the Norwegian mathematics curriculum, such as "What can TIMSS and PISA tell us about Norwegian pupils' competence in daily life mathematics in relation to more traditional pure mathematics taught in school?" This includes research on basic knowledge and skills and their role in school mathematics, which have been especially in focus in the most recent change of the mathematics curriculum in Norway.

Other central research questions that relate to how Norwegian pupils' competence and attitudes in mathematics have changed from 1995 to 2007 are discussed from several perspectives. Research on changes in the Norwegian school curriculum and possible impact of these on pupils' attitudes and achievement, as well as changes in organization and instruction in schools, is addressed. Attention is given to the possible impact of such changes on the roles of teachers and pupils. Popular slogans about "pupils' responsibility for their own learning" and "teachers as facilitators" and their possible impact on pupils' learning of mathematics are central to this discussion. Also, pupils' right to get instruction in accordance with their capacity and potential for learning (referring to the Norwegian curriculum) is part of this. Focus is on how successful the Norwegian school system seems to be in giving adequate instruction to the very able pupils, as well as to those who have learning problems in mathematics.

This chapter refers especially to national and international books and reports and to articles that use data from large-scale international studies such as TIMSS and PISA. Possible ways of developing research in this field—for example, combining data from TIMSS and PISA with data from other large-scale international studies such as TIMSS Advanced (specialists in mathematics in the last year of upper secondary school) and TEDS-M (college or university students being educated to become mathematics teachers in compulsory school) to conduct a type of "longitudinal" study through different school levels from primary till university—are mentioned. The need for additional research using a greater range of methods, such as classroom observation and video studies of instruction and communication in the classroom in order to shed light on results from large scale international studies, is part of this.

In the last part of the chapter, questions such as "What types of analyses and new research have been conducted related to large-scale international studies?" and "What are the opportunities for researchers to do analyses based on data from these studies?" are in focus. It is an aim to see research based on data from large scale international studies in the context of educational research from a more general perspective.

TIMSS and PISA

The first time Norway participated in a large-scale international study in mathematics was TIMSS in 1995, followed by PISA in 2000. This was followed up by both TIMSS and PISA in 2003, PISA in 2006, and TIMSS in 2007. This chapter is mainly based on primary and secondary analyses of data from these studies. TIMSS and PISA are large-scale international comparative trend surveys with many similarities in their methodology. They collect data from pupils using booklets with test items and questionnaires. In order to distinguish between mathematical test items and items in questionnaires about background, attitudes, and so forth, items measuring mathematical competence are referred to as *cognitive items*. There are rigorous rules for translation, adaptation, and implementation of the data collection procedures in these surveys. See http://timss.bc.edu/ for technical reports for all TIMSS studies, especially Olson, Martin, and Mullis (2008) for the TIMSS 2007 study. See http://www.pisa.oecd.org/ for technical reports for all PISA studies, especially OECD (2009) for the PISA 2006 study. Both studies use constructs or indexes based on different types of background data as basis for discussion about the results on pupils' achievement in mathematics.

As stated above, both studies have rigorous procedures for collecting the data; the procedures in the classrooms and at the schools have to be exactly the same in all participating countries. There are internationally appointed quality inspectors in all countries visiting at least 10% of the schools to observe if all procedures are followed. The national research center for the study also visits some schools for the same reason.

Both studies are based on a rotation system of booklets that are randomly distributed among pupils. The number of booklets has varied from 12 to 18 in different studies. This rotation is done in order to cover a broad range of mathematical topics, and it is acceptable because the aim of the studies is to measure and compare national and international levels, not individual pupils. There is also a rigorous system for exclusion of pupils, which is applied uniformly across participating countries.

The cognitive items are of different formats: Some are multiple choice items, and others are constructed response items that have to be scored according to a special scoring guide and procedure that is common to all countries. An international meeting is held to educate those in charge of the scoring in each participating country. About 25–30% of the constructed response items are scored independently by two different scorers, making it possible to estimate the inter-scorer reliability in each country. In addition, all countries have to score a number of pupils' responses in English, to estimate inter-country scoring reliabilities. Norway has always been able to meet all such reliability requirements, by means of a very firm and efficient organization of the scoring process. For more on this, see Grønmo and Onstad (2009) and Kjærnsli, Lie, Olsen, Roe, and Turmo (2007).

There are also some important differences between TIMSS and PISA that one should be aware of. Both are sample-based studies of clearly defined populations. However, while the definition of the populations in TIMSS 2003 and 2007 were pupils in Grades 8 and 4, the population in PISA 2003 and 2006 was defined as pupils who were 15 years old (in Norway that meant students mostly in Grade 10, the last year of compulsory school). Both studies randomly sampled a number of schools. Within each school, however, TIMSS sampled whole classes, while PISA sampled 30 pupils out of the age cohort.

Most important, however, is the difference between the frameworks for the two studies. A framework defines the type of mathematical competence that is assessed. The TIMSS framework is developed on the basis of the mathematical curricula in the participating countries, resulting in an international consensus on an "international curriculum" for the study. The framework of PISA, on the other hand, does not refer to the curriculum in any country, but is developed on the basis of what an expert group defines as mathematical knowledge all citizens may need in their future daily life, in PISA referred to as "mathematical literacy." While TIMSS (http://timss.bc.edu/timss2003i/frameworks.html; Mullis et al., 2003) gives a fine-grained description of central aspects of school mathematics within traditional school content areas like number, algebra, and geometry, PISA (http://www.pisa.oecd.org/; OECD, 2003) is based on a detailed description of contexts and phenomena related to daily life where mathematical competence may be needed. PISA emphasizes the use of mathematics in "a real-world context" in the form of numbers, tables, and graphs, which may be referred to as applied mathematics. TIMSS tests pupils to a great extent in more traditional school mathematics. The cognitive items in TIMSS cover both pure and applied mathematics, with a lot of emphasis on traditional pure mathematics. For more about similarities and differences between TIMSS and PISA, see Grønmo and Olsen (2006), Grønmo (2005a), Grønmo, Bergem, Kjærnsli, Lie, and Turmo (2004), and Kjærnsli et al. (2004). As is demonstrated later in this chapter, these differences between the frameworks offer unique opportunities to study a number of interesting research questions by combining data from both studies.

The case of Norway is central in this chapter. Nevertheless, it must be recognized that the overall relevance and purpose of what is analyzed goes beyond a national context (Grønmo & Olsen, 2006; Grønmo, Bergem, et al., 2004; Kjærnsli et al., 2004). Unless otherwise clearly stated in this text, the discussion refers to the performance and responses of Norwegian pupils.

Both studies are based on representative samples of pupils. Special software weighting the data according to the probabilities for pupils to be in the sample has been developed for the analysis. This ensures representativeness and accurate measurement of errors. In both studies, Item Response Theory is used as basis for the analyses. There are rigorous rules for participation rates for countries in order to be included in international reports. Norway has met all these requirements in both TIMSS and PISA. See Tables 4.1 and 4.2 for an overview of the sample sizes in the TIMSS and PISA studies.

The results for TIMSS 2003 and PISA 2003 revealed a decrease in pupils' achievement. This was especially the case in TIMSS, which compared with pupil achievement in 1995. These results initiated a new project called PISA+, with the aim of shedding more light on, and gettting a better understanding of, why Norwegian pupil performance was that low. Reference is made to results from PISA+ in mathematics later in this chapter. An overview of the sample for

TABLE 4.1 Samples in TIMSS

	Year of study	1995	2003	2007
Grade 4	Number of pupils	4476	4342	4108
	Number of schools	140	139	145
Grade 4	Number of teachers	351	228	310
Grade 8	Number of pupils	5758	4133	4627
	Number of schools	252	138	139
Grade 8	Number of teachers	269	180	264

TABLE 4.2 Samples in PISA

	Year of study	2000	2003	2006
Grade 10	Number of pupils	4147	4046	4700
	Number of schools	176	182	203

TABLE 4.3 Samples in PISA+

Number of schools	Number of teachers	Number of pupils
6	14	150

that study is given in Table 4.3; for more information see Bergem (2009).

To be able to measure trends, in TIMSS and PISA there are several identical items used in successive studies. About half of the items are released after each study; the remainder are kept secret to be used next time for trend analyses. For more about this, see Grønmo, Bergem, et al. (2004) and Kjærnsli et al. (2004). There has been some criticism of these studies because a number of items are not released. This is done, however, because otherwise it would not be possible to measure trends in achievement reliably. On the other hand, all items are open for researchers to do their analyses, but they cannot be published.

PROFILES IN GROUPS OF COUNTRIES RELATED TO PERFORMANCE IN CONTENT AREAS IN MATHEMATICS

International assessment surveys such as TIMSS and PISA aim at establishing reliable and valid scores for achievement that can be compared across countries or across groups of pupils within countries, and to relate achievement to various background and context variables that may give ideas about possible indicators for characterization of high performance in mathematics. These studies also offer opportunities for secondary analyses to answer a number of other research questions. This section aims at providing an answer to the following question: Is it reasonable to distinguish between different profiles of mathematics education in various countries or groups of countries? This creates a context for presenting results in achievement, and it has played an important role in the Norwegian national TIMSS reports from the studies in 2003 and 2007.

The method of analysis is a type of cluster analysis of so-called "item-by-country interactions" in order to investigate similarities and differences between countries or groups of countries across cognitive items. This has been conducted using data from both TIMSS and PISA. It has to be recognized that in these analyses, one is talking about relative performance. This cluster analysis displays groupings of countries according to similarities in relative response patterns across items. Countries in one group tend to have relative strengths and weaknesses in the same items. In the discussion of grouping of countries, the focus is on groups that are meaningful from a geographical, cultural, or political point of view. A possible Nordic profile in mathematics education plays a central role in the discussion, as well as what seems to characterize such a profile by contrasting it to other profiles.

This cluster analysis, obviously, depends on which countries participated in the respective study. Since there have been different countries participating in different studies, this influences which countries cluster together. Data from TIMSS 1995 were used to point out for the first time that is was reasonable to speak of a specific Nordic profile in mathematics education (Grønmo, Kjærnsli, & Lie, 2004). In the following, attention is also given to the so-called English-speaking profile, which is closely related to the Nordic profile. The difference between profiles is used to enlighten what is typical for the Nordic countries, with the aim of displaying typical characteristics of the Nordic profile (Grønmo, Kjærnsli, & Lie, 2004; Grønmo & Olsen, 2006; Olsen & Grønmo, 2006).

Figure 4.1 presents the results of the cluster analysis based on TIMSS 1995 data in mathematics in Grade 8. Figure 4.2 presents the results of the same type of analysis based on data from PISA 2003. The lines in these dendrograms show how countries are linked together. The closer to the left countries are linked together, the more pronounced are the similarities between the countries' relative performances on cognitive items. The further to the right countries or groups of countries are linked together, the more pronounced are the differences in relative performance on cognitive items. It should be noticed that in the end all countries are linked together. Grønmo, Kjærnsli, and Lie (2004) concluded that the following clusters of countries formed meaningful profiles from a geographical, cultural, or political point of view: English-speaking, German-speaking, East European, Nordic, and East Asian countries.

The dendrogram in Figure 4.1 clearly displays that the English-speaking countries form groups. The most pronounced group consists of England, Scotland, and New Zealand, clustering close to the left side. Australia is linked to this group a little further to the right. Canada and U.S. form their own rather strong North American group, with Ireland linked to this group further to the right. Among the Nordic countries, Norway, Sweden, and Iceland form a rather strong group, with Denmark linked to this group a little later. This Nordic group is then somewhat interwoven with the group of English-speaking countries. Other countries that form close groups are Hong Kong and Singapore, and Japan and Korea before these four Asian countries form an East Asian group. This group seems

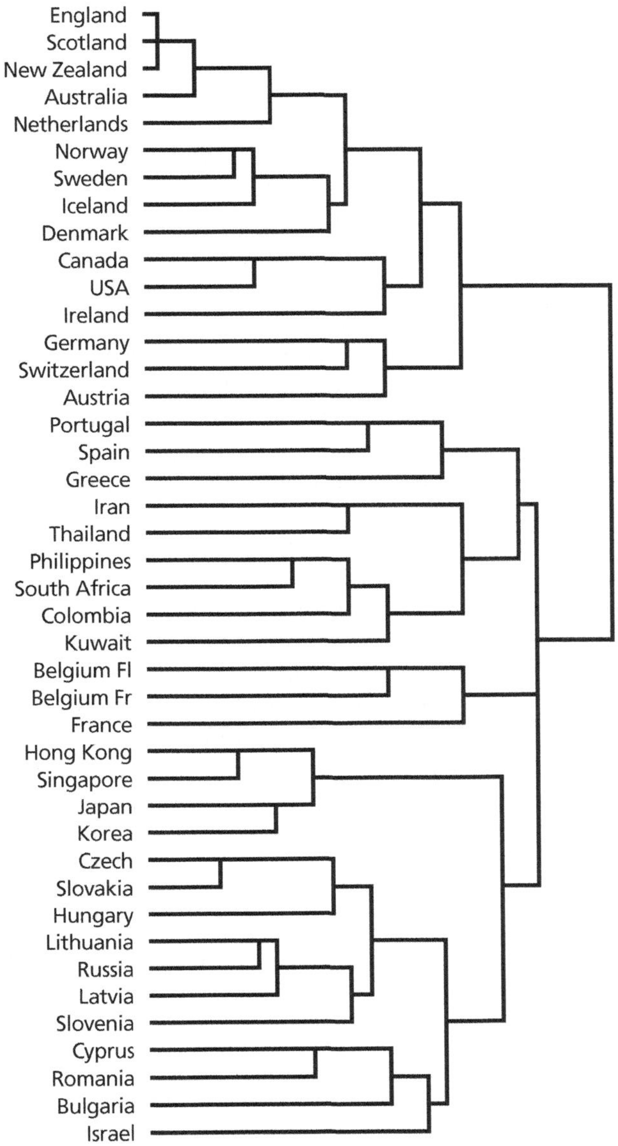

Figure 4.1 Dendrogram for clustering of countries according to similarities in patterns across mathematics items in TIMSS 1995, Grade 8 (From Grønmo, Kjærnsli & Lie, 2004).

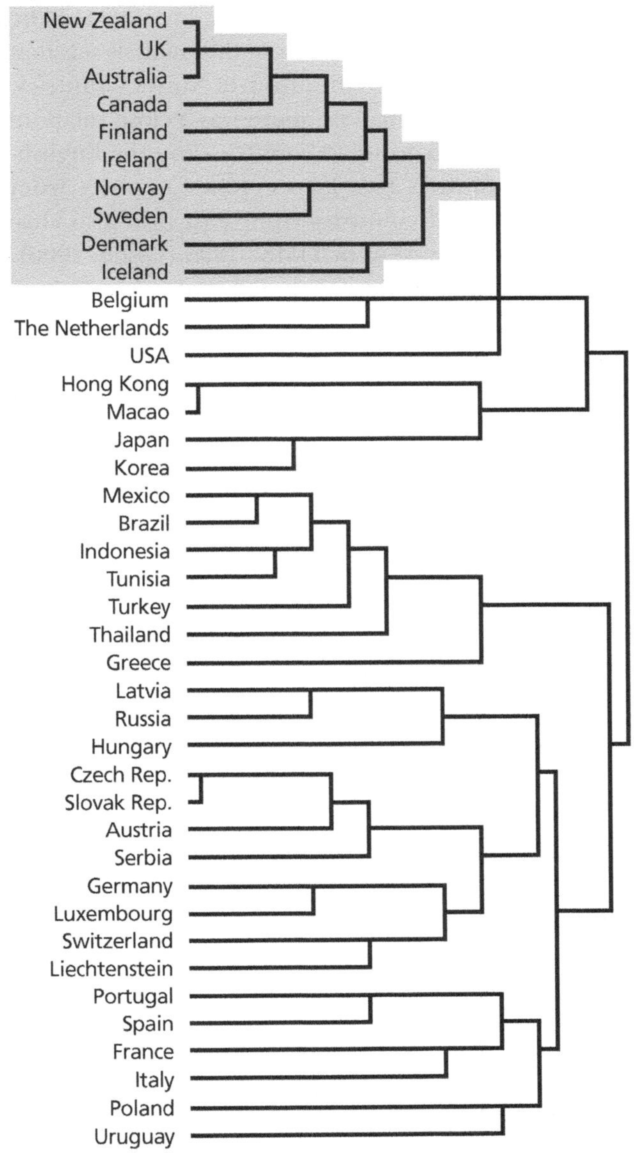

Figure 4.2 Dendrogram of the cluster analysis of countries in PISA 2003. The Nordic-English cluster is highlighted. (From Olsen, 2006)

to be distinguished from all the other countries, since it is rather late in the process that this group is linked to the East European group and then to the rest of the world. The Czech Republic and Slovakia are closer than most other countries, with Hungary joining this group later in the process. Also Lithuania, Russia, and Latvia form a strong group, with Slovenia linked later. Most of the East European countries are then linked together as one group. (It should be noted here that the label "East European" is strictly speaking not quite correct, but it still seems to be the most meaningful term for this group.)

The process of forming groups among many of the Western European countries is rather complicated, and will not be commented upon since it does not play a major role in the discussion. Attention will be given to the English-speaking profile, which is closely related to the Nordic profile, and to profiles that give important information by contrasting them to the Nordic and English-speaking profiles, as the East European and East Asian profiles.

Finland did not participate in TIMSS 1995 (the only Nordic country missing). However, all the Nordic countries participated in PISA 2003, where mathematics was the main school subject tested. Conclusions from an analysis of the Nordic profile based on data from this study (Olsen, 2006) were predominantly consistent with the corresponding conclusions from

TIMSS 1995 (Grønmo, Kjærnsli, & Lie, 2004). The important new information from this analysis was that Finland was included in the group of Nordic countries. As in the previous analysis, there was a close relationship between this group of countries and the English-speaking countries (see Figure 4.2). Countries from East Europe and countries from East Asia also clustered in similar ways as in TIMSS 1995 (Olsen, 2006). In both analyses it was concluded that it was reasonable to talk about a Nordic profile in mathematics education. Consistent in the analyses of TIMSS 1995 and PISA 2003 was that the closest related countries within the Nordic profile were Sweden and Norway.

It is worth noticing that a high-achieving country like Finland—in contrast to all the other Nordic countries—nevertheless clustered with the other Nordic countries in this type of analysis. Since these cluster analyses are based on the relative performance of countries on all items in the test, they can be interpreted like this: Pupils in Finland and the Nordic countries performed more strongly on the same items in relation to international levels of performance, but in general the Finnish pupils scored higher than their counterparts in the other Nordic countries.

Characteristics of the Nordic Profile in Mathematics

In the previous section it was concluded that cluster analyses based on data from TIMSS and PISA made it sensible to refer to a Nordic profile in mathematics education in compulsory school, and that this profile included all the Nordic countries. The profile appeared stable over time and across different studies with different definitions of mathematical competence. Figure 4.3 displays mean achievement for groups of countries within different profiles and in different content areas of mathematics in TIMSS 1995 (Grønmo, Kjærnsli, & Lie, 2004). It must be noted that this figure does not refer to relative achievement, as was the case in the cluster analysis presented in the previous section. This comparison of achievement gives an indication of the characteristics of the profiles over *content areas* in the study, as well as of *the achievement level* in each group of countries. The content areas refer to the categories used in the international achievement report for TIMSS 1995 (Beaton et al., 1996). For simplicity, a short form is used for two of the categories. "Fractions and number sense" is referred to simply as "Number sense." "Data representation, analysis and probability" is referred to by using

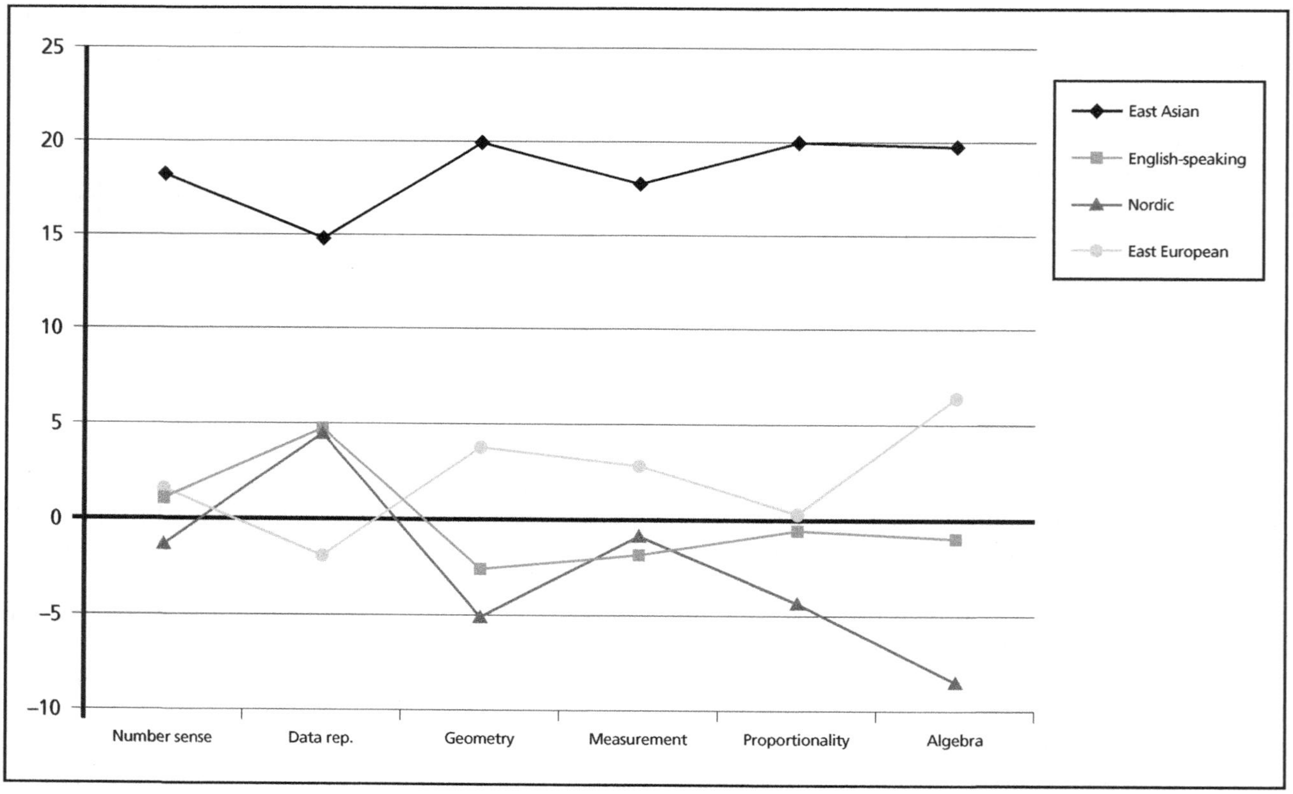

Figure 4.3 Mean performance in content areas of mathematics for countries in TIMSS 1995 forming different profiles in mathematics education. The graphs show score points above or below the international mean, which is set to zero in each content area (from Grønmo, Kjærnsli & Lie, 2004).

"Data representation" or only "Data" as a short label for this category.

The figure reveals that there are two very different types of profiles, one consisting of East Asia and East Europe, the other consisting of the English-speaking and the Nordic countries. Even if there are clear similarities between the groups in these two types of profiles, it is also possible to see distinct differences within each type.

The East Asian and the East European profiles are quite similar according to relative achievement in the various categories, but the East Asian group has much higher overall performance. Both groups performed relatively better in traditional mathematical content areas like geometry and algebra than they did in mathematics that related more closely to daily life such as data representation and probability. When it comes to proportionality, however, there is a clear difference between these two groups, with the Asian group performing better (relatively) than the East European group. This result is consistent with in the interpretation of the dendrogram: two distinct groups, one in East Asia and one in East Europe, which are linked together in a way that indicates a closer relationship between these two groups of countries forming separate profiles in mathematics education than with other profiles such as the Nordic and the English-speaking.

Also, countries in the English-speaking and the Nordic profiles reveal similarities as well as some distinct differences. Both these groups of countries performed relatively better on data representation and probability than was the case for countries forming the East Asian and the East European profiles. They also performed relatively better in number sense than in geometry, and relatively low in algebra compared to number sense and data representation, the latter being most pronounced for the group of Nordic countries. Both groups performed lowest in algebra, geometry, measurement and proportionality. The patterns revealed in Figure 4.3 are consistent with the result from the cluster analysis revealed in the dendrogram (Figure 4.1), and can be used as validation for which of the countries that first forms groups, close to the left side in the dendrogram, and which groups or countries that are joined together closer to the right side, which is interpreted as later in the clustering process.

Grønmo, Kjærnsli, and Lie (2004) also used the residuals in the matrix, which was the basis for the cluster analysis in the previous section in order to identify items for which a certain group of countries achieved particularly well or badly. They concluded that typical for the items where East European and East Asian countries seemed to perform relatively best was that they all are about classical, pure, and abstract mathematics. Contrary to this finding, the Nordic group as well as the English-speaking group performed relatively better on items closer to daily life mathematics like estimation and rounding of numbers. They also scored relatively low on items dealing with more classical abstract mathematics like fractions and algebra.

An analysis of curricula in different countries supported the view that what was seen as important in mathematics in school differs from one group of countries to another. This also seems to validate the results from the cluster analysis of relative performance on cognitive items (Grønmo, Kjærnsli, & Lie, 2004).

Olsen and Grønmo (2006) also looked for characteristics of the Nordic profile by contrasting it with profiles for other groups of countries based on data from PISA 2003. They developed a specific classification system for their analysis in which the items in PISA were reclassified according to how close they were to "real world" or "daily life" mathematics. The findings revealed that the profiles of the Nordic and the English-speaking countries were mainly accounted for by this variable. This variable had a higher degree of explanatory power than the aspects described in the framework of PISA when it came to understanding the clustering of countries in different profiles. The profile for the Nordic countries was strongly characterized by relatively high performance on items involving some sort of real-world mathematics, and the same was true for the English-speaking countries. The East Asian and East European countries, however, achieved relatively lower on items categorized as some sort of real-world mathematics. This result, consistent with Grønmo, Kjærnsli, and Lie's findings (2004), made it reasonable to conclude that real-world mathematics has been a driving force for school mathematics in the Nordic and in the English-speaking countries, in contrast to countries in East Europe and East Asia.

There was a tendency for the pupils in the Nordic countries to perform relatively best on easier items, and with a non-significant tendency to achieve lower on items requiring accuracy in calculations. For the English-speaking countries, the need for calculation was a significant factor indicating low relative achievement. This suggests that accuracy in calculations is not seen as that important in the Nordic and the English-speaking countries. To what extent an increased focus on daily life mathematics in Norway over the last decades has resulted in little attention being given to accuracy in calculation will be discussed later in this chapter, in connection with the relation between pure and applied mathematics. This has also been discussed in several other articles based on data from PISA and TIMSS

(Bergem, Grønmo, & Olsen, 2005; Grønmo, 2005a; Grønmo & Olsen, 2006; Olsen & Grønmo, 2006).

Even if the Nordic countries have many similarities when compared to other countries participating in PISA, there were also some differences regarding their relative achievement profiles. For example, the Danish pupils did not share the Nordic characteristic of performing relatively better on items including graphical representations of quantities. The profile for Finland did not deviate very clearly from the average Nordic profile, the largest deviation being a non-significant tendency for Finnish pupils to perform relatively better on items in a context more distant from the pupils' lives. For items in the topic Uncertainty, there was a tendency of relatively high performance for pupils from Norway and Iceland. Danish and Norwegian pupils tend to do better on easier items, while pupils in Iceland seem to be less affected by the item difficulty than pupils in the other Nordic countries. In spite of these differences, it is necessary to keep in mind that the similarities are more pronounced than the differences within the cluster of Nordic countries.

Olsen and Grønmo (2006) conclude that, on average, the pupils in the Nordic countries performed relatively better on items with a realistic context, on items that included some sort of graphical material, on low difficulty items, and on items that did not include explicit algebraic expressions. Pupils in the Nordic countries also performed relatively better on items relating to probabilities and statistics in a daily life context (classified as Uncertainty in the PISA framework), and on items that tended to be of a more qualitative type that did not require any accuracy in performance of calculations.

These findings are consistent with those of other researchers in mathematics education who have suggested that the mathematics curricula in the Nordic countries, as well as in the English-speaking countries, have been heavily influenced by an emphasis on real-world mathematics and a daily life perspective on mathematics in compulsory school (De Lange, 1996; Gardiner, 2004; Niss, 1996). This aspect of mathematics is very close to PISA's definition of mathematical competence, referred to as Mathematical Literacy.

An important driving force underlying changes in curriculum over the last decades has been an emphasis on everyday applications of mathematics (Mosvold, this volume, Chapter 11). The needs of mathematics for pupils in their daily lives have received more curricular attention than before, while more formal aspects of mathematics, such as algebra, have been reduced. Since the mid-1980s, there has been a lot of discussion about the tendency to give more attention to daily life mathematics; see for example De Lange (1996) and Kilpatrick, Hoyles, Skovsmose, and Valero (2005). Analysis of data from TIMSS 1995 and PISA 2003 seems to confirm that everyday applications of mathematics has been especially influential in the Nordic and in the English-speaking countries.

TRENDS IN MATHEMATICS ACHIEVEMENT FROM 1995 TO 2003

TIMSS and PISA have given Norwegian researchers an opportunity to study Norwegian pupils' competence and attitudes, and how these have changed from the first international comparative study in mathematics in which Norway participated, TIMSS 1995 (Lie, Kjærnsli, & Brekke, 1997a, 1997b). TIMSS 2003 showed a clear decrease in achievement for Norwegian pupils in mathematics from 1995 to 2003. This was the case both in Grade 8 and in Grade 4 (Grønmo, Bergem, et al., 2004). Figure 4.4 displays this for Grade 8. The results for Grade 10 in PISA 2003 pointed in the same

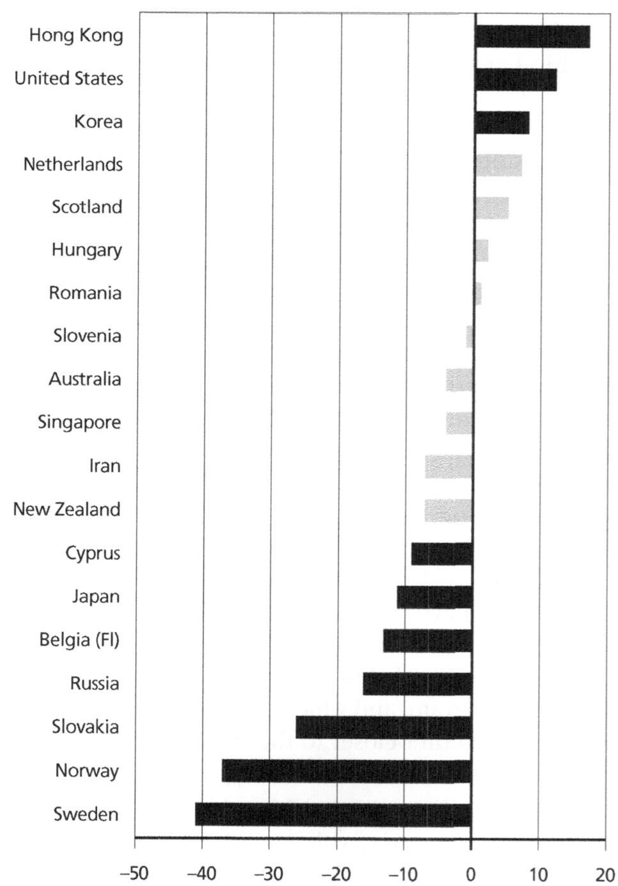

Figure 4.4 Changes in the 8th grade mathematics scores from TIMSS 1995 to TIMSS 2003 for countries where comparison is adequate (from Grønmo, Bergem, 2004). Dark color columns show significant changes.

direction, with indications of decrease from the first PISA study in 2000 (Kjærnsli et al., 2004).[2]

In 1997 a new Norwegian curriculum was implemented. This introduced a younger starting age for school. Earlier, Norwegian children had entered school in the middle of the year they turned seven years old. From 1997, they have started school one year earlier, that is, the year they turn six. The reference to grade level in Norway is therefore a little complicated. When pupils' achievement is compared in this chapter, it is based on pupils at the same age. But because of the change in school start, the grade levels for pupils at the same age will be different in 1995 than in later studies. This chapter refers to the names of the grade levels introduced with the curriculum change in 1997, even if those were not the correct names of the respective grades in 1995.

It was possible for TIMSS to provide reliable calculations for determining how achievements had changed and in which direction, based on identical items included in the tests in both 1995 and 2003. Improvements from 1995 in achievement are presented as a column pointing to the right, while a column pointing to the left represents a decline. Dark color for columns is used to show that the difference is significant. The margins of error are between 5 and 10 points. Countries that tested pupils with more than a half-year difference in the average age from 1995 to 2003 are not presented.

In 1995, two consecutive grade levels participated in the study in each population. It was therefore possible to measure the mean increase in achievement between two grade levels being roughly 40 points. Based on that, it was possible to relate a decrease or increase in achievement to years of schooling. It revealed that Norwegian pupils in grade 8 in TIMSS 2003 achieved about one year lower than pupils at the same age in 1995, while pupils in grade 4 were placed about half a year behind the 1995 level. Taking into account that pupils in grade 4 had one more year of formal schooling in 2003 than in 1995, these results were alarming. Possible reasons for the decrease in mathematical competence in TIMSS and PISA were discussed in the national reports published in 2004 from both projects (Grønmo, Bergem, et al., 2004; Kjærnsli et al., 2004; for English short version see http://www.timss.no/publications/eng2003_web.pdf).

Possible Explanations for the Decrease in Achievement from 1995 to 2003

One of the most important reasons pointed out was that too little attention might have been given to developing a necessary foundation of facts and skills in pure mathematics. This is important both for pupils in their daily lives as well as for those likely to study mathematics further after compulsory school. The importance of such a base of knowledge and skills is discussed in more detail later in this chapter, in the section about the relationship between pure and applied mathematics. Other explanations that were presented in the national reports from TIMSS and PISA in 2004 were changes in instruction and organization in school as well as in the roles of teachers and pupils. These conclusions were built on the results from both PISA and TIMSS, as well as on careful reading of national reports evaluating the implementation of the new curriculum from 1997 (Alseth, Brekke, & Breiteig, 2003; Klette, 2003).

It was argued that the Norwegian curriculum from 1997 took a constructivist perspective as important for pupils' learning of mathematics, the main point being that learners had to actively construct new knowledge themselves. A possible misinterpretation of this constructivist perspective on learning and the implications of this for the implemented curriculum in the schools may have contributed to more attention given to activities in themselves than to the goals for learning (Grønmo, Bergem, et al., 2004; Kjærnsli et al., 2004). There is no direct line from constructivist theory to what methods should be used in classroom instruction, as influential theorists in constructivism have pointed out (Ernest, 2004). Active learning being the basis for knowledge construction in a constructivist epistemology refers to an activity that takes place in the brain, and activity in the brain is not dependent on a particular method of instruction in school. The important point is whether the learning methods are able to initiate active mental processes within the learner. To what extent the pupils are active in a more colloquial sense is totally irrelevant, and no types of methods are to be rejected in principle based on a constructivist view of learning (Grønmo, Bergem, et al., 2004; Kjærnsli et al., 2004).

The researchers responsible for the TIMSS and PISA reports pointed out that the following excerpt from the curriculum—"the pupils shall be active, proactive and independent. They shall learn by doing, investigating and testing new concepts actively against their prior knowledge and understanding" (L97, 1996, p. 75)—may have been interpreted as referring to all kinds of activities, such as working in groups, project-based work, playing games, and experiments. It also seems reasonable to believe that such strong emphasis on specialized working methods might have led to the learning objectives of the subject being given lower priority. This is also in accordance with other national reports based on observation in school:

> The general impression is little systematic and summarized reflection around the learning potential in the different activities. Pupils are not given the opportunities to accumulate knowledge based on systematic experiences. The fact that little time is used for closure and summarizing of the different activities contributes to a situation where the different activities and intentions become diffuse for the pupils and a weak relation between doing and learning is established. (Klette, 2003, p. 73; short report in English on TIMSS 2003 in Norway, see http://www.timss.no/publications/eng2003_web.pdf)

Good instruction relies heavily on the teachers' competence: "A variety of recently published documents support the notion that the key to increasing students mathematical knowledge and to closing the achievement gap is to put knowledgeable teachers in every classroom" (Sowder, 2007, p. 157). The TIMSS 2003 survey revealed that Norwegian teachers had a high general competence, but this was not at all the case when it came to their competence in mathematics and science. Norwegian teachers had specialized in a school subject as part of their teacher education, but even if they taught mathematics and science, neither of these were usually the subject in which they had specialized (Grønmo, Bergem et al., 2004). In this regard, mathematics teachers in Norway were remarkably different from mathematics teachers in other countries (Grønmo, Bergem et al., 2004; Mullis et al., 2004).

Recent changes in the roles of pupils and teachers were also pointed out as possible reasons for low achievement in Norwegian schools (Grønmo, Bergem, et al., 2004; Kjærnsli et al., 2004). The democratization process in schools has resulted in pupils who know their rights and make demands, but they are not necessarily similarly aware of their responsibilities. Changes in the pupils' roles have of course also influenced the role of the teacher. It seems that to a certain extent, the role of the teacher has changed into that of a facilitator. This may have led to teacher uncertainty about what their responsibilities in the learning process are.

> A lot of teachers express an uncertainty about how they are suppose[d] to behave as guides to enhance student learning and teachers who earlier had their strength in lecturing, are now unsure of to what extent they can go through the curriculum in lectures, without being seen as old-fashioned and out of date. (Kjærnsli et al., 2004, p. 255; see English short version http://www.timss.no/publications/eng2003_web.pdf, p. 41)

The new role of the pupils poses greater demands on the pupils' self-regulation of learning. Placing more responsibility on the pupils for their own learning may favor the pupils who come from an academic background, as parents with a higher level of education are themselves probably more able to support their children with school work. There was a weak tendency in TIMSS pointing to a stronger connection between pupils' background and their performance in school in 2003 as compared to 1995 (Grønmo, Bergem, et al., 2004).

It also seemed to be the case that Norwegian classrooms were characterized by problems of noise and disturbances. This was reported from pupils at all levels, as well as from school leaders and teachers (Grønmo, Bergem, 2004; Kjærnsli et al., 2004; see English short version http://www.timss.no/publications/eng2003_web.pdf).

In December 2007, the results from PISA 2006 were announced (Kjærnsli et al., 2007). Science was the main school subject for this study; only a small part of mathematics was tested. Being aware of this limitation, analysis of the data once more displayed a tendency of decline in Norwegian pupils' achievement in mathematics. As in former studies, Finland outperformed all the other Nordic countries. The students from Iceland, Denmark, and Sweden performed higher than the OECD average, but only the Danish students scored significantly better. The Norwegian results appear disappointing, since the Norwegian students scored significantly lower than the OECD average for the first time in PISA. (OECD, 2007; http://www.pisa.oecd.org/).

FROM DECLINE TO PROGRESS IN MATHEMATICS ACHIEVEMENT IN NORWAY

In December 2008, the results from TIMSS 2007 were announced (Grønmo & Onstad, 2009). This was the first time an international comparative study showed an increase in Norwegian pupils' performance in mathematics. This increase was significant both in grade 4 and grade 8. Figure 4.5 gives an overview of the changes in mathematics achievement in Norway in TIMSS from 1995 to 2007.

This positive change in trends in mathematics achievement in compulsory school in Norway from 2003 to 2007 received a lot of attention in the media, at all levels in school, as well as among politicians. The national report from TIMSS 2007 had two main foci for understanding these results. One was to point out possible explanations for the increase in achievement from 2003 to 2007, the other to understand why the achievement of Norwegian pupils still was relatively weak from an international perspective. Norwegian

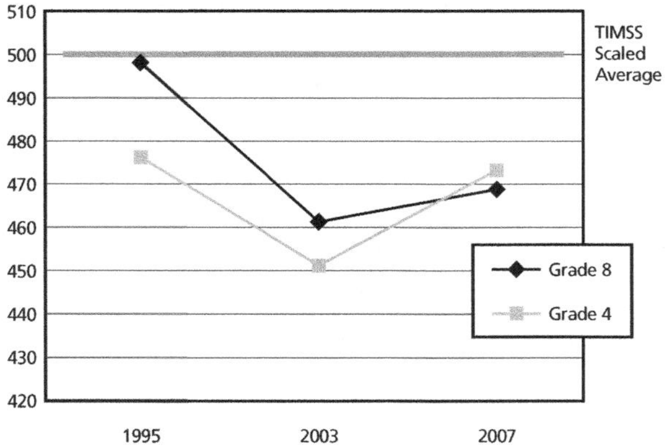

Figure 4.5 TIMSS Scaled Average is based on the international results from TIMSS 1995, standardised to average 500 and st.dev. 100. This is used as a fixed scale to measure and compare achievement levels from study to study.

pupils still performed low compared to the international mean as well as compared to selected countries it was reasonable to make comparisons with.

Possible reasons given for the increase in achievement in Norway were the following:

- The weak results in both TIMSS and PISA in 2003 attracted a lot of attention. This led to a broad political agreement on emphasizing knowledge in school.
- Several parliamentary white papers addressed the situation and stated the need for more emphasis on content matter in the curriculum.
- National tests in mathematics, Norwegian, and English were accepted as one way of emphasizing the importance of subject knowledge for pupils in schools.
- A new K–13 curriculum for Norwegian schools was implemented in 2006, under the name of "Knowledge Promotion."
- The new curriculum emphasized basic skills in arithmetic. Weak achievement in arithmetic had been pointed out as a main issue among pupils, based on the results from the TIMSS and PISA studies in 2003.
- As part of the new curriculum, the numbers of mathematics lessons were increased in primary school.

In addition to these more general changes in the society in favor of knowledge in school, the report also pointed out some other possible reasons for the improved achievement in mathematics based on responses in the TIMSS teacher questionnaire:

- There was an increase in the percentage of Norwegian mathematics teachers who reported that they followed up and reviewed their pupils' homework in TIMSS 2007 compared with TIMSS 2003.
- A higher percentage of Norwegian teachers participated in relevant professional development in 2007 than in 2003, referring to courses especially designed for teachers in mathematics.

Even if there was a clear increase for both these factors according to TIMSS data from 2003 and 2007, it is worth noticing that the values reported by the Norwegian teachers were still low compared to the international means for these questions. This was supported by the OECD study TALIS published in 2009 (Vibe, Aamodt, & Carlsten, 2009). In the national TIMSS report, these factors were therefore pointed out both as possible explanations for a national increase in achievement in mathematics as well as possible causes of the still weak results when compared with other countries in the study.

A closer analysis showed that it was especially in what may be referred to as formal mathematical knowledge that the Norwegian pupils' performance was weak. This was particularly pronounced in algebra in grade 8, and in arithmetic skills in grade 4. Figure 4.6 shows the Norwegian pupils' achievement in grade 8 in different areas of mathematical content. In this figure, performance of Norwegian pupils is compared to that of pupils in a number of selected countries. The selection of countries is based on the earlier analysis of countries having different profiles in mathematics education (see earlier in this chapter). Each country represents a different profile. In addition, it was taken into account that pupils in the selected countries should be of about the same age as pupils in Norway. This is the case for all the countries except for Japan. All the East Asian countries had pupils a little older, the Japanese pupils being about 8 months older than the Norwegian. However, this does not influence any conclusions made in this chapter. Adjusting the data for the Japanese pupils' achievement by age, these pupils will still perform much better than pupils in the other countries.

The tendency of low achievement in formal, pure mathematics is discussed in more depth in the section about the relationship between pure and applied mathematics later in this chapter. This was also mentioned earlier as a possible reason for why Norway performed low in PISA as well, pointing to a lack in basic skills (Grønmo, 2005a; Grønmo & Olsen, 2006).

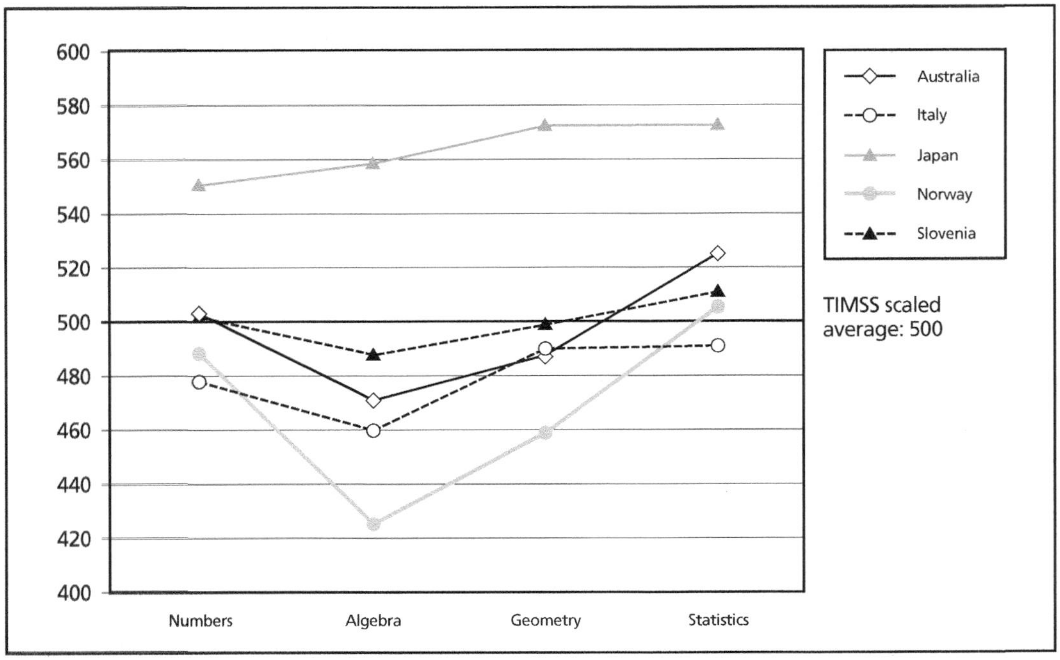

Figure 4.6 Pupils' performance in different content areas of mathematics in TIMSS 2007 (From Grønmo & Onstad, 2009).

TIMSS 2007 defined four benchmark levels for pupils' achievement in mathematics, Advanced level, High level, Intermediate level, and Low level of competence. Pupils at Low level have some basic knowledge about numbers and simple graphs, while pupils reaching Intermediate level can use basic knowledge in different content areas in simple one-step situations. Pupils at High level have a better understanding of concepts, and can use their skills and knowledge in more complex situations, including problems that need to be solved with more than one step. The highest competence level, Advanced level, requires that pupils can solve a number of complex non-routine problems; in Grade 4 they have to explain their solutions, in Grade 8 also to generalise and justify their answers. For a more detailed description of these competence levels, see Mullis, Martin, and Foy (2008) and Grønmo and Onstad (2009). Figure 4.7 displays the distribution over benchmark levels at grade 8 in TIMSS 2007 for Norway and the chosen reference countries.

Figure 4.7 reveals that Norway has more pupils than any of these countries at Low level and below, and fewer pupils at the two highest competence levels. As can be seen in the figure, none (less than 0.5 %) of the Norwegian pupils reached Advanced level in mathematics in grade 8. In 1995, 4% of the pupils in Norway reached that level, but none in 2003 and 2007. Results from the latest PISA study (Kjærnsli et al., 2007) also support the fact that it seems to be a special problem for Norwegian schools that there are fewer students at the highest achievement levels in mathematics than other countries relevant to compare with. This led to a question posed in the latest TIMSS national report about what the Norwegian schools offer to their best pupils—these being the pupils most likely to be recruited to studies and occupations requiring a high level of competence in mathematics. Recruitment to such studies and occupations has been pointed out several times as a serious problem in Norway (Norwegian Ministry of Education and Research, 2006; OECD, 2007). Hence, this was posed as an important question to policy makers and school teachers in the national TIMSS report (Grønmo & Onstad, 2009). At the same time, it is worth noticing that schools do not provide well for those pupils who have problems either. It seems that the Norwegian school is not doing well, either for the brightest pupils, or for the weak ones.

Pupils' and teachers' responses to questionnaires raise some important issues that seem to be typical for instruction in Norwegian schools. When pupils are asked about the use of different methods in mathematics lessons, Norwegian pupils score low compared to pupils in other countries in memorization and drill, as well as in explanation and reflection on mathematics. Both these methods may be seen as important and necessary in order to learn mathematics well, constituting mathematical competence as basic facts and skills combined with development of concepts. What seems to be a major strategy in Norway is pupils working individually—and according to teachers' answers

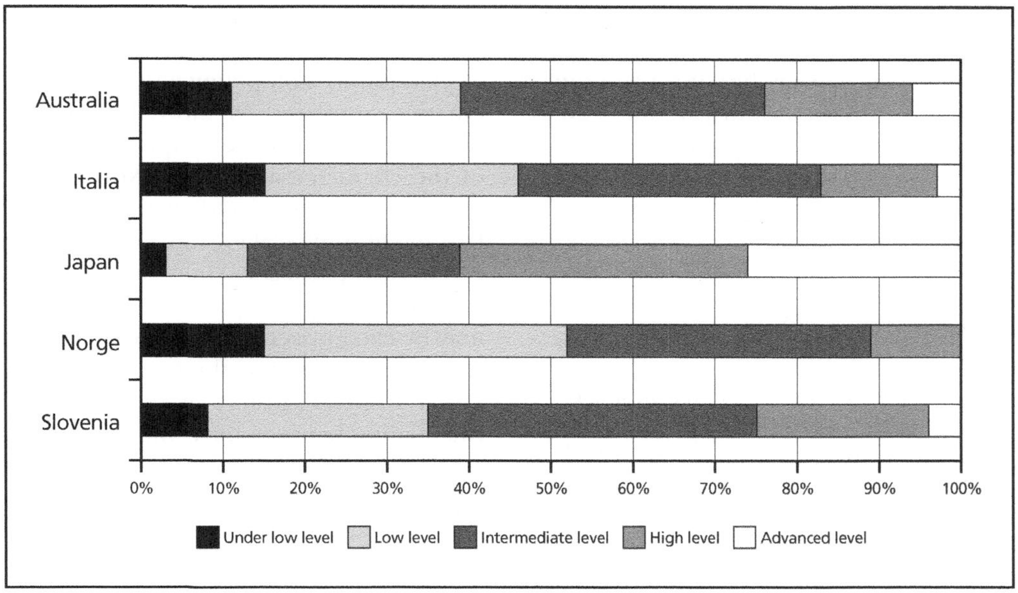

Figure 4.7 Distribution over benchmark levels at Grade 8 in Mathematics compared to reference countries in TIMSS 2007 (From Grønmo & Onstad, 2009).

to a greater extent than in other countries—without guidance from teachers (Grønmo & Onstad, 2009).

Both TIMSS and PISA are large-scale studies based on quantitative data and analysis. After these studies in 2003, the Ministry of Education initiated a project especially aimed at getting a deeper understanding of what happens in Norwegian classrooms. This project, called PISA+, was a study primarily based on qualitative research methods as observation and videostudies in classrooms and interviews with teachers and pupils after lessons. The motivation for this project was to shed light on some PISA and TIMSS findings from 2003 that were regarded as problematic, and to contribute to empirical school and classroom research in Norway. Results from this study were used in the national TIMSS 2007 report (Grønmo & Onstad, 2009) and made a substantial contribution to understanding the situation in Norwegian mathematics classrooms.

CLASSROOM-BASED RESEARCH—OBSERVATIONS AND VIDEO STUDIES

TIMSS 1995 conducted a video study in the U.S., Japan, and Germany. The video study explored a number of classrooms in each of these three countries, and compared the approaches to instruction and the discussions that followed between pupils and teachers. This study displayed some significant differences in teaching approaches and consequently pupils' experiences (Stigler & Hiebert, 1999). The main difference was not about whether the pupils listen to the teacher or not, but rather about what types of reflection their teachers required from the pupils, and how the objectives of the activities were integrated into the lesson. There were also clear differences in how the teachers treated the arguments of the pupils in the classroom.

Findings about mathematics from the PISA+ study indicated what characterised Norwegian classrooms (Bergem, 2009; Grønmo & Onstad, 2009):

- Interaction in Norwegian classrooms between pupils and teachers was characterized by a good and inclusive atmosphere, and there was considerable room for pupil initiative.
- Most classrooms were characterized by a high level of activity and pupil involvement, but activities were often isolated events, not related to subject content and aims. Intentions and goals behind the learning activities were seldom summarized and seen in a broader context.
- Learning was to a large extent privatized and left to the individual pupil. This effect was enhanced by the extensive use of individual work plans, which were used in all classes. In mathematics, there was an extensive use of individual tasks given to pupils as part of their working plans.

The sum of these factors makes learning a responsibility of the individual pupil. Learning a subject becomes a concern between the pupil, the text and the work plan, with the teacher in a peripheral role. These findings were consistent with results from TIMSS and

PISA, in the sense that a lot of attention is on individual work, and less focus on common classroom activities. This tendency in Norway has also been questioned by researchers who are not involved in TIMSS or PISA, as shown in the following quotes:

> ... a primitive rejection of community learning as reprehensible "Dissemination pedagogy" which has got nothing to do with our school of today. Instead, the classroom of pupil activities is about to reach single-tracked individualism. (Imsen, 2004, p. 70)

> It may therefore be questioned if research a bit single-mindedly has required adapted education to be taken as individualization in the form of individual work, individual guidance and pupil options, with the absence of full class instruction, or community learning. (Bachmann & Haug, 2006, p. 48)

Summarizing, it seems that the weak achievement by Norwegian pupils to some extent can be understood in the perspective that too much focus has been on individual responsibility given to the pupils, and too little attention has been given to the learning goals in mathematics. The latter especially seems to be the case when it comes to helping pupils to develop basic skills in mathematics. This is a main conclusion based on data from international comparative studies such as TIMSS and PISA, as well as from other studies. The other country having a lot in common with Norway in these respects is Sweden. And contrary to Norway, Swedish grade 8 pupils still showed a decrease in achievement from 2003 to 2007. It might be productive to have a discussion of these results including representatives from both Norway and Sweden. This discussion might focus on questions such as: What has been going on in our schools from 1995 to 2007? What types of changes have taken place, and what kinds of consequences have they had on pupils' performance in mathematics?

PURE AND APPLIED MATHEMATICS AND THE NORWEGIAN CURRICULUM

As already pointed out at the beginning of this chapter, the differences between frameworks, aims, and populations of TIMSS and PISA provide a great opportunity to gain knowledge and insight into the educational systems of different countries. In recent changes of the mathematics curriculum in Norway, discussions about applied mathematics as opposed to basic knowledge and skills in pure mathematics have played an important role. From that perspective it was of special interest to analyze data and see what kinds of information TIMSS and PISA can give about Norwegian pupils' competence in daily life applied mathematics compared to more traditional pure mathematics. Earlier in the chapter, results from analyses of the characteristics of the Nordic profile in mathematics were presented, with the main conclusion that the most typical for the Nordic profile—as for the group of English-speaking countries—being that the Nordic pupils performed relatively best on items that may be categorized as daily life or applied mathematics. The opposite was the case for the East Asian and East European clusters of countries. In this section, the relation between pure and applied mathematics is discussed a little more closely, and results of analyses comparing countries' levels of achievement in TIMSS and PISA are presented. The main focus is on results from TIMSS and PISA in 2003. The reason for this is that mathematics was the main subject in PISA 2003, so that a special opportunity was offered to compare results from both studies that year.

Grønmo (2005b) and Grønmo and Olsen (2006) have discussed the relationship between pure and applied mathematics in school, with focus on the situation for mathematics in compulsory school in Norway. Their discussion is based on the fact that pupils in some countries perform better in TIMSS while pupils in other countries perform better in PISA. Comparing the achievement levels in different countries in the two studies shows that some Nordic and English-speaking countries performed better in PISA than in TIMSS, in contrast to East European countries that performed relatively better in TIMSS. In their analyses, Grønmo (2005b) and Grønmo and Olsen (2006) referred to the differences in how mathematics competence is defined in the frameworks of the two studies (as described earlier in this chapter) as a background to understanding these shifts in rank. The findings of their analyses were that the assessment frameworks were formulated from largely different perspectives on mathematics, and that the differences in "operationalizations of mathematics in TIMSS and PISA reflect, to a large degree, this discussion about the relationship between mastery of basic facts and procedures in mathematics, on the one hand, and the ability to relate mathematics to real-world phenomena on the other" (Grønmo & Olsen, 2006, p. 3).

The discussion about what should constitute mathematics in compulsory school may be understood in the light of the considerable efforts and use of resources to develop education for all citizens in Western societies (Ernest, 1991; Skovsmose, 1994). It has been argued that the distinction between pure and applied mathematics may not be very well founded, referring to the fact that from a historical point of view, some of the main contributors in mathematics—Newton, Fermat, Descartes, and Gauss, among others—would probably not have recognized the distinction being made today between pure and applied mathematics, indicating that mathematics should be taught as a whole (Kline, 1972). However, in a discussion about what should be the content of mathematics in school, this distinction seems to be relevant and fruitful, as illustrated in the former analyses of different profiles in mathematics education. It also seems relevant in discussions of curriculum and curriculum changes for mathematics. As some have argued, an increasing focus on applied mathematics seems to have resulted in too little attention given to what we may call pure mathematics (Gardiner, 2004). Grønmo (2005b) and Grønmo and Olsen (2006) have pointed at problems created by underestimating the importance of pure mathematics and that only emphasizing applied mathematics may be one possible reason for the low performance of Norwegian pupils in both TIMSS and PISA.

Figure 4.8 presents a commonly accepted model of the relationship between pure and applied mathematics taken from an influential United States policy document on standards in mathematics (National Council of Teachers of Mathematics, 1989). PISA uses a slightly different form of this model (OECD, 2003). The right-hand side of the figure represents the mathematical world (what we may refer to as pure mathematics)—an abstract world with well-defined symbols and rules. The left-hand side represents the real, concrete world, containing an infinite number of different contexts and situations. The context or situation presented may either be scientific or what might be called daily life, the latter being the most relevant one for compulsory

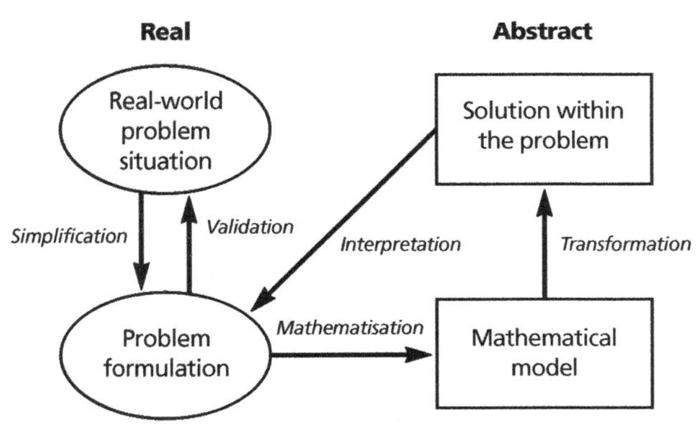

Figure 4.8 The Mathematization Cycle. *Source:* National Council of Teachers of Mathematics (1989). *Curriculum and evaluation standards for school mathematics.* Reston: Author.

school. Working with pure mathematics, such as numbers or algebra out of any context, means working only on the right-hand side of the model. In applied mathematics, the starting point is intended to be a problem from the real world, which first has to be simplified, and then mathematized into a model representing the problem. School mathematics rarely starts with a real problem. What is presented as a problem for pupils has in almost every case already been simplified to make it accessible to them.

For any type of applied mathematics, the pupils need to have some knowledge of pure mathematics to find a correct mathematical solution. Applied mathematics can therefore be seen as more complex than pure mathematics, if the same mathematics is involved in the two cases. Gardiner (2004) argues extensively that even if the ability to use mathematics to solve daily life problems is a main goal for school mathematics, this cannot be seen as an alternative to basic knowledge and skills in pure mathematics. It may rather underline the pupils' need to be able to orient themselves in the world of pure mathematics as a necessary prerequisite to solving real-world problems.

PISA aims at embedding all items in a context as close to a real-life situation as possible, while most items in TIMSS are pure mathematical items with no context, or items with a simplified context, as has long been the tradition in school mathematics. TIMSS therefore gives extensive information about pupils' knowledge in pure mathematics—or what may be called traditional school mathematics—while PISA mainly displays pupils' competence in solving items in a daily life context with the use of some mathematical knowledge—what may be referred to as applied mathematics.

Many countries have as a goal that, on leaving compulsory school, all pupils should have a type of competence that makes them well prepared to solve daily life problems using mathematics. This has been seen as important for active citizens in a modern society and has by some been referred to as functional numeracy (Niss, 1994; Niss, 2003; De Lange, 1996). The aim of PISA is to test pupils in this type of mathematical competence, defined in the study as Mathematical Literacy.

If a country or a group of countries achieves better in PISA than in TIMSS, the reason for this may be that pupils have not acquired a firm basis in pure mathematics. This lack of knowledge in pure mathematics may be general, or it may be in specific topics in mathematics. If the opposite is the case, that is, a country is achieving better in TIMSS than in PISA, it may be because pupils do not experience the full mathematization cycle for applied mathematics as illustrated in Figure 4.8.

Countries representing the East European profile performed relatively better in TIMSS than in PISA in 2003 (Grønmo & Olsen, 2006). From the simple proposition suggested above, the interpretation would be that most of the East European countries give little attention to the left-hand side of the mathematization cycle. The general message that this example serves to communicate is that concentrating only on pure mathematics in school may not be the best if the aim is to foster pupils who are mathematically literate—pupils who can use mathematics to solve the daily life problems they are likely to be exposed to. In contrast to the East European countries, countries representing the East Asian profile (i.e., Japan) are high achieving in both TIMSS and PISA. This may indicate that pure mathematics is emphasized in the mathematics curriculum in Japan, at the same time as attention is given to the full cycle of applied mathematics (see Figure 4.8).

A European country such as the Netherlands, also high-achieving in both studies in 2003, revealed some clear differences from Japan on performance levels in different topics in TIMSS. Comparing achievement in Japan and the Netherlands in grade 8, the countries achieved equally well in the topics Number, Measurement, and Data, while there were clear differences between these countries in their achievement levels in Algebra and Geometry. This indicates that even high-achieving countries may have pronounced differences in what they emphasize in their curriculum. Algebra and Geometry seem to be much more in focus in Japan than in the Netherlands. But when it comes to achieving well in mathematical literacy, as tested in PISA, the Netherlands is doing just as well as Japan. Grønmo and Olsen (2006) took this as an indication that the "basics" of most importance for daily life mathematics are the fundamental concepts of number and operations with numbers. A comparison of the Norwegian pupils' achievement in TIMSS with what the Norwegian mathematics curriculum focuses on suggests that the most problematic topic for instruction in schools in Norway is Number. Despite the fact that Number has been and is an extensive part of the curriculum all through compulsory school, Norwegian pupils tend to perform especially low in this content area in mathematics.

That the achievement in Algebra was even lower than in Number for pupils in Norway, in TIMSS 2003 as in 2007, may be easy to understand when taking into account that the Norwegian curriculum has not focused very much on this content area. Even if this is not the most important content for applying mathematics to solve daily life problems, this content knowledge is highly relevant for those going into studies and professions in need of more mathematical competence. The conclusions pointed out in the discussion of Figure 4.8 about the need for basic knowledge and skills to be able to apply mathematics is just as relevant for applying algebra as it is for applying number in mathematics problem solving. This aspect received more attention in the TIMSS 2007 report (Grønmo & Onstad, 2009), referring to the problems in Norway in recruiting pupils to educational programmes and professions requiring knowledge in algebra. In connection to this, the report also posed a critical question about what the Norwegian compulsory school offers to their most talented pupils in mathematics, the pupils who are most likely to be recruited to studies and professions in need of this type of mathematical knowledge.

One consequence of a growing focus on applied mathematics may be that problems arise if too little attention is given to pure mathematics. If pupils lack elementary knowledge and skills with numbers, this is important also for their later possibility to learn algebra. It has been pointed out that problems pupils have learning algebra in many cases are caused by a too weak basis in arithmetic (Brekke, Grønmo, & Rosén, 2000). And as already underlined, if talented pupils are not given the opportunity to learn basic concepts and skills in algebra, it will probably lead to later problems in recruiting them to studies and professions in need of such knowledge.

The problematic issues for Norway seem to be that too little focus has been on pure mathematics. This conclusion is based on results from a number of studies, as TIMSS 2003 and 2007, as well as from PISA 2003. In the following, a few items from TIMSS 2007 are presented to exemplify and discuss this in relation to pure and applied mathematics. Basic knowledge about number is related to two items from grade 4 in TIMSS 2007. Figure 4.9 and Figure 4.10 show two items together with the corresponding results in the

Multiply: 53×26

Answer: __________

Country	%
Australia	9
Italy	64
Japan	67
Norway	**2**
Slovenia	12
Int. Average	41

Figure 4.9 Item from the content category Number in Grade 4. The table shows percent correct for each country (p-values).

Al wanted to find how much his cat weighed. He weighed himself and noted the scale read 57 kg. He then stepped on the scale holding his cat and found that it read 62 kg.

What was the weight of the cat in kilograms?

Answer: ________ kilograms

Country	%
Australia	61
Italy	69
Japan	83
Norway	**67**
Slovenia	69
Int. Average	60

Figure 4.10 Item from the content category Number in Grade 4. The table shows percent correct for each country (p-values).

countries that were chosen as reference countries in the 2007 report for Norway.

The results for this item (Figure 4.9) exemplify that formal mathematical knowledge in pure mathematics, like the algorithm for multiplication, is not emphasized in Norwegian schools. It may be an issue that it is not obvious whether or not this item is part of the Norwegian curriculum at that level, even if the curriculum states that students have to be able to do multiplication with two-digit numbers in grade 4 in practical or daily life contexts. This item is not in a practical or daily life context, but referring to the former discussion about the need for pure mathematics to handle applied mathematics, it is not likely that Norwegian pupils would do better if the same numbers were presented in such a context. This is only given as an example of pupils' performance in Norway on arithmetical tasks in TIMSS 2007. The results point in the same direction for other items requiring basic numerical algorithms—for example, for subtraction or division. There is a consistent tendency that grade 4 pupils in Norway performed low on items requiring formal algorithms for calculation. Also Australian pupils performed low on this item, which is consistent with earlier results from TIMSS and PISA that there are many similarities between the Nordic and the English-speaking profiles in mathematics education, with more emphasis on applied and less on formal pure mathematics. The tendency, however, is most pronounced in the Nordic countries. This conclusion is also supported by analyses based on data from PISA 2003; a main problem seems to be that Norwegian pupils performed especially poorly on items requiring any type of calculation requiring an exact answer (Grønmo, 2005a, 2005b; Olsen & Grønmo, 2006). A relevant question seems to be why less attention is given to formal skills with numbers in the Nordic countries than seems to be the case in other parts of the world.

Norwegian pupils performed well on the item in Figure 4.10 compared both to the chosen reference countries and to the international average. This item, however, in contrast to the item in Figure 4.9, does not require any formal arithmetical skills. The item may be solved by counting to reach the answer. The results on these two items exemplify what seems to be typical for Norwegian pupils. They perform low at items requiring formal mathematical knowledge, but better on items that may be solved by some sort of reasoning without such knowledge.

The same tendency was revealed in pupils' achievement in grade 8. At this level it was not possible to draw any clear conclusions about their competence in arithmetical calculations, since pupils were free to use calculators for all items. However, at this level a similar tendency was clear in relation to items requiring formal algebra. These results indicate that little attention is given to formal mathematical knowledge in mathematics in Norway, interpreted as number in grade 4 and algebra in grade 8.

The item in Figure 4.11 was classified as an item at the highest competence level: Advanced International Benchmark. This item has to be solved in two steps, also requiring that pupils are able to conduct some reasoning to solve the problem. This, as well as the results internationally, exemplify that it is a rather difficult item for grade 8 pupils. The results on this item showed that Norwegian pupils performed at the same level as the international mean. However, it is worth noticing that this item can be solved without any formal competence in algebra. It may be solved by a strategy of trial and improvement, this being the main strategy for Norwegian pupils. Less than 2% of Norwegian pupils used any type of formal algebra to solve this item.

Figure 4.12 shows the results for another algebra item in TIMSS 2007. This item was classified as an easier item internationally than the item in Figure 4.11, being mirrored by the better international aver-

Joe knows that a pen costs 1 zed more than a pencil. His friend bought 2 pens and 3 pencils for 17 zeds. How many zeds will Joe need to buy 1 pen and 2 pencils?

Show your work.

Country	%
Australia	36
Italy	19
Japan	42
Norway	**18**
Slovenia	30
Int. Average	18

Figure 4.11 Item from the content category Algebra in Grade 8. The table shows percent correct for each country (p-values).

In Zedland, total shipping charges to ship an item are given by the equation $y = 4x + 30$, where x is the weight in grams and y is the cost in zeds. If you have 150 zeds, how many grams can you ship?

Ⓐ 630
Ⓑ 150
Ⓒ 120
Ⓓ 30

Country	%
Australia	26
Italy	24
Japan	65
Norway	**10**
Slovenia	37
Int. Average	34

Figure 4.12 Item from the content category Algebra in Grade 8. The table shows percent correct for each country (p-values).

age as well as the level of performance for the chosen reference countries. Nevertheless, the Norwegian pupils' performance on this item is strikingly lower compared to the international mean and to the reference countries (Grønmo & Onstad, 2009) than it was for the former item. A main difference between these two items, both classified as algebra items in TIMSS, is the unavoidable need of some formal mathematical competence in algebra in order to solve the last item. This result is a good illustration of how performance by pupils in Norway is strikingly low in Grade 8 compared to an international level when any formal knowledge of algebra is required.

Learning Goals and Learning Strategies

There has been an increasing emphasis on learning strategies in mathematics education (Grønmo & Trondsen, 2006). To some extent, it seems that the main focus has been on metacognition and self-regulated learning as a learning strategy. This strategy should not be underestimated, but as Grønmo and Trondsen (2006) argue, there are other learning strategies that are just as important. Learning strategies have to be chosen based on the mathematical competence pupils are supposed to develop, and it is important to recognize that the constructivist view of learning does not rule out any teaching techniques in principle (Ernest, 2004). Unfortunately, it seems that some learning strategies have been rejected by referring to constructivism (Grønmo et al., 2004b). "It is important that children should practice routine manipulations until they can be done with an appropriate degree of fluency" (D.E.S., 1982, Chapter 5). Routine practice may be the most relevant strategy for the learning of basic facts and skills, while probably *not* that relevant for concept building and developing strategies in problem solving (D.E.S., 1982; Schoenfeld, 1992). However, basic facts and skills learned this way can become richly connected into conceptual structures if subsequent links are made. Based on research presented in this chapter, a main problem for Norwegian pupils in compulsory school seems to be their lack of basic formal knowledge and skills in mathematics. Taking this into consideration, it seems relevant to argue for the use of a variety of learning strategies, also including different types of routine practice in Norwegian schools. The national report from PISA 2003 seems to support such a view. Schools in PISA 2003 where pupils' achievement was higher than expected, taking pupils' background into consideration, were characterised by more use of such learning strategies (Kjærnsli et al., 2004). That basic skills in elementary mathematics seem to be a necessary condition for doing well in applied mathematics is also supported in TIMSS: The countries that did well on items in problem solving also gained high scores on the more elementary items (Mullis, Martin, Gonzalez, & Chrostowski, 2004).

Examples of basic knowledge and skills will typically be number algorithms, knowledge of the multiplication table, and the four operations for calculating with natural numbers. This will probably be what most people have in mind when talking about basic skills in mathematics. This has also been in focus in this chapter, but it has to be remembered that even if this covers, reasonably well, the basic skills developed in compulsory school, there are basic skills that need to be developed and carried out automatically[3] at all levels in mathematics, such as manipulation with symbols in algebra and operations with complex numbers and matrices at higher levels in education. The need for well-mastered routines is necessary to free conscious attention as much as possible so that it can focus on aspects of a task which are novel or problematic. The mathematics curriculum in compulsory school in Norway emphasizes competence with numbers; nevertheless, some expressions in the curriculum from 1997 may contribute to underestimating the need for automatic recall of facts—for example, the multiplication table: *"Pupils who have difficulties with memorising the basic multiplication facts must nevertheless be free to proceed to concepts and tasks involving the multiplication concept. Understanding multiplication and how to use it is more vital than memorising the tables."* (L97, p. 166) This may give the impression that it is not very important to memorize the multiplication table. Also, the fact that the first main area in the curriculum from 1997 was "Mathematics in daily life"—apparently given greater status than areas such as number, geometry and algebra—may also contribute to giving the impression that pure mathematics is not so important. As already mentioned, some have argued hard against underestimating the importance of pure mathematics for being able to use mathematics to solve daily life problems (Gardiner, 2004). PISA may be good for measuring whether Norwegian pupils have the kind of competence we may refer to as mathematical literacy, while the TIMSS study can give more information about the elementary basic knowledge they may need to be able to apply mathematics in daily life problems (Grønmo, 2005; Grønmo & Olsen, 2006).

Data from TIMSS 2007 showed that Norwegian pupils reported low use of strategies in the classroom such as memorizing formulas and procedures, as well as minimal reflection about problem solving (Grønmo & Onstad, 2009). Both are important strategies for learning mathematics. This may indicate that it is reasonable to give more attention in schools to different learning strategies in mathematics. For more about learning strategies in mathematics, see Grønmo and Trondsen (2006).

CONCLUSIONS

Large-scale international studies of pupils' achievement are frequently criticized for only being used to rank countries. Several of the analyses presented in this chapter demonstrate that data from TIMSS and PISA can be used and analyzed from many different perspectives. Studies such as TIMSS and PISA are not to be seen as surveys only, but rather as data-banks offering researchers opportunities to access a large number of interesting first- and second-hand analyses of data of the highest quality, the product of rigorous systems for developing instruments, data collection, and handling developed in these studies. Some additional ways of secondary analysis have been presented in this chapter, but there are numerous unexplored possibilities for analysis that can be made on these data. About half a year after the first international results from these studies have been presented, the data are available for all researchers in the world on international websites for the studies (for TIMSS, see http://timss.bc.edu/; for PISA, see http://www.pisa.oecd.org/).

These types of second-hand analyses can be a resource for valuable educational research. IEA have, for that reason, initiated special conferences for secondary analyses of data from their studies (see IRC-conferences http://www.iea.nl/iea_irc.html). Some of the papers referred to in this chapter have been presented in such conferences. OECD has started initiating similar conferences for secondary analyses of PISA data (http://www.pisaresconf09.org/). It is also worth noting that follow-up studies like PISA+, using more qualitative methods, can contribute to a deeper understanding of results from international comparative studies. Starting with results from analyses of data from international comparative studies, and going further with research of different types using quantitative or qualitative methods have for example resulted in a number of master's and doctoral theses.

Two further international comparative studies have been completed and the first results are due to be published late in 2009 or in 2010. One of these is TIMSS Advanced, a study of specialists in mathematics and physics in the final year of upper secondary school. A similar study was part of the ordinary TIMSS in 1995, so this is a trend study designed to compare how pupils' achievement and attitudes at this level have changed in these two subjects from the mid nineties till 2008. The first results from this study were released in December 2009.[4] It is remarkable how consistent the results from this study are with the results from the international comparative studies in compulsory school presented in this article. "The decline in performance in mathematics for both Norwegian and Swedish students since 1995/1998 corresponds to the results of TIMSS surveys of students in primary and lower secondary school" (Grønmo et al., 2009, p. 7). The national report for mathematics in TIMSS Advanced (Grønmo, Onstad & Pedersen, 2010) emphasize consistency with results in primary and lower secondary school on a number of other factors as well. For example that solving items individually is more pronounced in Norway than in most other countries, while methods like discussing and arguing about problem solving and strategies is not that commonly used. Analysis of TIMSS Advanced data for Norway reveal that more use of methods as discussing and arguing are positively correlated with higher level of achievement in a class. Also amount of homework Norwegian teachers assign is positively correlated with level of achievement for classes.

The other study, results due to be published in 2010, is TEDS-M, which is the first international comparative study of students at university level. This is a survey of the education of future mathematics teachers and their knowledge and attitudes in sixteen countries around the world. This study also offers a number of new possibilities for different types of analyses. For example, can analyses of mathematics in school be linked to analyses of data from students to become future teachers in mathematics?

NOTES

1. Compulsory school in Norway (Grunnskole) covers primary grades 1 to 7, and lower secondary grades 8 to 10. Children start grade 1 at age 6. In addition, the national aim is to achieve 100% provision for children in kindergarten (ages 1–6 years). There is a national framework for early learning that applies to kindergarten.
2. PISA focuses on one school subject in each study: reading in 2000, mathematics in 2003, and science in 2006. Subjects not in focus will only be compared for smaller parts of the content. Trends in achievement in PISA in three-year cycles will only give some indications; measuring trends in each school subject is more to be seen as nine-year cycles.
3. The words "automatical" or "automatic" are used to cover any type of routine practice or manipulation that can be done with an appropriate degree of fluency.
4. For English short version see http://www.timss.no/rapporter%202008/Kortrapport%20TIMSS%20Advanced%202008%20engelsk.pdf

REFERENCES

Alseth, B., Brekke, G., & Breiteig, T. (2003). *Evaluering av Reform 97. Endringer og utvikling ved L97 som bakgrunn for videre planlegging og justering – matematikkfaget som kasus.* Notodden: Telemarksforsking.

Bachmann, K. & Haug, P. (2006). *Forskning om tilpasset opplæring,* Forskingsrapport nr. 62, Høgskulen i Volda.

Beaton, A. E., Mullis, I. V. S., Martin, M. O., Gonzalez, E. J., Kelly, D. L., & Smith, T. A. (1996). *Mathematics achievement in the middle school years: IEA's Third International Mathematics and Science Study (TIMSS).* Chestnut Hill, MA: TIMSS International Study Center, Lynch School of Education, Boston College.

Bergem, O. K. (2009). *Individuelle versus kollektive arbeidsformer. En drøfting av aktuelle utfordringer i matematikkundervisningen i grunnskolen.* (*Individual Seat Work Versus Collaborative Practices. A Discussion of Current Challenges in the Teaching of Mathematics in Lower Secondary School*). PhD thesis, Department of Teacher Education and School Development, University of Oslo.

Bergem, O. K., Grønmo, L. S. & Olsen, R. V. (2005). PISA 2003 og TIMSS 2003: Hva forteller disse undersøkelsene om norske elevers kunnskaper og ferdigheter i matematikk. *Norsk Pedagogisk Tidsskrift, 89*(1), 31–44.

Brekke, G., Grønmo, L. S., & Rosén, B. (2000). *KIM (Kvalitet i matematikkundervisningen): Veiledning til algebra.* Oslo: Nasjonalt Læremiddelsenter.

De Lange, J. (1996). Using and applying mathematics in education. In A. J. Bishop, K. Clements, C. Keitel, J. Kilpatrick, & C. Laborde (Eds.), *International handbook of mathematics education,* Vols 1–2 (pp. 49–98). Dordrecht, Kluwer Academic Publishers.

D.E.S. (1982). *Mathematics counts: Report of the Committee of Inquiry into the Teaching of Mathematics in Schools under the chairmanship of Dr. W. H. Cockcroft.* London: HMSO.

Ernest, P. (1991). *The philosophy of mathematics education.* London: Falmer Press.

Ernest, P. (2004). *The psychology of learning mathematics.* Advanced Course Module. Special Field: Mathematics Education. University of Exeter.

Gardiner, A. (2004, July). *What is mathematical literacy?* Lecture given at the ICME-10 conference, Copenhagen, Denmark.

Grønmo, L. S. (2005a). Matematikkprestasjoner i TIMSS og PISA. *Nämnaren, 32*(3), 5–11.

Grønmo, L. S. (2005b). Ferdighetenes plass i matematikkundervisningen. *Nämnaren, 32*(4), 38–44.

Grønmo, L. S. & Olsen, R. V. (2006, November). *TIMSS versus PISA: The Case of Pure and Applied Mathematics.* Proceedings at the 2nd IEA International Research Conference, Washington DC.

Grønmo, L. S. & Onstad, T. (2009). *Tegn til bedring. Norske elevers prestasjoner i matematikk og naturfag i TIMSS 2007.* Oslo: Unipub.

Grønmo, L. S. & Trondsen, I. S. (2006). Læringsstrategier i matematikk. In A. Turmo & E. Elstad (Eds.), *Læringsstrategier – Søkelys på lærernes praksis* (pp. 178–195). Oslo: Universitetsforlaget.

Grønmo, L. S., Kjærnsli, M., & Lie, S. (2004, May). *Looking for cultural and geographical factors in patterns of responses to TIMSS items.* Paper presented at the 1st IEA International Research Conference, Lefkosia, Cyprus.

Grønmo, L. S., Bergem, O. K., Kjærnsli, M., Lie, S., & Turmo, A. (2004). *Hva i all verden har skjedd i realfagene? Norske elevers prestasjoner i matematikk og naturfag i TIMSS 2003.* Oslo: Department of Teacher Education and School Development, University of Oslo.

Grønmo, L.S., Onstad, T., Pedersen, I. F., Lie, S., Angell, C., & Rohatgi, A. (2009). *Mathematics and physics in upper secondary school. One step back.* Abridged report presenting main results from TIMSS Advanced 2008 in Norway. Department of Teacher Education and School Development, University of Oslo.

Grønmo, L. S., Onstad, T. & Pedersen, I. F. (in press). *Matematikk i motvind – TIMSS Advanced 2008 i videregående skole.* Oslo, Unipub.

Imsen, G. (2004). *Det ustyrlige klasserommet: Om styring, samarbeid og læringsmiljø i grunnskolen.* Oslo: Universitetsforlaget.

Kjærnsli, M., Lie, S., Olsen, R. V., & Roe, A. (2007). *Tid for tunge løft. Norske elevers kompetanse i naturfag, lesing og matematikk i PISA 2006.* Oslo: Universitetsforlaget.

Kjærnsli, M., Lie, S., Olsen, R. V., Roe, A., & Turmo, A. (2004). *Rett spor eller ville veier? Norske elevers prestasjoner i matematikk, naturfag og lesing i PISA 2003.* Oslo: Universitetsforlaget.

Kilpatrick, J., Hoyles, C., Skovsmose, O., & Valero, P. (2005). *Meaning in mathematics education.* New York: Springer.

Klette, K. (2003). *Klasserommets praksisformer etter L97. Evaluering av L 97.* Oslo: Unipub.

Kline, M. (1972). *Mathematical thought from ancient to modern times.* Oxford: Oxford University Press.

L97. The Royal Ministry of Education, Research and Church Affairs. (1996). *The Curriculum for the 10-year Compulsory School in Norway.* Oslo: National Centre for Educational Resources.

Lie, S., Kjærnsli, M., & Brekke, G. (1997a). *9-åringers kunnskaper og holdninger i realfag i et internasjonalt perspektiv.* TIMSS-rapport nr. 25, Department of Teacher Education and School Development, University of Oslo.

Lie, S., Kjærnsli, M. & Brekke, G. (1997b). *Hva i all verden skjer i realfagene: internasjonalt lys på trettenåringers kunnskaper, holdninger og undervisning i norsk skole.* Oslo, Department of Teacher Education and School Development, University of Oslo.

Mullis, I. V. S., Martin, M. O., Smith, T. A., Garden, R. A., Gregory, K. D., Gonzalez, E. J., Chrostowski, S. J., & O`Connor, K. M. (2003). *TIMSS assessment frameworks and specifications 2003.* Chestnut Hill, MA: TIMSS & PIRLS International Study Center, Lynch School of Education, Boston College.

Mullis, I. V. S., Martin, M. O., Gonzalez, E. J., & Chrostowski, S. J. (2004). *TIMSS 2003 International Mathematics Report. Findings from IEA's Trends in International Mathematics and Science Study at the Fourth and Eighth Grades.* Chestnut Hill, MA: TIMSS & PIRLS International Study Center, Lynch School of Education, Boston College.

Mullis, I. V. S., Martin, M. O. & Foy, P. (2008). *TIMSS 2007 International Mathematics Report. Findings from IEA's Trends in International Mathematics and Science Study at the Fourth and Eighth Grades.* Chestnut Hill, MA: TIMSS & PIRLS International Study Center, Lynch School of Education, Boston College.

National Council of Teachers of Mathematics (1989). *Curriculum and Evaluation Standards for School Mathematics.* Reston, VA: Author.

Norwegian Ministry of Education and Research (2006). *A Joint Promotion of Mathematics, Science and Technology—Strategy 2006–2009.* Oslo: The Ministry of Education and Research. Retrieved May 16, 2008 from http://www.regjeringen.no/upload/kilde/kd/pla/2006/0003/ddd/pdfv/294463-strategi_engelsk.pdf

Niss, M. (1994). Mathematics in society. In R. Biehler, R. W. Scholz, R. Straesser, & B. Winkelmann (Eds.), *The didactics of mathematics as a scientific discipline* (pp. 367–378). Dordrecht: Kluwer Academic Publishers.

Niss, M. (1996). Goals of mathematics teaching. In A. J. Bishop, K. Clements, C. Keitel, J. Kilpatrick, & C. Laborde (Eds.), *International handbook of mathematics education,* Vols. 1–2 (pp. 11–18). Dordrecht: Kluwer Academic Publishers.

Niss, M. (2003). Mål for matematikkundervisningen. In B. Grevholm (Ed.), *Matematikk for skolen* (pp. 288–334). Bergen: Fagbokforlaget.

OECD. (2003). *PISA 2003 Assessment Framework. Mathematics, Reading, Science and Problem Solving, Knowledge and Skills.* Paris: OECD Publications.

OECD. (2007). *Education at a Glance 2007: Highlights.* Paris: OECD Publications.

OECD. (2009). *PISA 2006 Technical Report.* Paris: OECD Publications.

Olsen, R. V. (2006). A Nordic profile of mathematics achievement: Myth or reality? In J. Mejding & A. Roe (Eds.), *Northern lights on PISA 2003—A reflection from the Nordic countries* (pp. 33–46). Oslo: Nordisk Ministerråd.

Olsen, R. V. & Grønmo, L. S. (2006). What are the characteristics of the Nordic profile in mathematical literacy? In J. Mejding & A. Roe (Eds.), *Northern Lights on PISA 2003 – a reflection from the Nordic countries* (pp. 288–334). Oslo: Nordisk Ministerråd.

Olson, J. F., Martin, M. O., & Mullis, I. V. S. (2008). *TIMSS 2007 Technical Report.* Chestnut Hill, MA: TIMSS & PIRLS International Study Center, Lynch School of Education, Boston College.

Schoenfeld, A. H. (1992). Learning to think mathematically: Problem solving, metacognition, and sense making in mathematics. In D. A. Grouws (Ed.), *Handbook of research on mathematics teaching and learning* (pp. 334–370). New York: MacMillan.

Skovsmose, O. (1994). *Towards a philosophy of critical mathematics education.* Dordrecht: Kluwer Academic Publishers.

Sowder, J. T. (2007). The mathematical education and development of teachers. In F. Lester, Jr. (Ed.), *Second handbook of research on mathematics teaching and learning. Volume 1* (pp. 157–223). Charlotte, NC: Information Age.

Stigler, J. W. & Hiebert, J. (1999). *The teaching gap.* New York: Free Press.

Vibe, N., Aamodt, P. O., & Carlsten, T. C. (2009). *Å være ungdomsskolelærer i Norge. Resultater fra OECDs internasjonale studie av undervisning og læring (TALIS).* Rapport 23, NIFU STEP, Oslo.

5

COLLABORATIVE INQUIRY IN DEVELOPING MATHEMATICS TEACHING IN NORWAY

Barbara Jaworski

UNIVERSITY OF AGDER, NORWAY AND LOUGHBOROUGH UNIVERSITY, UK

Abstract: This chapter focuses on a developmental research project between teachers from eight schools and didacticians from a university in the south of Norway. The LCM project (Learning Communities in Mathematics) set out to create communities of inquiry between didacticians and teachers to promote development of mathematics learning and teaching in school classrooms. The chapter has four main sections focusing on Developmental Research, Co-learning in Inquiry Communities, Mathematics for Learning and Teaching, and Development: Collaboration and Learning. These sections deal with the main theoretical perspectives in the project, those of communities of practice and inquiry and of developmental research, along with issues and examples from developmental practice. The final section looks towards a more global perspective on development using an activity theory frame.{\abs}

Keywords: Inquiry community, Co-learning partnership, Mathematics teaching, Developmental research, Activity theory

INTRODUCTION

This chapter focuses on developmental research in Norway, commencing in January, 2004 and continuing to 2010 and beyond, in which teachers in schools and didacticians[1] in a university have worked together to promote better teaching and learning of mathematics in school classrooms. Several projects have been and are involved.[2] The first to be initiated was the LCM project, *Learning Communities in Mathematics*, which set the scene theoretically. It set out to establish *inquiry communities* (Jaworski, 1998; Wells, 1999) between mathematics teachers and didacticians to inquire into aspects of teaching and learning and to enhance opportunities for pupils to learn mathematics. Briefly, the idea was that teachers and didacticians, each working within their own established communities and aligning with the norms of practice as socially and historically constituted, would bring an element of inquiry to their practice. This would allow the questioning of established norms, and the design and introduction of innovative approaches to teaching and learning. Systematic scrutiny of questioning and innovation would allow a charting of the developmental process and simultaneously promote development. Teachers and didacticians, working together and inquiring into their own practices, would be insider researchers conducting inquiry, learning from their inquiry processes, and promoting new knowledge in practice. In addition, didacticians acting as academic researchers would chart the developmental process from the outside. These ideas and theoretical

The First Sourcebook on Nordic Research in Mathematics Education, pages 71–90

perspectives on which they are based are developed further in the sections which follow.

At the same time, a second project, ICTML (*ICT in Mathematics Learning*—see endnote 2 and Fuglestad, Chapter 6, this volume) was established on the same philosophical basis as LCM, but with a principal focus on the use of ICT as a tool in teaching and learning mathematics. When funding from the research council was agreed, schools were recruited to both projects; each school volunteered to take part in a project and agreed to its basic principles as expressed and explained at the time. Meetings were held with teachers and school leaders to explain what was envisaged in terms of community and inquiry. Volunteering schools agreed to have a commitment to the projects at school leadership level, and to have a minimum of three teachers committed to one or both projects. In total there were eight schools ranging from lower primary to upper secondary levels of which, initially, seven were in LCM and five in ICTML. A small amount of money was provided to each school to support its work in the project.

The four-year LCM project included three phases of field work, each of one school year. During the three phases, 16 workshops, each of about three hours' duration, were held between didacticians and teachers. These workshops formed the backbone of the projects. Their function was to be effective agents in achieving the developmental and research goals by contributing to building community, promoting inquiry, presenting mathematical tasks, encouraging discussion on classroom issues, and generating new perspectives for both teachers and didacticians on developmental opportunities. In parallel, ICTML held 12 workshops focusing on ICT use. Following a workshop, the teacher team in a school was expected to generate activity in school according to the teachers' perceptions of what would be valuable for their own pupils. School activity was diverse across the eight schools. Didacticians visited schools to work with teachers according to teachers' expressed wishes. Innovative activity in classrooms was video-recorded and data were collected from all events in a suitable format (e.g., video, audio, field notes, documents, etc.). Longitudinal testing of pupils was carried out, to locate pupils' mathematical performance in relation to data from Norway more generally as well as internationally (c.f. Breiteig & Grevholm Chapter 8, this volume).

The three phases of field work focused respectively on community building, developing a common understanding of inquiry, and establishing goals for each school team. As part of this fieldwork, vast amounts of data were collected in a variety of formats, including audio and video recordings of meetings and events and quantitative data from the longitudinal tests. A database was designed to organize this data and provide access to it. As outsider researchers, didacticians formulated, over time, a large collection of research questions. Particular members of the team addressed different subsets of this collection. Analysis was undertaken by individuals and small groups according to their expressed interests. Five doctoral students worked within the project, working as didacticians in contributing to development and analyzing data according to their own research focus.

This chapter presents a historical dimension, charting activity from project design through several phases of developmental research to highlight issues and findings. Collaboration between teachers and didacticians is the key cohering element of the chapter with a focus on what is learned by each of these groups as they have worked both jointly and independently to develop mathematics teaching. Reporting of events takes a narrative form (Bruner, 1986; Burton, 1999) in which activity and findings are presented and exemplified.

The narrative form is consistent with the qualitative methods employed in documenting developmental research. Naturalistic data were collected systematically from all events: for example, recordings (video or audio) of meetings, workshops, small group activity, classroom lessons; documents from planning, and presentation in university and in schools. To support analysis of naturalistic data, supplementary data (e.g., from interviews, focus groups, classroom tests) were collected purposefully. All data were stored electronically in a central data bank with access open to all didacticians. Analysis, conducted by didacticians, was related to research questions that had been discussed and refined. First, a process of data reduction of audio or video data was carried out to summarize as factually as possible the content of recordings. It was then possible to search the data according to a specific research question. Transcriptions were made for a finer grain analysis as appropriate to the questions addressed. Researchers sought to understand events and the meanings and understandings of their participants. Data and theory were used dialectically to interpret, inform, and theorize. The process is critically interpretative taking account of all the available evidence. Thus, when reporting from an event below, the narrative presented draws on such analysis.

Following this introduction, the chapter has 5 main sections, focusing on (1) *Developmental Research*, (2) *Co-learning in Inquiry Communities*, (3) *Mathematics for Learning and Teaching*, (4) *Development: Collaboration and Learning* and ending with (5) *Discussion and Conclusions*. The linear presentation of these sections belies their deeply inter-related nature. Thus, issues arising in some sections are addressed in others and there is a cycling of ideas and issues as the sections progress. The chapter builds on a growing collection of papers (http://fag.hia.no/lcm/) and a book published in 2007 (Jaworski et al., 2007).

DEVELOPMENTAL RESEARCH

Developmental research, as the name suggests, is research that is associated with developmental aims. The term has been used variously in relation to developmental projects in education (e.g., Cochran Smith & Lytle, 1999; Van den Akker, 1999). Here I focus specifically on research and development in *mathematics education* and take the term to refer to research that *promotes*, as well as charts, developmental processes: thus developmental research is "research which both studies the developmental process and, simultaneously, promotes development through engagement and questioning.... Not only are research questions defined and explored... but the whole research process is subject to question and exploration. We look critically at our research activity while engaging in and with it" (Jaworski & Goodchild, 2006, p. 353).

In the projects, we have sought to develop mathematics teaching and learning in school classrooms and research has been an integral process. The main idea theoretically is that as practitioners engage in research, they gain insights into the practices and processes in which they participate, and such insight promotes new knowledge, understanding, and awareness of issues that lead to new ways of being and doing in practice (Jaworski, 2003, 2004). I take these ideas further in the next section.

Simon Goodchild (2008), writing also about these projects, draws on the work of Koeno Gravemeier (1994) in mathematics curriculum development to present a model (Figure 5.1) in which development and research are both distinct and also closely inter-related and dependent.

Goodchild reports on Gravemeijer's argument, with reference to Freudenthal (1991), that knowledge of the developmental process is essential if the outcome of the process is to be understood: "... developmental research means: experiencing the cyclic process of development and research so consciously, and reporting on it so candidly that it justifies itself, and that this experience can be transmitted to others to become like their own experience" (Freudenthal, 1991, p. 161 cited in Goodchild, 2008, p. 209).

In the LCM and ICTML projects, typically, teachers engaged in activity in their schools relating to experiences in a recent workshop. The original proposal had suggested that a teacher team would design mathematical tasks for the classroom, observe each other in the classroom, and reflect together on outcomes. It was envisaged that tasks would encourage inquiry-based activity that would take pupils deeply into mathematics and promote conceptual understanding as well as procedural proficiency. A project aim was that teachers would develop their design and analytical qualities where use of tasks was concerned, able to go beyond the textbook in creating valuable tasks for their pupils and able to judge the quality of outcome of these tasks (see also Jaworski, Goodchild, Daland, & Eriksen, in press).

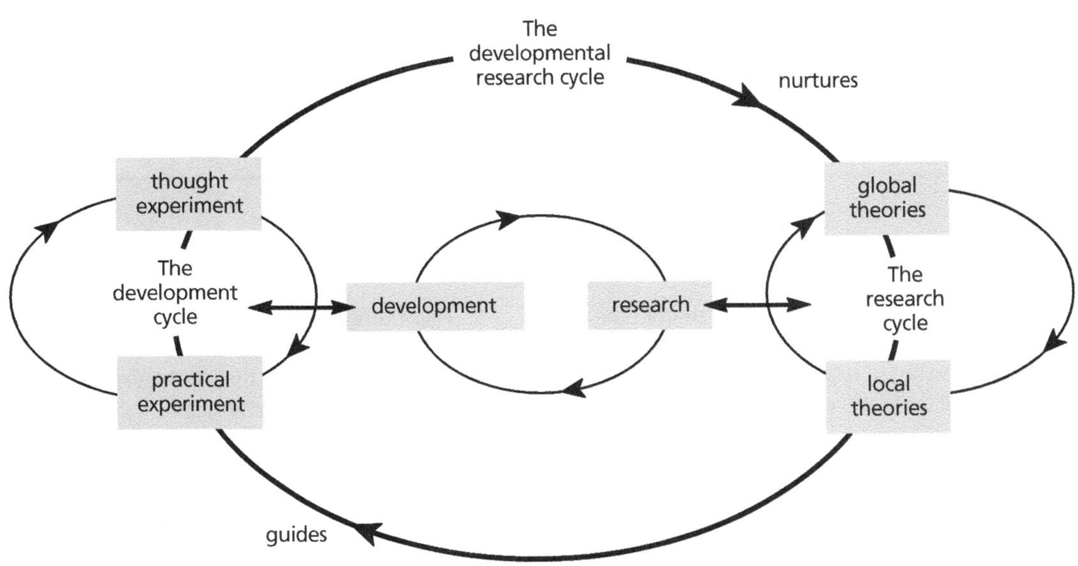

Figure 5.1 The development research cycle (Goodchild, 2008 p. 208).

Work in schools demonstrated a number of interpretations of these goals, not all in line with project proposals. One example, however, followed project goals quite closely. Early in the project (half way through Phase 1), three teachers in an upper secondary school decided they would like to design tasks related to the teaching of *linear functions* and invited didacticians to work with them. The teachers were concerned about the difficulties their students had experienced in learning about linear functions in the past. Two meetings were held, the first in their school and the second in the university. The teachers asked to meet in the university so that they could get away from the school environment to plan without interruption, which seemed an indication of commitment to the project, but also hinted that concentrated planning might be difficult within a school environment.

With reference to the textbook that the teachers normally used, teachers and didacticians discussed linear functions: the essential concepts that need to be understood, the particular difficulties for students that experience and research reveal, and the kinds of exploratory tasks with which students might engage. After this, the teachers, back in school, designed four tasks developing the ideas of linear functions and wrote these on cards for distribution to students. The four cards can be seen in Figure 5.2, translated from their original Norwegian. The ways the tasks are composed and written provide an indication of how teachers were interpreting the idea of inquiry-based tasks at this time. Each teacher taught one lesson based on the cards, and the lessons were video-recorded by didacticians.

Subsequently, teachers and didacticians met in an evening to reflect on the lessons while viewing segments of the video recording. The purpose of this meeting was twofold: first, to provide opportunity for teachers to reflect together and with didacticians, to allow issues to be made explicit and therefore, possibly, to be addressed; and second, to provide data as evidence for what was learned from these events.

At the next workshop, the teachers made a presentation of their work and reflections to the whole project community. In a project conference at the end of Phase 2, where each school made a presentation for the educational community more widely, this teacher team included their experience of designing the linear functions tasks. They also wrote a short chapter for

Card 1

X + Y = 7

1. Each pupil **write down** two whole numbers which add up to 7.
2. Can you find more such number pairs?
3. How many such number pairs can you find?
4. **Draw** these number pairs on millimeter paper.

Card 2

X + Y = 7

1. What happens if x = 2½?
2. What if y = 4,7?
3. How many different number pairs are possible now?
4. Draw these new number pairs together with those you have already drawn.
5. What happens if one of the numbers is 9?
6. What happens with the drawing now?
7. Find four new number pairs where one of them is greater than 9. Draw these also.
8. **Discuss** what you have found out. **Write** down at least two points.

Card 3

X + Y = 12

1. Find a number pair that fits and draw then in the same coordinate system as before.
2. How is this drawing in relation to the first?
3. Find similarities and differences. **Write these down.**
4. Repeat the points above with the expression:
5. **X + Y = 5**.
6. What do you see? **Write it down**.

Card 4

2X + Y = 9

1. Find number pairs that fit and draw these on a new piece of millimetre paper.
2. Draw number pairs that fit the expression **X + Y = 7** in this coordinate system
3. Find a number pair that fits both espressions. What have you done now?
4. Can you write **2X + Y = 9** in a different way such that it is easier to find the number pairs that fit?

Figure 5.2 The four cards.

the project book (Jaworski et al, 2007). Didacticians collected data from all parts of the described activity and analysis of this data led to a number of papers (Fuglestad, Goodchild, & Jaworski, 2006; Hundeland, Erfjord, Grevholm, & Breiteig, 2007; Jaworski, 2007a)

The events I have described in the previous paragraph provide an excellent example of developmental research. Teachers, with didacticians' support, designed tasks for their classroom, conducted the associated lessons, reflected afterwards on their teaching, presented their work for others in two events, and wrote an account of the events for publication. In the project we spoke of a design/inquiry cycle of

plan and replan
act and observe
reflect and analyze
feedback to the planning stage

Act refers to action in the classroom where teachers taught their lesson based on the cards. *Observation* in the classroom was done by video recording by didacticians. This provided a record that could then be drawn upon as a basis for reflection, developmental activity, and writing. Viewing the video allowed teachers to consolidate their thinking with respect to the planned activity. *Analysis* occurred as teachers reflected on the activity and constructed their two presentations and their book chapter. It also occurred as didacticians addressed the data in order to draw conclusions and write research articles as referred to above. Feedback to the planning stage occurred as teachers thought about what they might do differently another time. Although this was discussed, there was no opportunity to follow up with a subsequent activity based on the feedback. Teachers did not invite didacticians to observe activity a year later, and didacticians were reluctant to put pressure on teachers to fit with predetermined aims. This reflects a tension in the project, which is addressed further below.

Developmental findings here were important for both teachers and didacticians. At the outset of the project, the teachers in this school had expressed some reservations about joining the project and engaging with inquiry-based tasks in their classrooms. One reason given was the demanding nature of their upper secondary school curriculum and the shortage of time to cover its contents. They believed the project might be too demanding on their time, and that inquiry-based tasks would take up too much classroom time. Despite these reservations, they were the ones to suggest that didacticians might support them in their work on linear functions, and to carry this mini-project through. It certainly took time, which they had to find both within their busy school activity and in their own time out of school. Reflection showed that they considered the tasks had taken up more time in the classroom than a more traditional delivery of the same material. However, they believed that students had benefited from the exploratory activity that the tasks had promoted—requiring thinking about the nature of functions and their graphs. One teacher said that his students had appeared to be more attuned to the mathematical concepts than he remembered with previous cohorts of students. In their presentation they said that they could see value in the kind of tasks and the associated classroom activity in supporting students' conceptualization of mathematical ideas (here to do with linear functions—see Figure 5.2), and would consider doing this kind of work occasionally, but could not find time to do it often. There had not been time to observe each other in the classroom, although the video had allowed them to see each other in action.[3]

Didacticians were first of all delighted to see elements of project design being put into practice and to observe practical details and outcomes. The factor of time, well documented in the literature (e.g., Assude, 2005) was a key consideration in these events and throughout the project as a whole. Fitting into the school day and beyond extra elements of planning, design, and reflection placed a burden on teachers who considered themselves to be already over-committed in the demands of their school work. Didacticians had hoped that the stages of the inquiry cycle might fit into the normal events of a school day since teachers needed to plan, to teach, to reflect, and to replan as an ordinary part of their teaching. The constraints came in two ways (at least). The first related to the need to plan, and so on, *as a team*. The idea of *community* was an important part of project design—*inquiring together*. School structures did not allow time for a particular group of teachers (such as the LCM project team within a school) to meet in school hours unless that group was a designated group in the way the school was organized. Meeting outside school hours placed extra demands on teachers' personal and family life. So, the time factor was closely related to the school and societal systems in which teachers participated. The second constraint related to teachers' own perceptions of the use of inquiry-based tasks in their teaching. Particularly at upper secondary level, teachers resisted extra work. They perceived the use of inquiry tasks as peripheral to normal practice, as not clearly related to their curriculum, as something "fun" to engage with but a luxury considering curricular demands and therefore not warranting valuable curriculum time as they saw it. However, in one of the two upper secondary schools, teachers were interested in cooperative learning and designed their lessons to fit with a cooperative learning model. This involved design of tasks that would contribute to such a model. Over time and negotiation in the project, the kind of tasks they designed came to be recognized as having elements of inquiry as students and teachers talked together in exploration of mathematical ideas. These issues/constraints therefore led to differing perceptions of what was promoted in the projects according to teachers and didacticians. Examples of particular tasks discussed in the project and associated issues are addressed further below.

The observations and issues reported in the paragraph above are a result of didacticians' analysis of data collected during the activity described. Such analysis cannot be separated entirely from the more developmental analyses that happen as part of the inquiry cycle and lead to enhanced knowledge in practice. We see here cycles within cycles as emphasized by the diagram in Figure 5.1. In terms of developmental research, we see now, reflecting on the projects after some years of activity, the various cycles in action as *time-related.* Locally, we see cycles involved in the planning and acting described above; however, more globally there is growth of perception and awareness as both teachers and didacticians come to appreciate each others' ways of seeing and talking, the discourse in each of their practices related to the norms of practice, and concomitant growth of knowledge regarding theory-in-practice. I say more of this in the section on collaboration and learning later in the chapter.

CO-LEARNING IN INQUIRY COMMUNITIES

Learning Communities in Mathematics, as the title of the project implies, reflects two significant concepts on which the projects were based—*community* and *learning.* Rooted in Vygotskian theory, didacticians envisaged *learning as participation in social settings* with close collaboration between teachers and didacticians promoting *co-learning*: we should learn together through our joint participation in the projects, both in terms of each other's ways of doing and seeing and in terms of new insights into our own perspectives (Wagner, 1997). As indicated above, we drew particularly on social practice theory (Lave & Wenger, 1991; Wenger, 1998) and the concept of *community of practice.* From this perspective, established communities of school and university are communities of practice in which people work according to established norms and expectations and ways of being and doing. From our own established roots we looked at each other with perceptions of the other as seen from our own communities. To particularize rather crudely, teachers saw didacticians as the "experts" in terms of knowing what it means to teach and to learn, but perhaps as not having much idea about the needs and practices of working with students in school settings (Bjuland & Jaworski, 2009); didacticians saw teachers as professionals, knowledgeable in their school settings, but perhaps having little awareness of broader theoretical ideas and alternative associated practices (Jaworski, 2008). To illustrate, I quote the words of two teachers expressed in a focus group interview towards the end of Phase 2 of the project (cited from Bjuland & Jaworski, 2009, pp. 31 and 34, respectively):

Osvald: I maybe think that you didacticians have a rather naïve perception about our pupils... you have an ideal aim, while we have to give you some realism (laughter).

Agnes: In the beginning [of the project] I struggled, had a bit of a problem with this because then I thought very much about you should come and tell us how we should run the mathematics teaching. That was how I thought, you are the great teachers... but now I see that my view has gradually changed because I see that you are participants in this as much as we are even though it is you that organize. Nevertheless I experience that you are participating and are just as interested as we are to solve tasks on our level and find possibilities, find tasks, that may be appropriate for the pupils, and that I think is very nice. So I have changed my view during this time. And I think it's much better now, I feel myself much more comfortable, because now I feel that we are more equal than we were in the beginning from my point of view.

The idea of inquiry community was that together, through inquiry, we should challenge established norms and gain insights to ways of developing practice.

Wenger (1998) proposed that belonging to a community of practice involves *engagement, imagination,* and *alignment.* So, for example, teachers *engage* with school practices, ways of working with students, approaches to teaching in classrooms, and acting within a school's systems and structures. They use *imagination* in interpreting their role within these systems; imagination forms their worldview of teaching—for example, seeing teaching as one lesson followed by the next, or seeing it as a holistic approach to the education of young people. Teachers *align* themselves with the normal ways of doing and being within a school, so that their ways of engaging fit with established norms and structures. In the projects, our use of inquiry was designed to challenge existing norms while engaging in and with them, to expand imagination to allow us to see beyond what is established and find possibilities for development. We have used the term *critical alignment* to describe a process of looking critically at what we do while we do it—not just aligning with the status quo but looking consciously at the status quo with a view to identifying issues and a possibility of making changes (Jaworski, 2006). To inquire means to question, probe, investigate, and wonder; to identify problems and seek solutions; and to look critically at whatever it is we are inquiring into. Inquiry in the projects included inquiry into mathematics in workshops and classrooms. For example, participants in mathematical classrooms or workshops work on mathematical tasks or problems in a mode of open questioning that allows conjectures to be addressed and generalizations to be established (e.g., Mason, 2002). Teachers and didacticians inquire into the processes of designing tasks for learning and approaches to teaching. Through research, both groups inquire into the developmental processes in which they engage, and reciprocally promote development.

A focus on inquiry in the projects began in project workshops. In the beginning, mathematical problems were used to engage participants, both teachers and didacticians, in exploring together in mathematics. For example, one of the early problems was "Sum Factors":

> Any number can be expressed as a sum in a range of ways—for example, sums for 8 include 8 = 2 + 2 + 2 + 2; 8 = 3 + 2 + 1 + 1 + 1; 8 = 4 + 3 + 1. If we take the terms of the sum and make them factors of a product, we get a range of products—for example, 2.2.2.2 = 16; 3.2.1.1.1 = 6; 4.3.1 = 12. What is the largest product we can find for any starting number?

Inspection for the number 8 shows that the largest product is 3.3.2 = 18. How is 18 related to 8? Can we see a relationship with other numbers? Participants in an LCM workshop worked on this problem in small groups of four or five teachers from different levels of school plus one or two didacticians, resulting in differing degrees of resolution of the problem. A subsequent plenary session of all participants allowed findings to be discussed and further questions were raised. Participants could work further on the task in their own environment after the workshop. For the Sum Factors task, in a subsequent ICTML workshop, one teacher offered his resolution of the problem using a computer spreadsheet. Several teachers used this and other problems, suitably modified, in their classrooms with pupils at a range of levels. Such lessons were video-recorded allowing for developmental reflection, by teachers or didacticians, together or independently, on the lesson as recorded and subsequent analysis of the video data according to research questions.

A number of issues arose in relation to what I have described in the paragraphs above, for example:

1. Problems such as Sum Factors were seen by some teachers as peripheral to their curriculum, as mentioned earlier.
2. Working in small groups in workshops involving teachers from a range of levels of school was generally not liked by teachers who preferred to work with colleagues at their own level.
3. Didacticians were unsure of their role in working in the small groups.

In the first case, according to didacticians, Sum Factors could be seen to address number relationships, sums and products, and alternative factorizations. It provided opportunity to see and explore patterns, to make and test conjectures, and to seek some kinds of justification. It could lead to the exponential function. It therefore seemed to offer a worthwhile medium for exploring and establishing certain numerical relationships. Such problems were used fruitfully by some teachers in their classrooms, mainly with younger students (ages 6–14). Teachers of the older students asked for problems that they could see as being more clearly related to their curriculum, suggesting algebra, geometry, and probability as areas in which problems would be welcome. Didacticians responded by designing some workshops according to mathematical themes: for example, there were two workshops on probability and three on algebra. For these workshops, tasks were designed that were inquiry-based, but which didacticians felt were more clearly related to the designated area of the curriculum. However, the inquiry-based nature of the tasks still created problems for some teachers who found it hard to see how such tasks could be incorporated in their (largely textbook-based) work with students within the timescales they saw as being available to them.

An aim of the project was that teachers would design their own tasks, related to their own curriculum, and that the tasks in workshops would act as a stimulus and example for teachers' own design. Didacticians learned that this was an unrealistic aim for the early stages of engaging with inquiry. Such learning emerged through listening carefully to teachers talking about their experiences and feelings about the project, and through teachers' evaluations of the project that were sought purposefully at regular intervals. Although the meaning of "inquiry" was discussed from the very beginning of the projects and in every project meeting that was held, it took time for understanding of what inquiry-in-practice might mean. This was true for both teachers and didacticians, although in differing ways relating to their own established communities of practice. Teachers had to make sense of inquiry in relation to their students, to norms of classroom work, and to expectations related to their curriculum. Didacticians had to recognize that theoretical notions, however persuasive, are hard to translate into practice. Over time, their appreciation of issues moved from the design of "appropriate" tasks to a deeper awareness of the difficulties teachers experienced in conceptualizing the use of inquiry approaches as part of their normal practice—thus alignment with normal practice created obstacles that became the main focus of critical alignment.

We see here for both groups, teachers and didacticians, clear evidence of engagement, imagination, and alignment with respect to our own established communities of practice. A range of publications chart this process: see for example, Bjuland and Jaworski (2009); Cestari, Daland, Eriksen and Jaworski (2006); and Jaworski, Goodchild, Daland, and Eriksen (in press). Didacticians' expectations of inquiry practices came from alignment with theoretical notions of inquiry and its aims in relation to development in teaching and learning mathematics. Teachers' perceptions of what was possible in schools in relation to tasks within the project came from alignment with school practices and the norms and expectations of school and educational systems. Before teachers could design inquiry-based tasks for classrooms and work on them with pupils, teachers had to look critically at how they might modify established practices in schools. To understand the challenges faced by teachers, didacticians had to gain insights into practices in schools and look critically at their presumptions as to what could be possible. In neither case could the status quo be changed over night. Critical alignment meant a recognition of where our practices were aligned, a clearer perception of where we were aimed, and an overt inquiry into what might achieve these aims.

In the second case, concerning composition of small groups in workshops, critical alignment can be seen in action. Didacticians, according to theoretical perspectives, at first organized small groups in two ways: same-level groups in which teachers worked with colleagues from their own and other schools at the same level (e.g., lower secondary level, with students in grades 8–10) and cross-level groups. For example, one cross-level group working on the sum factors problem included four teachers from different schools, one from lower primary, two from lower secondary and one from upper secondary, plus one didactician. Discussion on the problem was led by one of the lower secondary teachers, but all members joined in the discussion, which focused both on aspects of sum factors and on issues relating to working with such a problem with students in the classroom. Analysis of such group discussion reveals teachers being introduced to and possibly starting to understand practices and expectations from teachers at different levels (Bjuland & Jaworski, 2009). Didacticians have argued that such cross-level understanding is valuable for teachers in working with the transition of students between school levels. However, teachers have been much more concerned by their experience that working on mathematics in cross-level groups can be uncomfortable, and not especially productive of material for their own classrooms. The upper secondary teachers wanted to be able to go more deeply into mathematical concepts, while teachers with less experience of mathematics felt threatened by mathematics which went beyond their familiarity.

Overwhelmingly, teachers preferred to work with colleagues who were teaching the same or near grades. In the second phase of the project, teachers requested strongly that opportunity could be provided in workshops to plan for the classroom since in Phase 1, little time had been found in school for planning. There is evidence here of teachers looking critically at what they perceived in the project to be demanding, and suggesting ways in which they could respond to these demands that were achievable within the school day and school patterns of activity. There is evidence too of didacticians gaining insights into teachers' possibilities for action within patterns of activity and overtly modifying project design to allow developmental progress. A challenge for didacticians was to maintain the theoretical integrity of project design while adapting to possibilities of practice. Thus, the project evolved from its initial expectations (expressed theoretically in the language of didacticians) to be accommodating of perceptions and needs from the school community (expressed by teachers with respect to school practice). Our data show examples of the results of critical alignment involving innovative activity in classrooms relating to planning in homogeneous small groups in workshops: for example, the use of probability and algebra tasks with students at lower primary level and teachers' invitations to didacticians to work with them on ideas for change within their own school environment. Many examples can be found in the book emerging from the projects (Jaworski et al., 2007).

The third issue, which relates directly to critical alignment by didacticians, concerns the roles of didacticians in working with teachers both in workshops and in schools. Throughout the project, didacticians met as a group (of 13 or 14 when all were present) to plan and to discuss issues. Early in the project, questions were raised as to what role a didactician should play. For example, in a small group in a workshop, should the didactician *lead* the group? We felt that leading, in the sense of directing activity, or guiding the mathematics was inappropriate. Didacticians also rejected the idea of being a *coordinator*; the word that was liked best was *facilitator*—enabling activity to take place according to what was planned, but being careful not to direct or condition the nature of activity (Cestari et al., 2006). How to engage suitably was something we all had to try out, experience, and discuss. Over time, didacticians became more aware of issues involved and clearer about choices that had to be made. When teachers engaged readily with the mathematical tasks suggested for group work, little facilitation was needed and a didactician could relax and contribute to the discussion. Sometimes facilitation involved ensuring that all participants in a group were able to have a voice. Sometimes it required difficult decisions about steering the group if the group seemed to be off-task. Such decisions were the focus of didacticians' meetings, although not overtly discussed with teachers. On one occasion, the facilitation of one didactician was recommended by another (who had also been a member of the group) as a good example of providing opportunity for all participants in the group to have a voice, and keeping the group to the agenda for the group session. The video recording of this group's activity was then used as a focus in a subsequent meeting, allowing didacticians to view their colleague in action and gain deeper insight to issues of facilitation. Thus didacticians learned from reflection and action in our own inquiry cycle. Mediation within the didacticians' community was central to this learning, as will be addressed in the final section of this chapter.

Similar issues arose with respect to activity in schools. Each school had a team of three didacticians associated with it. One didactician had the primary responsibility of acting as liaison with the school and communicating with the didactician team. In the early days it was suggested that teachers might invite didacticians to join them in school when there were issues to discuss, work in which to engage jointly, or video recording to be undertaken in classrooms. This worked well in some cases—for example, in the work on linear functions in an upper secondary school, described earlier. Teachers in different schools organized themselves in different ways, and didacticians over time responded and learned how best to work with each school. Many didactician meetings were devoted to discussing work in schools and the issues didacticians faced in this work—trying to encourage and facilitate and questioning how far to go in making suggestions or leading activity. When schools were very engaged in their activity and planning, didacticians could respond and did so in a variety of ways that we came to recognize and analyze. For example, in one school the teachers found it hard to work as a team since they taught at different levels within the school. Therefore, individual teachers engaged in innovative work with pupils in their own class and didacticians found themselves working closely with these individuals. Didacticians questioned whether there could be ways of helping these teachers to work as a team, but over time it became clear that the school dynamic was more powerful than the ideals of the project in this respect. Another school organized itself around *landscape teams*—for example, teachers teaching grades 4, 5, and 6 formed such a team. The team of teachers in the project crossed the landscape teams. Within the school, teachers in the project team discussed how they could make an input to their landscape team and thus influence their colleagues. Didacticians responded and supported, but here teachers took responsibility for activity. (See also Bjuland & Jaworski, 2009; Daland, 2007.)

A problem arose for didacticians when, over time, not much seemed to be happening within a school. In one case, after a period of time in which email messages and phone calls to teachers in the school produced little communication, a letter was written to the school by the directors of the LCM and ICTML projects. The letter indicated the difficulties experienced in achieving communication and raised questions about the school's participation in the project. Didacticians discussed these issues long and hard before writing such a letter, since it seemed to go against principles of facilitation and the desire that teachers should be responsible for forms of activity in their schools. However, if nothing happened in a school, then the project could achieve little. As a result of the letter, didacticians were invited to a meeting in the school at which the school team openly and frankly acknowledged their inaction and discussed reasons for it. From the meeting, a plan of action was initiated that was then taken further between teachers and the liaising didactician. The school became once again a central player in the project. However, despite considerable efforts between teachers and didacticians, one school decided to leave the project. The school had committed itself to too many initiatives, and teachers, despite continuing interest in the work of LCM and ICTML, found themselves unable to deal with the range of demands on their time.

Didacticians' roles in schools were diverse. In each case it was important for didacticians to learn about the school's systems, processes, and ways of being and thinking. Interpreting inquiry-based activity was always rooted in the specifics of these factors, although common issues emerged as the project progressed. Teachers expressed recognition of didacticians' endeavour to work with them according to their systems, wishes, and needs, and showed developing appreciation of project goals for shared responsibility and action. As understandings developed, a reciprocity began to emerge in which the aims of the project were transformed into workable forms while maintaining their essential ingredients. Goals of *community* and *inquiry* remained central to all activity, and we all learned how such goals can be addressed in practice. Didacticians turned to *activity theory* to make sense of the evolving growth and transmutation in the projects, as I will discuss in the final section of this chapter (see Jaworski & Goodchild, 2006; Jaworski, 2007b).

MATHEMATICS FOR LEARNING AND TEACHING

In these projects, we worked hard over three phases of activity to promote an inquiry community and shift towards inquiry communities in university and schools. It seems almost paradoxical that *mathematics* gets lost when we start discussing the main issues of the project; that didactical, pedagogic and developmental issues take the foreground and that mathematics takes relatively a background position. Ongoing research is addressing this issue. However, mathematics was at the roots of all our activity, sometimes in focus in its own right (as in working on a mathematics problem in workshop or classroom) or in meta-mathematical activity focusing on the didactics and pedagogy of mathematics in promoting learning and teaching. In this section, I try to bring mathematics to the foreground, to discuss its place in the project and the issues that are associated with it.

Every workshop had both a mathematical focus and meta-mathematical focuses. In the early stages, mathematical problems were chosen to stimulate mathematical inquiry and held to be a uniting factor that had potential for community building. Through our own engagement in these mathematical problems, our own inquiry, we (didacticians and teachers) gained insights into ways in which students in classrooms could work on such problems and appreciate mathematical relationships. As teachers requested a clearer curriculum focus, tasks and problems were designed to address specific topics. For example, in Phase 2, three workshops were devoted to algebra, building on input from one teacher who provided examples of problems that his students at upper secondary level were (still) having with basic algebraic relationships: an implication was that such problems should have been overcome at lower levels of schooling. His input provided an opportunity to address algebra across grade levels. To illustrate, one of the problems he highlighted was as shown in Figure 5.3.

$$\frac{x+4}{x} = \quad \longrightarrow \quad \frac{\cancel{x}+4}{\cancel{x}} = 4$$

Figure 5.3 Incorrect cancelling.

Given the fractional expression on the left, he indicated that students typically (and incorrectly) cancelled x in numerator and denominator to leave them with the answer 4. Using this example as a basis for exploration between teachers and didacticians in a workshop, we asked the question shown in Figure 5.4.

$$\frac{x+4}{x} = 4$$

What does this mean?
Is it true?
For what values of x

Figure 5.4 An inquiry task.

So for example, participants might try out different values of x, such as $x = 1$, or 3 or 9 (Figure 5.5).

$$\frac{1+4}{1}, \quad \frac{3+4}{3}, \quad \frac{9+4}{9}, \ldots \quad \neq 4$$

Figure 5.5 Unsystematic specializing.

None of these gives an answer of 4. Is it possible to find a value of x that gives an answer of 4? Well, it is hard to find this by a random process although a more systematic process, for example using trial and improvement, starting with 1. 2, 3/2, 4/3 . . . , might have more success. The power of algebra can be seen when we use it to solve the equation, as in Figure 5.6a.

$$\boxed{\frac{x+4}{x}=4} \longrightarrow \boxed{\begin{aligned} x+4&=4x\\ 4&=3x\\ 4/3&=x \end{aligned}}$$

(a)

$$\boxed{\begin{aligned} &\frac{4/3+4}{4/3}\\ &=\frac{16/3}{4/3}\\ &=\frac{16}{3}\cdot\frac{3}{4}\\ &=\frac{16}{4}\\ &=4 \end{aligned}}$$

(b)

Figure 5.6 An algebraic solution.

This solution (6a) suggests that when

$$x=4/3,\ \frac{x+4}{4}=4$$

We can test this out as shown in Figure 5.6b.

So, recognition of a student error, as presented by one teacher, provided a basis for mathematical exploration designed to highlight aspects of algebraic relationships and manipulation to find an answer to a starting question. Here we see an example of inquiry in mathematics with a focus on algebra. The task involved making sense of an algebraic expression and equation and using algebraic manipulation. It was designed to get teachers and didacticians to inquire into algebraic issues related to the kinds of student errors pointed out by teachers. Such inquiry by teachers could then inform their response to such an error when seen in the classroom. This was just one of a range of tasks which encouraged workshop participants to engage with algebra and explore algebraic relationships. Not all tasks could be taken directly into the classroom in the form presented in a workshop. However, there is considerable evidence of teachers adapting such tasks for classroom use (e,g., Jaworski et al., in press)

Some tasks were designed to encourage the use of algebra in making a mathematical formulation from a more general problem. One example was based on the idea of fitting a frame of width 1 cm around a square picture and considering the length of frame needed. Several arrangements might be envisaged, as seen in Figure 5.7.

For a square of side s: In the left diagram (7a) we see four pieces all the same and four corner pieces

$$\text{length of frame} = s+s+s+s+1+1+1+1 = 4s+4.$$

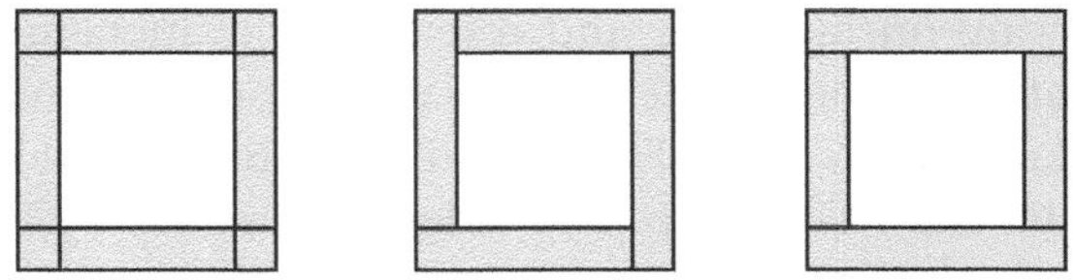

(a) (b) (c)

Figure 5.7 Mathematizing an everyday problem.

In the center (7b) we see four pieces of frame, all the same length,

$$\begin{aligned}\text{length of frame} &= (s+1)+(s+1)+(s+1)+(s+1)\\ &= 4(s+1) = 4s+4\end{aligned}$$

To the right (7c) we see 2 short and 2 long pieces,

$$\begin{aligned}\text{length of frame} &= s+s+(s+2)+(s+2)\\ &= 2s+2(s+2) = 4s+4\end{aligned}$$

Of course, regardless of formulation, the length of frame is the same for a given value of s, so the alternative formulations demonstrate the equivalence of the three algebraic expressions that can be manipulated according to the rules of algebra to give the same resulting formula ($4s+4$).

In this example, we see algebra emerging from the mathematization of an everyday problem leading to a process of algebraic generalization, while the previous example was set within algebra. In both cases, algebraic manipulation led to a solution to the posed problem with a didactical aim of contributing to meaning-making in algebra. At a meta-mathematical level, such examples contributed to an environment in which algebra could take on meaning rather than being seen just as a collection of meaningless symbols. Didacticians aimed that, by juxtaposing such examples in workshops, teachers and didacticians together could use algebra and work with it as a tool and environment for problem-solving and making mathematical sense. This workshop environment led to discussions about algebraic activity in classrooms and teachers' use of selected examples in inquiry mode with their students. For example, video from one grade 7 classroom shows students working on the *length of a frame* problem and the teacher's interactions with her students can be related directly to activity in the corresponding workshop.

The algebra workshops came in Phase 2 of the projects, by which time relationships within the project community had become more established. This was demonstrated by teachers' confident voice in engagement and discussion and didacticians feeling more relaxed to offer didactical or pedagogic input drawing on literature and research. The meta-mathematical theme in the algebra workshops was one of making mathematical sense and demystifying algebra. In the third workshop, one didactician made a plenary presentation, drawing together threads of algebraic use, shifting from pre-algebra involving pattern recognition and generalizations expressed in everyday language towards the introduction of symbolic forms and manipulation of algebraic expressions and equations. Evidence, largely in video form, from classrooms of teachers' confident use of algebraic ideas, showed that some meanings had been made and some transformation of activity emerged. This was particularly recognizable when certain teachers who had expressed considerable fear of algebra could be seen handling algebraic situations with students with greater ease and confidence.

Critical alignment in the case of mathematical development feels rather different from critical alignment as expressed in relation to group activity and planning for teaching, as discussed earlier. Above we saw development of *ways of doing* and *ways of thinking about doing* for both teachers and didacticians with perhaps the impetus coming from teachers and accommodation from didacticians. As a result, new *ways of thinking* resulted, particularly for didacticians. Where mathematics is concerned, *ways of thinking* about mathematics were central to possibilities for action. Thus, for teachers particularly, *ways of thinking about mathematics* needed to develop before *ways of doing* could be conceptualized. The mathematical tasks in workshops were central to encouraging new ways of thinking about mathematics that could lead to new possibilities for doing mathematics in classrooms (and to pupils' ways of thinking about mathematics).

Didacticians were involved fundamentally in mathematics task design that could open up awareness of mathematics learning and its promotion, for teachers, for their pupils, and, although not overt in the early stages, for ourselves. What does it mean for people to learn mathematics and in what ways do we see inquiry-based tasks as fostering learning? The process of design of tasks proved to be a central stimulus for didacticians' learning of didactics and pedagogy and, relatedly, of mathematics itself (Jaworski et al., in press). Teachers' engagement with the tasks brought teachers up against both concepts in mathematics and ways of doing and thinking mathematically that stimulated and excited them (as many expressed; see Bjuland & Jaworski, 2009). Through new ways of thinking for both groups, new possibilities opened up for alternative modes of practice (see Erfjord, 2008; Goodchild & Daland, 2009).

At the end of the three algebra workshops, one didactician invited teachers to engage in a research task with their students: to study one group of students to find out their understanding of or proficiency with algebra. In the succeeding workshop teachers were asked to bring some written account of their findings from their research, to present these to the assembled group, and to hang the written accounts onto a thread that was stretched across the room. Positioning of each item on the thread was intended to trace algebraic insight from being low in meaning making to demonstrating greater algebraic insight, regardless of the grade level. As teachers presented findings from their research, discussion focused on what it meant to have algebraic insight for students across the grade levels.

We might see this as a meta-mathematical task—firmly rooted in mathematics, but with meta-mathematical goals. The task proved to be inspirational, as reported by teachers. From their own enhanced perspectives on algebra, a requirement to focus on their students' perspectives provided a rare opportunity to see inside the teaching–learning interface. The zoom of the didactic lens into their students' algebraic thinking provided new and exciting insights into their activity and roles as teachers. Bjuland and Jaworski (2009) offer evidence for these claims in words expressed by teachers directly.

DEVELOPMENT: COLLABORATION AND LEARNING

The projects, funded for four years, completed three phases of field work which, for LCM, included 16 workshops and corresponding activity in schools. Development within the project could be traced through the data collected. Meetings between small groups of didacticians at the university; between didacticians, teachers, and school leaders; and between teachers and didacticians in a school were audio recorded and in some cases video recorded. Workshops were video recorded, with audio recording of some of the small group sessions. Innovative activity in classrooms was video recorded. Data from tests and surveys into students' achievement and students' and teachers' attitudes to mathematics learning and teaching provided a longitudinal cross-section on the projects.

Analysis of the vast collection of data presented a serious challenge. All didacticians had access to the data, and each individual or small group accessed data according to focus and interest. Thus, one group, addressing questions on dialogue and gesture in classrooms, selected from classroom video material (e.g., Bjuland, Cestari, & Borgersen, 2007). One doctoral student, interested in tasks, their design, and their use looked across the entire data set, drawing on data from meetings, workshops, and classrooms (e.g., Eriksen, 2007). One group focused on didactician learning and development (e.g., Cestari et al., 2006). Two doctoral students each focused on development within a particular school, or with particular teachers (Daland, 2007; Erfjord, 2008). Analysis of students' achievement in tests went on throughout the project and forms the basis of several master's dissertations (e.g., Andreassen, 2005; Espeland, 2006). Thus, data analysis within the projects is multifaceted and it is hard to consider its full complexity—although the book (Jaworski et al, 2007) tried to capture this complexity in a more holistic way. Interested readers can select from the many publications from the projects. Nevertheless, we want to be able to speak of development generally. Here I focus on two aspects cutting across the complexity of development—*collaboration* and *learning*.

As discussed above, teachers and didacticians were each committed to their own established communities, thinking and acting according to their own community norms and expectations. Within these norms and expectations were their perspectives of each other: teachers of didacticians and didacticians of teachers. Some teachers had taken part in other developmental projects and brought their experiences from those of working with didacticians. Most didacticians had worked with teachers in teacher education programs. The LCM and ICTML projects were unlike any of these other projects or programs. They proposed a partnership arrangement of working together in an inquiry community with both groups bringing knowledge and expertise and both groups promoting development. However, power in the project lay initially and overtly with didacticians who had written the proposal, had management of the funding, and led the way (at least initially) in planning. Nevertheless, as became clear gradually, schools and teachers held power over the projects, less overtly but influentially as I shall explain further below. The nature of the balance of power, of power sharing, and of ultimate reduction of imbalance can be seen as capturing the complexity of developmental progression within the projects.

The projects set out to develop inquiry communities between teachers and didacticians. Phase 1 was expressed as having the purpose of *building community*. Initial activity had to bring us together and provide opportunity to develop shared goals and ways of doing and being. Didacticians had the job of translating proposal principles into project activity. Their power over workshops, initially, was high. Teachers in early workshops were largely responsive, albeit willing to engage and respond. Relations from the beginning were extremely congenial, building on good will and expectations of what might be achieved. When it came to activity in schools, the schools and teachers were the most powerful. Didacticians might suggest ideas, ask for kinds of activity to take place, and assume that teachers would work in certain ways, but the established systems and processes of school functioning and a school's expectations of the teachers working there held sway. Didacticians, observing what happened and talking with teachers about what was possible, had, from very early days, to adjust to the school juggernaut. This constituted a considerable learning curve for the didacticians individually and as a group. Visions of project team activity in schools, teachers' joint planning for the classroom, and ways in which teachers would interpret workshop activity for their classrooms were all subject to challenge and reformulation. Meanwhile, teachers participated in workshop activity and expressed their enjoyment in engagement in workshops, tasks, and discussion. They met up with new terminology and questioned it. The word "inquiry" was itself a source of concern—what exactly do we mean by inquiry? What does it mean to engage in inquiry? Planning for the classroom meant different things for the two groups, seen from their own contexts. Didacticians had to learn from teachers what planning meant in their school context, before their visions of planning for students' engagement in mathematics could make sense for teachers. The didactical demands of planning for mathematics, as seen by didacticians, had no parallel in school contexts. Thus, the meta-mathematical goals of tasks designed for workshops were greatly challenged in early days of the project. Teachers tried to make sense of what they experienced and shifted from early expectations that didacticians would offer them tasks and teaching plans for direct use in the classroom towards didacticians' vision of a more reciprocal developmental relationship.[4]

The position sketched in the previous paragraph was the subtext of Phase 1, recognized through subsequent reflection and analysis. On the surface, didacticians and teachers met each other, engaged together in a range of activities, and negotiated didacticians' perceptions of activity in school. The word "negotiated" may imply an overt discussion of aims and actions, which of course are included in the term. However, negotiation was mainly perceptual: experienced and intuited rather than articulated and discussed. It involved an adjusting of awareness and expectation.

It involved learning to know the world of the other through engagement and interaction. In reporting and writing about the projects elsewhere, we have quoted extensively the words of Wagner (1997, p. 16) who wrote about co-learning:

> In a co-learning agreement, researchers and practitioners are both participants in processes of education and systems of schooling. Both are engaged in action and reflection. By working together, each might learn something about the world of the other. Of equal importance, however, each may learn something more about his or her own world and its connections to institutions and schooling. (Wagner, 1997, p. 16)

In our working together, developing principles of inquiry, working towards inquiry as a way of being and ultimately towards critical alignment, we were able to use what we learned of each others' worlds to look critically at our own and to adjust both thinking and action, as I shall now explain.

Phase 1 set up the possibility for what I have outlined above. An outcome of Phase 1, was that we gained insights to each other' expectations and possibilities in action. At the end of Phase 1, didacticians were aware that teachers held differing points of view on what was possible in the project, and a special meeting was held to which all participants were invited. Most came. Teachers were asked to express their views on Phase 1 activity and make suggestions for Phase 2. Most teachers exercised a voice, some passionately. The main outcomes of the meeting related to issues discussed above—relating tasks to curriculum, meeting in same-level groups, planning with colleagues in workshops for activity in classrooms. It seems important to recognize that while these proposals involved changes to project activity, they were changes that could be accomplished. No one was challenging the basic premises of the project, only the ways in which goals could be addressed. All of the teachers' suggestions were implemented, and teachers were able to see that their views were respected and that didacticians made a purposeful effort to make changes accordingly.

Phase 2 was intended (by didacticians) to develop notions of inquiry. In fact, it took on a focus of planning for the classroom. Although some overt planning had taken place in Phase 1, as discussed above, it was in Phase 2 that many project teachers, together with their colleagues in school, really began to take on the mantle of task design, albeit often based on tasks experienced in the workshops. An analysis of the nature and division of workshop time in Phase 2 shows that more than 25% of workshop time was devoted to planning for the classroom: "planlegge et opplegg" was the phrase used in Norwegian. As I understand it, this means the process of planning tasks and activity for work with pupils. A diagrammatic version of these data appears in Figure 5.8, in which the numbers correspond to those in Table 5.1.

Although we still had mathematical tasks in workshops with small group activity devoted to doing mathematics together, work on mathematics per se gave space to discussion and design of mathematical tasks for pupils. Such activity took place alongside the algebra work discussed above. Thus the activity was mathematical, meta-mathematical, and school process-focused. Largely this was done by teachers in the same levels of schooling, from which individual

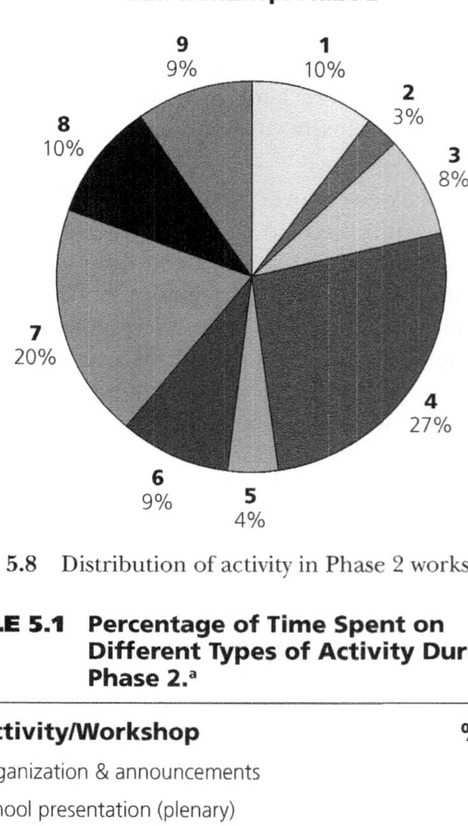

Figure 5.8 Distribution of activity in Phase 2 workshops..

TABLE 5.1 Percentage of Time Spent on Different Types of Activity During Phase 2.[a]

#	Activity/Workshop	% time
1	Organization & announcements	10.3
2	School presentation (plenary)	3.0
3	Reporting school activity in groups	8.2
4	Planning for the classroom ("planlegge et opplegg")	26.3
5	Didactical discussion in groups	4.3
6	Mathematical activity in groups	9.1
7	Reporting back from groups and plenary discussion	19.4
8	Mathematical input by didactician	9.9
9	Other input by didactician	9.5

Note: In small groups 89.2% of time teachers at same or near grades worked together. 10.8% of time groups mixed grades 1–13.

[a] I am grateful to Simon Goodchild for this analysis.

school groups could then modify for their own school environment. Although the major part of group work was in homogeneous groups, a small percentage was in mixed-grade groups to encourage cross-level discussion of issues. As can be seen in Table 5.1 and Figure 5.8, the second largest share of available time was given to feeding back from group work in plenary and discussing, in plenary, issues that had arisen.

At the end of Phase 2, a conference was held, inviting teachers and educators locally and from further afield in Norway. Teachers from each school made a presentation at this conference and, for most schools, planning for the conference occupied the latter part of Phase 2. It had a formative effect in bringing teachers and didacticians together with the classroom as a concentrated focus. During this time of planning for the classroom and preparation for the conference, *inquiry* started to have meaning beyond being just a term of reference. It started to enter into dialogue and teachers talked about inquiry in expressing their activity and thinking. For example, one teacher spoke as follows:

> Egil: Working in first and second grade is different, but the thinking behind is the same—the "inquiry"—it is nice to have that concept.... Before, I had an unreflected attitude to what I was doing, but I did it because I felt it was good.... [Now we are more attuned to be in a process], and the process I think has been very confirming... in respect of what you were doing as a teacher.... There is one thing that for me is the most important thing. That was at a meeting we had in December, when the school leaders were there, and we had a discussion about "inquiry," and when you [a didactician] said "inquiry as a way of being." That just put everything in place [for me]. [All words here are translated from the Norwegian except those in quotes which are verbatim.] (Bjuland & Jaworski, 2009, p. 36)

Towards the end of Phase 2, didacticians organized a focus group interview in each school with project teachers and the leaders of the school. Analysis of data from these meetings reveals considerable common understanding about project activity and its manifestation in schools. Teachers' words confirmed their understandings of project goals and their appreciation of didacticians' roles and intentions within the projects (Bjuland & Jaworski, 2009).

Phase 3 was a period of consolidation that focused on school goals. As a result of their presentations in the conference, school teams were asked to declare their particular goals for development of mathematics learning and teaching in their own schools. Considerable workshop time in small groups was given to discussing goals, and schools presented in plenary their particular planning, implications, and concerns. This enabled all participants to be introduced to a range of ideas and issues relating to differing grade levels but affecting every pupil who would pass through the system. Activity in school varied considerably from school to school, but all schools except one were engaged and working fruitfully with didacticians. During Phase 3, didacticians bid for a new project in response to the introduction of a new program by the Research Council of Norway. The new project (called Teaching Better Mathematics—TBM) would continue the work of LCM and ICTML with changes resulting from learning in these projects. The bid was successful, and didacticians invited current schools to consider continuing into the new project. Several schools expressed interest in doing so, although some had other developmental directions in mind.

THEORY, DISCUSSION AND CONCLUSIONS

The complexity acknowledged above presented problems in terms of reporting from the projects as a whole, and the range of papers testifies to ways in which we addressed these problems. An external evaluation of the projects was undertaken during Phase 3, and this sought to provide some overview of what we had achieved as well as issues arising (Säljö & Skovsmose, 2007). Some of these issues have been addressed in the text above. It is hard to present development simply. Often development can be "felt" rather than articulated clearly. We know we have moved on in our thinking and actions, and can articulate elements of this, but what does it mean in project terms, and more generally for learning and teaching mathematics in schools? The book from the first two projects (Jaworski et al., 2007) went some way towards expressing what we all learned. For example, the chapters written by teachers demonstrated real awareness of school concerns and developmental issues. The chapters written by didacticians showed a growth of theoretical understanding and a growing appreciation of what is possible in schools. This book was written in parallel with the final stages of the projects. The external evaluation took place at the same time. So, we now need some retrospective account of the contribution of the projects to development in mathematics learning and teaching across schools and university department.

Theory on which the projects were based has been expressed above in terms of communities of practice and inquiry and of co-learning between teachers and didacticians. The theoretical construct of *critical alignment* arose largely from these projects. We have written about how we see critical alignment fostering developmental processes in our engagement in the project (Goodchild, 2008; Jaworski, 2006, 2008). These areas of theory were valuable in giving us a theoretical basis for activity, in analyzing ways in which we thought about what we did and in characterizing activity in the projects. However, this theory proved to be too general to be useful where specific problems or issues were concerned. Yet to address complexity, we needed to be able to deal with specific issues and the learning which resulted from them. This reflects again the evolutionary nature of these projects in which emerging issues in practice influence ongoing developmental practice. Here we see them also influencing theory: When the theory we had worked with to this point proved not to be helpful in explaining emerging issues, new theory was sought.

In order to be able to make sense of the nature of learning from issues in the projects, and so to address complexity both locally and globally, we turned to activity theory, drawing on the work of Stieg Mellin-Olsen as well as that of Vygotsky, Leont'ev, and Engeström (Goodchild & Jaworski, 2005; Jaworski & Goodchild, 2006). Very briefly, activity theory is based in Vygotsky's concept of *mediation*. Learning in social settings is mediated by tools—physical, intellectual, and social—that are central to human interactions (Leont'ev, 1979; Vygotsky, 1978; Wartofsky, 1979). Engeström provides a specific expansion of the idea of mediation to emphasize the mediational nature of "rules," "community," and "division of labor" within an activity system. According to Leont'ev, all activity is necessarily *motivated* and can be analyzed in terms of goal-directed actions and related operations with conditions. Taking these basic theoretical elements into our projects that involve didacticians and teachers in school and university communities, we conceptualize activity in a variety of layers and levels. For example, we can see the school teaching of mathematics as motivated by a desire to educate young people. A school functions within an educational system in which we see norms and expectations—the "rules" of engagement. These include organization of students into classes, an academic timetable in which a curriculum is addressed, and forms of assessment to evaluate the outcomes of educational objectives. The school community includes teachers and students, head teachers, administrators, and so on, with close relations to society and culture—parents, civic leaders, and politicians, for example. We see clear division of labor within a school setting: students and teachers have long-established roles, they operate with respect to established conventions within the rules of the system. Thus, school systems are socio-historical in formation with complex activity structures that are often taken for granted and unchallenged. We can argue similarly for university systems.

Teachers and didacticians operate within these systems with goals which formulate action. One factor which became clear, almost from the beginning of the projects, was the different *worlds* from which teachers and didacticians came to the projects. In Engeström's (1998) terms, these worlds comprise differing communities, rules, and division of labor. Schools and universities manifest established activity systems in which ways of thinking, working, and being are part of the fabric, often unarticulated, but nevertheless acted on constantly from day to day. Partly because they are often unarticulated, they are hard to recognize overtly and even harder to change. Thus critical alignment through inquiry has been so important. It has brought with it, through the inquiry actions we have taken, first a possibility to *see* what we are doing and thinking and secondly an entry towards making a difference. This has been documented above for both teachers and didacticians. We can point to specific examples.

We might see this as action-development, where action here includes thought. Our thinking has changed along with our action as part of critical inquiry in engagement, imagination and alignment. What does this mean for the systems as a whole, and for the fledgling inquiry communities we have tried to create through the project? Goodchild, particularly (2007), has looked for "expansive development" as expressed by Engeström (1999). Briefly this means that perturbations within existing systems trigger for individuals or groups new ways of seeing and doing. Over time, through interactions between individuals and groups, these new ways expand to permeate the systems as a whole and we see what is described as *expansive development*. Goodchild found himself unable to trace expansive development within the inquiry communities. Rather he saw didacticians responding and reacting to teachers' expressed wishes and the evolution of the project through adaptation rather than the creative externalization of new forms of action.

Looking back from beyond the end of the projects and recognizing more at a distance what I see to have been achieved, I might express this differently. I recently attended a workshop of the project (TBM) that succeeded LCM and ICTML. Many of the same people were involved, didacticians and teachers, although there were new members in both groups. The workshop had a very familiar feel to it, and I realized that the style of activity, of working together in a workshop setting, has become established. The use of the term "inquiry" had become part of the fabric of established workshops. Didacticians and teachers greeted each other and worked together comfortably and with apparent confidence. Coming from the outside (I am no longer a central member of this project), I could bring a more critical eye and start to question some of the practices. For example, I wondered about the degree of mathematical engagement in activities that were designed to focus on mathematics and raise mathematical issues. Were we too ready to put aside the demands of the mathematics itself as we turned to a discussion of what pupils can do? I realized that this had happened in the LCM project, but had we made it a focus of critical inquiry? I suspect that, in LCM, we were too anxious about building community and finding a common understanding of inquiry processes to take mathematics itself as an issue.

As I discussed this with other didacticians, I could see that they recognized what I was talking about, but perhaps this had not been an issue at the forefront of their consciousness and concern. However, having been drawn into seeing it as an issue (perhaps?), might this be the beginning of an expansion? To understand this possibility we can look briefly at the issue in activity theory terms. Doing mathematics together had been a goal from the beginning of the LCM project. Firstly it was a *goal* for didacticians who subsequently designed mathematical activity (their *actions*) with the *object* of developing an inquiry community in mathematics and developing awareness of what inquiry in mathematics can mean. Teachers and didacticians participated in activity, bringing with them the rules and division of labor in their own established communities. The outcomes of such joint activity resulted in the growth of awareness and understanding and subsequent actions that have been described above. However, what I am now recognizing is what might be seen as avoidance of the overt challenges of mathematics, largely by diverting attention towards the pupils—a safer ground.

Now such recognition seems to me an important outcome of activity. We can see it arising in a tension (Engeström, 1999, speaks of "contradiction") between the *object* of activity from didacticians' perspectives and a twofold mediation. Firstly, teachers in their established setting are not used to doing mathematics except in their preparation for pupils' work in the classroom. It is not common for teachers to engage together in mathematics, except in relation to their pupils. Thus the tension is between teachers' rules, community, and division of labor and the object of the didacticians. Secondly, didacticians, intent on achieving a certain dimension of their overall object (largely about building an inquiry community) that is theoretically rooted in their academic understandings of pedagogy in mathematics (their own established university system), do not bring to attention the mathematical dichotomy: that the inquiry community is not really addressing mathematics and the mathematical issues that we need to be aware of together as educationists. The tension here is between didacticians' goals and the actions that are designed to lead to these goals. The activity that results is the focus of attention and its correspondence with what was envisaged through the planned actions. This obscures deeper issues that might be important to overall development.

An activity theory analysis both draws attention to tensions and illuminates such tensions. Tensions point to issues that need to be addressed as part of developmental progress. Two papers have pointed already to such issues that became evident and were addressed during the course of the projects (Goodchild & Jaworski, 2005; Jaworski & Goodchild, 2006). In developmental activity it is important to track action and outcome, and gain insights to development in thinking of individuals and groups involved. However, what the discussion here points to is the importance of becoming aware of and addressing issues that are (often deeply) embedded within activity and whose tackling leads to deeper knowledge and awareness and enhanced possibilities for practice.

For me it has been the insights we have gained through our activity that have been especially significant. Without such insights we cannot move forward. Recognition of issues, and tensions, can be painful or depressing as they come to light. However, as Engeström points out, the contradictions or tensions that are revealed become powerful forces for developmental progress—not only achieving our goals, but recognizing where our goals need refinement and where activity might usefully be redirected. I believe the years of the projects have been especially fruitful in this regard.

NOTES

1. In this context, didacticians are mathematics educators working in an academic setting with an interest in promoting learning and teaching of mathematics with both theoretical and practical perspectives. Many have been mathematics teachers or teach mathematics currently. Some are also teacher educators in a program for educating prospective teachers of mathematics.
2. In particular three projects LCM, ICTML and TBM. Historically, at the University of Agder, the project LCM, *Learning Communities in Mathematics* (Project No. 157949/S20) first gained funding in 2004, for 4 years, from the Research Council of Norway (RCN) followed shortly by ICTML, *ICT in Mathematics Learning* (Project No. 161955), both as part of the four-year KUL (Kunnskap Utdanning og Laering – *Knowledge, Education and Learning*) program of the RCN. Three years later, building on the work

of these two projects TBM, *Teaching Better Mathematics*, was also funded by the RCN, within the programme Praksis, Forskning og Utvidning (Praxis, research and development) (PraksisFoU) programme. Project No. 176442, again for four years (2006–2010). The TBM project, founded on the principles and findings to date of the other two projects, extended the research to five locations in Norway based around universities or colleges at Agder, Bergen, Bodø, Oslo, and Trondheim. This extension involved collaboration in a consortium between the University of Agder and four university colleges, and in each case a range of schools each linked to one college. Throughout these regions, teachers in schools from lower primary to upper secondary work, or have worked, with one or more of these projects. Depending on the type of school, the mathematical education and experience of the teachers in the project varies considerably.

3. One of our reviewers asked in what ways the developmental research described here differed from either *action research* or *lesson study*. While we can see elements of both of these in what is described, the methodology employed does not fit specifically with either an action research design, with the formality of well-defined action cycles, or a lesson study design with collaborative design, observation and modification of lessons. Design in developmental research here demonstrated a flexibility that allowed participants to act according to their own ideas and circumstances without the constraints of a formal methodology. Such an open approach had both advantages and disadvantages, as this chapter describes.
4. Throughout this paragraph, many references could be made to project papers, some of them already mentioned above—an up-to-date list can be found on the websites: http://fag.hia.no/lcm/LCM and http://fag.hia.no/iktml/ ICTML. Outcomes and issues from the projects fit within a wide literature in the area of teacher change (see, for example, Fullan & Hargreaves (1992). Providing a situated perspective relating to teacher change more widely is a project for the future.

REFERENCES

Andreassen, I. S. (2005). *Innsikt i elevers kompetanser som vises i skriftlige matematikktester. (Insights into pupils' competencies as shown in written mathematics tests.)* Mastersoppgave. Høgskulen i Agder, Kristiansand.

Assude, T. (2005). Time management in the work economy of a class, a case study: Integration of Cabri in primary school mathematics teaching. *Educational studies in mathematics, 59*, 183–203.

Bjuland, R. & Jaworski B. (2009). Teachers' perspectives on collaboration with didacticians to create an inquiry community. *Research in Mathematics Education, 11*(1), 21–38.

Bjuland, R., Cestari, M. L., & Borgersen, H. E. (2007). Pupils' mathematical reasoning expressed through gesture and discourse: A case-study from a sixth-grade lesson. In *Proceedings of the 5th Congress of the European Society for Research in Mathematics Education* (pp. 1129–1139). Cyprus: Department of Education, University of Cyprus

Bruner, J. (1986). *Actual minds, possible worlds*. London: Harvard University Press.

Burton, L. (1999). The implications of a narrative approach to the learning of mathematics. In L. Burton (Ed.), *Learning Mathematics: From hierarchies to networks*. London: Falmer Press.

Cestari, M. L., Daland, E., Eriksen, S., & Jaworski, B. (2006). Working in a developmental research paradigm: The role of didactician/researcher working with teachers to promote inquiry practices in developing mathematics learning and teaching. In M. Bosch (Ed.), *Proceedings of the 4th Congress of the European Society for Research in Mathematics Education (2005)* (pp. 1348–1357). Sant Feliu de Guíxols, Spain: Universitat Ramon Llull.

Cochran Smith, M., & Lytle, S. L. (1999). Relationships of knowledge and practice: Teacher learning in communities. *Review of Research in Education, 24*, 249–305.

Daland, E. (2007). School teams in mathematics, what are they good for? In B. Jaworski, A. B. Fuglestad, R. Bjuland, T. Breiteig, S. Goodchild, & B. Grevholm (Eds.), *Learning communities in mathematics* (pp. 161–174). Bergen, Norway: Caspar.

Engeström, Y. (1998). Reorganising the motivational sphere of classroom culture. In F. Seeger, J. Voigt, & U. Wascgescio (Eds.), *The culture of the mathematics classroom* (pp. 76–103). Cambridge, UK: Cambridge University Press.

Engeström, Y. (1999). Activity theory and individual and social transformation. In Y. Engeström, R. Miettinen, & R. -L. Punamäki (Eds.), *Perspectives on activity theory* (pp. 19–38). Cambridge, UK: Cambridge University Press.

Erfjord, I. (2008). *Teachers' implementation and orchestration of Cabri-use in mathematics teaching.* Doctoral Dissertation at University of Agder, Kristiansand, Norway.

Eriksen, S. (2007). Mathematical tasks and community building—"early days" in the project. In B. Jaworski, A. B. Fuglestad, R. Bjuland, T. Breiteig, S. Goodchild, & B. Grevholm (Eds.), *Learning communities in mathematics* (pp. 175–188). Bergen, Norway: Caspar.

Espeland, H. (2006). Elevers kunnskaper i tall og algebra: Hva har de med seg fra 9. og 11. trinn? (Pupils' knowledge in number and algebra. What do they bring from year 9 and 11?). Mastersoppgave. Høgskolen i Agder, Kristiansand.

Freudenthal, H. (1991). *Revisiting Mathematics Education: China Lectures*. Hingham, MA: Kluwer Academic Publishers.

Fuglestad, A. B., Goodchild, S., & Jaworski, B. (2006). Utvikling av inquiry fellesskap for å forbedre undervisning og læring i matematikk: Didaktikere og lærere arbeider sammen. [Development of inquiry communities to improve teaching and learning in mathematics]. In M. B. Postholm (Ed.), *Forsk med! Lærere og forskere i læringsarbeid undervisningsutvikling* [Research with us. Teachers and researchers in co-learning development of teaching] (pp. 34–73). Oslo: N W Damm & Søn.

Fullan, M. & Hargreaves, A. (Eds.). (1992). *Teacher development and educational change*. London: Routledge Falmer

Goodchild, S. (2007). Inside the outside: Seeking evidence of didacticians' learning by expansion. In B. Jaworski, A. B. Fuglestad, R. Bjuland, T. Breiteig, S. Goodchild, & B. Grevholm, (Eds.), *Learning communities in mathematics* (pp. 189–204). Bergen, Norway: Caspar.

Goodchild, S. (2008). A quest for 'good' research. The mathematics teacher educator as practitioner researcher in a community of inquiry. In B. Jaworski & T. Wood (Eds.), *The mathematics teacher educator as a developing professional. Volume 4 of the International Handbook of Mathematics Teacher*

Education (pp. 201–222). Rotterdam, Holland: Sense Publishers.

Goodchild, S. & Daland, E. (2009). Teaching development through discussion: a cultural-historical activity theory perspective. In C. Winslow (Ed.), *Nordic research in mathematics education* (pp. 151–158). Rotterdam, Holland: Sense Publishers.

Goodchild, S., & Jaworski, B. (2005). Identifying contradictions in a teaching and learning development project. In H. L. Chick & J. L. Vincent (Eds.), *Proceedings of the 29th Conference of the International Group for the Psychology of Mathematics Education* (Vol. 3, pp. 41–47). Melbourne, Australia: University of Melbourne.

Gravemeijer, K. (1994). Educational development and developmental research in mathematics education. *Journal for Research in Mathematics Education, 25*, 443–471.

Hundeland, P. S., Erfjord, I., Grevholm, B., & Breiteig, T. (2007). Teachers and researchers inquiring into mathematics teaching and learning: The case of linear functions. In C. Bergsten, B. Grevholm, H. S. Måsøval, & F. Rønning (Eds.), *Relating practice and research in mathematics education. Proceedings of Norma05, 4th Nordic Conference on Mathematics Education* (pp. 299–310). Trondheim, Norway: Tapir Akademisk Forlag.

Jaworski B. (1998). Mathematics Teacher Research: Process, practice and the development of teaching. *Journal of Mathematics Teacher Education*, *1*(1), 3–31.

Jaworski, B. (2003). Research Practice into/influencing Mathematics Teaching and Learning Development: Towards a Theoretical framework based on co-learning partnerships. *Educational Studies in Mathematics*, *54*(2/3), 249–282.

Jaworski, B. (2004). Grappling with complexity: Co-learning in inquiry communities in mathematics teaching development. In M. J. Høines & A. B. Fuglestad (Eds.), *Proceedings of the 28th Conference of the International Group for the Psychology of Mathematics Education* (Vol. 1, pp. 17–32). Bergen, Norway: Bergen University College.

Jaworski B. (2006). Theory and practice in mathematics teaching development: Critical inquiry as a mode of learning in teaching. *Journal of Mathematics Teacher Education, 9*, 187–211

Jaworski, B. (2007a). Learning communities in mathematics: Research and development in mathematics learning and teaching. In C. Bergsten, B. Grevholm, H. S. Måsøval, & F. Rønning (Eds.), *Relating practice and research in mathematics education. Proceedings of Norma05, 4th Nordic Conference on Mathematics Education* (pp. 299–310). Trondheim, Norway: Tapir Akademisk Forlag.

Jaworski, B. (2007b). Theoretical perspectives as a basis for research in LCM and ICTML. In B. Jaworski, A. B. Fuglestad, R. Bjuland, T. Breiteig, S. Goodchild, & B. Grevholm (Eds.), *Learning communities in mathematics* (pp. 189–204). Bergen, Norway: Caspar

Jaworski, B. (2008a). Development of mathematics teacher educators and its relation to teaching development. In B. Jaworski & T. Wood (Eds.), *The mathematics teacher educator as a developing professional. Volume 4 of the International Handbook of Mathematics Teacher Education* (pp. 335–361). Rotterdam, Holland: Sense Publishers.

Jaworski, B. (2008b). Building and sustaining inquiry communities in mathematics teaching development: Teachers and didacticians in collaboration. In K. Krainer & T. Wood (Eds.), *Participants in mathematics teacher education: Individuals, teams, communities and networks. Volume 3 of the International Handbook of Mathematics Teacher Education* (pp. 335–361). Rotterdam, Holland: Sense Publishers.

Jaworski, B., Fuglestad, A. B., Bjuland, R., Breiteig, T., Goodchild, S., & Grevholm, B. (Eds.). (2007). *Learning communities in mathematics.* Bergen, Norway: Caspar.

Jaworski, B., & Goodchild, S. (2006). Inquiry community in an activity theory frame. In J. Novotna, H. Moraova, M. Kratka, & N. Stelikova (Eds.), *Proceedings of the 30th Conference of the International Group for the Psychology of Mathematics Education* (Vol. 3, pp. 353–360). Prague, Czech Republic: Charles University.

Jaworski, B., Goodchild, S., Daland, E., & Eriksen S. (in press). Mediating mathematics teaching development and pupils' mathematics learning: the life cycle of a task. In O. Zaslavsky & P. Sullivan (Eds.), *Constructing knowledge for teaching secondary mathematics: Tasks to enhance prospective and practicing teacher learning.* New York: Springer.

Lave, J., & Wenger, E. (1991). *Situated learning: Legitimate peripheral participation.* Cambridge, UK: Cambridge University Press.

Leont'ev, A. N. (1979). The problem of activity in psychology. In J. V. Wertsch (Ed.), *The concept of activity in Soviet psychology* (pp. 37–71). New York: M. E. Sharpe.

Mason J. (2002). *Researching your own practice: The discipline of noticing.* London: Routledge/Falmer.

Säljö R. & Skovsmose, O. (2007). *Report on The KUL-projects: Learning Communities in Mathematics and ICT in Mathematics Learning.* Retrieved on March 30, 2009 from http://fag.hia.no/lcm/papers/

Van den Akker, J. (1999). Principles and methods of developmental research. In J. v. d. Akker, R. M. Branch, K. Gustafson, N. Nieveen, & T. Plomp (Eds.), *Design approaches and tools in education and training* (pp. 1–14). Dordrecht, The Netherlands: Kluwer.

Vygotsky, L. (1978). *Mind in society.* Cambridge, MA: Harvard University Press.

Wagner, J. (1997). The unavoidable intervention of educational research: A framework for reconsidering research-practitioner cooperation. *Educational Researcher, 26*(7), 13–22.

Wartofsky, M. (1979). *Models, representation and scientific understanding.* Boston: Reidel.

Wells, G. (1999). *Dialogic inquiry: Toward a sociocultural practice and theory of education.* Cambridge, UK: Cambridge University Press.

Wenger, E. (1998). *Communities of practice: Learning, meaning and identity.* Cambridge, UK: Cambridge University Press.

6

INQUIRY INTO MATHEMATICS TEACHING WITH ICT

Anne Berit Fuglestad

UNIVERSITY OF AGDER

Abstract: This chapter reports from a research and developmental project named *ICT and mathematics learning* (ICT-ML), in which teachers in schools and researchers at the university worked together on the use of ICT tools in schools to investigate how ICT can support an inquiry approach to teaching and learning mathematics. Background information about ICT in Norwegian schools is provided, and, following this, questions that have been explored concerning mathematics teaching with ICT are presented. A socio-cultural perspective of learning is adopted, with the key ideas of inquiry and learning community. In addition, elements from the theory of instrumental genesis concerned with understanding of ICT tools related to mathematics are applied. Questions for research are related to teachers' implementation of ICT tools in teaching and how the tool influences teaching and their didactical relationships with pupils. Although teachers from the project schools were part of the same developmental project together with researchers at the university, the findings reveal that teachers at three schools used different approaches and emphases in their teaching. Nevertheless, the emerging community of inquiry was seen to provide a strong support for their developmental work.

Keywords: ICT, inquiry, learning community, instrumental approach, research and development, lower secondary school.

INTRODUCTION

The implementation of ICT in mathematics teaching in schools appears to be a challenging task for teachers. The focus in this chapter is on the process of implementing ICT tools in Norwegian lower secondary schools (grades 8 to 10, pupils' ages 13 to 15 years) over three years with an inquiry approach to teaching and learning. I will briefly give some background information before I present the project and outcome.

Background of ICT in schools

Since 1984, Norwegian educational authorities have promoted the use of Information and Communication Technology (ICT) in schools by devising strategic long-term plans and plans of action[1] for introducing ICT in teaching (KUF, 1984; Krumsvik, 2007; UFD, 2004) and by including elements of computing or use of digital tools in curriculum guidelines. According to the plan of action from 1984, the reasons for introducing ICT in schools were first, expectations that use of ICT can support and improve learning; second, a recognition that ICT was becoming a common tool in workplaces and society; and third, the belief that ICT should be used in schools with the aim to give all pupils equal opportunities to access modern technology (KUF, 1984). More than twenty years ago, the curriculum guidelines from 1987 (KUF, 1990) mentioned modern technology including computers and media as teaching aids and listed computer

The First Sourcebook on Nordic Research in Mathematics Education, pages 91–108

studies (in Norwegian: datalære) as one of ten main topics in mathematics. This topic dealt with formulating ways of solving problems "where a computer is a suitable tool." In the following curriculum plan, introduced in 1997 (L97), the use of computers and calculators was regarded as important in mathematics, and it was asserted that ICT could provide new bases for work on tasks and problems in which experiments and investigations were emphasized (KUF, 1999). Additionally, elements of computer and calculator use were mentioned in several subtopics in the detailed plan for mathematics. Recent official plans for ICT in education and reports to the Norwegian parliament (Storting)[2] have made use of ICT compulsory (Krumsvik, 2007) and introduced the notions *digital competence* and use of *digital tools*. In the most recent curriculum guidelines, "The Knowledge Promotion" from 2006 (LK06), this intention is realized by including "facility with digital tools" as one of five basic competencies to be developed in all school subjects, with specific requirements given for their use in all aspects of teaching and learning of mathematics (KD, 2006). Nevertheless, as Krumsvik (2007) observes, despite the initiatives in official policy documents, the introduction of ICT in education has been slow, and teacher education is still challenged to provide teacher students with sufficient competence with ICT.

Access to ICT Tools

ICT equipment and computer software are now widely accessible and are used in schools in general. There are, on average, 4.7 pupils per computer in schools, and 90% of the computers have access to the Internet in primary and lower secondary schools. In upper secondary school the corresponding statistics are 1.8 pupils per computer and 98% with Internet access (Østerby, 2007). Learning Management Systems (e.g., Class Fronter and It's Learning) are widely used, with 58% and 98% used in lower and upper secondary schools, respectively. There have been further efforts for implementing the use of ICT in school subjects since the introduction of LK06. However, despite the efforts there is still only limited use of ICT in many school subjects, particularly in mathematics. An evaluation of the curriculum plan from 1997 revealed there was hardly any activity with ICT in mathematics teaching (Alseth, Breiteig, & Brekke, 2003), thus confirming the earlier findings reported as part of the TIMSS inquiry (Holmboe, 1997). Limited use of computers in mathematics teaching is also documented in biannual reviews of technology use in schools (Arnseth, Hatlevik, Kløvstad, Kristiansen, & Ottestad, 2007; Erstad, Kløvstad, Kristiansen, & Søbye, 2005) albeit revealing increased use from 2005 to 2007 in lower secondary schools. However, the situation is mixed, because some schools have classes that are using ICT extensively and moreover, in some classes all pupils have been provided with their own portable computers. An example where ICT use has developed can be found in a town close to University of Agder (UiA): In one upper secondary school pupils in some classes use computers extensively with hardly any use of paper and pencil in their mathematics lessons. Also in the same town, since 2007 all pupils in grade 8—that is, the first class in lower secondary school—have a personal laptop to use for their school work. An immediate consequence of these developments is the requirement that teachers develop their competence using computer software in teaching. In other schools in the region, a few teachers have, over some years, engaged in extensive developmental work with their own classes in utilizing various kinds of computer software for mathematics teaching.

QUESTIONS FOR INQUIRY

The provision of computers and software is just a first step. Substantial developmental work is needed to fully utilize the "affordances" (Greeno, 1994) opened up by the ICT hardware and software. In this chapter, questions for inquiry, related both to development and research concerning use of ICT tools in mathematics, are addressed. For those working with teachers' professional development, it is of concern to inquire into how teachers might be supported in using ICT and software in an effective and stimulating manner with their classes. There are also fundamental questions of interest to researchers in the field of mathematics and computer education. Among these I find the following to be particularly interesting and challenging: Does the use of ICT tools change the way a teacher approaches her or his work? How does the introduction of ICT affect teaching and learning and the didactical relationship between pupils and teacher?

The expressions ICT and digital tools encompass a variety of resources both hardware and software, including computers, generic programs such as Excel, specialized mathematics software such as Cabri or GeoGebra, simple and scientific or symbolic calculators, and other artifacts based on digital technology. Although the general background outlined above applies to a wide range of ICT tools, in this chapter I mainly consider computers together with software, and the examples I present come from my research in lower secondary schools.

My own experience, from more than 30 years of working with teachers and ICT, running professional developmental courses for teachers, and from my own use of computer software in mathematics teaching, indicates that implementing ICT in mathematics teaching is not straightforward. In some cases teachers lack the necessary technical knowledge of computers and knowledge and facility with relevant software. Also, teachers often express their need for help in finding good ideas about how to use ICT in the mathematics classroom. Similar observations are reported elsewhere with claims that many, but not all, teachers find that to utilize ICT in mathematics is difficult (Monaghan, 2004) and that to generate appropriate tasks for technology based lessons is demanding. Work by Hennesy, Ruthven and Brindley (2005) and others points to the UK government's effort to develop provision and use of ICT in schools, and the disappointingly slow implementation that has followed. One hindrance has been the lack of investment of time and resources for teachers to develop new approaches to teaching and learning with technology, and another is little research on how teachers can facilitate the pupils' learning with ICT. The review of the situation in the UK reveals a similar picture to what is experienced in Norway. ICT in mathematics teaching is an active field of research and there are many publications, both on research and about ideas and suggestions for use and their integration into schools and mathematics teaching. The reason for the slow integration is, according to Lagrange, Artigue, Laborde, and Trouche (2003), lack of precise understanding of what really happens; they point to the complexity of teaching and learning and that ICT adds even more complexity. A similar argument is presented in a review article on research on technologies for teaching algebra and calculus (Ferrara, Pratt, & Robutti, 2006).

THEORETICAL BACKGROUND—ICT ARTIFACTS AND INSTRUMENTS

These observations highlight the need to support teachers to understand how mathematics can be represented and communicated by the computer software, and to be able to work on these representations. Expressed in theoretical terms, the teachers need to develop ICT software as "instruments" for their work on mathematics (Trouche, 2004) and furthermore orchestrate their teaching for their pupils to support the processes involved in developing ICT software as instruments for mathematics. The notion of "instrument" in this connection relates to theory referred to as "instrumental approach," which has roots in the work of Rabardel and has been developed by several researchers, particularly in France (Artigue, 2002; Guin, Ruthven, & Trouche, 2005; Ruthven, 2002; Trouche, 2004; Vérillon & Rabardel, 1995). The process referred to as instrumental genesis concerns the development of an artifact[3] into an instrument for the learner.

"Instrument" refers to a psychological construct of the user that is personal for the user as he or she develops his or her mental schema associated with getting to know, understand, and use an artifact. This is a two way process between the individual and the artifact in which each influences and is influenced by the other. More detail about this process can be found in Trouche (2004, 2005a). In the following I limit the outline of the theory to what is relevant for this chapter. Developing an artifact into an instrument involves the development of utilization schemes that encompass usage schemes and instrumented action schemes. Usage schemes are directed towards managing the artifact by becoming familiar with the technical facilities or functionalities, using menus and knowing the basic concepts in the artifact. Developing a personal instrument also involves the development of instrumented action schemes, which are directed towards carrying out the mathematical tasks involved in the work (Trouche, 2004, 2005a); "instrumented action schemes can be regarded as a set of usage schemes" (Trouche, 2005a, p. 151). Using an instrumented action scheme then combines mathematical reasoning with the application of usage schemes in the context of a given task. An example could be to find a graphical solution for the intersection of functions by drawing the Cartesian graphs of the functions, zooming in or out, and adjusting the window or scales on axes to display the solution with the required accuracy. The instrumented action scheme includes the non-observable reasoning and decisions as well as the observed actions as an integrated unit. Such instrumented action schemes involve an understanding of how mathematics is represented by the software and how the necessary operations can be performed with the available tools embedded in the software.

Artifacts become part of cultural systems and are worked on in a learning community; teachers and pupils learn from each other, or from a wider community. The development of utilization schemes will be influenced by the mathematical culture, including tasks and investigations, in which the artifacts are introduced and used. Consequently, there are both social and individual aspects of the instrumental genesis. As observed by Hoyles and Noss (1996, p. 58), "Tools are not passive, they are active elements of the culture in which they are inserted." The teachers' "orchestration," enabling pupils to build their instruments and learn mathematics, is part of the culture. Furthermore, in order to act creatively using the instrument to solve mathematical problems, there is a need for a deep understanding of the mathematics, I suggest, beyond the creation of a personal instrument from the artifact.

Orchestration as a metaphor is inspired from music, to describe the articulation of planning; provision of artifacts such as ICT equipment, tasks and plans for the lesson, and so on; and the ongoing facilitation of the work. In teaching with ICT tools, the teachers are responsible for planning a suitable learning environment consisting of questions or tasks to work on, access to artifacts like computers and software, organizing the work, supporting and challenging the pupils, and so on. In this connection Kennewell (2001) uses the notion of orchestrating for the pupils' work. Trouche (2004) introduces the notion of instrumental orchestration for the necessity of steering or facilitating pupils' instrumental genesis. In a similar way, those didacticians working with the professional development of teachers have double challenges or goals, dealing both with instrumental genesis for teachers as users of ICT tools for working on mathematics, and additionally developing teachers' competence to use ICT as an instrument in their support for pupils learning mathematics combined with the pupils' instrumental genesis. Thus processes of instrumental genesis are going on in layers, for teachers and for pupils' own development, and across layers when teachers utilize ICT tools as an instrument for stimulating and supporting their pupils.

The view of ICT artifacts and how they might be utilized for mathematics will influence their implementation in schools, and, in turn, this might lead to new ways of looking at ICT tools and their influence on the learners and ways of working with mathematics. Several authors claim that a reorganization of thinking and ways of working with the subject is necessary when ICT tools are introduced. According to Pea (1987), Dörfler (1993), and others, the introduction of a new tool implies changes in the objects being worked on, partly introducing new objects and ways of working. An important aspect of this is that ICT software is not merely an "amplifier of operations" that performs basically the same operations as would be carried out without ICT, but can be seen, possibly, as a "reorganizer," implying the use of new objects to work on and new ways of working (Pea, 1987). Goos, Galbraith, Renshaw, and Geiger (2003, p. 74) view computers as "one of several types of cultural tools that not only amplify, but also re-organise, cognitive processes through their integration into the social and discursive practices of a knowledge community." Use of calculators and computers can be regarded as amplifying operations, doing many and rapid calculations, and plotting curves quickly, easily, and accurately. However, use of a graph plotter provides new ways of working on functions, such as zooming in and out, experimenting with parameters and families of functions, and in some software even dragging and manipulating the graph itself. The use of ICT tools might lead to a separation of planning from execution and implies thinking about the operations and shifting the work to a meta-level (Dörfler, 1993). Also, it is recognized that writing a computer program involves work on meta-level, with planning for the activity separated from the activity itself.

PERSONAL EXPERIENCE

The major portion of my research in the field of ICT started with research for a PhD degree, 1993–1996, where I collaborated with schools and classes of grades 5–9 focusing on the use of computers in mathematics teaching (Fuglestad, 1996). The field work was based in Norway, and the PhD degree taken at the University of Nottingham UK. During the years 2001–2004, I collaborated with an experienced teacher while leading a development and research project: *ICT competence in mathematics.*[4] The emphasis was on developing teaching approaches and pupils' competence with use of ICT tools so that they would be able to choose which tool would be suitable for a given task. In most classes, the ICT tools in use were computer software: spreadsheet, dynamic geometry, and a graph plotter, with the main emphasis on spreadsheets. In the ICT competence project, the main effort was focused towards development, but in the last half year, attention moved to research and evaluation of experiences. The remainder of this chapter focuses on the most recently completed ICT and mathematics learning (ICTML)[5] project based at the University of Agder (UiA) over the years 2004–2007. ICTML was implemented in close collaboration with another project at UiA, Learning Communities in Mathematics (LCM) (Jaworski, this volume, Chapter 5).

The ICTML Project

The ideas of inquiry and learning communities were fundamental principles of the ICTML project, as with the LCM project. In the ICTML project, didacticians worked with teachers in a learning community, with the aim of developing teaching approaches based on the belief that inquiry is a powerful tool for development and learning (Jaworski, this volume, Chapter 5). In the ICTML project the development and research aims were to investigate how ICT tools could support and stimulate inquiry approaches in teaching and learning, to support teachers' use of ICT software, and to stimulate the development of their teaching with ICT for inquiry approaches in mathematics.

In common with the LCM project, a socio-cultural view of learning was adopted as the theoretical framework of the ICTML project. From this perspective, learning is seen to be the result of participation in social practices and mediated by the social and cultural artifacts available or being developed in the community (Säljö, 2001). The theoretical framework and conceptualization of the project is built on common ground and in close collaboration with the project Learning Communities in Mathematics (LCM) and with some activities fully integrated. As the theoretical foundation and framework for the projects is described in another chapter in this book (Jaworski, this volume, Chapter 5) and elsewhere (e.g., Jaworski, 2007), I will only give a brief presentation of key ideas of the project. The methodology for the project can be characterized as developmental research, in which research and development are closely integrated. Development stimulates and informs research, and research guides development (Goodchild, 2008; Gravemeijer, 1994; Kelly, 2003).

At the start of the ICTML and LCM projects, schools were invited to take part, and school leaders were involved in discussions concerning the projects. Four schools agreed to join the ICTML project: Austpark, Vestpark, Fjellet, and Dalen. A small amount of money was allocated to each of the schools to cover expenses, such as travel costs. The schools could decide how this money would be used, but the amount was insufficient to release teachers from their classes to undertake additional planning, preparation, and familiarization. The teachers were expected to include work on the projects as part of their routine planning unless the school leader could allocate time for it.

Key activities in the project were workshops with teachers and didacticians, school meetings (mostly comprising teachers in their school teams but occasionally joined by a didactician), and meetings for didacticians, which were concerned with planning workshops, managing research and observations in schools, and developing the theoretical underpinning of the projects. Many of the teachers had limited experience of ICT tools, so the ICTML project had to include some competence development. Through activities in the project didacticians sought to stimulate development of learning communities with teachers working together in school teams within their schools, and in project workshops at the university together with other teachers and didacticians. Furthermore, didacticians together formed a learning community collaborating in planning for workshops and research, and discussing both developmental and research issues. The close collaboration between the projects made it possible for didacticians in LCM to participate in ICTML workshops, and the smaller ICTML team to be fully part of LCM. The ICTML team comprised three didacticians including the project leader, an experienced colleague who had been awarded a scholarship to work on a PhD within the project, and an experienced teacher who was engaged part time within the project for supporting the developmental work. In the remainder of this chapter I will normally refer to just the ICTML project; however, it is necessary to keep in mind that it was running alongside the LCM project with a strong symbiotic relationship between the two.

INQUIRY WITH ICT

A concern in the projects has been to stimulate inquiry as a tool—that is, stimulating pupils to find out about mathematical problems and issues by asking questions, wondering, investigating, and exploring mathematics. Furthermore, for the teachers the projects' concerns are to inquire into mathematics, the teaching of mathematics, and how teaching can be developed. Thus inquiry in ICTML can mean to question critically and reflect both on the mathematical problems, the way computer software can be used for mathematics and the approaches to teaching and learning mathematics with ICT. The intentions of the project were that didacticians, teachers, and pupils would be stimulated to ask questions and inquire by themselves instead of asking to be told the solutions, such as when pupils ask their teacher for an approach to work on a problem. The underlying assumption is that by developing the habits of inquiring into mathematics, ICT software, and its use for mathematics, the participants' attitudes towards teaching and learning will develop. Thus the learning communities can be seen as inquiry communities, where habits of inquiry develop into an attitude or "inquiry as a way of being" (Jaworski, 2004).

The developmental cycle can be described in five steps: plan, act, observe, reflect, and feed back, and as we worked in the projects, the cycle became a working model that we aimed to realize and sought to stimulate by providing ideas for initiating activity in the workshops. It was the intention that teachers could start from ideas presented in the workshops or from discussions in their school teams to develop their teaching approach with the use of ICT, developing lesson outlines or software applications for their classes. The implementation in classes was sometimes observed by a didactician or a colleague who later could take part in a discussion and feedback meeting to stimulate further reflection and development.

In the developmental work we deliberately did not want to "give" teachers prepared lesson outlines or tasks to use in their classes, but rather presented ideas as starting points and wanted to stimulate the teachers to engage in the development of tasks and ideas for pupils to work on. The activities in the workshops made a significant contribution, as did school meetings where the teachers could discuss both their experiences from the classes and further development of ideas for teaching. In the following I present some experiences from workshops and outline the work in schools to indicate how the project developed.

The pieces of software that were promoted in the workshops in the project were a spreadsheet (Excel), dynamic geometry software (Cabri), and a simple graph plotter developed in Norway called Grafbox. The basic idea was to use software open for investigations and experiments, and not with set tasks—in other words, software that is well suited for inquiry activities. The range of software was restricted with consideration to the grade levels of pupils involved in the project.

Alongside the developmental work in workshops and schools, didacticians followed the processes for the purpose of their own research (c.f. the research questions outlined earlier). For this purpose most meetings, including didacticians' meetings, and all workshops were audio or video recorded. When didacticians visited classes the activities were not only observed and video or audio recorded, but field notes were also written and some pupils' work collected, including copies of paper work or computer files. Video recordings provided opportunities also to discuss with teachers what happened in lessons and pupils' group activity as part of the developmental work. The recordings and field notes are the primary source of data upon which the following account of activity in schools is based.

KEY ACTIVITIES FOR DEVELOPING INQUIRY

Workshops at the university were the key developmental activities in the ICTML project for developing the learning community and teachers' competence with ICT, and to stimulate development of inquiry approaches using ICT software. Two workshops were held each semester. In addition, most of the participants took part in the LCM workshops in which the development of inquiry in mathematics teaching and ways of working was the main concern. From the outset, the plan for workshops was not fixed with themes and progression, except to develop competence using ICT tools for inquiry. The reason for this choice was the intention to pay attention to the participating teachers, to take their needs, wishes, and ideas about ICT use into account and furthermore, when appropriate, to follow up and develop ideas and themes from the LCM workshops for use with ICT tools.

The ICTML workshops followed a fairly regular pattern: an introduction to features of a piece of software with examples of applications in mathematics and then paired or individual activity on computers in which the specified software was experienced in the context of mathematical tasks. Participants were encouraged to utilize the presented ideas in the further development of teaching ideas with ICT support. This was followed by a plenary discussion in which participants were able to share the products of their computer-based activity. There were also opportunities for presentation and discussion of work in schools.

During the work at the computers, it became clear that many of the teachers did not have well developed competencies in using software. All had fairly basic knowledge of spreadsheets, and a few teachers had seen Cabri, but most of them had hardly any knowledge of the graph plotter and dynamic geometry software. This was confirmed in the first workshop and their answer to a short questionnaire. Many teachers needed to develop their own competencies and be given opportunities so they could incorporate the software as a personal instrument for their mathematical activities by developing utilization schemes, including usage schemes and instrumented action schemes related to the software and the mathematics. Tasks and problems to work with ICT are sources for the instrumental genesis. The mathematical environment or culture developed for teachers and pupils influences the kinds of instruments that they develop (Trouche, 2005b). Furthermore, there is also a need for teachers to develop ICT tools as instruments for teaching and for communicating mathematics with their pupils, as indicated by the term "professional instrumentation" (Abboud-Blanchard & Lagrange, 2006). An instrument for teaching might involve other features than just using the tool for solving problems. The teachers welcomed the time in the computer lab to work in small groups or alone at the computer, during which they had the opportunity to get hands-on experience with colleagues and didacticians at hand to support. The intention of the hands-on experiences with ICT tools just after the introduction was to support teachers developing their utilization schemes related to the software. Furthermore, supporting the development of inquiry approaches to teaching with ICT tools was the intention of the discussions that focused on the participants' own applications and related ideas for inquiry.

EXPERIENCES FROM WORKSHOPS IN THE ICTML PROJECT

The first workshop in September 2004 gave an introduction to the project, with examples of investigations and experiments using examples of the three ICT tools for the project: Excel, Cabri, and Grafbox. Two tasks were provided for work in groups followed by a plenary discussion on how the tasks could be solved with the three ICT tools. The work on computers and the following discussion in plenary revealed that the teachers' experiences varied a lot and few had any knowledge about Cabri and Grafbox. In the final session, four questions intended to inform future planning of the project were presented, such as what the teachers thought they needed to do at their own school, about their previous experiences with ICT, about their wishes for the workshops, and the contribution from didacticians at UiA. The answers confirmed first impressions of the teachers' knowledge of the ICT tools and that they wanted to have "examples, tasks, problems for starting and guidance on the way," and they welcomed workshops with practical work on computers, focused on the mathematics and methods of use, exchange of ideas and discussions. The experiences from the workshop, including the written answers to questions, made it clear that some systematic work on the ICT tools was wanted and necessary.

In the second workshop, the topic for investigation was number patterns on a spreadsheet. Otto, the teacher member of the project team, gave an introduction to the generation of number patterns using Excel. He presented tasks on spreadsheets with guiding text to make patterns of square numbers, rectangular numbers, and triangle numbers. The first task was partially constructed with some of the formulae prepared as guidance (See Figure 6.1). Other tasks and ideas that utilize the same principles were provided and, for a few teachers who had more experience with Excel, some ideas of using random numbers for simulation were suggested for investigation.

More than a year later, in an ICTML workshop in January 2006, didacticians aimed to follow up and utilize a spreadsheet for some algebra ideas that had emerged some weeks earlier in an LCM workshop, in which the focus had been on number patterns and

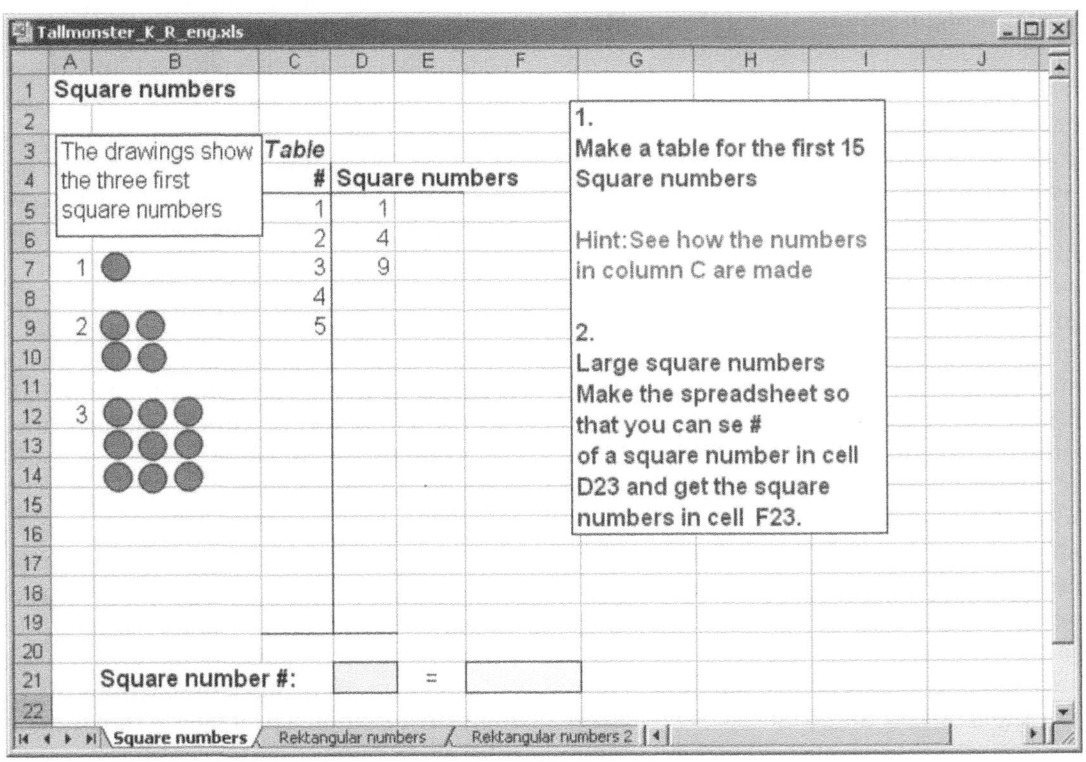

Figure 6.1 Spreadsheet layout for square numbers (no formulae in cells D5–D7).

how to express functional relationships (Fuglestad, 2007). At the start, Otto gave an introduction to how number patterns in columns or triangles can be produced. This included explanations about how to name a cell and make connections between cells, formatting issues, and how to hide and protect formulae. Thus, he demonstrated how it is possible to set up a sequence on a spreadsheet that can be explored by pupils without revealing the formula from which the sequence has been generated. The first challenge for participants was to try out and use the features of the spreadsheet that had been presented—that is, to build their usage schemes of the spreadsheet features. The second challenge was to build and understand an application, and thus, by working on this, to further develop their instrumented action schemes. And the third challenge was to utilize the spreadsheet features by envisaging and creating suitable tasks for pupils' inquiry into mathematics using a spreadsheet.

Several groups of teachers followed up by working on similar number patterns, in columns and triangles. This work provided the teachers with opportunities to develop their understanding of the spreadsheet, but the ideas for pupils' work in the examples mainly focused on exploring number patterns and their mathematical relations; the use of the ICT tool was hidden for pupils. However, a pair of teachers discussed whether it would be possible to have one triangle showing formulae and another revealing the numerical results. The discussion brought up interesting ideas and exposed limitations of the spreadsheet features.

In the following ICTML workshop, the introduction was devoted to presenting Grafbox,[6] a graph-plotting program (Fuglestad, 2007). It turned out that several teachers had not used such a piece of software before. Therefore, they needed experience with it to develop the graph-plotting software as an instrument, developing the necessary personal utilization schemes involving tasks such as entering a function to draw the graph, setting appropriate scales on the axes, and zooming in and out. Trouche (2004) points to an important idea: "the user's conception of the instrument is formed through use" (p. 295). The work on computers during the workshop provided the teachers with some opportunities for this. A challenge in this workshop was to solve the same task using different software and compare the results. Three pieces of software were available: a spreadsheet, a graph plotter and dynamic geometry. For this case a few mathematical tasks were provided, one entailed planning a rectangular sports field within the frame of a right-angled triangular space with shorter sides 30 and 40 metres. The task is to make the largest possible rectangular sports arena within the triangular field. What would be the best placement?

The tasks required the participants to have some user knowledge of the software, and at least partly developed usage schemes corresponding to the different programs. In order to be able to solve the problem with the three pieces of software, there was a need for instrumented action schemes corresponding to the mathematics in question together with mathematical insight with which they could implement suitable models for the triangle with the rectangle inside to be able to explore the case.

Two didacticians solved the problem by plotting functions representing the two ways of placing the rectangle, but apparently with the same result for the area. This turned out to be a surprise for them. They checked by using dynamic geometry constructing the rectangle inside the triangle, dragging, and observing the result. Their results were later presented in plenary and discussed. The reflection and discussions which followed provoked new questions: Could we consider perimeter instead of area for the rectangle? What if the triangle were not right angled? Perhaps paper folding could solve the same problem? The use of ICT tools in this case provided new approaches for inquiry into the task by comparing and experimenting with the different representations of the mathematics content. I suggest that the example provides an indication of how the approach to the mathematics might be reorganized for the benefit of pupils' learning.

The workshops played a key role in developing an inquiry community (Jaworski, this volume, Chapter 5). The teachers and didacticians worked together and supported each other, asked questions, wondered about possible solutions, discussed, and in so doing stimulated an inquiry approach in their work with ICT. Both didacticians and teachers revealed incomplete knowledge of the computer software and some mathematical features, and evidently experienced some surprises as they worked on the tasks. In the relaxed context of the workshop it was possible for all participants to ask each other for help and to stimulate further inquiry by raising new questions.

School Team Meetings

School team discussions and planning were other key activities in the project. At the start of the project it was the intention to have at least three teachers participating from each school. This would make it possible for teachers to discuss and support each other during the work between workshops. However, the regularity and effect of the work in school teams varied, and some organizational problems within the schools were difficult to resolve. Further descriptions of the work in school teams is given for each of three of the four schools participating in ICTML, followed by a discussion of some factors affecting the work, including references to the fourth school in the project.

Dalen School

Three teachers—Rikard, Viktor, and Otto—participated in ICTM. They were teaching in grades 8, 9, and 10. Otto had several years of experience with ICT in mathematics and had collaborated with me in a previous project, *ICT competence in mathematics*. Otto, who was introduced earlier in this chapter, was also employed part time in the ICTML project to contribute in the ICTML workshops and provide support for the school teams. From the start of the project the teachers at Dalen established regular meetings of about 1.5 hours on alternate Monday afternoons. They set as a goal for their work to develop tasks for investigations on a spreadsheet and to build a bank of spreadsheet tasks for various topics in the syllabus, coordinated with the textbook used and starting with grade eight, which was Rikard's class that year. The school team considered this work to be their main goal for the project and wanted to leave graph plotter and dynamic geometry for later development. During the school team meetings the teachers inquired into how a spreadsheet could be used for their purpose of stimulating pupils' inquiry into and discovery of mathematical relations. Many of these spreadsheet tasks were used in Rikard's class, followed by discussion in the school team (Fuglestad, 2008). Later some were also used in Victor's class and have since been developed further.

The inquiry cycle of plan, act, observe, reflect, and feedback was evident in the events that took place at Dalen. I will illustrate this briefly by presenting some episodes that took place over approximately two weeks when the teacher team discussed and developed tasks for work on decimal numbers, fractions, and percent (Fuglestad, 2008). The planning started with teachers' discussion of ideas in the school team. In the first attempt to create a task, the pupils would be required to compare three numerical representations and investigate the connections between them. Pupils would need to type in a value for a percent in a cell and observe the results in the corresponding cells, which displayed a decimal number and a fraction. The layout of the spreadsheet had space for the pupils to write comments and explanation.

When the task was used in a lesson with Rikard's class, I attended as participating observer together with Otto and Victor, who came in for part of the lesson to observe and talk with pupils as they worked at the computers. We were interested to get an impression of how the pupils experienced the tasks. Later the same day, a discussion for reflection and feedback took place. The school team discussed how the pupils worked on the task and suggested possible modifications and further ideas. Thus, the team discussed both didactical and technical issues related to the experiences and further development. Rikard was excited to follow up: "This is great fun. I will enjoy working more on it. I will make something. It is always too difficult in the start." He expressed his inspiration from the experiences with the class and following discussion. During the coming week he modified and expanded the task and implemented new ideas that were subsequently used in his class. These actions and observations in class stimulated the next steps, through reflection, feedback, into new planning, and hence a new cycle of inquiry.

Teachers' and didacticians' learning became evident during the work. Several of the tasks utilized the spreadsheet features to hide and protect formulae so that pupils could experiment with various numbers in cells without being required to make the layout themselves. Some challenges with implementing ideas in the spreadsheet came up during the work and were solved partly during the school team discussions, with suggestions from a didactician and partly by consulting another colleague at Dalen School. For example, the spreadsheet design to compare equivalent fractions (Figure 6.2) needed particular attention.

Thus the development of the spreadsheet tasks by Rikard stimulated his development, building a deeper knowledge of the spreadsheet and expanding his spreadsheet instrument by adding to his utilization schemes. There were also some more open tasks where the pupils had to plan and figure out what formulae to put into cells.

The collection of spreadsheet tasks was organized on several sheets in an Excel document. They were planned to support pupils' progression from guided investigations to tasks where the pupils had to fill in parts of the layout with formulae on the spreadsheet, and onward to tasks that offered ideas and challenges for pupils to create a full spreadsheet design. In this way some work leading to development of usage schemes was built into the tasks. Furthermore, in some cases there was a challenge for pupils to develop a more composite layout leading towards instrumented action schemes. These sheets challenged the pupils to develop their competence of using the spreadsheet for solving problems.

Rikard's lesson outlines typically included a plenary introduction in which the spreadsheet layout and the tasks were presented. Following this, pupils were organized to work in pairs or threes at the computers.

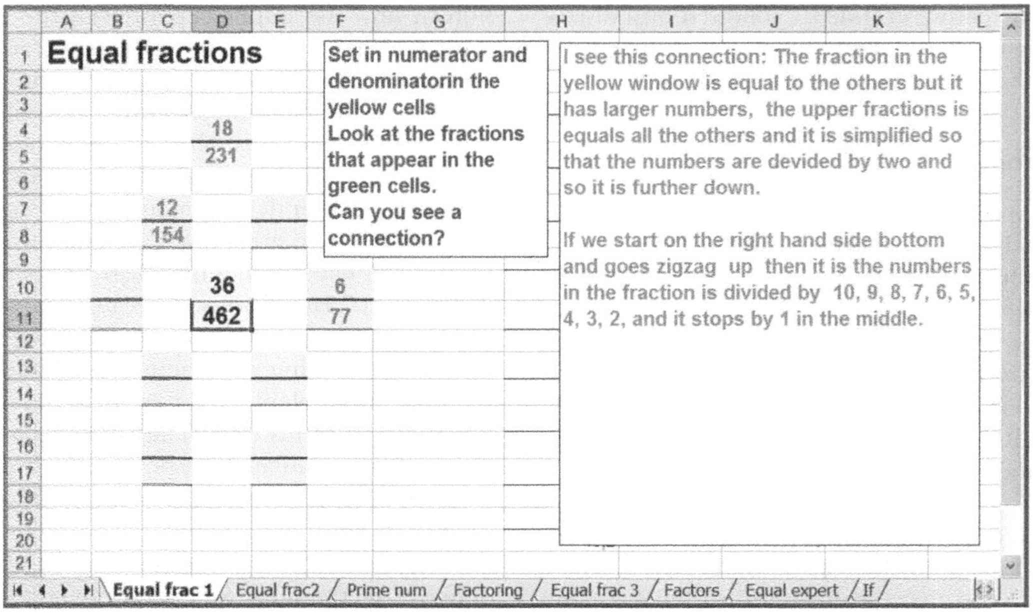

Figure 6.2 Spreadsheet task on comparing equivalent fractions, with a pupils' explanation included.

Rikard closed lessons by conducting a plenary discussion in the class, presenting the pupils' work accompanied by pupils' explanation of what they found and supported by their written comment on the spreadsheet. The didacticians responsible for the ICTML project (Aud and Kai) took part in some of the school team meetings and on several occasions observed work in classes. As in the first lesson described earlier, Otto and Victor occasionally joined Rikard's class for parts of lessons. In some cases a short meeting of the school team was held the same day or the observations and experiences were discussed in the following Monday school team meeting. Discussions often resulted in further work on the spreadsheet tasks, such as implementing a simplification, a further development, or a new idea, with new cycles of inquiry. It is evident that both the spreadsheet tasks and Rikard's lesson outline had a strong framework, but still with room for pupils to investigate particular properties designed in the task, and with emphasis on questions, explanations, and reasoning in the summary of the lessons. The tasks alone might seem limited as a source for inquiry, but seen as part of a lesson design, they presented opportunities to inquire into particular topic details of mathematics and how to explain and express relationships.

The teachers asked themselves or sometimes asked the didacticians for an idea for a specific topic and what could be done. Some ideas arose during workshops during the project. Ideas were discussed in the school team and discarded if the teachers did not regard them as suitable, but often ideas were initiated in a school team meeting. Following this Rikard would be inspired to start implementing ideas on a new spreadsheet, often using late evenings because he enjoyed the challenge and was keen to try new ideas in his class. The joint effort of the school team appeared as a strength and inspiration in their activity. A further source of inspiration and motivation was the close contact with didacticians and other teachers in workshops. The teachers maintained contact with didacticians by emails, and didacticians were invited to observe lessons. After observing work in the classes, the team followed up with discussions in meetings, informal conversations, and occasional e-mails. This gave ample opportunity for feedback and suggestions for new ideas.

The teachers at Dalen appeared to align with the key ideas of the ICTML with an inquiry approach to teaching and work in a learning community. They worked regularly with project tasks and supported each other. In developing tasks they aimed to provide situations for pupils' inquiry, using investigations and partly open tasks and with emphasis on discussing pupils' findings. The developmental work was mainly focused on inquiry into mathematical connections, such as in the example comparing decimals, fractions, and percent. To some extent the work was also concerned with the development of pupils' competencies with the ICT software, and some elements of this were built into the collection of tasks for various topics. Furthermore, their view of learning was influenced by a socio-cultural view. On several occasions when I observed his class, Rikard commented he could "see Vygotsky" in the way the pupils worked, referring to the zone of proximal development. He noticed the way pupils could do more together with a partner than alone, or from discussion with a helping teacher. On a later occasion Rikard, in an interview, commented on and confirmed that inquiry approaches to teaching fit well with his personal theory of learning (Fuglestad & Goodchild, 2008). His socio-cultural view of learning was developed from previous experiences with pupils and his reading. Rikard and Victor had discussed their thinking about learning and had developed their own principles for teaching. Their view of learning appears to support and provide a rationale for their strong commitment to and engagement with the project.

Austpark School

Three teachers, Frode, Harald, and Gunnar, took part in ICTML at Austpark School from the start. Austpark School is larger than Dalen, and has four classes in each grade, whereas Dalens has one. A school team was established, but the teachers experienced internal constraints, limited access to computer rooms, and disruption of the school's activities, which interfered with their project activity.

However, during the autumn of the second year, the teachers talked about using Cabri from the start of the spring term. Otto encouraged them by offering support in the form of a one-day visit to help with planning for Cabri use. Their choice was to have a day course for the teachers, getting to know the facilities of the software and in this way initiating the development of their Cabri instrument. Furthermore, the team was strengthened by the inclusion of a new recently qualified teacher, Jacob, who had just completed his master's degree in mathematics education and had experience with ICT and dynamic geometry. This appeared to provide the impetus they needed to start using Cabri. Thus, in the spring term all the teachers in grade 8 and two in grade 9 started using Cabri with their pupils. Only the grade 8 teachers, Jacob and Frode, were observed as part of ICTML.

Contrary to the work at Dalen, the teachers at Austpark planning for Cabri use did not include the development of their own teaching material. The teachers at Austpark chose to use some worksheets that had been developed by a teacher[7] working in another school and who was known by Jacob. The material consisted of seven two-page worksheets. The teachers found this material suitable as it covered a substantial part of the syllabus for grade 8, it was ready to be used, and it matched the way they wanted to work with Cabri in parallel with the traditional Euclidean geometry using ruler and compasses (Erfjord, 2008). The material was structured with explanations of main concepts and detailed instructions for the pupils how to construct basic geometrical figures in Cabri, for example, constructing angles and modifying sizes, construction of angles of 60 and 90 degrees, and perpendicular bisection of straight lines using circles and intersection points. These well-developed instruction sheets with step-by-step guidance were features that the teachers found useful for their teaching. The teaching approach, characterized by direct instruction and supportive comments, was also observed in their classes, when pupils were asked to perform constructions using Cabri following teachers' guidance (Erfjord, 2008,). In this way the teaching approach provided by the worksheets was close to the established tradition of constructing geometrical figures with use of compasses and ruler, but with additional possibilities of dragging points or other objects on the screen and changing the size and shape of figures in order to check that the properties of the constructions were preserved.

In conversations with a didactician, the teachers explained that they found the worksheets appropriate and helpful for their teaching and felt confident to use them since the methods were close to how they regularly work. They claimed it was difficult to be more innovative and make their own tasks. Also, the teachers at Austpark who did not take part in ICTML used the same material for introducing Cabri in their teaching. Writing about the worksheets used at Austpark, Erfjord (2008) points to a *double innovation* embedded in the project goal: first, implementation of computer software in teaching, and second, the adoption of an inquiry teaching approach. To aim for both is demanding for the teachers, and the fact that they needed to take into account their colleagues' working plans added further constraints to their work. Jacob and Frode argued for not using inquiry approach in the early phase and pointed to time as a critical factor. During a discussion in Workshop 11 in LCM, the teachers commented that their way of implementing Cabri did not include as much inquiry as they thought was expected from the project. As Frode expressed it: "...we ran very little of inquiry as you are saying. Because I ran it very basic; in the classroom with compasses: Look here. See how I do, and then you are allowed to do it afterwards." Erfjord (2008, p. 233). In this context it is reasonable also to see their choice in the context of time constraints, workload, and the fact that they collaborated with teachers who did not take part in the ICTML project. The observations reveal that there was strong emphasis on the "correct" or best way for specific operations with Cabri, with focus on figures "hanging together." This can be characterized as an emphasis on usage schemes, and to some extent instrumented action schemes. Nevertheless, a development appeared to start, and at the end of the projects the teachers reported strong engagement in discussions of mathematical subject matters with their colleagues and a wish to continue working on further development.

Fjellet School

At Fjellet school two teachers, Trude and Markus, who taught lower secondary grades, participated in ICTML. A third teacher, Ludvig, also teaching lower secondary grades intended to take part in ICTML, but was often prevented by other demands on his time. Three other teachers working at the primary level took part in LCM. The school has one class in each grade level.

In the spring term the second year Trude decided to implement Cabri in her teaching over some weeks in her grade 8 class. Her teaching style during this period was quite different from what was observed both at Dalen and Austpark. Trude did not use any prewritten material and gave only a few instructions at the start. Her approach was to give the pupils opportunity to explore to find out about different features of Cabri during their work on some geometrical tasks. The pupils were asked to experiment to find out the meaning of the menu items and thus develop usage schemes, simple one-step operations, for basic constructions. In the same way the pupils had to work on their instrumented action schemes, but with the opportunity to learn from each other and find out about various facilities of the software. Trude supported the pupils as they worked on Cabri, and in this way helped pupils to develop their usage schemes. Her approach was to present a problem written on a flip-chart (Figure 6.3) and ask the pupils to investigate the software and find out how to solve the problem. During the work they were also encouraged to discuss and help each other. Thus she stimulated the pupils themselves to investigate Cabri and solve the problems—in other words, to build their instrumented action schemes, combining various usage schemes and investigate further facilities provided in the menus of Cabri. In this way she tried to initiate inquiry in the work with Cabri and was not directive in the way she instructed the pupils. Half the class worked on Cabri while the other half had other tasks, and vice versa. She emphasized the use of dragging some object of the figure to test whether the figure "hangs" together, which is a significant feature of dynamic geometry. This requirement is specified in the first of the tasks, constructing a snowman.

From her presentation and comments during workshops in the project, it was clear that Trude had not used Cabri before. She worked a lot on her own to learn how to use Cabri and prepare herself for her pupils' questions. Trude talked about the necessity to learn Cabri herself enough "to dare to use Cabri in the class." I will interpret her comments to say she felt the need to get to know the tool, to develop Cabri as an instrument for herself before using it with the class. She described her enjoyment arising from working on

1) Snow man with a stick, hanging together

2) Make a figure by reflection
   a) Reflection in a line
   b) Reflection on a point

3) ΔABC AB = 5,0 cm
BC = 4,5 cm
AC = 2,5 cm

4) Make the circumscribed circle for a triangle

Figure 6.3 Trudes' tasks with Cabri, with English translation

Cabri, but also the stress she experienced. Trude considered it necessary to solve all the tasks herself before presenting them to the class. Given that she worked mainly alone as she introduced Cabri in her teaching, this is understandable.

In spite of being part of a school team, there appeared to be little support from the others during her work on Cabri, mainly because the teachers worked with classes of different grades and did not find time to get involved in each others' work. This is in contrast to the team work at Dalen, where the three teachers all engaged in discussion and development of tasks as a part of building their bank of tasks, which was regarded as a benefit for the school, not just for the single teacher. Nevertheless, Trude reported she had enjoyed using Cabri and although she regarded it as very demanding, she expressed satisfaction because she had been able to carry through the work.

FACTORS INFLUENCING THE WORK

During the project, the teachers' work was influenced by factors mainly external to the project, such as issues arising from the schools' environments and locality, size and organization of the schools, access to computers and software, personnel, and time constraints. Austpark School experienced disruption in their routines when a significant part of the school was badly damaged by a fire, and this strongly affected their engagement in the spring term of the first year of the project. Fjellet School is situated in a neighboring town, and this entailed longer distances to travel both for teachers and didacticians. This might have affected the frequency of attendance to workshops and didacticians' visits to the school.

Austpark is a large school with four parallel classes at each grade level, and not all the mathematics teachers were part of the two projects. This caused some problems of planning because the teachers of classes at the same grade were required to follow a common teaching plan and use the same internal tests in mathematics during the year. In addition, the school had planning meetings for teams across subjects at each grade level, in which mathematics teachers were also expected to participate. On the other hand, Dalen and Fjellet were small schools with only one class at each grade level; thus teachers taught classes at different grade levels. In order to achieve good progress in development, it was necessary to run the project over several years and have some stability in the school teams. However, occasionally teachers move to other schools and for some school teams this can have substantial impact. This happened in ICTML as with Vestpark, the fourth school within the project, one of the teachers in ICTML moved, and at Austpark a new teacher joined the school team. At Dalen, Otto moved to other work a long distance from UiA, but he continued to work with preparation of workshops, although obviously his school colleagues missed having his support.

Access to computer equipment and software for the pupils was essential for the work in the ICTML project. At Austpark a problem the teachers faced in the first year was access to the computer room because it was booked by other courses as the school was running a substantial number of short courses for all classes. The leader of the school was made aware of the problems regarding access to the computer room and time available for teachers to meet in their school team, and the problems were later resolved. At Vestpark School the teachers experienced problems with getting some software to run properly, caused by the way their computers were set up using a central server and thin clients.[8] During the first year, a spreadsheet was mainly in use, but efforts to get Cabri to run via the school network did not succeed. There was some effort to solve this problem, but little progress was made.

Teachers often experience time constraints. The meetings for various purposes, including meetings in school teams for the projects, added to the time constraints of the teachers' work. On the other hand, working in a large school gave opportunities for several teachers to work together and support each other. The project work itself added to the demands even if it was also regarded as valuable and supported teachers own competence building. Ludvig at Fjellet School hardly managed to attend any workshops due to other obligations and demands on his time. The complex organization and many meetings in a large school added to the demands on time, as did also commitment to more than one project. Regular meetings in school teams were intended and regarded as an essential as part of the developmental work. But the school team meetings appeared at times not to be as regular and frequent as intended and there was little initiative from some of the schools to invite didacticians to their lessons and school meetings. Therefore, from three of the four schools, didacticians sometimes had little information about what was going on related to the ICTML project. Some activities were going on, but didacticians were either not informed at all or were given insufficient notice to take part in the activities. There were exceptions, with Dalen from the start having a regular school team meeting every second week.

In late September of the second year of the project, the didacticians had little recent information about developmental activities at Austpark and felt it was necessary to address the problem of fulfilling the projects' intentions, to observe lessons and attend meetings at Austpark. The project leader (Aud) initiated a meeting with the teachers and school leaders to see if it would be possible to make better arrangements. This resulted in an improvement in the activities, and the school leaders and teachers declared their motivation continue with the project. The situation turned out to be similar at Fjellet and Vestpark concerning school team meetings and implementation of activities with ICT in the classes. Their development in the ICTML project was slower than expected and with scarce contact with the didacticians. As with Austpark, the didacticians initiated meetings with teachers and school leaders at Fjellet and Vestpark in order to strengthen the project work and offer support from Otto or other didacticians. During the final year of the project, there were no observations of activities with ICT, and after some weeks Vestpark School withdrew from both projects. The reason given was partly lack of economic support for work in the LCM project, and generally a demanding working situation at the school, but perhaps there were other reasons such as some teachers on leave or moving to another school. The same happened at Fjellet, but one of the teachers still attended workshops.

To summarize some key factors: The school leaders' support was necessary for the successful implementation of the project in the schools. The schools needed to provide organizational support; access to equipment and available time for the teachers' collaboration were considered essential in order to opportunities for development. It is unavoidable that teachers move to other schools and new teachers join during the project. A reasonable size of the school team at a project school is necessary to stabilize the work. Furthermore, the engagement and interest of individual teachers and stability of participation in the work in project activities are important factors concerning progress in the development and learning from the project for the teachers and the schools.

SUMMARY AND DISCUSSION

The philosophy of the ICTML project and ways to utilize ICT tools were presented in the workshops, in plenary sessions, discussions, and through practical activity with software. However, teachers had to decide themselves how to implement the project ideas in their classes. It turned out that teachers at three schools, Dalen, Austpark, and Fjellet, used different approaches, which led to different characteristics of their implementation of ICT in their classes. Here I will summarize some issues: (1) choosing a focus for ICT and limitations for its use, (2) style of teaching and how inquiry could take place with the teachers learning and orchestrating pupils' learning, and finally (3) issues related to learning community in workshops and school teams.

At Austpark the teachers chose to support the implementation of Cabri by using externally prepared teaching material, a set of worksheets, and they wanted to coordinate and compare the use of Cabri with the traditional constructions using compass and ruler. This suited their way of teaching and was close to the curriculum. The material was available, and it was helpful as they collaborated with teachers who did not participate in ICTML. The teachers at Dalen set their limits by focusing on using spreadsheets as their tool and the development of a series of structured tasks for investigations and experiments to find relations between numbers in various contexts. They chose to put less emphasis on other software. At Fjellet, Trude worked alone, and chose not to use or develop specific material except for planning tasks that the pupils were asked to work on.

The teaching style at Austpark followed a pattern with which the teachers were familiar, with fewer opportunities for inquiry. The teachers judged their own implementation as having less of an inquiry approach than was intended in the project. The demand of double innovation, both to use a new piece of software and a less familiar teaching approach with ICT, seemed too great. However, there was some emphasis on comparing Cabri constructions with traditional compass and ruler methods and the use the dragging feature of Cabri to test out their constructions, and this might appear to have an inquiry style. The tasks developed at Dalen were designed for pupils to experiment and investigate relations between numerical representations, find the connections, and describe and explain what they found. The investigations with the spreadsheet tasks were followed up by discussion in the class, with inquiry built into the discussion. Contrary to the other schools, at Fjellet, Trude's approach was quite open: She encouraged pupils to inquire into facilities of Cabri, explore, and utilize the constructions to solve problems. She supported their work at computers and by stimulating discussions with their peers.

At Dalen the school team had regular meetings that were established from the first autumn of the project, and these apparently were helpful for their joint development of the series of tasks using a spreadsheet. Problems were solved with joint effort from colleagues and sometimes suggestions from didacticians, and the work progressed. At the other schools, teachers experienced difficulties in arranging regular meetings. This perhaps was partly why some schools did not make so much progress and the didacticians had difficulty getting appointments for school visits.

The workshops were appreciated. In ICTML, as well as in the LCM project, the workshops proved to be a good means of support for building an inquiry community and to support teachers' development. Quite often teachers expressed satisfaction and commented that they appreciated the time devoted to working on computers with assistance from other teachers and didacticians and opportunities to discuss and ask questions. The teachers' reported experiences from working in a learning community in workshops and developmental work in school teams suggest that a learning community can be a strong support for their development. Furthermore, the didacticians' role, mainly supporting and not *telling* what teachers had to do in their classes, appeared to motivate some teachers, whereas others needed more intervention and suggestions in order to move on.

For the teachers involved, there were efforts to learn to use the necessary computer software and to implement what they wanted for the class. At Dalen the teachers investigated how their tasks could be implemented in Excel using various facilities. It was evident in several school team meetings that learning took place; teachers developed both knowledge of specific facilities and how to build applications combining operations. This implies developing Excel as an instrument for preparing tasks. Furthermore, work related to orchestrating use of the files, tasks, and other material required learning to use such facilities. The teachers at Austpark can be regarded as restricted by the resource material used, but on the other hand, they had to learn how to perform the constructions in Cabri and compare with traditional methods. At Fjellet, Trude worked alone, and commented that she had to sit alone at home to find out and learn Cabri herself to dare to use it. She had to work both on her own instrumental genesis and consider how she could help her pupils, planning instrumental orchestration. For all, I suggest, there was learning related to the didactical issues as they experienced the use of ICT in their class.

From all three schools, it is evident that the introduction of computer software for mathematics is demanding and needs substantial effort in competence development. Being able to work on a piece of software in teaching demands more than just working on some features of the software for teachers' own use. I suggest the teachers first have to develop the ICT artifacts as instruments for their own use with mathematics in order to be able to orchestrate for their pupils' development. This implies the development of utilization schemes that are concerned with getting to know the technical facilities or functionalities of the software—the usage schemes—and to developing instrumented action schemes related to the mathematical tasks involved in the work. The deep knowledge of the ICT artifact as it is becoming a personal instrument forms a foundation for investigative use—as an instrument for inquiry.

Furthermore, it is a demanding task to develop one's own material. It has been evident both in ICTML and LCM that teachers choose to use tasks from workshops or from other sources in their classes and to a lesser extent develop their own material. The development at Dalen was individual in style, and it seems that it was possible because of the close collaboration between the teachers and support from others in addition to the huge effort developing applications on spreadsheets by one of the teachers. This highlights the value of the community of inquiry that was one of the key ideas underpinning the project.

CONCLUSIONS AND A WAY FORWARD

Finally, what can be learned from the ICTML project, and how might this knowledge inform practice? Reporting from only three schools, a few classes and teachers, it is impossible to generalize. However, the research reported here confirms to some extent other findings. The summary I present here can be seen partly as conclusions and partly as suggestions for further inquiry. The first comments relate to the question of how teachers can be supported in using ICT software in an effective and stimulating manner.

The introduction of computer software in mathematics teaching takes time and is a demanding process for the teachers (Monaghan, 2004). Furthermore, to utilize computers with an inquiry approach implies a double innovation in teachers' work (Erfjord, 2008). Working with an inquiry approach demands use of software that gives opportunities for investigation and experiment. I suggest that such open software is more demanding than using task-specific software for drill and practice.

Even with more than basic knowledge of computer software, teachers might need to develop their competence further together with insight into how the software can represent mathematics and provide an environment for investigations and experiments with mathematics. In other words, teachers need to develop the various pieces of software as instruments for mathematics, not just for their own work on mathematics but for the orchestration of pupils' instrumental genesis with the software. Further research on orchestration both for teachers' and pupils' development of their instruments could prove valuable.

A strong affordance of the project is the learning community, or inquiry community comprising teachers and didacticians. The support from other teachers locally through work in school teams made development possible within a school. The support and insight from school leaders were crucial to overcoming some obstacles, as was demonstrated in the way the situation was handled at Austpark, for example: Hindrances were dealt with, and the support stimulated further development.

The question about whether ICT changes the way teachers approach their work is difficult to answer definitely from the evidence accumulated in this project. As seen at Austpark, the teachers chose to use an approach that seemed close to their familiar practice. However, comparing Cabri and compasses constructions together with using drag mode in Cabri did influence the teaching, perhaps more than using Cabri as an amplifier, since moving objects on the screen by dragging implies some conceptual changes (Dörfler, 1993), implying some reorganizing of objects worked on and associated thinking. On the other hand, accepting the challenge, at Dalen a new element came in, demanding changes in the orchestration of pupils' work to include use of computers for pupils and a system for administration of pupils' files, and not least development of their own spreadsheets for pupils to explore. With Trude, perhaps even more changes were implemented because, together with the start of using Cabri, she also sought to use an inquiry approach. Thus it appears that ICT tools have the potential to change the approach to mathematics, and perhaps they ought to. However, the extent to which it will happen depends largely on how the teacher follows up the potential.

The final question here is related to how ICT affects the didactical relationship between pupils and teachers. It is only possible to give a modest response to this, and it is very much dependent on how challenges in the previous question are met by the teachers. Using an inquiry approach with ICT tools as instruments for teaching will provide new arenas for work in which the teachers do not always know the answer, nor is there just one answer. An inquiry approach might imply that teachers and pupils together investigate and inquire into mathematics, and ICT tools can provide interesting situations for this.

Looking ahead, recent developments with digital competence as part of the Norwegian school curriculum for all subjects demand further developmental work with teachers and teaching approaches in order to implement the plans. The demands for digital competence in the curriculum plan and better provision of computers for schools make this need urgent. Providing computers and software is just a start; it is essential for the outcome in learning mathematics to consider how they are put to use. The work in ICTML has provided some insights that can be utilized for further development, and it is intended to follow up the work with further developmental research projects. Finally, development of teaching and teachers' competence with ICT takes time, and a period of three years is too short to fully see the effect of the work.

NOTES

1. The first plan of action "Handlingsplan for datateknologi i skolen" was introduced in Report no 39 to the Storting (1983–84), Chapter 6.
2. The Storting is the Norwegian parliament. Reports to the Storting are official policy documents from the government as part of the handling of items in the Storting.
3. Artifact—something created by humans, physical or abstract
4. ICT competence in mathematics was financed by the Department of Education with strategic funding for ICT in education.
5. ICTML and LCM were funded by the Research Council of Norway through the KUL program. KUL stands for Kunnskap, Utdanning og Læring–Knowledge, education and learning.
6. Norwegian software, prepared by Numerica
7. The material was developed by Henning Bueie, who was a colleague of Jacob.
8. Thin clients are machines used only as terminals, running the software on a central server.

REFERENCES

Abboud-Blanchard, M., & Lagrange, J.-B. (2006). Uses of ICT by pre-service mathematics teachers: Torward a professional instrumentation? *International Journal of Computers for Mathematical Learning, 13,* 183–190.

Alseth, B., Breiteig, T., & Brekke, G. (2003). *Endringer og utvikling ved L97 som bakgrunn for videre planlegging og justering—Matematikkfaget som kasus.* Notodden: Telemarksforsking.

Arnseth, H. K., Hatlevik, O., Kløvstad, V., Kristiansen, T., & Ottestad, G. (2007). *ITU monitor 2007 Skolens digitale tilstand 2007.* Oslo: Universitetsforlaget.

Artigue, M. (2002). Learning mathematics in a CAS environment: The genesis of a reflection about instrumentation and the dialectics between technical and conceptual work.

International Journal of Computers for Mathematical Learning, 7, 245–274.

Dörfler, W. (1993). Computer use and the views of the mind. In C.Keitel & K. Ruthven (Eds.), *Learning from computers: Mathematics education and technology* (pp. 159–186). Dusseldorf, Germany: Springer.

Erfjord, I. (2008). *Teachers' implementation and orchestration of Cabri-use in mathematics teaching.* Kristiansand, Norway: University of Agder.

Erstad, O., Kløvstad, V., Kristiansen, T., & Søbye, M. (2005). *ITU Monitor 2005. På vei mot digital kompetanse i grunnopplæringen.* Oslo: Universitetsforlaget.

Ferrara, F., Pratt, D., & Robutti, O. (2006). Technologies for teaching algebra and calculus. In A.Guitérrez & P. Boero (Eds.), *Handbook of research on the psychology of mathematics education past, present and future* (pp. 237–273). Rotterdam, The Netherlands: Sense Publishers.

Fuglestad, A. B. (1996). *Computers and the understanding of mathematics. A study of teaching decimal numbers. PhD thesis.* Kristiansand: Agder College, Research Series no 6, 1998.

Fuglestad, A. B. (2007). Teaching and teachers' competence with ICT in mathematics in a community of inquiry. In J.-H.Woo, H.-C. Lew, P. Kyi-Sik, & S. Don-Yeop (Eds.), *Proceedings of the 31st Conference of the International Group for the Psychology of Mathematics Education* (pp. 2-249–2-258). Seoul: The Korea Society of Educational Studies in Mathematics.

Fuglestad, A. B. (2008). Developing tasks and teaching with ICT in mathematics in an inquiry community. In D.Pitta-Pantazi & G. Philippou (Eds.), *Proceedings of the Fifth Congress of the European Society for Research in Mathematics Education (Cerme5)* (pp. 1409–1418). Larnaca, Cyprus: Department of Education, University of Cyprus.

Fuglestad, A. B. & Goodchild, S. (2008). Affordances of inquiry: The case of one teacher. In O.Figueras, J. L. Cortina, S. Alatorre, T. Rojano, & A. Sepúlveda (Eds.), *Proceedings of the joint meeting of PME 32 and PME-NA XXX* (pp. 3-49–3-56). Morella, Mexico: Cinvestav-UMSNH.

Goodchild, S. (2008). A quest for good research. The mathematics teacher educator as practicioner researcher in a community of inquiry. In B.Jaworski & T. Wood (Eds.), *International Handbook on Mathematics Teacher Education: Vol. 4. The Mathematics Teacher Educator as a Developing Professional: Individuals, teams, communities and networks* (pp. 201–220). Rotterdam, Holland: Sense Publisher.

Goos, M., Galbraith, P., Renshaw, P., & Geiger, V. (2003). Perspectives on technology mediated learning in secondary school mathematics classrooms. *Journal of Mathematical Behaviour, 22,* 73–89.

Gravemeijer, K. (1994). Educational development and developmental research un mathematics education. *Journal for Research in Mathematics Education, 25,* 443–471.

Greeno, J. G. (1994). Gibsons affordances. *Psychological Review, 101,* 336–342.

Guin, D., Ruthven, K., & Trouche, L. (2005). *The didactical challenge of symbolic calculators.* Dusseldorf, Germany: Springer.

Hennesey, S., Ruthven, K., & Brindley, S. (2005). Teacher perspectives on integrating ICT into subject teaching: Commitment, constraints, caution, and change. *Journal of Curriculum Studies, 37,* 155–192.

Holmboe, C. (1997). *"Je bruker itte IT" En rapport om informasjonsteknologi i norsk skole basert på resultater fra TIMSS undersøkelsen.* Oslo: Universitetet i Oslo.

Jaworski, B. (2004). Grappling with complexity: Co-learning in inquiry communities in mathematics teaching development. In M.Johnsen Høines & A. B. Fuglestad (Eds.), *Proceedings of the 28th Conference of the International Group for the Psychology of Mathematics Education* (pp. 17–36). Bergen, Norge: Bergen University College.

Jaworski, B. (2007). Theoretical perspectives as a basis for research in LCM and ICTML. In B.Jaworski, A. B. Fuglestad, R. Bjuland, S. Goodchild, & B. Grevholm (Eds.), *Læringsfellesskap i matematikk Learning Communities in Mathematics* (pp. 121–137). Bergen: Caspar Forlag.

KD. (2006). Mathematics subject curriculum. In KD (Ed.), *Knowledge Promotion.* Oslo: Kunnskapsdepartementet. Retrieved on February 15, 2008 from http://www.udir.no

Kelly, A. E. (2003). Research as design. *Educational Researcher, 1,* 3–4.

Kennewell, S. (2001). Using affordances and constraints to evaluate the use of information and communication technology in teaching and learning. *Journal of Information Technology for Teacher Education, 10,* 101–116.

Krumsvik, R. (2007). *Skulen og den digitale læringsrevolusjonen.* Oslo: Universitetsforlaget.

KUF. (1984). *St.mld. nr 39 Datateknologi i skolen.* Oslo: Kirke og undervisningsdepartementet.

KUF. (1990). *Curriculum guidelines for compulsory education in Norway (M87).* Aschehoug: The Ministry of Education and Research.

KUF. (1999). *The curriculum for the 10-year compulsory school in Norway (L97).* Oslo: The Royal Ministry of Education, Research and Church Affairs.

Lagrange, J.-B., Artigue, M., Laborde, C., & Trouche, L. (2003). Technology and mathematics education: A multidimensional study of the evolution of research and innovation. In A.J.Bishop, M. A. Clements, C. Keitel, J. Kilpatrick, & F. K. S. Leung (Eds.), *Second International Handbook of Mathematics Education Part One* (pp. 237–269). Dordrecht, The Nethelands: Kluwer Academic Publishers.

Monaghan, J. (2004). Teachers' activities in technology-based mathematics lessons. *International Journal of Computers for Mathematical Learning, 9,* 327–357.

Noss, R. & Hoyles, C. (1996). *Windows on Mathematical Meaning. Learning Cultures and Computers.* Dordrecht: Kluwer Academic Publishers.

Pea, R. D. (1987). Cognitive technologies for mathematics education. In A.Schoenfeld (Ed.), *Cognitive Science and Mathematics Education* (pp. 89–122). Hillsdale, NJ: Lawrence Erlbaum Associates.

Ruthven, K. (2002). Instrumenting mathematical activity: Reflections on key studies of the educational use of computer algebra systems. *International Journal of Computers for Mathematical Learning, 7,* 275–291.

Säljö, R. (2001). *Læring i praksis—Et sosiokulturelt perspektiv.* Oslo: Cappelen akademiske forlag.

Trouche, L. (2004). Managing the complexity of human/machine interactions in computerized learning environments: Guiding students' command process through instrumental orchestrations. *International Journal of Computers for Mathematical Learning, 9,* 281–307.

Trouche, L. (2005a). An instrumental approach to mathematics learning in symbolic calculator environments. In D.Guin, K. Ruthven, & L. Trouche (Eds.), *The didactical challenge of symbolic calculators* (pp. 137–160). Dusseldorf, Germany: Springer.

Trouche, L. (2005b). Instrumental genesis, individual and social aspects. In D.Guin, K. Ruthven, & L. Trouche (Eds.), *The didactical challenge of symbolic calculators* (pp. 197–230). Dusseldorf, Germany: Springer.

UFD. (2004). *Program for digital kompetanse 2004–2008.* Oslo: Undervisnings- og forskningsdepartementet. Retrieved on May 30, 2008 from http://www.udir.no/upload/Satsningsomraader/Program_for_digital_kompetanse_2004-2008.pdf

Vérillon, P. & Rabardel, P. (1995). Cognition and artefacts: A contribution to the study of thought in relation to instrumented activity. *European Journal of Psychology for Education, 10,* 77–101.

Østerby, S. (2007). *Utstyrs-og driftssituasjonen i grunnopplæringen 2006–2007 En gjentatt kartlegging og en sammenlikning med 2004–2005.* Oslo: Utdanningsdirektoratet.

7

INTERPRETATIVE RESEARCH AS COLLABORATIVE INQUIRY

Marit Johnsen-Høines

BERGEN UNIVERSITY COLLEGE

Abstract: The Learning Conversation in Mathematics Practice (LCMP) research project has engaged student teachers, schoolteachers, and didacticians in a collaborative study of the communicative conditions for learning. The research focus is twofold, investigating the role of communication in children's learning of mathematics and the professional development of mathematics teachers through participation in research. The aim of the project is to develop the concept of learning communication and to explore its use as a didactical tool; the method employed is collaborative inquiry. The LCMP approach has been developed in cooperation with the participating schools, based on studies of practical school activities organized by Bergen University College (BUC). This chapter outlines the development of the collaborative design that has made it possible to navigate among the different contexts in which communication and learning are studied.

Keywords: community of inquiry, collaborative research, dialogue and learning, innovation competence, teacher education in mathematics.

INTRODUCTION

This chapter relates to an ongoing research project that focuses on communication and learning in the field of mathematics, Learning Conversation in Mathematics Practice (LCMP).[1] The project goal is to develop the notion of the *learning conversation* as a didactical concept and tool for describing and facilitating learning processes. In order to gain insight into the communicative conditions for learning, it is first necessary to investigate and document the communication patterns, learning processes, and learning outcomes involved. The project builds upon the framework developed in *Dialogue and Learning* by Alrø & Skovsmose (2002)—more specifically, on their inquiry co-operation model for identifying and describing elements that constitute dialogical learning interactions, and on their discussions of the dialogical conditions for developing mathematical insight in correspondence with *mathemacy*.[2] It also builds upon the *flexible spaces of language*[3] approach developed by Johnsen-Høines (2002), based on a text-theoretical and dialogical perspective inspired by Bakhtin (1981, 1998) and Lotman (1988). According to this approach, insights[4] develop when one insight confronts another. New understanding is constituted based on different understandings (such as different ways of understanding a phenomenon, a structure, relations, etc.). In the implementation of the project, the perspective of *subject-oriented reflective discussion* is employed; this is a perspective that has been developed in the course of a study of communication in practice teaching[5] (Johnsen-Høines & Lode, 2006; Johnsen-Høines, 2009a). It is the aim of LCMP to develop *learning conversation*[6] as a concept by employing collaborative inquiry processes involving schoolteachers, student teachers, and didacticians. In addition, this concept will be developed as a didactical tool for facilitating *collaborative inquiry*, as described in this chapter. The LCMP project analyzes

The First Sourcebook on Nordic Research in Mathematics Education, pages 109–123

the communicative characteristics prevailing in the field of mathematics teaching and learning, and in the professional development of mathematics teachers. The ultimate intention is to strengthen the teaching and learning of mathematics in schools—that is, to strengthen the pupils' ability to participate actively in the learning process, using and communicating mathematical principles.

The LCMP project is a four-year project that, at the time of writing, was in its first year. Thus, this chapter does not present research results or conclusions; it will rather elaborate the research design and aims, and the preliminary research activities.

METHODOLOGICAL APPROACHES

Structured as Layers of Development

The LCMP project can be described as an inquiry into successive layers of development. A lower layer is represented by a school development initiative, which is intended to achieve what has been described by teachers and school authorities as "Real-Life Education,"[7] and is based on a teaching approach that was developed by a local teacher.[8] This approach forged a link between groups of lower-secondary pupils and local workplaces. The teacher wanted to stimulate learning in mathematics and science through dialogue with workers in trade and industry. He wanted to stimulate entrepreneurialism as an educational approach, in order to provide pupils with a basis for their education and career decisions. The approach was intended to improve and maintain links between education and industry. This work established the foundation for developmental initiatives that include all schools in Sund, Øygarden, and Fjell, a region of western Norway. This education–industry partnership agreement stated that (mathematics) teaching and learning should be organized sequentially at trade and/or industrial workplaces and in the schools. Experience gained from workplaces was expected to have an impact on the pupils' attitudes towards their learning and on the use they make of mathematics. It is conjectured that workplace experiences will also have an impact on the way mathematics is communicated (and thereby on what is learned). An important feature of the school development initiative, therefore, is the inclusion of discussions about the stated aim—that is, discussions about what it means to strengthen the quality of the pupils' mathematical knowledge in the context of "real-life education."

Researchers and teacher educators from Bergen University College (BUC) (hereafter referred to as didacticians) were invited to participate in these discussions at two schools. This was facilitated by the framework underlying teacher education in mathematics at BUC, which corresponds well with the real-life education initiative, particularly the concepts of learning rationale, culture production, empowerment, and knowledge control (Mellin-Olsen, 1987, 1989). Given the resultant focus on pupil communication and learning, it was evident that the benefits would be mutual if we invited the classroom teachers to participate in our discussions and explorations. The aim is that these discussions should be of equal interest to all participants and should take the form of collaborative inquiry processes. The first research focus in LCMP is the pupils' ability to communicate and to learn mathematics.[9]

A middle layer is represented by the professional development component. One of the 60-ECT[10] courses[11] in mathematics/mathematics education at BUC places teacher-training student (hereafter referred to as students[12]) in these schools for practice teaching. These students take part in the developmental initiative entitled Real-Life Education. As part of their practice teaching, they are challenged to develop new and alternative designs for teaching in collaboration with teachers, didacticians,[13] and employees from the designated workplaces. The planning, implementation, reflection, and evaluation of teaching and learning processes are to take place in an experimental mode, which facilitates the "trying out" of new ideas. The students join in collaborative conversations that inquire into the way in which children learn mathematics and the communicative preconditions for this learning. The data for the teacher-training study are drawn from these conversations. The second research focus, therefore, is on the communication that develops in the context of the students' learning—that is, communication related to their professional development as mathematics teachers.

The fact that practice teaching is linked to a school development initiative makes it possible to develop a new framework for collaboration between didacticians, schoolteachers and students. In the upper layer the schoolteachers, didacticians, and students are challenged to work as a learning community (Jaworski, 2007; Wenger, 1998). The emphasis is on inquiry and dialogical processes (Alrø & Skovsmose, 2001; Johnsen-Høines, 2002, 2009a; Johnsen-Høines & Lode, 2006), in order to develop new teaching and learning situations in the field of mathematics and mathematics education. The third research focus, therefore, is on the communicative learning processes that develop between the members of the learning community (Johnsen-Høines, 2009b).

Interpretative Collaborative Research

The research approach shares many characteristics with developmental research (Jaworski, 2006; Goodchild, 2008) in that processes are initiated, followed up, and investigated not just by the researchers but by all the participants in collaboration. In this approach, communicative and educational developments are analyzed critically and discussed; on the basis of these findings, conclusions are drawn that will provide a basis for planning and implementation of new actions which, in their turn, will be studied.

The school authority and some teachers initiated Real-Life Education as a common approach for all the schools in the municipality. However, the teachers that joined the project claim that the notion Real-Life Education is not clearly articulated. This is so in spite of the fact that the approach builds upon work that has been developed by a teacher and his pupils, and even though the school authority has provided courses, examples and advice, the teachers ask repeatedly: "What does Real-Life Education imply?" This lack of clarity is frustrating for the teachers but it also provokes them to develop and implement their own ideas about what they think should or could be implied. Many questions have to be answered: How can pupils' mathematical competency be colored by practice experiences at the work places? What are the preconditions necessary if the desired learning is to take place? (And what is required?) How can we organize courses to allow for such learning? How can what is happening be documented? These are just some of the questions for which teachers do not have *the* answers. None of us have the answers. Nor do teachers have concrete models or teaching plans. They are still at the stage of discussing ideas and "trying out" didactical approaches. The fact that they do not consider the initiative to be "clearly articulated" and easy to implement shapes the discussions with the students and didacticians who have been invited to join in. The teachers claim that it has been a challenge to handle simultaneously the poorly articulated initiative and the arguments raised by students and didacticians.

Another form of action is illustrated by the provision made for the establishment of a new practice-teaching initiative in which students shall participate in developing their own teaching practice. They are to play an active role in their teaching practice, employing an investigative approach in which teaching methods are co-created, rather than being provided with a guided approach in which they are told "how to do" or "what should be done." It is necessary to question the established student and tutor roles in teaching practice in order to realize such an approach. The tutors are challenged to take on an additional role as collaborator. The students are cast in the role of cooperating teachers, rather than their traditional trainee role. These challenges imply that it is necessary to develop both social and individual ownership of the processes, as described by Mellin-Olsen (1987) and to establish social and personal relationships, as stressed by McNiff and Whitehead (2006). The latter authors emphasize the differences between developmental processes based on what they describe as a *we* perspective, a *we–they* perspective and an *I* perspective. A w*e–they* perspective implies that someone (we) is in charge and responsible, and that others (they) are observed and assessed, and expected to, allowed to, or told to act in specific ways. This would be a natural approach in a research project in which the researchers are studying developmental work without intervening. However, a *we–they* perspective often proves to be counter-productive if the intention is to develop reflective collaboration. The fact that someone is regarding or observing the others as *they*, and that the latter regard themselves as having a *they* role will not stimulate a dialogical approach in which the participants take a mutual responsibility for the processes (Alrø & Skovsmose, 2002). Such relationships will influence the opportunities available for gaining analytical insights from the community. A *we* perspective differs from a *we–they* perspective because it implies common goals and collaboration underpinned by equality.[14] The *we* take part in the inquiry processes together, so the insight gained becomes *our* common insight. However, according to McNiff and Whitehead, the *we* collaboration may lack the necessary commitment, since participants do not readily take charge of, or responsibility for, progress. They may depend upon the fact that the processes are to be developed and followed up by *us*, not necessarily by *me*. In contrast, McNiff and Whitehead describe the *I* perspective as committed, responsible, and social. Such an *I* perspective corresponds with the kind of ownership that is considered ideal for the LCMP project and constitutes a developmental goal for major parts of the project. Thus, the project leaders have viewed the *empowerment* of the participants as crucial and have included this as a key theme for discussion in the groups. Such meta-discussions have been found to be important in the context of empowerment for the emancipation of the participants (for instance the students). Moreover, empowerment must be kept in mind when designing the project; how this becomes realized within the project will be considered later in the chapter.

It has been assumed in the LCMP project that approaches based on these three perspectives influence the communicative patterns within the group. While the resultant relations within a community are not evaluated as good or bad, the focus offered by McNiff and Whitehead is employed to distinguish the inter-participant relations that we want, and/or are able, to develop. This varies in relation to the project phase and context: At some points in the process, we strive to develop an *I* perspective, while at other points we strive for a *we* perspective as a transitional stage in the development of an *I* perspective. In addition, a *we–they* perspective also has to be included in our research processes, especially when, as researchers, we distance ourselves in order to undertake a more abstract and conclusive analyses (described as the level AL3 later in this section).

In the further development of this chapter much use is made of the pronoun "we" as a conscious attempt to signal the inclusive and subjective stance taken within the research. The group to whom *we* refers is not stable and thus it will be necessary to indicate occasionally the participants who are included.

The implementation of yet another action by the didacticians is evident in the introduction of *communication and learning* as a framework for the didactical discussions involving all the participants. This issue is considered to be of research interest because we believe that it will make the learning processes both fruitful and inspiring. We expect that the whole group will enter into the discussions, recognizing that the theme is useful and interesting. It is important that the participants consider the theme to be fruitful not only for this project, but also more generally for learning processes.

The empirical data for research is being (and will be) collected from communication taking place in the learning contexts embedded in the three layers described at the beginning of the chapter. The research approach is interpretive[15] in the sense that communication (communicative interactions recorded on videotapes and in transcripts) is interpreted and analyzed. The students, schoolteachers, and didacticians whose interactions are being recorded are engaging in a collaborative inquiry in which empirical data are generated and investigated. This research approach implies that all participants are expected to engage in the entire process: developing and sharing insights, expounding perspectives, exploring issues, and searching for didactical possibilities, as well as critically evaluating the various possibilities and formulating the analyses. This analytical approach builds upon the dialogical approach developed by Alrø and Skovsmose (2002), in which they stress that dialogue is a conversation of inquiry:

> Entering an inquiry means taking control of the activity in terms of ownership. The inquiry participants own their activity and they are responsible for the way it develops and what they can learn from it. The elements of shared ownership distinguish a dialogue-as-inquiry from many other forms of inquiry where, for instance, an authority sets the agenda for the investigation and the conversation. (p. 119)

This methodological approach has implications for the way analysis is undertaken jointly and has provided the framework for the analytical model developed and described in Johnsen-Høines and Lode (2007). It is upon this model that the following description is based.

Empirical data are collected from teaching and learning sequences in the mathematics classroom and in post-classroom discussions. Conversations in which mathematical and didactical issues are discussed are recorded and transcribed. These conversations are actually analytical processes in the sense that the participants are investigating how they understand and develop specific concepts or issues. The analytical processes are investigative and questioning processes, therefore, and they are developed on three different analytical levels (AL):

- AL1: The data are collected from conversations that are somewhat analytical by nature. This is also true on the lower layer where the pupils' communication and learning is studied. Participants at AL1 will be students, teachers, didacticians, and pupils. The pupils' conversations that are recorded as empirical data focus on their reflections on mathematical concepts and operations, and on how mathematics can be understood, used, and communicated. Implicit in these conversations are traces of analysis that can initiate valuable research inquiries. The fact that the pupils have been informed about the research focus is important not only for ethical reasons, but also to establish a common understanding of these issues. Such an understanding is essential in order to encourage pupil participation. Moreover, pupils meta-comments about their activity provide a starting point in the analytical processes. The data collected in the classroom provide a basis for conversations involving the students, teachers, and didacticians on the middle and upper layers: conversations that investigate the practices one is trying to

establish in order to lay the foundation for the pupils' learning, as well as for professional development in the field of mathematics teaching. These discussions are intended to be collaborative and inquiring, and to take place in an experimental or open atmosphere in which there is a willingness to share and investigate ideas and to try out suggestions for further didactical development. Issues related to the planning, implementation, and evaluation of teaching/learning situations serve as a basis for the conversations. Participants may draw upon previous experiences and issues raised in other phases of the study, as well as theoretical approaches.
- AL2: The project leaders initiate conversations about the conversations themselves in order to study these. This is done by referring to earlier conversations, and the participants, comprising students, teachers, and didacticians, use transcripts, tapes, or video recordings as a basis for the discussions. In dialogue we investigate how the conversations can be described and how they can be developed. We develop criteria for identifying and describing various aspects of or types of didactical conversations. Thus, the processes of investigation and of generating data are interwoven. The empirical material is generated from AL1 and AL2.
- AL3: At this level, analysis of the empirical data from AL2 (and AL1) is undertaken. While the didacticians have the overall responsibility for the analytical work—which implies responsibility for presentation of analyses, summing up of conclusions, and discussion of implementations—other participants may also be involved. This analytical work (AL3) builds on discussions with fellow researchers—for instance, in the Teaching Better Mathematics forum.[16]

The three analytical levels (AL1, AL2 and AL3) are not separate from each other, and more than one level may be represented as part of the same conversation. The conversations can be described as containing three descriptive and interpretative levels:

- events are described (these might be events that have been observed or activities that one plans to implement)
- these descriptions (and events) are the focus of reflection and discussion
- an effort is made to synthesize key ideas derived from the reflection in order to facilitate the operationalization of the insight gained at a later time.

The above model implies that the participants through inquiry and collaboration are engaged in analytical processes of description, reflection, and synthesis, and their analytical focus is on conversations. Their contributions thus constitute both the means and the content of the analysis.

The description of this analytical model can be read as a description of a research method. In addition, I have summarized key features of the whole project in the description of these analytical levels, and elaborate on the aims of the project as well. It is important to gain greater insight into these analytical, collaborative, and investigative conversations—into the content and into the way the interactions are organized. This project will enable us to better understand the way in which the participants go about analyzing the data and the conclusions they come to. Moreover, I see that this approach gives us the opportunity to learn about and stimulate the analytical competencies and potential of pupils and students in particular. As the LCMP project contains several research projects, the different projects will articulate their individual and different units of analyses; these will, however, be based on dialogue taking place within the 'learning loops' outlined below—these combine the various socio-cultural practices in which the interlocutors participate.

THE ORGANIZATIONAL STRUCTURE

Foundation for Structuring

The model and aims referred to above have to rest upon methodological and structural foundations. As stated earlier, the project leaders have been faced with the challenge of developing a design in which the conversations described can be materialized and stimulated. Burton (2001) stresses the importance of consistency when organising the study; the importance of keeping the basis, the aims and the contexts of the study in mind from the very beginning. This is to be achieved through the investigative processes. The integrity of each study rests on its foundations, theoretically and contextually. According to Burton, researchers face a challenge: The methodological choices that we make must be founded in the integrity of the study, and this integrity is further strengthened by our choices.

Thus, it is our aim in the LCMP project to develop a learning community (Wenger, 1998; Wells, 1999) as a basis for achieving Real-Life Education. The learning community, which is based on principles of inquiry and collaboration referred to in the introduction, is intended to strengthen and guide the professional development of all the project participants (students, schoolteachers, and didacticians). The aim is to develop a community of inquiry (Jaworski, 2007) as an emergent rather than established form of practice. Jaworski emphaszes the notion of *critical alignment*, which distinguish a community of inquiry from a community of practice, and states that "in order to move... from a community of practice to a community of inquiry, participants will engage in existing practices, aligning to some extent with those practices, but in a questioning or inquiry mode" (p. 129).

On the one hand, the students can benefit from the schoolteachers' experience and knowledge. On the other hand, these teachers openly admit that they too can benefit from the discussions and the collaboration, which supports them as they try out or implement a new approach that they have neither experienced nor had explained. They stress the importance of the group's support when discussing, planning, trying out and evaluating new teaching approaches. Their sceptical and hesitant attitude has proved to be a fruitful point of departure for establishing a learning community based on investigative dialogue (Alrø & Skovsmose, 2002). The participants differ as to whether, and to what degree, they are interested in gaining new insight through the conversations; they also position themselves according to whether, and to what degree, they consider that these processes will provide them with a better basis for teaching and learning. These attitudes influence the roles that the participants play in the collaborative process. In order to develop a teaching approach which none of the participants know "how to do," it has been necessary to stimulate mutual relationships between all the participants for the sharing of knowledge. The development of these relationships appears even more complex because in this project we are challenging the established tutor–student–didactician roles. The following learning loops provide a framework for the investigative processes we sought to develop.

The Pupils' Learning Loop

The education–industry partnership agreement is established on the belief that the pupils will learn mathematics in workplaces as well as in classrooms. As the Figure 7.1 illustrates, the pupils' learning is organized within these contexts and they move physically between these contexts. However, this does not mean that the learning periods in these locations are sequentially organized.

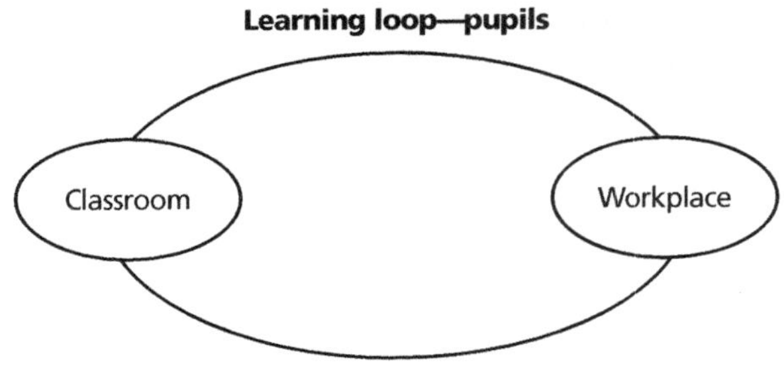

Figure 7.1

The LCMP project builds upon a socio-cultural perspective that considers learning a matter of personal cultural development—a matter of social and individual processes. Learning is understood in the context of the social and individual activities in which persons participate (Vygotsky, 1978). The project's focus on participation, particularly on the participants' ownership of and responsibility for the activities, further highlights the relevance of Vygotsky's perspective. Considerable emphasis has been placed on the social practice approach developed by Lave and Wenger (1991); in their view, learning is participation in the practice and is not limited to the individual consciousness of the participants. Lave (1999) and Dreier (1999) argue that learning is development through participating *in* and *between* different practices. The sketch in Figure 7.1 is also intended to illustrate that learning is taking place *in* as well as *between* the different contexts.[17] Different learning activities are organized for and related to the two contexts (classroom and workplace). Another sketch could be drawn to illustrate several learning activities. The theoretical approach referred to here gives rise to several questions: How can pupils' learning be organized and studied as the pupils participate *in* different learning activities and move *between* learning activities? Will the pupils see the different activities as support for and background for each other, or will they experience them as simply unrelated or vaguely related "happenings"? Will the different activities correspond with different practices as developed as basis for community of practice by Lave and Wenger? Will the pupils' collaboration and learning in different contexts (i.e., classroom and workplace) appear as participation in different learning communities (Wenger, 1998)? These are recurrent and ongoing questions that are part of the developmental processes.

The following example illustrates the kind of learning sequences to which this research project is connected, as well as the approach on which our inquiries are based. The pupils are 15 years old. They have visited an industrial workplace that provides the oil industry with a variety of off-shore supplies, ranging from huge mechanical components for installations, to more trivial support for daily household-type activities. The pupils have learned about containers of different sorts for storing supplies, and about vehicles for transporting and transferring huge quantities of heavy goods. They have studied some vehicles, containers, and buildings more closely than others. Pictures have been taken. In school they have made drawings and small preliminary models, and finally they have made models on a given scale using different materials such as metal, string, wood, polystyrene, papier mâché, cardboard, and paint. Some of the work was carried out in the school workshop, and they used different sorts of technical tools to shape their models. The teachers and the students have cooperated in drafting the "teaching plan," which may be considered the pupils' learning loop. Measurement and scale were chosen as the main mathematical theme. However, the students have discussed the fact that several other mathematical themes were also important, particularly a number of competencies in the field of geometry. This became evident, for instance, when the pupils decided to make a vehicle with a metal surface, wooden wheels, metal axles, and a moveable wooden hoisting crane. It was challenging for the pupils to cut the metal surface as a single piece that could be bent sufficiently to cover the vehicle, and to attach the axles in a way that enabled all the four wheels to touch the ground. In addition, they had to create a design to which a moveable hoisting crane could be attached.[18]

As a group comprising students, teachers, and didacticians, we discussed the processes that had been observed. Among other issues, we discussed the extent to which we could observe what we called the "mathematical conversations" included in the processes.[19] We discussed the preconditions for promoting such conversations. We observed that the episodes that demonstrated the strongest engagement were those in which the pupils were concentrated and eager to succeed in making their model function as they had planned, and concluded that these episodes included conversations in which the pupils seemed to feel a need to express *precisely* what they wanted to achieve or what hypothesis they had regarding how their design could be made workable. They used few words, short sentences. In these phases, words were seldom used that did not serve the purpose of getting the process to work. We interpreted this as an indication that the pupils were concerned about expressing exactly what they wanted, or questioning details of importance for the process. Their explanations most often worked; that is, the pupils seemed to understand each other; they gave instructions, asked relevant questions, and received answers. We interpreted these conversations as tools for supporting and structuring the collaborative processes. It was important for the pupils to help each other to produce precisely what was needed or do what had been planned: to hold, to bend, to make one more component exactly the same size, and so on. These conversations were based on a *focus on functionality*.

Another kind of conversation that was used as tool for goal-directed investigations was the questioning approach. This was evident in utterances such as: What if we use a nut here, would that help to make it flexible? Which size do we need if...? A distinctive mode of inquiry was evident in these sequences. The pupils appeared to be searching for words, and the utterances could be hesitant or incomplete, as if they were talking while thinking. Conflicting ideas could be introduced and reflected upon. More questions were often asked because they did not understand each other at once. Pauses were usual. Such conversations appeared to be goal-directed. The pupils did not start to talk about other things. If they interrupted each other, this was often because they had other ideas or saw alternative solutions that might work. They repeated utterances. They referred to earlier assumptions. These conversational interactions are in accordance with the Inquiry Co-operative Model developed by Alrø and Skovsmose (2002). Such interactions were based on a *focus on inquiry*, being referential and forthcoming.

A third kind of conversation was identified, which developed when the pupils had found some solutions and were in the process of doing the more trivial tasks that had to be accomplished—for instance, when painting the model or making patterns. In such exchanges they could engage in a kind of "social talk." They might talk about other pupils, about social events, computer games, and the like. This social talk might also include the students, teachers, and didacticians as well. During such exchanges, we could even ask about what they had done and what plans they had for their model. The key features of social talk included a less structured communicative pattern, lack of follow-up of utterances, interruptions, and follow-up of earlier utterances. These conversations can be interpreted as richer and more open to alternative ideas than those previously discussed, in that different kinds of initiative could be taken; however, they could also be described as less well informed, in that many initiatives were interrupted, not heard, or not followed up. We observed that these conversations were less goal-oriented, less structured, and more incidental. Nevertheless, it was also evident that they were socially important. References for ideas and investigations were established through these social conversations.

The above example serves to illustrate the field of enquiry that the LCMP research will explore in greater depth and detail. The preliminary analyses above are intended to illustrate how different sorts of conversations can be identified, and also how the interplay between these conversations, both sequentially and simultaneously, can be seen as constitutive of overarching conversations (Johnsen-Høines, 2002).

The group of students, teachers, and didacticians discussed whether we could identify the mathematical content to be implicated in such conversations and how we could do this, as well as what opportunities we had for developing a greater mathematical focus. However, we agreed that it would not be acceptable to intervene in ongoing conversations in order to guide them towards a more explicit mathematical discussion, or to a greater degree to implicate a mathematical register (Pimm, 1987). In our discussions we emphasized the importance of identifying examples of conversations in which we saw that the mathematics was implied without being explicitly expressed. In addition, we examined the language through which this mathematics was expressed. We agreed that our role was to support the ongoing processes, and thereby the conversations, to move in the direction pupils were heading. However, a number of concerns were expressed: *We* saw a lot of mathematics in use, but did the pupils see these mathematical traces and learn from them? Could we strengthen the ongoing and future processes in order that the pupils' learning could be supported by this work? Such questions recurred at intervals, and the students participated actively in these discussions. Together we study how such discussions influence the students' learning.

To return to the example, when the project finished, the pupils' products were demonstrated and discussed with the whole class. Also present were the students, didacticians, and teachers, as well as some teachers who had not been involved in the project. In the following excerpt, one of the pupils was holding the vehicle that was described earlier in her hand and described the working process:

A student: Do you see that you have worked on mathematics during this process?

The pupil (Elise): Yes... I have... lots of mathematics.... [She looks at the vehicle in a thoughtful way as if she is searching for ways of expressing more exactly what she means.]

A didactician: What do you think about this mathematics compared with the mathematics you work on using pencil, paper, ruler, and textbook?

The pupil (Elise): The mathematics we are used to is easier. Geometry is easier using pencil, ruler, and compass, and solving tasks. The mathematics we worked on here is really difficult. It was difficult to get it to work... to manage to make all the components fit. It is... complicated.... [She tried to envisage the complexity by showing the bending of a surface with her hands.] This was fun.

Earlier in the process the group of students, teachers, and didacticians had discussed whether, by making mathematics concrete and relevant for the pupils, we also made mathematics easier for them to understand. We discussed *how* Real-Life Education could serve as a positive approach for learning, and to what degree making mathematics easier could be seen as *the* option. Elise's observations failed to confirm these hypotheses. We interpret her statements as implying that she was excited by the complexity, and that their usual way of working with mathematics actually seemed like a simplification that she might find trivial and less interesting. Her perspective added another dimension to our further discussions.

This example serves to illustrate the ongoing nature of the analytical processes that constitute the LCMP project. Interactions that have been experienced, discussions that have been held, and findings that have been confirmed provide the basis for further collaborative investigations. Such processes generate the empirical data and provide the basis for this analytical approach.

The Students' Learning Loop

The above-mentioned theoretical and methodological perspectives underpin our work and have implications for the way we[20] organize pupils' learning. However, these perspectives are equally important for establishing a learning community as a basis for the professional development of teacher-training students, as well as that of teachers and the didacticians. As mentioned earlier when describing the middle layer, the kind of participation that we are trying to achieve in teaching practice may appear contradictory to the tradition with which the participants are familiar. Building upon Johnsen-Høines and Lode (2007), as a community of learning, we emphasize the importance of empowering the participants, especially the students and the teachers. The issue of empowerment became a methodological question: How can we organize for the project so that the participants develop a sense of ownership and a perception of themselves as valued and responsible participants in the discussions and the decision-making processes? How can we organize the project so that the participants can acknowledge their own and each other's analytical competencies and can perceive the potential relevance of these? How can we move in a direction of an *I* perspective, as elaborated by McNiff and Whitehead (2006)? How can we design a research model that will prove a basis on which research questions can be identified, data collected, and analytical discussions held?

Level 1 Learning Loop

Some activities are located at BUC where the 12–16 student participants are studying mathematics/mathematics education full time together with about 10 other students, and where the didacticians are teachers and researchers within the same field. The students are in their third or fourth year of teacher training and will have 90 ECT in mathematics/mathematics education by the end of the year.[21]

Other activities are located in public schools. Still others are located in workplaces of participating companies. The involvement of a variety of locations draws more people into participation (see Figure 7.2). This is particularly true for the activities based in the workplaces, where employees take part in the communication of knowledge and influence the learning processes. The three main contexts or locations included here are physically distinct because the buildings and spaces are different, having been constructed to serve different functions. In these spaces people move and (inter)act differently, and have developed different cultural practices. It is our hypothesis that the variety of participants engaging in the activities situated in these different contexts and the movement between these will influence the pupils' learning and the learning community. One of the aims of the LCMP study, therefore, is to shed light on these issues.

Level 2 Learning Loop

Figure 7.3 illustrates the ongoing activities. This diagram of the students' learning loop enables us to

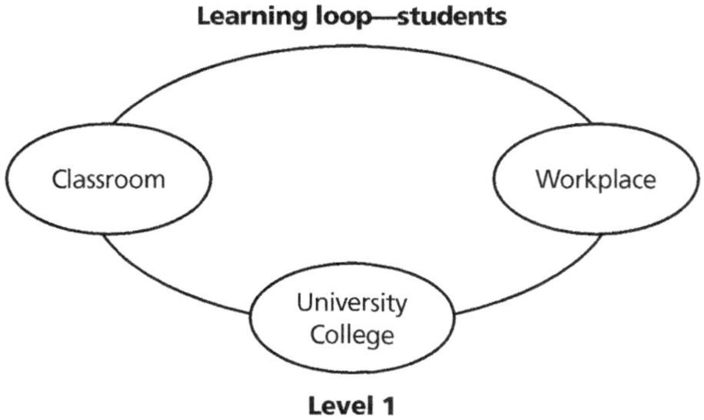

Figure 7.2

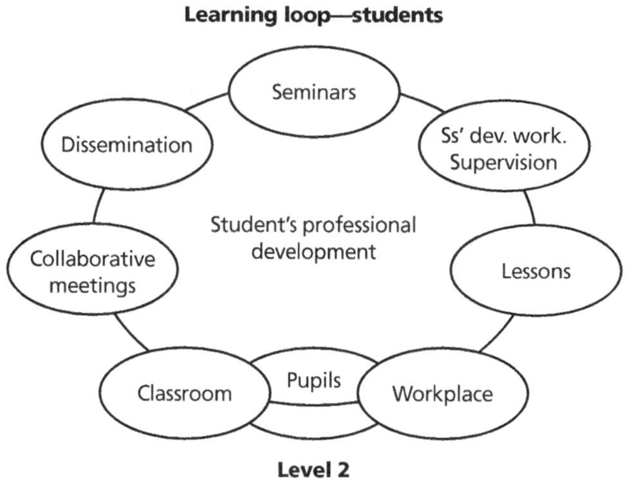

Figure 7.3

describe the different learning activities in which the students participate, as well as the learning community. The loop indicates where the students' learning is situated, both *in* the activities and *between* the activities. The activities are not sequentially organized, and they may develop at different rates and to different extents. The idea is that each activity is supposed to influence the others. Level 2 complements level 1. The two sketches can be seen as describing two separate loops that are organized on two levels and linked to one another. Thus, the level 2 activities are linked to the three contexts at level 1. However, the loops are not synchronized; the same activity may be located at BUC on one occasion and at one of the other locations on another occasion.

The various activities represented in this diagram warrant some explanation:

Pupil's learning. The students are actively planning as well as implementing and evaluating the teaching/learning activities—in cooperation with the teachers and the didacticians. All these activities are related to the pupils' learning of mathematics. When the group of students, trainers, and didacticians focus on the students' learning, the focus is on their learning in the context of teaching/learning mathematics in schools, using the current teaching/learning activities as a focus for collaborative investigations. The example referred to earlier illustrates how activities involving pupils are in focus and provide a basis for the students' (and the other participants') discussions in the context of their own learning.

Collaborative meetings. These meetings are intended to establish a close relationship between the students, teachers, and didacticians in order to support subject-based discussions of mathematics didactics in relation to the current teaching/learning activities. Building upon the framework developed by Johnsen-Høines and Lode (2007) and Johnsen-Høines (2009a), we seek to develop a subject-based and ongoing discourse in which evaluative perspectives are minimized. It is our intention to develop a form of discourse that is collaborative, investigative, and dialogical. Meta-level discussions about the conversations in which we engage are considered to support the learning processes; our shared goal is to engage in this conversation (between the teacher students, their trainers, and the didacticians), and the conversation itself is a topic of conversation. The collaborative meetings are intended to stimulate discussions about planning, and the following questions might be asked:

- What do we want to try out?
- Why do we want to organize the activities in just these sequences?
- What are we going to focus on?
- What do we want to observe in order to learn more about the issues on which we are focusing?

The collaborative meetings may be held just before the teacher student's teaching is to take place and after a teaching period is over. The conversation extract from the example indicates the sort of topics that may be considered relevant to raise in the collaborative meetings. The question of "what *is* Real-Life Education" (what do *we*, as a learning community, agree that this implies?) is one recurrent issue. Elise's observations challenged some of the earlier assumptions made by the group. As a result, discussions were initiated to consider the statement that making mathematics concrete and making claims about functionality makes mathematics more complex, more difficult, and more interesting. Another issue concerned the extent to which we could distinguish mathematical conversations from other conversations. This gave rise to several questions: What do we mean by "mathematical conversation"? What are the preconditions under which we as adults are allowed to interfere? Can we influence the discourse, and if so, how? Elise's comments stimulated still further discussion by documenting that she had actually worked on more mathematics than we had observed. Once again, her input challenged our earlier assumptions, and we interpreted this as an indication that our decision to further study this kind of learning episode in detail and in depth was justified. In addition, we discuss our discussions; for instance, how the interplay between teaching experiences influences our discussions, how we make use of earlier discussions, the degree to which we are *investigating* the content that we commonly include and develop, and how we can identify and characterize investigative conversations. Since these conversations are data, all the collaborative meetings are recorded. All the participants (students, teachers and didacticians) have access to these data and can use it in their professional development work and in their work as co-researchers.

The collaborative meetings were intended to replace the regular teacher-student supervisional meetings. However, this has proved to be difficult because the meetings could not be arranged often enough. Thus, the supervisional meetings and the collaborative meetings work in a complementary fashion.

Supervision related to the students' professional development work. The students are required to write a report (30 ECT) as part of their course work;[22] this is supervised by a didactician and may be submitted as individual or collaborative texts. The thesis must be a theory-based and analytical text, and is generally linked to the teaching practice and the LCMP discussions. The aim is for the students to develop an independent and professional attitude. The data for these theses is most often obtained during practice teaching, and the students discuss their developmental work with the schoolteachers during planning and implementation phases. However, the teachers do not have any responsibility as supervisors in this context.

Seminars. The seminars are a compulsory component of the course. Students (as well as didacticians, teachers, and other invited persons) are responsible for introducing and leading the seminars. Students may present work they have done in order to obtain feedback, to stimulate discussion, or to ask for advice regarding future directions. In addition, the seminars may include the presentation of an article, for instance, in order to explore mathematical and/or didactical (mathematics education) issues that have been raised. Students who are not in the LCMP project may also participate so that the whole group may learn from the LCMP discussions. This means that the LCMP discussions will be stimulated by the questions and ideas of these other students, who are also required to complete their teaching practice and to write a report based on what they learn from this experience. It is intended that the seminars provide an active, investigative forum in which the members share the responsibility making it function. In addition, the aim is to develop an *I* perspective, as previously described. Experience has shown that issues raised at the collaborative meetings are often discussed further in seminars, and that discussion points raised in seminars are introduced in and influence the discussion at the collaborative meetings. As with the meetings, the seminars are recorded as empirical data.

Lessons. These are the lessons relevant to teaching practice in the 60-ECT mathematics/mathematics education course. For instance, linear algebra is taught in the context of the computer programs that are used by the company for constructing mechanical models. Similarly, didactical issues such as "geometry and design" or discussions about "the functionality of knowledge" can be related to concrete discussions or situations related to the company. The intention is that the subject-based[23] course content should bear the imprint of the activities in the learning loop. The focus here is on the academic content of students' learning. The project teachers are invited to join these lessons in order to establish a common base and interests.

Dissemination. Experience has shown that the dissemination processes or "knowledge sharing" processes are important. The students give lectures or seminar presentations internally for the group and externally for other students and teacher educators at the university college. These activities have been effective both in initiating and inspiring discussions and in empowering the students. Their competence is demonstrated when they articulate their ideas. Our study of these processes has indicated that, after such presentations, the students are taken more seriously and listened to with greater respect by the other participants. Furthermore, the dissemination processes seems to have a positive influence on the development of the presenters' autonomy and authority. Ongoing research into this phenomenon is currently at the data analysis phase, and the intention is to document the processes in which dissemination has played an important role. While dissemination as part of the students' learning loop will not be further elaborated in this chapter, the pupils' dissemination processes will be discussed.

The students are challenged to investigate mathematical issues—to read about, practice, and discuss investigative dialogical approaches—which imply ownership, empowerment, and critique.[24] The activities in the students' learning loop, their practical implementation, and their content are intended to strengthen and be strengthened by these theoretical perspectives.

One important reason for illustrating the learning community through the participants' learning loops is to reveal the meaning of one activity in the light of all the others, or to demonstrate how the learning in one activity develops as a continuation of the others. We intend to investigate both ongoing and referential learning processes, and the sketches of the learning loops are intended to illustrate these relations.

COLLABORATIVE INQUIRY—TWO ISSUES FOR INVESTIGATION

One aim of the LCMP project is to provide insight into communicative preconditions for learning. It is for this reason that the notion of "the learning conversation" is to be developed as a concept and as a didactical tool. However, this is not to imply that we intend to develop models or clear-cut recommendations for practice. Rather, we hope that new insight into communicative phenomena or issues will help us to explore and develop reflective communication in the learning context. We hope to provide an analytical synthesis of our findings that will, when published, stimulate wider discussion of these issues. In the following sections, two of the issues that will be investigated are briefly presented.

Dissemination as Part of the Pupils' Learning Loop

Various sorts of dissemination are often included as part of teaching and learning sequences in schools. This is also the case in the classes participating in the LCMP project. When the pupils finish a project, the results are presented orally, illustrated, for example, by posters or power-point slides. The pupils in the LCMP project are asked to make presentations for several reasons. First, the presentations are considered to document the insight and competence that the pupils have employed and developed in completing the task. Second, a presentation is considered informative for the audience, which may include fellow pupils, teachers, and parents, as well as people from the workplace. Thirdly, the presentations may facilitate future discussions and learning processes, both for the presenters and for the audience. Finally, dissemination can be viewed as an important element in the collaborative learning processes, in that the preparatory activities have the presentation as their end product, and that the follow-up activities have it as their starting point.

The example presented earlier illustrates the way in which the work done has been partly directed towards the presentations. The pupils knew that they were to show their models and tell about what they had done, and that the students would encourage them to engage in conversations with the audience. The pupils were rather shy and gave the impression that they were not used to this kind of conversation; they did not speak loudly, formulate their answers clearly, or reveal much self-confidence. Nevertheless, their message was communicated very strongly. This was perhaps due to the quiet and caring way in which the students posed their questions. They did not comment on the form of the presentation, and did not ask the pupils to express themselves more clearly or behave in specific ways. The students expressed their interest in the content, and I observed that the audience found it easy to follow their style of communication. This provided us with another example that we could discuss at the collaborative meetings:

- How can we identify and characterize the kind of conversation that was taking place here?
- What are the preconditions for the development of such conversations?

Such episodes are not considered ideal models of any sort, but rather examples from which the participants can learn about communicative preconditions. Another group within LCMP is also focusing on the impact of dissemination activities on communication in the context of the pupils' learning. The students stressed that there should be a *need* for the pupils' work, not only from the perspective of the pupils' learning, but also from that of the industry. They put following question to the workplace: "There must be something that the pupils can investigate that you need information about?" One manager challenged the pupils to investigate internal health and safety measures, to study stock lists, and even to research overseas competitors to obtain information that the workplace needed. The results were presented critically, and representatives from the industry were present in the audience. The communicative processes were studied in the context of the learning that developed. The focus on these dissemination activities reflects the pupils' communication skills and learning; additionally, it is connected to the challenge given by the manager and the fact that they were expected to articulate their result (lower layer).

Dialogical and Conflicting Processes

The ability to bring about change is emphasized in the Norwegian teacher training curriculum. The practice teaching initiative (middle layer) must be understood in the context of this goal. Teacher trainers are educating teachers for the school of the future—a school none of us know. Thus, the students of today will be in the forefront of this development, shaping the new institutions of education. It is in the context of such innovation that LCMP's focus on empowerment and critical reflection is important, providing essential competencies for future school leaders. Also important are the investigative and "trying out" approaches described above, and the collaborative approach that underpins the entire project. In order to develop these competencies in the students, the redefinition of the student, teacher, and didactician roles becomes an issue. The teachers and students (and also the didacticians) understand their roles as part of a practice-teaching tradition in which the students are used to being supervised by the teacher-tutors. The teachers are expected to serve as experienced and "good" role models, demonstrating suitable teaching methods. The didacticians are to serve as advisors in matters related to mathematics. Although these descriptions are oversimplified and superficial, they reveal a basic contradiction to the collaborative approach. The LCMP project has challenged the established role patterns by proposing a collaborative investigative community, characterized by equality, respect, and curiosity. However, experience has shown that interaction patterns are not easily changed. This is a recurrent issue and has been openly and thoroughly discussed. The project leaders try to clarify what the new role patterns imply. As researchers, we identify the tension between various ways of understanding personal positioning parting the context of the investigative collaboration. These discussions are also recorded and provide an additional source of data.

For instance, it is possible to distinguish between the traditional teacher's voice as traditionally characterized and the collaborative and inquiring teacher's voice as they are employed by the same person.[25] A professional teacher who was actively discussing alternative teaching and learning possibilities in a collaborative mode stopped, wondered, and questioned whether he was doing his job as he *should* be: "Am I sure that these students are able to take leadership in ordinary classrooms?" He observed himself from a position that he earlier had been used to and moved away from. He described the insecurity and conflict that he felt by referring to the standards that he "had left."

In another case, an inquiring student who was investigating several options turned and asked the didacticians and teachers to decide between the alternatives, without joining in the discussion as critical and responsible collaborator. The student later commented: "I did not have the courage; I turned . . . and I do not know if the teacher really wanted me to take such chances. It is easier just to take a traditional stand."

These two examples illustrate the kind of shifts and contradictions that must be investigated in order to gain insight into the process of professional development. The shifts between different interaction patterns that are evident in these utterances can be identified in shorter sequences, as well as traced as they development over time. By analyzing such conversations, it becomes possible to identify and investigate various patterns collectively, employing the LCMP analytical model discussed earlier in the chapter.

Thus, insight is gained into professional development as a change process by identifying the conflicting "traditional" and "collaborative" patterns. The impact of meta-conversations and the reflections developed through these is discussed, and these conversations are, in turn, investigated as means of developing professional consciousness (Johnsen-Høines & Lode, 2005).

The LCMP approach is based on reflective and ongoing subject-related conversations, and critical and inquiring collaboration provides the foundation for all the activities. This collaboration is intended to encourage a willingness to share and investigate ideas and to try out suggestions. Taking a Bakhtinian perspective, understanding is considered to be constituted of different understandings that confront each other. Learning is seen as dialogical and is underpinned by the struggle between understandings, both interpersonal and intra-personal. These conflicts, and thereby change processes, characterize learning processes (Johnsen-Høines, 2002, 2004). Thus, the investigation of communicative processes in the context of learning processes includes identifying such conflicts and change processes. Similarly, the investigation of communication in the context of learning implies the identification of various ways of understanding and the analysis of the conflict issues distinguishing them. Thus, the conflicting content and the conflicting ways in which the conversations deal with this content becomes an issue for investigation in LCMP.

CLOSING COMMENTS

I set out to provide an overview of the Learning Conversation in Mathematics Practice (LCMP) project as interpretative and collaborative research. This has been done by elaborating on the underlying theoretical positions and on how the methodological design has grown out of a desire to gain deeper insight into the communicative preconditions for learning in the field of mathematics. The LCMP project is in an early phase, and this chapter provides an overview of the field in which our research will explore in greater depth and detail. It elaborates on those methodological considerations that are made in context of what we hope to learn from the project. By studying pupils' conversations related to learning mathematics, with other children and with adults, in different learning activities situated in classrooms and in workplaces, the hope is to identify how communication differs and discuss their characteristic in the context of the different learning activities and the quality of learning. The project design is developed in order to develop an inquiry community in which teacher students, their trainers, and didacticians collaborate in order to get insight into the communicative learning processes of the pupils, and discuss how to support communicative learning processes to develop. It is hoped that learning communication will develop as a concept and as a didactical approach, by emphasizing processes of "trying out" and analyzing what is going on in practice and what one wants to achieve. Thus the focus is on the pupils' learning of mathematics throughout the whole project. However, since this is part of the teacher students' practice teaching period, the emphasis is on their learning as well; the focus will be on how their communication and practice influence their becoming mathematics teachers. The teacher students' development as critical, independent, responsible, and competent professionals will be given additional attention, and it is hoped that knowledge will be gained about communicative conditions for their learning in the context of such professional development. Finally it is anticipated that knowledge will be gained about the complexity of building inquiry community including teacher students, their trainers, and teacher educators in mathematics, as well as researchers, also cooperating with people from working places. We further hope to learn about communicative conditions for realizing developmental practice supporting pupils' learning and the learning for all the participants.

NOTES

1. LCMP is financed by The Research Council of Norway (NFR) and Bergen University College. Marit Johnsen-Høines is the research leader. LCMP is member of the research consortium Teaching Better Mathematics developed in cooperation between mathematics educators from University of Agder, Bodø University College, Oslo University College and Sør-Trøndelag University College also partly financed by NFR. See Læringssamtalen i matematikkfagets praksis (LIMP) http://www.hib.no/fou/limp
2. Mathemacy refers to Skovsmose (2005, p. 46): "The notion of mathemacy signifies competencies related to mathematics, similar to the notion of literacy, as developed by Paolo Freire.... [M]athemacy refers to different competencies. One of these is to deal with mathematical notions, a second one is to apply such notions in different contexts, and a third is to reflect on such applications. This reflective component is crucial for the competence of mathemacy." This corresponds with Mellin-Olsen's (1989) discussions about *functionality of knowledge*, about being able to use mathematics critically for developing the society and ones own life.

3. From Norwegian: *Fleksible språkrom. Matematikklæring som tekstutvikling.*
4. Insight and understanding are used generally here—it might be insight into mathematical concepts or relations, into issues relevant for mathematics education, or other issues one seeks insight into.
5. The notion *practice teaching* is in this chapter referring to the practicum that is included as part of the teacher training study.
6. *Learning conversation* is used as the translation of the Norwegian *Læringssamtalen.*
7. In Norwegian *praksisnær undervisning.*
8. Gode sirklar AS has initiated an educational development which starts in kindergarten and ends at the university level. The aim is to establish an arena of competence where schools, parents, and companies collaborate through an involving and motivational education, and to improve the recruitment and career choices in the region by enhancing the relationship between the society and the local industries. A local teacher, Frede Thorsheim, commenced the forerunner to the project Real-Life Education in 1999. He has received several international awards for his teaching methods. See also http://www.godesirklar.no/index.php?l=en&m1=5&m2=14
9. The pupils are 11–16 years old.
10. ECT: European Credit Transfer
11. Equivalent to one year of full-time study
12. Children in school are referred to as pupils and teacher students as students in this chapter.
13. The schoolteachers that participate in LCMP are the children's teachers and the tutors (trainers) of the student teachers as well. The didacticians are teacher educators in mathematics and researchers.
14. "Equality" is here used in the sense that the contributions by each person are regarded as relevant and important as part of the collaboration. Differences are regarded as underpinning the equality.
15. The research is "interpretive" in the sense that it seeks improved understanding through the interpretation of the recordings made; it is not seeking to expose causal relationships or "laws."
16. See note 1.
17. The notion of "learning as taking place in and between different practices" refers to the fact that learning is going on when the organized activities are going on, but afterwards as well, and also when the learners are on their way into a organised learning activity. This issue is elaborated on in Johnsen-Høines (2009b).
18. These teaching sessions appeared quite interdisciplinary; more time than the mathematics lessons was used—the activity could be characterized as technology, art, and mathematics.
19. The notion of "mathematical conversations" was used rather informally by the participants. Particularly one of the teachers challenged the students and all of us to reflect on which conversations we would understand as "mathematical conversations" and what we would argue qualified them as such. Later this term was used as part of the continuing processes, to develop awareness about what characterized the mathematical content and the communicative actions.
20. "We" refers to the group of teacher students, teachers, and didacticians.
21. 60 ECT are equivalent to one year of full-time study, 90 ECT are one and a half year. These students have 30 ECT when they start this one year full-time study.
22. The students' study combines two 30 ECT courses; each course requires a 15 ECT report.
23. *Subject-based* or *subject-oriented* is connected to mathematics and mathematics education; to subjects as academic fields. *Discipline-based* may be a close translation. In Norwegian: *faglig.* (It does not refer to a subject-object discussion.)
24. Theoretical frameworks developed by Mellin-Olsen (1987, 1989), Skovsmose (1994, 2003), Alrø and Skovsmose (2002, 2003), Jaworski (2007), Goodchild (2001, 2007), and Johnsen-Høines and Lode (2007), all support the learning community we seek to establish.
25. The notion of voice is inspired by Bakhtin (1998) and developed further in Johnsen-Høines (2002).

REFERENCES

Alrø, H. & Skovsmose, O. (2002). *Dialogue and learning in mathematics education. Intention, reflection and critique.* Dortrecht, The Netherlands: Kluwer Academic Publishers.

Alrø, H. & Skovsmose, O. (2003). Læring gennem samtale. In O. Skovsmose & M. Blomhøj (Eds.), *Kan det virkelig passe?* (pp. 25–38). København: LrUddannelse.

Bakhtin, M. M. (1981). *The dialogic imagination* (M. Holquist, Ed.). Austin: University of Texas Press.

Bakhtin, M. M. (1998). *Spørsmålet om talegenrane.* Oversatt av Slaattelid. Bergen: Adriane Forlag.

Burton, L. (2001). Methodology and methods in mathematics education research: Where is "the why"? In S. Goodchild & L. English (Eds.), *Researching mathematics classrooms. A critical examination of methodology* (pp. 1–10). London: Praeger.

Dreier, O. (1999). Læring som endring av personlig deltakelse i sosiale kontekster. In K. Nielsen & S. Kvale (Eds.), *Mesterlære. Læring som sosial praksis* (pp. 196–214). Oslo: Ad Notam Gyldendal.

Goodchild, S. (2001). *Students' goals—A case study of activity in a mathematics classroom.* Bergen: Caspar Publishing Company.

Goodchild, S. (2007). Inside the outside: Seeking evidence of didacticians' learning by expansion. In B. Jaworski et al. (Eds.), *Læringsfellesskap i matematikk [Learning Communities in Mathematics]* (pp. 189–203). Bergen: Caspar Publishing Company.

Goodchild, S. (2008). A quest for good research. In B. Jaworski & T.Wood (Eds.), *International handbook on mathematics teacher education: Vol. 4. The mathematics teacher dducators developing a professional: Individuals, teams, communities, and networks* (pp. 201–220) Rotterdam: Sense Publishers.

Jaworski, B. (2004). Grappling with complexity: Co-learning in inquiry communities in mathematics teaching environment. In M. Johnsen-Høines & A. B. Fuglestad (Eds.), *Proceedings of the 28th Conference of the International Group for the Psychology of Mathematics Education.* (Vol. 1, pp. 17–33). Bergen: PME.

Jaworski, B. (2006). Developmental research in mathematics teaching and learning: Developing learning communities based on inquiry and design. In P. Liljedal (Ed.), *Proceedings Canadian Mathematics Education Study Group. Act 2006 Annual Meeting* (pp. 3–16). Calgary: University of Calgary.

Jaworski, B. (2007). Theoretical perspective as a basis for research in LCM and ICTML. In B. Jaworski et al. (Eds.), *Learning community in mathematics* (pp. 121–139). Bergen: Caspar Forlag.

Johnsen-Høines, M. (2002). *Fleksible språkrom. Matematikklæring som tekstutvikling.* Thesis Dr Philos. Bergen: University of Bergen.

Johnsen-Høines, M. (2004). Understanding constituted by understandings. *Nordic Studies of Mathematics Education, NOMAD, 9*(2), 63–81.

Johnsen-Høines, M. (2009a). Dialogical inquiry in practice teaching. *Nordic Studies in Mathematics Education, NOMAD, 14*(1), 39–59.

Johnsen-Høines, M. (2009b). Learning dialogue in mathematics practice. In C. Winsløw (Ed.), *Nordic research in Mathematics education . Proceedings from NORMA08 in Copenhagen, April 21–25, 2008* (pp. 135–142). Rotterdam: Sense Publishers.

Johnsen-Høines, M. & Lode, B. (2006). Positioning of a subject based and investigative dialogue in practice teaching. In J. Novotná et al. (eds.) *Proceedings of the 39th Conference of the International Group for the Psychology of Mathematics Education, PME30.* (Vol. 3, pp. 369–376). ISSN 0771-100X. Prague: Charles University.

Johnsen-Høines, M. & Lode, B. (2007). Meta-level mathematics discussions in teaching practice: An investigative approach. In C. Bergsten et al. (Eds.), *Relating practice and research in mathematics education. Proceedings of Norma05, Fourth Nordic Conference on Mathematics Education* (pp. 311–324). Trondheim: Tapir Akademisk Forlag.

Lave, J. (1999). Læring, mesterlære, sosial praksis. In K. Nielsen & S. Kvale (Eds.), *Mesterlære. Læring som sosial praksis* (pp. 35–52). Oslo: Ad Notam Gyldendal.

Lave, J. & Wenger, E. (1991). *Situated learning legitimate peripheral participation.* Cambridge: Cambridge University Press.

Lotman, J. (1988). Text within a text. *Soviet Psychology, 24*(3), 33–37.

McNiff, J. & Whitehead, J. (2006). *All you need to know about action research.* London: Sage Publications.

Mellin-Olsen, S. (1987). *The politics of mathematics education.* Dortrecht, The Netherlands: Kluwer Academic Publishers.

Mellin-Olsen, S. (1989). *Kunnskapsformidling. Virksomhetsteoretiske perspektiver.* Bergen: Caspar Forlag.

Pimm, D. (1987). *Speaking mathematically.* London: Routledge.

Skovsmose, O. (1994). *Towards a philosophy of critical mathematics education.* Dortrecht, The Netherlands: Kluwer Academic Publishers.

Skovsmose, O. (2003). Undersøgelseslandskaber. In O. Skovsmose & M. Blomhøj (Eds.), *Kan det virkelig passe?* (pp. 143–158). København: LrUddannelse.

Skovsmose, O. (2005). *Travelling through education. Uncertainty, mathematics, responsibility.* Rotterdam: Sense Publishers.

Vygotsky, L. S. (1978). *Mind in society: The development of higher psychological processes.* Cambridge, MA: MIT Press.

Wells, G. (1999). *Dialogic inquiry. Toward a sociocultural practice and theory of education.* Cambridge, UK: Cambridge University Press.

Wenger, E. (1998). *Community of practice. Learning, meaning, and identity.* Cambridge, UK: Cambridge University Press.

8

LONGITUDINAL STUDY AS AN INSTRUMENT FOR DEVELOPMENT IN MATHEMATICS TEACHING AND MATHEMATICS EDUCATION RESEARCH

Trygve Breiteig and Barbro Grevholm

UNIVERSITY OF AGDER

Abstract: A longitudinal study is designed to expose features of pupils' understanding of mathematics and contribute to a mathematics teaching developmental research project. The first part of the chapter describes the project context, the original goals of the longitudinal study, the methodology, and implementation of the study. The chapter proceeds to explain how the study evolved into an instrument for researcher development at a variety of levels—teacher research, master's and doctoral research, and mathematics teacher education research. The authors argue that within developmental research it is not so much the research outcomes that have an impact on practice; it is more the active engagement in research that is effective as an agent of change.

Keywords: Longitudinal study, mathematics education research, researcher training, pupils' mathematical conceptions, Norwegian compulsory school grades 1 to 13.

INTRODUCTION

This chapter describes the use of a longitudinal study (hereafter referred to as "the Study") within mathematics teaching and research development. The Study was designed, at the outset, to be a component of a mathematics teaching developmental research project, Learning Communities in Mathematics (LCM).[1] The LCM project and the broader methodology of teaching developmental research are described in detail by Jaworski (this volume, Chapter 5). The discussion in this chapter focuses specifically on the goals, methodology, implementation, and use of the Study, both within the LCM project as originally intended and beyond. The purpose of this chapter is, then, to explain how the Study has contributed to mathematics teaching development, researcher development, mathematics teacher education, and knowledge of teaching and pupils' conceptions within mathematics.

The chapter is organized to set out, initially, the original intention for the Study within the LCM project, the methodology of the Study, and an account of how it was implemented in schools and within the project. Following this are separate sections outlining the use of the Study in mathematics teaching development, researcher training, teacher education, and knowledge creation. The chapter concludes with a brief discussion of the role of research in mathematics teaching development.

The First Sourcebook on Nordic Research in Mathematics Education, pages 125–138

THE LONGITUDINAL STUDY AS A COMPONENT OF A MATHEMATICS TEACHING DEVELOPMENT PROJECT

The Study was originally proposed as an element of a large mathematics teaching developmental research project. The LCM project was founded on "a basic premise that people learn through responsible participation in a collaborative enterprise" (LCM, 2003, p. 3). Teachers and didacticians, in partnership, were intended to engage in research, or inquiry, into learning mathematics, teaching mathematics, and, for didacticians, the processes of mathematics teaching development. The project was to focus on:

- *Students' Learning*—students engaging in mathematical inquiry related to the mathematics curriculum with teachers and didacticians fostering and researching inquiry approaches
- *Teachers' Learning*—teachers engaging in inquiry into processes of learning and teaching as they interpret the curriculum in collaboration with their didactician colleagues
- *Researchers' Learning*—researchers (including teachers and didacticians) developing research knowledge and capacity through clearer understandings of approaches and issues (LCM, 2003, p. 6).

The Study was intended to contribute to the teaching development goal as well as contributing to the overall research enterprise of the project. The study focused on eliciting evidence that would contribute towards addressing a number of research questions, first concerning mathematics learning:

- What are the pupils' mathematical conceptions in agreed curriculum areas?
- What are the factors contributing to pupils' mathematical progress with respect to the current Norwegian curriculum in mathematics?

Second, in relation to mathematics teaching:

- What knowledge do teachers bring to mathematics teaching at agreed levels and what approaches are used?
- What is needed for teaching to develop in order to address the questions above about the pupils?

Further supplementary questions were proposed:

- What is the role of assessment in this relation between teachers' teaching and pupils' learning? What is the relation between teachers' teaching and pupils' learning?
- What is it that guides teachers when they decide about their teaching? How is our cooperative work with teachers helping them to develop skills as inquirers? (LCM, 2003, p. 6)

The methodology of the LCM project was based on the evolution of both research and development and, as will become clear in this chapter, the Study developed further significance, unanticipated in the original proposal. In this chapter we focus discussion on the way the Study evolved and contributed to teaching and researcher development in addition to describing how master's and doctoral students contributed to the effort to address these initial questions.

METHODOLOGY

The major component of the Study was comprised of mathematics tests to be completed by pupils in grades 4, 7, 9 and 11; these were chosen to provide complementary information alongside the Norwegian national tests. The tests were made up of a combination of multiple choice and short open answer items. It was decided to give the tests three years in succession during the period in which the LCM project was active with teachers in order to be able to study the consistency of pupils' responses over time. Thus the tests were to be given within a few weeks of the project's launch, and they had to be constructed during the six-month preparation phase. Consequently, it was not possible to develop and test new items.

A qualitative investigation of the curriculum was carried out (by the authors), and this was used to provide information about the types of tasks that were central in the different areas of mathematics addressed in Norwegian schools. Two test papers were prepared for each level, one on number and algebra, and the other on shape, space, statistics, and probability. A first critical evaluation of the tests was carried out together with an experienced teacher, who intended to use the tests as a basis for her master's research. The tests were shared with teachers in the project schools and piloted in one class at each grade level to check the usefulness and size. Following discussions with teachers in lower and upper secondary schools, grade 9 and 11, it was decided that only one of the tests be used with grade 9 and 11 pupils, and the number and algebra test was selected. Further, the proposed test was shortened substantially as the teachers would agree to use only one lesson for the test.

The items were taken or adapted from tests developed in research in which project members had previously participated, in particular the KIM project (KIM—Kvalitet i matematikkundervisninga [Quality in mathematics teaching]—Brekke, 1995; Breiteig & Brekke, 1993); the Exeter-Kassel project (Burghes, 1999; Hinna, 1996); and Evaluation of Reform 97 (Alseth, Breiteig, & Brekke, 2003). There were a number of advantages to be gained from using items from previous studies: it addressed issues of reliability and validity; pupils' responses and item facilities in the Study could be easily compared with larger previous samples; and as noted above, the items would not need extensive piloting in advance. Furthermore, items could be selected that had proven effective in exposing features of pupils' mathematics conceptions. There were some items that were common to tests for adjacent grades that aimed to provide an indication of pupils' progress in mathematical understanding and proficiency. Examples of items will be provided later in the chapter. It was decided that electronic calculators and other aids would not be used in the tests.

It is widely recognized that written tests in mathematics give only a limited picture of pupils' mathematical knowledge. Not all mathematical competencies can be captured by a written test, and one cannot be sure that pupils will correctly interpret an item and thus bring into play the required knowledge, which they command. Nevertheless, tests can provide a picture that provokes further inquiry about pupils' learning of mathematics and, related to that, teachers' teaching of mathematics. Test results can also indicate where there is potential for improved learning. This was the view that underpinned the Study and had previously been applied to explain the rationale for the KIM project, which was intended as an integrated test and professional development package for teachers. The LCM project differed from the KIM project in taking an alternative approach to teaching development, that is through the creation of learning communities and by developing "inquiry as a way of being" (Jaworski, 2004) and by didacticians working together with teachers in order to develop teaching.

IMPLEMENTATION

The initial proposal for the LCM project was based upon testing pupils from a random sample of 24 schools (12 primary and 12 secondary). However, the ambitious plan had to be curtailed due to limitations of the grant from the Research Council of Norway, and only pupils from those schools included within the LCM project and a partner project within the same Research Council program (ICT and Mathematics Learning, see Fuglestad, this volume, Chapter 6) participated in the Study. Thus eight schools participated in the Study: two primary schools (grades 1 to 7), two lower secondary schools (grades 8 to 10), two combined primary and lower secondary schools (grades 1 to 10), and two upper secondary schools (grades 11 to 13). In each of the grades that were to participate in the tests it was intended that pupils would be given the same test paper on two separate occasions, first in the autumn near the beginning of the school year and again in the spring towards the end of their year in that grade. It was not intended to follow pupils through the grades or across schools, and in fact, the measures to ensure pupils' anonymity made such tracking impossible.

The schools and pupils participating in the Study cannot be considered representative of Norway, or even the local region. The teachers had chosen to be part of the teaching development project, and the schools represented just part of the population of two communities in the south of Norway. It was never the intention to use the tests to make general statements about the population as a whole, but, as noted above, the selection of items enabled the comparison of project pupils with larger representative samples. Table 8.1 provides the numbers of pupils participating in each test on each occasion.

An intention of the Study was to give all the project's participants the best possible information that would represent the status of the mathematical attainment and competencies of pupils in the project schools. It was thus decided to include all classes in project schools in the chosen grades and to ask the teachers to implement the tests during regular mathematics lessons. Thus pupils would not experience

TABLE 8.1 Number of Participants in the Tests, Indicated for Each Project School

	Grade[a]			
	4	**7**	**9**	**11**
Autumn 2004	104 + 106	114 + 115	90	236
Spring 2005	61 + 64	67 + 64	78	218
Autumn 2005	113 + 109	130 + 135	167	227
Spring 2006	57 + 63	104 + 103	94	126
Autumn 2006	111 + 113	111 + 109	183	222
Spring 2007	106 + 109	111 + 111	80	207

[a] At grades 4 and 7 two tests were set, the numbers show the participation in each.

more stress than necessary, taking the test in a familiar location with a teacher that was well known to them. In order not to burden teachers unnecessarily and to ensure consistent evaluation of pupils' responses, the didacticians would take responsibility for marking and analysis. The class teachers would receive a full report of the results of their pupils. Teachers in Norway are familiar with the administration of externally managed assessments and summarized analyses of results, and thus the procedures did not make unusual demands that would require special training sessions to ensure proper implementation. To ensure pupils' anonymity, each was identified by a code number, used for both test occasions. Researchers received the pupils' answers identified by the code; only the class teacher had access to the information linking pupils' names to the code number. It was also agreed, and pupils were informed, that the test results would not be used by the teachers as part of the summative evaluation of the pupils. No comparison should be made between classes or schools. This created a risk that pupils would not take the tests seriously and not attempt to do their best, and indeed this was a significant concern as national tests had been undermined by pupils' refusal to participate and absenting themselves from school on the test day. The successful administration of the Study tests relied in large measure on the teachers convincing their pupils of the rationale for the Study, and that the pupils' cooperation would provide valuable information for the teacher and subsequently the quality of teaching. Clear instructions were given to teachers about the procedure for informing and carrying out the tests in class. Each school had a named didactician contact who provided schools and teachers with all necessary information.

A coding scheme for marking pupils' responses, based on experience gained in previous studies such as KIM, was developed. We, the authors together with a master's student, coded pupils' responses from the first round 2004; pupils' responses that were not directly classifiable within the coding scheme were subject to thorough discussion and subsequent agreement of all three coders. A statistics software package was used to provide summary statistics of the coded data. Using all the material available—that is, the raw data, coded data, and summary statistics—the master's student produced reports for each teacher and class and subsequently used the data as a starting point for her own dissertation. In the next round in 2005, two new master's students joined the Study team and built on the foundations laid in the first round, using the same coding scheme and approach to basic summary analysis. Later two more master's students joined the team and worked with the data from later rounds. Each master's student reported the outcome of the whole material for one semester (2004) or one year (2004/2005) and in addition used the Study data to address their own specific research questions. The resulting master's dissertations include the reports to schools and overviews of the outcome in addition to the investigation done by each student. The master's dissertations are published on the web pages of the university[2] (Andreassen, 2005; Espeland, 2006; Log, 2007; Tesaker, 2008; Tonheim, 2006). This collection of dissertations and reports enabled students who entered the Study later and other researchers to build on the knowledge being accumulated.

Each teacher received a comprehensive report of the responses made by pupils in his or her class; the reports included coding principles, lists of scores for each pupil, and means and standard-deviation for each item. In some cases didactician members of the Study team visited schools and presented the outcomes of the tests to groups of teachers. Further, in two of the LCM project workshops, we and a master's student presented reports of selected items and pupils' responses to both inform and stimulate discussion between teachers and didacticians.

The Tests in Teaching Development

As explained above, the Study was carried out as part of an embracing teaching development project, and it has been described how summary results were provided for all teachers whose classes participated in the tests. In some cases the reports were presented in person by didacticians who specifically motivated discussion with teachers on issues arising from the pupils' responses. Also, as noted above, pupils' responses to the test items were used within project workshops to stimulate discussion or as illustrations of possible pupil conceptions. The extent to which teachers were proactive in responding to the reports they received varied, and this was expected given that the project was based on co-learning with teachers rather than didacticians telling teachers what they should do. In this section we report one project teacher's response as exemplary of what was intended.

One teacher used a specific item included in the tests for grades 4 and 7 pupils; she believed this would prove challenging for her grade 6 pupils and created a workshop lesson for them. The task, reproduced in Figure 8.1, was originally used in the KIM study (the original concept of the task comes from resources produced by the Shell Centre for Mathematical Education, 1985). Pupils' responses to the task were presented and discussed in a project workshop; the grade 4 pupils had performed better on the task than the

(Norwegian in the diagram:
"Alder" means Age, "Høyde" means Height)

Write below which person corresponds to each of the points in the coordinate system.

Liv corresponds to the point __________

Gry corresponds to the point __________

Ole corresponds to the point __________

Hans corresponds to the point __________

Figure 8.1 From Gjone (1997), p. 6.

grade 7 pupils, thus attracting attention to the task. Within a project workshop when the item was discussed, several hypotheses were speculated: Could it be the choice of variables on the two axes? Does the task mislead pupils into error? Many hypotheses were put forward, among them that pupils in grade 4 did not so easily take for granted that the horizontal axis (by convention—as the independent variable) most often is used to represent the time axis.

The curiosity of the teacher was aroused, and she wanted to inquire further into this task with her pupils. After the teacher presented the task in her class, the pupils solved the problem in groups. A didactician was present in this lesson and made video recordings of both teacher and pupils. Subsequently the teacher came to the university to view the video with didacticians to discuss the pupils' activity that had been recorded.

Later further careful study by didacticians showed how teacher and pupils worked on the different representations, especially the visual representation—how the pupils coordinated the variables of height and age, their dialogue, and their use of physical gesture in communicating their reasoning. Also in the teacher's communication and dialogue, visualization and gesture were used to develop mathematical knowledge (Bjuland, Cestari, & Borgersen, 2007, 2008; this volume, Chapter 9). Thus, a new theoretical perspective, on the empirical data from the classroom created by the teacher's inquiry, led to new findings for the learning community of both teachers and didacticians. Data from the tests inspired practice, which in turn nurtured the theoretical inquiry.

That the tests motivated questions rather than provided answers was welcomed within a project established on a principle of inquiry as a developmental tool. In this respect the tests have had a wider reach than just the project schools. In one instance some items were discussed in a regular university course including teachers and researchers. In these items there was a statistically significant difference (small to medium effect size) in performance between girls and boys, with the boys outperforming the girls. In the last PISA[3] study that focused on mathematics (Organisation for Economic Co-operation and Development, 2004), Norwegian boys scored higher than girls, although the difference is not statistically significant. Thus the Study results appeared to contrast with the larger representative sample used in the PISA study. We note, however, the PISA test instrument is rather different to that used in the Study, and thus a direct comparison between the two studies is not possible. One lower secondary teacher present in the course found the results surprising and believed that in the mathematics classes she taught the girls would outperform the boys. However, when she tried the tasks with her classes, the Study's results were reproduced. Curiosity aroused, the teacher then set up a further study to explore differences between gender performances in mathematics items with her classes.

It is thus evident that the tests, or rather pupil responses to the test items, were able to provoke curiosity, stimulate inquiry and challenge teachers' (and didacticians') beliefs about pupils' conceptions.

The Tests in Researcher Training

It can reasonably be claimed that any research activity is an opportunity to learn more about research, including methodology, methods, and ethical considerations. This section is concerned with the development of those entering research activity through master's and doctoral programs. Four doctoral students participated actively within the LCM and ICTML projects, and their collaboration within the team of didacticians formed a significant part of their researcher education. In addition, one other doctoral student collaborated with the project through an attitude survey of pupils who took part in the Study reported in this chapter. The attitude survey provided an opportunity for a pilot study for the doctoral student in addition to making an important contribution to the projects.[4] As has been indicated above, a number of master's students made important contributions to the Study as they used it within their dissertation work. In this section we report in detail the way the Study contributed to the research of one doctoral student and provide more information about the contribution of the master's students.

Per Sigurd Hundeland's doctoral research focuses on teachers' justification of their teaching and assessment in mathematics. With Hundeland, we conducted a pilot study in autumn 2004, in which two teachers—one primary school teacher and the other an upper secondary school teacher—were interviewed (Hundeland, Grevholm, & Breiteig, 2005). Some of the questions in the interview were based on items used in the Study. Teachers were asked to comment on the items and evaluate the degree to which pupils would be able to solve them. The responses to these and other questions provided evidence of teachers' conceptions about mathematics teaching. The interviews were analyzed in relation to a modified description of Alba Thompson's levels concerning the development of teachers' conceptions about mathematics teaching (Pehkonen, 2003; Thompson, 1991).[5]

The interviews exposed differences between teachers' reasoning. This led to a positioning of the teachers in a preliminary analysis, where they were placed at different ends of a scale based on Thompson's levels. The primary teacher (at the time teaching a grade 4 class) expressed how it was important to him that the pupils would understand, that they were able to develop strategies for tackling mathematical tasks, and that they would appreciate a connection between language and mathematics. The teacher from the upper secondary school was more occupied by the mathematics as a product and conceived teaching as dissemination. He observed that there was a huge difference between the tasks pupils work with normally and the items that were presented in the Study tests. In the interviews it was discussed to what degree the items in the Study test could be integrated in regular teaching and used within an inquiry-based teaching plan in order that the items could then contribute more to the development of teaching. The teacher, on several occasions, expressed his scepticism to this possibility and pointed to the factors that constrained his teaching. He especially pointed to the lack of time, the demands of a content-laden curriculum, and the regulations for examination, which led to the situation where he had to go through the curriculum in "an efficient way." He referred to a situation where inquiry based planning and emphasis on inquiry had to be toned down.

Hundeland built on his experience from the pilot study and focused on three teachers who were interviewed three times each and once together over a period of 12 months. All three teachers took part in the LCM project and worked at the same upper secondary school, and all three of them used the Study tests. Hundeland attempted to explore the nature of the teachers' perspectives on effective learning and teaching of mathematics. Also in this continuation study, the teachers were presented with some of the items in the tests. They were asked to say something about what expectations they had for pupils' performance on the items. The questions elicited responses that clearly indicated that the teachers have a broad overview of the mathematics included in the curriculum and the degree to which pupils are expected to become competent in this mathematics. Further, the teachers revealed didactical competence in being able to explain, for the different tasks, what kind of inadequate conceptions, problems, or difficulties would emerge when pupils work with the items.

The Study tests were a resource external to the immediate research inquiry of Hundeland; they provided a common ground for the interviews which Hundeland conducted with the teachers. Thus the tests contributed to his research, but perhaps more significantly they provided a context in which he was able to collaborate with experienced researchers. In the language of Wenger (1998) Hundeland was on an *inward trajectory* as he participated in the practice of mathematics education research.

The research of master's students has made an important contribution to the development of mathematics education research in Norway, which is evident in many of the chapters in this Norwegian part of the Sourcebook—especially in the chapter by Borgersen, Cestari and Bjuland (this volume, Chapter 9). It has already been noted that master's students played a valuable role in the Study reported here; their interpretation of the data emerging from the tests have contributed crucially to addressing the initial research questions listed above. The remainder of this section offers a brief summary of the work of two of them. It must be appreciated that within a single chapter it is not possible to offer a fair representation of the extent and quality of their research. The account here is intended to serve two purposes: to record how the Study contributed to their researcher training, and to report some of the knowledge outcomes arising from the Study.

The first master's student, Irene Skoland Andreassen, developed the foundations of the Study laid by the didacticians by designing a scheme for evaluating, marking, and coding pupils' responses to items and reporting the results to schools. In this way her work built a framework within which the work of the students who joined the Study later could be structured. Andreassen (2005) chose to inquire into how pupils' responses to the test items can provide an insight into pupils' mathematical comprehension and competencies, and to reveal where they may need support. As an experienced teacher, Andreassen was curious about how tests in mathematics may be used with a positive effect on pupils' learning outcomes and contribute to raising pupils' attainment.

To offer an insight into Andreassen's work, we take one example of an item that was used. It is an item from the number and algebra test set for grade 7 pupils (translation by the authors, note that in Norway the decimal marker is a comma, the division symbol a colon and multiplication symbol is a dot):

> Draw a circle around the correct answer(s):
>
> Cakes are packed in boxes with 0,75 kg in each box. How many boxes do we need for 6 kg cakes?
>
> $6 \cdot 0{,}75 \quad 6 : 0{,}75 \quad 0{,}75 : 6 \quad 0{,}75 \cdot 6$
>
> $6 - 0{,}75 \quad 6 + 0{,}75$

The analysis of pupils' responses to this item revealed in Table 8.2.

TABLE 8.2 Distribution of Responses in Percentage

6 : 0,75 (Correct)	42
6 · 0,75 and 0,75 · 6	16
6 · 0,75 or 0,75 · 6	19
Other answers	19
No answer	4

Reflecting on the responses to this task, Andreassen observes:

> This is a division task where the divisor is less than 1. The task requires a conceptual model for quotient/quotition. One might regard it as repeated subtraction. We have 6 kg and take away 0,75 kg, which we pack in a box, then a new 0,75 kg in next box and so on until we have taken all the cakes. The task is also given in the KIM project, and Brekke (1995) remarks that this model is as simple to grasp as the model for sharing, where the number of parts is given. The difficulty here is that the measure is given as a decimal number. 35% think they must use multiplication. The reason may be that they see a grouping of 0,75 and interpret it as a repeated addition 6 times. Brekke (1995) concludes that more serious work should be devoted to building up conceptual models matching situations where quotition division is needed. Our results confirm this. (2005, p. 62. Authors' translation)

Andreassen chose to elaborate a small number of items in depth, such as interpretation of diagrams, understanding of decimal numbers, choices of operations in situations involving ratio and proportionality, and algebra to express generality. Here she could relate her findings to earlier research literature, as well as reflect more deeply on her own findings. She expressed the following conclusions as a result of her study:

> When written tests in mathematics are carefully designed, with selected diagnostic tasks included, such tests are a useful tool to get valuable knowledge of what pupils can do, and in which tasks they need help. This can be revealed by going deeply into the pupils' responses. That means among other aspects, to study alternative incorrect answers, map possible misconceptions, study correct or partially correct responses, study pupils' strategies, and try to identify on which thinking processes the responses are based. This increased insight may be used to enhance pupils' learning outcomes, by using this new knowledge actively in the teaching. In this way the tests will be a pedagogic tool that may be used in different ways to help pupils overcome obstacles that are hindering learning and a more deep understanding in mathematics. It is important that teachers very explicitly use the knowledge, which the test results and the evaluation of these may give. (Andreassen, 2005, p. iv)

It is important to keep in mind that this is the reflection of a master's student who is taking her first steps into doing research. Andreassen's work with the Study team has equipped her to adopt a more critical and inquiring stance as a teacher–researcher, and thus, we believe, she is better positioned to work on the quality of teaching and learning in her mathematics classes. Andreassen describes her own learning in the conclusion of her dissertation and sees the importance of diagnostic tests and evaluation as an integrated part of teaching and learning. Thus tests become a pedagogical tool to use in different ways to enable pupils to overcome the obstacles in their learning and gain improved understanding in mathematics. It is important that teachers apply the knowledge that is produced from test results and evaluation. Andreassen notes that earlier research has not led to improved results and points to the need to emphasize important knowledge gained earlier. She hopes that teachers in the future will make better use of the knowledge we have in order to let pupils be nurtured from that. The implication she suggests for teaching is that the tasks in the test can be used as starting points for diagnostic teaching. Given the widespread concerns that research in education has had little impact on teaching and learning (see for example the discussion in Bishop, 1998; Borgen, Gunnes, & Sivertsen, 2007; Pring, 2004), Andreassen's hopes may be rather optimistic. Andreassen has returned to her work in school, presumably a much more aware teacher who will use what she learned in her master's studies. Our contention in this chapter is that collaboration between experienced didactician–researchers and teachers does have an impact on teachers' perceptions of research and research findings.

In the second year, Hildegunn Espeland joined the Study.[6] Espeland had extensive experience as a mathematics teacher in upper secondary school when she embarked on the master's program. She claims that she is driven by curiosity arising from her teaching experience of pupils learning algebra. Consequently Espeland focused on the algebra items in the tests and interviewed pupils to confirm her initial conjectures about the reasons for their responses to the test items. Espeland found the synergy between practice and research so challenging that she decided to carry on with doctoral study directly after her master's dissertation. She is now working in a follow-up project based on the LCM and ICTML projects. She has also supervised two master students working with data from the Study or later follow-up.

An example from Espeland's inquiry relates to an item in the grade 11 test (Espeland, 2006, pp. 101–102) (authors' translation), as shown in Figure 8.2; results are shown in Table 8.3.

Espeland (2006, p. 102) makes the following interpretation of these outcomes:

> This task is difficult for most pupils. However, it is remarkable that only 6 percent leave it unanswered. This can indicate that they think they know the answer. Of course they may guess, but several students have solved it like this:
>
> $$\frac{\cancel{2x}+1}{\cancel{2x}+1+5}=\frac{1}{6}$$
>
> and tick for always true. This shows a missing conceptual understanding of fraction and division. They have learned to reduce a fraction, but this is extended to what seems equal above and below the fraction bar. Obviously reducing a fraction is a concept that is not connected to knowledge of multiplication and factorizing. From my school practice it is clear that this is complicated. It demands that the pupils have a fundamental and good conceptual understanding. The outcome shows lack of conceptual understanding when it comes to fractions and division.

Task 10c

$$\frac{2x+1}{2x+1+5}=\frac{1}{6}$$

This...

☐ is always true ☐ is never true

☐ *can* be true, namely when ______________

Figure 8.2

TABLE 8.3 Distribution of Responses in Percentage

	LCM grade 11		
Task 10c	**Group MX 2005**	**Group MY 2005**	**Total 2005**
Can be true, namely when $x = 0$	16	12	15
Can be true, but without comments	2	6	3
Always true	28	28	28
Never true	44	43	44
Can be true, but with wrong condition	5	2	4
No answer	5	9	6

Further, she relates the finding to the teaching. The purpose of the research is to improve the teaching:

> For a teacher on this level, it is useful to know how widespread such a problem is, as the curriculum has high ambitions that the pupils during this year in school should have a competency in algebra. Further work with functions, equations and algebra has as prerequisite that this kind of tasks should be simple and that students have a good understanding of letters as general numbers. (Espeland, 2006, p. 101, authors' translation)

The developmental aspect of research is important to Espeland, and she understands the need for teachers to know about research results.

An issue several of the master's students investigated concerned the progress made by pupils over the course of a school year and how progress varied across pupils. Espeland divided the cohort into three different subgroups according to their performance in the first test. Each subgroup was analyzed with respect to progress during the school year. Espeland found that in grade 9 the cohort with highest achievement at the beginning also made highest gains in scores on the tests over the year (Espeland, 2006).

Table 8.4 reveals that the upper third of the pupils shows a progress of 5.5 score points, the middle 3.6, and the lower third 3.3. The differences between the groups of pupils increase. The results from grade 9 provoke concerns about whether instruction is disproportionately directed towards higher-achieving pupils. If this is the case, it is a challenge that must be faced, especially since the curriculum in Norway emphasizes the right for every pupil to receive instruction matched to their level of attainment. The tendency that Espeland exposes here, that the difference between low-achieving and high-achieving pupils is growing over time, is well known internationally but had not been revealed previously for mathematics and pupils in Norway.

The Tests in Teacher Education

Teacher education in Norway is expected to be research-based, and students' participation in research and development work is one suggestion made by the Ministry of Education to realize this goal (Det Kongelige Kunnskapsdepartement, 2009). The research projects running in the department at UiA thus provided the opportunities sought (Grevholm, 2004a,b), and the Study provided an ideal context. We decided to offer students an opportunity to work as research assistants and, as part of their practicum, to conduct interviews with pupils. The Study thus evolved to incorporate this contribution of a group of student teachers who were taking an optional mathematics course in their third year (of a four-year course). They interviewed pupils regarding the items in the Study and attempted to explore deeper into pupils' thinking. By transcribing their interviews, analyzing the data, and relating it to theory, the student teachers gained experience that gave them a substantial insight into research activity and its potential to reveal pupils' knowledge and understanding. In addition, the data were made available and used by others in the research study. The didacticians' research questions in this "teacher education-based" part study were (1) In what ways is it meaningful to involve student teachers during practice in a research project? and (2) What kind of elements of learning do the students perceive in such a situation? The study has been reported by Grevholm, Berg, & Johnsen (2006, 2007).

The teacher education students became aware, among other things, of pupils' premature or partially formed conceptions that appeared to resist or present an obstacle to resolving problems in the test items. In the following extract from a student's report, the student sets out to explore the impact of cognitive conflict in response to pupils' errors. In this extract of the report, the student is analyzing the interview related

TABLE 8.4

Pupil group	*N*	Test year	Minimum	Maximum	Average	Standard dev.
All	74	2004	1	49	21.1	10.9
		2005	0	53	25.2	13.5
Lowest 1/3	25	2004	1	16	9.6	4.5
		2005	0	26	12.9	7.6
Medium 1/3	25	2004	16	27	20.6	3.0
		2005	12	41	24.2	7.9
Highest 1/3	24	2004	27	49	33.6	6.2
		2005	20	53	39.1	9.3

to the task: "24 necklaces weigh 3 kilograms. How will you find the weight of one necklace?"

> The pupils should not calculate the answer, but only find the operation they should use. However, when this pupil is firm on taking 24 and divide by 3, I challenge her to find the answer. We then go into point 2 in Brekke's template on diagnostic teaching.[7] A cognitive conflict is created.

Student teacher: If you take twenty-four and divide by three, what do you get?
Pupil: You get eight.
ST: If twenty-four necklaces weigh three kilos, then can one necklace weigh eight?
P: Eight kilos? No. So this obviously becomes gram?
ST: Perhaps eight grams?
P: (pause for some seconds). No, no, that is a little small.
ST: What do you think now?
P: No, I do not know.

> The pupil sees that this cannot be correct. It can neither be eight kilos nor eight grams that be the answer on the task. What we then did was to try to proceed to point 3 in the template mentioned, namely to try to solve the cognitive conflict. We tried to draw the task to make it more concrete. We did advance some steps but did not reach the goal.

ST: What have we done with the three kilos?
P: You divide them up.
ST: In how many parts have we divided them?
P: In twenty-four.
ST: And what did we find when dividing them into twenty-four. Then we found the one?
P: Yes. It weighs ca...
ST: We never mind how much it weighs, but what operation have you used for the calculation? This one? (points at 24 : 3)
P: Divided... yes.
ST: Or is it this one? (points at 3 : 24)
P: It is this one. (points at 24 : 3) Don't know...

> (From a student report reproduced with permission, name withheld by agreement. Authors' translation)

The resistance in pupils against dividing a number by a bigger one is exposed, as are the problems with decimal numbers smaller than one.

Several issues were raised for the student teachers such as: What are the roots of premature conceptions? How can pupils build more stable concepts? In the context here, we want to draw attention to the developmental power of these experiences as the student teachers engage in research within their practice, in addition to the students' development of research skills of use in their own teaching practice. Grevholm, Berg, and Johnsen (2006, 2007) concluded that through participation in the research project, by working as research assistants and carrying out interviews, the teacher education students experienced meaningful learning. The students observed that this learning is different in character to their other experiences in the teacher education program. We see the opportunities provided through the activity as potentially contributing to all six of the teacher competencies identified by Niss (2004). Niss described a model for mathematics teachers' specific competencies, four of which are connected to work with pupils: curriculum competency, teaching competency, uncovering of learning competency, and assessment competency. The remaining two are connected to the teacher's position in professional environments, namely: collaboration competency and professional development competency. We consider these six competencies to be a more precise version of the five competencies given in the curriculum for teacher education (Rammeplan for lærerutdanningen, 2003) and thus used Niss' model in order to interpret the data from student teachers' practice. Additionally, the students had the opportunity to develop a professional language through both the learning and the writing processes. The development of a professional language is not an evident part of most teacher education programs (Grevholm, 2004c). The students also experienced the writing process as a powerful learning opportunity. The data collected by the students are available for use by other researchers and master's students in the project and thus have the potential to make further contributions to the projects.

As students have no prior experience of conducting interviews, some of them found it a daunting prospect. A video was produced to demonstrate how to conduct a clinical interview in mathematics, which could be used by the course tutors in lectures; this was combined with the literature to facilitate discussion about how to interview effectively. Also produced was a scheme for observing in class, which would enable the students to link their class observations to the theories articulated in the course literature. Thus, students' knowledge gained from theoretical papers was connected to the work they did during their practicum. Students were provided with a template for the report they were to produce, which supported both the layout and the structure for their writing, as many of them were not aware of how to design a research paper. Again, this was a way to link their writing practice to didacticians' knowledge. Many of the students had never written a scientific report before and therefore both oral and written guidance was needed. The evaluation of the course revealed that all students except one were positive regarding this innovation within their practice. Students made several valuable suggestions for improvement of the course, which were implemented with the following cohort. In particular, they suggested an earlier start to the activity in the semester in order to allow more time to carry out the work in school and to transcribe and write the report. Also, despite all the measures taken to guide their activity they suggested providing more clarity in the information about the aims of the activity and the content and demands of the course.

Students indicated that mathematics didactics is the most valuable part of their education to become mathematics teachers. We conclude that the inclusion of research experience is valuable in order to make the practicum in mathematics teacher education research-based. Further, mathematics didactics research has an important part to play in the development of teacher competence (Grevholm, 2006). All the students' suggestions were implemented in the following year, and after a further year the course was revised and incorporated the innovation as a fixed feature.

The Tests as Research

So far this chapter has reported the research outcomes from the activity of those engaging in the Study as part of their research training. As the Study evolved, research training became an important feature of the work, and the students have made significant contributions to the Study's output. Alongside the students' work, routine inquiry based on the Study has been pursued. Here we note just two such examples.

One inquiry was pursued with the purpose of studying the development of algebraic thinking at the upper secondary level (Breiteig & Grevholm, 2006). We classified steps in the developing process and related it to an adapted APOS (action, process, object, schema) theory developed by Dubinsky and collaborators (Czarnocha, Dubinsky, Prabhu, & Vidakovic, 1999). The test item studied was the classical problem rooted in the work of Diophantus: Given the sum (s) and difference (d) of two numbers, find the numbers. In a specific given case ($s = 19$ and $d = 5$), the majority of respondents would find the numbers numerically using a guess and check method or by constructing a table. Three out of four pupils solved the item with such an approach. However, it is much more demanding, cognitively and metacognitively, to generalize and justify the method, and explain why it always is possible to solve the problem, given s and d. In the Study different levels of explanations were identified. Pupils might say, for example: $19 - 5 = 14$; $14/2 = 7$; $7 + 5 = 12$, where they may perceive the numbers as random representatives, such as in a generic example. Verbal rhetorical algebra might appear in reasoning, such as: "I found the numbers by taking the total sum–difference and divided by two, then I found the smallest number. The other I got by adding the smallest to 5." Only a small fraction of the students constructed and used equations, and reasoned from the structure of the problem. Our inquiry revealed that the essence of algebraic thinking develops slowly, and more slowly than skills in algebraic symbol manipulation. The APOS model was found to be useful for the researchers, and also seemed helpful for teachers in designing activities and guiding students to develop algebraic thinking.

Another inquiry emerged from an elementary statistical analysis of item responses in which differences in performance between grade 11 boys and girls were explored. Inquiring into where the differences were most pronounced, and then an analysis of pupils' errors suggested different approaches and possible attitudes to taking risks in mathematical problem solving (Goodchild & Grevholm, 2009).

DISCUSSION

The outcomes of the longitudinal study and subsequent analyses of pupil responses and related interviews hold lessons for both didacticians and teachers. We are convinced that we have not yet applied these lessons to achieve their full potential. The master's students recognized that the analysis of pupils' responses held important knowledge about pupils' conceptions, about which all mathematics teachers should be aware. Not only should teachers know about the (surprisingly high) proportion of pupils having common clearly defined problems, but also they should know how to study their pupils' responses in depth. The master's students themselves have been deeply impressed by the knowledge they gained about the situation they experienced in schools. They now make decisions about their own professional practice based on what they have learned through their investigations in the Study. For their part, the student teachers claim that they learned much about pupils' knowledge and learning from the interviews based on tasks from the tests.

Our intention is to continue to encourage teachers to reflect more closely on pupils' thinking and learning in similar ways as we have within the projects. We believe we need to continue to share these experiences, for example interviews with pupils, conversations between pupils about a problem, pupils' reactions to a given solution of a problem, pupils' explanations to peers about some mathematical task and so on. We need to learn together with teachers how such conversations and interviews can also contribute to the development of teaching. Concrete but limited pieces of evidence about pupils' knowledge and learning can have transformative power in a teacher's practice. A deep inquiry into limited pieces of evidence can provide much inspiration and knowledge about the learning outcomes of the pupils. Such work could give indication about where, in what mathematical areas, one needs to make more effort to offer pupils richer learning opportunities.

Through the projects a community has been developed in which didacticians and teachers are secure to share—both knowledge and uncertainties. In this milieu it is possible to continue collecting evidence and trying to understand through inquiry more about teaching and learning mathematics based on this evidence. We do not need to include new elements in our work; at the present it is necessary to refine and develop what we are already doing. We still need to consider what it means to develop teaching and to improve pupils' learning, and the relation between teachers' teaching and pupils' learning must come into focus. We know little about how the specific teaching acts of a teacher influence pupils' learning, and what is effective in the teaching and how this varies across pupils. One significant reason for the present uncertainty is that the teaching takes place within in a specific situation in time and space, but the learning can occur anywhere, at any time, and maybe long after the teaching that possibly caused it (Grevholm, 2007).

We would like to go on with the tests and the interviews connected to the tests in order to see if we can uncover any changes over time. So far the indication is that pupils' knowledge, as it is measured here, has not improved since the 1995 measurements that were made within the KIM project. Some recent results from TIMSS indicate a slight improvement by Norwegian students (Grønmo & Onstad, 2009). Of course we need to continue to report on the LCM project and to assist and inspire the teachers and student teachers to do so, too. We see an inbuilt paradox in developmental research. Teachers and didacticians who participate in the studies are learning and developing themselves through their participation in learning communities. But how can we share this learning and development with others outside the community? We recognize the difficulties faced by others—teachers and researchers—to learn from what we write about the projects because the knowledge is gained *in practice*. It is the participation in the research that leads to learning and development.

SUMMARY AND CONCLUSION

At the outset when it was proposed in the project to include the longitudinal study, it was not foreseen how the study would evolve and contribute so widely in other areas of our (and our colleagues' in teacher education) work. In this chapter our goal has been to explain how the study contributed to the development of pupils' mathematical competencies through the analysis of pupils' responses, reporting to teachers, and subsequently the teachers' inquiry that was motivated. We have offered examples of how the reporting of test outcomes in workshops led to development of teaching when teachers were inspired to use items from the tests as the basis for activities in class. We have reported about how the tests were used in the teacher education program and how the student teachers were invited to work as research assistants and use items from the tests as a basis for interviews with pupils. We have reported on the work of master's students who based their different studies in data collected in the tests and how this led to their development as teachers or teacher educators. Also, we have offered insight into how doctoral students used data from the longitudinal tests as part of their studies and by this work complement our knowledge about what teachers think in relation to tasks used in the tests and to pupils' attitudes and beliefs in relation to mathematics.

We have also noted follow-up studies where researchers have been able to pose fruitful questions starting from item responses, which have led to further investigation and theory building. In fact almost all of the participants in the learning community of didacticians have been involved in work related to the longitudinal study in some way or other. All of these aspects of the Study have added to our knowledge about mathematics teaching and learning in schools and mathematics teaching development. Thus, what from the outset seemed to be a rather conventional pencil-and-paper test instrument has opened opportunities for different kinds of developmental activities building on the pupils' responses to test items.

NOTES

1. The LCM project was supported by a grant from The Research Council of Norway, project number 157949/S20.
2. http://www.uia.no/no/portaler/om_universitetet/teknologi_og_realfag/matematiske_fag/mastmat5/masteroppgaver
3. Programme for International Student Assessment, Organisation for Economic Co-operation and Development, 2004

4. This work is reported by Kislenko (2007a,b) and Kislenko, Grevholm, and Lepik (2007)
5. Thompson (1991) identifies three levels (0, 1, and 2) of teachers' conceptions of mathematics teaching based on five categories (the nature of mathematics, learning mathematics, the content of teaching mathematics, teachers and students' roles, and evidence of students' knowledge of mathematics).
6. In fact two students joined the study, the other being Odd Helge Mjellem Tonheim. Tonheim also made an important contribution, but here we only report details from Espeland's study.
7. Reference is to Brekke, 1995.

ACKNOWLEDGEMENT

The authors wish to thank Hildegunn Espeland, Per Sigurd Hundeland, and Kirsti Kislenko for constructive contributions to the writing of this chapter. We also thank the master's students for the use of their dissertations as basis for the chapter and the student teachers for use of their reports. The Study was part of the research projects LCM (project number 157949/S20) and ICTML (project number 161955), both funded by the Research Council of Norway.

REFERENCES

Andreassen, I. S. (2005). *Innsikt i elevers kompetanser som vises i skriftlige matematikktester* [Insights into pupils' competencies as shown in written mathematics tests]. Masteroppgave. Kristiansand, Norway: Høgskolen i Agder.

Alseth, B., Breiteig, T., & Brekke, G. (2003). *Endringer og utvikling ved R97 som bakgrunn for videre planlegging og justering: Matematikkfaget som kasus* [Changes and development with R97 as a background for further planning and revision: The case of Mathematics]. Notodden, Norway: Telemarksforsking.

Bishop, A. J. (1998). Research, effectiveness, and the practitioners' world. In A. Sierpinska & J. Kilpatrick (Eds.), *Mathematics education as a research domain: A search for identity* (pp. 33–45). Dordrecht, The Netherlands: Kluwer.

Borgen, J. S., Gunnes, H., & Sivertsen, G. (2007). *Kartlegging av norsk utdanningsforskning* [Survey of Norwegian education research]. NIFU STEP Studier av innovasjon, forskning og utdanning [Studies of innovation, research and education] RAPPORT 32/2007. Retrieved August 8, 2009 from http://www.nifustep.no/norsk/publikasjoner/kartlegging_av_norsk_utdanningsforskning

Bjuland, R., Cestari, M. L., & Borgersen, H. E. (2007). *Pupils' mathematical reasoning expressed through gesture and discourse: A case-study from a sixth-grade lesson.* Paper presented at CERME 5, Larnaca, Cyprus.

Bjuland, R., Cestari, M. L., & Borgersen, H. E. (2008). The interplay between gesture and discourse as mediating devices in collaborative mathematical reasoning: A multimodal approach. *Mathematical Thinking and Learning, 10*(3), 271–292.

Breiteig, T., & Brekke, G. (1993). Ansvar for egen læring i matematikk [Responsibility for own learning in mathematics]. Report in *Program for utdanningsforskning,* Norges forskningsråd. Oslo, Norway: Norges forskningsråd.

Breiteig, T., & Grevholm, B. (2006). The transition from arithmetic to algebra: To reason, explain, argue, generalize and justify. In J. Novotná, H. Moraová, M. Krátká, & N. Stehlíková (Eds.), *Proceedings of the 30th Conference of the International Group for the Psychology of Mathematics Education,* Vol 2, (pp. 225–232). Prague: PME.

Brekke, G. (1995). *Introduksjon til diagnostisk undervisning i matematikk.* Oslo, Norway: Læringssenteret, Utdanningsdirektoratet. Retrieved August 6, 2009 from http://bestilling.utdanningsdirektoratet.no/Bestillingstorg/PDF/59447_KAR_MAT_007_innmat.pdf.

Burghes, D. (1999). The Kassel Project: An international longitudinal comparative project in secondary mathematics. In B. Jaworski & D. Phillips (Eds.), *Comparing standards internationally* (pp 135–153). Oxford: Symposium Books.

Czarnocha, B., Dubinsky, E., Prabhu, V., & Vidakovic, D. (1999). One theoretical perspective in undergraduate mathematics education research. In O. Zaslavsky (Ed.), *Proceedings of the 23rd conference of the International Group for the Psychology of Mathematics Education* (vol 1.) (pp. 95–110). Haifa, Israel: Israel Institute of Technology.

Det Kongelige Kunnskapsdepartement (2009). St.meld. nr.11 (2008–2009) *Læreren: Rollen og utdanningen* [Government white paper 11 (2008–2009) The teacher: The role and the education]. Oslo, Norway: Avdeling for offentlige publikasjoner. Retrieved August 10, 2009 from http://www.regjeringen.no/nb/dep/kd/dok/regpubl/stmeld/2008-2009/stmeld-nr-11-2008-2009-.html?id=544920

Espeland, H. (2006). *Elevers kunnskaper i tall og algebra: Hva har de med seg fra 9. og 11. trinn?* [Pupils' knowledge in number and algebra. What do they bring from year 9 and 11?]. Masteroppgave. Kristiansand, Norway: Høgskolen i Agder.

Gjone, G. (1997). *Rettleiing til funksjonar.* [Guide to functions]. Oslo: Nasjonalt læremiddelsenter, Utdanningsdirektoratet. Retrieved June 7 2010 from http://bestilling.utdanningsdirektoratet.no/Bestillingstorg/product.asp?catalog_name=LS&category_name=&product_id=KAR-MAT-018

Goodchild, S., & Grevholm, B. (2009). An exploratory study of mathematics test results: What is the gender effect? *International Journal in Science and Mathematics Education, Vol. 9*(2), 161–182.

Grevholm, B. (2004a). A research based vision of mathematics teacher education. In M. Lepik et al (Eds.), *Teaching mathematics: Retrospective and perspectives. Proceedings from Tallin conference, May 24–25, 2003,* (pp. 25–29). Tallin, Estonia: Tallin University.

Grevholm, B. (2004b). What does it mean for mathematics teacher education to be research based? In R. Strässer, G. Brandell, B. Grevholm, & O. Helenius (Eds.), *Educating for the future. Proceedings of an international symposium on mathematics teacher education* (pp. 119–134). Stockholm: The Royal Swedish Academy of Sciences.

Grevholm, B. (2004c). Mathematics worth knowing for a prospective teacher. In B. Clarke, D. Clarke, G. Emanuelsson. B. Johansson, F. K. Lester, D. V. Lambdin, A. Wallby, & K. Wallby (Eds.), *International perspectives on learning and teach-*

ing mathematics. (pp. 519–536). Göteborg, Sweden: National Center for Mathematics Education.

Grevholm, B. (2006). Matematikdidaktikens möjligheter i en forskningsbaserad lärarutbildning [The opportunities for didactics of mathematics in a research based teacher education]. In S. Ongstad (Ed.), *Fag og didaktikk i lærerutdanning. Kunnskap i grenseland* (pp. 183–206). Oslo: Universitetsforlaget.

Grevholm, B. (2007). What is quality in mathematics teaching and learning? In C. Bergsten & B. Grevholm (Eds.), *Developing and researching quality in mathematics teaching and learning. Proceedings of the Fifth Mathematics Education Research Conference, Madif5, in Malmö,* January 24–25, 2006, (pp. 17–30). Linköping: SMDFs skriftserie.

Grevholm, B., Berg, C., & Johnsen, V. (2006). Student teachers' participation in a research project in mathematics education. In E. Abel, R. Kudzma, M. Lepik, T. Lepmann, J. Mencis, M. M. Ivanov. & M. Tamm (Eds.), *Teaching mathematics: Retrospective and perspectives, 7th international conference May 12-13, 2006* (pp. 61–68). Tartu, Estonia: Tartu Ulikool.

Grevholm, B., Berg, C., & Johnsen, V. (2007). Forskning under lärarpraktik i matematik. [Research during school practice in teacher education]. In O. Eskilsson & A. Redfors (Eds.), *Ämnesdidaktik ur ett nationellt och internationellt perspektiv. Rikskonferensen i ämnesdidaktik 2006* (pp. 121–132). Kristianstad, Sweden: Kristianstad University Press.

Grønmo. L. S., & Onstad T. (Eds.). (2009). *Tegn til bedring: norske elevers prestasjoner i matematikk og naturfag i TIMSS 2007* [Signs of improvement: Norwegian pupils performance in mathematics and science in TIMSS 2007]. Oslo, Norway: Unipub forlag.

Hinna, K. (1996). *Elevers kunnskaper og fremgang i matematikk. En toårig undersøkelse av 300 norske ungdomsskoleelever.* [Pupils' knowledge and progress in mathematics. A two-year investigation of 300 Norwegian secondary school pupils]. Hovedoppgave i matematikkdidaktikk. Kristiansand, Norway: Høgskolen i Agder.

Hundeland, P. S., Grevholm, B. & Breiteig, T. (2005). Læreres oppfatninger om matematikkundervisning [Teachers' conceptions of mathematics teaching]. In I. Stedøy (Ed.), *Vurdering i matematikk–Hvorfor og hvordan? Fra småskole til voksenopplæring. Konferanserapport No. 3, 2005* (pp. 59–70). Trondheim, Norway: Nasjonalt Senter for Matematikk i Opplæringen.

Jaworski, B. (2004). Grappling with complexity: Co-learning in inquiry communities in mathematics teaching development. In M. J. Høines & A. B. Fuglestad (Eds.), *Proceedings of the 28th Conference of the International Group for the Psychology of Mathematics Education* (Vol. 1, pp. 17–32). Bergen, Norway: Bergen University College.

Kislenko, K. (2007a). Structuring students' beliefs in mathematics: A Norwegian case. In K. Hoskonen & M. S. Hannula (Eds.), *Current state of research on mathematical beliefs XII. Proceedings of the MAVI-12 Workshop, May 25–28, 2006* (pp. 47–57). Helsinki, Finland: University of Helsinki. Department of Applied Sciences of Education. Research Report.

Kislenko, K. (2007b). A study of Norwegian students' beliefs in mathematics. In B. Jaworski, A. B. Fuglestad, R. Bjuland, T. Breiteig, & B. Grevholm (Eds.), *Læringsfellesskap i matematikk Learning Communities in Mathematics* (pp. 215–227). Bergen, Norway: Caspar Forlag.

Kislenko, K., Grevholm, B., & Lepik, M. (2007). "Mathematics is important but boring": Students' beliefs and attitudes towards mathematics. In C. Bergsten, B. Grevholm, H. Måsøval, & F. Rønning (Eds.), *Relating practice and research in mathematics education. Proceedings of NORMA05, Fourth Nordic Conference on Mathematics Education* (pp. 349–360). Trondheim, Norway: Tapir Academic Press.

LCM (2003). Learning Communities in Mathematics: Enhancing knowledge of mathematics, and of mathematics learning and teaching through a collaborative inquiry model. Original proposal to The Research Council of Norway in the Kunnskap, utdanning and Læring (Knowledge, education and learning) programme. Retrieved on August 6, 2009 from http://fag.hia.no/lcm/papers/KUL%20febr%2003%20Research%20proposal.pdf

Log, E. L. (2007). *Achievement of 9th and 11th grade students in algebra and numbers: research based on data from the KUL–LCM project.* Masteroppgave i matematikkdidaktikk. Høgskolen i Agder, Kristiansand.

Niss, M. (2004). The Danish KOM-project and possible consequences for teacher education. In R. Strässer, G. Brandell, B. Grevholm, & O. Helenius (Eds.), *Educating for the future. Proceedings of an international symposium on mathematics teacher education.* (pp. 179–190). Stockholm, Sweden: The Royal Swedish Academy of Sciences.

Organisation for Economic Co-operation and Development. (2004). *Learning for tomorrow's world: First results from PISA 2003.* Paris: OECD Publications. Retrieved April 4, 2006, from http://www.pisa.oecd.org.htm

Pehkonen, E. (2003). Lærere og elevers oppfatninger som en skjult faktor i matematikkundervisningen. [Teachers' and pupils' conceptions as a hidden factor in mathematics teaching]. In B. Grevholm (Ed.), *Matematikk for skolen* [Mathematics for school] (pp. 154–181). Bergen, Norway: Fagbokforlaget.

Pring, R. (2004). *Philosophy of educational research.* (2nd ed.). London: Continuum.

Rammeplan for allmennlærerutdanningen. (2003). [Curriculum for teacher education]. Oslo: Utdannings-og forskningsdepartementet. Retrieved May 18 2009 from http://www.regjeringen.no/upload/kilde/kd/pla/2006/0002/ddd/pdfv/175666-rammeplan_2003_allmennlaererutd.pdf

Shell Centre for Mathematical Education. (1985). *The language of functions and graphs.* Nottingham, UK: Nottingham University.

Tesaker, E. M. (2008). *Minoritetsspråklige elever og matematikk. En komparativ studie av skriftlige tester på 9. og 11. trinn.* [Minority language pupils and mathematics. A comparative study of written tests in grade 9 and 11]. Master's thesis, University of Agder, Kristiansand, Norway.

Thompson, A. G. (1991). The development of teachers' conception of mathematics teaching. In R. G. Underhill (Ed.), *Proceedings of the PME-NA 13, volume 2* (pp. 8–14). Blacksburg, VA: Virginia Tech. Retrieved August 8, 2009 from http://www.eric.ed.gov/ERICDocs/data/ericdocs2sql/content_storage_01/0000019b/80/13/00/96.pdf

Tonheim, O. H. M. (2006). *Innsikt i elevar si utvikling i matematikk over eit år. Bruk av skriftlege testar.* [Insights into pupils' development in mathematics over a year. Use of written tests]. Masteroppgåve. Kristiansand, Norway: Høgskolen i Agder.

Wenger, E. (1998). *Communities of practice: Learning, meaning, and identity.* Cambridge, UK: Cambridge University Press.

9

AN OVERVIEW OF THE DEVELOPMENT OF RESEARCH ON COLLABORATIVE PROBLEM SOLVING IN MATHEMATICS AT THE UNIVERSITY OF AGDER

Hans Erik Borgersen[1]
Maria Luiza Cestari
UNIVERSITY OF AGDER

Raymond Bjuland
UNIVERSITY OF STAVANGER

Abstract: Research on collaborative problem solving at the University of Agder has developed over more than 35 years. In this chapter we present the background for this development and show how three phases of methodological approaches have resulted in substantial research findings. The first phase focuses on students' written group reports, describing the procedures used during their problem-solving processes. The second phase brings a shift in the methodological perspective by focusing on a dialogical approach to analyzing communication and cognition. More specifically, this approach has been used to analyze the dynamics of conversation during pupils'/students' collaborative problem solving in small groups. More recently, the third phase focuses on a multimodal approach, which takes into account different ways to express pupils'/students' mathematical reasoning. By analyzing videos from mathematics lessons and working groups, we have started to pay attention to the modalities discourse and gesture, and the interplay between them.

Keywords: problem solving, mathematics, small groups, dialogical and multimodal analysis

INTRODUCTION

This chapter aims to present an overview of the development of collaborative problem-solving research carried out at the University of Agder (UiA) from 1972 to the present. To do so, we need to go back to the establishment of a new type of a decentralized university-like institution, Agder Regional University College, or simply Agder College (AC) [Agder Distriktshøgskole-ADH], in Kristiansand in 1969. The region had already a teacher training college, Kristiansand College of Education (KCE) [Kristiansand Lærerhøgskole-KLH], established in 1839, which

The First Sourcebook on Nordic Research in Mathematics Education, pages 139–153

was responsible for educating teachers for work in elementary and lower secondary schools. From the very beginning, AC started to educate lower and upper secondary mathematics teachers. So, these two institutions represented the two main traditions in teacher education offered by the teacher training colleges and the universities in Norway at the time. Since 1972, the departments of mathematics at AC and KCE started to cooperate on building mathematics study programs for teachers. The aim was to combine the best in the two traditions represented by AC and KCE, emphasizing both subject matter and didactics of mathematics.

Agder College and Kristiansand College of Education were formally united in a merging process of several colleges under the name of Agder University College (AUC) in 1994, and became a fully accredited university, the University of Agder in 2007. The institution, represented by the department of mathematics, was accredited to offer a master's program in 1993 and a doctoral program in Mathematics Education in 2002 (see the chapter by Breiteig and Goodchild in this volume). From being five mathematicians in total who started cooperating in 1972, the department has about thirty faculty members at the time of writing this chapter; almost twenty of these are researchers and teachers in the different programs in mathematics teacher education.

In this chapter we present the introduction of small-group teaching in the 1970s, collaborative open-ended problem solving in the 1980s, the introduction of new methodologies—dialogical approaches to analysing communication and cognition—in the 1990s, and examples of collaborative problem-solving research throughout the whole period. We include research that has contributed to the production of masters- and doctoral dissertations, and, towards the end of the chapter, we present some of our recent work.

SMALL-GROUP TEACHING AND LEARNING

When Agder College was established, a revision of traditional university teaching methods was encouraged, among other things, by introducing collaborative work in small groups alongside ordinary lecturing. This way of organizing the teaching, *small-group teaching*, developed mainly through the students' initiative. The college provided suitable rooms to meet the students' need for discussion groups to solve problems on their own as a supplement to large classes in statistics and mathematics. In a short time, this became the most frequent way the courses at AC were taught (Dahl, 1994, 1995), including the courses created through cooperation between AC and KCE.

In 1977, the mathematics departments at AC and KCE were asked by the Norwegian Mathematics Council (NMC) to report on mathematics as a subject for teaching with emphasis on mathematics didactics and its place in mathematics education. The report (Bekken, Borgersen, & Breiteig, 1978) advocated integration of subject matter and mathematics didactics [matematikk fagmetodikk] to include the methods and ways of thinking in mathematics through problem solving, modelling, and the historical and cultural development of mathematics. Polya's recommendations for teaching mathematics through problem-solving activities were emphasized (Polya, 1945, 1954, 1962, 1965), and the report offered a list of references to international literature on teaching and learning mathematics. A closer cooperation between all responsible parties in mathematics education, across faculties and institutions, was also recommended. The NMC report stimulated further cooperation and effort to improve teaching at both AC and KCE, and it opened opportunities for starting research on teaching and learning of mathematics at the college/university level (and later at other school levels). One such focus was to document the small-group teaching in mathematics described above as a combination of lecturing and small-group work (Dahl, 1994, 1995).

The problems used in small-group teaching are mostly regular textbook problems linked to specific topics taught in the lesson. In mathematics, new problems are given each week and the students have one week to work individually with the problems and the related theory before meeting their co-students in the small groups. The students are responsible for structuring the activities in the groups: how they organize their work, which problems they choose, and how to engage in the work. The teachers are available for discussing with the students if called upon, and most teachers visit the groups to get feedback about the students' understanding of the mathematical concepts involved in the problem-solving process.

This way of teaching is based on a collective student-teacher commitment to improve teaching and learning within the given resource frames (budgets, physical space, etc.), and it has developed to be part of the institution's teaching–learning identity. The emphasis on collaborative learning is built on the positive experience students and teachers have had with this way of teaching. It is interesting to look back and see the development in light of the expansions of the dominant theoretical perspectives on teaching and learning over the last forty years, from a cognitive Piagetian perspective to a collaborative Vygotskian or socio-cultural one. The small-group teaching at UiA developed at first through practice and later as a result of the influence and awareness of theoretical perspectives. Davidson, Reynolds, and Rogers (2001) cite UiA, among other institutions, when they observe that this is today one of several examples of large-scale implementation of cooperative learning in undergraduate mathematics.

The research presented in this chapter can be conceived in three phases related to different theoretical and methodological perspectives: First, the focus is on written group reports describing the procedures used during the problem-solving processes. In the second phase the focus is on spoken language used in interaction, when the communication produced during the discussions was recorded. At this point a dialogical approach to analyzing communication and cognition has been introduced. In the third phase, we (Bjuland, Cestari, & Borgersen, 2008b) take into account different ways to express mathematical reasoning (multimodal approach). The main focus is on the modalities discourse and gesture used to express meaning making. It is important to make clear that we consider these two modalities not only as *expressive* but also as *constitutive* of mathematical reasoning.

FIRST PHASE: COLLABORATIVE OPEN-ENDED PROBLEM SOLVING IN GEOMETRY

Initiating in 1986, a supplement to the small-group teaching described above, called *collaborative open-ended problem solving in geometry*, has been implemented in a project within a geometry course (Borgersen, 1994, 2004). In this, students were given open problems from Euclidean geometry. They worked in small collaborative groups over a month, without teacher intervention. The project aimed at trying to meet two basic challenges:

> One is to have students experience what mathematics is (that is to give a true picture of mathematics and of doing mathematics) including to experience mathematics as a process giving full respect to the students' way of thinking and need for time.... The other challenge is to create a learning environment of acceptance, safety and trust, which encourages the students to take responsibility for their own learning, for their common learning environment and for them to use their full potential in the learning process. (Borgersen, 1994, p. 7)

The reports from 1986 to 1994 constituted the data for the first research study at UiA on students' small-group collaborative problem solving in geometry, in which the focus was on how and why the project was initiated, how it developed, and what came out of it (Borgersen, 1994).

The project started out and it developed through cooperation with professors Paul J. Kelly and Julian Weissglass at the University of California, Santa Barbara. It was based on the experiences with small-group teaching at UiA and on the philosophy expressed in Polya's and his followers' work on problem solving (Lester, 1994; Polya, 1945, 1954, 1962, 1965; Schoenfeld, 1985, 1992; Silver, 1985; Stanic & Kilpatrick, 1989). The research literature on cooperation in small groups by Johnson and Johnson (1990), Davidson (1990a, 1990b), and Weissglass (1993) became central in the development of the project. For example, Borgersen (1994) offers a problem-solving model in seven steps as an extension of Polya's four-stage model, and the emphasis on the students' process writing, proof writing, and self-evaluation can be related to the aspects of complex intellectual activity involved in problem solving identified by Schoenfeld (1985, 1992). The focus on processing during the project period (both in the groups and in plenum) was inspired by Julian Weissglass. The selection and formulation of geometry problems were collected from different sources (journals, books, and generous colleagues). Paul J. Kelly contributed with many problems, and some were developed through research carried out by Borgersen (1984a, 1984b, 1987, 1993).

In preparing for the project, the students were introduced to Borgersen's (1994) problem-solving model and given recommendations about working together in small groups similar to the "basic principles for cooperative learning" as later articulated by Johnson and Johnson (1990). They were also given a short introduction to the computer program Cabri as a tool for geometry activity. The students made their own decisions about how to organize their meetings, how to cooperate, and how to present their final reports.

The students normally worked in groups of five, which they formed without teacher intervention. The projects lasted for four weeks, with groups meeting at least once a week. In the group reports, the students were expected to set out the schedule for their meetings and their way of organizing the group work (*log*). They were expected to write about their problem-solving process (*process writing*) and to write down and provide an argument for their solutions (*proof writing*). They were also expected to evaluate their own group work and the project in general (*self-evaluation*). The analysis of the group reports was based on the text written by the students about their problem-solving process, their solutions, their processing of ideas, strategies and feelings, and their evaluation of the group work. The teacher of the geometry course combined the roles of teaching and researching within the project.

Borgersen (1994) showed that collaborative open-ended problem solving in small groups is one way of meeting the two challenges mentioned in the quotation above. It confirmed that it is possible for university students to experience the whole process of doing mathematics, from carrying out experiments and making conjectures via attempts at proving the conjectures to generalizing and formulating new problems. The study showed that elementary geometry is a rich resource for finding appropriate problems, and that it is possible to achieve interesting results even though students have limited mathematical background and experience. The study also confirmed that it is possible to create learning environments where students cooperate and give each other support to go through the process together. There is a great need for encouragement and support in the process of doing mathematics. The emphasis on processing and self-evaluation is claimed to have stimulated the students' motivation and understanding. Collaborative small-group work is highly recommended by the students as a way of learning mathematics.

A later study (Borgersen, 2004) confirmed these findings even when the geometry course from 1994 to 1996 changed from being a medium-sized course with about thirty students within the same study program to a large course of nearly one hundred students from three different study programs. Some more detailed findings have been identified, such as: properly demanding and thoughtfully formulated problems are important for creating interest and inspiration. Explicit questions in the formulation of a problem are important for motivating the students to try out all steps in problem solving, also generalizing or formulating new problems. Cooperation in small groups made it possible for students to be actively involved in the tasks. They emphasized the importance of all group members to be active and willing to help each other. Students experienced the importance of good humor. Cooperative small-group work helped the students get started on problem solving and helped them to persevere if they ran into difficulties. The students experienced how ideas developed gradually and how cooperative work contributed to deeper understanding of mathematical concepts. Borgersen is thus led to observe that "it appeared as though having students cooperating in small groups focusing on the process of problem solving, promoted reflection on their own learning. It might help the students to become aware of their cognitive resources and affective reactions and to develop their metacognitive abilities and social skills" (2004, p. 63).

Inspired by the small-group teaching approach and the collaborative open-ended problem-solving research, several theses from the very beginning of the MA program in Mathematics Education at AUC focused on different aspects of collaborative problem solving in small groups at different grade levels in primary and secondary school and in teacher education (see below). This trend was developed by the introduction of new methodological approaches to the analysis of students' conversations when solving problems.

SECOND PHASE: DIALOGICAL APPROACH—OPENING A NEW METHODOLOGICAL PERSPECTIVE

While Polya (1945) based his work mainly on his own knowledge and reflections, Schoenfeld (1985) went well beyond that and based his analysis on the so-called *speak-aloud protocols.* His data consisted of recordings of informants speaking out their reasoning to a researcher as they tried to solve given problems. In the collaborative problem-solving research at AUC before 1994, the data consisted of students' *written group reports* from their work in the small groups. One important element that was not exposed in those reports was the dynamic of conversation that took place during the resolution of problems. So, when we started to ask about the communication among members of these groups, about the meaning-making process, we started to use an analytical approach that focused on the dynamic of the participants' contributions. Looking at transcriptions of recordings of conversations between students during the small-group sessions, new research question arose: How do pupils/students collaborate and express their solutions in conversations? What kinds of difficulties do pupils/students meet when solving mathematical problems? What kinds of mistakes are produced in the process of solution? Concerning the third question, the aim is indeed to investigate all possible ways the pupils/students use to solve the problem proposed, including those that do not correspond to an adequate solution to the problem.

In order to answer these questions, we needed to use methodological approaches that could take into account the sequentiality of the verbal contributions of the group members when elaborating the solution for the proposed problems. As a consequence, we also started to use new procedures to collect data, audio-recording the conversations and analyzing the transcripts according to the *dialogical approach* to communication and cognition (Bjuland, 2002, 2005; Cestari, 1997, 1998, 2000; Linell, 1995, 1998).

The main dialogical principles are, according to Linell (1998), the *sequentiality*, *joint construction*, and *act–activity interdependency*. Concerning the *sequential organization* of discourse, "each constituent action, contribution or sequence, gets significant parts of its meaning from the position in a sequence. That means that one can never fully understand an utterance or an extract, if taken out of the sequence which provides its context" (Linell, 2008, p. 85). Focusing on this principle we can identify, for example, monitoring moves (Bjuland, 2002; Lambdin, 1993) that are responsible for the introduction of elements that provoke developments in the meaning making of a problem posed and advances for possible solutions. The *joint construction* principle is based on the assumption that actions and meanings are jointly constructed by the participants of a group, for example, by collaborative completion. Those who understand the ways to solve a problem, when expressing their reasoning, open up possible ways of understanding for the other participants. The contributions can be considered as building blocks in the process of meaning making and solution. The *act–activity interdependence* is a principle emerging in task-oriented activities. Those activities are, in general, part of a special genre of discourse, in our case, part of the mathematical discourse. In fact it is the mathematical activity that enhances particular kinds of speech acts.

According to Linell (1995), in the sequence of the conversation, the unit of analysis is defined as an *inter-act*. In this way, each contribution is considered as a *unit*, defined in relation to the *previous* contribution as well as to the *subsequent* one. This kind of analysis allows us to take into account one of the basic properties of dialogue: the sequentiality, which is one of the main characteristics of human interaction, constituted by speech acts socially co-produced. The use of this analytical approach to interpret conversations in mathematics classrooms has been very fruitful. For example, some central features identified in the communication between teacher and pupils in lessons are the *conflict of perspectives* in the process of attributing meaning to the mathematical task. Another feature is the *co-production of mistakes*, which occurs when the question of the teacher is complex, containing many referents, provoking in this way, different answers, some of which are inadequate (Cestari, 1997). As a consequence, there has been more concern in relation to the teachers' formulation of questions instead of blaming the pupils for their wrong answers.

The dialogical approach to communication and cognition used to analyze classroom conversations has also been implemented in the master's program at UiA through the MERG project (Mathematics Education Research Group) developed in the course of Teaching and Learning of Mathematics. It started in 1995 and has been developed regularly; it is now in its fifteenth edition. Ninety research papers have been written based on the analysis of classroom observation, teacher and pupil interviews, and tasks solved by the pupils. A description of this project has been given in Cestari, (2000), and presented at ICME 10 (Cestari, 2008).

In addition to the studies related to problem solving and communication in mathematics classrooms, the dialogical approach has also been used in master's studies to analyze data within research topics related to mathematics used in everyday life of enterprises, laboratory, and industry (Bradal, 1997); mathematics in house construction and measurements (Samland, 1998); learning disabilities (Daland, 2001); multicultural classrooms (Ringøen, 2001); motivational and emotional dimensions of learning (Fiskerstrand, 2005); semiotic analysis of textbooks (Askevold, 2005); activities related to solution of algebraic expressions (Vinje-Christensen, 2005); addition and subtraction (Simensen, 2006); mathematics teaching in vocational school (Martin, 2006); and multiplication in fifth grade (Engeland, 2007).

THIRD PHASE: MULTIMODAL APPROACH

The *multimodal approach* is based particularly on the identification of pupils' reasoning, focusing on the two modalities *discourse* and *gesture*. Both modalities are important semiotic resources in the constitution and expression of mathematical knowledge. We started to pay attention to them as well to the interplay between them when analyzing videos from mathematics lessons and working groups. We have also observed the modality of *writing* (written solutions on a worksheet) while pupils have worked on a mathematical task.

Inspired by the pioneering research of David Efron in the 1940s, who focused on spontaneous gestures accompanying speech, McNeill (1992) was concerned with narrative discourse, conjecturing that "speech and gestures are elements of a single integrated process of utterance formation in which there is a synthesis of opposite modes of thought" (p. 35). Gestures can be defined as "movements of the arms and hands... closely synchronized with the flow of speech" (p. 11).

McNeill (1992) classified gestures in four major categories: iconic, metaphoric, deictic, and beat. Gestures are classified as *iconic* if they present imagery, and they refer to a concrete object, event, or action that is expressed simultaneously in speech. *Metaphoric gestures* also present imagery similar to iconic, but they "present an image of an abstract concept, such as knowledge, language itself etc." (p. 80). These gestures are therefore more complex since they depict two aspects: the concrete entity or action presented in the gesture and the concept of, for instance, a question or an answer. McNeill emphasizes that mathematicians' gestures are often metaphoric since the gestures cover single mathematical concepts throughout the discourse. *Deictic gestures* are defined as "pointing movements, which are prototypically performed with the pointing finger" (op. cit., p. 80). Vygotsky (1978) describes how the development of pointing can be identified from the perspective of a child's activity. Initially, this gesture is only an unsuccessful attempt to grasp an object, and the grasping movement designates forthcoming activity. The situation changes dramatically when the mother discovers the child's movement since the pointing becomes a gesture for another person and in this way, a communicative tool. This becomes a true gesture when it manifests all the functions of pointing. According to Roth (2001), there is a development from concrete pointing, appearing when the child is about ten months old, to abstract pointing, indicating a non-present referent, when the child is about ten years old. This example, related to the development of pointing from child's activity, illustrates how gestures are embedded into the role of intentionality of actions. The *beat gesture* does not present a discernible meaning, but it indexes the word or phrase it accompanies as being significant. According to Roth (op. cit., p. 370), "beats are simple, non-pictorial gestures that can include the up and down flick of a hand, or the tapping motions used to emphasize certain utterances."

McNeill (2005) has developed his gesture framework from categories to dimensions by proposing that a given gesture may have aspects of more than one category. He therefore refers to dimensions of *iconicity, metaphoricity, deixis, "temporal highlighting"* (beat), and *social interactivity.* This suggests that a gesture could be loaded by different dimensions. In one of our recent studies (Bjuland et al., 2008b), we have taken into account the developed gesture framework proposed by McNeill (2005). We have focused on sixth-grade pupils' reasoning through gesture and speech while working on a task that involves movement between three different semiotic representations (*figure, diagram,* and *written text*). In our study there is a particular focus on the gesture dimension of *deixis* since the pupils point to different objects within these semiotic representations. They point to the figure of four persons, they point to coordinate axes or points in the diagram, or they use pointing to move between the semiotic representations. We are aware from McNeill's framework that there are many dimensions displayed when the pupils move between or within semiotic representations. However, we have restricted our analyses to the *deixis* since this dimension has been frequently identified in different moments of the pupils' activity with the solution of the mathematical task.

According to Radford (2003), the notion of *semiotic means of objectification* can be conceived as "objects, tools, linguistic devices, and signs that individuals intentionally use in social meaning-making processes to achieve a stable form of awareness, to make apparent their intentions, and to carry out their actions to attain the goal of their activities" (p. 41). Radford claims that gestures are part of the *semiotic means of objectification* that allow students to objectify knowledge. In addition to gestures, these *means* include "signs, graphs, formulas, tables, drawings, words, calculators, rules, and so on" (Radford, 2005, p. 143).

It is important to mention that we use the concept of semiotic means of objectification from Radford (2005) not only in the sense of gesturing as a form of externalization of knowledge, but also in the sense of *constitution of meaning* as a process of internalization of semiotics resources. For example, when children are producing or reproducing gestures and discourses, they are incorporating new meanings to their repertoire of knowledge, and at the same time, expressing how they understand it. This means that semiotics has two functions in learning processes: constitution and expression of mathematical knowledge.

MA AND PHD THESES ON COLLABORATIVE PROBLEM SOLVING

The two topics of communication in mathematics classrooms (Cestari, 1997, 1998) and collaborative problem solving in geometry (Borgersen, 1994, 2004) have been the basis for several MA and PhD studies focusing on problem solving and cooperation in small groups at different grade levels. The method used in these studies was based mainly on microanalysis of the dialogue in small groups. The dialogues were audio taped and transcribed for analysis. Since 2002, most studies have used both audio and video recordings. Thus, the analysis introduced and developed by Cestari (1997, 1998, 2000) has been extended into a multimodal approach (Bjuland et al., 2008b).

MA Studies

In the following we will present a sample of master's theses ordered chronologically—first a short listing of seven theses with their overall focus and then each study in more detail.

Bjuland (1997) studied problem-solving processes in geometry as part of teacher education when preservice teacher students cooperated in small groups. The study was done in the first semester at a teacher training college. Mydland (1998) studied how pupils took responsibility for their own learning by focusing on problem solving in geometry in small groups in a ninth-grade class. Eriksen (1999) and Damsgaard (1999) studied open problems and communication in small groups in a first-year class of upper secondary school. Carlsen (2002) studied pupils' sharing of responsibility and development of pupils' understanding of a mathematical concept (scalar product) through the dialogue in a small group of high-ability final year upper secondary school pupils. In a multigrade classroom (classes with mixed ages), Wrånes (2003) studied problem-solving strategies used in a sixth- and seventh-grade class solving problems in geometry. Kibsgaard (2005) studied final year senior high school pupils' questions and explanations in their collaborative small-group work on series.

Bjuland's study was designed according to the collaborative open-ended problem-solving project at UiA. The others were designed more like the small-group teaching at UiA.

Both in design and content some of these studies built on each other. For example, Wrånes followed Bjuland's recommendation focusing on heuristic strategies used by pupils aged 11–13 solving open-ended geometry problems collaboratively, and one of the problems she used was created by Damsgaard and Eriksen.

Bjuland (1997, 1998, 1999) studied one small group of pre-service teacher students' problem-solving processes to capture elements of their mathematical thinking and the development of their mathematical language and collaborative skills. This was done as part of a larger project on collaborative open-ended problem solving in geometry with all students in their first semester at a teacher training college. The students had one month of training in the working approaches that would be used in the project before they started their group collaboration on solving problems without teacher intervention (four meetings during three weeks). The design of the project and the problems used were inspired by the open-ended geometry project at UiA described earlier in this chapter.

Bjuland claims that the students developed their mathematical thinking, their social skills, and their ability to communicate about mathematics during the project period. The group discussions showed how one group member generated an idea or a question that was further developed by another student. In this zig-zag communication, the students discussed different concepts. Even if they did not succeed in solving all the problems, it seemed as if they were gradually developing their understanding of mathematical concepts such as perpendicular, distance from a point to a line, cyclic quadrilateral, and similar quadrilaterals. The students were uncertain about what a proof really is, and they did not spend much time on generalizing and posing their own problems. The solution process varied between analysis, implementation, and reflection. Examples of problem-solving strategies that could be identified in the discussion were looking back on the solution process (in five different ways), using analogous problems, and going to extremes. The students also developed their metacognitive skills, and they reflected on their actions at the end of the group meetings and evaluated their own control behavior.

Mydland (1998) investigated to what degree a model of learning (emphasizing holistic learning goals, pupils' freedom of choice, autonomy, metacognition, and self-assessment) can contribute to lower secondary pupils' increased responsibility for their own learning of mathematics. The model is based on differentiated and pupil-centred teaching material, designed for self-study and small-group problem solving in geometry. The pupils, the teacher, and the parents were all in focus in the project.

The pupils revealed great involvement. They especially appreciated the problem solving in small groups and the opportunity they had to find out solutions on their own. Mydland claims that the pupils became more conscious of their own learning; their log books showed a development towards enhanced metacognition, and several pupils considered this way of working as contributing more to their understanding than regular teaching in mathematics. The way they organized and carried out their work indicated responsibility for their own learning. The teacher was inspired by the pupils' engagement, especially the pupils' curiosity and reflective questions. The pupils' logs gave valuable information about their thinking. The lessons were dynamic, and the teacher had more time for each individual pupil. Parents were also positive about the project. Several of them found that their children were more involved in mathematics than usual, especially by telling more and discussing the subject at home.

Damsgaard (1999) and Eriksen (1999) explored cooperative learning by studying pupil dialogues in small groups working on open-ended problems. The problems were developed in cooperation with the teacher in the beginning of the first year of an upper secondary class, covering the first chapters of the textbook in use. They collected and transcribed the data together, but had different focuses for their studies.

Damsgaard focused on mutuality in the pupils' mathematical discourse. His study developed an analytical method with a main purpose to capture the sequentiality of contributions. From the dialogical approach to conversation, it became possible to look closely at the principles of how meaning is developed in social contexts. The role of other members influenced the individual's learning process. Some contributions were given attention, whereas others were not. To explain this observation, Damsgaard focused on group norms that developed through conversation. These norms showed that a group member's status and the shaping of contributions were related to the other group members' attention to the contribution.

Eriksen tried to find characteristics of an open task that produce a dialogue that supports the construction of knowledge by the individual pupil. He found that characteristics of open tasks that positively affect the pupils' discourse seem to be that: the tasks require some sort of result, the tasks have alternative solutions, several methods can lead to solutions of the tasks, start conditions are based on the pupils' mathematical understanding, the entry of the task is not too difficult, and the topic of the task is new to the pupils. If the pupils have previous knowledge of the topic, it seems as if the task should approach the topic in a new way.

Carlsen (2002, 2005) studied cooperative learning in one small group of high-ability final year upper secondary pupils. The teacher, who was experienced in this approach, had organized the classroom very much like our small-group teaching at UiA as a blend of lectures and work with exercises and problems in homogeneous small groups (in separate rooms). Carlsen wanted to investigate whether cooperative group work can be a tool for sharing responsibility and for differentiation in teaching mathematics. In addition, he wanted to investigate the pupils' mathematical development and conceptual understanding of the scalar product.

The pupils developed skills related to cooperative learning of mathematics. Argumentation and vocabulary used became more accurate and clear, and some pupils acquired mathematical reasoning and strategy skills. The group members revealed a fairly clear development when it came to conceptual understanding of scalar product, and they took responsibility for each other's social and mathematical involvement in the group. Carlsen suggests the group could have emphasized more certain group norms, such as being on time, how to organize the work (repetition of theory, preparation at home, and choice of coordinator), being focused (use of the blackboard), and processing their work. The pupils took different complementary roles during the group work—roles such as initiator, coordinator, and explainer. Carlsen claims that by using small-group teaching, the pupils' effort and commitment increase in such a way that they take responsibility for their own and other's learning. In addition, this way of working may enhance more adapted teaching[2] to establish well-founded conceptual understanding.

Wrånes (2003) studied which problem-solving strategies pupils used and which strategies were most successful for solving open-ended geometry problems collaboratively. The data were collected twice in a combined grade 6 and 7 classroom of five pupils in the first year and four of the same pupils one year later. Wrånes was both class teacher and researcher. As teacher she tried to intervene as little as possible in order to see how the group was able to work out the problem solving on its own.

In the analysis, the problem-solving process was divided into three phases: sense making, trying to find a solution, and looking back. The pupils used a wide and varied range of problem-solving strategies even if they were not familiar with either problem solving or group collaboration. The strategies that were used appeared to be intuitive and based on common sense. About half of the strategies appeared to be important for the pupils' understanding and for initiating progress in the problem-solving process. Wrånes claims that the pupils in her study reveal a rich potential and a strong basis on which to develop problem-solving skills and group collaborative skills. To be able to monitor their own problem-solving process, the pupils have to be aware of their own behavior. They have to develop their metacognitive skills to assess their own choice of strategies and to improve their ability to communicate their own mathematical thinking. The findings appeared to demonstrate that collaborative problem solving can open for otherwise untapped resources in the pupils.

Kibsgaard (2005) conducted a video analysis of pupils' questions and explanations expressed orally (in the dialogue) and non-orally (by gestures and gazes) to their peers in a small group, working on problems in arithmetic and geometric series. The pupils were in their final year of upper secondary school. They were offered small-group teaching. In addition to the teacher's expositions to the full class, they mostly worked on their own in separate rooms, and the teacher circulated from group to group.

The analysis was divided into three parts: identification and categorization of questions/utterances that provoked an explanation, identification and categorization of the mathematical explanations, and a search for relations between questions and explanations. Kibsgaard found that the pupils used a great variety of both questions/utterances and explanations. She categorized eleven types of questions/utterances that provoked an explanation and nine types of explanations. About 60% of the questions/utterances had a direction towards a goal, 40% were more open. The pupils questioned and explained by use of language, gesture, and gaze. They showed willingness and ability to explain such that their co-pupils could understand, they combined different types of explanations, and it seems that the pupils understood each other.

PhD Studies

Two of the master's students have developed and extended their studies into PhD theses (Bjuland, 2002; Carlsen, 2008). Carlsen's work is presented separately in this volume (Chapter 15).

Bjuland (2002) analyzed the dialogues of student teachers working collaboratively in small groups in a problem-solving context. In his analysis of the conversation from two groups of students with different mathematical backgrounds working on two problems of classical geometry without teacher intervention, he identified the students' heuristic strategies and observed how elements of reasoning were expressed in their mathematical discussions. Reasoning was defined as five interrelated processes of mathematical thinking categorized as *sense-making, conjecturing, convincing, reflecting,* and *generalizing.* These categories were derived from three models of problem solving (Borgersen, 1994; Mason, Burton, & Stacey, 1982; Mason & Davis, 1991; Polya, 1945), and from the contributions of Schoenfeld's (1985, 1992) framework for analyzing problem-solving behavior. The strategies of *posing questions, monitoring* and *looking back,* and *visualizing* were the most prominent, used by both groups in order to succeed in solving the problems. Bjuland particularly identified different categories of questions used in the solution process: *looking back, request for agreement, why, monitoring, conjecturing,* and *generalizing.* Many of these questions stimulated the establishment of a conjecturing atmosphere in the group dialogues. The study revealed that it is possible for students with limited mathematical backgrounds to succeed in solving unfamiliar problems without teacher involvement if they are given sufficient time to work on them. Further analysis (Bjuland, 2007b) of the student teachers' dialogues from their collaborative small-group work on the two classical problems in geometry (Bjuland, 2002) revealed four main categories of strategies: Visualizing strategy (VS), Monitoring strategy (MS), Questioning strategy (QS), and Logical strategy (LS). These analyses have demonstrated the importance of identifying student teachers' heuristic strategies used in group dialogues while solving mathematical problems.

The question of teacher involvement in group dialogues was further developed in Bjuland (2007a). An episode of the dialogue from the group of students with limited mathematical backgrounds revealed that the students transferred the special condition for similar triangles to similar quadrilaterals by just focusing on corresponding angles. This observation challenges the argumentation of working in groups without teacher involvement. When students make an overgeneralization of a special mathematical concept and draw an incorrect conclusion, "should they remain in this situation without being helped by a teacher?" (Bjuland, 2007a, p. 185). It is natural that a teacher, monitoring a group, could correct partial conceptions and stimulate students' solution process. On the other hand, it is important that students are not too often interrupted in their discussion. A monitoring teacher could observe and identify students' difficulties in the solution process. These observations could then be brought into an elaborated discussion with the whole class when the collaborative small-group work is finished.

Bjuland (2004) reported research that focused on student teachers' reflections on their collaborative problem-solving process. The students in one group with limited mathematical background realized that they were making the problems too difficult, and they perceived differences in their participation in the solution process. In order to provoke participation, the students reflected on how they could introduce particular ideas or give hints before presenting a solution. The reflections from the students' difficulties in the solution process led to elaborated reflections in order to improve their cooperation on the solution process. Particular group reflections were raised to a meta-level in which they reflected on their learning process. The students reflected on their preparation as future teachers of mathematics. Based on the experience of getting stuck, the students suggested that this experience could help them to better understand when pupils have difficulties in solving problems in school.

Recently, research in the didactics of mathematics has emphasized the importance of identifying mathematically productive discourses (MPD) among students (Kieran, 2001; Ryve, 2004, 2006). The identification of heuristic strategies in a collaborative problem-solving context (Bjuland, 2002, 2007b) and students' reflections on their solution process (Bjuland, 2004) are promising attempts at finding indicators of mathematically productive discourses (MPD). Bjuland (2007c) claims that even though research has recently focused on this issue, "there is a need for a careful discussion of how MPD can be defined and identified in group discussions" (pp. 33–34). Bjuland introduced a definition of MPD, involving five criteria in which a monitoring utterance is a crucial element. The definition is used as an analytical tool to illustrate mathematically non-productive and productive discourses respectively. Even though MPD is defined and identified in group discussions, it is difficult to conclude when a discourse is productive or not. This is particularly difficult when students are working on unfamiliar problems in which they are challenged to experience the entire problem-solving process (Borgersen, 1994).

RECENT RESEARCH

From 2004 to 2007, we participated in the research and developmental project Learning Communities in Mathematics[3] (LCM), at the University of Agder (see Jaworski, this volume, Chapter 5). The project, comprising teachers from eight schools and a didactician team from UiA, aimed to "create *inquiry communities* of teachers and didacticians to both develop, and explore the development of mathematics teaching and learning" (Jaworski et al., 2007, p. 7); see also Cestari, Daland, Eriksen, and Jaworski (2005). In spring 2006, didacticians led focus group interviews with teacher teams to gain insights into how teachers reflected on the activities of the project, including workshops. These interviews also illustrated some of the teachers' perspectives on collaboration with didacticians to create an inquiry community (Bjuland & Jaworski, 2009).

Sixteen workshops took place throughout the project period, enabling the participants to collaborate on mathematical tasks and to discuss issues in didactics of mathematics. A substantial amount of video and audio data were collected from workshops, classrooms, planning meetings, and the like. Bjuland, Cestari, and Borgersen (2007) reported how one of the teachers in the project transposed ideas from workshops at UiA to her own sixth-grade classroom. More particularly, we illustrate how this teacher brought a mathematical task (we will refer to this as "the diagram task") that involves movement between three different semiotic representations (*figure*, *diagram*, and *written text*) into her classroom. *Figure* refers to the representation of four persons, *diagram* refers to four points plotted against rectangular axes representing age (vertical) and height (horizontal), and *written text* refers to the task questions. The pupils worked on the task in collaborative groups of two or three.

Researchers in mathematics education have recently focused on the importance of gestures as a function of communication and reasoning in the construction of mathematical concepts and procedures (Edwards, 2009; Núñez, 2004; Reynolds & Reeve, 2002). There has also been a growing interest in this issue among some researchers at UiA. We have identified, from two groups of pupils working on the diagram task, the pupils' gestures, their discourse strategies, and their difficulties in the solution process (Bjuland, Cestari, & Borgersen, 2007, 2008a.). These studies revealed that gesture and speech occurred simultaneously in the pupils' mathematical reasoning. The gestures, *pointing* and *sliding*, play an important role when they make the transition between the two semiotic representations, figure and diagram.

We have made more detailed and elaborated analyses of the dialogues of the same two groups of sixth-grade pupils, expressing their mathematical reasoning through speech and gesture while working on the diagram task (Bjuland et al., 2008b). Inspired by Radford (2003) and Arzarello, Ferrara, Robutti, Paola, and Sabena (2005), we used a multimodal approach in order to focus on the interplay between gesture and discourse as mediating devices in collaborative mathematical reasoning. The pupils used the gesture dimension of deixis (McNeill, 2005) exemplified through pointing gestures (*pointing, held-point, linear point-slide, circular point-slide,* and *repeated pointing*). Our study has revealed a close relationship between these pointing gestures and "the reasoning strategies: comparison of persons in the figure, coordination of two dimensions in the diagram, recapitulation and going to an extreme location" (Bjuland et al., 2008b, p. 271). The interplay between gesture and discourse mediate mathematical reasoning among the pupils. These modalities are displayed when the pupils make moves between figure and diagram. Using the analytical framework of Arzarello and colleagues (2005), the pupils' learning situation includes both a synchronic process (to make sense of the task) and a serial process "where the pupils' learning situation has been developed by their use of pointing gestures to highlight different features of the semiotic representations, from one axis to both axes, suggesting an increasingly sophisticated understanding of the task" (Bjuland et al., 2008b, p. 289).

We have, in two studies (Bjuland, Cestari, & Borgersen, 2008c, 2010), focused on the teacher's use of gesture and discourse as communicative strategies: (1) by her presentation of the diagram task, (2) in her dialogue with two girls working on the task in a small group, and (3) when summing up in full class after the small-group work. Based on the analysis of the teacher–pupil dialogue when the diagram task was presented in full class (Bjuland et al., 2008c), three of the main communicative strategies used by the teacher have been identified: First, she used the verb, *see* and the personal pronoun, *you*[4] (singular you), making individual pupils aware of any differences about the persons in the figure. Second, the teacher posed open questions (*questioning*), introducing the two variables, height and age; and third, she used gestures (four consecutive pointings in the diagram and circular sliding), making a connection between persons in the figure and points in the diagram. Related to these strategies, the teacher was contextualizing the diagram task by likening it to a familiar situation. She was also giving advice for small-group work in order to establish the problem-solving context.

In the teacher's dialogue with the two girls in their attempt at solving the task, she used the strategy of posing questions. She also made explanations in combination with pointing gestures (*linear point-slide*) in order to show three connections between figure and diagram (Bjuland et al., 2008c).

In the second study (Bjuland et al., 2010) the teacher's most used strategy in summing up the activity was questioning (who, how, why, other suggestions). Her use of gesture was more restricted than in the presentation and during the supervision. The teacher seemed to make it a priority for the pupils to articulate their solutions.

From this microanalysis, we have identified the teacher's communicative strategies by using gestures and verbal explanations, both in the introduction of "the diagram task" in full class, in dialogue with the two girls, and in summing up at the end. The teacher started to pose questions and to use gestures, then shifted to explanations in combination with gestures, and ended up by mostly posing questions, playing the role of presenter, supervisor, and coordinator, respectively. In this way, the mathematical activity defines a collaborative community interacting with different ways of knowing and being mutually attuned to each other.

FINAL REMARKS

The research reported above has given more detailed insight about "what is going on" in mathematics classrooms at different levels of Norwegian schools over recent years. To a certain extent, this knowledge has already been implemented in the school system to improve teaching and learning. In fact, many of our master's and doctoral students are today in permanent teaching positions and concerned with how to transform mathematics teaching, not only at the primary and secondary schools, but also at universities and teacher training colleges.

The knowledge about communication in classrooms and collaborative problem solving in mathematics published over the last years was based on development over a period of more than twenty years. Mathematicians affiliated with Agder College and Kristiansand College of Education were determined to integrate the best in the two traditions of teacher education in Norway at the time with the goal of improving tertiary teaching and learning. They developed small-group teaching in their own practice and, among other things, tried out projects on collaborative open-ended problem solving in geometry. This experience, competence, and knowledge have been the basis for the later research in this area.

The initial research work, which here has been labeled the first phase, was based on the analysis of students' written reports of their problem-solving processes. Through work with elementary geometry problems, the students had the opportunity to experience what it means to do mathematics, and in this way getting a grasp of mathematical ways of thinking. The students started to take more responsibility for their own learning and for cooperation with colleagues, based in a supportive and trusting atmosphere of work. From the institutional point of view, a more active learning environment has been created that today has been implemented in our academic context.

In the second phase, we described how the need to understand the constitution of the mathematical reasoning during the small-group work stimulated the introduction of the dialogical approach to analyzing communication and cognition. It became necessary to capture the students' reasoning through the dialogues and conversations about how to solve problems. This has been a clear methodological shift from the analysis of text written by the students about their problem-solving process to the focus on the dynamics of conversation in the process of meaning making. These contributions from social sciences, particularly from conversation analysis and social psychology, were the basis for this shift. The new approach has been used in many master's and doctoral theses as well as in the project MERG, illuminating the so-called hidden dimensions of mathematics lessons. New findings have been identified from those analyses: for instance, crucial problem-solving strategies, monitoring moves among participants in the groups, and the conflict of perspectives and co-production of mistakes in the communication among teachers and pupils/students. By focusing on different school levels, several master's and doctoral studies have confirmed the main findings from the first phase.

In the third phase of this developing research, we continued with the methods used earlier and added a new dimension brought by the collection of data through the use of video recordings. At this point we introduced a multimodal approach to analyze interactions, particularly focusing on the use of gestures considered as expression and constitution of mathematical reasoning. In terms of future research, new questions have been raised—for example, those related to the semiotics of written texts including signs, formulas, operations, drawings, graphs, tables, and words. So, a third modality of expression will be added to our analysis. An important issue is to keep the focus on the sequentiality of actions and activities when analyzing solutions to mathematical tasks.

Our research experiences and findings have the potential for creating a more effective pupil/student and teacher engagement, promoting learning environments of acceptance and trust. It has potential for improving pupils'/students' learning processes, for instance, by meeting the challenge of giving them sufficient time to discuss and solve problems, and for the teacher to get suitable feedback.

It is very helpful to discuss episodes from dialogues in classrooms or in small groups in courses at teacher-training colleges and in in-service training of teachers. Furthermore, we see multimodality not only being about developing a tool for researchers, but also as providing a tool for teachers. How the knowledge of multimodality might be made available and useful for teachers of mathematics is a challenge for further investigation.

The theoretical methodological approaches used in the studies here have contributed to make learning and teaching more visible and accessible for researchers, and in this way, promoting a better understanding of pupils/students and teachers doing mathematics in the classroom.

NOTES

1. Universitetet i Agder, Institutt for matematiske fag, Serviceboks 422, N-4604 Kristiansand, Norway
2. Meeting individual needs [In Norwegian: tilpasset opplæring]
3. The LCM project was supported by the Research Council of Norway (Norges Forskningsråd): Project number 157949/S20.
4. The Norwegian language makes a distinction in the second person pronoun—singular "du" and plural "dere."

REFERENCES

Arzarello, F., Ferrara, F., Robutti, O., Paola, D., & Sabena, C. (2005). Shaping a multi-dimensional analysis of signs. In H. L. Chick & J. L. Vincent (Eds.), *Proceedings of the 29th Conference of the International Group for the Psychology of Mathematics Education*, Vol. 1 (pp. 127–131). Melbourne: Melbourne University, Australia.

Askevold, G. A. (2005). *Matematikklæreboka som semiotisk verktøy. Eit studie av VGS elevar som løyser oppgåver i smågrupper: Ei dialogisk tilnærming* [The mathematics textbook as a semiotic tool. A study of upper secondary pupils' problem solving in small groups: A dialogical approach]. Unpublished master's thesis, Agder University College, Kristiansand, Norway.

Bekken, O. B., Borgersen, H. E., & Breiteig, T. (1978). *Matematikk som undervisningsfag. Rapport fra Norsk Matematikkråds arbeidsgruppe i fagmetodikk* [Mathematics as a subject for teaching. Report from Norwegian Mathematics Council's working group on mathematics education]. Skrifter 1978:2, Fagseksjon for matematikk. Kristiansand: Agder College, Norway.

Bjuland, R. (1997). *Problemløsningsprosesser i geometri. Lærerstudenters samarbeid i smågrupper. En dialogisk tilnærming* [Problem-solving processes in geometry. Teacher students'collaboration in small groups: A dialogical approach]. Unpublished master's thesis, Agder University College, Kristiansand, Norway.

Bjuland, R. (1998). Lærerstudenters matematiske tenkning og utvikling i en sosial kontekst. Problemløsning i smågrupper [Teacher students' mathematical thinking and development in a social context. Problem solving in small groups]. *Nordisk matematikkdidaktikk, NOMAD, 6*(2), 41–66.

Bjuland, R. (1999). Problem-solving processes in geometry. Teacher students' cooperation in small groups: A dialogical approach. In O. Zaslavsky (Ed.), *Proceedings of the 23rd PME* (Vol. 2, pp. 121–128). Haifa, Israel: Israel Institute of Technology.

Bjuland, R. (2002). *Problem solving in geometry. Reasoning processes of student teachers working in small groups: A dialogical approach.* Unpublished doctoral dissertation, University of Bergen, Norway.

Bjuland, R. (2004). Student teachers' reflections on their learning process through collaborative problem solving in geometry. *Educational Studies in Mathematics, 55*, 199–225.

Bjuland, R. (2005). Dialogiske tilnærminger i klasserommet: Utvikling av matematiske begreper gjennom lærer-elev dialog i klasserommet eller gjennom elevsamarbeid i smågrupper [Dialogical approaches in the classrrom. Development of mathematical concepts through pupil-teacher dialogue in the classroom or through pupil collaboration in small groups]. In K. Hirsch (red.), *Ulike perspektiv på læring* [Different perspectives on learning] (pp. 67–76). Trondheim, Norway: Senter for voksenopplæring.

Bjuland, R. (2007a). Collaborative problem solving in geometry without teacher intervention. In C. Bergsten, B. Grevholm, H. S. Måsøval, & F. Rønning (Eds.), *Relating practice and research in mathematics education.* Proceedings of NORMA 05, Fourth Nordic Conference on Mathematics Education (pp. 177–188). Trondheim: Tapir Academic Press, Norway.

Bjuland, R. (2007b). Adult students' reasoning in geometry: Teaching mathematics through collaborative problem solving in teacher education. *The Montana Mathematics Enthusiast, 4*(1), 1–30.

Bjuland, R. (2007c). Mathematically productive discourses among student teachers. *Nordic Studies in Mathematics Education, 12*(2), 33–55.

Bjuland, R., Cestari, M. L., & Borgersen, H. E. (2007). En lærers bruk av problemløsningsverksted: Fra LCM til klasserommet [A teacher's use of problem-solving workshop: From LCM to the classroom]. In B. Jaworski, A. B. Fuglestad, R. Bjuland, T. Breiteig, S. Goodchild, & B. Grevholm (Eds.), *Læringsfelleskap i Matematikk/Learning Communities in Mathematics* (pp. 151–160). Bergen, Norway: Caspar.

Bjuland, R., Cestari, M. L., & Borgersen, H. E. (2008a). Pupils' mathematical reasoning expressed through gesture and discourse: A case study from a sixth-grade lesson. In D. Pitta-Pantazi & G. Philippou (Eds.), *Proceedings of the Fifth Congress of the European Society for Research in Mathematics Education* (CERME 5, Larnaca, Cyprus, pp. 1129–1139). Department of Education, University of Cyprus.

Bjuland, R., Borgersen, H. E., & Cestari, M. L. (2008b). The interplay between gesture and discourse as mediating devices in collaborative mathematical reasoning. A multimodal approach. *Mathematical Thinking and Learning, 10*(3), 271–292.

Bjuland, R., Cestari, M. L., & Borgersen, H. E. (2008c). A teacher's use of gesture and discourse as communicative strategies in the presentation of a mathematical task. In O. Figueras, J. L. Cortina, S. Alatorre, T. Rojano & A. Sepúlveda (Eds.), *Proceedings of the 32th Conference of the International Group for the Psychology of Mathematics Education,* (PME 32). Vol. 2 (pp. 185–192). Morelia, Mexico: Universidad Michoacana de san Nicolás de Hidalgo.

Bjuland, R., Cestari, M. L., & Borgersen, H. E. (2010). A teachers' use of gesture and discourse as communicative strategies in concluding a mathematical task. In V. Durand-Guerrier, S. Soury-Lavergne, & F. Arzarello (Eds.), *Proceedings of the Sixth Congress of the European Society for Research in Mathematics Education* (CERME 6, Lyon, France, pp. 884–893). Universitè de Lyon 1.

Bjuland, R. & Jaworski, B. (2009). Teachers' perspectives on collaboration with didacticians to create an inquiry community. Research in Mathematics Education, 11(1), 21–38.

Borgersen, H. E. (1984a). *Discovery and investigation—starting with a problem in plane geometry.* Skrifter 1984:1, Department of Mathematics. Kristiansand, Norway: Agder College.

Borgersen, H. E. (1984b). *The maximal T-triangle of a given triangle.* Skrifter 1984:2, Department of Mathematics. Kristiansand, Norway: Agder College.

Borgersen, H. E. (1987). Synspunkter på geometri i lærerutdanning [Geometry in teacher education]. *Nordisk Matematisk Tidskrift, 37*(3), 57–72.

Borgersen, H. E. (1993). *A family of associated triangles and Fermat points.* ADH-serien no. 58. Kristiansand, Norway: Agder College.

Borgersen, H.E. (1994). Open ended problem solving in geometry. *Nordic Studies in Mathematics Education, NOMAD, 2*(2), 6–35.

Borgersen, H. E. (2004). Open ended problem solving in geometry revisited. *Nordic Studies in Mathematics Education, NOMAD, 9*(3), 35–66.

Bradal, R. (1997). *Matematikk i arbeidslivet. Implikasjoner for skoleverket* [Mathematics in the work place. Implication for schools]. Unpublished master's thesis, Agder University College, Kristiansand, Norway.

Carlsen, M. (2002). *Smågruppeundervisning i matematikk. Ei dialogisk tilnærming til samarbeid i ei gruppe av faglig sterke elever i videregående skole* [Small group teaching in mathematics. A dialogical approach to collaboration in a small group of high achieving upper secondary pupils]. Unpublished master's thesis. Kristiansand, Norway: Agder University College.

Carlsen, M. (2005). Conceptual understanding of the dot product. *Nordic Studies in Mathematics Education, NOMAD, 10*(3/4), 3–28.

Carlsen, M. (2008). *Appropriating mathematical tools through problem solving in collaborative small-group settings.* Unpublished doctoral dissertation, University of Agder, Kristiansand, Norway.

Cestari, M. L. (1997). *Communication in mathematics classroom. A dialogical approach.* Unpublished doctoral dissertation, University of Oslo, Norway.

Cestari, M. L. (1998). Teacher–student communication in traditional and constructivist approaches to teaching. In M. B. Bussi, A. Sierpinska & H. Steinbring (Eds.), *Language and communication in the mathematics classroom* (pp. 155–166). Reston, VA: National Council for Teachers of Mathematics, NCTM.

Cestari, M. L. (2000). Opening the door of the mathematics classroom: Students wondering and reflecting. In B. K. Selvik, C. Kirfel, & M. Johnsen Høines (Eds.), *Matematikk i samfunnsmessig, historisk og kulturelt perspektiv* [Mathematics in a social, historical, and cultural perspective] (pp. 187–193). Skriftserien. Rapport nr. 2/2000. Bergen, Norway: Bergen College.

Cestari, M. L. (2008). From the mathematics classrooms: Dialogues and tasks under analysis. Returning to teacher autonomy. In M. Niss (Ed.), *Proceedings of the ICME 10; Regular Lecture* (CD, pp. 1–10), Roskilde: Roskilde University, Denmark.

Cestari, M. L., Daland, E., Eriksen, S., & Jaworski, B. (2005). Working in a developmental research paradigm: The role of didactician/researcher working with teachers to promote inquiry practices in developing mathematics learning and teaching. In M. Bosch (Ed.), *Proceedings of the Fourth Congress of the European Society for Research in Mathematics Education* (CERME 4, Saint Feliu de Guixóls, Spain) (pp. 1348–1358). Universitatat Ramon Llull.

Dahl, H. (1994). *Teaching small groups. Experiences at Agder College 1973–1993.* ADH serien no. 72. Kristiansand, Norway: Agder College.

Dahl, H. (1995). Teaching small groups at Agder College. Teaching and learning undergraduate mathematics. *Newsletter no. 2 from the Mathematical Association, UK.*

Daland, E. (2001). *Dialogisk tilnærming til kommunikasjon og matematikkvansker. En vei til bedre forståelse og økt engasjemaent i læring og undervisning av matematikk* [Dialogical approach to communication and pupils' difficulties in mathematics.

A way to better understand and to improve engagement in learning and teaching of mathematics]. Unpublished master's thesis, Agder University College, Kristiansand, Norway.

Damsgaard, K. (1999). *Læring av matematikk i sosiale sammenhenger* [Learning of mathematics in social connections]. Unpublished master's thesis, Agder University College, Kristiansand, Norway.

Davidson, N. A. (Ed.). (1990a). *Cooperative learning in mathematics. A handbook for teachers.* Menlo Park, CA: Addison-Wesley.

Davidson, N. A. (1990b). Small-group cooperative learning in mathematics. In T. J. Cooney & C. J. Hirsch (Eds.), *Teaching and learning in the 1990s* (pp. 52–61), 1990 Yearbook. Reston, VA: NCTM.

Davidson, N. A, Reynolds, B. E. & Rogers, E.C. (2001). Introduction to cooperative learning in undergraduate mathematics. In E. C. Rogers, B. E. Reynolds, N. A Davidson, & A. D. Thomas (Eds.), *Cooperative learning in undergraduate mathematics: Issues that matter and strategies that work* (pp. 1–11). Washington, DC: The Mathematical Association of America.

Edwards, L. (2009). Gestures and conceptual integration in mathematical talk. *Educational Studies in Mathematics, 70,* 127–141.

Engeland, T. (2007). Elevers strategier for å løse multiplikasjon i 5. trinn—Lesson Study for lærere: En case studie [Pupils' strategies for solving multiplication in 5th. Grade—Lesson Study for teachers: A case-study]. Unpublished master's thesis, University of Agder, Kristiansand, Norway.

Eriksen, S. (1999). *Åpne oppgaver i skolen. Elevers dialog og læring i smågruppe-undervisning* [Open problems in school. Pupils' dialogue and learning in small group teaching]. Unpublished master's thesis, Agder University College, Kristiansand, Norway.

Fiskerstrand, A. (2005). *Motivasjon for aktivitet i matematikk. Om hvordan ulike oppgaver motiverer elever i matematikk, og kva som påvirker elevers motivasjon* [Motivation for activity in mathematics. How different problems motivate pupils in mathematics, and what influence pupils' motivation]. Unpublished master's thesis, Agder University College, Kristiansand, Norway.

Jaworski, B. (2007). Introducing LCM—Learning communities in mathematics. In B. Jaworski, A. B. Fuglestad, R. Bjuland, T. Breiteig, S. Goodchild, & B. Grevholm (Eds.), *Læringsfelleskap i matematikk/Learning Communities in Mathematics* (pp. 13–25). Bergen, Norway: Caspar.

Jaworski, B., Fuglestad, A. B., Bjuland, R., Breiteig, T., Goodchild, S., & Grevholm, B. (Eds.). (2007). *Læringsfelleskap i matematikk/Learning Communities in Mathematics.* Bergen, Norway: Caspar.

Johnson, D. W. & Johnson, R. T. (1990). Using cooperative learning in math. In N. Davidson (Ed.), *Cooperative learning in mathematics. A handbook for teachers* (pp. 103–125). Menlo Park, CA: Addison Wesley.

Kibsgaard, E. (2005). *Elevers matematiske spørsmål og forklaringer. Smågruppesamarbeid om rekkeoppgaver i videregående skole: En dialogisk tilnærming* [Pupils' mathematical questions and explanations. Collaborative small group work on series in upper secondary school: A dialogical approach]. Unpublished master's thesis, University of Agder, Kristiansand, Norway.

Kieran, C. (2001). The mathematical discourse of 13-year-old partnered problem solving and its relation to the mathematics that emerges. *Educational Studies in Mathematics, 46,* 187–228.

Lambdin, D. V. (1993). Monitoring moves and roles in cooperative mathematical problem solving. *Focus on Learning Problems in Mathematics, 15,* 48–64.

Lester, F. K. (1994). Musings about mathematical problem-solving research 1970–1994. *Journal for Research in Mathematics Education, 25*(16), 660–675.

Linell, P. (1995). Troubles with mutualities: Towards a dialogical theory of misunderstanding and miscommunication. In I. Markova, C. F. Graumann, & K. Foppa (Eds.), *Mutualities in dialogue* (pp. 176–213). Cambridge: Cambridge University Press.

Linell, P. (1998). *Approaching dialogue. Talk, interaction and contexts in dialogical perspectives.* Amsterdam: John Benjamins.

Martin, S. P. K. (2006). *Matematikk i yrkesutdanningen. En studie av lærerens valg av mål for matematikkundervisningen ved en yrkesfaglig videregående skole* [Mathematics in vocational education. A study of the teacher's choice of goals for the mathematics teaching in an upper secondary vocational school]. Unpublished master's thesis, Agder University College, Kristiansand, Norway.

Mason, J., Burton, L., & Stacey, K. (1982). *Thinking mathematically.* London: Addison-Wesley.

Mason, J. & Davis, J. (1991). *Fostering and sustaining mathematics thinking through problem solving.* Victoria: Deakin University Press.

McNeill, D. (1992). *Hand and mind: What gestures reveal about thought.* Chicago: Chicago University Press.

McNeill, D. (2005). *Gesture and thought.* Chicago: Chicago University Press.

Mydland, O. (1998). *Ansvar for egen læring i matematikk. En studie av en helhetlig læringsmodell brukt på ungdomsskoleelevers arbeid i geometri* [Responsibility for own learning in mathematics. A study of a holistic learning model used by lower secondary pupils working with geometry]. Unpublished master's thesis, Agder University College, Kristiansand, Norway.

Núñez, R. (2004). Do real numbers really move? Language, thought, and gesture: The embodied cognitive foundations of mathematics. In F. Iida et al. (Eds.), *Embodied artificial intelligence* (pp. 54–73). New York: Springer.

Polya, G. (1945). *How to solve it.* Princeton, NJ: Princeton University Press.

Polya. G. (1954). *Mathematics and plausible reasoning. I. Induction and analogy in mathematics. II. Patterns of plausible inference.* (Vols. 1, 2). Princeton: Princeton University Press.

Polya. G. (1962). *Mathematical discovery: On understanding, learning and teaching problem solving.* (Vol. 1). New York: John Wiley and Sons.

Polya. G. (1965). *Mathematical discovery: On understanding, learning and teaching problem solving.* (Vol. 2). New York: John Wiley and Sons.

Radford, L. (2003). Gestures, speech and the sprouting of signs. *Mathematical Thinking and Learning, 5*(1), 37–70.

Radford, L. (2005). Why do gestures matter? Gestures as semiotic means of objectification. In H. L. Chick & J. L. Vincent (Eds.), *Proceedings of the 29th Conference of the International Group for the Psychology of Mathematics Education,* Vol. 1 (pp. 143–145). Melbourne: Melbourne University.

Reynolds, F. J. & Reeve, R. A. (2002). Gesture in collaborative mathematics problem-solving. *Journal of Mathematical Behavior, 20,* 447–460.

Ringøen, S. I. (2001). *Matematikk—et språklig og kulturelt fag? En empirisk undersøkelse av funksjonelle tospråklige elever i videregående skole* [Mathematics—a language and a cultural subject? An empirical study of functional bilingual pupils in upper secondary school]. Unpublished master's thesis, Agder University College, Kristiansand, Norway.

Roth, W. M. (2001). Gestures: Their role in teaching and learning. *Review of Educational Research, 71*(3), 365–392.

Ryve, A. (2004). Can collaborative concept mapping create mathematically productive discourses? *Educational Studies in Mathematics, 56*, 157–177.

Ryve, A. (2006). Making explicit the analysis of students' mathematical discourses—Revisiting a newly developed methodological framework. *Educational Studies in Mathematics, 62*, 191–209.

Samland, G. M. (1998). *Matematikk i huset. Et prosjekt om målestokk, omkrins og areal* [Mathematics in the house. A project about scale, perimeter and area]. Unpublished master's thesis, Agder University College, Kristiansand, Norway.

Schoenfeld, A. H. (1985). *Mathematical problem solving.* Orlando, FL: Academic Press.

Schoenfeld, A. H. (1992). Learning to think mathematically: Problem solving, metacognition and sense making in mathematics. In D. A. Grouws (Ed.), *Handbook for research on mathematics teaching and learning* (pp. 334–370). New York: Macmillan.

Silver, E. A. (Ed.). (1985). *Teaching and learning mathematical problem solving: Multiple research perspectives.* Hillsdale, NJ: Lawrence Erlbaum Associates.

Simensen, A. M. (2006). *Utforskende aktiviteter som utgangspunkt for addisjon og subtraksjon i en andreklasse: En kasusstudie* [Explorative activities as a starting point for addition and subtraction in a second degree class: A case study]. Unpublished master's thesis, Agder University College, Kristiansand, Norway.

Stanic, G. M. A. & Kilpatrick, J. (1989). Historical perspectives on problem solving in mathematics curriculum. In R. T. Charles & E. A. Silver (Eds), *The teaching and assessing of mathematical problem solving* (pp. 1–22). Reston, VA: NCTM.

Vinje-Christensen, P. (2005). *$5x \cdot 2x = x^{10}$. 9. klasseelevers forståelse av algebra. En dialogisk tilnærming til elevers arbeid med algebra* [$5x \cdot 2x = x^{10}$. Ninth grade pupils' understanding of algebra. A dialogical approach to pupils' work with algebra]. Unpublished master's thesis, Agder University College, Kristiansand, Norway.

Vygotsky, L. S. (1978). *Mind in society: The development of higher psychological processes.* Cambridge, MA.: Harvard University Press.

Weissglass, J. (1993). Small-group learning. *American Mathematical Monthly, 100*(7), 662– 668.

Wrånes, B. (2003). *Problemløsningsstrategier i geometri. Samarbeid mellom elever på 6. –8. klassetrinn i en fådelt skole. En dialogisk tilnærming* [Problem-solving strategies in geometry. Cooperation among pupils in grade 6–8 in a small school with multigrade classes]. Unpublished master's thesis, Agder University College, Kristiansand, Norway.

10

EDUCATIONAL REFORMS—HOW ARE THEY IMPLEMENTED?

Bodil Kleve

OSLO UNIVERSITY COLLEGE

Abstract: This chapter reports how three mathematics teachers responded to the curriculum reform L97 in Norway and reveals that although one curriculum was intended, three enacted curricula were observed. Possible constraints (*cultural factors influencing teachers' beliefs and teaching practice which may conflict with the recommendations of the curriculum*) explaining why the curriculum was not implemented as intended were identified. The constraints were found at different levels: Some resting in the teacher's beliefs about good mathematics teaching; some are identified between the teacher's beliefs about good teaching and his/her classroom practice, such as time pressure and parents' expectations; and some become most evident in the teacher's classroom practice and reveal that a mathematics teacher's vision about good mathematics teaching is not so easily transformed into the classroom practice.{\abs}

Keywords: Curriculum reform, enacted curriculum, socio-cultural constraints, mathematics teaching practice, teachers' beliefs

INTRODUCTION

This chapter reports a study of three mathematics teachers and how they responded to the curriculum reform of 1997, L97, in Norway (Kleve, 2007). In the next section about background and context, I present some issues that have been of great concern in Norway related to how Norwegian students have performed in international studies such as TIMSS and PISA. Following this, I present the curriculum L97; it was the teachers' interpretation of the L97 curriculum upon which the study focused. A theoretical analysis of L97 is also included. I include a brief overview of research literature that relates to how teachers in other countries have responded to curriculum reforms and thus relate the findings from my study of the mathematics teachers' teaching practice and how they responded to a new curriculum reform to international research more widely.

To study the mathematics teachers and their teaching practices, I used research methods commonly associated with ethnographic approaches. Obstacles, issues, and constraints influencing the teachers' teaching practices were identified and discussed and related to the three teachers' expressed beliefs about good mathematics teaching. Reflections around factors that are of decisive importance for mathematics teachers' teaching practice are presented in the final part of the chapter.

BACKGROUND AND CONTEXT

Mathematics in school and students' learning outcomes in mathematics have been widely discussed in Norway; this has included both political debate and scientific inquiry. Norwegian students have been included in large-scale international comparative stud-

The First Sourcebook on Nordic Research in Mathematics Education, pages 155–168

ies that have taken place over the last two or three decades. One of these, the Trends in International Mathematics and Science Study, TIMSS, which was carried out in 1995, revealed that Norwegian students performed relatively poorly, especially in mathematics, where performance was significantly lower than the mean of all participating countries (Lie, Brekke, & Kjærnsli, 1997). Again, in TIMSS 2003, Norwegian students performed significantly lower than the international mean. However, the group of countries that participated in 2003 was not the same as in 1995, so it is not easy to compare directly. Particular concern arises as the results from these two studies are compared because, on the same tasks, Norwegian students' performance in 2003 was lower than in 1995 (Grønmo, 2004).

Norway also participated in PISA, Programme for International Student Assessment, another international comparative study, organized by OECD in 2000 (Lie, 2001) in 2003 (Kjærnsli, Lie, Olsen, Roe, & Turmo, 2004) and in 2006 (Kjærnsli, Lie, Olsen, & Roe, 2007). The students participating in PISA were 15 years old, and thus in lower secondary school in Norway. In these studies Norwegian students performed at an average level from an OECD perspective. However, studying the results from a Nordic perspective, Norway performed significantly lower than the other Nordic countries (Finland, Denmark, Iceland, and Sweden). Comparing the Norwegian results in 2000 and 2003, the decline in absolute performance in Norway is modest, but compared to the other Nordic countries, the gap in performance is considerably higher in 2003 than in 2000 (Kjærnsli et al., 2004). Also the results from PISA 2006 show a decline from 2003 in performance in mathematics among Norwegian 15 year old students. Norwegian students perform lowest in mathematics compared to the other Nordic countries, (Kjærnsli et al., 2007). The greatest concern with regard to the results from PISA 2006 is that Norwegian students still perform as poorly in numeracy as they did in TIMSS 2003 (Grønmo, 2004; Grønmo, this volume, Chapter 4).

There is a national concern to find reasons for the poor and declining performance of Norwegian students. One suggested factor is the frequent changes of curricula in Norway that have occurred over the last five decades (see Breiteig & Goodchild, this volume, Chapter 2). Frequent shifts in the political stance of government have led to successive changes in curricula. L97, on which this chapter focuses, was operative for nine years only (ten for one class). Looking back over the history of curriculum reform in Norway, that is the shortest time for any curriculum. Stigler and Hiebert (1999) suggest that frequent reforms can make the teachers grow weary. Teachers are asked to engage in change over and over again, and nothing happens.

EDUCATIONAL REFORMS, R97 AND THE CURRICULUM L97

Reform 97 (R97) was a nationwide educational reform that took place in Norway in 1997. Part of the reform started some years earlier, in 1993 at upper secondary school level; however, I will not describe this in detail because I focus on teaching in lower secondary schools. Before I present the reform, there are two issues related to the background of R97 that need to be explained because I believe them to be essential in understanding how teachers respond to a new curriculum.

First, R97 was a "top-down reform" (Gjone, 2003), meaning that it was the outcome of political initiative rather than a professionally motivated reform sought by teachers. Hence teachers perceived it as a directive from above. Second, as noted above, in Norway there had been several changes in government within a relatively short period resulting in reforms that were initiated by one government being implemented by another. Hence, R97, which was prepared by a left-wing government, was implemented by a minority coalition government made up by center and right-wing politicians. During their time in power, the minority coalition government started their work with yet another reform, "Kunnskaps-løftet" (Knowledge Promotion), LK06 (Kunnskapsdepartementet, 2006; Utdannings- og forskningsdepartementet, 2005) and introduced national tests in Norwegian, English, and Mathematics, which were carried out in 2004 and 2005. In my work with in-service training courses with teachers, I experienced that the introduction of these national tests led teachers to concentrate more upon meeting the assessment criteria of the tests rather than upon the working methods encouraged in the curriculum.

One of the most important changes in the way education was organized to come out of Reform 97 was that pupils should begin school at the age of six instead of seven. Thus there would be 10 years of compulsory schooling instead of 9. Furthermore, as part of the more wide-ranging Reform 97, which affected the whole of the compulsory education system, a new curriculum was implemented in August 1997. This curriculum, for grades 1–10 (age six to age sixteen), is referred to as L97 (following the Norwegian title "Læreplanverket for den 10-årige grunnskolen"), and throughout this chapter I refer to the English version of the curriculum, "The Curriculum for the 10-years compulsory school in Norway" (Hagness & Veiteberg,

1999). The curriculum intended by planners and Government to be implemented in classrooms is thus L97. In the curriculum, learners in school are referred to as pupils and, to be consistent with other chapters in this book I follow that convention here despite the custom in international research in mathematics education in which it appears that the word "students" is preferred.[1]

The curriculum for the 10-years compulsory school in Norway, L97, consists of three main parts: a general part or the Core Curriculum (which applies to compulsory, upper secondary and adult education), Principles and Guidelines for compulsory education, and Subject syllabuses, under which the mathematical section belongs. In the introduction to the mathematics section, emphasis is given to the practical application of mathematics, and on the links between school mathematics and mathematics in the outside world (see Mosvold, this volume, Chapter 11). The curriculum emphasizes that the working methods chosen should ensure equal opportunities for all pupils to develop favorable attitudes that are considered to be essential for the learning of the subject. Elements such as reasoning, imagination, and experience are identified as important, and the subject syllabus stresses that pupils should be active, enterprising, and independent and that they should learn by doing, exploring, and experimenting. Mathematics should be looked upon as a science, an art, a craft, a language, and a tool, and the learning of mathematics should be looked upon as a process.

A Theoretical Analysis of L97

I interviewed the leader of the committee responsible for developing the mathematics section of the curriculum. He informed me that the committee was inspired by constructivist learning theory in its work. This is confirmed by several statements in L97 that emphasized the value of an individual approach to learning, reflecting notions of assimilation and accommodation, which are central in Piaget's development of a constructivist theory of cognition (Piaget, 1970). The curriculum encouraged rooting mathematics in practical matters and the use of play and games. Pupils' own experience was held to play an important role in the teaching–learning process, and it was asserted that pupils should build up their knowledge largely by themselves and that their experience and previous knowledge are important in the learning process. Moreover, L97 put weight on learners' own construction of mathematical concepts and on the importance of reflection and discussion in the learning process, as the following indicates: "Learners construct their own mathematical concepts. In that connection it is important to emphasize discussion and reflections. The starting point should be a meaningful situation, and tasks and problems should be realistic in order to motivate pupils" (L97, p. 167).

L97 also paid attention to the social aspect of learning and the central role of mathematics in society and social and everyday activities. The curriculum encouraged taking factors such as pupils' different social and cultural backgrounds into account when choosing examples and explanations; it was intended that mathematics should make sense for all pupils. The mathematics section also drew attention to the importance of classroom discourse; "In *a confident and constructive atmosphere* such matters [i.e. incomplete concepts, misconceptions, and occasional mistakes] are grounds for further learning" (L97, p. 167, my emphasis), and later, "mathematics teaching must at all levels provide pupils with opportunities to... work cooperatively on assignments and problems" (L97, p. 168). Collaboration, adult guidance, social practice, and the role of the teacher in the classroom were recognized as having an important role to play. According to my interpretation of L97, the teacher's role was to provide guidance and to support pupils in becoming participants in classroom activities and to provide pupils opportunities to learn. Teachers were encouraged by L97 to help, maintain, develop, and systematize the mathematical concepts pupils have formed before they start school.

Thompson (1992) suggested that a common view of what knowing mathematics entails is being skilful in performing procedures and identifying basic mathematical concepts. This is often referred to as a traditional view on mathematics. L97 contrasted with this common or traditional view and focused more on conceptual knowledge and less on procedural and factual knowledge than previous Norwegian curricula. The connections between development of procedural and conceptual knowledge, what is most important or what is a reasonable balance between them, have been discussed within research in mathematics education over the years (Hiebert & Lefevre, 1986). L97 emphasized the contradiction between rote learning and conceptual understanding in mathematics and put weight on the latter: "Pupils who have difficulties with memorizing the basic multiplication facts must nevertheless be free to proceed to concepts and tasks involving the multiplication concept. Understanding multiplication and how to use it is more vital than memorizing the tables" (L97, p. 166).

An *investigative approach* might be taken as an appropriate collective term for what L97 presented as being important in the teaching of mathematics. I base this assertion on the descriptions of investigative mathematics, reform mathematics and teaching experiments presented in the literature of mathematics education (Cobb, Wood, & Yackel, 1990; Ernest, 1998; Jaworski, 1994; Norton, McRobbie, & Cooper, 2002; Smith Senger, 1998/1999; Wood, Cobb, & Yackel, 1995). L97 stressed that pupils' own activities are of the greatest importance in their study and learning of mathematics. Pupils' activities within the compass of a collaborative investigative approach generate understanding. The curriculum thus acknowledged both the importance of pupils' active construction of knowledge and their interaction with others in a classroom.

MATHEMATICS TEACHERS' TEACHING PRACTICE

Teaching, in actual practice, has been an important subject of research in investigating how teachers have responded to curriculum reforms. This aspect of the curriculum, the enacted curriculum—that is, that which takes place in the classroom—describes what is jointly constructed by the teacher and the pupils and the teaching material (Remillard, 1999, 2005; Ross, 2003; Spielman & Lloyd, 2004; Tarr, Chávez, Reys, & Reys, 2006). The literature on studies of teaching practices in relation to curriculum reforms reports both qualitative and quantitative studies and research methods varying from large-scale surveys to small case studies. The studies describe curriculum reforms and identify obstacles, constraints, and issues such as lack of time, parents' expectations, and the school's discourse in teachers' decision making when implementing a reform. Studies also point to variation among teachers in how they respond to new reforms and implications for teachers' development.

The international TIMSS study, which is the largest international comparative study carried out in mathematics, created the background for large-scale studies about teachers' reactions to reform movements. Based on the material collected in the TIMSS' video study of mathematics lessons in Japan, Germany, and the U.S. in 1995, Stigler and Hiebert analyzed and compared mathematics teaching in the three countries. They revealed that there were significant cultural differences in teaching practices within each country (Stigler & Hiebert, 1999). When the teachers in the U.S. were asked about the reform document published by the National Council of Teachers of Mathematics, they claimed to have read it and that they used the reform ideas in their teaching. However, studies of the videos from 1995 revealed that there was great unevenness in how reform ideas were interpreted among the teachers, and little evidence was found that the teachers' classroom practices reflected the goals of the reforms (Stigler & Hiebert, 2004). This was also emphasized by Jacobs, Hiebert, Givvin, Hollingsworth, and Wearne (2006). They used a questionnaire to inquire into the practices of the U.S. teachers who participated in the TIMSS video studies in 1995 and 1999. The results from the questionnaires gave the impression that the majority of the teachers were teaching in line with the ideas in the NCTM standards (NCTM, 2000). The teachers also reported that the videotaped lessons illustrated this. However, the researchers found that classroom practice was not consistent with the Standards and that "the typical eighth-grade classroom displays teaching at odds in many respects with the recommendations" (Jacobs et al., p. 28). Thus these studies demonstrate findings that teachers' practices were different from what the teachers had reported.

In research, attention has been drawn to many constraints and issues between the curriculum as intended and the curriculum as enacted. Even when teachers have reported their agreement with the principles upon which reform is based, the actual classroom practice, the enacted curriculum, has turned out to be traditional in style[2] (Broadhead, 2001; Norton et al., 2002). Traditional beliefs and practices regarding school mathematics are challenged by reform-oriented curricula, and teachers' deeply held beliefs can serve as obstacles in implementing new reforms (Lloyd, 2002; Manouchehri & Goodman, 1998; Reys, Reys, Barnes, Beem, & Papick, 1998; Smith Senger, 1998/1999). Also, obstacles and constraints are reported as lying outside the teacher's scope of influence—for example, other colleagues, the school's discourse, and parents' expectations, perceptions, and concerns. Furthermore, the classroom context and the pupils' contributions during the lessons are of decisive importance for how the enacted curriculum turns out (Remillard, 1999; Skott, 2001a, 2004).

There have also been reported varying degrees of consistency between teachers' conceptions of mathematics and their instructional practice (Thompson, 1992). There appears to be a higher degree of consistency when teachers report traditional conceptions about mathematics and its teaching than when teachers report a more reform-oriented view. The importance of the relationship between teachers' teaching practice and teachers' conceptions about mathematics and mathematics teaching in connection with the implementation of a curriculum reform was also emphasized by Cooney (2001). He viewed teachers' teaching practices as highly influenced by their views about mathematics and mathematics teaching.

Skott (2001b) challenged much of the underlying rationale and premises lying behind research about teachers' beliefs, and he questioned research that has an implicit premise that a teacher's beliefs can serve as explanatory principles for practice. Being inclined to take more cultural factors into account, he claimed that what the teacher does in the classroom makes sense for the teacher based on the multiple motives for the present action, even if those actions may seem inconsistent to an observer. "Students' and teachers' actions do make sense…teachers cannot be inconsistent" (Skott, 2001b, pp. 6–7). Instead of assuming that there is something lying behind a teacher's practice which is called a teacher's beliefs, he looked upon the motives determining a teacher's practice not as predetermined beliefs but rather as entities emerging from the interactions with the students in the classroom. This illuminates the manner in which the complexity of the classroom plays a role in research about teachers' beliefs.

Taking Skott's (2001b) claim that "inconsistency is an observer's perspective," Leatham (2006) accounted for the problem of consistency or inconsistency between a teacher's beliefs and practice by introducing the notion of beliefs as a "sensible system." Taking this view of teachers' beliefs about teaching and learning, teachers' beliefs may be seen not as inconsistent with their actions in the classroom, but rather as systems where certain beliefs have more influence over actions than others. Leatham (2006) exemplified this perspective by accounting for how a teacher's beliefs about classroom management had more influence over her classroom practice than that of her beliefs about the effect of group work. Therefore, the pupils did not work in groups, although the teacher had expressed her beliefs about group work as an effective learning activity.

The above mentioned research provides the backdrop for my study into Norwegian teachers' practices in implementing the intended national curriculum L97. The research above both informs the methodology and provides a framework for the analysis and interpretation.

THE STUDY

One perspective from which to study a teacher's interpretations of a curriculum is what teachers *say* about the curriculum; about their teaching; about their pupils, their abilities, and how they learn; and about how they prepare their lessons. Another perspective is to observe the mathematics teacher's teaching practice in the classroom. I have looked at both as important components of a teacher's interpretation of the curriculum. Through the analysis of the data collected about what they say and what they do, together with an analysis of the relation between the two, characteristics of teaching are identified that in each case provide an example of how a teacher interprets L97. It was the interpretation of the curriculum that was the focus in my study, not the teachers themselves.

In this study I used the term *belief*, and I looked upon teachers' beliefs about teaching and learning mathematics and about L97 as cognitive constructions highly influenced by socio-cultural factors such as teachers' own experience and the school context, and also influenced by the teachers' knowledge of mathematics and about mathematics teaching. The insight I could get into teachers' beliefs was through what the teachers said and wrote and through my interpretations of what I observed in their classrooms. I did not look upon beliefs as something that can be directly observed. My conceptions about teachers' beliefs had to be inferred from what they said about what they were doing in the classroom; what they thought about their practice; what they thought was good mathematics teaching; and about the curriculum L97. A teacher's understanding of the curriculum develops over time as the teacher returns to the curriculum for guidance, clarification, or perhaps merely to refresh her/his memory. Thus there is a reflexive process between L97 and teachers' beliefs. I recognize that what can be observed about a teacher's practice is influenced by pupils' contributions and their interactions with each other and with the teacher.

It was important for me both to study teachers' beliefs about teaching and learning mathematics and also what I could observe them doing in their classrooms. Thompson (1992) observed that in order to understand teachers' practices from teachers' own perspectives, it is important to understand the beliefs with which they understand their own work. I do not accept that a teacher's beliefs and his or her practice exist in some cause–effect relationship, but rather as a reflexive process. A teacher's beliefs are not only influenced by the curriculum but also by his or her experience, and interactions in the classroom are, in turn, influenced by the teacher's beliefs. Thus a teacher's practice can both act as a reinforcement of his or her beliefs and as an incitement for change.

I adopted a qualitative research design that "embodies a view of social reality as a constantly shifting emergent property of individuals' creation" (Bryman, 2001, p. 20), because I see the classroom as a *constantly shifting emergent property of individuals' creation* rather than as an *external objective reality*.

The research focused on the practice of four teachers using a case study approach (Bassey, 1999). Although I collected data from four teachers, only data from three of the teachers have been analyzed in detail and thus reported in this chapter. Teaching mathematics is part of a complex school context. The complexity of the classroom is widely recognized; it is physically complex, with a variety of furniture sometimes in some degree of disarray, pupils often off-task, and perhaps without the intended resources and occupying irregular positions. Moreover, the large variety of actions that might take place in any one lesson leads to further complexity in making sense of both episodes and sequencing. Not only physically, but also on the mental plane, the classroom is complex and characterized by a youth culture (Cobb, 2001; Daniels, 2001; Lerman, 2000). Concurrent data-gathering methods gave me the opportunity to grasp this complex reality. I observed the teachers one lesson each week over a period of three months, and I had conversations with them before and/or after the lessons. Focus group interviews with the teachers, together with the individual conversations I had with them, provided information about how their teaching practices were related to their beliefs about teaching and learning mathematics. All of this was audio taped, transcribed, and analyzed. For the purpose of minimal disturbance when entering the classroom as a researcher, I decided not to use video recording. I also obtained information from the teachers through questionnaires, a self-estimation form, and their writings about what they looked upon as "ideal" teaching of mathematics. In the self-estimation form I drew on the same three aspects as Pehkonen and Törner (2004). They used Dionne's (1984) perspectives of mathematics (traditional, formalist, and constructivist perspectives) in terms of a toolbox metaphor—T (mathematics is a toolbox; doing mathematics means working with figures, applying rules and procedures, and using formulae); a system metaphor—S (mathematics is a formal, rigorous system; doing mathematics means providing evidence, arguing with clear and concise language and working with concrete problems); and a process metaphor—P (mathematics is a constructive process; doing mathematics means learning to think, deriving formulae, applying reality to mathematics, and working with concrete problems). The teachers were asked to distribute 30 points or "credits" corresponding to their estimation of the factors T, P, and S in which they should value their own teaching, their opinion about ideal teaching, and how they saw the these aspects reflected in L97.

Coding and Categorizing the Data

A general issue when working with coding of data is whether categories or codes are derived from the data or if they are brought to the data a priori (Wellington, 2000). I did not start my study and I did not read through the transcripts with a blank mind. Before I started the coding process, I had had conversations with the teachers, I had studied L97, and I had been sitting in the classroom. All this was part of the analytical process. Also, my own experience as a teacher and teacher educator influenced what categories I used. Some of the categories were derived from past research done by other researchers that I have studied. In analyzing the data collected, I did not look upon the data as an objective reality to be discovered, but rather as an interactive process between me and the data. Thus the codes and categories I used in the analysis emerged from the data through the way I dealt with them. I used a grounded theory approach in a sense offered by Charmaz (2000), who suggests a constructivist approach to grounded theory that "recognised that the categories, concepts, and theoretical level of an analysis emerge from the researcher's interactions within the field and questions about the data" (p. 522).

In analyzing the data from conversations and focus groups, I read through the transcripts and listened to the recordings several times to get ideas of what we were talking about, and codes and categories emerged through this work. To ensure reliability of this work I followed one of Bryman's (2001) recommendations, which was to do the coding several times, and I also imported the data into NVivo, a qualitative research program for the purpose of analyzing a large amount of qualitative data.

The recorded classroom observations provided me with a lot of data, principally in the form of audio-recordings. Processes for handling this quantity of data became an issue while working with it. To get a holistic overview, I made a table with all lessons observed. I also transcribed everything as uniform text. However, this was not easy to analyze because it did not give me the desired overview, so I organized the transcriptions in tables. Furthermore, I considered it useful to make some kind of data reduction and wrote narrative summaries of each lesson. These different ways of dealing with different sets of data can be looked upon as a possible condition of trustworthiness. I dipped in and out of the data, immersing myself in them by listening to the recordings, transcribing, using my field-notes to support the transcriptions, organizing the data in tables, and writing summaries and overviews of lessons. One of Bassey's (1999) recommendations to ensure trustworthiness is "prolonged engagement" (p. 76) in the data, which is about immersing in its issues as described above. Another recommendation is to keep going back to the raw data, which I also did.

THE TEACHERS AND THEIR TEACHING PRACTICE

The three teachers Bent, Cecilie, and David, were mathematics teachers in lower secondary school (pupils in grades 8–10). I begin by offering a general presentation of each of the teachers' practices underpinned by extracts from the data collected, before comparing their practices and explicitly pointing out constraints that I noted were influencing their mathematics teaching practices.

David

David expressed professional pride and he demonstrated a high level of professional confidence and addressed his pupils in a friendly way. Based on what he said in our conversations, he liked teaching mathematics, he looked upon himself as doing a good job, and he enjoyed the company of his pupils. The following quotation from one of our conversations (which was an answer to my question about whether my presence in the classroom would make him feel uncomfortable or disturbed) emphasizes his level of personal and professional confidence:

> No, no. But of course I am not perfect, nobody is, but I have the feeling that my pupils throughout years have been content with what I have been doing. I have received clear feedback for that. Not only what I am doing from the board as a math teacher, but also my behaviour and attitude towards the pupils. So I am quite confident that I am doing well.[3]

David has more than 30 years experience as a teacher; he said explicitly that he did not relate to L97, and this attitude characterized his teaching. With regard to the mathematical content to be taught, he relied on the textbooks and on his own experience. He expressed disappointment that algebra had been toned down in L97 and, according to what he said, he put more weight on algebra, functions, and equations than L97 suggested. Critical comments about L97 were presented in focus groups and in our conversations. During a conversation before a lesson, David showed me some transparencies with statistics that he had taken from magazines that he was going to present in class to illustrate use and misuse of statistics. Since statistics is a topic on which more weight was put in L97 than in prior curricula, I questioned this.

BK: Did you use L97 when you . . . ( )
D: (interrupting) No, I never leaf through that book
BK: You never leaf through it?
D: No. I can't stand it.
BK: But this was not in the prior curriculum, M87?
D: That's right, but I do this rather because it is fun and not because it is in L97

This emphasized the impression of his reluctance to follow L97. Even when some of his opinions about mathematics teaching and learning coincided with L97, he did not base his rationale on L97 but rather on the fact that it was reasonable from his point of view. He said he should have used more play and exploring activity—however, not because L97 encouraged that. David said that he challenged his pupils according to their abilities, but not because L97 said so.

David's teaching style might be described as transmission of knowledge, and he presented detailed explanations of how to solve mathematical exercises focusing on mastering methods, exercising skills, and procedures. In a conversation, he said: "I like being a conveying teacher. That is not politically correct now, but I think that is fun and I feel that I succeed in it and that I get the pupils to follow me." This was also emphasized in his lessons. For example, a pupil had found $x = 2$ in the set of equations:

$$x + y = 4$$
$$4x - 3y = 2$$

and did not know how to proceed further.

Pupil: What shall I do now?
D: Let me see. You have multiplied that one by four, and then you are half way through. If x is 2, now you have to find y. And that equation is easiest. And there it says $x + y = 4$, you see? And since x is two we exchange that with the number two plus y is four. What do we do to find y?
Pupil: Must divide?
D: No, that is when it says multiply, when it is nothing in between. When it says plus in between we move it over. So y is four minus two. Makes two and then you have finished.

David had a great concern for pupils' different abilities in mathematics and when presenting a new mathematical topic, he forged links with pupils' prior knowledge by asking questions about what they were expected to have learned earlier.

David thought carefully about his way of teaching, which had developed through many years of teaching experience and work with mathematics. With regard to the characterization of good teaching, he expressed a greater belief and faith in his own judgment than the recommendations of L97. His beliefs about teaching and learning mathematics were thus socioculturally rooted in his own educational background, in his own experience as a teacher, and in the school context. He had experienced that his way of teaching mathematics had worked well; he had experienced success as a teacher; his pupils performed well in examinations and according to him, pupils and parents liked his way of teaching. David appeared confident in his knowledge of mathematics, and he always had an answer ready to present when a pupil asked for help. He also strongly advised the pupils about what to have in their rulebooks[4] and he handed out photocopies he had made for them to paste in. Their use of the rulebooks in the examination had been seen to work well. Based on this there was no reason for David to consider changing his way of teaching. Why should he?

Cecilie

Cecilie had been teaching for eight years but she had worked as an engineer for many years before taking teacher education; she was the least experienced of the three teachers described here. Cecilie expressed a genuine interest for mathematics as a science and for mathematics history—an interest she shared with her pupils. She linked school mathematics to pupils' everyday life, to the real world, and to social science. Cecilie rarely used textbooks, and many topics in her teaching were not typically school mathematics. She expressed a special concern for clever[5] pupils and thought it was important to give them sufficient challenge. Unlike David, Cecilie admitted that she liked L97 and that she used it in planning her lessons. She was the only teacher who said that she actively used L97 in her teaching and that she rarely used textbooks. In the same way as David "used his own head" (David's own expression) rather than L97 in his teaching of mathematics, Cecilie used her own ideas and ideas picked from other literature rather than the textbook. She said:

> In addition [to studying L97] I read a lot. And I make notes in a separate book. And if I find something on the internet I write it down and sort it out in a file. One file is called prime numbers, another is called [something else]. If not I would never have found back the exciting things just when I needed it. And then I leaf through my notebook to see if I can use something, so it turns out being quite varied.{\ext}

Thus both Cecilie and David can be seen as teachers who had faith in what they were doing and who had made their own judgments about how to teach and on which aspects of mathematics to focus. They had constructed their own conceptions of good mathematics teaching based on their own ideas and experience. Cecilie believed that pupils learn best from exploring things themselves and that they then learn some mathematics they would not learn by only using ready-made formulae. She therefore prepared "exploring activities." However, the way it turned out in the classroom, in the enacted lessons that I observed, it was Cecilie who became *the teacher* and did the exploring, and the pupils were led through the activity by answering straightforward, manageable, closed questions. The pattern of discourse was often a typical triadic I-R-E (Mehan, 1979) as in the following sequence from a lesson for which Cecilie had prepared to explore Pythagorean Triples:

Cecilie: I hope you've got your calculators. You'll need them in doing this task.
What kind of triangle do we have there, Mikkel?

Mikkel: Right angled triangle.

Cecilie: Then we know the lengths of two sides. And now I don't use unit. We are only interested in the numbers. How can I find the third side, Leif?

Leif: You have to use Pythagoras

Cecilie: Yes. Have to use Pythagoras. Let us try to do that with this triangle. If we call this side x, Leif?

Leif: Must take $x^2 = 3.6^2 + 4.8^2$

Cecilie: Yes, let's calculate that. Three point six squared is?

Pupils: 12.96

Cecilie: Four point eight squared is?

Pupils: 23.04

Cecilie: Twenty three point o four (*she writes it on the board*). The sum of these numbers is?

Pupils: Thirty six

Cecilie: It is thirty six

Pupil: It makes six

Cecilie: Yes. Okay. It became 6 long. This was lots of calculations. If we look at the numbers here we might have simplified it. Is it like, here I have added one point two, and if I add another one point two I'll get the third side? *Is that a rule which always works?* Let us take another example. New triangle (She draws a new triangle on the board with sides 7.5 and 10). If that is 7.5 and that is 10, will that one be 12.5? Can you check if it works?

Baard: Yes.

Cecilie: That worked as well. Your exploratory task is now: *Does it always work? Does it work for any length?* Find new lengths on the smaller sides of a triangle and check if it works on any side.

Morten: But I didn't understand what you did!

After this comment from Morten, Cecilie carried on with a similar example:

> Okay, once more, if we see that the third side is unknown. We don't know that one yet, but we claim that it is 12.5, because that one is 7.5, I add 2.5, and I claim that I can add 2.5 there and get the hypotenuse. And that is easier than taking that one squared and that one squared and add and take the square root. What is 7.5 squared?

However, again Morten asserted that he still did not understand, and surrendered. I observed that he started talking to his classmates. Baard and a few other pupils in Cecilie's class were the only ones who kept following Cecilie's "exploring activity." That leads us to another significant aspect in the course of her lessons, which was that many pupils lost track through her exploration and stopped paying attention. Higher-attaining pupils followed her and contributed with comments and suggestions. This indicates that factors such as having pupils with different mathematical abilities and different interest for the subject in the same classroom and the complexity of the classroom where some pupils were very interested and captured the teacher's attention while others talked to their classmates, influenced the enactment of the lesson. Thus Cecilie's vision about doing exploring activities was not so easily translated into her classroom practice.

Bent

The third teacher, Bent, was the youngest in the study, with 10 years of teaching experience. He focused on the conceptual aspect of mathematics and explained that he wanted the pupils to be active in the learning process. He wanted to do more exploring activities, and thought this should be the case as recommended in L97. He thus expressed more uncertainty about his own teaching than the other two teachers. Contrary to Cecilie, who perceived herself able to prepare exploring activities, Bent indicated that he was not sure how to do that. In the first of our conversations he said, "I have to improve myself where exploring mathematics is concerned," and in the second, when talking about exploring activities, he said, "there I have a way to go myself." What Bent said in our last conversation when we talked about exploring activities in mathematics two months later emphasizes this:

> I think I agree more with L97 than I practice. It is a challenge to find good methods. I feel I don't have the methods to be able to do that kind of process. What methods shall I use and what kinds of tasks shall I give to get into that process?

Bent reflected more than the other two teachers on how he saw himself as not yet sufficiently accomplished as a teacher. Furthermore, he demonstrated a more inquiring attitude towards his practice than the other teachers. In addition to admitting that he did not know how to do exploring activities, Bent suggested other reasons for not responding adequately (as he saw it) to L97. He often mentioned time pressure as one such reason:

> I notice if the pupils shall be more exploring like L97 encourages, I think it will take more time. However, I believe they learn better that way. So I haven't found a good way to make it effective for them to learn better. That doesn't mean that my opinion is that teaching from the board is the best way, but if I have to spend so much time doing exploring activities, will we then make it?, I'm thinking. Possibly we will, maybe.

Another issue Bent said he had to compromise on was parents' expectations. He referred to parent–teacher meetings where parents had been saying they wanted him to teach more from the board. He had tried to argue for not teaching so much from the board by saying that some of the pupils would not understand, for some it would be too easy and that only about one third of the pupils would benefit from it. He had told the parents that he believed that pupils learn better if they find out things themselves. Despite his arguments, they still wanted him to teach more from the board, and he responded to these demands by doing it "at least for a while," he said, to "calm them down." Also, pupils wanted him to teach from the board as they kept blaming the teacher for not having taught them if it was something they did not manage. They expected him to give examples of tasks on the board before they started working on similar tasks in their own books. These views are culturally rooted and mirror a view on mathematics consistent with a transmission or absolutist view on knowledge (Reys et al.1998) or they reflect practices that have been common in mathematics teaching for generations.

One outcome of the analysis of Bent's lessons that I observed was how he dealt with a highly complex classroom with many management issues responding to demanding pupils. For example, at one point a pupil, Siv, interrupted into a dialogue between Bent and Einar. Einar had been asked to tell the whole class how to derive the formula for the surface of a prism. He had started saying that two and two sides have the same size and suggested $2lb$ for two times length times breadth, which Bent had written on the board, when Siv raised her hand and called on Bent's attention:

Bent: Is it a comment on what he is doing now?
Siv: It is about the formula.
Bent: Yes, but let him finish first.
Einar: Then I have done the same for the sides.
Bent: Have you put any sign between here?
Einar: Yes, plus. And then I have done the same so it makes two $l\,h$
Siv: Can I just ask a question?
Bent: Yes?
Siv: I would have put l multiplied by b in brackets.
Bent: Ok, brackets like that? I can take that afterwards because there are several possible ways to put brackets here.

After this episode, they spent 20 minutes on a discussion about brackets in which several pupils took part. This illuminates that Bent took pupils' contributions into account which in turn changed the direction of the lesson. Siv's contribution caused both a shift in discourse (from deriving a formula to probing pupils' thinking) and also a shift in mathematical focus.

During individual seatwork, Bent challenged and structured the pupils' thinking. These were also elements of teaching mathematics reflecting the requirements of L97, thus apparently meeting one of the challenges of the curriculum.

I noticed, through the analysis, Bent's reflections on how much time to spend on developing pupils' conceptual understanding as opposed to a method of mastering a procedure and that some pupils are happy just knowing the rule and using it. He demonstrated that he had reflected on how pupils' different abilities in mathematics influence their learning of mathematics. David also expressed the same kind of awareness with regard to the relation between pupils' abilities and working methods. There was a difference, however, in how the awareness became evident. Whereas David expressed a certainty and demonstrated an acceptance that the weak pupils would manage in the classroom while exercising procedures, but would forget later because, as he said, "they won't digest it," Bent offered a reflection on and expressed an uncertainty about how much weight to put on computational methods as opposed to relational understanding.

One Curriculum Intended, Three Enacted

Earlier in the chapter I referred to research that revealed great differences in teaching practices, both within the same nation (Stigler & Hiebert, 1999) and also within the same school (Kilpatrick, 2003; Stigler & Hiebert, 2004; Tarr et al., 2006). Several explanations have been given why teaching style both across schools and within the same school vary (Collopy, 2003; Rowan, Harrison, & Hayes, 2004). Hence the enacted curriculum has been the subject of research when investigating how teachers respond to curriculum reforms.

In my study I saw socio-cultural factors such as teachers' own educational background, their experience as teachers, and experienced or fear of time pressure, parents' and pupils' expectations, and the school context as constraints influencing both the teachers' beliefs and their teaching practice. Thus I define *constraints* in this study as *cultural factors influencing teachers' beliefs and practice that may conflict with the recommendations of the curriculum and internally with each other.*

I studied one curriculum, L97, and three teachers, and I will now discuss the differences in teaching styles among the teachers I found, and hence three different enacted curricula.

For the purpose of contrasting and comparing the three teachers, I have made an overview in Table 10.1. In the first column I quote an expression from each teacher about L97; in the next I have chosen a quotation that, according to the analysis of each teacher, characterizes his or her beliefs about teaching and learning mathematics; in the third column I present the most significant teaching strategies based on the analysis of classroom observations. This overview allows me to compare the teachers across what they said about L97, across their beliefs, and across their teaching strategies. The last column deals with "constraints" and issues of which a discussion will follow.

Based on the foregoing, I note three types of teaching in this study:

- According to David, the best way for pupils to learn mathematics is to have it well explained. The mathematical focus in his lessons was procedural and the discourse in the lessons was characterized by him showing and telling as if

TABLE 10.1

Teacher	Quote about L97	Expressed thoughts about teaching and learning mathematics	Teaching strategies observed in the classroom	Constraints and issues
David	"I never leaf through that book" (from a conversation)	"I think pupils learn best and quickest if they get it clearly and well explained" (from a Focus group conversation)	He explained by showing and telling and exercising skills and procedures	There was lack of agreement between his beliefs about good teaching and L97's recommendations
Cecilie	"I liked it [L97]", "when I start a new topic, I read L97" (from a conversation)	"pupils are supposed to explore things themselves,[ ] it becomes more exciting that way and I believe they will learn some mathematics they won't learn by learning formulae by heart" (from a conversation)	She prepared for exploring activities and linked school mathematics to everyday life and to mathematics history	Her visions about good teaching and L97's recommendations were not so easily transformed into the classroom. Mismatch between mathematical focus and pupils' abilities
Bent	"If the pupils shall be more exploring like L97 encourages, I think it will take more time. However, I believe they learn better that way" (from a conversation)	"I think pupils learn best by exploring things themselves" (from writings about ideal teaching)	He probed pupils' thinking through the way he asked questions and responded to their questions	There were constraints between his beliefs about good teaching which are in line with L97, and his classroom practice

mathematics were being transmitted from the teacher to the pupils.

- Cecilie prepared exploring activities. She expressed a belief that pupils learn best by exploring things themselves, and that they will then discover mathematics that cannot be learned from only using readymade formulae. From my perspective, the lessons turned out differently from what Cecilie (according to what she said) had intended. It was Cecilie herself who carried out the exploring activities through which the pupils were channelled by easy manageable questions.
- Bent focused on pupils' conceptual understanding. In his teaching he challenged pupils' thinking and encouraged them to see connections between different mathematical entities. He was thus "bridging" between previous and new knowledge.

Earlier in the chapter I commented on obstacles, constraints, and issues in teachers' decision making. The comments were illuminated by mathematics educational research. As an outcome of the analysis of the three teachers in my study, I was able to identify three types or levels of constraints influencing the different stages in teachers' implementation of a curriculum. Based on what the teachers said about L97 and about their own teaching related to L97, I made an informed interpretation about the beliefs each teacher had about L97. I see these expressed beliefs, which are highly influenced by socio-cultural factors as discussed earlier (the teacher's own educational background, teaching experience, and the school context), as one level of possible constraints that prevent the teacher from implementing a reform curriculum. If a teacher does not believe in the reform, if s/he does not want to teach according to it, if s/he believes that the way of teaching mathematics s/he has always done is the best way, then one cannot expect that s/he will implement the curriculum as intended. I look upon this as one type of constraint. These constraints that prevent the teachers from implementing the reform rest in the teacher's beliefs. This is the level of constraints that I found most evident in David's teaching.

The second type of possible constraint that influences the teacher in another stage is evident when the teacher expresses a wish to implement the reform. A teacher believes in the reform, s/he believes that L97's recommendations enhance pupils' possibilities for learning mathematics, but does not teach according to this to the extent s/he wishes because factors such as parents' expectations, pupils' demands, the work plan, and time pressure are constraints that prevent him or her from doing it. These constraints are lying between the teacher's beliefs and his or her teaching practice in the classroom, and they influence the extent to which the teacher teaches according to his or her beliefs. This was where I found the constraints in Bent's teaching most visible.

The third type of possible constraints is seen when the teacher believes in the reform, prepares the lessons according to it by choosing exploring activities and thus an investigative approach to teaching as L97 recommends. However, the way it turns out in the classroom becomes quite traditional. The constraints—such as having pupils with different abilities and interests in the same classroom and the fact that, in Norway, it is not lawful to group pupils according to abilities—are in the school context. The teacher's classroom practices together with the complexity of the classroom are the constraints; they are lying in the activities jointly constructed by the teacher, the pupils, and the teaching material used, in the enacted curriculum. This was most visible within Cecilie's teaching.

This "conclusion" may raise the question: If I had studied four teachers in detail, would I have seen four types of constraints? And with five teachers? I studied three teachers in detail and I have reported the constraints I found most visible for each teacher. However, there were features of all types of constraints in all three teachers' interpretation of the curriculum. If I had studied more teachers, I suggest I would have found features of all constraints in the teachers' interpretation of a curriculum, and each type of the constraints would have been more or less visible for each teacher.

CURRICULA COME, CURRICULA GO—WHAT ABOUT MATHEMATICS TEACHERS' TEACHING PRACTICE?

In Norwegian there is a saying that, applied to curricula, says: "curricula come, curricula go, but the classroom remains" [*Læreplaner kommer, læreplaner går, klasserommet består*]—thus recognizing that classroom practices remain the same despite changes in curriculum. As I indicated in the beginning of this chapter, frequent shifts in government have led to frequent changes in curricula. L97 was operative for nine years only (ten for one class). That is the shortest time for any curriculum in Norway.

As part of a wide-ranging evaluation of the reform, R97, the research project "Change and development with R97 as a basis for further planning and adjustment—the case of mathematics" suggested that the mathematical part of L97 was not implemented as intended (Alseth, Brekke, & Breiteig, 2003). Similar findings in the U.S., which reveal that features from the NCTM standards were implemented only at the margin of the teaching rather than at its core, were presented by Jacobs and colleagues (2006). I have also pointed out that pupils' performances on similar tasks (the attained curriculum) were lower after the implementation of L97 than before. The latter can be explained by Stigler and Hiebert's (1999) argument that only individual features of the curriculum have been changed and that has been "downright risky" (p. 100).

WHY IS IT IMPORTANT TO IDENTIFY CONSTRAINTS?

In this study I have pointed to possible reasons, or constraints, for why the curriculum, L97, was not implemented as intended:

- Elements of teachers' belief systems form a basis for a choice not wanting to do what the curriculum says
- There are socio-cultural issues preventing them from implementing it
- Teachers' visions are not easily translated into classroom practice.

I have learned that teachers responded differently to the curriculum L97. I found both similarities and differences in their teaching practice; they expressed different beliefs about teaching and learning mathematics, and I have identified constraints and issues influencing and, as I see it, partly preventing the teachers from implementing the curriculum. Hence, constraints—*cultural factors influencing teachers' beliefs and teaching practice that may conflict with the recommendations of the curriculum and internally with each other*—are important aspects of how teachers respond to a curriculum.

For a teacher educator with the task of in-service training, knowledge about constraints is valuable. I suggest that enhancement in a teacher's practice has a greater possibility of occurring when factors constraining the practice are identified and accompanied by a deliberate plan to address the constraints. I have suggested constraints as lying in the complexity of the classroom, in the difficulties of communicating vision about good mathematics teaching into practice; between the teacher's beliefs and practice, and in the socio-cultural environment as society's and parents' expectations and the school context. Being conscious of such factors, which to a certain extent can be dealt with, can thus open up possibilities for professional development of mathematics teacher educators' and teachers' teaching practice. However, constraints lying in the teacher's beliefs are more difficult to deal with, because only the teacher can change his or her own beliefs. Mason (2002) writes: "I cannot change others, but I can work at changing myself" (p. xii). I suggest that collaboration between teachers and with teacher educators can influence beliefs so the teachers and teacher educators can work at changing themselves.

NOTES

1. Editor's note: The convention in Norway is to refer to learners in school as "pupils" and in higher education as "students." The Norwegian chapters in this book have been written to reflect the Norwegian convention. This should also be helpful to readers as it makes clear the distinction between pupils, students in teacher education courses, and teachers.
2. Focus on factual knowledge, drill and practice
3. Here and elsewhere, teacher's words have been translated into English by the author.
4. According to the L97 reform, students should produce their own rulebook where they could write down all solved tasks, rules, formulae, methods, etc. they wished. Students may bring their own rulebooks and use them in the final exam in mathematics.
5. Cecilie used the Norwegian expression "flinke elever" which is usually translated as "clever pupils." "Flink" can also be referred to as able, skillful, or gifted. In Norwegian being "flink" does not necessarily mean hardworking. Those students getting good grades are called "flinke elever."

REFERENCES

Alseth, B., Brekke, G., & Breiteig, T. (2003). *Endringer og utvikling ved R97 som bakgrunn for videre planlegging og justering: matematikkfaget som kasus.* Notodden: Telemarksforsking.

Bassey, M. (1999). *Case study research in educational settings.* Buckingham: Open University Press.

Broadhead, P. (2001). Curriculum Change in Norway: thematic approaches, active learning and pupil cooperation—from curriculum design to classroom implementation. *Scandinavian Journal of Educational Research, 45*(1), 19–36.

Bryman, A. (2001). *Social research methods.* Oxford: Oxford University Press.

Charmaz, K. (2000). Grounded theory: Objectivist and constructivist methods. In N. K. Denzin & Y. S. Lincoln (Eds.), *Handbook of qualitative research* (2nd ed., pp. 509–535). Thousand Oaks, CA: Sage.

Cobb, P. (2001). Supporting the improvement of learning and teaching in social and institutional context. In S. Carver & D. Klahr (Eds.), *Cognition and instruction: Twenty-five years of Progress* (pp. 455–478). Mahwah, NJ: Lawrence Erlbaum Associates.

Cobb, P., Wood, T., & Yackel, E. (1990). Classrooms as learning environment for teachers and researchers. In R. B. Davis, C. A. Maher, & N. Noddings (Eds.), *Constructivist views on the teaching and learning of mathematics* (pp. 125–146). Reston, VA: National Council of Teachers of Mathematics.

Collopy, R. (2003). Curriculum materials as a professional development tool: How a mathematics textbook affected two teachers' learning. *Elementary School Journal, 103*(3), 287.

Cooney, T. J. (2001). Considering the paradoxes, perils and purposes of conceptualizing teacher development. In F.-L. Lin & T. J. Cooney (Eds.), *Making sense of mathematics teacher education* (pp. 9–31). Dordrecht, The Netherlands: Klüwer.

Daniels, H. (2001). *Vygotsky and pedagogy.* London: Routledge-Falmer.

Dionne, J. (1984, October). *The perception of mathematics among elementary school teachers.* Paper presented at the North American chapter of the International group of the psychology of mathematics education (PME), Madison, WI: University of Wisconsin.

Ernest, P. (1998). The culture of the mathematics classroom and the relation between personal and public knowledge: An epistemological perspective. In J. Voigt, F. Seeger, & U. Waschescio (Eds.), *The culture of the mathematics classroom* (pp. 245–267). Cambridge: Cambridge University Press.

Gjone, G. (2003). Læreplaner og læreplanutvikling i matematikk. In B. Grevholm (Ed.), *Matematikk for skolen* (pp. 261–287). Bergen: Fagbokforlaget.

Grønmo, L. S. (2004). *Hva i all verden har skjedd i realfagene?: Norske elevers prestasjoner i matematikk og naturfag i TIMSS 2003.* Oslo: Institutt for lærerutdanning og skoleutvikling, Universitetet i Oslo.

Hagness, R., & Veiteberg, J. (1999). *The curriculum for the 10-year compulsory school in Norway.* Oslo: Nasjonalt læremiddelsenter, Kirke- utdannings- og forskningsdepartementet, National Centre for Educational Resources.

Hiebert, J., & Lefevre, P. (1986). Conceptual and procedural knowledge in mathematics: An introductory analysis. In J. Hiebert (Ed.), *Conceptual and procedural knowledge: The case of mathematics* (pp. 1–27). Hillsdale, NJ: Erlbaum.

Jacobs, J. K., Hiebert, J., Givvin, K. B., Hollingsworth, H., & Wearne, D. (2006). Does eighth-grade mathematics teaching in the United States align with the NCTM standards? Results from the TIMSS 1995 and 1999 Video Studies. *Journal for Research in Mathematics Education, 37*(1), 5–32.

Jaworski, B. (1994). *Investigating mathematics teaching: A constructivist enquiry.* Washington, DC: Falmer Press.

Kilpatrick, J. (2003). What works? In S. L. Senk & D. R. Thompson (Eds.), *Standards-based school mathematics curricula: What are they? What do students learn?* (pp. 471–493). Mahwah, NJ: Lawrence Erlbaum Associates.

Kjærnsli, M., Lie, S., Olsen, R. V., & Roe, A. (2007). *Tid for tunge løft norske elevers kompetanse i naturfag, lesing og matematikk i PISA 2006.* Oslo: Universitetsforl.

Kjærnsli, M., Lie, S., Olsen, R. V., Roe, A., & Turmo, A. (2004). *Rett spor eller ville veier? Norske elevers prestasjoner i matematikk, naturfag og lesing i PISA 2003.* Oslo: Universitetsforlaget.

Kleve, B. (2007). *Mathematics teachers' interpretation of the curriculum reform, L97, in Norway.* Kristiansand, Norway: Høgskolen i Agder, Kristiansand.

Kunnskapsdepartementet. (2006). *Læreplanverket for Kunnskapsløftet: Midlertidig utgave.* Oslo: Utdanningsdirektoratet.

L97. The Royal Ministry of Education, Research and Church Affairs. (1996). *The Curriculum for the 10-year Compulsory School in Norway.* Oslo: National Centre for Educational Resources.

Leatham, K. (2006). Viewing mathematics teachers' beliefs as sensible systems. *Journal of Mathematics Teacher Education, 9*(1), 91–102.

Lerman, S. (2000). Getting used to mathematics: Alternative ways of speaking about becoming mathematical. *Ways of Knowing, 1*(1), 47–52.

Lie, S. (2001). *Godt rustet for framtida?: Norske 15-åringers kompetanse i lesing og realfag i et internasjonalt perspektiv.* Oslo: Institutt for lærerutdanning og skoleutvikling, Universitetet i Oslo.

Lie, S., Brekke, G., & Kjærnsli, M. (1997). *Hva i all verden skjer i realfagene?: Internasjonalt lys på trettenåringers kunnskaper, holdninger og undervisning i norsk skole.* Oslo: Institutt for lærerutdanning og skoleutvikling, Universitetet i Oslo, Third International Mathematics and Science Study.

Lloyd, G. M. (2002). Mathematics teachers' beliefs and experiences with innovative curriculum materials. In C. G. Leder, E. Pehkonen, & G. Törner, (Eds.), *Beliefs: A hidden variable in mathematics education?* (pp. 149–159). Dordrecht, The Netherlands: Kluwer Academic Publishers.

Manouchehri, A., & Goodman, T. (1998). Mathematics curriculum reform and teachers: Understanding the connections. *Journal of Educational Research, 92*(1), 27.

Mason, J. (2002). *Researching your own practice the discipline of noticing.* London/New York: RoutledgeFalmer.

Mehan, H. (1979). *Learning lessons: Social organization in the classroom.* Cambridge, MA: Harvard University Press.

NCTM. (2000). *Principles and standards for school mathematics.* Reston, VA: National Council of Teachers of Mathematics.

Norton, S. J., McRobbie, C., & Cooper, T. (2002). Teachers' responses to an investigative mathematics syllabus: Their goals and practices. *Mathematics Education Research Journal, 14*(1), 37–59.

Pehkonen, E., & Törner, G. (2004). Methodological considerations on investigating teachers' beliefs of mathematics and its teaching. *NOMAD, 9*(1), 21–50.

Piaget, J. (1970). Piaget's theory. In P. H. Mussen (Ed.), *Carmichael's manual of child psychology* (3rd. ed.) (pp. 703–733). New York: Wiley.

Remillard, J. T. (1999). Curriculum materials in mathematics education reform: A framework for examining teachers. *Curriculum Inquiry, 29*(3), 315–342.

Remillard, J. T. (2005). Examining key concepts in research on teachers' use of mathematics curricula. *Review of Educational Research, 75*(2), 211–246.

Reys, R. E., Reys, B., Barnes, D., Beem, J., & Papick, I. (1998). What is standing in the way of Middle School Mathematics Curriculum Reform? *Middle School Journal, 30*(2), 42–48.

Ross, V. (2003). Walking around the curriculum tree: an analysis of a third/fourth-grade mathematics lesson. *Journal of Curriculum Studies, 35*(5), 567–584.

Rowan, B., Harrison, D. M., & Hayes, A. (2004). Using instructional logs to study mathematics curriculum and teaching in the early grades. *Elementary School Journal, 105*(1), 103–127.

Skott, J. (2001a). The emergent practices of a novice teacher: The roles of his school mathematics images. *Journal of Mathematics Teacher Education, 4*, 3–28.

Skott, J. (2001b, June 8–12). *Why belief research raises the right question but provides the wrong type of answer.* Paper presented at the 3rd Nordic Conference on Mathematics Education, Kristianstad, Sweden.

Skott, J. (2004). The forced autonomy of mathematics teachers. *Educational Studies in Mathematics, 55*(1–3), 227–257.

Smith Senger, E. (1998/1999). Reflective reform in mathematics: The recursive nature of teacher change. *Educational Studies in Mathematics, 37*(3), 199–221.

Spielman, L. J., & Lloyd, G. M. (2004). The impact of enacted mathematics curriculum models on prospective elementary teachers' course perceptions and beliefs. *School Science & Mathematics, 104*(1), 32–44.

Stigler, J. W., & Hiebert, J. (1999). *The teaching gap: Best ideas from the world's teachers for improving education in the classroom.* New York: Free Press.

Stigler, J. W., & Hiebert, J. (2004). Improving mathematics teaching. *Educational Leadership, 61*(5), 12–17.

Tarr, J. E., Chávez, Ó., Reys, R. E., & Reys, B. (2006). From the written to the enacted curricula: The intermediary role of middle school mathematics teachers in shaping students' opportunity to learn. *School Science & Mathematics, 106*(4), 191–201.

Thompson, A. G. (1992). Teachers' beliefs and conceptions: A synthesis of the research. In D. A. Grouws (Ed.), *Handbook of research on mathematics learning and teaching* (pp. 127–146). New York: Macmillan.

Utdannings- og forskningsdepartementet. (2005). *Kunnskapsløftet: læreplaner for gjennomgående fag i grunnskolen og videregående opplæring : læreplaner for grunnskolen* (Midlertidig trykt utg. ed.). Oslo: Utdanningsdirektoratet.

Wellington, J. J. (2000). *Educational research contemporary issues and practical approaches.* London: Continuum.

Wood, T., Cobb, P., & Yackel, E. (1995). Reflections on learning and teaching mathematics in elementary school. In L. P. Steffe & J. E. Gale (Eds.), *Constructivism in Education* (pp. 401–423). Hillsdale, NJ: Lawrence Erlbaum.

11

TEACHERS' USE OF PROJECTS AND TEXTBOOK TASKS TO CONNECT MATHEMATICS WITH EVERYDAY LIFE

Reidar Mosvold

UNIVERSITY OF STAVANGER

Abstract: Many pupils do not find mathematics meaningful. In an attempt to address this issue, recent curriculum reforms have made an effort to forge a stronger connection between school mathematics and the everyday life of pupils. In the Norwegian curriculum reform of 1997, "mathematics in everyday life" became one of five main areas along with number and algebra, geometry, handling data, and graphs and functions. This chapter reports a multiple case study of three Norwegian teachers' beliefs and practices related to the intentions of the curriculum in this area, with a particular focus on how they use projects and textbook tasks. The data included field notes and transcripts from classroom observations, interviews, and questionnaires. The chapter also provides a brief socio-historical account of the background for the development of the Norwegian curriculum on this area, as well as the significant Scandinavian contributions to this field.

Keywords: everyday life, teacher beliefs, teaching practice, mathematics

INTRODUCTION

This chapter focuses on how teachers connect the mathematics they teach in school with everyday life, and it is thereby related to the Norwegian curriculum development, where "mathematics in everyday life" became a distinct area of the mathematics framework in a reform in 1997 (RMERC, 1999). I present results from a case study of three experienced teachers, which focuses on their beliefs and practices concerning this issue. In the first part of the chapter, I focus on the Norwegian curriculum development and theories that appeared to influence the development of the curriculum reform in 1997 (often referred to as L97). Most of the literature presented in the section referred to as "international research" were chosen in response to my own conversations with the leader of the group that was responsible for the mathematics framework in L97. The particular influence of international research traditions on this area of the curriculum has not been thoroughly reported elsewhere. I also present some of the main contributions from the Scandinavian research in this area, although it would be inaccurate to claim that this research had a direct influence on the formation of the Norwegian curriculum. I then present some results from my own study in order to illustrate some of the issues that arose. (For a more thorough presentation and discussion of results, see Mosvold, 2006.)

In the field of mathematics education research, there are several expressions that appear to be related to mathematics in everyday life. "Everyday mathematics" is an expression used by several researchers (e.g., Brenner & Moschkovich, 2002; Wistedt, 1993), and it might include mathematics that we attain or need in our daily lives. The TIMSS 1999 Video Study (Hiebert et al., 2003) used the term "real-life connections,"

The First Sourcebook on Nordic Research in Mathematics Education, pages 169–180

which referred to a connection with the physical world outside the classroom. They distinguished between real-life connections in problem settings and real-life connections in non-problem settings. Both these expressions appear to be similar or at least related to the one found in L97: mathematics in everyday life. The Dutch tradition, on the other hand, often uses the word "realistic." This word has a distinct meaning in Dutch, and it might be defined as something that students can imagine, or something that is meaningful to students. Dutch classrooms and Dutch textbooks have a large proportion of real-life connections, but the contexts of problems do not have to be real-world situations as long as they are "real" or imaginable to the students (van den Heuvel-Panhuizen, 2003).

Although L97 had a strong focus on mathematics in everyday life, the report from the evaluation of the reform concluded that to a large extent the teaching remained unchanged and traditional (Alseth, Breiteig, & Brekke, 2003). Nevertheless, the report noted that the teachers' knowledge about the content of the curriculum was good.

THEORETICAL BACKGROUND

Norwegian Curriculum Development

Mathematics was mentioned as a distinct school subject in Norway as early as 1604, and there have been national regulations of Norwegian schools since the middle of the 18th century (Frøyland, 1965). The first school law was passed in 1739, and during the 19th century, there was a development towards what we now refer to as a national curriculum for Norwegian schools. A plan for the district schools was published by the department in 1890, and this is regarded as the first national curriculum in Norway (Dokka, 1988).

In the 20th century, there were national curriculum reforms in 1922/25, 1939, 1974, 1987, and 1997. In all these curriculum reforms, the plans for mathematics included some kind of reference to a connection between school mathematics and life outside of school. Some of the plans did not emphasize this in a strong way, but there has always been some kind of statement such as: "The children should learn to solve the kind of tasks that will be of use to them in life" (KUD, 1925, p. 21). Even the plan from 1974 (Norwegian curriculum development was then strongly influenced by the New Math reforms taking place in the U.S. and elsewhere) had a focus on the connection with everyday life, and the following was stated as one of the main goals for mathematics: "The aim of the teaching of mathematics is to exercise the pupils in the application of mathematics to problems from daily life and other subjects" (KUD, 1974, p. 132).

In the national curriculum of 1987, problem solving had a focus, and already in the introductory section, a connection with everyday life was emphasized:

> Mathematics is a necessary tool within technology and science and other areas of life in society. Knowledge of mathematics is also part of our culture. Mathematics can be used to convey precise information, and such information presupposes that the recipient has some understanding and knowledge of the subject. We need mathematical knowledge and skills in order to solve many everyday tasks, and to take care of personal interests and duties. For this reason, all pupils in the compulsory school receive instruction in mathematics. (MER, 1990, p. 210)

When the national curriculum of 1997 finally presented "mathematics in everyday life" as a new *main* area of the mathematics curriculum (the other areas were number and algebra, geometry, handling data, and graphs and functions), the focus on a connection between school mathematics and life outside of school was not new at all. Still, this area had an important role in the syllabus, and it had a somewhat different character than the other main areas: "The first area of the syllabus, mathematics in everyday life, establishes the subject in a social and cultural context, and it is especially oriented towards users. The further areas of the syllabus are based on the main areas of mathematics" (RMERC, 1999, p. 168).

Mathematics in everyday life (in the Norwegian curriculum) was therefore not an area of mathematics itself, but it was supposed to be a more overarching topic that could show pupils how mathematics can be placed in a social and cultural context. As for the other main areas, lists of content were presented that the pupils should work with in all compulsory school grades 1–10 (ages 6–16). However, several textbooks that were published in response to the new curriculum included separate chapters on mathematics in everyday life, and it thereby appeared to be a completely new content area rather than a new main focus.

Previous curricula also included some references to a connection with everyday life, but there was an intentional refocusing worked out in the curriculum of 1997. In all previous curricula there had been a focus on how pupils should learn the mathematics that they could eventually apply in everyday life. In L97, teachers were supposed to start with situations from the pupils' everyday life and then let the pupils reinvent the mathematical knowledge embedded in these situations. These ideas were also emphasized in the report from the evaluation of the mathematics part of the curriculum reform (Alseth et al., 2003). Although the teachers appeared to know about the content of the curriculum, they continued to teach in a traditional way:

> The teaching normally still takes place by the teacher starting the class with an introduction where homework assignments are reviewed and new content is presented. This presentation normally ends up in an explanation of how a certain kind of problem is to be solved. After this, the pupils work individually on solving such tasks from their textbooks. (Alseth et al., 2003, p. 117)

International Research

In the TIMSS 1999 Video Study (Hiebert et al., 2003), mathematics in the Dutch classrooms had by far the largest proportion of real-life connections. This might be related to the fact that the framework of Realistic Mathematics Education (RME) has a strong influence in the Netherlands. It is presented as the main theoretical framework for developmental research within the Freudenthal Institute, and Freudenthal himself was interested in the connection between mathematics and everyday life. One of his main ideas was to teach mathematics so as to be useful (Freudenthal, 1968).

The idea of mathematizing is central in RME, and according to Freudenthal (1991) this implies mathematizing reality or pieces of reality. There are as many everydays as there are people, so the concept of reality is not a simple one. Nevertheless, Freudenthal indicates that "as soon as mathematizing is didactically translated into reinventing, the reality to be mathematized is that of the learner, the reality into which the learner has been guided, and mathematizing is the learner's own activity" (Freudenthal, 1991, p. 67).

The Dutch tradition influenced the Norwegian curriculum of 1997, and there are several passages that are related to the ideas of RME. One example can be found in the section called "Approaches to the study of mathematics": "Learners construct their own mathematical concepts. In that connection it is important to emphasise discussion and reflection. The starting point should be a meaningful situation, and tasks and problems should be realistic in order to motivate pupils" (RMERC, 1999, p. 167).

Norwegian society is influenced by the U.S. in many respects, and in this context we can find links between the Norwegian curriculum and the NCTM Standards (Alseth, 2004). In the Standards of 1989 (NCTM, 1989), mathematical knowledge was emphasized as important for gaining an understanding of the world. There is a need to understand and be able to use mathematics in everyday life, and this need appears to be stronger than ever: "The mathematical abilities needed for everyday life and for effective citizenship have changed dramatically over the last decade and are no longer provided by a computation-based general mathematics program" (NCTM, 1989, p. 130).

This was also noted in the Standards from 2000 in which four categories were distinguished: "Mathematics for life, Mathematics as a part of cultural heritage, Mathematics for the workplace and Mathematics for the scientific and technical community" (NCTM, 2000, p. 4).

The idea of mathematics being useful for everyday life was also recognized in the Cockroft report, which had a strong influence on the development of mathematics as a school subject in the UK:

> The usefulness of mathematics is perceived in different ways. For many it is seen in terms of the arithmetic skills which are needed for use at home or in the office or workshop; some see mathematics as the basis of scientific development and modern technology; some emphasise the increasing use of mathematical techniques as a management tool in commerce and industry. (Cockroft, 1982, p. 1)

The Cockroft report summarized the requirements of mathematical knowledge for adult life as "a feeling for number" and "a feeling for measurement." From this set of requirements, it appears evident that school mathematics should not only have a focus on the mathematics that is needed in everyday life. Mathematics would then be a rather narrow and small subject.

Scandinavian Research

Several Scandinavian researchers have been involved with issues concerning the connection of mathematics with everyday life in the past decades. One of the early projects—everyday life knowledge and school mathematics—was initiated by Swedish professor Inger Wistedt in the early 1990s. One of the suggestions here was that "School mathematics, they say, should become connected to the children's everyday life experiences, and collect material from the environment that surrounds the pupils" (Wistedt, 1990, p. 2).

Wistedt used the term "everyday mathematics," and she thereby referred to a distinction between the mathematics children attain in their everyday lives, and the mathematics that is needed in everyday life.

Wistedt's project showed that the use of contexts that were well known for the children did not automatically help them gain a better understanding of mathematics. Using such contexts might actually "increase the risk of the pupils missing the mathematical ideas in the problem" (Wistedt, 1993, p. 3). Wistedt claimed that it was important for the children to be able to distinguish between the everyday understanding of a problem and the mathematical one.

The use of realistic contexts in mathematical problems was further studied by another Swedish researcher, Torulf Palm. He had a focus on the authenticity of the contexts, and he concluded, "The results of the study show that authenticity, even under the restrictive constraints of normal classroom resources, can affect students' tendencies to effectively use their real world knowledge in the solutions to word problems" (Palm, 2002b, p. 31). Students still present unrealistic solutions to realistic tasks, and Palm (2002b) suggests that the students carry beliefs that do not require a consistency between school mathematics and life outside of school. In school mathematics, all tasks have a solution, and that solution is normally represented by a single number (Palm, 2002a).

Although many students seem to provide unrealistic answers to realistic questions, Palm found that some of these students had actually made use of their understanding of the problem context. Their understanding was consistent with their beliefs concerning the socio-cultural aspects of the context, but their understanding still differed considerably from that of many teachers and researchers (Palm, 2008).

The discussion about connections with everyday life in textbook tasks has been going on for a long time. As early as 1868, the Swedish scholar A.T. Bergius wrote an article where he claimed that the textbooks were poorly organized because they had little connection with the pupils' everyday life. One of his four criteria for good teaching of mathematics was to "connect with practical problems in the pupils' present or future everyday life" (Prytz, 2003, p. 48).

Swedish researchers such as Palm (2008) and Lithner (2000) have focused on authenticity and realism in the contexts of word problems. The realism of contexts has also been a main focus by two Danish scholars who study adults' learning of mathematics. One of these, Tine Wedege, builds on the theories of situated learning when she claims that mathematics itself is an activity that is bound to context, and "that there is no such thing as a context-free mathematics" (Wedege, 1999, p. 206). Wedege regards adults' learning of mathematics as a complex issue, and she presents three conditions that reveal this:

1. Knowing mathematics is a contextually determined and thus relational concept in the adult world. Whether or not an adult knows mathematics can only be answered after the questions such as who, where, when, what and related to what.
2. The situation for learning mathematics depends on the experience of the individual adult with mathematics in school and everyday practice and their individual perspectives for learning. Emotional factors are just as important as cognitive ones in the psychological learning process.
3. Adults have a range of different opportunities to learn (or not to learn) maths in adult or further educational systems and/or in everyday practice. There is a large variety of formal adult education programs where mathematics is taught as a separate subject or integrated in other subjects, and where the objectives are either mathematical, general, or specific vocational competence (Wedege, 1999, p. 206).

The context is vital in the teaching and learning of mathematics, and Wedege makes a distinction between "task context" and "situation context." The latter refers to the context of mathematics education in a more political sense, and this relates to the works of Mellin-Olsen (1987) and others. The former is strongly connected with the focus on realistic contexts that can be found in Freudenthal's (e.g., 1991) work and the Dutch tradition of RME in general.

The focus on realistic contexts and mathematizing was also central in the master thesis of Bjørnar Sæterås (2001). He combined the theoretical framework of RME with social-constructivist theories and situated views on learning in his study of students in the first year of upper secondary school. Sæterås found that the students did not take advantage of the realistic contexts of the problems, but rather fell back on beliefs that were created on the basis of previous experience with mathematical problems in school. This project had a stronger connection with the framework of RME, but it appears that Sæterås reached some of the same conclusions as Palm (2002a) did in his research.

Whereas Sæterås, Palm, and others have been concerned with the context of word problems, Fyhn (2008) had a stronger focus on the context of activities. Like Sæterås, she also drew upon ideas from RME in her study. She used the activity of climbing as a context and starting point for further classroom activities that involved mathematizing and reinventing the angle concept. Her research points towards the use of analytical drawings when working with analytical texts in geometry (see Chapter 13, this volume).

The focus of the research project described in this chapter was not on the problem contexts as such. Although it refers to RME, I did not use the theoretical framework of this Dutch tradition as a point of departure. My focus was to analyze the beliefs and practices that some experienced teachers had concerning the connection of mathematics with everyday life (Mosvold, 2006). Two research questions formed a starting point:

1. What are the teachers' beliefs about connecting school mathematics and everyday life?
2. What ideas are carried out in their teaching practice?

METHODS

The three teachers in the study are referred to as Harry, Karin, and Ann. They are all experienced teachers working in schools in the same town in Norway. At the time of the study, Karin and Ann are colleagues in a smaller school in the outskirts of the town, whereas Harry is teaching in a larger school close to the town center. Harry and Ann teach pupils in grade 9 at the time of the study, whereas Karin's pupils are in grade 8. Harry has approximately 16 years of teaching experience, and Karin has 18 years of experience as a teacher at the time of the study, the largest part of it in lower secondary school. Ann is also an experienced teacher, and she has 17 years of experience at the same lower secondary school where she is now.

I collected data in the school of Karin and Ann in February and March 2003, and followed up with studies in Harry's school in April and May the same year. The entire project followed the same three-phase model:

1. Before the study started, I conducted a planning meeting with the teachers and their respective principals. In this meeting, my aims were presented, and we discussed the details of the classroom study.
2. In the main phase of the study, I observed the lessons as a non-participant observer. Data were collected through audio-recordings of teacher talk and classroom conversations, field-notes, collection of handouts, etc. In the beginning of this period, the teachers answered a questionnaire.
3. After the observation period, the teachers were interviewed individually. These semi-structured interviews were audio-recorded.

The methods and order of these three phases of the project were chosen because I wanted to make sure that the informants had an idea of my intentions and how I wanted to carry out the study. The planning meeting was thus held for a practical reason. I decided to collect data from different sources, because the connection between teachers' beliefs and practices is complex (cf. Leatham, 2006; Skott, 2001), and researchers therefore need to gather information from different sources. My rationale for including classroom observations was based on my belief that the classroom context sometimes influences classroom practice more than the teachers' beliefs. Some researchers even put a lot of energy into discussing what appear to be inconsistencies between teachers' beliefs and practice (see Cooney, 1985; Raymond, 1997). Research on beliefs and practice is strongly influenced by the psychological tradition, and there are many examples of research where questionnaires are the only source of data in a study of teachers' beliefs and (self-reported) practices, and this is an approach I think should be reviewed. I decided not to analyze the questionnaire results before the observation period was finished, because I did not want these results to influence my observations.

The questionnaire consisted of a collection of questions with Likert-scale responses, a few comment-on questions, and some list questions. The questionnaire was tested and discussed with some fellow doctoral students and some experienced scholars prior to the study.

During the classroom observations, I made extensive field-notes from all mathematics lessons during the four-week periods I spent with each teacher. In one of the weeks, all lessons were audio-recorded in their entirety. The recorded lessons were transcribed, and so were the teacher interviews. In addition to this, the teachers' textbooks were also analyzed, but this part of the study will not be reported further here.

The data material was analyzed in two phases. In the first phase, the codes from the TIMSS 1999 Video Study (Hiebert et al., 2003) were used as a starting point. Their initial distinction between real-life connections in problem and non-problem situations was expanded, and it resulted in a more complex coding scheme in three levels (see Mosvold, 2006 for further elaboration). In the second phase of analysis, this process was revised and refined. The analysis was then mainly carried out following the four-stage process suggested by Stake (1995):

- Categorical aggregation
- Direct interpretation
- Establishing patterns
- Development of naturalistic generalisations

The analysis of the questionnaire results was used as the basis for an aggregation of ten categories divided in three main themes:

- Beliefs concerning activities and organization
  - Cooperative learning
  - Reinvention
  - Projects
  - Repetitions and hard work
- Beliefs concerning content and sources
  - Textbooks
  - Curriculum
  - Other sources
- Beliefs concerning practice
  - Teaching and learning
  - Vocational relevance
  - Connections with everyday life

These categories emerged from the data and my analysis of the data. The categories were used to create profile tables of each of the three teachers, and these profile tables were analyzed. In the following, I will present some data that are related to one category from each main theme: projects, textbooks, and connections with everyday life.

RESULTS

Beliefs Concerning Projects

In respect of the ways in which teaching might be organized, L97 had a sharp focus on projects. Mini-projects were mentioned explicitly as a way of connecting school mathematics with everyday life. When Ann was asked about using projects in the interview, she replied:

> Yes, very good. But as I said: since I started off this autumn, I have felt that there have been too many projects, and too many loose things. And as I said, in this class, when you do that, everything will float. I am in need of putting things into some kind of system. And when you in a way have got the system and are in control, then you can start loosening up. (Teacher interview)[1]

Ann thought of large projects that involved several subjects, and her experience was that mathematics often played a minor role in such projects. She was, however, strongly in favor of organizing the pupils in groups in order to facilitate rich discussions:

> That they can . . . discuss and . . . help each other and . . . no, there is a lot of learning in that! And then you can both . . . that is, and put together pupils that are on the same level. You do that from time to time. And that can become very successful! But you might also put together pupils that you believe might complement one another. And then you have a stronger pupil who can help one who is not so strong. And I believe that both will benefit a lot from it! But again, so . . . the pupils must have confidence in each other. I have experienced here that they sit side by side, but they kind of sit and do individual work. (Teacher interview)

Karin viewed herself as a conservative and rather old-fashioned teacher. Her teaching was traditional, and she believed that the endorsement of projects and similar methods was exaggerated. She explained that she was not opposed to the idea of projects in principle, but she believed it was difficult to manage in a school system where everyone is included. She said that it was not her job only to put the knowledge into the pupils' heads:

> [I]t is important that I don't just put things into their heads, and then believe that it will become theirs. I don't believe in that! They have to apply it. So, it is important for me not to spend too much of the lesson teaching. But on the other hand it is . . . because there are pupils who are uncertain, who will look at the neighbor, who will look in the answers, so it isn't good enough just throwing them out into it. (Teacher interview)

Harry also emphasized that it is necessary to ensure pupils' active engagement in mathematics, but he appeared to have a different interpretation of this. Harry explained that his thinking is influenced by the social learning theories of Stieg Mellin-Olsen (see Mellin-Olsen, 1984, 1987). He used a lot of "mini-projects" in his teaching, and a large part of the pupils' activities were organized as themes or projects.

> Very often, I say that today we will, or for the next two weeks we will work with. . . . And then I present, call it a plan or an angle of incidence. And then the pupils work on that. But my aim then is that I believe that they will do many mathematical calculations through that. My experience is that the pupils find it interesting. Then there are periods where I don't follow the book much. Like now, we are going to work with the bicycle as a geometrical phenomenon, and with this Pythagoras—so I believe that I get the pupils to work differently. And I force them to relate to mathematics in a completely different way. (Teacher interview 2)

Harry's projects normally included some kind of activities that the pupils would actually have to carry out in practice. In these activities he tried to connect with everyday life as much as possible. This was the case with the bicycle assignment that he gave his pupils during the classroom observations.

> Yes, I gave an example earlier today, where I did some investigations with one of the girls—it was in connection with the geometry they had last year, in grade 8—how such a Tipp-Ex was formed [referring to marks painted on the cogs to indicate a particular alignment] on the inside with two cogwheels, and the gear relation between them. And then we tried how many times the one rotated for the two marks to meet again and likewise. And then I asked the girl: "what might be the reason that the cogwheels don't have the same size?" And then we had a quiet and calm situation, so I got a very good discussion with her about that. So, it doesn't have so much to do with the pupils' everyday lives, but it has to do with technology. So, I believe that the pupils' everyday life is very seldom incorporated into school mathematics. But I believe that the teachers use it when they come up with it. And some also have really good examples. (Teacher interview 2)

Beliefs Concerning Textbooks

Karin often let the pupils solve tasks from the textbook. This statement was supported by the observations of her teaching. Other sources than the textbook were only used occasionally, and in the classroom observations few examples of this were noted. Her teaching was traditional in the sense that she often started off with a presentation of the theory, and then let her pupils practice solving problems from the textbook. The pupils' rulebook (in Norwegian called "elevbok" or simply "regelbok") was used a lot in her class, and she normally told the pupils explicitly what to write in this book.

> I believe in a way that they get more of an ownership towards what they themselves have written in a book from an example, than from an example that is printed in the textbook. So, they return to something they feel more related to. So in that way I believe, I am quite convinced that this is a way of making things more clear, emphasizing and engraving something into their brains. (Teacher interview)

Contrary to Karin, Harry believed that the textbook was cramping the teachers' style, and he therefore did not use it much. He mainly used it as a source for the pupils' homework, but in his regular teaching he tried to use other sources than the textbook.

Ann was somewhere between Karin and Harry as far as her beliefs concerning textbooks were concerned. She felt dependent on the textbook, and it appeared to be something she was not overly happy with.

> No, I have said that I use it [the textbook]. It is what directs/controls my mathematics! And again, the exam at the other end directs me. So I think that if one is to change the teaching a lot, one also has to change the final exam. If that had been done, it is absolutely certain: I would have changed at once! I believe so. (Teacher interview)

In the questionnaire, she replied that she believed it would make mathematics more interesting to the pupils if the teacher managed to put away the textbook more often. Still, she claimed that her pupils very often solved textbook tasks.

Both Ann and Karin replied that they sometimes use other sources than the textbook, but they did not clarify what these sources might be.

Beliefs Concerning Mathematics in Everyday Life

The three teachers had quite different opinions about the connection of mathematics with everyday life. Karin described herself as a traditional teacher, and she did not emphasize connections with the pupils' everyday lives a lot. In the questionnaire she claimed that she sometimes connected mathematics with everyday life. In the interview, however, she expressed that it was not a good idea:

> Yes, well, I must say that I am not very fond of the idea. Like, at least not when it comes to having a direct connection. Just as I believe that when a teacher of sports makes them do the high-jump, it is not because they are going to do the high-jump later in life. It is plainly to train the body. And I actually have that idea about most of the things we do here, that—at least when it comes to mathematics, what the mathematics offer of training—is logical thinking, being able to transfer principles from one... or, yeah, transfer something to something else, rules and principles. Plainly to use your head. And... because they need this! No matter what kind of job. So, right, you have x and y-s and a-s and b-s and lots of things they don't have to do for the rest of their lives.... But this is my motivation for the pupils. (Teacher interview)

Harry, on the other hand, explained that this was an important issue for him. He focused on activating the pupils, as he believed that active pupils would learn more. He replied in the questionnaire that he often emphasized the connection of school mathematics with everyday life. The notion of "school mathematics" was something he found problematic, though.

> Yes, well, I believe that the main problem lies exactly in the expression "school mathematics." Because, I said

> in a meeting we had here, that there is a big difference between school mathematics and mathematics. Where school mathematics is mainly calculations on a piece of paper, as I see it. What the teachers do, I don't know, but it is certainly difficult to draw upon the pupils' everyday lives. Not least because the pupils in a way have a school tradition where they are removed from everyday life . . . in school, so the pupils' everyday life is not a part of school. (Teacher interview 2)

Like Harry, Ann was also in favor of connecting school mathematics with everyday life, and this was something she emphasized in the interview. She had, however, experienced many practical and organizational problems in this relation, and she felt that she had not succeeded in implementing these ideas.

> No, well that is the very problem: being able to relate mathematics to their everyday lives. I think so. I get this question very often. Why do we have to learn this? What is the point? So, that's where the challenge lies. No, when it comes to me, I am probably not good enough at this at all, to connect with that. Because I am too bound by the textbook that is! Just like that. I have no problems admitting that. But again, the final exam is what directs me. (Teacher interview)

Still she thought it was important to start with the pupils' everyday life, and that this could help her initiating a discussion with the class. This oral approach was important for her, and she believed that a teacher should be able to pose the right questions.

Into the Classrooms

I now present some examples from the teachers' classroom practices. The examples presented here are chosen in order to provide a basis for discussing the issues raised above. I do not want to imply that they are representative of any generality of teaching practices; the examples are intended to illustrate the characteristics of the teaching approaches used by these three teachers.

The Bicycle Assignment

Harry expressed a strong belief in using projects, and in particular what he called "mini-projects" in his teaching. He believed that such projects would hold opportunities for doing many mathematical calculations. He further said that he would normally use the textbook as a source of exercises for the homework assignment, rather than as a main tool in his teaching. Still, he found it difficult to connect mathematics with everyday life, and he believed that a major reason for this was that the pupils' everyday life is normally not a part of school. This following relates to an activity that is typical of Harry's use of mini-projects, and it is also an example of how he used a context from everyday life to facilitate mathematical activities.

Harry introduced his class with the bicycle assignment in this way:

> When you are now going to work with the bicycle assignment, you are going to draw the bicycle on the scale of 1:5. You are going to collect as accurate measurements and angles as possible! And I just said: draw a sketch today, a rough draft! So that you can sit down later and make an accurate drawing. And then you will pick out as many geometrical shapes as possible. (Transcriptions 130503)

In the introductory part of the lesson, Harry also brought up some issues concerning the construction of bicycles, like how the brakes work, the digital speedometer, etc. Harry then let the pupils start measuring some bicycles that were brought into the classroom. The pupils decided for themselves whether they would like to work in groups or not. Most pupils ended up working in groups or pairs, and some of the pupils decided to work on the assignment outside the classroom.

A problem that arose quite soon concerned how large the drawing should be in order to fit on a page. This initiated an interesting discussion between Harry and some pupils about proportions and scales.

Harry: Yes. Perhaps we should say something about how big this bicycle will be. Is there room for you to draw it on a page?
Pupil: No!
Harry: If it is 1 metre high, how big will it be in your book then?
Pupil: 5, 50, no 5 centimetres.
Harry: Remember to draw on a scale of 1:5.
Pupil: I don't have a clue. 5, 15, something else, 5 then! I haven't got a clue! Asking me about these things . . .
Harry: No. No. Do you follow, Sandra? If that one is 1 metre high, if we say that the bicycle is 1 metre high now. How high will it be in the book then?
Pupil: Yes of course. You just divide by 5!
Harry: Yes, of course.
Pupil: But I don't know what that is. Fourteen.
Harry: 20 centimeters. (Transcriptions 130503)

Several issues came up during their involvement with the assignment, and the pupils were given the opportunity to solve different kinds of realistic problems. They also had the opportunity to work with measurement in a realistic context, and they encountered challenges that motivated the introduction of a tool such as the slide calliper. Being a science teacher as well as a mathematics teacher, Harry took the opportunity and went into discussions related to physics and technology.

Harry: For instance—you are going to find as many geometrical figures as possible—and for instance, I would have included the length of that [the pedal]. Yes, so you must write the length of it. But why is it necessary to include the length of it, why is that a point?

Pupil: Yes, because you must see how much force . . . no, I don't know.

Harry: Yes, you are on the right track. If it were shorter, what would it be like to cycle then?

Pupil: Hard. (Transcriptions 130503)

This mini-project was quite typical for Harry's teaching, and we can see how he made links to other subjects like science. The example also illustrates how complex such an activity can be. The discussions that come up can be difficult to prepare for, and it is challenging for the teacher to come up with the good ideas for these projects.

Area of Figures

Ann believed in connecting mathematics with everyday life, but she found it hard to do in practice. A main reason appeared to be a strong dependence on the textbook. She was also positive towards projects as an approach, but her experience was that mathematics often had a minor role in such projects.

The following example is from one of Ann's geometry lessons. She started off presenting the pupils with a piece of paper that contained different geometric figures. Some were common geometric shapes (e.g., triangles and rectangles), but there were also other figures. All figures were inscribed in a square grid. The pupils were asked to work in pairs and find the area of the figures. There were different possible approaches to reach a solution, and Ann asked the pupils to discuss and decide what approach they would use.

When the pupils had worked with this task for a while, their solutions were discussed in class. During this discussion both Ann and the pupils brought up several connections with everyday life.

Ann: But, . . . eh, you can buy boxes of mineral water, right? (the teacher draws on the chalkboard) One, two, three, four, five . . . Imagine you're having a birthday party, and you buy mineral water. Eh, and you buy a box, and you are to count the number of bottles in the box, how would you do that? Put your hands up.

Pupil: You count the vertical row and the horizontal.

Ann: Are there other ways of doing it?

Pupil: Look at the receipt, and see the number that is printed there.

Ann: Yes . . . the receipt says so? But, as we heard before, instead of counting them all, you can count the vertical row and the horizontal. And then you could say that the area of this box of mineral water is actually 24 bottles then!

Ann suggested that a similar method could also be used when we try to figure out the area of a room, and she then went into a discussion about units of measurement.

If I Go Shopping

Karin clearly expressed a disbelief in connecting mathematics with everyday life. In her view, school mathematics was supposed to be more of an opportunity for the pupils to exercise their brains. Similarly, she thought the idea of using projects had been exaggerated, and she considered herself to be a traditional teacher with a strong focus on using the textbook.

In a lesson where she introduced the pupils to algebra, Karin gave the following example:

> If I bring 100 kroner to the store, then buy some item at a cost of 20 kroner and then something for 10, how much do I have left?

She used this example to introduce arithmetical operations, and she was interested in how the pupils calculated when trying to solve this task. Karin also presented other similar examples to the pupils, and they were asked to write down these examples in their rule books. Karin let the pupils present their own solutions and suggestions for ways of solving the tasks, and she encouraged them to participate in a discussion about what methods to use. Here is an example of the discussion concerning a different example:

(Karin writes on the blackboard: 3 + 4 * 2)

Karin: What is the answer, and how do you get it?

Pupil: You take four and multiply by two, and then you add.

Karin: Would anyone do it differently?

Pupil2: 3 plus 4 and multiply with 2.

Karin: Here we see that we have to choose, because we can only have one answer. Which is correct?

Pupil: 11.

Karin: Can we make a rule?

Pupil: No parenthesis.

Karin was opposed to connecting mathematics with everyday life, and although this example includes something that might be categorized as a real-life connection, the following discussion indicates that this connection is of less importance. Karin's main focus is on the mathematical contents and the rules involved.

DISCUSSION

The discussion of how teachers connect mathematics with everyday life is related to a number of issues. During the analysis of data, a list of categories emerged with three main headings: (1) activities and organization, (2) content and sources, and (3) beliefs concerning practice. My assumption is that when we try to analyze how teachers connect mathematics with everyday life, this is related to how they organize their class, what teaching methods they use, how they use the textbook, what other sources than the textbook they use, their beliefs concerning teaching and learning, and so on.

The use of projects and mini-projects are explicitly mentioned in the Norwegian curriculum (RMERC, 1999) in relation to mathematics in everyday life. Pupils in grade 9, for example, should have the opportunity to "use mathematics to describe and process some more complex situations and small projects" (RMERC, 1999, p. 180).

Harry appeared to focus more on projects than Ann and Karin did, and the data from the questionnaire, the interview, and the observations all supported this assumption. It appeared, however, that they thought of different things when they referred to projects. When Harry talked about projects, he did not follow a more formal project ideology. He talked about mini-projects, activities, and themes, and these concepts appeared to relate to the same kind of activity. Ann, on the other hand, clearly thought about a more formal framework for project work. She experienced that mathematics often played a minor part in such projects, and this appeared to be part of the reason why she was reluctant to use them. In the evaluation of L97, Alseth, Breiteig and Brekke (2003) found that regular teaching more or less remained unchanged, whereas there were several examples of connections with everyday life in more *extensive activities.* The projects referred to here fit with this description of "extensive activities." Mini-projects are not used by all teachers in this study, although it is suggested explicitly in the curriculum. The case of Harry offers some interesting indications about how this could be done.

Textbooks are still a main source of influence for mathematics teachers (see Alseth et al., 2003). This appeared to be the case in my study as well. Again Harry was the exception, and he claimed to seldom start his teaching activities with a focus on solving textbook tasks. Karin and Ann both had a strong emphasis on solving textbook tasks, but Ann expressed some concern about her dependence on the textbook. Palm (2008) investigates the impact that authentic contexts have on pupils' problem solving processes. The example from Karin's shopping story illustrates how an apparent real-life context could be presented without having any practical influence on the way pupils solve the task. The task was: How much do I have left? In a real-world setting this would have been easily solved by counting the remaining coins, but this was never an option in this context. As Palm (2008) points out, this might be related to the didactical contracts that are predominant in many classrooms. These kind of tasks are probably seen as 'school problems' by the pupils, and this influences the way in which the problems are solved. In the mathematics problems found in textbooks one can hardly ever solve the problem by counting or measuring. Instead, one is expected to uncover some 'mysterious' method to reach the correct answer, and there is always one and only one correct answer.

The idea of giving the pupils an opportunity to construct their own mathematical concepts is apparent in L97, and the intentions in the Norwegian curriculum strongly resemble the idea of (guided) reinvention that can be found in the Dutch tradition of RME. These ideas are not easily carried out in the traditional approach of starting with a presentation of theory and then letting the pupils solve routine tasks form the textbook. As I remarked earlier, in the TIMSS 1999 Video Study, the Netherlands had by far the highest amount of real-life connections (Hiebert et al., 2003). When examined more closely, however, it appears that most of these instances are related to textbook tasks (Mosvold, 2008).

Harry was positive towards connecting school mathematics with everyday life, although he had some difficulties with the concept of "school mathematics," and this became apparent in his teaching as well. Karin, on the other hand, was more negative towards the entire idea of connecting with everyday life, and her teaching reflected this. Somewhere in between these two "extreme" points of view was Ann. She claimed to be positive towards connecting mathematics with everyday life, but she experienced problems and constraints in her attempts to carry out these ideas. My observations in her classroom only revealed a few connections with everyday life, and this did not appear to be a main focus in her teaching. I have included one example where she referred to bottles to illustrate the concept of area, and this context was used to initiate further discussion around the issue. Still, it would have been easy to suggest that there was a conflict between Ann's beliefs and her teaching practice. This is not my intention, and I find it more interesting to learn about the complexities of the teachers' belief systems, and also about the complexities of the mathematics classrooms. Ann had struggled a lot with this class, so there were other and more vital issues that had to be addressed first. When it came to making connections with everyday life, this was something she believed to be important, but she acknowledged that she had not managed to put this into action successfully—perhaps because she was so closely attached to following the textbook.

It is my belief that Ann's challenge is fairly typical of that faced by many teachers, in Norway and elsewhere. It also illuminates some of the complexities and challenges of being a teacher, and it indicates that the road from curriculum intentions to teaching practice is not always straightforward.

CONCLUDING THOUGHTS

In my study, I intended to analyze the ways in which some experienced teachers connected mathematics with everyday life. I also wanted to learn more about their professed beliefs concerning this issue. The results of this study cannot be generalized to the entire population of Norwegian teachers. The study does, however, provide some interesting information about some experienced teachers' beliefs and practice, and this information provides us with some important insights. First, it appears that even the most experienced teachers (such as Ann) might struggle when it comes to putting the curriculum intentions into action. The complexities of the classroom context might be a serious hurdle, and it appears to be somewhat simplistic only to point to what appears to be a discrepancy between teachers' beliefs and practice. One of the teachers (Karin) was opposed to the idea of connecting school mathematics with everyday life as such, and such opinions should also be brought into the discussion. Similar findings are reported by Kleve (2007). She identified several factors that constrained the mathematics teachers' implementation of the intentions of L97 (See Kleve, Chapter 10, this volume).

Even some of the most experienced teachers appear to rely on the textbook, and issues related to how textbooks make connections with everyday life need to be investigated further. Palm (2002) and others have already focused on this, but other perspectives concerning this issue might be further investigated. Even an innovative teacher like Harry had problems finding other sources than the textbook, and this might be a significant issue in this context.

In 2006, a new national curriculum was introduced in Norwegian schools, and the area of mathematics in everyday life was taken out entirely. The new curriculum has a stronger focus on basic skills, and it also omits making suggestions about methods of organization as L97 did. Still, there are many issues concerning a connection of mathematics with everyday life that need to be discussed, and the issues discussed in this chapter remain important, although there is no longer an explicit demand for such a focus from the curriculum.

NOTE

Transcriptions from the teacher interviews have been translated from Norwegian into English by the author.

REFERENCES

Alseth, B., Breiteig, T., & Brekke, G. (2003). *Evaluering av Reform 97—Endringer og utvikling ved R97 som bakgrunn for videre planlegging og justering–matematikkfaget som kasus.* Notodden: TFN-rapport 02/2003.

Alseth, B. (2004). Endret læreplan = endret matematikkundervisning? Matematikkopplæringen på småskoletrinnet etter R97. In K. Klette (Ed.), *Fag og arbeidsmåter i endring? Tidsbilder fra norsk grunnskole* (pp. 38–54). Oslo: Universitetsforlaget.

Brenner, M. E., & Moschkovich, J. N. (2002). *Everyday and academic mathematics in the classroom.* Reston, VA: National Council of Teachers of Mathematics.

Cockroft, W. (1982). *Mathematics counts: Report of the committee of inquiry into the teaching of mathematics in schools under the chairmanship of W.H. Cockcroft.* London: HMSO Books.

Cooney, T.J. (1985). A beginning teacher's view on problem solving. *Journal for Research in Mathematics Education, 16,* 324–336.

Dokka, H., & Bruhn, V. (1988). *En skole gjennom 250 år: den norske allmueskole, folkeskole, grunnskole 1739.* Oslo: NKS-Forlaget.

Freudenthal, H. (1968). Why to teach mathematics so as to be useful. *Educational Studies in Mathematics, 1*(1, 2), 3–8.

Freudenthal, H. (1991). *Revisiting mathematics education: China lectures.* Boston: Kluwer Academic Publishers.

Frøyland, E. (1965). *Matematikk i skolen. Hvorfor—hva og hvordan?* Oslo: Hovedoppgave i pedagogikk.

Fyhn, A.B. (2008). A climbing class' reinvention of angles. *Educational Studies in Mathematics, 67*(1), 19–35.

Hiebert, J., et al. (2003). *Teaching mathematics in seven countries: Results from the TIMSS 1999 video study.* Washington, DC: National Center for Education Statistics.

Kleve, B. (2007). *Mathematics teachers' interpretation of the curriculum reform, L97, in Norway.* Unpublished doctoral thesis, Agder University College, Kristiansand, Norway.

KUD. (1925). *Normalplan for byfolkeskolen.* Oslo: Stenersen.

KUD. (1974). *Mønsterplan for grunnskolen: Bokmål.* Oslo: Aschehoug.

Leatham, K. (2006). Viewing mathematics teachers' beliefs as sensible systems. *Journal of Mathematics Teacher Education, 9*(1), 91–102.

Lithner, J. (2000). Mathematical reasoning in school tasks. *Educational Studies in Mathematics, 41*(2), 165–190.

Mellin-Olsen, S. (1984). *Eleven, matematikken og samfunnet. En undervisningslære.* Bekkestua, Norge: NKI Forlaget.

Mellin-Olsen, S. (1987). *The politics of mathematics education.* Dordrecht, The Netherlands: Kluwer Academic.

MER. (1990). *Curriculum guidelines for compulsory education in Norway.* Oslo: Aschehoug.

Mosvold, R. (2006). *Mathematics in everyday life: A study of Norwegian teachers' beliefs and actions concerning the connection with mathematics and everyday life.* Unpublished doctoral thesis, University of Bergen.

Mosvold, R. (2008). Real-life connections in Japan and the Netherlands: National teaching patterns and cultural beliefs. *International Journal of Mathematics Teaching and Learning.* Published online, July 3, 2008. Retrieved May 28, 2010, from http://www.cimt.plymouth.ac.uk/journal/mosvold.pdf

NCTM. (1989). *Curriculum and evaluation standards for school mathematics.* Reston, VA: National Council of Teachers of Mathematics.

NCTM. (2000). *Principles and standards for school mathematics.* Reston, VA: National Council of Teachers of Mathematics.

Palm, T. (2002). *The realism of mathematical school tasks—Features and consequences.* Doctoral Thesis No 24. Umeå: Umeå University.

Palm, T. (2002b). *Impact of authenticity on sense making in word problem solving* (Research report, No 2, in Mathematics Education). Umeå, Sweden: Umeå University, Department of Mathematics.

Palm, T. (2008). Impact of authenticity on sense making in word problem solving. *Educational Studies in Mathematics, 76*(1), 37–58.

Prytz, J. (2003). "Moderna" ideer från förr och nu. *Nämnaren 30*(1), 42–47.

Raymond, A.M. (1997). Inconsistency between a beginning elementary school teacher's mathematics beliefs and teaching practice. *Journal for Research in Mathematics Education, 28*, 550–576.

RMERC. (1999). *The curriculum for the 10-year compulsory school in Norway.* Oslo: The Royal Ministry of Education, Research and Church Affairs.

Skott, J. (2001). The emerging practices of a novice teacher: the roles of his school mathematics images. *Journal of Mathematics Teacher Education, 4*(1), 3–28.

Stake, R.E. (1995). *The art of case study research.* Thousand Oaks, CA: Sage Publications.

Sæterås, B. (2001). *Matematikkundervisning som vektlegg matematisering, med utgangspunkt i reelle situasjonar. Utprøving og vurdering av to undervisningsopplegg.* Unpublished master's thesis, Agder University College, Kristiansand, Norway.

van den Heuvel-Panhuizen, M. (2003). The didactical use of models in realistic mathematics education: An example from a longitudinal trajectory on percentage. *Educational Studies in Mathematics, 54*(1), 9–35.

Wedege, T. (1999). To know or not to know—Mathematics, that is a question of context. *Educational Studies in Mathematics, 39*(1), 205–227.

Wistedt, I. (1990). *Vardagskunskaper och skolmatematik—Några utgångspunkter för en empirisk studie.* Stockholm: Stockholms Universitet, Pedagogiska Institutionen

Wistedt, I. (1993). *Att använda barns informella kunskaper i matematikundervisningen–slutrapport från ett uppföljningsprojekt.* Stockholm: Stockholms Universitet, Pedagogiska Institutionen.

12

MATHEMATICAL ARCHAEOLOGY ON PUPILS' MATHEMATICAL TEXTS

Un-Earthing of Mathematical Structures

Ole Einar Torkildsen[1]

VOLDA UNIVERSITY COLLEGE

Abstract: This chapter reports research falling within the naturalistic inquiry paradigm; it is a study of pupils' collaborative solutions of some open-ended tasks. The purpose of the analysis is to make explicit the mathematical structures inherent in the solutions rather than arithmetical errors or inferred cognitive processes. The method entails decoding the learners' language into the formal mathematical language. The analysis reveals that the pupils in their solving procedures relied heavily on a few fundamental mathematical structures. In each solution at least one relationship was stated; some of the relationships had an explicit structure, while others had a recursive structure. It is argued that the mathematical structures inherent in the solution procedures can be localized at two levels. A top level that embraces the mathematical structures that can be identified in the relationships or rules that is clearly stated in the answers, and an underlying level that consists of mathematical structures applied in the solving procedures, but not emphasized or (directly) focused in the stated relationships.

Keywords: Mathematical archaeology, mathematical structures, open-ended tasks, control of knowledge, decoding

INTRODUCTION

The aim of this chapter is to present research that focuses on unearthing mathematical structures mathematical texts, written by pupils age 6 to 16, in Norwegian compulsory schools. This unearthing process I have characterized as mathematical archaeology.

The analyzed texts are pupils' collaborative solutions on some investigational tasks presented in a competition organized by a Norwegian journal for mathematics teachers in the beginning of the 1990s. The same tasks were set for all ages.

Together with this introduction, the chapter is structured in seven sections. The next section provides a brief outline of the contextual background for the study—in particular, the founding of the journal and the competition for school classes. The contribution of Stieg Mellin-Olsen and mathematical archaeology is the focus for the third section. A theoretical framework is described in the fourth section; this is based on a brief explanation of mathematical structures and education, and open-ended tasks. The fifth section explains the methodology of the research presented here, and gives a short description of the data collection and an in-depth discussion of the method of the analysis. The purpose of the analysis is to make explicit the mathematical structures inherent in the pupils' solutions. The only concern of the analysis is the uncovering of mathematical structures; it is not aimed at exposing possible arithmetical errors or inferring cognitive processes. Examples are provided to demonstrate how the analysis method is applied. The conclusion, which summarizes the main findings of the study, is given in section seven.

The First Sourcebook on Nordic Research in Mathematics Education, pages 181–195

BACKGROUND

The Founding of Tangenten and the Competition for School Classes

Research perspectives shift. During the 1960s and 1970s, a large proportion of the research in mathematics education focused on the individual child learning mathematical concepts where the context was the child in an isolated situation—that is, a "laboratory" rather than a classroom. Gradually during the 1980s an increasing body of the research focused on children's learning of mathematics or mathematical thinking in the context of regular classrooms. The shift in research perspective was from the individual child in an isolated situation to children in a social situation. During that period, in 1990, in the mathematics didactics milieu in Bergen where Professor Stieg Mellin-Olsen was at the forefront, the journal, *Tangenten—Tidsskrift for matematikk i grunnskolen* [*The TANGENT—Journal for mathematics in compulsory school*], was founded. In texts the abbreviated title *Tangenten* [*The Tangent*] is the most frequently used, and it, or the English translation, will be used here. The subtitle, *Journal for mathematics in compulsory school*, indicated that *The Tangent* was intended for teachers in primary and lower secondary school. (In 1996 the subtitle changed to its present form: *Journal for mathematics teaching*.) The first issue of the journal, which was edited by Mellin-Olsen, came out in September 1990.

The founders of the journal decided that each issue of *The Tangent* should include a section in the form of a competition for school classes. I was asked to edit this section, which meant responsibility for finding suitable competition tasks, for evaluating the pupils' answers, and for commenting on their work in the journal (Torkildsen, 1991a, 1991b, 1991c, 1991d, 1992a, 1992b, 1992c, 1993). In total eight tasks were constructed, and a total of 25 solutions were received. The number of competing classes was never high, and as a result, after just eight rounds, the competition came to an end in September 1992.[2]

The evaluation of the pupils' responses revealed that some of the solutions were quite unexpected and surprising. The unexpected features were manifested in different ways. Some of the work revealed an originality and inventiveness in the way the pupils attacked the tasks, and hence their solutions, others showed originality in the way they argued. This variation was found both between the solutions on the same task and between the tasks. It was observed that the use of mathematical symbols, to a large extent, varied from solution to solution. There were instances where few mathematical symbols were used and others where several symbols were used. For a few of the solutions there appeared to be no correspondence between the use of mathematical symbols and the mathematical richness of the solution. The use of few mathematical symbols did not necessarily imply that a solution could be classified as mathematically "meagre" or "poor." Disregarding whether the pupils in their solutions had used few or many mathematical symbols, the solutions were the result of an organizing and structuring activity, *mathematizing*.

Mathematizing is here used in the way defined by Treffers (1987). According to Treffers (1987), mathematizing has both a *vertical* and a *horizontal* component. The horizontal component relates to "transforming a problem field into a mathematical problem" (Treffers, 1987, p. 247) or, in Freudenthal's words, "from the world of *life* to the world of *symbols*" (Freudenthal, 1991, p. 41). The vertical component relates to "processing within the mathematical system" (Treffers, 1987, p. 247), or to stay in "the world of symbols" where "symbols are shaped, reshaped, and manipulated, mechanically, comprehendingly, reflectingly" (Freudenthal, 1991, pp. 41–42). However, Freudenthal (1991) stressed also that there is no clear distinction between these two directions, the distinction is dependent on several factors such as the specific situation, the person involved, and the environment. The mathematical richness in the pupils' work could be classified according to those two directions, the vertical and/or the horizontal. The horizontal component, which transforms a problem into a mathematical problem, lays the foundation for using a mathematical language, and hence the use of mathematical symbols. The vertical component elaborates and stays inside the "world of symbols." This means that the horizontal component ends with or inside "the world of mathematical symbols" and the vertical component is inside that world.

A Pedagogical Philosophy and the Contribution of Mellin-Olsen

Even though Mellin-Olsen was not responsible for construction of the competition tasks, he had a very strong and indirect influence through both the competition rules, and the guidelines concerning the tasks, which he communicated orally to me as the one responsible for running the competition.

At the first Political Dimensions of Mathematics Education (PDME) conference in London (1990) in a paper entitled "Liberation of knowledge," Mellin-Olsen spoke about the notion *control of knowledge*. However, he first introduced and developed the notion in Norwegian in his book *Kunnskapsformidling* [Knowledge dissemination] (Mellin-Olsen, 1989). Mellin-Olsen considered knowledge as tool for a goal-oriented activity. He exemplified with a teaching project where mathematics was to be used, and this raised many questions:

- Who takes the initiative in the project? The teacher or the learners?
- Is the project idea discussed between the teacher and the learners?
- Do the learners have any impact on the direction of the project?
- What about the tools? Who decides which tools to use—the teacher or the learners?
- Is it expected or required that particular parts of the mathematical syllabus should be applied?

The overall questions are:

- To what extent does the teacher support the learners?
- To what extent does the teacher scaffold the learners work in the project/project work?

These questions all focus on the relation between the teacher and the learners: "who is in the control of ______________?" In that respect Mellin-Olsen used the word *independence*. If a learner can use or apply her knowledge on her own, independent of the teacher, the learner controls her knowledge. The teacher is no longer scaffolding the learner. The learner can manage her knowledge on her own, without assistance of the teacher. However, Mellin-Olsen stresses that control of knowledge also has an aspect of meta-knowledge. To have control of knowledge implies that the learner has realized that she can apply the knowledge when it is relevant, and more, that she is aware it is her right to use the knowledge.

Suppose a didactical activity where the passing on of knowledge plays an important role. In such an activity Mellin-Olsen identifies three different levels where a learner and a teacher can negotiate about the control of knowledge.

Goal level: A didactical activity has a purpose; some goals are supposed to be achieved. Who is to control these goals? The teacher? The learners? Is there a negotiation process? When the activity develops, do the goals change? If so, who is controlling the development of the activity? Will the focus or direction change when the activity is developing?

Choice level: Knowledge is a tool. Who controls the choice of suitable tools when a learner is working with or on a problem? The teacher or the learners?

If the solving of the problem calls for adjustments of the available tools, or even construction of a new and appropriate tool, who is deciding what to do when the solving process is running?

Tool level: The learners work with or on a problem. Who is controlling the use of the tool? The learners or the teacher? Is it expected that the learners shall use the tool in a particular way?

Related to the solutions analyzed in this study, I need to point out: When the pupils started the problem-solving process, the goal was to solve the given task. That meant that the pupils did not control the goal level when they started the solving procedure. However, the decision to engage in the task was voluntary, which meant that the engagement in the solving process was in the hands of the pupils. The two other levels were controlled by the pupils. Mellin-Olsen (1990, 1993) argues that if the pupils controlled these three levels the intensity of their engagement will increase.

At the end of the 1980s and the beginning of the 1990s, Mellin-Olsen worked on his book *Hvordan tenker lærere om matematikkundervisning?* [*How do teachers think about mathematics teaching?*] (Mellin-Olsen, 1991). In this book he discusses the role that tasks play in the teaching of mathematics. One of his findings was that the teachers believe that the learning of mathematics in the mathematics lessons is predominantly linked to the solving of closed exercises—that is, exercises that have a beginning and an end. The pupils were not invited to discuss or reflect or to come up with their own problems. The teaching of mathematics was classified as a journey through that type of exercises where the goal was an examination or the next grade. For that he introduced the notion *the discourse of exercise*. Mellin-Olsen used the Norwegian word *Oppgavediskursen*. There are at least two possible translations into English: the discourse of exercise or the discourse of task. Skovsmose (2005) used the first of these two. On the journey through the exercises the stops are few, and the time spent at each stop is limited. The discussion can thus be poor since there is nothing to discuss. Therefore Mellin-Olsen maintained that this type of exercise could increase the instrumentalism (1991). However, he also discerned among the teachers another discourse, a discourse that invited pupils to reflect and discuss. There can be no doubt that Mellin-Olsen advocated this alternative discourse. He maintained that it is possible to construct tasks rooted in significant features of the pupils' environment—that is, tasks that could lead the pupils into mathematics and that also challenged their problem-solving capacity and creativity. He was also of the opinion that it would be possible to construct tasks rooted in pure mathematics that would have the same quality. My own experience as a mathematics teacher in teacher education confirmed that it was possible to find such tasks, and that students enjoy working with such tasks.

In the decade before the competition began there had been a growing interest in mathematics education for promoting collaboration and communication in mathematics teaching (cf. Christiansen, 1990c; Christiansen & Walther, 1986; Dekker, 1987; Freudenthal, 1973, 1978, 1991; Hoyles, 1985; Schoenfeld, 1992; Steen, 1988). Also, collaboration and communication were two central pedagogical philosophical features of Mellin-Olsen's philosophy of mathematics education and mathematics teaching (Mellin-Olsen, 1987, 1989, 1991, 1996). It is therefore quite natural that the competition rules reflected these main pedagogical philosophical principles.

The first rule required pupils to collaborate on solving a task, and as a consequence it was assumed that the group of pupils, and the class as a whole, could handle more complex tasks than pupils working individually. The second rule gave strong guidelines for what may be called the technical standards of the tasks. According to that rule, and the personal information from Mellin-Olsen, it was one task for all grades, which meant it should be possible for the pupils in all grades to work on the same task—this would imply that solutions could be on different levels. Concerning the construction of the tasks, this rule was primarily influential in two respects: the numbers that could be used in the tasks, and the arithmetic operations performed on those numbers.

In addition to the written competition rules, Mellin-Olsen orally provided further guidelines. It would not be accurate to imply that these guidelines had the form of an instruction; it would be better to characterize them as strong wishes from Mellin-Olsen. He wanted the tasks to have a broad perspective, which meant that it should be possible to extend the tasks beyond what was asked for in the competition, and if possible, to extend in several directions. He also wanted to promote the alternative discourse of reflection and discussion. With respect to the tasks this was a particularly important guideline, and a fundamental requirement was that the tasks should be *open-ended.*

These guidelines together with the competition rules and a wish to emphasize the underlying pedagogical philosophy of the competition—collaboration and communication—made rather stringent demands for the construction of the tasks. It was desirable that a task could be solved in different ways—that is, that there was not one and only one way to solve a task, and a task should give room for solutions at different levels. These requirements, to be more precise, wishes, were also in line with my own personal pedagogical principles, and they pointed towards open-ended tasks.

Related to the context, the tasks can be sorted into two categories. Some tasks had their context inside the "world of mathematics"; others draw on contexts outside that world. In the *Tangent* articles that commented on the pupils' solutions (Torkildsen, 1991a, 1991b, 1991c, 1991d, 1992a, 1992b, 1992c, 1993), these two directions, the horizontal and the vertical, were, to a greater or lesser degree, present. Which component, and the degree of presence of these two components, depended mainly on the task. For the tasks of the inside category only, the articles focused on the vertical component. Both components, the horizontal and the vertical, were present in articles that commented on tasks from the outside category.

The structure in the *Tangent* articles was more or less the same. For the actual task at least one formal solution was given, for some tasks several solutions were presented. The number of solutions given was dependent on both the task and the number of different solutions given by the classes. In these articles the pupils' solutions were compared or related to the(se) formal solution(s). In order to simplify this comparison, sections of pupils' solutions were translated or decoded into a formal mathematical symbolic language. The result of this decoding process was a product in which the mathematics or the mathematical richness used by the pupils in their solution procedures became more visible than in the original work. The mathematical symbolic language unearthed or brought to light, to a certain extent, the mathematics used by the pupils in their solution procedures. This decoding process could also be characterised as *mathematical archaeology.*

Mathematical Archaeology

The notion *mathematical archaeology* is relatively new. To my knowledge it was introduced by Ole Skovsmose in his book, *Towards a Philosophy of Critical Mathematics Education* (1994). In this he asserts that mathematical archaeology can be interpreted in two ways, one being *global,* and the other *educational.* The global interpretation he applies to excavating or bringing to light the mathematics hidden and used in "social structures and routines" (p. 95). In a later paper, Skovsmose described mathematical archaeology as "the process of excavating mathematics which might be encapsulated in certain political arguments, technologies or administrative routines" (1999, p. 14). A process of clarifying or "bringing to light" the mathematics that in many instances in our society has become a nearly tacit and invisible actor in decision-making processes, is an example of mathematical archaeology. The global interpretation of mathematical archaeology can be characterised as reflecting on the sociopolitical dimensions of mathematical models being developed or applied.

Skovsmose (1994) stresses that it will make sense to carry through a mathematical archaeology in an educational setting. The educational interpretation rests on the fact that even in educational settings, mathematics can be hidden or invisible to both pupils and teachers. Mathematics can be so integrated in the pupils' activities that neither pupils nor their teacher realizes that the pupils are doing mathematics when the activity is performed. After engaging in the activity, pupils and teacher are not able to identify the mathematics carried out. Skovsmose (1994) points out that it therefore will make sense to carry through a mathematical archaeology in an educational setting in order to be able to identify the mathematics in such contexts. Mathematical archaeology in the classroom should focus on the fact "that some of the activities carried out in the classroom... are in fact mathematics" (Skovsmose, 1994, p. 96).

Outside the classroom, ethno-mathematical research programs in many instances exemplify a mathematical archaeology focusing on the pure mathematics inherent in activities carried out in non-mathematical contexts—such as street mathematics (Nunes, Schliemann, & Carraher, 1993), mathematics found in decorations (Gerdes, 1995), and mathematics found in handicrafts such as the weaving of baskets (Gerdes, 1986) and in knitting (Mellin-Olsen, 1989). A more comprehensive overview of ethno-mathematical studies is found in Gerdes (1996).

However, an archaeological excavation can also be made in a mathematical and educational context. A mathematical archaeology in a mathematical and educational context can be done within at least two different arenas. It can be used in order to uncover the pure mathematics that is hidden or embedded in pupils' mathematical activity, and it can also be used for analyzing teaching elements constructed by educators. It is the first of these two arenas, the pupils' activity, upon which the study I report here focuses.

The extent to which the classes applied mathematical symbols in their solutions varied. In some instances it appeared that there was no correspondence between the use of mathematical symbols and the mathematical richness of the solution. The *explicit making* of the mathematics inherent in these works was the goal of the research, and the question I was grappling with was: *How can the mathematics unearthed from the pupils' solution procedures be characterized?*

This question was also discussed with Mellin-Olsen, and the outcome was that mathematical structures could be an appropriate theoretical framework. With this as a background, the following research question was formulated: *What kind of mathematical structures are inherent in the solution procedures that pupils have applied in some of the open-ended tasks used in the* Tangent *competition?*

THEORETICAL FRAMEWORK

Mathematical Structures and Education

Mathematical structures had a very strong position in the modern mathematics reform programs. After the "demise" of modern mathematics in the 1970s it seems that the notion of mathematical structures more or less was largely avoided by the researchers in mathematics education (Breiteig & Goodchild, this volume, Chapter 2). Gradually, from the beginning of the 1980s, the term *mathematical structures* began to regain currency among researchers in mathematics education. It was used by different authors and in a broad spectrum of areas. One highly influential mathematics educator who actively used this term in the 1980s was Hans Freudenthal. He considered the meaning of the term mathematical structures and discussed it on a broad basis and in an educational setting (Freudenthal, 1983, 1991). He offers many examples of his conception of mathematical structures (i.e., length, natural numbers, graphs, functions, etc.). My interpretation of Freudenthal's definition of mathematical structures, as explained in Freudenthal (1983, 1991), is: A mathematical structure consists of a set and at least one relation on that set, and these relations (or eventually the relation) are not chosen at random, but have to meet some given postulates or axioms.

There are some requirements a mathematical structure should meet. Since it is indisputable that variables are needed in mathematics, it must be possible to handle variables and/or constants in a mathematical structure. Another requirement is that it should be possible to construct or make expressions consisting of variables and/or constants, and thirdly it should be possible to make statements. This leads to the following definition of a mathematical structure.

A mathematical structure consists of

1. a set
2. one or more operations
3. one or more relations

This definition, which I adopted in my study, does not differ markedly from the definitions found in the literature; see for example Freudenthal (1983).

The definition has a number of implications: The set ensures the existence of variables and/or constants. The operation(s) make it possible to construct expressions, and the relation(s) imply that statements can be made. For example, the natural numbers, the operations of addition and multiplication, and the identity relation constitute a mathematical structure. Other examples are functions and equivalence relations.

However, other mathematical educators who have used the notion of mathematical structures have not agreed on the meaning of the term (cf. Bell, 1976; Christiansen, 1969; 1990a, 1990b; Dreyfus & Hoch, 2004; Hoch & Dreyfus, 2004; Kieran, 1992; Mamona-Downs & Downs, 2002; Schoenfeld, 1992; Steen, 1990; Van Dormolen, 1986). It is realized by these educators, as observed by Hoch and Dreyfus (2004, p. 50), that "[the term] is used in the field of mathematics education to cover various different meanings."

Open-Ended Tasks

During the 1960s, *investigational* activity was introduced in mathematics teaching in Britain. This became enshrined, for a while, in the national rhetoric by the Cockroft (1982) report, which recognized that investigations play an important role in mathematical activity in school. The use of investigational tasks or open problems in mathematics education was widespread in the 1980s, and the research on the use of such tasks increased (cf. Boaler, 1998; Pehkonen, 1995).

A task has both a starting situation and a goal situation. If a situation, either starting or goal, can be exactly explained, the situation is called closed; if not the situation is called open (see Table 12.1). It is usual to characterize type A as a closed task.

A categorization of tasks was done in the AMI project—*Applying Mathematics International.* The AMI project operated also with four different types or categories of tasks: *word problems, problem solving, investigations* and *mathematization* (Van den Heuvel-Panhuizen & Van Rooijen, 1997). Each of these types was classified or characterized by a set of characteristic qualities, and for each type an example was given. Related to the starting situation and goal situation, the following characteristics are found (van den Heuvel-Panhuizen & van Rooijen, 1997).

TABLE 12.1 Different Types of Tasks

Task type	Starting situation	Goal situation
A	Closed	Closed
B	Closed	Open
C	Open	Closed
D	Open	Open

Word problems:

- Operations can in most cases be deduced from the text
- There is nearly always one and only one good answer

Example: Sporting club Atilla had, on the 1st of January, 234 members. Over the course of the year, 58 members left the club, and 113 new became members.

How many members did the sporting club have at the end of the year?

Problem solving:

- The solving strategies cannot be deduced directly from the problem text
- In many cases to a trial-and-error attempt is resorted to
- The solution has an all-or-none character; semi-good solutions do not in most cases exist
- In most cases there is always one good solution

Example: Here is a sketch of land belonging to farmer Jenkins. He wants to divide this land into four parts that are alike. Each part has identical form and surface.

Find out how this can be done.

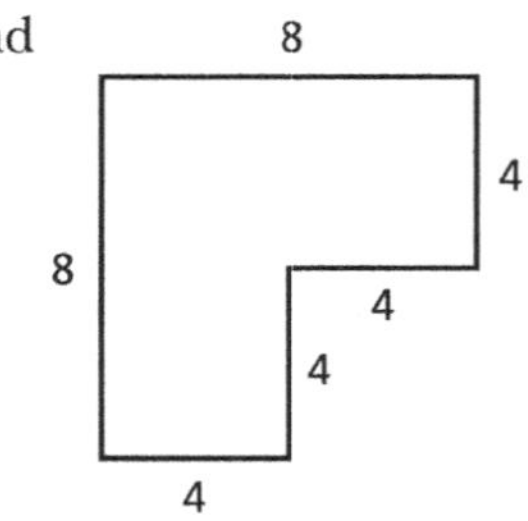

Investigations:

- In most cases these are descriptive and directed towards a goal
- In many cases they are infinitely open (the Dutch text: onbegrensd open)
- Often one must choose descriptive goals; that is, more than one correct solution is possible

- Often it is not possible to decide what a good solution is.

Example: Find out the approximate distance covered by your walking during one day.

Mathematizations:

- In many cases it is a question of adapting personal goals
- The information given in many cases has to be organized and/or completed and/or omitted
- Often it is possible to solve with more than one good solution
- In most cases it is not possible to tell what a good solution is
- The operations cannot be deduced directly from the problem text.

Example: How big is this flag in reality?

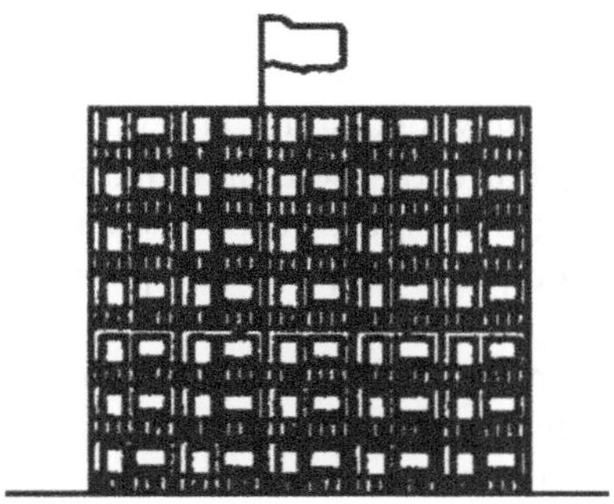

Since some of the characteristics are used for more than one category it should be noticed that there is no clear distinction between the categories.

The links or connections between the description in Table 12.1 and the AMI project are: Word problems and problem solving are identified as type A; investigations are classified as type B, and mathematizations as type D. Earlier I explained that the competition tasks should be open-ended, which means that the goal situation is open—that is, tasks of type B or/and type D. In order to separate these two from each other it is convenient to name each type, and I found it appropriate to adopt the AMI classification. The tasks used in the competition were all of type B—investigations.

The reason for using this type of task in the competition was pragmatic. Through my teaching experience as a teacher educator I had observed that investigations had the qualities that met the pedagogical philosophy manifested in the competition rules and additional guidelines set out by Mellin-Olsen. In the context of mathematics classrooms in Norwegian compulsory schools in the beginning of the 1990s, it was believed that the use of mathematizations would be exceptional. For most of the pupils and the teachers, this type of task would be unfamiliar. A task with a closed starting situation would, on the other hand, be more familiar to both pupils and teachers. The number of competitors would most likely be higher if the tasks used in the competition were investigations and not mathematizations.

The structure of the tasks as presented did not vary greatly. A context was given that could be purely mathematical or an approximation of an everyday phenomenon that could be mathematized. Blum's characterization of one type of problem or situation applied to mathematical modeling is an appropriate description of the latter context: "an artificial dressing-up of a mathematical problem" (Blum, 1993, p. 4). In some instances the context was illuminated by an example. An example is the task "A Set of Ten Cubes," used in *The Tangent 1*(2):

> A child has a set of ten cubes.
>
> One cube has 1 cm edge
>
> One cube has 2 cm edge
>
> One cube has 3 cm edge
>
> And so on, until the largest cube has 10 cm edge.
>
> The child tries to build two towers of equal height.
>
> Is it possible for the child to do that construction if she/he wants to use all the cubes?
>
> Show how, or explain why it can not be done.
>
> If the largest cube, the cube with edge 10 cm, is removed, is it then possible to build two towers of equal height (still using all the remaining cubes)?
>
> What if also the cube with 9 cm edge is removed?
>
> What about the building of two towers of equal height if, in addition to the ten original cubes, a cube with edge 11 cm is added?
>
> Investigate this problem for other numbers and try to find a rule/ a pattern.
>
> Use the rule/pattern and solve this problem:
>
> If we have a set of twenty different cubes with edges from 1 cm to 20 cm, is it possible to build two towers of equal height, using all the cubes?
>
> Try to explain why your rule is true.
>
> An extension of the problem could be:
>
> If it is possible to build two towers of equal height with a set of 11 different cubes [with edges from 1 cm to 11 cm], find as many ways as possible to do this construction.

From the given starting point, the pupils were invited to work through investigations. Based on these investigations, they were challenged to formulate relationships or state one or more rules. For all the tasks, what may be called a *guided path* conducted the pupils in their investigational activity. One purpose behind this guided path was that it should be possible for the pupils from grade one to grade nine to enter the task, and start working on it. The intention was that the threshold to step into the task should be sufficiently low to enable pupils to produce or generate data that could act as a basis for an exploration fairly quickly and that led to a statement of relationships or rules. How to organize or to systematize these data was, however, an open question—that is, the pupils were responsible for devising their own means of organization or systematization of the data. For most of the solutions offered, the organization took the form of tables. For some of the tasks, the pupils were also challenged to provide a proof or argue for the validity of the stated relationship(s) or rule(s).

METHODOLOGY

Data Collection

The data corpus consists of pupils' solutions to the first six competition tasks. This comprised 23 solutions altogether. The number of solutions per task varied from one, the task set in *The Tangent 2*(3), to seven, the task presented in *The Tangent 1*(2).

All solutions were carried out in the pupils' ordinary class context, which meant that I was not present when the pupils worked on their solutions. With this as a background, significant questions concerning the *independence* of the data arise. Independence in this context means that the pupils really produced the data, and that they had worked without their teachers' intervention—that is, that the data were the pupils' self-obtained solutions. From the competition rules it was emphasized that the teacher should play a passive role when the pupils were working on a task, a role that may be characterized as the role of a non-participant observer. The rules stated that the teacher was only allowed to assist the pupils in interpreting the problem text. For all of the solutions, with one or possibly two exceptions, it can be reasonably assumed that the solutions were produced and written by the pupils. In the exceptional case the solutions have the form of a report written by the teacher(s) (Torkildsen, 2006).

Analysis Method

The purpose of the analysis was to make explicit the mathematical structures inherent in the pupils' solutions of the open and investigational tasks set in the competition. Since the interest is in analyzing the produced solutions of the pupils in terms of mathematical structures, the analysis can be viewed as a decoding process. A brief description of this method could be analyzing the pupils' solutions through the lens of a mathematician. The method is thus one of translating or decoding the artifacts, which are manifest in the learners' language, into a formal mathematical language.

A first level of this analysis method is illustrated in Figure 12.1. In this context the term "mathematical language" means a formal mathematical language—a language that uses mathematical symbols/notations. The forms of solutions could be characterized using Pimm's (1987) classification of verbal, mixed, and symbolic styles. For some of the solutions or parts of a solution, this decoding process was relatively simple; the mathematical meaning of the text was clear, and a *direct* decoding was possible. This was the case when the pupils had already used mathematical symbols in their solution; that is, they used a *symbolic* style of recording. Direct decoding was also possible when the pupils used a *mixed* style of recording in their solutions, which means that the pupils used their natural language together with some mathematical symbols. A direct decoding can, however, also be possible and can be utilized if the pupils use a *verbal* style of recording; that is, very few or no mathematical symbols are used by the pupils in their solutions but the natural language is used in such a manner that the "mathematical language" can be easily inferred.

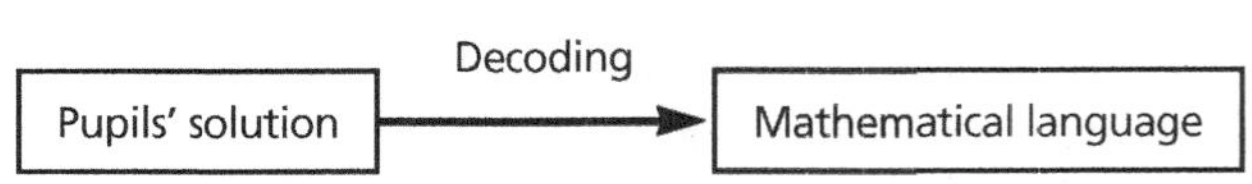

Figure 12.1 Direct decoding process.

In some cases where direct decoding is not straightforward, the decoding process leads to a product with gaps. This situation is shown in Figure 12.2. This will usually be the situation if a solution is ambiguous—that is, for the whole solution or parts of the solution there are several decoding possibilities. Missing or inaccurate use of words and symbols and/or unexpected use of or combinations of words could cause such ambiguity.

When a group of people work on a particular problem or task for a period of time, the members of the group obtain specific knowledge about the problem, results, a solution procedure, and a solution. For the members of the group some of this knowledge may be understood tacitly. A consequence of this could be that the written solution is comprehensible to the members of the group, but could be hard to comprehend or even incomprehensible for a person outside the group, and hence the decoding process is better represented by Figure 12.2.

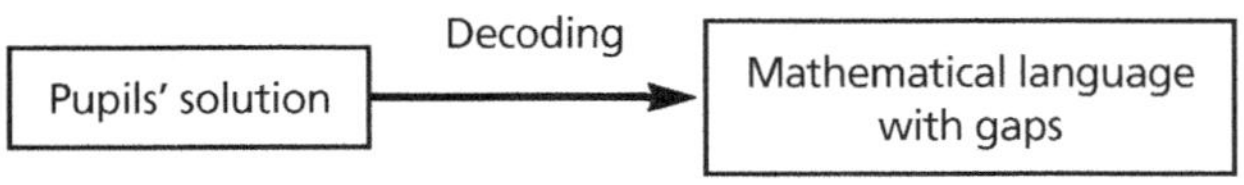

Figure 12.2 Incomplete decoding process.

The goal for the decoding process is the presentation of a solution in a mathematical language without gaps, as illustrated in Figure 12.1. When the decoding process ends with a product dressed up in a mathematical language with gaps, is it both necessary and required to fill the gaps in order to get a solution written in a mathematical language. In doing so, an emendation process is applied. This situation is illustrated in Figure 12.3.

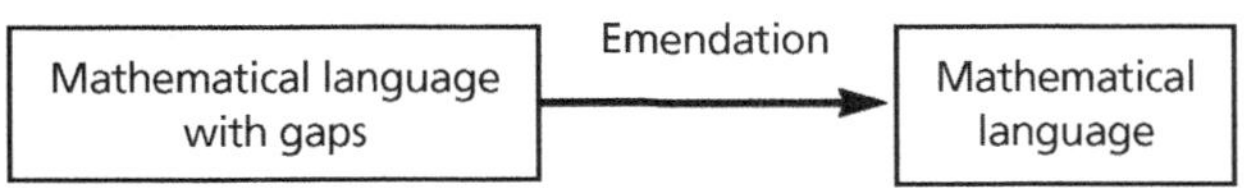

Figure 12.3 Emendation process.

It is not possible, in advance, to describe the outcome of the emendation process. This process is dependent of several factors. The main factor is the text under consideration, and another factor is the nature of the gaps. During the emendation process it is important that the focus on the text is not too narrow. It is necessary to focus on a particular gap and the corresponding part of the text, but at the same time a holistic view of the text could be of great importance. Text written elsewhere in the document could provide the key to a successful emendation process. There is also another aspect that can be taken into consideration: the gap or the gaps mean that some text is missing. Reflection or speculation over why it might be missing can be useful.

It is by no means certain that an emendation process is unambiguous. This process could end up with more than one possible mathematical statement. In this case it will often only be possible to argue which of the statements are the most likely. There is also the possibility that the outcome of the emendation process is "empty"—that is, that it is found to be very difficult or impossible to fill the gaps.

During the decoding and/or the emendation processes, there is a risk of wishful thinking, to over-interpret the pupils' solutions, or to discover or expose more mathematics than is actually embedded in the solutions. In order to eliminate or reduce eventual over-interpretation, it is necessary to "look back." This means that after finishing the decoding process and the possible emendation process, it is necessary to appraise the results against the original text. In order to minimize the wishful thinking, the outcome of the decoding process should be evaluated and/or discussed with knowledgeable persons.

All of the preceding figures have a linear form, but that does not imply that the analytical process is linear. A triangular illustration (Figure 12.4) of the complete analytical process may be more appropriate. Compared with the earlier figures, this has double arrows, indicating that a "looking back" process, an evaluation, takes place not only at the end of the decoding process or the emendation process. The looking back processes are intertwined with both the decoding process and the emendation process; they occur concurrently and continuously.

To summarize, the approach taken is one in which a knowledgeable person analyzes the pupils' self-generated mathematical artifacts in terms of mathematics. If this process leads with some certainty to a known formal mathematical product, then this product is declared the most probable one that conforms to the learner-generated product. Regarding the methodological demands of research, the trustworthiness of the analysis lies in descriptions that are so factual and convincing that others can in some way re-walk the analysis. Thus both the activity and the decoding process should be open for scrutiny by others.

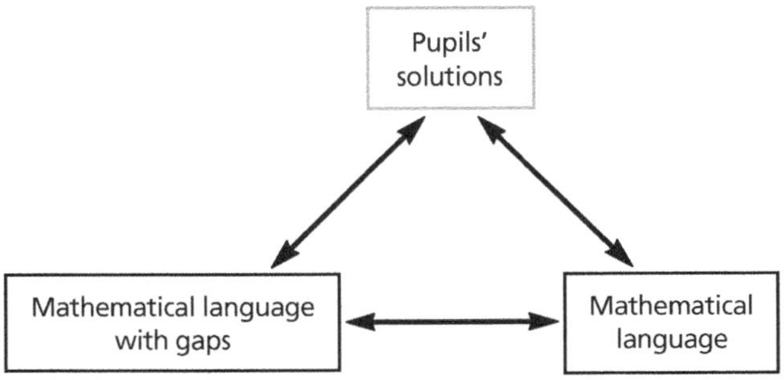

Figure 12.4 Complete decoding process.

AN EXAMPLE OF THE ANALYSIS

The Hundred Square (Tangenten 2(1))

1	2	3	4	5	6	7	8	9	10
11	12	13	14	15	15	17	18	19	20
21	22	23	24	25	26	27	28	29	30
31	32	33	34	35	36	37	38	39	40
41	42	43	44	45	46	47	48	49	50
51	52	53	54	55	56	57	58	59	60
61	62	63	64	65	66	67	68	69	70
71	72	73	74	75	76	77	78	79	80
81	82	83	84	85	86	87	88	89	90
91	92	93	94	95	96	97	98	99	100

If we pick out a 3×2-rectangle from the table, for example

72	73	74
82	83	84

and multiply the numbers in opposite (diagonally) corners, $72 \cdot 84 = 6048$ and $74 \cdot 82 = 6068$, we find that the difference of their products is 20.

a. Pick out other 3×2 rectangles and carry through the same procedure.
   What are your findings?
b. Explore this procedure for different type of rectangles, e.g.,
   i. 4×2
   ii. 5×2
   iii. 4×3
   What are your findings in each of the cases?
c. Try to formulate a rule or a pattern (relation)!
d. Can you explain (prove) why your rule for the 3×2 rectangle is true?

It should be noticed that the symbol $\times$ does not mean or indicate the multiplication operator. In Norway the symbol for the multiplication operator is a dot ($\cdot$). The symbol, $\times$, is frequently, or usually, used to indicate the size or the dimension of a rectangle: *length* $\times$ *height.*

It was intended that the text of the task should be organized in such a way that it guided the pupils to carry through systematic investigations. These investigations had to be carried out in several directions, which means that several independent variable quantities became an integral part of the task—quantities such as:

- the position of the figures—that is, where in the table (hundred-square) the rectangles have been located
- the size (dimension) of the rectangles

The sizes of the rectangles depend on two independent variables, or one variable if the rectangle is a square.

It is appropriate to introduce some symbols and terms. The complete process, picking a rectangle from the given table and calculating according to the given algorithm, is called the *rectangle algorithm* or shortly the *algorithm.* For this algorithm I use the symbol $R_{m \times n}(x)$, where m and n respectively mean length and height of the rectangle, and x is the number that is located in the upper left corner of the rectangle. The variable x is then a variable for the position to the rectangle, and m and n for the size.

The following solution was submitted by a grade 3 class (pupils aged 9–10 years).

The outcome of the decoding process, based on their investigations of the rectangles, can be summarized in the statements: $R_{3\times 2}(x) = 20$, $R_{4\times 2}(x) = 30$, $R_{5\times 2}(x) = 40$, and $R_{4\times 3}(x) = 60$.

Each of these four statements was illustrated or supported by three examples. That the class just included three examples for each of the rectangular sizes may look like it was "typographically" motivated. Each of the three examples fitted nicely on a page in their written answer. The solution was a collaborative work made by the class. It was the end-product of the collaboration, it is therefore very likely to suppose that the statements, mentioned above, rest on far more examples than the ones written in their solution. On the basis of their findings the class formulated the relationship in Figure 12.5.

As seen from the quotation the class has written $3 \cdot 10 = 20$. Compared with what they wrote earlier, where they stated that $R_{4\times 2}(x) = 30$, this is most likely a writing error, and as such, an error of no significance for my research. It is noticed that the class first stated a general rule, which is explicitly formulated, and then the rule is illustrated with an example.

In the stated rule, they have not explicitly emphasized that "rectangles of equal size gives identical answers"; that is, the position of the rectangle in the table does not have any significance for the answer, for the outcome of the algorithm. Most likely that is tacitly understood and accepted, since they claim that the result of the algorithm only depends on the size of

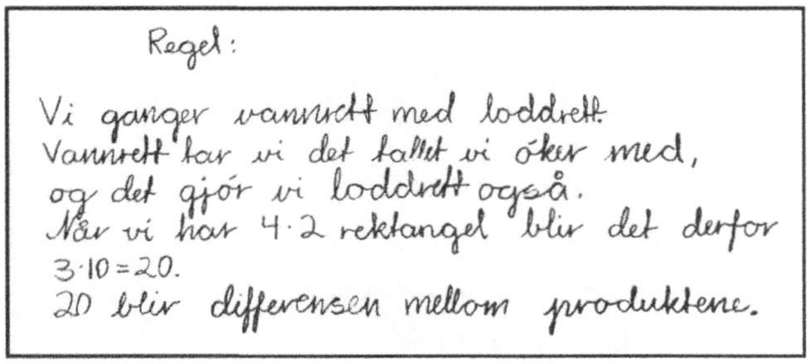
Regel:
Vi ganger vannrett med loddrett.
Vannrett tar vi det tallet vi øker med,
og det gjør vi loddrett også.
Når vi har 4·2 rektangel blir det derfor
3·10=20.
20 blir differensen mellom produktene.

The Norwegian text reads:

Rule:

We multiply horizontally by vertically.
Horizontally we take the number we increase with,
And that we also do vertically.
When we have a 4 · 2 rectangle it therefore becomes
3 · 10 = 20
20 is the difference between the products.

Figure 12.5 Facsimile of the solution to the class (from p. 3 in the solution).

the rectangle, and not on its position in the table. The size of the rectangle is described: "Horizontally . . . the number we increase with, and . . . also vertically" (see stated rule, above), respectively length and height in the given rectangle. A direct decoding process is thus possible. Using the mathematical symbols, the rule stated by the class can be presented as:

> The number we increase with in the horizontal direction is $m-1$, in the vertical direction the increase is $10(n-1)$. Multiplying these two numbers gives $R_{m\times n}(x) = (m-1) \cdot 10(n-1)$.

This relationship is mathematically correct. This interpretation is supported by the example presented in the solution. For a 4×2 rectangle, the increase in the horizontal direction is 3, which is $4-1$, and the increase in the vertical direction is 10, which is $10 \cdot (2-1)$. Related to mathematical structures, this is a function in three independent variables, x, m and n, or what may be more in accordance with the mathematical terminology, an explicit function in one independent variable, x, and two parameters m and n.

A proof of the rule for a 3×2 rectangle was also given by this class. Even though they used fixed number symbols, the line of argumentation carried out in this proof has a general nature (character). In their argumentation the class utilised only what may be called general qualities (characteristics) of the symbols. The proof was carried out as shown in Figure 12.6.

Another solution also given by a class of 9–10-year-old pupils stated: "When the rectangle increases by one e.g., from $3 \cdot 2$ to $4 \cdot 2$ the difference increases with 10."

The stated relationship is not unambiguous or unique; it opens up for different interpretations, and a direct decoding process is not possible. There are two terms in the relationship that have to be considered and interpreted. These terms are "the rectangle increases by one," and "difference." According to the degree of stress that is put on the number example included in the statement, the interpretation leads to statements that can be classified as more or less general.

I first address the interpretation of the word "difference" and argue that the term is used as a name for $R_{m\times n}(x)$. There are several indicatorss that support this interpretation. First, the last operation that is carried out in the algorithm is subtraction, which implies that the answer of the algorithm $R_{m\times n}(x)$ is expressed as a difference, and hence it is practical to call the answer a difference. In addition, the word is used in the text of the task where it is stated "that the difference of their products is 20." Also, this interpretation is confirmed by what the class has written elsewhere in their solution: "All differences within the same type of rectangle are identical."

What then with the term "the rectangle increases by one"? Isolated from this context the term "the rectangle increases by one" is without any meaning, but in this context it has several possible interpretations. What shall be *increased by one*? The *length*, the *height*, or *both*? The *both* alternative, that both the length and height in the rectangle increase by one at the same time, I believe to be very unlikely. The reason for this conclusion is mainly based on the method used when the pupils wrote the results concerning the rectangles; when going from one rectangle to the next, only one of these two variables was increased.

If no attention is attached to the given number example, it means that the statement is valid for all types of rectangles when either length or height is increased by one. If so, the class has formulated a general statement for all the rectangles. Using the earlier introduced symbols this interpretation implies that the relationship can be decoded:

If the length of the rectangle is increased by one then:

$$R_{(m+1)\times n}(x) = R_{m\times n}(x) + 10 \qquad (1)$$

or if the height of the rectangle is increased by one then

$$R_{m\times(n+1)}(x) = R_{m\times n}(x) + 10 \qquad (2)$$

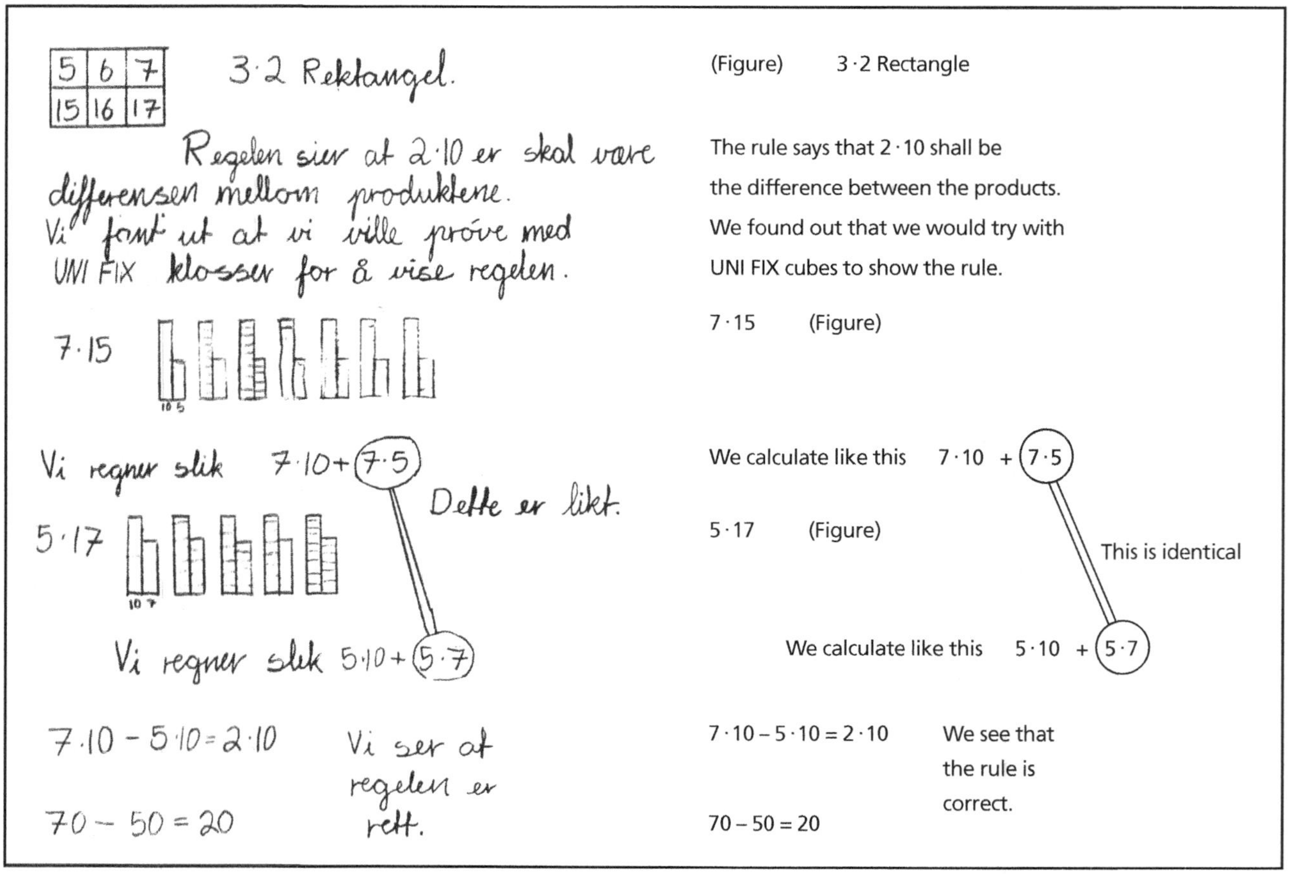

Figure 12.6 Facsimile of the proof to the class. (Norwegian text translated is shown on the right.)

If more weight is given to the number example (i.e., the statement is only valid for rectangles with length 2), the interpretation of the statement is less general. In this case the relationship can be symbolized:

$$R_{(m+1)\times 2}(x) = R_{m\times 2}(x) + 10 \qquad (3)$$

I believe interpretation (3) is most likely one on the basis of the following reasoning: Interpretation (2) is very unlikely since the class has observed that $R_{4\times 2}(x) = 30$ and $R_{4\times 3}(x) = 60$. The increase in the differences is in this case 30. Interpretation (1) is possible. From the pupils' answer it is not clear if they explored any other rectangular types than the four mentioned earlier. If they, for example, also had investigated $R_{3\times 3}(x)$ or $R_{4\times 3}(x)$, they would most likely have observed or discovered that this interpretation was incorrect. However, on the basis of the presented number examples, the interpretation (1) cannot be rejected, but I believe it to be less likely than interpretation (3). The reason for this is that in three out of four presented number examples, the rectangles were of the type $m \times 2$, the same type as exemplified in their relationship. Regardless of which of these interpretations is the correct one, what is common for all of them is the recursive structure.

CONCLUSIONS

As could be expected, the analysis confirmed that the pupils applied different solving procedures. The analysis uncovered that each of the answers stated one or more relationship or rule, and that the solving procedures relied heavily on some fundamental mathematical structures. Some of these statements have an explicit structure, or explicit statements, while others have a recursive structure, or recursive statements. In the solutions submitted, the number of explicit statements was higher than the number of recursive statements. Explicit statements are found in nearly all the solutions. Recursive statements are found in the solutions to four of the six tasks. Mathematically, the stated relationships or rules can be identified as functions, and in some instances functions with more than one independent variable. All the classes used functions in their solving procedures.

Another mathematical structure identified in some of the solution procedures is equivalence relations. This mathematical structure appears in the following way: the data produced by the pupils, which for these tasks were sets of numbers, were separated into disjoint subsets where the elements in each of the subsets had an identical property. This means that the subsets can be identified as equivalence classes, and based on the study or inspection of these equivalence classes, a relationship or a rule was conjectured.

A third mathematical structure identified in the solution procedures is the use of figurative numbers: odd/even, triangular numbers, and square numbers. To be mathematically precise, the odd/even can be identified as the (mod 2) relation, which is an equivalence relation; however, even numbers can also be given as the function $f\colon n \to 2n$ where $n \in$ N, and the odd numbers as the function $g\colon n \to 2n-1$ where $n \in$ N. The triangular numbers and square numbers can be located in the function structure, the triangular numbers with the function

$$f : n \to T_n = \frac{n(n+1)}{2}$$

where $n \in$ N, and the square numbers with the function $g\colon n \to S_n = n^2$ where $n \in$ N.

The mathematical structures inherent in the solution procedures can be localized at two levels, a top level and an underlying level. The top level embraces the mathematical structures that can be identified in the relationships or rules that the pupils have stated in their answers. At this level the mathematical structures are clearly visible. The underlying level consists of mathematical structures that are applied in the solving procedures but not emphasized or (directly) focused on by the pupils in their stated relationships or rules. The mathematical structures found in the underlying level are to a certain extent hidden or concealed.

The function structure is present at both levels. At the top level, functions are identified as directly represented by tables or through the stated rules or relationships. The function structure at the underlying level can be inferred through the solving procedure. For most of the classes, the pupils assumed a stated relationship or rule to be true if they had tested it and found it correct for some examples. In some instances the pupils emphasized that they tested or evaluated their conjecture applying an alternative procedure. In a few cases, pupils argued or even gave a formal proof for the stated conjecture. However, it is necessary to point out that not all the tasks encouraged or challenged the pupils to prove or argue for the correctness of their stated relationships or rules. That proofs occurred in only some of the solutions does not necessarily mean, therefore, that the pupils were not able to carry out a proof. It was not only the oldest pupils that proved or argued for the correctness for the statements; classes at lower grades also did this.

The archaeology carried through in my study revealed that the pupils in their collaborative solutions in some cases used mathematical structures that are not usually taught in school. Most likely, the pupils will first be taught some of these mathematical structures in mathematics courses at university or college level. An example is equivalence relations.

A mathematical archaeology uncovers the past; it gives knowledge about the mathematics identified in works that are finished. This knowledge can be interesting, but not necessarily useful. The usefulness of this knowledge is related to what extent it can be used in order to improve the mathematics teaching.

The explicit-making of the mathematics identified in pupils' mathematical texts has implications in relation to both a *local* and a *global* perspective. The local perspective is related to the practice at the classroom level and concerns the learning process in the individual classroom. It has implications for the teaching of the subject—how to plan and to carry out the teaching, and how to revise and adjust the teaching. This arena is the arena for mathematics teachers, and it is also a field where researchers in mathematics education should play important roles. These roles should be played in close collaboration with the mathematics teachers. The global perspective concerns the school at large, such as development of curricula, the design and development of suitable tasks and teaching elements, education of mathematics teachers, and so on. This arena, the global level, is the field for researchers in mathematics education.

ACKNOWLEDGEMENTS

I would like to thank Cyril Julie for fruitful discussions, comments and suggestions.

NOTES

1. Volda University College, Postboks 500, N-6101 Volda, Norway. oet@hivolda.no
2. *The Tangent* continues to thrive to this day and is the leading Norwegian language printed professional journal for mathematics teachers.

REFERENCES

Bell, A. W. (1976). *The learning of general mathematical strategies. A developmental study of process attainments in mathematics, including the construction and investigation of a process-oriented curriculum for the first secondary year.* Unpublished doctoral thesis, University of Nottingham.

Blum, W. (1993). Mathematical modelling in mathematics education and instruction. In T. Breiteig, I. Huntley, & G. Kaiser-Messmer (Eds.), *Teaching and learning mathematics in context* (pp. 3–14). New York: Ellis Horwood.

Boaler, J. (1998). Open and closed mathematics: Student experiences and understandings. *Journal for Research in Mathematics Education, 29*(1), 41–62.

Christiansen, B. (1969). Induction and deduction in the learning of mathematics and in mathematical instruction. *Educational Studies in Mathematics, 2,* 139–157. (Also in *Proceedings of the first international congress on mathematical education* (pp. 7–27). Dordrecht, Holland: D. Reidel Publishing.)

Christiansen, B. (1990a, September). *Interactive aspects of mathematics teaching and learning.* Paper presented at the 4th SCTP conference Brakel, Germany. Copenhagen: Danmarks Lærerhøjskole (Royal Danish School of Education).

Christiansen, B. (1990b). *Gymnasiets matematikundervisning set i fagdidaktiske perspektiver* [The teaching of mathematics in secondary school from (mathematical) didactical perspectives]. København: Danmarks Lærerhøjskole.

Christiansen, B. (1990c). *Synspunkter vedrørende kvalitet i matematikundervisningen* [Standpoints concerning the quality on teaching mathematics]. København: Danmarks Lærerhøjskole (Royal Danish School of Education). Also in *Tangenten, 3*(1), 10–12 , 21–22, and *3*(2), 10–12, 21–23.

Christiansen, B. & Walther, G (1986). Task and activity. In B. Christiansen, A. G. Howson, & M. Otte (Eds.), *Perspectives on mathematics education* (pp. 243–307). Dordrecht, Holland: D. Reidel Publishing Company.

Cockroft, W. H. (Chairman). (1982). *Mathematics counts: Report of the committee of inquiry into the teaching of mathematics in schools.* London: Her Majesty's Stationary Office.

Dekker, R. (1987). Roos and José, two children in a mixed ability group. *Educational Studies in Mathematics, 18,* 317–324.

Dreyfus, T. & Hoch, M. (2004). Equations—A structural approach. In M. J. Høines & A. B. Fuglestad (Eds.), *Proceedings of the 28th conference of the international group for the psychology of mathematics education Volume 1* (pp. 1-152–1-155). Bergen: Bergen University College.

Freudenthal, H. (1973). *Mathematics as an educational task.* Dordrecht, Holland: D. Reidel Publishing Company.

Freudenthal, H. (1978). *Weeding and sowing* (2nd ed.). Dordrecht, Holland: D. Reidel Publishing Company.

Freudenthal, H. (1983). *Didactical phenomenology of mathematical structures.* Dordrecht, Holland: D. Reidel Publishing Company.

Freudenthal, H. (1991). *Revisiting mathematics education. China Lectures.* Dordrecht, Holland: Kluwer Academic Publishers.

Gerdes, P. (1986). On culture. Mathematics and curriculum development in Mozambique. In M. J. Høines & S. Mellin-Olsen (Eds.), *Mathematics and culture: a seminar report* (pp. 15–41). Rådal (Bergen): Caspar forlag & Bergen Lærerhøgskole.

Gerdes, P. (1995). *Woman and geometry in southern Africa.* Maputo, Mozambique: Instituto Superior Pedagógico Moçambique.

Gerdes, P (1996). Ethnomathematics and mathematics education. In A. J. Bishop, K. Clements, C. Keitel, J. Kilpatrick, & C. Laborde (Eds.), *International handbook of mathematical education. Part one* (pp. 909–943). Dordrecht, Holland: Kluwer Academic Publishers.

Hoch, M. & Dreyfus, T. (2004). Structure sense in high school algebra: the effect of brackets. In M. J. Høines & A. B. Fuglestad (Eds.), *Proceedings of the 28th conference of the international group for the psychology of mathematics education Volume 3* (pp. 49–56). Bergen: Bergen University College.

Hoyles, C. (1985). What is the point of group discussion in mathematics? *Educational Studies in Mathematics, 16,* 205–214.

Kieran, C. (1992). The learning and teaching of school algebra. In D. A. Grouws (Ed.), *Handbook of research on mathematics teaching and learning* (pp. 390–419). New York: Macmillan.

Mamona-Downs, J. & Downs, M. (2002). Advanced mathematical thinking with a special reference to reflection on mathematical structure. In L. D. English (Ed.), *Handbook of international research in mathematics education.* (pp. 165–195). Mahwah, NJ: Lawrence Erlbaum Associates.

Mellin-Olsen, S. (1987). *The politics of mathematics education.* Dordrecht, Holland: D. Reidel Publishing Company.

Mellin-Olsen, S. (1989). *Kunnskapsformidling. Virksomhetsteoretiske perspektiver.* [Knowledge dissemination. Activiy theoretical perspectives] (2nd ed.). Bergen: Caspar Forlag.

Mellin-Olsen, S. (1990). Liberation of knowledge. In R. Noss, A. Brown, P. Dowling, P. Drake, M. Harris, C. Hoyles, & S. Mellin-Olsen (Eds.), *Political dimensions of mathematics education action & critique. Proceedings of the first International Conference.* Revised edition. (pp. 173–185). London: Department of Mathematics, Statistics & Computing, Institute of Education, University of London.

Mellin-Olsen, S. (1991). *Hvordan tenker lærere om mateatikkundervisning?* [How do teachers think about mathematics teaching?]. Bergen (Landås): Bergen Lærerhøgskole.

Mellin-Olsen, S. (1993). Dialogue as a tool to handle various forms of knowledge. In C. Julie, D. Angelis & Z. Davis (Eds.), *Political Dimensions of Mathematics Education 2, PDME Curriculum Reconstruction for Society in Transition* (pp. 243–252). Cape Town: NECC Mathematics Commission (Maskew Miller Longman).

Mellin-Olsen, S. (1996). *Samtalen som forskningsmetode.* [Conversation as research method]. Bergen: Caspar Forlag.

Nunes, T., Schliemann, A. D., & Carraher, D. W. (1993). *Street mathematics and school mathematics.* Cambridge: Cambridge University Press.

Pehkonen, E. (1995). On pupils' reactions to the use of open-ended problems in mathematics. *Nordic Studies in Mathematics Education (NOMAD), 3*(4), 43–57.

Pimm, D. (1987). *Speaking mathematically.* London: Routledge.

Schoenfeld, A. (1992). Learning to think mathematically: Problem solving, metacognition, and sense making in mathematics. In D. A.Grouws (Ed.), *Handbook of research on mathematics teaching and learning* (pp. 334–370). New York: Macmillan.

Skovsmose, O. (1994). *Towards a philosophy of critical mathematics education.* Dordrecht, Holland: Kluwer Academic Publishers.

Skovsmose, O. (1999). *Linking mathematics education and democracy: Citizenship, mathematical arckaeology, mathematcy and deliberative interactions*. Publication no. 2. Roskilde: Centre for Research in Learning Mathematics.

Skovsmose, O. (2005). *Travelling through education. Uncertainty, mathematics, responsibility.* Rotterdam: Sense Publishers.

Steen, L. A. (1988). Celebrating mathematics. *American Mathematics Monthly, 95*, 414–427.

Steen, L. A. (1990). Pattern. In L. A. Steen (Ed.), *On the shoulders of giants. New approaches to numeracy* (pp. 1–10). Washington DC: National Academy Press.

Torkildsen, O. E. (1991a). Resultater av konkurransen i nr. 1-1991 [Results from the competition no. 1-1991]. *Tangenten, 2*(1), 1–11.

Torkildsen, O. E. (1991b). Matematikken bak oppgaven i nr. 2-1990 [The mathematics behind the problem in no. 2-1990]. *Tangenten, 2*(2), 6.

Torkildsen, O. E. (1991c). Stadig nye prisvinnere! [All the time new prizewinners]. *Tangenten, 2*(3), 20–21.

Torkildsen, O. E. (1991d). Nå kommer svenskene! [Now comes the Swedes!]. *Tangenten, 2*(4), 20–21.

Torkildsen, O. E. (1992a). En eneste deltaker! [A single competitor!]. *Tangenten, 3*(1), 23–25.

Torkildsen, O. E. (1992b).Veteranklasser i matematikk vinner! [Veteran-classes in mathematics wins!]. *Tangenten, 3*(2), 19–20.

Torkildsen, O. E. (1992c). Presterud skole igjen! [Presterud school again!]. *Tangenten, 3*(4), 18–19.

Torkildsen, O. E. (1993). Resultater konkurransen i nr. 3 1992 [Results from the competition in no. 3 1992]. *Tangenten, 4*(2), 29–30.

Torkildsen, O. E. (2006). *Mathematical archaeology on pupils' mathematical texts. Un-earthing of mathematical structures.* Unpublished doctoral thesis, Faculty of Education, University of Oslo. Retrieved on June 7, 2010 from http://www.ils.uio.no/forskning/pdh-drgrad/doktoravhandlinger/docs/Avhandling%20Ole%20Einar%20Torkildsen.pdf

Treffers, A. (1987). *Three dimensions. A model of goal and theory description in mathematics instruction—The wiskobas project.* Dordrecht, Holland: D. Reidel Publishing Company

Van den Heuvel-Panhuizen, M. & van Rooijen, C. (1997). *Toelichting AMI-vragenlist. Toepassen als doel van reken-wiskunde-onderwijs* [Introduction to the AMI-questionnaire Applying mathematics as a goal for mathematics education]. Utrecht: Freudenthal Institute.

Van Dormolen, J. (1986). Textual analysis. In B. Christiansen, A. G. .Howson & M. Otte (Eds.), *Perspectives on mathematics education* (pp. 141–171). Dordrecht, Holland: D. Reidel Publishing Company.

13

THE CLIMBING EXPERIMENT

From Doing to Thinking

Anne Birgitte Fyhn[1]

DEPARTMENT OF EDUCATION, UNIVERSITY OF TROMSØ

Abstract: Analyses of the twelve-year-old Mary's narrative from a climbing trip (Fyhn, 2006) reveal the notion of angle to be an emerging object in her climbing discourse. In a second study, Mary's class is introduced to climbing as basis for the teaching of angle. The analysis of the study indicates that the pupils' analytical drawings (where issues are abstracted from their context) may function as basis for their further work with geometry (Fyhn, 2008). The single case study of Mary and the study of her class constitute the two first phases in a design research study. Retrospective analyses of these two phases constitute a third phase, where examination of validity plays an important role. The analysis of phase three shows that angles are identified and described in phase two, but they do not emerge as thoroughly as they do in phase one. This indicates a need for new kinds of data, method, and theoretical framework in the teaching experiment. In a redesigned teaching experiment, the data will consist of the pupils' narrative texts from climbing events that they can imagine to have taken place.

Keywords: Angle, climbing, validity, embodied mathematics, design research

INTRODUCTION

Research into Norwegian primary school pupils' understanding of angles reveals that they find it difficult both to understand what an angle is and how to compare angles of different sizes (Fyhn, 2007; Gjone & Nortvedt, 2001; Johnsen, 1996). In this chapter I present the "climbing experiment," which has the purpose of enabling pupils to establish a conceptual understanding of angles based on their own bodily experiences from climbing.

Climbing is an activity that involves one's complete body and its actions in a three-dimensional world. The joints in the climber's own body shape different angles, and these angles are crucial for whether the climber will succeed or not. Thus this chapter starts with an introduction to the terms "different conceptions of space," "embodied geometry," and "angle," followed by a survey of how angles are treated in Norwegian schools. One idea of the study is to follow Freudenthal's (1983) suggestion: not to teach abstractions by concretizing them, but rather to go for the converse approach—to start with the climbing activity and then search for the mathematics in it. This chapter includes a very brief outline of Realistic Mathematics Education, which has been developed from Freudenthal's ideas.

The climbing experiment constitutes the three first phases in a design research study (Cobb, 2001; Gravemeijer & Cobb, 2006). Phase one is a single case study of the twelve-year-old Mary's narrative from a climbing trip. I guide her to discover several angles in her own story (Fyhn, 2006). Beforehand, Mary claimed that her narrative concerned neither angles nor geometry, but several angles emerge through the analyses.

The First Sourcebook on Nordic Research in Mathematics Education, pages 197–209

The second phase is a multiple case study of a teaching experiment in Mary's class, where the teaching of angles is based on the pupils' climbing experiences. The analyses in phase two are based on the pupils' writings and drawings from the teaching experiment (Fyhn, 2008), and many angles occur in this phase. The boys' drawings present fragments of a situation with focus on angles; their drawings are interpreted as analytical. The girls who succeeded best in mathematics chose to make narrative drawings; that is, overview drawings with much detail.

The purpose of the first two first phases was to "measure" the occurrences of angles in the informants' utterances. The third phase, however, focused on whether I really measured what I intended to measure. The analyses in phase three concern the quality of the different findings, theoretical frameworks, and choices of data in the first two phases. In the climbing experiment, examination of validity turns out to become an instrument that influences the choice of data, method, and theoretical framework, and thus it plays an important role in the refinement of the design. This chapter pays most attention to the retrospective analyses in phase three.

DIFFERENT CONCEPTIONS OF SPACE

According to Lakoff and Núñez (2000), space has been conceptualized in two different ways in the history of mathematics; either naturally continuous or as a set of points. Most people normally think of space as naturally continuous. "It arises because we have a body and a brain and we function in the everyday world. It is unconscious and automatic" (Lakoff & Núñez, 2000, p. 265). The metaphor invented by Descartes' "Numbers are points on a line" (Lakoff & Núñez, 2000, p. 260) leads to the conceptualization of space as a set of points (SSP), and "the Set-of-Points conception is the one taken for granted throughout contemporary mathematics" (p. 265). Climbing takes place in the naturally continuous space (NCS), where coordinates and axes do not exist. Primary school geometry takes place in the NCS as well as in the SSP.

Berthoz (2000) subdivides space into three: "personal" space, "extrapersonal" space, and "far" space, where personal space in principle is located within the limits of a person's own body. The brain uses two different frames of reference for representing the position of objects in a room; either an "egocentric" encoding, by relating everything to oneself, or "allocentric," related to a frame of reference that is external to one's body. "Moreover, allocentric encoding is constant with respect to a person's own movement; thus it is well suited to internal mental simulation of displacements" (Lakoff & Núñez, 2000, p. 100). Children first relate space to their own bodies, and the ability of allocentric encoding appears later (Lakoff & Núñez,).

Climbing is a step-by-step problem solving process, where every move must be carried out carefully to prevent falling. The position of the different parts of the climber's body is important; if she manages to move the weight of her body carefully from foot to foot, the possibility of success will become stronger. In order to manage a long reach, the climber must be aware that the diagonal from her left hand to her right foot is longer than the distance between the hand and the foot on the same side of one's body (see Figure 13.1)

When trying to ascend a passage of a climbing route, the climber encodes the actual passage egocentrically within her personal space. But when the climber stands below a climbing route considering whether or how to ascend it, she exercises allocentric encoding in extrapersonal space by considering how the route's different elements and her body relate to each other. Thus climbing can offer pupils good opportunities

Figure 13.1 Climber and observer: This climber's challenge is to move only one limb at a time in order to prevent falling. The observer may suggest ideas about how to solve the climber's challenge.

for moving back and forth between egocentric and allocentric representations. The climbing experiment intends to guide pupils to reflect upon these experiences in drawings and written texts.

EMBODIED GEOMETRY

Lakoff and Núñez (2000, p. 365) assert that "human mathematics is embodied, it is grounded in bodily experience in the world." The term "embodied" here is interpreted to mean based on experiences from actions that can involve the entire body. This requires a mastery of a relation with space that in Norway traditionally is focused on in gymnastics lessons.

Several researchers (e.g., Nemirovsky, Borba, & Dimattia, 2004) appear to confine the term *embodied* to experiences from using one's hands in manipulating objects. Watson and Tall (2002) refer to Lakoff and Núñez' (2000) terms, but their examples are confined to the manipulation of objects by one's hands. Gravemeijer (1997) includes this visual image in such a way that the action and the visual image are partly treated as a whole. A person's shadow is a visual mapping of the person's movement with her body. She can immediately observe the result of the manipulation.

WHAT IS AN ANGLE?

Euclid defines angle as an inclination of lines (Freudenthal, 1973; Henderson & Taimina, 2005), but "Euclid does not know zero angles, nor straight and bigger than straight angles.... Nevertheless one feels that Euclid's angle concept is consistent" (Freudenthal, 1973, pp. 476–477).

When one creates angles by bending different parts of one's body, one makes neither zero angles nor angles bigger than straight angles. "Dihedral angle" means the "union of a line and two half-planes which have this line as a common edge. This line is the edge of the dihedral angle and the union of the line and one of the planes is a face" (James & Beckenbach, 1976, p. 112–113). In climbing terminology, the word *dihedral* refers to a *shape*—a visual gestalt. A dihedral is an "open book" rock formation, and not the angle between the two planes. The dihedral formation has special properties; it is a corner with two walls, and the dihedral angle of this formation is bigger than zero and smaller than a straight line. The technique for climbing a dihedral is for the climber to stem her feet (and sometimes her hands, too) towards the walls in order to move upwards.

Henderson and Taimina (2005) define angle from three different perspectives: angle as geometric *shape*, as a *dynamic* notion (angle as movement) and angle as *measure*. A dynamic notion includes an action, a turning point and a rotation. Freudenthal (1983) claims that the angle as turn, the process of change of direction, "is the most natural, the most instinctive aspect of angle" (p. 328). However, the passing from the turn angle to the angle determined by two sides is abstract and difficult to attain (Freudenthal, 1983).

According to Van den Heuvel-Panhuizen (2005), measurement and geometry are two domains, each with their own characteristics. The issue of geometry is to understand the space around us while measurement is aimed at quantifying our physical environment.

In order to measure drawn angles and to draw angles of a given measure, the protractor is the classical instrument. But Freudenthal warns against the use of a protractor; he describes it as "a trap the learner should have walked into once in order to avoid it in the future" (Freudenthal, 1983, p. 363). This is because when one measures angles by a protractor, one needs to know that there is no correlation between the angle Φ and the length of the arc subtended by Φ (Lakoff & Núñez, 2000): "Angles themselves are nonnumerical. In plane geometry there are angles but no numbers. If we are going to do arithmetic calculation on angles, we have to get angles to *be* numbers. That is a job for a metaphor—what we call the Trigonometry metaphor" (p. 387).

The Trigonometry metaphor is based on the unit circle (Lakoff & Núñez, 2000). An implicit understanding of the unit circle is not part of the curriculum for the pupils in primary school. Angle as measure is categorized as belonging to SSP, while dynamic angles and angle as shape are found in NCS too. It takes special training to think in terms of the SSP (Lakoff & Núñez, 2000). This supports Freudenthal's warning against introducing angles by the measure approach: "angles will be distinguished as different or equal before measuring angles is discussed" (Freudenthal, 1983, p. 323). Several years ago, Johnsen (1996) claimed that measurement was the most typical approach to angles in Norwegian schools. Following the defense of my doctoral dissertation, I pointed out in interviews that my focus on angles was not measurement. Nevertheless journalists insisted on using headlines such as "The body as angle measurer" (Christensen, 2007, my translation).

THE TEACHING OF ANGLES IN NORWAY

The national curriculum of 1987 (KUD, 1987) introduces angles as a concept, and the 1997 curriculum

(KUF, 1996) focuses on a dynamic approach. The 2006 curriculum (KD, 2006a; KD, 2006c) to a large extent places the matter of focus in the hands of the teacher.

Johnsen (1996) describes how the pupils in one particular Norwegian class develop their understanding of angles, and she categorizes angles as static, dynamic, or analyzed by measurement. The teacher focused on angles in different contexts and not just angle as measurement. The teacher also spent much time discussing the different examples with the pupils. Johnsen (1996) concludes that measuring angles is difficult for the pupils, and along with Freudenthal (1983), she claims that a large proportion of pupils use the protractor in a wrong way. In the curriculum of 1987 (KUD, 1987), neither measuring angles nor the protractor are mentioned explicitly. Her study was conducted in a private school that is headed by a religious community, although Norwegian children usually go to public schools. This particular school has many enthusiastic teachers who know most of the parents in their private lives. Thus the class may not be representative for Norwegian classes of the same age.

The 1997 curriculum (KUF 1996) requires the introduction of angle using a dynamic approach. Simmons and Cope (1993) point out that several researchers have examined how pupils explored the concepts of rotation and angle using the graphics component of the programming language Logo. A large proportion of these pupils seemed to gain only instrumental understanding of both the programming concepts and the geometric concepts. Freudenthal (1983) warns that the progression from angle as turn to angles determined by two sides is difficult, and that teachers meet difficulties when trying to explain the turn concept of angle. However, a dynamic approach to angles includes more than just the use of computers. Freudenthal (1983) recommends several examples of angles in contexts well known to children: the hands of a clock, scissor blades, the steepness of a ladder, and the opening of a door. At the Nordic preconference to ICME 10, I introduced the angle as the position of one's skis in uphill skiing—the fishbone technique (Fyhn, 2003). These are all examples of dynamic angles determined by two sides. The dynamic skiing angles are used in analytic reasoning as a basis for measurement; in the actual contexts, the point is what happens when the angle increases or decreases its size. All these dynamic angles can be categorized as belonging to NCS.

In the curriculum of 2006 (KD, 2006a; KD, 2006c) for primary school, the word *angle* occurs only twice, and that is under the heading "competence aims" in the subject area "measuring" for pupils at the fourth and seventh grades: "The aims for the education are that the pupil shall be able to... make estimates of and measure... angles" (KD, 2006c, p. 5). Angles occur only implicitly elsewhere as part of geometry for the lower grades, for instance after the second grade the learning goals concern "simple two- and three-dimensional figures in connection with corners, edges and surfaces" (KD, 2006c, p. 4).

THE DUTCH APPROACH: RME

The Dutch Realistic Mathematics Education, RME (Van den Heuvel-Panhuizen, 2003), which is based on Freudenthal's ideas, constituted the basis for the experiment in phase two. In RME, "students should learn mathematics by developing and applying mathematical concepts and tools in daily-life problem situations that make sense to them" (Van den Heuvel-Panhuizen, 2003, p. 9). A central characteristic of RME is the "activity of organizing matter from reality or mathematical matter—which he [Freudenthal] called 'mathematization'" (ibid., p. 11).

Treffers (1987) distinguishes between *horizontal* and *vertical* mathematizing. "Horizontal mathematising leads from the world of life to the world of symbols" (Freudenthal, 1991, p. 41), while vertical mathematizing is the more or less sophisticated process that goes on within the mathematical system; "vertical mathematising is the most likely part of the learning process for the bonds with reality to be loosened and eventually cut" (Treffers, 1987, p. 68). The distinction between horizontal and vertical mathematizing is relative to the involved person, the situation, and the environments: "For the expert mathematician, mathematical objects can be part of his life in quite a different way but (sic.) for the novice" (Treffers, 1987, p. 42).

Another characteristic of RME is that "students pass through different levels of understanding on which mathematizing can take place" (Van den Heuvel-Panhuizen, 2003, p. 12). According to Van Hiele (1986), it is possible to discern between five different levels of thinking in mathematics. People approaching the first level are guided by intuition, while the second level concerns descriptions. At the third level people are able to perform informal deduction, while formal deduction does not appear before level four. Level five concerns the nature of logical laws. Van Hiele (1986) points out that the fifth level has only theoretical value in school; when our pupils do not understand us, we need to focus on the instruction given at the lower levels. In addition to informal deduction, the third van Hiele level also concerns congruency. The climbing experiment focuses on angles, and because two equal angles are congruent whatever length of their sides, the climbing experiment level three (CE3) does not focus on congruency. Except for this, the levels match the van Hiele levels.

Typical for RME is *reinvention*; "that is recreating mathematical concepts and structures on the basis of intuitive notions..." (Treffers, 1987, p. 241). Freudenthal (1991) describes the term "guided reinvention" as

> striking a subtle balance between the freedom of inventing and the force of guiding, between allowing the learner to please himself and asking him to please the teacher. Moreover, the learner's free choice is already restricted by the "re" of "reinvention." The learner shall invent something that is new to him but well-known to the guide. (p. 48)

A careful choice of context can prevent learners from being left with what Freudenthal (1991) denotes as "heaps of loose ends"; mathematizing of pupils' body movement with respect to angles is an activity meant for pupils in primary school while mathematizing of pupils' body movement with respect to vectors is an activity they can perform years later in high school (Fyhn, unpublished).

MATHEMATICAL ARCHAEOLOGY

Mathematics can be integrated into an activity to such a degree that it disappears for both the children and the teachers. Skovsmose (1994) claims, "Mathematics has to be recognised and named, that is the task of a mathematical archaeology" (p. 94). It makes a difference whether the teaching is built upon situations that contain possibilities for the application of mathematics or just for descriptive purposes. Many sorts of descriptive uses of mathematics can be possible as well as appropriate through mathematical archaeology; mathematics can be treated as an emerging subject (Skovsmose, 1994). For a more extensive discussion of mathematical archaeology see Torkildsen, this volume, Chapter 12.

Within a project where mathematics is an implicit element, it is important to spend some time on mathematical archaeology. Skovsmose further claims: "If it is important to draw attention to the fact that mathematics is part of our daily life, then it also becomes important to provide children with a means for identifying and expressing this phenomenon" (1994, p. 95).

THE TWO FIRST PHASES IN A DESIGN RESEARCH CYCLE

Design research involves both instructional design and classroom-based research (Cobb, 2001). Doorman (2005) points out that design research has both a hypothetical and a reflective side, and this leads to "a delicate and iterative process of testing, reflection and redesign" (p. 68). The climbing experiment is based on the idea that the pupils may be guided to see lots of apparently invisible angles in their climbing contexts. Phase one focuses on whether it is possible for one single pupil to see angles as objects in her climbing discourse, and phase two intends to carry out this focus in one entire class. The third phase concerns retrospective analyses of the two first phases, and together, these three phases constitute one turn in a design research cycle.

In design research, the teachers usually perform the teaching while the research team makes "guidelines for the teacher" (Doorman, 2005, p. 76). I chose to perform the teaching myself in order to get as close as possible to my intentions with the design. This provided me better possibilities for testing out how the intended teaching works, but on the other hand, my possibilities of observation were limited. Video recording of the lessons might have provided better insight into what actually happened throughout the lessons, but then a larger research team would have been needed. Since the climbing experiment is a low-cost, one-person research project, video recording was not practicable.

Phase One: Preparation. One Girl's Narrative

This preparation phase consists of a single case study of the twelve-year-old Mary's narrative. The analysis (Fyhn, 2006) indicates that the "shape approach" to angles in a climbing context could be worth trying with an entire class. Early in my doctoral studies, I visited some friends just after they had made a successful climbing trip; two friends of mine had ascended a classic climbing route in the nearby region together with their two young daughters. I wrote down the girl Mary's narrative from this trip, and afterwards I guided her to discover several angles in her own story. The ascent of a climbing route can be interpreted to be what Freudenthal (1973) denotes as geometry; as grasping the space in which you live, breath, and move. Thus the main research question in this phase was (Fyhn, 2006): Is a climbing discourse a possible resource for a school geometry discourse? This closed question is of interest because before telling her story,

Mary claimed that it concerned neither angles nor geometry. In order to legitimate the teaching experiment in phase two, it is necessary to make sure that "no" is not a certain answer to the above question.

Mary's use of "angle" is analyzed within two frameworks: Foucault's (2004) discourse theory and Lakoff and Núñez' (2000) theory of NCS and SSP. The latter theory was considered in an earlier section, so here I concentrate on the relevant elements of Foucault's work. Foucault (1972/2004) refers to a discourse object's surface of emergence. At the beginning Mary found no angles in her climbing discourse, and thus the discourse object "angle" appeared with a larger surface of emergence in my climbing discourse than in Mary's. The different angles in Mary's story were categorized into four different visibility levels according to their emergence. The analysis shows how she forms and deforms the object angle in her climbing discourse, through her discussions with me a lot of angles emerged and developed. As Foucault (1972/2004) says, "discursive practice is a place in which... objects is formed and deformed" (p. 53). Phase one concluded that a climbing discourse could be a possible discourse for the teaching of angles.

Phase Two: The Teaching Experiment

One intention of phase two was to get further into the question of how pupils describe and explain angles in a climbing context (Fyhn, 2008).
The analysis of Mary's narrative (Fyhn, 2006) combined with my fascination with Freudenthal's (1973, 1983, 1991) theories and the Dutch RME (Van den Heuvel-Panhuizen, 2003) gave direct rise to a two-day climbing-and-angles teaching experiment for Mary's class. As for the pupils' reasoning and reflections in mathematics, the RME theories were supposed to offer a broader insight than Foucault's (1972/2004) discourse theory.

The first day was a climbing day, while day two was a follow-up day at school. The pupils completed a pre-test and a post-test as part of the experiment. The tests included tasks that dealt with identifying the smallest and the largest in a given group of angles (Fyhn, 2008). The pre-test was taken from the national quality in mathematics (KIM) project (Brekke, 1995; Gjone & Nortvedt, 2001), which among other things aims to identify and survey pupils' most common misconceptions. The post-test I designed for the task, specifically in order to get as close as possible to the pre-test without being identical. The aim of phase two is to develop theory that can lead to an improved design of the teaching experiment. My description of phase two (Fyhn, 2008) focuses on the participating pupils during these two days of work. The research question is: How do pupils describe and explain angles in drawings and written text when they mathematise climbing with respect to angles?

At the end of day two, the pupils were to make a drawing from the climbing walls in their own way and point at something concerning angles in their figure. These drawings were categorized as narrative or analytical, and the terms *narrative* and *analytical* are used similarly as by Murphy and Elwood (1998) in their categorization of texts. A narrative drawing is an overview of a climbing situation that includes details that are irrelevant for the mathematics focus. Examples of such details are the color of the climber's shoes, how the rope is folded on the floor, the climber's hair, unused climbing holds on the wall, and clasps on the climbers' harnesses. Analytical drawings extract angle(s) from the context and describe or explain these angles, as shown in Figure 13.2.

All the narrative drawings turned out to be made by girls—actually both Mary and the other girl who had top scores on the pre-test made narrative overviews that were rich with details. Some of the other girls made analytical drawings. The narrative drawings turned out to

Figure 13.2 Analytical drawings which focus on angles. These drawings do not tell stories from a climbing context. The angles in the figures are extracted from the context by leaving out respectively the body's upper part, the person's clothes and the unimportant part of the rope. In addition the pupils have marked the angles. The Norwegian "spiss vinkel" means "acute angle."

show either horizontal mathematizing (Freudenthal, 1991; Treffers, 1987) or no mathematizing at all, while five of the six analytical drawings were interpreted to show both horizontal and vertical mathematizing. The pupils' writings were analyzed with respect to different levels of understanding. The pupils who succeeded with the angle tasks on the pre-test consequently referred to angle as measure in their drawings and written texts, and none of them produced anything that was categorized as the CE3 level.

Phase Three: The Retrospective Analyses

Regarding design research, Gravemeijer and Cobb (2006) claim that a theoretical analysis is the result of a purposeful and complex problem-solving process, and thus different researchers are not expected to develop identical theoretical constructs from analyzing the same set of data. Russel Hanson (1958) imagined Tycho Brahe and Johannes Kepler watching the sunrise together. Most likely they would make the same observation, but their interpretations would differ: Kepler would regard the sun as fixed; it was the earth that moved. According to Brahe the earth would be fixed and all other celestial bodies moved around it. Hanson asked if Kepler and Brahe did see the same thing in the east at dawn. They watched the same visual object, but most likely they interpreted it differently because of different theoretical frameworks.

Method

The analyses in phase three point towards a change of data which in turn leads to new needs regarding the theoretical framework. This influences the design of the teaching experiment, which constitutes the core of the climbing experiment. I questioned myself: Are my findings "sufficiently authentic (isomorphic to some reality, trustworthy, related to the way others construct their social worlds) that I may trust myself in acting on their implications?" (Guba & Lincoln, 2005, p. 205). The research question in the third phase is: When retrospective analyses are carried out with respect to validity, then how may these analyses influence the choice of theoretical framework?

Lincoln and Guba (2000) denote naturalist and interpretive researchers as new-paradigm practitioners, as opposed to positivists. In new-paradigm inquiry, both method as well as the process of interpretation will deliver some set of local or text-grounded truths, and this point raises the following questions: "Are we interpretively rigorous?" (Lincoln & Guba, 2000, p. 179). Lincoln and Guba (2000) furthermore point out that validity is neither easily dismissed nor readily configured by new-paradigm practitioners.

Validity—Four Criteria

"Validity is generally concerned with the extent to which an instrument measures what it is supposed to measure" (Ary, Jacobs, & Razavieh, 1996, p. 262), while reliability concerns how consistently we are performing the measurement (ibid.). The "valid" rifle is adjusted to hit the target, while the rifle that is both "valid" and "reliable" also hits the bull's eye consistently. The rifle that hits very precisely, but always above and to the right of the target, is reliable but not valid. According to Lesh, Lovitts, and Kelly (2000), new problems and new theoretical perspectives require us to re-examine how we interpret constructs as validity, objectivity, and reliability, and they further claim that maybe new standards of quality are needed. In the past, credibility was closely linked to objectivity. However, throughout the natural and social sciences there is a growing recognition of the importance of distinguishing signal from noise while shaping the interpretations of the phenomena under investigation (Lesh, Lovitts, & Kelley, 2000). Lincoln and Guba (1985, p. 328) ask "Credibility for whom?" This is interpreted to mean that the researcher as well as the consumer needs to use evidence and lines of argument that are auditable and credible to a sceptic.

Doorman (2005) addresses the demands from Lesh, Lovitts, and Kelly (2000) and from Lincoln and Guba (2000) in his description of four criteria that can validate the scientific basis for design research: (1) *testable conjectures* about the pupils' development with respect to the created educational environment; (2) the *theoretical foundation* of the interpretative framework that guides reduction and interpretations of the data; (3) the *credibility* of the instructional design, the data interpretation, and the researcher's arguments; and (4) the *engagement* of teachers and pupils during the teaching experiments (Doorman, 2005). These four aspects of validity will be used as the basis for the climbing experiment analyses in order to improve the design of the teaching experiment.

Flow

Doorman's (2005) fourth criterion concerns the participants' engagement. Csikszentmihalyi (2000) introduced the concept of *flow*, described as the holistic sensation that people feel when they act with total involvement. People seek flow for itself and not for any extrinsic rewards that might occur from it: "Achievement of a goal is important to mark one's performance but is not in itself satisfying. What keeps one going is the experience of acting outside the parameters of worry and boredom: the experience of flow" (Csikszentmihalyi, 2000, p. 38). One quality of flow experiences is that they provide clear unambiguous feedback to the person's actions. Csikszentmihalyi (2000) points at climbing, chess, research, and mathematics as activities that may result in flow experiences. However, there are certainly people who claim they experience flow from one of these subjects but not from the others. The climbing experiment intends to let the pupils experience that mathematics concerns flow activities.

Hanson (1958) claims that "seeing is not only the having of a visual experience; it is also the way in which the visual experience is had" (p. 15). The introduction of angles in a context that most pupils experience as exciting and fun prepares for a positive attitude towards mathematics: "If the students experience the process of reinventing mathematics as expanding common sense, then they will experience no dichotomy between everyday life experience and mathematics. Both will be part of the same reality" (Gravemeijer & Doorman, 1999, p. 127). The pupils' attitudes towards climbing and towards their participation in the research may be of importance for whether the study measures what it intends to measure. Those who are not fascinated by climbing, who dislike mathematics, and in addition are skeptical towards participating in research, will probably not feel comfortable with taking part in a project like this.

Relevance

Mellin-Olsen (1987) claims that whether a topic in a curriculum is regarded as important by a pupil, depends on how she relates it to her overall life situation. For the teacher this means to focus on the questions, "how does the lesson relate to her pupils' conception of the important totalities of their world, and how can this lesson eventually transform this totality?" (Mellin-Olsen, 1987, p. 33). In November 2002, the Norwegian Ministry of Education and Research introduced a strategy plan for strengthening the subjects of mathematics, science, and technology in the educational system, in working life and in the general public (KD, 2006b). Once a year the cabinet minister releases a revised version of this plan. One goal of the plan is to develop arenas for experiencing and learning mathematics outside school (KD, 2006b). Indoor climbing walls might be a suitable arena for the teaching of angles for pupils who claim that climbing is part of the important totalities of their world. Pupils who are not familiar with climbing might be introduced to it as part of their mathematics lessons.

Replicability

The rifle that hits the bull's eye can usually not be successfully used by other hunters without some adjustment. Despite the fact that the gun is very reliable, other hunters most likely will miss the bull's eye because there are hardly two identical hunters; their cheeks may be of different shape or the length of their arms may vary, and so on. Phase one (Fyhn, 2006) has quite good possibilities for replication. But there is no guarantee that the details in another climbing narrative will match this, even if it should be a story from a climbing route with similar angle possibilities as in Mary's case. The second phase (Fyhn, 2008) is hardly replicable because it involves teachers and teaching: "Teachers are professionals who continuously adjust their plans on the basis of ongoing assessments of their students' mathematical understanding" (Gravemeijer & Cobb, 2006, p. 44). Furthermore, it is highly unlikely that there exist two identical groups of pupils or identical sets of conditions.

Ethical Considerations

One important ethical aspect of doing research with teachers and pupils is the use of time; to do classroom research it is necessary to use some of the teachers' and pupils' time, and there is no guarantee that their time is used in a way they will feel comfortable with. The qualitative approach in this study intends to give the participants a more or less direct response, but the final analyses have to be carried out before the result can reach the participants. This means that the participating pupils will not benefit directly from the results of this study. The research cannot be characterized as a waste of time for the pupils if they declare they enjoy taking part in the experiment. However, in carrying out flow activities (Csikszentmihalyi, 2000) as part of the teaching, there is an extra need to control the pupils' learning; the moral aspect of occupying the pupils' mathematics lessons must not depart from the researcher's focus. A really problematic situation occurs if the pupils experience enjoyment from the activity while I can see that despite my intentions only a very small amount of mathematics emerges from the fun situations.

The Data in the Three Phases

The data in phase one is my written version of one pupil's narrative. The data in phase two come from the pupils in one class. It is their written expectations and experiences from two days at school, their drawings at the end of the second day, and their answers to a pre-test and a post-test. The research in phases one and two are presented in two scientific papers (Fyhn, 2006; Fyhn, 2008). These papers constitute the data for the retrospective analyses in phase three. The retrospective analyses will focus on the different findings, theoretical frameworks, and choices of data in the two first phases. Validity is used as instrument in analyzing the choice of theoretical framework

Analyses

Because I did not take part in the climbing trip in phase one, there is limited evidence for whether the written narrative is factual. Even if I had video-recorded the described climbing trip myself, it is unlikely that Mary and I would have told identical narratives from the trip. The question of factual accounts of experiences here is a question of interpretation. My written version of Mary's narrative is only an image of the actual climbing event: "Images, reflexions, pictures and maps in fact copy originals with different degrees of strictness" (Hanson, 1958, p. 28). Mary's climbing discourse and its practice both differ from mine (Fyhn, 2006). Thus I expected to benefit from taking part in the pupils' climbing event.

Four Validity Criteria

The second phase fulfils Doorman's (2005) four validity criteria in design research. First, the formulation and verification of testable conjectures: The pupils' performed a pre-test and a post-test in connection to the climbing event. Second, the theoretical foundation of the interpretative framework guides reduction and interpretation of data: Freudenthal's (1973) idea of mathematization is not to teach abstractions by concretizing them. He advises to use the converse approach, to start off by searching for a phenomenon that might lead the learner to constitute the understanding. The teaching experiment starts with the phenomenon of climbing. The pupils' drawings are analyzed with respect to a framework based on the two categories of analytical and narrative as used by Murphy and Elwood (1998), and Treffers' (1987) horizontal and vertical mathematizing. The analyses of the texts are based on a framework quite similar to van Hiele's (1986) level theory; I noted earlier that the level CE3 differs to some extent from the third van Hiele level. Third, the credibility of the instructional design, the data interpretation, and the researcher's arguments: The fact that a report from this research is peer-reviewed and published (Fyhn, 2008) supports its credibility. Fourth, the engagement of teachers and pupils during the teaching experiment: The pupils were accustomed to receiving visitors in their class, and they said that they saw guests as a welcome interruption to their ordinary teaching. Climbing is a flow activity (Csikszentmihalyi, 2000), and the first part of the experiment took place at the climbing wall at the community center quite near the school. This indicates that the participating teachers and pupils acted with engagement. In addition, the participating teachers were familiar with climbing and enjoyed taking part in it.

The Findings in Phase Two

Five of the eighteen pupils in the class did not take part in the climbing. Two of these did not wish to participate in the research and three others were absent on the day in which the climbing activity took place. This could be due to a negative attitude towards taking part in research, or it could be just a coincidence. A follow-up study at two schools with 105 registered sixth-grade pupils indicates that climbing appeals to pupils at this age: 98 pupils were present at a climbing day, and all of them participated in the climbing. There were none who declined to take part in the project, which in this case was defined as developmental work and not as research.

Most of the angles from the pupils' written and drawn works are identified as belonging to level one. Maybe that is because the van Hiele (1986) level theory is not appropriate as a theoretical foundation for the interpretative framework. As for Mary, the analysis in phase two does not reveal anything similar to the richness of emerging angles in phase one. The analyses in phase one made clear how angles emerged through different levels of visibility. But the analyses of the written and drawn works in phase two did not present any emerging angles. One reason for this could be the choice of theoretical framework: The term *mathematization* (Freudenthal, 1973, 1983; Treffers, 1987) does not point explicitly at "mathematics as an emerging subject."

Two of the pupils who took part in the second phase made oral statements that were interpreted to belong to the CE3 level. The angles in these statements were not categorized as measurement. One of these statements was made by Mary during phase one: "If you stand on your toe with your heel low, then it is harder to raise oneself than if you have the heel a little higher. The angle between the leg and the foot should be large" (Fyhn, 2006, p. 99). The other statement was made by George, who receives special lessons because of his reading and writing difficulties. I was able to guide him to make a logical mathematical statement. George had ascended two climbing routes on different walls, and he applied angles and parallelism in an informal proof for deciding which one was the highest (Fyhn, 2008).

One idea of the climbing experiment is to base the teaching on the pupils' talk about climbing. I believed that I had good opportunities of getting valid data in the second phase because I could control and influence the climbing event in which the pupils were taking part. The goal is that the pupils build up their understanding of angles as a tool for improved climbing technique. The point of guiding the pupils in their reinvention of angles is to guide them on their way from doing to thinking. This was an important issue in the design of the research. The problem is, however, that it was not easy to collect valid and reliable data about what the pupils actually did, and so the data in phase two turned out to be the pupils' own written and drawn material.

DISCUSSION

According to Guba and Lincoln (2005), the process of uncovering forms of inscription is archaeological. In describing his term *mathematical archaeology*, Skovsmose (1994) refers to a study where the teacher claims that the pupils did not see any mathematics. This could be because "The idea of mathematics as an emerging subject had not been realised" (Skovsmose, 1994, p. 90). Skovsmose further distinguishes between application of mathematics and descriptive use of mathematics. The climbing experiment focuses on descriptive use of mathematics, and one difference between phase one and phase two is that angles emerge only in phase one. In phase two angles occur and are described, but they do not emerge. The retrospective view of the climbing experiment indicates a need for a more appropriate theoretical foundation for the interpretative framework in phase two. This means that Doorman's (2005) second validity criterion is not fulfilled.

Foucault (1972/2004) discusses the different discourse objects' surface of emergence. A combination of Skovsmose's (1994) and Foucault's (1972/2004) term "emergence" turns out to become central for the refined version of the climbing experiment's interpretive framework.

Throughout phase one, the girl Mary was guided to unearth angles in her climbing discourse, and Foucault's (1972/2004) discourse theory was central in these analyses. When the pupils made their drawings and written texts during the second phase, no angles emerged except from the situation where I guided a boy to use an informal geometrical proof to decide which was the highest of two climbing routes. The climbing experiment intends to realize the idea of mathematics as an emerging subject; to start with a climbing activity in which the pupils do not see any mathematics at all, and then the pupils are guided to *unearth* (Torkildsen, 2006) the angles that appears to them in this discourse. A change of theoretical foundation for the interpretative framework from mathematization to mathematical archaeology leads from guided re-invention (Freudenthal, 1991) to *guided unearthing* of angles in climbing. In a refined teaching experiment, the pupils' work will be denoted as unearthing of climbing with respect to angles.

The validity of my data needs to be questioned. The data in the first phase represented what one pupil could tell when she was encouraged to tell about her experiences. In this case, the data did not represent what actually happened, but what Mary referred to when she recalled the event. The focus in this phase was the girl's narrative. My observations comprised the account that she told me. I wrote down her words, subsequently corrected by her. Because Mary took an active part in the analysis, she helped me to interpret her correctly. In phase two, my intentions were that the data should represent what the pupils actually did and their immediate ideas about it. My observations in this phase were the pupils' interactions with each other and the teachers, their climbing, what the pupils did while I could see them, their response to my presence, and so on. I ended up analyzing the pupils' drawings and written texts, and thus I had to admit that there was a distance between what I intended to do and what I actually did. A retrospective view leads me to recognize that the data appears more valid in the first phase than in the second.

In an improved climbing experiment, the pupils might have been asked to spend one hour at school in writing down their narratives from the climbing day. A focus on these narratives would probably give more valid data than my attempt to focus on what actually happened. A focus on the pupils' narratives is supported by Hanson's question: "For what is it for things to make sense other than for descriptions of them to be composed of meaningful sentences?" (1958, p. 25).

Lincoln and Guba (2000) argue that qualitative methods are appropriate as instrument in naturalistic inquiries. They do not adopt an anti-quantitative posture or the exclusivity of methods, but rather a focus on the philosophies of which paradigms are constructed. Guba and Lincoln (2005) as well as Lincoln and Guba (2000) claim that validity points to the crucial question about whether the findings are sufficiently authentic for the researcher to trust her- or himself in acting on their implications. However, Bassey (1999) suggests that validity is not a vital concept in case study research and cites Lincoln and Guba to support his assertion: "As an alternative to reliability and validity, Lincoln and Guba put forward the concept of trustworthiness in 1985 in *Naturalistic Inquiry*, and in my view this successfully illuminates the ethic of respect for truth in case study research" (p. 75). Nevertheless, as I observed above, Guba and Lincoln (2005) do claim validity to be a useful term in qualitative research. The climbing experiment is a multiple case study on the pupils who participate in a teaching experiment in which there are no clear borders between the experiment and the case study. The examination of validity has emerged as a useful instrument in improving the experiment's quality.

CONCLUSION

The analysis in phase one (Fyhn, 2006) points at angles as an emerging object in Mary's climbing discourse. In phase two, her class is introduced to climbing as basis for the teaching of angles (Fyhn, 2008), and the pupils' use of drawing and text in describing and explaining angles in a climbing context is analyzed. The analyses indicate that the teaching should point out differences between narrative and analytical text and drawing, and then the pupils' analytical drawings can function as a basis for their further work with geometry. But despite an adapted version of the van Hiele (1986) levels, there is no sign of emerging angles in phase two. Angles occur; they are identified and described. But they do not emerge.

The idea of the researcher being as close as possible to the actual climbing event might lead to a blind track. The point is that the pupils' narratives concern something they can imagine has taken place in their own climbing discourses. The word "realistic" in the Dutch RME does not refer to "realness" or authenticity of problems, but it refers to the Dutch verb *zich realiseren*, which means "to imagine" in English (Van den Heuvel-Panhuizen, 2003). This underlines the closeness between RME and the climbing experiment's theoretical framework.

A difference between the pupils' climbing experiences and their narratives from climbing is not any problem. However, the point is that the researcher needs to be aware of similarities and differences between her climbing discourse and that of the pupils, and thus it might be important for the researcher to be familiar with the actual climbing arena.

When the two first phases are analyzed with respect to validity, it turns out that despite the quality of the data and the interpretative framework in phase two, these were not appropriate to a satisfactory level. This indicates a need for new data and some more appropriate theoretical framework that incorporates theory where mathematics as an emerging subject is taken into account.

In a redesigned teaching experiment, the pupils' narratives and drawings from their climbing events constitute the data. These data will represent events that the pupils can imagine to have taken place. The pupils will take part in the analyses, which will be based on Foucault's (1972/2004) discourse analyses as in phase one (Fyhn, 2006). The intention is to focus on mathematics as an emerging subject (Skovsmose, 1994). Through this descriptive use of mathematics, the pupils are guided in unearthing angles from their own oral narratives. The redesigned climbing experiment will base the interpretative framework on Skovsmose's (1994) mathematical archaeology as a revised version of the Dutch "mathematization" (Freudenthal, 1973).

A refined research question for the experimental phase could be "How do pupils unearth angles from their own narratives and drawings?" Then the pupils' path from doing to thinking may be described by the way angles emerge from their climbing discourses: "We must explore the gulf between pictures and language, between sketching and describing, drawing and reporting... picturing and speaking are different... and brought together they must be if observations are to be significant and noteworthy" (Hanson, 1958, p. 25).

ACKNOWLEDGEMENTS

I would like to thank PhD student Elin Damsgård for her participation in our fruitful discussions regarding quality criterions in research, and for her comments to my manuscript. Thanks to Odd Valdermo for his valuable comments, and Linn Sollied Madsen who checked the language of the text.

NOTE

1. anne.fyhn@uit.no Associate Professor, Department of Education, Faculty of Social Science, Teorifagbygget 1.592, N-9037 University of Tromsø, Norway. Phone +47 7764 6120. home: http://www2.uit.no/www/ansatte/organisasjon/ansatte/person?p_document_id=42365&p_dimension_id=30943

REFERENCES

Ary, D., Jacobs, L. C. & Razavieh, A. (1996). *Introduction to research in education.* Orlando, FL: Harcourt Brace College Publishers.

Bassey, M. (1999). *Case study research in educational settings.* Berkshire, UK: Open University Press.

Berthoz, A. (2000). *The brain's sense of movement.* Cambridge, MA: Harvard University Press. (Originally published as *Le sens du mouvement,* 1997)

Brekke, G. (1995). *Introduksjon til diagnostisk undervisning i matematikk. (Introduction to diagnostic teaching in mathematics).* Oslo: Læringssenteret. Retrieved August 23, 2008, from http://www.hint.no/~toh/m19900/introduk.pdf

Cobb, P. (2001). Supporting the improvement of learning and teaching in social and institutional context. In S. M. Carver & D. Klahr (Eds.), *Cognition and instruction: Twenty-five years of progress* (pp. 455–478). Mahwah, NJ: Lawrence Erlbaum

Christensen, K. M. (2007, July 11). Kroppen som vinkelmåler. (Measuring angles by the body). *forskning.no.* Retrieved August 15, 2008, from http://www.forskning.no/artikler/2007/juli/1183965404.31

Csikszentmihalyi, M. (2000). *Beyond boredom and anxiety. Experiencing flow in work and play,* San Francisco: Jossey-Bass. (Original work published in 1975)

Doorman, L. M. (2005). *Modelling motion: from trace graphs to instantaneous change.* Utrecht, The Netherlands: CD-ß Press, Center for science and mathematics education.

Foucault, M. (2004). *The archaeology of knowledge.* London and New York: Routledge Classics. (Original work published 1972)

Freudenthal, H. (1973). *Mathematics as an educational task.* Dordrecht, The Netherlands: D. Reidel.

Freudenthal, H. (1983). *Didactical phenomenology of mathematical structures.* Dordrecht, The Netherlands: D. Reidel.

Freudenthal, H. (1991). *Revisiting mathematics education. China lectures.* Dordrecht, The Netherlands: Kluwer Academic Publishers.

Fyhn, A (2003). *Learning of geometry via physical activities in snow.* Paper presented at group 5, *PICME 10, Nordic pre-conference to ICME10,* Växjö University, Sweden. Retrieved August 23, 2008, from http://www.vxu.se/msi/picme10/F5FAB.pdf

Fyhn, A. B. (2006). A climbing girl's reflection about angles. *The Journal of Mathematical Behavior, 25,* 91–102.

Fyhn, A. B. (2007). *Angles as tool for grasping space: Teaching of angles based on students' experiences with physical activities and body movement.* Unpublished doctoral dissertation, University of Tromsø, Department of Mathematics and Statistics, Faculty of Science.

Fyhn, A. B. (2008). A climbing class' reinvention of angles. *Educational Studies in Mathematics, 67,* 19–35.

Fyhn, A. B. (unpublished). Hvordan en klatrefilm belyser noen grunnleggende elementer i vektorregning. Fokus på filmen Vektorer i klatring. (How a climbing video may enlighten some basic elements in vector calculus. A focus on the video Vectors in climbing.)

Gjone, G. & Nortvedt, G. (2001). *Kartlegging av matematikkforståelse. Veiledning til geometri F og I, (Identification of mathematical understanding. A guide to geometry F and I).* Oslo: Directorate for Primary and Secondary Education.

Gravemeijer, K. P. (1997). From a different perspective: Building on students' informal knowledge. In R. Lehrer & D. Chazan (Eds.), *Designing learning environments for developing understanding of geometry and space* (pp. 45–66). Mahwah, NJ: Lawrence Erlbaum Associates.

Gravemeijer, K. & Cobb, P. (2006). Design research from a learning design perspective. In J. van den Akker, K. Gravemeijer, S. McKenny & N. Nieveen (Eds.), *Educational design research* (pp. 17–51). Oxon, UK: Ruthledge.

Gravemeijer, K. & Doorman, M. (1999). Context problems in realistic mathematics education: A calculus course as an example. *Educational Studies in Mathematics 39,* 111–129.

Guba, E. G. & Lincoln, Y. S. (2005). Paradigmatic controversies, contradictions, and emerging confluences. In N. K. Denzin & Y. S. Lincoln (Eds.), *The Sage handbook of qualitative research* (3rd ed.). Thousand Oaks, CA: Sage Publications.

Hanson, N. R. (1958). Observation. In N.R. Hanson, *Patterns of discovery: An inquiry into the conceptual foundations of science* (pp. 4–30). Cambridge, UK: Cambridge University Press.

Henderson, D. W. & Taimina, D. (2005). *Experiencing geometry. Euclidean and non-Euclidean with history.* Ithaca, NY: Cornell University.

James, R. C. & Beckenbach, E. F. (1976). *Mathematics dictionary: James and James* (4th Ed.). New York: Van Nostrand Reinhold.

Johnsen, V. (1996) Hva er en vinkel? (What is an angle?). *NOMAD Nordisk Matematikkdidaktikk Nordic Studies in Mathematics Education, 4*(1), 25–49.

KD, Ministry of Education and Research. (2006a). *Kunnskapsløftet. (Knowledge promotion).* Oslo: Utdanningsdirektoratet/ Norwegian Directorate for Education and Training. Retrieved June 09, 2008, from http://www.udir.no/templates/udir/TM_UtdProgrFag.aspx?id=2103

KD, Ministry of Education and Research. (2006b). *Strategy 2006–2009. A joint promotion of mathematics, science and technology (MST).* Oslo: KD. Retrieved June 10, 2008, from http://www.regjeringen.no/en/dep/kd/Selected-topics/compulsory-education/strategic-plans/a-joint-promotion-of-mathematics-science.html?id=279649

KD, Ministry of Education (2006c). *Knowledge promotion. Mathematics subject curriculum.* Oslo: Utdanningsdirektoratet. Retrieved July 11, 2008, from http://udir.no/upload/larerplaner/Fastsatte_lareplaner_for_Kunnskapsloeftet/english/Mathematics_subject_curriculum.rtf

KUD, Ministry of Church and Education. (1987). *Mønsterplan for grunnskolen.* Oslo: Kirke- og undervisingsdepartementet og Aschehoug & Co.

KUF, Ministry of Church, Education and Research. (1996). *The curriculum for the 10-year compulsory school.* Oslo: Nasjonalt læremiddelsenter. Retrieved June 09, 2008, from http://www.udir.no/L97/L97_eng/index.html

Lakoff, G. & Núñez, R. (2000). *Where mathematics comes from. How the embodied mind brings mathematics into being.* New York: Basic Books.

Lesh, R., Lovitts, B. & Kelly, A. E. (2000). Purposes and assumptions of this book. In A. E. Kelly & R. A. Lesh (Eds.), *Handbook of Research Design in Mathematics and Science Education* (pp. 28–31). Mahwah, NJ: Lawrence Erlbaum Associates.

Lincoln, Y. S. & Guba, E. G. (1985). *Naturalistic inquiry.* Thousand Oaks, CA: Sage Publications.

Lincoln, Y. S. & Guba, E. G. (2000). Paradigmatic controversies, contradictions, and emerging confluences. In N. K. Denzin & Y. S. Lincoln (Eds.), *Handbook of qualitative research.* (2nd ed.). Thousand Oaks, CA: Sage Publications.

Mellin-Olsen, S. (1987). *The politics of mathematics education.* Dordrecht, The Netherlands: Kluwer Academic Publishers.

Murphy, P. & Elwood, J. (1998). Gendered learning outside and inside school: Influences on achievement. In D. Epstein, J. Elwood, V. Hey & J. Maw (Eds.), *Failing boys?: Issues in gender and achievement* (pp. 162–181). Buckingham, UK: Open University Press.

Nemirovsky, R., Borba, M., & Dimattia, C. (2004). PME special issue: Bodily activity and imagination in mathematics learning. *Educational Studies in Mathematics, 57*, 303–321.

Simmons, M. & Cope, P. (1993). Angle and rotation: Effects of different types of feedback on the quality of response. *Educational Studies in Mathematics, 24*, 163–176.

Skovsmose, O. (1994). *Towards a philosophy of critical mathematics education.* Dordrecht, The Netherlands: Kluwer Academic Publishers.

Torkildsen, O.E. (2006). *Mathematical archaeology on pupils' mathematical texts. Un-earthing of mathematical structures.* Unpublished doctoral dissertation, University of Oslo, Faculty of Education.

Treffers, A. (1987). *Three dimensions. A model of goal and theory description in mathematics instruction—The Wiskobas Project.* Dordrecht, The Netherlands: D. Reidel.

Van den Heuvel-Panhuizen, M. (2003). The didactical use of models in realistic mathematics education: An example from a longitudinal trajectory on percentage. *Educational Studies in Mathematics, 54*, 9–35.

Van den Heuvel-Panhuizen, M. (2005). Measurement and geometry in line. In M. van den Heuvel-Panhuizen & K. Buys (Eds.), *Young children learn measurement and geometry. A learning–teaching trajectory with intermediate attainment targets for the lower grades in primary school. Tal project* (pp. 9–14). Utrecht, The Netherlands: Freudenthal Institute (FI), Utrecht University and National Institute for Curriculum Development (SLO).

Van Hiele, P. M. (1986). *Structure and insight. A theory of mathematics education.* Orlando, FL: Academic Press.

Watson, A. & Tall, D. (2002). Embodied action, effect and symbol in mathematical growth. In A. Cockburn & E. Nardi (Eds.), *Proceedings of the 26th Conference of the international group for the Psychology of Mathematics Education, University of East Anglia* (pp. 370–377). UK: Norwich.

14

PUPILS' PARTICIPATION IN THE CLASSROOM DISCOURSE OF MATHEMATICS

Åse Streitlien

TELEMARK EDUCATIONAL RESEARCH

Abstract: This chapter describes findings of a study of communication patterns of mathematics classrooms in primary schools, focusing on how young pupils participate in the classroom discourse. Classroom discourse is a form of institutional talk, and as such has certain characteristics. There is a complex relationship between form and function in the teacher's instructions in mathematics lessons. The data illustrate that pupils have to recognize both the mathematical demands as well as the social demands in the teacher instruction. It becomes apparent that various discourse patterns create different premises for pupil participation. Sometimes there is a discrepancy between what the teacher asks about and what young pupils are able to contribute with. The feedback from the teacher determines whether pupils' responses are included and viewed as valid knowledge in the discourse of mathematics. When pupils have to struggle in concentrating on the rules of participation, less attention is given to the mathematical content. The chapter is based on qualitative analysis of discourse excerpts, using transcribed audio recordings and field notes.

Keywords: communication skills, dialogical analysis, discourse patterns, primary school, pupil-teacher interaction, social positioning, sociocultural perspective, whole class teaching

INTRODUCTION

There has been a growing interest over the last decades in the study of communication in education, based on the assumption that successful learning might depend as much on the type of interaction that takes place in the classroom as on the learning approaches used (Alrø & Skovsmose, 2004; Bjuland, 2002; Cestari, 1997). It is widely recognized that pupils' participation, initiatives, and involvement in the negotiation of meaning represent an important aspect of classroom interaction (Mellin-Olsen, 1987, 1989). At the University of Agder in Norway, a development in methodology has taken place from the 1990s—from using a dialogical approach, to analyzing communication and cognition, to combining this with a multimodal approach based particularly on the identification of pupils' reasoning focusing on the two modalities *discourse* and *gesture*. (This is reported by Bjuland, Cestari, & Borgersen, this volume, Chapter 9.)

In her studies of classroom discourse, Cazden (2001) considers classroom talk as a communication system with special rules and participation structures that children have to recognize: "In classrooms, spoken language is the medium by which much teaching takes place and in which pupils demonstrate to teachers much of what they have learned" (Cazden, 2001, p. 2). The meaning of classroom events emerges for each participant, children as well as teacher, through the interactive processes in which both actors take part. In accordance with this perspective, pupils both experience and influence the classroom context.

The First Sourcebook on Nordic Research in Mathematics Education, pages 211–222

Children entering school have some ideas about what they are going to learn. They come to the classroom with preconceptions about how the world works (Fosse, 2004). They also have some experiences with mathematics. According to Donovan and Bransford (2005), they may fail to grasp the new concepts and information if their initial understanding is not engaged. The language pupils use about mathematical phenomena diverges from the teacher's language. Children's utterances are characterized by pictorial language and experiences with concrete objects from everyday life. The teacher's presentation of mathematics topics is assumed to bridge the gap between the known and the unknown, between the children's prior knowledge and "new" mathematical knowledge.

The aim of this chapter is to explore the mediational role of discourse in the development of knowledge and understanding of mathematics. The research focus is the interaction between the teacher and the pupils in mathematics lessons and how discourse patterns affect pupils' participation in learning processes. Central questions concern the feedback pupils are given when they try to express themselves orally and arguing for their solutions in the classroom discourse. What happens when the teacher and the pupils misunderstand each other regarding the focus in the ongoing dialogue? How are children's meanings included in the classroom discourse?

CLASSROOM DISCOURSE

Pedagogical situations have rules for speaking and participation (Edwards & Mercer, 1987; Edwards & Westgate, 1994). Firstly, classroom discourse is oriented to pedagogical goals, and the participants are interacting for the specific purpose of learning. Secondly, the participants take the roles of "instructor" and "instructed," and therefore have unequal rights of participation. Finally, there is a certain amount of centrally focused attention with basic rules of participation; either one person speaks at a time, or multiple speakers say more or less the same thing. Because of these characteristics, it is generally the teacher who initiates the interaction, introduces the topic, and decides who can talk and when talk is welcome. Instruction is often characterized by a three-part dialogue—described as the IRF-structure (Sinclair & Coulthard, 1975): The teacher initiates (I), a pupil responds (R), following which the teacher gives feedback (F).

Mehan's (1979) study of classroom communication identified also a three-part dialogue in the classroom, but he found that the pupils have the right to take initiatives on their own. However, they have to recognize when such initiatives are welcome. According to Mehan (1979), lessons provide a hierarchical and sequential structure where teachers pursue academic goals while maintaining social control. To participate as a competent class member, a pupil must use normative social conventions of the classroom and supply correct academic information.

The description of the three part discourse pattern as ritualized and strictly controlled is clearly seen in the research of Nystrand and his co-researchers (1997). However, they made a distinction between what they call "monologically organized instruction and dialogically organised instruction" (p. 19; see Table 14.1).

The differences between these two types of instruction, according to Nystrand (1997), emerge when it comes to epistemology, communication model, and source of valued knowledge and texture of the lesson. In the case of monologically organized instruction, knowledge is given, and its source is the teacher or the textbook, not the pupils. The teacher prescripts and monitors the answers. The teacher initiates all topics of discussion and determines what is worth knowing. Knowledge is treated as fixed, objective, and autonomous. Recitation involves interaction, but this is superficial and procedural, prescribed by the teacher.

In the case of dialogically organized instruction, pupils' interpretations and personal experiences are included. Dialogic methods put a premium on close teacher–pupil interaction, a high degree of individualization, peer groups, open-ended discussion, and

TABLE 14.1 Key Features of Monologically and Dialogically Organized Instruction

	Monologically Organized Instruction	Dialogically Organised Instruction
Paradigm	Recitation	Discussion
Communication model	Transmission of knowledge	Transformation of understandings
Epistemology	Objectivism: Knowledge is a given	Dialogism: Knowledge emerges from interaction of voices
Source of valued knowledge	Teacher, textbook authorities: Excludes students	Includes students' interpretation and personal experience
Texture	Choppy	Coherent

Source: Nystrand, 1997, p. 19

curricula and lesson plans that are not completely planned in advance. Knowledge is something generated and constructed. Both teacher and pupils participate in this construction. Pupils figure out, not just commit to memory.

It is thus evident that different research perspectives on classroom interaction have evolved describing participation structure and language use in school lessons. With the background of the research referred to here, I wanted to explore how young pupils participate in mathematics classroom discourse. The episodes of interaction presented here are part of the data connected to a doctoral study (Streitlien, 2006). The aim of the study was to analyze the opportunities for participation in the learning processes that pupils encounter, how they handle communicative and mathematical challenges, and how they are positioning themselves in the discourse.

THEORETICAL APPROACH

My assumptions in this exploratory study are that discourse patterns and participation structure affect what counts as mathematical knowledge and the pupils' beliefs about mathematics. Attending two classrooms as an observer in grade one and grade two provided the opportunity to view the meanings teachers and pupils constructed through formal and informal processes. Before discussing the findings, I will give a short description of my approach to learning and the how the data collected were analysed.

Traditionally, research on learning of mathematics has focused on the individual learner and how the individual constructs knowledge. According to Sfard (1997), there has been a shift from a cognitive approach to a discursive approach in mathematics educational research.

From the 1990s on, we can see a tendency to look at learning processes as more complex than viewing the individual child constructing knowledge by her/himself without being influenced from other factors (Rogoff, 1990; Schoultz, 1998). The perspective has changed from an individual-centered view to look at cognitive processes as dependent on social, cultural, and historical factors (Mellin-Olsen, 1989). This means more attention must be paid to communication and classroom activities. In the perspective of learning that emerges from this discursive turn, the permanence of having (a conception, a scheme) gives way to a constant flux of doing (participating in an ongoing activity) (Sfard, 1997). The ongoing mathematical activities are never considered separately from the context within which they take place. Methodological consequences of this theoretical perspective were new procedures to collect data, audio-recording the conversations, and analyzing the transcripts according to a dialogical approach to communication and cognition (Bjuland, 2005; Cestari, 1997, 1998, 2000; Linell, 1995, 1998). As Bjuland, Borgersen, and Cestari write in Chapter 9, there has been more concern in relation to the teachers' formulation of questions instead of blaming the pupils for their wrong answers.

Perspectives on Learning

From the perspective of learning presented above, children learn by participating in a socio-cultural context and in socially situated practices (Lave & Wenger, 1991; Wertsch, 1991). Negotiation of meaning and learning processes of mathematics are integrated in a social, cultural, and material context. The socio-cultural view of learning in this chapter is mainly grounded on the theories of Vygotsky (1978, 1987) and Bakhtin (1981, 1986).

Vygotsky's theory is based on the idea that human beings develop cultural identity by participating in cultural activities and, by doing this, acquire the instituted meaning of the activity. Knowledge is socially constructed, and language is the medium for such construction. Language arises as a way for the child to communicate with others, and more sophisticated mental processes like thinking in concepts will develop as a transformation of semiotic activity. The semiotic activity will start as an external social activity and then will be reconstructed mentally. The distance between the actual developmental level of the learner as determined by independent problem solving and the level of potential development as determined through problem solving under adult guidance or in collaboration with more capable peers is described by Vygotsky as the "zone of proximal development." According to Wells (1999), Vygotsky stressed the role of more expert members of the culture in providing the guidance and assistance that enables the learner to become increasingly competent and autonomous participants in their activities. The form such guidance should take was a matter to which Vygotsky returned several times (Wertsch & Stone, 1985). To be effective, he argued, this guidance or instruction must always be in advance of development. For a learner in any situation, here is a zone of proximal development—a window of potential learning—that lies between what he or she can manage to do unaided and what he or she can achieve with help.

Wood, Bruner, and Ross (1976) describe the adult's support for children to build their own understanding as "scaffolding." This is a metaphoric term describing the way the teacher makes it possible for the child to internalize interpersonal knowledge and convert it into a tool for conscious control. The teacher supports the pupils' learning by means of cultural tools (linguistic and non-linguistic) (Lindén, 1992). Scaffolding is understood as "assisted performance." Much of the description of scaffolding in the literature, however, deals with mother/child or expert/novice dyads and practical activities (Rogoff, 1990). Mercer (1995) suggests that scaffolding can offer a metaphor for the teacher's participation in pupils' learning, but then scaffolding should be reinterpreted if it is to be useful in analyzing classroom practice. As Roehler and Cantlon (1997) write: "Scaffolding characterizes the social interaction among students and teachers that precedes internalization of the knowledge, skills and dispositions deemed valuable and useful for the learners" (p. 9). Scaffolding may involve helping pupils to apply frames of reference that they only partially grasp and are inexperienced in applying. It also needs to be a process that involves more than two people to be generally applicable to a classroom context. It seems that scaffolding in a whole class situation might be a complex enterprise (Bråten, 1999).

When studying *learning processes* in mathematics classrooms from a socio-cultural perspective, the focus is on interaction and dialogue. According to Bakhtin (1986), the utterance is the fundamental unit of investigation for anyone studying communication. And in every utterance there are two building blocks—*theme* and *meaning*. Any true understanding is dialogical in nature. Meaning is the effect of interaction between speaker and listener produced via the material of a particular sound complex. Bakhtin asserts that meaning and therefore learning—understood as the expansion of personally coherent interpretation of information and events—are actively constructed and negotiated through language use. Learning is a dialogical event—part the instructor's and part the learner's. If events and information are to acquire meaning and pupils are to learn, the teacher must think of curriculum not just in terms of points to be made, information to be conveyed, and abstract skills to be mastered. Rather she or he must engage pupils in activities and projects that advance the purposes of their pupils and the goals of instruction. Children do not learn merely by being spoken to or written to, but by participating in communicative exchanges (Johnsen-Høines, 1998; Mellin-Olsen, 1989).

METHODOLOGY

This perspective of learning is the background for how the research design was developed. For me, a crucial question was to find methods that could grasp the dynamics of interaction and how pupils were engaged in creating mathematical meaning. The dynamics of interaction and the development of the thematic content of the subject are, in principle, separable, but in practice they are independent aspects in communication (Edwards & Mercer, 1987; Lemke, 1989).

The study relies on naturally occurring data. My work involves detailed and close analysis of the text of lesson transcripts combined with field notes. My observations took place in two classes at two primary schools in Norway, Lindeskog and Furuli.[1] Furuli is a small school, situated in a rural district, and Lindeskog is in the suburb of a city. The rationale for the selection of schools was to find out more about to what extent the discourse patterns and participation structure in mathematics education were independent of time, place, school size, and number of pupils. The data were collected in mathematics lessons from the second semester of the pupils' first school year (10 observations) and continued until the end of their second year (30 observations). All excerpts in this chapter are taken from mathematics lessons in the second grade.

These data provided me with the opportunity to view the meanings teachers and pupils constructed through formal and informal processes. I made extensive field notes and audio recordings of classroom talk in order to capture the dialogue between the teachers and pupils and the activities going on. Transcription of accounts in the classroom required a notation system[2] for "transforming" speech into written texts, here presented as a simplified version of the Jefferson system (Sacks, Schegloff, & Jefferson, 1974).

Significant episodes were selected by looking for interesting events, such as disturbance or a tendency to a breakdown of the ongoing dialogue (Jørgensen & Phillips, 1999). Other interesting episodes for my study were when pupils extended or changed the content of the dialogue (Wells, 1999). The sequences of interesting events were thereafter deconstructed into interaction format. Conflict settings are good moments to see pupils' ideas, their version of educational topics, the discursive resources used to state and defend a position, and, thus, their subjectively constituted roles in the interaction (Edwards & Mercer, 1987).

The Focus of the Analysis

I analyzed the ways pupils and teachers communicated and the ways the content of the lesson developed. Young pupils have other experiences than the teacher, and misunderstandings occur in the classroom dialogue. Sometimes pupils' contributions to the issue are as expected; sometimes their interpretation of the social rules and the mathematical problems differ from the teacher's curriculum.

From the perspective of discourse analysis (Jørgensen & Phillips, 1999; Potter, 1997), asymmetry is one of the interactional features revealed in discourse. Asymmetrical aspects are not introduced into the analysis as part of pre-established reality, but as features of the relationships displayed and oriented to as such by the participants (Marková & Foppa, 1991). Pupils' engagement in the discourse affects more or less the triadic dialogue, and sometimes this will change the power relation between the teacher and the pupils. According to Candela (1999), asymmetric changes occur in situations where the issue is important for the pupils. Examples of such changes are when pupils ask critical questions or evaluate the teacher's instruction in a critical way.

EPISODES FROM MATHEMATICS CLASSROOMS

The following analysis is based on transcriptions of audio tape-recordings from mathematics lessons in the two research classes. The episodes take place in the pupils' second-grade year; episode I takes place at Furuli school, and episode II and III at Lindeskog school.

Episode I: Working with Dice

The class is working on addition (the numbers between zero and ten). The eight pupils present are seated in two groups of four. In the first part of the lesson, the pupils have been working on the positions of numbers on the number line (the ordinal meaning of numbers). The teacher draws two big dice on the blackboard and says they are going to do some arithmetic.

Excerpt 1

74 **T:** Now you are going to work together two and two / later with // dice // how many dots do you see here Chris? [pointing at one of the drawn dice on the blackboard]

75 **Chris:** Three

76 **T:** Then I write the number / three // how many dots do I have here? [pointing at another die on the blackboard]

77 **Jack:** [shouts] Si-i-ix

78 **T:** Six / then I write the number six

79 **Jack:** [banging at his desk]

80 **T:** Now this is important that you follow and that you learn how the die / the dots on the die look like.

81 **Rich:** I know that.

82 **T:** Well, not all pupils know / you see Rich / now I will show how we do it if we are going to draw two / dots / if we are going to draw / two dots let's have a look.

83 **Rich:** We already know that.

84 **T:** [drawing] We have to practice a little.

85 **Rich:** All of us already know.

The teacher (74) tells the pupils what they are going to do, but first she poses some questions to the pupils. She nominates the pupil who is going to answer without waiting for the pupils to put their hands up. Chris answers "three," and the teacher says she is writing three (76). She then asks a new question and points at the dots on a die she has drawn. Jack shouts six. The teacher does not question why Jack is shouting. She repeats the word six and says she is writing six (78).

Jack makes noise by hitting his fist on his desk. It becomes unclear what she emphasizes as important; if it is what the die looks like or what the dots on the die look like (80). At first the teacher says that it is important that "*you*" (Norwegian plural "dere") pay attention, but then continues that "*you*" (Norwegian singular "du") learn this. Probably the last "you" is directed towards Jack, who is making a noise. Then she says that it is important to learn what the *die* looks like, but then corrects herself immediately to what the *dots* on the die look like.

The teacher says that not everybody knows and adds, "you see Rich" (82). The last "you see" could be to indicate that she resents the interruption or it might be because she wants to remind Rich that he should pay respect to his fellow pupils in the class.

There appears a tension or discrepancy in the interaction between the teacher and two of the pupils. We may ask if this is due to a fixed structural pattern (pedagogic genre) or to the subject-matter content (mediation aspects). Looking back to the interaction between Rich and the teacher, they seem to disagree on what they are talking about when it comes to the die. Rich, on his side, does not give in and repeats his claim, but now by including the whole class: "we already know that" (83). The teacher is talking about "*we*" as well (84). She does not answer Rich's statement directly. Rich then tries again and claims that *all of us* already know, and of course he does, if we look at what the teacher said (80). Any pupil in grade two would claim that he/she knows what the dots on a die look like. The meaning of the teacher's utterance is more likely to be related to how the dots are positioned on the die, what Sfard (2001) describes as the attended focus. Rick does not want to be subordinate to the teacher's orientation. He argues for his interpretation of the teacher's statement. Even when attending to the same object, interactants, as here the teacher and a pupil, interpret it from different perspectives as a result of their previous experience and current concerns. This means that, for dialogue to proceed satisfactorily, participants have to make a persistent attempt to understand each other's perspectives—to achieve a state of "intersubjectivity."

The teacher seems not to be willing to discuss Rich's statement that "I know that" or ask him to explain what he means.

How to Use the Dice

The teacher hands out dice to the groups, four to one group, but then she discovers that she has insufficient dice. Accordingly, one of the groups gets only two dice. The teacher shows on the blackboard how to use the dice.

The teacher tells the pupils to work together and work out what "becomes two" by adding two dice. Mary answers there are "one there and one there," pointing towards the two dice on the blackboard. The teacher repeats Mary's answer and writes on the blackboard.

Excerpt 2

86 **T:** Now you should try to collaborate / now you are going to make an arithmetic problem for me // with the dice / it might well happen that it will turn out differently for each group / and / the number that the die adds up to // that is four / please do collaborate / agree about how you will set the problem / when you have finished you put your hands up.
87 **Jack:** I don't have any die.
88 **T:** You have two dice / and you should cooperate / unfortunately I do not have enough dice for everyone so we have to work together four and four / but that will be all right / then I ask / Jim's group.
89 **Jim:** Two plus two.
90 **T:** Two dots [draws] // and two dots become / two plus two becomes four // do you have any other suggestion?
91 **Kim:** Yes
92 **T:** Yes / let me hear.
93 **Kim:** Four plus zero
94 **T:** Four // but you should use two dice.
95 **Kim:** Eh
96 **T:** But it will pass / but four plus zero it / just now we should use the dice you know two dice // there is no zero / there is no side of the dice without a dot on / on all sides.
97 **Kim:** One plus three

In excerpt 2 it can be observed that the teacher opens up for suggestions and says that there might be various answers. She emphasizes that the pupils should collaborate (86). Jack says that he has no die. He is in the group that was given only two dice. The teacher answers that they have to work together four by four, without giving any explanation how to do it and goes on asking Jim's group (88). The teacher asks for other suggestions and Kim suggests four plus zero equals four (93). He has probably forgotten that they were supposed to use the dice and is just thinking of what becomes four. The teacher repeats Kim's answer, but then reminds him of using the two dice. She brings him back into the discourse (where dice should be used). Kim seems confused, and the teacher then hurries up saying that this will pass, but not just now, because of the dice (96). Before the teacher has finished her utterance, Kim comes with a new suggestion that fits with the look of the dice (97).

Kim manages what is in fact an important aim in arithmetic; that is to say adding up "in his head" without using concrete aids. It is highly probable that he can add up without the dice, also with respect to larger numbers. He realizes, however, that this is not what the teacher wants him to do. The dice represent some limits to what he is allowed to suggest.

Collaboration Problems

Something that seems to disturb and distress the activity focus is the lack of one die in one of the groups. The dice seem to become a problem rather than a support. The pupils pay more attention to the dice than the content of the teacher's question.

Excerpt 3

98 **T.** One dot /plus / pay attention back there Mary / do you turn around One dot [counting while drawing] / two three / that becomes / one plus // three equal
99 **Rich:** Four
100 **T:** Rich
101 **Jim:** Yes but may I say something?
102 **T:** Yes
103 **Jim:** Certainly we cannot use two dice when we just have two.
104 **T:** Well yes / because you collaborate all four of you / you four work together for the first time you are supposed to work together a little.
105 **Jim:** xx

106 **T:** We try again /two dice are now going to equal // the number // seven.

Rich answers before the teacher nominates him (99). Still the teacher asks him after he has given his answer. Perhaps she wants Rich to repeat the answer in order to maintain the structure of the conversation or to be sure that the other pupils should hear his answer. The next thing to be noticed is Jim's polite question[3] whether he is allowed to say something (101) and how the teacher answers him. For Jim it is important to clear up the problem with the lack of dice. Instead of asking directly, he asks if he is allowed to say something. He argues that four pupils are not able to operate with two dice as they only have two. The teacher again refers to collaboration—all four of them should collaborate. It seems to be a real problem for this group to do this in practice. They are not able to handle the task they are given. The teacher does not seem to notice that the group has a real problem. She continues her questioning, and the pupils at the group with two dice give up using them.

Some of the pupils certainly can solve the arithmetical problems without the dice, but one of the groups become frustrated because they are not able to do the operations they are asked to do. Probably they find this situation quite unfair when they compare themselves with the other group, where they have got four dice.

Dice are a common tool that children know, although from other contexts such as playing games. In this lesson the dice are intended to be used in a new situation for the children—the addition of numbers. In the mathematical activities of this lesson, the pupils are intended to handle various forms of visual and mental representations of numbers:

- Pictorial (e.g., drawings of dice)
- Iconic (e.g., drawings of the dots on the dice)
- Symbolic (the formal number symbols)

Visualization by means of concrete tools does not work automatically. The pupils have to interpret and structure them by themselves and be responsible for the abstractions by reading new, but invisible relations and structures into the visualizing material (Bruner, 1964). A crucial question is whether, for example, dice used in this context are of any help for the pupils[4].

Dice used as teaching aid in primary school mathematics are recommended in the pupils' textbooks. The idea behind this kind of material tool is that this will support the child's mental strategies (Vygotsky, 1978). Since the teacher cannot transmit her knowledge of arithmetic directly to the child, she must try to base it on something well known to the children, on something that can be verbalized and visualized in a way that the children can understand. In connection with addition, the material tools are meant to be a support for children who are not yet able to add mentally. This does not mean that they do not have mental strategies, but rather that the children ascribe the object they are counting with a particular meaning. The teacher can choose between common tools (money, calendars, rulers, dice) or specially developed educational materials such as Multilink and Unifix cubes. Teaching aids are part of the children's real world, but at the same time they are artificial objects intended to express certain theoretical relationships. Finding suitable tools and then using them in a way that helps pupils to understand how they are related to the concepts, procedures, and other aspects of mathematics is a didactical challenge.

Episode II: The Rules of Communication

The pupils, aged six and seven in this study, are very eager to participate. In different ways they try to get the teacher's attention by waving their hands and other non-verbal signs. Before the following episode took place, the teacher reminded them about raising hands and not saying anything before they are asked. Raising hands in the classroom is a way of displaying to the teacher and to other pupils that you know the answer. When a pupil is showing a hand, the teacher may assume that the pupil knows the answer. Thereby, there is less risk for asking someone who does not know the answer.

The class is working with numbers between ten and twenty. The teacher draws two columns on the blackboard where the tens and units should be marked. Then she raises a question.

Excerpt 4

01 **T:** How many squares do we need to color on the other column in order to get fifteen together?
02 **A pupil:** Hh-mm
03 **T:** How many squares do we have to color here? [points at column number 2] / Per?
04 **Per:** Eh //
05 **T:** How many squares do we have to color on the other column in order to / it is fifteen we should get...
06 **Per:** Is it / I think it is / five / five or something
07 **Tor:** [shouting] I know!
08 **T:** There is ten / here

09 **Tor:** I know!

10 **T:** Yes / that's good / I can see that you have raised your hand // there are many more besides // how many squares do I have to color on the other column in order to get fifteen // then we should listen to someone else / Kari?

11 **Kari:** Ten and five

12 **T:** Ten and five / do you agree with Kari?

13 **Pupils:** Yes / yes / yes

In this episode the teacher asks Per, who hesitates (04). The teacher then repeats her question, and Per gives his answer, but he is not sure—he thinks it is "five or something" (06). Tor then shouts out that he knows. For Tor it might be difficult to understand why the teacher does not give a turn to him since Per seems so doubtful about his answer. In other situations, Tor's action would be valued as a helping hand, but not so in classroom discourse.

The episode shows a typical communication pattern: the teacher asks questions, and the pupils are expected to raise their hands. The teacher's questions are linked to what the pupils are intended to learn in mathematics. Interesting dialogues emerge in this episode between the use of language to develop the thematic content of the mathematics lesson and its use to produce an accepted social behavior, which the teacher can control. In utterances 04 and 06, Per answers doubtfully—he thinks the answer might be five. In utterance 10, expectation of curricular knowledge is mixed with what the teacher expects of social behavior.

On one hand, the pupils have to find the right answer very quickly in competition with fellow pupils. On the other hand, they have to control their behavior, keeping quiet about what they know. There is a complex relation between form and function in language use. Tor does not understand why his knowledge is not valued. He seems to be confused by the hidden rules. Why should not the teacher give turn to him since Per is so doubtful about his answer? In other situations Tor's initiative would be valued as a helping hand, but not so in classroom discourse. When he says he knows, the teacher answers that many others also know.

Excerpt 5

Later on, one of the pupils, Eva, is sighing deeply, and the teacher remarks:

78 **T:** Yes / was this a little bit difficult

79 **Pupils:** No / no

80 **Eva:** Beate [the teacher's name] / I did not get my turn in spite of raising my hand constantly

81 **T:** Should we take another one // then I will ask what is // what equals sixteen // Eva

82 **Eva:** That is six / six / ten and six

83 **T:** [writes on the blackboard] so let us take one more question / what is the neighbor number to ten/ do you know what a neighbor is?

84 **Pupils:** Yes

85 **Eva:** They are nine and eleven

86 **T:** Well / yes

87 **Eva:** [laughs]

88 **L:** Eva / you may try to say it once more then

89 **Eva:** Nine and eleven

If we take a look at Eva's situation, she has followed the rule that Tor broke. She has done exactly what the teacher had asked her to do—but in spite of that, she was not given the opportunity to answer. The teacher expresses the rule of hand raising openly, but the relation between form and function in the teacher's message is not obvious for the children. From the children's perspective, this rule might seem quite unfair. In spite of raising hands, many of them will not get their turn to answer.

The analysis of the episode reveals that what seems obvious for adults is not always comprehensible for children. Tor and Eva have raised their hands, but the underlying meaning is that raising hand is no guarantee for being the chosen one. Eva has done exactly what the teacher had asked her to do—but in spite of that, she did not get the opportunity to answer (80). However, she argues for her right to get turn and has success. She then wants to go on answering (85), but seems to realize she has broken the rules of participation and starts to laugh (87). The teacher repeats the question and gives the turn to Eva. The discourse pattern is then re-established.

Episode III: Extending the Answer

The following episode illustrates the type of effect the teacher's questions may have on interaction. The pupils are grouping wooden sticks into tens and then adding them. The teacher has drawn three figures named Eli, Johan, and Karen on the blackboard. She has explained for the pupils that the figures are intended to take part in a game and each of them has been awarded points in this game.

Excerpt 6

The teacher then asks:

01 **T:** How many points did Eli get / Daniel
02 **Daniel:** / ninety
03 **T:** Ninety /do you all agree?
04 **Pupils:** Yes/ yes
05 **T:** How did you do that?
06 **Daniel:** // counted in my head
07 **T:** You counted in your head / you counted fifty / plus / twenty / plus [twenty
08 **Daniel:** [No I didn't / I thought that four plus five is nine
09 **T:** You thought that four plus five is nine / yes
10 **Daniel:** Yes
11 **T:** And then you added zero / like this that's smart [writes at the blackboard] how many points has Tor got // how many points has Tor // Marian?
12 **Marian:** Eighty
13 **T:** /are you quite sure
14 **Marian:** Yes
15 **T:** // look here once more / you were quite close
16 **Marian:** ... [N]
17 **T:** What do you think, Susan?
18 **Susan:** Seventy
19 **T:** Seventy / why do you think so?
20 **Susan:** Fifty and twenty / that's seventy
21 **T:** Fifty plus twenty equal seventy / mm [writing] can you see it Marian?
22 **Marian:** Yes

The teacher is asking her pupils to explain how they "thought." Daniel tells that he counted in his head. He has developed his own strategy for adding forty and fifty (06, 08). Marian is not quite sure of what the answer will be, and the teacher gives her time for reflection before she gives turn to Susan. As in excerpt 4, the interaction pattern in the excerpt above is a three-part structure, controlled by the teacher.

The difference between the communication pattern in the dialogues seems trivial at first sight. According to Nystrand (1997), the contrast between "monologic" and "dialogic" lessons is clarified when we examine the roles of the conversants in the discourse structure and each exchange. The interaction pattern of excerpt 4 is reminiscent of the monologically organized instruction; no pupil ideas are elaborated, and the teacher moves on to the next question just as soon as a pupil has demonstrated what she knows. The teacher is responsible for both topic and elaborations. The resources of knowledge are the teacher and the textbook. Wells' (1986) study of children's talk experience at home and in school showed that teachers were far less likely to use interactions to include children's meanings into their own talk, either by elaborating upon their contributions or by asking the pupils to extend their thinking. Not only do children almost cease to ask real questions at school, but teachers rarely ask them to express and explain their beliefs and opinions—at least with respect to the official curriculum.

The general emphasis on the importance of good lesson planning and teacher control tends to put pressure on following lesson plans as closely as possible and avoiding any unplanned learning activities. Consequently, the discourse structure becomes less flexible, and the teacher and the textbook are resources of knowledge. The aim of the dialogue is then to find the "right" answer. In case of such discourse patterns, there will be less time for discussion and negotiation of meaning. This is in conflict with the view that pupil initiative, participation, and involvement in instruction represent an important aspect of learning processes. How the teacher gives feedback might change their role of participation.

The pupils in excerpt 6 are encouraged to extend their answers and explain how they come to a solution (Wells, 1999). The teacher is uptaking[5] the pupils' responses and establishes by doing this the dialogic character of the interaction. These behaviors are a form of "fine tuning" because the teacher gives attention to what the pupils are able to communicate at this stage. Both teacher and pupils take part in knowledge construction.

DISCUSSION

The analysis of the transcripts shows pupils' version of the educational topics presented, their discursive resources used to state and defend a position, and their subjectively constituted roles in the interaction. The analysis also shows that classroom discourse is a complex mediator of learning where the participants refer to different subject matters and rules of participation in the communication. Sometimes, the result will be communication breakdown, sometimes asymmetric changes in the dialogue.

The teacher's feedback and support are significant for pupils' possibilities to contribute in the discourse, balancing challenges and support. According to current theories of pedagogy (Barnes & Todd, 1977), the practice of fine-tuning is crucial in the process of learning, because it pushes the pupils' understanding a bit further than its current limit (Vygotsky, 1978). As Nystrand writes, "When teachers ask questions about what students are thinking, and when they ask questions about students' previous answers, they promote fundamental expectations for learning by seriously treating students as thinkers, that is, by indicating that what students think is important and worth examining" (1997, p. 28).

The analysis shows that the classroom discourse regulates who is going to talk, how to talk, and what to talk about. Such regulations affect how the mathematical topic is presented. If mathematical knowledge is viewed as given (Alrø & Skovsmose, 2004), pupils are supposed to discover the given knowledge, not participate in creating knowledge. In the mathematics lessons referred to in this chapter, the teaching of mathematics emphasized procedures—how to do it. According to Edwards and Mercer (1987), institutionally produced knowledge, rather than involving conceptual understanding of the "principles" of educational knowledge—or *principled* knowledge—often subordinates this understanding to the power relations of the classroom interaction and to the constraints of the lesson. Thus a procedural type of knowledge is emphasized—a knowledge of knowing how to execute classroom activities. As Edwards and Mercer (1987) observe, procedural knowledge will, when not accompanied by an understanding of the principles of the activity, often give rise to *ritual* knowledge—that is, knowledge that does not connect the pedagogic activity with explanations and the pupils' own ways of understanding.

In the transcribed dialogues there seems to be a tendency to take pupils' self-selected initiatives as a disturbance and interruption of the turn-taking structure. The children referred to in the study are young pupils, 6 and 7 years old. They have limited experience of being in school. Nevertheless, their first meeting with the discourse of mathematics will affect their attitudes and beliefs of mathematics for years to come. If pupils have to struggle with concentrating on the rules of participation, trying to find the "right" answer, less attention will be given to the mathematical reasoning. If pupils gain several experiences of being rejected by the teacher, it is reasonable to assume that they will stop taking initiative on their own. As Wells (1999) sees it, there is knowledge that pupils need to take over from previous generations, and monologic, direct instruction is sometimes the best way of providing the necessary opportunity for such learning. However, in the light of a sociocultural perspective on learning, monologic instruction alone is not sufficient. Pupils do not always understand what they are told and so need to engage in clarifying dialogue to get the meaning of the task and what they are expected to do. Pupils may also have alternative perspectives on a topic that need to be brought into the communication and explored in more symmetric dialogues.

According to Myhill (2002), interactive learning through whole class teaching places considerable emphasis on the skills of the teacher; the teacher has to "orchestrate" verbal encounters, keep control of the turn-taking in the discourse and the subject matter, and at the same time, enhance the learning process of her pupils. Teachers might feel the pressure to get through the mathematical content planned for the lesson, and time pressure gives little space for pupils' initiatives (Chilver & Gould, 1982). The discourse structure then becomes less flexible, and the teacher and the textbook are the resources of knowledge (Nystrand, 1997). The aim of the dialogue is to find the "right" answer. In the case of such discourse patterns, there will be less time for discussion and negotiation of meaning. This is in conflict with the view that pupil initiative, participation, and involvement in instruction represent an important aspect of learning processes.

Mathematics education seems to be an area where insight into discourse patterns and rules of interaction is highly relevant. In my opinion, we need more knowledge of how the actual subject matter content of lessons affect and is affected by the interaction patterns of the classroom. The instructional discourse opens up particular epistemic roles for the participants. What the pupils learn will depend on how their teacher responds to their responses and the opportunities given them in the negotiation of mathematical meaning. By exploring the role of language use and communicative actions in the construction of meaning, things taken for granted are much more tentative and open to question than one might assume. A deep micro-level analysis, in which a dialogical approach to communication and cognition is used, indicates a complex dynamic relationship between teaching and learning and between the participants in the classroom, the act of learning, the language, and the object of learning.

The young pupils need space and time for expressing what they know and can manage on their own. The challenge for teachers will be balancing the conflicting demands of, on the one hand, a child-centred ideology of learning and, on the other hand, an essentially socializing role as the society's agents of cultural transmission in the context of a system of compulsory education (Edwards & Mercer, 1987). The pupils need to learn how to communicate mathematics, and they need to practice these skills.

NOTES

1. For reasons of anonymity, the names of the school, the teachers and the pupils are fictional.
2. Pause < 2 secs. = /
   Pause > 2 secs = //
   Longer pause = ... (N)
   Emphasis on a word =___
   Overlapping speech = [
   T indicates teacher's utterance
   xx indicate unintelligible speech
3. Schegloff (1980) describes utterances like "Can I ask you something?" as preliminary to preliminaries.
4. I will here refer to the ideas of Bruner (*The Course of Cognitive Growth*, 1964), who formulated his theory of three different types of representation of human knowledge: enactive, which is a physical process, while the other two, iconic and symbolic, are drawn or written.
5. Nystrand refers to Collins (1982) in order to explain "uptake." Collins says that "uptake" occurs when one conversant—for example, a teacher—asks someone else—for example, a student—about something the other person said previously.

REFERENCES

Alrø, H. & Skovsmose, O. (2004). *Dialogue and learning in mathematics education: Intention, reflection, critique.* Dordrecht, The Netherlands: Kluwer Academic Publishers.

Bakhtin, M.M. (1981). *The dialogic imagination: Four essays.* M. Holquist (Ed.), C. Emerson & M. Holquist (Trans.). Austin: University of Texas Press.

Bakhtin, M.M. (1986). *Speech genres and other late essays.* C. Emerson & M. Holquist (Eds.). Austin, Texas: University of Texas Press.

Barnes, D. (1976). *From communication to curriculum.* Harmindsworth, UK: Penguin.

Barnes, D. & Todd, F. (1977). *Communication and learning in small groups.* London: Routledge & Keagan Paul.

Bjuland, R. (2002). *Problem solving in geometry: Reasoning processes of student teachers working in small groups: A dialogical approach.* Unpublished doctoral dissertation, Department of Applied Education, Faculty of Psychology, University of Bergen.

Bjuland, R. (2005). Dialogiske tilnærminger i klasserommet: Utvikling av matematiske begreper gjennom lærer-elev dialog i klasserommet eller gjennom elevsamarbeid i 2 smågrupper. In K. Hirsch (Ed.), *Ulike perspektiv på læring* (s. 67–76). Senter for voksenopplæring, Trondheim.

Bruner, J. (1964). The course of cognitive growth. *American Psychologist, 19,* 1–15.

Bråten, I. (1999). Selvregulert læring. (Self-regulated learning). *Bedre skole, 3,* 25–30.

Candela, A. (1999). Students' power in classroom discourse. *Linguistics and Education,* 10, 139–163.

Cazden, C.B. (2001). *Classroom discourse: The language of teaching and learning* (2nd ed.). Portsmouth: NH: Heineman.

Cestari, M. L. (1997). *Communication in mathematics classrooms.* Unpublished doctoral dissertation, Agder College,

Cestari, M.L. (1998). Teacher–student communication in traditional and constructivist approaches to teaching. In H. Steinbring, M.G.B. Bussi, & A. Sierpinska (Eds.), *Language and communication in the mathematics classroom* (pp. 155–166). Reston, VA: National Council of Teachers of Mathematics.

Cestari, M. L. (2000). Opening the door of the mathematics classroom: Students wondering and reflecting. In B. K. Selvik, C. Kirfel & M. Johnsen-Høines (Eds.), *Matematikk i samfunnsmessig, historisk og kulturelt perspektiv* (pp. 187–193). Skriftserien. Rapport nr.2/2000. Bergen: Høgskolen i Bergen.

Chilver, P. & Gould, G. (1982). *Learning and language in the classroom: Discursive talking and writing across the curriculum.* Oxford, UK: Pergamon Press.

Collins, J. (1982). Discourse style, classroom interaction and differential treatment. *Journal of Reading Behavior, 14,* 429–437.

Donovan, S. M. & Bransford, J. D. (Eds.). (2005). *How students learn: Mathematics in the classroom.* Committee on How People Learn: A Targeted Report for Teachers. Washington, DC: National Academies Press.

Edwards, D. & Mercer, N. (1987). *Common knowledge.* London: Routledge.

Edwards, A.D. & Westgate, D.P.G. (1994). *Investigating classroom talk* (2nd ed., rev. and extended). London: Falmer Press.

Fosse, T. (2004). *Skolestart: en studie av 6-åringers forventninger til skolen med særlig vekt på matematikkundervisningen.* (*Starting school: A study of the expectations of 6 year-old pupils, with a special focus on mathematics education*). Hovedoppgave i pedagogikk. Universitetet i Bergen.

Jørgensen, M.W. & Phillips, L. (1999). *Diskursanalyse som teori og metode.* Frederiksberg: Roskilde Universitetsforlag.

Johnsen-Høines, M. (1998). *Begynneropplæringen: fagdidaktikk for barnetrinnets matematikkundervisning.* Landås: Caspar forlag.

Lave, J. & Wenger, E. (1991). *Situated learning: Legitimate peripheral participation.* Cambridge, MA: Cambridge University Press.

Lemke, J.L. (1989). The language of science teaching. In C. Emihovich (Ed.), *Locating learning: Ethnographic perspectives on classroom research* (pp. 216–239). Norwood, NJ: Ablex Publishing Corporation.

Lindén, N. (1992). *Stillaser om barns læring.* Bergen: Caspar forlag.

Linell, P. (1995). Troubles with mutualities: towards a dialogical theory of misunderstanding and miscommunication. In I. Markova, C. F. Graumann & K. Foppa (Eds.), *Mutualities in dialogue* (pp. 176–213). Cambridge: Cambridge University Press.

Linell, P. (1998). *Approaching dialogue. Talk, interaction and contexts in dialogical perspectives.* Amsterdam: John Benjamins Publishing Company.

Marková, I. & Foppa, K. (Eds.). (1991). *Asymmetries in dialogue.* Hemel Hempstead, UK: Harvester Wheatsheaf.

Mehan, H. (1979). *Learning lessons: Social organization in the classroom.* Cambridge: Harvard University Press.

Mellin-Olsen, S. (1987). *The politics of mathematics education.* Dortrecht, The Netherlands: Kluwer Academic Publishers.

Mellin-Olsen, S. (1989). *Kunnskapsformidling. Virksomhetsteoretiske perspektiver.* Bergen: Caspar Forlag.

Mercer, N. (1995). *The guided construction of knowledge: Talk among teachers and learners.* Cleveden, UK: Multilingual Matters.

Myhill, D. (2002). Bad boys and good girls? Patterns of interaction and response in whole class teaching. *British Educational Journal, 28*(3), 341–352.

Nystrand, M. et al. (1997). *Opening dialogue: Understanding the dynamics of language and learning in the English classroom.* New York: Teachers College Press.

Potter, J. (1997). Discourse analysis as a way of analysing naturally occurring talk. In D. Silverman (Ed.), *Qualitative research: Theory, method and practice.* London: Sage.

Roehler, L.R. & Cantlon, D.J. (1997). Scaffolding: A powerful tool in social constructivist classrooms. In K. Hogan & M. Pressley (Eds.), *Scaffolding student learning: Instructional approaches & issues.* Cambridge, MA: Brookline Books.

Rogoff, B. (1990). *Apprenticeship in thinking: Cognitive development in social context.* Oxford: Oxford University Press.

Sacks, H., Schegloff, E.A., & Jefferson, G. (1974). A simplest systematic for the organisation of turn-taking for conversation. *Language, 50,* 696–735.

Schegloff, E. A. (1980). Preliminaries to preliminaries: "Can I ask you a question?" *Sociological Inquiry, 50,*104–152.

Schoultz, J. (1998). *Kommunikation, kontext och artefact—studier av elevers behärskning av naturvetenskapliga diskurser.* Linköping: Linköpings Universitet, Institutionen för pedagogik och psykologi.

Sfard, A. (1997). From acquistionist to participationist framework: Putting discourse at the heart of research on learning mathematics. In T. Lingefjärd & G. Dahland (Eds.), *Research in Mathematics Education. A Report from a Follow-Up Conference after PME 1997* (pp. 109–136). Sweden: Gothenburg University.

Sfard, A. (2001). There is more to discourse than meets the ear: Looking at thinking as communicating to learn more about mathematical learning. *Mathematics, 46*(1/3), 13–17.

Sinclair, J. & Coulthard, R. (1975). *Towards an analysis of discourse: The English used by teachers and pupils.* Oxford: Oxford University Press.

Streitlien, Å. (2006). *Rom for deltakelse—en studie av kommunikasjon og interaksjon i matematikkundervisningen* (*Room for participation—a study of communication and interaction in mathematics classrooms*). Series of dissertations submitted to the Faculty of Education, University of Oslo. Oslo: Faculty of Education.

Vygotsky, L. S (1978). *Mind in society. The development of higher psychological processes.* Cambridge, MA: Harvard University Press.

Vygotsky, L. S. (1987). *Thought and language.* Cambridge, MA: Harvard University Press.

Wells, G. (1986). *The meaning makers: Children learning language and using language to learn.* Portsmouth, NH: Heinemann.

Wells, G. (1999). *Dialogic inquiry—Towards a socio-cultural practice and theory of education.* Cambridge, UK: Cambridge University Press.

Wertsch, J. V. (1991). *Voices of the mind: A sociocultural approach to mediated action.* Cambridge MA: Harvard University Press.

Wertsch, J.V. & Stone, A. (1985). The concept of internalization in Vygotsky's account of the genesis of higher mental functions. In J.V. Wertsch (Ed.), *Culture, communication and cognition. Vygotskian Perspectives.* Cambridge, UK: Cambridge University Press.

Wood, D., Bruner, J. & Ross, G. (1976). The role of tutoring and problem solving. *Journal of Child Psychology and Psychiatry. 17,* 89–100.

15

UPPER SECONDARY STUDENTS' APPROPRIATION OF MATHEMATICAL TOOLS THROUGH COLLABORATIVE PROBLEM SOLVING IN SMALL GROUPS

Martin Carlsen[1]

UNIVERSITY OF AGDER

Abstract: This chapter reports a case study conducted to explore how upper secondary students appropriate mathematical concepts or tools such as scalar or dot product of vectors and geometric series. A sociocultural perspective on learning and development is adopted. Recordings of naturally occurring talk-in-interaction in small-group problem-solving settings were collected, with the intention of scrutinizing the subtleties of the appropriation process in mathematics. The students were found to: (1) steadily improve their accuracy in use of concept-related terminology; (2) experience difficulties in coordinating, on the one hand, their knowing of definitions, theorems, and formulae and, on the other hand, the utilization of these in situ; (3) appropriate the mathematical tools through the use of semiotic means of objectification such as linguistic devices and inscriptions; and (4) be able to explain the mathematical components of the tools to their fellow students, and thus to have come far in their process of appropriating these tools. The analyses also show how resistance manifests itself in the appropriation process.

Keywords: Appropriation, Case study, Dot product, Geometric series, Problem solving

INTRODUCTION

The aim of this chapter is to offer an account of my research which focused on Norwegian students' learning of mathematics in upper secondary school (Carlsen, 2008).[2] The chapter reports the theoretical framework adopted, the research methods used, the data analyses, and the results of my study. Some of these results have been published elsewhere (Carlsen, 2005, 2007). Analyses were made of the subtleties of how students reason and interact to appropriate mathematical concepts and related tools and actions such as definitions, theorems, formulae, and procedures.

One aim of the research was to make theoretical and empirical contributions to understanding the appropriation processes in mathematics learning of students at upper secondary school. Additionally, my study was an attempt to address the lack of in-depth analyses of how upper secondary students appropriate mathematical tools through reasoning. This is achieved by focusing on the nature of students' participation in processes of appropriating the mathematical concepts of dot product and geometric series.

The reasons for researching these issues are also located in my interest in studying the nature of mathematical thinking and learning. According to Lerman (2000), this is a topic that mathematics educa-

The First Sourcebook on Nordic Research in Mathematics Education, pages 223–238

tion research ought to study in depth. As a mathematics student and participant in collaborative problem solving in mathematics, I started to wonder about the nature of the learning process. My subsequent experience as a mathematics teacher further developed my interest in studying how students engage in collaborative problem solving in mathematics and how students appropriate mathematical concepts through reasoning and interaction. Moreover, reasons for my topic of research were found in my reading of the work of Cestari (1997) and Bjuland (2002), in addition to my own experience arising from a related research study, which I had conducted (cf. Carlsen, 2002).

In studying how students, 17–18 years of age, appropriate mathematical tools such as the dot product[3] and geometric series, I adopted a sociocultural perspective of learning and development. The reason for adopting this perspective when analyzing mathematical learning through problem solving in small-group settings was that this framework focuses particularly on the role of communication and use of psychological and physical tools in collective activities (cf. Säljö, 2001; Vygotsky, 1986). When studying the appropriation process, one notices that students are, to a significant degree, dependent on a range of cultural tools such as language, mathematical terminology and notation, paper and pencil, and the calculator (Säljö, 2005; Wertsch, 1998). The sociocultural perspective is a theoretical position in which the role of tools in learning and development is argued to be fundamental. This perspective has its origin in the work of Vygotsky (Vygotsky, 1978, 1986) and his followers Leont'ev and Luria. Following Rogoff (1990, 1995), Säljö (2001, 2005), and Wertsch (1991, 1998), I used the appropriation metaphor of learning[4] to analyze how students participate in and contribute to the ongoing collaborative problem-solving activity in small groups (Cole, 1996; Dysthe, 2001, Mercer, 2000; Wells, 1999). The term appropriation will be further discussed later. My research focus on collaborative learning in small-group settings was chosen because this is a way to access how students reason mathematically.

With this as a background, the following two general research questions, or rather issues of research, were formulated for the study:

1. How do students' utterances and actions in joint mathematical reasoning contribute to their appropriation of specific mathematical tools?
2. How can interaction between students in joint mathematical reasoning episodes be analyzed so as to provide a comprehensive understanding of processes of appropriation of mathematical reasoning?

These broad and general issues were refined into three more specific research questions:

1. What are the characteristics of students' appropriation of the cultural tool of dot product?
2. What are the characteristics of students' appropriation of the cultural tool of geometric series?
3. What are the features of students' use of inscriptions[5] in the appropriation process?

THEORETICAL FRAMEWORK

Some basic ideas arising from the sociocultural perspective on learning and development require clarification. In the following I will first give a brief historical background of research on mathematical problem solving together with clarifications with respect to the theoretical terms used in the study. Second, I discuss the role of tools in mathematical problem solving. This will be followed by a discussion of the sociocultural view of learning as an individual process of appropriation. At the end of this section I will focus on the mathematical substance of my study and relevant research.

Collaborative Problem Solving

Interest in how people learn mathematics through problem solving has roots that are usually traced back to the work of Polya (1957; cf. Lesh & Zawojewski, 2007) and his four-step model of the problem-solving process (Understanding the problem, Devising a plan, Carrying out the plan, Looking back). Schoenfeld (1992) asserts that the term *problem solving* in the context of mathematics learning has been described by terms as diverse as "working rote exercises" and "doing mathematics as a professional" (p. 334). According to him, problem solving is also a term used to describe what mathematicians do as well as a way of learning mathematics. Furthermore, Stanic and Kilpatrick (1989) claim that the term problem solving is used in three different ways: as a context in which mathematics is learned, as a skill where techniques and heuristics are learned, and as the art of mathematicians. In my study, following Bjuland (2002), I view mathematical problem solving as a "cognitive, meta-cognitive, socio-cultural and affective process of figuring out how to solve a mathematical problem"

(Bjuland, 2002, p. 9). With respect to the notion of a *mathematical problem*, I follow Schoenfeld's (1993) definition: "For any student, a mathematical problem is a task (a) in which the student is interested and engaged and for which he wishes to obtain a resolution, and (b) for which the student does not have a readily accessible mathematical means by which to achieve that resolution" (p. 71).

Researchers such as Lester (1994) and Wyndhamn and Säljö (1997) have offered insights into how research on mathematical problem-solving activity has evolved during the last three decades. Lester (1994) has divided the period from 1970 to 1994 into different epochs. According to him, research has in this period focused on identification of determinants with respect to the possible difficulties in problems, comparisons of unsuccessful and successful problem solvers, and metacognitive issues. According to Wyndhamn and Säljö (1997), problem-solving activities in the 1980s were explained by referring to mental structures. In the 1990s the research shifted towards a situated perspective in which problem solving was treated as a contextually dependent and situated activity. Schroeder and Lester (1989) introduced the phrases teaching and learning *for*, *about*,[6] and *via* problem solving. Following Schroeder and Lester (1989), Wyndhamn (1998) used these categories as analytical tools in his analyses of Swedish curricula. However, I view these phrases as being representative with respect to the issue of mathematical problem solving in Norwegian schools as well. It seems reasonable to utilize the *for* and *via* as describing how the students engaged with the process of problem solving in my study. The mathematics teacher, with whom I collaborated, taught mathematics *for* and *via* problem solving and the students learned mathematics *via* or through problem solving. Moreover, following the example of Springer, Stanne, and Donovan (1999), students' efforts in appropriating the mathematical tools of dot product and geometric series are labeled *collaborative learning*—that is, a team-based learning approach characterized by a relatively unstructured process where goals are negotiated, problems are defined, procedures are developed, and socially shared knowing is achieved (Springer et al., 1999).[7]

As indicated in the title of this study, collaboration in small-group settings is the context in which my study has been conducted. In *group collaboration*, a term used by Bjuland (2002) to describe problem-solving interaction among three or more students, students get opportunities to share ideas, ask each other questions, and bring their perspectives to the mathematical discussion. The students in my study, I assert, work collaboratively as a group when they actively participate in and contribute with ideas and arguments to their joint problem solving. Hiebert (1992, pp. 443–444) maintains that by "expressing ideas publicly, by defining them in the face of others' questions, and by questioning others' ideas, students are forced to deal with incongruities." By working with such challenges, the students are encouraged to establish a shared focus of attention and to achieve shared meanings for concepts through elaborating and clarifying their thinking. As will be seen later, my study revealed that these aspects of communication and interaction play significant roles in students' learning of mathematical tools. Group collaboration is hence confirmed to be fruitful for mathematics learning to occur.

The Role of Tools

The mathematical learning process is an extremely complex phenomenon, and to study the subtleties of this process one has to attend to several different but co-existing aspects of communication and interaction. Duval (2006) claims that one reason for this, among others, is that mathematics is different from other scientific domains. According to Duval, three differences have been found:

1. Semiotic representations are of paramount importance in mathematics. The role played by signs is not only to designate mathematical objects or to communicate, but also to work on and with mathematical objects.
2. Mathematical objects are not accessible by perception or by instruments, but only through the use of signs and semiotic systems.
3. The large variety of semiotic representations used in mathematics is not found in other sciences.

It is worth mentioning that these differences relate to mathematics as a subject matter. With regard to the mathematical learning process, several other aspects have to be acknowledged as well, such as differences between learners, the social and cultural context of learning, teachers, and available resources (text books, calculator, etc.). The arguments of Duval further pinpoint the important role of cultural tools in mathematics learning and interaction, such as the semiotic representations signs and inscriptions. With respect to the learning of mathematics, examples of tools are the graphical calculator, the textbook, and paper and pencil. When students use such kinds of physical tools in reasoning and problem solving in various settings, these tools become useful and significant psychological tools in the problem-solving process (cf. Säljö, 2001). Traditionally, the literature has distinguished between psychological or intellectual tools on the one side and physical or material tools on the other side. Related to this, Vygotsky discriminated between the terms sign (psychological) and tool (physical). A sign is an auxiliary means or instrument to solve psychological problems "in a manner analogous to the role of a tool in labor" (Vygotsky, 1978, p. 52). Tools' function is to act on and change the object of activity; they are "*externally* oriented" (op. cit., p. 55, emphasis in original). Signs' function, however, does not change the object of psychological activity; they are "*internally* oriented" (op. cit., p. 55, emphasis in original). Nevertheless, when analyzing social interaction in practice, these functions merge (Resnick, Pontecorvo, & Säljö, 1997), and in such circumstances the classical dichotomy between psychological signs and physical tools is unnecessary. In my opinion, signs are tools as well as the compasses and the ruler. The psychological tools such as language, ideas, and thoughts are developed in close relationship with the development of physical artifacts and vice versa. Hence, in order to understand the role played by tools in interaction, it is more fruitful to consider them as containing both psychological *and* physical aspects (Säljö, 2005). Figure 15.1 (Carlsen, 2008, p. 38) tries to illustrate these relationships.

Figure 15.1 is meant to communicate that the sociocultural notion of tools might be divided into two parts: psychological and physical tools. The vertical double-headed arrows indicate, however, that in the literature these notions are used interchangeably with intellectual and material tools, respectively. One particular kind of tool is what is referred to as an *inscription*, where the intellectual and material aspects co-exist. This is why this notion is placed between them. Inscriptions are considered as powerful tools in collaborative problem solving in mathematics—for example, graphs, algorithms, tables, and drawings. In student reasoning, these inscriptions are significant elements (Latour, 1987; Lehrer, Schauble, Carpenter, & Penner, 2000; Lehrer, Schauble, & Petrosino, 2001) through which thinking is externalized and made explicit (Cobb, 2002; Säljö, 2005; Sfard, 2002). *Reasoning* is seen as a process encapsulating both mathematical thinking and externalizations of this mathematical thinking, through argumentation and interaction in collaborative problem solving. This significant role of inscriptions in student reasoning is based on a premise of active participants who are able to make sense when interacting with them, looking at them, and using them. Through mediating tools such as inscriptions and language, students objectify their knowing (cf. Radford, 2002, 2003) and establish shared focus of attention and shared meanings for concepts (cf. Moschkovich, 2004; Rogoff, 1990).

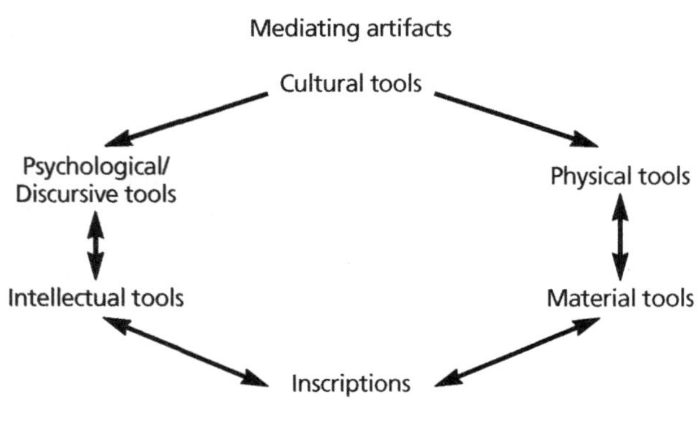

Figure 15.1 Mediating artifacts and cultural tools—an overview.

Learning as Appropriation

The metaphor of appropriation is used with regard to learning, as a process resulting from interactional and social engagement in which people are active participants and contribute with ideas, questions, and arguments (Rogoff, 1990, Säljö, 2001). Hence, learning and knowing are initiated through interactions and collaborative efforts, that is, collective thinking (Mercer, 2000) between people. Secondly, this knowing and experience become individually appropriated, that is, taken as the possession of the one who appropriates (Säljö, 2005; Vygotsky, 1986; Wertsch, 1998).

Following Bakhtin (1981) and Wertsch (1998), appropriation is viewed as a process of "taking something that belongs to others and making it one's own" (Wertsch, 1998, p. 53). Included in this process is the irreducible triad of agent (students), tool (e.g., concepts, inscriptions), and context (collaborative small-group problem solving) (Wells, 1999; Wertsch, 1998). The process of appropriating mathematical concepts thus concerns a process of becoming a cultural knower of these concepts together with related tools and actions. In this process of appropriation, it is important to emphazise the role of tools (Cobb, 2002; Cole, 1996; Säljö, 2001). A basic assumption is that people purposely utilize tools in interaction and problem solving (Resnick et al., 1997; Rogoff, 1990; Vygotsky, 1979, Wertsch & Rupert, 1993). In this sense, appropriation implies that students increase their knowing of and ability to interact with the culture called mathematics and all its concepts, theorems, procedures, and formulae—knowing and abilities that might not be that relevant within other contexts (Säljö, 2005). An aspect of the learning process from a sociocultural position is what Wertsch (1998) calls mastery. Mastery is a term used equivalently to knowing how to use tools in concrete situations—for example, multiplication of multi-digit numbers and evaluating the sum of a geometric series. Students might master the algorithm or calculator to solve such problems without completely appropriating the mathematical tool of multiplication or geometric series (Wertsch, 1993). In that sense, mastery of a mathematical tool is a necessary but not sufficient part of the process of appropriating that tool.

To achieve shared meanings for concepts is the core of the process of appropriation; that is, it is when meanings are shared that students take over what someone else produces during joint activity and make it their own. Appropriation thus "occurs in the process of participation, as the individual changes through involvement in the situation at hand" (Rogoff, 1995, p. 153). Participation here means to take part in, engage with, and contribute to the collaborative problem-solving efforts in the small group through verbal and nonverbal actions.

Language, and communication more precisely, is thus also a significant constituent of the appropriation process (cf. Mercer, 2000). According to Vygotsky (1986), language is the most important mediating tool humans have at their disposal. Alongside language, Radford (2002, 2003) uses what he calls *semiotic means of objectification* as analytical tools when studying the appropriation process. According to Radford (2003), these semiotic means are "objects, tools, linguistic devices, and signs that individuals intentionally use in social meaning-making processes to achieve a stable form of awareness, to make apparent their intensions, and to carry out their actions to attain the goal of their activities" (p. 41). Furthermore, Radford describes the process of appropriating mathematical tools and actions by utilizing the concept of objectification, which is "a process aimed at bringing something in front of someone's attention or view" (2002, p. 14). This concept highlights a significant aspect of a sociocultural perspective on learning, to be able to externalize and make explicit one's thinking and arguments in interaction and collaboration, in order for the participants to make sense of each others' contributions. The objectified thinking contributes at the interpsychological, social level—thoughts and ideas that later will appear on the individual, intrapsychological level, to use Vygotskian parlance. In this study, I have drawn on the work of Radford and analyzed the students' utilizations of semiotic means and their role within an objectification process. Additionally, the students' sensemaking through relating earlier understandings and experience with the mathematical problem at hand, and externalizations of thinking through explaining and reasoning, are emphasized as constituents of the objectification process and hence of appropriation.

I have found the terms *interthinking* and *cumulative talk* used by Mercer (2000) helpful when it comes to the appropriation process. Interthinking is, in Mercer's terminology, a joint intellectual activity through which students' participation and actions are coordinated. Moreover, through this mode of thinking together, students build on each others' contributions, that is, their externalized thinking and arguments. The term cumulative talk is closely related to interthinking. This kind of talk is characterized by participants who are mutually establishing joint knowing and meanings and are adding individually new information to their discussion. Cumulative talk is hence a kind of dialogue that is characterized by interthinking, and this shared intellectual activity supports the interlocutors' individual processes of appropriating tools and actions. Through participating in such a joint activity, the collaborating students achieve shared foci and taken as shared meanings (Rogoff, 1990; Wertsch, 1998).

Resistance is a significant issue of the appropriation process of mathematical tools (Säljö, 2005). As a theoretical term, resistance denotes the obstacles or tensions that emerge when students use tools in context, such as difficulties in calculations, in recalling exponents, or in how to use concept-related terminology. According to Wertsch (1998), the process of appropriation always involves resistance of some sort. This resistance he metaphorically describes as "'friction' between mediational means and unique use in mediated action" (p. 54). This resistance manifests itself as problems between cultural tools per se and their use in situ—that is, the obstacles that emerge between knowing constituent parts of a concept and how to coordinate and utilize this knowing in practice. With the words of Säljö (2005), the "cultural tool mediates a certain meaning that is not immediately or intuitively available to the individual" (p. 231, my translation). An example of the issue of resistance from mathematical problem solving would be that to be able to define what a geometric series or dot product are at a general level, is not the same as being able to productively utilize such tools in specific situations. Individuals' appropriation of mediational means is tied to the context in which it takes place. In my work I view resistance slightly differently than Wertsch (1998). The resistance involved in the appropriation process is not seen as an active position taken by the learner against what to be learned. Resistance, in the way I use the term, is an inevitable element of the learning process, when encountering and using tools and actions for the first time and in later use. The microdynamics of resistance are thus "shaped by factors in the immediate spatial and social context" (Wertsch, 1998, p. 168).

This resistance fades and disappears when individuals have come farther in their process of appropriation. When students have come farther in their appropriation of mathematical tools such as dot product and geometric series, original conceptually subordinate difficulties such as recalling exponents and mathematical calculations related to specific procedures are no longer experienced as difficulties. At this point the students are so used to operating with the mathematical tools that their meaning in a specific situation is evident. The tool is transparent to the user—its meaning and its use are obvious. The resistance that emerged in the gap between the tool and its applications has faded. Furthermore, it is the overcoming of this resistance that is crucial in appropriating mathematical tools. However, in general not all aspects of a tool are appropriated. The learner may know how to use the tool in practices where she has encountered it, but generally there are applications that the learner may not have encountered as yet. Hence new and different manifestations of resistance may occur later in the appropriation process. The model of Säljö (2005, p. 230, my translation) shown in Figure 15.2 summarizes the sociocultural perspective on learning as a process of appropriation.

Research literature relating to the appropriation process (e.g., Moschkovich, 2004; Radford, 2002, 2003; Rogoff, 1990; Säljö, 2005) identifies five crucial aspects of the process. To become a cultural knower of a tool an individual has to:

1. participate in a joint activity
2. establish shared foci of attention with collaborators. Students need to develop a working consensus of the significant issues to which they need to pay attention in mathematical tasks.
3. develop shared meanings for concepts and terminology (i.e., meanings in accordance with the mathematics community built up through participating in mutual decision-making processes)
4. be actively involved in a process of transforming. Through this process the individual appropriates utterances and actions made by fellow students in their collaborative problem solving and utilizes these in ongoing activities.
5. attend to the relationship between sense and meaning of mathematical tools. The relations between the students' joint activity in the small group and the pre-existing established mathematical knowledge in the classroom have to be identified.

As mentioned in the introduction, the two main research issues are formulated so as to inform these

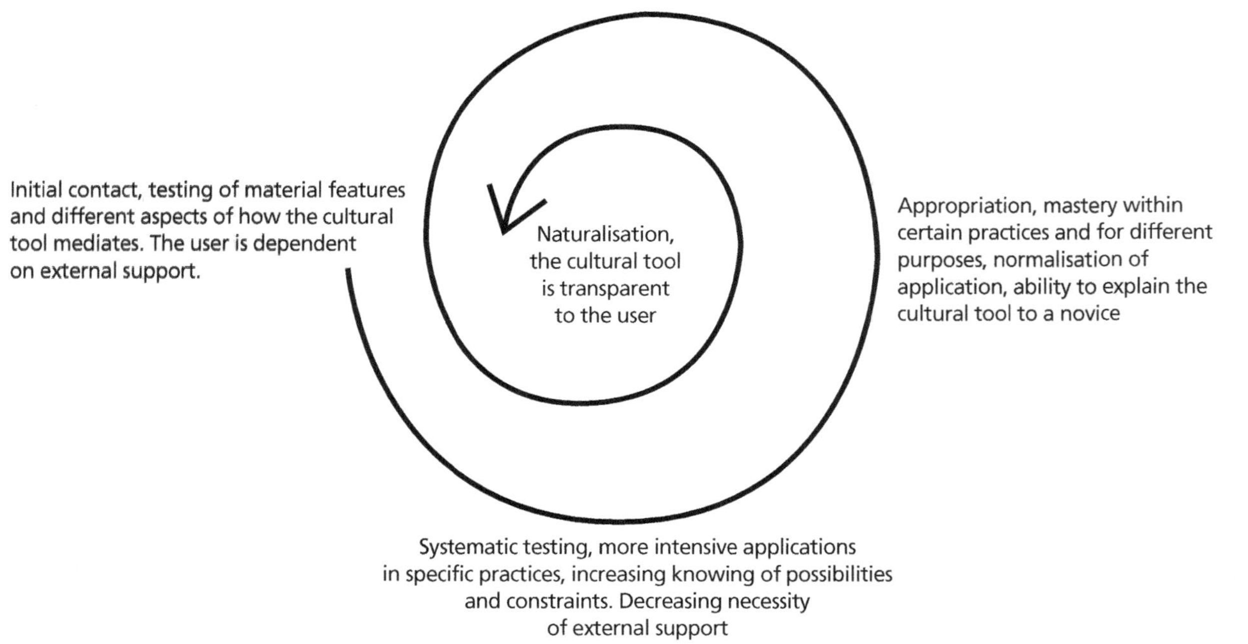

Figure 15.2 Model of the process of appropriation.

five aspects with respect to how to become a cultural knower of mathematical tools. By analyzing the students' collaborative problem-solving efforts, I aim to elucidate their process of appropriating the mathematical tools of dot product and geometric series. It is my intention to contribute to the collective knowing of how this process evolves and how it can be analyzed. To summarize, I have in this section tried to elaborate a sociocultural perspective on learning and development and, in particular, those aspects of the theoretical position relevant for the conducted study. The terms *appropriation* and *tool* are thus central. The present study is about analyzing how students make mathematical tools such as the dot product and geometric series their own, in the context of problem solving in small-group settings. Hence, the features of the appropriation process have been theoretically discussed. In the following, I will turn to the empirical foundation for my research, the methods used to collect empirical material, how I analyzed the data to address the research questions, and the background and context of the study. However, before this I will briefly outline the mathematical content upon which students and teacher focus within the study.

The Mathematical Substance

In this study I explore the subtleties of how upper secondary students appropriate the concepts of dot product and geometric series. The focused mathematical concepts were chosen for different reasons. The focus on dot product emerged from the data. The students' efforts to appropriate this tool appeared to be interesting subsequent to the data collection. Moreover, this gave guidance for focusing on geometric series prior to the second data collection. An additional reason for focusing on these topics was an assumption that it might be favorable to study the students' processes of appropriating mathematical concepts that they had not previously met in their school experience. Furthermore, it appeared that students' efforts to appropriate these topics have not been particularly focused upon in the mathematics education research. The research literature regarding issues with respect to appropriating these concepts is not extensive (cf. Moschkovich, 2004).

In $\mathbf{R}^3$, the concept of dot product has two definitions ((1) and (2)) and a relevant theorem (3). From a sociocultural point of view, the definitions can be regarded as two different inscriptions or semiotic means of objectification (Radford, 2002, 2003), which mediate the idea of dot product slightly differently. Both the definitions and the theorem are used interchangeably by the students according to the information given in the task text:

1. $\vec{a}\cdot\vec{b}=|\vec{a}|\cdot|\vec{b}|\cdot\cos\alpha$, where $\vec{a}$ and $\vec{b}$ are oriented line segments, $|\vec{a}|$ and $|\vec{b}|$ are the magnitude of the vectors $\vec{a}$ and $\vec{b}$ respectively, and α is the angle between the vectors $\vec{a}$ and $\vec{b}$.
2. $\begin{bmatrix} a_1 & a_2 & a_3 \end{bmatrix}\cdot\begin{bmatrix} b_1 & b_2 & b_3 \end{bmatrix}=a_1b_1+a_2b_2+a_3b_3$, where a_i and b_i are real numbers for $i=\{1, 2, 3\}$
3. $\vec{a}\perp\vec{b}\Leftrightarrow\vec{a}\cdot\vec{b}=0$

The concept of dot product is mathematically related to algebra and vectors. With regard to the learning of algebra and algebraic thinking, a rich field of research is found within mathematics education. In school the process of learning algebra often concerns manipulation of variables, simplifying expressions, and solving equations (Kieran, 1992, 1996). In researching this process, the focus has been, among other things, on the role of interaction and communication (Sfard & Kieran, 2001) and how students make sense of algebraic expressions. Regarding the latter, students have been found to view algebraic expressions as strings of symbols (an algebraic object), as representing a certain number, as a computational process, or as a function (Sfard & Linchevski, 1994). Nevertheless, I have found only the studies of Wiliam (1986) and Manes (1996) immediately relevant when it comes to research whose object of study is appropriation of the concept of dot product. According to Wiliam (1986), a geometrical approach (i.e., definition (1) above) is preferable over an analytical approach (i.e., definition (2) above) to appropriate the concept of dot product, because with respect to the latter, students often become occupied with practicing techniques. However, as in Wiliam's study, when students are introduced to the geometrical approach, he found that they were using the dot product as a tool for expressing geometrical insights, and the students rarely experienced difficulties in practicing techniques and procedures. Manes (1996) found that students were able to reason using the dot product definitions (cf. (1) and (2) above). From task-based interviews, Manes found that students were reasoning about the tasks using their knowing of the concept of dot product, but she also found students who generalized incorrectly from working with unique cases.

With respect to the concept of geometric series, the research literature is richer (Alcock & Simpson, 2004; Eade, 2003; Stergiou & Patronis, 2002). Mathematically, issues concerning geometric series are related to issues of limits and infinity, topics that have been widely studied (cf. Fischbein, Tirosh, & Melamed, 1981; Sierpinska, 1994; Tall, 1992; Tall &

Tirosh, 2001). All these studies analyze how students work with tasks involving the issue of infinity, which might be solved by using the tool of geometric series. These studies' documentation of student difficulties may have to do with the constituent concepts of limit and the process of infinity, which are "intuitively contradictory" (Fischbein et al., 1981, p. 506). Thus, the difficulties the students' experienced did not necessarily have to do with the process of appropriating the mathematical tool of geometric series per se.

The studies of Alcock and Simpson (2004) and Eade (2003) used task-based interviews to analyze university-level students' understanding of convergence of series and sequences. Alcock and Simpson (2004) found that some students were able to recall definitions and to categorize series. These students also gave relevant arguments that comprised the definitions. Other students were neither able to recall definitions nor to categorize the series. Eade (2003) reports, in accordance with the latter group of students, that the researched trainee teachers in his study showed limited knowledge of the concept of convergence. However, Eade argues that it is problematic to tell whether the trainee teachers' difficulties in appropriating the concept of convergence were due to "semantics" or "basic ideas in analysis" (2003, p. 381). For instance, the interviewees struggled with the meaning of the word "bounded." In my opinion, and in accordance with a sociocultural position, mathematics might be viewed as semantics in several aspects. To appropriate mathematical tools is very much an enterprise of making meaning, of signs, concepts, and words. Mathematics might be viewed as language of a particular kind, used to communicate and solve problems in everyday activities as well as in academic settings. It is thus inadequate to argue whether students' problems in appropriating mathematical tools relate to issues of semantics or issues of mathematics, since these aspects are seen as intertwined.

Mathematically, the concept of geometric series is defined as a sum where every term is the square root of the product of the two neighboring terms, or, alternatively, as a sum where each term is equal to the preceding term multiplied by a constant number, the ratio (here written as k). Symbolically it might be denoted as

$$a_1 + a_1k + a_1k^2 + a_1k^3 + a_1k^4 + \dots$$

where a_1, the first term of the series and k, the ratio between the terms, are real numbers.

Additionally, the students potentially had three inscriptions at their disposal, two inscriptions for the sum of a geometric series (finite (4) and infinite convergent (5)) and one inscription for the nth term of a geometric series (6):

4. $S_n = \dfrac{a_1 \cdot (k^n - 1)}{k - 1}$

   S_n is the sum of the n first terms in the geometric series.

5. $S = \dfrac{a_1}{1 - k}, k \in \langle -1, 1 \rangle$

   S is the sum of an infinite convergent geometric series.

6. $a_n = a_1 \cdot k^{n-1}$

   a_n, for all $n \geq 1$, is the nth term of a geometric series

RESEARCH METHODS

The research conducted stems from my interest in how students appropriate mathematical tools and actions through interaction in the classroom. In taking a sociocultural perspective on learning and development as a given premise, the methodical approach used seeks to document and analyze learning practices as they emerge in the classroom. I have thus analyzed naturally occurring talk-in-interaction in two small groups, fairly homogeneously composed both with respect to gender and achievement (cf. Cohen, 1997; Good & Marshall, 1997; Johnson & Johnson, 1985; Linchevski & Kutscher, 1998; Webb, 1991).

Research Paradigm and Design

The study is framed within a naturalistic research paradigm (Carr & Kemmis, 1986; Lincoln & Guba, 1985). Situating my research within this research paradigm gives guidance for the work with respect to what is important, legitimate, and reasonable, and it gives direction for the practical research (Macbeth, 2003). Ernest (1995) asserts the basic constituents of a research paradigm to be ontology, epistemology, and methodology. In accordance with a naturalistic research paradigm, I view ontological, epistemological, and methodological issues as follows: Ontologically, realities of the world are constructed and hence inseparable from persons and objects. Students are actors in a social world, and they make sense of and interpret the activities in which they participate and act accordingly. The epistemological stance I take is to view the individual as to appropriate knowing in thinking and actions, which subsequently is inextricably tied to the cultural tools utilized in communicative settings. This is the position taken, revealing how people achieve their knowing and how they know in general. Methodologically, my research is drawing on "purposive sampling focusing on qualitative investigations of humans' interaction in a natural setting" (Carlsen, 2008, p. 54).

The research was organized through the use of a case study design (Bassey, 1999; Stake, 2000) within a qualitative research strategy (Bryman, 2001, Silverman, 1993). In designing my research as a case study, detailed and intensive analyses of the two small groups' collaborative problem solving have been made. My study is hence an empirical and contextual inquiry in a real-life setting.

Data Collection

I took an ethnographic approach to the data collection (Eisenhart, 1988; Tedlock, 2000); that is, I visited a mathematics classroom and studied two small groups. The gathering of empirical material was done through audio recordings (cf. Carlsen, 2002, 2005), video recordings, field notes, and a research diary of the students' collaborative problem solving in separate rooms with chalkboards (cf. Carlsen, 2008). I took an overt but passive researcher role in my collection of naturally occurring talk in the small groups. Hence, I did not explicitly intervene in the students' mathematical problem solving. Moreover, it was my intention that the students were totally aware of my status as a researcher.

More than other research methods, according to Tedlock (2000), the ethnographic approach makes possible an uncovering of cultural aspects of the group, such as ways of communicating and interacting in mathematical small-group problem solving. Thus, I chose an ethnographic approach in order to scrutinize the students' appropriation processes of mathematical tools. According to Bryman (2001), by taking a case study design the ethnographic approach is also adequate, usual, and suitable in seeking to grasp and portray the students' collaborative learning process. Furthermore, following Chambers (2000), the ethnographic approach is characterized by "long-term and relative intimate acquaintance" (p. 862) with the persons studied. Through this approach it is thus possible to gain contextual information that may increase the depth of the researcher's insight and knowledge as regards these persons.

Data Analysis

I used a dialogical approach to analyze the data. The dialogical approach is an operationalized analytical tool drawing on dialogism and dialogical principles such as *sequentiality*, *joint construction*, and *act-activity interdependence* (Linell, 1998). Dialogism is an epistemological framework that is particularly interested in how people discuss, communicate, and negotiate in collective settings (Bakhtin, 1981, 1986; Linell, 2006; Markovà, 1990a, 1990b). According to Cestari (1997), the dialogical approach "allows one to analyse the co-construction of formal knowledge among participants in a defined situation" (p. 41). Thus, dialogism is claimed to be compatible with a sociocultural perspective on learning and development (cf. Wertsch, 1990). Following this analytical framework, to analyze student dialogues one has to use the analytical principle of double dialogicality (Linell, 2006). The term dialogicality denotes that utterances, both spoken and written, have to be analyzed with respect to their interrelationships with prior and subsequent utterances. As regards double dialogicality, one as a researcher has to consider the relationship between the dialogues' embeddedness in a global institutionalized setting *and* their embeddedness in a local small-group context.

In the analysis three dialogical principles have been considered. The principle of *sequentiality* means that every contribution to a dialogue is interpreted and gets its meaning from its positioning in a sequence, i.e. from what has already been said and what is said in the consecutive utterances. The principle of *joint construction* means that dialogues are seen as jointly made. The dialogues are made up, experienced, and possessed collaboratively by the participants. The principle of *act-activity interdependence* is equivalent to the notion of double dialogicality (see above). Drawing on both theoretical and methodological aspects, the following unit of analysis was formulated for the study: *individuals-using-tools-in-cultural-practice.*

Context and Participants

The data corpus originated from two data collection periods, autumn 2001 and the academic year 2004/2005, at the same urban upper secondary school in the south of Norway. In both cases the research participants were taking an optional mathematics course[8] intended, in part, to prepare for mathematics studies at university level, and it was the most theoretical and demanding mathematics course at upper secondary school. The students were 17–18 years of age and in their final year of upper secondary school. In both data collection periods I collaborated with the same mathematics teacher, and the mathematics teaching was organized similarly. In particular, the teaching included whole-class exposition and small-group sessions. In these sessions, in small rooms with chalkboards, the students were expected to engage in collaborative problem solving. During the group sessions, the mathematics teacher was available for support if needed by the groups.

In both periods of data collection I followed the mathematics class for several weeks. In the first period of data collection I observed one group consisting of six students—five boys and one girl. All six students were recognized as high-attainers in the prior (optional) course (achieving highest or next to highest grades[9]). In the second period of data collection, two groups were followed. However, in the analysis only one of them has been particularly in focus in my study, for both pragmatic and theoretical reasons. One group was chosen to delimit my study, and I chose the group because the students in this group externalized their thinking more extensively, and arguments and ideas came to the fore more often than in the other group. My dialogical approach to analysis of learning activities in small-group settings is evidently dependent on externalizations of thinking. This group consisted of six students—three boys (two so-called high-achievers and one so-called average-achiever) and three girls (two high-achievers and one average-achiever). It was totally the mathematics teacher's enterprise to compose the small groups, in both periods of data collection.

DISCUSSION OF THE RESEARCH RESULTS

The analyses of the students' dialogues show that they gradually improved the accuracy in their use of concept-related terminology regarding the process of appropriating the mathematical tool of dot product. At early stages in their process of appropriating the tool, the students used inaccurate mathematical formulations such as: "Eh, yes, how do you multiply two such parameters?", "If that is parallel to that, then they are," and "AB multiplied by AD" (Carlsen, 2005). Particularly, the students said "multiply" and "multiplied with" instead of "dot" and "dotted with," which were the teacher's phrases; "parameters" instead of "vectors"; and "parallel" instead of "perpendicular." However, relevant terminology, such as "dotted with" and "perpendicular," was used successfully at later stages. Nevertheless, these terminological discrepancies did not seem to cause substantial difficulties with respect to the students' collaborative reasoning and problem solving. I argue this finding to illustrate one aspect of the students' process of appropriating the concept of dot product. The students are in a process of making the terminology their own. The analyses also revealed evidence that one of the students made significant progress in making the tool his own. He argued in a concrete case by utilising both an inscription as well as a theorem in his reasoning (cf. Excerpt A below, Carlsen, 2005, p. 18). By his argument, he explains the tool of dot product to the other group members:

Excerpt A

103	**Jon:**	((Jon writes on the black board AB · AD = 0)) If that's equal to zero then they are perpendicular, that is they are 90 degrees. =
104	**Tor:**	Yes, that is actually right
105	**Jon:**	= You're following right? Well then you just write(.) the coordinates to *that*, =
106	**Tor:**	Yes A
107	**Jon:**	= AB, the coordinates to AB, =
108	**Tor:**	Yes yes, just yes
109	**Jon:**	= and then multiply them and then add them the three coordinates, =
110	**Tim:**	Okay
111	**Jon:**	= and then you get like that is 28, I got. And then AB multiplied by AD actually is 28 and then they aren't 90 degrees
112	**Tor:**	Totally correct

The issue of resistance manifests itself in appropriating tools, and it becomes evident in a variety of ways, such as terminological confusion and inaccurate use of relevant terminology (as exemplified above), within details in the solution process such as directions of vectors and estimation of parameters, through students' difficulties in calculations and in mathematical symbolism. The students in this study first met the concept of dot product from what Wiliam (1986) calls a geometrical approach. The analyses resonate with his findings in that the students were able to reason by utilizing the concept's related definition ((1) above) as a tool to express geometrical insights. However, the students also experienced difficulties in coordinating, on one hand, their knowing of definitions and the related theorem and, on the other hand, the utilization of these in situ (cf. Manes, 1996). Nevertheless, the students were able to collaborate and to achieve shared meanings regardless of inaccuracy (Webb, 1991).

With respect to the concept of geometric series, the students were found to utilize the tool and its theoretical components as problem-solving tools in various situations. Furthermore, the analyses show that the students are able to explain the mathematical components of the tool to their fellow students. Hence, following Figure 15.2 above (cf. Säljö, 2005), the students are argued to have come far in their process of appropriating the tool of geometric series. Further evidence of this was found in that the students objectified their knowing when making their thinking explicit (cf. Radford, 2002, 2003). By calling for help, by externalizing their thinking, and by asking questions, learning opportunities were created.

In their process of appropriating the mathematical tools, the students in my study made use of semiotic means of objectification (Radford, 2002, 2003). Throughout their involvement in joint activity, the students used mathematical terminology such as parameter letters a_1 and k and a written class note[10] to objectify their knowing with respect to the concept of geometric series. Moreover, linguistic devices such as "the nth term" (related to geometric series), "the usual formula," "the good old formula," and "the other formula" (all three expressions related to the concept of dot product) were also used to objectify their knowing and to establish shared meanings. The reason why the use of mathematical terminology and linguistic devices are found to be significant when it comes to the students' appropriation process is that both terminology and devices were used without any conceptual elaborations. The students seemed to be establishing shared meanings for these constituents of the concepts since the students were neither concrete with regard to how the formula is expressed (cf. the definitions (1) and (2) above) nor explicit about which formula they were thinking of.

The analysis also exposed the significant role of inscriptions in the students' processes of appropriating the mathematical tool of geometric series. These inscriptions were also among the semiotic means of objectification used by the students (cf. Radford, 2002, 2003). However, the inscriptions were found to be a particular kind of semiotic means. Thus they have become a particular focus in a separate study (Carlsen, 2009). The students were externalizing their thinking and making it explicit through utilizing inscriptions, and these served as concretizations, clarifications, and as anchor points in their reasoning. Moreover, I argue that the inscriptions played significant roles in the students' appropriation process, although some of the inscriptions in fact were misleading and mathematizing the problems incorrectly. The nature of geometric series was clarified among the students through the utilizations of inscription such as drawings and graphs. The problem-solving process, with arguing, interthinking, and cumulative talk (cf. Mercer, 2000), and the utilization of inscriptions were intertwined. The inscriptions thus served as intermediaries (cf. Latour, 1987) in the process of mathematizing the tasks given in the textbook. The inscriptions were used to transform the original problem, which might be seen as diffuse and ambiguous, into something more distinct and concrete. The students' collaborative problem solving in small groups as well as the co-thinking of the kind seen in the study would have been impossible without the presence of such devices. The various contributions to interthinking and cumulative talk among the students are argued to be conducive to their process of appropriating the mathematical tool of geometric series.

Through the analyses of dialogues with respect to geometric series, I documented how the students took various roles in their collaborative problem-solving process. Particularly, two roles were identified as "questioner" and "explainer." A girl, Eli, took the role of questioner, and she eagerly asked questions throughout the discussions. By doing that she made the other students elaborate, explain, and share their knowing. A boy, Jan, took the role of explainer. When his fellow students asked questions, he took responsibility and sought to establish shared meanings (cf. Moschkovich, 2004; Rogoff, 1990). Additionally, by taking such roles these students actively contributed to perseverance and thoroughness in the group's problem-solving process. Related to this, objections might be raised in that the students' role-taking was a result of them being high-achievers in mathematics—that is, that high-attainers may be more eager to ask questions and explain for each other than other categories of students. That is, however, a contradiction to the findings of Webb (1991). She summarizes the results of a range of studies, and she found that homogeneously composed small groups were not preferable because they took a lot for granted. Consequently, the amount of explanation was thus low. I argue that this role-taking, exposed in my study, is more the effect of the working environment within the group. Student personality and ability to create a supportive learning environment are significant factors when it comes to small group interaction (cf. Webb, 1991).

As mentioned above, from a sociocultural position, mastery is an important aspect of learning (cf. Wertsch, 1998). The analysis reveals that on several occasions the students show mastery in their solving of mathematical problems. For instance, with respect to evaluating the sum, S, and the general term, a_n, of a geometric series (cf. the inscriptions (5) and (6) above), the students showed mastery of various components of the tool.[11] As I have argued elsewhere (Carlsen, 2007), mastery or knowing how is shown by one student when solving the task: Calculate the sum $1 + 3 + 9 + 27 + \ldots + 729$. This student knew how to use inscription (6) above to evaluate the number of terms through the use of a self-made inscription. It is also worth mentioning that the ellipsis represented an obstacle in the task and became a source of resistance in the students' efforts to estimate parameters and to calculate the sum.

Both with respect to the concept of dot product and to the concept of geometric series, the analyses show the significance of the temporal dimension in the process of appropriating mathematical tools. The issue of time is relevant because extensive work is needed to experience and to become familiar with the various contexts in which the mathematical tools might be utilized. The appropriation process never ends (cf. the process modeled as a spiral in Figure 15.2), and thus students obviously need time in experiencing this process. On a general level, these mathematical concepts are abstract tools that are not immediately applicable in students' life out of school. To appropriate and experience the use of these scientific concepts (cf. Vygotsky, 1986) is thus a time-consuming enterprise that could be made via problem-solving activities in school (cf. Schroeder & Lester, 1989). Moreover, the appropriation of mathematical tools is a cumulative process since the concepts are frequently hierarchically interrelated. To compare and discriminate between them with respect to content and structure takes time.

With respect to the first and the second specific research question formulated for the study—(1) What are the characteristics of students' appropriation of the cultural tool of dot product? and (2) What are the characteristics of students' appropriation of the cultural tool of geometric series?—these results give important answers. The students' individual processes of appropriating the tools of dot product and geometric series are characterized by resistance and the use of semiotic means of objectification. The resistance manifested itself as inaccurate and confusing use of concept-related terms as well as obstacles with respect to estimation of parameters and directions of vectors. Moreover, the students were also found to be bogged down with mathematical symbolism and difficulties in calculations. Among the semiotic means the students used in their collaborative problem solving were parameter letters, a class note, and linguistic devices. All these means served to establish shared foci of attention and shared meanings for concepts were achieved. With regard to the third research question—What are the features of students' use of inscriptions in the appropriation process?—a substantial finding in my study is also that the students were using inscriptions in their problem-solving efforts and interaction. Through these inscriptions, thinking was made explicit, explanations were given, and anchors for their reasoning were externalized.

Although Cobb (2007) argues that the educational and instructional values of sociocultural studies are low, I still claim my study offers interesting implications for mathematics teaching and instruction in the classroom. The study shows the complexities that lie behind what may seem trivial at first sight. Learning mathematics is a time-consuming enterprise, and this relates to the delicate dilemma of pace and progress versus thoroughness and distinctness in mathematics teaching. Mathematics teaching has to arrange for the students to develop eagerness, thoroughness, and perseverance in their engaging with mathematical problem solving to achieve the desirable outcomes. Furthermore, establishing coherence between the various mathematical tools is not straightforward.

Another issue that is raised in the analyses is the relation between knowing the mathematical definition of a tool and knowing how to use the tool in situ. Mathematics teaching has to emphasize that it is one thing to appropriate a tool to know the formulae and definitions involved and quite another thing to know how to use them in problem-solving activities. Related to this is also the issue of student training in mathematics teaching as well as the issue of resistance in the learning process. Students need different amounts of training to make mathematical tools their own—that is, they overcome the resistance involved in the appropriation process at different paces. Training is hence absolutely necessary for students to achieve shared meanings for concepts. Training here means that students need to experience how to apply mathematical tools in various problem-solving situations, practicing their knowing in various contexts.

The role of mathematical tasks is also significant in student reasoning. What might be called the learning potential of a task is not pre-given. Significant efforts ought to be put into the mathematics teacher's enterprise of choosing and formulating mathematics tasks for the students to work with. The students should experience how to mathematize, and thus "close the gap," between the problem situation and the mathematical situation. The tasks should carry potential for the students to actively participate and contribute with ideas and arguments that are shown to be substantial parts of the appropriation process. However, I realize that making or finding such mathematical tasks is certainly not an easy matter. This calls for collaboration among the mathematics teachers in a school, and also collaboration between mathematics teachers, mathematicians, and didacticians.[12]

More research is needed to further investigate the subtleties of the appropriation process. To possibly complement and inform the analyses made here, more studies would be necessary. In my study I have focused particularly on upper secondary students and the concept of dot product and geometric series. Analyses of mastery and appropriation of other mathematical tools and other categories of students may add to the picture of students' processes of appropriation drawn here. Additionally, my study suggests further studies of how people are dependent on and use semiotic means of objectification in interaction and problem solving.

To summarize, the process of appropriating mathematical tools such as dot product and geometric series I argue to be closely related to the use of external means such as mathematical language, arguments, and inscriptions (Lehrer et al., 2000; Lehrer et al., 2001; Radford, 2002, 2003). This argument "is an antidote against a rationalistic view upon learning as an internal cognitive endeavour" (Carlsen, 2008, p. 87). By focusing on student appropriation of mathematical tools, I have attempted to offer empirical and theoretical perspectives to explain how the subtleties of this process might be characterized, as this was a deliberate aim of my study. Moreover, I will claim that the results of the research add to the pool of knowledge in mathematics education with respect to how upper secondary students, in collaborative problem-solving settings, appropriate the mathematical concepts of dot product and geometric series.

Nevertheless, I am aware that the outcomes of my study might arise from the researched students being "special cases." Objections might be raised with respect to them being high-achievers, and therefore they are able to externalize their thinking, overcome the resistance, and achieve shared meanings. This is a relevant critique of the study. However, I am convinced, in accordance with both Bassey (1999) and Niss (2004), that since it happened in this case it might as well happen in other cases.

ACKNOWLEDGEMENTS

I want to give thanks to the two reviewers of this chapter, for extremely helpful comments and suggestions. It is my opinion that the content of this research report has improved significantly due to their contributions.

NOTES

1. Address: Service box 422, 4604 Kristiansand, Norway. E-mail: Martin.Carlsen@uia.no
2. This study was originally made for a PhD degree in mathematics education.
3. Throughout this chapter the term *dot product* is used, while acknowledging the equivalent term *scalar product*. In Norwegian both terms are used.
4. Vygotsky (1978) used the concept of *internalization* when describing the individuals' learning process. There is, however, a terminological problem due to its connotations. Metaphorically, internalization seems to differentiate between the internal (i.e., thinking) and the external (i.e., communication between people). Socioculturalists (cf. Rogoff, 1990, 1995; Säljö, 2001, 2005; Wertsch, 1993, 1998) thus find the concept to be unfortunate. In this study I thus choose to use mastery and appropriation about the individual learning process.
5. This term will be discussed later, in the section called "The Role of Tools."
6. Teaching and learning about problem solving is a phrase that puts emphasis on the teaching of strategies, heuristics, and sequences of operations.
7. The term *collaborative learning* is used in contrast to the term *cooperative learning*, which Springer et al. (1999) describe as a structured small-group learning approach. Cooperative learning is characterized, among other ways, by assigning individual tasks and roles to the group members and offering rewards for achieving goals.
8. The course was referred to as "3MX" (following the introduction of a new curriculum in 2006 the present equivalent course is called R2). The course 3MX followed on from 2MX (at present called R1), which was the first optional mathematics course.
9. In upper secondary school in Norway, the grades in the assessment scale run from six (highest) to zero. Students that were categorized as high-attainers had the grades 6 or 5 in the mathematics course 2MX.
10. The written class note was made by the mathematics teacher, and it was basically a summarizing document of a lesson given prior to a group session.
11. The identified mathematical components of the tool geometric series were: (1) estimating of parameters, (2) establishing of the general term, (3) composing of the sum, and (4) deciding on convergence.
12. The notion didactician is referring to a person who conducts research within mathematics education.

REFERENCES

Alcock, L., & Simpson, A. (2004). Convergence of sequences and series: Interactions between visual reasoning and the learner's beliefs about their own role. *Educational Studies in Mathematics, 57*, 1–32.

Bakhtin, M. M. (1981). *The dialogic imagination: Four essays*. Austin, TX: University of Texas Press.

Bakhtin, M. M. (1986). *Speech genres and other late essays*. Austin, TX: University of Texas Press.

Bassey, M. (1999). *Case study research in educational settings*. Buckingham, UK: Open University Press.

Bjuland, R. (2002). *Problem solving in geometry. Reasoning processes of student teachers working in small groups. A dialogical approach*. Published doctoral dissertation. Bergen, Norway: University of Bergen.

Bryman, A. (2001). *Social research methods.* Oxford, UK: Oxford University Press.

Carlsen, M. (2002). *Smågruppeundervisning i matematikk. Ei dialogisk tilnærming til samarbeid i ei gruppe av faglig sterke elever i videregående skole.* [Small group teaching in mathematics. A dialogical approach to collaboration in a group of high-achieving upper secondary students in school]. Unpublished master's thesis, Agder University College, Kristiansand, Norway.

Carlsen, M. (2005). Conceptual understanding of the dot product. *Nordic Studies of Mathematics Education, 10*(3–4), 3–28.

Carlsen, M. (2007). Geometric series and mathematical mastery and appropriation. In C. Bergsten, B. Grevholm, H. S. Måsøval, & F. Rønning (Eds.), *Relating practice and research in mathematics education. Proceedings of Norma 05, Fourth Nordic Conference on Mathematics Education* (pp. 203–215). Trondheim, Norway: Tapir Academic Press.

Carlsen, M. (2008). *Appropriating mathematical tools through problem solving in collaborative small-group settings.* Published doctoral dissertation. Kristiansand, Norway: University of Agder.

Carlsen, M. (2009). Reasoning with paper and pencil: The role of inscriptions in student learning of geometric series. *Mathematics Education Research Journal, 21,* 54–84.

Carr, W., & Kemmis, S. (1986). *Becoming critical. Education, knowledge and action research.* London: Routledge Falmer.

Cestari, M. L. (1997). *Communication in mathematics classrooms. A dialogical approach.* Unpublished doctoral dissertation, University of Oslo.

Chambers, E. (2000). Applied ethnography. In N. K. Denzin & Y. Lincoln (Eds.), *Handbook of qualitative research (2nd ed.)* (pp. 851–869). Thousand Oaks, CA: Sage.

Cobb, P. (2002). Reasoning with tools and inscriptions. *The Journal of the Learning Sciences, 11,* 187–215.

Cobb, P. (2007). Putting philosophy to work. Coping with multiple research perspectives. In F. K. Lester (Ed.), *Second handbook of research on mathematics teaching and learning* (pp. 3–37). Charlotte, NC: Information Age Publishing.

Cohen, E. G. (1997). Restructuring the classroom: Conditions for productive small groups. In E. Dubinsky, D. Mathews, & B. E. Reynolds (Eds.), *Readings in cooperative learning for undergraduate mathematics* (pp. 135–156). Washington, DC: The Mathematical Association of America.

Cole, M. (1996). *Cultural psychology: A once and future discipline.* Cambridge, MA: The Belknap Press.

Duval, R. (2006). A cognitive analysis of problems of comprehension in a learning of mathematics. *Educational Studies in Mathematics, 61,* 103–131.

Dysthe, O. (2001). Sosiokulturelle teoriperspektiv på kunnskap og læring. [Sociocultural perspectives on knowing and learning]. In O. Dysthe (Ed.), *Dialog, samspel og læring* [Dialogue, interaction and learning] (pp. 33–72). Oslo: Abstrakt.

Eade, F. (2003). Secondary trainee teachers' understanding of convergence and continuity. *International Journal of Mathematical Education in Science and Technology, 34,* 371–384.

Eisenhart, M. A. (1988). The ethnographic research tradition and mathematics education research. *Journal for Research in Mathematics Education, 19,* 99–114.

Ernest, P. (1995). The one and the many. In L. P. Steffe & J. Gale (Eds.), *Constructivism in education* (pp. 459–486). Hillsdale, NJ: Lawrence Erlbaum.

Fischbein, E., Tirosh, D., & Melamed, U. (1981). Is it possible to measure the intuitive acceptance of a mathematical statement? *Educational Studies in Mathematics, 12,* 491–512.

Good, T. L., & Marshall, S. (1997). Do students learn more in heterogeneous or homogeneous groups? In E. Dubinsky, D. Mathews, & B. E. Reynolds (Eds.), *Readings in cooperative learning for undergraduate mathematics* (pp. 93v106). Washington, DC: The Mathematical Association of America.

Hiebert, J. (1992). Reflection and communication: Cognitive considerations in school mathematics reform. *International Journal of Educational Research, 17,* 439–456.

Johnson, D. W., & Johnson, R. T. (1985). The internal dynamics of cooperative learning groups. In R. Slavin, S. Sharan, S. Kagan, R. Hertz-Lazarowitz, C. Webb, & R. Schmuck (Eds.), *Learning to cooperate, cooperating to learn* (pp. 103–124). New York: Plenum Press.

Kieran, C. (1992). The learning and teaching of school algebra. In D. A. Grouws (Ed.), *Handbook of research on mathematics teaching and learning* (pp. 390–419). New York: Macmillan.

Kieran, C. (1996). The changing face of school algebra. In C. Alsina, J. M. Alvarez, B. Hodgson, C. Laborde, & A. Pérez (Eds.), *8th international congress on mathematical education. Selected lectures* (pp. 271–290). Sevilla, Spain: S.A.E.M. 'THALES'.

Latour, B. (1987). *Science in action: How to follow scientists and engineers through society.* Cambridge, MA: Harvard University Press.

Lehrer, R., Schauble, L., Carpenter, S., & Penner, D. (2000). The inter-related development of inscriptions and conceptual understanding. In P. Cobb, E. Yackel, & K. McClain (Eds.), *Symbolizing and communicating in mathematics classrooms. Perspectives on discourse, tools, and instructional design* (pp. 325–360). Mahwah, NJ: Lawrence Erlbaum.

Lehrer, R., Schauble, L., & Petrosino, A. J. (2001). Reconsidering the role of experiment in science education. In K. Crowley, C. D. Schunn, & T. Okada (Eds.), *Designing for science: Implications from everyday, classroom, and professional settings* (pp. 251–278). Mahwah, NJ: Lawrence Erlbaum.

Lerman, S. (2000). Some problems of socio-cultural research in mathematics teaching and learning. *Nordic Studies in Mathematics Education, 8,* 55–71.

Lesh, R., & Zawojewski, J. (2007). Problem solving and modelling. In F. K. Lester (Ed.), *Second handbook of research on mathematics teaching and learning (2nd ed.)* (pp. 763–804). Charlotte, NC: Information Age.

Lester, F. K. (1994). Musings about mathematical problem-solving research: 1970–1994. *Journal for Research in Mathematics Education, 25,* 660–675.

Linchevski, L., & Kutscher, B. (1998). Tell me with whom you're learning, and I'll tell you how much you've learned: Mixed-ability versus same-ability grouping in mathematics. *Journal for Research in Mathematics Education, 29,* 533–554.

Lincoln, Y., & Guba, E. (1985). *Naturalistic inquiry.* Beverly Hills, CA: Sage.

Linell, P. (1998). *Approaching dialogue. Talk, interaction and contexts in dialogical perspectives.* Amsterdam: John Benjamins.

Linell, P. (2006). *Essentials of dialogism. Aspects and elements of a dialogical approach to language, communication and cognition.* Retrieved April 14, 2006 from http://www.tema.liu.se/temak/personal/perli/

Macbeth, D. (2003). Hugh Mehan's Learning Lessons reconsidered: On the differences between the naturalistic and critical analysis of classroom discourse. *American Educational Research Journal, 40,* 239–280.

Manes, C. (1996). Student learning in linear algebra: The gateways to advance mathematical thinking project. In *Proceedings of fifth conference on the teaching of mathematics.* Retrieved January 13, 2006 from http://archives.math.utk.edu/CTM/5th.html

Markovà, I. (1990a). Introduction. In I. Markovà & K. Foppa (Eds.), *The dynamics of dialogue* (pp. 1–22). New York: Springer-Verlag.

Markovà, I. (1990b). A three-step process as a unit of analysis in dialogue. In I. Markovà & K. Foppa (Eds.), *The dynamics of dialogue* (pp. 129–146). New York: Springer-Verlag.

Mercer, N. (2000). *Words & minds. How we use language to think together.* London: Routledge.

Moschkovich, J. N. (2004). Appropriating mathematical practices: A case study of learning to use and explore functions through inter-action with a tutor. *Educational Studies in Mathematics, 55,* 49–80.

Niss, M. (2004). Key issues and trends in research on mathematical education. In H. Fujita, Y. Hashimoto, B. R. Hodgson, P. Y. Lee, S. Lerman, & T. Sawada (Eds.), *Proceedings of the ninth international congress on mathematical education* (pp. 37–57). Norwell, MA: Kluwer.

Polya, G. (1957). *How to solve it.* Princeton, NJ: Princeton University Press. (Original work published 1945)

Radford, L. (2002). The seen, the spoken and the written: A semiotic approach to the problem of objectification of mathematical knowledge. *For the Learning of Mathematics, 22,* 14–23.

Radford, L. (2003). Gestures, speech, and the sprouting of signs: A semiotic-cultural approach to students' types of generalization. *Mathematical Thinking and Learning, 5,* 37–70.

Resnick, L. B., Pontecorvo, C., & Säljö, R. (1997). Discourse, tools, and reasoning. Essays on situated cognition. In L. B. Resnick, R. Säljö, C. Pontecorvo, & B. Burge (Eds.), *Discourse, tools, and reasoning. Essays on situated cognition* (pp. 1–20). Berlin: Springer-Verlag.

Rogoff, B. (1990). *Apprenticeship in thinking. Cognitive development in social context.* New York: Oxford University Press.

Rogoff, B. (1995). Observing sociocultural activity on three planes: participatory appropriation, guided participation, and apprenticeship. In J. V. Wertsch, P. del Río, & A. Alvarez (Eds.), *Sociocultural studies of mind* (pp. 139–164). Cambridge, MA: Cambridge University Press.

Säljö, R. (2001). *Læring i praksis. Et sosiokulturelt perspektiv* [Learning in practice. A sociocultural perspective], Oslo, Norway: Cappelen.

Säljö, R. (2005). *Lärande & kulturella redskap. Om lärprocesser och det kollektiva minnet* [Learning & cultural tools: On processes of learning and the collective memory], Stockholm: Norstedts Akademiska Förlag.

Schoenfeld, A. H. (1992). Learning to think mathematically: Problem solving, metacognition, and sense making in mathematics. In D. A. Grouws (Ed.), *Handbook of research on mathematics teaching and learning* (pp. 334–370). New York: Macmillan.

Schoenfeld, A. H. (1993). Teaching mathematical thinking and problem solving. *Rapport fra en konferanse om matematikkdidaktikk og kvinner i matematiske fag* [Proceedings of the conference Mathematics education and women in mathematical sciences] (pp. 67–89). Oslo, Norway: Norges forskningsråd.

Schroeder, T. & Lester, F. (1989). Developing understanding in mathematics via problem

Solving. In: P. Trafton and A. Shulte, (Eds.), *New directions for elementary school mathematics (1989 Yearbook)* (pp. 31–42). Reston, VA. NCTM.

Sfard, A. (2002). The interplay of intimations and implementations: Generating new discourse with new symbolic tools. *The Journal of the Learning Sciences, 11,* 319–357.

Sfard, A., & Kieran, C. (2001). Cognition as communication: Rethinking learning-by-talking through multi-faceted analysis of students' mathematical interactions. *Mind, Culture, and Activity, 8,* 42–76.

Sfard, A., & Linchevski, L. (1994). The gains and the pitfalls of reification—The case of algebra. *Educational Studies in Mathematics, 26,* 191–228.

Sierpinska, A. (1994). *Understanding in mathematics.* London: The Falmer Press.

Silverman, D. (1993). *Interpreting qualitative data: Methods for analysing talk, text and interaction.* London: Sage.

Springer, L., Stanne, M. E., & Donovan, S. S. (1999). Effects of small-group learning on undergraduates in science, mathematics, engineering, and technology: A meta-analysis. *Review of Educational Research, 69,* 21–51.

Stake, R. E. (2000). Case studies. In N. K. Denzin & Y. Lincoln (Eds.), *Handbook of qualitative research (2nd ed.),* (pp. 435–454). Thousand Oaks, CA: Sage.

Stanic, G. M. A., & Kilpatrick, J. (1989). Historical perspectives on problem solving in the mathematics curriculum. In R. I. Charles & E. A. Silver (Eds.), *The teaching and assessing of mathematical problem solving* (pp. 1–22). Reston, VA: NCTM.

Stergiou, V., & Patronis, T. (2002). Exploring a modern path to infinitesimals: Mathematics students' understanding of the concept of rate of convergence. *International Journal of Mathematical Education in Science and Technology, 33,* 651–659.

Tall, D. (1992). The transition to advanced mathematical thinking: Functions, limits, infinity, and proof. In D. A. Grouws (Ed.), *Handbook of research on mathematics teaching and learning* (pp. 495–511). New York: Macmillan.

Tall, D., & Tirosh, D. (2001). Infinity—The never-ending struggle. *Educational Studies in Mathematics, 48,* 129–136.

Tedlock, B. (2000). Ethnography and ethnographic representation. In N. K. Denzin & Y. S. Lincoln (Eds.), *Handbook of qualitative research (2nd ed.),* (pp. 455–486). Thousand Oaks, CA: Sage.

Vygotsky, L. S. (1978). *Mind in society. The development of higher psychological processes.* Cambridge, MA: Harvard University Press.

Vygotsky, L. S. (1979). The instrumental method in psychology. In J. V. Wertsch (Ed.), *The concept of activity in Soviet psychology* (pp. 134–143). Armonk, NY: M. E. Sharpe.

Vygotsky, L. S. (1986). *Thought and language.* Cambridge, MA: The M. I. T. Press.

Webb, N. M. (1991). Task-related verbal interaction and mathematics learning in small groups. *Journal for Research in Mathematics Education, 22,* 366–389.

Wells, G. (1999). *Dialogic inquiry: Towards a sociocultural practice and theory of education.* Cambridge, MA: Cambridge University Press.

Wertsch, J. V. (1990). Dialogue and dialogism in a socio-cultural approach to mind. In I. Markovà & K. Foppa (Eds.), *The dynamics of dialogue* (pp. 62–82). New York: Springer-Verlag.

Wertsch, J. V. (1991). *Voices of the mind: A sociocultural approach to mediated action.* Cambridge, MA: Harvard University Press.

Wertsch, J. V. (1993). Commentary. *Human Development, 36,* 168–171.

Wertsch, J. V. (1998). *Mind as action.* New York: Oxford University Press.

Wertsch, J. V., & Rupert, L. J. (1993). The authority of cultural tools in a sociocultural approach to mediated agency. *Cognition and Instruction, 11,* 227–239.

Wiliam, D. (1986). A "black box" approach to scalar and vector products. *The Mathematical Gazette, 70,* 197–200.

Wyndhamn, J. (1998). Problem solving as a metaphor and a practice. Assumptions and conceptions of problem solving and their consequences for activities in the mathematical classroom. In T. Breiteig & G. Brekke (Eds.), *Theory into practice in mathematics education. Proceedings of Norma 98, the Second Nordic Conference on Mathematics Education* (pp. 45–58). Kristiansand, Norway: Agder College.

Wyndhamn, J., & Säljö, R. (1997). Word problems and mathematical reasoning—A study of children's mastery of reference and meaning in textual realities. *Learning and Instruction, 7,* 361–382.

16

PUPILS' MOTIVATION FOR LEARNING MATHEMATICS

Research in Norway

Kjersti Wæge

NORWEGIAN UNIVERSITY OF SCIENCE AND TECHNOLOGY

Abstract: This chapter presents an overview of Norwegian research on students' motivation for learning mathematics. The focus is on three contributions to the international body of research into motivation in mathematics: (1) Mellin-Olsen's articulation of rationales for learning mathematics; (2) Skaalvik and colleagues' studies on students' math self-concept; students' motivation in mathematics; and relations between achievement, self-concept and motivation in mathematics; and (3) the author's study on students' motivation in terms of needs and goals, and in terms of intrinsic and extrinsic motivation. The relation between the concept of intrinsic motivation and extrinsic motivation and rationales for learning mathematics is discussed. The three perspectives presented in this chapter provide different insights into the research field of motivation in mathematics, and complementary additions to our knowledge.

Keywords: motivation, rationales, self-concept, needs, goals

INTRODUCTION

Affective factors play an important role in mathematics learning and teaching. They exercise a determining function in the learning process, both as constituent elements of the learning process and in close interaction with (meta)cognitive factors; as Op't Eynde, De Corte and Verschaffel (2002) remark:

> Motivation and volition (i.e. the conative factors) are no longer seen as just the fuel or the engine of the learning process, but are perceived as fundamentally determining the quality of learning. In a similar way, self-confidence and positive emotions (affective factors) are no longer considered as just positive side effects of learning, but become important constituent elements of learning and problem solving. (p. 14)

The learning process encompasses constructs such as motivation and self-confidence. In this chapter I present an overview of Norwegian research on students' motivation in mathematics. I focus on three contributions to the international body of research into motivation in mathematics, namely the work of Mellin-Olsen (1987), Skaalvik and colleagues (e.g., Skaalvik & Valås, 1999; S. Skaalvik & Skaalvik, 2005), and my own work Wæge (2007a). There are a great many ways to study motivation, and the three Norwegian contributions to the research field differ from each other in the approaches they take. Therefore, the three perspectives presented in this chapter provide different insights into the research field of motivation in mathematics, and complementary additions to our knowledge. Mellin-Olsen (1987) argues that if pupils are going to learn mathematics, they must approach their activity with

The First Sourcebook on Nordic Research in Mathematics Education, pages 239–258

a rationale for learning. He identifies two possible rationales for learning mathematics in school: an S-rationale (social rationale) and an I-rationale (instrumental rationale). Coming from a different research tradition, educational psychology, Skaalvik and colleagues (e.g., Skaalvik & Valås, 1999; S. Skaalvik & Skaalvik, 2005) have conducted several quantitative research studies on pupils' motivation in mathematics Their research focuses on mathematics self-concept. In my own research (Wæge, 2007a), I explore how pupils' motivation for learning mathematics can develop when they experience an inquiry approach taken in mathematical teaching. I analyze the pupils' motivation in terms of needs and goals and focus on the basic psychological needs for competence and autonomy. I also analyze the pupils' motivation in terms of intrinsic and extrinsic motivation.

First in this chapter, I present Mellin-Olsen's articulation of rationales for learning mathematics. Second, the results of the studies made by Skaalvik and colleagues are described. Third, my own analytical framework for analyzing the students' motivation for learning mathematics and the results of the study are presented. Finally, I discuss the relation between the concepts of intrinsic and extrinsic motivation and rationales for learning mathematics.

MOTIVATION

Within psychology, one important feature of the theorization of motivation has been to distinguish between intrinsic and extrinsic motivation (Deci & Ryan, 1985; White, 1959). Another approach to motivation is to distinguish motivational orientations in educational settings: learning (or mastery or task) orientation and performance (or ego) orientation (e.g., Dweck, 1986; Linnenbrink & Pintrich, 2000; Nicholls, 1984). Other prominent social cognitive conceptualizations of motivation are self-efficacy (e.g., Bandura, 1994), expectancy–value theory (e.g., Wigfield & Eccles, 2002), and attribution theory (e.g., Graham & Weiner, 1996; Urdan & Turner, 2005).

Only a few researchers in mathematics education have distinguished between intrinsic and extrinsic motivation in mathematics. In Middleton and Spanias' (1999) review of research in the area of motivation in mathematics education, *intrinsic motivation* is defined as the student's desire to engage in learning for its own sake. According to them, *extrinsically motivated* students are doing something to obtain rewards, such as good grades or approval, or to avoid punishment. Some mathematics educators distinguish task orientation and ego orientation. According to Nicholls, Cobb, Wood, Yackel, and Patashnick (1990) *task-oriented* students have a goal of gaining insight or developing skills or accomplishing something that is personally challenging. *Ego orientation* means that a student is concerned about being superior to others. In the review of research I referred to above, Middleton and Spanias (1999) distinguish three motivational orientations: mastery orientation, ego orientation, and work avoidance. Their definitions of mastery and ego orientation are similar to those of Nicholls et al. (1990). *Work avoidance* is defined as a goal pattern in which working hard is not valued. Goodchild (2001) relates *extrinsic* and *intrinsic motivation* to ego and task orientation, and to performance and learning goals. Evans and Wedege (2004) consider people's motivation and resistance to learning mathematics as interrelated phenomena. They present and discuss a number of meanings of these two terms as used in mathematics education and adult education. Hannula (2004) developed a theoretical foundation of motivation as a structure of needs and goals. This will be elaborated upon more thoroughly when I present my own theoretical framework, which builds upon Hannula's conceptualization of motivation. Finally I will mention that some mathematics educators have discussed students' motivation under the terms of motivational beliefs (Kloosterman, 1996; Op't Eynde et al., 2002) and interest (e.g., Köller, Baumert, & Schnabel, 2001; Schiefele & Csikszentmihalyi, 1995). (For an overview, see McLeod, 1992; Middleton & Spanias, 1999). The aim of this chapter is to present an overview of Norwegian research on pupils' motivation for learning mathematics, in particular the work of Mellin-Olsen, Skaalvik and colleagues, and my own.

ACTIVITY THEORY AND RATIONALES FOR LEARNING

Mellin-Olsen (1987) argues that if pupils are going to learn mathematics, they must approach their activity with a rationale for learning. He develops his work through the synthesis of independent, nevertheless related, stands of theory—in particular, activity theory, building upon especially the work of A. N. Leont'ev (e.g., 1981), the social psychology of G. H. Mead (e.g., 1934), and the work of the social anthropologist G. Bateson (1973). My presentation of Mellin-Olsen's work in this section will consider his use of activity theory and Mead's theory of the "generalized other" (Mead, 1934).

Activity Theory

Mellin-Olsen develops activity theory, a socio-cultural theory of learning rooted in Marxist philosophy. Within Activity theory the person and society are considered a unity. "the individual acts on her society at the same time as she is being socialised to it" (Mellin-Olsen, 1987, p. 33). A person is considered a political individual within society, and that means that the theory recognises and conceptualises the person as having responsibility for his or her own life situation and for the society. Activity belongs to the individual and is a way of describing the complete life of him or her: "In the broad sense Activity is the way Man acts in his world, transforms it, and is being transformed himself in a variety of ways. Such transformation takes place in environments which are primarily social" (Mellin-Olsen, 1987, p. 38).

Mellin-Olsen uses an upper case "A" relating to a person's activity to signal distinction between "activity theory" as articulated by Soviet psychologists and his own notion of "Activity." He argues that activity theory (without the upper case "A"), does not take into consideration major social structures fundamental to an understanding of Norwegian classrooms such as oppression, resistance, and culture. Mellin-Olsen develops a theory of Activity that includes these fundamental concepts as a basis for a theory of mathematics education. Within Activity theory, the individual experiences himself through others. At this point Mellin-Olsen (1987) develops his account of rationales from Mead's (1934) notion of the *generalized other*. Mead's concept of generalized other relates to an individual's personal schema of all the social groups in the environment that have influence on her or his own life. It includes the common attitudes, expectations, and reactions as experienced by an individual and which function as a referent for her or his actions. Usually the person has a system of generalized others—for example, friends, family, neighbors, and school. According to Mellin-Olsen, the individual may be exposed to various generalized others at the same time, and the generalized others may communicate different views.

Rationales for Learning

To understand the way pupils act and respond to learning situations, Mellin-Olsen introduces the concept of *rationales for learning*: "The individual's rationale belongs to the individual. It is the way he "chooses" to act in his world under the material and social conditions under which he lives" (Mellin-Olsen, 1987, p. 156).

An individual's rationale is the way she or he "chooses" to act within the material and social conditions of her or his life. According to Mellin-Olsen's explanation, the individual's rationale is a dialectical concept. "The rationale belongs to the person, but it is a product of the individual's relations to his system of GOs [generalized others]" (Mellin-Olsen, 1987, p. 157). He claims that several rationales may be present for behavior and identifies two rationales important for learning mathematics in school. Mellin-Olsen calls one of the rationales an *instrumental rationale (I-rationale)*, indicating that it has a role as an instrument for the pupils. This rationale is related to the school's influence of the pupils' future, by the contribution of formal qualifications to further studies or professions: "In its purest form the I-rationale will tell the pupil that he has to learn, because it will pay out in terms of marks, exams, certificates and so forth (Mellin-Olsen, 1987, p. 157). For example, a pupil learns mathematics because she is planning to be a doctor and to get into the medical school she needs a good grade in mathematics, or she learns mathematics because she is planning to be a hairdresser and therefore needs to pass the mathematics exam.

Mellin-Olsen uses the concept of a *social rationale (S-rationale)* as a category of pupils' learning behavior when the pupil recognizes that the knowledge to be acquired has a value beyond exams and grades. This rationale includes everything that makes the knowledge so important and interesting for the pupils that they want to acquire it for its own sake. The S indicates that the pupil evaluates knowledge through a reference to her or his generalized others that goes beyond the I-rationale. The I- and S-rationale operate together, and the pupil's rationale for learning is considered a result of both rationales. The two rationales mutually influence each other, and the pupil's evaluation of I- and S-knowledge may change over time.

Empirical Basis

Mellin-Olsen supports his theoretical development empirically. By observing mathematics classrooms in lower secondary schools Mellin-Olsen (1984) and colleagues found that although mathematics teachers emphasized understanding of mathematics rules, the pupils were nevertheless mainly concerned about how to use these rules. These observations led to a research study on pupils' strategies for reasoning about mathematics within several Norwegian curricula (Mellin-Olsen, 1984, 1987). In one study Norwegian pupils (15 years) were required to produce explanations of 10–12 given rules. They were asked to explain why the rules were true. After they had done this, the pupils were asked to associate their mathematical explanations to a mathematical principle according to their satisfaction with it. The pupils' explanations were clearly separated into two main groups: (1) explanations that show how to use the rule in practice, and (2) explanations that focused on the mathematical structure. There was small variation in the results in relation to different curricula. Mellin-Olsen's interpretation of these results is that there is a common rationale for pupils' learning in mathematics, which is independent of the curricula within which they are working. In another study Mellin-Olsen and the English psychologist Skemp (Mellin-Olsen, 1984, 1987) carried out some research in this area within several curricula in Norway and England. They developed a test, with both a Norwegian and an English version. In this test pupils were given a set of mathematical explanations from which they had to choose. The criterion for selection was that the pupils had to choose explanations that demonstrated why a mathematical principle or rule was true. The pupils had to choose between two explanations that focused on how to use the rules, two explanations that focused on mathematical structure, and one explanation that was "pure nonsense"—that is, it was not an explanation at all. This differs from the previous study where the pupils had to produce their own explanations. There was some variation in the results in relation to different curricula, but it was so small that it could not reveal anything about qualitative differences in the pupils' learning. Mellin-Olsen and Skemp's interpretation of the results was that they point to the existence of a common rationale for learning mathematics.

Mellin-Olsen also worked with the development of pupil projects as a means to present mathematics as a useful tool. Teachers participating in an in-service course developed and tested projects in their mathematics classes. One important aim of the projects was to contribute to the pupils' S-rationale. In the classes where the projects were carried out, Mellin-Olsen and colleagues (Mellin-Olsen, 1981) observed a change of rationale for learning from the I-rationale to the S-rationale. This can be illustrated by a project called "Stability of Residence," which was about how often the pupils' in the class (8–9 years of age) had moved since they were born, what reasons they had for moving, and what effects this had on the their lives. The pupils collected some data, which they analyzed using simple statistics. Finally they had a discussion in relation to the interpretation of the data: "Why do we move?" Each pupil in the mathematics class possessed data that was her/his personal contribution to the investigation. The data related to important events in the pupils' lives. The change of rationale from I- to S-rationale was demonstrated by the pupils' comments and the discussions among the pupils in relation to the data treatment. The quality of the discussion in the class emerged from the fact that the pupils could relate their own experiences to the actual topic and from the fact that they had different opinions on this matter: "We moved because we wanted higher wages" and "We were forced to move" (Mellin-Olsen, 1981, p. 365).

Later in this chapter I will consider whether Mellin-Olsen's S-rationale and I-rationale can be directly mapped onto the concepts of intrinsic and extrinsic motivation as defined by Self Determination Theory (Ryan & Deci, 2002) which is common practice within mathematics education in Norway (e.g., Goodchild, 2001; Holden, 2003).

PUPILS' SELF-CONCEPT AND MOTIVATION IN MATHEMATICS

Coming from a different research tradition, educational psychologists E. M. Skaalvik and colleagues have conducted several studies on Norwegian pupils' motivation in mathematics using quantitative methodology. Their research has centered on mathematics self-concept. Here I present first Skaalvik and colleagues' research on relations between achievement, self-concept, and motivation in mathematics, and second, their development of an "Internal–External frame of reference model" that was first proposed by Marsh (1986). The model is central in research on the relation between pupils' achievement and self-concept in mathematics and verbal arts. Note that Skaalvik and his colleagues' definitions of the theoretical concepts in question relate to the methodology they use.

Methodology

The data in Skaalvik and colleagues' studies were analyzed using structural equation modeling (SEM) and LISREL. The data were collected by the means of questionnaires and mathematics and verbal tests. Mathematics and verbal self-concept were measured by a modified version of the Self-Description Questionnaire (SDQ). Examples of mathematics items are "I always do well in mathematics," "Mathematics is difficult for me," and "I do well on tests in mathematics" (Skaalvik & Valås, 1999). Mathematics and verbal achievement were measured by asking pupils to perform a mathematics test and a verbal test (Skaalvik & Rankin, 1995), or with both achievement tests and teacher ratings. The teacher ratings were provided by the mathematics teacher and the Norwegian teacher.

They were asked to rate a pupil's level of achievement in mathematics and in reading/writing on a six point scale (Skaalvik & Valås, 1999; Skaalvik & Valås, 2001). Intrinsic motivation, interest, investment, and general motivation for schoolwork were measured with 5-point scales designed in the same way as the Self-Description Questionnaire. Examples of items are "I like mathematics" (Interest or intrinsic motivation subscale), "I give up quickly if I get a difficult mathematics task" (investment subscale), and "I like all school subjects" (general motivation for schoolwork). The two dimensions of ego orientation and task orientation were measured by scales consisting of four items each. Examples of items are "When I am in math classes it is important for me to avoid looking stupid" (self-defeating ego orientation), "I always try to do better that other pupils in mathematics (self-enhancing ego orientation), and "In math classes it is important for me to learn something new" (task orientation).

Relations Between Achievement, Self-Concept and Motivation in Mathematics

Relations between mathematics achievement, self-concept, and different measures of motivation have been examined in several studies conducted by Skaalvik and colleagues (Skaalvik & Rankin, 1990, 1995; Skaalvik & Valås, 1999; Skaalvik & Skaalvik, 2005). They define *mathematics self-concept* as self-perceived abilities or the feeling of doing well or poorly in mathematics (Skaalvik & Skaalvik, 2004b). The Norwegian research on the relation between mathematics achievement and self-concept (Skaalvik & Skaalvik, 2005; Skaalvik & Rankin, 1990; Skaalvik & Rankin, 1995) corresponds with a large body of research in the area, showing moderate to strong relations between mathematics achievement and mathematics self-concept. Skaalvik and Valås (1999) pointed out that although the relationship between (general) academic self-concept and achievement is well established in the research literature, there is no agreement about the causal ordering of those constructs. Skaalvik and Valås (1999, 2001) examined, taking a longitudinal perspective, the relations between achievement and self-concept in mathematics. Three cohorts of Norwegian primary and lower secondary school pupils participated in the study. The pupils in the three cohorts were in 3rd, 6th, and 8th grades (9–10, 12–13, and 14–15 years of age) at the time of the first data collection. The second data collection took place one academic year later, when the pupils attended 4th, 7th, and 9th grades. According to Skaalvik and Valås, previous research supports a skill-development model of the relation (achievement affects self-concept) during the early elementary school years and reciprocal influence of achievement and self-concept during upper secondary school years. The results of Skaalvik and Valås' study were consistent with a skill-development model, but the authors emphasized that no firm conclusions should be drawn from these results. They further argued that there is need for more research in this area.

Quantitative studies have revealed strong relations between pupils' academic self-concept and a variety of motivational variables (for an overview, see Bong & Skaalvik, 2003). Skaalvik and Rankin (1995) examined the relations between achievement, self-concept, and several motivational variables: intrinsic motivation, self-perceived effort, anxiety, and general motivation for schoolwork. *Intrinsic motivation* in mathematics was defined as interest in working or liking to work with mathematics tasks in school or in future education and jobs. *General motivation for schoolwork* was defined as liking to work with school tasks in general (Skaalvik & Rankin, 1995). Norwegian pupils in 6th grade and 9th grade (12–13 and 15–16 years, respectively) participated in the study. Data were analyzed using path analysis modeling. The results showed moderate-to-strong relations between achievement and motivation. Moreover, the study showed strong relations between pupils' mathematics self-concept and intrinsic motivation, effort, and anxiety in mathematics. The study also indicated that the effect of mathematics achievement on motivation is mediated through mathematics self-concept.

In the longitudinal study of Norwegian elementary and lower secondary school pupils mentioned above, relations among mathematics achievement, self-concept, and motivation were also examined (Skaalvik & Valås, 1999). In this study, interest and investment were used as indicators for motivation. The definition of interest was similar to the definition of intrinsic motivation in the former study; that is, *interest* was defined as interest in working, or liking to work, with mathematics tasks in school. *Investment* was defined as making an effort and not giving up easily when working with mathematics tasks. The results showed a relation between motivation and achievement in the two oldest cohorts. Contrary to expectations, they found no evidence that motivation was affected by self-concept.

One may also expect pupils' goal orientation in mathematics to be influenced by their mathematics self-concept. Skaalvik and Skaalvik (2005) differentiated between three different goal orientations: task orientation and two dimensions of ego orientation—self-enhancing ego orientation and self-defeating ego orientation. *Task orientation* concerns developing skills, learning, and understanding (Nicholls, 1984). Pupils with a *self-enhancing ego orientation* are more concerned about outperforming others or to demonstrate superior ability. Pupils with a *self-defeating ego orientation* are more occupied with not looking "foolish" or avoiding being poorest (Skaalvik & Skaalvik, 1998; S. Skaalvik & Skaalvik, 2005). Skaalvik and Skaalvik (2005) explored relations between mathematics achievement, self-concept, goal orientation, intrinsic motivation, and help-seeking behavior among Norwegian adults studying for a high school diploma (the age varied from 18 to 52 years). *Intrinsic motivation* was defined as interest in working with or liking to work with mathematics. The results revealed mathematics self-concept to be systematically related to all dimensions of goal orientation. Task orientation and self-enhancing ego orientation related positively to self-concept, whereas self-defeating ego orientation related negatively to self-concept; that is, high score on one variable was associated with low score on the other variable. Mathematics self-concept was also related positively to intrinsic motivation. The researchers claim the results demonstrate that self-concept is a key variable predicting pupils' motivational orientation.

Skaalvik and Skaalvik have also examined gender differences in mathematics self-concept, performance expectations, intrinsic motivation, and goal orientation of Norwegian pupils in 6th grade (11 years), 9th grade (14 years), 11th grade (16 years), and adult students (age varied between 18 and 52) in first year of high school (Skaalvik & Skaalvik, 2004b). Males had higher self-concept, performance expectations, and intrinsic motivation than did females. The results show that gender did not predict task orientation or self-defeating ego orientation in mathematics, but males had higher self-enhancing ego orientation in mathematics than did females. According to the authors, the findings indicated that gender differences in mathematics continue to exist and that the differences are consistent with gender stereotypes (Skaalvik & Skaalvik, 2004b). In an earlier study (Skaalvik & Rankin, 1990), where *mathematics self-concept* was defined as the pupils' success expectations on defined and familiar mathematics tasks, the results showed no differences between girls and boys (12 years) in mathematics self-concept.

The I–E Model

The relation between pupils' achievement and self-concept in mathematics and verbal arts has received much attention by researchers, including Skaalvik and colleagues. In Skaalvik and colleagues' studies (e.g., Skaalvik & Rankin, 1995; Skaalvik & Skaalvik, 2002) *verbal arts* refers to reading and writing skills in Norwegian. *Verbal self-concept* is defined as self-perceived abilities or the feeling of doing well or poorly in verbal arts. The main body of research reveals high correlations between achievement in the mathematics and verbal areas (e.g., Skaalvik & Rankin, 1990). Surprisingly, data from several studies show a near-zero correlation between mathematics and verbal self-concept (Skaalvik & Rankin, 1995; Skaalvik & Valås, 2001). These unexpected results led to the development of the Internal–External Frame of Reference model (I–E model) proposed by Marsh (1986). Marsh developed the model as an attempt to theorize (or explain) the influence of both external and internal frames of references or comparisons to pupils' academic self-concept, for instance mathematics self-concept: "*External comparison* is a process by which a student compares his or her own performance with the perceived performance of another, which may be a comparison group or a comparison person" (Skaalvik & Skaalvik, 2002, p. 234).

In Norwegian primary and lower secondary schools the pupils attend the same class with the same classmates in all school subjects. Skaalvik and colleagues argue that the school class is the most important comparison group because pupils directly interact and compare subject evaluations with classmates (S. Skaalvik & Skaalvik, 2004b). They have therefore defined *external comparison* as comparison with other pupils in class (e.g., Skaalvik & Rankin, 1995). "*Internal comparison* refers to students comparing their perceived abilities or achievements in one area (e.g., mathematics) with their perceived abilities or achievements in other areas (e.g., verbal skills) independent of how these self-perceived abilities compare with those of other students" (Skaalvik & Skaalvik, 2002, p. 238).

These internal impressions constitute a second basis of self-concept. According to the I–E model, the joint effect of external and internal comparisons is assumed to balance the effect of each process. The result is a near-zero correlation between mathematics and verbal self-concept. The I–E model predicts both (1) a near-zero correlation between mathematics and verbal self-concept, and (2) negative direct effect of mathematics achievement on verbal self-concept and of verbal achievement on mathematics self-concept (Skaalvik & Skaalvik, 2002). These predictions are examined in several studies of Norwegian pupils conducted by Skaalvik and colleagues (Skaalvik & Rankin, 1990, 1995; Skaalvik & Valås, 2001). Skaalvik and Valås (2001), in a longitudinal study of 3rd , 6th and 8th grade Norwegian pupils (8–9, 11–12, and 13–14 years of age), found strong support for the I–E model in the oldest cohorts. However, the analyses provided only partial support for the youngest cohort. Longitudinal analyses proved only partial support for the model in all three cohorts. Skaalvik and Valås pointed out that the inconsistency of the results demonstrates that there is a need for a better theoretical understanding of the psychological processes underlying the I–E model.

In the article "Internal and External Frames of Reference for Academic Self-Concept," Skaalvik and Skaalvik (2002) analyze the frames of references pupils use in making self-judgments in mathematics and verbal arts. The authors claim that external and internal frames of reference are not restricted to the comparisons outlined above and that has been included in the I–E model. They argue that pupils use multiple frames of references as well as multiple sources of information. By analysing studies on pupils' self-concept, they identify four possible frames of reference related to external comparisons: (1) other pupils in school, (2) other pupils in class, (3) selected pupils in class, and (4) selected pupils outside of class. They propose that pupils use several, possible all, of the four frames of reference in forming self-judgments. Based on analyses of several studies, Skaalvik and Skaalvik also propose four types of internal comparisons related to schoolwork: (1) comparison of achievement in different school subjects at a given time, (2) comparison of achievement by the same subject across time, (3) comparison of achievement with goals and aspirations, and (4) comparison of achievement in a subject matter area with applied effort in that area. The authors argue that the impact of external and internal comparisons depend on which frames of references are most salient for the pupil, on which sources of information that are available to the pupils, and which sources the pupil uses for seeking self-relevant information (Skaalvik & Skaalvik, 2002).

In a study of Norwegian pupils in 6th , 9th, 11th grade (10–11, 13–14, 15–16 years), and adult students (the age varied between 18 and 52) attending senior high school Skaalvik and Skaalvik (2004b) used measures of these eight frame-specific self-evaluations of ability in mathematics to predict general mathematics self-concept and self-efficacy. *Mathematics self-efficacy* refers to the pupils' beliefs about their capabilities to successfully perform given mathematics tasks at designated levels (Bandura, 1994; Bong & Skaalvik, 2003). In this study, mathematics self-efficacy was measured at a domain-specific level focusing on pupils' grades. The results show that comparison with others in the class (an external frame of reference) and comparison of mathematics achievement with achievement in other school subjects (an internal frame of reference) are robust predictors of both mathematics self-concept and self-efficacy. The results further suggest that learners are using multiple external and internal frames of reference in their self-evaluation of mathematics ability. The authors emphasize that there is a need in future research to extend the I–E model with multiple frames of reference.

In this section I have presented several of Skaalvik and colleagues studies on mathematics self-concept. Some of their studies in Norwegian classrooms support a large body of international research in this area, while other studies give new insight into the area; however, the researchers argue that there is a need for more research in the area of mathematics self-concept.

PUPILS' MOTIVATION FOR LEARNING MATHEMATICS AND AN INQUIRY-BASED MATHEMATICAL TEACHING APPROACH

In my own research, I explore how pupils' motivation for learning mathematics can develop when pupils experience an inquiry-based mathematical teaching approach (Wæge, 2007a). The study I report addresses the following research question: "How can pupils' motivation for learning mathematics develop when they experience an inquiry-based mathematics teaching approach?" In this section I suggest a framework for analyzing pupils' motivation for learning mathematics. The pupils' motivation is examined in terms of needs and goals, and the emphasis is on the psychological needs for competence and autonomy. The usefulness of the analytical framework will be illustrated with an exemplary case. To obtain another perspective, I also analyze the pupils' motivation in terms of intrinsic and extrinsic motivation. I place the pupils' motivation in mathematics on the Self-Determination continuum. Here I will give a short introduction of the continuum and shortly discuss the exemplary case in relation to it.

Motivation is defined in different ways in the literature of (achievement) motivation, and I have chosen to use the following definition: "Motivation is a potential to direct behaviour that is built into the system that controls emotion. This potential may be manifested in cognition, emotion and/or behaviour" (Hannula, 2004, p. 3). Motivation is considered as a potential to direct behavior, and my focus is on the orientation of motivation. This concerns the underlying goals that give rise to an action—that is, it concerns the why of actions. Why does a particular student choose to put a lot of effort into activities A and B and not into activity C? According to Hannula (2004), human beings are always in some emotional state. Emotions are not restricted to hot, intense states of arousal (Op't Eynde,

De Corte, & Verschaffel, 2006). This is important in understanding the definition. According to the definition, students' motivation may be manifested in cognition, emotion, and/or behavior. For example, a student's motivation to get a good grade in mathematics may be manifested in happiness (emotion) if he or she scores high on a test. It may also be manifested in studying for a test (behavior) and in new conceptual learning (cognition) when studying for the test. According to Hannula, motivation, as a potential to direct behavior, is realized through needs and goals (Hannula, 2004).

Self-Determination Theory and Needs

Self-Determination Theory (SDT) is a general theory of motivation that focuses on psychological needs, and I have chosen to use Ryan and Deci's (2002) definition of needs. Before presenting the definition, I will give a short presentation of the theory. SDT is founded on the assumption that human beings have an innate tendency to integrate. *Integrating* means to forge interconnections among aspects of one's own psyche as well as with other individuals and groups in his or her social world: "all individuals have natural, innate, and constructive tendencies to develop an even more elaborated and unified sense of self" (Ryan & Deci, 2002, p. 5).

This assumption of active, integrative tendencies in development is not unique to SDT. Specific to SDT is the proposition that this evolved integrative tendency cannot be taken for granted. According to SDT, social-contextual factors may facilitate and enable the integration tendency, or they may thwart this fundamental process of the human nature: "SDT posits that there are clear and specifiable social-contextual factors that support this innate tendency, and that there are other specifiable factors that thwart or hinder this fundamental process of human nature" (Ryan & Deci, 2002, p. 5).

In other words, according to SDT, there is a dialectic between an active organism and a dynamic environment (social context), such that the environment not only acts on the individual, but is also shaped by the individual. To describe and organize the environment as supporting versus thwarting the integrative process, the concept of needs is used. *Needs* are defined through optimal functioning (growth and well-being), and I have chosen to use the following definition: "There are necessary conditions for the growth and well-being of people's personalities and cognitive structures, just as there are for their physical development and functioning. These nutriments are referred to within SDT as basic psychological needs" (Ryan & Deci, 2002, p. 7).

Looking back at Hannula's definition, psychological needs are specified instances of the general potential to direct behaviour. SDT posits that human beings have three basic psychological needs, the needs for competence, relatedness and autonomy (Deci & Ryan, 2000; Deci, Vallerand, Pelletier, & Ryan, 1991; Ryan & Deci, 2000a, 2000b). Within SDT, competence, autonomy, and relatedness are defined in the following way:

> *Competence* refers to feeling effective in one's ongoing interactions with the social environment and experiencing opportunities to exercise and express one's capacities (Ryan & Deci, 2002, p. 7). *Relatedness* refers to feeling connected to others, to caring for and being cared for by others, to having a sense of belongingness both with other individuals and with one's community (Ryan & Deci, 2002, p. 7). *Autonomy* refers to being the perceived origin or source of one's own behaviour. (Ryan & Deci, 2002, p. 8; emphasis mine in all three quotations)

According to the definition, competence is not an attained skill or capacity, but it is a felt sense of confidence and effectiveness (effectance) in action. The individual feels and experiences competence in the specific situation; it is not a product that shall be used (Wæge, 2007a). In that case it is different from the way it is used by Niss and Jensen (2002) and in the Norwegian curriculum "Kunnskapsløftet" (LK06). When individuals are autonomous, they experience themselves as volitional initiators of their own actions. Cobb and colleagues (Cobb, 2000; Cobb, Gravemeijer, Yackel, McClain, & Whitenack, 1997; diSessa & Cobb, 2004) use the concept of intellectual autonomy as a characteristic of a student's way of participating in the practices of a classroom community. They speak of the students' awareness and willingness to draw on their own intellectual capabilities when making mathematical decisions and judgments as they participate in mathematics activities. Relatedness, in the definition above, refers to the psychological feeling of being together with other persons in a secure community or unity.

The concept of needs is useful because it allows the specification of the social-contextual conditions that will facilitate motivation. According to SDT, students' motivation will be maximized within social contexts that provide them with the opportunity to satisfy their basic psychological needs for competence, autonomy, and relatedness. I have chosen to use Ryan and Deci's definitions of these three psychological needs.[1] In my study I focus on the pupils' needs for competence and autonomy. The data in the study do not give a basis for detailed analyses of the pupils' needs for relatedness and the goals they have in relation to this need. Therefore, the need for relatedness is not a focus in my study.

Needs and Goals Structures

Hannula's definition of motivation (above) purports that the potential to direct behavior is structured through needs and goals. Needs and goals are specified instances of the potential to direct behavior. According to Hannula, goals are derived from needs, and the difference between needs and goals is their different level of specificity. A need may be directed toward a relatively large category of objects, while a goal is directed toward a specific object (Hannula, 2004). For example, in my study, a pupil translated her need for competence into the more specific goal of gaining a good grade. A pupil may translate her need for autonomy into the more specific goal of developing her own ideas, independently of the teacher. The pupil's social need might be translated into the more specific goal of contributing to group work.

Five Motivation Variables

There is a serious methodological problem with research on a mental construct such as motivation. Pupils' motivation is not directly observable, and thus measured, and it needs to be reconstructed through interpretation of the observable. I have developed instruments to assess pupils' motivation for learning mathematics in terms of cognition, emotion, and behavior. In doing this I focus on the five sets of motivational variables that Stipek, Salmon, Givvin & Kazemi (1998) used in their study entitled: "The value (and convergence) of practices suggested by motivation research and promoted by mathematics education reformers."[2] These are the pupils':

1. focus on learning and understanding mathematics concepts as well as on getting right answers
2. enjoyment in engaging in mathematics activities
3. related positive (or negative) feelings about mathematics
4. willingness to take risks and to approach challenging tasks
5. self-confidence as mathematics learners

All these motivation variables figure prominently in the achievement motivation literature and in the mathematics reform literature (e.g., Cobb, Wood, Yackel, & McNeal, 1993; Dweck, 1986; McLeod, 1992; Middleton, 1995; Nicholls, 1984; Resnick, Bill, Lesgold, & Leer, 1991). In analyzing the data, I assess these five motivation variables and I analyze the needs and goals of the pupils in relation to these specific motivational orientations. Furthermore, my emphasis is on the pupils' need for autonomy and competence.

Methodology

The study I report focuses on the development of Norwegian upper secondary school pupils' (age 16) motivation for learning mathematics when they experience a mathematics teaching approach that invites the pupils to be active and to explore. It is a design study in that it involves both instructional design and classroom-based research (Cobb, 2001). I will give a very short presentation of the teaching approach in the next section. As mentioned above, it is not possible to directly assess pupils' goals. Some of the goals need to be inferred through interpretation of the data (Hannula, 2002). I collected a large and varied corpus of data (participant observation, semi-structured interviews, conversations with the teacher, pupils' diaries) on seven of the pupils. The focus of this section is the development of theory, and a case of one pupil is presented as an example of the application of the analytical framework that I am presenting.

Teaching Approach

The teaching approach in the study was intended to give more space for the pupils to satisfy their needs for competence, autonomy and relatedness, than possible with teacher-centered and teacher-controlled approaches. In the study attention was given to the development of pupils' mathematical thinking and reasoning. Our task (the teacher's and mine) was to create instructional activities that supported the development of both collective mathematical meanings evolving in the classroom community and the mathematical understanding of individual pupils. We tried to support "the collective learning of the classroom community, during which taken-as-shared mathematical meanings emerge as the teacher and pupils negotiate interpretations and solutions" (Gravemeijer, Cobb, Bowers, & Whitenack, 2000, p. 226).

The teacher regularly asked the pupils "What did you think when you solved this problem? What strategies did you use?" In the written tasks we developed, the pupils were frequently asked to explain their solutions and strategies, and the pupils were invited to find several solution strategies to a problem. The teacher tried to promote a classroom microculture

(Cobb, Boufi, McClain, & Whitenack, 1997) where active participation and encouragement to understand were emphasized. In some of the instructional activities, the pupils had to develop their own ideas, apply the mathematics in realistic situations, and draw their own conclusions. Collaboration was an important feature of the approach we took. When the pupils were given problems they were not familiar with, we wanted the pupils to collaborate. The pupils had an opportunity to experience themselves and their peers as active participants in creating mathematical insight. Every pupil brought a personal contribution at his or her level. These elements of our design study were suitable for meeting the pupils need for competence, autonomy and relatedness.

Analyses of an Interview

I chose to use interviews in order to gain insight into the complexity of the pupils' motivation, which was one of the goals for the research. The purpose of the interview was to gain insight into pupils' motivation for learning mathematics at a given time. The questions in the interview were developed with the intention to produce qualitative descriptions of the pupils' motivation in relation to the five motivation variables:

- focus on learning and understanding mathematics concepts as well as on getting right answers
- enjoyment in engaging in mathematics activities
- related positive (or negative) feelings about mathematics
- willingness to take risks and to approach challenging tasks
- self-confidence as mathematics learners

In the analysis of the interviews, I used a matrix as a working tool. When I analyzed the data I was moving from the left side of the matrix to the right side (see Table 16.1). That is, first I analyzed the interview according to the five motivation variables. This analysis was further used as a basis for analyzing the pupils' needs and goals.

Here I will present part of an analysis of a semi-structured interview of a pupil, Berit, whose motivation for learning mathematics changed within a few months.[3] The interview is the first out of two interviews I conducted with Berit, and this interview took place at the end of the first semester.[4] I have divided the analysis into two parts. First I present part of the analysis of Berit according to the five motivation variables and I include some of the quotations I used to make my interpretations. In the second part, I present the analysis of Berit's needs and goals in relation to these specific motivational orientations. Here I refer to quotes and interpretations I presented in the first part.[5]

Motivation Variables

Focus on Learning and Understanding Mathematics Concepts as Well as on Getting Right Answers

Berit says that in the beginning of the school year, she thought that the new kind of teaching approach was difficult, but now she prefers this kind of teaching approach. It is better to try to understand and think for yourself than remembering rules without understanding them, even if the grades get lower. She emphasizes that she believes it is possible to get better grades when she gets used to the new way of thinking:

1 **I:** Could you describe the mathematics lessons the first semester?

2 **B:** [...] I think it has been pretty enjoyable. In the beginning I thought it was a bit difficult [I: mm] because I was not used to this kind of teaching approach. And maybe it got kind of, we didn't have any homework and [inaudible] then you didn't do any homework either. They only said "You can do what you want to do," and generally you are busy. And it becomes like "No, you don't have to do the mathematics." So, I think it is good that we have got more homework now [I: mm]. Things like that. [...] I think this mathematical approach is much better. The full-day test[6] was pretty special this time, because

TABLE 16.1 Working Tool

Interview (quotations)	Motivation variables	Needs	Goals

usually I didn't quite understand what I was doing [inaudible]. Do this, follow rules and things like that. This time I thought that I understood everything and I thought the test went very well. And then I get a 4[7] and when I didn't understand it I used to get a 5. But I almost think it's better to try to understand a little more and nevertheless get a lower grade. Anyhow, I think it is possible to increase the grade. It's only a new way of thinking. It's quite interesting, I think [laughing] strange, yes.

Enjoyment and Related Positive or Negative Feelings about Mathematics

Berit says that the first semester has been quite enjoyable (Statement 2). She mentions specifically an activity in the beginning of the school year. She says it was enjoyable because they had to find their own strategies, they were working in groups and collaborated with each other, and they could be social with each other:

9 **I:** Is there something special you want to comment on? Something that was particularly good or bad or anything you just want to say something about?

10 **B:** [...] I especially remember the project. It was pretty enjoyable, because we had to use our own thoughts. It was as if it were real, we had to build a garden pavilion. It was pretty enjoyable, to work in groups and collaborate. And of course it wasn't all about mathematics, we could be social with one another and then I, it's more fun, right?...

On several occasions during the interview, Berit mentions that the all-day test in mathematics was enjoyable, because she was able to find solutions to problems by being patient and thinking about ways to solve them (also see statement 2):

12 **B:** I think we've had many challenging tasks like "Thinking opposite" and not getting the solution before you have found it yourself and things like that. That has been ok, but difficult. But it helps. I noticed that working with the all-day test. I thought it was much more fun working with the problems, and I was very happy when I solved a problem. Previously, if I didn't manage to solve a problem, I didn't manage to solve it. I wasn't able to reason or anything. I wasn't able to do that before, but now I managed to be persistent and to think about ways to solve them, and to get it right. And then you get "pretty pleased," right? When you are able to do that.

She says that she enjoys being able to have an opinion about mathematics and to express it:

30 **I:** How do you interpret the word "interesting"?

31 **B:** Enjoyable. [I: mm] It is, you are simply motivated, and it is enjoyable to see that you are able to do it yourself, actually. That it is possible for you to develop something, to make something, to have your own thoughts and ideas, at least. [I: mm] [...] It is enjoyable to notice that you are able to express something about other areas, like mathematics, for example, that you can have a mathematical opinion and might be right [inaudible]. That is interesting.

Willingness to Take Risks and to Approach Challenging Tasks

She claims that she likes challenging tasks when she is working in groups, collaborating with peer pupils. Berit says that she does not like challenging tasks when she is working individually:

57 **I:** Do you like challenging tasks?

58 **B:** Yes, sometimes [laughing]. It depends on in what way they are challenging. I have noticed that I like them better if we are working in groups. But when I am working individually, then it becomes kind of [inaudible] frustrated if you are not able to find the solution. Then it becomes like "No, I think I will work with the easier tasks instead." [inaudible] But if you are working in groups, then you can discuss the problem and figure it out together "Is this correct?" and "Yes, I think it is". [...]

Self-Confidence

Berit says that when she is working individually and having difficulty solving the mathematics tasks, she easily gives up and asks for help. She claims that she does not see things very easily in mathematics, and that discussing and collaborating with peer pupils helps her in understanding the mathematics (statement 58). Berit reports that on the full-day test she was willing to approach the tasks, she exerted effort in doing that, and she was happy when she succeeded (statements 2, 12).

Berit says she does not think she is good at mathematics anymore, because there are so many clever (high-achieving) pupils in the class. She claims that she is an average mathematics pupil, and refers to her grades:

59 **I:** Do you think you are good in mathematics?

60 **B:** Not any more [laughing]. In secondary school I thought I was good. That was because I was kind of. In this class all the pupils who have been the best pupils in their classes in secondary school, are gathered in one class, and then [.]. It is always someone who is better than me. In one subject or another. I don't think I am very good, no. I think I am normal, actually. My grades are 4s (fours) and 5s (fives), but I think it is OK.

Needs and Goals

The foregoing sets out my classification of Berit's responses in relation to the five motivation variables. Now I will present my interpretation of what the data from this interview reveal about Berit's motivation for learning mathematics, in terms of needs and goals. The first analysis of Berit's motivation for learning mathematics according to the five variables is used as a basis for analyzing her goals in relation to her psychological needs for competence and autonomy. In the analyses a general goal of mastery in mathematics refers to understanding, learning, developing skills, and mastering tasks in mathematics. This is based on the distinction between mastery goal and performance goal (see Dweck, 1986; Lemos, 1999; Linnenbrink & Pintrich, 2000; Nicholls, 1984).

Need for Competence

Berit's need for competence is translated into several specific goals. The data from the interview indicate that Berit has a general goal of mastery in mathematics, and that she has a specific goal of relational understanding in mathematics. She is concerned about knowing what to do and why (Skemp, 1976). Berit's feeling of mastery and of succeeding in mathematics is stronger when she feels that she really understands the problems than when she is remembering rules and formulas without knowing the reasons. Relational understanding in mathematics is Berit's main goal now (statements 2, 12).

The data further indicate that Berit has a goal of succeeding in tests (statement 12), and that she also has a goal of influence and authority in group work, whole class discussions, or in more general situations outside the classroom (statement 31). Grades are important to Berit, and she has a performance goal of getting a good grade in mathematics. Her goals of understanding mathematics and getting good grades are goals that mutually support each other (statement 2).

To summarize, Berit's need for competence is translated as a general goal of mastery in mathematics. Her need for competence in mathematics is translated into the more specific goals of relational understanding in mathematics, of succeeding on tests, and of influence and authority in different social situations where the focus is on mathematics.

Need for Autonomy

The data from the interview indicate that Berit has a goal of using her own thoughts and ideas in mathematics and finding her own solutions. This goal is closely connected to the goal of understanding mathematics (statements 10, 12, 31). Her need for autonomy is closely related to her need for competence and the more specific goals of relational understanding in mathematics.

Two Key Points

There are two key points to be made at this stage. First is that the proposed framework for analysing motivation is useful in describing the pupils' goals and changes in goals in detail. Second is that the pupils' motivation for learning mathematics, although it is considered relatively stable, can be influenced by changes in the teaching approach. The detailed analysis of the one case illustrates how the proposed framework can be used. As the example of Berit illustrates, the framework is useful in clarifying pupils' notion of what it might mean to understand in mathematics. This case study shows that for this particular pupil, Berit, to understand means to know what to do and why. We may also understand the relations between different goals through the use of this framework. For Berit, there was a strong connection between her goal of relational understanding in mathematics and her goal of finding her own solutions. In accordance with Hannula's results (Hannula, 2004, 2006), this case illustrates that performance goals, here a goal of getting good grades in mathematics, and mastery goals, here a goal of relational understanding in mathematics, should not be seen as mutually exclusive. For Berit these goals mutually supported each other.

Skaalvik and Skaalvik (2004a) did not examine the relation between task orientation and ego orientation (Hannula, 2004). They probably expected these orientations to be mutually exclusive. Neither did they differentiate between pupils' focus on relational and instrumental understanding. Their instruments did not capture the complexity of the pupils' motivational orientations.

As some researchers in mathematics education have emphasised, the example with Berit shows that pupils' motivation for learning mathematics may be influenced by changes in the teaching approach (Hannula, 2004, 2006; Middleton & Spanias, 1999). The case of Berit illustrates that pupils' motivation for learning mathematics might change in a relatively short time. Within the first semester of the school year, Berit changed her goal of instrumental understanding in mathematics to a goal of relational understanding in mathematics. We may also understand the relations between different aspects of the instructional designs and the pupils' motivation for learning mathematics in terms of needs and goals through this framework. This case shows that the teaching approach in the study gave the pupil opportunities to satisfy her need for competence and autonomy.

Results of the Study

In the previous section I presented part of an analysis of Berit, one of the seven informants in the study. Here I will give a short summary of the results of the study, that is, the results based on the analysis of the seven pupils included within the study. The detailed analyses of the seven pupils' motivation contribute to enhanced insight into the pupils' motivation for learning mathematics. The study shows that satisfaction of the need for competence, in terms of understanding, is a central aspect of the pupils' motivation. The results indicate that the pupils' feeling of competence is higher when they experience a relational understanding in mathematics than when they experience an instrumental understanding (Skemp, 1976). The data show that there is a close relation between the pupils' enjoyment in engaging in mathematical activities and their feeling of competence. The pupils think it is enjoyable and interesting to work with mathematical activities when they experience understanding or mastery during action. These results are illustrated with the example of Berit.

The study indicates that there have been some changes in the pupils' goals in relation to the need for competence in mathematics. Most of the pupils changed their focus from instrumental understanding to relational understanding, or their focus on relational understanding was enhanced during the school year. The study further indicates that there is a close relation between the need for competence and the need for autonomy in mathematics, as the example of Berit illustrates. The pupils' goal of finding their own strategies for solving problems is closely related to their feeling of learning and of relational understanding in mathematics.

The results indicate that the teaching approach provided the pupils with opportunities to satisfy their needs for competence and autonomy. There are particularly three aspects of the teaching approach that are conducive to pupils' feelings of competence and autonomy during action: (1) the instructional design, (2) the pupils' collaboration with each other, and (3) encouragement and acceptance of the pupils' own strategies for solving problems. The three factors are closely related to each other.[8] I am referring to instructional designs such as projects, problem solving activities, mathematical modelling (Blomhøj & Jensen, 2003), mathematical games, and real-life problems. Many of the instructional designs involve the use of manipulative materials, and the pupils are engaged in collaborative small-group activity where they try to find their own strategies for solving problems.

Pupils' Intrinsic and Extrinsic Motivation in Mathematics

To obtain another perspective on the pupils' motivation I also analyzed the pupils' motivation in terms of intrinsic and extrinsic motivation. Self-Determination Theory, as many other motivational theories, distinguishes between intrinsic and extrinsic motivation.[9] Within SDT, "intrinsic motivation is defined as the doing of an activity for its inherent satisfactions rather than for some separable consequence" (Ryan & Deci, 2000a, p. 56).

Intrinsic motivation reflects the inherent tendency of human nature to engage in activities that are novel and challenging and that result in learning and development (Ryan & Deci, 2000a). According to SDT, if active engagement or intrinsically motivated behaviors are to be maintained, they require satisfaction of the needs for competence, autonomy, and relatedness (Ryan & Deci, 2000b, 2002).

Different Types of Extrinsic Motivation

Extrinsic motivation refers to the performance of an activity in order to maintain some outcome that is separated from the activity, that is, a person is doing an activity for its instrumental value rather than because she/he finds it interesting. According to Self Determination Theory, extrinsic motivation can vary greatly in its relative autonomy. *Internalisation* and *integration* are important concepts in describing the different types of extrinsic motivation, and they are defined in the following way: "Internalisation is the process of taking in a value or regulation, and integration is the process by which the individuals more fully transform the regulation into their own so that it will emanate from their sense of self" (Ryan & Deci, 2000a, p. 60).

According to SDT, internalization is a natural process where the individual tries to transform social practices, values or regulations into personally endorsed values and self-regulation (Deci & Ryan, 2000; Ryan & Deci, 2000a). I interpret regulation to be about what regulates, orients, or determines behavior, or in other words, what causes behavior. SDT assumes the following:

> [I]f external prompts are used by significant others or salient reference groups to encourage people to do an uninteresting activity—an activity for which they are not intrinsically motivated—the individuals will tend to internalize the activity's initially external regulation. That is, people will tend to take in the regulation and integrate it with their sense of self. (Ryan & Deci, 2002, p. 15)

In SDT, internalization is seen as a continuum. It describes how people's motivation varies from amotivation to self-regulated motivation. How autonomous an individual is when acting depends on the extent to which the regulation of the extrinsically motivated behavior is internalized. SDT identifies four types of extrinsic motivation: external regulation, introjected regulation, identified regulation, and integrated regulation (Deci & Ryan, 2000; Ryan & Deci, 2000a, 2002). The four types of motivation are organized in a taxonomy that reflects their differing degree of autonomy. Figure 16.1 shows the different types of motivation, arranged from left to right in terms of the extent to which the motivation is autonomous or self-determined.

At the left end of the continuum in Figure 16.1 is *amotivation.* When people are amotivated they lack an intention to act, and either they do not act at all or they act passively. This happens when they are not valuing the activity or the outcomes it would yield, or when they are not feeling competent to do it (Ryan & Deci, 2000a, 2000b, 2002). At the right end of the continuum is intrinsic motivation. Intrinsically motivated behaviors are the prototype of autonomous or self-determined behavior, because these behaviors are interesting and enjoyable, and are performed volitionally.

External Regulation

External regulation is the least autonomous type of motivation. It is the classic case of extrinsic motivation in which people's behavior is externally regulated by, for example, tangible rewards or threats (Deci & Ryan, 2000; Ryan & Deci, 2000a, 2002).

Introjected Regulation

Another type of extrinsic motivation is introjected regulation. The regulation is partially internalized by the individual but not accepted as one's own. These behaviors are performed with a sense of pressure to avoid guilt and shame and to attain a feeling of pride or worth (Deci & Ryan, 2000; Ryan & Deci, 2000a, 2002).

Identified Regulation

Identified regulation is a more autonomous or self-regulated type of extrinsic motivation. If a regulation or goal is personally valued by the individual, and is consciously accepted as one's own goal, the regulation is identified (Deci & Ryan, 2000; Ryan & Deci, 2000a, 2002).

Integrated Regulation

Integrated regulation is the most autonomous type of extrinsic motivation. It not only involves identifying with the importance of the behavior, but the regulation is evaluated and brought into harmony with the individual's own personal values, goals, and needs that are already a part of the self (Ryan & Deci, 2002). The external regulation is completely internalized to self-regulating, and the result is self-determined extrinsic motivation (Deci & Ryan, 2000; Ryan & Deci, 2002). According to SDT, integrated regulation has many of

Type of Motivation	Amotivasjon	Extrinsic Motivation				Intrinsic Motivation
Type of Regulation	Non-Regulation	External Regulation	Introjected Regulation	Identified Regulation	Integrated Regulation	Intrinsic Regulation
Quality of the behavior	Nonself-Determined					Self-Determination

Figure 16.1 The self-determination continuum showing different types of motivation and regulations (Ryan & Deci, 2002, p. 16)

the same qualities as intrinsic motivation, but there is one essential difference. When people's behavior is governed by integrated regulations, they are performed volitionally to attain personally important outcomes rather than because the activity is inherently interesting or enjoyable. The behaviors are performed to attain a separate outcome where the value of the outcome is well integrated with the self. According to SDT, integrated regulation has an internally perceived locus of causality (Deci & Ryan, 2000; Ryan & Deci, 2000a, 2002). An example to illustrate this could be Anna, in my study. Anna is very clever in mathematics. The analyses indicate that doing well in mathematics is personally important to her. She is working hard with mathematics and focuses on conceptual understanding. The regulation is integrated and her behavior is self-regulated, but she is not intrinsically motivated to learn mathematics. She does not think mathematics is enjoyable or interesting.

The continuum illustrated in Figure 16.1 is purely descriptive. According to SDT, the individual does not need to progress through each stage of internalization. Internalization of a new regulation of behavior may happen at any point along this continuum (Deci & Ryan, 2000; Ryan & Deci, 2002).

Pupils' Motivation in Terms of Intrinsic and Extrinsic Motivation

In my study I placed the pupils' type of motivation on the Self-Determination continuum. These analyses were based on both interviews with the pupils. Due to space constraints, it is not possible for me to present the evidence to support my analysis. The statements I have chosen to illustrate the analysis of Berit are from the second interview with her, and this interview took place at the end of the second semester.[10]

The analyses of Berit's motivation indicated that her motivation for learning mathematics depends on what kind of mathematical activities she is working with:

42 **B:** [. . .] I think it is much more enjoyable to work with practical tasks, I think. And, for example, put into, like that, that you get kind of, like the [. . .] preparing task. That you get a formula, and then make your own task.

43 **I:** Mmm

44 **B:** I, at least, learn much more out of those things, when I can choose, use a little fantasy and some things like that. Then it gets more enjoyable.

45 **I:** Mmm

46 **B:** I think [. . .]. But I also need to have theoretical tasks. To understand it eventually.

The data indicate that she is intrinsically motivated to work with practical tasks and open problems. A common feature with these activities is that the pupils are working in groups, collaborating with each other. By working with these kinds of activities, Berit's need for competence is satisfied through the attainment of the goals of understanding and mastering the mathematics problems, and to have influence and authority in the group. Her need for autonomy is satisfied through the goals of using her own thoughts and ideas in mathematics.

The data indicate that Berit is not intrinsically motivated to work with theoretical tasks. Her kind of motivation for working with theoretical tasks can be placed as integrated regulation on the continuum. She realizes that it is necessary to work with these kinds of tasks to develop understanding in mathematics, which is her main goal, but she is not intrinsically motivated because theoretical tasks do not satisfy her need for autonomy and the goals of developing own strategies and using own ideas.

125 **B:** Like Barbie and things like that.[11] That I think is enjoyable. I think it is a little boring to do exercises. Even if I need it, to kind of understand it, so I know that I need to do the exercises. But it is much more enjoyable to do the practical things. [. . .]

As the example of Berit illustrates, the results from the study show that the pupils' motivation for learning mathematics may vary from activity to activity, depending on their feeling of competence or sense of autonomy during action.

RELATION BETWEEN THE CONCEPTS OF INTRINSIC MOTIVATION AND EXTRINSIC MOTIVATION AND RATIONALES FOR LEARNING MATHEMATICS

I will discuss and make conjectures about the relation between the concepts of intrinsic motivation and extrinsic motivation as defined by Self Determination Theory and rationales for learning mathematics as defined by Mellin-Olsen. As a starting point for my discussion I will present an example described by Mellin-Olsen:

> Jarle is a son of a doctor. From aunts, uncles, neighbours and teachers Jarle understands at an early age

> that he most likely is going to study, a thought which he also makes his own. For this reason the S-rationale nevertheless works for Jarle, when the teacher presents equations and the class does not understand at all how this can be useful for them. He knows that some time he will meet equations again, maybe as a necessity to become a doctor. For this reason, Jarle has no difficulty with accepting equations and how to solve them as a meaningful activity in school. (Mellin-Olsen, 1984, p. 39, my translation)

This example illustrates that the pupil's S-rationale can work even though the pupil does not experience the activity as enjoyable, interesting, or challenging. Jarle considers equations to be important in relation to future studies. A pupil who is acting from the S-rationale has considered the knowledge, through a reference to his generalized others, to be important and interesting. According to Self-Determination Theory, a pupil will internalize the external regulation of a learning activity if significant other or salient reference groups encourage the pupil to do an uninteresting activity. Within SDT, Jarle's type of motivation will be described as identified regulation. Jarle has recognized and accepted the underlying value of learning equations. He has identified with the value of the learning activity, and accepted it as his own. The behavior is still extrinsically motivated, because he is not doing the activity for its own sake. Another type of extrinsic motivation that also falls under the S-rationale is integrated regulation, which is the most autonomous type of extrinsic motivation. When a pupil has integrated a regulation, the behavior is performed to attain goals that are brought into harmony with the pupil's values, goals, or needs. The two least autonomous types of motivation in SDT's model are external regulation and introjected regulation. I consider these two types of extrinsic motivation to be similar to the I-rationale, because in the former case the behaviors are regulated by rewards such as grades and exams, and in the latter case to avoid guilt and shame and to attain a feeling of pride or worth.

The Self-Determination Theory model of the different types of motivation makes it possible for me to place the rationales in relation to intrinsic and extrinsic motivation. The model also makes it possible to be critical of the direct translation from S-rationale and I-rationale to intrinsic and extrinsic motivation as, for example, Goodchild (2001) and Holden (2003) do. Further, the SDT model made it possible to discuss the relation between the frameworks of intrinsic and extrinsic motivation versus social and instrumental rationales for learning.

SUMMARY

First in this section I consider the international impact of the work of the three researchers/research groups presented in this chapter. Second I will give an idea about its value to teachers and teacher educators. Finally I will give a view of future research questions within the research field of motivation in mathematics.

International Impact of the Norwegian Research Studies

Mellin-Olsen (1987) develops Activity theory which has its root in Soviet activity theory. He argues that if pupils are going to learn mathematics, they must approach their activity with a rationale for learning. He identifies two possible rationales for learning mathematics in school: an S-rationale (social rationale) and an I-rationale (instrumental rationale). Mellin-Olsen's theoretical development is based on results of his and Skemp's (Mellin-Olsen, 1984, 1987) research on pupils' choice of mathematical explanations within several curricula in Norway and England. The international impact of the work of Mellin-Olsen can be illustrated by a study conducted by Goodchild (2001). To explore pupils' activity in the mathematics classroom, Goodchild develops a theory of goals based on the works of Mellin-Olsen and Doyle. To complement Mellin-Olsen's S- and I-rationale for learning mathematics, he introduces another rationale, a *P-rationale*: the pupils engage in classroom activity because that is what is expected of them. He labels the rationale "P" for practice, because it is about adopting the behavior of the practice. The P-rationale is different from the S- and I-rationale because it is not a rationale for learning. Goodchild uses the constructs of S-, I-, and P- rationales to explore pupils' activity in a mathematics classroom. So how does the P-rationale fit in the Self-Determination continuum? A pupil with a P-rationale does what is expected of her in the classroom. Her behavior is externally regulated and I consider the P-rationale to be similar to introjected behavior. She is extrinsically motivated, not for learning mathematics, but for fitting into the classroom practice.

In the last two decades, Skaalvik and colleagues have conducted several studies on pupils' mathematics self-concept and motivation by using quantitative methodologies. Their research indicates a relation between mathematics achievement and self-concept, but Skaalvik and colleagues point out that there is a need for more research to make any conclusions about the causal ordering of those constructs. Based on the results of their studies, Skaalvik and colleagues claim that self-concept is a key variable in predicting pupils' motivational orientation. The relation between pupils' achievement and self-concept in mathematics and verbal arts has also received much attention by Skaalvik and colleagues. They extended the I–E model with multiple frames of reference, and the extended model was tested out in a study of Norwegian pupils. The authors emphasize that there is a need for future research to extend the I–E model with multiple frames of reference. Skaalvik and his colleagues' work on mathematics self-concept and motivation has certainly had impact on the international field of educational psychology. Their work has been published in several international journals over the last two decades and other (international) researchers are referring to Skaalvik and his colleagues work in their studies (e.g., Marsh & Yeung, 1997; Pajares & Graham, 1999).

My own work was published recently, so it is relevant to consider what contributions the research study makes to the field of mathematics education. I explore how pupils' motivation for learning mathematics can develop when they experience an inquiry mathematical teaching approach. I analyze the pupils' motivation in terms of needs and goals, and the emphasis is on the basic psychological needs for competence and autonomy. Hannula (2006) criticizes studies on pupils' motivation for learning mathematics for not giving detailed descriptions of the quality of the pupils' motivation. My contribution was to develop an analytical framework that captures the complexity of the pupils' motivation for learning mathematics. The analytical tool made it possible to suggest detailed descriptions of the pupils' motivation. The study provides a deeper insight into the pupils' motivation for learning mathematics. I have also brought Self-Determination Theory, which is a general theory of motivation, into the research field of mathematics education. First I developed a theoretical framework that combines Hannula's theory and definition of motivation with Self-Determination Theory. To obtain another perspective on the pupils' motivation, I placed the pupils' motivation in mathematics on the self-determination continuum. Finally I discussed the relation between intrinsic and extrinsic motivation, as defined by Self-Determination Theory and Mellin-Olsen's concept of rationale for learning.

Possible Value for Teachers

The knowledge I produced in my qualitative study is connected to a certain classroom and a certain period of time. Nevertheless, the contextual knowledge could be adapted to other settings. Gravemeijer and Cobb (2006) use the concept of *ecological validity* about this kind of generalization. They argue that design research aims for ecological validity; that is, the description of the results from a study should provide the reader with a basis for adaption to other situations. They emphasize that "the conception of teachers as professionals who continually adjust their plans on the basis of ongoing assessment of their students' mathematical understanding in fact suggests that complete replicability is neither desirable or, perhaps, possible" (Gravemeijer & Cobb, 2006, p. 45). By reading the report of my study, teachers and teacher educators can understand more about pupils' motivation for learning mathematics. They can get a deeper insight into pupils' goals in mathematics and the different types of extrinsic motivation. One result teachers should be aware of is that the pupils' goals of good grades in mathematics and their goals of relational understanding are not necessarily mutually exclusive. The results of Hannula's (2004) and my own study demonstrate that these goals may support each other. Teachers could use the theoretical framework I developed as a tool when they are planning and carrying out the instructional activities. Both the theoretical framework and the results of the study increase knowledge about how to provide more opportunities for the pupils to satisfy their psychological needs for competence and autonomy. The theoretical framework in my study is based on the work of Hannula. He claims that the specificity in his, and therefore also my own, approach to motivation is that it focuses on what the subject wants to do and why. He further points out that this becomes very important when we want to find implications for practice. The pupils' basic psychological needs for competence, autonomy, and relatedness can all be met in a mathematics classroom that emphasizes exploration, understanding, and communication. These claims are in accordance with the results of my study.

Most of the mathematics instruction in primary and lower secondary schools in Norway follows a traditional pattern consisting of teacher presentation of the problems from the pupils' homework and of new content to be covered followed by the pupils' individual work on textbook tasks (Alseth, Breiteg, & Brekke, 2003). My study shows that it is possible to implement an inquiry-based mathematics teaching approach within the Norwegian curriculum (R94) in upper secondary school. I have shown that it is possible to introduce an inquiry-based curriculum in a mathematics class (with 30 students) of above-average-attaining 16 year olds. The study contributes to increased insight into the possibilities teachers in Norwegian upper secondary schools have when it comes to instructional methods. The study does not provide evidence that the approach can be successfully implemented with younger children in classes with teachers who are not mathematics specialists. There is a need for more research in this area.

Mellin-Olsen's conceptualization of motivation differs from mine. By reading his work, teachers can get a deeper insight into another perspective on pupils' motivation for learning mathematics, where the pupil is considered in relation to the social groups in which he or she relates. This knowledge could help teachers to be aware of and thus take into consideration the pupils' Activities when designing instructional activities. This could further lead to changes in the students' rationales for learning mathematics from the I-rationale to the S-rationale.

Skaalvik and colleagues' quantitative studies could contribute to teachers' increased knowledge about pupils' self-concept and about relations between achievement, self-concept, and different motivational constructs. One particular finding teachers should be aware of is that there exist gender differences in mathematics and that the differences are consistent with gender stereotypes.

Future Research Questions

The three approaches to research on motivation in mathematics presented in this chapter differ from each other and give different knowledge into the actual research field. There is certainly a need for more research in this field, both in general and specifically in Norway. In my opinion there is a need for more research on the relation between pupils' motivation in mathematics and their learning of mathematics. My own study focuses on the pupils' sense of learning mathematics (their feeling of competence), but it does not say anything about the pupils' learning process and if they actually learn anything. Another interesting and important question concerns the relation between pupils' motivation for learning mathematics and different teaching approaches. Personally, I have not yet followed up my research study, but what I would like to do next is to further develop the theoretical framework so that it could be used in quantitative studies. The aim would be to develop a questionnaire that could give careful analyses of pupils' motivation in mathematics. Quantitative methods based on questionnaires make it possible to study a larger sample of pupils and to conduct comparative studies on pupils' motivation in mathematics.

More research and knowledge of pupils' motivation in mathematics could contribute to improving Norwegian pupils' performance in and enjoyment of mathematics, and thus improve the recruitment to mathematics, science, and technology in all levels of the educational system.

ACKNOWLEDGEMENTS

Sincere thanks are due to my colleague, Trond Solhaug, for helpful comments and suggestions. Also the referees are thanked for their valuable suggestions concerning a number of specific points.

NOTES

1. See Wæge (2007a) for a detailed description of my interpretation of the definitions.
2. I have changed the original order of the motivation variables due to the analysis of the data. For pragmatic reasons I have chosen not to differentiate between enjoyment and related positive feelings about mathematics in my analysis of the data.
3. See Wæge (2007a) for the entire analyses of the interview.
4. See Wæge (2007a) for a detailed description of the interview guide.
5. Key to transcripts: [. . .] extracts edited out of transcript for sake of clarity; [inaudible] unclear words; [text] comments about context or emotional behavior like laughing; [.] 1 sec pause, [..] 2 sec pause, and so on.

   The interviews took place in Norwegian. I have tried to translate from colloquial Norwegian to colloquial English, but it does not give an exact word for word translation. My analysis took place without any translation; that is, I analyzed the transcripts in the original language.
6. At the end of each semester, the pupils have an all-day test in mathematics.
7. 1 is the lowest grade and 6 is the highest.
8. See Wæge (2007a) for more detailed descriptions and discussions about these results.
9. See Wæge (2007b) for a more detailed description of intrinsic and extrinsic motivation as defined in Self-Determination Theory.
10. See Wæge (2007a) for the entire analyses of the two interviews.
11. Barbie refers to a modelling task called "Barbie is Bungy jumping."

REFERENCES

Alseth, B., Breiteg, T., & Brekke, G. (2003). *Endringer og utvikling ved R97 som bakgrunn for videre planlegging og justering.* Notodden: Telemarksforskning.

Bandura, A. (1994). Self-efficacy. *Enclyclopedia of human behavior, 4,* 71–81.

Bateson, G. (1973). *Steps to an ecology of mind.* Herts: Paladin Books.

Blomhøj, M., & Jensen, T. H. (2003). Developing mathematical modelling competence: Conceptual clarification and educational planning. *Teaching mathematics and its applications, 22*(3), 123–139.

Bong, M., & Skaalvik, E. (2003). Academic self-concept and self-efficacy: How different are they really? *Academic Psychology Review, 15*(1), 1–40.

Cobb, P. (2000). The importance of a situated view of learning to the design of research and instruction. In J. Boaler (Ed.), *Multiple perspectives on mathematics teaching and learning* (pp. 45–82). Stamford, CT: Ablex.

Cobb, P. (2001). Supporting the improvement of learning and teaching in social and institutional context. In S. Carver & D. Klahr (Eds.), *Cognition and instruction: Twenty-five years of progress* (pp. 455–478). Mahwah, NJ: Lawrence Erlbaum Associates.

Cobb, P., Boufi, A., McClain, K., & Whitenack, J. (1997). Reflective discourse and collective reflection. *Journal for Research in Mathematics Education, 28*(3), 258–277.

Cobb, P., Gravemeijer, K., Yackel, E., McClain, K., & Whitenack, J. (1997). Mathematizing and symbolizing: The emergence of chains of significance in one first-grade classroom. In D. Kirshner & J. A. Whitson (Eds.), *Situated cognition. Social, semiotic, and psychological perspectives* (pp. 151–235). London: Routledge.

Cobb, P., Wood, T., Yackel, E., & McNeal, E. (1993). Mathematics as procedural instructions and mathematics as meaningful activity. The reality of teaching for understanding. In R. B. Davis & C. A. Maher (Eds.), *Schools, mathematics, and the world of reality* (pp. 119–133). Boston: Allyn and Bacon.

Deci, E. L., & Ryan, R. M. (1985). An Introduction. In *Intrinsic motivation and self-determination in human behavior* (pp. 3–40). New York: Plenum.

Deci, E. L., & Ryan, R. M. (2000). The "what" and "why" of goal pursuits: Human needs and the self-determination of behavior. *Psychological Inquiry, 11*(4), 227–268.

Deci, E. L., Vallerand, R. J., Pelletier, L. G., & Ryan, R. M. (1991). Motivation and education: The self-determination perspective. *Educational Psychologist, 26*(3&4), 325–345.

diSessa, A. A., & Cobb, P. (2004). Ontological innovation and the role of theory in design experiments. *The Journal of the Learning Sciences, 13*(1), 77–103.

Dweck, C. S. (1986). Motivational processes affecting learning. *American Psychologist, 41*(10), 1040–1048.

Evans, J., & Wedege, T. (2004). *Motivation and resistance to learning mathematics in a lifelong perspective.* Paper presented at the 10th International Congress on Mathematical Education, http://www.icme10.dk/, TSG 6, Copenhagen, Denmark.

Goodchild, S. (2001). *Students' goals. A case study of activity in a mathematics classroom.* Norway: Caspar Forlag.

Graham, S., & Weiner, B. (1996). Theories and principles of motivation. In D. C. Berliner & R. C. Calfee (Eds.), *Handbook of educational psychology* (pp. 63–84). New York: Simon & Schuster/Macmillan.

Gravemeijer, K., & Cobb, P. (2006). Design research from a learning design perspective. In v. d. Akker, K. Gravemeijer, S. McKenney, & N. Nieveen (Eds.), *Educational design research* (pp. 17–51). Oxon: Routledge.

Gravemeijer, K., Cobb, P., Bowers, J., & Whitenack, J. (2000). Symbolizing modelling and instructional design. In P. Cobb, E. Yackel & K. McClain (Eds.), *Symbolizing and communicating in mathematics classrooms. Perspectives on discourse, tools, and instructional design* (pp. 225–273). Mahwah, NJ: Lawrence Erlbaum Associates.

Hannula, M. S. (2002). Goal regulation: Needs, beliefs, and emotions. In A. D. Cockburn & E. Nardi (Eds.), *Proceedings of the 26th Conference of the International group for the Psychology of Mathematics Education* (Vol. 4, pp. 73–80). Norwich, UK: University of East Anglia.

Hannula, M. S. (2004). *Regulation motivation in mathematics.* Paper presented at the 10th International Congress on Mathematical Education, http://www.icme10.dk/, TSG 24, Copenhagen, Denmark.

Hannula, M. S. (2006). Motivation in mathematics: Goals reflected in emotions. *Educational Studies in Mathematics, 63*, 165–178.

Holden, I. M. (2003). Matematikk blir gøy—gjennom et viktig samspill mellom ytre og indre motivasjon. In B. Grevholm (Ed.), *Matematikk for skolen* (pp. 27–50). Bergen: Fagbokforlaget.

Kloosterman, P. (1996). Students' beliefs about knowing and learning mathematics: Implications for motivation. In M. Carr (Ed.), *Motivation in Mathematics* (pp. 131–156). Cresskill, NJ: Hampton Press.

Köller, O., Baumert, J., & Schnabel, K. (2001). Does interest matter? The relationship between academic interest and achievement in mathematics. *Journal for Research in Mathematics Education, 32*(5), 448–470.

Lemos, M. S. (1999). Students' goals and self-regulation in the classroom. *International Journal of Educational Research, 31*, 471–485.

Leont'ev, A. N. (1981). The problem of activity in psychology. In J. V. Wertsch (Ed.), *The concept of activity in soviet psychology* (pp. 37–71). New York: M. E. Sharpe.

Linnenbrink, E. A., & Pintrich, P. R. (2000). Multiple pathways to learning and achievement: The role of goal orientation in fostering adaptive motivation, affect, and cognition. In C. Sansone & J. M. Harackiewicz (Eds.), *Intrinsic and extrinsic motivation. The search for optimal motivation and performance* (pp. 195–227). San Diego: Academic Press.

Marsh, H. W. (1986). Verbal and math self-concepts: An internal/external frame of reference model. *American Educational Research Journal, 23*(1), 129–149.

Marsh, H. W., & Yeung, A. S. (1997). Causal effects of academic self-concept on academic achivement: Structural equation models of longitudinal data. *Journal of Educational Psychology, 89*, 41–54.

McLeod, D. B. (1992). Research on affect in mathematics education: A reconceptualization. In D. A. Grouws (Ed.), *Handbbok of research on mathematics teaching and learning* (pp. 575–596). New York: MacMillan Publishing Company.

Mead, G. H. (1934). *Mind, self and society.* Chicago: The University of Chicago Press.

Mellin-Olsen, S. (1981). Instrumentalism as an educational concept. *Educational Studies in Mathematics, 12*, 351–367.

Mellin-Olsen, S. (1984). *Eleven, matematikken og samfunnet. En undervisningslære.* Bekkestua, Norge: NKI Forlaget.

Mellin-Olsen, S. (1987). *The politics of mathematics education.* Dordrecht: Kluwer Academic Publisher.

Middleton, J. A. (1995). A study of intrinsic motivation in the mathematics classroom: A personal constructs approach. *Journal for Research in Mathematics Education, 26*(1), 254–279.

Middleton, J. A., & Spanias, P. A. (1999). Motivation for achievement in mathematics: Findings, generalizations, and criticism of the research. *Journal for Research in Mathematics Education, 30*(1), 65–88.

Nicholls, J. G. (1984). Achievement motivation: Conceptions of ability, subjective experience, task choice, and performance. *Psychological Review, 91*(3), 328–346.

Nicholls, J. G., Cobb, P., Wood, T., Yackel, E., & Patashnick, M. (1990). Assessing students' theories of success in mathematics: Individual and classroom differences. *Journal for Research in Mathematics Education, 21*, 109–122.

Niss, M., & Jensen, T. H. (2002). *Kompetencer og matematiklæring. Ideer og inspiration til udvikling af matematikundervisning i Danmark.* København: Undervisningsministeriet, Uddannelsesstyrelsen temahæfteserie nr. 18–2002.

Op't Eynde, P., De Corte, E., & Verschaffel, L. (2002). Framing students' mathematics-related beliefs. In G. C. Leder, E. Pehkonen & G. Törner (Eds.), *Beliefs: A hidden variable in mathematics education?* (pp. 13–37). Dordrecht: Kluwer Academic Publishers.

Op't Eynde, P., De Corte, E., & Verschaffel, L. (2006). "Accepting emotional complexity": A socio-constructivist perspective on the role of emotions in the mathematics classroom. *Educational Studies in Mathematics, 63*, 193–207.

Pajares, F., & Graham, L. (1999). Self-efficacy, motivation constructs, and mathematics performance of entering middle school students. *Contemporary Educational Psychology, 24*, 124–139.

Resnick, L. B., Bill, V. L., Lesgold, S. B., & Leer, M. N. (1991). Thinking in arithmetic class. In B. Means, C. Chelemer & M. S. Knapp (Eds.), *Teaching advanced skills to at-risk students: Views from research and practice* (pp. 27–67). San Fransisco: Jossey-Bass.

Ryan, R. M., & Deci, E. L. (2000a). Intrinsic and extrinsic motivations: Classic definitions and new directions. *Contemporary Educational Psychology, 25*, 54–67.

Ryan, R. M., & Deci, e. L. (2000b). Self-determination theory and the facilitation of intrinsic motivation, social development, and well-being. *American Psychologist, 55*(1), 68–78.

Ryan, R. M., & Deci, E. L. (2002). Overview of self-determination theory: An organismic dialectical perspective. In E. L. Deci & R. M. Ryan (Eds.), *Handbook of self-determination research* (pp. 3–33). New York: The University of Rochester Press.

Schiefele, U., & Csikszentmihalyi, M. (1995). Motivation and ability as factors in mathematics experience and achievement. *Journal for Research in Mathematics Education, 26*(2), 163–181.

Skemp, R. R. (1976). Relational and instrumental understanding. *Mathematics Teaching, Bulletin of the Association of Teachers of Mathematics, 77*, 20–26.

Skaalvik, E. M., & Rankin, R. J. (1990). Math, verbal, and general academic self-concept: The internal/external frame of reference model and gender differences in self-concept structure. *Journal of Educational Psychology, 82*(3), 546–554.

Skaalvik, E. M., & Rankin, R. J. (1995). A test of the internal/external frame of reference model at different levels of math and verbal self-perception. *American Educational Research Journal, 32*(1), 161–184.

Skaalvik, E. M., & Skaalvik, S. (1998). *Selvoppfatning, motivasjon og læringsmiljø.* Norway: TANO.

Skaalvik, E. M., & Skaalvik, S. (2002). Internal and external frames of reference for academic self-concept. *Educational Psychologist 37*(4), 233–244.

Skaalvik, E. M., & Valås, H. (1999). Relations among achievement, self-concept, and motivation in mathematics and language arts: A longitudinal study. *Journal of Experimental Education, 67*(2), 135–149.

Skaalvik, E. M., & Valås, H. (2001). Achievement and self-concept in mathematics and verbal arts: A study of relations. In R. J. Riding & S. G. Rayner (Eds.), *International perspectives on individual differences* (Vol. 2: Self perception, pp. 221–238). London: Ablex Publishing.

Skaalvik, S., & Skaalvik, E. M. (2004a). Frames of reference for self-evaluation of ability in mathematics. *Psychological Reports, 94*, 619–632.

Skaalvik, S., & Skaalvik, E. M. (2004b). Gender differences in math and verbal self-concept, performance expectations, and motivation. *Sex Roles, 50*(3/4), 241–252.

Skaalvik, S., & Skaalvik, E. M. (2005). Self-concept, motivational orientation, and help-seeking behavior in mathematics: A study of adults returning to high school. *Social Psychology of Education, 8*, 285–302.

Stipek, D., Salmon, J. M., Givvin, K. B., & Kazemi, E. (1998). The value (and convergence) of practices suggested by motivation research and promoted by mathematics education reformers. *Journal for Research in Mathematics Education, 29*(4), 465–488.

Urdan, T., & Turner, J. C. (2005). Competence motivation in the classroom. In A. J. Elliot & C. S. Dweck (Eds.), *Handbook of competence and motivation* (pp. 297–317). New York: The Guilford Press.

White, R. W. (1959). Motivation reconsidered: The concept of competence. *Psychological Review, 66*(5), 297–333.

Wigfield, A., & Eccles, J. S. (2002). The development of competence beliefs, expectancies for success, and achievement values from childhood through adolescence. In A. Wigfield & J. S. Eccles (Eds.), *Development of Achievement Motivation* (pp. 91–120). San Diego: Academic Press.

Wæge, K. (2007a). *Elevenes motivasjon for å lære matematikk og undersøkende matematikkundervisning.* Trondheim: Norwegian University of Science and Technology.

Wæge, K. (2007b, February). *Intrinsic and extrinsic motivation versus social and instrumental rationale for learning mathematics.* Paper presented at the Fifth Congress of the European Society for Research in mathematics Education (CERME 5), Cyprus.

17

THE FUTURE OF MATHEMATICS EDUCATION RESEARCH IN NORWAY

Some Indicators

Simon Goodchild

UNIVERSITY OF AGDER

Liv Sissel Grønmo

ILS, UNIVERSITY OF OSLO

Abstract: An attempt is made to consider possible developments and directions for mathematics education research in Norway. Uncertainties at the time of writing make meaningful prediction impossible. This view of the future is informed by the current position, as outlined in the other chapters in this section of the Sourcebook and the relevant reports policy and statements coming from government and the Department of Education. The discussion continues on the basis of the authors' knowledge, experience, values, and beliefs upon which their hope for a lively, purposeful, and effective future of the research field in Norway is founded.

Keywords: Developments in research, Norwegian national policy, Impact of research, Continuing professional development, International benchmarks

INTRODUCTION

Research in mathematics education is a relatively young scientific field, starting out internationally about 1960. This is a rather short time for establishing a new scientific field, and there is still some way to go before it can be considered as a mature, full-fledged, and well-defined discipline. In Norway, the research field is even younger; most development occurring since the beginning of the 1990s. In this final chapter of the Norwegian section of the Sourcebook, we give a brief summary of our perspective of some characteristics of Norwegian research in mathematics education, pointing to challenges that must be faced and reflecting upon possible directions for future research in the field. It is likely that anyone faced with writing a chapter that attempts to make reasonable predictions about the future would be confronted by the uncertainties of the present. Nevertheless, reflection on current opportunities and challenges, and their implications for the future, is one approach to planning and developing research in our field. As we write this chapter, we are aware of the hopes invested in current proposals to the Research Council of Norway for funding new research projects. We are also aware of current national policies and initiatives that will have differing impacts depending on the way they are interpreted and implemented.

Developments in the field of mathematics education research since 1990 have established a solid base for further growth and expansion, which will be realized if conditions are favorable. However, there are a number of challenges that confront the Norwegian community of researchers in mathematics education. Based on current research activity, we will point to some strengths and challenges to be faced without

The First Sourcebook on Nordic Research in Mathematics Education, pages 259–266

making any claims to be able to predict how the conditions for research and research activity will change or stabilize.

FOUNDATIONS

It was noted in the introduction to the Norwegian chapters that there has been a considerable expansion in research activity in the field from the beginning of the 1990s. This can be attributed to a number of factors that have emerged in this section of the Sourcebook.

- The first master's degrees in mathematics education in Norway were completed in the beginning of the 1990s at the University of Oslo and Agder University College, followed by master's and doctoral degrees at a number of Norwegian universities.
- Norway's participation in international comparative studies in mathematics starting with TIMSS 1995 (Grønmo, this volume, Chapter 4): Pupils' achievement, attitudes as well as teachers' view of mathematics have been among several important issues investigated at all levels in school—from primary through lower and upper secondary schools. Since then, Norway has participated on a regular basis in both PISA and TIMSS and other related studies. This has led to national reports focusing on mathematics education as well as articles based on secondary analyses of data, which have contributed to national concerns about raising the quality of mathematics education.
- The development of Agder University College and its transformation into the University of Agder in autumn of 2007 was based in part, first, on the introduction of a masters program in mathematics education in the early 1990s, and later, on the doctoral program (Breiteig & Goodchild, this volume, Chapter 2).
- The Nordic Graduate School in Mathematics Education, hosted at the University of Agder, was established in 2004, ending in 2009 (Goodchild, this volume, Chapter 1). In 2007, the University of Oslo set out to establish the National Graduate School in Mathematics and Science Education (RDID skolen) with funding guaranteed through 2011. These two universities are planning to unite their efforts and collaborate in a Graduate School in Science and Mathematics Education.
- Doctoral research being undertaken at the universities of Oslo, Tromsø, NTNU and Agder (chapters 10–16 in part 3 of this volume, and Breiteig & Goodchild, Chapter 2)
- Programs financed through the Research Council of Norway (KUL, PraksisFoU) that have supported research in mathematics education (this volume: Jaworski, Chapter 5; Fuglestad, Chapter 6; Johnsen-Høines, Chapter 7).

In addition, we should pay tribute to the small but influential group of mathematics educators whose drive and energy inspired policy makers to invest in research in the field in the early days of its development. All of the above have contributed to the current position in which there is a new generation of researchers in the field, which is steadily growing as the doctoral programs mature. Also of significance is the now well-developed national network of researchers and the opportunities for further collaboration this opens up.

The research reported in the chapters in this section of the Sourcebook is wide ranging and reveals the development of a rich and varied community of researchers who are concerned with:

- Teaching and learning mathematics in schools at all levels
- Quantitative studies of pupils' performance, attitudes, and experiences with instruction in school
- Trends in students' achievement over time
- Mathematics teaching development
- Discourse and multi-modal approaches
- Socio-cultural studies
- Ethnomathematics
- Problem solving and meta-cognition
- Learning strategies in mathematics
- Design research
- Students' motivational goals
- ICT and mathematics education
- Use of assessment as a tool for development and learning
- Video studies of instruction and practice in the classroom
- Pupils with learning difficulties

There are other active research interests related to these that have not been included, for example research into teachers' knowledge, pupils with special talent for mathematics, and recruitment of students and motivation for studying mathematics at university.

NATIONAL INDICATORS

Quality and Characteristics of National Research

A review of Norwegian educational research resulted in a rather critical report (NIFU STEP: Borgen, Gunnes, & Sivertsen, 2007) that appears to be influential in guiding national policy. The report concluded that Norwegian education research was characterized by researchers' substantial freedom of choice for research themes, that largely it is researchers who have responsibility to evaluate the social relevance of their studies, and third, that a large proportion of research was published in reports rather than papers in peer-reviewed journals that would provide an evaluation of scientific quality. The report was also critical of the national processes for defining research programs in that they did not reach to all areas in which research was needed.

Programs of the Research Council of Norway

The first practical outworking of the NIFU STEP report was the program Utdanning 2020 (U2020) [Norwegian Educational Research towards 2020], which was announced by the Research Council of Norway in December 2008, and first call for proposals made in April 2009. The program calls for multidisciplinary research, including economics and social sciences, and seeks the formation of larger groups of researchers with strong international links. It is the intention that "the programme will also strengthen social science research on the outcomes of education including the effects of educational interventions, as well as research based on longitudinal studies and empirical research, including research based on quantitative data" (RCN, 2009, p. 8).

The program identifies four "thematic priority areas":

A. Educational objectives, content, and teaching and working methods
B. Assessment forms, learning processes and learning outcomes in education
C. Management, administration, and organization of educational and research institutions
D. Education and society (RCN, 2009, p. 14).

On the surface, these would appear to give plenty of scope for mathematics education research. However, the funds made available for the program, at the time of writing, are disappointingly small, and the competition, which includes all areas of educational research, is intense. Recently, in May, 2009, the Norwegian Ministry of Education (KKD, 2009b) have indicated an intention to continue support of programs such as PraksisFoU (Praksis, Forskning og Utvikling—Praxis, research and development), which is currently funding mathematics education projects at several universities and university colleges.

One element of the call for proposals within the U2020 programme was aimed at establishing a national unit for psychometric testing. This, taken together with the quotation above drawing attention to effects and longitudinal and quantitative studies, suggests that greater emphasis will be placed in this area than is represented by most of the chapters in this Norwegian part of the Sourcebook, which, apart from Grønmo's chapter, report qualitative studies.

Teacher Education

The Norwegian Government white paper on teacher education (KKD, 2009a) listed a number of reasons for developing basic teacher education for all into a five-year program with a master's level qualification at the end. At the same time, the white paper recognizes that the ambition will be difficult to meet in the short term because of individual teacher education institutions' inadequate capacity to provide master's level education. Another recommendation of the same white paper is the establishment of a National Graduate School in Teacher Education that would support doctoral education. These two issues have clear implications for education in mathematics education research. First, the strengthening of institutions' capacity to offer master's level education and the development of a graduate school implies doctoral programs will continue to be supported and will possibly increase in size and number. An increase in doctoral education, to be meaningful, will need to be accompanied by an increase in the number of senior active researchers who will be competent to supervise doctoral research. Thus, it may be reasonably expected that research in mathematics education will develop in volume. The national ambition, as expressed in the white paper (KKD, 2009b) is that research output will also develop in quality and significance to achieve internationally recognized standards of excellence. However, if this Norwegian part of the Sourcebook is as representative as claimed, and the signals regarding the national focus on quantitative methods have been correctly interpreted, there will need to be some adjustment to the focus of research.

Strengthening research such as longitudinal studies and empirical research, including research based on quantitative data (RCN, 2009) may perhaps also be a good starting point for combining quantitative approaches with qualitative studies. One example is given in this book, where PISA+ was initiated to shed more light on results from PISA and TIMSS 2003 results, by using more qualitative approaches such as classroom based case studies (Grønmo, this volume, Chapter 4).

To develop and take this further, we see the need for closer cooperation between researchers with different backgrounds and between different institutions. The current developments based on collaboration between researchers at different institutions will facilitate this. This will include collaboration between researchers in mathematics education and researchers in pure mathematics, as already initiated in several institutions. This collaboration is especially for research in mathematics education in upper secondary school and in teacher education.

Norway has participated in the first international comparative study that takes place at university and college level, Teacher Education and Development Study in Mathematics Teacher Education, TEDS-M 2008 (www.teds.no). The plan is to release international results in 2010. An important perspective for the national report will be to reflect upon and discuss the knowledge *in mathematics* and *mathematics education of* students on different educational tracks, and also their motivation and view of mathematics. It appears that the national discussion about teacher education to this point has been dominated by issues relating to the length and organization of courses, rather than on approaches to improve students' content knowledge. The TEDS-M study could be a good starting point for further research using multiple approaches.

Continuing Education and Training, In-Service Training for Teachers in Mathematics

Education to become a teacher is important, but a teacher's professional development is a continuous process. Continuing professional education and training needs to be an integrated part of what it means to be a teacher. Research in mathematics education points to how pupils learn mathematics and how to improve textbooks and assessment, and it can make an important contribution to the construction of new curriculum reforms for the school, among other things. This research may have the highest level of quality, be published in international journals, and be cited by politicians; however, unless research outcomes are communicated to teachers in school, the impact on pupils' experience will be very small. In Norway there has been an increase in funding for continuing education and training for teachers, but as shown by TIMSS, Norwegian teachers' participation in courses relevant for their teaching of mathematics is lower than that of teachers in most other countries (Grønmo, this volume, chapter 4). There may be several reasons for this. One reason may be that much of the continuing professional education offered is too general and not closely related to content knowledge in a specific school subject such as mathematics. Another reason may be that research institutions do not have the capacity to develop and offer this kind of education. Resources for continuing professional development have mainly been devolved to local communities and schools for them to purchase further professional development opportunities for their teachers. However, if insufficient funds are devolved to research institutions to develop the type of programs required, it does not help to have the resources devolved to schools and communities. This seems to have been a major problem in Norway. A further problem may be that too much responsibility has been placed on too few researchers in the field. To develop good and adequate continuing professional education for teachers, it may be more effective to devolve more resources for this purpose to the institutions in which initial teacher education and research in mathematics education takes place. Most Researchers recognize that dissemination of research findings to the professional community is an important part of their work, although at the moment it is not included among the criteria for evaluating a researcher's performance, and thus is not prioritized. Research in mathematics education that is not communicated to the schools is not likely to make any significant contributions for better instruction in school. The significance of teachers' continuing professional education and training could also be recognised more formally through the award of academic credit, and developments in teachers' career and salary structures.

Institutional Development

The Norwegian Ministry of Education set up a committee under the chairmanship of Professor S. Stjernø to inquire into the development of higher education in Norway. The committee reported its findings in January 2008 (NOU, 2008). The committee recommended, among other things, a consolidation of higher education by the amalgamation of universities and university colleges into fewer, larger units. The suggestion was not well received by many of the institutions that would lose identity and autonomy, and it is not clear whether the government will pursue this recommendation. Nevertheless, there is evidence of institutions moving closer together and establishing partnerships in selected research fields and niches and exploratory talks between institutions' leadership exploring the benefits of closer collaboration. Such coming together could benefit research communities such as in mathematics education because the country and milieu is too small to be able to sustain or benefit from strong internal competition. Moreover, the national community of researchers has a sense of identity that would probably feel more at ease and be more productive in an environment of collaboration and mutual support than it would be motivated by strong competitive pressures.

The Stjernø committee also noted the lack of research in higher education that could be useful to inform discussion on policy and practice. The report also repeats the criticism of the NIFU STEP 2007 report noting the fragmentation and limited impact of Norwegian research that failed to achieve international levels of excellence. At the same time consideration was given to international research in higher education and noted that at compulsory school level education research tended to focus more on pedagogy whereas at higher education it focused more on political science, economics and sociology. RCN's most recent education research programme U2020, emphasises research in higher education. The chapters in this Sourcebook are consistent with the criticisms expressed in the Stjernø committee's report. Mathematics education researchers might be well advised to set their focus on research on higher education, and perhaps upper secondary schools. This would be an interesting twist given that the early research at Agder University College (AUC) focused especially on pedagogical and didactical developments in teaching mathematics at the colleges that came together to form AUC (see chapters by Beiteig & Goodchild, and Bjuland, Cestari & Borgersen). It is only more recently that research has tended to concentrate on compulsory and upper secondary school didactics.

International comparative studies have mainly focused on pupils' achievement and attitudes in compulsory schools (Grades 1 to 10), except for TIMSS 1995 in which students at the end of upper secondary school were also included. With the TIMSS Advanced study conducted in 2008, the first results from a trend study in upper secondary school will be presented internationally. The Department of Education gave funding for the study, and researchers in Oslo were heavily involved in developing instruments for the study. Also this study may form a good starting point for further research using multiple approaches, and several master students and doctoral students have started to use data from the study in their research.

International Comparison and Trend Studies

Evidence from international comparative studies initiated by IEA or OECD such as TIMSS and PISA have attracted an influential audience of politicians, policy makers, curriculum developers as well as researchers in mathematics education. These studies can give insights into pupils' attainment, attitudes and experiences in school in different countries, and how they have changed over time nationally as well as internationally. The results of analyses of data from international comparative studies have to be interpreted according to the national needs and goals in education in Norway, and used to support reflection on the Norwegian school system from an international perspective (Grønmo, this volume, Chapter 4). In Norway, the leadership for international studies in mathematics education is placed at ILS, the University of Oslo. These studies should not be seen only in terms of the reports that draw attention to the international comparisons, the studies also provide a significant advanced database that can support a number of secondary analyses, including master and doctoral research in the field. Given the practice of many countries where national representatives in the international comparative studies' working groups are drawn from non-academic fields, or non-mathematics education research groups, the importance of having these studies placed at a university is not to be underestimated.

Dualism in Research in Mathematics Education

Researchers in mathematics education face a special challenge in what Niss refers to as the dualism between descriptive/explanatory and normative dimensions of research.

> In other words, in addition to its descriptive/explanatory dimension, the didactics of mathematics also contains a *normative* dimension, in which the genetic questions are 'what *ought* to be the case?' and '*why* should this be so?' It may come as a surprise to some that issues such as these are considered part of a scholarly and scientific discourse and are claimed to belong to the scope of research. But this is unavoidable in the same way as it is unavoidable to operate with the notion of 'good health' and 'sound treat-

> ment' in much medical research, or 'satisfactory functioning' of devices constructed in engineering. (Niss, 1998, pp. 5–6)

As researchers, we may feel more comfortable with descriptive/explanatory research. As soon as the normative dimension is included, one is more likely to meet criticism. Moreover, we are aware of the challenges of practice that arise from the unique identities of every pupil, teacher, and class—we know there are no universal recipes. Further, we recognize differences in the expertise and responsibilities of researchers, teachers, and policy makers. Media and politicians appear to be most often responsible for the interpretation of the research and making suggestions about action to be taken. We contend that if research is to have an impact on policy and practice, researchers must recognize that they share responsibility for correctly interpreting the research implications and suggesting ways that policy and practice might address the implications. One consequence of this may be that those embarking on research training in the field would be required to have greater experience in policy and or practice than is presently the case.

Local Conditions

In the introduction to the Norwegian section of the Sourcebook, mention was made of the important contribution made by the Nordic Graduate School in Mathematics Education (NoGSME) in building up the community and network of researchers. It was further noted that as the funding for NoGSME has ceased, a new body—the Nordic society for Research in Mathematics Education (NoRME)—has been established to seek funding and coordinate activities that NoGSME had managed so successfully. The first evidence that this is an achievable goal has been the awarding of funds to run a doctoral summer school in 2010. Within Norway, it is unfortunate that an application for a national graduate school in teacher education led by a partnership of the natural sciences and mathematics education group at the University of Oslo with the mathematics education group at the University of Agder did not succeed. The impact of this outcome is yet to be felt but it is anticipated that it will have significant consequences for the continued development of the research milieu.

Breiteig and Goodchild (this volume, Chapter 2) suggested that the development of the mathematics education master's program and later doctoral program at the University of Agder has been significant in the development of the research field in Norway. The institution is now developing other doctoral programs, and mathematics education research is losing its favored status. The mathematics education group has been successful in the past to draw in research funds from external sources, but efforts to sustain this flow will need to be sustained if the milieu is to continue to flourish.

The University of Oslo initiated collaboration between four teacher education institutions in the international comparative study TEDS-M 2008. The cooperation is between the two universities in Oslo and Agder and the two university colleges in Bergen and Trondheim. There are master's and doctoral students involved in the study, and the plan is to develop this further. It has been an interesting and promising experience for all researchers involved to be part of this cooperation, with plans for further development in writing reports and scientific articles based on analyses of data from the study.

The research group at Agder initiated a successful collaboration between five institutions as part of the Teaching Better Mathematics (TBM) project. The collaboration within the project consortium, including Bodø, Sør Trøndelag, Bergen, and Oslo University Colleges, has provided an effective forum for sharing research and developing researcher competencies. The funding for this collaboration comes to an end in 2010, but the possible extension of funds within the PraksisFoU program noted above offers some hope that the activity of TBM may be continued.

Several senior researchers are due for retirement within the next few years. The hope is that some of the new generation of Norwegian researchers will be ready to fill their positions. At the moment, evidence suggests that Norway lacks national capacity to meet the national need for senior researchers in mathematics education. Recent announcements of positions for professorial appointments in mathematics education have had disappointing responses. Most institutions are prepared to appoint a well qualified person who is fluent in English, and certainly prepared to appoint a well qualified person fluent in at least one of the Scandinavian languages. This broadens the field somewhat, but it is still a big step for a person to move to Norway with a spouse and family. Even within Norway there are very strong regional and local identities, and the culture of mobility is not well developed, which has an impact both on appointments to permanent positions and temporary researcher posts, such as doctoral fellowships and post-doctoral positions.

CONTINUATION OF RESEARCH

The foregoing outline of the national indicators regarding research suggests that the future of mathematics education research is secure, provided that researchers are willing and prepared to adapt to national priorities. The drive for quality and internationalization also promises the development of the field, given that researchers are willing to make the effort to lift their work to the required levels. For these developments, a great deal rests on the enthusiasm of the younger generation of researchers who have either recently completed their PhDs or will soon do so. Concurrent with developments in the national context, there are local changes and indicators that provide further challenges.

The local conditions as outlined above may give rise to some uncertainties, but the chances are that the new generation of researchers now established will continue to be active for many years to come. The chapters in this Sourcebook have not represented the research activity relating to teacher knowledge that researchers at the University of Stavanger have established in collaboration with researchers at the University of Michigan. Presumably, it is likely that the group at the University of Agder will continue to engage in mathematics teaching developmental research and that research applying discourse and multimodal approaches will continue at Bergen University College and the University of Agder. Researchers at the University of Oslo will continue to engage in research relating to the large international comparative studies. Work related to the use of digital technologies and mathematics will continue at Agder, Norwegian University of Science and Technology, Trondheim, and ethnomathematics at Tromsø. New research interests will also emerge as individuals and groups respond to changes in the national priorities.

FUNDING RESEARCH

Norway ranks amongst the wealthiest, per capita, countries in the world and is developing high-tech industries that demand a highly qualified and mathematically competent workforce. The research presented in this volume has had many different sources for funding:

- Norway's participation in international comparative studies get funding from The Department of Education, with additional funding from the university where these studies take place
- Programs of the Research Council of Norway (NCR) have founded some of the project presented in this Sourcebook
- Doctoral stipends funded by the Department of Education
- Other research project has been funded from the universities or institution where it is conducted, or from local sources

Funding of research is an important issue. Research presented in this Sourcebook has used different combinations of sources for funding their projects.

It is not only funding of research project that is important, but also the opportunities given to researchers to do research of high international quality. Currently there is much debate among researchers about ways that institutions organize the opportunities for research—for example, for researchers to have sabbatical periods in which they can concentrate on field work, analysis, and writing. If research in Norway is to develop to play a more significant role internationally, it seems reasonable to relate such sabbaticals to claims about publishing in international journals of high quality.

LEADERSHIP

The research community is making its presence felt at international conferences but has yet to make a significant impression through publication in first-rank international journals. The Norwegian society of researchers in mathematics education is challenged to make more substantial contributions to research internationally.

Foreign researchers of high international reputation have worked in Norway for fairly short periods—for example, Barbara Jaworski (UK) and Cyril Julie (South Africa). Also we note Paul Ernest's (UK) positions as adjunct professor at Sør-Trøndelag University College and the University of Oslo. In this respect it is appropriate to mention Barbro Grevholm (Sweden) who settled for a prolonged engagement at the University of Agder. Her contribution both in leading the doctoral program at the University of Agder and her role in leading the Nordic Graduate School in Mathematics Education has to be recognised. Also, we note, the appointment of Professor Birgit Pepin at Sør Trondelag University College.

There are opportunities for developing international links and collaboration; indeed, this is strongly encouraged within the country and an important criterion used to evaluate research proposals. Nevertheless, strong leadership needs to come from within the Norwegian community; we cannot look for it from outside. Leadership is not merely about providing a sense of direction that can be based upon a deep knowledge of the scientific field and the local, national context. Leadership is also about setting an example regarding scientific activity and output and demonstrating the commitment to the scientific enterprise. Furthermore, good leadership will be astute in spotting fresh talent and nurturing that talent by seeking out opportunities for development. To create a strong leadership for research in mathematics education in Norway, one of the challenges seems to be to find effective ways for researchers and institutions to cooperate. With a large number of fairly small institutions, it may be appropriate for each institution to specialize in some part of research in the field, such as employing different types of methodology, focusing on different levels in school, or taking different perspectives on practice.

It has been noted that many of the senior researchers who are currently active will be due for retirement within a few years. Will those who have recently completed their doctoral research be ready to fill the places left? There are some opportunities for post-doctoral research positions that open the possibility for rapid development to levels of international excellence, but, as in most educational research environments, people carry substantial teaching and administrative responsibilities. The problem may be heightened because so much reliance is made on foreign researchers who are unable to make a realistic contribution to initial teacher education because of the language; thus more demands are made on the upcoming Norwegian generation of researchers.

As observed above, mathematics education research in Norway is still in a developmental phase and has yet to make a significant impact internationally. This present volume provides an opportunity to present research activity to the international community, but we are aware that we still have some way to travel. There are some individuals who have made or are presently making a contribution to mathematics education research on an international scale—some of these have contributed to this section of the Sourcebook or have been referred to in some detail. The hope is that research in mathematics education will continue to develop and begin to be recognized as significant within the international community. It is a problem that international research in the field is heavily influenced by English-speaking countries, and that researchers in these countries appear to neglect research from other countries (Sowder, 2007). However, the main responsibility is on Norwegian researchers to develop the national output from mathematics education research by lifting the quality of research and publishing more articles internationally.

We want to end on an upbeat note. The community of mathematics education researchers is inspired by the effort of the generation of practitioners and researchers who have built up the field—a generation now nearing or recently past retirement. Those of us who continue in the field have a responsibility to build on the secure foundation laid for us, and we will! Norwegian research in mathematics education will continue to be a developing field, and there is every reason to be optimistic.

REFERENCES

Borgen, J. S., Gunnes, H., & Sivertsen, G. (2007). Kartlegging av norsk utdanningsforskning [Survey of Norwegain Education Research]. NIFU STEP Rapport 32/2007. Oslo, Norway: NIFU STEP Norsk institutt for studier av innovasjon, forskning og utdanning http://www.nifustep.no/norsk/publikasjoner/kartlegging_av_norsk_utdanningsforskning

KKD. (2009a). *Læreren Rollen og utdanningen: St.meld.nr.11* [The teacher the role and the education: Government white paper nr. 11] (2008–2009). Oslo, Norway: Det Kongelige Kunnskapsdepartement. http://www.regjeringen.no/nb/dep/kd/dok/regpubl/stmeld/2008-2009/stmeld-nr-11-2008-2009-.html?id=544920

KKD. (2009b). *Klima for forskning: St.meld.nr.30* [Climate for research: Government white paper nr. 30] (2008–2009). Oslo, Norway: Det Kongelige Kunnskapsdepartement. http://www.regjeringen.no/nb/dep/kd/dok/regpubl/stmeld/2008-2009/stmeld-nr-30-2008-2009.html?id=556563

NOU (2008). *Sett under ett: Ny struktur i høre utdanning* (Stjernø report) [Seen as one: New structure in higher education]. Oslo, Norway: Kunnskapsdepartentet. http://www.regjeringen.no/nb/dep/kd/dok/nouer/2008/NOU-2008-3.html?id=497182

RCN (2009). Work programme: Norwegian educational research towards 2020 (Utdanning2020). The Research Council of Norway. http://www.forskningsradet.no/servlet/Satellite?c=Page&cid=1224697819042&p=1224697819042&pagename=utdanning%2FHovedsidemal

Niss, M (1998). *Aspects of the nature and state of research in mathematics education.* Tekster fra IMFUFA. Roskilde Universitetscenter. http://dspace.ruc.dk/bitstream/1800/3396/1/IMFUFA_351.pdf

Sowder, J. T. (2007). The mathematical education and development of teachers. In F. Lester Jr (Ed.), *Second handbook of research on mathematics teaching and learning. Volume 1* (pp. 157–223). Charlotte, NC: Information Age.

SECTION II

Swedish Research in Mathematics Education

Christer Bergsten, *Section Editor*

18

MATHEMATICS EDUCATION RESEARCH IN SWEDEN

An Introduction

Christer Bergsten

LINKÖPING UNIVERSITY

Keywords: Mathematics education, Swedish research, overview, theoretical orientations, research areas

Any survey of research, claiming to present the "state of the art," remains by necessity a contemporary document and a victim of selection principles. This is especially important to point out in a period of rapid expansion of the field in focus, which indeed is the case with mathematics education research in Sweden. During the time of writing the chapters in this section, new developments of research have widened the range of theoretical approaches as well as areas of interest. In this introduction to the Swedish section, therefore, space will be given not only to introduce the different chapters but also to account for some of these new developments that promise to significantly strengthen the Swedish landscape of mathematics education research, as well as to some previous research not covered by the chapters. First, a short glimpse of the development of school mathematics in Sweden is given, with occasional reference to the emergence of mathematics education research. The three subsequent sections are organized by the same principle as the chapters on Swedish research presented in this volume: some of the theoretical orientations that are employed by groups of Swedish researchers, some of the major areas of research, and some examples of recent studies. Naturally, the content of these sections overlaps, and the headings should not be read as analytical categories for classifying research.

THE DEVELOPMENT OF SCHOOL MATHEMATICS IN SWEDEN

Mathematics teaching is commonly categorized, in different kinds of publications, as "traditional" or "innovative," often without offering very clear descriptions of what is actually happening in the classrooms in terms of responsibilities for knowledge development, social relations, and the structuring of content in the activities. It has thus been claimed that before the time of the New Math there had been very little change in the way mathematics was taught at schools in Sweden (Magne, 1986). The teacher showed the students a mathematical technique (such as two-digit multiplication), using explanations and visual tools, and the students worked, under the guidance and help from the teacher, through a number of tasks to practice this technique from a book of exercises, which was the only textbook. This "teaching by rules" was considered the most efficient method for mathematics teaching in the upper grades of compulsory

The First Sourcebook on Nordic Research in Mathematics Education, pages 269–282

school (Magne, 1986). In the lower grades, the "heuristic method" was considered to be more adapted to the level of thinking of young children: the students should first get tasks sufficiently simple to enable them to solve them without instruction, with their own methods, then, by systematic training through tasks with increasing difficulty, develop mathematical skills (Bergsten, 1939). Both these methods were rooted in the strong Swedish tradition of elementary mathematics teaching to keep the learning activities close to the reality of the young child, with an emphasis on visual perspicuity by using concrete materials and visual tools to make children develop mathematical knowledge through their own activities (see, e.g., Wigforss, 1952). The ideas behind the New Math were therefore strongly criticized (Lindström, 1968) already before its curriculum was implemented in 1969. The national curriculum from 1962 had introduced a nine years compulsory school common for all. Widespread, large-scale projects with an emphasis on individualized teaching within the class that took place in the 1960s added to the many changes in school mathematics during this period (Larsson, 1973). The unsatisfactory Swedish mathematics results, as measured in comparative international achievement studies by the Association for the Evaluation of Educational Achievement in 1980 (Murray & Liljefors, 1983), fostered intense attention on continuing education of mathematics teachers in compulsory school. The emergence of the electronic calculator during this time contributed to the discussion about students' apparently weak computational skills (Hedrén & Köhlin, 1983). In parallel to these curriculum reforms, and the subsequent counter-reforms in 1980 (related to the "back to basics" movement) and 1994 (with more emphasis on mathematical reasoning and communication), research in mathematics education has seen a strong increase in quantity and scope. Looking back on the development of mathematics education in classrooms in Sweden during the last four decades, one gets the impression of a search for a lost identity, witnessed by survey reports describing the dominating practice in Swedish classrooms as weakly framed and textbook governed, leaving an increased responsibility to the student to recognize criteria for producing legitimate knowledge (e.g., Kling Sackerud, 2009; Skolverket, 2003). A consequence of this practice seems to be an increased stratification of achievement (Skolverket, 2009a) that could be interpreted as a stabilization of schooling as an instrument for social reproduction. The recently initiated and ongoing knowledge-centered curriculum reform in Sweden, with an elaborated assessment and grading system, only supports this impression. During the last years, large state resources have been spent on intense activities with continuing education of mathematics teachers in school.

To understand the role of research for the development of educational practice in mathematics in Sweden, one needs to consider the distribution of labor between the different actors involved in mathematics education. There is a comparatively well established relation between the community of researchers and the school authority in Sweden. Skolverket (National Agency of Education) has issued several reports on a range of topics concerning teaching, learning, and evaluation of education, some with a focus on mathematics (Skolverket, 2003, 2009a, 2009b, 2009c). Skolverket is an independent executive state institution directed through regulations from the Ministry of Education. About 10 years ago, a National Center for Mathematics Education (NCM) was established by the government. It is affiliated with Göteborg University, with its main task to support, as a resource center, the development of Swedish mathematics education in preschool, school, and adult education. It is only in recent years that the field of mathematics education as a research domain has been established at several universities (see also below). In an exploration based on Bernstein's (2000) conceptualization of pedagogic identities, Johansson (2010) describes positions that influence, to different degrees, the official and unofficial pedagogic discourse in the recent reform initiatives in Sweden, on for example what "should be done" with mathematics to improve performance measures. The power or the empowerment of an institution, within the reform, is determined by the state or, indirectly, through other positions in the official pedagogic arena. A reform is then seen as a result of the struggle in the arena in which the different pedagogic identities try to turn their bias and focus into state policy and practice. In the diverse enterprises that constitute the current mathematics reform in Sweden, several identities are influential. Skolverket, even though a state agency, works as a player in the market and chooses affiliates (institutions or individuals) who are then empowered and influence the reform. The main "co-drivers" and "performers" in the reform are the colleges, the universities, and the national resource centers (for example, the NCM). By giving support and advice to the authorities, in the next phase, they could be commissioned to act in the suggested direction. In the current reform initiatives, these institutions seem to have to compete with each other to get their share in the official pedagogic arena. Consequently, the role of academic research in the process is ambivalent.

THEORETICAL ORIENTATIONS

As a field of research, mathematics education has become inhabited by a plethora of theories of different origin, scope, range, and purpose. This has been described in terms of an increased plurality or diversity of theories seen as a problem, a resource, or a challenge to the field (Lerman, 2006; Prediger, Bikner-Ahsbahs, & Arzarello, 2008). A key issue of concern in this discussion is the notion of theory itself. A static definition, as for example the one given by Niss (2007), describing a theory as a hierarchically ordered structure of well-defined concepts and claims about some domain, is weak in that it does not account for the dynamics of the research process and the relation to the empirical. These critical dynamic aspects of a theory are captured more adequately by Radford's (2008) proposal to conceive of a theory as a triplet of three interrelated components: a system of basic principles, a methodology, and a set of paradigmatic research questions. In order to distinguish a theory from a systematic empirical description, Moore (2006), in elaboration of Bernstein, requires from the former to be able to produce descriptions of instances of a category of phenomena not yet observed.

Thus, when researchers employ terms such as "a theory of" or "a theoretical model of," they do not always refer to the same conceptualization of what constitutes a theory, and it is up to the reader to critically judge if and in what sense the claims made deserve to be called theoretical. Such an endeavour would be facilitated by explicitly embedding the work presented in its intellectual and theoretical roots, or by contrasting it to alternative languages of description.

Considering mathematics education research as a field of knowledge production, it has, according to Lerman (2006), a "horizontal knowledge structure": "horizontal knowledge structures consist of a series of specialised languages with specialised modes of interrogation and criteria for the construction and circulation of texts" (Bernstein, 1999, p. 162). Rather than by way of an integrated progression of accumulated knowledge, knowledge expands by inserting new domains alongside existing ones without replacing these (for a discussion, see Jablonka & Bergsten, 2010). Such a development is also visible in the Swedish community of mathematics education research, where one can find groups of researchers working within different theoretical orientations of research, more or less isolated from each other, and contributing predominantly to the development of the group's shared approach, as well as individuals adhering to other frameworks or working within a range of theoretical orientations. For example, one can find groups of researchers that work within phenomenography and variation theory, with a focus on how a mathematical "learning object" is experienced by the students in its different varying manifestations, in particular regarding which aspects are discerned and which are not (see Chapter 20, this volume, by Runesson and Kullberg); within the theory of contextualization, employing intentional analysis, attempting to take conceptual, situational, and cultural aspects of students' conceptions of a given task into account (see Chapter 21, this volume, by Ryve, Scheja, and Wistedt); and within the framework of "imitative and creative reasoning" (see Chapter 19 in this volume by Lithner and Palm). The latter can also be seen as research on students' learning difficulties but categorizes itself as a research framework that has been applied to task analysis, students' task solving, textbook analysis, and assessment. Recently, research drawing on sociological theorizing in education, using frameworks that allow an integrated study of macro and micro levels of mathematics education, set out to analyze the emergence of disparity in achievement in mathematics classrooms, in particular with a focus on how teachers facilitate the students' access to the knowledge criteria (e.g., Jablonka, Johansson, & Rohdin, 2010). This research fits well into an established tradition of empirical classroom research in Sweden (see below).

One also finds individual researchers using a more traditional cognitively oriented psychological approach as well as research with socio-cultural and semiotic orientations (as examples of these approaches, see the chapters by Eriksson, Hemmi, Juter, and Österholm). Several researchers have used Vygotskyan approaches, for example, to study problem solving (Wyndhamn, 1993), students' context-dependence in task solving in classrooms (Säljö & Wyndhamn, 1990), changes between everyday and specialized discourses in the primary mathematics classroom (Riesbeck, 2008), and the use of computers in lower secondary mathematics education (Lantz-Andersson, 2009).

Some Swedish researchers have also pursued the issue of theorizing in mathematics education (Bergsten, 2008; Bergsten & Jablonka, 2009; Jablonka & Bergsten, 2010a, 2010b) and historical-critical analyses of the development, character, and role of (school) mathematics in society (Lundin, 2008; Mouwitz, 2006).

AREAS OF RESEARCH

In this section, some major areas of Swedish mathematics education research are briefly reviewed, organized under the following subheadings: low achievement, assessment, textbooks, classrooms, teacher education, ICT, tertiary mathematics education, and "other areas." These areas naturally cut across the different theoretical orientations discussed in the previous section. Some extra space will here be devoted to research on the use of ICT in mathematics education, as this area is not well represented in the chapters.

Low Achievement

One important role of research is to explain differences in achievement in school mathematics. This amounts to finding appropriate approaches to support high achievement for all students as well as identifying those differentiation mechanisms that influence students' study trajectories that determine their future profession and position in society. For example, there is still a considerable influence of the educational level of their parents on students' grades in compulsory school in kernel subjects such as mathematics, amounting to differential opportunities for entering secondary education (Skolverket, 2009a). Also at upper secondary level, a strong influence of social background factors on student achievement was observed in the TIMSS Advanced 2008 study (Skolverket, 2009b). According to these data, there was in addition a change in mathematics achievement over time with significantly lower results in 2008 as compared to 1995. That such negative developments had not been observed in the national tests during the same period, despite the good match of the TIMSS tasks with the national curriculum and tests (Skolverket, 2009c), points to the necessity of analyzing the underlying assumptions and constructions of tests and grading in both the international and the national testing "packages" (see, for example, Clarke, 2003; Jahnke & Meyerhöfer, 2007; Keitel & Kilpatrick, 1998).

The study by Sjöberg (2006) highlights the complexity of the problem of low achievement in school mathematics in general, and of the notion of "dyscalculia" in particular. Data from an extensive empirical longitudinal study show that the educational setting and environment, including the critical role of the teacher, lack of students' engagement in the classroom, large classes, stress and anxiety prior to tests, and gender issues, have to be seen as exerting a stronger influence on achievement than individual student characteristics. In the chapter by Eriksson (see also below), a psychological approach was used for studying low-achieving children.

The chapter by Engström and Magne provides an overview of Swedish research on low achievement in mathematics in compulsory school, from the first systematic studies of arithmetic disorder by Henschen, who coined the term "acalculia" and demonstrated the complexity of mathematics as a cognitive construct, to the Middletown Studies carried out almost 90 years later. The overview shows a change in focus from students' "disabilities" to low achievement in mathematics. Magne's extensive Gothenburg Studies in the 1950s, with 6000 students involved, were followed by The Mathematics Clinics project, testing special education methods in mathematics. The recent Middletown Study is an investigation of students' mathematical achievement in the compulsory school system of a Swedish municipality during a period of 25 years.

Assessment

In Sweden, national assessment in school mathematics has a strong tradition since the seminal work of Wigforss, who introduced standardized tests to facilitate the work of teachers to give comparable grades in mathematics to students all over the country (see Kilpatrick & Johansson, 1994). National testing still prevails in the new criterion-referenced system that in the curriculum reform in 1994 replaced the previous norm-referenced grading. The purpose to increase the comparability of grades across the country still guides the work, but results are also used to evaluate the educational outcomes nationwide, as well as locally (for example, to compare schools). The tests for the upper secondary level have been shown to have a lower influence on the practice of teachers' own (local) test constructions than expected, the latter remaining more "traditional," not testing the wide range of mathematical competencies described in the official curriculum and operationalized in the national tests (Boesen, 2006). It should be noted that national tests have been obligatory only at the end of compulsory school (one test in grade 9) and at the end of the first and last mathematics course taken at upper secondary school. However, in recent reforms one can observe a change towards an increased number of national tests in mathematics and several other school subjects.

Recent Swedish research on assessment in mathematics has been carried out mainly at the two institutions responsible for the production and processing of the national tests: the Department of Educational Measurement at Umeå University and the PRIM-gruppen (PRIM stands for "Prov i matematik," which means "Testing in Mathematics") at Stockholm University. Studies at Umeå University have focused on secondary school mathematics assessment and competency frameworks, in addition to analyses of TIMSS results (e.g., Skolverket, 2009b). The chapter by Pettersson and Björklund Boistrup reports from the work by the PRIM group, focusing on both formative and summative assessment; providing the students with opportunities to show as many aspects as possible of their mathematical knowledge has been a major concern of the group. As an example of a tool to also support teachers in their work, a task-specific assessment matrix has been developed for oral group tasks, where three different qualitative levels of knowledge are described for the dimensions of understanding, language, and participation, aimed to provide a holistic assessment of the students' performance on the task. Seeing assessment as communication, a tool for the analysis of tasks that draws on a multimodal social semiotic framework uses Halliday's distinction between interpersonal, ideational, and textual meaning, as well as their combination in the mathematical discourse (Björklund Boistrup, 2007). The chapter also points out that work on national assessment is not isolated from other research in mathematics education, but provides important contributions, for example, to the study of how students solve mathematical tasks.

Textbooks

Students' individual work with textbook tasks is a dominating practice in mathematics classrooms in primary and secondary education in Sweden (Kling Sackerud, 2009; Skolverket, 2003), with only little space given to teacher interventions, group work, student presentations, and other forms of interaction. Considering the goals for the study of mathematics as described in the present national curriculum, this fact is surprising. It has also been suggested that this mode of working has contributed to the perceived decline in mathematics performance. Particularly from this perspective, studies on mathematics textbooks and their use in classrooms are highly relevant. The chapter by Jablonka and Johansson offers a structured overview of Swedish textbook research by looking at textbooks as a window into classroom practice and as a window into students' struggle with the construction of mathematical meaning. The research points out that textbooks and their use have a constraining influence on reforms, views of mathematics as an academic discipline, transition processes, students' opportunities to engage in more demanding tasks, and the possibility for identity construction for specific groups of students.

Classrooms

In Sweden there exists a considerable tradition of empirical classroom research with an interest in the social constitution of classroom learning. Within this tradition, mathematics classrooms were occasionally the focus of interest. Some more recent studies include a comparative perspective in studies of Swedish mathematics classrooms. These studies are also characterized by the employment of more sophisticated technology in data generation (e.g., Emanuelsson & Sahlström, 2006).

Some early studies attempted to link linguistic analysis of classroom interaction on the micro level with curriculum analyses with an interest in the reproduction of social inequalities (Lundgren, 1972). Gustafsson (1977) investigated the constitution of students' pedagogical identities. She found that independently of their perceptions of the students' experiences and pre-requisites, the teachers generally interacted with all students in the same way. Her empirical material included mathematics lessons in the middle grades.

More recent studies convey a more dynamical view of classroom interaction that leaves more space for students' contributions. The more open framing cannot always be seen as productive in terms of the development of mathematical knowledge (Emanuelsson & Sahlström, 2008; Liljestrand, 2002; Sahlström, 2008). Classroom studies have also been conducted, for example, from the perspective of variation theory (see the chapter by Runesson and Kullberg). Based on a socio-cultural perspective, Säljö and Wyndhamn (1990) carried out an often-cited experiment in which students had to deal with a contextualized mathematics task involving a table with postal charges in a mathematics classroom and in a social science classroom. The students reacted differently to the task in those two settings.

As an example of an ongoing classroom study, which is part of the international project "The emergence of disparity in mathematics performance" that investigates how discursive and interactional mechanisms affect stratifications of achievement within the classroom (see web page at www.acadiau.ca/~cknippin/sd), Jablonka, Johansson, and Rohdin (2010) employ Bernstein's (e.g., 1990) concepts of classification and framing for analyzing the practices during the very first lessons of two upper secondary mathematics classes. To respond in a legitimate way to the tasks given, the students need to have access to the instructional and regulative classroom rules. Differences in these rules can be described as variations of the type of knowledge that can be accessed and of the ways this knowledge is made accessible. The underlying principles can be captured by the strength of external and internal classification and framing. A strong or weak classification refers to the strength of the boundaries established between, for example, an everyday and a highly specialized mathematical discourse, while framing captures how strongly the transmission and the acquisition of this discourse are regulated in the pedagogic context. The first results from the study point to a weak framing and an implicitness of the classification criteria in both classrooms. The instructional discourse seems to be delegated to the textbooks and the teachers act mainly as a coach. As a consequence, students' success "is likely to depend largely on study habits and behaviour" (Jablonka, Johansson, & Rohdin, 2010, p. 10). In addition, the textbooks employed do not seem to facilitate students' access to specialized mathematical discourse, but largely operate in a "hybrid" discourse of contextualized mathematics tasks. This result echoes some outcomes of earlier empirical classroom studies in Sweden with a similar theoretical orientation.

Teacher Education

Teacher education in Sweden has undergone a number of reforms during the recent decades, partly driven by a need to better integrate subject knowledge, pedagogical knowledge, didactical knowledge, and teaching practice in the education, in order to overcome the "didactic divide" between these components (Bergsten & Grevholm, 2004). With this background, the chapter by Grevholm explores research on mathematics teacher education in Sweden and raises questions about research funding and how it has influenced research agendas. The fragmented picture of the reviewed research outcomes calls for a more systematic program for research on mathematics teacher education.

The ongoing debate on the quality of teacher education has initiated several studies on student teachers' knowledge and the effect of the education and so on. For example, E. Persson (2009) studied student teachers before and after their graduation in mathematics teacher education and observed that one year after graduation these teachers felt well prepared for teaching mathematics in preschool and primary school, in contrast to their dissatisfaction with the limited emphasis on specific recommendations provided by their education, as they expressed it immediately after graduation. This result points to the importance of longitudinal studies to evaluate teacher education.

Another example is Nilsson's (2005) study on student teachers' activities with laboratory work in compulsory school geometry. He found that such work amounted to increased knowledge and interest in geometry with many of these students, who also increased their ability to explain geometrical notions and properties to their school students.

Information and Communication Technologies—ICT

The rapid developments of digital technologies available for use in mathematics teaching and learning have been paralleled by a large body of international research on the gains and pitfalls of employing these tools in the mathematics classrooms at all educational levels, as well as more theoretically inspired analyses of the cognitive, social, and didactic dimensions of their use. In Sweden there has been a research interest in these issues since the introduction of electronic calculators in the 1970s. One can note, however, that there is comparatively little research carried out in Sweden in this area given the importance of technical aids put forward in the curriculum documents and the large financial investments over the years in computer equipment and hand-held calculators for schools. In addition, the fact that teachers generally have been hesitant to integrate modern technologies in their everyday mathematics classroom practices (Dahland, 1998) seems to support the need for developmental work and research on ICT.

In Sweden some early studies investigated on a broad basis the effects of the use of the electronic calculator in elementary mathematics teaching. In a study during the years 1977–1982, called the RIMM project (Hedrén & Köhlin, 1983), seven classes were studied during their years in the Swedish "middle grades" (grades 4–6). Using a design involving experimental and control groups, it was clearly shown that the experimental groups, where the calculator was consequently used, showed the same ability as the control groups to carry out arithmetic calculations, mentally as well as with pencil and paper, but showed significantly better results on number estimation, on choosing the correct operation, and on using relevant information when solving word problems. The explanation given for these results was that the calculator reduced the time needed for performing the necessary calculations and increased the amount of time spent on problem solving. This idea also formed the basis of the large-scale longitudinal developmental project ADM ("Analys av Datorns konsekvenser för Matematikundervisningen," or "Analysis of the consequences of the computer for mathematics teaching"), investigating the use of calculators, computers, and, later, graphic calculators in secondary school (Björk & Brolin, 1995). The project started in 1976 with a study of the use of calculators. When compared to a control group, the classes involved in the project showed significantly better results on conceptual understanding and problem solving, while no difference was reported for basic routine skills.

The influence of the use of calculators in compulsory school on students' problem solving abilities was investigated earlier by Ekenstam and Greger (1983), who emphasized the importance of what they called "non-algorithmic basic skills" in mathematics learning. They claimed that purely algorithmic routine skills could be done by the use of calculators but noted that some skills learned through performing routine skills may then be lost. The study by Dahland and Lingefjärd (1996) on the use of graphing calculators during exams pointed to the need of technical insight and good mathematical understanding to be able to take advantage of the tool. Students' use of graphic calculators to graph a quadratic function along with the two linear functions of its factorization was studied by Bergqvist (1999) to investigate the benefits of this activity for the understanding of the factor theorem. In addition to some observed positive benefits, through analyzing students' investigations by means of the triplet "visualisation–conjecture–verification," the study also showed the students' improved understanding of the relations between graphical and algebraic representations of functions. In a recent meta-study, P-E. Persson (2009) surveyed research literature on the educational use of hand-held calculators, with a specific focus on the impact of algebra learning, and concluded, though acknowledging that things are not unproblematic, that "calculators can be powerful cognitive tools for the enhancing of students' skills and understanding of mathematics, and in particular algebra" (p. 70). Instead of causing decreased manual or mental skills, they activate students and support motivation.

At the school level, some research on the use of specific computer software has been carried out in Sweden. Using an experimental design to investigate middle grade students' work with Logo in small groups, Hedrén (1990) found that such investigative activities supported the students' conceptual learning both in arithmetic and geometry. In addition, the students developed a more open attitude to mathematical problem solving, provided that the teacher, instead of being too instructional, allowed the pupils the freedom to explore.

Lingefjärd (2000) investigated student teachers' understanding of mathematical modeling when they had access to computer technology during their problem-solving activities. He found, among other things, that technology acted as a tool as well as a source of authority in the mathematical modeling process. During the first weeks of the course, by a kind of transformation of authority, the students became uncritical of the computer and graphing calculator results despite the emphasis made during the course to be critical when using the software.

Using classroom observations and interviews, Engström (2006) studied how teachers in Sweden and Switzerland used the dynamic geometry software Cabri. She concluded that in addition to good knowledge of mathematics and the potentials of the software, the ability to pose problems that are both open-ended and thought-provoking is needed in order to influence students' learning through working with the software. The use of dynamic geometry software in education has also been studied by Lingefjärd and Holmquist (2003).

Samuelsson (2003) discusses the influence of the computer as a change agent on the conditions, methods, and results in mathematics teaching. By participating in all computer-aided lessons given by two teachers during one year and interviewing 18 teachers in lower secondary school, the author found that the computer was assimilated into traditional views of methods and content by different types of drill programs where students can learn mathematical procedures. For laboratory work, with the intention to learn mathematical concepts, the speed of the computer was seen as an important condition. This way of teaching mathematics thus seems to have such a strong tradition that the influence of the computer as a change agent is relatively weak.

The use of ICT in tertiary mathematics education has not been well researched in Sweden. That the use was widespread by the end of the 1990s was evidenced by a national questionnaire survey by Bergsten (1999a), showing that computer support was involved in about half of all mathematics courses at Swedish universities, mainly as computer labs added to the traditional teaching formats. Computers were used in the courses mainly because of a combination of pedagogic and pragmatic reasons, and for informing about available software, while arguments given against the use included lack of resources, lack of suitable software and working formats, and ineffectiveness. The most commonly used software was MATLAB® and computer algebra systems (CAS). At a national conference, organized by the National Agency for Higher Education, a large variety of ICT use was presented (see Bergsten, 1999b). In an early study on CAS, Bergsten (1993) found that students in a first-year calculus course appreciated the graphing tool linked to the algebraic representations—for example, to visualize potentials and limitations of Taylor polynomials of increasing degree, such as to approximate trigonometric and logarithmic functions over different intervals—but were more hesitant to employ the automatic algebraic commands for differentiation, integration, or equation solving. They saw it as the computer doing the job the students themselves should do.

Tertiary Mathematics Education

One of the outcomes of the National Graduate School for mathematics education during the years 2001–2006 (see the chapter by Brandell) is an increased interest in research dealing with mathematics in higher education. While rare in Sweden during the 20th century, a number of recent studies have dealt with some problematic areas of tertiary mathematics education. Having always been a topic of research interest, the secondary–tertiary mathematics transition gained public attention during the 1990s due to a significant lowering of pass rates nationwide in introductory mathematics courses in engineering education, marked by a conference organized by the National Agency of Higher Education in 1999 (HSV, 1999). Brandell, Hemmi, and Thunberg (2008) studied the transition problem by analyzing the mismatch in mathematical content and approach in upper secondary and tertiary mathematics courses, with a particular focus on how the notion of proof is handled, relating their results to the school reforms and the expansion of higher education. In a recent doctoral study, Stadler (2009) reconstructed the student perspective in analyzing classroom activities and interviews of six student teachers in introductory mathematics courses in terms of three categories emerging from the data in a grounded theory approach. These were "mathematical learning objects" (learning content and goals as conceptualized by the students), "mathematical resources" (such as teachers and textbooks), and the "student-acting-as-learner." The results describe the transition as an increasing mismatch during their studies between the first two categories.

Other recent studies on tertiary mathematics education have employed cognitive, contextual, socio-cultural, and epistemological approaches, investigating students' conceptions of the notion of limits of functions (see Juter, this volume, Chapter 29; Bergsten, 2008), student teachers' understanding of the function concept (Hansson, 2006), students' discourse in a linear algebra concept mapping exercise (Ryve, 2004), the development of understanding calculus in an algorithmic context (Pettersson & Scheja, 2008), the issue of quality of undergraduate mathematics lectures (Bergsten, 2007), and proof (see Hemmi, this volume, Chapter 28). In a study of the didactic transposition of proof with a focus on the Fundamental Theorem of Calculus, including a historical study as well as interviews with mathematicians, Klisinska (2009) showed how the "adaptation" of mathematical knowledge for the purpose of teaching also had an influence on the field of production of knowledge, as it fostered standardization of the integrated body of knowledge called calculus. The mathematicians' views on what counts as a proof and how to teach this topic to different groups of students was influenced by their research and teaching background. Thus, the didactic transposition of a specific body of knowledge cannot be described as a linear process from scholarly to taught knowledge.

Some Other Researched Areas

There has been a continuing interest in Sweden in research on early mathematics learning, discussed in this volume by Eriksson and shortly in the chapters by Runesson and Kullberg. As another example, the study by Ahlberg (1997) shows the complexity of early number experience. Preschool children were interviewed on everyday problems, decomposition problems, and contextual problems, not allowed to use any manipulatives. Interview outcomes were classified into a number of main categories, in line with the phenomenographic approach, and were analyzed under the main headings "ways of handling numbers" and "ways of experiencing numbers." The result showed that there was no simple relationship between the way children handle numbers and the way they experience them. The identified categories of experiencing numbers all came into play, more or less, when children handled numbers by counting, structuring, and using known facts. The manifold interplay of different dimensions deepens the understanding of the complexity of early numeracy development.

As examples of studies on components of mathematical knowledge, one finds the powerful factorial study by Werdelin (1958) on "mathematical ability" based on intelligence factors, and the more recent work by Bergsten (1990) on "mathematical operativity," a construct describing the dynamic interplay between four dimensions of mathematics: form and content, on the one hand, and structure and operation, on the other. One also finds research on mathematical modeling, often connected to the use of ICT, such as the studies by Lingefjärd (2000, 2006), Holmquist and Lingefjärd (2003), and Wikström (1997). Årlebäck (2009) studied students' solving of Fermi problems as a way to introduce mathematical modeling, and developed a "modeling activity diagram" to analyze video-recorded empirical data on students' modeling subactivities. A questionnaire study on 400 upper secondary students' conceptions of mathematical modeling showed that the students were not very familiar with this notion (Frejd & Ärlebäck, 2010). Modeling was connected mainly with aspects of problem solving and with applying mathematics in extra-mathematical situations, and seen as a constructive and creative activity. Mathematical models were conceptualized mainly in terms of mathematical formulas.

SOME RECENT STUDIES

As indicated above, there has been a quantitative "explosion" of research studies in mathematics education in Sweden during the last two decades compared to the very few studies that were carried out earlier. Instead of attempting to cover this present flow in a comprehensive overview, this volume includes four elaborated descriptions as examples of recent studies carried out by Swedish researchers. In addition, the chapter by Brandell presents an outline of the development and the role of a National Graduate School, which was in place during the years 2001–2006 and provided a significant input for mathematics education research in Sweden for the new millennium. According to Brandell, this accounts for a widening of traditional research areas and institutional engagement as well as an increased international visibility of the research.

With the aim to examine children's conceptual development in early arithmetic, Eriksson conducted investigations in a compulsory school and in a special school for students with special needs. Multiple clinical interviews, based on the epistemology of radical constructivism, were used. Various behavior patterns that emerged when the participants solved tasks in their own ways are exemplified in the chapter. An analysis of the underlying conceptions and a comparison shows that the patterns are compatible between children at the same conceptual level, irrespective of age or type of school.

In her chapter, Hemmi presents a framework for the study of proof in mathematics education. The framework combines a socio-cultural perspective and the social practice theories of Lave and Wenger with theories and main themes in mathematics education research concerning the teaching and learning of proof. The components of intuition/formality, visibility/invisibility, and deduction/induction are seen as properties of proof and how to approach proof, and illuminate the dynamic character of proof as reification, as both a process and a product. The framework was developed and tested on data from students' encounters with proof at a university in Sweden.

With a focus on cognitive aspects, Juter studied how students of undergraduate mathematics perceive limits of functions and what pre-knowledge they bring to university courses in calculus. Several of the students' concept images were incoherent, often due to theoretical misconceptions. Intuitive perceptions influenced, in some cases, students' pre-knowledge and fostered establishment of invalid links between concepts.

Österholm examines in his chapter the roles of different types of prior knowledge for the comprehension process when reading mathematical texts. For three related aspects (i.e., cognitive structure, cognitive process, and metacognition) questions arise regarding relationships between general and content-specific types of prior knowledge. In the empirical studies reported, different types of texts were used, and the lack of correlation found between reading comprehension for different types of texts shows that reading ability does not seem to be a homogenous construct. The main difference does not appear to be between texts from different domains, but other properties of texts, such as the use of symbols in mathematical texts, seem more relevant.

DEVELOPMENTS

With only a few active senior researchers in mathematics education until recently, much of the research in Sweden has been, and still is being done, as doctoral work. To describe Swedish research, an overview of the doctoral theses produced in Sweden has therefore to be considered as a significant part of the full picture. The first and most striking observation from such an overview refers to the quantitative aspect: before 1970 there were only five theses in the field of mathematics education presented in Sweden, while there was a strong increase from 1990 onwards, with 56% of all theses presented only during the last decade (see Table 18.1).

This recent "explosion" within mathematics education is stronger than the general increase in the number of PhD theses in Sweden over all topics (a little more than doubled from 1990 to 2005; see, e.g., Bergsten, 2006) and must be seen in relation to the high visibility in the public educational and political debate of the role and state of mathematics in education and society (see also the chapter by Brandell about the possible influence of the National Graduate School on

TABLE 18.1 Number of Theses in Mathematics Education in Sweden per Decade During the Last Hundred Years

Years	Number of PhDs
1910–1919	1
1920–1929	0
1930–1939	0
1940–1949	0
1950–1959	2
1960–1969	3
1970–1979	5
1980–1989	5
1990–1999	18
2000–2009	43

Source: list of doctoral theses provided by the Swedish Center of Mathematics Education at their web site at <ncm.gu.se>

this increase). That there has been and still is a perceived need of personnel with a research background in, for example, teacher education (see the chapter by Grevholm) and at educational authorities, as well as in schools, is another explanation for this "inflation," which may evoke a critical discussion about quality criteria in mathematics education research (see, e.g., Editorial, 2009). It should be noted here that chairs in mathematics education at Swedish universities have been appointed only during the last decade and only at five universities, contributing to the establishment of the field as a new academic discipline.

Table 18.2 outlines an attempt to provide an overview about the content of the doctoral theses, adding to earlier reviews by Björkqvist (2003) and Bergsten (2002, 2006). The categorizations of the content areas that appear in the table were made by the author by simply grouping the main topics of the theses (four of the theses are missing). Though several issues are often raised in a thesis, each thesis was classified as belonging to only one category, judged to be its main focus.

The table shows that there is not one or only a few dominant levels of education or content areas that have been the main focus of Swedish research in mathematics education; a rather even distribution over most of the categories listed can be observed (apart from adult education and a low representation on belief research). However, when adding the time dimension, one finds a focus on higher educational levels as well as on mathematical content (MC) during the last decade (cf. the chapter by Brandell), categories that were only rarely represented earlier. A deeper analysis would be needed to see possible trends regarding for example the distribution of different theoretical and methodological approaches used (for comments on these issues, see Björkqvist, 2003; Bergsten, 2002, 2006).

While reading about these research studies based on doctoral work, it should be kept in mind that recently there has also been a considerable increase in post-doctoral research as well as an initiative from the state to provide research education to practising teachers. In addition, several larger research projects funded by Vetenskapsrådet (Swedish Research Council), and other funds are currently conducted or are in the process of starting. The increase of senior research parallel to the one at the level of PhD studies indicates that mathematics education is becoming an increasingly more mature field of academic research in Sweden, even though research at an international level is still sparse.

In addition to academic work, much developmental work, policy overviews, and surveys of students' achievement and attitudes have been carried out by or through state educational authorities such as

TABLE 18.2 Educational Level and Content Area of Doctoral Theses in Mathematics Education Presented in Sweden 1919–2009 (number of theses of each category)

Level/Content area[a]	MT	PS	ICT	MC	C	B	CR	T/A	Total
Primary	2	3	2	6	1	0	1	1	**16**
Lower Secondary	3	3	1	1	2	0	3	2	**15**
Upper Secondary	1	1	3	1	0	1	1	1	**9**
Tertiary	3	0	0	5	1	1	0	1	**11**
Teacher Education	0	1	2	2	1	2	1	0	**9**
Adult Education	0	0	0	0	0	0	1	0	**1**
Mixed	1	0	0	0	5	0	1	5	**12**
Total	**10**	**8**	**8**	**15**	**10**	**4**	**8**	**10**	**73**

[a] MT = Mathematical Thinking; PS = Problem Solving; ICT = Information and Communication technology; MC = Mathematical Content; C = Curriculum; B = Beliefs (in a wide sense); CR = Classroom; T/A = Textbooks, Assessment

Skolverket (National Agency for Education; see web page at skolverket.se) and Högskoleverket (National Agency for Higher Education; see web page at hsv.se). Considering the small quantity of academic research done in Sweden before 1990, these often large studies, most likely, have had a much stronger influence on the development of educational practice than national (local) research, in addition to the influence of international trends in educational policy and results from comparative achievement studies. The provision of large financial resources by the government for small but numerous research and developmental studies placed directly within schools adds to this picture. It therefore remains to see what effect the recent expansion of academic research carried out at universities may have on present and future curriculum policies and on the practices of mathematics teaching in schools and on mathematics education at teacher education institutions.

ACKNOWLEDGEMENT

The author wishes to thank Eva Jablonka, who contributed with her knowledge in the field and with critical comments during the completion of this text.

REFERENCES

Ahlberg, A. (1997). *Children's ways of handling and experiencing numbers.* Göteborg Studies in Educational Sciences 113, Göteborgs universitet.

Ärlebäck, J. B. (2009). On the use of realistic Fermi problems for introducing mathematical modelling in school. *The Montana Mathematics Enthusiast, 6*(3), 331–364.

Bergqvist, T. (1999). Gymnasieelever undersöker ett matematiskt begrepp med grafräknare [Upper secondary school students investigate a mathematical concept by using graphing calculators]. *Nordic Studies in Mathematics Education, 7,* 35–60.

Bergsten, A. (1939). *Folkskolans räkneundervisning* [Arithmetic teaching in elementary school]. Pedagogiska Skrifter, Häfte 163. Sveriges Allmänna Folkskollärarförenings Litteratursällskap.

Bergsten, C. (1990). *Matematisk operativitet : en analys av relationen mellan form och innehåll i skolmatematik* [Mathematical operativity: an analysis of the form-content relationship in school mathematics]. Doctoral thesis. Linköping: Linköping University.

Bergsten, C. (1993, September). On analysis computer labs. In B. Jaworski (Ed.), *Proceedings of the Conference Technology in Mathematics Teaching (TMT93)* (pp. 133–140). Birmingham: University of Birmingham.

Bergsten, C. (1999a). Datoranvändning i högskolans matematikundervisning. Resultat av en enkätundersökning [Computers in tertiary mathematics education: results from a survey study]. In C. Bergsten (Ed.), *Datorstödd eller datorstörd matematikundervisning?* (pp. 21–47) [Computer supported or computer distorted mathematics teaching?]. Högskoleverkets skriftserie 1999:4 S, Stockholm.

Bergsten, C. (1999b). *Datorstödd eller datorstörd matematikundervisning?* [Computer supported or computer distorted mathematics teaching?]. Högskoleverkets skriftserie 1999:4 S, Stockholm.

Bergsten, C. (2002). Faces of Swedish research in mathematics education. In C. Bergsten et al. (Eds.), *Action and research in the mathematics classroom. Proceedings from Madif2* (pp. 21–35). Linköping: SMDF.

Bergsten, C. (2006, October). Matematikdidaktikens vägar (The routes of mathematics education]. Paper presented at 'Forskarskolans Jubileumskonferens', Linköping University. Retrieved on December 1, 2009 from http://www.mai.liu.se/jubileumskonferens/11.html]

Bergsten, C. (2007). Investigating quality of undergraduate mathematics lectures. *Mathematics Education Research Journal, 19,* 48–72.

Bergsten, C. (2008). On the influence of theory on research in mathematics education: the case of teaching and learning limits of functions. *ZDM–The International Journal of Mathematics Education, 40*(2), 189–199.

Bergsten, C. & Grevholm, B. (2004). The didactic divide and the education of mathematics teachers in Sweden. *Nordic Studies in Mathematics Education, 9*(2), 123–144.

Bergsten, C. & Jablonka, E. (2010). Interpreting students' reasoning through the lens of two different languages of description: integration or juxtaposition? In V. Durand-Guerrier, S. Soury-Lavergne, & F. Arzarello (Eds.), *Proceedings of the Sixth Congress of the European Society for Research in Mathematics Education. January 28th - February 1st 2009, Lyon* (pp. 1555–1564). Lyon: Institut National de Recherche Pédagogique.

Bernstein, B. (1990). *The structuring of pedagogical discourse.* Cass, codes and control, volume IV. London: Routledge.

Bernstein, B. (1999). Vertical and horizontal discourse: an essay. *British Journal of Sociology of Education, 20*(2), 157–173.

Bernstein, B. (2000). *Pedagogy, symbolic control and identity. Theory, research and critique* (rev. ed.). Oxford: Rowman & Littlefield.

Björk, L-E. & Brolin, H. (1995). Using new technology as a tool to increase student understanding of calculus. In T.Scott (Ed.), *Proceedings of ICTMT 1995, International Conference of Technology in Mathematics Teaching.* Edinburgh: Napier University.

Björklund Boistrup, L. (2007). Assessment in the mathematics classroom. Studies of interaction between teacher and pupil using a multimodal approach. In D, Pitta-Pantazi. & G, Philippou (Eds.), *European Research in Mathematics Education V. Proceedings of the Fifth Congress of the European Society for Research in Mathematics Education* (pp. 1160–1169). Nicosia, Cyprus: University of Cyprus.

Björkqvist, O. (2003). *Matematikdidaktiken i Sverige—en lägesbeskrivning av forskningen och utvecklingsarbetet* [Mathematics education in Sweden—a description of present research and developmental work]. Göteborg: NCM.

Boesen, J. (2006). *Assessing mathematical creativity: comparing national and teacher made tests, explaining differences and examining impact.* Doctoral thesis. Umeå: Umeå University.

Brandell, G., Hemmi, K., & Thunberg, H. (2008). The widening gap—A Swedish perspective. *Mathematics Education Research Journal, 20*(2), 38–56.

Clarke, D. (2003). International comparative research in mathematics education. In A. Bishop, M. Clements, C. Keitel, J. Kilpatrick, and F. Leung (Eds.), *Second international handbook of mathematics education* (pp. 143–184). Dordrecht: Kluwer.

Dahland, G. (1998). *Matematikundervisning i 1990-talets gymnasieskola* [Mathematics teaching in upper secondary mathematics teaching during the 1990s]. Volym I+II, Rapport nr. 1998:5. Institutionen för pedagogik, Göteborgs universitet.

Dahland, G., & Lingefjärd, T. (1996). Graphing calculators and students' interpretations of results: A study of four upper secondary classes in Sweden. *Nordisk matematikkdidaktikk: Nordic Studies in Mathematics Education, 4*(2/3), 31–50.

Editorial (2009). Quality in mathematics education research. *Nordic Studies in Mathematics Education, 14*(1), 1–5.

Ekenstam, A. & Greger, K., (1983). Some aspects of children's ability to solve mathematical problems. *Educational Studies in Mathematics, 14*(4), 369–384.

Emanuelsson, J., & Sahlström, F. (2006). Same from the outside—different on the inside. Swedish mathematics classrooms from students' points of view. In C. Clarke, C. Keitel & Y. Shimizu (Eds.), *Mathematics classrooms in 12 countries: The insider's perspective* (pp. 307–322). Dordrecht: Sense Publishers.

Emanuelsson, J. & Sahlström, F. (2008). The price of participation. Teacher control versus student participation in classroom interaction. *Scandinavian Journal of Educational Research, 52*(2), 20–223.

Engström, L. (2006). *Möjligheter till lärande i matematik. Lärares problemformuleringar och dynamisk programvara* [Teaching mathematics posing problems using dynamic geometry software]. Doctoral thesis. Stockholm: HLS förlag, Stockholm Institute of Education.

Frejd, P. & Ärlebäck, J. B. (2010, January). *On Swedish upper secondary students' descriptions of the notion of mathematical models and modelling.* Paper presented at MADIF7, The 7th Swedish Mathematics Education Seminar, Stockholm.

Gustafsson, C. (1977). *Classroom interaction. A study of pedagogical roles in the teaching process.* Studies in Education and Psychology Bd. 1. Lund: Liber.

Hansson, Ö. (2006). *Studying the views of preservice teachers on the concept of function.* Doctoral thesis. Luleå: Luleå University of Technology, Department of Mathematics.

Hedrén, R. (1990). *Logoprogrammering på mellanstadiet: en studie av fördelar och nackdelar med användning av Logo i matematikundervisningen under årskurserna 5 och 6 i grundskolan* [Logo-programming in the middle grades: a study of advantages and disadvantages of using Logo in mathematics teaching in grades 5 and 6 in compuksory school]. Doctoral thesis. Linköping Studies in Education, 28. Linköping: Linköping University.

Hedrén, R. & Köhlin, K. (1983). *Miniräknaren på mellanstadiet. Slutrapport från RIMM-guppens arbete inom Skolöverstyrelsens ARK-projekt* [The pocket calculator in the middle grades. Final report from the RIMM project within the ARK project of the National School Agency]. Rapport 1983:5, Högskolan i Falun-Borlänge.

Holmquist, M., & Lingefjärd, T. (2003). Mathematical modeling in teacher education. In Q. Ye, W. Blum, S. K. Houston, & Q. Jiang (Eds.), *Mathematical modeling in education and culture, ICTMA 10: Applications in Science and Technology* (pp. 197–208). Chichester: Horwood.

HSV (1999). *Bra start i matematik* [A good start in mathematics]. Sigtuna 11-12 November 1999. Rådet för högskoleutbildning, Högskoleverket.

Jablonka, E. & Bergsten, C. (2010a). Commentary on Theories of mathematics education: Is plurality a problem? In B. Sriraman, B. & L. English (Eds.), *Theories of mathematics education: Seeking new frontiers* (pp. 111–120). New York: Springer.

Jablonka, E. & Bergsten, C. (2010b). Theorising in mathematics education research: differences in modes and quality. *Nordic Studies in Mathematics Education, 15*(1), 25–52.

Jablonka, E., Johansson, M., & Rohdin, M. (2010, January). *Achievement as a matter of choice?* Paper presented at MADIF7, The 7th Swedish Mathematics Education Seminar, Stockholm.

Jahnke, T. & Meyerhöfer, W. (Hrsg.) (2007). *Pisa & Co. Kritik enies Programms* (2., erweiterte Auflage). Berlin: Verlag Franzbecker.

Johansson, M. (2010, March). *Pedagogic identities in the reform of school mathematics.* Paper presented at the 6th Mathematics and Society Conference, Berlin, Germany.

Keitel, C. & Kilpatrick, J. (1998). The rationality and irrationality of international comparative studies. In G. Kaiser, I. Huntley, & E. Luna (Eds.), *International comparisons in mathematics education* (pp. 241–257). London: Falmer Press.

Kilpatrick, J. & Johansson, B. (1994). Standardized mathematics testing in Sweden: The legacy of Frits Wigforss. *Nordic Studies in Mathematics Education, 2*(1), 6–30.

Kling Sackerud, L-A. (2009). *Elevers möjligheter att ta ansvar för sitt lärande i matematik: En skolstudie i postmodern tid* [Students opportunities to assume responsibility for their own learning with regard to mathematics: A classroom-based study in a postmodern era]. Doctoral thesis. Umeå: Umeå University.

Klisinska, A. (2009). *The fundamental theorem of calculus. A case study into the didactic transposition of proof.* Doctoral thesis. Luleå: Luleå University of Technology.

Lantz-Andersson, A. (2009). *Framing in educational practices: learning activity, digital technology and the logic of situated action.* Göteborg Studies in Educational Sciences, 278. Göteborg: Acta Universitatis Gothoburgensis.

Larsson, I. (1973). *Individualized mathematics teaching. Results from the IMU project in Sweden.* Studie Psychologica et Paedagogica, no 21. Lund: CWK Gleerup.

Lerman, S. (2006). Theories of mathematics education: Is plurality a problem? *Zentralblatt für Didaktik der Mathematik, 38*(1), 8–13.

Liljestrand, J. (2002). *Klassrummet som diskussionsarena* [The classroom as an arena for discussions]. Örebro Studies in Education 6, Örebro University.

Lindström, S. (1968). *Överteoretiseringen av den elementära matematikundervisningen* [The overtheorizing of elementary mathematics teaching]. Stockholm: Almqvist & Wiksell.

Lingefjärd, T. (2000). *Mathematical modeling by prospective teachers.* Electronically published doctoral dissertation, University of Georgia.

Lingefjärd, T. (2006). *Faces of mathematical modeling.* Zentralblatt für Didaktik der Mathematik, *38*(2), 96–112.

Lingefjärd, T., & Holmquist, M. (2003). Learning mathematics using dynamic geometry tools. In S. J. Lamon, W. A. Parker, & S. K. Houston (Eds.), *Mathematical modelling: A way of life, ICTMA 11* (pp. 119–126). Chichester: Horwood.

Lundgren, U. P. (1972). *Frame factors and the teaching process. A contribution to curriculum theory and theory on teaching.* Stockholm: Almqvist & Wiksell.

Lundin, S. (2008). *Skolans matematik: en kritisk analys av den svenska skolmatematikens förhistoria, uppkomst och utveckling* [The Mathematics of Schooling. A critical analysis of the prehistory, birth and development of Swedish Mathematics Education]. Doctoral thesis. Uppsala: Uppsala University.

Magne, O. (1986). *Teorier för folkundervisningen i matematik.* [Theories for general education in mathematics]. Notiser och rapporter från institutionen i pedagogik, nr 461. Lärarhögskolan i Malmö.

Moore, R. (2006). Knowledge structures and intellectual fields: Basil Bernstein and the sociology of knowledge. In R. Moore, M. Arnot, J. Beck, & H. Daniels (Eds.), *Knowledge, power and educational reform: Applying the sociology of Basil Bernstein* (pp. 28–43). New York: Routledge, Taylor & Francis Group.

Mouwitz, L. (2006). Matematik och bildning: berättelse, gräns, tystnad [Mathematics and Liberal Education: narrative, limit, tacitness]. Doctoral thesis. Stockholm: KTH Royal Institute of Technology.

Murray, Å. & Liljefors, R. (1983). *Matematik i svensk skola* [Mathematics in the Swedish school]. FoU-rapport 46. Stockholm: SÖ och Liber Utbildningsförlaget.

Nilsson, G. (2005). Att äga π. Praxisnära studier av lärarstudenters arbete med geometrilaborationer [To own Pi. A study on student teachers' practical work with geometry]. Göteborg studies in educational sciences, 228. Göteborg: Acta Universitatis Gothoburgensis.

Niss, M. (2007). Reflections on the state of and trends in research in mathematics teaching and learning. From here to Utopia. In F. K. Lester (Ed.), *Second handbook of research on mathematics teaching and learning* (pp. 1293–1312). Reston: NCTM.

Persson, E. (2009). *Det kommer med tiden: Från lärarstudent till matematiklärare* [It will come with time. From student teacher to mathematics teacher]. Doctoral thesis. Stockholm: Stockholm University.

Persson, P-E. (2009). Handheld calculatros as tools for students' learning of algebra. *Nordic Studies in Mathematics Education, 14*(2), 49–77.

Pettersson, K. & Scheja, M. (2008). Algorithmic contexts and learning potentiality: A case study of students' understanding of calculus. *International Journal of Mathematical Education in Science and Technology, 39*(6), 767–784.

Prediger, S., Bikner-Ahsbahs, A., & Arzarello, F. (2008). Networking strategies and methods for connecting theoretical approaches: first step towards a conceptual framework. *ZDM – The International Journal on Mathematics Education, 40*(2), 165–178.

Radford, L. (2008). Connecting theories in mathematics education: Challenges and possibilities. *ZDM—The International Journal on Mathematics Education, 40*(2), 317–327.

Riesbeck, E. (2008). *På tal om matematik: matematiken, vardagen och den matematikdidaktiska diskursen* [Speaking of mathematics–Mathematics, every-day life and school mathematics discourse]. Linköping Studies in Behavioural Sciences, 129. Linköping: Linköpings universitet.

Ryve, A. (2004). Can collaborative concept mapping create mathematically productive discourses? *Educational Studies in Mathematics, 56*(2–3), 157–177.

Sahlström, F. (2008). *Från lärare till elever, från undervisning till lärande. Några utvecklingslinjer i klassrumsforskningen.* [From teachers to students, from teaching to learning. Lines of development in classroom research]. Vetenskapsrådets rapportserie nr 9:2008. Stockholm: Vetenskapsrådet.

Samuelsson, J. (2003). *Nytt, på nytt sätt? En studie över datorn som förändringsagent av matematikundervisningens villkor, metoder och resultat i skolår 7-9* [New, in a new way? A study of the computer as an agent of change of conditions, methods and results in mathematics teaching in grades 7-9]. Doctoral thesis. Uppsala: Uppsala universitet, Pedagogiska institutionen.

Säljö, R. & Wyndhamn, J. (1990). Problem-solving, academic performance and situated reasoning. A study of joint cognitive activity in the formal setting. *British Journal of Educational Psychology, 60,* 245–254.

Sjöberg, G. (2006). *Om det inte är dyskalkyli—vad är det då? En multimetodstudie av eleven i matematikproblem ur ett longitudinellt perspektiv* [If it isn't dyscalculia—then what is it? A multi-method study of the pupil with mathematics problems from a longitudinal perspective]. Doctoral thesis. Umeå: Department of Mathematics Technology and Science Education, Umeå Univeristy.

Skolverket (2003). *Lusten att lära—med fokus på matematik. Nationella kvalitetsgranskningar 2001-2002* [The desire to learn—with a focus on mathematics. National quality surveys 2001-2002]. Rapport 221. Stockholm: Skolverket.

Skolverket (2009a). *Vad påverkar resultaten i svensk grundskola? Kunskapsöversikt om betydelsen av olika faktorer. Sammanfattande analys.* [What influences the results in Swedish compulsory school? A synthesis of the impact of different factors. A summary analysis]. Stockholm: Skolverket.

Skolverket (2009b). *TIMSS Advanced 2008. Svenska gymnasieelevers kunskaper i avancerad matematik och fysik i ett internationellt perspektiv* [Swedish upper secondary school students' knowledge in advanced mathematics and physics in an international perspective]. Skolverkets Rapport 336. Stockholm: Skolverket.

Skolverket (2009c). *Hur samstämmiga är svenska styrdokument och nationella prov med ramverk och uppgifter i TIMSS Advanced 2008?* [How do the Swedish curriculum and the national tests match the framework and the tasks in TIMSS Advanved 2008?]. Stockholm: Skolverket.

Stadler, E. (2009). *Stadieövergången mellan gymnasiet och universitetet. Matematik och lärande ur ett studerandeperspektiv* [The transition from secondary school to university: Mathematics education as seen from a student perspective]. Dissertation. Växjö: Växjö University Press.

Werdelin, I. (1958). *The mathematical ability. Experimental and factorial studies.* Studia psychologica et paedagogica, 9. Lund: Lund University.

Wigforss, F. (1952). *Den grundläggande matematikundervisningen* [Elementary mathematics teaching]. Stockholm: AB Magnus Bergvalls förlag.

Wikström, H. (1997). *Att förstå förändring* [To understand change]. Göteborg studies in educational sciences no 114. Gothenburg: Acta Universitatis Gothoburgensis:

Wyndhamn, J. (1993). *Problem-solving revisited. On school mathematics as a situated practice.* Doctoral thesis. Linköping Studies in Arts and Science, 98. Linköping: Linköping University.

19

LEARNING DIFFICULTIES AND MATHEMATICAL REASONING

Johan Lithner and Torulf Palm

UMEÅ UNIVERSITY

Abstract: This chapter presents a synthesis of a series of studies aimed at clarifying characteristics and causes of learning difficulties in mathematics. A particular focus is on the distinction between imitative reasoning that is related to learning by rote, and more creative mathematically founded reasoning that helps develop non-routine problem-solving competencies and deeper conceptual understanding. In the investigation of the characteristics of learning difficulties, students' performance in both experimental situations and in genuine classroom settings was studied. When searching for causes behind the difficulties, genuine teaching situations, textbooks, and both teacher-made tests and official national tests were analyzed. In essentially all aspects, imitative reasoning dominates, and although this may be efficient in a very short perspective, it is devastating for long-term learning and therefore a main cause behind learning difficulties. The studies build on earlier international research in related areas, and add to this research by clarifying how rote learning through imitative reasoning takes place and by formulating a research framework for analyzing imitative and creative mathematically founded reasoning.

Keywords: Creative reasoning, imitative reasoning, learning difficulties, mathematical reasoning, problem solving

INTRODUCTION

During the last ten years a series of empirical studies related to learning difficulties and mathematical reasoning have been conducted by researchers from Umeå Mathematics Education Research Centre (UMERC), based in Umeå, Sweden. Some of the studies are published, some are submitted for publication, and some are in preparation. In order to structure the analyses and communicate the findings a research framework has gradually been formed, focusing the relation between *Creative mathematically founded reasoning* and *Imitative reasoning* (Lithner, 2008). This chapter will start by presenting the background to the research; it will then describe the research framework and summaries of some empirical studies, and end with a discussion.

Background

The number of students entering mathematical studies at university level in Sweden has increased dramatically in the last decades. In a longer perspective one can say that undergraduate mathematics was an élite education, but today it is a mass education. About 15% of each age group study undergraduate mathematics, mainly as a service subject within programs in technology, natural science, and computer science, but also in mathematics and teacher programs. As many as 98% of each Swedish age group (about 100,000 persons) study mathematics in upper secondary school for one to three years, and mathematics is one of three core subjects (together with Swedish and English) for all students in the compulsory school (grades 1–9). A severe problem for us engaged in teaching mathematics at any level and at

The First Sourcebook on Nordic Research in Mathematics Education, pages 283–298

any place in the world is that we are not able to help sufficient numbers of students reach a desired level of mathematical competence. There is a large and increasing demand from society for persons with different kinds of mathematically intense educations, but our educational system is neither able to be sufficiently attractive (too few students apply for mathematically intense programs) nor sufficiently efficient (too many drop out). Even among the majority who pass there are clear signs of severe weakness in competence, and it also seems clear that it is an international problem (see the references below). Central questions in relation to all this are:

Q1. What are the *characteristics* of the students' main learning and achievement difficulties?

Q2. What are the main *reasons* behind these difficulties?

The main purpose behind Q1 and Q2 is to provide a foundation for answering the following question, which has not (yet) primarily been treated within the studies:

Q3. What *measures* should be taken in order to reduce the difficulties?

The Role of Research

In the literature and in discussions with students, parents, teachers, researchers, administrators, and so on, there is a great variety in the proposed answers to the questions above. For example: The students are lazy, unintelligent; they do not learn anything sensible at the earlier educational levels. Or the mathematics learning environments provided by schools and universities are badly adapted to the real situation: There is too much, or too little, tests, grades, algorithms, calculators, lectures, small-group work, projects, exploratory work, individualization, real-life math, hard exercises, or easy exercises. Or the problems are caused by mass education, budget reductions, changed attitudes, social factors, and the like. This variety in the proposed answers is related to two hypothetical conclusions about the above research questions:

1. The answers are actually very complex. This assertion is strengthened by Niss (1999, 2007), when summarizing the results of research in mathematics education, concluding that mathematical learning in general is astonishingly complex and affected by many factors.
2. Although many researchers have described the severe learning and achievement difficulties of large groups of students, often in relatively elementary situations (Harel & Dubinsky, 1992; Harel & Trgalová, 1996; Hiebert, 2003; Schoenfeld, 1985; Tall, 1996; Thompson, 1994), there is much more that we do not know than what we know (Niss, 2007). Thousands of teachers and researchers all over the world have worked hard with these questions for a long time but many difficulties remain.

Thus it seems far from possible to provide complete answers to the complex questions Q1, Q2, and Q3, but the aim of the projects presented in this chapter is to treat parts of the answers of decisive importance. The projects focus on what we claim are central roles in students' ways of reasoning in learning and achievement situations.

A RESEARCH FRAMEWORK FOR ANALYZING IMITATIVE AND CREATIVE MATHEMATICALLY FOUNDED REASONING

This section is largely a summary of parts of the framework presented by Lithner (2008). A problem in mathematics education is that although problem solving has been, and still is, a central learning goal and that years of research and reform have gone by, many students still try to solve problems by faulty or inefficient rote learned methods (Hiebert, 2003). Rote thinking is a main factor behind learning and achievement difficulties and has been analyzed in a series of empirical studies presented below. This section presents the framework that has gradually been formed to specify and communicate the findings of these studies. To learn by rote is seen as a mechanical repetition of a fact or method, without an effort of understanding. The concept of understanding is very complex (Sierpinska, 1996), and will not be pursued here beyond noting that several of the theoretical constructs concern relations between rote learning and deeper understanding (see below). The notion "conceptual understanding" will thus be used in a relatively intuitive way, referring to insights in the origin, motivation, meaning, and use (Brousseau, 1997) of a mathematical method or fact.

There exist several frameworks for analyses of aspects related to rote learning. Some are not primarily designed as research tools and encompass all central goals of mathematics education (National Council of Teachers of Mathematics, 2000; Niss, 2003), while others propose aspects of understanding. Skemp (1978) distinguishes between instrumental and relational understanding of mathematical procedures. The former can be apprehended as 'true' (relational) understanding, but is only the mastering of a rule or procedure without any insight in the reasons that make it work. A distinction between "action" and "process" is made by Asiala and colleagues (1996), and Hiebert and Lefevre (1986) distinguish between conceptual and procedural understanding. Sfard (1991), Asiala et al. (1996), and Tall (2004) analyze how development of understanding of objects and processes interact, and Pirie and Kieren (1994) describe levels of understanding. Some define categories of cognitive achievement, like the Bloom (1956) and van Hiele (1986) taxonomies. Other frameworks structure particular competencies, such as for non-routine problem solving (Schoenfeld, 1985). However, there are not many that aim at characterizing the reasoning as such. One example is given by Balacheff (1988), but as in most reasoning frameworks it concerns proofs that is too narrow for the scope of this paper, even if proof is considered in a wider sense as justification (Ball & Bass, 2003). The purpose of the framework presented in this chapter is to characterise Imitative reasoning and Creative mathematically founded reasoning, and explain the origins and consequences of different reasoning types (see Figure 19.1).

A basic idea of this framework is that rote learning reasoning is *imitative*, while another important type of reasoning is *reasoning that is creative*. When trying to explain origins of different reasoning types, the approach (Figure 19.1) is to separate the *reasoning sequence* from the *thinking process* that created it. Thus, the reasoning that we see as data in these empirical studies can be seen as a trace of the students' thoughts. In order to understand why different thinking processes are activated or not, this framework sees them as guided and limited by the *student competencies* that are formed in the learning milieu.

Reasoning

The series of studies mentioned above started as an attempt to understand what made students fail or succeed in frequent occurring task solving and test tasks. Soon the focus came to be on the disadvantages of being restricted to imitative reasoning. Consider the following example.

The task is to find the largest and smallest values of $y = 7 + 3x - x^2$ on the interval $[-1,5]$. Ben, a skilled tenth-grade natural science student, solves it by a familiar method: Maxima and minima can be found where $y'(x) = 0$, where $y'(x)$ is undefined, and at the endpoints of the interval. Ben knows that for polynomials, $y'(x)$ is defined for all x, and differentiation and calculating function values are trivial for him.

This example of reasoning is structurally simple, but raises some fundamental questions. In the situation above (a laboratory test), what are the reasons that Ben succeeds and others fail? If this was a practice task in a textbook, what is learned by solving it? If it was included in a high-stake assessment, what is tested? In order to address such questions, one aim is to be relatively specific in defining both general characteristics of reasoning and particular reasoning types.

Ball and Bass (2003, p. 28) claim that "mathematical reasoning is no less than a basic skill," but still, the term "reasoning" is mostly used among mathematics educators without defining it, under the implicit assumption that there is a universal agreement on its meaning. *Reasoning* is defined in this chapter as the line of thought adopted to produce assertions and reach conclusions in task solving. It is not necessarily based on formal logic, thus not restricted to proof, and may even produce an incorrect result as long as there are some kinds of sensible (to the reasoner) reasons backing it. Reasoning can be seen as thinking processes, as the product of these processes, or both. What we can mainly see in data is behavior, while we can only speculate about the underlying thought processes (Vinner, 1997). Since one purpose of this framework is to characterize data, the choice is to see reasoning as a *product*. It is the line of thought that is visible to an observer in behavior data, such as written solutions and oral utterances during task solving and interviews. Reasoning during task solving starts in a task and ends in an answer. *Tasks* include most of the assignments given to students in classrooms, such as exercises, tests, group work, and so on. An *answer* provides the information requested. A *solution* is an answer and an explanation why the answer is correct. This solution often does not display the actual reasoning used to reach the answer, but an idealized summary of it. The term "problem" has been used in the literature with many different meanings but denotes

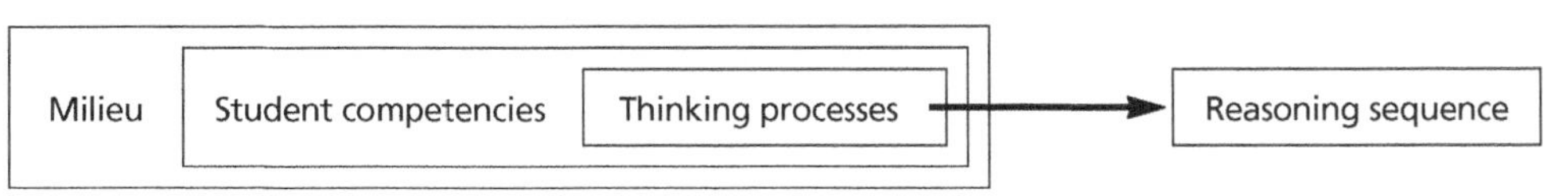

Figure 19.1 The origin of reasoning.

in this paper a task that is intellectually difficult for an individual (Schoenfeld, 1985).

Solving a task can be seen as carrying out the following four steps (Lithner, 2008):

1. A (sub)task is met, which is denoted as a *problematic situation* if it is not obvious how to proceed.
2. A *strategy choice* is made, where "strategy" ranges from local procedures to general approaches and 'choice' is seen in a wide sense (choose, recall, construct, discover, guess, etc.). It can be supported by *predictive argumentation*: This argumentation includes arguments that are future-oriented, that is, arguments about a predicted outcome. The arguments concern the reasons for the choice of a certain strategy. Why will the strategy solve the task?
3. The *strategy* is *implemented*, which can be supported by *verificative argumentation*: This argumentation includes arguments that are past-oriented. It includes the students' reasons for thinking that an appropriate solution has been found (or not). Why did the strategy solve the task, and why is the solution appropriate?
4. A *conclusion* is obtained.

The reasoning sequence can be represented by a path in a directed graph (see Figure 19.2). A vertex v_n represents both a momentary state of knowledge and a (sub)task. The reasoner makes a strategy choice among the edges leading from v_n. The strategy implementation is represented by a transition edge $e_{n,m}$. Here knowledge not already accessed in v_n is recalled or constructed and added up to form the new knowledge state in v_m, where the task is partially resolved and therefore a new task state is formulated. A *reason* is the justification supporting transitions between vertices. There is always a reason for attempting a particular transition, even if it is vague or superficial. The analogy of a directed graph is not intended to imply that a solution process is always linear. It is only a simplified model of a solution process, highlighting important parts of the process that are central in the framework. Sometimes a part of the reasoning sequence will lead the task solver farther away from a successful solution, but the partitioning of task-solving reasoning into sequences of reasoning solving different subtasks has been an efficient way of analyzing student data. The analyses in our studies of empirical reasoning data were typically done in several steps, including transcription of the reasoning, interpretation (a completion, e.g., by interviews, giving a more coherent picture of the actual reasoning), identification of the main problematic situations, and characterization of the argumentation.

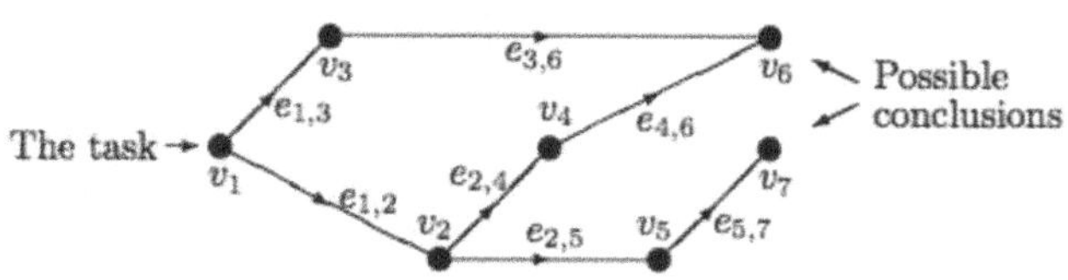

Figure 19.2 The reasoning sequence.

Imitative Reasoning

A key characteristic of Ben's solution above is that the path is laid out from the start. He imitates a solution procedure memorized from the textbook. The empirical studies behind this framework have identified two main types of imitative reasoning; memorized reasoning and algorithmic reasoning.

Memorized Reasoning (MR)

Memorized reasoning fulfils the following conditions:

1. The strategy choice is founded on recalling a complete answer.
2. The strategy implementation consists only of writing it down.

All task solving builds partly on recollection, but as an overall strategy it is useful only in a few task types like those asking for facts ("How many cm³ is a litre?"), definitions ("What is a polynomial?"), and proofs. An undergraduate examination task was "State and prove the Fundamental Theorem of Calculus." The answers by almost all of the 50% of the 150 students who got full credit were copies of the two-page textbook proof. Most of the faulty answers contained large parts of the same proof, but with some parts missing or in reversed order resulting in broken logical chains. A post-test asked the students to explain a sequence of six equalities included in the proof, most of them elementary in relation to the proof's complexity:

$$F'(x) = \lim_{h\to 0}\frac{F(x+h)-F(x)}{h}$$

$$= \lim_{h\to 0}\frac{1}{h}\left(\int_a^{x+h} f(t)\,dt - \int_a^{x} f(t)\,dt\right.$$

$$= \lim_{h\to 0}\frac{1}{h}\int_x^{x+h} f(t)\,dt$$

$$= \lim_{h\to 0}\frac{1}{h} f(c)$$

$$= \lim_{c\to x} f(c)$$

$$= f(x)$$

Most students were able to explain only a few of them; thus they did not understand the long proof they managed to memorize.

Another type of MR influence is through established experiences from the learning environment, including apprehensions of facts, concept images, and beliefs. One example is when Jed calculates $k = 6$ as the correct value of the slope of a linear function, but dismisses it by the argument that "It shouldn't be that big, k, the line is too steep.... It is always a smaller number like 1 or –2." Such arguments often override more mathematically founded reasoning.

Algorithmic Reasoning (AR)

School tasks normally ask for calculations where it is more appropriate to recall not the answer but an algorithm. "An algorithm is a finite sequence of executable instructions which allows one to find a definite result for a given class of problems" (Brousseau, 1997, p. 129). Its importance is that it can be determined in advance. The nth transition doesn't depend on any circumstance unforeseen in the n-1st transition—not on finding new information, any new decision, any interpretation, and thus on any meaning that one could attribute to them (ibid.).

In this framework, "algorithm" includes all prespecified procedures (not only calculations)—for example, to find the zeros of a function by zooming in its intersections with the x-axis on a graphing calculator. *Algorithmic reasoning* (AR) fulfils the following two conditions:

1. The strategy choice is to recall a solution algorithm. The predictive argumentation may be of different kind (see below for examples), but there is no need to create a new solution.
2. The remaining reasoning parts of the strategy implementation are trivial for the reasoner; only a careless mistake can prevent an answer from being reached.

This notion of "algorithm" is much wider than the tendency (which seems fairly common) to see an algorithm only as something that is explicitly taught in the form of a chain of executable instructions (for example, the division algorithm). The main point with this wider notion is that all parts that are conceptually difficult are taken care of by the algorithm, and only the easy parts are left to the student, which may limit the learning effect. We have easily taught normally talented seven-year old children to differentiate simple polynomials. They of course had absolutely no conceptual insight into what they were doing, but their performance would grant a few credit points in common upper secondary mathematics examinations. Thus, AR may be carried out with limited as well as full understanding (insights in the origin, motivation, meaning, and use) of the procedure.

Ben's implicit argument in the subsection "Reasoning," above, was that he knew that the task type could be solved by a particular algorithm. He was in no problematic situation, and neither did he have to consider the mathematical meaning of the components of the task nor provide any predictive or verificative arguments. He understood the mathematical foundation of the solution method, but his reasoning did not refer to this understanding and did not go beyond the reasoning of the seven-year-old children differentiating polynomials.

One may question whether the term "reasoning" is proper for a solution without explicit arguments, but there are always at least implicit reasons for the strategy choices. In order to clarify what these reasons may be, it is necessary to first clarify the concept of "argument."

Argumentation

Reasoning can have many functions in mathematics, including verification, explanation, systematization, discovery, communication, construction of theory, and exploration (Yackel & Hanna, 2003). Our framework proposes a focus relating to the phases of problem solving formulated by Pólya and elaborated by Schoenfeld (1985): reading the task (including noting conditions and goals), analysis (to understand, select perspective, and perhaps reformulate), exploration (a broader and less structured search for information), planning, implementation (including evaluation of progress), and verification. We suggest that analysis, exploration, and planning are primarily supported by predictive argumentation, and implementing and verification by verificative argumentation. In the analysis phase, one argues why properties of the task components have certain consequences. In the exploration phase, one considers why some of the outcomes may be useful. Planning concerns why certain approaches may better lead towards a solution. When implementing, one can have metacognitive arguments concerning why it proceeds in a proper way, and perhaps why the strategy needs to be reconsidered. Verification is an explanation of why a solution is actually reached. If understanding of the task is immediate in the reading phase, no argumentation is needed. Otherwise understanding is sought by analysis or exploration.

The quality of an argument in task solving is here seen as being determined by three factors: validity, ability to convince, and constructiveness. There is a distinction between the more general "argument" and "valid reasoning," where the latter is based on the organization of several propositions into a deductive step and the organization of several steps into a proof (Duval, 2002). Harel (2006) argues that if this distinction is not understood, we may advance mostly argumentation skills and no mathematical reasoning. The meaning of a statement is based on its content, status (premise, conclusion, theorem), logical value, and epistemic value, where the latter is the degree of trust (absurd, unreal, possible, likely, obvious) that a person has in a statement as soon as she understands its content. Logical value is independent of the person and characterizes a proof (Duval, 2002). Outside mathematics, valid reasoning is not the main concern but the strength of an argument in the sense that it is convincing. Regardless of whether it is valid or not, reasoning that does not convince fails (Duval, 2002).

Anchoring in Mathematical Properties

An acceptable justification is based on mathematics rather than social status—for example, the authority of the teacher or the smartness of a peer (Yackel & Cobb, 1996). But what does it mean for the warrant to be based on mathematics? Schoenfeld (1985) found that novices used naive empiricism to judge that geometrical constructions are correct if they "look good," and experts used more relevant properties (e.g., congruence). Thus content is important: What are the arguments about?

In order to address this question, we introduce the notion of *anchoring*, which refers not to the logical coherence of the warrant but to its fastening in relevant mathematical properties of the *components* one is reasoning about. The components are objects, transformations, and concepts. The *object* is the fundamental entity, the "thing" that one is doing something with (e.g., numbers, variables, functions, diagrams, etc.). A *transformation* is what is being done to an object, and the outcome is another object. A sequence of transformations—for example, finding polynomial maxima—is a *procedure*. A *concept* is a central mathematical idea built on a set of objects, transformations, and their properties—for example, the function or infinity concept. The status of a component depends on the situation: $f(x) = x^3$ can be seen as a transformation of, for example, the input object 2 into the output object 8. If f is differentiated, then the differentiation is the transformation, $f(x)$ is encapsulated (Tall, 1991) into an input object, and $f'(x)$ is the output object.

It is not sufficient to distinguish between properties that are mathematical and properties that are not. A property may be correctly apprehended but still irrelevant when solving a particular task. The relevancy of a mathematical property thus depends on the mathematical task. In deciding if 9/15 or 2/3 is largest, the size of the numbers (9, 15, 2, 3) is a *surface* (irrelevant in this case) property that is insufficient to consider, while the quotient captures the *intrinsic* (central in this case) property. Another example (Lithner, 2003) is a student trying to determine if the same test for absolute convergence is applicable to the series

$$\sum_{n=1}^{\infty} \frac{\cos n\pi}{(n+1)\ln(n+1)} \text{ and } \sum_{n=1}^{\infty} \frac{n \cos n\pi}{2^n}$$

He decides it is, based on the surface property that $\cos n\pi$ appears in both numerators, while the intrinsic property in this comparison lies in other parts of the fractions. The intrinsic–surface notion was introduced since one of the reasons behind students' difficulties was the anchoring of arguments in surface properties (Lithner, 2003).

Variants of Algorithmic Reasoning

In AR, the main difficulty is to identify a suitable algorithm. Three common ways are found, which may be based on surface property considerations only and are then mathematically unstable.

Familiar AR

The elementary Keyword strategy is determined by the appearance of keywords, for example "more" or "less" that correspond to the addition and subtraction algorithms, respectively (Hegarty, Meyer, & Monk, 1995). It may be generalized to more complex situations, and the reasoning in a task solution attempt fulfilling the following we call *Familiar AR*.

1. The reason for the strategy choice is that the task is seen as being of a familiar type that can be solved by a corresponding known algorithm.
2. The algorithm is implemented.

The (implicit) argument that convinces the reasoner of the strategy choice is often based, through similar practice tasks, on the established experience that a task with certain textual, graphical, and/or symbolical features is related to a corresponding algorithm—for example, as Ben connects a differentiation algorithm to a task type asking for the largest value of a function without considering the meaning of differentiation. Since an algorithm is "designed to avoid meaning" (Brousseau, 1997, p. 130) there is no need for any argumentation once the algorithm is identified.

However, the validity of superficial Familiar AR is not based on any anchoring in intrinsic mathematical properties and is therefore not reliable in problematic situations.

Familiar AR is common. For example, Lithner (2000a) found that 33 of 46 university mathematics students made a superficial interpretation of the task of finding the number of produced units x that maximizes the profit if the production cost is $f(x) = -2x^2 + 2000x - 42000$ kr/unit, the sale price is $g(x) = -x^2 + 700x$ kr/unit, and $400 \le x \le 600$. They reached a clear but faulty image that it was of a more familiar type, where the difference $g(x) - f(x)$ is maximized (only 5 of 46 made a correct interpretation, to maximize $x(g(x) - f(x))$).

Delimiting AR

Sally's approach to the same task as Ben is also algorithmic, but in a quite different way (Bergqvist, Lithner, & Sumpter, 2008). She differentiates y, solves $y'(x) = 0$ $(x = 1.5)$, and evaluates $y(1.5) = 9.25$. She hesitates: "I think I should have got two values [meaning max and min], and I don't know what I did wrong." She abandons this method only because the algorithm gave one value instead of two, and she does not reflect on why. Sally instead tries her calculator's minimum function, and in order to use it she has to enter a rough guess of the minimum's location. She does not understand why she cannot see the minimum on the graph (y has only one critical point, a maximum) to estimate its location, and again abandons the method without reflection. Sally then uses the calculator's table function to evaluate y for $x = -1, 0, \ldots 5$. She finds min = −3 and max = 9, but is troubled since she found a larger value, $y = 9.25$, earlier. She does not understand why the table method does not work and abandons it without reflection. She finally solves $7 + 3x - x^2 = 0$ and gets two values, $x_1 \approx 4.54$ and $x_2 \approx 4.54$. This method does not produce a correct result, but she sees it as the solution.

In Familiar AR, the vertices and edges of Figure 19.2 are all apprehended (wrongly or rightly) to be specified by a familiar algorithm, but this task is not sufficiently familiar to Sally. However, though there is no attempt at planning or analysis based on the intrinsic properties of the task, her choice of algorithm is not random. This would be unrealistic since she knows hundreds of algorithms, so Sally delimits the set of potentially useful ones by trying only those that have some connection to the task. "Largest and smallest" is related to the first two methods, functions (represented by the polynomial) to tables, and the polynomial to the quadratic equation algorithm. The initial drawback with this approach is that these connections are based only on surface property considerations, so she is unable to predict where the transition will take her. There is nothing wrong in exploring different algorithms with weak predictive argumentation, but in order to be constructive this has to be compensated by valid evaluations of the outcomes. Sally's evaluations are based on surface considerations that do not take her far enough. She could, for example, have tried to understand why the first method only gave one value, why she could not see a minimum on the graphing calculator, why the table gave a smaller maximum, or how equation solutions are related to maxima and minima.

Sally's reasoning is an example of *Delimiting AR*, defined as follows:

1. An algorithm is chosen from a set that is delimited by the reasoner through the algorithms' surface relations to the task. The outcome is not predicted.
2. The verificative argumentation is based on surface considerations that are related only to the reasoner's expectations of the requested answer or solution. If the implementation does not lead to a (to the reasoner) reasonable conclusion, it is simply terminated without evaluation, and another algorithm may be chosen from the delimited set.

According to our data, Delimiting AR is the main AR approach in problematic situations where Familiar AR does not work and no guidance source (see below) is at hand. Often it is done in one step, when the reasoner only knows one related algorithm or when the first attempt yields an acceptable conclusion. This can be seen in the perspective that justifying a successful solution by simply describing the algorithm seems to be an accepted sociomathematical classroom norm that may influence most practice and test situations studied.

Guided AR

When Familiar or Delimiting AR does not work, one may aim for external guidance. In *Text-guided AR* the following conditions hold:

1. The strategy choice concerns identifying surface similarities between the task and an example, definition, theorem, rule, or some other situation in a text source.
2. The algorithm is implemented without verificative argumentation.

Seventy percent of the exercises in some common American calculus textbooks may be done with Text-guided AR, 20% by essentially copying a solution but with a minor modification by Local mathematically founded creative reasoning (see next subsection), and 10% require Global mathematically founded creative reasoning (Lithner, 2004). The latter are generally the most difficult exercises at the end of each exercise section, which few students ever attempt. Similar distributions are found in ongoing studies of Swedish grade 5–12 textbooks. Text-guided AR is the dominating type of student reasoning found in our studies of individual or small-group learning situations (Lithner, 2003; Långström & Lithner, 2010). Ben's and Sally's task can be solved by copying the following standard textbook example:

Find the maxima and minima of $f(x) = x^2 + 4x + 1$ if $x \in [-3, 0]$.

Solution: $f'(x) = 2x + 4$. Solve $f'(x) = 0$.
$2x + 4 = 0 \Rightarrow x = -2$.
Since $f'(x)$ exists for all x and $x = -2$ is its only zero the function is evaluated there and at the endpoints of the interval. $f(-3) = -2$, $f(-2) = -3$, and $f(0) = 1$. Thus we have maximum $f(0) = 1$ and minimum $f(-2) = -3$.

The first step is to identify this as a proper example to copy. In the school textbooks analyzed, most exercises are preceded by only one example, and then the identification is trivial. At higher educational levels, the theory sections preceding the exercises become longer and may, in a university calculus textbook, contain ten pages including some ten solved examples. However, the identification in the 70% Text-guided AR tasks mentioned above can still be based on surface considerations alone. It is sufficient to know that the expression "largest and smallest value" in Ben's task is equivalent to "maxima and minima" in the example above, and to note that both tasks contain quadratic equations on closed intervals. In the strategy implementation (consisting of subalgorithms), it is necessary to differentiate and evaluate a quadratic polynomial, and solve a linear equation. Thus a task that is supposed to concern extreme values is reduced to a practice in basic algebra. The reasoning has low logical value since there are no arguments anchored in intrinsic properties of the function, the derivative, or extreme values. Still, the epistemic value is normally regarded as high since, by the didactic contract (Brousseau, 1997), the role of examples is seen as providing complete templates and the authority of the book is seen as guaranteeing that the algorithm will provide a correct solution.

The other common version of guidance is denoted *Person-guided AR*.

1. All strategy choices that are problematic for the reasoner are made by a guide, who provides no predictive argumentation.
2. The strategy implementation follows the guidance and executes the remaining routine transformations without verificative argumentation.

To exemplify, the ninth grade student Moa asks the teacher for help to calculate 15% of 90. The teacher starts by writing 0.15 multiplied by 90 in Moa's notebook, and the complete dialogue is as follows (Lithner, 2002):

T: What is 5 · 0? **M:** 0 [T writes 0]
T: What is 5 · 9? **M:** 45 [T writes 45]
T: What is 1 · 0? **M:** 0 [T writes 0]
T: What is 1 · 9? **M:** 9 [T writes 9]
T: What is 5 + 0? **M:** 5 [T writes 5]
T: What is 4 + 9? **M:** 13 [T writes 13]
T: Where shall the decimal point be placed?
M: [Silence]
T: [Marks the decimal point at the correct place and leaves].

Moa participates in no mathematical activities beyond one-digit arithmetic, which is not her difficulty. There is no argumentation, and no intrinsic properties of the problematic components (percentage, the multiplication algorithm, and its relation to the task) or strategy choices are considered. In the graph analogy, the teacher takes the student by the hand and chooses the transition edge at each vertex. As in Text-guided AR, validity is not an issue if the guidance is provided by an authority that is seen as guaranteeing a correct solution.

A study of students' collaborative reasoning in their ordinary classroom setting included Person-guided AR by peers that had approximately the same level of knowledge as the guided student (Långström & Lithner, 2010). The guidance was often unstable and erroneous, but predictive or verificative argumentation was still rarely requested or provided. The students' reasoning was extremely focused on text-guided AR, mainly based on fragmentary searches for surface clues, and their dialogues were seldom mathematically meaningful.

Teacher presentations of task solutions rarely contain true problematic situations, since the teacher normally has prepared them. The framework was used to study if teachers simulate creative mathematically founded reasoning (see next subsection) in 23 ordinary teaching situations from grade 7 to university (Bergqvist & Lithner, 2005). The teaching mainly focused on algorithmic guidance in three versions, of which the first two are person-guided AR: without comments (as above), with descriptions of the strategy choices, and with explanations of the strategy choices.

Creative Mathematically Founded Reasoning

Per is given the graph $f(x)$ in Figure 19.3a (Lithner, 2000b). His task is to draw a primitive function $g(x)$ to $f(x)$.

Per's strategy choice, which correctly implemented would solve a part of the task, is to combine sequences of Familiar AR: to find an algebraic expression for the leftmost linear part of $f(x)$ on the interval $[-3,-1]$, integrate it to obtain an algebraic expression for $g(x)$, and finally draw $g(x)$. However, he makes several careless mistakes in the strategy implementation. He reaches after 20 minutes of laborious work, the faulty graph $g(x)$ in Figure 19.3b and considers himself finished with the interval $[-3,-1]$. He immediately turns to the next interval, $[-1,0]$, but the observer interrupts.

Observer: Is the derivative of your $g(x)$ described by the graph of $f(x)$?

Per: No, it isn't. My answer is wrong. The derivative of $g(x)$ is positive all the way, but $f(x)$ is not.

Observer: Can you from this reasoning make a rough sketch of what the function should look like?

Per: The derivative is first positive, then zero, and then negative.

Per skillfully makes a rough but reasonable sketch (Figure 19.3c) by reading the slope of $g(x)$ point-wise as the function value of $f(x)$. After Per's initial unsuccessful AR attempt, when "mildly guided," his reasoning is non-imitative in three central aspects.

1. Per constructs novel reasoning sequences, first when evaluating the graph b and then when forming the graph c. He has not practiced any tasks that are similar to the questions the observer asks him (which is clear from the post-interview), and though his basic knowledge is used, it does not equip him with complete answers or solution algorithms as in MR or AR.
2. There are arguments supporting the logical validity of the reasoning. It is justified why the first solution (b) is inconsistent with the premises and then what the graph should look like. These arguments are not just explaining something he knew beforehand, but guide the strategy choices and implementations and therefore the constructiveness of the reasoning that results in new knowledge (to Per).
3. Per's arguments are mathematically founded and provide validity through anchoring in intrinsic properties of the components in the reasoning: the relation between the function value and the graph of the function, the relation between the slope and the graph, and that point-wise the slope of $f(x)$ at $x = a$ is equal to the derivative $f'(a)$.

Before defining the reasoning illustrated by Per's solution, a short discussion on logical rigour is required. To some authors 'mathematical reasoning' seems equivalent to the strict proof (Duval, 2002; Harel, 2006) while others include other types of less strict reasoning (NCTM, 2000; Ball & Bass, 2003). School tasks normally differ from the tasks met by mathematicians, engineers and economists, etc. Within the didactic contract (Brousseau, 1997) of school it is allowed, and sometimes encouraged, to guess, take chances, and use reasoning with considerably reduced requirements of logical rigour. Even in examinations it is accepted to have only, for example, 50% of the answers correct, while it is absurd if mathematicians, engineers, or economists are correct only in 50% of their conclusions. This framework proposes a wider conception of mathemati-

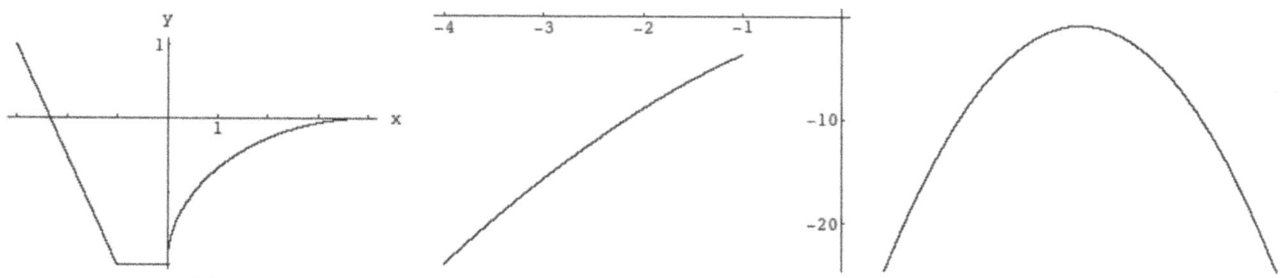

(a) (b) (c)

Figure 19.3 a) $f(x)$ b) Per's first, faulty, $g(x)$ c) Per's second, fairly correct, $g(x)$

cal reasoning inspired by Pólya (1954): "In strict reasoning the principal thing is to distinguish a proof from a guess.... In plausible reasoning the principal thing is to distinguish a guess from a guess, a more reasonable guess from a less reasonable guess." In proof, the value of the reasoning relies on its correctness. Analogously, plausible reasoning is in this framework not necessarily logically strict but constructive through the support of plausible arguments. The stronger the logical value, the more plausible it is.

A reasoning is called *creative mathematically founded reasoning* (CMR) if it fulfills all of the following criteria:

1. Novelty. A new (to the reasoner) reasoning sequence is created, or a forgotten one is recreated.
2. Plausibility. There are arguments supporting the strategy choice and/or strategy implementation motivating why the conclusions are true or plausible.
3. Mathematical foundation. The arguments are anchored in intrinsic mathematical properties of the components involved in the reasoning.

Using the definitions of AR and MR, two categories of CMR can be defined: Reasoning that is mainly based on MR or AR but contains minor, local elements of CMR will be called *local CMR*, while reasoning with large CMR elements is called *global CMR*. The latter may still include large elements of MR or AR. The local aspect was introduced in (Lithner, 2004) to capture the substantial empirically established differences between local and global CMR. CMR does not, as problem solving, have to be a challenge. The definition also includes elementary reasoning. Still, in most studies, CMR is rare and AR is dominating. The only exceptions so far are two studies on reasoning requirements in assessment. The official national tests had large proportions of tasks judged to require CMR, while such tasks were scarce in teacher-made tests (Palm, Boesen, & Lithner, 2005).

Summarizing by relating back to Figure 19.2, the transition in CMR is created. In AR it follows a known path, and is in MR immediate through recollection. Furthermore, in CMR the epistemic value lies in the plausibility and logical value of the reasoning; in MR and AR it is determined by the authority of the source of the imitated information.

SUMMARIES OF THE EMPIRICAL STUDIES

This section contains summaries of some of the empirical studies that constitute the overarching project presented above in the Introduction. The research methods include, for example, observation of task solving in classrooms and video-recordings, and analyses of think-aloud protocols and interviews. The analyses of empirical reasoning data were typically done in several steps, including transcription of the reasoning and interpretation (a completion, e.g., by interviews, giving a more coherent picture of the actual reasoning), identification of the main problematic situations, and characterization of the argumentation. The formation of the reasoning framework has been an iterative process, including analyzing and trying to explain and understand data from each study, and summarizing and structuring these results at an overall level. For a more detailed description of how these research methods were used we refer to each study (see below).

University Students' Learning and Achievement Difficulties

Four studies on university students' work in experimental situations constitute the basis of a doctoral dissertation by Lithner (2001). These studies are also published elsewhere in revised versions (Lithner, 2000a,b; 2003; 2004).

Students' General Difficulties in Task Solving

In the paper "Mathematical Reasoning and Familiar Procedures" (Lithner, 2000a), four first-year undergraduate students are working with two tasks. The underlying question treated is, "What are the characteristics and background causes of their difficulties when trying to solve these tasks?" The purpose is to give a general survey of their main difficulties, rather than to go deeply into details. One of the common characteristics is that the students are more focused on what is familiar and remembered than on (even elementary) creative mathematically founded reasoning and accuracy. See Per's example above.

Students' Reasoning in Task Solving

The research question in the study above concerns students' general difficulties, and the results point to the particular role of reasoning. Therefore, it was followed up by the study "Mathematical Reasoning in Task Solving" (Lithner, 2000b), which focuses on, and extends, the part of the earlier study that concerns task-solving strategies. The results indicate that focusing on what is familiar and remembered at a superficial level is dominant over reasoning based on mathematical properties of the components involved, even when the latter could lead to considerable progress. The students did not use the potential of constructive creative reasoning as a means to make progress in problematic task solving situations, except in a few cases when guided by the observer (see Per's solution in Figure 19.3). Another study (Bergqvist et al., 2008) with similar focus and method was conducted on upper secondary students, with similar results (see subsection "Delimiting AR" above for an example).

The main difference between the two studies (Lithner, 2000a,b) is that the former is "wider" (all their main difficulties) and the latter is more limited to treating, on a firmer theoretical foundation, certain types of mathematical reasoning. In the latter, the first embryo to the framework presented in this chapter was introduced. These two studies indicate (together with studies of research literature) that students focus on routines and superficial reasoning, combined with their inability and/or reluctance to to consider the intrinsic properties involved in the reasoning, is one of the main reason behind their difficulties. The two studies below aim at searching for possible reasons behind these indications.

Strategies and Reasoning Possible to Use When Solving Textbook Exercises

The aim of the paper "Mathematical Reasoning in Calculus Textbook Exercises" (Lithner, 2004) is to study some of the strategies that it is possible to use in order to solve the exercises in undergraduate calculus textbooks. It is described how most exercises may be solved by mathematically superficial strategies. Strategy choices and implementations can usually be based on identifying similar solved examples and copying, or sometimes locally modifying, given solution procedures. One consequence is that exercises may often be solved without actually considering the core mathematics of the book section in question. See also the subsection "Guided AR," above.

Strategies and Reasoning Applied by Students When Solving Textbook Exercises

The study "Students' Mathematical Reasoning in University Textbook Exercises" (Lithner, 2003) investigates the ways students conduct their study work, in particular their mathematical reasoning when working with textbook exercises (see the subsection "Anchoring in Mathematical Properties" for an example). The results indicate that: (1) most strategy choices and implementations are carried out without considering the intrinsic properties of the components involved in the solution work, which in turn leads to different difficulties; (2) it is crucial for these students to find solution procedures to copy; (3) there are extensive attempts, often successful, to understand each step of the copied solution procedures, but only locally; (4) the students make almost no attempts to construct their own solution reasoning, not even locally; (5) the main situations where the students' work is not simply a straightforward implementation of provided solution procedures are where mistakes are made in minor local solution steps.

To summarize the four studies, the research methods used in the completed studies above are mainly qualitative: relatively fine-grained analyses of a small number of students' reasoning characteristics in limited task solving situations, including the development of analytical frameworks. These types of analyses can (1) show the existence of some reasoning types, and (2) indicate plausible characteristics of larger student groups. The latter may also be supported by studying similar or related aspects from other theoretical perspectives or by other methods, hereby finding reasonable and general explanations behind the indicated behavior. One example of this is a study (Lithner, 2004) that is partly quantitative (600 textbook exercises were classified), where possible reasons behind the students' behavior in the other studies is investigated.

Though the studies above only treat limited aspects of students' competence and limited aspects of the learning environment, and though the work of rather few students is investigated, the overall picture emerging is coherent: It seems that the students are founding their work mainly on superficial reasoning, and that the reasons behind this originate to a large extent from the learning environment provided by the educational system. This is (at least partly) already known in the literature, and it also seems to be experience-based knowledge familiar to many teachers. The motivation for carrying out as research the studies above is:

1. The studies explicitly and primarily address the ways that the students' reasoning is based on mathematical properties or not, something that is not done by many other studies. The studies (Lithner, 2000a,b) indicated that the domination of "non-mathematical" reasoning is one of the main causes behind task-solving difficulties.
2. The reasoning is studied in rather fine-grained detail (a framework for this type of studies is one of the outcomes of the studies (Lithner, 2008).
3. There are few studies on textbook structure (Love & Pimm, 1996), especially from the perspective mentioned in (1) above and on students' learning strategies and textbook usage.
4. The achievement difficulties of mathematics students at all levels have been known

> for many years, but the difficulties mainly remain (Hiebert, 2003). Extensive research on characteristics and causes of learning difficulties is still required in order to be able to construct well-founded measures for improvement of the learning environment.

Simulated CMR in Teaching

Real teacher presentations rarely contain true problematic situations, since the teacher is normally prepared for or at least familiar with the tasks presented at the board. Bergqvist and Lithner (2005) present a modification of the framework in order to study in what sense teachers simulate CMR in ordinary presentations. Twenty-three ordinary teaching situations from grades 7–12 and first-semester university mathematics were analyzed, and the teachers' reasoning was categorized through four aspects or criteria: mathematical foundation, creativity, reflection, and focus. In the situations analyzed, the criteria were sometimes fulfilled, but in rather limited and modest ways. The teaching was mainly focused on presenting algorithmic methods by solving routine tasks. See the subsection "Guided AR" above for an example of a teacher helping an individual student.

Reasoning Characteristics in Student Collaboration

It is often advocated that student collaboration will enhance learning. There are many aspects of collaboration to consider, and one is the quality of the reasoning according to the framework of this paper. The study (Långström & Lithner, 2010) analyzes students' collaborative reasoning in their ordinary classroom setting, and the framework of Section 2 is extended to encompass aspects of help-seeking and collaboration. In short, the students' reasoning in this study is extremely focused on text-guided AR. Their collaboration is mainly based on a fragmentary search for clues and, with few exceptions, without intrinsic property considerations, and thus of very low mathematical quality. It seems that collaboration works poorly since the tasks and dialogues are not mathematically meaningful to the students.

Affect, Beliefs and Reasoning

Sumpter (2007) studied the influence affect and beliefs have on the central reasoning decisions that students make in problematic task solving situations. Three major themes stand out: expectations (e.g., superficial experiences lead to expectations on the character of an answer to a mathematical task), security (that some approaches feel safer), and motivation (including motivational beliefs and active goals). These beliefs are quite influential and often lead to mathematically superficial strategy choices. For example, imitative reasoning feels familiar and therefore safer than creative mathematically founded reasoning, even when the latter can lead to better progress. In an ongoing follow-up study, Sumpter investigates in what ways reasoning is gendered, and the results so far shows that at least teachers see different types of reasoning as gender-related.

Reasoning Requirements in Tests

The assessments that the students encounter is another important part of the learning environment that influences student learning. There are several ways in which assessments can influence learning, and the tasks that make up the assessments are crucial for this influence. Assessments have a formative function when assessment information is primarily used for helping students learn (Wiliam, 2007). In such assessments, the kind of reasoning that the tasks elicit is important for the students' learning since this is the reasoning that the teachers' feedback can address. But the characteristics of assessment tasks may also be important for students' learning even when the assessment is not having a formative function. The types of tasks, and thus the competencies that are valued, in externally mandated tests such as national tests can under some circumstances influence the work of teachers and textbook writers (Barnes, Clarke, & Stephens, 2000), which in turn influence students' learning. Teacher-made classroom assessments may influence students' learning focus more directly. Thus, in the endeavor to search for conditions that might provide some explanations for students' emphasis on imitative reasoning, it is appropriate to investigate the characteristics of the assessment tasks that students encounter.

However, before analyzing the characteristics of assessment tasks administered to students, we conducted a study in an authentic national test situation that investigated the relation between task characteristics and the mathematical reasoning the students use when they try to solve the tasks (Boesen, Lithner, & Palm, in press). The results show that when confronted with test tasks that share important properties (described in the study) with tasks in the students' textbook, they solve the test tasks using imitative reasoning by trying to recall facts or algorithms. Such test tasks did not require conceptual understanding. In contrast, test tasks that did not share important properties with tasks in the textbook mostly elicited creative mathematically founded reasoning. In addition, most successful solutions to such tasks were based on this type of reasoning. The results of the study may be used in the development and analysis of assessment tasks and in the interpretations of assessment results; they have already been used in two studies analyzing the kind of reasoning emphasized in assessments administered to students (Palm et al., 2005; Bergqvist, 2007).

Palm and colleagues (2005) investigate the mathematical reasoning required to solve the tasks in Swedish national tests and a random selection of Swedish teacher-made tests at upper-secondary level. The results show that only a small proportion of the tasks in the teacher-made tests require the students to produce new reasoning and to consider the intrinsic mathematical properties involved in the tasks. In contrast, the results also show that the national tests include a large proportion of tasks for which memorization of facts and procedures is not sufficient. In the study by Bergqvist (2007), a modified version of the framework developed by Boesen et al. (in press) was used in examining what types of reasoning students taking introductory calculus courses are required to perform. Tasks from 16 exams produced at four different Swedish universities were analyzed. This study, investigating university teachers' assessment tasks, showed similar results as the study involving upper-secondary teachers assessment development. Nearly 70% of the tasks were solvable by imitative reasoning and another 22% were solvable by local CMR. In addition, 15 of the 16 exams could be passed using only imitative reasoning. In a follow-up study to the one by Palm et al. (2005), Boesen (2006) interviewed teachers in an attempt to find reasons for the upper-secondary teachers' focus on imitative reasoning in their tests. The following three factors were identified as contributing to this emphasis: (1) limited awareness of differences in reasoning requirements, (2) low expectations of students' abilities, and (3) the desire to get students passing the tests, which was believed easier when excluding creative reasoning from the tests. In a similar study by Bergqvist (2006), factors (2) and (3) were also found to contribute to the university teachers' focus on imitative reasoning in their tests.

So, in the development of their classroom assessments, the upper-secondary teachers did not seem to be much influenced by the national tests regarding the type of reasoning focused upon. Differences in the conditions and constraints under which the test development takes place for upper secondary teachers and national test developers indicate some of the reasons for the discrepancy between their assessments. Some of these differences were the available time and resources to acquire the knowledge that may be useful to interpret shortly written curriculum documents and to develop new kinds of tasks (Palm et al., 2005).

In summary, these studies show that many of the teacher-made tests the students meet are possible to solve, and are likely to be solved, by imitative reasoning. They do not require conceptual understanding and so join the other parts of the learning environment where imitative reasoning dominates and the learning of creative mathematically founded reasoning is not facilitated.

DISCUSSION

The main components of the mathematical learning environment, at least of the part organized by the educational system, can be seen as students' activities, teaching, textbooks, and tests. The framework and the empirical studies above make the imitative character of all four components transparent in the analyses, and indicate that the students' IR focus is to a large extent caused by insufficiencies in the learning environment. The IR focus will lead to short-term gains like managing textbook exercises and passing exams that are adapted to IR, but to long-term losses like weak concept understanding and weak problem-solving competence. It is possible to be quite successful by superficial IR strategies in exercises (Lithner, 2003, 2004) and in tests (Palm et al., 2005). This is in line with a large part of the research results on reasons behind learning and achievement difficulties, which can be characterized as an unwarranted and too extensive reduction of ambiguity, risk, and complexity of mathematical concepts and processes by teachers, textbook writers, and students trying to cope with curricular goals that are hard to reach (Doyle, 1988; Schoenfeld, 1991). McGinty, VanBeynen and Zalewski (1986) found that the number of word problems in American grade 5 textbooks from 1924, 1944, and 1984 had decreased and they had become less rich, while the number of drill tasks had increased. In Australia, algebra instructions and exercises were reduced to easier arithmetic (Stacey & MacGregor, 1999), and in Sweden (using American textbooks), university calculus was reduced to grade 10 algebra (Lithner, 2003; 2004).

The results presented above go beyond identifying the fact that students are making mistakes, and also beyond characterizing what their reasoning lacks: It emphasizes what they actually do and why. Analyses through the framework confirm, but may also further clarify and explain, assertions proposed by other authors. For example, when Tall (1999) claims that it is often possible to pick up marks on procedural exams while having difficulties in considering definitions and proofs, this claim can be further understood when analyzing the reasoning requirements of the tests and the actual competencies of the students (Boesen et al., in press).

In Sweden in the last 15 years there has been a strong trend towards individualization in the sense that the teacher spends most or all classroom time in helping students individually. The advantage is that the teacher may better meet each student at her or his level, but there are at least two disadvantages: Firstly, the teacher has very little time with each student and may be confined to Guided AR. Secondly, there is a pressure on textbooks to be self-contained in the sense that students do not have to ask the teacher so many questions.

The easiest way to achieve this is to provide guiding examples to most exercises, which is done (see the subsection "Guided AR" above). Therefore students may develop strategy choice approaches where the question "what method should be applied?" is immediately asked, instead of first trying to reach a qualitative representation of the task and base the solution attempt on intrinsic mathematical properties. When the textbook is at hand, this question leads to text-guided AR. When the book is not there, like in tests, the same question leads to a memory search for earlier situations that are experienced as similar. If the student then has difficulties in understanding the concepts involved and no real practice in creative problem solving, the only alternative may be to search for superficial familiarities. The main alternative seems to be to ask the teacher for help and perhaps receive it in the form of teacher-guided AR. The help students obtain from their peers is also dominated by direct algorithm descriptions, but then often in incorrect ways. These connections may partly explain the students' focus on superficial reasoning: One memorizes algorithms through guided AR and, conversely, one does not learn so much beyond familiar and delimiting AR. Indeed, the most frequent type of reduction of complexity seems to be to focus on actions in the form of routine procedures (Tall, 1996; Vinner, 1997) that may prevent the development of conceptual and relational understanding (Skemp, 1978; Hiebert & Carpenter, 1992). In a research literature survey, Hiebert (2003) finds massive amounts of converging data showing that students know some basic elementary skills but there is not much depth and understanding: "One can model their behaviour and predict the errors they will make by looking only at the symbol manipulation rules they have been taught and pretending that they are following these rules like robots with poor memories" (p. 12). Memorized operations without structural conceptions can only be stored in unstructured, sequential cognitive schemata, which are inadequate for the modest dimensions of human working memory (Sfard, 1991). The reduction of complexity may thus go so far that the imitative thinking is detached from its mathematical meaning, as in the MR and AR examples above. In this case it is not constructive to analyse students' reasoning only by characterizing its mathematical coherence.

The framework can also be used to relate phenomena across educational levels. One could, for example, hypothesize that there is a progression in superficial reasoning from elementary to more complex mathematical situations: In primary school there are normally few algorithms to choose from, and the key-word strategy works well. In secondary school, mathematics becomes more complex and there are more algorithms. It becomes necessary to limit the amount of algorithms to try, and delimiting AR may be a way to cope if familiar AR does not work. As the student progresses through secondary and perhaps into tertiary mathematics, the mere amount of algorithms may at some stage be insurmountable. This may be one of the causes behind the learning collapse we see when students who have managed fairly well seem to suddenly and completely lose contact with mathematics and drop out.

The ways that students meet mathematics form their beliefs about what mathematics is and what they are supposed to do when learning it (Schoenfeld, 1985). Mathematics is often seen as a meaningless, fragmentary collection of algorithms to be memorized (Hiebert, 2003). The aspects that can make mathematics meaningful to the students like conceptual understanding, creative reasoning, and insights in the central roles of mathematics in our society are not enhanced by rote learning. In a sense, the main problem with an AR focus is not that students do not learn CR but that they do not develop the conceptual understanding that may be necessary for learning many other aspects of mathematics (Lithner, 2004). These complicated phenomena have to be communicated, but this is usually done with vague and ambiguous terminology. This may well be one of the main obstacles for educational development, in particular with respect to turning research findings into practice. Just to say that students need to practice more creative mathematically founded reasoning will not help much if we can not communicate what creative mathematically founded reasoning is and construct tasks that will promote the learning of it. The above presented research program is an attempt to contribute to such communication.

REFERENCES

Asiala, M., Brown, A., DeVries, D., Dubinsky, E., Mathews, D., & Thomas, K. (1996). A framework for research and curriculum development in undergraduate mathematics education. In A. Schoenfeld, J. Kaput, & E. Dubinsky (Eds.), *Research in collegiate mathematics education II, CBMS Issues in mathematics education* (pp. 1–32). Providence, RI: American Mathematical Society.

Balacheff, N. (1988). Aspects of proof in pupils' practice of school mathematics. In D. Pimm (Ed.), *Mathematics, teachers and children* (pp. 216–235). London: Hodder and Stoughton.

Ball, D. & Bass, H. (2003). Making mathematics reasonable in school. In J. Kilpatrick, G. Martin, & D. Schifter (Eds.), *A research companion to principles and standards for school mathematics* (pp. 27–44). Reston, VA: National Council of Teachers of Mathematics.

Barnes, M., Clarke, D., & Stephens, M. (2000). Assessment: The engine of systemic curricular reform? *Journal of Curriculum Studies, 32*(5), 623–650.

Bergqvist, E. (2006). *Mathematics and mathematics education: Two sides of the same coin.* Unpublished dissertation, Department of Mathematics and Mathematical Statistics, Umeå University.

Bergqvist, E. (2007). Types of reasoning required in university exams in mathematics. *Journal of Mathematical Behavior, 26,* 348–370.

Bergqvist, T., & Lithner, J. (2005). *Simulating creative reasoning in mathematics teaching.* Research reports in mathematics education 2, Department of Mathematics, Umeå University.

Bergqvist, T., Lithner, J., & Sumpter, L. (2008). Upper secondary students task reasoning. *International Journal of Mathematical Education in Science and Technology, 39,* 1–12.

Bloom, B.S. (Ed.). (1956). *Taxonomy of educational objectives, Part 1: The cognitive domain.* New York: David McKay.

Boesen, J. (2006). *Assessing mathematical creativity: comparing national and teacher-made tests, explaining differences and examining impact.* Doctoral thesis. Umeå: Department of Mathematics and Mathematical Statistics, Umeå University.

Boesen, J., Lithner, J., & Palm, T. (in press). The relation between types of assessment tasks and the mathematical reasoning students use. Accepted for publication in *Educational Studies in Mathematics.*

Brousseau, G. (1997). *Theory of didactical situations in mathematics.* Dordrecht: Kluwer.

Doyle, W. (1988). Work in mathematics classes: The context of students thinking during instruction. *Educational Psychologist, 23,* 167–180.

Duval, R. (2002). Proof understanding in mathematics: What ways for students? *In Proceedings of the 2002 International Conference on Mathematics: Understanding Proving and Proving to Understand* (pp. 23–44). Taipei: National Taiwan Normal University, Department of Mathematics.

Harel, G. (2006). Mathematics education research, its nature and its purpose: A discussion of Lester's paper. *Zentralblatt fuer Didaktik der Mathematik, 38*(1), 58–62.

Harel, G., & Dubinsky, E. (Eds.). (1992). *The concept of function: Aspects of epistemology and pedagogy.* Washington, DC: Mathematical Association of America.

Harel, G., & Trgalov´a, J. (1996). Higher mathematics education. In A. Bishop, K. Clements, C. Keitel, J. Kilpatrick, & C. Laborde (Eds.), *International handbook of mathematics education* (pp. 675–700). Dordrecht: Kluwer.

Hegarty, M., Mayer, R., & Monk. C. (1995). Comprehension of arithmetic word problems: a comparison of successful and unsuccessful problem solvers. *Journal of Educational Psychology, 87*(1), 18–32.

Hiebert, J. (2003). What research says about the NCTM standards. In J. Kilpatrick, G. Martin, & D. Schifter (Eds.), *A research companion to principles and standards for school mathematics* (pp. 5–26). Reston, Va.: National Council of Teachers of Mathematics.

Hiebert, J., & Carpenter, T. (1992). Learning and teaching with understanding. In D. Grouws (Ed.), *Handbook for research on mathematics teaching and learning* (pp. 65–97). New York: Macmillan.

Hiebert, J. & Lefevre, P. (1986). Conceptual and procedural knowledge in mathematics: An introductory analysis. In J. Hiebert (Ed.), *Conceptual and procedural knowledge* (pp. 1–27). Hillsdale, NJ: Lawrence Erlbaum Associates.

Långström, P., & Lithner, J. (2010). Students' learning strategies and mathematical reasoning in textbook-related metacognitive processes. In preparation.

Lithner, J. (2000a). Mathematical reasoning and familiar procedures. *International Journal of Mathematical Education in Science and Technology, 31,* 83–95.

Lithner, J. (2000b). Mathematical reasoning in task solving. *Educational Studies in Mathematics, 41,* 165–190.

Lithner, J. (2001). *Undergraduate learning difficulties and mathematical reasoning.* Doctoral thesis. Roskilde: Roskilde University.

Lithner, J. (2002). *Lusten att lära, Osby* (The motivation to learn, Osby). Skolverkets nationella kvalitetsgranskningar (Quality inspections of the Swedish National Agency for Education), (in Swedish).

Lithner, J. (2003). Students' mathematical reasoning in university textbook exercises. *Educational Studies in Mathematics, 52,* 29–55.

Lithner, J. (2004). Mathematical reasoning in calculus textbook exercises. *Journal of Mathematical Behavior, 23,* 405–427.

Lithner, J. (2008). A research framework for creative and imitative reasoning. *Educational Studies in Mathematics, 67,* 255–276.

Love, E., & Pimm, D. (1996). "This is so": A text on texts. In A. Bishop, K. Clements, C. Keitel, J. Kilpatrick, & C. Laborde (Eds.), *International handbook of mathematics education* (pp. 371–409). Dordrecht: Kluwer.

McGinty, R., VanBeynen, J., & Zalewski, D. (1986). Do our mathematics textbooks reflect what we preach? *School Science and Mathematics, 86,* 591–596.

National Council of Teachers of Mathematics (2000). *Principles and standards for school mathematics.* Reston, VA: National Council of Teachers of Mathematics.

Niss, M. (1999). Aspects of the nature and state of research in mathematics education. *Educational Studies in Mathematics, 40,* 1–24.

Niss, M. (2003, January). *Mathematical competencies and the learning of mathematics: The Danish KOM project.* Paper presented at the Third Mediterranean Conference on Mathematics Education, Athens.

Niss, M. (2007). Reactions on the state and trends in research on mathematics teaching and learning: From here to utopia. In F. Lester (Ed.), *Second handbook of research on math-*

ematics teaching and learning (pp. 1293–1312). Charlotte, NC: Information Age Publishing.

Palm, T., Boesen, J., & Lithner, J. (2005). The requirements of mathematical reasoning in upper secondary level assessments. *Research Reports in Mathematics Education 5,* Department of Mathematics, Umeå University.

Pirie, S., & Kieren, T. (1994). Growth in mathematical understanding: How can we characterise it and how can we represent it? *Educational Studies in Mathematics, 26,* 165–190.

Pólya, G. (1954). *Mathematics and plausible reasoning, Vol. I and II.* Princeton, NJ: Princeton University Press.

Schoenfeld, A. (1985). *Mathematical problem solving.* Orlando, FL: Academic Press.

Schoenfeld, A. (1991). On mathematics as sense-making: An informal attack on the unfortunate divorce of formal and informal mathematics. In J. Voss, D. Perkins, & J. Segal (Eds.), *Informal reasoning and education* (pp. 311–344). Hillsdale, NJ: Erlbaum.

Sfard, A. (1991). On the dual nature of mathematical conceptions: Reflections on processes and objects as different sides of the same coin. *Educational Studies in Mathematics, 22,* 1–36.

Sierpinska, A. (1996). *Understanding in mathematics.* Routledge: Falmer.

Skemp, R. (1978). Relational understanding and instrumental understanding. *Arithmetic Teacher, 26*(3), 9–15.

Stacey, K., & MacGregor, M. (1999). Taking the algebraic thinking out of algebra. *Mathematics Education Research Journal, 11,* 24–38.

Sumpter, L. (2007). *Affect and reasoning.* Licentiate thesis. Umeå: Department of Mathematics and mathematical statistics, Umeå University.

Tall, D. (1991). Reflections. In D. Tall (Ed.), *Advanced mathematical thinking* (pp. 251–259). Dordrecht: Kluwer.

Tall, D. (1996). Functions and calculus. In A. Bishop, K. Clements, C. Keitel, J. Kilpatrick, & C. Laborde (Eds.), *International handbook of mathematics education* (pp. 289–325). Dordrecht: Kluwer.

Tall, D. (1999). The chasm between thought experiment and formal proof. In G. Kadunz, G. Ossimitz, W. Peschek, E. Schneider, & B. Winkelmann (Eds.), *Mathematical literacy and new technologies* (pp. 319–343). Conference: 8 internationales Symposium zur Didaktik der Mathematik, Klagenfurt (Austria).

Tall, D. (2004). Thinking through three worlds of mathematics. In M. Johnsen Høines & A. B. Fugelstad (Eds.), *Proceedings of the 28th Conference of the International Group for the Psychology of Mathematics Education* (pp. 281–288). Bergen, Norway: Bergen University.

Thompson, P. (1994). Students, functions, and the undergraduate curriculum. In E. Dubinsky, A. Schoenfeld, & J. Kaput (Eds.), *Research in collegiate mathematics education I, CBMS Issues in mathematics education* (pp. 21–44). American Mathematical Society.

van Hiele, P. M. (1986). *Structure and insight. A theory of mathematics education.* Orlando: Academic Press.

Vinner, S (1997). The pseudo-conceptual and the pseudo-analytical thought processes in mathematics learning. *Educational Studies in Mathematics, 34,* 97–129.

Wiliam, D (2007). Keeping learning on track: Classroom assessment and the regulation of learning. In F. Lester (Ed.), *Second handbook of research on mathematics teaching and learning* (pp. 1051–1089). Charlotte, NC: Information Age Publishing.

Yackel, E. & Cobb, P. (1996). Sociomathematical norms, argumentation, and autonomy in mathematics. *Journal for Research in Mathematics Education, 27*(4), 458–477.

Yackel, E. & Hanna, G (2003). Reasoning and proof. In J. Kilpatrick, G. Martin, & D. Schifter (Eds.), *A research companion to principles and standards for school mathematics* (pp. 227–236). Reston, VA: National Council of Teachers of Mathematics.

CHAPTER 20

LEARNING FROM VARIATION

Differences in Learners' Ways of Experiencing Differences

Ulla Runesson

UNIVERSITY OF GOTHENBURG AND UNIVERSITY OF SKÖVDE

Angelika Kullberg

UNIVERSITY OF GOTHENBURG

Abstract: Our aim with this chapter is to present some studies in mathematics education framed by a specific research approach: phenomenography and variation theory. The theme of the chapter is "variation" and how that is related to learning. We describe how variation in how something is experienced is central to this research approach. We report on some empirical studies of mathematics learning in the early grades and account for what has been learned by such descriptions from a theoretical and practical point of view. We suggest that studies of variation in learning have contributed to a better understanding of the nature of learning, namely that an experienced variation is a necessary condition for learning. Finally, we will report on studies where variation is used deliberately for enhancing learning in mathematics.

Keywords: variation theory, phenomenography, learning study

INTRODUCTION

Why are some better learners than others? What is necessary to take into consideration when planning for learning? How could learning be improved? These questions could be answered differently depending on what perspective on learning and teaching that is taken. From the point of view of the research tradition presented in this chapter, we would reply to such questions by asking: "better at learning *what*?" The assumption is that it is only when we know *what* learners are expected to learn and *what* they actually learn in those situations that we can ask the initial questions. Learning is always learning of something; it has an object. In this chapter we present some studies of mathematics teaching and learning where the object of learning is central. These are framed within a tradition, phenomenography and variation theory, which started some thirty years ago at the University of Gothenburg, Department of Education, as a research approach for addressing certain educational issues. It has been adopted worldwide by researchers in different research fields of education and has recently developed into an emerging theoretical framework.

Initially, we report studies of mathematics learning within a phenomenographic research approach from which certain theoretical conclusions could

The First Sourcebook on Nordic Research in Mathematics Education, pages 299–317

have been drawn. Next, we describe how a theoretical framework, variation theory, has emerged on the basis of a number of phenomenographic studies and how it has been applied in research of classroom teaching and learning. We conclude the chapter with a report of learning study, an approach for promoting learning that is a teaching intervention framed by the variation theory framework.

VARIATION IN WAYS OF EXPERIENCING

Phenomenography developed during the early 1970s from an interest to understand better the nature of different learning outcomes by studying how learners experience the act of learning as well as *what* they had learned (Marton, 1981). It is best described as a research approach with some methodological implications and a particular research interest. The aim of phenomenographic research is to describe the qualitative differences in how *the same thing* (a phenomenon, a situation, or a problem) could be experienced. In this tradition, "experience" has a particular meaning. Experience is seen as an internal relation between the object and the subject; thus they are inseparable. What is experienced is neither a feature within the subject nor within the object, but constituted as a relation between them both, and hence, learning is seen as a change in this relation. A phenomenographic study takes a second-order perspective. The researcher is oriented to "the others'" ways of comprehending the world. As a consequence of the relational view of experiencing, it is not an individual's ways of experiencing that is accounted for, but *the different ways the subject–object relation could be constituted.* Differences in experiencing could be a difference within the individual or between individuals: "The objective of a study is to reveal the variation, captured in qualitatively distinct categories, of ways of experiencing the phenomenon in question, regardless of whether the differences are differences between individuals or within the individuals" (Marton & Booth, 1997, p. 124).

More than 30 years of phenomenographic research have provided insights, not just into the different ways something can be experienced, but also into the nature of and structures of ways of experiencing or understanding. To illustrate this, we will start with reporting on some studies of mathematics learning in the early grades.

Experiencing Numbers

One of several phenomenographic studies in mathematics education is Neuman's study (1987) of different strategies that young children use in the beginning of their first school year. One hundred and five children were interviewed about their line of reasoning in relation to specific tasks concerning counting objects as well as simple addition and subtraction tasks. Neuman's aim was to trace difficulties with learning simple arithmetic to the initial learning in early years. She had noticed that some students in high school had the same difficulties in arithmetic as young pupils have. Why had they not learned although they had been taught for many years? In her study she found that pupils' difficulties with early arithmetic could be traced to their ways of experiencing numbers. The different ways of experiencing numbers had impact on their capability of solving mathematical problems. There is an extensive literature of children's learning of simple arithmetic, and there is, to a great extent, also a consensus that modeling and counting strategies are the development path to learning arithmetic. Neuman came up with an alternative view, that the experiencing of structure—not counting—is the origin of arithmetic skills. Some of the children who experienced numbers as "finger numbers" solved arithmetic problems by using their fingers—not for counting—but for "seeing" the number. When one uses fingers for counting, the fingers are used for "keeping track." When one uses "finger numbers," the number is structured as a part–whole relation. The fingers (hands) are used to structure the numbers (0–10) and make it possible to visually experience the number structure. For instance, the "sevenness" of seven becomes visible as a structure by visually seeing five (undivided) fingers on one hand and two on the other. This offers an opportunity to develop a part–whole relationship. Structuring numbers promotes number sense, whereas counting single objects could lead to math difficulties, Neuman concludes.

Hence, Neuman found five different ways of experiencing number among the pupils: "numbers facts," "finger numbers," "counted numbers," "number as extents," and "numbers as names" (Figure 20.1). But what does "different ways of experiencing" really entail? The categories in Neuman's study are ordered in a hierarchy, starting from the level of experiencing numbers only as names to the "higher" level of experiencing numbers as number facts.

If we take a closer look at what distinguishes the different ways of experiencing, we can conclude that the difference between ways of experiencing numbers has to do with what features of numbers that are focused on simultaneously. Starting with analyzing "numbers as names" and "numbers as extents" from the point of view of what aspects are focused on, only one of the two aspects, the ordinal or the cardi-

number facts	Reflects a complete experience of numbers and their part-whole relationships. A simultaneous experience of cardinal and ordinal aspects is reflected by the way problems are handled, even if there is no conscious awareness of them.
finger numbers	Numbers are laid on fingers and made visible. By arranging the largest part first there is usually a pattern of a full hand (the largest undivided five) plus odd fingers. Numerosity (cardinal aspect) and ordered items (ordinal aspect) can be experienced simultaneously, as well as parts and whole.
counted numbers	Ordinal and cardinal aspects of numbers are present in parallel, side by side and separately. Track is kept of numerals that are spoken by counting, either by speaking other numerals or by setting up fingers.
numbers as extents	Focus is only on the cardinal aspect of number. Children experience wholes and parts as vague approximate numerosities without experiencing the units within the parts.
numbers as names	Focus is on the ordinal aspect of number. Children are aware of the part-whole relationships. They are also aware of the ordered set of units, and the focus on the last units in each of the two parts.

Figure 20.1 Five ways of experiencing numbers (from Marton & Booth, 1997).

nal aspect, is attended to or discerned. In "counted numbers" and "ordinal numbers," both the ordinal and the cardinal aspect are focused on. In "number facts" (note, it is not about knowing the addition/subtraction table by heart), the ordinal and the cardinal aspects are focused on simultaneously with the part–whole relationship.

Neuman suggested that to be able to develop good concepts of numbers, the learner has to structure the numbers and discern their part–whole relationship simultaneously with the ordinal and the cardinal aspects. When experiencing numbers as "finger numbers," the number is experienced as a structure. This makes it possible to visually see ordinal and cardinal features at the same time.

Experiencing Arithmetic Problem Solving

From Neuman's study it could be concluded that the aspects that are focused on simultaneously is important for how something is experienced or understood. Another phenomenographic study about children's mathematical learning, carried out by Ahlberg (1992), demonstrated the same thing. She conducted an intervention study with a designed mathematical problem-solving sequence in primary school. The study also included interviews with 38 primary school children. The intervention study had a quasi-experimental design that included five different types of mathematical problems. The designed problem sequence included story problems with no arithmetic content, story problems with arithmetic content, numerical calculations that elaborated on the students' own stories, story problem with estimation, and textbook problems. During these problem-solving activities, the pupils were encouraged to use communicative skills such as drawing, writing, and talking to their peers. A sub-goal for the problem solving was to see if different methods of solutions could contribute to understanding. Three classes and a control group participated in the study. The analyses of the pre- and posttests showed that the participating classes solved the problem to a higher degree than the control group.

Four different approaches were found in the pupils' problem solving. The first approach (gestalt approach) was signified by pupils having an open orientation towards all the parts of the content and the internal relationship of different parts of the mathematical problem. The second approach (hypothesis approach) described an open orientation towards numbers, operation, the problems' content, and their internal relationships, but pupils do not relate to the relevant parts. In the third category (procedure approach) pupils focused solely on numbers and operations in comparison to the fourth identified approach (operand approach), in which pupils focus only on the numbers (Ahlberg, 1992). The different approaches had an impact on the pupils' success with solving mathematical problems. Ahlberg states that to enable the pupils to shift from one approach to the other, teaching should take the point of departure in children's own way of experiencing the problem and the problem-solving process. Ahlberg found that different forms of communication while doing mathematics (drawing, writing, talking) helped the pupils differentiate the content of the problem and helped to focus on different parts of the problems. However, it does not determine the success of problem solving. Alhberg (in press) concludes that "the outcome of the problem solving depends instead on the extent to which the pupils simultaneously focus on different aspects of the content and integrate them to a composite entity" (p. 113). Alhberg's work demonstrates that it is not the methods used that make the difference for students' learning but rather what the pupils pay attention to and discern in the problem-solving process. In this way Ahlberg's study supports Neuman's results about the importance of what features are focused upon by students.

WHAT WAS LEARNED FROM DESCRIPTIONS OF VARIATION IN LEARNING? THE THEORETICAL TURN

The ultimate goal of phenomenographic research is to come up with a description of different ways of experiencing something. Initially, this approach was inductive and empirically driven. During the last decade, a more theoretical orientation within this approach has emerged. Based on a synthesis of hundreds of phenomenographic studies, an embryo to a theoretical framework—variation theory—was presented by Marton and Booth (1997) and Bowden and Marton (1998). Hence, variation theory is grounded in empirical studies of learning and implies a theoretical turn of the research approach. Variation theory is a theoretical framework of learning and experience, but with practical implications. In a series of studies, variation theory has been used for analyzing classroom learning from the point of view of what is made possible to learn, what is actually learned, and how to provide better learning. Some examples of this will be given in the following sections.

Variation theory is a conceptual framework for understanding the nature and structure of learning. It is based on theories of the nature of human awareness, that when we direct our awareness to something, some aspects become more in focus than others, which are un-discerned or not attended to (Gurwitsch, 1964).[1] As was demonstrated in the studies referred to above, what aspects are discerned simultaneously is of critical difference for how something is experienced. Thus, "a way of experiencing" implies a description of discerned aspects present in the awareness simultaneously. The variation between different ways of experiencing the same thing "derives from the fact that different aspects or different parts of the whole may or may not be discerned and be objects of focal awareness simultaneously" (Marton & Booth, 1997). From this it follows that a certain way of understanding takes the simultaneous discernment of certain critical aspects. For instance, as has been discussed in the previous section, in order to understand the first ten numbers, it is necessary to discern the cardinal, the ordinal, and the part–whole relationship aspects simultaneously (Neuman, 1987). The same thing is the case for arithmetic problem solving: Pupils were more successful if they simultaneously focused on aspects of the content and the structure when solving mathematical problems. Discerning content and structure at the same time was thus critical for successful problem solving.

VARIATION IN POSSIBILITIES TO LEARN IN MATHEMATICS CLASSROOMS—REVEALING VARIATION IN DIMENSIONS OF VARIATION

The main idea of variation theory, that the discernment of aspects is critical for learning, has been used as an analytical tool in several studies of the mathematics classroom (Emanuelsson, 2001; Emanuelsson & Sahlström, 2008; Häggström, 2008; Liljestrand & Runesson, 2006; Olteanu, 2007; Runesson, 1999; Runesson & Mok, 2005). In these studies, two or more classrooms where the same topic was taught have been studied. The lessons have been video-recorded, and in some cases, students' learning outcomes have been studied by interviews or tests. The point of departure is that *what* is made possible to learn in the mathematics classroom is significant. From other perspectives it may be argued that teaching arrangement, organization, or the possibility to communicate and interact are critical for learning. Taking a variation theory perspective, it is not denied that this is of significant importance. However, it is assumed that in communication, for instance, there is always *something* communicated. When a learning situation is arranged in a certain way, it is arranged in order to learn *something*. The goal for learning is the development of a certain capability. There is something beyond communication and interaction (Runesson, 2005).

However, the object of learning is not equivalent to learning objectives. It refers to a certain capability that the students are expected to develop during a lesson, or during a limited sequence of lessons (Marton & Pang, 2006). Capabilities are not confined to theory or concepts only. They can also be associated with skills, values or attitudes. Every capability has two aspects, one general and one specific (Marton & Booth, 1997, p. 84). The general aspect refers to the nature of the capability—for instance, "understanding," "calculating," "explaining," or "describing." The specific aspect refers to what is acted upon, hence the subject matter, such as multiplication with two digits, linear equations, or Pythagoras' theorem. To be able to afford vital learning possibilities for the learner, it is probably important for a teacher to focus on both aspects. Hence, teachers should be aware not only of what the students are trying to learn, but also of the way the students master that which is learned. How the object of learning is enacted in the classroom and experienced by learners has been the focus of research in several studies in mathematics education, some of which we will describe in the following sections.

To reveal and describe the object of learning, principles from variation theory have been applied when analyzing classroom data. From a variation theory perspective, learning implies awareness of certain aspects of what is learned. From this principle of discernment, it follows that if different ways of experiencing derive from differences in discernment of aspects of that which is learned, then differences in what aspects are presented simultaneously (and thus made possible to experience) in the learning situation must be important for what is made possible to learn. Thus, in order to describe what is made possible to learn, the aspects present and made possible for the learners to discern must be identified.

However, variation theory also says that an aspect can only be discerned if it is discerned as a dimension of variation (Bowden & Marton, 1998). In order to know what something is, one must know what it *is not*. To know, for instance, what "seven" is, you must have experienced other numbers. In the same way, the meaning of positive numbers emerges in the contrast of negative numbers, for example. One must have experienced a variation of numbers or number systems. To understand the idea of proof it is not sufficient to have experienced just one proof, since the understanding of a general idea springs from having experienced different examples of the idea.[2] So, to make the discernment of an aspect possible, the aspect must be opened up as a dimension of variation. This could be made spontaneously by the learner herself or with the help of the teacher (Runesson, 2006). Therefore, the variation that the learners are exposed to is of prime interest when studying possibilities to learn in the classroom.

Variation in Possibilities to Learn Fractions

Runesson (1999) studied five experienced teachers teaching fractions and percentage in grades six and seven in Swedish compulsory school. The aim was to reveal differences in how *the same* topic was handled when discussed and interacted about in the classroom. This was done by analyzing what aspects the teacher (or the learners as well) opens up as dimensions of variation and thus draws the learners' attention to. In this way, different patterns of variation and invariance constituted in the classrooms were identified. For instance, she found that in some cases *different ways of solving the same problem* (e.g., ¾ of 12) were presented to the learners. Thus, the method of calculating was opened up as a dimension of variation, whereas the example was invariant. In other cases, it was the opposite: Only one method was presented (invariant) and applied to different examples (varied), and thus the example was opened up as a dimension of variation. From a variation theory perspective, this is of importance, since what is opened up as a dimension of variation is likely to be discerned. When the method varies and the example remains invariant, it is possible to learn how to solve the same example with different methods. In the opposite case—the method is invariant and the example varies—it is made possible to learn how to solve different examples with the same method. In the same way, how other features of fractions and percentages were presented in terms of patterns of variation was investigated. From this it was concluded that what was made possible to learn about fractions and percentages was different in the five classrooms. These differences existed even though the same topic and teaching unit was taught, the same assignment was talked about, and the lessons were arranged in the same way.

Runesson also found that how the content was handled, in terms of patterns of variation and invariance, was to a great extent unreflected upon by the teachers. When they talked bout their teaching (before and after the lessons) in the interviews, they merely focused on teaching arrangements, organization, or what hands-on-material to use. However, Runesson found that in the enactment, they constituted a space of learning (Marton, Runesson, & Tsui, 2004) by opening up for variation in some dimensions of the mathematical content, which they seemed not to have reflected about.

Data from recorded lessons could be analyzed in different ways depending on the purpose and theoretical assumptions taken. Runesson's study implied something new in the research of the mathematics classroom. In her study, the same theoretical concepts were used for analyzing teaching as well as learning. In this way, learning and teaching were brought closer together by describing them in commensurable ways. Furthermore, the aim of the study was not to study possibilities to learn in a general sense, but possibilities to learn the same topic. It was demonstrated that it was possible to reveal variation in what was made possible to learn about a mathematical concept or procedure in different classrooms.

Drawing on the conclusion that what is varied and what is invariant create both constraints and possibilities for learning, mathematics classrooms in different school cultures have been studied with a similar design as was used in Runesson's study—that is, the same topic taught in different classrooms (e.g., Emanuelsson & Sahlström, 2008; Liljestrand & Runesson, 2006). In order to better understand their own mathematics school culture—not to make generalizations about different school cultures—Runesson and Mok (2005) studied mathematics lessons in Hong Kong and Sweden. How the same topic—comparison of fractions with different denominators—was taught was the object of research. The aim was to identify differences in the two cultures by studying how the object of learning was enacted as the learners and the teacher interacted in the classroom. In comparative studies like this, several cultural differences can be identified, of course. For instance, the curricula in the two countries were very different. In Hong Kong this topic is introduced in primary four (age 10), and in Sweden usually later (grade six and seven). Since the study drew on data from the above reported study by Runesson, the pupils were older than the pupils in Hong Kong.

Runesson and Mok (2005) identified a striking difference with respect to the construal of fractions. The Hong Kong lesson showed a pattern of variation consisting of many dimensions of variation. To mention some, in the Hong Kong lesson, alternative methods for producing an equivalent fraction with a larger denominator was presented. The pupils were encouraged to find a method for the task given: "___ is larger than ___ by ___." The different methods (diagrammatic and amplification) suggested by the pupils were discussed in the lesson. The method was opened as a dimension of variation. Another variation opened up was the variation of the fractional parts of different wholes. Once the pupils had been exposed to different methods to compare—for example, 5/8 and 1/2—and had concluded that 5/8 is greater than 1/2, the question: "Is it possible for 5/8 to be smaller than 1/2?" was raised by the teacher. Seeing 5/8 as numerical values, 5/8 is greater than 1/2. In order to see it in another way, it is necessary to consider the fractions as a fractional part of a whole. By guiding the pupils to think of pizzas of different sizes, they were able to see a different meaning of fraction, a fractional part of a whole. In this way it was made possible for the learners to focus on various aspects of comparison of fractions *simultaneously.* This simultaneous opening of dimensions of aspects (i.e., several aspects of fractions was possible to experience in the same lesson) was characteristic for the Hong Kong lesson, whereas in the Swedish data a more sequential opening of dimensions of variation was identified. For instance, that different fractions can have the same value was talked about in lesson one. The teacher showed a chocolate bar (squared 3×4) and explained that three twelfths equals one quarter. This idea was brought up again in the next lesson. However, this time a new teaching aid was used for presenting a diagrammatic method. Now the task was to mark 1/4 of a circle partitioned into eights in order to realize that 1/4 = 2/8. In the following lesson (lesson 3), again two fractions with different denominators were compared, this time with yet another diagrammatic representation used. So, in this case only a variation of diagrammatic methods was opened and made possible for the pupils to experience. Furthermore, no connections were made between the methods of comparison presented in the different lessons. Runesson and Mok found other instances in the Swedish data showing how aspects were presented as disjoint instances without any connections or references to what the pupils had experienced previously. In this way, the enacted object of learning appeared to be more sequential in the Swedish classroom as compared to Hong Kong. The Hong Kong lessons were condensed, in that they showed a pattern of variation of many juxtaposed dimensions of variation of different aspects of comparing fractions. In contrast, the Swedish learners were exposed to a different pattern of variation with a sequential, widespread character with little connections between the concepts and skills between the lessons. The authors concluded:

> Therefore, it is unlikely for students to discern how these are related. In the Hong Kong class however, the students may experience the intertwined relationship between these dimensions of variation that form a special arrangement or simultaneity of variation. Such experiences give the students a chance to consider several aspects of the object of learning simultaneously and to see a connection between the critical aspects of the mathematical object—where such an opportunity is less likely in the Swedish case. Therefore, the possibility of making connection makes a significant difference. (Runesson & Mok, 2005, p. 12)

Variation in Possibilities to Learn Equations

In a more extensive study of how the same topic is dealt with in different school cultures, Häggström (2008) found similar differences. Sixteen lessons from a total of six classes in Sweden, Hong Kong, and Shanghai were video recorded,[3] analyzed, and compared to find how systems of linear equations in two unknowns were taught. A wider and more complex space of opened dimensions of variation was found in the Chinese classrooms compared to the Swedish classrooms. Häggström explains this as a consequence of a deliberate and systematic approach among the teachers to direct the learners' awareness to important features of mathematics. Fewer aspects of the topic were taken for granted in the Chinese classrooms. For instance, the concept of systems of linear equations was not introduced as *the* system. It was not taken for granted that there are different types of systems. Häggström found this a bit paradoxical. When teaching mathematics it is always necessary to take something for granted, you cannot "start from scratch" every time. The more advanced and familiar with the topic you assume the learners are, the more you can reasonably take for granted that they are familiar with. In Häggström's study, the Chinese students were already familiar with the topic, so it would have been possible to take many aspects for granted. However, the opposite was found. Many dimensions of variation were opened and few aspects were taken for granted.

What is then the danger of taking something for granted and hence not open up for variation in that dimension? Häggström gives an example: If a student has not experienced variation in the kind of letter used and has only met the letters *x* and *y*, would it be possible for her to understand and solve a system of equations where the unknown are *u* and *v*? There is no direct answer to this, of course. However, it is more likely that she will be able to cope with such equations, if she has this experience. A classroom situation is complex and, although one aspect may remain invariant in the lesson, the students themselves can open dimensions of variation on their own, or a dimension may be opened but still not discerned. What can be said is *whether it was possible to discern certain critical features or not.*

From the described studies it has been suggested that different patterns of variation, in different combinations and structures, create different learning opportunities. Do such subtle differences in how the object of learning is handled matter? Do they have an impact on students' learning outcomes? In a similar way as Häggström (2008) did, Olteanu (2007) studied differences between how teachers in upper secondary school handled the same content (second-degree equations and functions). To be able to draw conclusions about how the content was handled and the students' learning she also collected data from four tests given to all participating students. In the study, Olteanu observed two teachers and conducted interviews with them and with eight students, video recorded lessons, and recorded individual sessions in the lesson. The teachers used the same mathematics textbook. In the analysis, Olteanu focused on which aspects of the content the teacher, the students, and the textbook brought up during the lessons and compared them with what the students focused on when they solved exercises. Also, aspects ignored (or taken for granted) by the teacher and the textbook (cf. Häggström, 2008) were reported. She found that one of the two teachers brought up more features of second-degree equations and functions than the other teacher (and the textbook), in both implicit and explicit ways. This had an impact on students' ways of solving tasks. The way the content was handled in terms of aspects focused on was reflected in what students focused on when solving the corresponding tasks.

DESIGNING LEARNING FROM VARIATION

The development of variation theory, in this chapter previously described as a "the theoretical turn" of phenomenography, made it possible to analyze classrooms and teacher practice. But was it possible to use the theory to enhance practice too? Runesson (1999) found that a pattern of variance and invariance is always present and possible to notice in a lesson. However, teachers did not seem to be aware of this, nor did they deliberately try to make use of variation as a means to promote learning. Could variation theory be a tool for teachers to plan and analyze their practice? The research presented in this section answers these questions and elicits the relationship between teaching and learning in classrooms. Teaching and learning are not described in general, but instead very specifically and on a detail level, using variation theory as an analytical tool. The focus is on the capabilities that the students are intended to develop, and the aim of enhancing and promoting student learning is central.

The background to a new era of intervention studies, conducted within framework of variation theory, can be found in results from the international TIMSS (Trends in Mathematics and Science Study) video study presented in the book *The Teaching Gap* (Stigler & Hiebert, 1999). In the analysis of the study, differences between the ways teachers teach mathematics in different countries were found. The Asian teachers handled the content in the lessons in a different and, it seemed, a more powerful way than their Western colleagues (cf. Ma, 1999). In the Asian countries, students were also more successful in mathematics performance. One explanation for this finding was the Japanese model for teachers' in-service training and development called *lesson study* (Lewis, 2002; Yoshida,

1999). In a lesson study a group of teachers collaboratively develop a lesson, implement it, and make revisions of the lesson in an iterative process. The collaborative effort of working together with planning and discussing the content of the lesson developed the teachers' teaching skills in a way that enhanced student learning (Stigler & Hiebert, 1999).

Learning Study

In the early 2000s, Ference Marton developed an approach, *learning study*, together with researchers at Hong Kong University and Hong Kong Institute of Education (Lo, Pong, & Chik, 2005; Marton & Tsui, 2004; Runesson, 2008). Learning study was inspired by the above described lesson study model but has a stronger theoretical underpinning. The use of theory is in a similar way emphasized as in design experiments (Brown, 1992; Cobb, Confrey, di Sessa, Lehrer, & Schauble, 2003; Collins, 1992), both as a means to use and develop theory in intervention studies. Over a 120 learning studies have been conducted in a project (VITAL project) over three years in Hong Kong. In Sweden, learning study was implemented by a research project[4] in 2003 (Gustavsson, 2008; Holmqvist, 2006; Holmqvist, Gustavsson, & Wernberg, 2008; Kullberg, 2004, 2006; Kullberg & Runesson, 2006; Runesson, 2007, 2008) in mathematics, English, and Swedish.

A learning study (LS) is an intervention model in which teachers work together in teams designing lessons on a specific topic (often in collaboration with a researcher) with the aim to promote students' learning. It is a cyclic process of planning and evaluating one lesson (or a series of lessons) and follows the following steps (see Figure 20.2):

1. The cycle starts with choosing an *object of learning*. An object of learning is a certain capability or understanding that the teachers want the students to develop—for example, students' understanding of the equal sign. In the process of LS, the teachers are discussing the nature of a capability: What does it imply to understand the equal sign?
2. The second step is about finding out the students' pre-knowledge. This is done by designing a test that would give information about students' difficulties in the selected content area. The test is implemented in all the participating classes and is analyzed by the teachers.
3. The third step involves a collaborative planning of the lesson. The teachers discuss and plan in what way the object of learning should be brought up during the lesson. Also, methods and tasks to be used are discussed, though the main focus is on the object of learning: How does the object of learning connect with other knowledge? What features must the students discern? How can these features be brought out in the learning situation (by variation)?
4. After careful planning of the lesson, it is implemented by one of the teachers in his or her class. The lesson is video recorded.
5. The next step after the implemented lesson is the post-test. A few days after the lesson, the students who participated in the lesson take a test (in most cases the same as the pre-test). The post-test gives the teachers valuable information about how the students responded to the actual lesson and what they learned.
6. At the post-lesson meeting, the video-recorded lesson is analyzed together with the pre- and post-tests. Conclusions drawn from the results and the analysis contribute to the planning of Lesson 2. The cyclic process takes us back to step 4 in the model (see Figure 20.2). Lesson 2 is implemented by another teacher in his or her class (with a new group of students). After the analysis of Lesson 2, the lesson is revised again, and Lesson 3 is planned, implemented (with a new group of students), and analyzed.
7. An important last step in the LS cycle is to report findings to other teachers.

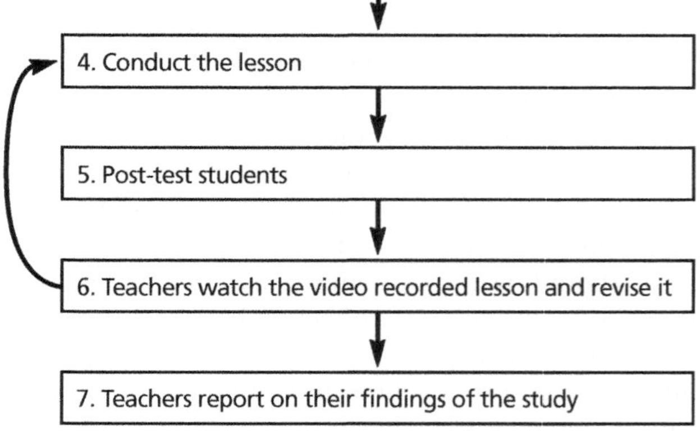

Figure 20.2 The learning study cycle.

Identifying Critical Aspects

In a learning study, teachers and researchers are involved in the production of knowledge. They start with identifying the object of learning (a certain capability they want the students to develop). They work systematically with a particular method and with a theory as a guiding principle, gathering and evaluating data about learning and teaching in their own practice in order to find the critical aspects/features for an object of learning.

The notion of "critical aspect" was previously introduced in the section about studies in early mathematics. We described how certain ways of experiencing entailed the discernment of certain aspects, while other ways of experiencing did not (Figure 20.1).

Variation in discerned aspects is the critical difference between the different ways the same thing may be experienced. A critical aspect/feature (Marton et al., 2004) could be seen as a necessary condition for learning something particular. The critical features are not critical in general; they are critical for the group investigated and probably for similar groups within the same age and experience.

If you, for example, want your students to learn what an angle is, what features of angles is it important that the students discern? What do they need to see and understand in order to grasp what an angle is? In a learning study about angles (Runesson, 2007), the teachers, due to the iterative analysis, evaluation, and re-designing of a lesson, found critical features for student learning about angles. The object of learning concerned the capability to be able to recognize the 180-degree angle, to differentiate an angle from a non-angle, and to see that the size of the angle is independent of the length of its "legs" (sides). For example, to be able to see that the length of the legs is not a part of the actual size of the angle, the teachers and researcher planned for this feature to be brought up in Lesson 1. During the analysis of the first (video-recorded) lesson in the cycle and the pre- and post-test, the teachers found that they had not succeeded in making the students discern this feature since only 33% of the pupils showed on the post-test that legs/sides are not part of the angle itself. Consequently, the pupils who did not answer this gave the answer b (see Figure 20.3) to the task of which of the two angles was the biggest.

Further, in the analysis of Lesson 1, the teachers experienced that they needed to make this feature of the angle as a turn more explicit in Lesson 2. In Lesson 1 the feature of the length of the legs was brought up in terms of the teacher telling the pupils and repeating such statements as "an angle is two lines (sides) and a point" and "the lengths of the legs do not have to be equal in size":

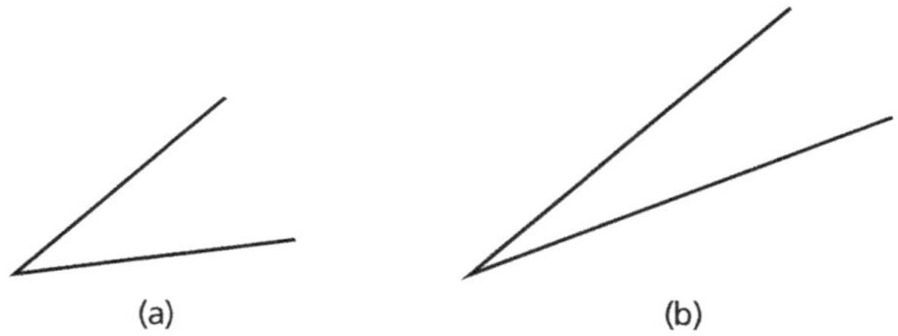

Figure 20.3 Compare the angles. Which one is the biggest?

> For example, they came to the conclusion that the pupils see the sides as constituent part of the angle, and this way of understanding is reasonable and logical from the point of view of how the definition of an angle was given in the lesson. So, they decided to come up with another definition in the next lesson and bring out the idea of angle as a turning more clearly. (Runesson, 2007 p. 12)

It was found that that this decision of bringing out the concept of angle as a turn (dynamic) was more successful than the definition used in Lesson 1 (static). The result on the post-test after Lesson 2 showed that 95% of the pupils could differentiate between angles with different sizes and different lengths in legs (see Figure 20.3). Comparing angles (on different sizes of clock-faces) and making the angles smaller and bigger by "turning" the legs, as in Lesson 2, were necessary conditions for learning what was intended. A critical feature for learning was, for this group of students, to see that the size of the angle is dependent on the turning of the legs. This feature made it possible to separate the legs from the turning. Hence the legs were constant and did not vary, but instead it was the turning that made the angle vary in size. The turning was seen as critical for learning and was even brought out in Lesson 3 with similar result on student learning, where 78% of the pupils answered correctly on the items about comparing angle size (see Figure 20.3).

Marton and colleagues (2004) argue that "the critical features have, at least in part, to be found empirically—for instance through interviews with learners and through the analysis of what is happening in the classroom—and they also have to be found for every object of learning specifically, because the critical features are critical features of specific objects of learning" (p. 24).

It is thus not favoring learning to simply tell the students the critical features; these must be discerned. There is a significant difference between "being told" something and "experiencing a variation of different aspects" of an object of learning. This implies that the teacher must focus the students' attention on these critical features/aspects.

> What is of decisive importance for the students, is what actually comes to the fore of their attention, that is, what aspects of the situation they discern and focus on. In the best case, they focus on the critical aspects of the object of learning, and by doing so they learn what the teacher intended. But they may also fail to discern and focus on some of the critical aspects, or they may discern and focus on other aspects. (Marton et al., 2004, p. 5)

In a LS, variation theory is used as a means to plan and analyze the lessons. Once the teachers have an idea about what may be critical aspects, they plan a pattern of variation and invariance to implement in the classroom in order to make that aspect noticeable for the learners. For instance, to help the pupils to experience the change of the angle as a change of a turning (not a change of the lengths of the arms), the pattern planned was a systematic variation of the size of the angle while keeping the length of the arms invariant, and vice versa.

The Need for Collaborative Work

Learning study may be described as a form of action research made by teachers with the specific goal of finding out what matters for student learning of a particular capability, such as understanding the nature of the size of an angle in an adequate way. Our experience of LS is that it is an effective way to investigate and come up with descriptions of what matters for learning. In a learning study, three to five teachers with subject matter knowledge and teaching experience meet, often in collaboration with a consultant from the university, and share and develop their expertise over a long period of time. Although the LS takes a lot of effort from the participants, it is rewarding in terms of new knowledge and professional skill. Our experience of working with teachers suggests that, by deeply investigating the particular, they can gain knowledge that is of general character also. So in that sense, the benefit from a learning study might have an effect on teachers' work in general, just as in lesson studies (Lewis, Perry, & Murata, 2006; Yoshida, 1999).

A Learning Study about the Density of Rational Numbers

In this section we report more in detail on the analysis of a learning study about decimal numbers conducted in three classes in grade six, with 12-year-old students (Kullberg, 2004, 2007). In this study the students were supposed to understand that there is an infinite amount of numbers between two decimal numbers. The teachers chose this object of learning since they all thought that this was a topic that was hard for students to understand and hard to teach in a good way (cf. Fischbein, Tirosch, & Hess, 1979; Monaghan, 2001; Moskal & Magone, 2000; Vosniadou, Vamvakoussi, & Skopeliti, 2008). Three teachers and a researcher worked together with the planning of the lessons and to implement them in different classes (see LS cycle, Figure 20.2). It is important to note that the teachers had the ownership of the study. The role of the researcher was merely supportive, and she encouraged the teachers to come up with ideas of their own and try them out. So the way the lessons in this study were designed and enacted was principally based on the teachers' own ideas.

A pre-test was given the three participating classes. One of the tasks on the test was about whether there were one or many numbers in an interval between two decimal numbers:

Ann says there is a number between 0.97 and 0.98. John says there is no such number. Who is right and why?

In the analysis of the pre-test the following categories of answers to the task were found (all classes);

a. there is one number 0.97.5
b. there are many numbers between 0.97 and 0.98
c. there are ten numbers between 0.97 and 0.98
d. there is one number 0.975
e. there are no numbers between 0.97 and 0.98

The answers to the task varied to some extent between the three classes. In Class A (Lesson 1), only one student on the pre-test showed that there were many/infinite numbers in the interval (answer b), as opposed to four students in Class B (Lesson 2) and two in Class C (Lesson 3).

Lesson 1

In the first lesson, the teachers wanted the students to learn the object of learning by letting the students discuss how many numbers there were in the interval between two decimal numbers, in this case 0.97 and 0.98. This task, originated from the pre-test, the teachers found very useful because of the diversity of answers it generated. The plan for Lesson 1 was to let the students discuss the different answers that the students had given in the pre-test, in smaller groups. The groups would draw their answers to the task on a big poster, as a number line with the numbers between 0.97 and 0.98. In the end of the lesson they planned a discussion about the different answers to the task. The rationale for giving the students the categories of different answers was that the students themselves would pick the correct answer and develop an understanding of the density of rational numbers through the discussion.

However, the lesson did not generate the expected learning outcome since on the post-test only 21% of the students showed that there were very many or infinite numbers in an interval between two decimal numbers. From the pre-test there was an increase of three students who learned what was intended. In fact, about half of the students in the class thought after the conducted lesson that there were ten numbers and only ten in an interval as given in the task. This was an interesting fact, since none of the students showed that on the pre-test. In this case, half of the class had learned something that was not intended by the teacher. In conclusion, even if the density of rational numbers was brought up and discussed, it was not possible for most students to discern the infinite divisibility of decimal numbers during the lesson.

The teachers watched the video-recorded lesson and analyzed the post-tests. It is important to remember that the lesson was carried out according to the plan that the team had made. It was therefore the whole team's lesson that was evaluated and analyzed. *How was the content handled in the lesson? How was the student learning reflected in the tests? What was possible to learn during the lesson? What could be critical features?* In the analysis of the lesson, the teachers found that the teacher did not contribute with something that could have helped the students to enhance their understanding before group-work. The solutions to the task that different groups presented in front of the class did not develop an understanding of the infinity of decimal numbers automatically. Therefore, they decided to bring up connections to other rational numbers (fractions, percentages) in Lesson 2.

Lesson 2

Lesson 2 in the LS cycle was planned to bring out the connection between the rational numbers in an explicit way. The plan for the lesson was, in the beginning of the lesson, to introduce other rational numbers such as fractions and percentage and the part–whole relationship of a decimal number. The rationale for this was an idea that this would have an impact on the result of the group assignment and the students' oral account of the task. The rational numbers were introduced through a question about the decimal numbers 0.97 and 0.98. "Could 0.97 be said and interpreted in another way?" The students contributed with many different answers such as "97 hundredths," "970 thousandths," and "97 percent" that *represented the same number.* The part–whole relationship was elicited by representing 0.97 as fractions and percentage *at the same time.* This was also shown when the teacher asked the students if it was possible to take 97 hundredths of something. In this case the students contributed with suggestions like a ruler, a person, a pen, and a human being. It was shown that it is possible to take 0.97 of a whole, a ruler, with 0.97 in this case being seen as 97 centimeters of the one-meter ruler. After this introduction, the students worked in groups with the same assignment as in Lesson 1. In the oral account of the assignment, in the end of the lesson the teacher had the students reflect upon aspects of decimal numbers. For example, the teacher showed *the same number* on several rulers (Figure 20.4), *with different number of parts* (ten parts, hundred parts, thousand parts, etc.). By doing this, the teacher gave the students many examples of how the same number could be expressed in different number of parts and that this implied that the ruler could be "cut up" in smaller and smaller parts.

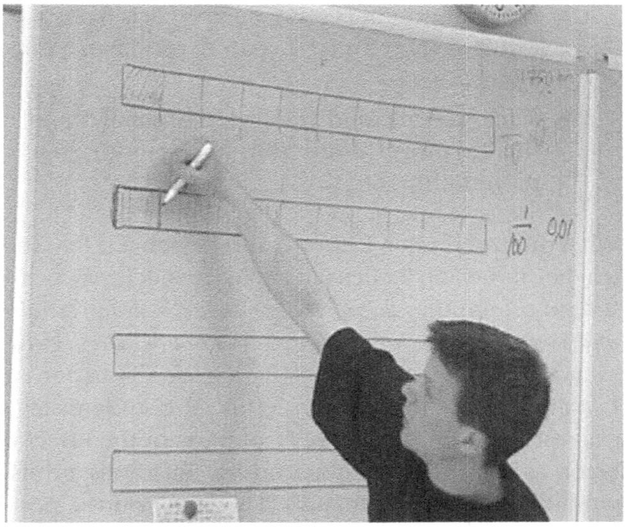

Figure 20.4 The picture is showing the teacher dividing several rulers in different number of parts.

The post-test showed that 94% of the students in Lesson 2 (all students except one) answered correctly. The teachers were pleased with the result and made no major revisions for Lesson 3.

Lesson 3

Lesson 3 was designed to bring up the features that seemed to be critical for student learning based on the analyses from Lesson 2. Theses features were

- Decimal numbers, seen as a form of rational numbers (fractions, percentages)

- Decimal numbers seen as part/parts from a whole (e.g., 0.97 as 97 centimeters from a one-meter ruler)
- The interval could be divided in smaller and smaller parts (hundredths, thousandths, ten thousandths, etc.).

What differs between Lesson 2 and Lesson 3 is that the teacher in Lesson 2 elicits in a more explicit way (with several rulers) that *the interval could be divided in smaller and smaller parts (hundredths, thousandths, etc.)*. However, this was elicited in the whole class discussion in Lesson 3, in another way, by letting the students give examples of different number of parts in the interval (such as hundredths, thousandths, ten thousandths, ten billionths, infinite many). The result on the post-test after Lesson 3 was similar to Lesson 2.

Conclusions and Further Analysis of the Lessons

One reason it was not made possible to learn (for most students) that there was an infinite number of numbers in the interval, in Lesson 1, could be that the numbers were treated as countable numbers on a number line. The group with the most advanced solution to the task, in the oral account, showed that there were nine numbers in the interval. One group said that there was one number, and three groups that said there were no numbers in the interval between the two decimal numbers. What the teacher and the students contributed during the lesson was obviously not enough to develop students' understanding. The team discovered, in the analysis of the lesson, that during the lesson the decimal numbers were treated as numbers on a number line, and no connection to other rational numbers (e.g., fractions or percentage) was made. In conclusion, in Lesson 1 decimal numbers were treated as countable numbers in an interval (0.971, 0.972, 0.973 etc.), as opposed to Lesson 2 and 3 where decimal numbers were treated as different numbers of parts in an interval (Figure 20.5). The different ways of handling the content contributed to different possibilities for learning the infinity of decimal numbers. On one hand, the variation that was possible for the students to discern was in Lesson 1, a variation of numbers in the interval. On the other hand, in Lesson 2 and Lesson 3 it was possible to discern different number of parts in the interval. This way of experiencing decimal numbers was more successful in terms of student learning.

The use of fractions contributed to a clearer conception of the decimal numeration system for the students. It has been found in previous studies that students often fail to establish the connection between decimals and fractions, and this seems to be critical for learning the density of rational numbers (cf. Moskal & Magone, 2000; Smith, Solomon, & Carey, 2005).

Further Analysis of the Learning Outcomes

The result from the post-test showed that the students in Lesson 2 and 3 improved the most in this study. The explanation for this was that critical features for learning were brought up during the lessons and had an impact on the student learning. One question to ask is whether all students benefit from a learning study. A recent study about the effect of learning studies for low-achievers (Gaversjö & Lööf Widing, 2008) found that in all five learning studies that were studied, the low achievers benefited the most.

Figure 20.6 demonstrates the learning progress in the study about decimal numbers referred to above. For each class/lesson, the five lowest scores on the pre-test were identified as the "low group," and the rest as "high group." As can be seen from Figure 20.6, in Lesson 1 the average score on the pre-test in the low group was 28% correct answers (on the whole test), and for the high group it was 50%. On the post-

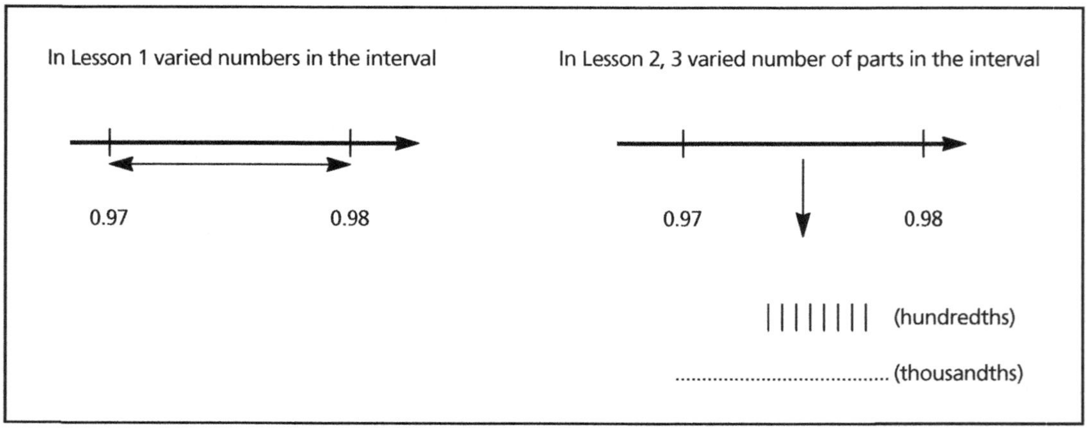

Figure 20.5 Two dimensions of variation for experiencing decimal numbers.

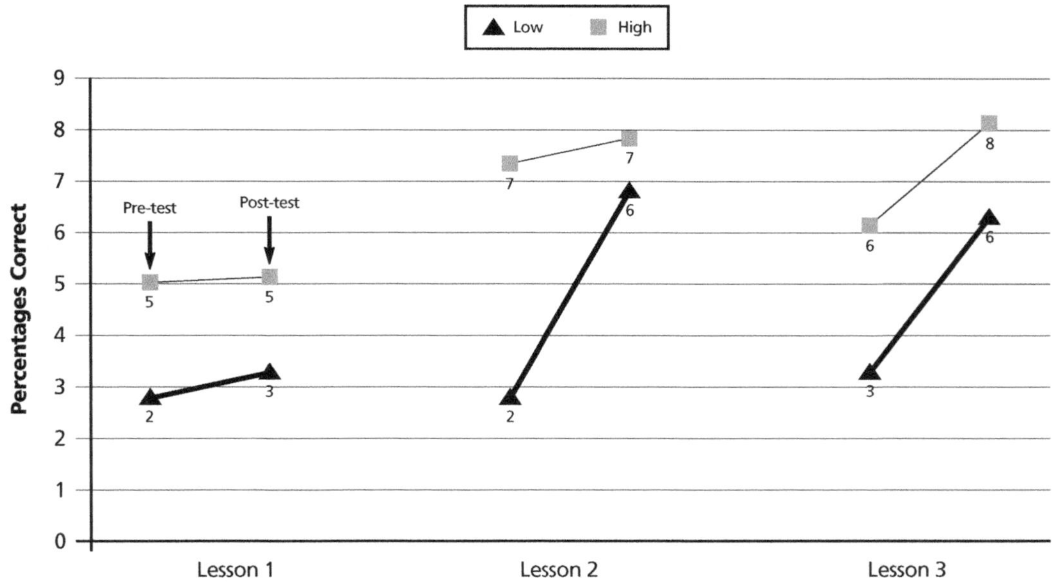

Figure 20.6 Results on pre- and post-test about decimal numbers (whole test) for high and low achievement groups respectively before and after lesson/class 1–3 (Gaversjö & Lööf Widing, 2008).

test (two days after the lesson), it was 33 and 51%, respectively. Thus, the gap between the two groups was almost the same after the lesson. However, for Class/Lesson 2, the low group had the same score as in Class 1, whereas the high group performed better on the pre-test compared to Class/Lesson 1. After the lesson, the low group had increased the average score from 28% to 68% correct answers. The high group had a fairly small increase of correct answers. In Lesson 2, the difference between the high and the low group score was remarkably decreased. A similar tendency was found in Lesson 3/Class 3.

Similar results were found in other studies: LS promotes student learning. In almost all of the approximately 20 studies made within the project, the results showed that the students' learning improved significantly (cf. Lo et al., 2005). The biggest difference between the lessons was often found between the first and the second lesson in the cycle. In most cases, the last lesson in the cycle generated the best learning outcomes. So the systematic collaborative work seems to be beneficial for student learning.

The Next Step: Implementing and Validating Variation

Analyses of learning studies raised questions about whether the good results concerning student learning could in any way be applicable for other groups of students. Although it has been argued that teacher action research, where teachers experiment with and gather data about their classroom performance, is a model that is likely to have an impact on improving teachers' classroom practice (Hill et al., 2008; Nuthall, 2004), it has been criticized for its context-specific nature (Hiebert, Gallimore, & Stigler, 2002) and its lack of generalizable knowledge that can be transferred from classroom to classroom (Nuthall, 2004). We took this questioning as a challenge. Studies that could validate the research results were needed. The next step was thus to implement and validate the results from LS in a form we call "design studies." In the design studies, the researchers reproduce lesson designs from learning studies with new teachers and students (Kullberg, 2007a, 2007b). The purpose of these studies differs from the learning studies: Whereas the aim with a learning study is to find the critical features for student learning through a systematic approach of analyzing teaching and student learning, the design studies aim to validate learning study findings. The design studies also aim at exploring in what way the results could be communicated to other teachers. *How could a lesson design be communicated and transformed in a way that is possible for other teachers to make use of?*

In design studies, new teachers are informed about the critical features identified in previous learning studies. They make use of such experiences, use them as resources, and try to make the critical features visible and possible to discern for new groups of students. The lessons are video recorded and analyzed together with pre- and post-tests. Variation theory is used in the analysis as a means to determine whether it is possible for the students to discern the critical features of the object of learning: Which patterns of variation are used to make the student discern the object of learning, and how is that variation brought about? In the analyses, the researchers try to relate how the mathematical content is related in the lesson to the students' learning, thus to study how the lesson design is reflected in the students' learning. It must be noted that when we talk about "lesson design," we do not mean "lesson plan." A lesson design is not a description of tasks, questions to ask, or teaching arrangements, for instance. It is a description of those aspects that have been identified as necessary for a particular group of learners, in a particular situation to discern in order to learn something in a certain way. It does not prescribe a certain way of interaction with the students. Rather, it does imply that variation (and invariance) is used to bring out the critical features in the learning situation. The tentative results from these studies show that it is possible to some extent to predict learning when the critical features for learning are brought up during lessons in a special way.

A Design Study About the Density of Rational Numbers

One recent study, based on the results from the learning study about density of rational numbers, is presented in this section (Kullberg, 2010). In this study, critical features found in the LS are reproduced. The *object of learning* was therefore the same, namely that the students should understand that there were infinitely many decimal numbers in an interval between two decimal numbers, though in this study there were two lesson designs. The lesson designs (LDs) are characterized by their content-specific critical features. The rationale for using two designs was to make it possible to compare their outcome in the analysis. The LD used in the reported study contains the following critical features:

Lesson Design 1 (LD1)

One critical feature of infinity of decimal numbers to be brought up during Lesson Design 1:

- Decimal numbers, seen as numbers (on a number line)

Lesson Design 2 (LD2)

Critical features of infinity of decimal numbers to be brought up during Lesson Design 2:

- Decimal numbers, seen as numbers (on a number line)
- Decimal numbers, seen as form of rational numbers (fractions, percentages)
- Decimal numbers seen as part/parts of a whole (e.g., 0.97 as 97 centimeters from a one meter ruler)
- The interval can be divided in smaller and smaller parts (hundredths, thousandths, etc.)

In this study, four teachers implemented two lessons each, with different lesson designs (Table 20.1). Lesson Design 1 (LD1) and Lesson Design 2 (LD2) were intended to bring out different critical features of infinity of decimal numbers (see above).

The Outcome of the Lessons

In this study, the researcher worked together with two teams of teachers. It should be noted that in this study the teachers, in each team (lessons 1–4 and 5–8), jointly planned the lessons together with the researcher. The teachers had therefore the same intentions and tasks planned for the lessons. The teachers watched and discussed the video-recorded lessons from the learning study about density of decimal numbers together with the researcher as a means to better understand the critical features. Note that it was not the entire learning study lesson that was to be reproduced, but only the critical features.

Each teacher taught two lessons, one of each design (see Table 20.2). The lessons were implemented in classrooms, with grade five (Y-school) and six (X-school) students (11–12 years old), and video recorded. One question this study addresses was whether or not it was possible to reproduce and "take over" a lesson design. *Could teachers implement a lesson design developed in another setting in their classroom? and what were the implications for the students' learning?*

The analysis of the video-recorded lessons showed that in some cases the teachers succeeded in reproducing the critical features in the intended way (lesson 1, 2, 4, 5, 6, 7), but in some they did not. In two of the lessons (3 and 8), it was intended that the critical features in LD 2 should be brought out, but this was not enacted in the classroom. The reason for this could be that the teacher forgot, or did not interpret the critical features in the intended way.

TABLE 20.1 The Eight Implemented Lessons in the Study about Infinity of Decimal Numbers

Teacher	Lesson Design 1	Lesson Design 2
Teacher A (school X)	Lesson 1 (Class 1)	Lesson 3 (Class 2)
Teacher B (school X)	Lesson 2 (Class 3)	Lesson 4 (Class 4)
Teacher C (school Y)	Lesson 5 (Class 5)	Lesson 7 (Class 6)
Teacher D (school Y)	Lesson 6 (Class 7)	Lesson 8 (Class 8)

TABLE 20.2 Percentage of Correct Answers on the Post-Test

Task 3: *Are there numbers between 0.5 and 0.6?* and

Task 5: *Anne claims that there is a number between 0.97 and 0.98. John says there is no such number. Who is right and why?*

Lesson Design	Lesson	Task 3		Task 5	
Lesson design 1	Lesson 1, $n = 19$	84%	(+26%)[a]	74%	(+16%)
	Lesson 2, $n = 14$	64%	(+21%)	64%	(+36%)
	Lesson 5, $n = 13$	54%	(+23%)	62%	(+31%)
	Lesson 6, $n = 15$	33%	(+14%)	47%	(+33%)
Lesson design 2	Lesson 4, $n = 13$	100%	(+62%)	100%	(+62%)
	Lesson 7, $n = 11$	91%	(+64%)	91%	(+55%)
Intended Lesson design 2	Lesson 3, $n = 13$	69%	(+8%)	69%	(+0%)
	Lesson 8, $n = 13$	77%	(+23%)	62%	(+31%)

[a] Change from the pre-test

The Learning Outcome

Besides the video-recorded lessons, individual pre- and post-tests were used to capture students' learning outcomes. The pre- and post-tests were the same, and several items tested the same ability. The pre-test was given a week before the research lesson, and the post-test was given two days after the lesson. The two tasks (*"Anne claims that there is a number between 0.97 and 0.98. John says there is no such number. Who is right and why?"* and *"Are there numbers between 0.5 and 0.6?"*) were used in the pre- and post-tests. In the students' written explanations, it was possible to analyze whether they thought that there was an infinite number (expressed by the students in some cases as "millions," "many, many," or "infinite many") of decimal numbers. Students who did not answer correctly on these items answered, for example, that there were no numbers or ten numbers in the intervals.

The result from the post-test showed that the lessons/classes with LD2 (lesson 4 and 7) were more successful on the post-test than LD1 (see Table 20.2). In these classes almost 100% of the students, in some way, expressed that there was an infinite number of decimal numbers in an interval between two decimal numbers. The result on the test showed that the Lessons 3 and 8, which were intended to bring out the critical features but did not succeeded in that, had improved less than LD2. Instead of the intended LD2, the lessons were enacted in the classroom in a similar manner as LD1. In the lessons with LD1, the students' results improved from the pre-test, but not as much as LD2. After these lessons (1, 2, 5, and 6) the increase was about 25%, compared to LD2 (lessons 4 and 7) with an increase over 60%. In some of the lessons, the teacher was new to the classes. This fact, it seems, did not affect the result and outcome of the lessons on a general level, in either a positive or negative way, since the results improved the most and the least (lesson 3 and 7) during these lessons.

All the lessons (1 to 8) in this study were organized the same way: They started with an introduction, followed by a group/pair activity, and ended with an oral account and discussion of the activity. However, the results on the post-test were different, and the analysis of the lessons showed that how the content was handled afforded different possibilities to learn. The analysis showed that during Lessons 4 and 7 (both LD2), the teachers in a systematic way brought up critical features for learning. In the next section we will go deeper into the analysis of these two lessons.

The Critical Features for Learning

This section will show how the teachers brought up critical features for learning in Lessons 4 and 7. In Lesson 4, the teacher started the lesson with the question about how many numbers there were between 0 and 1. In this part the students contributed with many answers, such as 0.4 and 0.99999. By doing that, it was possible for the students to experience a variation of numbers between 0 and 1. In the next part of the lesson, the relationship between fractions and the number of parts in the interval was made clear. This was done by letting the students name the same number 0.30 with different number of parts in the fractions, 3/10 and 30/100. Also, 0.29 was brought up the same way and compared with 0.30. The teacher used a drawn cucumber to illustrate the part–whole relationship of a rational number. In this case 0.29 and 0.30 were seen as parts of a cucumber. In a comparison of the numbers, the space between 0.29 and 0.30 was expressed by the students as different amount of parts in the interval such as 1/100, 10/1000, 100/10000, and so on (see

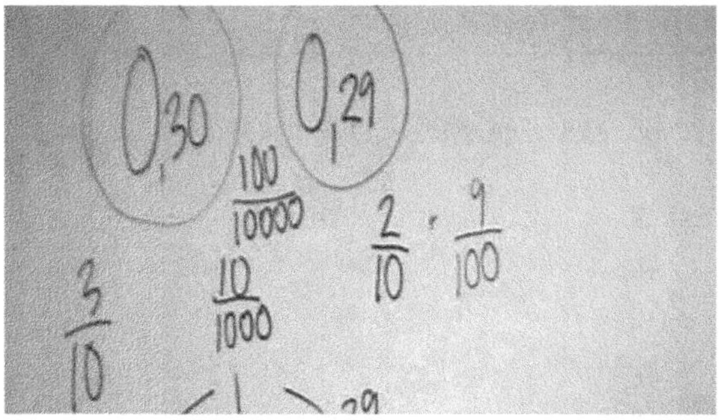

Figure 20.7 A picture of different ways of representing the decimal numbers 0.29 and 0.30 and the part in the interval between the two numbers shown on the whiteboard.

Figure 20.7). The same interval was in this case seen as smaller and smaller parts of the cucumber. Thereafter, the teacher introduced the task/group activity*: How many numbers can you find between 0.17 and 0.18? Draw a picture to show your answer/thinking.*

In the oral account of the task, a student repeated that 1/100 is the same as 10/1000 and so on. Students' understanding of the divisibility of number seems to have had a great impact since they wanted to show more ways of showing the number 1/100. The account of the students' group activity showed that two of three groups used fractions (in some way) to explain their solutions. The fact that the students used fractions and showed the connection between fractions and decimals in the account of the group activity in whole class probably contributed to the good result for all students in Lesson 4, where 100% of the students on the post-test answered correctly on tasks about density of rational numbers. In Lesson 7, another teacher brought up the same critical features, in a different way than in Lesson 4. As in Lesson 4, the introduction of the lesson started with a discussion about the number line, decimal numbers, and fractions. Also as in lesson 4, the same decimal number was expressed in many ways, including fractions.

A decimal number (0.5) was seen as a part from a whole on a pizza (5 times 1/10). The number line was also used to elicit the part-whole relationship as well as how the same interval could be split in different number of pieces. In the end of the lesson the teacher concluded the discussion about number of parts, in this case hundredths and thousandths and its relationship with the number of parts in the interval (see excerpt 1). In this excerpt the teacher make a connection between the fractions and its number of parts.

Excerpt 1:

Teacher: If it was the case that we had split it in hundredths, the whole ruler, and started counting here and came all the way up to here, then we would had found 70 such hundredths. Do you agree with me? That's why it says so. We could have split it in thousand parts. If we have split the ruler in thousand parts... That we had found 550 thousandths... if we had split the ruler in thousandths.. had we found 550 thousands at this part between the zero and this part then?

Student: Yes

Teacher: Yes, that is why it is called that.

Both lessons (4 and 7) brought up the feature of different number of parts in the interval between two decimal numbers. This had implications on the result of the post-tests. The critical features that were intended to be brought up were brought up by using variation (and invariance). One example of this is shown in Lesson 7, where the decimal number 0.25 is represented in many different ways:

$$0.25, \quad \frac{25}{100}, \quad 0.250, \quad \frac{250}{1000}, \quad 0.2500, \quad \frac{1}{4}$$

This opened up for experiencing a variation in representation of the decimal numbers which makes it possible to discern divisibility of number and connections between decimal numbers and fractions.

What implications could these findings have? Could teachers implement a lesson design in their own classroom, and what are the implications for the students learning? Tentative conclusions drawn from this study suggest that it is possible for teachers to "take over" a lesson design and implement it in their own classrooms, using other methods, materials, and organization than the original lesson. In this study it was possible to reproduce the LD in six of eight cases. The ways the critical features are brought up in the lesson have an impact on the student learning. It is therefore not enough that the critical features are present in the lesson—the students must be given the possibility to discern them.

CONCLUSIONS AND REFLECTIONS

In this section we will sum up some of the points made in this chapter. First, we will emphasize that there is an explicit interest in the phenomenography and variation theory tradition: to enhance and improve learning. Second, the fundamental idea is that what aspects that are experienced as dimensions of variation by the learner is of decisive importance for learning. Variation has been a theme in this chapter. Phenomenographic studies aim at describing the variation in how the same phenomenon, situation, or problem is experienced. We would suggest that phenomenographic studies, like those reported in this chapter about children's ways of experiencing simple arithmetic, can contribute to the understanding of why some learners are more successful than others.

From such studies it has been concluded that learning requires the simultaneous discernment of the critical features of that which is learned. The difference between different ways of experiencing has to do with differences in aspects discerned or paid attention to at the same time. To enhance and make learning possible it is necessary to help the learners to be aware of such aspects.

However, it is necessary to find out what the critical aspects are. This is an aim in the iterative processes of collaborative work among teachers, described as *learning study*. In order to enhance student learning, questions such as the following are raised: What does it take to learn something? What are the critical features in relation to learning a certain capability? How is it possible to create the necessary conditions for learning?

The awareness of an aspect presupposes an experienced variation of the aspect in question. Discernment is a function of an experienced variation—that is, we attend to or discern certain aspects of a situation when things stand out, and when they change or vary against a stable background they are likely being discerned. Hence, to make discernment of critical features noticeable, it is necessary to create a pattern of variation and invariance in the classroom.

How the content is handled by the teacher and the students, in terms of what aspects are brought out and opened as dimensions of variation during the learning situation, has been demonstrated to be of importance for learning. We have given several examples of how differences between how the same topic is taught in different classrooms affects what is made possible to learn. We have also demonstrated how such differences are reflected in students' learning. Hopefully it has been apparent to the reader that in the phenomenographic and variation theory tradition, the object of learning, thus *what is learned* and *what is made possible to learn*, is in focus. Hence, there has been a distinct and very specific object of analysis within this approach. It could be argued that this is a limited way of looking upon teaching and learning. What is made possible to learn in a classroom situation or why learners differ as regards learning outcomes depends on several factors, proximal as well as distal (Hattie, 1992; Pong & Morris, 2002). And from the point of view of the socio-linguistic turn in education, one could argue that it not possible to study peoples' ways of conceiving the world, only their participation in social practices and appropriation of linguistic tools. However, it must be noted that phenomenographic research does not aim at "looking into peoples' heads," but to identify critical differences between different ways of how something can be understood or experienced in order to better understand and also enhance learning. It is likely of significance for a teacher to know about what differs between one way and another way of experiencing that which is supposed to be learned when planning for and evaluating learning, for instance. The focus on "ways of experiencing" does not imply a denial of the social and semantic dimension of learning (Marton & Tsui, 2004). However, in the social interaction there is always something interacted about and something possible to learn. We would suggest that the inclusion of what is learned in the social interaction brings an important dimension to the understanding of learning (Runesson, 2005). In our view, variation theory is a complement to other theoretical perspectives due to its power to reveal constraints on what is possible to learn and what is actually learned.

We also agree with those who claim that there is not a one-to-one correspondence between teaching and learning (e.g., Nuthall, 2004). Therefore, it is important to be aware of the risk of using variation theory in an instrumental way when designing classroom learning. For instance, we do not think that it is possible to implement results from learning studies gained in one socio-cultural context to another *directly*. What we believe is that insights about aspects critical for learning, gained and documented by teachers in their specific context, could be used as resources by other teachers if those are adapted and developed to their situation and their learners.

NOTES

1. From a variation theory perspective, learning is seen as a process of differentiation, rather than enrichment. This is in line with psychologists like Gibson and Gibson (1955).
2. Earlier educationalists have also emphasized the importance of contrast and variation. For instance, Maria Montessori stressed that the training of the senses is done by means of systematic variation in different sense modalities against a background that is invariant (Marton & Signert, 2005). A similar rationale was proposed by Dienes (1960) in his theory of mathematics learning. A concept must be taught by varying essential features of the concept, he argues. Watson and Mason (2006) have suggested similar ideas in their notions "range of permissible change" and "dimensions of possible variation."

3. The study draws on data from the Learners' Perspective Study (LPS) (Clarke, Keitel, & Shimizu, 2006).
4. The research project *The Pedagogy of Learning* was funded by the Swedish National Council of Educational Research year 2003 to 2006.

REFERENCES

Ahlberg, A. (1992). *Att möta matematiska problem. En belysning av barns lärande [Meeting mathematical problems. An illumination of children's learning; in Swedish].* Göteborg studies in educational sciences 87. Göteborg: Acta Univerisitatis Gothoburgensis.

Ahlberg, A. (in press). Communicating mathematics in Primary school. In J. Emanuelsson, L. Fainsilber, J. Häggström, A. Kullberg, B. Lindström, & M. Löwing (Eds.), *Voices on learning and instruction in mathematics.* Göteborg: National Centre of Mathematics Education, University of Gothenburg.

Bowden, J., & Marton, F. (1998). *University of learning.* London: Cogan.

Brown, A. L. (1992). Design experiments: Theoretical and methodological challenges in creating complex interventions in classroom settings. *The Journal of the Learning Sciences, 2*(2), 141–178.

Clarke, D., Keitel, C., & Shimizu, Y. (2006). *Mathematics classroom in 12 countries: The insider's perspective.* Taipei: Sense Publishers.

Cobb, P., Confrey, J., di Sessa, A., Lehrer, R., & Schauble, L. (2003). Design experiments in educational research. *Educational researcher, 32*(1), 9-13.

Collins, A. (1992). Toward a design science of education. In E. Scanlon & T. O'Shea (Eds.), *New directions in educational technology.* Berlin: Springer.

Dienes, Z. P. (1960). *Building up mathematics.* London: Hutchinson Educational.

Emanuelsson. J. (2001). *En fråga om frågor. Hur lärares frågor i klassrummet gör det möjligt att få reda på elevernas sätt att förstå det som undervisningen behandlar i matematik och naturvetenskap [A question about questions. How teachers' questioning makes it possible to learn about students' ways of understanding the content taught in mathematics and science; in Swedish].* Göteborg studies in educational sciences 168. Göteborg: Acta Universitatis Gothoburgensis.

Emanuelsson, J., & Sahlström, F. (2008). The price of participation—Teacher control versus student participation in classroom interaction. *Scandinavian Journal of Educational Research, 52*(2), 205–223.

Fischbein, E., Tirosch, D., & Hess, P. (1979). The intuition of infinity. *Educational Studies in Mathematics, 10,* 3–40.

Gaversjö, A., & Lööf Widing, H. (2008). *Kan man lära alla allt? En studie i matematik skolår 4-8 som tar sin utgångspunkt i Learning study* [Is it possible for everyone to learn everything? A study in mathematics in 4th to 8th grade about Learning study]. Göteborg: Institutionen för Pedagogik och didaktik.

Gibson, J. J., & Gibson, E. J. (1955). Perceptual learning: Differentiation or enrichment? . *Psychological reweiv, 62*(1), 32–41.

Gurwitsch, A. (1964). *The field of consciousness.* Pittsburgh: Duquense University Press.

Gustavsson, L. (2008). *Att bli bättre lärare. Hur undervisningsinnehållets behandling blir till samtalsämne lärare emellan* [Becoming a better teacher. Ways of dealing with the content made a topic of conversation among teachers; in Swedish]. Umeå University, Umeå.

Hattie, J. (1992). Measuring the effects of schooling. *Australien Journal of Educational Research, 11,* 155–164.

Hiebert, J., Gallimore, R., & Stigler, J. W. (2002). A knowledge base for the teaching profession: What would it look like an how can we get one? *Educational Researcher, 31,* 3–15.

Hill, H. C., Blunck, M. L., Charalambous, C., Lewis, J. M., Phelbs, G. C., Sleep, L., et al. (2008). Mathematical knowledge for teaching and the mathematics quality of instruction: An exploratory study. *Cognition and Instruction, 26*(4), 430–511.

Holmqvist, M. (Ed.). (2006). *Lärande i skolan. Learning study som skolutvecklingsmodell* [Learning in school. Learning study as a model for school development]. Lund: Studentlitteratur.

Holmqvist, M., Gustavsson, L., & Wernberg, A. (2008). Variation theory: An organizing principle to guide design reserach in education. In A. E. Kelly, J. Y. Baek & R. A. Lesh (Eds.), *Handbook of design research methods in education.* Mahwah, NJ: Erlbaum.

Häggström, J. (2008). *Teaching systems of linear equations in Sweden and China: What is made possible to learn?* Göteborg studies in educational sciences 262. Göteborg: Acta Universitatis Gothoburgensis

Kullberg, A. (2004). *Tal, delar och oändlighet. En studie om avgörande skillnader i undervisning och lärande om decimaltal.* [Numbers, parts and infinity. A study about differences in teaching and learning about decimal numbers]. Göteborg: Institutionen för pedagogik och didaktik.

Kullberg, A. (2006, December). *Teaching for understanding infinity of rational numbers.* Paper presented at the 2nd Annual Conference on Learning Study, The Hong Kong Institute of Education, Hong Kong.

Kullberg, A. (2007a). Can lessons be replicated? In J. H. Woo, H. C. Lew, K. S. Park, & S. Dong-Yeop (Eds.), *Proceedings of the 31st Conference of the International Group for the Psychology of Mathematics Education* (Vol. 3, pp. 121–128). Seoul: PME

Kullberg, A. (2007b, November–December). *Students opening up dimensions of variation for learning negative numbers.* Paper presented at the World Association of Lesson Studies International Conference, Hong Kong.

Kullberg, A. (2010). *What is taught and what is learned. Professional insights gained and shared by teachers of mathematics.* Göteborg studies in educational sciences 293. Göteborg: Acta Universitatis Gothoburgensis.

Kullberg, A., & Runesson, U. (2006). Exploring teaching and learning of letters in algebra: A report from a Learning study In J. Novotná, H. Moraová, M. Krátká & N. Stehlíkova (Eds.), *Proceedings of the 30th Conference of the International Group for the Psychology of Mathematics Education* (Vol. 1, p. 273). Prague: PME.

Lewis, C. (2002). *Lesson study: A handbook of teacher-led instructional change.* Philadelphia: Research for Better Schools.

Lewis, C., Perry, R., & Murata, A. (2006). How should research contribute to instructional improvement? The case of lesson study. *Educational Researcher, 35*(3), 3–14.

Liljestrand, J., & Runesson, U. (2006). Interaction, organisation, tasks and possibilities for learning about mathematical relationships: A Swedish classroom compared with a

US classroom. In D. Clarke, J. Emanuelsson, E. Jablonka & I. A. C. Mok (Eds.), *Making connections: Comparing mathematics classroom around the world* (pp. 165–183). Rotterdam: Sense Publishers.

Lo, M. L., Pong, W. Y., & Chik, C. P. M. (2005). *For each and everyone. Catering for individual differences through learning studies.* Hong Kong: Hong Kong University Press.

Ma, L. (1999). *Knowing and teaching elementary mathematics.* Mahwah, NJ: Erlbaum.

Marton, F. (1981). Phenomenography—Describing conceptions of the world around us. *Instructional Science, 10,* 177–200.

Marton, F., & Booth, S. (1997). *Learning and awareness.* Mahwah, NJ: Erlbaum.

Marton, F., & Pang, M. F. (2006). On some necessary conditions of learning. *Journal of the Learning Sciences,, 15*(2), 193–220.

Marton, F., Runesson, U., & Tsui, A. B. (2004). The space of learning. In F. Marton & A. B. Tsui (Eds.), *Classroom discourse and the space of learning* (pp. 3–40). Mahwah, NJ: Erlbaum.

Marton, F., & Signert, K. (2005, August). *Affordances for learning. Studying teaching in terms of the learning that the inherent pattern of variation and invariance makes possible: Maria Montessori´s pedagogy as an example.* Paper presented at the 11th Biannual Conference of the Association for Research on Learning and Instruction Nicosia, Cyprus.

Marton, F., & Tsui, A. B. (2004). *Classroom discourse and the space of learning.* Mahwah, NJ: Erlbaum.

Monaghan, J. (2001). Young peoples' ideas of infinity. *Educational Studies in Mathematics, 48,* 239–257.

Moskal, B. M., & Magone, M. M. (2000). Making sense of what students know: Examining the referents, relationships and modes students displayed in response to a decimal task. *Educational Studies in Mathematics, 43*(3), 313–335.

Neuman, D. (1987). *The origin of arithmetic skills: A phenomenographic approach.* Göteborg studies in educational sciences 62. Göteborg: Acta Universitatis Gothoburgensis.

Nuthall, G. (2004). Relating classroom teaching to student learning: A critical analysis of why research has failed to bridge the theory–practice gap. *Harvard Educational Review, 74*(3), 273–306.

Olteanu, C. (2007). *Vad skulle x kunna vara? Andragradsekvation och andragradsfunktion som objekt för lärande* [What could x be? Second degree equations and second degree functions as objects for learning]. Dissertation, Högskolan Kristianstad.

Pong, W. Y., & Morris, P. (2002). Accounting for differences in achievement. In F. Marton & P. Morris (Eds.), *What matters? Discovering critical conditions of classroom learning* (pp. 9–18). Göteborg studies in educational sciences 181. Göteborg: Acta Universitatis Gothoburgensis. Gothenburg: Acta Universitatis Gothoburgensis.

Runesson, U. (1999). *Variationens pedagogik. Skilda sätt att behandla ett matematiskt innehåll [The pedagogy of variation. Different ways of handling the same topic; in Swedish].* Göteborg studies in educational sciences 129. Göteborg: Acta Universitatis Gothoburgensis.

Runesson, U. (2005). Beyound discourse and interaction. Variation: a critical aspect for teaching and learning mathematics. *Cambridge journal of education, 35*(1), 69–87.

Runesson, U. (2006). What is it possible to learn? On variation as a necessary condition for learning. *Scandinavian Journal of Educational Research, 50*(4), 397–410.

Runesson, U. (2007). A collective enquiry into critical aspects of teaching the concept of angles. *Nordic Studies in Mathematics Education, 12*(4), 7–24.

Runesson, U. (2008). Learning to design for learning. The potential of learning study to enhance learning on two levels: Teachers' and students' learning. In T. Wood & P. Sullivan (Eds.), *Knowledge and beliefs in mathematics and teaching development* (pp. 153–172). Rotterdam: Sense Publishers.

Runesson, U., & Mok, I. (2005). The teaching of fractions: A comparative study of a Swedish and a Hong Kong classroom. *Nordic Studies in Mathematics Education, 8*(2), 1–15.

Smith, C., Solomon, G., & Carey, S. (2005). Never getting to zero: Elementary school students' understanding of the infinite divisibility of number and matter. *Cognitive psychology, 51,* 101–140.

Stigler, J. & Hiebert, J. (1999). *The teaching gap: Best ideas from the world's teachers for improving education in the classroom.* New York: The Free Press.

Watson, A., & Mason, J. (2006). Seeing an exercise as a single mathematical object: Using variation to structure sense-making. *Mathematical Thinking and Learning, 8*(2), 91–111.

Vosniadou, S., Vamvakoussi, X., & Skopeliti, I. (2008). The framework theory approach to the problem of conceptual change. In S. Vosniadou (Ed.), *International handbook of research on conceptual change* (pp. 3–34). New York: Routledge.

Yoshida, M. (1999). *Lesson study: A case study of a Japanese approach to improving instruction through school-based teacher development.* Champaign, IL: University of Illinois.

21

CONTEXTUALIZING MATHEMATICAL TEACHING AND LEARNING

Andreas Ryve

STOCKHOLM UNIVERSITY AND MÄLARDALEN UNIVERSITY

Inger Wistedt and Max Scheja

STOCKHOLM UNIVERSITY

Abstract: The social turn in mathematics education has implied that researchers wish to include collaborative and social aspects when accounting for students' learning in mathematics. To accomplish this task, many theoretical frameworks have been introduced, and some of these are also used in the field of mathematics education. Several publications produced by Swedish researchers in mathematics education have used the theory of contextualization and the related analytical principles of intentional analysis to conceptualize and analyze mathematical teaching and learning. The theory of contextualization is situated in constructivist ontological and epistemological principles and is operationalized to include conceptual, situational, and cultural aspects of students' conceptions of a given task. In this chapter we elaborate on the theory of contextualization and how it may be used to study mathematics learning and teaching by bringing together five examples of students´ learning of mathematics in compulsory school and higher education previously separately presented in international publications in mathematics education.

Keywords: classroom discourse, context, contextualization, intentional analysis

INTRODUCTION

The social turn in mathematics education (cf. Lerman, 2000) is a product of, and has resulted in, researchers trying to include social and collaborative aspects when accounting for mathematical teaching and learning. One way of conceptualizing these efforts is to distinguish between two theoretical positions: (1) constructivist frameworks incorporating social aspects in their theoretical approaches, and (2) socio-culturally inspired approaches such as situated cognition (Kirshner & Whitson, 1997) and activity theory (Engeström, Miettinen, & Punamäki, 1999). In both cases, the theoretical approaches must be developed to meet the demands of mathematics education (Bartolini Bussi, 1998; Cobb, 2007). In this chapter we take a closer look at how one approach of the former kind, the theory of contextualization, has been used and developed in mathematics education research.

When characterizing the theory of contextualization (TC) and the research carried out from this theoretical position, three dialogical relations may serve as a framework for pointing out the specifics of TC. These three relations could all be seen as parts of the general scientific criteria for research in mathematics education (Lester & Lambdin, 1998), including: (1) the relation to other theoretical approaches, (2) the

The First Sourcebook on Nordic Research in Mathematics Education, pages 319–332

relation to the research subjects, and (3) the relation to researchers from other scientific disciplines. The first dialogical relation concerns how TC has been developed in relation to other theories of learning—for example, conceptual change theory, the alternative frameworks movement, socio-cultural theory—as well as to theories developed within the field of mathematics education. The second relation refers to the theoretical sensitivity to students' interpretations of learning situations and to their learning potential rather than to their misconceptions of the tasks presented or concepts encountered. The third relation refers to the multidisciplinary and collaborative efforts common among researchers using TC, and how such joint efforts characterize the development of TC.

All three dialogical relations may seem self-evident in relation to the criteria used to discuss quality in mathematics education research (cf. Lester & Lambdin, 1998). However, as Niss (2007a, b) argues and evaluations of educational research in general show (e.g., Swedish Research Council, 2007), many publications in educational sciences fail to meet such criteria. Hence we regard it as worthwhile to situate TC in relation to the three dialogical relations by drawing on theoretical discussions and several empirical examples previously presented in international journals and anthologies in mathematics education.

The chapter consists of three parts. First, we give a short theoretical introduction to TC where we discuss ontological and epistemological assumptions as well as analytical constructs used and how TC relates to other theoretical approaches. Second, research from several publications is presented to illustrate the dialogical qualities of TC research. We conclude this chapter by using the three dialogical relations described above to make explicit what characterizes TC as a research approach to understanding mathematics teaching and learning.

THEORETICAL BACKGROUND

The research presented in this chapter has its roots in constructivist research on student learning and conceptual change in school and higher education. Out of a series of early research studies (Halldén, 1982; Wistedt, 1987) that had observed difficulties that students had in solving tasks assigned to them by teachers in various subject areas, grew an interest in coming to understand what the students were actually reading into these tasks. Students who were confronted with a particular task would often end up approaching this task in very different ways—so diverse that teachers sometimes would have great difficulty in recognizing the problems that they had intended the students to work on. Thorough qualitative analyses of how students tackled various tasks made it clear that the difficulties initially observed among students had to do with the ways in which the students interpreted the tasks to begin with. This conclusion led Halldén (1988) to introduce an explicit distinction between *task* and *problem*, defining "task" as "what is presented to the [students] by the teacher with the intention that they are to do something and/or that they are to learn something" (p. 125), and "problem" as the student's "personal interpretation of the task given" (p. 125).

This distinction had methodological implications, as it involved shifting from a teacher-oriented to a student-oriented perspective on the learning activities that occur in different educational settings (Entwistle, 1996; Entwistle & Ramsden, 1983). It also brought with it the important recognition that although students may be given the same learning task, they may very well end up working on different problems, depending on the interpretation of the task. The central quest for the research thus became to try to understand what sort of activities students do in fact engage in when they are confronted with a learning task in a particular setting, and what these activities imply in terms of opportunities for learning.

Focusing on the Intentionality of Action

A fundamental epistemological theme running through this research strand has been that learning is to be seen as an *intentional activity* where students' intentions and personal beliefs about their own learning capabilities in relation to perceived demands of the learning environment, strongly influence the ways in which they take on a particular learning task. This approach falls back on the philosophy of action developed by von Wright (1971, 1979, 1980) suggesting that it is possible to not only understand but also to logically explain human conduct through teleological arguments of the following kind (practical syllogism describing the explanation of action, modified from von Wright, 1971, 1979, 1980):

(i) A person P intends to bring about x
(ii) P believes that bringing about x would require the fulfillment of y
(iii) Hence: P does y (in order to bring about x)

Using this syllogism as a tool for thinking around human activity yields the following structure for rationalizing an observed piece of action: What we see as observers is P doing y, and it is by ascribing to P's doing y

the intention of bringing about x that explains why *P* did *y* in the circumstances. In von Wright's words:

> Behavior gets its intentional character from being seen by the agent himself or by an outside observer in a wider perspective, from being set in a context of aims and cognitions. This is what happens when we construe a practical inference to match it, as premises match a given conclusion. (von Wright, 1971, p. 115; cf. von Wright, 1979, 1980)

As a piece of logic, this argument makes perfect sense. However, in real-life settings, there are many factors that can make it difficult to say for sure why a particular action was undertaken. Nevertheless, despite these obvious difficulties, the ideas expressed in von Wright's model have reached beyond the immediate scope of academic philosophy and have been introduced into the field of student learning research as a heuristic tool for analyzing learning activities that individuals engage in when they are confronted with learning tasks in various instructional settings (e.g., Halldén, 1999; Scheja, 2002). Often, the application of this intentional model to a piece of empirical material has been referred to in terms of *intentional analysis.* Such an analysis applies both to analyses of physical actions and to utterances placed in a dialogue between interlocutors, the central assumption being that all acts can be regarded as intentional—that is, can be rendered meaningful with references to objectives on the part of the acting individual (cf. Downes, 1998).

Since the inception of intentional analysis into educational research (Booth, Wistedt, Halldén, Martinsson, & Marton, 1999), a range of studies have been carried out on student learning in a variety of subject areas, such as mathematics (Nilsson, 2007; Pettersson & Scheja, 2008; Ryve, 2006; Wistedt & Brattström, 2004), science (Halldén, Haglund, & Strömdahl, 2007), history (Halldén, 1999), environmental education (Lundholm, 2005), biology (Caravita & Halldén, 1994), and engineering (Scheja, 2002, 2006), all aiming to further our understanding of how students understand and deal with different learning tasks and learning environments. Out of these studies, several of which are described in more detail in this chapter, has grown a theoretical framework for analyzing student learning that pays particular attention to the contexts within which students operate in dealing with learning tasks. *Context* here should be understood, not as the physical situation occupied by the learner, but rather as the cognitive context within which learners think about learning material and the learning environment as a whole (Halldén, 1999; Scheja, 2002; Wistedt, 1993). This constructivist conceptualization of context has proved particularly helpful in explicating students' interpretations of learning tasks and in discussing educational implications of the varying contextualizations through which students' come to an understanding of learning tasks that they are confronted with.

Understanding the interpretive process through which students arrive at an understanding of learning material is of pivotal importance in any research aiming to explicate students' learning processes, and so more recent research has qualified the analytic use of the concept of context in terms of a *theory of contextualization* (e.g., Ryve 2006; cf. Halldén, Scheja, & Haglund, 2008), offering a basis for conceptualizing learning from the viewpoints of the students themselves. In essence, this theory contends that students who are confronted with a learning task engage in an interpretive process through which they strive to put the task in a personal context to make sense of it for themselves. From such a perspective, learning can be described as a process of approximation and feedback in which students try out contextualizations of learning material and, on the basis of the response they get from teachers and other significant influences in the learning environment, develop conceptions of the learning material and of what it means to study and learn a particular subject.

The theory of contextualization claims that it is through such complex communicative processes that students gradually learn to differentiate the preconditions for developing conceptual understanding in various subject areas (Halldén et al., 2002; Halldén et al., 2008; Petersson, 2002). Importantly, and in contrast to more mentalistic theories of the past, the constructivist theory of contextualization does not consider students' conceptions in cognitive isolation. Rather it views these conceptions as outcomes of the individual learner's efforts to cope with both intra-individual cognitive constraints and the perceived socio-cultural requirements of the learning environment (Scheja, 2002). The theory of contextualization thus takes into consideration both the cognitive and the socio-cultural dimensions of learning, obviating the traditional socio-cultural critique of constructivist research for not paying sufficient attention to the discursive and situated aspects of learning (e.g., Säljö, 1994, 1997; Säljö, Schoultz, & Wyndhamn, 2001; Säljö & Wyndhamn, 1990).

EXAMPLES OF CONTEXTUALIZATION IN MATHEMATICS EDUCATION

The social turn in mathematics education does not only imply a growing interest in studying student learning in collaborative and social settings. In a wider perspective it also includes the researchers themselves calling for a change in perspectives on how research is done and communicated in the social context of mathematics education as a research field. The theory of contextualization has been developed and refined in dialogue with researchers in mathematics (e.g., Wistedt, 1994a, b; Wistedt & Martinsson, 1996; Wistedt & Brattström, 2004), and in other academic subjects (e.g., Caravita & Halldén, 1994; Halldén, 1999; Halldén, Hansson, & Skog, 1994; Molander, Halldén, & Pedersen, 2001; Wistedt, 1998), often in joint projects where researchers from different disciplines or with different research agendas have come together to discuss the strengths and restraints of the respective approaches with the aim of contributing to the understanding of student learning in context (e.g., Booth, Wistedt, Martinsson, Halldén, & Marton, 1999). In this section we will give some examples of such collaborative projects and discuss their contributions to the development of the theory of contextualization and Intentional analysis as means to capture the dynamics of the learning processes which take form when students jointly solve mathematical tasks.

Abandoning Hierarchies, Abandoning Dichotomies is the title of an anthology edited by Leone Burton (1999). The quest calls us to stop looking for a "right" or "best" theory that will uncover every aspect of student learning. Rather, mathematics education needs a variety of theoretical perspectives in order to understand the learning of mathematics as it takes place in and out of school. Abandoning hierarchies and dichotomies can also mean that we should not primarily look for a "right" or a "best" understanding of mathematics in students' grappling with mathematics tasks. In the perspective of contextualization we do not view students' conceptions of a certain subject matter as hierarchically ordered, but as simultaneously existing, each of them possible to bring to the fore depending on what is relevant to a particular situation. This means that "learning in context" must be taken seriously, as a process highly dependent on students' aims and goals and their understanding of what it takes to reach these goals in a particular situation as they understand it. Below we will elaborate on these characteristics and others based on examples from studies that have been carried out in collaboration with fellow researchers.

Example 1: A Piece of Wood

In a study published in the anthology mentioned above (Burton, 1999), Shirley Booth and Inger Wistedt worked closely together on a common set of data consisting of audio and video recordings of eleven-year old students' work on a mathematics task confronting them with the notion of infinity (Booth et al., 1999). Data had previously been analyzed from one of these perspectives only (Wistedt, 1996). In the referred article, Wistedt used the concept of *reflective variation* introduced by Ference Marton in a comparative study of Swedish and Hungarian students' interpretations of a Kafka essay (Marton, Asplund-Carlsson, & Halász, 1992)—a concept that was later developed by Marton in the *Theory of Variation* (Marton & Booth, 1997; Marton & Tsui, 2004; see Runesson & Kullberg, this volume). In a joint venture, Booth and Wistedt revisited the data from Wistedt (1996), with the aim of investigating the possibilities to expand the interpretative frameworks of *phenomenography* and *contextualization.* Thus, the reason for opening such a dialogue was not to blend perspectives but to articulate the respective theoretical stances. In order to elicit the perspective of contextualization, we will discuss what characterizes such a perspective by using examples from that joint study. Since the aim of this section is to give examples that show the characteristics of the theory of contextualization and intentional analysis, we will restrict the presentation of data material to short excerpts (for a more elaborate account, see Wistedt & Martinsson, 1996; Booth et al., 1999).

In a junior school in Sweden, four groups of 11-year-old pupils, three in each group, were pondering a task carefully designed, in cooperation with mathematicians, in such way that it would subtly lead the children to reflect on the problem of infinity:

> During a craft lesson Charlie decides to make Christmas presents for his family. He finds a piece of wood that would be exactly suited to making two advent candlesticks, or perhaps four small candlesticks or maybe three bookstands. Charlie ponders the matter.

The task was presented in the form of a small booklet where three consecutive questions with accompanying illustrations were given, one on each page. The children were asked to answer the question *exactly how big a part* is needed to make (a) the two advent candlesticks, (b) the four small candlesticks, and (c) the target question, the three bookends, and to give their answers in decimal form. However, none of the children knew how to write decimal numbers so they asked the observer if they were allowed to write their answers in percentages. Since such a restriction would also lead them to an infinite decimal expression, the observer allowed them to do so.

Open-Ended Tasks Allow for a Variety of Contextualization

The children started out with task (a) trying to figure out exactly how big a part is needed to make two candlesticks. This is what the *task* requires, but what is the *problem* to be solved? In the discussion among the children we found that the task could be interpreted in different ways and put into different contexts. In one of the groups a girl, Maria, stated, after pondering the question (b): *But how can you know how big an advent candlestick is, then?* She read the text anew: *How big a part is needed for each of the candlesticks? Half.* In her first reading she puts the task in the context of the physical world where candlesticks have a certain length, which in this case could not be measured since we have no information about scale. In her second statement she puts the task in a different, virtual context, where it is regarded as an example of fractions, an interpretation that is further elaborated upon when she walks to the blackboard and writes ½.

In all four groups, these two competing interpretations of the task were found. In a perspective that focuses on people's contextualizations of tasks given to them, the object of study is how people put given tasks into context depending on what goals they have when acting (1) in the context of their *conceptual knowledge* about the content matter to be addressed, in this case knowledge about numbers and how to represent them; (2) in the context of the observed *demands and restrictions of the situation in which they act*, in this case the booklet with written instructions, the other students present and their utterances, the observer, and the instructions she is giving them, access to physical material such as pencils, paper, blackboard, and so on; and (3) in the *context of the conventional rules* guiding their interpretations of how the situation is to be characterized: as a test, as a mathematics lesson, or, as the phrasing of the task seems to suggest, as a crafts problem. The theory of contextualization focuses on how these three aspects of students' contextualizations of a task are brought to the fore and varied in a learning situation.

Since we are interested in studying peoples' potential to act in resourceful ways, in this case how 11-year-old students can use their current mathematical knowledge to master a task which will take them far beyond the boundaries of what is commonly expected of them in school, we prefer to use tasks which are open-ended and which will allow for a variety of contextualizations.

Probing Students' Learning Potential

Varying interpretations of a given task or situation also means that the full potential of students' present knowledge is activated as competing interpretations are contrasted and evaluated in relation to their respective relevance to the situation at hand. In the next excerpt, we will move to the students' discussion of the target question (c). The first two questions were handled quite swiftly, and two possible interpretations occurred in all of the four groups: the task interpreted as a problem situated in a physical context or in a virtual, mathematical one. Since the answers to the two first tasks (a) and (b) did not present any difficulties in reconciling these two competing interpretations and the answers—(a) ½ or 50% and (b) ¼ or 25%—did not stir up any heated discussion among the children about what contextualization to prefer, they moved on to task (c) quite confidently. The three children, Maria, Christian, and Sofia, used the strategy that they had used effectively in the two former tasks when dividing 100 into two and four equal parts, respectively:

Maria: A hundred divided by three... hang on... I'm going to have to think about this one. If you say... oh blinking heck... you can't do it.

How could this problem be resolved? Maria suggests a solution:

Maria: But you can't get them the same size. There has to be a little bit over. [turning to the observer, asks pleadingly] Can't there be a little bit of wood over?

However, she does not take her own suggestion seriously and together with the two other group members she starts exploring the problem 100/3 by trying out different solutions: *35 and a half, 34 and a half, times three*, but they all lead to sums bigger than 100.

Maria: Take 33 then
Christian: ... point three maybe

As we can see, the children oscillate between the two interpretations of the task, as contextualized within the physical world where you may make one bookstand a bit shorter than the other two, and the virtual world of numbers where you have to come up with a solution to the mathematical problem of dividing 100 by three by giving an exact answer in percentages. The sum on the blackboard on which the students are writing is filled with decimals, 33.3333 times three equals 99.9999, but adding decimals does not seem to help them in reaching an even 100. Christian nearly loses his hope:

– *It can't be done. There is always going to be that little thing left* (in the case above 0.0001).
– *It's impossible, Maria agrees. You can't divide 100 by 3. It's impossible.*

The pupils conclude that there will always be a remainder even if infinitely small. Christian suggests that they could disregard the problem "since it's such a little bit of wood that you could hardly see it. " But Maria does not want to give up. She goes to the blackboard and starts drawing 100 apples of which the three children would get 33 apples each:

> **Maria:** We will get 33 apples each and this apple is over. Then we divide that. It could be done: 33 and one third, plus 33 and one third, plus 33 and one third, make one hundred!

She writes the numbers on the blackboard 33.3 + 33.3 + 33.3 (where .3 supposedly stands for one third). However, when she adds them up the sum turns out to be 99.9 and now she gets really angry:

> **Maria:** I don't understand a thing! Why could it be done a minute ago and not when you write it with numbers instead of apples?
>
> **Christian:** You cannot do it with apples either: One of the three will get an extra kernel
>
> **Maria:** Well then you cannot divide it into two equal parts either.
>
> **Christian:** Yes you could [remembering how the children successfully solved subtask (A)]
>
> **Maria:** No, not two and not three either, not *exactly*. But *here* it could be done! This is exact: 33 and one third, plus 33 and one third, plus 33 and one third make one hundred *exactly*. It could be done, Christian!

Contextual Variation

As we can see, the children have little knowledge about fractions and decimal numbers and how to write them. Nevertheless the children successfully oscillate between three different mathematical contexts for interpreting the equation 3x=100: the set of whole numbers, the set of rational numbers expressed as finite decimal expansion, and the set of whole numbers with halves, as in "34 and a half," which seems to hold an intermediate position between the two former ones. These three contexts render different meaning to the task of solving the equation and the possibilities of satisfying it. In the last excerpt above, Maria shows a certain *contextual awareness* (Wistedt, 1993), that is, the ability to consciously reflect on competing contextualizations of a task and their respective relevance to the current situation, as she urges Christian to appreciate the difference in the meaning of the word "dividing" something in the real world and in the mathematical world: between physically cutting something in pieces where someone will get "an extra kernel" and performing the mental operation of dividing 100 by 3, where the word "here" indicates the contextual shift.

Intentional Action

Above we have discussed the children's contextualizations of the task given, but what about their intentions when solving the task? It is obvious that the children will not settle with a solution to the task as a crafts problem. They try out such interpretations when they run into difficulties, but they never seem to accept them as relevant to the situation as they understand it. Suggestions such as disregarding the missing part (*such a little bit of wood that you could hardly see it*) are not accepted by the group members in any of the four groups.

In the task the children are asked to answer the question "*exactly* how big a part is needed..." and by varying contexts they find that the word *exactly* takes on different meaning in a crafts context and in a mathematical context. In the physical world of crafting candle sticks and book stands, the word *exactly* means "equal enough so that your eyes will not detect the difference," but in a mathematical context, *exactly* means that the parts should be "numerically equal." It seems as if the children find the latter meaning more in tune with the task requirements. This interpretation is further supported by the fact that the students in all four groups look for a mathematical expression that will satisfy the quest for a numerically exact answer. Maria is very disappointed when she cannot find a way to express "33 and one third times three" in percentages, although it could be done when she used apples. Why could it not be expressed in decimal numbers? On pondering the problem, she comes up with a solution that shows that she has realized that the number system has (unforgivable) limitations in expressing infinite quantities: "Maybe the numbers don't get it. It's because you only divide *one*. The numbers think that's queer and they just don't bother to learn how it's done," she says in a stern voice.

In another group, a girl tries a solution which will save the face of the decimal numbers:

> "I've got it!" she exclaims. "Now I know how it can be done! We could write 99.9 and then plus that little something. You could sort of write... this is a bit complicated, but it's better than nothing. Of course it will still be 99, sort of, but plus that something, she says and writes 33.33 + 33.33 + 33.33 = 99.99 + 0.01: And then it will come out even," she proudly states. "There will always be a small error that you will have to add."

Intentional analysis requires a close reading of data. Of particular interest is the subtle shift in pronoun in the first two utterances: from *I* to *we* and to *you*. First it tells us that the speaker views herself as involved in a personal task of understanding the puzzling problem of dividing 100 by three (*I*'ve got it. Now *I* know...). The pronoun *I* changes into *we* as she invites her fellow students to take her argument into consideration (*We* could write...) and finally into *you* as she states her general claim (*You* could sort of...). In a comment to the study cited above, Renshaw and Brown (1998) point to this shift of pronoun in the children's expressions: "Note the student's use of the pronoun 'you' which indicates that she is speaking in the general voice of the mathematician—that is the procedures and outcomes she is reporting are not personal opinions but general claims about the nature of such problems" (p. 472).

Such shifts also tell us that understanding does not require a replacement of personal or concrete concepts (what *I* know...) for more abstract and general ones (what *you* could...). These different forms of understanding exist parallel to each other and a full conceptual understanding requires the integration of the two forms rather than the replacement of one form by the other, where the collaborative situation, in which claims are made and challenged by appeal to reason, plays a major role in helping the students to make their understanding accessible to themselves and others. Following their aim of exploring the problem of expressing one third as a decimal number, the students opt for the general solution as the most relevant to the situation at hand and to their own commonly shared and developing understanding of the number system.

Example 2: Limits and Integrals

Another example of research using intentional analysis to capture the dynamics of the learning processes that unfold when students tackle a mathematical task is provided in a recent study of higher education students' conceptual understanding of calculus (Pettersson & Scheja, 2008). While much of the previous research (see, e.g., Davis & Vinner, 1986; Driver & Erickson, 1983, Duit, 2006; Nussbaum & Novick, 1982; Tall, 1993) has been putting effort into investigating students' misunderstandings and shortcomings in coming to understand scientific concepts, this study set out to look for the *potentiality* in the conceptual understanding that a group of undergraduate students showed as they struggled with the task of explaining the meaning of two central calculus concepts: limit and integral.

As with most studies using intentional analysis this research involved close collaboration between researchers specializing in education and mathematics respectively. Data were collected among undergraduate engineering students at a medium sized Swedish university college. The students (N = 20) were, at the time of the data collection, taking a basic course in calculus that introduced typical calculus concepts such as limit, derivative, and integral. At the end of this course, just before the exam, the students were asked to reflect in writing on their understandings of the concepts of limit and integral. Specifically, the students were given two open questions that asked of them to "explain as clearly as possible the meaning of the mathematical concept [limit/integral]." In addition, the students were asked to rate their own understandings of these concepts in terms of "poor," "less well," "neither poor nor good," "good," or "very good."

The students were also asked whether they would be willing to participate in a subsequent individual interview to discuss their understandings of the two concepts. Five of the students volunteered for interviews and ultimately four students—two women and two men—were interviewed a few days after the writing session. Both researchers involved in the study were actively involved in the interviewing of students. One of the interviewers—with formal training in mathematics—focused on questions pertaining to the mathematical concepts and the students' understandings of those, and the other researcher—specializing in educational psychology—focused on questions relating to how the students usually set about their everyday studies. This design illustrates the dialogic nature of research using intentional analysis, combining the researchers' different competencies to ensure proper coverage of both the mathematical content and the students' conceptions of the wider teaching–learning environment. Such close collaboration at the data collection stage also has the advantage of preparing the researchers for the strenuous task of jointly analyzing the interview transcripts in search of findings of relevance of the research questions posed.

Fragmentation, Algorithmic Contexts and Potentiality for Conceptual Reflection

Focusing on how the students explained the concepts of integral and limit, the analysis of the interview transcripts at first revealed students' understandings to be fragmented and scattered. Typically the students seemed to rely on ways of thinking that focused on different conceptual aspects in isolation rather than on possible interconnections and the integration between different aspects. Although students would stress that they tried to understand the "connected whole" of what they were learning, their accounts would testify to a rather patchy view of connections between the concepts involved and other related mathematical concepts. The way in which students explained their approaches to their studies also revealed a strong preference for procedural understanding and learning how to solve problems rather than thinking about the function and nature of concepts, or, as one of the students put it: "I [focus on] how to solve problems.... I try to understand and apply. I don't go around mulling over why it's one way or the other."

However, as the analysis progressed in the dialogue between the researchers involved in the study, it was clear that this initial picture of the students' understandings as exclusively fragmented needed to be revised. A closer look at the interview data, and in particular at how the students had dealt with the questions asked in the interview, revealed instances in the interviews where the students, prompted by probing questions from the interviewers, touched on issues that seemed to have led them to reflect on mathematical topics and connections that they had not previously considered. For example, one of the students, Philip, was asked whether all integrals could be seen as areas and all areas as integrals.

I: Are all integrals areas and are all areas integrals?

Philip: No.

I: Can you say a bit more about it? When, for instance, is it impossible to use an integral to compute the area...?

Philip: It's when the integral... to take an example, if you... or yeah, maybe so. If you have a... if you have, like, a broken graph or something... But then again, you could calculate with that... and find the different areas and then add them together. I have to think about whether you can do that for all integrals. I've never thought about that before...

In analyzing this piece of interview dialogue it is clear that the topic brought up by the interviewer sparked ideas about the relationship between area and integral that seemingly Philip had not considered before and that make him reflect on this issue from a new angle. At first Philip firmly replies that integral and area is not the same thing. His utterance "It's when the integral..." suggests the notion that not all areas can be computed with the help of integrals. Philip goes on to exemplify: "take an example, if you... or yeah, maybe so. If you have a broken graph or something... But then again, you could calculate with that...," and from the interruptions in his chains of thought it is clear that Philip, unsuccessfully, tries to find examples illustrating his point. After a couple of attempts he suggests the notion of a "broken graph," but in arguing that the integral of such a function cannot be computed, Philip stumbles across a counter argument: "you could calculate with that... and find the different areas and then add them together." So Philip moves from resolute to hesitant in his view on the relationship between area and integral. He realizes that he has not previously considered the possibilities and limitations of the use of integrals. The interviewers' probing questions challenge Philip's procedural understanding of the concept, and he finds himself pondering an issue that he has evidently not considered a topic for conceptual reflection. Moreover, Philip's way of phrasing the final sentence in the extract above indicates that he is projecting his thinking, not just towards the realm of calculus but towards a mathematical way of thinking, resting on logical conditions of the form "for all x...": "I have to think about whether you can do that *for all integrals.* I've never thought about that before..." (emphasis added), thus suggesting a more profound change in viewpoints.

Previous research on student learning in mathematics and other subject areas (e.g., Halldén, 1999; Wistedt & Brattström, 2004) has shown that students often find ways of contextualizing learning material that may at first appear inconsistent and flawed but which may nevertheless enable the students to cope quite successfully with learning tasks confronting them in different settings. In line with these results, Petterson and Scheja (2008) argue that while students' understandings may at first appear fragmented and inconsistent, a closer look at the students' ways of reasoning suggests that they were in fact viewing limit and integral within an *algorithmic context,* in which the very "operations" of these concepts were seen as defining features and as a basis for understanding these concepts. Furthermore, the analysis of how students dealt with probing questions posed in the interview also indicated clear potentiality for conceptual understanding, as students—prompted by the interviewers' challenging questions—were encouraged to venture outside their algorithmic contextualizations of the concepts in hand, to think about calculus in terms suggesting that the students were experiencing an emerging awareness or a conceptual context inviting elaboration and conceptual reflection.

Example 3: Concepts Maps in Linear Algebra

Ryve (2006) used the theory of contextualization to amend and strengthen the methodological framework of Kieran (2001), Sfard (2001), and Sfard and Kieran (2001). Sfard and Kieran have developed a methodological framework that aims at characterizing the students' mathematical discourses while they are working in groups. The framework deals with two types of analyses: *focal* and *preoccupational* analysis. In short, the focal analysis addresses the question of whether the students are using the articulated concepts in reasonably similar ways in the discourse. In the preoccupational analysis, investigations are performed of whether the interlocutors, when solving mathematical tasks together, are engaged in communicating with each other or not and whether they are communicating about mathematics or not. A specially designed tool, an interactive flowchart, is developed for conducting the preoccupational analysis.

The interest in amending and strengthening the methodological framework of Sfard and Kieran stemmed from an earlier article by Ryve (2004), which used that methodological framework to analyze groups of engineering students constructing concept maps in a linear algebra course. Ryve (2004) found that the preoccupational analysis and the interactive flowcharts were weak in basically two aspects. First, there existed limited opportunities to distinguish between different types of mathematical turns. Second, to determine whether the students were engaged in communicating with each other or not, the preoccupational analysis offered no analytical constructs enabling the analyst to relate a single turn to the whole interaction and, thereby, substantiate the analyst's interpretations of the communicative intentions of the students.

Ryve (2006) shows how the first weakness related to differentiation between types of mathematical turns could be solved using Lithner's (2003) framework of different kinds of mathematical reasoning, and the second weakness by using the theory of contextualization. Since the latter additional analysis is most relevant for this chapter, we will now focus on it.

The theory of contextualization was helpful in producing solid arguments for understanding whether the students are talking to each other or to themselves. For instance, it is shown that the three participants in a group contextualize the task of constructing a concept map in linear algebra differently and how these different contextualizations affect the participants' types of engagements in the interaction. Ryve (2006) shows one student, called S, contextualizing the task of constructing the concept map as a problem of connecting many concepts to each other. Turns such as: "I think it is important that we continue now!", "Write inverse so we can move on!", "I don't care anymore, just write gives or gives or something," and "Are we finished now?" served as indication of the above ascribed contextualization.

In contrast to S, another participant, called G, is much more interested in discussing in which ways the concepts are related to each other. G makes jokes and asks conceptually focused questions such as "The dimension defines the rank, or is it the rank that defines the dimension?" Hence, G contextualizes the task of constructing a concept map in linear algebra as a problem of discussing the relations between concepts. The third participant, K, takes the role of a secretary and only to a little extent engages in mathematical discussion. So, since S is in a hurry to connect many concepts to each other, he is not really interested in being engaged in deep conceptual discussions initiated by G. This results in a conversation in which S avoids responding to conceptually focused turns introduced by G and that clearly affect the interactional patterns and the mathematical productivity of the interaction.

To conclude, the important thing to notice in relation to this chapter is that the theory of contextualization could be used to strengthen interactional frameworks developed within mathematical education by highlighting how participants contextualize mathematical tasks and therefore act in specific ways in relation to each other and the mathematical content.

Example 4: Problem Solving in Teacher Education

Ryve (2007) uses the theory of contextualization to serve as a complement to the dialogical approach (Linell, 1998), a theory originally developed for studying face-to-face interaction. The dialogical approach has been used by several scholars in mathematical education (e.g., Bjuland, 2002, 2004, 2007; Carlsen, 2005; Cestari, 1997; Cestari, Daland, Eriksen, & Jaworski, 2005) even though it is not especially developed for the purposes of mathematical education. Therefore, the dialogical approach must be complemented by theories of mathematical knowledge and learning (see Ryve, 2008). Ryve (2007) shows how the theory of contextualization could help the analysts to make sense of the mathematical whole-class discourse of prospective teachers taking a problem solving course. More precisely, it is shown not only that different students contextualize the same task differently but also how those contextualizations affect the characteristics of the whole-class discourse.

For instance, the prospective teachers were working with the task: "In a charity meeting money was collected for welfare purposes. It was decided that each and every person should give as much money as the number of participants attending the meeting. In that way 1600 Skr [Swedish crowns] was collected. How many participants attended the meeting?"

While some students contextualize the task as a problem of solving it by only using mathematical ideas from middle-school mathematics, other students contextualize it as a problem where you are allowed to use any mathematical idea. Let us illustrate this by introducing a section from the classroom discourse in which the teacher educator and the prospective teachers discuss different ways of solving the above introduced task.

Here we enter the discussion when the teacher educator (TE1) initiates the communicative project (Linell, 1998) of discussing, with the prospective teacher at the chalkboard (PT1) and the audience (A1, A2, A3...), alternative ways of solving the task.

[22] **TE1:** By the way, have you got any alternative solution for this?

[23] **A1:** I just think that it became a little bit clearer when I used *x* for the number of participants and *y* for the amount of money they should pay and after a while I realized that *x* and *y* must be the same.

[24] **TE1:** Two unknowns instead of one makes it even more sophisticated.

[25] **A2:** I thought that one could draw a graph also.

[26] **TE1:** Yes.

[27] **PT1:** I was also thinking in that direction.

[28] **TE1:** Yes, mhm, then you are working with *x* squared.

[29] **A3:** Write a table with the number of people and money then *x* times *x*.

[30] **TE1:** By the way, when is *x* introduced?

[31] **PT1:** Yes and graphs, they did not exist.

[32] **TE1:** No, but *x*, when does one start to use one unknown?

[33] **A4:** Secondary school.

[34] **TE1:** In secondary school they work with pretty simple problems and the question is if *x* squared comes in to play.

[35] **A3:** That is the reason why you should use this table.

[36] **TE1:** Yes, yes that's right.

[37] **A3:** It will be easier for the children to understand.

[38] **TE1:** Mhm.

[39] **PT1:** I think it is easier to take a chance, perhaps it is not so good but it feels like it is, first one does not know which two numbers one will have and then it will be terribly complicated to name them *x* and *y*, I think that one should just take a chance.

The prospective teacher at the chalkboard (PT1) indicates by [31] and [39] that he contextualizes the task as a problem of finding a solution that adapts to middle-school students' knowledge of mathematics and in [39] he explicitly questions the suggestion by A1 of introducing two unknowns. That is, [23] indicates that A1 contextualizes the task as a problem in which any kind of mathematical idea could be used for solving the problem. Hence, individuals' different contextualizations affect which kind of mathematics is discussed.

An important underlying principle of the theory of contextualization is that individuals could be regarded as rational agents. In the case of the task discussed above, the interpretations of PT1 and A1 could be seen as sensible in relation to the contexts in which they situate the task. That is, PT1 contextualizes the task in relation to middle-school students, which seems rational in light of the fact that the task is originally designed for middle-school students and that PT1 and his fellow students are prospective teachers. On the other hand, the course is given in a mathematics department, and there are no clear instructions on what kinds of mathematical ideas the prospective teachers are allowed to use, and both these facts make A1's contextualization reasonable.

To sum up, Ryve (2007) shows how the theory of contextualization is used to make explicit the individuals' contributions to the interaction and, through that, support Linell's (1998) dialogical approach. It must here be stressed that neither Linell's dialogical approach nor the theory of contextualization is specifically developed for studying mathematical interactions, so both of them must be complemented by theories of mathematical teaching and learning. Ryve (2007) accomplished this by introducing problem-solving frameworks from Schoenfeld (1992), Wyndhamn, Riesbeck, and Schoultz (2000), Pòlya (1945), and Mason and Davis (1991).

DISCUSSION

In concluding this chapter we make explicit characteristics of the theory of contextualization by relating it to the three dialogical relations introduced in the beginning of this chapter. That is, we argue that the theory of contextualization could productively be characterized as dialogically connected to other theories, research agents, and other researchers. We use the theoretical discussion and the examples from international publications to support this claim.

Let us start by discussing the dialogical relation between the theory of contextualization and other perspectives and theories. Two different strands of this relation are highlighted. First, as the theoretical section and the discussion connected to the first empirical example made clear, the theory of contextualization has been developed in relation to other theoretical perspectives such as sociocultural theory, the theory of variation, and phenomenography. In line with the alternative frameworks' tradition (see Duit, 2006 for an overview), analysts using the theory of contextualization search for research subjects' ways of making sense of the world, rather than their misconceptions of it. However, cognitively oriented researchers have been criticized by sociocultural researchers (e.g., Säljö, 1991) for not paying attention to the situational character of students' conceptions of learning tasks. Empirical studies (e.g., Caravita & Halldén, 1994; Halldén, 1999) carried out within the theory of contextualization tradition support such critique, and analytical constructs have been developed to account for both situational and cultural aspects of students' sense-making processes.

The second strand highlights the theory of contextualization as it has been used to amend and complement other theories in understanding mathematical teaching and learning, and specifically the relation between the two (Wistedt 1994). Ryve (2006, 2007) used it to amend the methodological framework of Sfard and Kieran and as a complement to the dialogical approach (Linell, 1998). Petterson and Scheja (2008) used it to question the thesis that students' strong preferences for procedural understanding of a mathematical content will prevent future conceptual understanding of the content in question, and Nilsson (2007) used it to complement and amend theories of probability reasoning (e.g., Fischbein, Nello, & Marino, 1991; Polaki, 2005; Pratt, 2000). In essence, the theory of contextualization is constantly being explicitly developed and related to other theoretical approaches.

The second dialogical relation that characterizes the theory of contextualization concerns the research approach towards the agents/students being studied. Much educational research could be characterized in terms of a "negative scholarly rationale" (Rommetveit, 1978) by which "children (and adults) are constantly portrayed as lacking in abilities and situationally appropriate and perfectly rational modes of quantifying and handling problems... are marginalized in the experiment (or in the formal setting)" (Säljö, 1991, p. 123). The nature of the theory of contextualization prevents researchers from falling into such a negative rationalism by engaging them in the investigation of how research agents understand and rationalize assignments in learning settings and what potential their reasoning entails (see, e.g., Pettersson and Scheja, 2008). Such a focus is also productive in relation to students working with larger, open-ended problems. As all the empirical examples given above show, mathematical problems could be contextualized in several different ways that affect how the students orient towards each other as well as towards the mathematical aspects of the assignments. In essence, the theory of contextualization facilitates the process of engaging in students' ways of interpreting learning material and classroom discussions rather than ascribing to the students a lack of abilities.

A third dialogical relation connected to the theory of contextualization is that of collaborations among researchers from different research disciplines. As Lester and Lambdin (1998) make clear, "it may often be the case that mathematics educators need to collaborate with researchers in other fields" (p. 421), and such projects characterize research conducted within the research programs using the theory of contextualization. As the discussion of the first empirical example illustrates, Wistedt, a professor in education, has collaborated with mathematicians such as Martinsson and Brattström, and such collaboration between researchers from different disciplines is common among researchers from the contextualization tradition (see, e.g., Caravita & Halldén, 1994; Halldén, Haglund, & Strömdahl, 2007). Further, Pettersson, a doctor of mathematics with a special interest in mathematics education, collaborated with Scheja, a doctor of education, in a study of university students' potential in understanding the concepts of limit and integral. Hence, research projects using the theory of contextualization strive to develop relations among researchers holding different competencies in analyzing and understanding mathematical teaching and learning.

Even though the theory of contextualization has been developed over more than two decades, researchers using it are still very much engaged in continuing, refining, and amending it. Such projects are typically connected to empirical studies in collaboration among researchers from different research disciplines, and we want to conclude this chapter by suggesting ways of developing the theory of contextualization to make it even more useful to understanding mathematics teaching and learning. We wish to highlight two ongoing projects focusing both on understanding empirical data better and on further developing the theoretical approach. In an ongoing project the empirical data used in Nilsson (2007) is reanalyzed in an investigation of how interactional theories, such as conversation analysis (Schegloff, 2007), could serve as a complement to the theory of contextualization for better understanding how students interact and work in groups with complex mathematical assignments. The theory of contextualization is also used in relation to unusual empirical material. In an ongoing project on gifted education in mathematics (Edfeldt & Wistedt, 2008), it is used in collaboration with researchers in informatics and mathematics with the purpose of conceptualizing how students with special abilities in mathematics collaborate when solving mathematical tasks.

REFERENCES

Bartolini Bussi, M. G. (1998). Verbal interaction in the mathematics classroom: A Vygotskian analysis. In H. Steinbring, M. G. Bartolini Bussi, & A. Sierpinska (Eds.), *Language and communication in the mathematics classroom* (pp. 65–84). Reston, VA: National Council of Teachers of Mathematics.

Bjuland, R. (2002). *Problem solving in geometry. Reasoning processes of student teachers working in small groups: a dialogical approach.* Unpublished doctoral dissertation. Department of Applied Education, University of Bergen.

Bjuland, R. (2004). Student teachers' reflections on their learning process through collaborative problem solving in geometry. *Educational Studies in Mathematics, 55*, 199–225.

Bjuland, R. (2007). Mathematically productive discourses among student teachers working on problem solving in geometry. *Nordic Studies in Mathematics Education, 12*(2), 33–55.

Booth, S., Wistedt, I., Halldén, O., Martinsson, M., & Marton, F. (1999). Paths of learning. The joint constitution of insights. In L. Burton (Ed.), *Learning mathematics: From hierarchies to networks* (pp. 62–82). London: Falmer Press.

Burton, L. (1999) *Learning mathematics: From hierarchies to networks.* London: Falmer Press.

Caravita, S., & Halldén, O. (1994). Re-framing the problem of conceptual change. *Learning and Instruction, 4*, 89–111.

Carlsen, M. (2005). Conceptual understanding of the dot product. *Nordic Studies in Mathematics Education, 10*(3/4), 3–28.

Cestari, M. L. (1997). *Communication in mathematics classroom: a dialogical approach.* Unpublished doctoral dissertation. Faculty of Education, University of Oslo.

Cestari, M. L., Daland, E., Eriksen, S., & Jaworski. B. (2005). Working in a developmental research paradigm: The role of didactician/researcher working with teachers to promote inquiry practices in developing mathematics learning and teaching. In M. Bosch (Ed.), *European Research in Mathematics Education IV, Proeedings of CERME4* (pp. 1348–1358). Barcelona: Fundemi IQS-Universitat Ramon Llull.

Cobb, P. (2007). Putting philosophy to work: Coping with multiple theoretical perspectives. In F. K. Lester (Ed.), *Second handbook of research on mathematics teaching and learning: A project of the National Council of Teachers of Mathematics* (pp. 3–38). Charlotte, NC: Information Age Publishing.

Davis, R. B., & Vinner, S. (1986). The notion of limit: Some seemingly unavoidable misconception stages. *Journal of Mathematical Behavior, 5*(3), 281–303

Downes, W. (1998). *Language and society.* Cambridge, UK: Cambridge University Press.

Driver, R., & Erickson, G. (1983). Theories in action. Some theoretical and empirical issues in the study of students' conceptual frameworks in science. *Studies in Science Education, 10*, 37–60.

Duit, R. (2006). Bibliography – STCSE. Students' and teachers' conceptions and science education. Leibniz Institute for Science Education at the University of Kiel. Retrieved on November 6, 2009 from http://www.ipn.uni-kiel.de/aktuell/stcse/stcse.html

Edfeldt, Å.W., & Wistedt, I. (2008). High ability education in Sweden—The Swedish model. In T. Balchin, B. Hymer, & D. Matthews (Eds.), *The Routledge companion to gifted education* (pp. 76–83). London & New York: Routledge.

Engeström, Y., Miettinen, R., & Punamäki, R.-L. (1999). *Perspectives on activity theory.* New York: Cambridge University Press.

Entwistle, N. (1996). Recent research on student learning and the learning environment. In J. Tait & P. Knight (Eds.), *The management of independent learning* (pp. 97–112). London: Kogan Page.

Entwistle, N., & Ramsden, P. (1983). *Understanding student learning.* London: Croom Helm.

Fischbein, E., Nello, M. S., & Marino, M. S. (1991). Factors affecting probabilistic judgements in children and adolescents. *Educational Studies in Mathematics, 22*, 523–549.

Halldén, O. (1982). *Elevernas tolkning av skoluppgiften* [*The learners' interpretation of the school task*]. Unpublished doctoral dissertation. Stockholm University, Department of Education.

Halldén, O. (1988). The evolution of the species: Pupil perspectives and school perspectives. *International Journal of Education, 10*(5), 541–552.

Halldén, O. (1999). Conceptual change and contextualisation. In W. Schnotz, S. Vosniadou, & M. Carretero (Eds.), *New perspectives on conceptual change* (pp. 53–65). London: Pergamon.

Halldén, O., Haglund, L., & Strömdahl, H. (2007). Conceptions and contexts. On the interpretation of interviews and observational data. *Educational Psychologist. 42*, 25–40.

Halldén, O., Hansson, G., & Skog, G. (1994). *Evolutionary reasoning in answers to two questions used to measure the development of university students' understanding of evolutionary theory.* Working Paper Series, 4. Stockholm University: Department of Education.

Halldén, O., Petersson, G., Scheja, M., Ehrlén, K., Haglund, L., Österlind, K., & Stenlund, A. (2002). Situating the concept of conceptual change. In M. Limon, & L. Mason (Eds.), *Reconsidering conceptual change—Issues in theory and practice* (pp. 137–148). Dordrecht, The Netherlands: Kluwer Academic Publishers.

Halldén, O., Scheja, M., & Haglund, L. (2008). The contextuality of knowledge. An intentional approach to meaning making and conceptual change. In S. Vosniadou (Ed.), *International handbook of conceptual change* (pp. 509–532). London: Taylor & Francis Group.

Kieran, C. (2001).The mathematical discourse of 13-year-old partnered problem solving and its relation to the mathematics that emerges. *Educational Studies in Mathematics, 46*, 187–228.

Kirshner, D., & Whitson, J. A. (1997). *Situated cognition: Social, semiotic, and psychological perspectives.* Mahwah, NJ: Lawrence Erlbaum Associates.

Lerman, S. (2000). The social turn in mathematics education research. In J. Boaler (Ed.), *Multiple perspectives on mathematics teaching and learning* (pp. 19–44). Greenwich, CT: Ablex Publishing.

Lester, F. K., & Lambdin, D. V. (1998). The ship of Theseus and other metaphors for deciding what we value in mathematics education research. In J. Kilpatrick, & A. Sierpinska (Eds.), *Mathematics education as a research domain: A search for identity* (pp. 415–425). Dordrecht, The Netherlands: Kluwer.

Linell, P. (1998). *Approaching dialogue: Talk, interaction and contexts in dialogical perspectives.* Amsterdam: John Benjamins Publishing.

Lithner, J. (2003). Students' mathematical reasoning in university textbook exercises. *Educational Studies in Mathematics, 52,* 29–55.

Lundholm, C. (2005). Learning about environmental issues: Postgraduate and undergraduate students' interpretations of environmental contents in education. *International Journal of Sustainability in Higher Education, 6,* 242–253.

Marton, F., Asplund-Carlsson, M., & Halász, L. (1992). Differences in understudying and the use of reflective variation in reading. *British Journal of Educational Psychology, 62,* 1–16.

Marton, F., & Booth, S. (1997). *Learning and awareness.* Mahwah, NJ: Lawrence Erlbaum.

Marton, F., & Tsui, A. B. M. (Eds.), (2004), *Classroom discourse and the space of learning.* Mahwah, NJ: Lawrence Erlbaum.

Mason, J., & Davis, J. (1991). *Fostering and sustaining mathematical thinking through problem solving.* Greelong, Australia: Deakin University Press.

Molander, B-O., Halldén, O., & Pedersen, S. (2001). Understanding a phenomenon in two domains as a result of contextualization. *Scandinavian Journal of Educational Research, 45*(2), 115–123.

Nilsson, P. (2007). Different ways in which students handle chance encounters in the explorative setting of a dice game. *Educational Studies in Mathematics, 66,* 293–315.

Niss, M. (2007a). Reflections of the state of and trends in research on mathematics teaching and learning: From here to utopia. In F. K. Lester (Ed.), *Second handbook of research on mathematics teaching and learning* (pp. 1293–1312). Charlotte, NC: Information Age Publishing.

Niss, M. (2007b). The concept and role of theory in mathematics education. In C. Bergsten, B. Grevholm, H., Måsøval, & F. Rönning (Eds.), *Relating practice and theory in mathematics education, proceedings of Norma 05* (pp. 97–109). Trondheim, Norway: Tapir Academic Press.

Nussbaum, J., & Novick, S. (1982). Alternative frameworks, conceptual conflict and accommodation: towards a principled teaching strategy. *Instructional Science, 11,* 183–200.

Petersson, G. (2002). *Begreppsutveckling genom imitation, generell assimilation och differentiering: hur studenters uppfattningar förändras över tid i en akademisk utbildning* [Conceptual development through imitation, general assimilation and differentiation: how students conceptions change over time in an academic program]. Unpublished doctoral dissertation. Department of Education, Stockholm University.

Pettersson, K., & Scheja, M. (2008). Algorithmic contexts and learning potentiality: A case study of students' understanding of calculus. *International Journal of Mathematical Education in Science and Technology, 39*(6), 767–784.

Polaki, M. V. (2005). Dealing with compound events. In G. A. Jones (Ed.), *Exploring probability in school: Challenges for teaching and learning* (pp. 191–214). New York: Springer.

Pòlya, G. (1945). *How to solve it: A new aspect of mathematical method.* Princeton, NJ: Princeton University Press.

Pratt, D. (2000). Making sense of the total of two dice. *Journal for Research in Mathematics Education, 31,* 602–625.

Renshaw, P. D., & Brown, R. A. J. (1998). Orchestrating different voices in student talk about infinity: Theoretical and empirical analyses. In C. Kanes, M. Goos, & E. Warren (Eds.), *Teaching mathematics in New Times: Proceedings of the twenty-first annual conference of the Mathematics Education Research Group of Australasia* (Vol. 2, pp. 468–475). Gold Coast, Australia: MERGA.

Rommetveit, R. (1978). On the negative rationalism in scholarly studies of verbal communications and dynamic residuals in the construction of human intersubjectivety. In M. Brenner, P. Marsh, & M. Brenner (Eds.), *The social contexts of method* (pp. 16–32). London: Croom Helm.

Ryve, A. (2004). Can collaborative concept mapping create mathematically productive discourses? *Educational Studies in Mathematics, 56,* 157–177.

Ryve, A. (2006). Making explicit the analysis of students' mathematical discourses—Revisiting a newly developed methodological framework. *Educational Studies in Mathematics, 62,* 191–209.

Ryve, A. (2007). What is actually discussed in problem solving courses for prospective teachers? *Journal of Mathematics Teacher Education, 10,* 43–61.

Ryve, A. (2008). What analyzing mathematical classroom discourse: initiating elaborations on the usefulness of the dialogical approach. *Nordic Studies in Mathematics Education, 13, 7–29.*

Säljö, R. (1991). Piagetian controversies, cognitive competence, and assumptions about human cognition. *Educational Psychology Review, 3,* 117–126.

Säljö, R. (1994). Minding action: conceiving of the world versus participating in cultural practices. *Journal of Nordic Educational Research, 14,* 71–80.

Säljö, R. (1997). Talk as data and practice — a critical look at phenomenographic inquiry and the appeal to experience. *Higher Education Research & Development, 16,* 173–190.

Säljö, R., Schoultz, J., & Wyndhamn, J. (2001). Heavenly talk. A discursive approach to conceptual knowledge and conceptual change in children's understanding of elementary astronomy. *Human Development, 44,* 103–118.

Säljö, R., & Wyndhamn, J. (1990). Problem-solving, academic performance, and situated reasoning. A study of joint cognitive activity in the formal setting. *British Journal of Educational Psychology, 60,* 245–254.

Schegloff, E. A. (2007). *Sequence organization in interaction.* Cambridge: Cambridge University Press.

Scheja, M. (2002). *Contextualising studies in higher education: First-year experiences of studying and learning in engineering.* Unpublished doctoral dissertation. Stockholm University, Department of Education.

Scheja, M. (2006, May). *Contextual variation and conceptual understanding in higher education.* Paper presented at the SIG Symposium on Conceptual Change, Stockholm, Sweden.

Schoenfeld, A. H. (1992). Learning to think mathematically: Problem solving, metacognition, and sense making in mathematics. In D. Grouws (Ed.), *Handbook of research in mathematics teaching and learning* (pp. 127–146). New York: MacMillan.

Sfard, A. (2001). There is more to the discourse than meets the ears: Looking at thinking as communication to learn more about mathematical learning. *Educational Studies in Mathematics, 46,* 13–57.

Sfard, A., & Kieran, C. (2001). Cognition as communication: Rethinking learning-by-talking through multi-faceted analysis of students' mathematical interactions. *Mind, Culture, and Activity, 8,* 42–76.

Swedish Research Council. (2007). *International evaluation in educational sciences.* Report 4:2007. Stockholm: Swedish Research Council.

Tall, D. (1993). *Students' difficulties in calculus.* Proceedings of Working Group 3 on Students' difficulties in calculus, ICME-7, Québec, Canada, 1992 (pp. 13–28). ISBN 2-920916-23.

von Wright, G. H. (1971). *Explanation and understanding.* Ithaca, NY: Cornell University Press.

von Wright, G. H. (1979). Reasons, action, and experience. In H. Kohlenberger (Ed.), *Essays in honour of Raymond Klibansky* (pp. 107–119). Hamburg: Felix Meiner Verlag.

von Wright, G. H. (1980). Freedom and determination. *Acta Philosophica Fennica, 1*(31), 1–8.

Wistedt, I. (1987). *Rum för lärande* [*Latitude for learning*]. Unpublished doctoral dissertation. Stockholm University, Department of Education.

Wistedt, I. (1993). Elevers svårigheter att formulera matematiska problem [Pupils' difficulties in contextualizing mathematical problems]. *Nordic Studies in Mathematics Education, 1*(1), 40–54.

Wistedt, I. (1994a). Everyday common sense and school mathematics. *European Journal of Psychology of Education, 9,* 139–147.

Wistedt, I. (1994b). Reflection, communication, and learning mathematics: A case study. *Learning and Instruction, 4,* 123–138.

Wistedt, I. (1996). *Gender-inclusive higher education in mathematics, physics and technology. Five Swedish development projects.* Stockholm: National Agency for Higher Education.

Wistedt, I. (1998). Assessing student learning in gender inclusive tertiary mathematics and physics education. *Education and Programme Planning, 21,* 143–153.

Wistedt, I. & Brattström, G. (2004). Mathematical induction in a co-operative setting. Merits and limitations of classroom communication among peers. In A. Chronaki & M. Christiansen (Eds.), *Challenging perspectives on mathematical classroom communication* (pp. 173–203). Charlotte, NC: Information Age Publishing Inc.

Wistedt, I. & Martinsson, M. (1996). Orchestrating a mathematical theme. Eleven-year-olds discuss the problem of infinity. *Learning and Instruction, 6,* 173–185.

Wyndhamn, J., Riesbeck, E., & Schoultz, J. (2000). *Problemlösning som metafor och praktik* [Problem solving as a metaphor and practice]. Linköping: Linköpings Universitet.

22

FROM HENSCHEN TO MIDDLETOWN MATHEMATICS

Swedish Research on Low Achievement in Mathematics

Arne Engström

LINKÖPING UNIVERSITY

Olof Magne

MALMÖ

Abstract: This chapter presents a summary of Swedish research on low achievement in mathematics from the first systematic studies of arithmetic disorder done by Salomon Eberhard Henschen to the Middletown studies carried out almost 90 years later. Henschen's case studies challenged the prevalent localization of math ability to a single brain region. He coins the term acalculia and demonstrated the complexity of mathematics as a cognitive construct. During the 1950s, Olof Magne introduced Swedish research on low achievement in mathematics with his extensive Gothenburg Studies, with 6,000 students involved. The Mathematics Clinics project tested special education methods in mathematics. The Middletown study is an investigation of students' mathematical achievements in the compulsory school system of a Swedish municipality during a period of 25 years.

Keywords: Error analysis, learning disability, low achievement, mathematics education, special needs.

INTRODUCTION

In this chapter a summary of Swedish research on low achievement is presented, from the first systematic studies on arithmetic aphasias done by Salomon Eberhard Henschen to the Middletown studies carried out almost 90 years later. This overview also shows a change in focus from students' disabilities to low achievements in mathematics. Henschen's seminal case studies focused on arithmetical aphasias whereas the Middletown studies took their stance in the normal variability in students' behavior and aptitudes defined as low achievement in mathematics.

Individual differences in mathematics are very marked, and some students really have to struggle to make any progress in mathematics. Whether such difficulties are attributable to an underlying disability is still an open question in the research community. There is still much research to do on clarifying and defining Mathematical Learning Disabilities (MLD) as a theoretical construct (see, e.g., Berch & Mazzocco, 2007).

During the last 10–15 years the concept of (Mathematics) Learning Disability (LD) has been questioned and criticized. LD existed from the beginning as a disorder that was difficult to define. Later on it was defined as a heterogeneous group of disorders with a common marker of individual variability (i.e.,

The First Sourcebook on Nordic Research in Mathematics Education, pages 333–345

unexpectedness) often considered to represent a discrepancy between IQ and achievement. The primary approach to identification involved a search for intra-individual variability as a marker for the "unexpectedness" of learning disabilities, along with an emphasis on the exclusion of other causes of underachievement that would be "expected" to produce underachievement. Now it is recognized that this model for intra-individual differences lacks empirical support and should be abandoned (Fletcher, Morris, & Lyon, 2003).

There seems to be strong empirical support for seeing most of the students with low mathematics attainment as representing the lower end of a continuum (Dowker, 2005). This does not mean that they do not have a genuine problem with learning mathematics. Complaints that are found in these students can be observed among most students to a varying degree.

LOW ACHIEVEMENT IN MATHEMATICS

There are different reasons why an individual in school is assessed as a low achiever in mathematics. Usually, the reasons are related to education, although some students are aphasics who display low arithmetical skills due to lesions to the nervous system. Still other individuals are not very competent to master mathematical tasks in their trade or leisure time activities.

Obviously, definitions and standards vary from one complaint to another or from one nation to another. Low achievement seems to be assessed according to different criteria in and outside the school systems. In Sweden, an official report on conditionally able-bodied workers in and outside the labor market stresses the relativity of the concept of working capacity (SOU, 1977). The report explains that "Disability connotes a deviation from a common societal norm" (p. 87).

In the EU countries, the expression *special educational needs in mathematics* (SEM) is often used to signify that a student fails in his or her efforts to master one or several topic strands in the mathematics syllabus according to set standards. Special educational needs in mathematics are not looked upon as a fact but a human interpretation of the relation between mathematics, the individual, and his or her environment (MIE). We will use the expression SEM for low-achieving students. Terms such as acalculia, dyscalculia, and mathematical disabilities are in vogue in the United States in addition to other expressions, but we avoid them because of their depreciative meaning.

If an individual claims to have a condition of some kind or if somebody proposes to assume a condition in an individual, it is common practice to look upon the case objectively and find operational symptoms. In Sweden, the public opinion views many symptoms of low achievement in children as a "failure-to-match" in the interaction between the child and the system that surrounds him or her. Such a viewpoint assumes that children are inseparable parts of a mini social system and that in order to be maximally effective, interventions must address themselves to the realities of those systems. We see low achievement in mathematics as a social construct but, obviously, one that is possible to study from various standpoints. We will suggest five standpoints, following the arguments of Tredgold (1952) and Wallin (1949). Each standpoint adds an important facet to the total picture. Definitions must accordingly vary with the basic point of view. The following approaches seem to be valid. Examples will then be given from Swedish research.

A. Anatomical (neurophysiological) definition
B. Psychological definition
C. Educational definition
D. Socio-occupational definition
E. Eugenic definition.

If we consider current international research, approaches A, B and C are mainly discussed in research on low achievement. Approach D attracts a growing interest, but approach E seems to lose popularity in recent research.

Anatomical and Neurophysiological Studies

S. E. Henschen, Professor of medicine at Karolinska Institutet, Stockholm, started Swedish studies of low performance in arithmetic in the beginning of the twentieth century and published two magistral volumes in the year of 1920. He investigated arithmetical aphasias associated to lesions to the neural central system (Henschen, 1920). His studies are based on *the anatomical definition*. Accordingly, he presumed a one-to-one correspondence between arithmetical areas and cortical areas. Henschen developed the fundamentals of *the anatomical theory*. His main term was *acalculia*—absence of arithmetical skill. Although Henschen was a pioneer in this field, there are no follow-up studies in Sweden.

The anatomical (neurophysiological) approach is represented in the National Board of Health and Welfare "Classification of Disease and Health Problems" in its 2008 edition (Socialstyrelsen, 2008). It includes "specific arithmetic difficulty" as a specific impairment of mathematical skills that cannot be blamed on mental developmental disturbance or insufficient school attendance. Specific arithmetic difficulty is defined as a deficient ability in basic arithmetic skills in the four rules rather than in the abstract proficiency called for in algebra, complex computations, and so on. Acalculia, arithmetic impairment, and Gerstmann's syndrome are classified as anatomically conditioned. Thus, the usual anatomical/neurophysiological approach is based on individual impairment.

Psychological Views

Although research from a psychological point of view is mainly lacking in Sweden, many implicit meanings circulate, often imported from the international debate. School psychologists and hospitals practice psychological testing and guidance of low achievers in mathematics.

In Sweden, representatives can be found in Ahlberg and Csocsán (1999), a study on blind students' mathematics; Foisack (2003), a study on deaf students' concept formation in mathematics; and Andersson (2008, 2009) and Andersson and Lyxell (2007) on working memory deficits in students with mathematics difficulties. Johansson (2005) has studied students' arithmetic solutions procedures.

Educational Studies

Magne (1958, 1960) introduced Swedish research on low achievement in schools. Low achievement in mathematics was defined as a student's retention of mathematical teaching below a minimum cut-off level. Thus, for school systems, Magne suggested *an educational definition.* The primary subject of these studies was low achievement in school mathematics. The project comprised two extensive investigations on low achievement in a sample of the compulsory schools in Gothenburg, Sweden (7–16 years). The project took place in 1953 and 1954–1955. Additional studies aimed at mapping various presumed secondary symptoms, such as learning capacity, perception, ability, working capacity, emotional traits, motivation, and achievement in other school subjects.

Another experiment was called the Mathematics Clinics project (1963 to about 1985). It was introduced by Magne, supported by the government, and entrusted to the local authorities. Education of the eligible students, aged 7 to 16 years in the compulsory school, took place according to directions in the curriculum. The teaching methods should be modernized. According to several evaluations, the students were able to acquire up to 1.5 school years' worth of knowledge in one year´s attendance.

The Middletown Mathematics study (Magne, 1990, Engström & Magne, 2003, 2006, 2008) belongs here. Other recent educational studies are a phenomenographic study on students' development of number (Ahlberg, 1997) and a longitudinal study on students who were assessed to have dyscalculia (Sjöberg, 2006). Recently, Bentley (2008) has presented error analyses of students' calculations of various arithmetical tasks, based on material from the international evaluation project TIMSS 2007.

Studies of Socio-Occupational Character

Sociological and ecological studies of low achievement in mathematics were neglected for many years. Gradually social workers are now becoming aware of the fact that students often meet barriers of a social nature when entering school. It is also a fact that the means of mathematics performance vary from one school to another and that various ethnic groups differ considerably in their mathematics attainment. This phenomenon is characterized by the expression *the achievement gap.* A search in databases discloses the information that there are hundreds of thousands of contributions in books, articles, and the press on this phenomenon.

There are some Swedish studies related to occupation and labor market, for instance the official report from SOU (1977) on "Betingat arbetsföra" (Conditional able-bodiedness). Malmqvist (2007) has conducted a long series of experimental studies on motor impairments.

Eugenic Conditions

In Sweden, there seems to be scarcity of relevant genetic research on genom studies associated with achievement in mathematics. According to older traditions, there was general agreement regarding a strong relationship between heredity and mathematical ability. Eugenics was introduced about 1900. It aimed at methods to neutralize effects of unfavorable or undesired heredity dispositions in the population. Following a period around the middle of the twentieth century, sterilization was rejected as a eugenic practice.

SOME MAIN FINDINGS OF THE GOTHENBURG PROJECTS 1953 AND 1954–1955 ON MATHEMATICAL ACCOMPLISHMENTS IN SCHOOLS

The project in 1953, as well as in 1954–1955 (Magne, 1958, 1960), defined SEM as the achievement level at which the student did not get a pass in mathematics according to the grading system. The 1953 project comprised all the 6,000 students of three random school districts of the compulsory school system in the city of Gothenburg. First, the mathematical performance was mapped in the respective districts. Secondly, various psychological, social, and medical conditions were investigated as well as family circumstances.

In the Gothenburg studies the following prevalence frequencies were found: A minority of about 15% of the population in the compulsory school could be described as SEM students—in other words, those who got marks in mathematics below the pass standard according to the current marking system. In this case, the low achievement is often considered *general*, relating to the whole set of mathematical topic strands or even to all other school subjects as well. The low achievement sometimes is *partial*, meaning that the student manages a subset of the specifications in the mathematical curriculum; this refers to knowledge of some mathematical topics but not to others. In exceptional cases the special need is tied to mathematics only but to no other school subjects (often termed *specific*), a case which in the United States is sometimes called underachievement or specific educational needs in mathematics (Lewis, Hitch, & Walker, 1994). In the two latter cases, the Germans speak of *Teilleistungsschwäche*. Contrary to prevailing views, there are very few students with *specific* educational needs. At Gothenburg there were not more than about two tenths of a percent (.2%) prevalence of specific SEM students. Later, similar estimations have corroborated the Gothenburg frequencies in various Swedish investigations.

The 1954–1955 project contained a detailed analysis of the students' mathematical behavior. The SEM students (15% of the student population) were responsible for more than 50% of the mathematical errors. They displayed particularly low results in problem solving, geometry, and equations. SEM students were best in subtraction and least successful in division. The achievement in rational numbers was very low (fraction, decimal form, and percent). We made a thorough error analysis. From a comparison between the SEM students and the five percent highest performing students, we found that the SEM students mainly committed errors when trying to solve mathematical tasks because they committed logical mistakes ("thinking errors"). The top 5% performers only seldom committed thinking errors. Many observations made us believe that the teachers' instruction for the SEM students had aimed at mechanical training rather than rational insight.

The following conclusions were made: Prevalence of low achievement in mathematics was found to be low in preschools, at least as a registered entry in population statistics. The frequency increases during compulsory school age and amounts to 15 to 20% in Sweden. For adult persons, prevalence is unknown, probably very low.

Secondary Complaints

As accomplishment in mathematics was treated as the primary complaint, other symptoms were considered secondary. Secondary complaints were numerous in the SEM students, although the majority of them did not reveal behavior problems. Some secondary symptoms were of little value. The following observations were comparatively frequent:

1. Low intelligence (learning capacity) is assessed to be as frequent as 95 to 98% among the SEM students.
2. A symptom assessed as reduced "working capacity" is as frequent as about 75%. These students are averse to trying hard, to exerting themselves to the utmost when necessary, to strain every nerve. We interpret that this is a crucial symptom of low achievement.
3. Failing concentration/staying power has a frequency of about 75%.
4. Failing motivation totals nearly 50%.
5. Mathematical phobia or anxiety amounts to 20 to 25%, with the majority of cases in the later classes.
6. Ill-health and illness, including symptoms of nervous disorders, are restricted to less than 20% in SEM students but are not intensively explored.
7. Low socio-economic status and poverty does not exceed 10 to 15%.
8. The so-called specific SEM is considered overrated as to frequency and importance by the international literature. Its prevalence is assessed to be less than five per thousand (<.5%).

Summarizing the Gothenburg studies, the SEM students appear as individuals within normal socio-biological variance of the human population. For specific information, we refer to the original reports.

These investigations have not been replicated in Sweden or elsewhere. Hannele Ikäheimo (1989, 1991) performed two parallel investigations in Finnish-speaking and Swedish-speaking school systems in Helsinki. On the whole, Swedish science seems to contain rather few detailed investigations of low achievement in mathematics, compared with corresponding studies of low achievement in language skills.

THE MATHEMATICS CLINICS PROJECT—A SEM TEACHING AND LEARNING PROJECT

Magne's Gothenburg investigations (1958) gave impulses to didactic research and development projects in the Nordic countries. The two Swedish cities of Karlskrona and Arboga decided to organize small-group education for the *folkskola* (primary school) students who failed in mathematics. For tackling the problem, Magne worked out a design for a nationwide experimentation together with the school authorities at Karlskrona. In 1963, the Swedish government published directions on "Education at Mathematics Clinics," a name that was gradually replaced with the colloquialism "Math land," originally introduced in Helsinki, Finland.

The basic educational philosophy has its origin in the international reform movements in the middle of the 20th century. We chose the concept called *the rational model* as a foundation of the project's didactical aims (see Nuthall & Snook, 1973). The expression "rational" relates to the intention to find a philosophy and theory of reasoning power as the main generator of knowledge instead of psychology. The rational model can fit well to the context of mathematics education. It is the striving after logical meaning in knowledge, which is conceived as the central element in knowledge. The model assumes that man is a reasoning creature with judgment, logic, and language. Mathematics education involves active learning of the students and differentiated tuition of the teacher. Among other things, the model presupposes that the student's learning depends on his or her reasoning power to establish long-range retention in mathematics. As to the role of the teacher, the progress of the student would depend on the teacher´s trust in the student's gift for actively exploring and educing mental processes. Deliberate co-operation between teachers and students on rational reasoning should be considered crucial for the genesis of knowledge. Thus, we look at mathematics learning in terms of logical rather than psychological relevance.

According to the operational design of the experiment, the special education should be carried out according to the general directions in the general curriculum passed by the government. The curriculum, however, grants the teachers the liberty of instructing their students in various ways. While the subject content was kept within the frame of the official syllabus, the teaching and learning should be influenced by ideas of the reform movement. The method was considered laboratory, rational, and constructivistic. *Laboratory* means to start from authentic everyday situations and find mathematical solutions of problems by planning and reasoning with the help of the mathematical language. *Rational* implies logical reasoning. *Constructivism* signifies a differentiated view on learning strategies for a learner who meets problem situations.

We worked out tutorials for students according to these aims. In our didactics we tried to tutor the student to a spontaneous planning skill. This education gravitated towards applied mathematics (everyday math). With the preconception that only the active learner learns mathematics, we wanted the student to develop an inquiring attitude to reasoning in order to build up mathematical concepts and relations. The stress on reasoning does not mean neglect of practicing and rehearsing mathematical operations. There are two sides to the matter. If the students are motivated, they choose themselves to train and drill their skills. This is called productive learning. One method was to act on open tasks in order to enable the students to work at the subject matter in various ways.

To be eligible for the Math land program, the student would have marks that indicated that he had failed in mathematics but had at least average success in other academic subjects. The program got state subsidies. A total of about 200 students aged 7 to 16 were educated yearly. Several guidebooks and reports were published (e.g., reports by Magne, 1973; a guidebook by Magne, Bengtsson, & Carleke, 1973). Reiterated observations and interviews with the teachers in experimental and control groups indicated that the teachers accepted and followed the directions.

The outcome has been summarized as follows. The efficacy was positive, resulting in an average raise of the student's achievement from a half to a whole grading step after one year's training. The SEM students with a good working capacity answered for this progress. Those who succeeded were shown to be energetic, well motivated and mentally healthy. However, for some students the result was negative. Various kinds of maladjustment created barriers to learn mathematics. Poverty and ethnic or language disadvantage were as common as up to three fourths of the SEM-population. Some students had low working pace (or working passivity). When assessed to belong to the lowest working pace level in mathematics the students remained stagnant on their bottom level or even got lowered grades (Magne, 1973).

During the 1960s and 1970s, the project was gradually expanding until the subsidy unexpectedly was stopped around 1980.

Programs for low achievers seem to be rare in Sweden nowadays. When the State Commission *Matematikdelegationen* (Mathematics Delegation) in 2004 summarized their propositions for improving mathematics education (SOU, 2004) there were no comments on the critical situation for students with low achievement. There are more efforts to construct intervention programs targeting children who are perceived to belong to social high-risk conditions—for instance, children living in poverty and immigrant children. There is also more enterprizing power and promising development work in the research actions to promote progress in teaching and learning for dyslexic students.

THE MIDDLETOWN MATHEMATICS PROJECT

Middletown Mathematics Project is one of the largest empirical investigations on changes of mathematical skills and knowledge in the compulsory, comprehensive school (*Grundskola*) in Sweden. Middletown (a ficticious name, Swedish *Medelsta*) is a municipality of 25,000 inhabitants, selected on statistical criteria to represent an average municipality of Sweden. The Middletown Grundskola (compulsory school) comprises approximately 2,000 students.

Although the main focus in the Middletown Mathematics Project was on changes in students' achievement in mathematics over time, special attention was paid to low achievement in mathematics.

In 1977, all students from grades 1 to 9 were tested in March–April. The assessment materials used were developed as part of the project. The Middletown Diagnoses, as they were called later, were developed in order to measure the individual mathematical achievement. The comprehensive character of the account is a reason why we have not been able to include significant statistical specifications. We refer the reader to the original reports.

The original Middletown Mathematics study was replicated twice, the first time in 1986 and secondly in 2002. During this span of 25 years, different educational reforms and curriculum changes took place. The project was designed to make it possible to compare the outcomes of the mathematics education according to successive curricula. These curricula were based on rather different ideological, educational, and operational assumptions.

Method

Design

Replication is an essential part of the scientific method. But still, replication studies are not common, particularly not in mathematics education. One of the most well-known and discussed replications of community studies in social sciences may be the Middletown Study conducted during the 1920s (Lynd & Lynd, 1929). It was replicated twice, the first time ten years later (Lynd & Lynd, 1937), and the second time in a follow-up study fifty years later through the Middletown III Project (Bahr, 1980; Caplow & Bahr, 1979; Caplow, Bahr, Chadwick, Hill, & Williamson, 1982).

Replication has two main purposes in social sciences (Bahr, Caplow, & Chadwick, 1983): to check the findings of an earlier study (the reliability test), and to assess whether they hold under altered conditions (the generality test). The essence of a replication is in some way to test a specific previous program of research, so it must be linked together with the previous study and cannot appear without an identifiable baseline study. A replication may differ in some way in its major features of the baseline study: time, place, subjects, methods, and researchers' purpose.

In the Middletown Mathematics Project, the two replications were made under the same conditions according to the major features of the first study. Our main purpose of the replications was not to test the reliability or generality, but to the study hypotheses about changes in teaching and learning that educational and curricular reforms may bring about through the years. In this sense the Middletown Mathematics Project seems to be rather unique. To our knowledge, no similar educational studies on mathematical achievement over time in Sweden have been made as a replication case study.

Construction of the Middletown Diagnoses

The assessment materials used were developed as part of the project by one of the authors (Magne) together with the teachers in Middletown. In order to measure the individual mathematical achievement a series of eleven test instruments were developed. With only one exception the diagnoses were used in more than one grade. With this set of diagnoses it was possible to compare achievements of the Middletown students in many ways, not only between different grades, different individuals or groups of individuals, but also between different curricula and main areas of mathematics.

Preliminary versions of the diagnoses were tested several times for the Middletown Project. Items of various topic strands and difficulty were tested, whereupon they were modified or gradually discarded due to insufficient discrimination power. Validity and reliability were gradually improved. Item selection techniques and statistical information are described by Magne (1990) and Engström and Magne (2003).

Basic Criteria for the Items and Main Areas

The items were chosen from five basic criteria:

- *Grade-typical criteria.* Grade-typical items refer to the grade-level specifications assumed in the curriculum to start learning a specified topic strand (type of item or task).

- *Mathematical main areas.* To distribute simple and various complex topic strands among curricular main areas.
- *Mathematical complexity levels.* Complexity in the item factorial structure is supposed to cause decreasing achievement across main areas and age levels.
- *Retention criteria.* For elementary learning topic strands, we assume a retention level near 100% and lower percentages for complex areas. For the simplest topic strands, 100% retention is considered necessary and attainable.
- *Downward sloping of retention.* We assume a gradual decrease of achievement in grade-typical topic strands for successive grades.

The diagnoses were constructed to cover the elementary parts of the curriculum and were categorized in six main areas:

- *P* for problem solving and mathematics language
- *N* for number and number sense
- *G* for form and geometry
- *ASMD* for arithmetic (the four rules)
- *E* for elementary statistics
- *F* for functions

Results

Stability or Change in Achievement at Middletown

The most striking result was probably the stability of students' achievement over years. Despite educational reforms and curricular changes, the outcome of mathematics education was almost the same during the period from 1977 to 2002.

Achievement in Different Main Areas

On the whole, if a student had low item frequencies in one main area, the results in other main areas were equally low. But there are differences in attainment between some main areas. Occasionally, the inference intimates that there are unexplored regions on the map. This is the case for geometry in all grades (the G area). Number sense seems to be weak, particularly in rational numbers (the N area). Problems (the P area) displays low item frequencies.

The ASMD area seems to have too many low item frequencies from a practical point of view. The young students in grade 1, 2, and 3 were quite successful in elementary algebra, but in the upper grades, both power of comprehension and skill were extremely low in the more complex algebra topic strands.

One result from all three investigations in 1977, in 1986, and in 2002 is that means of the scores gradually increase from grade to grade although the increment becomes less for each additional grade. Also, the variability increases for each successive grade, above all for the low achievers. Most of the content in the upper grades (7 to 9) appears to be directed to achievement levels far beyond the present number competence taught to low achievers. It seems open to doubt whether the lowest achievers are getting appropriate tuition in mathematics.

Downward Slope for Grade-Typical Items

Grade-typical items relate to given objectives of a defined grade in the Swedish Standards—for instance, to use the concept of probability in grade 9. For scores of the grade-typical items, there is a downward slope with higher grades. The grade-typical items score highest in the lowest grades. In grades 1 and 2 (8 to 9 years of age, respectively), they are particularly high, at 82 and 86%, respectively. The scores of grade-typical items then drop steadily through the grades until they reach the low mean score of 68% in grade 9.

On the other hand, Figure 22.1 indicates that the Middletown students' 2002 scores increase for some

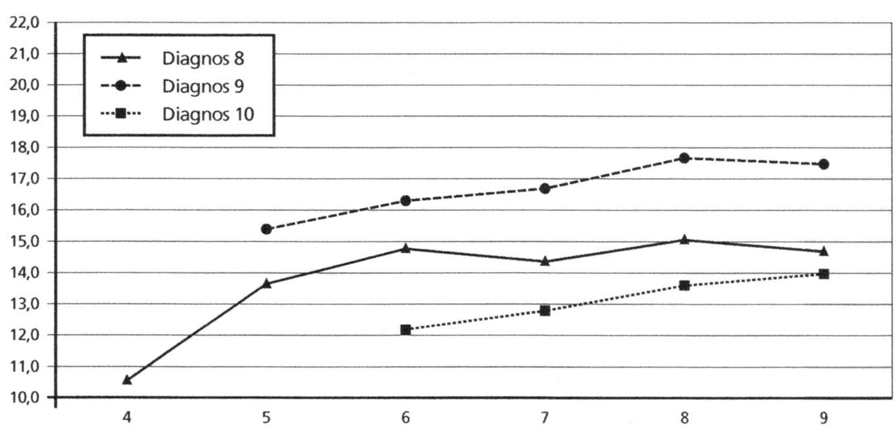

Figure 22.1 Score development from grade 4 to 9 for diagnoses 8, 9 and 10 for Middletown 2002.

TABLE 22.1 Percentages of Correct Responses (%) for the 15% Lowest Achievers in Grade 9 in Comparison to Students in Other Grades for Diagnoses 8–11

	Grade				
Diagnosis	**4**	**5**	**6**	**7**	**SEM9**
8	59	76	82	80	64
9		70	74	76	59
10			55	58	29
11				47	12

of the diagnoses. The same pattern of score development was found in 1977 and 1986.

The downward slope of scores in the higher grades proved to be particularly serious for the 15% lowest achievers. Low achievers scored far below the general population of the Middletown school. Table 22.1 shows a comparison between the 15% lowest achievers in grade 9 and students in other grades for diagnoses 8–11.

Low Achievement in Mathematics in the Middletown Project

The achievement of low-achieving students is described and analyzed in all three investigations. Some of the general observations are the following:

- There is the same proportion of low-achieving students in all three investigations—no better, nor worse.
- The low-achieving students seem to commit the same types of errors in all investigations.
- We find no valid gender differences in frequency, nor in category.
- Apparently all three curricula were "unfair" to low-achieving students. In grade 9 these students could only reach the objectives of grade 4.

These quantitative results were confirmed in qualitative analyses. We performed qualitative specifications of the SEM students' work in every grade from the age of 7 up to the age of 16.

The Attainment Levels of the SEM Students in Middletown

As we have shown earlier, most diagnoses were administered to more than one class. Thus diagnosis 5 was administered to grades 2 and 3, and diagnosis 7 to grades 3, 4, and 5. A special case is diagnosis 8: We have used it in grades 4 through 9. This makes it possible to test the outcome of instruction in the two grades 2 and 3 with diagnosis 5. We can also study the six grades 4, 5, 6, 7, 8 and 9 using diagnosis 8.

TABLE 22.2 A Comparison Between the Means of SEM Students and the Means of the Whole Population of Students in the Grades Referred to for Diagnosis 8 in Grades 4 to 9 in the 2002 Investigation

Grade	SEM students (M)	Whole population (M)
4	5.3	10.6
5	8.4	13.7
6	11.0	14.8
7	10.1	14.4
8	11.1	15.1
9	11.5	14.7

How low is the achievement of the 15% lowest achievers in comparison to the other students? Let us take the students in the top grade (9). For diagnosis 8, we have made a comparison between the means of the these students and the means of the whole population in the Grades referred to. Table 22.2 shows the result of the comparisons.

The first, and perhaps the most important, observation we make in Table 22.2 is that scores stagnate starting in grade 6. Neither the total population nor the low-achieving students increase their scores after grade 5. Low-achieving students seem to fall far behind the general population. The Middletown Mathematics Project results indicate that an average low attainer, aged 16, only reached the average level of grade 4 at the school-leaving age. More generally speaking, Figure 22.2 demonstrates the average levels of the 15% lowest attaining students (SEM students) in every age group compared with the mean attainment of the remainding population.

As can be seen in Table 22.3, the low-achieving students have about the same percentage of correct solutions in 1977, 1986, and 2002. There are very few exceptions. Particularly remarkable is the decrease in diagnosis 11 from 1977 and 1986 to 2002.

Variability

One last issue concerns the variabilities in skills (see Figures 22.3–22.5). In most tests some students come near the maximum while others get no more than one or two items correct. But the background of these students varies too. The same pattern of variability could be seen in all the three investigations. The variability increases during the school grades, to reach a maximum in grade 9 with several years.

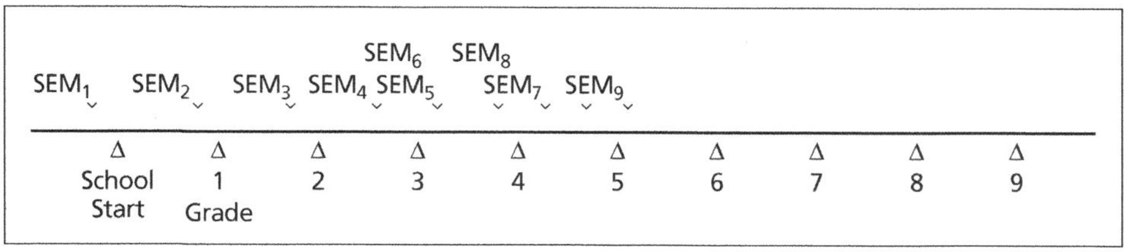

Key to the signs: SEM_1, SEM_2, etc. stand for the various performance means of the SUM students in relation to the means of the average students.

The symbol ˅ marks the position for the mean of the performance of the SEM-students at the end of each Grade.

The symbol Δ marks the beginning and the end of each school year.

SEM_1	Mean after the first school year is lower than the mean of the school beginners
SEM_2	The mean is equivalent to the knowledge level after the first half school year.
SEM_3	The mean has reached the level at the beginning of second school year.
SEM_4	The mean has almost arrived at the finishing tape of grade 2.
SEM_5	Manages the first half of grade 3.
SEM_6	Reaches the goal of grade 3.
SEM_7	Has reached the beginning of grade 4.
SEM_8	Near the target of grade 4
SEM_9	Has just passed the transition from grade 4 to grade 5.

Figure 22.2 Diagram illustrating the performance level of the SEM students at Medelsta at the end of the spring term 2002 for the nine grades of the Grundskola in relation to average performance level in each Grade. (Revised from Engström & Magne, 2006).

TABLE 22.3 Percentages of Correct Solutions for the SEM Students and the Total Population of Grade 9 in 1977, 1986, and 2002

	Year					
	1977		1986		2002	
Diagnosis	SEM	Total	SEM	Total	SEM	Total
8	68	83	67	87	64	84
9	53	79	54	85	59	81
10	23	68	22	71	29	67
11	24	59	26	65	12	52

After three years of schooling, the 15% lowest achievers reach the average level of grade 1. Three years later they reach the average level of grade 3. After nine years of compulsory schooling, the 15% group reach the average level of a student in grade 4.

Error Analyses

In the Middletown Project the classification system for error analyses starts from assumptions of the rational model (Nuthall & Snook, 1973; Scheffler, 1960, 1965), followed up by empirical case and mapping studies (Engström & Magne, 2008; Magne 1990; Magne & Thörn, 1987).

Let us look at the high and low achievers. We use the terms "students with special need for supplementary mathematics" (SUM students—15% of the population) and "students with special educational need" (SEM students—15% as well). The test forms of SUM and SEM students were carefully scrutinized in order to check the solutions. We treated each note and calculation the students wrote alongside an item as a proposed solution. When the student made no such note whatsoever, the student was thought to have refrained from solving the item. Thus, we get two alternatives: solution or no solution. If the student refrained, we can do no more. But otherwise we are able to decipher a meaning of the writings and assign them to our various categories. Thus, the student presents various attempted solutions and we classify them according to the following main categories: (a) correct solution, (b) refrained, (c) not interpretable attempts, and (d) errors. The main categories (a) to (c) refer to one single type of assessment. Main category (d) consisted of several subcategories, which in the final account were brought together into categories such as "oversight errors" and "logical errors." We exemplify our analysis by comparing ten-year-old SUM and SEM students who were taught in grade 3.

Table 22.4 illustrates our method of categorizing the answers of the 15% highest and 15% lowest achiev-

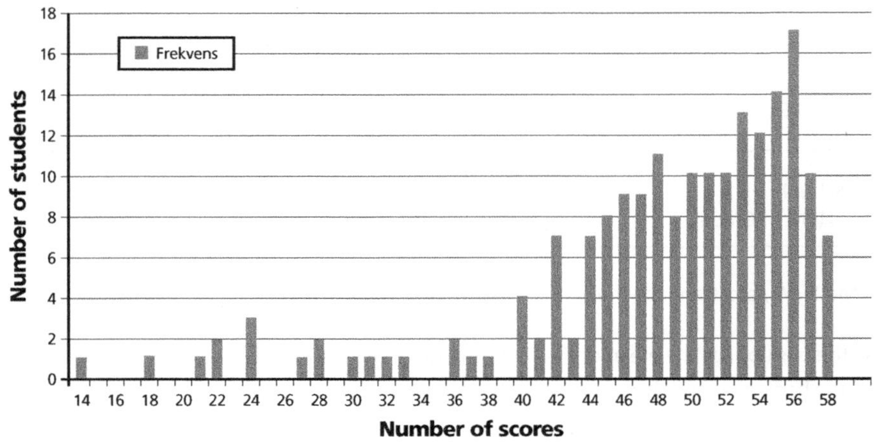

Figure 22.3 Distribution of scores for grade 1, Middletown, 2002.

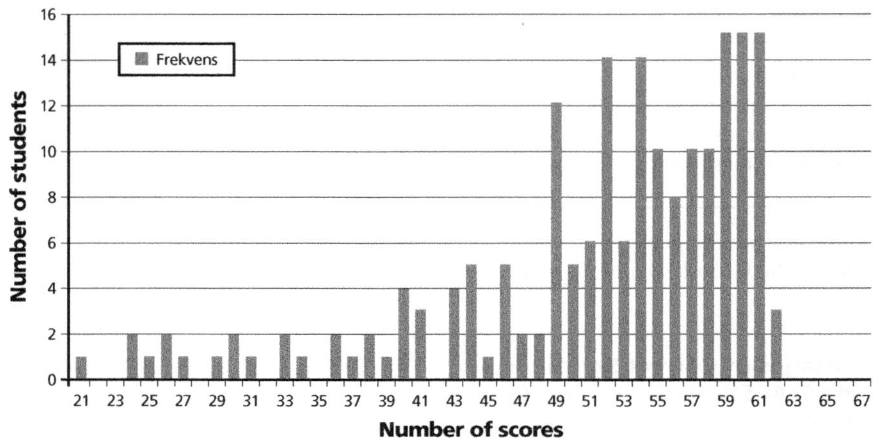

Figure 22.4 Distribution of scores for grade 5, Middletown, 2002.

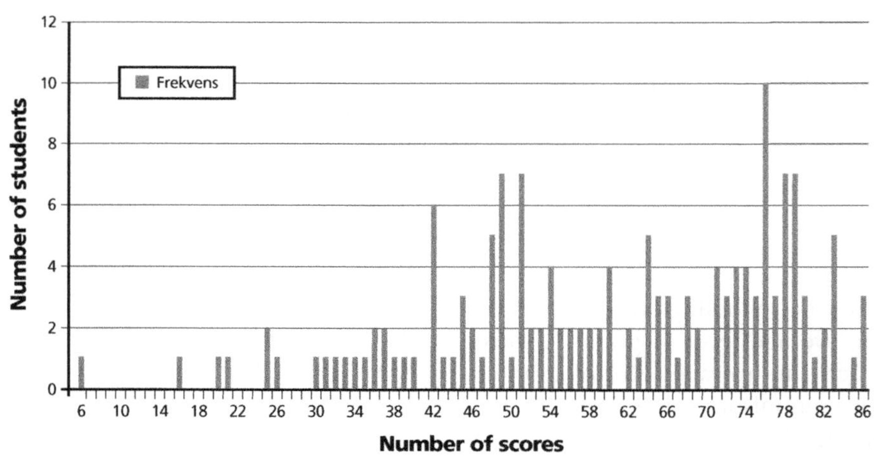

Figure 22.5 Distribution of scores for grade 9, Middletown, 2002.

ers in grade 3. Each grade 3 student was presented 76 items. As the SUM students numbered 31 and the SEM students 30, we have studied a total of 4,536 items in this partial analysis of mathematics performance at Middletown. The SUM students solved nearly 90% of the items correctly, seldom refrained from attempting to solve an item (about 1%), made few oversight errors (about 3%), but made quite a few logical errors (about 7%). The SEM students solved roughly one third of the items, refrained from attempting to solve

TABLE 22.4 Comparisons Between the 15% Highest (SUM) and the 15% Lowest Achieving Students (SEM) in Grade 3, Middletown, in 2002, Displaying the Different Solution Strategies Used in the Diagnoses 5, 6 and 7. Frequencies in Percent

Solution categories	D 5		D 6		D 7	
	SEM	SUM	SEM	SUM	SEM	SUM
Correctly solved	79	92	34	89	20	84
Refrained solution	8	0	29	0	43	3
Oversight error	1	5	5	3	7	2
Logical error	11	3	31	8	30	11
Not interpretable	1	0	1	0	0	0

another third of the items, made few errors of oversight (roughly 4%), but committed logical errors in nearly one third of the items.

Linguistic influence on solution strategies in mathematics appeared in various ways. We noticed many different types of linguistic errors assuming the following categories, (a) mathematical errors due to linguistic mistakes, and (b) errors due to mistakes in reading or writing.

The SEM students' correct solutions fall very much below the correct solutions of the SUM students. Logical errors are few among the SUM students but common among the SEM students. The SEM students refrain from solution in up to a third of all items. This is unusual among the SUM students. Errors due to oversight have about the same frequencies for both SUM and SEM students.

We suggest that the SUM students tend to solve the items by thinking logically more often than SEM students. Other observations seem to indicate that the crucial condition that makes SEM students succeed is when they recognize the content as (1) very familiar, (2) very elementary, or (2) having an everyday character. Analyses of students' mathematical solutions have been explored for every age group in Middletown. Engström and Magne (2006, 2008) have hinted at some hypotheses of why SEM students do what they do. The following survey of the analyses summarizes these studies.

Our interpretation is the following. Basically, success depends on mathematical knowledge. Success may occur when a task is familiar to the students. A boy or a girl may find solving a problem easy if it is about things that remind them of their well-known environment. In this case, it is a question of authentic problems the children have met in the family, in the candy shop, in the playground, or in talk among adults as well as children of their own age.

DISCUSSION

The idea of mathematics for all has been in focus for reforms and developmental work in mathematics education during the last decades. But in spite of waves of educational reforms and curricular changes, the impact on classroom practice seems to be remarkably low. The elementary mathematical skills have remained nearly the same through the 25-year period of the investigations in Middletown independent of the reforms passed by the Swedish Parliament.

Low-achieving students continue to make progress through the grades, but we have observed an increase of variability in mathematical knowledge and skills—the achievement gap—in Middletown. Similar results have been shown by Bodovski and Farkas (2007). They used a large sample of representative data to examine the pattern and determinants of mathematics achievement growth for U.S. elementary school students. Students who began with the lowest knowledge also showed the smallest gains. The beginning level of students' mathematical knowledge is associated with the progress students make subsequently.

This increasing inequality during the school grades has also been presented by other studies (e.g., Alexander, Entwisle, & Horsey, 1997). The students who are falling behind come predominantly from high-poverty and high-minority areas (Balfanz & Byrnes, 2006).

The possibility of closing or reducing the achievement gap seems to be one of the greatest educational challenges of today. In a study by Balfanz and Byrnes (2006), high-poverty students in the middle school followed one of two quite different paths. A significant number of high-poverty students closed the achievement gap and made large achievement gains, which shows that closing the gap can be achieved. But the majority of the students entered the middle grades behind grade level and left even further behind. More attention on the situation for low-achieving students seems urgent. We need a better understanding of the factors that constrain achievement growth.

One interpretation of the Middletown project is that for some students, the school system itself produces special educational needs in mathematics. Students in upper grades score lower on grade-typical items than students in lower grades on their grade-typical items. We suggest that this downward slope of grade-typical items, according to what we have found, is due to an increase of complex items in the syllabus of the higher grades—the complexity model. Low-achieving students often find that degree of complexity is an adverse condition that undermines achievement.

Students with special educational needs seem to be unfairly treated. It is open to debate whether the lowest achieving students are getting appropriate tuition in mathematics. Most of the content taught in the upper grades (7 to 9) appears to be directed to levels far beyond their present number competence. In the upper grades they are virtually excluded from learning mathematics of their own achievement level. Some critics have questioned whether this school system is consistent with the overall educational goal of a "school for all." There will always be some students who experience barriers when trying to reach the prescribed binding goals; that is, due to various interfering social circumstances, some individuals are doomed to fail in such a school system. It seems urgent to carry out reforms of teaching and learning mathematics, also for students with special educational needs. Many students with special educational needs may achieve the goals to be attained when the teaching and learning conditions are changed both for students individually and for the grade as a whole.

One could probably distinguish between three different levels of mathematical attainment among students with special educational needs:.

1. For most SEM students, the so-called social and life-skills mathematics defines the essential aims for their mathematical education. Every student develops mathematics competence depending on his or her ability, motivation, energy, and the conditions in his or her social networks.
2. For the lowest achievers the mathematics teaching and learning might be organized according to the social and life skills mathematics. For these students, the more complex specifications of the syllabus seem to be unrealistic and could be said to fall outside these students' ad hoc competence field. The more complex strands of present mathematics education seem to be inappropriate for some students.
3. There are hard-working and energetic SEM students who are assessed to display personality traits that favor acquisition of the elementary topics of school mathematics. By sensible tuition, they can reach the elementary targets of the syllabus.

The Middletown Mathematics Project has, of course, its restrictions. Important educational dimensions such as students' social and ethnic backgrounds were not investigated. The mathematical knowledge and skills that were measured do not cover all parts of the syllabus. Another objection concerns the question of generalizing the results. What was found in this community study cannot straight off be said to be valid for a whole country. Further replication studies are needed in order to falsify or confirm its results. For the time being, the results from the Middletown Mathematics Project are the most extensive available.

REFERENCES

Ahlberg, A. (1997). *Children's ways of handling and experiencing numbers.* Göteborg: Acta universitatis Gothoburgensis.

Ahlberg, A. & Csocsán, E. (1999). How children who are blind experience numbers. *Journal of Visual Impairment and Blindness, 93*(9), 549–560.

Alexander, K., Entwisle, D., & Horsey, C. (1997). From first grade forward: Early foundation of high school dropout. *Sociology of Education, 70,* 87–107.

Andersson, U. (2009). Skill development in different components of arithmetic and basic cognitive functions: Findings from a three-year longitudinal study of children with different types of learning difficulties. Unpublished manuscript.

Andersson, U. (2008). Mathematical competencies in children with different types of learning difficulties. *Journal of Educational Psychology, 100,* 48–66.

Andersson, U. & Lyxell, B. (2007). Working Memory deficits in children with mathematical difficulties: A general or specific deficit? *Journal of Experimental Child Psychology, 96,* 197–228.

Bahr, H. (1980). Changes in family life in Middletown, 1924–77. *Public Opinion Quarterly, 44*(Spring), 35–52.

Bahr, H., Caplow, T., & Chadwick, B. A. (1983). Middletown III. Problems of replication, longitudinal measurement, and triangulation. *Annual Review of Sociology, 9,* 243–264.

Balfanz, R. & Byrnes, V. (2006). Closing the mathematics achievement gap in high-poverty middle schools: Enablers and constraints. *Journal of Education for Students Placed at Risk, 11*(2), 143–159.

Bentley, P-O. (2008). *Svenska elevers matematikkunskaper i TIMSS 2007. En djupanalys av hur eleverna förstår centrala matematiska begrepp och tillämpar beräkningsprocedurer.* Stockholm: Skolverket.

Berch, D. B. & Mazzocco, M. (Eds.). (2007). *Why is math so hard for some children? The nature and origins of mathematical learning difficulties and disabilities.* Baltimore: Paul Brookes Publishing Co.

Bodovski, K. & Farkas, G. (2007). Mathematics growth in early elementary school: The roles of beginning knowledge, student engagement, and instruction. *The Elementary School Journal, 108*(2), 115–130.

Caplow, T. & Bahr, H. M. (1979). Half a century of change in adolescence attitudes: Replication of a Middletown survey by the Lynds. *Public Opinion Quarterly, 43*(Winter), 1–17.

Caplow, T., Bahr, H. M., Chadwick, B. A., Hill, R., & Williamson, H. M. (1982). *Middletown families: Fifty years of change and continuity.* Minneapolis: University of Minnesota Press.

Dowker, A. (2005). Individual differences in arithmetic. Implications for psychology, neuroscience and education. New York: Psychology Press.

Engström, A. & Magne, O. (2003). *Medelsta-matematik. Hur väl behärskar grundskolans elever lärostoffet enligt Lgr 69, Lgr 80 och Lpo 94?* (Rapporter från Pedagogiska institutionen 4). Örebro: Örebro universitet.

Engström, A. & Magne, O. (2006). *Medelsta-matematik III. Eleverna räknar.* Rapporter från Pedagogiska institutionen, 12. Örebro: Örebro University.

Engström, A. & Magne, O. (2008). Medelsta-matematik IV. En empirisk analys av Skolverkets förslag till mål att uppnå i matematik för årskurs 3. LiU-PEK-R-248. Linköping: Linköping University.

Fletcher, J. M., Morris, R. D., & Reid Lyon, G. (2003). Classification and definition of learning disabilities: An integrative perspective. In H. Lee Swanson, K. R. Harris, & S. Graham (Eds.), *Handbook of learning disabilities* (pp. 30–56). New York: Guilford Press.

Foisak, E. (2003). *Döva barns begreppsbildning i matematik.* Malmö Studies in Educational Sciences, 7. Malmö: Malmö University.

Henschen, S. E. (1920). *Klinische und anatomische Beiträge zur Pathologie des Gehirns. 5. Teil. Über Aphasie und Akalkulie.* Stockholm: Nordiska bokhandeln.

Ikäheimo, H. (1989). *Matematiikan keskeisen oppiaineksen hallinta Helsinkin periskouluissa.* Helsinki: Helsinkin kaupungin kouluviraston julkaisusarja A4:1989.

Ikäheimo, H. (1991). *Hur väl behärskar eleverna i Helsingfors svenska skolor det Centrala lärostoffet i matematik?* Helsingfors: Helsingfors stads publikationsserie, A3.

Johansson, B. S. (2005). *The development from informal to formal arithmetic solution procedures: Gradual or continuous?* Slutrapport. Stockholm: Vetenskapsrådet.

Lewis, C., Hitch, G. J., & Walker, P. (1994). The prevalence of specific arithmetic difficulties and specific reading difficulties in 9- to 10-year-old boys and girls. *Journal of Child Psychology and Psychiatry, 35*(2), 283–292.

Lynd, R. S. & Lynd, H. M. (1929). *Middletown.* New York: Harcourt, Brace, and World.

Lynd, R. S. & Lynd, H. M. (1937). *Middletown in transition.* New York: Harcourt, Brace, Jovanich.

Magne, O. (1958). *Dyskalkyli bland folkskoleelever.* Göteborg: Göteborgs Universitet, Pedagogiska institutionen.

Magne, O. (1960). *Räknesvårigheter i folkskolan* (Pedagogiska skrifter 229). Stockholm: Svensk Läraretidnings Förlag.

Magne, O. (1973). *Matematiksvårigheter* (Pedagogiska skrifter 253). Stockholm: Sveriges Lärarförbund.

Magne, O. (1990). *Medelsta-matematik. Hur väl behärskar grundskolans elever lärostoffet enligt Lgr 69 och Lgr 80?* (Pedagogiska Problem 539). Malmö: Lärarhögskolan.

Magne, O. & Thörn, K. (1987). *En kognitiv taxonomi för matematikundervisningen. Del 1–2.* (Pedagogiska Problem 471). Malmö: Lärarhögskolan.

Magne, O., Bengtsson, M., & Carleke, I. (1973). *Hur man undervisar elever med matematiksvårigheter.* Stockholm: Esselte Stuidum.

Malmqvist, J. (2007). *Gymnasieval för ungdomar med svåra rörelsehinder: Slutrapport från regeringsuppdrag.* Stockholm: Socialstyrelsen.

Nuthall, G. & Snook, I. (1973). Contemporary models of teaching. In R. M. W. Travers (Ed.) *Second handbook of research on teaching* (pp. 47–76). Chicago: Rand McNally College.

Scheffler, I. (1960). *The language of education.* Springfield. IL: Charles C. Thomas.

Scheffler, I. (1965). *Conditions of knowledge: An introduction to epistemology and education.* Chicago: Scott Foresman.

Sjöberg, G. (2006). *Om det nu inte är dyskalkyli–vad är det då? En multimetodstudie av eleven i matematikproblem ur ett longitudinellt perspektiv.* Doktorsavhandling i Pedagogiskt Arbete, 7. Umeå: Umeå University.

Socialstyrelsen (2008). *Klassifikation av sjukdomar och hälsoproblem 1997 - version 2008.* Stockholm: Socialstyrelsen.

SOU. (1977). *Betingat arbetsföra* (Rapport av sysselsättningsutredningen No. 89). Stockholm: Arbetsmarknadsdepartementet.

SOU. (2004). *Att lyfta matematiken—intresse, lärande, kompetens. Betänkande av Matematikdelegationen.* Rapport No. 97. Stockholm: Utbildningsdepartementet.

Tredgold, A. F. (1952). *Mental deficiency.* New York: William Wood.

Wallin, J. E. W. (1949). *Children with mental and physical handicaps.* New York: Prentice-Hall.

23

RESEARCH ON MATHEMATICS TEACHER EDUCATION IN SWEDEN

Barbro Grevholm

UNIVERSITY OF AGDER & LULEÅ UNIVERSITY OF TECHNOLOGY

Abstract: The chapter explores research on mathematics teacher education in Sweden carried out by mathematics teacher educators over a period of three decades. Questions about research funding and how it has influenced the research are discussed. An overview of the development of teacher education in Sweden forms a background for the chapter. Studies carried out by experienced researchers or doctoral students are presented explaining who did the study, what it is about and the results. Conclusions based on the studies are drawn, and some suggestions are given for future research on mathematics teacher education.

Keywords: Mathematics teacher education research, teacher educators, doctoral studies

INTRODUCTION

This chapter includes historical and structural aspects on mathematics teacher education research and development in Sweden. Reasons why so few resources for research were available and the policy to support development projects are discussed, as well as consequences of this educational policy in Sweden. The question about similarities and differences between international studies and studies carried out in Sweden is also raised. At ICME10 in 2004, an international survey on research on mathematics teacher education was reported (Adler, Ball, Krainer, Lin, & Novotna, 2005). I will here relate to this survey, partly using their methods and categories. International research on mathematics teacher education has expanded immensely during the 1990s and later (Even & Ball, 2009; Jaworski & Wood, 2008).

The chapter starts with two sections intended to give some background to the problem. The method used to structure the chapter is described, followed by a section about the funding situation for research on mathematics teacher education. Then the presentation deals chronologically with studies published before 1988, between 1988 and 2001, and after 2001. Finally a discussion about the questions raised is presented, and a comparison is made with the international situation. The chapter concludes with a reflection on what I find to be important questions for future research on mathematics teacher education and how results from such studies could be used in order to develop the education and help resolve some of the critical issues that are discussed at the moment.

WHY IS RESEARCH ON MATHEMATICS TEACHER EDUCATION IMPORTANT AND INTERESTING?

Mathematics is a subject that arouses strong feelings in almost everybody. It can either create joy to learn or anxiety and unpleasant feelings (Skolverket, 2003; Kislenko, Grevholm, & Lepik, 2005). The teacher is

The First Sourcebook on Nordic Research in Mathematics Education, pages 347–362

of crucial importance for the motivation and progress of pupils' learning. Thus the education of good teachers of mathematics is a matter of concern for society but also for both the pupils who succeed in and those who reject mathematics. In the efforts of society to educate teachers of mathematics I discern problems with several aspects. The access to well-educated mathematics teachers is rarely sufficient, the recruitment to the education is problematic, and the effect of the education is often questioned. Repeated evaluations of teacher education in Sweden claim that it is not research-based enough and the quality of the education is too low (HSV, 1996, 2005, 2008). Society has reacted to this situation by implementing iteratively new reforms in teacher education. Why do these problems exist, and what are the characteristics of them? What kinds of research have been undertaken on this often debated area of society? Have research results guided the reforms of teacher education or the creation of the teacher education programs? In the following sections I will elaborate on these and related questions.

BACKGROUND WITH HISTORICAL AND ORGANIZATIONAL SITUATION FOR MATHEMATICS TEACHER EDUCATION

The organization and structure of teacher education is adapted to suit the school system and thus normally changed according to that, although often with some delay. In Sweden a nine-year compulsory school was introduced in 1962 as *enhetsskolan*, and in 1969 it was renamed *grundskolan*. Starting in 1968, prospective teachers for school years 1–3 and 4–6 were educated at teacher training colleges in class teacher programs (Prop. 1967:4). Before that, the education took place in so-called seminars and aimed for school years 1–2 (*småskola*) and 3–7 (*folkskola*). Teachers for older pupils took a degree at university.

In the 1960s, the education of class teachers and subject teachers was collected in special teacher training colleges (*Lärarhögskolor*) with the National Board of Education as steering authority. Primary teacher education was integrated into university studies in 1977 when the special teacher training colleges became part of the universities or university colleges (UHÄ, 1992). Two different educational traditions were in this way brought together—the teacher training college tradition (the seminar tradition) and the university tradition—causing some problems in the years to follow. Education was arranged in two separate programs for teachers for the age groups 7–9 and 10–12. Teachers for age group 13–15 and for upper secondary school were given a common education with three years of subject matter studies at university, most often in two school subjects, followed by one year of practical teacher education with pedagogy (general education theory), methodology, and teacher practice and training in schools.

Through the university reform of 1977, the demand that teacher education must be research-based was raised. This had consequences for both the content of the education and the competence of the teacher educators, as will be illustrated below.

During the 1980s teacher education was discussed politically and criticized because of claimed low results. The results from the IEA tests, among other things, started a thorough investigation into mathematics education. A report from the Ministry of Education was presented with several suggestions in order to improve mathematics teaching and learning in Sweden at all levels (Ds U 1986:5). It was claimed that teachers were not specialized enough in their subjects and it was wished that teachers follow their pupils through a longer time period. The result was a reformed teacher education in 1988 where teachers should be trained for school year 1–7 or 4–9, either in mother tongue and social science or in mathematics and science. The intention was to give the compulsory school its own specialized teachers. The educational reform was never followed up, however, by an intended reform of the construction of positions in school (for years 1–7 or 4–9). Teachers for upper secondary school received separate education from teachers for the compulsory school. For the former, the education program was almost the same as described above except that the practical pedagogical year was interspersed in two parts in between the academic subject study semesters. These new education programs were evaluated in the mid 1990s, and critique was exposed (Gran, 1995; HSV, 1996; Ds, 1996).

As a consequence of the evaluations and other changes in the educational system, the government appointed in 1997 the Teacher Education Committee (LUK97) consisting of politicians from all parties in the parliament. The government instructed LUK97 to set forward a proposition to renew teacher education on the basis of recent changes in society, school, and university. Teacher education was one of the most important missions in universities, with around 11% of the total number of enrolled students. The government decides which programs should exist and the goals for them, while the universities themselves decide what to offer and the content and organization of the education inside the national frame. Schools have undergone substantial change with new curriculum in 1994 (Lpo94, 1994; Lpf94, 1994) and are goal- and achievement-directed.

The teacher profession has undergone changes due to the fact that pupils are now seen as active, responsible for their own learning, and part of the creative work in school rather than as objects for teaching. The organization of work in school demands leadership, social competence, and awareness of and insight into how individuals and groups function. The teacher profession also demands more theoretical competence. The curriculum reform and the new system of governing schools demand from teachers an ability to analyze, argue for, and mediate a way of thinking about content, planning, and organization of work in school (Dir 1997:54). The final document from LUK97 (SOU 1999:63) led to the government proposition in 2000 (Prop. 1999/2000:135).

The structure was new and radical, characterized by an integration of three areas: a general education domain, directions, and specializations. The first area contains common teacher knowledge such as learning, socialization, and interdisciplinary studies. The "direction" consists of the subject or subject fields in which the teacher intends to work (e.g., mathematics and science). Parts of the direction are school-based, consisting of teaching practice and practice-related course tasks. A compulsory diploma thesis is written related to the profession of teaching. All teachers receive the same academic degree but with specialization for primary, lower secondary or upper secondary school. Important aims with the new structure were to create openness and flexibility and let students find individual combinations and specialization. This flexibility led to a critique of the new structure as it was found that some prospective teachers were lacking important parts of teachers' professional knowledge, such as insights about early learning of reading and mathematics or about assessment. Two evaluations took place (HSV, 2005, 2008), shortcomings were found, and this led to a new committee for design of a new teacher education. Thus again the demand was raised for a new teacher education structure and a suggestion was proposed in spring 2009 for a new education program to start in 2011 (Prop. 2009/2010: 89). In a parallel process, new curricula for compulsory school and upper secondary school are being developed and intended to be implemented in 2011.

Based on this historical account, Table 23.1 lists the time epochs given by the restructuring of the teacher education programs.

METHOD

The opportunities for doing research in and on mathematics teacher education have depended on the organizational place and structure of the teacher education. Thus I will use the chronological structure outlined above as the overarching structure for the analysis. I claim that the chronological structural aspect has guided opportunities for financing of research and development, the kinds of studies done, the position of the person who could carry out the study, and the questions, methods, and results of the study. I use a model including these aspects as basis for the presentation below.

Arguments for why these aspects are important seem obvious. In each epoch of time, the financing opportunities have differed depending on how the importance of research was seen in relation to teacher education. The level of experience of the researcher has depended on whom the government wanted to support to do research and where the money came from. When the claim was raised for the need for more teacher educators with a research degree, financing was made available for those teacher educators without a research degree, while little funding could be found for experienced researchers. This fact influences what kind of study can be carried out, questions, and methods. If the teacher educator has to become a doctoral student supervised by a pedagogue (for one who wants to do a doctorate in mathematics education the only available doctoral programs at that time were in pedagogy), the study will be framed by this fact (time-limited, with a pedagogical content focus).

By these methodological decisions, this study links closely to the methods used in the international survey (Adler et al., 2005). The unit of analysis for both studies is research and development studies (in

TABLE 23.1 Time Epochs

Time	Structure and Organizational Place for Teacher Education
Pre 1968	Year 1–2 and year 3–7 in seminars, year 8–12 at University
1968–1977	Year 1–3 and year 4–6 in teacher education (seminars), year 7–12 at university
1977–1988	Year 1–3 and year 4–6 in teacher education (colleges), year 7–12 at university
1988–2001	Year 1–7 and year 4–9 in teacher education (colleges), year 10–12 at university
2001–2010	One teacher education degree with specialization, university or university colleges
2010–	Toward a sustainable teacher education? University or university colleges

a broad sense), and what is observed is "the who (who was writing/doing the research, and from where), the how (what methods were used) and the what (what was being studied, theoretical orientation, assumptions and outcomes)" (Adler et al., 2005, p. 365). Adler and colleagues (2005) present four claims:

1. Small-scale, qualitative research predominates.
2. Most teacher education research is conducted by teacher educators studying the teachers with whom they are working.
3. Research in countries where English is the national language dominates the literature.
4. Some questions have been studied, not exhaustively but extensively, while other important questions remain unexamined.

I will include comments relating to these claims to enable comparison between the international and the Swedish situations for research on mathematics teacher education.

As I worked as a mathematics teacher educator in Sweden between 1973 and 2003, my own lived professional experience provides me with first-hand knowledge. This has then been quality assured by checking other sources such as governmental documents, regulations, journals, papers, books, evaluations, and reports. The sources that have been searched are government bills about teacher education, official evaluations of teacher education, *Educational Studies of Mathematics* (as the earliest scientific journal in mathematics education and covering the period I investigate), *Nordic Studies in Mathematics Education* (the only scientific journal focusing on mathematics education in the Nordic countries), the proceedings from the six research seminars MADIF of the Swedish Society for Research in Mathematics Education, and, from the Nordic Research conferences NORMA (1994, 1998, 2001, 2005, and 2008), doctoral theses in Sweden with a focus on didactics of mathematics done by mathematics teacher educators or dealing with mathematics teacher education issues. All research studies found will be recognized in the chapter. A mathematics teacher educator here means a professional teacher who works in a teacher education program with teaching of mathematics, teaching methods in mathematics, or didactics of mathematics.

THE FUNDING SITUATION FOR RESEARCH ON MATHEMATICS TEACHER EDUCATION IN SWEDEN

The main criticisms over the years toward teacher education in Sweden have pointed at the lack of research carried out within it and on it, the low number of teacher educators with research education, and the lack of research-based education for the students. For example, the National Swedish Board for Universities and Colleges (UHÄ) writes (in an evaluation of the teacher education from 1988) that the realization of the aim that all university education shall rest on scientific foundations depends on access to relevant research and the scientific competence of the teacher educators (UHÄ 1992:21, p 36). They suggested that special resources be given to teacher education programs over a longer period of time for research and development and for competence development of teacher educators. Although some resources were given, the situation is still not satisfactory when it comes to the scientific foundation of teacher education. In fact, after the evaluation in 2008, a number of teacher education programs are threatened with being closed down if the proportion of research educated teacher educators is not raised (the lowest in 2008 was 25%, which was the normal situation in the 1990s). Actually, teacher education differs from other university education programs by the low number of senior lecturers and professors.

From the time of the teacher seminars before 1968, it is possible to trace some research and development that has been done by mathematics teacher educators. As one illustration, we can take the work by Fritz Wigforss on national standardized tests (Kilpatrick & Johansson, 1994). His work took place systematically over a long period of time and had much influence on the Swedish educational system. The standardized tests were used over many years and later replaced by an updated version of tests, the national tests in the 1990s. It seems as if there were no special resources for research in and on teacher education before 1960, but some teacher educators carried out research and development work as part of their duty. In the 1960s, the seminars were replaced by schools of teacher education (*Lärarutbildningar*) and, linked to them, research and development activity existed. Such activity was probably financed directly by each of these research and development centers. One example of a study carried out in that way is the work by the two mathematics teacher educators, Ekenstam and Nilsson (1977), on pupils' learning algebra, which was highly relevant for student teachers, and the results were used in teacher education. This study exemplifies early studies that were published internationally (Ekenstam & Nilsson, 1979).

One way of trying to increase the number of teacher educators with research education was to finance the further education of *adjunkter* (a position for academic teachers without research education degree). In the 1980s and later, this was a source of some research education studies going on in teacher education. Below, examples will be given of such studies in the area of mathematics education. As no research education programs existed in subject didactics at that time, the studies were carried out in pedagogy (general education), although in some cases the studies belong to the area of didactics of mathematics.

Another positive change related to the financing of research is that in the mid 1990s the demand for teacher education programs to include research time in the normal work of teacher educators with research degrees (senior lecturers and professors) was raised, and in at least in some of the universities, 20–50% of working time could be devoted to research and development. Financing was given by the government to the teacher education institutions to be distributed locally among the research-active teacher educators.

In the 1990s, the government suggested an educational science committee (Prop. 1999/2000:135) linked to the Research Council, with the task to announce and distribute research financing for teacher education. Thus, not until the end of the1990s was financing of research possible for educational science in similar ways as for other fields of research. Among the studies given support by this new committee, we find rather few concerning didactics of mathematics, which is not so surprising when most of the experts evaluating the applications are professors of pedagogy, not of subject didactics.

The amount of money for research in and on teacher education has been limited compared to any other academic field. Most of the senior lecturers do not have research time as part of their positions. Only professors can be sure to have time for research, and the professor/student ratio is surprisingly small for teacher education as compared to psychology studies (1/3100 compared to 1/240) (SOU 1999:63). Professorships in mathematics education did not exist in Sweden until 2001, and the number of professors in this field is still very low. One of the results of that fact is that the changes carried out in mathematics teacher education on a national level have not been based in mathematics education research results from Sweden (SOU 1999:63; SOU 2008:109). In the following sections I will investigate and illustrate the examples of research on mathematics teacher education that I found.

STUDIES BEFORE 1988

The first signs of interest for research on mathematics teacher education came in the late 1970s and dealt with student teachers' level of mathematical knowledge. These two linked studies (in Swedish) from the 1970s and 1980s were carried out on mathematics teacher education but by researchers in pedagogy. One is the report on written tests and students' mistakes during problem solving (Lindblad, 1978). This study caused the discussion on student teachers' pre-knowledge in mathematics, which later led to changes that meant that the level of prerequisites was raised (Ds U, 1986). A similar study was conducted nine years later, with some test problems in common with the first investigation (Ljung, 1987). The later study shows that only 12% of the candidates reach a level in the test that he claims to be reasonable (more than 57 correct out of 65 tasks), and almost 17% solve fewer than half of the tasks correctly. Ljung claims that questions about teacher competence and the professionalism of teachers cannot be isolated to details about subject knowledge. It should be obvious that such knowledge is necessary but not sufficient to create a successful teacher. Higher demands on prerequisites for teacher education had been put forward. A substantial part of the education time should be spent on the content the future teachers are going to teach. A coordinated and common approach is necessary from subject theory, pedagogic, methodological, and practical–pedagogical education. The subject studies must give a substantially deeper knowledge on the areas of mathematics that are taught in school (Ljung, 1987).

In the 1980s, many mathematics teacher educators devoted much of their time to development work with computers and calculators. The new technology was more in focus than mathematics as such. The government invested substantial amounts of money to support teachers in developing computer competency and software. Much emphasis was on offering competence development for teachers and integrating knowledge about ICT in teacher education. New courses for in-service and pre-service teachers were developed. As one illustration of this I can use an example I was involved in myself. Financed by the teacher education at Lund University, were I worked at the time, I was asked to take a course in 1986 on how to develop software for teaching mathematics. The outcome of this was a design of a computer program "The Counting Cave," for which we got a good deal of funding from the Nordic Council of Ministers to develop it and have it programmed by a professional programmer. The aim of the program was to assist pupils in developing number sense (Grevholm et al., 1989; Grevholm, 1990). Much research and development work was spent in similar ways and implemented in mathematics teacher education. For a more extensive overview of the projects in DOS (the Computer in School), see Riis (1991).

Another typical feature of the 1980s was the development work done to offer competence development to teachers in pre-service and in-service education. Many mathematics teacher educators were involved in the development and production of literature for such teacher courses. A series of booklets should be mentioned called *Täljaren*, which consisted of a handful of texts of about 50 pages on, for example, number and number sense, counting stories, geometry in our surroundings, part of a whole, basic skills, and letters to count with (Täljarenserien, 1988–1989). The material was also used in pre-service education. The production was financed by a special resource for teacher competence development, initiated thanks to the departmental report in 1986 (DsU, 1986) and worked out by the development and reference groups at the National Board of Education. The aim was to produce basic material for competence development of teachers, and all the authors were mathematics teacher educators. Most of the texts do not refer to research literature, but lists of literature are provided that in some of the booklets contain some international research reports. Earlier books used for mathematics teaching methods courses in teacher education do not refer in the actual text to mathematics education research literature at all. One of the later books lists one page of reports from Swedish research and development work, which can be seen as further evidence of the shifting role of research in relation to mathematics teacher education. Thus the government, by offering directed resources in such a way for several years, steered the development work done by many mathematics teacher educators.

STUDIES AFTER 1988 AND BEFORE 2001, THE FIRST ERA OF INTEGRATED COMPULSORY SCHOOL TEACHER EDUCATION

In the preparations for the new teacher education implemented in 1988, the content of teacher education changed in character. Up to then it had consisted of four parts, labeled subject studies (mathematics), methods, pedagogy, and practice. In the mid 1980s, subject didactics entered the discussion, becoming an important part of teacher education and in some ways replacing part of both teaching methods and pedagogy. At the same time, introductory research education courses in mathematics didactics were created and offered, for example, at University of Gothenburg and at Lund University. This became an invitation to mathematics teacher educators to enter formally into research education. Research seminar series were created at several universities, and groups of teacher educators met regularly to discuss research in didactics of mathematics. The so-called LUMA conferences, yearly national meetings with mathematics teacher educators that started in 1982, became more research-based (Dahland & Emanuelsson, 1999). At Linköping University, a new doctoral program in pedagogy with a direction toward subject didactics of mathematics was opened, and in this and one other program, three PhD theses dealing fundamentally with school mathematics relevant for teacher education were produced (Bergsten, 1990; Hedrén, 1990; Wyndhamn, 1993). At Umeå University, a doctoral program in didactics of mathematics started in 1995 with five doctoral students, among them several experienced teacher educators. In the sections below, the studies on mathematics teacher education from this epoch will be presented.

A Cross-Sectional Study on Problem Solving

The Swedish National Agency for Education funded a three-year research project during 1997–1999 called "Problem Solving as Metaphor and Practice." One part of this study inquired into student teachers' conceptions of problem solving. The study was carried out by an experienced teacher educator and researcher, Jan Wyndham, involving his own students as research objects. What student teachers see as problem solving was examined in a cross- sectional study. Fifty-five novice students at the first lecture in mathematics and mathematics didactics were given the task of planning a lesson for pupils in the compulsory school about problem solving. Individually, in writing, they were asked to answer and discuss the questions of "what, how, and why." Another group of 64 student teachers, at the end of their education in mathematics, were given the task of drawing a mind-map of problem solving and, guided by that, constructing proper tasks in a test or for discussion about problem solving in compulsory school. In the first group, most students planned a lesson in arithmetic. Thirty-one students suggested an experimental approach or use of concrete material, and 24 students suggested the use of the blackboard. Some arguments for "why" problem solving were that the design will offer understanding, pupils see connections, pupils will be allowed to work practically, problems are linked to reality, and it is better that the teacher explains. In the group of older students, arithmetic is still dominating among the suggested tasks, followed by geometry, measurement, and combinatorics.

For a deeper analysis, a model was used with two dimensions, one from learning to application, and the other from way of working to way of thinking. Using the four quadrants created by these axes, it is shown that the novice students took most notice of application/way of working and way of working/learning. In the second group of students, more stress is put on the way of thinking/application and way of thinking/learning quadrants. For them, problem solving is more than just searching for an answer. They widen the scope of problem solving compared to the novices. For the older students it is more important to discuss the way to a solution than to get a right answer. In the second study, the students stated that mathematics is about abstract things and relationships and is founded on logic rather than on observations as truth criteria. Experiments are means to discover new relationships. Five main categories for interpreting problem solving as a metaphor and practice were identified (Wyndham, 1998): Problem solving is to rationally use your acquired competence in mathematics in a new situation; to work with quantitative questions close to everyday life; to discuss and interpret information, facts, and different relationships in order to learn something new in mathematics; to make and evaluate intentional intellectual experiments in mathematics; and to experience and develop a sense of "can do" in mathematics. The full study was reported only in Swedish (Wyndham, Riesbeck, & Schoulz, 2000); for a presentation in English see Wyndhamn (1998).

A Doctoral Study Carried Out in the U.S. by a Swedish Teacher Educator

The doctoral study by Lingefjärd (2000), carried out on student teachers at Gothenburg University, focused on research questions concerning how preservice teachers relate mathematical models to reality when using software tools to generate the models and what conceptions and misconceptions lie behind the decision to believe more in a mathematical model than in real-world phenomena. Being an experienced mathematics teacher educator in Sweden, he was formally enrolled as a doctoral student in the U.S. The data used consisted of questionnaires, videotaped interviews, observations of lecture sessions and the labs, students' lab reports, students' written assignments, and their final exams. One major conclusion is that a normal prospective teacher can, during a ten-weeks course, change his source of authority and become "seduced" by technology. The data show that students' mathematical knowledge can be fragile and fragmented, and to compensate for this, different learning experiences are recommended, such as reasoning and constructing mathematical models and developing, comparing, and evaluating alternative solution processes. A discussion about the problems of being at the same time the teacher educator and the researcher of your own students is presented (Lingefjärd, 2000). One obvious effect of the study was that the results changed the local course in which the students took part. The first of the three-part case studies led to more emphasis on the validation part of the modeling process. After the second part of the study, the course lectures emphasized both validation and common misconceptions. The third part of the study led to additional topics in the course: responsibility, authority, and peer tutoring. Thus there was an immediate local implementation of the research results. The study has also been reported in English in an international research seminar (Lingefjärd, 2002).

A Longitudinal Study of one Swedish Mathematics Teacher Education

In Sweden few research studies have been undertaken concerning the learning of mathematics for student teachers, and no earlier longitudinal study is reported before the one by Grevholm (Grevholm, 1998, 2003a, 2003b, 2004b, 2005a, 2005b, 2008; Hansson & Grevholm, 2003). The study was carried out as part of the normal work as senior lecturer and teacher educator, while I was at times a teacher for the students. I followed a group of prospective mathematics teachers for grades 4–9 for eight years in 1996–2003. The studied group consisted of 48 students at the beginning of the program, and after 4.5 years when the education was finished, 25 of them took their degree as teachers. Data consisted of interviews, audio-recordings, video-recordings, mathematical tests, essays, exam papers, concept maps, field notes, and classroom observations. As can be seen from the numbers mentioned above, the proportion of students who were able to complete the education in the expected time is low (50%). The student teachers' conceptual development in mathematics and didactics of mathematics was especially in focus, and it was revealed that initially their concept images of central mathematical concepts (such as numbers, fractions, algebra, equation, function) were vague and meagre and somewhat confused. Repeated observations of the development of mathematical concepts indicate that over a longer period of time, students' conceptions grew more adequate and became richer, even in periods where the students did not study mathematics (Grevholm, 1998, 2003b, 2005a, 2005b, 2008). Many student teachers had problems with the mathematics courses and failed several times in the examinations. Mathematics and mathematics education take up only around 20% of the study time in the program, and still it seems as if the learning outcome in mathematics creates significant problems (Grevholm, 2003b). Thus mathematics comes to play the role of sorting instrument of the students. The student teachers, who feel less successful in mathematics, carry this experience with them out into the classrooms, which might influence their own pupils in less favorable ways (Pehkonen, 2001).

Once the student teachers graduate and start working in a classroom, they find that their professional language is too limited (Grevholm, 2004b). When they try to assist pupils during lessons, they have problems expressing themselves in ways that pupils understand. The teachers also find it difficult to vary their explanations if the pupil does not follow the first explanation (Grevholm, 2003a). As a result of this longitudinal study, changes took place in the local teacher education program in order to better meet the student teachers' demands to develop a professional language. Natural study groups were introduced where student teachers were given appropriate tasks to discuss in groups without a teacher present. This way they could practice using a professional language and developing it (Gellert, 2009; Grevholm, 2004b; Grevholm & Bergsten, 2005).

Classroom management is crucial for beginning teachers. Video-recordings from the study referred to here give evidence of different kinds of problems (Grevholm, 2003a). Teacher competences such as being able to help all pupils that need help at the same time proved to be difficult to achieve when students were not skilled in using many different ways of working. Similarly, to find appropriate tasks on different levels of difficulty adequate for each pupil was demanding and would have required access to and knowledge about different materials and methods, which not all students had. Pupils who got bored because of inadequate tasks created disciplinary problems, and the student teachers who were newcomers in the teaching profession had difficulties handling such situations (Grevholm, 2003a).

The outcome of Grevholm's study was implemented in the local teacher education, and some of the research features were later used by younger researchers such as Örjan Hansson, who partly replicated the study by Grevholm and by Andreas Ryve, who used concept maps as introduced by Grevholm (1998, 2005a, 2005b). More about these studies follows below.

At the end of this epoch, a crucial change took place for research in didactics of mathematics in Sweden. Financial resources were given to a National Graduate School in didactics of mathematics for a period of five years, and extensive preparations took place during 2000–2001. In the autumn of 2001, the program could accommodate 21 doctoral students to start their studies in mathematics education (Leder, Brandell & Grevholm, 2004; see also the chapter by Brandell in this volume). Some of these students who investigated issues related to mathematics teacher education will be mentioned in the next sections. Through the creation of the National Graduate School, the conditions for doing research in mathematics education were completely changed. These doctoral students could make their studies on full-time basis with a full salary, and with all the resources from a graduate school in the form of supervisors, mentors, guidance from the board, a group of fellow research students, research seminars, and academic courses. Never before were the conditions so supportive.

STUDIES AFTER 2001 CONCERNING MATHEMATICS TEACHER EDUCATION

Another Longitudinal Study on Student Teachers

In her doctoral work, Bjerneby Häll (2002, 2006) followed a group of student teachers during their teacher education and the first three years after their debut as mathematics teachers in school during the years 1993 to 2001. She was at that time working as an experienced teacher educator. The aim of her study was to describe and analyze arguments for mathematics in compulsory school and to understand why and how the official arguments change. During their education, the student teachers develop a view on mathematics and mathematics education that is in harmony with the goals of mathematics in the national syllabus. The student teachers express ten main arguments for mathematics in school: to manage one's everyday life, for future education and occupation, to take care of one's own interests in society, because society needs and demands this knowledge, to develop one's thinking skills, because it is funny and will increase one's self confidence, because it is needed in many other subjects, because it is an important body of knowledge and a language, because it is part of our culture and common knowledge, and because there will be a test.

As novice teachers, many experience a conflict between different goals in the national curriculum and the course syllabus for mathematics. Those who teach in the later part of compulsory school experience a pressure to fulfill the course content before the national test in mathematics. Their goal in teaching becomes to prepare pupils for the test. A tension is that pupils need to have an acceptable score in order to be allowed to continue to the national programs in upper secondary school. The conditions and reality for teachers and the mathematics subject in compulsory school are influenced by how schools organize teachers in working teams and the teachers' combinations of teaching subjects. Her conclusion for teacher education is that during their education the student teachers must get the opportunity to develop their competence to evaluate textbook series. Teacher education must give a basic competence and then the continued education and competence development must take place in school and at the work place. Communities and schools must take their responsibility for teachers' competence development and school development. Bjerneby Häll points to the need for research on the important competence development and in service training of mathematics teachers. The thesis is written in Swedish, but the work has later been presented in English (Bjerneby Häll, 2009).

A Doctoral Study on Pre-Service Teachers' Concept of Function

In his doctoral work, Hansson started by replicating a study by Grevholm (Hansson & Grevholm, 2003) on student teachers' conceptions of functions and equations. He was at the time relatively new mathematics teacher educator, and the study was conducted on his own students. One of his main instruments was concept maps, as used earlier by Grevholm, in addition to interviews with students. Questions were posed concerning pre-service teachers' conceptions of the significance of functions in mathematics and the presence of functions in school mathematics. One result is that the pre-service teachers, in their reasoning, show that their knowledge is fragmented, which Hansson sees as an issue of concern in the students' development of conceptual frameworks that are rich in meaningful connections. He indicates that this fact can influence their ability to work in a constructivist environment and reason with their pupils from different points of view. The main findings in Hansson's thesis (2006), which is written in English, are that the student teachers' view of the function concept is represented by a weakly developed knowledge structure. The findings indicate that the prospective teachers need additional support to develop their view of functions and understanding of mathematics. This leads to consequences for mathematical courses in teacher education. Among the suggestions are that students encounter functions in a variety of contexts, and are offered opportunities to experience different properties of functions. Hansson has also reported his work in English in other publications (2007, 2009).

A Doctoral Study on Mathematical Discourse

Another doctoral study in the graduate school was done by Ryve (2006), who partly studied student teachers' learning of mathematics. The study, related to student teachers (Ryve, 2007), characterizes classroom discourse in two problem-solving courses in teacher education. It also discusses the possibilities of developing two analytical approaches for studying mathematical discourse, the communicational approach to cognition and the dialogical approach. Ryve was not the teacher in the courses, and he had limited experience as a teacher educator at the time. He investigated the existence of problem-solving courses in 28 institutions of teacher education in Sweden and found such courses in six cases. From these six he decided to collect data in actual classes in two of the institutions. Data were field notes and 10 hours of audiotapes, recorded while he was present in the classrooms. Ryve describes the analysis of data as a cyclic process between transcripts and field notes, audio recordings, and theoretical and methodological constructs. He used the constructs of turns, topical episodes, and task presentation episodes for structuring the accounts of the mathematical discourses. The concepts of contextualization (Halldén, 1999), communicative projects, and communicative activities (Linell, 1998) were used to ascribe meaning and thus giving accounts for the mathematical discourses (see the chapter by Ryve, Wistedt, and Scheja in this volume). The principles of intentional analysis are used when he moves from descriptions to interpretations and from accounts to accounts. He offers three transcripts to illustrate the analysis.

The results show that the courses differ widely when it comes to the kind of communicative projects. While course 1 has 92% of subject-oriented communicative projects, there are none in course 2. The didactically oriented projects are one out of 25 in course 1, and 6 out of 18 in course 2; and, finally, the problem-solving-oriented communicative projects are one out of 25 in course 1, and 12 out of 18 in course 2. Ryve concludes with a number of reasons for why the results are of interest. One is the ongoing discussion about how much mathematics should be included in mathematics teacher education, a discussion that needs to be complemented by questions about quality of the content included. Another reason is that teacher educators planning to arrange problem-solving courses could learn from it. Should the emphasis of the course be on mathematical problem solving or didactical problem solving? He concludes that it is important to discuss the role of problem solving in the classroom and suggests that the work of Wyndhamn and colleagues (2000) could serve as a conceptual framework for discussions about the use of mathematical problems in teaching.

Another Doctoral Study on Communication in Mathematics

In her doctoral thesis, Riesbeck (2008) studied interaction and problem solving in primary school mathematics using a socio-cultural framework. Her data concern both pupils and student teachers. Her findings show that teachers use two different discourses when they interact with pupils: everyday discourse and a mathematical discourse. She concludes that when teachers and pupils become aware of crossing the discursive boundaries in mathematics, an understanding of mathematical phenomena can start to develop. Teachers and pupils can construct a meta-language leading to new knowledge and new learning in mathematics. The consequences for student teachers are obviously that they should be offered deeper insight in the classroom discourse and become aware of the teachers' opportunities of guiding the communication. Riesbeck points out that the student teacher/teacher must observe how one in a conscious way can pass discursive borders, change between discourses adapted to different situations, and choose and use concepts and notions for a specific reasoning situation. Riesbeck worked for many years as mathematics teacher educator at Linköping University, specializing towards teaching for school years 1–5. Riesbeck's thesis is written in Swedish.

An Ongoing Study on Student Teachers' Learning of Geometry

In an ongoing doctoral work, Holmquist (2004) is carrying out a study on student teachers' learning in geometry. He has solid experience of being a mathematics teacher educator at Gothenburg University, and the study is conducted on his own students. Using an epistemological tetrahedron model with the corners concept, term, referent, and definition he analyzes the answers from 213 students to a questionnaire on the concepts diagonal, congruence, parabola, rhomb, and cycloid. The results show that the concept images of the prospective teachers are based mainly on referents with weak connections to the concept definition. The visual aspect played a fundamental role in the students' ways of presenting their explanations. Holmquist suggests that it is appropriate to believe that the importance of the concept images when developing understanding of a concept is underestimated in mathematics education. Metaphorically speaking, he says, that to reach good understanding we have to find different ways in the epistemological tetrahedron model. He also suggests that the presented models may be used as starting point for how a teacher can analyze a student's written production as expression for understanding of mathematical content.

Studies on the Content and Form of Mathematics Teacher Education

Bergsten and Grevholm have explored several areas of importance for mathematics teacher education together. One is what they termed the "didactic divide" (Bergsten & Grevholm, 2004, 2005); another is the professional language of mathematics teachers (Grevholm & Bergsten, 2005); and a third is the relation and linking practices between theory and practice in teacher education (Bergsten & Grevholm, 2008; Bergsten, Favilli, & Grevholm, 2009). The issues have had importance for the development of the local teacher education programs and through publications internationally (Even & Ball, 2009). The conception of a didactic divide refers to a lack of or weak connection between discipline and pedagogical knowledge, which can appear at a cognitive as well as at an organizational level. This notion is used as an analytical tool to describe the rationale behind the reforms of teacher education. The didactic divide is claimed to be at the core of several critical domains, such as tensions between tradition and renewal, experience-based and research-based knowledge, and theoretical and practical components of teacher education.

Three critical issues are mentioned in relation to the development of the structure and content of mathematics teacher education: views on what should constitute teacher knowledge, back-up research, and the creation of communities of learning (Bergsten & Grevholm, 2004). An empirical study is used to illustrate how teacher educators work to overcome the didactic divide. Based on an analysis of the effect of the split knowledge structure displayed in teacher education practices of the weakly organized body of pedagogical knowledge combined with the split institutional organization of teacher education, Bergsten and Grevholm (2005) claim that one of the goals for research in mathematics education can be described as the development of an integrated body of *education-*

al knowledge in mathematics. The interest in bridging the didactic divide in teacher education practices is also observed in the outcome from the ICMI study on mathematics teacher education (Even & Ball, 2009), while Bergsten, Favilli, and Grevholm (2009) explore the content and results of a number of international studies related to the role of practicum and how it can be expanded to become an integrated part of teacher education.

In the same vein, Bergsten and Grevholm (2008) explore the issue of knowledgeable teacher educators and linking practices. That handbook chapter discusses knowledge of mathematics teacher educators in relation to their work on practices with student teachers and beginning teachers—in particular, those practices aiming at linking theoretical course work and teaching in schools. An overview of practices used by mathematics teacher educators is given, as well as some more detailed accounts of documented practices that focus on links between theoretical course work and teaching practice. Such linking practices have the potential to bridge the didactic divide between disciplinary and pedagogical knowledge, often separated by the organization of teacher education programs. The conclusion reached is that mathematics teacher educators not only need knowledge and experience in critical competence areas and activities involved in the work of mathematics teachers, but also to be knowledgeable persons in the activities they design, which comprises a critical awareness of the activity and of those with whom the teacher works as well as a will to move forward to the best of the others (Bergsten & Grevholm, 2008). The efforts to build on research in mathematics education to develop mathematics teacher education permeate the work discussed in this section.

DISCUSSION

The new teacher education intended to be implemented in 2010 is called a sustainable one (SOU, 2008), which indicates that even the educational authorities consider the earlier changes too many and are aiming for a more long-term solution. It has been claimed that the many new programs created instability in teacher education and experiences gained never had a chance to influence the development before a completely new scheme was introduced. A question is whether the results from research in mathematics education will have a potential to develop mathematics teacher education. Teacher education programs, which exist at 27 different universities and university colleges in Sweden, have great independence in designing the studies for prospective teachers. Thus it is not easy for locally developed research results to gain acceptance in all these environments. From the account above it is clear that most of the research studies on mathematics teacher education have had an influence on the local program or some of the outsider programs. A question is how long-lasting this influence can have been, considering the many changes in structure that have taken place.

From the historical development as sketched out above, it is clear that research has come to play a more important role over time. Before the university reform in 1977, where teacher education became a university education, research was not emphasised. Teacher educators in the teacher seminars and *Lärarhögskolor* (teacher education colleges) could not apply for research money, as there were no sources. From 1977 on, the importance of research has been more and more pointed out, and new steps to achieve a research-based education and research-trained teacher educators have been taken (Bergsten & Grevholm, 2004; Grevholm, 2004a). The repeated evaluations of teacher education (HSV, 1996, 2005; 2008) have shown some impatience with the slow development of the scientific foundation of teacher education. As mentioned above, even in 2009 about 10 of the teacher education programs were threatened to be closed down partly because of the low number of research-trained teacher educators. In several cases, mathematics was one of the subjects under discussion. The situation is presently in a process of change, and some of the new doctors from the graduate school have been hired as mathematics teacher educators.

Before 1977 there were no opportunities to apply for research funding, but tasks given by the National Board of Education could open for research and development studies. A good deal of funding has over the years been given to the departments of pedagogy in order to allow teacher educators without a research degree to qualify for a competence development. As can be understood from the above, this has assisted many mathematics teacher educators in getting a doctoral degree, although this has often been achieved very late in their careers and in some cases even after retirement, so one could debate how much impact this opportunity has had.

Most of these earlier doctoral dissertations are written in Swedish, and some of them have never been presented internationally. However, the doctoral students in the national graduate school wrote mainly in English, making their studies more easily available to a wider audience, and some of them have been published in international scientific journals (Brandell, Grevholm, Wallby & Wallin, 2009).

The outline above raises questions on the reasons for giving so few resources for research to mathematics education, and specifically mathematics teacher education, and what determineded the policy to support development projects rather than research studies. One hypothesis is that teacher education is seen as a practical professional education. Well-founded experience from practice was more valued than the more theoretical results from research (Bergsten & Grevholm, 2004). Another hypothesis could be that the professors of pedagogy were successful in claiming the whole research resource for general education, leaving little or no space for subject didactics. It is obvious that mathematics didactics took a long time to gain ground in Sweden, with the first professorship only in 2001 and still rather few of them. The first professor of pedagogy in Sweden came already in the beginning of 1900, and many have been involved in teacher education since then. The consequences of this educational policy have been that general education issues have dominated over subject didactics. A fight about resources has been more or less visible at many teacher education institutions (HSV, 1996).

The outline above has pointed to problems that have been studied in research on mathematics teacher education, along with some answers to them. Several of the studies have dealt with issues about what school mathematics is and how to justify the teaching of mathematics in school, while other studies investigated the changes in school mathematics based on recent technological developments. A few studies inquired specifically into student teachers' learning in mathematics and how it could be improved. The professional language of teachers has been emphasized in a few studies. This is a new area of interest, which has had very little visibility in the curriculum plans for teacher education. Mathematical problem solving has been of interest for a longer period with a changing character from analyses of pupils' solutions to the interactions between pupils and teachers. A few researchers have focused on the content, form and design of mathematics teacher education. The problem with repeated changes of structure and content of teacher education might have been that research results from one era could not be implemented before the education had already gone into a new structure and design. A consequence is that the long-term effect of one specific area of teacher education has not been investigated. Both these facts might have prevented a development of mathematics teacher education based on research results.

The many evaluations of mathematics teacher education have generally expressed much criticism. One exception is the evaluation from 1992 (UHÄ, 1992) where the conclusion is that the form given to teacher education for compulsory school is overall suitable, and the evaluators saw no need for more substantial change. Still they write that it is not possible to link teacher education to research unless the teacher education programs are given resources for such steps. Such resources do exist in other areas of study within the fixed faculty allowance. Thus, teacher education must be given such resources (UHÄ, 1992).

The International Situation and Possible Comparisons in Relation to Studies on Mathematics Teacher Education

As mentioned in the introduction, international research on mathematics teacher education has expanded during the 1990's and later (Adler et al., 2005; Even & Ball, 2009; Jaworski & Wood, 2008; Ponte & Chapman, 2006). To discuss similarities and differences between international studies and the studies in Sweden, I now go back to the claims made by Adler and colleagues (2005) presented earlier in the chapter. Claim number 1 is that "Small scale qualitative research predominates." In Sweden we find two longitudinal studies, but most of the studies are small scale and qualitative in nature. Although it is possible to trace some links between the studies, they mainly appear as fragmented.

Claim number 2 states, "Most teacher education research is conducted by teacher educators studying the teachers with whom they are working." Also, this seems to be the case in Sweden, although there are exceptions. A natural explanation is that teacher educators can easily get access to research data inside teacher education. A number of the researchers have raised some concern about the ethical problems about researching one's own students (Lingefjärd, 2000; Grevholm, 2001).

Claim number 3 is that "Research in countries where English is the national language dominates the literature." It is obvious that the early studies in Sweden are written mainly in Swedish, and sometimes no follow-up papers in English have been published. Thus the studies are difficult to access for non-Swedish speaking researchers. But starting around 2000, several studies made in Sweden have been published in English, and they are easy to access for international researchers. Whether these studies will be taken into account by an international audience is still to be seen. None of the early studies have been presented internationally, with the exception of Ekenstam and Nilsson (1979). Studies from the 1990s are the first ones found in international proceedings and journals, and then later studies and the studies carried out by students in the national graduate school have become visible (e.g., Hansson & Grevholm, 2003; Bergsten & Grevholm, 2004; Hansson, 2007; Ryve, 2007; Bjerneby Häll, 2009).

Claim number 4, that "Some questions have been studied, not exhaustively but extensively, while other important questions remain unexamined," certainly is the case also for Sweden; many important questions are left unexamined. I will return to this issue in the next section.

The ICMI15 study on teacher education (Even & Ball, 2009) presents an overview of the structure of teacher education programs in different countries. This overview shows how difficult it is to compare because of the varying structures. In Sweden such variation is present already through its 27 different teacher education institutions with the present open program structure.

Regarding other issues to be noted in an international comparison of studies, an OECD report (OECD, 2005) claims that the demands on schools and teachers are becoming more complex. The same arguments were used in the Swedish evaluations to initiate change and implementation of new programs. A search for research that could help and guide work to improve schools in order to meet the higher expectations will give little result. Askling, who led one of the evaluations in Sweden, writes that the need for subject didactical research in Sweden has been emphasized and admits that this has been a neglected area (Askling, 2006). She continues to say that it has not been easy to decide what research has relevance for teacher education, when orienting away from traditional pedagogical research. Many name variants for such research have been tried out, such as pedagogical work, learning, innovative learning, educational science, general didactics, subject didactics, and so on. I have claimed elsewhere that mathematics didactics is the relevant research area for mathematics teachers (Grevholm, 2006). But for Sweden, it remains to be seen if this emerging area of research will survive and grow strong as it has done internationally (Ponte & Chapman, 2006; Even & Ball, 2009).

CONCLUSIONS

This survey of research in mathematics teacher education in Sweden reveals that such studies began to appear at the end of the 1990s. The two smaller studies before that time were quantitative studies of results from students' mathematical tests, carried out in order to measure whether prospective teachers had adequate mathematical knowledge in areas that their pupils should study.

The studies presented before 2001 focus on what is seen as problem solving, students' views on mathematical modeling and use of ICT, and student teachers' conceptual development in mathematics and didactics of mathematics. Problem solving had been an area of interest in Sweden since the 1980s and Wyndhamn (1998) points to the seemingly inconsistent fact that there still are a multitude of questions to be studied, developed, and written about. The interest for teacher education courses on problem solving is taken up again by Ryve ten years later, and he then suggests that "the work of Wyndhamn et al. (2000) could serve as a conceptual framework for introducing discussions of how to use mathematical problems in teaching" (Ryve, 2007, p. 58). Ryve studied mathematical problem solving from another perspective than Wyndhamn and carried out microanalyses of the communication in the classroom during problem solving. Both studies have implications for teacher education and raise issues about what content to include in teacher education to prepare students in the area of problem solving.

Mathematical modeling and the use of ICT in teacher education had not been studied in Sweden before Lingefjärd carried out his study. Mathematical modeling was appearing as an interest in education later than problem solving, but researchers have indicated a link between problem solving and modeling (e.g., Mousolides, Sriraman, & Christou, 2007). Modeling has had a limited place in Swedish curricula, and Lingefjärd's study explores unique courses in teacher education, which may inspire other teacher educators to include new content areas for prospective teachers. The study indicates that to be fruitful, such a step must be taken with much consideration about how it is done.

Behind Grevholm's study, with its main focus on student teachers' conceptual development in mathematics, lies a concern about the long-term effect of mathematics teacher education. Among teacher educators there is a deeply felt need to learn more about the ways in which teacher education can influence prospective teachers. Such insights could guide the future development of contents and structure in teacher education. A longitudinal design is then necessary. This has also been a feature of one of the studies after 2001, the one by Bjerneby Häll (2006). While Grevholm investigated cognitive aspects of development, Bjernby Häll searched attitudinal aspects. In both cases the development over many years was revealed and disputes claims that no learning or development takes place during teacher education. However, the change is slow and therefore hard to prove.

Hansson's study (2006) also explores conceptual development, mainly of the function concept during teacher education. He shows that there is a need to carefully design the teaching in analysis, especially about the function concept and allow students a much more varied experience of the properties of functions. His study complements Grevholm's and offers implications for the design of analysis courses in teacher education.

Ryve (2007) and Riesbeck (2008) both engaged in studies about communication in mathematics teaching. Riesbeck discerns an everyday discourse and a mathematical discourse in the primary classroom, and Ryve categorizes communicative projects as subject-oriented, didactically oriented, or problem-solving oriented. The prospective teachers' awareness of the professional language and how it can be analyzed and steered is studied here and links to Grevholm's findings about professional language. This is an area of learning that has had little visibility in curricula for teacher education.

The areas of mathematics that have been touched upon in these studies are analysis by Grevholm and Hansson, number theory and algebra by Grevholm, geometry by Holmquist (2004), and then more general mathematical problem solving by Wyndhamn and Ryve and mathematical modeling and ICT by Lingefjärd. One study focuses on attitudes and belief— the one by Bjerneby Häll. Communication and language has been of interest in several of the studies. More overarching issues about how to achieve a consistent and holistic teacher education have been studied by Bergsten and Grevholm, such as working formats and links between theory and practice and between subject knowledge and teaching practice. Obviously the conclusion must be that there is space for many more studies on mathematics teacher education in Sweden in order to achieve a more coherent picture of student teachers' learning and development during the education, as well as issues concerning its content and organization.

After didactics of mathematics entered as an area of learning in teacher education in the mid 1980s, the interest in research on questions about teaching methods has been low. The "what" and "why" questions have been studied more than the "how" question. A similar case seems to be emerging when it comes to studies on teacher education. Only a few studies have as their main focus questions about how to design teacher education (e.g., Bergsten & Grevholm, 2004). But many of the studies could be used to raise questions about how to teach, for example, about equations, functions, geometry, problem solving, modeling, a professional language, communication in the classroom, and views upon what mathematics is.

I thus want to conclude, based on these observations of the Swedish studies, that future research is needed with a focus on

- Research-based design of the preservice mathematics teacher education, its content and form
- development of research-based competence development programs for teachers
- development of research-based textbooks for pupils and handbooks for teachers
- the factors that influence teachers' working conditions and professional opportunities
- teachers' development of a professional identity and its long-term influence on their practice

These are a few of the areas of inquiry that I see as being the most urgent for teachers' professional life and practice. Studies carried out need to be coordinated in order not to give more of the kind of fragmented research outcomes we have had up to this point (Adler et al., 2004; Even & Ball, 2009). A full program for research on teacher education needs to be set up and given resources to be carried out and implemented.

REFERENCES

Adler, J., Ball, D., Krainer, K., Lin, F-L., & Novotna, J. (2005). Reflection on an emerging field: Researching mathematics teacher education. *Educational Studies in Mathematics*, 60, 359–381.

Askling, B. (2006). En förnyad svensk lärarutbildning. In K. Skagen (Ed.), *Lærerutdannelsen i Norden* (Teacher education in the Nordic countries) (pp. 91–108). Kristiansand: Norwegian Academic Press.

Bergsten C. (1990). *Matematisk operativitet. En analys av relationen mellan form och innehåll i skolmatematik.* Dissertation. Linköping: Linköping University.

Bergsten, C. & Grevholm, B. (2004). The didactic divide and education of teachers of mathematics in Sweden. *Nordic Studies in Mathematics Education*, 9(2), 123–144.

Bergsten, C. & Grevholm, B. (2005, May). *The didactic divide and educational change.* Paper presented at the ICMI15 Study Conference in Aguas de Lindoia, Brazil. Retrieved June 3, 2010 from http://stwww.weizmann.ac.il/G-math/ICMI/log_in.html and as an electronic publication on CD.

Bergsten, C. & Grevholm, B. (2008). Knowledgeable teacher educators and linking practices. In B. Jaworski & T. Wood (Eds.), *The mathematics teacher educator as a developing professional* (pp. 223–246). Rotterdam: Sense Publishers.

Bergsten, C. & Grevholm, B. (2009). Form or content of mathematics teacher education—What matters most? In C. Winsløw (Ed.), *Nordic Research in Mathematics Education. Proceedings from NORMA 08* (pp. 205–209). Rotterdam: Sense Publishers.

Bergsten, C., Favilli, F., & Grevholm, B. (2009). Learning to teach mathematics: Expanding the role of the practicum as an integrated part in a teacher education programme. In R. Even & D. L. Ball (Eds.), *The professional education and development of teachers of mathematics: The 15th ICMI Study* (pp. 57–70). Berlin: Springer.

Bjerneby Häll, M. (2002). *Varför undervisning i matematik? FiF-avhandling 60/02,* Licentiatavhandling. Institutionen för beteendevetenskap, Linköpings universitet.

Bjerneby Häll, M. (2006). *Allt har förändrats och allt är sig likt. En longitudienll studie av argument för grundskolans matematikundervisning* (Dissertation). Studies in Education and Psychology, no 110, Linköping University.

Bjerneby Häll, M. (2009). Everything has changed and all remains the same—a longitudinal study of arguments for mathematics teaching in compulsory school. In C. Winsløw (Ed.), *Nordic Research in Mathematics Education. Proceedings from NORMA08* (pp. 345–352). Rotterdam: Sense Publishers.

Brandell, G., Grevholm, B., Wallby, K., & Wallin, H. (Eds.). (2009). *Matematikdidaktiska frågor—Resultat från en forskarskola.* Göteborg, Sweden: SMDF & NCM.

Dahland, G. & Emanuelsson, G. (1999). *LUMA 1998. Konferens för lärarutbildare i matematik 23-25 september 1998. Dokumentation.* Rapporter från institutionen för ämnesdidaktik. Rapport 1998: 03. Göteborgs universitet.

Dir 1997:54. (1997). *Lärarutbildningen. Kommittédirektiv.* (Teacher education. Committee guiding principles). Stockholm: Utbildningsdepartementet.

DsU 1986:5. (1986). *Matematik i skolan.* Stockholm: Utbildningsdepartementet.

Ds 1996:16. (1996). *Lärarutbildning i förändring* (Teacher education in transition). Stockholm: Utbildningsdepartementet.

Ekenstam, A. af, & Nilsson, M. (1977). *Om behållingen av grundskolans matematikkurser.* Försöks- och demonstrationsverksamheten, Lärarutbildningen i Linköping. Nr 7/1977.

Ekenstam, A. af, & Nilsson, M. (1979). A new approach to the assessment of children's mathematical competence. *Educational Studies in Mathematics, 10*(1), 41–66.

Even, R., & Ball, D. L. (Eds.) (2009). *The professional education and development of teachers of mathematics: The 15th ICMI Study.* Berlin: Springer.

Gellert, U. (2009). Practising mathematics teacher education: Expanding the realm of possibilities. In R. Even & D. L. Ball (Eds.), *The professional education and development of teachers of mathematics: The 15th ICMI Study* (pp. 35–55). Berlin: Springer.

Gran, B. (1995). *Professionella lärare? Lärarförbundets utvärdering av grundskollärarutbildningen.* Unpublished. Stockholm: Lärarförbundet.

Grevholm, B. (1990). Räknegrottan. *Datorn i utbildningen, nr 1.*

Grevholm, B. (1998). Teacher students' development of concepts in mathematics and mathematics education. In G. Brekke & T. Breiteig (Eds.), *Theory into practice in mathematics education. Proceedings from NORMA 98* (pp. 139–146). Kristiansand: Agder College.

Grevholm, B. (2001). Läraren som forskare i matematikens didaktik. Några exempel och reflektioner. In B. Grevholm (Ed.), *Matematikdidaktik—ett nordiskt perspektiv* (pp. 257–274). Lund: Studentlitteratur.

Grevholm, B. (2003a). Teachers' work in the mathematics classroom and their education. What is the connection? In C. Bergsten & B. Grevholm (Eds.), *Proceedings from Madif3, 2002* (pp. 96–106). Linköping: SMDFs skriftserie.

Grevholm, B. (2003b, May). *Student teachers' conceptions of equations and inequalities.* Paper presented at the Pre ICME10 conference, Växjö, Sweden. Retrieved February 2006 from http:// www.vxu.se/msi/picme10/L4GB.pdf

Grevholm, B. (2004a). What does it mean for mathematics teacher education to be research based? In R. Strässer, G. Brandell, B. Grevholm, O. Helenius (Eds.), *Educating for the future. Proceedings of an International Symposium on Mathematics Teacher Education* (pp. 119—134). Stockholm: The Royal Swedish Academy of Sciences.

Grevholm, B. (2004b). Mathematics worth knowing for a prospective teacher. In B. Clarke et al. (Eds.), *International perspectives on learning and teaching mathematics* (pp. 519–536) Göteborg: National Center for Mathematics Education.

Grevholm, B. (2005a). Research on student teachers' learning in mathematics and mathematics education. In H. Fujita, Y. Hashimoto, B. R. Hodgson, P. Y. Lee, S. Lerman & T. Sawada (Eds.), *Proceedings of the Ninth International Congress on Mathematics Education* (pp. 131–132 & CD). Dordrecht: Kluwer Academic Publishers.

Grevholm, B. (2005b). Concept maps as a tool in research on student teachers' learning in mathematics and mathematics education. I C. Bergsten & B. Grevholm (Eds.), *Conceptions of mathematics, Proceedings of Norma01, Third Nordic Conference on mathematics Education* (pp. 127–139). Linköping: SMDF.

Grevholm, B. (2006). Matematikdidaktikens möjligheter i en forskningsbaserad lärarutbildning. In S. Ongstad (Eds.), *Fag og didaktikk i lærerutdanning. Kunnskap i grenseland* (pp. 183–206). Oslo: Universitetsforlaget.

Grevholm, B. (2008). Concept maps as research tool in mathematics education. In A. J. Cañas, P. Reiska, M. Åhlberg & J. D. Novak, (Eds.), *Concept mapping: Connecting educators. Proceedings of the Third International Conference on Concept Mapping. Tallinn, Estonia & Helsinki, Finland 2008* (pp. 290–297). Tallin: Tallin University. Available at http://cmc.ihmc.us/cmc2008papers/cmc2008-p301.pdf

Grevholm, B., et al. (1989). *Räknegrottan.* Datorprogram i matematik för åk 1–12. Gävle: Gävle Pedagogiska centrum.

Grevholm, B. & Bergsten, C. (2005, May). *The development of a professional language in mathematics teacher education.* Demonstration at the ICMI15-study conference in Aguas de Lindoia, Brazil. Retrieved June 3, 2010 from http://stwww.weizmann.ac.il/G-math/ICMI/log_in.html and electronic publication on CD.

Halldén, O. (1999). Conceptual change and contextualisation. In W. Schnotz, S. Vosniadou & M. Carretero (Eds.), *New perspectives on conceptual change* (pp. 53–65). London: Pergamon.

Hansson, Ö. (2006). *Studying the views of preservice teachers on the concept of function.* Doctoral thesis. Luleå: Luleå University of Technology.

Hansson, Ö. (2007). Preservice teachers' conceptions of the function concept and its significance in mathematics. In C. Bergsten, B. Grevholm, H. Måsøval, & F. Rønning (Eds.), *Relating practice and research in mathematics education. Proceedings of Norma05, Fourth Nordic Conference on Mathematics Education* (pp. 271–285). Trondheim: Tapir Academic Press.

Hansson, Ö. (2009). Examining prospective teachers' reasoning of functions: a feedback perspective. In C. Winsløw (Ed.),

Nordic research in mathematics education. Proceedings from NORMA 08 (pp. 175–183). Rotterdam: Sense Publishers.

Hansson, Ö. & Grevholm, B. (2003). Preservice teachers' conceptions about y=x+5: Do they see a function? In N. A. Pateman, B. J. Dougherty, J. Zilliox (Eds.), *Proceedings of 2003 Joint Meeting of PME and PME-NA* (Vol. 3, pp. 25–32). Hawaii: University of Hawaii.

Hedrén, R. (1990). *Logoprogrammering på mellanstadiet.* Dissertation. Linköping: Linköping University.

Holmquist, M. (2004). Prospective mathematics teachers' learning in geometry. In C. Bergsten & B. Grevholm (Eds.), *Mathematics and language. Proceedings of Madif 4* (pp. 122–132). Linköping: SMDF.

HSV Högskoleverket. (1996). *Grundskollärarutbildningen 1995—en utvärdering.* HSVs rapportserie 1996:1 R. Stockholm: Högskoleverket.

HSV Högskoleverket. (2005). *Utvärdering av den nya lärarutbildningen vid svenska universitet och högskolor.* Högskoleverkets rapportserie 2005:17 R. Stockholm: Högskoleverket.

HSV Högskoleverket. (2008). *Uppföljande utvärdering av lärarutbildningen.* Högskoleverkets rapportserie 2008:8 R. Stockholm: Högskoleverket.

Jaworski, B. & Wood, T. (2008). *The international handbook of mathematics teacher education, Volume 4: The mathematics teacher educator as a developing professional.* Rotterdam: Sense Publishers.

Kilpatrick, J. & Johansson, B. (1994). Standardized mathematics teaching in Sweden: The legacy of Frits Wigforss. *Nordic Studies in Mathematics Education, 2*(1), 6–30.

Kislenko, K., Grevholm, B., & Lepik, M. (2007). "Mathematics is important but boring": Students' beliefs and attitudes towards mathematics. In C. Bergsten, B. Grevholm, H. Måsøval , & F. Rønning (Eds.), *Relating practice and research in mathematics education. Proceedings of Norma05, Fourth Nordic Conference on Mathematics Education* (pp. 349–360). Trondheim: Tapir Academic Press.

Leder, G. C., Brandell, G., & Grevholm, B. (2004). The Swedish graduate school in mathematics education: Conception, birth and development of a new doctoral programme. *Nordic Studies in Mathematics Education, 9*(2), 165–182.

Lindblad, T. (1978). *Blivande klasslärares ingångskunskaper i matematik.* Institutionen för praktisk pedagogik, Mölndal: Göteborgs Universitet.

Linell, P. (1998). *Approaching dialogue: Talk, interaction and contexts in dialogical perspectives.* Amsterdam: John Benjamins Publishing.

Lingefjärd, T. H. (2000). *Mathematics modelling by prospective teachers using technology.* Unpublished doctoral thesis, University of Georgia.

Lingefjärd, T. (2002). Mathematical modelling and prospective teachers. In C. Bergsten, G. Dahland, & B. Grevholm (Eds.), *Research and action in the mathematics classroom, Proceedings of MADIF2, The 2nd Swedish Mathematics Education research Seminar* (pp. 131–137). Linköping: Skrifter från SMDF.

Ljung, B-O. (1987). *Klasslärarkandidaters räkneförmåga.* Stockholm: Högskolan för lärarutbildning.

Lpf 94. (1994). Läroplaner för de frivilliga skolformerna (Curriculum for the upper secondary school system). SKOLFS 1994:2. Stockholm: Skolverket.

Lpo94. (1994). Läroplaner för de obligatoriska skolformerna (Curriculum for the compulsory school system). SKOLFS 1994:1. Stockholm: Skolverket.

Mousolides, N., Sriraman, B., & Christou, C. (2007). From problem solving to modeling, the mergence of models and modeling perspectives. *Nordic Studies in Mathematics Education, 12*(1), 23–47.

OECD. (2005). *Teachers matter. Attracting, developing and retaining effective teachers.* Paris: Author.

Ponte, J. P. & Chapman, O. (2006). Mathematics teachers' knowledge and practices. In A. Gutierrez & P. Boero (Eds.), *Handbook of research on the psychology of mathematics education: Past, present and future* (pp. 461–191). Rotterdam: Sense Publishers.

Pehkonen, E. (2001). Lärares och elevers uppfattningar som en dold factor i matematik-undervisningen. In B. Grevholm (Ed.), *Matematikdidaktik—ett nordiskt perspektiv* (pp. 230–253). Lund, Sweden: Studentlitteratur.

Proposition 1967:4 (1967). *Kungl. Maj:ts proposition till riksdagen angående reformerad lärarutbildning* [Government bill]. Stockholm: Riksdagens tryck.

Proposition 1999/2000:135. (1999/2000). *En förnyad lärarutbildning.* Stockholm: Regeringen.

Riesbeck, E. (2008). På tal om matematik. Matematiken, vardagen och den matematik-didaktiska diskursen. *Linköping Studies in Behavioural Science No. 129.* Doctoral thesis. Linköpings: Linköpings Universitet.

Riis, U. (1991). *Skolan och datorn.* Huvudrapport och sammanfattning av utvärderingen av treårssatsningen 1988–1991 på datateknikanvändning i skolan. Arbetsnotat 88 Tema T. Universitetet i Linköping.

Ryve, A. (2006). *Approching mathematical discourse. Two analytical frameworks and their relation to problemsolving interactions.* Unpublished doctoral dissertation, Mälardalen University.

Ryve, A. (2007). What is actually discussed in problem-solving courses for prospective teachers? *Journal of Mathematics Teacher Education,* 10, 43–61.

Skolverket. (2003). *Lusten att lära—med fokus på matematik: Nationella kvalitets-granskningar 2001–2002* (Skolverkets rapport nr 222). Stockholm: Skolverket.

SOU 1999:63. (1999). *Att lära och att leda. En lärarutbildning för samverkan och utveckling* (To learn and to lead. A teacher education for co-operation and development). Stockholm: Utbildningsdepartementet.

SOU 2008:109. (2008). *En hållbar lärarutbildning* (A sustainable teacher education). Stockholm: Statens Offentliga Utredningar.

Täljarenserien. (1988–89). *Studiematerial i matematik.* Stockholm: Utbildningsförlaget.

UHÄ. Universitets- och högskoleämbetet. (1992). *En grund att bygga på* (A foundation to build upon), UHÄ-rapport 1992:21. Stockholm: Author.

Wyndhamn, J. (1993). *Problem-solving revisited. On school mathematics as a situated practice* [Doctoral thesis]. Linköping Studies in Arts and Science No 98. Linköping: Linköping University.

Wyndhamn, J. (1998). Problem solving as a metaphor and a practice. Assumptions and conceptions of problem solving and their consequences for activities in the mathematical classroom. In G. Brekke & T. Breiteig (Eds.), *Theory into practice in mathematics education. Proceedings from NORMA 98* (pp. 45–58). Kristiansand: Agder College.

Wyndham, J., Riesbeck, E., & Schoulz, J. (2000). *Problemlösning som metafor och praktik.* Institutionen för tillämpad lärarkunskap, Linköpings universitet.

24

USING TEXTS AND TASKS

Swedish Studies on Mathematics Textbooks

Eva Jablonka and Monica Johansson

LULEÅ UNIVERSITY OF TECHNOLOGY

Abstract: During the last decades, a considerable number of scholars in Sweden have been concerned with the use of textbooks as a major resource for mathematics teaching. The chapter surveys Swedish research on textbooks. After an introduction into the Swedish educational context, the first section reports studies that aim at reconstructing the implemented curriculum through investigating the structure and representation of mathematical topics in textbooks, including historical analyses that portray changes over time in the Swedish curriculum. The next section surveys studies that infer hidden assumptions about the acquisition of mathematical knowledge and hidden messages about the nature and relevance of mathematics from an analysis of textbook tasks. Finally, research is reviewed that investigates how teachers and students use textbooks. The chapter documents an interesting variety of research on textbooks and tasks that have been carried out in Sweden.

Keywords: Textbooks, curriculum, tasks

BACKGROUND: THE SWEDISH CONTEXT

About ten years ago, in a cross-national study of mathematics textbooks and their use in sixteen countries in grade four and eight, textbooks were found to be a major resource used by teachers (Foxman, 1999). Even though there is no newer international survey on the impact of textbooks based on reports from teachers and students, more recent studies from single countries or small-scale comparative studies suggest that this practice has not changed in most places (e.g., Christou, Eliophotou-Menon, & Philippou, 2004; Fujita & Jones, 2002; Haggarty & Pepin, 2002; Newton & Newton, 2006; Ponte, 2007). Extensive use of textbooks still is a prevalent practice in Swedish mathematics classrooms. The textbooks are commercially produced, mostly authored by a team including experienced teachers, and available to all students free of charge. There is open competition on the market, and since 1991 there is no formal procedure of evaluation and approbation by the school authority. Abolishment of state control came along with decentralization, handing over more and more responsibility for schools to the municipalities. The choices for using a particular textbook are made at the school level or by individual teachers.

In Sweden, the classic model of progressive welfare state education was at its peak in the 1970s, giving birth to the comprehensive school (*grundskolan*) and to an expansion of secondary education and subsequently of tertiary education. Lindblad (2006) characterizes present trends by describing some major transformations this model has undergone since then. The current curriculum prescriptions for the comprehensive school are outcome-based; at present formulated for the end of year three, five, and nine; and more general compared to earlier; there are no

The First Sourcebook on Nordic Research in Mathematics Education, pages 363–372

recommendations concerning the order of topics and methods. Students control to a higher degree the pacing and sequencing of their work, compared to lessons in 1970, the classroom interaction being less dominated by the teacher. This points to a transition from teaching to "studenting." Swedish classrooms very often show a high proportion of lessons devoted to individualized work—students working at their own pace, occasionally scaffolded by the teacher (Petterson & Kjellström, 2005).

Both the focus on outcomes in curriculum prescriptions and a preference for individualized instruction that might be due to the heterogeneity of the classrooms in the comprehensive school are likely to give rise to extensive use of textbooks containing collections of tasks that are sometimes grouped according to levels of difficulty.

After nine years of education in a system without organizational differentiation, students can chose between different tracks in upper secondary school (*gymnasieskolan*). The number and the level of obligatory mathematics courses vary across these tracks. The textbooks refer to the different courses, which are named *A*, *B*, *C*, *D* and *E*, the order defining the level, although the content is not completely vertically structured. The completion of each course culminates with a national test, the results of which are meant to inform the teacher's evaluation of the students' attainment.

Also in undergraduate mathematics education at Swedish universities, the use of commercially produced textbooks is a widespread practice. In many places, those are books produced for another market, often the U.S.; consequently, students are expected to work with books written in English.

Textbooks have attracted considerable attention of researchers in mathematics education in Sweden in the last two decades (for a short history see Bremler, 2003, pp. 16–18, and Johansson, 2003, pp. 41–43). Studies on textbooks refer to all levels of education and comprise analyses of how mathematical topics in textbooks are portrayed and structured, investigations with a historical focus, analyses of particular types of tasks found in textbooks, and investigations of textbook use in classroom practice.

TEXTBOOKS AS A WINDOW INTO CLASSROOM PRACTICE: THE "POTENTIALLY IMPLEMENTED CURRICULUM"

In a practice that highly depends on textbook use, much of the curriculum is mediated through these books. Consequently, textbook studies inform about the curriculum and provide a window into the particular school mathematical knowledge that is likely to be developed by teachers and students in classrooms. Textbooks can be viewed as the "potentially implemented curricula" (Foxman, 1999; Schmidt, McKnight, Valverde, Houang, & Wiley, 1997; Valverde, Bianchi, Wolfe, Schmidt, & Houang, 2002). This leaves the question open whether they also mediate the official curriculum. For lack of large-scale classroom observation studies, textbooks might indeed be the only source of information about the type of tasks and problems students are working with; about how mathematics is portrayed in terms of the topics, methods, and principles chosen (or omitted); and in terms of the messages about its nature inherent in the particular representations and inner- and extra-mathematical relationships.

Changes Over Time: Different Visions of School Mathematics—Different Versions of Textbooks?

Studies of different editions of textbooks as well as of new launches, in juxtaposition with changing curriculum prescriptions, make it possible to reconstruct the development of particular areas of school mathematics, to identify periods of reform and restoration, and to investigate the role of textbooks within these processes as a progressive or as a conservative force.

Johansson (2003) investigates the alignment of five textbooks with changing portrays of school mathematics in curriculum prescriptions. She sees a major difference in the ways mathematics is portrayed in how it is supposed to be linked to out-of-school practices. In comparison with the curriculum prescription from 1969, the later version from 1980 shows a vision of a more "practical" mathematics that is linked to the students' everyday life. For her case study, she investigated the chapters about percentages in different editions of one textbook as well as in one more recent textbook. By drawing on a classification system developed by Palm (2002; see below), all word problems dealing with percentages were categorized according to three dimensions: (1) the match in the description of a situation in the school task with an imaginable corresponding situation, (2) the match in the question posed in the school task and the question(s) that might be posed in a corresponding situation, and (3) the match in the availability of the relevant information/data in the school task with its availability in the simulated situation. The outcome of the analysis is summarized in Table 24.1. It shows the number of problems that showed a high degree of match in terms of these dimensions. There is no indication of

TABLE 24.1 Realism of Word Problems about Percentages in Textbooks from 1979 until 2001

	Undvall et al. 1979[a]	Undvall et al. 1979[b]	Undvall et al. 1985[a]	Undvall et al. 1985[b]	Undvall et al. 2001
Total number of word problems	47	87	43	74	48
Event	47 (100%)	87 (100%)	43 (100%)	72 (97%)	46 (96%)
Question	46 (98%)	83 (95%)	42 (98%)	70 (95%)	42 (88%)
Existence	26 (55%)	62 (71%)	25 (58%)	55 (74%)	26 (54%)

an increase in "realism" of the tasks over the years. The textbooks studied were also not aligned to the national curriculum guidelines in several other aspects.

The change in general features of the same textbooks was investigated by Johansson (2003). Apart from some special units (such as problem solving and thematic work) introduced in the most recent book, the texts turned out to be comparable. If these additional units are not taken into account, the number and the type of tasks did not change dramatically. Adapting a textbook to curriculum changes only by adding new units affords a practice of simply omitting the additional parts, in this case those dedicated to problem solving and other enrichments. Students basically work with the same type of exercises as the students did in the beginning of the 1980s.

Another study that aims at describing how a distinct area of mathematics is portrayed in textbooks over time, is concerned with the development of school algebra in the Swedish upper secondary school. Jakobsson-Åhl (2006) attempts to reconstruct the "experience-of-the-learner-as-intended-by-the-author-of-textbook" in her analysis of a selection of textbooks and their revisions from the period between 1960 and 2000. The study includes two commonly used sets of mathematics textbooks from the period (altogether eleven books) for the upper-secondary school courses (courses *A* and *B*) for the natural science program in Sweden. The results of the study are hypotheses about the teachers' and students' conceptions of school algebra, that is, their hypothetical answers to the question "What is (school) algebra?" One outcome of the analysis is a system of classification that can be used to describe what upper secondary students were supposed to experience as school algebra, according to how the textbook authors portray the subject. The classification includes 19 categories that are to describe (1) the "objects of study," (2) the introduction and motivation for studying these objects, (3) the tasks for practicing algebraic skills, (4) the characteristics of algebraic word problems, and (5) the use of symbols or numerical examples. The results suggest that the specific portrayals of algebra in the textbooks reflect the trends that, at a specific time, pervade mathematics education.

The change of geometry instruction in elementary school (*folkskolan*) and lower secondary school (*realskolan*) in Sweden is the focus of a study carried out by Prytz (2007). Textbooks from the period between 1905 and 1962 provided a major source of data for the study. The author discusses the findings of his textbook analysis in the light of the goals of geometry instruction put forward in the debate by "curriculum experts" and in relation to the tasks used in the final examination after secondary education. One prevalent objective for lower secondary education was the training in reasoning through geometry instruction, a goal that appears not to have been relevant for textbook authors. In general, an axiomatic approach was kept. On the other hand, the final examination contained tasks that could be successfully solved by heuristic approaches. However, Prytz also notices changes throughout the period studied. Some alternative (non-axiomatic) textbooks began to be produced. Those became dominant in the 1930s. In the debate about geometry instruction in elementary school, visualizability (*åskådlighet*) became an important concept in methodological discussions. Prytz obeserves that almost all of the textbook authors started to design their textbooks in accordance with the arguments on visualizability. In the textbooks investigated, the "experimental routine" was applied, a routine that was described as linked to ideas about visualizability.

An evaluation of a mathematics textbook series designed for grade seven to nine is reported by Areskoug and Grevholm (1987). For the survey, three aspects were taken into account: the content, the formats for working with the tasks, and the ways in which the readers are guided through the book. The content was analyzed from different perspectives, such as legibility and choice of images. Some more implicit aspects related to the values transmitted through the texts were included in the evaluation. For example: Are pupils from different social environments likely to recognize themselves? Is there a gender bias? Are questions of solidarity and equity dealt with? In the report, the committee expresses a rather harsh critique of the textbook series. It states that the textbooks examined are monotonous, characterless, and uninteresting, and the authors see an improvement of textbooks as a major contribution to the development of mathematics education. The gender imbalance found in the texts (also transmitted through illustrations) was the most sensational result of the evaluation and became a focus of the subsequent discussion.

Another evaluation report of mathematics textbooks with a historical focus appeared in the late 1980s (Grevholm, Nilsson, & Bratt, 1988). Textbooks from the period between 1940s and 1980s were studied in terms of language (in particular how the reader is addressed), of images, the "realism" in applied tasks, and the complexity of the tasks (in terms of "steps"). The students' right to have their own books, so-called gift books, is also discussed in the report.

Johansson (2006) points out that while textbooks certainly offer a window into the enacted curriculum, textbooks cannot simply be seen as mediators of the curriculum as intended by school authorities and documented in official curriculum prescriptions. There are additional or conflicting rationales involved in commercial production.

Differential Portrayals of School Mathematics: Proof and Derivative

School textbooks are not explicit about the rationale behind the choice for a specific organization of the mathematical content. The most common model of text organization exhibits a repeated cycle of "exposition–examples–exercises" (Love & Pimm, 1996, p. 386). This is also a prevalent type of text organization in Swedish textbooks. As these textbooks are not written following a "Euclidean rhetoric" (cf. Fauvel, 1988), we can expect proofs or other explicit demonstrations of the truth of statements to be rare.

In Sweden, the current official curriculum prescription for upper secondary education does not emphasize proof. However, the marking criteria state that students are expected to be able to differentiate between guesses and assumptions from given facts, as well as deduction and proof, even for passing the courses with the lowest mark, and, in order to pass with distinction, to be able to provide a mathematical proof. Starting from this observation, Nordström and Löfwall (2006) investigate how mathematical proof is portrayed in two sets of commonly used Swedish textbooks for upper secondary school. The study shows that the distinction between examples and general proofs is not always made visible. A proving task could, for instance, start with "Explain that..." instead of "Prove that...," even though the generic example already could be seen as an explanation. Deductive proofs are not always labeled as proofs in the textbooks. In general, meta-mathematical terms such as "proof," "definition," "assumption," and "conjecture" are avoided. In one of the books, the concept "definition" is defined in a chapter summary for geometry, but the term is not used throughout the chapter. Where tasks are grouped according to differing levels of difficulty, the proving tasks are placed in the set with higher levels of difficulty. The authors interpret the invisibility and implicitness of proofs in the textbooks as a reaction to the earlier, very formal way of presenting theorems and proofs, a reaction that is in line with the official curriculum from 1994 (in which the term "proof" did not appear).

The development of the concept of derivative in Swedish upper secondary school (*gymnasium*) is investigated by Bremler (2003). Fifty-eight textbooks from the period between 1967 and 2002 are analyzed. A change could be expected after 1994, when a new curriculum steering document appeared. Starting with a discursive approach, the author puts forward some theoretical arguments for the role of textbooks in establishing the discourse of school mathematics—a discourse that is situated in the borderland between everyday and mathematical discourse (cf. Wyndhamn, 1993). Textbooks are mediating artifacts that can be seen as both a semiotic tool and as an "actor" who is "explaining," or "showing" how to solve examples, "answering" questions. The textbook and the teacher mutually support each other and thus create a stable teaching and learning situation in which the participants recognize themselves. The book gives trustworthiness to the teacher and the teacher sanctions the book.

The choice for studying the concept of derivative is based on its exemplary role as a new mathematical topic in relation to what the students have previously encountered in their school career, as well as on the fact that there is a long tradition in teaching it in upper secondary schools. The study includes questions about the explicitness of the goals; the role of historical references; geometrical interpretations of the derivative; the status of proof (for the rule for calculating derivatives of polynomials); the role of worked examples as models for solution procedures; the variation of the voices in the text; the role of images, realism, and gender in applied tasks; and information about the authors of the textbooks (age and gender). The analysis is mostly quantitative and based on identifying, for each issue, a small number of categories, respectively noticing their absence, in the texts (such as "historical reference without mathematics," "historical reference including mathematics," or "no historical reference"). It turns out that historical references became more common, but still remain cursory. Explicit statements of goals for studying derivatives remain rare; if mentioned they refer to the importance of the knowledge for higher mathematics education and/or for studies in other subjects. The number of textbooks that neither show a proof for the rule for polynomials nor mention that there is one, has doubled since 1994. However, some of the textbooks have preserved the proof but as an exercise with hints, or hint at the rule by an example. Worked examples continue to play an important role. It is possible to "copy" the solution procedure for solving a vast majority of the exercises, at least in the chapters about derivatives of polynomials that the author studied in more detail. Bremler (2003) finds gender role stereotyping even in the more recent textbooks, a result that he did not expect. The authors of the textbooks are mostly male, and "a small group of veterans seem to have dominated the book production" (p. 108).

TEXTBOOKS AS A WINDOW INTO STUDENTS' STRUGGLE WITH THE CONSTRUCTION OF MATHEMATICAL MEANING

Curriculum prescriptions, be it as general outcomes, as process skills, or in the form of lists of topics, by nature leave an interpretational space for textbook authors. This space gives room for ideologies to play, and it can be filled with assumptions about how students best learn, assumptions about levels of difficulty in relation to students' achievement, hidden messages about the nature and the relevance of mathematics, and assumptions about the ideal student as the ideal reader of the text who successfully accomplishes the tasks while working through the book.

Differing Levels of "Cognitive Demand"

Encapsulating mathematical knowledge as a series of tasks (be it as worked examples, exercises, applications, mixed tasks, etc.) is a traditional and typical didactical practice in the context of school mathematics teaching. This is reflected in school mathematics textbooks, which often contain a high proportion of text composed of tasks and not much of other types of text (such as narratives about famous mathematicians, popular science descriptions of the development of a mathematical theorem, formal proofs, or other elements found in non-school mathematics texts).

Love and Pimm (1996) observe that exercises in textbooks are most likely to be arranged in sequences of progressive difficulty either in a single set or in parallel sets of differing levels. In Swedish textbooks for grades 7 to 9 in the comprehensive school, tasks are often differentiated by "level" in order to support the selection of tasks according to the students' attainment. We find two or three parallel strands defining different pathways for the students working with the same textbook. Usually the books contain some diagnostic test for assigning groups of students to the strands. As the total number of tasks is much higher than what a student can work through, it can be assumed that teachers or students in fact make selections. Using the diagnostic tests is a convenient way for doing so.

Brändström (2005) analyzes the levels of difficulty embodied in the tasks in three Swedish textbooks commonly used in grade 7. She conceptualizes these levels in terms of the type of cognitive activities that are required for solving the tasks. The analysis also includes some other features of the texts, such as connectivity and illustrations. The outcome suggests that a taxonomy that conceptualizes mathematics learning as progression through a hierarchy of cognitive demands seems to be the hidden agenda of the construction of the sequences of tasks. Typically, in hierarchical taxonomies, reproduction is seen as the lowest level, and some more reflective and analytic activities (such as "creating," "evaluating," and "doing mathematics") are seen as the highest level of mathematical performance. Brändström argues that sequencing and grouping tasks according to such taxonomy is likely to limit the possibilities of the students who are already identified as low attainers by teachers who actually group their students for allocating them to the pre-defined strands of tasks. Brändström's analysis shows that the tasks for the low-attaining students in fact do not at all or only occasionally contain tasks that can be classified as high level in a hierarchical taxonomy. Consequently, even if students become proficient in solving the "lower" strand tasks, they would not have got much opportunity to encounter the type of tasks contained in the "higher" strands. However, even in the strands for the more advanced students, the proportion of intellectually demanding tasks turned out to be very low. However, there was more variation in those.

Making Sense of Word Problems

It is well documented that students face a challenge when asked to solve the typical word problems found in textbooks (for an overview see Verschaffel, Greer, & De Corte, 2000; Verschaffel, Greer, B., Van Dooren, W., & Mukhopadhyay, 2009; for a study on algebraic word problems carried out in Estonia see Lepik, 1990). These problems describe the substratum of a situation, sometimes as a little narrative. The situation is supposed to be familiar to the students. In the most artificial versions both the problem and the answer are well defined, no redundant information is given, and no data are missing; the missing quantity can be calculated without consideration of situational constraints or issues of validity and implementation of the solution. The functioning of such problems for establishing mathematical meaning has been disputed. Many critics of the artificial nature of typical textbook problems have pointed out that their unrealistic nature might be the reason for the students' difficulties.

Palm (2002) starts from the assumption that the degree of "realism" of the out-of-school situations that are indexed by word problems influence students' appropriate use of everyday knowledge for making sense of the problems and solutions. Based on an analysis of word problems in textbooks and of assessment tasks, he develops a framework for evaluating the "realism" of school tasks that he sees as simulations of out-of-school situations. It includes a couple of aspects of those situations that are reasoned to be important to be considered in their simulation. These aspects are "essential in the sense that their simulations clearly can affect the possibilities for the students' to engage in the same mathematical activities in the school tasks as in the corresponding out-of-school situation" (Palm, 2003, p. 3). The proposed aspects are event, question, purpose, information/data, presentation, solution strategies, circumstances and solution requirements. The category "question" describes whether the question posed in the task would be reasonably posed in the corresponding out-of-school situation. However, one major difference between out-of-school and school mathematical practices in fact concerns the 'question'. In domestic practices, often an efficient strategy relies on changing the constraints of the problem and thereby changing the question. So even if the same question as in the school task might be posed in the corresponding out-of-school situation, this would not amount to more realism in terms of the solution of the school task, if the question cannot be changed.

Empirical studies about the ways in which a change in any of those aspects affects students' solution strategies are still to be provided. The tension between the practice of school mathematics and out-of-school situations from domestic practices, in which school mathematical knowledge is supposed to be applied, cannot easily be resolved. In domestic practices, the situations are organized as specialized segments of everyday life. The acquisition of knowledge is likely to be a local activity embedded in that segmentation, and the strategies are often tacitly acquired. Consequently, making use of everyday knowledge might be impossible, inappropriate, or both in a school mathematical context, a context that aims at transforming everyday knowledge.

Restrictive Hypothetical Solution Strategies for Exercises in Calculus Textbooks

Lithner (2004) tries to reconstruct the solution strategies that could be employed by students solving exercises in calculus textbooks. Seeing this textbook as a typical text, in particular in terms of the content and organization of the worked examples and exercises, Lithner uses Adams (2003) for developing different categories of solution strategies. For an investigation of the distribution of different types of exercises that afford particular solution strategies, three textbooks were selected (Adams, 2003; Edwards & Penny, 2002; Salas, Hille, & Etgen, 1999).

The analysis draws on a distinction between solution strategies that are grounded in considering the mathematical properties of the objects involved, and strategies based on searching the textbook for similar worked examples and applying the same solution procedure (see also the notions of pseudo-conceptual and the pseudo-analytical thought processes discussed by Vinner, 1997).

In the approach based on identifying similarities, the textbook is searched for passages that show some resemblances in the components of the exercise to be solved (such as a similar worked example, procedure, definition, or theorem). The approach involves a transfer and/or combination of the procedure found. The author terms the type of similarities that allow for such a strategy "surface mathematical properties" of the components involved. The solution strategies based on such an approach are labeled "imitative reasoning" (Lithner, 2004).

A qualitatively different approach to solving the textbook exercises, labeled as "plausible reasoning," draws on some mathematical properties of the components of the exercise to be solved (the "intrinsic mathematical properties"). The author distinguishes between more "local" and more "global" versions of it. This distinction seems particularly necessary because the components might differ in some aspects, which complicates the identification of a "similarity." In principle, all textbook exercises could be solved by "global plausible reasoning," but this might not be necessary for successful completion.

It turns out that it is theoretically possible to complete a high proportion of the exercises from the three books analyzed (about 70%) by searching the worked examples for similarities. Only a few examples, such as those included in sections headed "challenging problems," can only be solved by "global plausible reasoning."

The question of the extent to which students actually employ these hypothetical strategies when completing the exercises in the textbooks remains open. There might well be quite different ways of dealing with the tasks. It can be assumed that neither the textbook authors nor the lecturers of the calculus courses who employ these textbooks aim at fostering imitative and local solution strategies, but rather aim at eliciting more demanding strategies from the students who work through (the piles of exercises in) the books. As long as the exercises contained in typical textbooks do not afford solution strategies that are based on drawing conclusions from the mathematical properties of the components, it should not be surprising that these turn out to be more demanding, because the students lack the possibility of encountering this type of tasks. The purpose of producing and using those texts then remains unclear.

Hidden Messages in Textbooks from Different Levels of Academic Mathematics

Raman (2001) sees a major problem for students' difficulties in the transition from pre-calculus to calculus in the opportunities students get for coordinating informal and formal aspects of mathematical meaning. She analyzes two popular U.S. textbooks, one pre-calculus and one calculus text (Demana, Waits, & Clemens, 1993; Stewart, 1995), in order to identify the emphasis given to formal and informal aspects. The analysis focuses on the exercises, worked examples, and definitions and theorems that are marked as important. These are the parts students tend to focus on. The texts show two polarized ways of separating the formal from the informal—that is, emphasizing one at the expense of the other. The separation coincides with students' difficulties, which are also studied. In a related study (Raman, 2004) that focuses on the concept of continuity, a popular analysis textbook is also included (Rudin, 1976). The texts from the three different levels seem to convey incongruent or conflicting "epistemological messages" about the status and purpose of definitions. There are distinct ways in which informal and formal aspects of the definitions interact, and different relative emphases on syntax (manipulation of symbols without recourse to meaning) or semantics (focus on the meaning or conceptual understanding) in the worked solutions. As there are different epistemological messages at different levels, students can be expected to have typical problems with the transition from high school to lower-division and from lower-division to upper-division tertiary level mathematics. Raman sees a major problem in the implicitness of the messages in the textbooks.

These studies can also be interpreted as evidence that different mathematical discourses are established at different levels of mathematics education and that textbooks have a considerable impact because they convey epistemological claims about the status of mathematical knowledge (cf. Jablonka, 2000, 2001). As the differences remain largely implicit, students repeatedly face the problem of understanding hidden messages when moving through a range of (hierarchically structured) mathematical discourses. Many get lost along the way.

The Struggle with Technical Language

A prominent feature of mathematical texts is the use of highly condensed symbolism accompanied by low redundancy, a sparseness of modifiers, use of agentless sentences (impersonal or personal passives, impersonal pronouns), and avoidance of deixis. These distinctive features of mathematical texts can be expected to cause a challenge for inexperienced readers.

Based on a general model for text comprehension (Kintsch, 1998), Österholm (2004) produced data of individuals' success in extracting information from different types of texts in order to find out whether the differences affect the readers' understanding. The three texts used in the study are about one-page length each. One is a text containing some facts about the Russian revolution, the other two are mathematical texts about the definition of a group. The two mathematical texts only differ in the use of symbols, while not changing all other features of a typical mathematical text. This choice was based on the assumption in the model that the mental representation of information operates with the syntax of natural language and that algebraic terms differ in this respect from natural language. The preknowledge of the readers is considered to be another important factor. In the two mathematical texts, for example, the version with symbols contains the following passage (translated from Swedish):

> A group is a mathematical system, that is, a set G together with a combination rule ~ that, in addition, has the following four properties:
>
> 1) If a and b belong to G, then a~b also belongs to G. This is to say that the system is closed. (Österholm, 2004, p. 96)

In the version without symbols, the same passage reads as follow (translated from Swedish):

> A group is a mathematical system, that is, a set together with a combination rule that, in addition, has the following four properties:
>
> 1) If two objects belong to the set, then the combination of those also belongs to the set. This is to say that the system is closed. (Österholm, 2004, p. 97)

The study included 106 readers (69 upper secondary school students and 37 university students). All participants were reading the history text and one of the two mathematical texts about groups, and afterwards had to answer questions about the information in the text. The pre-knowledge of the readers was tested and they were informed that they were expected to answer some questions after reading the texts. The quantitative analysis of the data revealed the same coupling of the pre-knowledge of the readers with answering the text-related questions for all three texts. However, the text-based questions were less successfully answered by those who were confronted with the mathematical text containing symbols. Some similarities between the reading of the history text and the mathematical text without symbols were revealed.

HOW TEACHERS AND STUDENTS USE TEXTBOOKS

Without information about how teachers and students actually use a textbook in classroom practice, the conclusions drawn about the enacted curriculum remain limited. On the other hand, it can be argued that textbooks are usually not criticized by their users. In particular, exercises contained in textbooks, on which a couple of the studies reviewed above focus, directly address the students and usually do not convey an invitation to change or deconstruct them. Consequently, analyzing textbooks without at the same time studying their classroom use remains an untapped area of research. However, the extent to which and the ways in which textbooks influence the content, level, and organization of classroom practice as well as the differences in individual teachers' or students' textbook use are a matter of empirical classroom research.

Andersson and Håkansson (2006) carried out a small study about textbook use in Swedish primary school by asking a group of 46 teachers working in preschool through year 3 in primary school about textbook use in mathematics. They found that the teachers use textbooks more often in year 1 than in year 3. Teachers reported that they use the textbook very frequently in the first year. The textbook use drops in year 2, but there still seems to be more than half of the time in the lessons spent with it, and less than half in year 3. In an interview, one of the teachers argued for this practice by stating that reading and writing are the main focus when the children begin school. It is then "comfortable" to let the students work by themselves for a while with the textbook. From the interviews with two teachers, the authors conclude that Swedish language has a higher status than mathematics in the early school years.

Gidlund (2005) examined to what degree teachers in upper secondary school use other material than the textbook. Table 24.2 shows the result of this small investigation.

Johansson (2006) carried out an empirical study of the use of textbooks in year 8 classrooms by analyzing 13 videotaped lessons from three classrooms.[1] As to classroom organization, the results coincide with earlier studies (cf. the ones mentioned in the introduction). In large portions of the lessons, students are working individually solving tasks from the textbook, while the teacher is walking between the desks, offering individual assistance and support (for an analysis of this form of interaction in different countries, see O'Keefe, Xu, & Clarke, 2006). In the lessons analyzed by Johansson, students are exclusively working with tasks from the textbook during these periods, which on average cover more than half the time of a lesson. Also, the examples the teachers present on the board, the kind of concepts introduced, and the way this is done reflect the content and text organization of the textbook. Only one teacher more frequently uses different examples by drawing on his experience as a teacher of physics. The portrayal of mathematics as a scientific discipline mirrors the approach in the textbook. In 100 out of 119 occasions when the teachers made mathematical generalizations or uttered statements, these were comparable to or identical with those in the textbook. This means that hardly any other definitions, conventions, or rules than what the textbook offers are presented to the students. Two of the three teachers studied use the textbook also as the main source for background and motivational dis-

TABLE 24.2 The Extent of Using Other Material During the Last School Year as Reported by Teachers (in Umeå, Sweden; N = 25)

	Number	Percent
Always	4	6
1–2 times per period	24	36
1–2 times per month	27	41
1–2 times per week	11	17
	66	100

cussions. Homework is not assigned on a regular basis. However, when the teachers do give assignments, students are supposed to work with tasks from the textbooks. These results are based on a quantitative analysis of lesson episodes that were coded according to the categories mentioned above.

Johansson (2006) also examined in detail the interaction between the teacher and the students during some critical incidents in the classroom. The analysis reveals that the construction of the textbook tasks has an effect on the teacher–student interaction. In the "standard" pattern of interaction, the teacher interacts with the students in a confident way, scaffolding the students' attempt to solve the tasks from the textbook. However, the situation changes when there is a discrepancy between the solution in the textbook and what the teacher thinks is a correct solution. In such an episode, the teacher becomes ambiguous but does not argue against the textbook. In another incident, the same teacher dismisses the authority of the textbook and proposes suggestions that deviate from the text. As a consequence of a question raised by a student working on a specific task, the teacher opens a general and public discussion. These examples show that the teacher establishes a critical reading of the text and sanctions the text, but does not always decide to change the students' perception of the text as the institutionalized mathematical meaning in the classroom.

A SHORT SUMMARY

The studies reviewed point out some limiting effects of textbooks and their use, be it as obstacle to reform, as conveying an inappropriate picture of mathematics as an academic discipline, as causing problems of transition, as restricting the students' opportunities for engaging in intellectually demanding tasks, or as not providing sufficient possibilities for identification for distinct groups of students (such as females).

In fact, the identity of the textbook authors, who often are experienced teachers, might counteract attempts at change within an agenda of reform. Acquaintance with classroom practice suggests the construction textbooks that are "easy to work with" for the teachers. This is in accordance with the criteria of an open market. "Easy to work" means that there is no disruption in the everyday work of the classroom. In many classrooms in Sweden this implies, for example, spending much time with individualized work with textbook exercises. Consequently there is no need to make explicit the values and norms that govern current classroom practice, which would be a prerequisite for change.

The recurrent and prevalent practice of using textbooks as a major resource—despite all criticism—has to be seen as both a consequence and an outcome of standardization and institutionalization of the mathematics curriculum.

CONCLUDING REMARK

The summaries of research presented in this chapter reflect the authors' reading of the studies quoted. For more and unbiased information readers should feel free to study the original works.

NOTE

The data are from *The Learner's Perspective Study* (http://extranet.edfac.unimelb.edu.au/DSME/lps/).

REFERENCES

Adams, R. (2003). *Calculus: a complete course* (5th edition). Toronto: Addison-Wesley.

Andersson, K., & Håkansson, C. (2006). *Matematikundervisning för de tidigare åren—en kartläggning av lärares uppfattningar om arbetsform och innehåll.* Unpublished graduation thesis (examensarbete). Växjö: Växjö University.

Areskoug, M ., & Grevholm, B. (1987). *Matematikgranskning* (No. 1987:3). Stockholm: Statens Institut för Läromedel.

Bremler, N. (2003). *Matteboken som redskap och aktör. En studie av hur derivatan introduceras i svenska läroböcker.* Unpublished licentiate thesis. Lärarhögskolan, Stockholm.

Brändström, A. (2005). *Differentiated tasks in mathematics textbooks: An analysis of the levels of difficulty.* Licentiate thesis. Luleå: Luleå University of Technology.

Christou, C., Eliophotou-Menon, M., & Philippou, G. (2004). Teachers' concerns regarding the adoption of a new mathematics curriculum, *Educational Studies in Mathematics, 57,* 157–176.

Demana, F., Waits, B., & Clemens, S. (1993). *Precalculus: Functions and graphs.* Reading, MA: Addison-Wesley.

Edwards, C.H. & Penney, D. (2002). *Calculus* (6th edition). Upper Saddle River, NJ: Prentice Hall.

Fauvel, J. (1988). Cartesian and Euclidean rethoric. *For the Learning of Mathematics, 8,* 25–29.

Foxman, D. (1999). *Mathematics textbooks across the world: some evidence from the Third International Mathematics and Science Study* (TIMSS). Slough: NFER.

Fujita, T. & Jones, K. (2002). Opportunities for the development of geometrical reasoning in current textbooks in the UK and Japan, *Proceedings of the British Society for Research into Learning Mathematics, 22*(3), 79–84.

Gidlund, M. (2005). *Variationsrik matematikundervisning?* Graduation thesis (examensarbete). Umeå: Umeå University.

Grevholm, B., Nilsson, M., & Bratt, H. (1988). Läroböcker i matematik. In *Skolböcker 2. Den (o)möjliga läroboken. Rapport från Läromedelsöversynen* (Ds 1988:24). Stockholm: Allmänna Förlaget.

Haggarty, L., & Pepin, B. (2002). An investigation of mathematics textbooks and their use in English, French and German Classrooms: who gets an opportunity to learn what?, *British Educational Research Journal, 28*(4), 567–590.

Jablonka, E. (2000). Perceptions of mathematics and reality in TIMSS mathematics items. In A. Ahmed, J.M. Kraemer & H. Williams (Eds.), *Cultural diversity in mathematics (education)* (pp. 127–132). Chichester: Ellis Horwood.

Jablonka, E. (2001). Modelling: Good problems—not only a question of (good) taste. In J.F. Matos, W. Blum, S. Carreira, & K., Houston (Eds.), *Modelling, applications and mathematics education – Trends and issues* (pp. 305–316). Chichester, UK: Ellis Horwood.

Jakobsson-Åhl, T. (2006). *Algebra in upper secondary mathematics: a study of a selection of textbooks used in the years 1960-2000 in Sweden.* Licentiate thesis. Luleå: Luleå University of Technology.

Johansson, M. (2003). *Textbooks in mathematics education: a study of textbooks as the potentially implemented curriculum.* Licentiate thesis. Luleå: Luleå University of Technology.

Johansson, M. (2006). *Teaching mathematics with textbooks: a classroom and curricular perspective.* Unpublished doctoral dissertation. Luleå University of Technology

Kintsch, W. (1998). *Comprehension: A paradigm for cognition.* Cambridge: Cambridge University Press.

Lepik, M. (1990). Algebraic word problems: Role of linguistic and structural variables. *Educational Studies in Mathematics, 21*(1), 83–90.

Lindblad, S. (2006). Sweden. In D. Clarke, C. Keitel & Y. Shimizu (Eds.), *Mathematics classrooms in twelve countries: the insider's perspective* (pp. 369–372). Rotterdam: Sense Publishers.

Lithner, J. (2004). Mathematical reasoning in calculus textbook exercises. *Journal of Mathematical Behaviour, 23,* 405–427.

Love, E., & Pimm, D. (1996) "This is so": a text on texts. In A. Bishop et al. (Eds.), *International Handbook of Mathematics Education* (pp. 371–409). Dordrecht: Kluwer Academic Publishers.

Newton, D. & Newton, L. (2006). Could elementary mathematics textbooks help give attention to reasons in the classroom? *Educational Studies in Mathematics, 64,* 69–84.

Nordström (Hemmi), K. & Löfwall, C. (2006). Proof in Swedish upper secondary school mathematics textbooks: The issue of transparency. In M. Bosch (Ed.), *European research in mathematics education IV. Proceedings of CERME4, San Feliu de Guı´xols, Spain 2005* (pp. 448–457). Barcelona: Universitat Ramon Llull.

O'Keefe, C., Xu. L.H., & Clarke, D. (2006). Kikan-Shido: Between desks instruction. In D. Clarke, J. Emanuelsson, E. Jablonka, & I. Mok (Eds.), *Making connections. Comparing mathematics classrooms around the world* (pp. 73–106). Rotterdam: Sense Publishers.

Österholm, M. (2004). *Läsa matematiska texter: Förståelse och lärande i läsprocessen.* Licentiate thesis. Linköping: Linköping University.

Palm, T. (2002). *The realism of mathematical school tasks. Features and consequences.* Doctoral thesis. Umeå: Umeå University.

Pettersson, A. & Kjellström, K. (2005). Matematiken och ungdomskulturen—är det förenligt? I *Grundskolans ämnen i ljuset av den nationella utvärderingen 2003.* Stockholm: Skolverket.

Ponte, da J.P. (2007). Proportion in school mathematics textbooks: A comparative study. In D. Pitta-Pantazi & G. Philippou (Eds.), *European research in mathematics education V, Proceedings of the Fifth Congress of the European Society for Research in Mathematics Education* (pp. 2443–2452). Cyprus: ERME & University of Cyprus.

Prytz, J. (2007). *Speaking of geometry: A study of geometry textbooks and literature on geometry instruction for elementary and lower secondary levels in Sweden, 1905–1962, with a special focus on professional debates.* Doctoral thesis. Uppsala: Uppsala University.

Raman, M. (2001). Coordinating informal and formal aspects of mathematics: Student behavior and textbook messages. *Research Reports, No 7, 2001 in Mathematics Education.* Umeå: Department of Mathematics, Umeå University.

Raman, M. (2004). Epistemological messages conveyed by three high school and college mathematics textbooks. *Journal of Mathematical Behavior,* 23, 389–404.

Rudin, W. (1976). *Principles of mathematical analysis* (3rd edition). New York: McGraw-Hill.

Salas, S., Hille, E., & Etgen, G. (1999). *Calculus: One and several variables* (8th ed.). New York: Wiley.

Schmidt, W. H., McKnight, C. C., Valverde, G. A., Houang, R. T., & Wiley, D. E. (1997). *Many visions, many aims: a cross-national investigation of curricular intentions in school mathematics* (Vol. 1). Dordrecht, The Netherlands: Kluwer.

Stewart, J. (1995). *Calculus: Early transcendentals* (3rd edition). Pacific Grove, CA: Brooks/Cole.

Undvall, L., Olofsson, K.-G., & Forsberg, S. (1979a). *Matematikboken ak 7.* Stockholm: Esselte Studium.

Undvall, L., Olofsson, K.-G., & Forsberg, S. (1979b). *Matematikboken sk 7.* Stockholm: Esselte Studium.

Undvall, L., Olofsson, K.-G., & Forsberg, S. (1985a). *Matematikboken 7A.* Stockholm: Esselte Studium.

Undvall, L., Olofsson, K.-G., & Forsberg, S. (1985b). *Matematikboken 7S.* Stockholm: Esselte Studium.

Undvall, L., Olofsson, K.-G., & Forsberg, S. (2001). *Matematikboken X.* Stockholm: Liber.

Valverde, G. A., Bianchi, L. J., Wolfe, R. G., Schmidt, W. H., & Houang, R. T. (2002). *According to the book. Using TIMSS to investigate the translation of policy into practice through the world of textbooks.* Dordrecht, The Netherlands: Kluwer.

Verschaffel, L., Greer, B., & De Corte, E. (Eds.). (2000). *Making sense of word problems.* Lisse, The Netherlands: Swets & Zeitlinger.

Verschaffel, L., Greer, B., Van Dooren, W., & Mukhopadhyay, S. (2009). *Words and worlds: Modelling verbal descriptions of situations.* Rotterdam: Sense Publishers.

Vinner, S. (1997). The pseudo-conceptual and the pseudo-analytical thought processes in mathematics learning, *Educational Studies in Mathematics, 34,* 97–129.

Wyndhamn, J. (1993). *Problem solving revisited: On school mathematics as a situated practice* (TEMA Communication Studies 98). Doctoral thesis. Linköping: Linköping University.

25

NATIONAL ASSESSMENT IN SWEDISH COMPULSORY SCHOOL

Astrid Pettersson and Lisa Björklund Boistrup

STOCKHOLM UNIVERSITY

Abstract: In this chapter we describe assessment in mathematics in Sweden at a national level in the compulsory school. In doing this we address important aspects of assessment in mathematics. The basis for the chapter is the work of the *PRIM-gruppen* (PRIM = Assessment In Mathematics) at Stockholm University. The PRIM-gruppen does research on assessment and has been active for several years. We discuss the history of the national test system and research performed in connection to national tests and national evaluation. The chapter also presents analyses of students' answers to test tasks and analyses of tasks.

Keywords: national test, national evaluation, diagnostic materials, longitudinal studies, compulsory school

INTRODUCTION

This chapter reports on national assessment in mathematics in Swedish compulsory school. We present some of the theoretical considerations that have an impact on our work during the construction of materials for assessment. After this we give a brief presentation of the various materials for assessment in mathematics, national tests and diagnostic materials, and national evaluation, of which we have been part of the construction. Furthermore, we present examples of analyses of students' answers and of tasks that have been performed in relation to the different materials and their use. These analyses are, alongside the theoretical considerations and steering documents, a base that strongly influences the development of our work.

BACKGROUND

Before we describe the Swedish national mathematics tests of today, we give a short historical background of the national test system and how it has developed over time. In the early twentieth century, students in Sweden had to pass an entrance examination before entering lower secondary school. This changed in 1940 and one of the reasons was that the increasing number of applicants made the examinations too expensive and complicated. Another reason was that it was thought that a single day of a student's life should not decide whether the student could gain admittance to an education program or not. Instead, school marks were used as the instrument for selection. In order to make them comparable across the whole country, standardized tests were introduced (Kilpatrick & Johansson, 1994; SOU 1942:11; Wigforss, 1937, 1941). The theoretical underpinning of the tests was the bell curve: The idea was that the teacher should adapt the marks so that the distribution of marks in their class should follow the results of the standardized tests. This system was the basis of the relative grading system that was introduced into the comprehensive school. Starting in 1962, there were five grade levels from 1 to 5, where 5 was the highest mark.

The First Sourcebook on Nordic Research in Mathematics Education, pages 373–385

From the middle of the 1940s until the middle of the 1990s, the system of grades was norm-referenced in Sweden. Marks were based on students' performance in relation to each other and to the national average. Students' marks were mainly determined by the rank order in the class. The aim of the tests was to get maximum uniformity in teachers' marking throughout the country. The relative grading system was severely criticized. The teachers criticized it for intruding on their freedom in teaching. The test was so important that it steered teaching into too narrow a range. The students complained that it created a terrible competition, although cooperation among students should be encouraged according to the curriculum.

Since 1994, the school system in Sweden has been decentralized and goal-based. The main responsibility for education lies within the municipalities and authorities responsible for independent schools. The overall national goals are stipulated by the Swedish Parliament and government. The National Agency for Education is responsible for evaluating, following up on, and supervising the public school system. The agency is also responsible for the national testing system and the national evaluation system. Together with universities, the agency develops national tests and diagnostic and evaluation materials in order to ensure equality in the assessment of pupils. Since 1984, PRIM-gruppen (from 2008 at Stockholm university) has been engaged in constructing standardized tests and national tests, and it has also been involved in national evaluations. A very important part of our work is to analyze how the students solve the items and what kind of reasoning they show, which can be regarded as correct or not correct, in their solutions. We also perform analyses of the tasks, regarding content, language, context, and form.

In the Swedish National Syllabus of Mathematics, since 1994, the mathematical content in the teaching and learning of mathematics is formulated under the headings "Aim of the subject and its role in education" and "Structure and nature of the subject" (Swedish National Agency of Education, 2001). There are two kinds of goals in the syllabus. One is "goals to aim for." The purpose for these is to "express the direction the subject should take in terms of developing pupils' knowledge. They clarify the quality of knowledge which is essential in the subject" (SNAE, 2001, p. 5). The other goals are "goals to attain." They define the minimum knowledge to be attained by all pupils in the third, fifth and ninth year of school.

THEORETICAL CONSIDERATIONS

Here we describe aspects concerning the relationship between assessment and the learning process. We also address relational and communicative aspects that are a part of the theoretical foundation on which our work is based.

The Role of Assessment in the Learning Process

People learn in many different ways. Learning is dependent on various factors but also on how and on what aspects one is assessed. Students focus their learning according to the content and methods used in the assessment. How the subject is presented to the students in the assessment will therefore influence students' experiences, beliefs, and views in regard to the subject (see, e.g., Black, 2001; Black & Wiliam, 1998; Gipps, 1994; Leder, 1992; Shephard, 2000). From our own experience, we know that not all teaching and assessment are stimulating and supportive for learning. Assessment is not only a "receipt" for knowledge displayed, but it also influences an individual's learning, his or her self-esteem, and confidence in his or her knowledge. Assessment can provide great potential for learning. But what does assessment mean for the individual? The consequences of assessment can be illustrated by Figure 25.1.

An assessment that supports and stimulates learning means that a student's knowledge is analyzed and evaluated in such a way that the student progresses in his or her learning and feels self-confidence in his

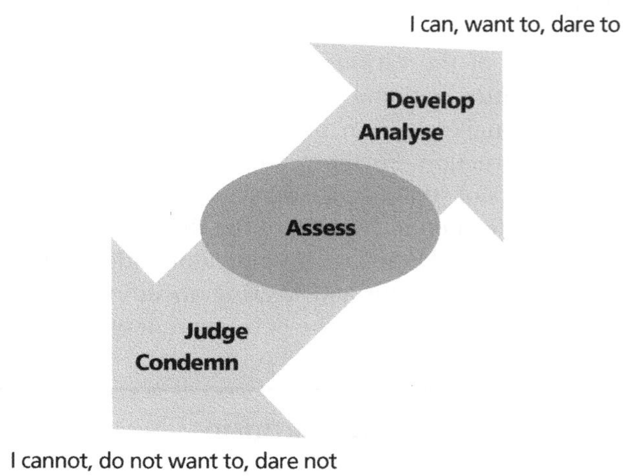

Figure 25.1 The consequences of assessment (Pettersson, 2004, p. 43).

or her own ability (I can, want to, dare to), instead of an assessment that leads to a judgment and perhaps a condemnation (I cannot, do not want to, dare not). For this students need to get feedback on what qualities their performances demonstrate and also on what they should focus their learning in the near future.

A central area of interaction processes in the classroom occurs when a student gets feedback on her or his learning processes. In some of the studies referred to in Black and Wiliam (1998), there is evidence that feedback to a pupil on the knowledge she or he has shown in certain tasks has impact on her or his interest and self-confidence, whereas feedback on what has to be learned has an impact on her or his learning. Unreflected feedback can have an impact on student's learning that is worse than points and marks on tests (Smith & Gorard, 2006). Thus, it is clear that *what* the teacher and student focus on during the student's learning process is a main aspect when it comes to feedback.

If the students are placed in a variety of situations, they will get more opportunity to show their knowledge in different ways. Consequently, the instruments for assessment must be characterized by great flexibility (Niss, 1993). In the past, solutions have been either right or wrong, and scoring has been based merely on the number of correct answers. If the result of a test only gives the number of correct answers, you lose detailed information about students' performances across and within domains (Black & Wiliam, 1998, Gipps, 1994).

Gipps (1994, 2001) uses the term "educational assessment," which means a wide variety of methods for testing, measuring, and evaluating students' achievements. An integral part of this is also students' self-assessment, assessment of peers, and assessment of the learning environment. Gipps argues that students ought to be active in the assessment process. The basic view is that the student is responsible for his or her own learning and thereby a participant in the assessment processes.

Assessment can be seen mainly from two perspectives. One is a summative perspective, to determine what knowledge and what level of knowledge an individual or group has shown on a given occasion. The level of performance is related either to stipulated criteria or to other students. The other perspective, the formative perspective, regards assessment as an integral part of the learning process for the individual (Erickson & Börjesson, 2001).

Relational and Communicational Aspects

Theoretical considerations, for our work with national tests, diagnostic materials, and evaluations, address relational and communicational aspects. We acknowledge that the knowledge demonstrated by a student is strongly related to the context. What knowledge a student has opportunity to show is connected to the student's social life, the student's history, and the actual situation. Four "corner pillars" in the work in schools are knowledge, learning, assessment, and opportunities to learn (Pettersson, 2007)—see Figure 25.2. These are strongly related to each other and important aspects of the classroom's micro-world, as well as to aspects in the macro-world. One example is that the work in school is highly affected by decisions on school, community, and national levels. Here, the national tests play an important role.

When addressing communicational aspects, we choose to view assessment as an act of language-making (language in a broad sense). This viewpoint is inspired by social semiotics (e.g., Kress, Jewitt, Ogborn, & Tsatsarelis, 2001). This means that it is only possible to assess knowledge that a student shows/communicates. Hence, there is no use for a statement such as "(s)he does not know this." Instead we focus on students' performances in different situations, on what knowledge students show through different modes of communication. The considerations in the curriculum emphasize the process of generating knowledge: a consequence of the formulations in the curriculum is that the students must be given opportunities to demonstrate their knowledge in various ways. In our work with the national tests and evaluations, we thus strive towards *making the essentials assessable and not making the most easily measured things the most essential.*

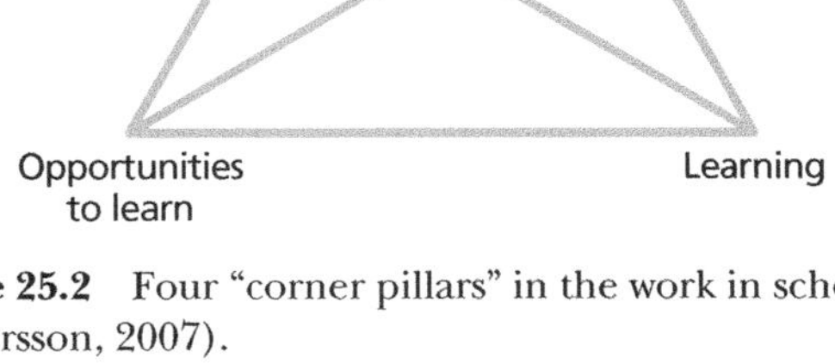

Figure 25.2 Four "corner pillars" in the work in schools (Pettersson, 2007).

An additional base in the construction of national tests is of course also the ongoing development that is taking place in mathematics teaching and in international and national research in this area.

NATIONAL TESTS AND DIAGNOSTIC MATERIALS

In this section we describe national tests and diagnostic materials that PRIM-gruppen has developed on behalf of the Swedish National Agency of Education. We also give example of analyses that have been performed.

Presentation of National Tests and Diagnostic Materials for Compulsory Schools

When we in PRIM-gruppen develop test material, an important base is our theoretical considerations as outlined in the previous section. Another point of departure is the view of knowledge stated in the curriculum, the view of the subject in the syllabus and, for grade 9, also the criteria for the different grades. A test consists of several different parts in order to give the students an opportunity to show as many aspects of their knowledge in mathematics as possible. We construct our tests in as balanced means as possible, with various kinds of tasks in a variety of contexts and a range of response formats. The tasks in the testing material should be designed in such a way that the student is given opportunities to demonstrate different areas of knowledge and different qualitative levels in mathematics.

The following schematic schedule (see Table 25.1) shows the different test materials in mathematics for compulsory school in Sweden constructed by PRIM-gruppen. Of course, there are also other test materials available for Swedish teachers. Many of the Swedish textbooks in mathematics consist of different types of diagnoses. The National Agency of Education has one more diagnostic material, called "Diamant," which is constructed by researchers at Gothenburg University (Skolverket, 2009).

The *diagnostic materials in mathematics* have a formative purpose and consist of two parts, one part for use up to grade 6 and one part for grades 6 to 9. Each part consists of an assessment scheme for analysis and a bank of diagnostic tasks. The purpose of the materials for analysis is to help teachers to analyze and document the students' shown knowledge in mathematics. The teacher can use the same scheme for students of different ages. By using this scheme of analysis, the development of a student's shown knowledge can be traced over several years.

The *national subject tests in mathematics for grade 3* were used for the first time in 2009, and they are compulsory. The purposes of this test are to support the formative assessment in the classroom, to support the teachers in assessing whether the student has reached the goals to attain for grade 3, *and* to be part of a national evaluation. The test is intended to be integrated into the ordinary teaching, and each student is given sufficient time to work on the various tasks in the test. The different parts of the test are framed by a story about two children and the adventures they encounter. The teacher reads one part of the story before each test part is carried out. To describe each student's shown mathematical knowledge the teachers can use a *Knowledge Profile.* This profile is similar to the one used in grade 5, which is described below. In the material there is also a part for self-evaluation. In this the student is asked to state whether the student feels certain, rather certain, or not certain in different mathematical situations. The student shows this through the use of different colors.

The *national subject tests in mathematics for grade 5* are compulsory starting in 2009. The tests are only one component in the entire assessment of the student's

TABLE 25.1 Test Materials in Mathematics for Compulsory School in Sweden Constructed by PRIM Gruppen

Diagnostic materials Not compulsory	Up to Grade 6, and from Grade 6 up to Grade 9 (started 1996)	Assessment scheme for analysis of students' competence in mathematics A bank of tasks.
Subject test Compulsory	Grade 3 (started 2009)	6-8 different small parts + group tasks and self-assessment
Subject test Not compulsory until 2008, compulsory from 2009	Grade 5 (started 1996)	4–5 different parts + group tasks and self-assessment
Subject test Compulsory	Grade 9 (started 1998)	2–3 different parts + group task or/and oral test

knowledge done by the teacher. The main purpose of the test is to assist the teachers in assessing whether the students have reached the goals to attain for school year 5 in the syllabi, but, as with the test for grade 3, the test also has a formative purpose, which includes analyzing how the students solve the problems and examining the quality of their work in order to identify what knowledge they show and what they need to learn in mathematics. The teachers are supposed to assess the test holistically and not use scoring points. The intention is that both the way in which the student has solved the problem and the answer will be taken into consideration. To describe each student's mathematical performance the teachers can use a knowledge profile (see Table 25.2), which is included in the material. Teachers are then expected to consider both their assessment of the student's work on the subject test as well as their overall assessment of the student's mathematical knowledge.

For each area of competence in the knowledge profile, the teacher can describe the assessment in words making use of the opportunity to comment on or to describe verbally their assessment. The example in Table 25.2 shows only two of the knowledge areas.

By the use of such profiles, teachers can hopefully gain a more balanced picture of their students' knowledge in mathematics. Use of the knowledge profile is optional for the teachers.

The curriculum states that each student should develop the ability to critically evaluate his or her own results and place his or her own results in relation to his or her own achievements (Skolverket, 2000). In the test for grade 5, there is a section entitled "You and mathematics," in which students are to judge how confident they feel in various situations where they use mathematics (for example: "In calculating 8 – ___ = 3," "In calculating with your calculator," "In looking at the newspaper and finding out how long a TV program lasts," "In working together with someone," etc.). The response alternatives are "certain," "quite certain," "uncertain," "very uncertain."

TABLE 25.2 Examples of Knowledge Areas in the Knowledge Profile

Knowledge areas	Teacher's comment (examples of comments are shown)
Basic spatial perceptions, shapes and patterns, scale	*Difficulty with the names of the shapes.*
Diagrams, tables, measures of central tendency	*Can handle simple tables and diagrams. Easily passes the goals for year 5.*

The students start with this part and are to evaluate themselves without referring to their answers to the testing materials. A comparison can then be made with their results on similar tasks on the test and thus provide a basis for judging the student's self-confidence in using mathematics and how realistic this degree of confidence is.

Many teachers use the section "You and mathematics" in conversation with students and even with parents in a so-called progress discussion. Self-assessment as an important part of the learning process for each student is emphasized in Sweden (Skolverket, 2003), and can be seen in the following quotation: "Students who think about and discuss their progress toward the achievement of class goals and the learning of mathematical concepts, on the basis of the evidence they see in their own work, will lead to better understanding and control of their own success" (Stenmark, 1991, p. 56).

The *national subject tests in mathematics for grade 9* are compulsory. The main purpose of the test is to help the teachers assess to what extent the students have reached the goals set up in the syllabus and to provide support for teachers in awarding marks. The test paper consists of different parts in order to give the student an opportunity to show as many aspects as possible of her or his knowledge in mathematics. The student can show this by different ways of working (individually or in pairs or groups) and by different ways of answering (orally, by written answers with or without explanations and calculations). The teacher should use different ways of marking and assessing (right/wrong, scoring rubric, and analytic scoring). Two parts of the test are more traditional, and in one of these the students are not allowed to use calculators. The third part consists of one more extensive task, and the fourth part is often an oral test.

A distinguishing feature of more extensive tasks, often placed in realistic contexts, is that they are assessed using methods different from those used for ordinary items. The teachers assess extensive tasks with the support of assessment matrices (see Table 25.3). More than other items these more extensive tasks are a measure of the ability of the students to do independent work, to be creative, to systematise, to carry out mathematical reasoning, to create mathematical models, to formulate and test assumptions, as well as to draw conclusions. It should be possible to assess the students' entire solution on different qualitative levels. A requirement for these tasks is that they give all students the opportunity of at least starting a solution, while at the same time the task should be sufficiently challenging that a good solution can show qualities at the highest grade level.

TABLE 25.3 Task-Specific Assessment Matrix for "Geometric Figures"

	Qualitative levels		
The assessment concerns	**Lower** ⟶		**Higher**
Understanding *To what extent the student shows an understanding for the task and motivates his or her conclusions* *To what extent the student uses relations and generalisations*	Shows understanding for either of the concepts perimeter and area, i.e. by describing how you can calculate perimeter or area for some of the figures.	Shows good understanding for both the concepts perimeter and area, i.e. by explaining how you can see that perimeter and area are the same/ not the same for the two figures.	Shows good understanding for both the concepts perimeter and area and relations between them i.e., by carrying out general reasoning and extending the problem.
Language *How clear and distinct the student's account is.* *To what extent the student uses relevant mathematical language and terminology*	Comprehensible and possible to follow.	Easy to follow. Uses acceptable mathematical language and terminology.	Well-structured and clear with relevant mathematical language and terminology.
Participation *To what extent the student participates in discussions with mathematically based ideas and explanations*	Accounts only for one's own figures.	Contributes with his or her own ideas and explanations after explanations of others or in the final discussion.	Is involved in the arguments of others. Carries the discussion forward.

Source: Kjellström & Pettersson, 2005, p. 316

The oral test for grade 9 is carried out in groups of 3–4 students who work well together. The teacher selects the groups. Our intention with this is that there should be a conversation between the students with as little interference as possible from the teacher. It is quite a challenge to construct problems and make assessment directives for the oral part. The aim is to construct problems that are more suitable for oral communication than for written solutions. The exercise consists of two sub-sections. First, every student gives an account of her or his problem to the other students in the group. Then the whole group discusses all of the problems.

> One of the oral tasks we have used: *Each one of you will get a sheet of paper with two geometrical figures. Your task is to describe similarities and differences between the figures e.g., concerning perimeter and area.* (Two examples of pairs of figures are shown in Figure 25.3.)

In this exercise the student is asked to describe and compare geometrical figures and to find similarities and differences between them. The exercise also tests to what extent the student can use mathematical language, see relations, argue, and accept arguments from others. Pairs of figures can be chosen with regard to the capacity of the student.

Every student gets a sheet of paper with one pair of figures and is then given approximately five minutes for preparation. The students report in turn to the other students in the group. To help the student to advance, the teacher might insert short questions. After each report, the other students in the group are invited to ask questions, make comments, and make contributions. Finally, all figures are put on the table and the students are invited to discuss together similarities and differences between as many figures as possible.

The different pairs of figures are of varying difficulty to compare. In assessing the student's performance one should therefore take into account what pair of figures the student has compared. The teacher should make a holistic assessment of a student's performance with the support of a task-specific assessment matrix, as shown in Table 25.3 (Kjellström, 2005). The matrix is constructed in line with the criteria for the different grades as stated in the National syllabus.

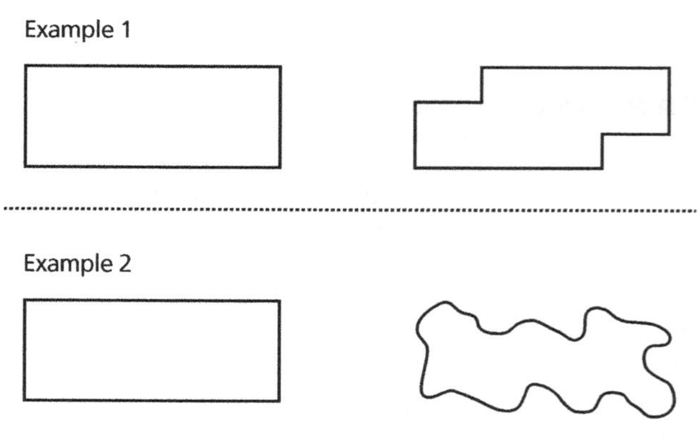

Figure 25.3 Geometric figures (two examples out of six).

Analysis of Students' Work

A key element in qualitatively assessing students' knowledge is analyzing, through various tasks and situations, how the students work with and master an area of mathematics. In our work with the national tests we examine whether the students have tried to find a solution to the task, how well they have understood the task, which concepts they have displayed, and in what ways the students have dealt with the task (Alm, 2004; Pettersson, 1994, 1997).

The students can work on tasks in many different ways. We have seen that those who arrive at correct results may have used different strategies, either ones that are dependent on the context, or ones that are more general. The students who arrive at incorrect results may display errors that are *accidental* in nature, that is, they are not found systematically but are of a more random character. There are also errors that are *systematic*, that is, they occur consistently (Brown & Burton, 1978). This often indicates deficiencies in grasping concepts. There are many possible reasons for reaching incorrect results. The student might have misunderstood the task, or might have understood the task but used an incorrect method, or might have understood the task, used an appropriate method, but made a miscalculation.

There are different components to be assessed in mathematics. We examine knowledge of facts, knowledge of skills, and knowledge of concepts.

One example of an outcome of our analysis of students' work is that we have found that quite a few students have difficulties with numbers in fractional and decimal form (Alm, 2004). These students cannot reach the goals that the students are to have attained by the end of grade 5: "Students should have a basic understanding of numbers, including natural numbers and simple numbers in fractional and decimal form" (Skolverket, 2000, p. 28).

In the test of year 2000 for grade 5, there are two tasks involving fractions. The first is: *Amir is in school for a quarter of a day and Linda for a third of a day. Who has the longest school day? Write down how you worked out your answer.* Seventy-five percent of the children in grade 5 solve this task correctly. The most common way they use to show their understanding is by pictures (38 %), then by calculations (29 %), and also by explanations with words (6 %). The most common misconception, for 8 % of the students, is that 1/4 is greater than 1/3 because "4 is greater than 3" or "4 is more than 3."

Analysis of Tasks in a Multimodal Social Semiotic Framework

Some of our analyses are based on multimodal social semiotics (Kress et al., 2001; see also Morgan, 2006). According to this, assessment can be seen as communication. In the multimodal approach, communication is not only considered from a linguistic perspective. Instead, all modes of communication are recognized. Modes can be, for example, speech, writing, gestures, and pictures. Each mode has its "affordance" for a person engaged in a communication (Kress et al., 2001)—that is, which mode that is chosen in a specific situation is not arbitrary; instead it is the best way for a person to communicate a certain message in a particular moment that leads to the choice of modes for this communication. The modes that are available are dictated and formed by the existing culture and context. Halliday's (2004) meaning-making principles have been elaborated within the multimodal approach. According to Halliday, there are three kinds of meaning that all communication is understood to reflect: interpersonal, ideational, and textual. The focus of the *interpersonal* meaning is on interactions and relations between people. *Ideational* meaning can reflect what is going on in the world, and *textual* meaning refers to formation of whole entities that are communicatively meaningful. Morgan (2006) uses these functions with a focus on linguistics and on the construction of the nature of mathematical school activity. They are also used by Björklund Boistrup (2007) to analyze processes of assessment in a mathematical classroom. These three functions are in our analysis combined with a focus on multimodality according to Kress et al. (2001), which we use in our analysis of diagnostic materials in mathematics. In social semiotics, signs of communication are a central part of communication (Kress et al., 2001). We use the meaning-making principles as an analytical tool for analysis of materials.

We have applied multimodal analysis to one of the diagnostics materials for grades 1–3 (Björklund Boistrup, Strandberg, Sjöholm, & Pettersson, 2007). We describe this analysis shortly as an example of how the multimodal social-semiotic framework can support analysis of test and diagnostic items.

Analytical Tools in Relation to Research Focus

As mentioned in the section on theoretical considerations, the focus of interpersonal meaning is interactions and relations between people. We look at how the student is addressed in the tasks, with focus on relational aspects. When we use the ideational meaning, the focus is on what is going on in the world, and in this case, the focus is on mathematics. We look for areas of mathematics in which the students are given opportunity to show signs of learning. When we focus on the textual meaning, we are interested in what possible and different communicational modes can be used for the assessment in mathematics. We analyze which modes the students are given the opportunity to use in the material and also to what extent the different modes are present.

We also focus on a combination of the three metafunctions. Mathematical discourse is one aspect of the ideational meaning (Morgan, 2006), and, since it is about a relation to this particular discourse, it can also be seen as an interpersonal meaning. This is expressed in different communicational modes, and therefore the textual meaning is also present. Our interpretation of this is that there are two ways of looking at the student in relation to the mathematical discourse: One is to focus on the context and the other is to what extent the tasks have open solutions and answers.

Interpersonal Meaning

For the interpersonal meaning, we analyzed how and to what extent the student is addressed directly in the material, that is, to what extent there is a relation to the student created in the formulation of the task. We found four ways for this. When the student is addressed with "you," we considered this as a direct addressing. This goes also for the use of "I" or "we." The third alternative is when there is a picture of a person looking "out" towards the reader, posing the task. The fourth alternative is when the students are expected to work in groups, since they do address each other directly in this work. In the grade 1–3 diagnostic materials the student is addressed directly, in one or more ways, in 94% of the tasks.

Ideational Meaning

For the ideational meaning, we compared the tasks with "goals to aim for" in the national syllabus (Swedish National Agency of Education, 2001). Looking first at the goals of competence, the results are presented in Table 25.4.

In the material, several goals of competence are represented. Looking at the goals of subject content knowledge, the picture becomes somewhat different, since there are differences in how different areas are represented (see Table 25.5).

The material focuses mainly on numbers and calculation. Nothing is focused on functions and probability. This analysis can be used for various purposes. One purpose might be for a future revision of the material. What should be addressed in that process, with relevance for this analysis, is to what extent more goals should be assessed in the new version of the material.

TABLE 25.4 Frequency in Percentage of Goals of Competence

Goals of competence	
Roles of mathematics in different contexts, development. Historical development of mathematics.	1%
Reflect on and appreciate the use of mathematical forms of expression	17%
Logic, Argumentation	66%
Problem-solving, Analyzing	47%
Models	46%
Calculators	22%

TABLE 25.5 Frequency in Percentage of Goals of Subject Content Knowledge

Goals of subject content knowledge	
Numbers and calculation	86%
Measurement	50%
Geometry	21%
Statistics	9%
Algebra	4%
Functions	0%
Probability	0%

Textual Meaning

For the textual meaning, we have analyzed which communicational modes (action, drawings, speech, words, and/or symbols) the student is given the possibility to use according to the task instruction, and also if there are complementary instructions that the teacher is told to tell the students orally. In Table 25.6 we show the frequency of each possible communicational mode.

In Table 25.7 we show the percentage share of the numbers of possible modes per task. With number of possible modes, we mean the total of the different modes the students are allowed to use when solving the task, according to the task instruction.

Symbols have a prominent role. For the material, however, drawing is even more common. The results show that there are many opportunities for the students to use different modes in the material.

The Mathematical Discourse

Finally, we have analyzed which kind of context that is used in the tasks and also the degree of openness (answer and solution) in the tasks. For the analy-

TABLE 25.6 Frequency in Percentage of Modes the Student is Given Possibility to Use

Modes	
Action	39%
Drawing	67%
Speech	18%
Writing	41%
Symbols	61%

TABLE 25.7 Frequency in Percentage of Numbers of Possible Modes per Task

Number of possible modes per task	Material
one	39%
two	17%
three	5%
four	12%
five or more	28%

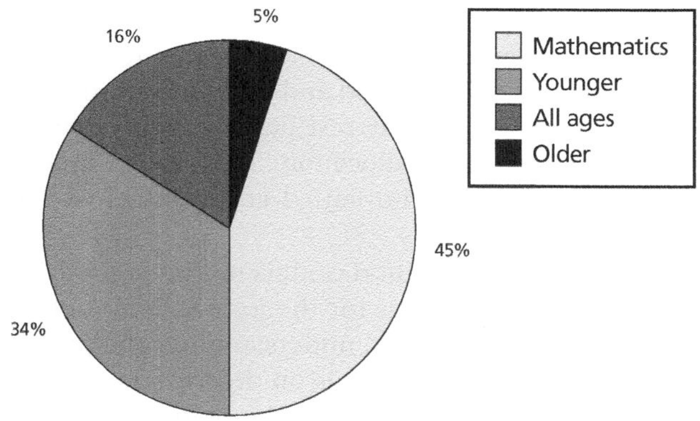

Figure 25.4 Kind of task contexts, distribution in percent.

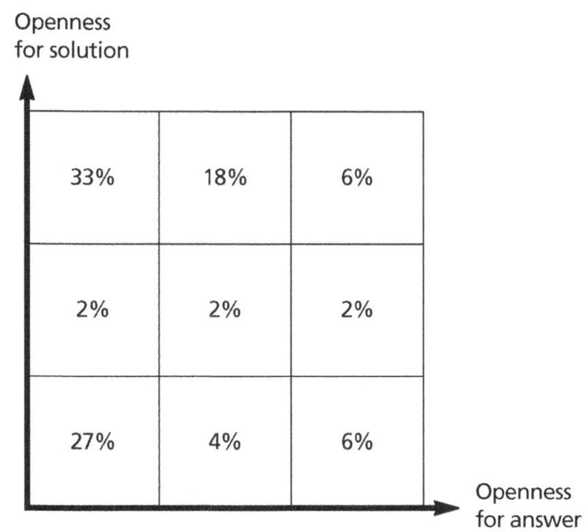

Figure 25.5 Openness of tasks, distribution in percent.

Note: In the first column from the left there is only one answer expected. In the middle column, tasks with two or three possible correct answers are placed, and in the third column, tasks with four or more possible correct answers are placed. In the lowest row the tasks with just one expected solution are placed. Also, tasks where the student is not asked to show the solution, but only the answer, are placed here. Tasks that have two or three possible solutions are placed in the middle row, and in the highest row, the ones with four or more possible solutions are placed.

sis of the context we looked for whether the context is "pure" mathematics (called "mathematics") or if it is a "real-world" task. The "real-world tasks" were divided into three groups: typical for the age of the students (called "younger"), for people in almost all ages (called "all ages") or for people older than the students (called "older"). The result is shown in Figure 25.4.

For the material 45% of the tasks are within an exclusive mathematical context. For the analysis of the openness we constructed a model. The results are shown in Figure 25.5.

The material contains a good many tasks where there is more than one possible answer and/or solution. Our conclusion is that these open tasks invite the student to participate in the mathematical discourse more than when there are only closed tasks, since there is less restriction on the student's ability to show his or her knowledge. One can say that the student gets the opportunity to be part of the assessment process as a subject and not only as an assessed object (Mellin-Olsen, 1993).

When using the multimodal approach within social semiotics as an analytical framework, we find that it helps to reveal important aspects of the diagnostic materials and tests. We are aware that our interpretation of the meaning-making principles is based on our experiences and research interests. We are also aware that there are other important aspects that we have not focused on. However, using the framework inspired us to analyze other aspects than if we had performed the analysis without this particular framework. One can say that the operationalization of the meta-functions helped us to take at least one more step in viewing aspects of the material, which can be vital in the future revision and construction process.

NATIONAL EVALUATION

Extensive evaluation programs in many countries have been established in order to measure the effectiveness of education and to control the outcomes of the money put into it. The dominating evaluation tradition is focused on student achievement, assessing students´ basic skills to read and write and solve arithmetic problems. The interest lies in central control for effective resource allocation. This type of evaluation, which is product-oriented and summative, has been questioned.

An alternative method centers on the potentiality of evaluation as a means for the local school itself to develop its own processes, understanding, and professional resources. The focus is on description and understanding. Evaluation of this type is focused on the process of education as well as on the achievement of the students. It is formative and process-oriented. The research interest has led to a quest for more complex information than is usually produced by the summative approach. A greater variety of test items is being developed, aimed at describing the process, the context, and the related outcomes.

Here we describe the background and some results of the national evaluation in Sweden. We continue by describing a longitudinal national project where extensive analyses have been performed.

The Swedish National Evaluation

A program for national evaluation was established in Sweden and put into effect in 1989 in grades 2 and 5. In 1992, a second data collection was carried out in grades 5 and 9 (Ljung & Pettersson, 1990; Pettersson, 1993a,b). In 1995, a third data collection was carried out, and in 2003 the fourth data collection, which was a repeat of the evaluation in 1992, was carried out (Skolverket, 2004).The program exists in a field of tension between the two traditions described above. It is centrally initiated and administrated. It aims at delivering results to central authorities on the standard of the Swedish schools. But it is much more complex than such programs usually are. It covers the whole range of the curriculum from, reading, writing, and arithmetic to history, technology, music, arts and crafts, and other subjects. The overriding goal of the program is to evaluate every aspect of the compulsory school. The program thus aims at describing the whole range of qualities, which means that, in addition to easily measured knowledge and skills, aspects like problem solving, understanding, valuing, oral skills, and so on are to be described.

The national evaluation is important both on a national and a local level. Feedback from results has been delivered directly to the schools concerned, to the National Agency of Education, and to the Ministry of Education. The assessment will also provide a platform for future development of the curriculum, teacher training, and in-service training as well as for research and debate, among other uses.

There were many aims for the national evaluation. From a mathematical viewpoint, we can describe the aims as

- to assess the students´ knowledge and skills in mathematics and their attitudes about mathematics as a subject and about mathematical education
- to describe the "mathematical situation" for students and their teachers in the schools

The evaluation in mathematics should give as complete a picture as possible of the students' knowledge and skills in mathematics and the students' attitudes about mathematics. But the evaluation in mathematics should also give a picture of the pedagogical process and the working methods.

Some results from the three latest national evaluations, in 1992, 1995, and 2003, are listed here (Pettersson, 1993a, 1993b, 1997; Skolverket, 2004). These results are chosen in order to show the many different aspects in the national evaluations.

- About 20% of the students in grade 5 and about 30% in grade 9, 1992, thought that much of what they learned about mathematics was unnecessary. In grade 9 in 2003, this opinion had increased to 55%.
- 30% in grade 9 in 1992 and 57% in 2003 were of the opinion that they had worked with too many difficult items in mathematics.
- Compared with 1992, a deterioration in performance can be observed in 2003, most clearly in grade 5.
- We get a picture of an increasingly individualized teaching—compared with both 1992 and 1995—here students work in isolation from the teacher and their class colleagues.

The results show different aspects of the national evaluation and the advantage of having the same questions for different years and different grades, which makes it possible to compare performances and attitudes over years and grades.

Analysis by Longitudinal Studies of Students' Mathematical Achievement

Many researchers have, by investigating students' written protocols, studied what types of errors students make when they solve mathematical tasks (e.g., Morton, 1924; Brueckner & Elwell, 1945; West, 1971). Most of the researchers have classified the errors according to how serious they are. This previous research has shown how pedagogically important it is to study how a student solves mathematical tasks and not only to note if the answers are correct or incorrect. There are, however, very few longitudinal studies of students' solution strategies. A study from the 1980s (Pettersson, 1990) investigates students' achievements in mathematics from grade 3 (age 9–10) to grade 6 (age 12–13). What strategies are used when they solved different mathematical items correctly, and what types of errors are involved when they performed incorrectly? The main question of the study was: "How do students with varying developmental trends differ in their solution strategies for the problems?" The students' strategies when they solved an item correctly have been studied, and the nature of the students' errors has been analyzed in detail. The errors have been classified into one of two categories: grave errors or simple errors. Examples of grave errors are: the students used incorrect rules of arithmetic, for example, addition instead of subtraction; they "borrowed a ten" incorrectly; they used 1 hour = 100 minutes. Simple errors are errors in calculation and slips. In the study, the achievements are considered as part of an interaction process, an interaction between the individual and his or her environment. For the theoretical consideration, the individual prerequisites, the learning, and the teaching are considered essential for how the student performs and how the student solves mathematical tasks (Pettersson, 1990).

The mathematical test, which was constructed for the longitudinal study, consisted of 15 items and was divided into two sections arithmetic and problem solving. Most of them were short answer items. Examples of items:

- *Calculate 263 + 654*
- *Eva and Mats got 60 SEK to split up. Eva should have twice as much as Mats. How much did Eva and Mats get?*

A random sample of 10,000 students in grade 3 in Sweden were given 15 mathematical tasks to solve in the spring of 1982. The same students solved the same tasks three years later. The students solved three more items correctly on the average in grade 6 compared to grade 3. Approximately 86% of the students achieved a better result in grade 6 than in grade 3. About 7% achieved the same result in grades 3 and 6, and about 7% achieved a result in grade 6 inferior to that in grade 3.

The study showed that when the students solved a task correctly, they frequently used the same standard algorithm. The students who achieved good results in both grades solved almost every item correctly. If they made a mistake it was almost always a simple error. The students who had poor results in both grades made a large number of grave errors in both grades. The grave errors persisted in grade 6. This group of students left items unanswered to a greater extent then the students in the other groups. The group who performed much better in grade 6 than in grade 3 made many grave errors in grade 3, but in grade 6 the errors were mostly simple errors.

In grade 9 many students have a standardized achievement test in mathematics. The group of students who achieved good results in both grades 3 and 6 achieved results that are superior to the students in the other groups. The group of students with poor results in both grades achieved inferior results on the standardized achievement test compared to the other groups. Another result is that a large number of the group of students which had poor results in both grades had difficulties in understanding the teachers´ explanations; they also wished to ask the teacher for help more often than they actually did.

CONCLUSION

In this chapter we have presented the background and framework for national assessment materials that are constructed within the PRIM-gruppen. We have briefly described different materials, and along with this we have described examples of analyses that have been performed.

What we have wanted to show is that assessment and the construction of assessment material is not a trivial pursuit. Instead it is a process that contains a lot of both theoretical and empirical considerations. The further development of our work with national tests and diagnostic materials relies heavily on our analyses of students' performances and of tasks.

Moreover, the work with national assessment also makes important contributions to the research in mathematical education—for example, in the sense of how students solve mathematical tasks. What strategies are used when they solve a problem correctly and what types of errors are involved when they solve a problem incorrectly? The work with the national tests and diagnostic material also contributes to the research of assessment both for summative and formative purposes.

ACKNOWLEDGEMENTS

To construct and develop national tests and evaluations and to do research in connection with this is not the work of only one or two people—many people are involved. We would like to thank Lena Alm, Katarina Kjellström, and Gunilla Olofsson, who in other situations have written about the topic of national tests (e.g., Alm, Björklund, Kjellström, Olofsson, & Pettersson, 2004; Kjellström & Pettersson, 2005). Their work has been of great use for us while writing this chapter. We would also like to thank Max Ljungberg and Kari Sjöholm for their analysis of the diagnostic materials.

REFERENCES

Alm, L. (2004). På upptäcktsfärd i elevernas värld av tal [On expedition of discovering the students' world of numbers]. In Skolverket (Eds.), *Att visa vad man kan, En samling artiklar om ämnesproven i år 5* [*To show what you know, A collection of articles about the national tests in grade 5*] (pp. 101–113). Stockholm: Skolverket.

Alm, L., Björklund, L., Kjellström, K., Olofsson, G., & Pettersson, A. (2004). Assessing students' knowledge. *SMDF Newsletter, 9,* 4–24.

Björklund Boistrup, L. (2007). Assessment in the mathematics classroom. Studies of interaction between teacher and pupil using a multimodal approach. In D. Pitta-Pantazi & G. Philippou (Eds.), *European research in mathematics education V. Proceedings of the Fifth Congress of the European Society for Research in Mathematics Education* (pp. 1160–1169). Cyprus: Larnaca.

Björklund Boistrup, L., Strandberg, M., Sjöholm, K., & Pettersson, A. (2007). *Multimodal aspects of assessment—Analysis of diagnostic materials in Sweden.* Paper presented at EARLI 2007, 12th Biennial Conference for Research on Learning and Instruction, Budapest, Hungary.

Black, P. (2001). Formative assessment and curriculum consequences. In D. Scott (Ed.), *Curriculum and assessment. International perspective on curriculum studies* (Vol. 1, pp. 7–24). London: Ablex.

Black, P. & Wiliam, D. (1998). Assessment and classroom learning. *Assessment in Education, 5*(1), 7–74.

Brown, J.S. & Burton, R.R. (1978). Diagnostic models for procedural bugs in basic mathematical skills. *Cognitive Science, 2,* 155–192

Brueckner, L.J. & Elwell, M. (1945). Reliability of diagnosis of error in multiplication of fractions. *Journal of Education Research, 26,* 175–185.

Erickson, G. & Börjesson, L. (2001). Bedömning av språkfärdigheter. I nationella prov och bedömningsmaterial [Assessment of language proficiency. In national tests and materials for assessment]. In P. Malmberg & R. Ferm (Eds.), *Språkboken—En antologi om språkundervisning och språkinlärning* [*The book of language—An anthology about language education and language learning*] (pp. 255–269). Stockholm: Skolverket.

Gipps, C. V. (1994). *Beyond testing. Towards a theory of educational assessment.* London: The Falmer Press.

Gipps, C V. (2001). Sociocultural aspects of asssessment. In G. Svingby & S. Svingby (Eds.), *Bedömning av kunskap och kompetens. Konferensrapport* [*Assessment of knowledge and competence*] (pp. 15–67). Rapport nr 18 från PRIM-gruppen vid Lärarhögskolan i Stockholm.

Halliday, M.A.K. (2004). *An introduction to functional grammar.* (Revised by Matthiessen, M.I.M.). New York: Oxford University Press.

Kilpatrick, J. & Johansson, B. (1994). Standardized mathematics testing in Sweden: The legacy of Frits Wigforss. *Nordisk Matematikkdidaktik, 2*(1), 6–30.

Kjellström, K. (2005). Bedömningsmatriser—en metod för analytisk bedömning [Assessment matrices—a method for analytical assessmen]. In L. Lindström & V. Lindberg (Eds.), *Pedagogisk bedömning. Om att dokumentera, bedöma och utveckla kunskap* [*Educational assessment. To document, assess and develop knowledge*] (pp. 193–217). Stockholm: HLS Förlag.

Kjellström, K. & Pettersson, A. (2005). The curriculum´s view of knowledge transferred to national tests in mathematics in Sweden. *ZDM, 3,* 308–317.

Kress, G., Jewitt., C., Ogborn., J., & Tsatsarelis, C. (2001). *Multimodal teaching and learning: The rhetorics of the science classroom.* London: Continuum.

Leder, G. (1992). Curriculum planning + assessment = learning? In G. Leder (Ed.), *Assessment and learning of mathematics* (pp. 313–146). Camberwell, Australia: ACER.

Ljung, B. & Pettersson, A. (1990). *Matematiken i nationell utvärdering. Kunskaper och färdigheter i årskurserna 2 och 5* [Mathematics in national evaluation. Knowledge and skills in grades 2 and 5]. Stockholm: PRIM-gruppen.

Mellin-Olsen, S. (1993). A critical view of assessment in mathematics education: Where is the student as a subject? In M. Niss (Ed.), *Investigations into assessment in mathematics education. An ICMI study* (pp. 143–156). London: Kluwer Academic Publishers.

Morgan, C. (2006). What does social semiotics have to offer mathematics education research? *Educational Studies in Mathematics, 61*(1), 219–245.

Morton, R. L. (1924). An analysis of pupils' errors in fractions. *Journal of Educational Research, 10,* 117–125.

Niss, M. (1993). Assessment in Mathematics Education and its Effects: an Introduction. In M. Niss, (Ed.), *Investigations into Assessment in Mathematics Education. An ICMI study* (pp. 1–30). London: Kluwer Academic Publishers.

Pettersson, A. (1990). *Att utvecklas i matematik. En studie av elever med olika prestations-utveckling* [To grow in mathematics. A study of pupils with different development of performance]. Doctoral thesis. Stockholm: Almqvist & Wiksell International.

Pettersson, A. (1993a). *Den nationella utvärderingen av grundskolan 1992. Matematik åk 5. Huvudrapport.* [The national evaluation of compulsory school in 1992. Mathematics grade 5. Main report]. Stockholm: Skolverket.

Pettersson, A. (1993b). *Den nationella utvärderingen av grundskolan 1992. Matematik åk 9. Huvudrapport* [The national evaluation of compulsory school in 1992. Mathematics grade 9. Main report]. Stockholm: Skolverket.

Pettersson, A. (1994). *Den nationella utvärderingen av grundskolan 1992. Hur löser elever uppgifter i matematik. Fördjupad analys.* [The national evaluation of compulsory school in 1992. How do students solve tasks in mathematics. Analysis in depth]. Stockholm: Skolverket.

Pettersson, A. (1997). *Matematiken i utvärdering av grundskolan 1995. Analys av elevernas arbeten med mer omfattande matematikuppgifter i åk 9.* [Mathematics in the evaluation of compulsory school in 1995. Analysis of students' solutions of more comprehensive mathematical tasks in grade 9]. Stockholm: Skolverket.

Pettersson, A. (2004). Assessing students' knowledge: Language in mathematics tests. In C. Bergsten & B. Grevholm (Eds.), *Mathematics and language. Proceedings of Madif 4* (pp. 35–44). Linköping: SMDF.

Pettersson, A. (2007). Pedagogisk bedömning—Bedömning för lärande [Educational assessment—Assessment for learning]. In M. Olofsson (Ed.), *Bedömning, flerspråkighet och lärande [Assessment, multilingualism and learning]*. Symposium 2006. Nationellt centrum för sfi och svenska som andraspråk (pp. 11–20). Stockholm: HLS Förlag.

Shepard, L. (2000). The role of assessment in learning culture. *Educational Researcher, 29,* 4–14.

Skolverket. (2000). *Grundskolan. Kursplaner och betygskriterier.* [Compulsory school. Syllabi and criteria for grading]. Stockholm: Skolverket.

Skolverket. (2003). *Lusten att lära—Med fokus på matematik. Nationella kvalitetsgranskningar 2001–2002.* [The desire to learn—With focus on mathematics. National quality reviews 2001–2002]. Stockholm: Skolverket.

Skolverket. (2004). *National evaluation of the compulsory school in 2003.* Stockholm: Author.

Skolverket. (2009). *Diamant. Diagnoser i matematik.* [Diamonds. Diagnoses in mathematics]. Retrieved on June 6, 2009 from http://www.skolverket.se/sb/d/260/a/14694

Smith, E. & Gorard, S. (2006). "They don't give us our marks": The role of formative feedback in student progress. *Assessment in Education, 12*(1), 21–38.

SOU 1942:11. (1942). *Betänkande om betygsättningen i folkskolan.* [Report of the grading in comprehensive school]. Stockholm.

Stenmark, J. K. (Ed.) (1991). *Mathematics assessment: myths, models, good questions and practical suggestions.* Reston, VA: National Council of Teachers of Mathematics.

Swedish National Agency of Education. (2001). *Compulsory school. Syllabuses.* Retrieved on June 1, 2009 from http://www3.skolverket.se/ki/eng/comp.pdf

West, T. A. (1971). Diagnosing pupil errors. *Arithmetic Teacher, 18,* 469–469.

Wigforss, F. (1937). *The entrance examination in view of later school performance.* Stockholm: P. A. Norstedts & Söners Förlag.

Wigforss, F. (1941). *Marks and certificates in the primary school.* Stockholm: P. A. Norstedts & Söners Förlag.

26

THE SWEDISH GRADUATE SCHOOL IN MATHEMATICS EDUCATION

An Effort to Promote an Interdisciplinary Research Field

Gerd Brandell

LUND UNIVERSITY

Abstract: During the years from 2001 to 2006, a national graduate school in Mathematics Education worked through a structured cooperation between ten universities in Sweden. The graduate school is the greatest single effort promoting research in mathematics education in Sweden so far. The funding from two of the main research financing bodies in Sweden was generous but limited in time. The purpose of the graduate school was twofold. The immediate goal was to give about twenty students the opportunity to join a doctoral program leading to a PhD in Mathematics Education within the frame of a graduate school. A more long-term goal was to support and develop environments for research education at a number of mathematics departments, some of them without any earlier such program. The chapter will briefly describe the goals, functioning, and outcomes of the graduate school. The choice of the doctoral students' research topics will be broadly positioned within the field. The possible impact of the graduate school on the Swedish research in mathematics education will be discussed with a focus both on the content of the research and on the organization of the research environments within faculties and departments.

Keywords: Doctoral program, graduate school, mathematics education

BACKGROUND

Teaching and learning of mathematics has been a theme in Swedish academic research in the field of education for the past ninety years. Bergsten (2002) identifies 23 doctoral dissertations in the area through 1999, the first published as far back as 1919, but the majority during the 1990s. A current list at the homepage of the National Center for Mathematics Education uses a slightly wider definition and contains 31 dissertations from the same period (National Center for Mathematics Education, 2009). With very few exceptions, the dissertations were presented and defended as PhD theses in general education (pedagogy). Beginning around 1980, subject didactics slowly transformed into an independent area, no longer so closely related to pedagogy. The concept of subject didactics spread within the academia during the 1980s, supported by the publication of a book in three volumes with the title *Fackdidaktik* (Subject didactics) edited by Ference Marton (1986). The area of didactics of mathematics developed during this period to become a fairly broad subject, both in teacher education and as a research area, treating not only teaching and learning of mathematics but also the place and role of mathematics and mathematics education in society and in the school system. During the same period, two PhD dissertations in pedagogy, treating didactics of mathematics, were presented within the

The First Sourcebook on Nordic Research in Mathematics Education, pages 387–396

frame of a collaborative program between education and mathematics at Linköping University (Bergsten, 1990; Hedrén, 1990). This program can be viewed as a forerunner of the graduate school in the sense that it included a mathematics department and contained a strong component of coursework in mathematics (Bergsten, 1996). A Swedish Society for Research in Mathematics Education (SMDF) was established in 1999 and has been very active since then with biannual conferences documented in proceedings.

In Sweden, chairs for professors in mathematics education or didactics of mathematics were created later than in the other Nordic countries. Iceland got its first professor in 1990, and Finland, Denmark, and Norway had professors in place during the 1990s. In Sweden the first permanent professor in the area was appointed at Luleå University of Technology in 2003. That same year, one senior lecturer was promoted to be a professor in mathematics education at the University College of Kristianstad. Since then, the development has accelerated, and by 2009, professors had been appointed or promoted in mathematics education at four other Swedish universities (Umeå, Linköping, Växjö, and Malmö Universities).

Programs for research education specific to the area of mathematics education were established during the 1990s at Umeå, Luleå, and Göteborg Universities. These programs were run by mathematicians and professors in education in cooperation, with researchers from other countries involved in the supervision. However, the funding for doctoral students was limited, and few students were enrolled. The first dissertation within this frame was presented in 1996 (Dunkels, 1996) and the second in 2001 (Bergqvist, 2001).

A great effort was made to improve the conditions for the research area in Sweden through an initiative in 2000 by the Bank of Sweden Tercentenary Foundation (RJ) to fund a graduate school in mathematics education. RJ is a major Swedish research foundation supporting humanities and social science.

A short outline of the Swedish research education system may be relevant. Two levels of research degrees exist—a licentiate degree and a doctoral degree. The latter corresponds approximately to a PhD or slightly more in the Anglo-Saxon system and requires four years of full-time study, including substantial coursework. The PhD thesis ought to report original and high-quality research work and may be in the form of a monograph, or as a combination of articles already published or submitted to journals. In the latter case there is also an introduction and overview that connects the research, the "kappa" (coat).

The licentiate degree is shorter and requires two years of full-time study. The requirements on the size and quality of the thesis are lower. The licentiate degree does not lead to an academic career but qualifies the holder for a position as lecturer (senior teacher) at the Swedish upper secondary school. Coursework is part of both exams but may not exceed what amounts to half of the total time, that is, not more than two years of full-time study for a doctoral degree and no more than one year of full-time study for a licentiate degree.

The history leading to the creation of the graduate school is interesting since it brings out some of the challenges the new graduate school had to overcome right at the beginning. In 1999 the Swedish National Committee for Mathematics at the Swedish Royal Academy of Sciences put forward a proposal to RJ. The mathematicians suggested a graduate school, not in mathematics education but in mathematics, specifically aimed at mathematics teachers in upper secondary school who were willing to qualify for a senior teaching position. Since 1980 very few lecturers in mathematics had been appointed in Swedish upper secondary schools, and at the time (in 1999), there were almost no such teachers still active. The proposal from the mathematicians for a national graduate school aimed to remedy or at least improve this situation. The focus would therefore be on the licentiate program in mathematics.

The RJ foundation chose another direction when the national graduate school was launched. Their analysis of the prolonged difficulties within mathematics education in the Swedish school led to the conclusion that research in this area would be the best way for RJ to support mathematics education in Sweden in the long run. However, the foundation did put an emphasis on the subject of mathematics when they decided to define the area for the graduate school as *mathematics, with specialization in didactics of mathematics* rather than mathematics education in a broad sense. As a consequence, the faculties and departments involved in the graduate school were mathematics departments and science faculties rather than educational departments and social science faculties. The foundation obviously wished that the mathematics departments should take more responsibility for developing the research area. The reasons for this wish will be discussed further on.

Ten departments at ten different universities (a couple of them formally university colleges) participated in the graduate school, most of them mathematics departments, but in some cases interdisciplinary departments with mathematics, applied mathematics, and other sciences. The universities were chosen through an application and review process, with international experts evaluating the quality of the graduate program and the research environment. Most of the departments created their doctoral program during the application process since their faculties had not earlier awarded PhD exams in the area. Hence, many departments without any tradition in the field and many supervisors new to the field participated in the graduate school.

Leder, Brandell, and Grevholm (2004) have described the start and the early development of the graduate school with special attention on the departments and the conditions for the supervision. Grevholm (2008) has positioned the graduate school within a Nordic context. Brandell (2009) describes and discusses the process behind the creation of the school, the organization and steering, the recruitment process, the students and the environments, courses and conferences, exams and scientific publications within the frame of the school, and evaluations and reports about the school in a comprehensive report in Swedish.

Obviously at least three major challenges faced the new graduate school at the start. Firstly, in more than half of the ten departments, the program was new. In these cases there was no research environment in mathematics education and no earlier experience of graduate students in the area to build on. Secondly, researchers (mathematicians) with little or no experience of supervision in mathematics education took on responsibility for the supervision in cooperation with experts in the area, often external and in some cases from other countries. This organization called for new and fruitful collaboration around the supervision. No professors in mathematics education in Sweden were involved for the obvious reason that there were no such positions at the time. Finally, a critical and even hostile reaction from some influential mathematicians met the graduate school during the preparation process and in the first years (Brandell, 2009).

Three questions will be addressed in this chapter, questions that are closely related to the challenges mentioned above. Firstly, did the graduate school reach its immediate goal to offer the graduate students reasonable possibilities to complete their PhD studies? Secondly, in what way did the graduate school broaden or change the earlier general picture of research topics in Swedish research in didactics of mathematics? Thirdly, did the graduate school succeed in supporting the creation and development of research environments for mathematics education/didactics of mathematics within the mathematics departments?

In the following two sections I will give a brief survey of the organization and the content of the graduate school and the outcomes of it. I will then describe the research topics and the environments in some detail. In the final parts of the chapter I will address the three questions above and discuss possible future implications for the area of mathematics education research in Sweden from the same perspectives. The answers to inquiries sent out to all the doctoral students and to all the departments in 2006 and 2008 will be used in the final part.

Goals and Organization of the Graduate School

The funding body, RJ, laid down two goals for the graduate school, one directly concerning the students, the other concerning the research environments:

- To provide teacher education institutions and upper secondary schools with a larger number of mathematics educators and teachers with a PhD degree in mathematics with specialization in didactics of mathematics
- To develop didactics of mathematics as a research area in Sweden.

The first goal required the school to guarantee good conditions for the doctoral students in order for them to complete their PhD degree in due time and get an education that prepared them for relevant jobs. Positions such as senior teachers in the upper secondary school, as mathematics educators, or as university teachers and researchers were first-hand alternatives. Some PhDs could also be expected to find positions in the local, regional, or national school administration where their qualification as researcher was required. Thus the prospective labor market was a fairly broad one. Secondly, the five-year-long duration of the graduate school also aimed for a development of the research environments in all participating departments. This last goal implied, among other things, that senior staff, acting as supervisors in the graduate school, should raise their competence via their experience from the graduate school.

The graduate school worked during five years as a network with around 20 doctoral students within the school, 10–15 other doctoral students taking part in some of the programs, and roughly 35 supervisors. The network included researchers from other countries, invited to contribute to the courses and conferences or acting as assistant supervisors. The members of the steering group took part in the conferences and acted as mentors for the students and thus expanded the network further.

The ten departments that participated in the graduate school were spread all over the country, from Kristianstad University College in the south to Luleå University of Technology in the north and from Göteborg University in the west to Stockholm University in the east. Each student in the graduate school was enrolled as doctoral student at his or her home university. In a few cases a student was inscribed as doctoral student at another university, owing to the fact that his or her own either did not have the rights to award doctoral degrees—this is the case for university colleges in Sweden—or to the choice of the particular department. In this way there was a special cooperation within the graduate school between some pairs of participating departments.[1]

All of the doctoral students were employed by her or his university. In most cases they held a position as doctoral student, but in a few cases as junior lecturers with the same duties and assignments as a doctoral student. The reason was that these students were already employed when they applied to the graduate school. A position as a doctoral student implies part-time (20%) teaching duties besides the research studies. Thus the time for completion of a doctoral degree prolongs to five years, which is also the time limit for the employment. All students but one started their doctoral studies in August 2001. Thus the formal study time lasted until 2006. For some of the students (six out of 24) the formal time extended due to maternity leave between one and three years.

The graduate school funded the departments to an extent that covered the salary and overhead costs for each doctoral position as well as about a month's salary for one supervisor per doctoral student per year. Besides the main funding from RJ, the Swedish Research Council supported the graduate school through two substantial grants that made it possible to increase the number of positions from originally 15 to 21.

The main activities of the graduate school were courses and conferences. The participating departments organized the courses in turn in cooperation with the graduate school. The main funding emanated from the graduate school. The courses were all directly related to research in mathematics education and all or nearly all students of the graduate school took all the common courses. The students complemented their coursework with doctoral courses in mathematics at their own department and in other areas at their home universities. The steering group of the graduate school and the administration planned and organized conferences for students and supervisors together. During the first years, special conferences for supervisors were also arranged.

The home department organized the supervision with support from the graduate school in some cases. Each student had two or more supervisors. Most often a mathematician was the main supervisor. An expert in mathematics education research took on the main responsibility for supervising the thesis work. Assistant supervisors often held positions at other departments and in several cases at universities abroad.

Outcomes of the School

In total, 24 students were enrolled in the graduate school; 21 students started in 2001 and three in 2003. As of December 2009, 17 students had finished their program with an exam, most of them during 2006. Fourteen students have successfully defended a PhD thesis in mathematics with specialization in didactics of mathematics. One more student plans to present her PhD thesis early in 2010. Three students chose to finish their graduate education by presenting a licentiate thesis and being awarded a licentiate degree. Three quit the graduate studies rather early (during the first or second year), and two have left the school and turned to other research areas. One of these presented a doctoral thesis in sociology of education in 2008.

The quantitative outcomes of the school can be compared to the national statistics for research education. In the official statistics it is possible to find the proportion of doctoral students awarded the PhD degree within eight years after the year of enrollment. For the graduate school this proportion is 63% (15 out of 24). In the subjects of mathematics and pedagogy the corresponding numbers are 74% and 59%, respectively, for students enrolled in 2000 and examined through 2008, which is the latest available information.[2]

However, national graduate schools have been established in Sweden in order to increase the efficiency and quality of research education (Swedish National Agency for Higher Education, 2000). It is therefore of interest to compare the outcome of the graduate school in mathematics education with other graduate schools. In 2001, 16 national graduate schools were launched, funded within a large governmental program (Swedish National Agency for Higher Education, 2008a). Two of these belong to educational science, one in didactics of science and engineering, and the other in pedagogy. There is also one national school in mathematics and computing. These three schools have much in common with the graduate school in mathematics education, since they are working in the larger fields of education or mathematics. The corresponding proportions of PhD exams awarded in these three schools through the end of 2009 are expected to be higher, about 65–75%.[3] The conclusion is that the output from the graduate school in mathematics education in the form of doctoral exams is slightly higher than for the PhD programs in pedagogy and somewhat lower than for mathematics and for programs within graduate schools in mathematics and in educational sciences. It is important to realize that the numbers for the four graduate schools are fairly small. Hence the comparisons may be influenced by random individual factors.

The new doctors have presented and published their research in many contexts such as national and international research conferences and national competence development programs for teachers. The dissertations—both PhD theses and licentiate theses—have been published before the defense according to the Swedish rules. In most cases the PhD theses were in the form of a combination of published articles. Some of the authors, whose theses were monographs, have also published articles in international research journals after the PhD exam. A list of publications from 2004 to 2008 extends to 21 papers in international research journals by eleven authors (Brandell, 2009). This is a sign of good quality research, judged by the standards of the international community of researchers in mathematics education.

All the doctors and licentiates from the graduate school have relevant positions at universities, schools, or local or national school authorities, most of them at universities. Positions as senior lecturers at teacher education departments were more or less waiting to be filled with experts in didactics of mathematics. Six persons have sought-after positions with favorable conditions for research (between 50% and full-time research) during three or four years. In such positions there are excellent opportunities to reach a further step in one's research career, namely to become a docent, which is an academic title mainly based on research competence beyond the PhD exam and awarded by the faculty after a review process. This qualification is a crucial merit when applying for a permanent position as professor.

Research Object

The graduate school did not define common research projects, and there were no requests on the research topics other than belonging within "mathematics with specialization in didactics of mathematics." This area was conceptualized as being part of *mathematics in a broad sense* in the strategy document formulated by the steering group for the graduate school before the start of the school (Ledningsgruppen, 2000). Mathematics viewed in this way would include, for instance, applied mathematics, parts of theoretical computer science, and optimization theory, as well as parts of the history and philosophy of mathematics and of mathematics education. However, other parts of mathematics education ought to be placed outside of mathematics, even if viewed in such a broad sense.

The choice of the content and design of the projects was entirely up to each student and her or his supervisors. But the very definition of the area as a sub-domain or specialization within mathematics suggested that mathematics would be a strong ingredient in the projects. An expectation from the steering group was that most of the research projects would have a clear connection to mathematics as a subject, which would make mathematics visible in the thesis in one way or the other. Focus was also expected, but not required, to be on higher levels and older students, in principle on upper secondary and tertiary level. Advanced mathematics had only been touched upon occasionally in earlier Swedish doctoral theses (through 2000).

There are of course many ways to characterize research studies. I have chosen to focus on two aspects, the topic or object of study and the age group of students involved directly or indirectly. Thus, in this categorization I have neither considered methods nor theoretical perspectives.

Research Topics

What is the object of study in the research didactics of mathematics? According to Mogens Niss the field extends beyond narrow questions about learning and teaching mathematics.

> **Subject:** The didactics of mathematics, alias the science of mathematics education, is the scientific and scholarly field of research and development which aims at *identifying, characterising,* and *understanding* of phenomena and processes actually or potentially involved in the *teaching* and *learning* of *mathematics at any educational level.*
>
> . . .
>
> **Approaches:** In pursuing these tasks the didactics of mathematics addresses *all matters* that are pertinent to the teaching and learning of mathematics, irrespective of which scientific, psychological, ideological, ethical, political, social, societal or other spheres this may involve. (Niss, 1999, p. 5)

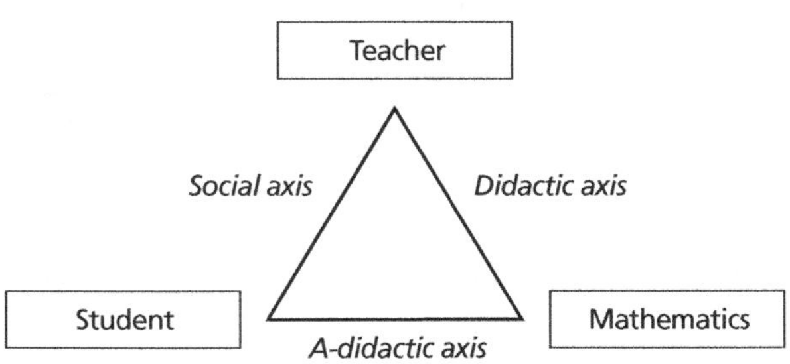

Figure 26.1 Didactical triangle.

One way of characterizing the topic of a research study is to locate it relative to the didactical triangle. The well-known didactical triangle (Figure 26.1), as introduced by Guy Brousseau, has the student, the teacher, and the subject (mathematics) in each of the corners (Brousseau, 1997).

The sides of the triangle are defined as the didactic axis, the a-didactic axis, and the social axis respectively. Many of the fundamental processes are enacted along these axes. The didactical triangle may be thought of as describing didactical situations delimited in time and place. But learning and teaching take place within larger contexts and are influenced by curriculum, steering documents, schools, families and local communities; assessment programs at local, national and international level; political steering systems of education, and so on. Historical perspectives are relevant across the field.

However, most of the dissertations of the graduate school may be located within the triangle with their main focus on one of the axes—here I allow myself to use the concept of the didactical triangle quite freely. There may be more than just one of the axes involved, so the categorization is fairly rude. Those studies that have the main focus on students' relation to mathematics are placed along the *a-didactic* axis. In these dissertations, the researcher treats topics such as students' perceptions of mathematical concepts; students' learning of certain areas of mathematics; long-time development of students' apprehension of mathematics; and problem solving, modeling, and communication within groups of students. The teacher may also be involved in such studies, for instance, as one part in a dialogue steered by the student(s) as long as the focus of the research remains on students' relation with mathematics, not on the teacher's choices, attitudes, or knowledge. A subgroup here refers to the use of technology or textbooks as a mediator for learning mathematics. Questions of communication with the teacher and working methods in the classroom as well as students' apprehension of mathematics learning are associated with the *social* axis. The *didactic* axis is not represented in Table 26.1, but is present in some theses as one aspect but not the main focus. In such cases the teachers' didactical transposition of the subject is studied. Finally, a few studies are located outside the didactical triangle.

TABLE 26.1 Topic of Research Studies in 14 PhD Theses in Didactics of Mathematics within the Graduate School until December 2009

Didactic triangle			System level—outside the didactic triangle
A-didactic axis		Social axis	
No focus on artefacts	Focus on artefacts		
6	2	1	5

Note: Each thesis is attributed to one category according to the main topic.

Obviously a great deal of the research within the graduate school is related to the a-didactic axis. Among the dissertations categorized along this axis, several focus on one specific mathematical object such as limits (Juter, 2006), functions (Hansson, 2006), proofs (Hemmi, 2006), and probability (Nilsson, 2006). Others concern problem solving (Pettersson, 2008; Ryve, 2006), textbooks (Johansson, 2006), and reading comprehension of mathematical texts (Österholm, 2006). Stadler's study about the transition from secondary to tertiary level is associated with the social axis (Stadler, 2009). The theses with their main focus located outside the didactical triangle concern national assessment (Boesen, 2006), beliefs and gender aspects (Sumpter, 2009), a historical study of mathematics education (Prytz, 2007), and studies of the historical development of certain mathematical concepts (Bråting, 2009; Pejlare, 2007). The three latter studies represent a renewed interest in the history of mathematics education in Sweden.

All students of the graduate school studied mathematics to a relatively high level both as undergraduates and graduates. Their strong interest in mathematics may be one explanation for the attraction to the "mathematics corner" of the didactics triangle. Another reason may be the opportunities offered within the environment of a mathematical department and the access to supervisors who were specialists in mathematics.

According to Bergsten (2002) and Björkqvist (2003), in overviews of Swedish research in mathematics education up until 2001, earlier and ongoing projects showed few studies with a starting point in a specific topic in mathematics or a certain procedure or method in mathematical work. Björkqvist judges this situation as a weakness. The graduate school has helped to improve the situation.

In an evaluation of ten licentiate theses from the graduate school in 2005, Skovsmose (2006) raises the question of the place of history in the graduate school. His view, from three licentiate theses, is that these are studies in history of mathematics rather than in mathematics education. On the other hand, he also points to possibilities to connect these historical studies to central issues in mathematics education research. One of the doctoral theses has such a clear connection by its very topic, history of mathematics education.

Age Groups

Many studies in didactics of mathematics are connected with a specific educational level. This is the case for all dissertations of the graduate school. In Table 26.2, the research studies have been categorized according to four groups of students: lower secondary level corresponds to students aged from 13 to 15 years; upper secondary level corresponds to school the age group 16 to 18 years; and university students are divided in two groups: student teachers and all other students.

According to the table, the majority of the 14 PhD studies in the graduate school treat problems related to university students, among them student teachers. According to Bergsten (2007), in an overview of nearly all doctoral theses presented in Sweden from 1919 to 2006, the great majority concerned students in school year K–12. Only ten theses outside the graduate school had their main focus on higher education in mathematics, teacher education, or teachers. The contrast becomes even more distinct when the theses presented during the decade 1990–2000 immediately preceding the creation of the graduate school are analyzed. This decade was an important period for research in mathematics education in Sweden, with a sharp increase in the production of new PhDs in the area. According to a list of all doctoral theses in mathematics education, 18 theses were presented during 1990–2000 in Sweden or at universities outside Sweden by researchers with positions at Swedish universities (National Centre for Mathematics Education). Only two of these 18 theses address questions about university students' mathematics education, compared to nine out of 14 for the graduate school.

Several aspects of the graduate school have certainly influenced the doctoral students' choice of educational level to address in their research studies. Firstly, the doctoral students' home departments were mathematics departments, involved in teaching mathematics students and, among them, prospective teachers for secondary level. As a consequence, students and courses at secondary and university level have been easy to access. Secondly, the graduate school students had the necessary mathematical preparation with a strong background at the start and mathematics courses at the doctoral level within the graduate school. Thirdly, some of them had earlier teaching experience from teacher education or other university programs. Finally, their mathematics supervisors may have felt more comfortable with a topic close to their own teaching experience.

TABLE 26.2 Student level Considered in 14 PhD Theses in Didactics of Mathematics Presented within the Graduate School 2006–2009

Lower secondary level (School year 7–9)	3
Upper secondary level (School year 10–12)	3
Teacher education	3
University other than teacher education	6

Note: In one case two age groups are equally involved (upper secondary and university level), making a total of 15 entries in the table.

Research Environments

A Changed Position for Didactics of Mathematics

Umeå University and Luleå University of Technology were the first Swedish universities to launch independent doctoral programs in didactics of mathematics in 1995, based only in the mathematics department. Didactics of mathematics formally became a specialization within mathematics. A year later, the first doctoral dissertation in the area was presented as a thesis in mathematics in Luleå (Dunkels, 1996). A close cooperation was developed with teacher education and pedagogy from the very beginning.

Inspired by the invitation from the graduate school, several similar doctoral programs were launched during 2001. Ten departments at ten universities, including the ones mentioned above, created a working environment for the doctoral students. Doctors and licentiates have been examined within the graduate school from eight doctoral programs based in mathematics departments and in most cases belonging to a science faculty.

There were several reasons why the universities and RJ wished to develop the area in these new environments presented during the discussion that led to the decision of a graduate school (Ledningsgruppen, 2000; NFR, 1999). First of all, the initiative was seen as a complement to the doctoral programs within pedagogy and subject didactics already existing. The hope was to attract more resources to the area and new groups of students, not to compete. The main reason was the reported problematic situation for mathematics education at school and universities. Upper secondary level and the first year of university studies were much in focus. The transition from secondary to tertiary level caused serious problems to many students. The earlier research addressing higher levels was judged to be far too rare. A close cooperation with mathematicians was expected to increase the interest and add to the quality of the research when more advanced mathematics was the topic. A more immediate but serious concern was the lack of competent academic staff for teacher education. The requirement is that all university teachers should have a PhD and be prepared to do research and supervision within a university position. This is still far from the reality in teacher education, where a majority of the teachers do not have a PhD (Swedish National Agency for Higher Education, 2008b). In many cases no qualified applicants were found when senior lecturers in mathematics education were sought. Finally, some mathematicians showed a strong interest to enter the area as researchers, an interest among individuals that was crucial for the positive outcome of the process. These reasons represent the most important arguments put forward in the discussion documents, even if the list is not complete.

According to Leder, Brandell, and Grevholm (2004), who discuss various types of motives for entering a new interdisciplinary field, different reasons were decisive for joining the program in different environments.

The Graduate School—Incentive or Sole Support?

At the closure of the graduate school in September 2006 and again two years later, a survey was sent to all former students and another to all environments of the graduate school. The ten environments/departments were represented by the leader of the research group in mathematics education during the period of the graduate school, who answered the survey as the representative of the environment.

All research leaders acknowledge the importance of the graduate school for the creation and development of the graduate program and the research environment at the department. They also wish for a continuation of the graduate school and for new doctoral students in the area. However, the conditions are very different at the ten departments.

The research environments of the graduate school may be divided into three groups: departments and faculties with programs already in place before applying to the graduate school and which have developed during the graduate school and after its closure; departments that developed their program in connection with the start of the school and have developed the environments further and still have activities two years after the closure; and finally, the third group of departments for which the graduate school represented the peak and which show little or no activity two years after the school finished. All but one environment can be placed in one of the categories based on the surveys in 2006 and 2008.

As mentioned earlier, Luleå University of Technology and Umeå University belong to the first group. Each has a professor in the field, other senior researchers are employed, and doctoral students and external money keep coming in. Umeå is the largest environment. Three universities clearly fall in the next group: Kristianstad University College, Linköping University, and Växjö University—the two latter also with a professor in the field. All have senior researchers, doctoral students, and viable research groups. In all five cases, the department in question has a large responsibility for teacher education and close cooperation with other departments involved in teacher education. In all cases the department is a mathematics department or an interdisciplinary department within which mathematics is a part.

The four largest universities all fall into the last group: Göteborg University, Royal Institute of Technology, Stockholm University, and Uppsala University. In the departments connected to the graduate school, the research area seems to be fading away. This fact may be related to the special difficulties of larger universities due to organizational matters. Traditions create more obstacles to structural changes. The cultural differences between departments of teacher education and traditional disciplinary departments are significant and so is the physical and organizational distance. One contributory factor may well be the doubts about the value of research in mathematics education demonstrated by some mathematicians during the early period of the graduate school. The criticism has been less apparent during the last years, but has not disappeared from the mathematical community.

It is not possible to draw too far-reaching conclusions about the future from the situation in 2008, but it is very likely that departments with low activity and no senior staff in 2008 will have great difficulties building up an environment in the near or not-so-near future. It is also expected that those departments that have a working environment with senior researchers and doctoral students will be able to develop further in the future, hopefully with the help of expanding resources for research connected to teacher education on the national level. It seems as if the long-term goal of the graduate school to build up and/or further develop research environments has been reached for five or six of the ten departments involved, but not for all.

CONCLUDING DISCUSSION

The impact of the graduate school on the research area of mathematics education in Sweden is substantial. New researchers have completed their exams in sufficient numbers that will have a decisive influence on the access of academic staff during a long time. The new PhDs are considerably younger than is otherwise common among new PhDs in educational sciences. They have a widespread network, both nationally through the graduate school and internationally through courses, conferences, and publications. Swedish research in the field of mathematics education is becoming more visible internationally through the new PhDs. Research about older students including student teachers and more advanced mathematics has been launched on a much larger scale than earlier in Sweden.

According to the survey of former students in 2008, they found the network that they built via the graduate school especially positive. For those who continue in the field, the network will be of great value in the future.

The graduate school has been reasonably successful when it comes to its direct results. Quantitatively, the comparison with other research education shows comparable or somewhat lower results in examination as other similar graduate schools and areas. Qualitatively, the many positions as researchers held by doctors from the graduate school, the many international publications, and the relevant positions now held by the doctors and licentiates, mostly in academia, indicate a high level of quality in the preparation of the students for research and professional work.

Several mathematics departments have been drawn into the area. Senior staff got more opportunity and time to work in the area through supervision and cooperation with graduate students in the graduate school. On the other hand, several of the participating departments did not continue after the closure of the graduate school. This may be considered a failure of the graduate school. It is quite serious, since these departments represent some of the largest Swedish universities with major influence over research policy and resources.

The vision of the graduate school was to build up research through the introduction of a graduate program in many departments that had no environment for the research area. The vision was a daring one. Normally, the development would be the other way around: first build up research in an area with senior researchers, and then start a graduate program and involve students. Against this background, the failure at some departments might have been anticipated and the good result for the other departments may instead be seen as a great success. Björkqvist (2003), in his analysis of the current state of Swedish research in mathematics education in 2001, stresses the importance of connecting to a tradition while expanding a research area into new environments:

> From what I have brought forward, it must be clear that I do not fully accept the idea of a build-up stage that implies starting from nothing. Scientific knowledge always builds on previous knowledge, and part of the academic culture is acknowledging the wisdom that we use in our own search. Swedish research in mathematics education needs to do that, too. (p. 14)

As positive as the cooperation with mathematicians is when it succeeds as in the graduate school in many cases, the tendency in some places to isolate from other research of importance for the field, such as didactics of science and technology, psychology, pedagogy, cognitive science, sociology, history, and other areas is just as risky. There is not just one place for research in mathematics education in relation to other areas. This calls for cooperation between researchers interested in the area but with background in different subjects and organized within different faculties and departments.

NOTES

1. Kristianstad University College had their students enrolled as doctoral students at Luleå Unversity of Technology. Mälardalen University College and the Royal Institute of Technology had their students enrolled at Stockholm University.
2. Personal communication, Andreas Frodell, Statistics Sweden, November 2009.
3. Personal communication, Ulf Lindström, Umeå and Helge Strömdahl, Linköping, September 2009. For the graduate school in mathematics and computing the sources are the homepage www.math.uu.se/fmb/ retrieved September 4, 2009 and the Swedish National Agency for Higher Education (2008a).

REFERENCES

Bergsten, C. (1990). *Matematisk operativitet: en analys av relationen mellan form och innehåll i skolmatematik* [Mathematical operativity: an analysis of the relation between form and content in school mathematics]. Doctoral thesis. Linköping: Department of Education and Psychology, Linköping University.

Bergsten, C. (1996). Is there a royal road to math ed? In J. Lithner & H. Wallin (Eds.), *Preparation of Researchers in Mathematics Education* (pp. 1–9). Research Reports in Mathematics Education, No 1, Umeå Universitet.

Bergsten, C. (2002). Faces of Swedish research in mathematics education. In C. Bergsten, G. Dahland, & B. Grevholm (Eds.), *Research and action in the mathematics classroom. Proceedings of MADIF 2* (pp. 21–36). Linköping, Sweden: SMDF.

Bergsten, C. (2007). Matematikdidaktikens vägar [The paths of mathematics education]. *SMDF Medlemsblad, 13,* 5–19.

Bergqvist, T. (2001). *To explore and verify in mathematics.* Doctoral thesis. Umeå: Department of Mathematics, Umeå University.

Björkqvist, O. (2003). Mathematics education in Sweden: A review of research and development work. In C. Bergsten & B. Grevholm (Eds.), *Challenges in mathematics education, Proceedings of MADIF 3* (pp 7–14). Linköping, Sweden: SMDF.

Boesen, J. (2006). *Assessing mathematical creativity. Comparing national and teacher-made tests, explaining differences and examining impact.* Doctoral thesis. Umeå: Department of Mathematics and Mathematical Statistics, Umeå University.

Brandell, G. (2009). *Forskarskola i matematik med ämnesdidaktisk inriktning 2001–2006* [The graduate school in mathematics with specialisation in didactics of mathematics 2001–2006] (No. 2009:1). Umeå: Department of Mathematics and Mathematical statistics, Umeå University.

Brousseau, G. (1997). *Theory of didactical situations in mathematics.* Dordrecht, The Netherlands: Kluwer Academic Publishers.

Bråting, K. (2009). *Studies in the conceptual development of mathematical analysis.* Dissertation. Uppsala University.

Dunkels, A. (1996). *Contributions to mathematical knowledge and its acquisition.* Doctoral thesis. Luleå: Department of Mathematics, Luleå University of Technology.

Grevholm, B. (2008). Nordic doctoral programs in mathematics education. In R. E. Reys & J. A. Dosey (Eds.), *U.S. Doctorates in mathematics education: Developing stewards of the discipline, Issues in Mathematics Education, Vol. 15* (pp. 189–194). CBMS. Providence, R.I.: American Mathematical Society, in cooperation with Mathematical Association of America.

Hansson, Ö. (2006). *Studying the views of preservice teachers on the concept of function.* Doctoral thesis. Luleå: Department of Mathematics, Luleå University of Technology.

Hedrén, R. (1990). *Logoprogrammering på mellanstadiet: en studie av fördelar och nackdelar av användningen av Logo i matematikundervisningen under årskurserna 5 och 6 i grundskolan.* [Programming in Logo in primary education: a study of the advantages and disadvantages of the use of Logo in mathematics teaching in school year 5 and 6]. Doctoral thesis. Linköping: Department of Education and Psychology, Linköping University.

Hemmi, K. (2006). *Approaching proof in a community of mathematical practice.* Doctoral thesis. Stockholm: Department of Mathematics, Stockholm University.

Swedish National Agency for Higher Education. (2000). *Forskarskolor, ett regeringsuppdrag* [Graduate schools, a commission from the government] (No. 2000:2 R). Stockholm: Högskoleverket.

Swedish National Agency for Higher Education. (2008a). *Utvärdering av 16 nationella forskarskolor* [Evaluation of 16 national graduate schools] (No. 2008:16 R). Stockholm: Högskoleverket.

Swedish National Agency for Higher Education. (2008b). *Uppföljande utvärdering av lärarutbildningen* [A follow-up evaluation of teacher education] (No2008:8 R). Stockholm: Högskoleverket.

Johansson, M. (2006). *Teaching mathematics with textbooks—A classroom and curricular perspective.* Doctoral thesis. Luleå: Department of Mathematics, Luleå University of Technology.

Juter, K. (2006). *Limits of functions—University students' concept development.* Doctoral thesis. Luleå: Department of Mathematics, Luleå University of Technology.

Leder, G., Brandell, G., & Grevholm, B. (2004). The Swedish graduate School in mathematics education: Conception, birth and development of a new doctoral programme. *Nordic Studies in Mathematics Education, 9*(2), 165–182.

Ledningsgruppen. (2000). Forskarskola i matematik med ämnesdidaktisk inriktning [Graduate school in Mathematics with specialization in didactics of mathematics]. Unpublished report. The Bank of Sweden Tercentenary Foundation.

Marton, F. (Ed.). (1986). *Fackdidaktik* [Subject didactics]. Lund, Sweden: Studentlitteratur.

National Centre for Mathematics Education. (2009). Doktorsavhandlingar (Doctoral dissertations). Retrieved September 1, 2009, from http://ncm.gu.se/node/171

NFR. (1999). *Matematikkommitténs positionsdokument* [Position document of the mathematics committee]. Stockholm: Naturvetenskapliga forskingsrådet.

Nilsson, P. (2006). *Exploring probabilistic reasoning. A study of how students contextualise compound chance encounters in explorative settings.* Doctoral thesis. Växjö: Växjö University Press.

Niss, M. (1999). Aspects of the nature and state of research in mathematics education. *Educational Studies of Mathematics, 40*(1), 1–24.

Österholm, M. (2006). *Kognitiva och metakognitiva perspektiv på läsförståelse inom matematik* (Cognitive and metacognitive perspectives on reading understanding in mathematics). Doctoral thesis. Linköping: Department of Mathematics, Linköping University.

Pejlare, J. (2007). *On axioms and images in the history of mathematics.* Doctoral thesis. Uppsala: Department of Mathematics, Uppsala University.

Pettersson, K. (2008). *Algoritmiska, intuitiva och formella aspekter av matematiken i dynamiskt samspel. En studie av hur studenter nyttjar sina begreppsuppfattningar inom matematisk analys.* [Algoritmic, intuitive and formal aspects of mathematics in a dynamic interplay. A study of how students use their conceptions in mathematical analysis]. Doctoral thesis. Göteborg: Department of Mathematical Sciences, Göteborg University.

Prytz, J. (2007). *Speaking of geometry. A study of geometry textbooks and literature on geometry instruction for elementary levels in Sweden, 1905–1962, with a special focus on professional debates.* Doctoral thesis. Uppsala: Department of Mathematics, Uppsala University.

Ryve, A. (2006). *Approaching mathematical discourse. Two analytical frameworks and their relation to problem solving interactions.* Doctoral thesis. Västerås: Department of Mathematics and Physics, Mälardalen University.

Skovsmose, O. (2006). *Graduate school in mathematics education* (No. 2006:1). Umeå: Department of Mathematics and Mathematical Statistics, Umeå University.

Stadler, E. (2009). *Stadieövergången mellan gymnasiet och universitetet: matematik och lärande ur ett studerandeperspektiv* [The transition from secondary school to university: mathematics education as seen from a student perspective]. Doctoral thesis. Acta Wexionensia No 195/2009. Växjö: Växjö University.

Sumpter, L. (2009). *On aspects of mathematical reasoning: Affect and gender.* Doctoral thesis. Umeå: Department of Mathematics and Mathematical Statistics, Umeå University.

27

CHILDREN'S KNOWLEDGE BUILDING IN ARITHMETIC IN THE COMPULSORY SCHOOL AND IN THE SPECIAL SCHOOL FOR INTELLECTUALLY DISABLED IN SWEDEN

Göta Eriksson

STOCKHOLM UNIVERSITY

Abstract: Investigations have been conducted in a compulsory school and in a special school for intellectually disabled students in Sweden to examine children's conceptual development in early arithmetic. The studies were based on the epistemology of radical constructivism and the methodology of multiple clinical interviews. A theoretical model elucidates the behavior patterns and the corresponding mental structures underlying them. The individual interviews were video recorded. In this chapter, results are shown, and various behavior patterns that emerge when the participants solve tasks in their own ways are exemplified, conceptually analyzed, and compared. The activated behavior patterns, which are responses to well-adapted contexts presented by the researcher, are compatible between children at the same conceptual level irrespective of age or type of school. The study gives substantial impetus to change the education in early arithmetic to harmonize with the children's own ways of operating.

Keywords: multiple clinical interviews, counting schemes, conceptual progression, early arithmetic, remedial teaching, intellectually disabled

INTRODUCTION

Changing the teaching of mathematics to better suit those who should learn this demanding subject is a worldwide issue. A lot of research has already been done to meet the demands, but still the need is acute (De Lange, 2001). Too many pupils experience failure too early in their years of schooling, and this fact has remained constant over the decades. Research shows that the gaps in children's mathematical knowledge increase up through the school years (Engström & Magne, 2004; Wright, 2000); evidently this is because children do not later learn what they fail to learn in the early school years. Thus, as the content in teaching increases in complexity year after year, these children lag behind more and more. Moreover, as research shows, the content of what is taught is not related to children's varied prior knowledge (Bentley, 2003), so the gap between what the child needs and what the curriculum demands in teaching is expanding.

The First Sourcebook on Nordic Research in Mathematics Education, pages 397–409

A critical point in mathematics education is what happens in teaching from the very beginning, that is, when children enter school. Children do not begin in the educational system as blank slates. To the contrary, children all over the world have been shown to have various degrees of developed knowledge in the area of mathematics (Ginsburg, Choi, Lopez, Netley, & Chao-Yan, 1997). This knowledge usually is informal, in contrast to the formal knowledge taught in school (Hughes, 2001), and it is not limited to the preschool level (Nunes Carraher, Carraher, & Schliemann, 1985). But generally, teachers do not seriously take children's self-constructed knowledge and their natural ways of learning in consideration in teaching arithmetic. Consequently, the children's right to get an education in relation to their prior knowledge can seldom be fulfilled. When teaching is not adapted to the children's prior knowledge in the early years, it becomes a problematic situation for many children, becoming even more critical later on (Ellemor-Collins & Wright, 2007).

Research describes how it is possible to interpret the children's own constructions in early arithmetic (Olive, 2001; Steffe, 1992a; Steffe, Cobb, & von Glasersfeld, 1988; Steffe, von Glasersfeld, Richards, & Cobb, 1983; Wright, 1989). Indeed a new subject field, "the arithmetic of the child" or "the child's arithmetic," has been constructed from longitudinal research by Leslie P. Steffe and collaborators. The children's early arithmetic is a logical as well as progressive construction of their own and "taken to be the results (or possible results) of operating in environments and, as such, essential in their mathematical progress, but these experiences cannot possibly be determined by the mathematical programs of the schools" (Steffe, 1992b, p. 97). Further, the arithmetic of the child is an adult construction based on children's innate learning processes (Steffe, 2004).

Investigations have been conducted with children in the compulsory school (CS) as well as in the special school for intellectually disabled (SID) in Sweden with the aim of examining the variation in children's conceptual development in arithmetic (Eriksson, 2008a,b). In this chapter the aim is to report some results and exemplify the various behavior patterns emerging when children solve tasks in their own ways. Based on conceptual analyses, comparisons are made between children's counting behavior in CS and in SID to examine essential differences related to types of schools and/or different ages, which can give implications for teaching in early arithmetic.

CONSIDERATIONS OF EPISTEMOLOGY

The study is based on the epistemology of radical constructivism (von Glasersfeld, 1995). Inspired by Piaget's well known and influential research, this theory views knowledge building as resulting from the cognitive system of self-organizing and self-regulation. In this sense, the children construct their own progressive arithmetic, that is, "the arithmetic of the child," through the apparatus of cognition. But the children's learning is dependent on four main factors: namely, maturation, social transmission, experience, and self-regulation (Steffe, 1996). These factors are interdependent; that is, none can be excluded. The following focuses on the process of cognition.

The interdependence of the abovementioned factors gives a crucial role to the cognitive process of self-regulation, as this process is viewed to result in learning. Learning as such is defined as modification of the available mental structures to a higher mental level (Steffe et al., 1988). Accordingly, the arithmetic of the child is a subject field in progression. The fundamental requirement for learning is the child's prior mental structure, but also crucial is the child's opportunities to use his or her mental structures to make experiences—that is, to interpret settings or tasks in accordance with his or her own cognitive capacity. This in turn results in an emerging activity of the child to solve the task in his or her own way. The solution is a consequence of the child's activated interpretative structure. Logically, if the child cannot interpret the task, he or she cannot manage it in his or her own way, either. This circumstance is important for education: In arithmetic, for example, what to teach must be determined in relation to the child's interpretative structures in order to bring forth his or her own pathways for solutions. When the arithmetic situation is not interpretable for the child, no solution is to be expected.

The behavior patterns involved in the children's solutions differ in accordance to the conceptual content of the task. This is due to the children's different assimilatory interpretative structures in "the arithmetic of the child" consisting of various pre-concepts or concepts. A pre-concept is defined as "any conceptual structure whose use requires sensory material of some type" (Steffe et al., 1988, p. 336), and a concept is "any conceptual structure that can be re-presented in the absence of perceptual material" (p. 335). The child's dynamic constructions of more and more complex concepts in arithmetic occur in a counting scheme. A scheme consists of three parts: namely, a situation to be interpreted, an activity accomplished in the child's own way, and a solution. Five different counting schemes have been distinguished in the child's early arithmetic (Eriksson, 2005; Olive, 2001; Steffe et. al., 1988; Wright, 1989). Thereafter follow schemes not included in early arithmetic. All of them have their own typical counting behavior related to the involved conceptual structures. Counting—that is, to coordinate number words to units of one kind or another with the aid of the number sequence—is one of the child's typical behaviors. The concepts in turn are not observable in the same explicit way as external behaviors. Thus, they have to be constructed by the observer in such a way that they show a logical dynamic sequence; that is, a concept typical for one scheme constitutes a logical basis for the construction of the concept in the subsequent scheme. The conceptual advancement becomes a consequence when the child has been active in using a counting scheme for a period, which enables her or him to take the result in the scheme as input in the next scheme.

The complexity of the involved concepts in the assimilatory interpretative structures of the children determines the complexity of the conceptual content in the problems. Thus, children at different levels of cognitive development demand varying content in teaching from a conceptual perspective. Children with simple arithmetical concepts need simpler problems, while those with more complex concepts demand more complicated ones for their learning to proceed. Logically, it becomes essential to learn the involved concepts in the children's diverse assimilatory structures and to understand the variation of the conceptual content in varying problems. This is essential, partly to assess the child's prior knowledge and partly to determine what to teach considering further development (Eriksson, 2004). Taken together, to learn "the arithmetic *of* the child" and to elaborate "the arithmetic *for* the child," (i.e., what to teach), requires conceptual analyses (Steffe & Kieren, 1994) of children's behavior patterns in their counting schemes as well as the content in teaching. In Table 27.1, the five steps in the conceptual progression in "the arithmetic of the child" earlier elaborated in research is presented (Eriksson, 2005; Olive, 2001; Steffe, 2004; Wright, 1989). The sixth step is also mentioned to give a hint of further progression.

According to Table 27.1, children's most basic counting scheme in the arithmetic of the child is the perceptual one. At this stage their counting activities are restricted to perceptual objects. Thus, when the

TABLE 27.1 A Model of the Conceptual Progression in the Arithmetic of the Child

Stage	Counting scheme	Counting behavior	Unit item	Result of the scheme
1	The perceptual counting scheme	Coordinate number words with objects	Perceptual unit item	Number as a perceptual collection
2	The figurative counting scheme	Coordinate number words with visualized objects Counts from "one" Keeping track	Figurative unit item	Numbers as a figurative collection
3	The initial number sequence	Counting on Counting backwards in take away tasks Counting on in unknown addend Regulate counting	Abstract unit item	Numbers as a numerical composite
4	The implicitly nested number sequence	Disembedding operations Counting up to Counting down to Counting off Coordinate and regulate composite units Segment and regulate composite units	Abstract unit item Composite unit during activity	Numbers as a composite Numbers as implicitly embedded
5	The explicitly nested number sequence	Part-whole operations Thinking strategies Transforming addition to subtraction Constructing algorithms Coordinate or segment composite units in regulation	Abstract unit item Composite unit The iterable unit of one	Numbers as iterated units of one Numbers as explicitly embedded
6	The generalized number sequence	Multiplication and division as inverse operations	The iterable composite unit	Numbers as iterated composite units

objects are hidden they cannot manage the situation. When the child can visualize hidden objects he/she is viewed to count figurative unit items at Stage 2. Children at these two stages are pre-numerical, and their results are experiential and tied to the here-and-now situation, and thus easily forgotten. The typical activity for pre-numerical children is their need to count from "one." At this stage, the children have to keep track of their counting to know when to stop. At Stage 2, this is managed with the aid of visualized number patterns or linear patterns. In situations where one addend is visible and the other hidden or both addends are hidden, there are limitations to the possible range of the hidden second addend.

At the three numerical stages (3–5), the unit items are abstract. Thus, children at Stage 3, the initial number sequence, have no need to visualize objects in counting. They just regulate their counting acts considering tasks of addition or unknown addends. In take-away tasks they count backwards; that is, they take away already counted counting acts. In all these activities they have to regulate their counting to know when to stop—they must think of how many are counted and how many are left to be counted. Thanks to this reflective behavior, the children from Stage 3 and further on are capable of explaining their solutions. The initial counting scheme produces numbers as a sequence of counting acts or a numerical composite. When this result can be taken for granted, the child has advanced to Stage 4, the implicitly nested number sequence, viewing numbers as implicitly interwoven in each other. The typical behavior at this stage is to count up to in subtraction tasks or in tasks with unknown subtrahends. They are also capable of coordinating or segmenting composite units, which is the precursor to multiplication and division. Just as at Stage 3, the children have to regulate their counting when they count one at a time or the number of constructed composite units.

When the children experience the relation between the parts and the whole using the part–whole operation, the explicitly nested number sequence at Stage 5 is realized. This operation, together with the constructed iterable unit of one, makes flexibility of the counting behavior possible in the four rules of arithmetic. For example, thinking strategies together with constructed number facts shorten the counting procedure as well as the constructions of algorithms when they count with tens and ones. At Stage 5 the children view a number as consisting of iterable units of one and are not yet capable of interpreting the place value system.

Thus, further development is required to fully understand this system from a conceptual perspective. At the next stage, the general number sequence, this advancement occurs when the iterable composite unit (i.e., the unit of units of units) is constructed (Steffe, 2001). Then children also realize the relation between multiplication and division. But by that time, the child has left the arithmetic of the child and proceeded to a more complex arithmetic. Further reorganization of the number sequence is needed to construct complex fractional schemes (Hackenberg & Tillema, 2009; Olive, 2006; Olive & Steffe, 2002).

METHODOLOGY

In the investigations the methodology of multiple clinical interviews was applied to assess the children's conceptual development. This is not a standardized methodology, and a similar teaching experiment methodology was used in Steffe and collaborators' pioneering model building of children's conceptual progression over extended periods (Steffe & Thompson, 2000). Multiple clinical interviews are composed of four main elements, namely the researcher as a teacher, pupils participating in teaching, a method to preserve the interviews, and a theoretical model of "the arithmetic of the child."

The role of the researcher is active and gives great freedom of action in the interviews, which also is true for the participating children. The essential element is to design different tasks for the children to bring forth the conceptual structures they have developed as well as the limit of their current competences. It is the researcher's responsibility to make the child's various counting behaviors visible, and this can only be realized when the child can interact with the tasks from a conceptual perspective using one of his or her interpretative assimilatory structures. When this occurs the child solves the tasks in his or her own way. From the emerging behavior pattern, the researcher can challenge the child further with a task that, for its solution, demands the assimilatory structures of the next stage. Thus, the procedure of assessment consists of tasks with increasing complexity up to the limit of the child's capacity. In the interviews, the researcher has to experience successful solutions of the child as well as tasks he or she cannot yet manage. As the researcher is aiming to establish interaction with the child at increased conceptual levels, the assessment can also cause learning. If so, it is hypothesized that the interactive communication occurs in the zone for potential construction that is sensitive to the child's learning (Norton & D'Ambrosio, 2008; Steffe & D'Ambrosio, 1995).

The researcher is also responsible for creating a permissive as well as a playful atmosphere for the child during the interviews. Thus, the tasks can consist of games where calculating is necessary. The researcher can give some degree of mild assistance; however, the form of assistance provided must be close to the researcher's modeling of the child's conceptual level, for the purpose of getting more information about the child's competence. The arithmetical situations are presented verbally in everyday language. At the higher stages this can be combined with figures too. As the researcher asks the child questions and also wants them to be answered, it becomes natural that the child is permitted to do so as well. Thus, the researcher also gives answers in a respectful way. This is important as the child's question is viewed to be a product of preceding cognitive activity. Indeed, these questions have a special value for the researcher as they give a splendid opportunity to establish interaction with the child, as the answer has to be formulated in such a way that it is interpretable to the child. Moreover, to understand the child's question and to give him or her an adequate answer, the researcher has to, metaphorically speaking, be within the child's symbolic world. This could be demanding, and if the researcher fails, the child can be confused, which can affect the outcome of the interview in a negative way. With this in mind, it becomes necessary to have more than one interview session for assessment. Another reason is to get a more trustworthy judgment: The results have to be viable.

The researcher's most important means of establishing interaction with the child is to design adapted tasks for him or her. Presenting prescribed tasks to challenge unknown children's arithmetical competence is not profitable, so the tasks are designed in a here-and-now situation. This is especially important in the first session of the interviews. In subsequent sessions, the circumstances are somewhat different. These can be arranged in accordance to what happened in the former session—for example, testing hypotheses that emerge in the analyses or collecting more data if the judgement of the counting behavior is uncertain. The researcher also has to be sensitive to the child's reactions in the sessions at all times and be ready to meet the unexpected and to make modifications of the task on the spot. The hypotheses of the child's knowledge continuously have to been made trustworthy during the interviews. Thanks to the preserved data, the material of the completed interviews with each child is scrutinized in retrospective analyses to strengthen the viability of each assessment as a whole. Thus, the conclusive assessment is based on the constructed data from the final revisions. Accordingly, the methodology demands acquaintance with the arithmetic of the child and how to design solvable tasks at different stages. To fulfill both of the claims, the model of the conceptual progression in Table 27.1 is an essential hypothetical principle. Moreover, teaching experience is a requirement.

All of the interviews are video recorded to make retrospective analyses possible and to make comparisons possible between different children's results or with a specific child's earlier achievement. The video recordings also are used to transcribe episodes. Further, on the basis of the recorded material, the research process could be documented with the aim of reporting both the results and the way they are created.

Description of the Study

The data used for the conceptual analyses of children's counting behaviors, which emerged in their spontaneous solutions of arithmetical tasks posed by the researcher, were generated in two different studies conducted in Sweden. Excerpts from these studies are used for the conceptual analyses.

A request was made to the head of a school in a smaller municipality about conducting research in the school. After a positive answer and an evident interest in the planned research, a request for participation was sent out to the parents of the beginners in the compulsory school (CS) and to parents of the children in two classes in the school for intellectually disabled children (SID). The children in CS and in SID attended the same school, so the two target groups were established from the affirmative answers to participation in the research. One of the groups consisted of 15 children (ten boys and five girls), out of a total of 26 beginners in CS (17 boys and nine girls). In Sweden, children begin their first school year when they become or have turned seven years old. The other group comprises of all children in two SID classes (eight boys and seven girls, 8–13 year old). Five of them have for various reasons been transferred from the compulsory school after 2–5½ years. They are in school year 3–6, while the other ten, who have belonged to the special school from the beginning, are in school year 2–7.

The aim of the study in CS was to follow the 15 children's conceptual progress after practicing school arithmetic and using arithmetic outside school during their early school years (Eriksson, 2008b). To assess their conceptual progress, this group participated in multiple clinical interviews in the beginning of their first, second, and third school year. In this study only data from the first interview conducted in autumn 2005 are taken. During the 2006–2007 school year, the fifteen children in SID also participated in interviews to assess their conceptual competences aimed at building a basis for a subsequent intervention study (Eriksson 2008c). The data used in this study are also from the SID children's first interview.

All of the interviews were carried out in a little room behind the children's ordinary classroom. The children participated in the interviews one at a time. Each interview for the CS group consists of up to five sessions; each lasted about 20 to 30 minutes. The sessions were dispersed with a few days between them. The eight children in one of the SID classes participated in up to five sessions in their interviews too (Eriksson, 2008a), while the seven children in the other SID class only experience one session each. The viability of the assessments was additionally tested in the follow-up interviews; for the CS group, this was the following year's interviews, and for the SID group it was the subsequent intervention study. All assessments showed viability.

The study also has ethical implications. This is especially important as children belonging to the special school and their parents are sensitive to assessment situations. All the children and their parents were thus fully informed of the purpose and process of the research. They had to consent to participation and could at any time withdraw from the investigation without any negative sanctions. One urgent issue is to give all of the children sufficient time and opportunity to demonstrate their knowledge in arithmetic. Moreover, to presuppose that children have a progressive as well as logical arithmetic of their own, being general for all of them and thus possible to build a model of, is indeed a bold starting point. From an ethical perspective it is less problematic to explain the children's knowledge from such a view than interpreting their symbolic understanding from a formal arithmetical basis. The hypothetical approaches, partly to the assessments and partly to the model, incorporate a more humble attitude to the research process.

The results have to be viewed as a co-construction of the researcher and the participating children. The hypothetical dimensions in the methodology of the research process have to be incorporated in the teaching process. Under these circumstances other researchers or readers have to take responsibility for their own interpretations of the current study while the researcher behind it is accountable for his/her own constructions.

RESULTS

Results from the first interviews with children in the compulsory school (CS) as well as in the special school for intellectually disabled (SID), with the aim to examine their conceptual development in arithmetic, are presented and compared. Examples of behavior patterns emerging when they solve problems in their own ways are given to illustrate the conceptual process and a conceptual analysis is applied.

Comparison of Stages for Children in CS and SID

In Table 27.2, the results from the assessment interviews with children at the compulsory school (CS) and in the school for the intellectually disabled (SID) are presented. In the interviews with the children in year 1 (age 6–7) in CS, it was possible to bring forth all of the various counting schemes in the children's arithmetic. Only three of the 15 children are assessed to be pre-numerical. Six of them have reached the first numerical stage, Stage 3, the initial number sequence. Further, two belong to Stage 4, and those who screen highest are four boys at Stage 5, the explicitly nested number sequence. It can be concluded that the beginner's knowledge in the arithmetic of the child differs to a great deal from a conceptual view.

The results from the interviews with the children at the special school for intellectually disabled (SID) are also shown in Table 27.2. Their ages vary from 8–13. In this target group there are ten children who have been registered in this type of school from the beginning. The lowest results are found in this group. One boy cannot yet be assigned to manage the first counting scheme in the model. He is also the youngest. Three are pre-numerical belonging to year 3, 6, and 7. But, the most surprisingly is that six are numerical: four at Stage 3 (two in year 5, and one each in year 6 and 7), and the remaining two are in year 5 and are assessed at Stage 4 and 5.

The other five children in SID have all begun in the compulsory school and after some years been moved to the school for intellectually disabled. All of these children are numerical. Three are at Stage

TABLE 27.2 Stage Affiliation in the Arithmetic of the Child for Pupils in CS and SID

	CS (year 1)		SID (year 2–7)	
Stage	**Boys**	**Girls**	**Boys**	**Girls**
1	—	—	1	1
2	1	2	1	—
3	4	2	4	3
4	1	1	—	1
5	4	—	1	2

3—one each in year 3, 4, and 5. Two girls, one each in year 4 and 6, are assessed at Stage 5. In this target group, many years at school do not imply a more advanced conceptual thinking than a few years. Comparing school forms, there are as many pre-numerical children in SID as in CS. Most of the numerical children in SID as well as in CS are assessed at Stage 3 (seven compared to six), while four in comparison to six have progressed to Stage 4 or 5.

Examples of the Pupils' Counting Behavior at Five Conceptual Levels

To illustrate the pupils' spontaneous activities, when they solve well-adapted problems presented by the researcher, examples are given at each stage. They are arranged in increasing complexity. For space consideration, only two examples from each stage are given, one similar task from each target group. The principles for selection are typical solutions for each stage or counting scheme, clear episodes, only one excerpt from the same pupil, and representation from both girls and boys. Episodes from the video recordings are transcribed and presented as "excerpt." The excerpts are numbered to make references possible later on. For the pupils in SID, the school year is mentioned, and for pupils in CS, the notation "beginner" is used. Thereafter the assumed name of the child is given. In the excerpt, "T" stands for the researcher as a teacher and the other capital letter for the initial letter in the child's assumed name.

Stage 1. *Excerpt 1.* David, SID, 7th year in school.

In this situation cases with a card inside each are used. Two numerals within the range 1–9 are printed on the case. Inside each case two groups of dots are applied correspondingly to the numerals.

T: [Puts a case with the numerals 3 and 4 on at the table.] In your math book there are numerals, like these [points at them]. Can you name them?
D: Three and four.
T: Mm. [Takes out the card from the case, points at the two groups of dots] three dots and four dots. How many are there together? [Puts the card in the case again]
D: [Utters number words; no other observable activity] three, four, five six [looks at the teacher].
T: [Takes out the card from the case, puts it on the table]
D: [Points and looks at each dot simultaneously with uttering number words] one, two, three, four, five, six, seven.

Excerpt 2. Arne, CS, beginner. In this excerpt we play with two dice and note down how many dots there are together.

A: [Throws two dice]
T: Six and four [points at each of the dice]. How many are there?
A: [Looks at the dice and counts sub-vocally. When he counts the six, he also points with a pencil at each dot.] Ten [looks at the teacher].
T: Yes.

In the situation with the numerals "3 and 4" and the corresponding dots hidden in excerpt 1, no counting scheme becomes activated for David. Thus, he only recites some number words. But in the task "three and four visible dots" the perceptual counting scheme emerges. This also happens in Arne's task "six and four visible dots." Both of the boys count dots, on a card and on dice. So doing, they coordinate number words to each of the dots, simultaneously looking and pointing at each. They use perceptual unit items. Arne, however, counts the four on the dice only by looking but has to point at each of the six dots. The behavior "to look at" and "to point at" the objects at Stage 1 is a way to keep track of the perceptual items. Both boys also succeed to extend their counting from one group to the next. David is assessed to Stage 1 as he fails to count when the objects were hidden. But, Arne managed to solve hidden tasks too. Thus, he has also constructed the next counting scheme, the figurative scheme at Stage 2. However, when the task is presented with visible objects, as in these excerpts, the perceptual counting scheme is sufficient for its solution.

Stage 2. *Excerpt 3.* Lars, SID, the 3rd year in school

T: [Puts five visible buttons and a closed little bag on the table]. There are four in the bag. How many are there together?
L: [Points and looks at each visible button simultaneously uttering number words] one, two, three, four, five [thereafter makes four points on the bag; the points form a domino-four pattern, simultaneously uttering number words] six, seven, eight, nine.
T: Check yourself.
L: [Opens the bag and takes out the four buttons and counts all of them, now totally visible].

Excerpt 4. Lena, CS, beginner

T: I want you to put down five buttons in the bag [holds up an open bag in front of Lena]

L: [Takes buttons from a pot, one at a time] one, two, three, four, five.
T: Now I close the bag [puts it on the table]. How many buttons are there? [points at the bag]
L: Five.
T: [Puts some visible buttons beside the bag] Five [points at the bag] and three buttons more [points at the visible buttons], how many buttons are there?
L: [After 18 seconds she begins to move her lips indicating counting sub-vocally] seven!
T: How did you work that out? You did something?
L: [Doesn't answer]
T: I think you count, but silently so I couldn't hear. I saw your lips. Can you do it once again and this time aloud so I can hear.
L: I count.
T: How do you count?
L: One, two, there are five [points on the bag] and [points on the bag again and whispers "five," then points at the visible buttons one at a time, simultaneously uttering number words] six, seven, eight [with emphasis; smiles at the teacher]
T: Yes, it makes eight. Now I know how you think.

In these two excerpts, some objects are hidden and some are visible. To count the visible group, the perceptual counting scheme is sufficient, but this scheme cannot be used to count hidden groups. Lars counts the visible buttons first, while Lena begins to sub-vocally count the hidden group. So doing, she coordinates number words to visualized buttons, using the figurative counting scheme and figurative unit items. When she arrives at "five," she stops and begins to count the visible group using the perceptual counting scheme. When the last visible button is coordinated with a number word, this is taken as the answer. As Lars begins to count the visible group first, he has to keep track of the second hidden addend in some way to know when all are counted in. For this reason he uses the elements in a domino-pattern for number four. He just visualized the pattern and coordinates all of its elements with one number word. The last number word uttered at the last element in the pattern is taken as the result. Lars also managed to solve tasks in Lena's way. When Lena uses Lars' behavior, she keeps track with a linear pattern for the hidden second addend.

The teacher asks Lena to explain her solution. However, pre-numerical children do not usually manage to reflect on recently performed activity. The teacher asks her to do it once again and to count aloud. This time she spontaneously uses overt behavior in her activity typical for the figurative counting scheme. Hence, Lena's and Lars' counting behaviors belong to Stage 2. They are also assessed at Stage 2.

Stage 3. *Excerpt 5.* Jane, SID, the 5th year in school

T: [Puts a little bag on the table] Before there were twenty-seven buttons in the bag, and while you kept your eyes closed I did something. So, now there are thirty-one there [points at the bag]. What did I do?
J: You took from the pot [points at the pot].
T: How many more did I add?
J: [Thinks silently; moves her fingers a little on her right hand] Four.
T: How did you work that out?
J: You count like this. [moves some fingers]
T: What did you say then?
J: Twenty-eight, twenty-nine, thirty, thirty-one [raises one finger at a time simultaneously with each uttered number words]
T: Well done, Jane!

Excerpt 6. Fabian, CS, beginner

T: Today we only pretend. Thirty-two [points at a cover with nothing under] and eight more [lifts a closed hand].
F: Thirty-two, thirty-three, thirty-four, thirty-five, thirty-six, thirty-seven, thirty-eight [looks at his raised fingers], thirty-nine [looks at his raised fingers again], thirty-ten [raises one finger at a time simultaneously with uttering number words, the first finger raised at thirty-three].
T: After thirty-nine comes?
F: Forty!
T: Very good!

In these two excerpts Jane "counts on" to find the missing addend, while Fabian "counts on" to include eight units to the first addend thirty-two. They both are careful to keep the first addend separate from the inclusion. Thus, Jane raises one finger first at "twenty-eight," and Fabian raises his first finger at "thirty-three." Jane knows in beforehand to stop when she arrives at thirty-one, while Fabian does not know the terminal point. On the other hand, Jane does not know how many she has to count while Fabian knows that. Jane constructs the unknown addend through monitoring and unitizing the counting acts "twenty-eight, twenty-nine, thirty, thirty-one," while Fabian regulates the inclusion of the eight units "thirty-three, thirty-four, thirty-five, thirty-six, thirty-seven, thirty-eight, thirty-nine, thirty-ten." His pauses after thirty-nine, indicating reflection of how many are already counted and how many are left to count. The behavior pattern in these two excerpts is typical for the first numerical counting scheme (i.e., the initial counting scheme at Stage 3). Both Jane and Fabian are assessed at this stage.

Stage 4. *Excerpt 7.* Ada, SID, the 5th year in school

> **T:** [Puts a strip on which some dots are applied on the table]. How many dots are there?
> **A:** [Counts the dots sub-vocally] sixteen.
> **T:** I will hide away some of them [partly inserts the strip in a case so only some of the dots are visible and puts it on the table again]. It was sixteen on the strip, but I have hidden some [points on the case]. How many have I hidden there [draws a finger along the case]?
> **A:** [Thinks silently, looking at the case and the visible dots. After 20 seconds the teacher interrupts].
> **T:** How many visible are there?
> **A.** [Counts aloud and points at each dots] one, two, three, four, five, six, seven. Seven and there were sixteen [thoughtfully scratches on her cheek].
> **T:** Which number lies there? [points on the case quite close to the seventh dot]
> **A:** [After six seconds] eight [with one finger of her left hand points to the case quite close to the seventh dot; after a little while straightens out her right thumb], nine [points to a place near the earlier with one finger of her left hand and straightens out her right index finger], ten [points to a place nearby the earlier with the same finger as before hand and straightens out her right middle finger], eleven [behaves as before and straightens out her right ring finger], twelve [behaves as before and straightens out her right little finger], thirteen [behaves as before and the teacher straightens out her left thumb at the side of Ada's five outstretched fingers; Ada looks at the teacher's thumb and continues], fourteen [points at the case; the teacher straightens out the index finger; Ada looks at the teacher's fingers and continues], fifteen [points at the case; the teacher straightens out the middle finger; Ada looks at the teacher's fingers and continues], sixteen [the teacher straightens out the ring finger; Ada looks at her own as well as the teachers fingers and answers with a smile] nine.
> **T:** Please check yourself!
> **A:** [Draws out the strip from the case and counts the earlier hidden dots] Nine. Yes!

Excerpt 8. Loke, CS, beginner

> **T:** There were fifteen [puts a little bag on the table], now there are only eleven, because I have taken some away. How many have I taken away?
> **L:** [After four seconds] How many were there? [points at the bag]
> **T:** Now there are eleven and from the beginning there were fifteen.
> **L:** Eleven, twelve, thirteen, fourteen, fifteen, [after eleven he straightens out one finger on the right hand after each uttered number word] four!
> **T:** Yes!

Both of the problems in these two excerpts consist of an unknown subtrahend. Ada is asked to count the minuend before she gets the task. In the case of Loke, he is not given such an opportunity. Further, Ada's subtrahend is of bigger value than Loke's. Both of them construct the subtrahend using the behavior pattern "counting up to." They view "seven" and "eleven," respectively, as well as the unknown subtrahend as embedded in the minuend "sixteen" and "fifteen." Thus, they keep "seven" and "eleven" apart, counting up to the minuend beginning from "eight" and "twelve" to find the subtrahend. They monitor the counting activity with the aid of their fingers, of which each represents one counted abstract unit, successively unitized to build up the unknown subtrahend. Ada is very concentrated when she points at the case with one finger on her left hand while she uses the fingers on her right hand to regulate the activity. When all fingers on her right hand are outstretched, she gets assistance from the teacher's fingers. She has control over the teacher's fingers as well as of her own the whole time.

To be sure, Loke asks the teacher of the number in the bag before he solves the problem. After the teacher's description of the task once again, he solved it accurately. As Ada thinks for a good while, the teacher gives a hint to her without being asked to. Whether her solution is successful owing to this, is hard to know. Despite this, the means to solve the problem is hers, and the aid of the teacher's fingers harmonized with that behavior. This aid was given with a view to facilitating her strategy. Despite some diversity in these two excerpts, they are equally good from a conceptual perspective. The strategy "counting up to" is brought forth in both excerpts, typical at Stage 4. Ada and Loke are also assessed at Stage 4.

Stage 5

In the next excerpt, bunches of five are used. Each bunch consists of five toothpicks tied up with an elastic band.

Excerpt 9. Mikael, SID, the 5th year in school

T: This is a five bunch [shows one bunch of five]
M: Like five in each.
T: Yes! Six bunches of five [puts six bunches in a box].
M: Okay, I think five is much better. Two are ten [straightens out two fingers on his left hand], three are fifteen [straightens out one more finger], twenty [straightens out one more finger], twenty-five [straightens out one more finger], thirty! [straightens out the sixth finger]
T: How many are there? [points to the box]
M: There are thirty.

Excerpt 10. Villy, CS, beginner

T: [Holds out some cases with strips with some groups of dots]. Pick one of these!
V: The biggest [puts the case at the table].
T: What's that?
V: [Draws out the strip a little bit] Six!
T: Is it?
V: Yes, [counts the dots] eight, oh I see!
T: There were eight! And there are many eights [draws the whole strip out from the case] Look! [Shows the whole strip]
V: Oh!
T: There are many. There are six groups of eight in each. [puts the strip in the case again] How many are there?
V: May I count on my fingers?
T: Yes!
V: One, two, three, four, five, six, seven, eight [simultaneously straightens out fingers one at a time], nine, ten, eleven, twelve, thirteen, fourteen, fifteen, sixteen [simultaneously straightens out fingers one at a time].
T: Then you have two groups of eight, haven't you?
V: Mm.
T: Then there are?
V: Sixteen! [Counts eight more, one at a time] twenty-four
T: Then you have?
V: Three! [Counts eight more, one at a time] thirty-two are four, [counts eight more again, one at a time] forty are five, [counts eight more, one at a time] forty-eight are six!
T: How much are six groups of eight?
V: Forty-eight.

Mikael and Villy have to coordinate six composite units of the value "five" and "eight," respectively, in these episodes. The composite units are represented as bunches of toothpicks and groups of dots on a strip, respectively. Thereafter ,all of the units are hidden in a box correspondingly. Mikael is satisfied with the value of his composite units because he knows the sequence for five. He regulates the number of counted composite units with the aid of outstretched fingers, one at a time, simultaneously counting on the number sequence for five at a time, uttering "five, ten, fifteen, twenty, twenty-five, thirty." In the case of Villy, he has more trouble as he does not know the number sequence for eight at a time. Thus, he counts in eight ones to construct each composite unit of eight. He uses his fingers, counts "one to eight" and thereafter from "nine to sixteen." The teacher hypothesizes that he uses double counting to regulate the number of eights thinking: eight is one, sixteen are two. Thus, the teacher wants a manifestation of his strategy and asks him "Then you have two groups of eight, haven't you?" Villy answers "Mm." The teacher also wants to know if he controls his way of operating, asking "Then there are?" He answers correctly and counts on to twenty-four. The teacher asks him once again "Then you have?" Villy directly answers "three." In the continuation he makes an explicit double counting without any further demands. Very likely, he has managed the situation with a silent regulation, too.

In Villy's episode the iterated unit of one is used when he unpacks each composite unit of eight into eight units of one with the intension to coordinate the six composite units of eight. Mikael never unpacks his composite units and knows the hallmarks beforehand, but he works on the same conceptual level as Villy (i.e., at Stage 5). Both of the boys are assessed at Stage 5.

DISCUSSION

The results show a wide range of conceptual development in early arithmetic for the 30 participants. Half were beginners in the compulsory school (CS) and half were different aged pupils in the school for the intellectually disabled (SID). Only a few of them were pre-numerical (three in each target group). The most striking result is that the majority of the pupils in SID were numerical. Considering the review of Baroody (1996), this is remarkable, as he states, "there is no evidence that children with mental retardation can devise a keeping-track or a counting-on process of their own" (p. 74). However, in this study, eleven pupils who all were numerical display those abilities—namely, to count on and regulate counting activities. Further, many years at school does not necessarily mean a higher conceptual level than those with only a few years of schooling. Despite attending seven years in school, they used their own constructed counting schemes to solve the problems. Only one boy in SID has not yet constructed the first scheme. All of the other fourteen children have their own ways of operating, owing to the researcher's well-adjusted tasks in accordance to the conceptual progression involved in the arithmetic of the child.

In the interviews with the pupils in the two types of school, there were no differences in bringing forth their conceptual competence. When the presented tasks harmonized with their constructed mental structures, they solved the problems spontaneously in their own ways, using the counting schemes compatible with Table 27.1. If the problem demanded a scheme other than what was available, it became unsolvable. The limit of the current ability then was reached. All of the pupils experienced satisfaction and a feeling of competence in the interviews, which is not obvious in the excerpts. Some of them also showed amazement over their own capabilities. This indicates that the feeling of competence seems to be tied in with the pupils' own pathways of operating.

In the ten excerpts, we can follow the conceptual progression in the arithmetic of the children (cf. Table 27.1) as well as the various developmental levels of the participants. In this way, the relations between available mental structures and solvable problems in the four rules in arithmetic are shown but not completely exemplified due to lack of space. At the two first stages, the adding schemes are pre-numerical (cf. excerpt 1–4) as the constructed unit items (perceptual and figurative unit items) are pre-numerical. The numerical adding scheme occurs first at Stage 3, when the items are abstract (cf. excerpt 5 and 6). At this stage, "take away" subtraction tasks also are solvable. This is not exemplified in the excerpts, but all of the participants at this stage show such capabilities. At Stage 4, the prerequisites for multiplication and division emerge when the pupils constructed composite units during their counting activity. The composite unit is also present when they count up to (or down to) in subtraction (cf. excerpt 7 and 8). On basis of these activities, explicit relations between numbers in part to whole operation become realized at Stage 5. With the available mental structures at this stage all of the four rules of arithmetic are managed in a more fluent and secure way than at Stage 4 (cf. excerpt 9 and 10). Operations with several figures are managed in self-constructed algorithms counting on or off, for example, with ten and ones. The procedure of counting is possible to curtail thanks to thinking strategies and number facts. The decimal system is not fully understandable until the generalized number sequence is constructed.

Comparisons between the beginners' solutions in CS and the pupils' of various ages in SID were made trough conceptual analyses. The main finding is the similarities of counting behavior at each stage disregarding the pupils' age or type of school. The dissimilarities are found between pupils' ways of operating at different conceptual levels.

CONCLUSION

This study gives a striking piece of evidence that the field of "child's arithmetic" is active for the beginners in the CS group as well as for the older pupils in the SID group. The investigation gives some essential implications for teaching early arithmetic and the necessity to change the process of teaching to better suit those who are at a disadvantage due to the gap between the child's operating scheme and the arithmetic teaching in school.

First of all, the participants' conceptual levels are not explicitly related either to age or type of school. Their natural ways of learning arithmetic (i.e., the child's arithmetic) appear to be similar in the sample. National research from other theoretical bases also reports how blind or deaf children solve problems quite comparable with the ways demonstrated in this study (Ahlberg, 2004; Foisack, 2003). Obviously, even pupils in special schools can have their own progressive ways of solving problems (Eriksson, 2008a). Not obvious in this study is that they can have capacity of doing conceptual advancement. too (Eriksson, 2008c).

Secondly, what the child is able to learn in arithmetic is dependent on his or her available conceptual structure and not his or her age. Logical assessment of the child's developmental readiness and teaching based on the child's condition demands knowledge of the child's arithmetic, which "is a construction of the observer, and it is a kind of mathematics that can be learned in the context of teaching children" (Steffe, 2004, p. 224). Further, to bring forth the child's current conceptual level demands posing tasks up to and next to that level. The tasks in turn are designed from and dependent on a particular model of a particular child's arithmetic, which is a teacher's construct.

Thirdly, the pupils' own constructed knowledge building in arithmetic is progressive. Logically, the content in teaching has to harmonize with and challenge the pupils' development from a conceptual perspective. Then the content has to harmonize with the pupils' own ways of operating. This is essential, as prior knowledge is a prerequisite for further learning. Consequently, differences in pupils' conceptual competences demand content adjusted to their own ways of operating. This in turn necessitates a good acquaintance with the pupils' current ways of operating and how to successively foster conceptual advancements.

Last but not least, we have to cope with a wide range of ability levels in teaching. This could hardly be successfully done giving the same content to an age group, for example, the same textbook to all pupils in a class. What children could learn by themselves ought to be facilitated by adequate teaching. This implies that the teacher has to be acquainted with the pupils' ways of operating and understand how to challenge their conceptual growth. Accordingly, the content in teaching has to be framed in relation to the children's prior knowledge and need of learning, and, when learning occurs, the content has to be adjusted to a higher conceptual level. This demands an individually adapted curriculum, an already legislated right in Sweden in the form of individual development plans for children in the compulsory school as well as in the special schools.

Considering previous national research in the area (e.g., Ahlberg, 1992; Neuman, 1987), this study displays an additional approach to research in early arithmetic. From the author's point of view, it is urgent that there be a good deal more research in the spirit of this study with the aim of teaching early arithmetic in harmony with children's conceptual prerequisites and need of further learning. This is especially important for remedial teaching and teaching in special schools. Regarding the latter group, the results at least raise the same hope as Baroody (1996) does in his study, namely that intellectually disabled children can also exhibit spontaneous and self-invented learning.

ACKNOWLEDGMENTS

The research on which this paper is based was funded by the Foundation of Clas Groschinsky's Memorial Fund, No. SF5 10 and SF7 95. The author thanks Professor Siv Fischbein, the Department of Special Education, Stockholm University, for invaluable feedback on the author's research. The author also wants to thank the children and their parents who made this research possible. All opinions expressed are solely those of the author.

REFERENCES

Ahlberg, A. (1992). Att möta matematiska problem. En belysning av barns lärande. [To meet mathematical problem. An illustration of children's learning]. Gothenburg: Acta Universitatis Gothoburgensis.

Ahlberg, A. (2004). Att förstå tal och lära sig räkna. Om hur barn som är blinda, barn som är döva och barn som ser och hör erfar talens innebörd. [To understand numbers and learn to count. About blind or deaf children and children who see and hear experience numbers.]. In A. Engström (Ed.), *Democracy and participation. A challenge for special needs education in mathematics. Proceedings of the 2nd Nordic Research Conference on Special Needs Education in Mathematics* (pp. 101–119). Report No. 7 from the Department of Education. Örebro: Örebro University.

Baroody, A. J. (1996). Self-invented addition strategies by children with mental retardation. *American Journal on Mental Retardation, 101*(1), 72–89.

Bentley, P-O. (2003). *Mathematics teachers and their teaching. A survey study.* Gothenburg: Acta Universitatis Gothoburgensis.

De Lange, J. (2001). The P in PME: Progress and problems. In M. van den Heuvel-Panhuizen (Ed.), *Proceedings of the 25th Conference of the International Group for the Psychology of Mathematics Education* (Vol.1, pp. 3–4). Utrecht: Utrecht University, The Netherlands.

Ellemor-Collins, D., & Wright, R. J. (2007). Assessing pupil knowledge of the sequential structure of number. *Educational and Child Psychology, 24*(2), 54–63.

Eriksson, G. (2004). *Assessing and influencing early arithmetic learning: A challenge to special education.* Paper presented at the 10th International Congress on Mathematical Education, July 4–11, 2004, Copenhagen, Denmark.

Eriksson, G. (2005). *Tidig aritmetisk kunskapsbildning. Ett radikalkonstruktivistiskt perspektiv.* [Early arithmetic knowledge building from the perspective of radical constructivism.] Doktorsavhandling [Doctoral dissertation]. Stockholm Institute of Education, Studies in Educational Sciences 76. Stockholm: HLS Förlag.

Eriksson, G. (2008a). Arithmetical thinking in children attending special schools for the intellectually disabled. *Journal of Mathematical Behavior, 27*(1), 1–10.

Eriksson, G. (2008b). Beginners' progress in early arithmetic in the Swedish compulsory school. *Journal of Mathematical Behavior, 27*(3), 177–187.

Eriksson, G. (2008c). Supervision of teachers based on adjusted arithmetic learning in special school. *Journal of Mathematical Behavior, 27*(4), 264–276.

Engström, E., & Magne, O. (2004). Middletown, 1977–1986–2002. In A. Engström (Ed.), *Democracy and participation. A challenge for special needs education in mathematics. Proceedings of the 2nd Nordic Research Conference on Special Needs Educa-*

tion in Mathematics (pp. 29–39). Reports No. 7 from the Department of Education. Örebro: Örebro University.

Foisack, E. (2003). *Döva barns begreppsutveckling i matematik.* [Deaf children's concept formation in mathematics.] Doktorsavhandling [Doctoral dissertation]. Malmö: Studies in Educational Science, No 7.

Ginsburg, H.P., Choi, E., Lopez, L.S., Netley, R., & Chao-Yan, C. (1997). Happy birthday to you: Early mathematics thinking in Asian, South American, and U.S. Children. In T. Nunes & P. Bryant (Eds.), *Learning and teaching mathematics: An international perspective* (pp. 163–207). East Sussex: Psychology Press.

Hackenberg, A. J., & Tillema, E. S. (2009). Students' whole number multiplicative concepts: A critical constructive resource for fraction composition schemes. *Journal of Mathematical Behavior, 28*(1), 1–18.

Hughes, M. (2001). Linking home and school mathematics. In M. van den Heuvel-Panhuizen (Ed.), *Proceedings of the 25th Conference of the International Group for the Psychology of Mathematics Education* (Vol.1, pp. 5–8). Utrecht: Utrecht University, The Netherlands.

Neuman, D. (1987). *The origin of arithmetic skills. A phenomenographic approach.* Acta Universitatis Gothoburgensis.

Norton, A., & D'Ambrosio, B. S. (2008). ZPC and ZPD: Zones of teaching and learning. *Journal for Research in Mathematics Education, 39*(3), 220–246.

Nunes Carraher, T., Carraher, D. W., & Schliemann, A. D. (1985). Mathematics in streets and in schools. *British Journal of Developmental Psychology, 3,* 21–29.

Olive, J. (2001). Children's number sequences: An explanation of Steffe's constructs and an extrapolation to rational numbers of arithmetic. *The Mathematics Educator, 11*(1), 4–9.

Olive, J. (2006). Making sense of instruction on fractions when a student lacks necessary fractional schemes: The case of Tim. *Journal of Mathematical Behavior, 25*(1), 18–45.

Olive, J., & Steffe, L.P. (2002). The construction of an iterative fractional scheme: The case of Joe. *Journal of Mathematical Behavior, 20,* 413–437.

Steffe, L. P. (1992a). Schemes of action and operation involving composite units. *Learning and Individual Differences, 4*(3), 259–309.

Steffe, L. P. (1992b). Learning stages in the construction of the number sequence. In J. Bideaud, C. Meljac, & J-P. Fischer (Eds.), *Pathways to number. Children's developing numerical abilities* (pp. 83–98). Hillsdale, NJ: Lawrence Erlbaum Associates.

Steffe, L. P. (1996). Social-cultural approaches in early childhood mathematics education: A discussion. In H. Mansfield, N. A. Pateman, & N. Bednarz (Eds.), *Mathematics for tomorrow's young children* (pp. 79–99). Dordrecht, The Netherlands: Kluwer Academic Publishers.

Steffe, L. P. (2001). A new hypothesis concerning children's fractional knowledge. *Journal of Mathematical Behavior, 20*(3), 267–307.

Steffe, L. P. (2004). PSSM from a constructivist perspective. In D. H. Clements, J. Sarama & A.-M. DiBiase (Eds.), *Engaging young children in mathematics. Standards for early childhood mathematics education* (pp. 221–252). Mahwah, NJ: Lawrence Erlbaum Associates.

Steffe, L. P., & D'Ambrosio, B. S. (1995). Toward a working model of constructivist teaching: A reaction to Simon. *Journal for Research in Mathematics Education, 26*(2), 146–159.

Steffe, L. P., & Kieren, T. (1994). Radical constructivism and mathematics education. *Journal for Research in Mathematics Education, 25*(6), 711–733.

Steffe, L. P., & Thompson, P. W. (2000). Teaching experiment methodology: Underlying principles and essential elements. In R. Lesh & A. E. Kelly (Eds.), *Research design in mathematics and science education* (pp. 267–309). Hillsdale, NJ: Erlbaum.

Steffe, L. P., Cobb, P., & von Glasersfeld, E. (1988). *Construction of arithmetical meanings and strategies.* New York: Springer-Verlag.

Steffe, L. P., von Glasersfeld, E., Richards, J., & Cobb, P. (1983). *Children's counting types. Philosophy, theory and application.* New York: Praeger Scientific.

von Glasersfeld, E. (1995). *Radical constructivism. A way of knowing and learning.* London: The Falmer Press.

Wright, R. J. (1989). *Numerical development in the kindergarten year: A teaching experiment.* Unpublished doctoral dissertation, University of Georgia.

Wright, R. J. (2000). Professional development in recovery education. In L. P. Steffe & P. W. Thompson (Eds.), *Radical constructivism in action. Building on the pioneering work of Ernst von Glasersfeld* (pp. 134–151). London: RoutledgeFalmer.

CHAPTER 28

A THEORETICAL FRAMEWORK FOR THE STUDY OF PROOF IN MATHEMATICS EDUCATION

Kirsti Hemmi

LINKÖPING UNIVERSITY

Abstract: The chapter presents a theoretical framework to study proof in mathematics education. The framework was developed and tested against the data during an extensive four-year study on students' encounters with proof at a university in Sweden and combines a socio-cultural perspective and Lave and Wenger's social practice theories with theories and main themes concerning the teaching and learning of proof. It offers a fresh locus to the issue and provides a coherent network about important aspects of proof and, therefore, suits well to be applied and developed in further studies.

Keywords: Proof, mathematical practice, artifact, transparency, transfer

INTRODUCTION

Proof is a current topic in mathematics education, and there is a renewed emphasis on proof in the ongoing revisions of school curricula in many countries (Mariotti, 2006). The research on proof in mathematics education has expanded during the last three decades, and, consequently, the 19th ICMI study in 2009 in Taiwan was devoted to proof and proving. Although the research in mathematics education during the last three decades has led to a rich variety of results, there have been some problems in bridging between research results obtained within different paradigms (e.g., Hoyles, 2001). Educational research has also been accused of not being cumulative (e.g., Wellington, 2000). The framework presented in this chapter is an attempt to bring together the main concerns from the earlier results in the field and create a structured network to study various aspects in the teaching and learning of proof.

The framework was developed and tested during a four-year study concerning students' encounters with mathematical proof at a mathematics department in Sweden (Hemmi, 2006).

The empirical data ranged from students' survey responses and transcripts of interviews with both mathematicians and students to field notes from observations of lectures as well as instructional material like textbooks and examinations. Both quantitative and qualitative methods were applied in the study.

In this chapter, I do not touch on the empirical results of my study. I have reported about them elsewhere (e.g., Brandell, Hemmi, & Thunberg, 2008; Hemmi, 2006, 2008). Instead, I make an attempt to illustrate solely the theoretical ideas developed within and for the study.

A lot of research on the teaching and learning of proof has been conducted within a cognitive re-

The First Sourcebook on Nordic Research in Mathematics Education, pages 411–418

search paradigm. The results from these studies shed light on the complexity of proof and the teaching and learning of proof from various points of view. These studies have often been in harmony with a constructivist view on learning and Piaget's theory about the developmental stages that an individual goes through. During the last decade, socio-cultural and sociological approaches to the study on proof have also appeared in the field.

The perspective of Lave and Wenger (1991) provides a bridge between cognitivist perspectives and sociological perspectives because their theory of social practice "emphasizes the relational interdependency of agent and world, activity, meaning, cognition, learning, and knowing" (p. 50). The world is seen to consist of objective forms and systems of activity, on the one hand, and agents' subjective and intersubjective understanding of them, on the other hand. These mutually constitute both the world and its experienced forms. Further, cognition and communication in, and with, the social world are situated in the historical development of ongoing activity.

Wenger's theory offered an appropriate network of concepts that helped me to examine and give structure to the object of my study, the practice of people exercising/sharing university mathematics at a mathematics department. I wanted to describe mathematicians' participation as well as students' participation in this practice with a special focus on students' access to proof. The notion of *community of practice of mathematics at a mathematics department* provided me with an appropriate level of analysis. The community of practice perspective allowed me to look at the mathematics department in a dynamic way, view the students as active participants in the practice, and define the mathematicians' activities in the practice also as learning. In my empirical study, I investigated the problems of drawing newcomers into the practice of proof. There is an affinity between my study and the one of Solomon (2006). She applies Wenger's theory about communities of practice in order to study the nature of the gap between expert practice and the student experience about proof at a mathematics department in the UK.

LEARNING AS INCREASING PARTICIPATION IN THE COMMUNITY OF PRACTICE

Wenger (1998) defines learning as increasing participation in a community of practice that leads to changing identities in relation to the practice. He also introduces the following four components of learning: *belonging—community*, *becoming—identity*, *experience—meaning*, and *doing—practice*. This definition of learning is very general and does not offer tools for analyzing the details, for example, in an individual students' learning of mathematics and questions like what qualities are demanded for participation; what is the practice; what are the emotional, cognitive aspects of the participation; and so on. However, we can find a lot of earlier pedagogical studies that illuminate these components.

A defining characteristic of participation is the possibility of developing an *identity of participation*. Building an identity consists of negotiating meanings of our experience in social communities. One's identity is always changing; it is a constant becoming. Students' possibility of participating in different kinds of activities around proof can develop their *identity of participation* in the mathematical practice if they can follow and experience meaning in the activities. Conversely, not being able to follow and find a meaning in the activities can lead to the development of an *identity of non-participation*, because the students already at the beginning of their studies view proof as a central part of mathematics (Hemmi, 2006).

Wenger stresses that structuring resources for learning come from a range of different sources, not only verbal interaction or intentional teaching of the teacher as a cultural mediator. Teachers, lectures, lessons, and instructional materials like textbooks and computer software become resources for learning in complex ways. Much of what students learn is not intended, and much of what teachers want to convey is not captured by students.

In my study, I contrasted the mathematicians' views and intentions with the students' learning experiences and observations of lectures in order to shed light on how the structuring resources become resources for learning. Next, I very briefly present the central notions in the social practice theories and relate proof and proving to these notions.

NEGOTIATION OF MEANING

Negotiation of meaning is seen to be a fundamental process on different levels and in different manners in all social practices. Negotiation constantly changes the situations to which it gives meaning and affects the participants. This process creates new circumstances for further negotiation and further meanings. The negotiation of meaning involves the interaction of two constituent processes—*participation* and *reification* (see Figure 28.1).

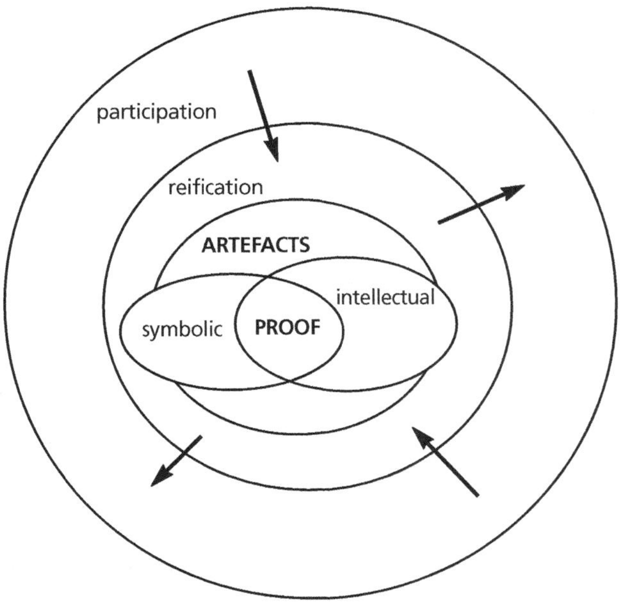

Figure 28.1 Negotiation of meaning involving proof

Participation refers to a process of taking part in interpreting proofs as well as constructing one's own proofs. When students grapple with their lecture notes and try to make sense of proofs and mathematical arguments, or when they discuss a new concept with some other students or listen to the lecturer, they participate in the practice. Negotiation does not necessarily refer to something going on between people but can as well be conceived as processes going on silently in one's head.

With *reification,* Wenger refers to the projection of our meanings into the world and then perceiving them as existing in the world and having a reality of their own. Wenger uses the concept of reification very generally to refer to the process of giving form to our experience by producing objects that "congeal" this experience into "thingness" that does not only refer to matters or material objects but also thoughts and ideas. In doing so we create *points of focus around which the negotiation of meaning becomes organized.*

In the practice of mathematics, for example, mathematical symbols, definitions, theories, and proofs can be seen as different kinds/levels of reifications. Proof as reification can refer both to a process of proving and its product, proof. Proof is not only formalizing mathematics and organizing it in a deductive manner but also creating conditions for new theorems and proofs and also a means of communication and thus production of a new context of both participation and reification. The newcomers in the mathematical practice have not designed the mathematical theories and proofs, yet they must absorb a part of them into their practice. The reifications coming from outside have to be *reappropriated* into a local process in order to become meaningful.

Participation and reification cannot be considered in isolation: They come as an interacting pair. Reification always rests on participation: For example, proofs always assume a history of participation, and, in turn, participation always organizes itself around reification because it always involves artifacts—that is, symbols and concepts that allow the negotiation of meaning to proceed.

PROOF AS AN ARTIFACT

One of the main arguments I put forward in my theoretical work is that from a socio-cultural/social practice perspective, proof can be seen as an *artifact,* a tool with a lot of functions in mathematical practice. I found support for this view both from the literature and from the interview data with the mathematicians who very clearly talked about proof as a tool in their practice. Artifacts are the concrete and abstract tools that *mediate* between the social and the individual. Vygotsky developed the concept of mediation in human–environment interaction to the use of signs as well as tools. Tool systems and sign systems like language, writing, and number systems have been created by societies during human history (Cole & Scribner, 1978). We *come to know* the world and the culture by *mediation* through artifacts: For example, meanings are known through language, which is also seen to be an artifact, and in mathematical practice, meanings are mediated through symbols and language—by a teacher, another student, or a textbook that can be the mediator.

There are several ways to classify artifacts (e.g., Wartofsky, 1979), but I found Säljö's (2005) classification most helpful to investigate the role of proof in mathematical practice. Säljö divides artifacts into two groups, *intellectual tools,* like discourses and systems of ideas, and *physical tools,* like texts, maps, and computers. He also classifies them into *primary tools* (e.g., a hammer) and *symbolic tools* (used for communicating ideas). He states that artifacts are carriers of information. As proof is a system of ideas and used for communicating ideas, it can be seen as an intellectual and symbolic tool (Figure 28.1).

Further, Säljö states that artifacts serve as tools for mediating in social practice, as they

1. stabilize human practice
2. facilitate continuities across generations
3. coordinate and discipline human reasoning by suggesting how to do things.

Proof stabilizes the practice of mathematics because it offers mathematicians common criteria for accepting and generating new mathematical knowledge and to connect the new theorems to the previous ones. Proof mediates methods of reasoning.

The systematization of mathematical results into a deductive system of axioms, definitions, and theorems unifies and simplifies mathematical theories by integrating unrelated statements, theorems, and concepts with one another, thus leading to an economical presentation of results (de Villiers, 1990). Hence, proof facilitates continuities across generations because the axiomatic deductive way of organizing mathematics makes it easier for new generations to reappropriate the mathematical knowledge obtained by the previous generations and allows new generations to further new problems in mathematical practice. The idea of mathematical proof has made it possible to create a body of knowledge, a core of mathematics that is relatively stable from generation to generation.

Finally, proof also coordinates and disciplines mathematical reasoning because of the severe demands it puts on precision in reasoning and justifying results. Proof helps with identification of inconsistencies, circular arguments, and hidden and not explicitly stated assumptions.

FUNCTIONS OF PROOF

Proof has several functions in the mathematical practice. Many researchers in the field of mathematics education have examined these functions and their significance for the teaching of mathematics. Bell (1976) was one of the first to deal with the nature and the role of proof in mathematics in relation to mathematics education. De Villiers (1990) presented the following model for the functions of proof ,which is an expansion of Bell's original distinction between the functions of *verification, illumination,* and *systematization.*

- Verification (concerned with the *truth* of a statement)
- Explanation (providing insight into *why* it is true)
- Discovery (the discovery or invention of *new* results)
- Systematization (the *organization* of various results into a deductive system of axioms, major concepts, and theorems)
- Communication (the *negotiation* of *meaning* and *transmission* of mathematical knowledge) (de Villiers, 1990, p. 18)

De Villiers' categories above have found a shared consensus among researchers in the mathematics education community and have been applied in many research projects and articles (e.g., Almeida, 2000; Knuth, 2002).

The functions of verification and explanation in connection to proof, in particular, have been discussed in mathematics education research (e.g., de Villiers, 1990; Hanna, 2000; Hersh, 1993), and these considerations have led to a lot of empirical studies. However, Weber (2002) states that besides proofs that convince or/and explain, there are proofs that justify the use of definitions or an axiomatic structure and proofs that illustrate technique.

Also, mathematicians in my study talked about proof as a tool in their mathematical practice. Besides the functions mentioned above they talked about *aesthetic, intellectual challenge,* and a function that I decided to call *transfer* (Hemmi, 2006; Hemmi & Löfwall, 2009). *Aesthetic* and *intellectual challenge* have been mentioned in earlier research (e.g., Burton, 2004; de Villiers, 1990), but the function of *transfer* has not been in the focus in mathematics education research community, and it has not been involved in the earlier models (Hemmi, 2006; Hanna & Barbeau, 2008). This function is close to and partly overlapping the aspect Weber (2002) describes, but not exactly the same.

With the function of transfer, I refer to the following: There are proofs that can offer new techniques, concepts, or ideas that are useful for other problems in other contexts of mathematics—for example, in problem solving. They can also offer understanding for something different from the original context. For example, in mathematics, Galois' result that the fifth degree equation cannot be solved by radicals has had much less importance to mathematics than his *proof* of the theorem, which opened a possibility to develop a new theory (see also Rav, 1999).

There are also examples in teaching contexts at an elementary level about the function of transfer. For example, the method of completing the square in deriving the formula for the solution of the second degree equation may be also used in problem solving in other mathematical contexts (Hemmi, 2006; Hanna & Barbeau, 2008).

Indeed, there is a need to study various proofs with respect to benefits to other contexts in mathematics such as problem solving (Hemmi & Löfwall, 2009). It seems clear that mathematical proofs are carriers of mathematical knowledge, and there are obviously various ways of enlightening this for students. However, we cannot look at the function of transfer mechanically, even if there are situations where it is possible to just copy a proof technique to another proving task.

What experts consider as evident connections may be difficult to see for a learner. Hence, it is important to study "the personal constructions of similarity" across proving and problem solving also from the learner's perspective (cf. Lobato, 2003).

All the functions of proof express something about the meaning of proof in mathematical practice, and therefore the knowledge about them is important for how students experience the practice. As mentioned above, the functions of conviction and explanation have particularly been in focus in mathematics education research, but the knowledge of the other functions should also enhance students' participation and their understanding of the role of proof in the practice.

TRANSPARENCY OF ARTIFACTS

Access to artifacts both through their use and through understanding their significance is crucial for students' access to the practice of mathematics. Lave and Wenger (1991) introduce the concept of *transparency* of the artifacts. They use it in connection to technology (physical and symbolic tool), but I have examined its strengths for describing the conditions of proof (intellectual and symbolic tool).

According to the theory of Lave and Wenger (1991), artifacts need to be both seen (visible) and to be used and seen through (invisible) in order to provide newcomers access to a practice. By the first characteristic, visibility, I refer to the different ways of focusing on the significance of proof in mathematical practice. How does one construct a proof? What are the main components in specific proofs? What is the logical structure of a proof? What is the historical role and function of proof in mathematics? How were proofs created for the first time? How is it possible to differentiate and define various proofs? What is the meaning of proof in mathematics?

With the second characteristic, invisibility, I refer to the inclusion of proof in the mathematical instruction in a way that does not provide extended information about various aspects of proof. For example, there is no focus on the logical structures of proofs, not proof techniques or information about how one constructs the proofs, but proof exists invisibly in the presentation of mathematics, derivations of formulas, explanations, or convincing arguments. There is no focus on the process of proving, but on the products, such as formulae and theorems and the justifying of the solutions of problems and mathematical tasks, without thinking of it as proving.

Both characteristics are needed and, according to the theory, they support each other. Combining these two aspects in the teaching of proof is called the *condition of transparency*. The condition of transparency refers to the intricate dilemma in the teaching of mathematics. It is impossible to focus on proof without some experience of "unproblematic" use of it. Conversely, it can be difficult for students to understand the meaning of proof or learn to produce their own proofs in the mathematical practice without any explicit focus on it.

There is extensive literature in mathematics education research that illuminates this dilemma from various points of view. The condition of transparency of proof proved to be a useful metaphor for shedding light on crucial aspects when studying the structuring recourses for learning and students' actual learning (see also Hemmi, 2008).

INTERACTING COMPONENTS OF PROOF

The framework developed in my study includes a model of interacting components of proof (Figure 28.2). In this model I summed up the main themes in mathematics education research that involve some tension/balance between two components. They can be seen as properties of proof and how to approach proof. They also illuminate the dynamic character of proof as reification, as both a process and a product, a view that allows one to describe the complex process of working with and creating proofs. There is an ongoing negotiation of meaning along with the interacting aspects of proof—for example, intuition/formality and induction/deduction in which both mathematicians and students participate. All these aspects are partly overlapping and intertwined, and they had an important role in the data analysis in my empirical study:

- Induction/Deduction
- Intuition/Formality
- Invisibility/Visibility

These interacting components are based on the literature concerning mathematical practice, mathematics education research, and empirical studies illuminating the concerns in the pedagogical debates. *Induction/deduction* refers both to the way in which the mathematician approaches proof in the teaching of mathematics and how the students are allowed to approach proof in various activities. There have been discussions in the field of mathematics education research about the advantages and disadvantages of

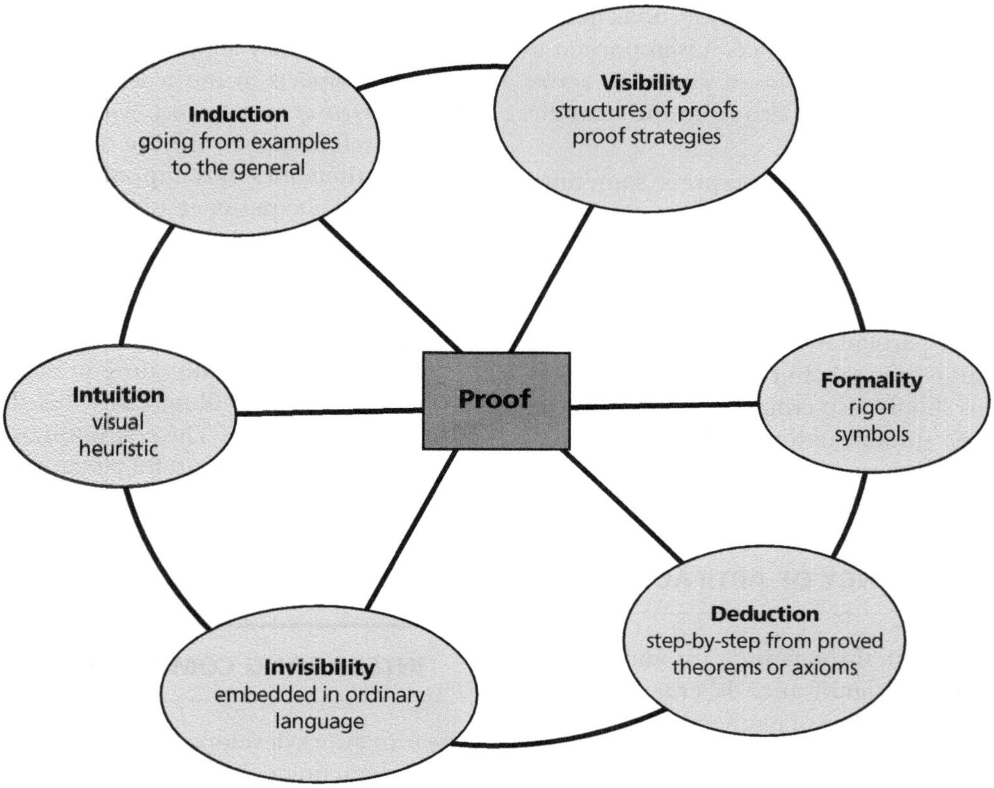

Figure 28.2 The interacting components of proof.

the axiomatic deductive approach versus a heuristic style (e.g., de Villiers, 1986; Lakatos, 1976; Lerman, 2000).The idea that students should not just meet "ready-made" proofs but also participate in constructing them by investigating, testing, conjecturing, and finally proving and disproving their hypotheses has led to a lot of empirical studies about new approaches to proof in the teaching of mathematics (e.g., Alibert, 1988; Almeida, 2003; Bell, 1976; de Villiers, 1991; Schalkwijk, Berge, & Rooij, 2001). These concerns can be related to the social practice theory in the following way: With inductive ways of working, and by conjecturing and trying to justify the conjectures, students participate in constructions of proofs and do not just meet the complete reifications. Of course they cannot create all the mathematics themselves, but they do have to participate in appropriating the reifications presented to them as complete deductive proofs. To study a "ready-made" deductive proof and make sense of it is also a form of negotiation of meaning that involves both participation and reification.

Intuition/formality overlaps with the aspect of induction/deduction in the sense that working on an investigative, inductive level is often associated with intuitive and informal ways of reasoning. Also, the analyses of the levels of students' argumentation involve the aspect of intuition/formality because the criteria for the different levels in some of the studies address the students' ability to produce deductive proofs with general symbols as well. This pair includes also the use of language and symbols. The relation between the formal and the intuitive has appealed to many researchers in the field of mathematics education, both in debates and in empirical studies (e.g., Chin & Tall, 2000; Fischbein, 1987; Moore, 1994; Viholainen, 2008). Fischbein (1987) argues that the educational problem is not the elimination of intuition but to "develop new, adequate intuitive interpretations as far as possible, together with developing the formal structures of logical reasoning" (p. 211). He criticizes the two opposite didactical strategies, the one which emphasizes the intuitive, pictorial components on one hand, and the one in which the body of knowledge is presented axiomatically and formally on the other hand.

There are parallels in Fischbein's considerations to those of Wenger (1998), who states that an excessive emphasis on formalism without corresponding levels of participation, or conversely a neglect of explanations and formal structure, can easily result in an experience of meaninglessness. Further, he states that explicit knowledge is not freed from the tacit, as formal processes are not freed from the informal. According to Wenger, a more abstract formulation, viewed as reification, will require more intense and specific participation to remain meaningful, not less. It could mean that, for example, to experience proof as meaningful, students would have to participate in different kinds of activities, involving the negotiation of meaning on various aspects of proof.

One can say that behind every abstract mathematical formula and proof, there is a lot of participation during the history of mathematics. But new generations do not need in every case to start from scratch. In the light of research (e.g., Lerman, 2000, p. 65), it can be questioned whether a higher level of abstraction, in general, needs to require more participation at an individual level compared with struggling with a concrete problem with a lot of details. Abstraction in mathematics can help us to see connections and structures, and in that way we can use them in many occasions without participating in the negotiation of meaning at all levels. These kinds of abstractions also offer us a means to understand problems in new contexts. I found support to this view also from the empirical data.

This questions Wenger's assumption that a more abstract reification in general would require more participation. Starting from the abstract and general in mathematics may sometimes require less participation than starting from the concrete.

The aspect of *invisibility/visibility* is described in section "Transparency of Artifacts."

The functions of proof are related to the interacting aspects in the model in various ways. For example, *explanation* may in certain situations be obtained by proceeding in an inductive manner, whereas in other situations a deductive presentation can offer a better explanation. Also, *communication* may sometimes be facilitated by emphasizing the logical structure of a proof (visibility), but in other cases it can be easier to communicate using ordinary language (invisibility).

SUMMARY AND DISCUSSION

I have presented a theoretical framework that was developed for and within an extensive study about students' encounters with mathematical proof at a mathematics department in Sweden. I combined socio-cultural perspective and a social practice approach with theories and main issues from the field of mathematics education research.

I looked at the mathematics department as a community of practice where the joint enterprise is the learning of mathematics. One of the main arguments I have put forward in my work is that from a socio-cultural perspective, proof can be seen as an artifact in mathematical practice, a tool with a lot of functions in this practice. I examined proof using Säljö's (2005) classification. In this way, proof is viewed as a symbolic and intellectual tool. As an artifact, proof is also a specific kind of reification in Wenger's (1998) terms. This allows a view of proof as both a process and a product that includes the complex process of working with and creating proofs. Proof is not only formalizing mathematics and organizing it in a deductive manner but also creating conditions for new theorems and proofs, and also a means of communication and thus production of a new context of both participation and reification, which are two constituent processes in negotiation of meaning. The didactical research on various functions of proof can be seen as an examination of proof as an artifact, whereas didactical research on various approaches/properties of proof, such as intuition/formality and induction/deduction, can be connected to proof as a process of reification.

Lave and Wenger (1991) introduce the metaphor of transparency of artifacts (invisibility/visibility), which I examined with respect to proof because it illustrates the intricate dilemma about how much to focus on various aspects of proof at a meta-level and how much to work with proof without a focus on it as a proof in the teaching of mathematics. The interacting aspects of proof as well as the functions of proof are based on the main issues in the mathematics education research. This is the way in which I have brought together a socio-cultural perspective, the social practice theories of Lave and Wenger (1991) and Wenger (1998), and didactical research on proof.

The theoretical framework developed for my study proved to be useful in analyzing and describing the mathematicians' pedagogical perspectives as well as students' experiences concerning proof. For example, a model with three different teaching styles was developed from the data with help of the aspects in the framework (Hemmi, 2006).

The view of proof as an artifact prompts further studies on various proofs with respect to what different functions and aspects specific proofs can illuminate in the didactic processes. Investigations of questions like: "In what phase of mathematical studies and in what ways do certain proofs benefit students' learning of mathematics?" and "What proofs are useful for the illumination of different functions of proof?" could help to improve the teaching of mathematics. Here, the framework could make a contribution by helping to discern and analyze proofs and the instructional settings from various points of view.

To learn more about the teaching and learning mathematics, it is not enough to just *describe* the educational practice. In order to study the learning outcomes, one has to study the organization of learning environments, for example, with respect to the choices of methods. All of them combined in various ways, together with instructional materials and together with teachers' intentions, create the structuring resources (the plans) for learning. The theoretical framework presented in this chapter could be applied and developed in such studies as it offers a fresh perspective to proof in mathematics and mathematics education and provides a structured network of results and issues from earlier research in the field.

REFERENCES

Alibert, D. (1988). Towards new customs in the classroom. *For the Learning of Mathematics, 8*(2), 31–43.

Almeida, D. (2000). A survey of mathematics undergraduates' interaction with proof: Some implications for mathematics education. *International Journal for Mathematics Education in Science & Technology, 31*(6), 869–890.

Almeida, D. (2003). Engendering proof attitudes: can the genesis of mathematical knowledge teach us anything? *International Journal of Mathematical Education in Science & Technology, 34*(4), 479–488.

Bell, A. W. (1976). A study of pupils' proof-explanations in mathematical situations. *Educational Studies in Mathematics, 7*(1), 23–40.

Brandell, G., Hemmi, K., & Thunberg, H. (2008). The widening gap—A Swedish perspective. *Mathematics Education Research Journal, 20*(2), 38–56.

Burton, L. (2004). *Mathematicians as enquirers. Learning about learning mathematics.* Boston: Kluwer Academic Publishers.

Chin, E. T., & Tall, D. O. (2000). Making, having and compressing formal mathematical concepts. In T. Nakahara & M. Koyama (Eds.), *Proceedings of the 24th Conference of the International Group for the Psychology of Mathematics Education* (Vol. 2, pp. 177–184). Hiroshima: PME.

Cole, M., & Scribner, S. (1978). *Introduction.* In L. Vygotsky, *Mind in Society* (pp. 1–14). Cambridge: Harvard University Press.

de Villiers, M. (1986). *The role of axiomatisation in mathematics and mathematics teaching.* Stellenbosch: University of Stellenbosch.

de Villiers, M. (1990). The role and function of proof in mathematics. *Pythagoras, 23,* 17–24.

Fischbein. (1987). *Intuition in science and mathematics.* Boston: Kluwer Academic Publishers.

Hanna, G. (2000). Proof, explanation and exploration: An overview. *Educational Studies in Mathematics, 44,* 5–23.

Hanna, G. & Barbeau, E. (2008). Proofs as bearers of mathematical knowledge. *ZDM—The International Journal on Mathematics Education, 40*(3), 345–353.

Hemmi, K. (2006). *Approaching proof in a community of mathematical practice.* Doctoral thesis. Stockholm: Department of Mathematics, Stockholm University.

Hemmi, K. (2008). Students' encounter with proof—The condition of transparency. *ZDM—The International Journal on Mathematics Education, 40*(3), 413–426.

Hemmi, K., & Löfwall, C. (2009). *Why do we need proof?* Paper presented at CERME 6, January 28–February 1, 2009, Lyon, France.

Hersh, R. (1993). Proving is convincing and explaining. *Educational Studies in Mathematics, 24,* 389–399.

Hoyles, C. (2001). From describing to designing mathematical activity: The next step in developing a social approach to research in mathematics education? *Educational Studies in Mathematics, 46,* 273–286.

Knuth, E. J. (2002). Teachers' conceptions of proof in the context of secondary school mathematics. *Journal of Mathematics Teacher Education, 5,* 61–82.

Lakatos, I. (1976). *Proofs and refutations.* Cambridge: Cambridge University Press.

Lave, J., & Wenger, E. (1991). *Situated learning: Legitimate peripheral participation.* Cambridge: Cambridge University Press.

Lerman, S. (2000). Some problems of socio-cultural research in mathematics teaching and learning. *Nordic Studies in Mathematics Education, 8*(3), 55–71.

Lobato, J. (2003). How design experiments can inform a rethinking of transfer and vice versa. *Educational Researcher, 32*(1), 17–20.

Mariotti, A. (2006). Proof and proving in mathematics education. In Gutiérrez, A. & Boero, P. (Eds.), *Handbook of research on the psychology of mathematics education: Past, present and future* (pp. 173–204). Rotterdam: Sense Publishers.

Moore, R. (1994). Making the transition to formal proof. *Educational Studies in Mathematics, 27,* 249–266.

Rav, Y. (1999). Why do we prove theorems? *Philosophia Mathematica, 7*(3), 5–41.

Schalkwijk, L. v., Berge, T., & Rooij, A. v. (2001). Learning to prove by investigations. *Educational Studies in Mathematics, 43*(3), 293–311.

Solomon, Y. (2006). Deficit or difference? The role of students' epistemologies of mathematics in their interactions with proof. *Educational Studies in Mathematics, 61,* 373–393.

Säljö, R. (2005). *Cultural tools and human knowledge: A sociocultural perspective on learning and the study of learning.* Lecture at the NoGSME Summer School (The Nordic Graduate School in Mathematics Education), August 8–14, 2005, in Jyväskylä, Finland.

Wartofsky, M.W. (1979). *Models: representation and the scientific understanding.* Dordrecht, The Netherlands: Reidel.

Weber, K. (2002). Beyond proving and explaining: Proofs that justify the use of definitions and axiomatic structures and proofs that illustrate technique. *For the Learning of Mathematics, 22*(3), 14–17.

Wenger, E. (1998). *Communities of practice.* Cambridge: Cambridge University Press.

Wellington, J. (2000). *Educational research: Contemporary issues and practical approaches.* New York: Continuum.

Viholainen, A. (2008). *Prospective mathematics teacher's informal and formal reasoning about the concepts of derivative and differentiability.* Unpublished doctoral dissertation. Faculty of Mathematics and Science, Jyväskylä University.

29

STUDENTS' PERCEPTIONS OF LIMITS

Kristina Juter

KRISTIANSTAD UNIVERSITY

Abstract: Learning calculus at university level requires skills such as linking concepts together, logical reasoning, and understanding abstract notions. Questions about how students perceive limits of functions and what pre-knowledge they bring to university courses in calculus have been investigated in two different studies. The students' perceptions were depicted in terms of concept images. Several of the students' concept images were incoherent, often due to theoretical misunderstandings. Students' pre-knowledge included, in some cases, invalid links between concepts that were influenced by intuitive perceptions.

Keywords: Conceptions, limits of functions, concept images, pre-knowledge, calculus

INTRODUCTION

Background and Questions

Mathematics has several characteristic properties, such as the use of models to analyze real-world phenomena, compact and unambiguous formulations for clear expositions, and deductive reasoning in proving and problem solving. Each property offers its own set of challenges for mathematics learners, often including a transition stage, for example, from a model to the real world, from everyday language to a mathematical expression, or from one step to another in a mathematical proof. Poincaré (1983), also identifying a transition, claimed that mathematics strives from the particular to the general through, for example, increasingly sophisticated deductive reasoning.

An initial course in mathematics at Swedish universities usually comprises algebra and calculus, including the notion of limits of functions, which has proven to be difficult for students to learn (Cornu, 1991; Juter, 2006c; Tall, 1991; Tall & Vinner, 1981). This chapter reports from two studies that were conducted at the basic university level at two different universities in Sweden. The research question in the first, and larger, study was: How do students deal with the concept of limits of functions? and in the second: What pre-knowledge do students have about functions, limits, continuity, derivatives, and integrals (when starting these courses)? Consequences of students' conceptions of limits from possible pre-knowledge in analysis are raised. Both studies were conducted over a period of time to grasp the students' development, with various foci added such as attitudes and confidence, complementing existing research on students' learning limits (e.g., Bezuidenhout, 2001; Cornu, 1991; Przenioslo, 2004; Tall, 1991; Tall & Vinner, 1981). This chapter describes and summarizes the results of the empirical studies after an outline of theoretical settings of individual concept formation. Nordic as well as international research results on students learning limits are discussed. Implications for teaching and learning conclude the chapter.

The First Sourcebook on Nordic Research in Mathematics Education, pages 419–430

THEORETICAL OVERVIEW

The learning and concept formation theories used in the studies are discussed in this section, followed by an analysis of the nature of the concept of limits.

Learning

From a constructivist point of view, students construct their own mental representations of concepts. A *concept image* is a model of an individual's total cognitive representation of a concept (Tall & Vinner, 1981; Vinner, 1991). The concept image comprises all representations from experiences linked to the concept, of which there may be several sets of representations constructed in different contexts that possibly merge as the individual becomes more mathematically mature. Multiple representations of the same concept can co-exist if the individual is unaware of the fact that they represent the same concept. Possible inconsistencies may remain unnoticed if the inconsistent parts are not evoked simultaneously. Parts of concept images, *concept definitions*, are the individual's definitions of the concept, learned by rote or self-constructed. A person's concept definition can be different from the *formal concept definition*, that is, an unambiguously stated stringent definition that is accepted (used) by the general mathematical community. Studying the mental networks or representations of a concept is not possible, but the students' performances are determined by their capacities (Star, 2000) represented in their minds, and we can therefore study the effects of their mental representations.

Tall (2004) has introduced a model with three worlds of mathematics to distinguish different modes of mathematical development. The theory of the three worlds emphasizes the construction of mental representations of concepts and has emerged from several theories on concept development, such as Piaget's abstraction theories and Sfard's (1991) work on encapsulation of processes to objects (Tall, 2004). The three worlds are somewhat hierarchical in the sense that there is a development from just perceiving a concept through actions to formal comprehension of the concept. In the first world, *the conceptual-embodied world*, individuals use their physical perceptions of the real world to perform mental experiments to build mental conceptions of mathematical concepts. The mental experiments can be children's categorizations of real-world objects, such as an odd number of items or, later, students' explorations of intuitive perceptions of limits of functions, that is, non-formalized descriptive representations. The second world is called the *proceptual-symbolic world*. Here individuals start with procedural actions on mental conceptions from the first world, as counting, which by using symbols become encapsulated as concepts. The symbols represent both processes and concepts, for example counting and number or addition and sum. The symbols, together with the processes and the concepts, are called procepts (Gray & Tall, 1994) and are used dually as processes and concepts depending on the context. The third world is called the *formal-axiomatic world*, and here properties are expressed with formal definitions as axioms. Individuals go between the worlds as their needs and experiences change and mental representations of concepts are formed and altered.

Limits can initially be handled through an explorative approach with tables of function values and graphs and later as symbolically expressed entities. Learning limits of functions demands leaping between operational and static perceptions (Cottrill, Dubinsky, Nichols, Schwingendorf & Vidakovic, 1996). There is a challenge in understanding the termination of an infinite procedure as a finite object, such as

$$\lim_{x\to 0}\frac{\sin x}{x}=1$$

It is necessary to reach all significant stages to comprehend the aforementioned duality, and be able to change between the different representations. Only then can an individual understand the concept if understanding of a mathematical concept is defined as Hiebert and Carpenter (1992) did—that is, as something an individual has achieved when he or she can handle the concept as part of a mental network. Each connection or creation of a mental representation can be conceived as an *act of understanding* (Sierpinska, 1990), and the strength or depth of this understanding of a concept can be measured in the quantity and quality of the acts of understanding for that particular concept (Dreyfus, 1991).

Limits of Functions in Analysis

The limit concept is an important part of the foundations of mathematical analysis, and not understanding it clearly might lead to problems when dealing with concepts such as continuity and derivatives. It is an elusive concept because it can be considered as both an infinite process and a rigid entity, depending on the situation. Infinity and this duality demands flexibility in the mental representations we use when thinking of limits. An inexperienced individual who does not possess enough representations or flexibility in the representations cannot understand the crucial similarities and differences for handling certain situations. In that case, knowing which operations and manipulations are allowed and which are not allowed is unclear (Juter, 2006a).

Hähkiöniemi (2006) has studied students learning the derivative and concluded that students could have a procedural understanding of the limit of the difference quotient, but still lack a conceptual understanding of the limit process. He stated that the students used a variety of different representations of limits in their learning of derivatives, but the students had difficulties linking them to formal mathematics. Pettersson's (2008) somewhat different results from a study on how students use their conceptions in analysis imply the possibility of starting in an algorithmic context and proceeding to a conceptual context to learn successfully. Viholainen (2006) also studied students learning analysis, and his results showed that the students in his study had problems using concepts in the conceptual-embodied world (Tall, 2004) in a constructive correct manner when they worked with continuity and differentiability. Students are supposed to understand crucial mathematical concepts at different levels of abstraction and in various contexts, which means that they have to have several ways to represent them, to link them together, and to label them. Hemmi (2006) studied students' work with proofs and found that the students thought that the mathematical language was hard to comprehend. The language in upper secondary school differed from that at the university level, and the students felt that they did not have enough practice from work with mathematical symbols. The high pace of the lectures further complicated their understanding of words and symbols. Language and algebraic experience influence students' abilities to learn, and Bergsten (2006) concluded that algebra works both as an aid and an obstacle to students handling limits of functions. Algebraic manipulations were often used, prompted by surface properties of the tasks rather than conceptual understanding.

Everyday language can have a slightly different meaning compared to the language used in mathematics, which can have an effect on students' perception of the concept (Monaghan, 1991). We use such words or phrases as convergence, arbitrarily close, tend to, and limit when we work with limits of functions. There is an ambiguity in the way the concept of limits can be perceived. One can focus on the process of approaching the limit and hence consider it a never-ending procedure. But one can also think of the limit as a static entity to which functions can be compared. This kind of ambiguity was exemplified in a study by Williams (2001), where two students' conceptions of limits were examined. Both of the students understood that continuous functions reach their limit values, but they were not sure if it was correct to say that the limit was reached in the limit process. Williams mentioned the complexity of the notion of infinity as a source for the problem in this situation, but it can also be that the students did not accept or understand the definition of limits. Przenioslo (2004) has done a study about limits of functions with university students in years three to five. She found that students' concept images differing from the formal concept definition could have their origin in an intuition that limits do not exist if the function is not continuous. The intuitive views come from secondary school work, according to Przenioslo.

To sum up, some students have an intuitive sometimes procedural conception of the concepts and need guidance to comprehend and formalise their knowledge.

THE EMPIRICAL STUDIES

One study (Juter, 2005a,b,c; 2006a,b,c; 2007) was conducted to reveal how students learn limits of functions and one about students' pre-knowledge in calculus (Juter, 2009). The students were enrolled in their first university-level mathematics courses. They had encountered the notion of limits of functions in upper secondary school, at least in connection with derivatives. The treatment of limits can be intuitive or peripheral at this level and students can hence have very different conceptions of limits upon arriving at university. Responses of the students to questionnaires and interviews and their actions are regarded as traces from their concept images and were used to assess their developments and abilities.

The First Study

Comprising 112 students in their first university mathematics semester, the study was carried out over 20 weeks divided in two 10-week courses. Students majoring in mathematics or science took the courses, but they were also given as separate courses. The first course covered calculus and algebra, and the second was a continuation course in calculus and algebra. Three questionnaires, E1 to E3, were handed out, and two sets of interviews, I1 and I2, were conducted at the times indicated by the timeline in Figure 29.1. Limits were introduced in the course after the first questionnaire. Field notes were taken throughout the course when limits were taught. Information from the field

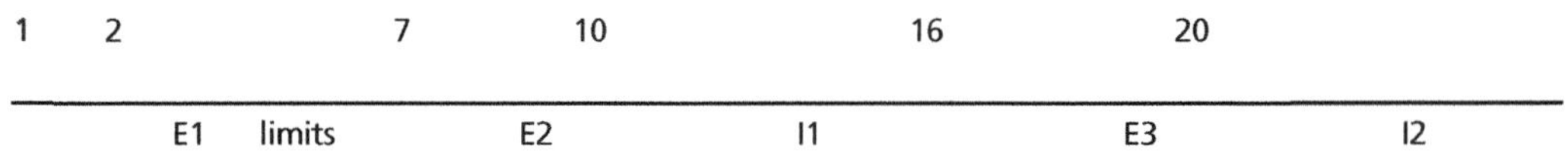

Figure 29.1 Timeline (in weeks) for the five times data was collected.

notes was used in the creation of tasks for questionnaires, in addition to being descriptions of the students' learning situations. The tasks and questions used were tested in a similar study with 143 students (Juter, 2006c).

The first questionnaire, E1, was designed to give information about the students' previous education, attitudes to mathematics and their knowledge of limits with a few open limit questions before limits were taught in the course. The work of Mohammad Yusof and Tall (1994) influenced the design of the attitudinal part. The intention with these questions was to get a picture of how well the students knew this concept and what other concepts they connected to it, to find out how developed their concept images were.

The second questionnaire, E2, was distributed to the students after they had seen the limit definition, proven some theorems, and solved some limit problems. The first part concerned how the students interpreted their learning situation, with a focus on their struggle to learn. The design of the second part of the questionnaire was partly influenced by Tall (1980) and Szydlik (2000). The students were asked to determine the limits (if they existed) in four problems and to explain what they had done as if they were explaining to a person who was just learning about limits. They were also asked if the functions in the problems could attain their limit values, if they exist, and to explain their claims. The aim of the four problems was for the students to show how well they could explain their calculations, and how functions and limits of functions were linked in their concept images. Task 1 is an example of these tasks:

Task 1:

a. Determine the limit:

$$\lim_{x\to\infty}\frac{x^3-2}{x^3+1}$$

b. Explanation.

c. Can the function

$$f(x)=\frac{x^3-2}{x^3+1}$$

attain the limit value in 1a?

d. Why?

Two problems were accompanied by solutions that were either wrong or incomplete, or both wrong and incomplete. Here, the aim was to see if the students could handle their knowledge about limits in a non-traditional situation and how they handled a given solution, by either accepting it or challenging it with arguments of their own.

The third questionnaire, called E3, was just one task where two fictitious students discussed a limit problem. One had an approach based on a conception compatible with the formal limit definition. The other had an approach based on a misinterpretation of the limit definition. The students in the study were asked to determine which fictitious student's arguments they preferred and why, and also to explain the misinterpretation in the other student's argumentation.

Eighteen students were selected for interviews, based on their responses to the first two questionnaires. Their answers were compared and sorted into groups with similar responses to the questionnaires. The students indicated in the second questionnaire if they were willing to participate in interviews. The selection was done based on these factors. If one type of answer was more frequent than another, then more students were chosen from the larger group to obtain a representative subset of the students. Two interviews were planned with each of the eighteen students. One was conducted a few weeks into the second course and the other was directly after the exams. Three students did not go through the second interview; hence, fifteen pairs of interviews were done.

The duration of the first interview was between 30 and 60 minutes, depending on the interviewed student. I questioned the responses to the questionnaires and assigned a few more limit tasks for the students to solve. One task comprised six statements about limits to respond to, such as: "A limit value is a number or a point beyond which a function cannot attain values." The statements were similar to those used by Williams (1991). The pre-formulated statements worked as stimuli for the students to describe their own conceptions. Other tasks were about the limit definition—for example, if the students could state the formal definition or recognize it next to an incorrect one. The graphs of two functions, both discontinuous at a point, were presented to the students to determine whether the functions had limits or not at the discontinuous points. The tasks in the interview were of different character to work as instruments to measure the students' perceptions of limits in different situations (Juter, 2005c; 2006c).

The second interview was about 15–20 minutes long. The main topic was the limit definition compared to the example with two fictitious students arguing about solutions in questionnaire E3. Both interviews were directed by a set of questions, with follow-up questions adjusted to the students' answers.

The Second Study

The second study, about pre-knowledge, took place at another university. The 15 students in the study were attending a calculus course with functions, limits, continuity, derivatives, and integrals as main topics. No further selection of students was done since there were 16 students taking the course and one student was absent when the data were collected. Most of the students took the course as part of their teacher education program, but it was also given outside the program.

At the beginning of the course the students filled out a questionnaire where they were asked to describe the five aforementioned concepts and also to write what the concepts are used for. The reason for this open approach was to prevent the students from becoming influenced by other formulations than their own. The tasks were: Describe the concept of function/limit/continuity/derivative/integral with your own words. What are functions/limits/derivatives/integrals used for? Continuity was omitted in the last set of tasks since it is a property of functions rather than an entity like the other concepts. The pre-knowledge study is part of a larger study about students' conceptual development over a year after the calculus course.

RESULTS

Categories of the limit tasks from the questionnaires were created from the students' responses. Some categories were created in advance, and some became apparent from the data after sorting the responses—for example, correct answers based on misconceptions. The analysis of the data was hence inspired by grounded theory (Strauss & Corbin, 1998). The categories were created to reveal as much as possible about the students' development in understanding of limits of functions through the duration of the semester with respect to a number of features of the concept, such as static or procedural perceptions, functions' abilities to attain limit values, perceptions of the limit definition, and strategies for problem solving in various contexts. The students' descriptions, with respect to these features, were then connected to Tall's (2004) three worlds to illustrate the different modes of development.

The categories in the pre-knowledge study were created from the collected data with respect to the students' responses in a similar way as in the first study.

Results of the First Study

The results from the first study are summed up in Table 29.1, where typical high achievers' and low achievers' respective developments during the semester are portrayed. High achievers had the highest

TABLE 29.1 Typical Developments for High-Achieving Students and Low-Achieving Students at the Five Data Collection Events Presented in Figure 29.1

Event	High achievers	Low achievers
E1	Links limits to prior studies. Solves easy tasks well.	Links limits to prior studies and other topics. Solves easy tasks well.
E2	Limits are attainable in problem solving. Solves tasks and explains well. Problems stating a limit definition.	Limits are attainable in problem solving. Solves routine tasks and explains with some flaws, other tasks problematic. Cannot state a limit definition.
I1	Limits are attainable in problem solving and in theory. Problems stating the definition. Can identify the definition with uncertainty about the quantifiers' meanings. Solves tasks fairly well.	Limits are attainable in problem solving but not in theory. Cannot state the definition. Problems identifying the definition, quantifiers not understood. Problems solving tasks. Not an actual limit if attainable by the function.
E3	Identifies the error.	Problems identifying the error.
I2	Can identify the definition. Can explain the quantifiers' meanings.	Problems identifying the definition. Cannot explain the quantifiers' meanings.

mark on the exams and low achievers had the lowest passing mark or failed the exams.

Most students could solve the routine tasks, but had trouble solving the non-routine tasks, which required them to understand specific features of the notion of limits (Juter, 2005a,c; 2006b; Juter & Grevholm, 2007)—for example, to decide whether a function can attain a limit value, or to determine what the different components in the definition represent. Algebra was more problematic than anticipated. Students failed in several cases of straightforward limit problem solving due to insufficiently developed algebraic skills. Manipulations with symbols were done in a manner implying that the students were unable to understand what the symbols represent or what the effects of the operations should be, leaving them unable to solve the limit problems. One example of such an algebraic calculation is:

$$\frac{x^3-2}{x^3+1}=\frac{\frac{x^3}{x^3}-\frac{2}{x^3}}{\frac{x^3}{x^3}+\frac{1}{x^3}}=\frac{2x^3}{x^3}$$

The student's explanation showed that the two equal fractions were omitted and then the student overlooked the minus sign. Quite a few students carried out this type of calculation (Juter, 2006b), and many failed to differentiate surface properties from intrinsic properties (Lithner, 2003), leading the students to select unsuitable solving methods. Examples are given, in the first interview, in the responses to the task of determining

$$\lim_{x\to 1}\frac{x^2-1}{x-1}$$

Several students recognized the rational function and, without considering intrinsic properties of the task (similar to Bergsten's (2006) results), tried to divide numerator and denominator with what would have been the dominant term if x had tended to infinity. They went for the solution they were used to from working with limits of rational functions, where x tends to infinity, and found it difficult to see other possible solution methods, implying a need for further development in the proceptual-symbolic world (Tall, 2004). The symbols in the task representing the limiting process did not evoke suitable parts of the concept images to enable the students to come up with a method to solve the problem, if such parts existed.

There was also some confusion of functions with limits of functions, which made the students give wrong answers to tasks or correct answers with wrong explanations often caused by misconceptions of infinity or the limit definition (Juter & Grevholm, 2007). Intuitive representations of the concepts, i.e. concepts not developed enough, in the conceptual-embodied world (Tall, 2004) caused some of the misconceptions. The intuitive representations interfered with the students' formal representations of limits. The meaning of the quantifiers in the formal limit definition seemed to be a mystery for many students throughout the entire semester (Juter, 2003; 2005a; 2006a). Few were able to state the definition or explain what it actually means, which is implying that they solved tasks without using the fundamental theory behind them. Perhaps the students did understand the concept from the problem solving in the course or through discussions with peers, but few students could actually connect tasks with the definition (Juter, 2005a; 2006a; 2007). They had serious problems in bridging the gap from intuitive or procedural perceptions to formal and static perceptions. Students' developments need to have reached the proceptual-symbolic world of Tall's three worlds of mathematics (Tall, 2004) for them to be able to perceive limits both as objects and processes, and then the formal-axiomatic world to be able to understand, in the aforementioned sense (Hiebert & Carpenter, 1992), what the components of the definition correspond to in the task. The transition to the third world of mathematics demands an ability of abstraction making the concept transparent to the student. This transparency means that students can switch between perceiving limits operationally and perceiving limits as entities (Cottrill et al., 1996; Sfard, 1991; Tall, 2004). Table 29.2 shows the students' inconsistencies in their conceptions of limits. Each column represents how many times the students confirmed regarding limits as, for example, attainable or an exact value.

Examples of students' statements in the different categories of Table 29.2:

Attain limit value, T: Functions can attain limit values in theory.

> Example: (Emma) "...the definition of limit values is like that it is possible, most or it depends, there are limit values which are attainable."

> Example: (Julia) "...the function can attain values beyond the limit value if it...if x is not, well for other x so to speak"; "Yes it is okay that it reaches its limit value, that is no problem."

Attain limit value, P: Functions can attain limit values in problem solving situations.

> Example: (Leo)... (Interviewer): "If it [the variable] had tended to minus one instead and the same function, what would that be?" (Leo): "Mm,

it becomes 0." (Interviewer): "Mm, can it attain that value?" (Leo): "Yes."

Example: (Martin) (Interviewer): "Can this function ever be equal to two?" (Martin): "Yes when it is one, then it becomes two."

Not attain limit value: Functions can not attain limit values.

Example: (Martin) "...you never reach the limit value, you approach the limit value."

Example: (Emma) "Visually it feels like... if one continues long enough, it will reach its limit value, but it does not since it does not reach the limit value, it is part of the definition, but..."

Distance between function and limit: Speaks of distance between limit value and function values.

Example: (David) "Yes, the difference between *f(x)* and A is less or equal to epsilon." [David speaks of "difference" as a distance as he explains how the function values are getting closer to the limit value.]

Border: Limits as borders.

Example: (Oliver) "You get close to another value as some kind of border, yes. [...] As a principle when it comes to limit values there can not be anything beyond."

Example: (John) "You think of some border of some kind, that you come to a border and beyond it you have decided something, or I think a lot about the body what you can do physically or something like that."

Approximation: Limits as approximations of function values.

Example: (Mikael) "We can do it as accurate as we want, but we can never get it exact."

Exact: Limits as exact values.

Example: (Dan) "...approximations can be done in several ways but a limit value is uniquely determined."

TABLE 29.2 Students' Conceptions of Limits of Functions

	Attain limit value, *T*	Attain limit value, *P*	Not attain limit value	Distance between function and limit	Border	Approximation	Exact
Filip	1	1	4	2	1		1
Martin		2	3				1
Tommy	1		3	2		1	
Anna		1	3	1			
John	2	1	1	1	5		
Frank		1	6				1
Louise	1		2	2	1	1	
Leo		1	3				
Dan	1	1	2	3			1
Mikael		1	6	2	1	1	
David	1		1	2	1		1
Julia	3			1		1	
Dennis	1				1	1	
Emma	4	1	3	3	1		1
Oliver		3	4		1		1

Note: The numbers show how many times each view was confirmed during the interviews. The students are listed from lowest marks to highest on the exams.

Attainability is undoubtedly an issue in need of attention since a clear majority of the students had conceptions from one of the two first categories as well as the third category in Table 29.2. The last two categories, on the other hand, neatly complement each other as there are no cases of inconsistencies regarding limits as approximations or as exact values. Three students did not think of limits as either an exact or an approximate value, though.

Six of the students stated that limits are unattainable and that limits are attainable in problem solving situations, but not in theoretical contexts. Theory and problem solving seem to be separate issues for these students. Three students revealed an opposite trace to the six students, where they claimed that limits are not attainable and that limits are attainable in theory, but they did not say that limits are attainable in problem solving. The theory used appears to be unclear for the students. Emma and Martin, as two examples, contradict themselves in the examples given above. The definition is causing Emma's divided perception, and Martin seems to evoke different parts of his concept image in the different situations.

Only two students, Julia and Dennis, showed traces of coherent concept images. They showed no traces of problems with functions attaining limit values, and they both explicitly stated that functions can attain limit values. The rest of their conceptions did not match each other, but there were no contradictions either. Students with high marks did not have very different structures of their perceptions compared to the other students. The one difference is that the two students with coherent traces of their concept images were among the students with the highest grades.

Limits of functions was considered the most important topic of the calculus part of the course, according to most students, and also the most difficult topic (Juter, 2005b). The first part probably inspires or encourages students to keep working despite their difficulties. The students' concept images were not necessarily coherent or in agreement with generally accepted definitions, though the students were not always aware of this. Some students actually showed great confidence in their own capability to handle limits, even after they had failed to solve problems or explain some theoretical property (Juter, 2003; 2005a). Successful routine problem solving can cause such reactions since the abilities of students become confirmed by their correct answers. One of the previously mentioned results is that many correct answers were founded on reasoning which suggested misinterpretations of concepts. This undermines the creation of useful concept images, and not many students in the study showed evidence of a concept image of limits that allowed them to solve non-routine tasks or explain essential features of the definition. Students with great confidence in their own abilities do not work further to adjust their understanding of a concept because nothing to adjust is perceived.

The field notes revealed almost no interaction between the students and lecturer during the lectures. There were more discussions at the problem-solving sessions, but few students were active in these discussions. Students who solved tasks at the blackboard were usually those who were active in the discussions. Some of the students who silently and inactively participated in the problem-solving sessions were among those interviewed, and they had various reasons for their behavior. One student did not want to go to the blackboard because he wanted to see other solution methods, preferably the teacher's. Some students were shy and did not want to speak in front of the other students, with the risk of embarrassment. The learning environment was consequently not easily adjusted to the students' unexposed levels of understanding.

Results of the Second Study

Parts of the study connected to limits are presented in this section. The students' responses to the second task, which was "Describe the concept of limit with your own words," are shown in Table 29.3.

Student answers in the first category reveal perceptions of limits as a value closest to the function at a certain point—for example, student C: "A limit is the closest value a for a given number in a." This student

TABLE 29.3 Categories for Students' Descriptions of Limits

Category	Students in each category
1. Closest value	C, O
2. Derivative	D
3. Interval within which function values can be	A, B, F, G
4. A number or infinity	H, K
5. Maximum or minimum for a function	I
6. Stabilizing point	J
7. Mean value	N
8. No answer	E, L, M

Note: Each student is denoted by a letter A–O.

confused the limit with an interval, which is a common way to see limits as the statements in category three reveal. Student F described limits: "Within what frame the answer can be." Students in the third category described limits as intervals, but their descriptions imply that they may have seen the endpoints of the intervals as different limits. Student I, who stated that a "limit is either max or minimum values in the function," had an intuitively logical interpretation of the word "limit" as something confining the function, but not a mathematically useful one.

Students in the fourth category thought of limits as the outcome, the value: "It can be a determined value or infinity" (student H). They perceived limits as objects in a problem solving perspective. The student in category two had a more theoretical view remembering limits from an evoked concept image of derivatives (Tall & Vinner, 1981), without being able to make sense of the notations.

Student N in category seven thought that a "limit is levelling out between values"—that is, some kind of mean value. This type of reasoning can be similar to the student's description in category six, of an oscillating function stabilizing at one value, if the function oscillates over and under the limit symmetrically.

Table 29.4 shows the students' responses to the task: *Describe the concept of derivative with your own words.*

A common way to describe derivatives was as the slope of a function. In category one, student D described derivatives as "functions' slope at different places/times." Student E in category three associated derivatives with values in intervals: "Determine many values and then check off in a certain interval." Student F had a similar pattern of several parts, but more like the pattern of a Riemann integral: "You divide the area within a graph in little parts to be able to determine the result at a certain value."

TABLE 29.4 Categories for Students' Descriptions of Derivatives

Category	Students in each category
1. Slope of a function	D, G, H
2. Rate of change	C
3. Values in an interval	E, F
4. Means to simplify functions/equations	A, B, N
5. Means to find where a function is zero/ max or min	I, K, O
6. Limit equals derivative	M
7. No answer	J, L

Note: Each student is denoted by a letter A–O.

The three students in category four saw derivatives as something to use to alter functions to more manageable functions. For example, student N wrote: "You add several *x*, etc. in equations and similar (depending on how the equation looked like from the start)," and then on the question on what derivatives are used for: "You derive equations to more easily be able to determine the variables in the equations." There seems to be some kind of confusion with finding roots to polynomial equations with double roots or higher.

The fifth category comprises student answers related to finding specific values, such as functions' extreme values. Student K wrote: "Derivative is when you derive a function and get, for example zero values."

Student M in category six wrote: "Isn't derivative and limit, in principle, the same thing?" The students' responses to the task "Describe the concept of integral with your own words," showed one connection to limits as student B simply wrote "Study of limits."

Table 29.5 displays responses to the question: "What does it mean for a function to be continuous?" The first category comprises student answers depicting continuous functions as a smoothly developing graph, such as student E: "It develops evenly all the time." The responses in this category are fairly intuitive.

Students in the second category described continuous functions stricter: "It is defined for all *x* and connected" (student H). Student I described a continuous function in the third category: "It moves the same way all the time, for example the sine curve." Student L wrote: "Every step at the *x*-axis gives the same difference at the *y*-axis." Both students' conceptions include a reoccurring pattern implying that the students' conceptions were influenced by examples with this property.

Several of the categories in Tables 29.3 to 29.5 show the students' conceptions of limits in the embodied world (Tall, 2004) with intuitive and explorative descriptions (e.g., categories 1 and 3 in Table 29.3, categories 1 and 3 in Table 29.4. and categories 1 and 3 in Table 29.5). Many descriptions were method-oriented (e.g., category 4 in Table 29.3 and categories 4 and 5 in Table 29.4) and showed a symbolic treatment in the proceptual world. There were indications of students working in the formal world (e.g., category 2 in Table 29.5) but the data at this point is not enough to say more.

TABLE 29.5 Categories for Students' Descriptions of Continuity

Category	Students in each category
1. Even rate of change/development	A, C, E
2. Function connected and defined for all *x*	H, K, M
3. Function changes the same way all over	I, L
4. The function has no limits	B
5. The function is not discrete	J
6. The function tends to infinity	F
7. No answer	D, G, N, O

Note: Each student is denoted by a letter A–O.

DISCUSSION AND CONCLUSIONS

Results of the studies are discussed in terms of conceptual development and teaching and learning followed by some concluding remarks.

Conceptual Development

The gaps mentioned in the results section between upper secondary school and university—limits as infinite and finite, intuitive and formal, and object and process—imply too many and too large discontinuities for the students to understand the full extent of limits. Students are expected to handle limits in the formal-axiomatic world, since textbook and classroom presentations are on a formal level, but the students themselves may only be experienced enough to handle limits in the proceptual-symbolic world (Tall, 2004).

Intuitive representations and language based in non-mathematical situations interfere with students' conceptual development (Monaghan, 1991), causing hesitation and confusion as in the cases reported by Prezenioslo (2004) and Viholainen (2006). The second study showed how intervals are used to picture the concepts of limits and derivatives, but instead of helping the students to understand, they sometimes lead to misinterpretations, as depicted in Tables 29.3 and 29.4. The results of the first study imply that this pre-knowledge remains and results in difficulties in understanding the intervals stated in the ε,δ- definition of limits with the consequence that many students never learn the meaning of the definition.

Students found it hard to see connections between formal representations and their other representations from, for example, problem solving. The formally stated parts were unsuccessfully integrated in their concept images (Tall & Vinner, 1981), leaving the students with parallel incoherent parts of the concept image, one for theory and one for problem solving. This confusing duality became apparent from a task similar to Task 1, with a linear function instead of a rational function, where several students knew that linear functions attain all values but thought that limits are unattainable according to the definition. In the interviews, the frustration with the incompatible parts resulted in doubt of their own conceptions and in some cases in annoyance to the task. Some students said that there is no point in talking about limits if they are attainable, contradicting Przenioslo's (2004) result that limits only exists for continuous functions. The students in Williams' (2001) study were in the same way hesitant to say that continuous functions reach their limits. The students appear, as I see it, to be confused by the context of the task. They know that continuous functions can reach their limit values, but their concept images of limit processes interfere with their concept images of continuous functions, causing this hesitation.

Implications for Teaching and Learning

Students' rather intuitive pre-knowledge can, as exemplified in the results section, be vague and hence interfere with acquisition of new knowledge. Understanding limits at a university level requires awareness of their own conceptions (Hiebert & Carpenter, 1992), including links from formally stated theory to the initial, sometimes vague, conceptions. Adding the challenge of perceiving a limit as a process and as an object at the same time (Tall, 2004) further stresses the importance of a varied selection of problems for students to work with. Students strategically dispose of their time depending on their purposes of their education. The questionnaires and interviews in my study and in that by Mohammad Yusof and Tall (1994) imply that it is probably typical exam problems the students work with to have any possibility of passing an exam. Creating more room for questions about the nature of limits may encourage students to study further and hence allow them to understand deeper. The students' struggles with theory, for example, about what consequences small changes in the formulation have on the meaning of the limit definition from the first interview (Juter, 2006c), reveal a need for working with limits in a wider manner. It could be that the symbols used as procepts (Gray & Tall, 1994), such as a limit L, are explicitly linked to representations from all of Tall's three worlds (Tall, 2004)—particular examples of limits explored with computer or calculator, limits determined by algebraic manipulations, all compared to the general expression $\lim_{x\to a} f(x) = L$, emphasizing the concept's dual nature. Exploring and comparing limits of functions in various cases, intuitive and formal, strengthens students' concept images with the added links between the representations.

Conclusions

The students readily solved routine tasks, although as soon as the tasks became slightly unfamiliar to them, many of them became confused about what they should do. The limit definition was often the reason for errors because it was wrongly interpreted. Connected concepts such as infinity, together with the students' algebraic inadequacy, also caused problems in their problem solving. Students who were unable to formally handle limits did not possess rich or strong enough concept images for them to handle the required abstraction levels. Students' existing concept images, when they come to university courses, often need adjustments, and this is an obstacle to overcome before further abstractions are possible. Deficient, and sometimes intuitive, conceptions of absolute values, inequalities, limits, and infinity were some of the students' impediments when they started their courses (Juter, 2006c).

Many students complete basic courses in calculus without ever being able to link procepts (Gray & Tall, 1994) of limits to embodied or formal representations (Tall, 2004), or even being aware of the missing links. Several disconnections in their learning developments have been detected in the present study—for example, between intuitive and formal perceptions, procedural and static perceptions, and the finite and the infinite, which are all cognitively demanding to link together. The lack of interaction between students and lecturers or teachers, reported from the field notes, may be a reason for the students' struggles to see connections of various kinds, such as those between everyday problem solving and the formal definition. Added to the disconnections are the students' inclinations to misjudge their own abilities or the accuracies of their mental representations. Students who are convinced that their perceptions of limits are coherent and correct do not feel any urge to recreate mental representations of the concept. Students need to become aware of their misrepresentations to adjust them appropriately. For that to happen, they must be in a situation where incoherent parts of their concept images are evoked simultaneously and the inconsistencies become apparent. Connectedness and continuity are essential features of teaching and learning limits.

REFERENCES

Bergsten, C. (2006). Trying to reach the limit—The role of algebra in mathematical reasoning. In J. Novotná, H. Moraová, M. Krátká, & N. Stehliková (Eds.), *Proceedings of the 30th Conference of the International Group for the Psychology of Mathematics Education* (Vol. 2, pp. 153–160). Prague: PME.

Bezuidenhout, J. (2001). Limits and continuity: Some conceptions of first-year students. *International Journal of Mathematics Education and Science Technology, 32*(4), 487–500.

Cornu, B. (1991). Limits. In D. Tall (Ed.), *Advanced mathematical thinking* (pp. 153–166). Dordrecht: Kluwer Academic Publishers.

Cottrill, J., Dubinsky, E., Nichols, D., Schwingendorf, K., & Vidakovic, D. (1996). Understanding the limit concept: Beginning with a coordinated process scheme. *Journal of Mathematical Behaviour, 15*, 167–192.

Dreyfus, T. (1991). Advanced mathematical thinking processes. In D. Tall (Ed.), *Advanced mathematical thinking* (pp. 25–41). Dordrecht, The Netherlands: Kluwer Academic Publishers.

Gray, E., & Tall, D. (1994). Duality, ambiguity and flexibility: A procedural view of simple arithmetic. *The Journal for Research in Mathematics Education, 26*(2), 115–141.

Hemmi, K. (2006). *Approaching proof in a community of mathematical practice.* Doctoral thesis. Stockholm: Stockholm University.

Hiebert, J., & Carpenter, T. P. (1992). Learning and teaching with understanding. In D. A. Grouws (Ed.), *Handbook of research on mathematics teaching and learning* (pp. 65–97). New York: Macmillan.

Hähkiöniemi, M. (2006). *The role of representations in learning the derivative.* Unpublished doctoral thesis. University of Jyväskylä, Finland.

Juter, K. (2005a). Limits of functions—How do students handle them? *Pythagoras, 61*, 11–20.

Juter, K. (2005b). Students' attitudes to mathematics and performance in limits of functions. *Mathematics Education Research Journal, 17*(2), 91–110.

Juter, K. (2005c). Limits of functions: Traces of students' concept images. *Nordic Studies in Mathematics Education, 10*(3/4), 65–82.

Juter, K. (2006a). Limits of functions as they developed through time and as students learn them today. *Mathematical Thinking and Learning, 8*(4), 407–431.

Juter, K. (2006b). Limits of functions—Students solving tasks. *Australian Senior Mathematics Journal 20*(1), 15–30.

Juter, K. (2006c). *Limits of functions—University students' concept development.* Doctoral thesis. Luleå: Luleå University of Technology, Department of Mathematics.

Juter, K. (2007). Students' conceptions of limits, high achievers versus low achievers. *The Montana Mathematics Enthusiast, 4*(1), 53–65.

Juter, K. (2009). Learning analysis: Students' starting point. In C. Winsløw (Ed.), *Nordic Research in Mathematics Education—Proceedings from NORMA08 in Copenhagen, April 21–April 25, 2008* (pp. 127–134). Rotterdam, The Netherlands: Sense Publishers.

Juter, K., & Grevholm, B. (2007). Limits and infinity: A study of university students' performance. In C. Bergsten, B. Grevholm, H. Måsøval, & F. Rønning (Eds.), *Relating practice and research in mathematics education. Proceedings of NORMA 05, Fourth Nordic Conference on Mathematics Education* (pp. 337–348). Trondheim, Sweden: Tapir Academic Press.

Lithner, J. (2003). Students' mathematical reasoning in university textbook exercises. *Educational Studies in Mathematics, 52*, 29–55.

Mohammad Yusof, Y., & Tall, D. (1994). Changing attitudes to university mathematics through problem solving. In J. Pedro da Ponte, & J. Filipe Matos (Eds.), *Proceedings of PME 18* (pp. 401–408). Lisbon, Portugal: University of Lisbon.

Monaghan, J. (1991). Problems with the language of limits. *For the Learning of Mathematics, 11*(3), 20–24.

Pettersson, K. (2008). *Algoritmiska, intuitiva och formella aspekter av matematiken i dynamiskt samspel* (Algorithmic, intuitive and formal aspects of the mathematics in dynamic interplay). Doctoral thesis. Göteborg: Göteborg University.

Poincaré, H. (1983). On the nature of mathematical reasoning. In P. Benacerraf, & H. Putnam (Eds.), *Philosophy of mathematics, selected readings* (pp. 394–402). Cambridge: University Press.

Przenioslo, M. (2004). Images of the limit of function formed in the course of mathematical studies at the university. *Educational Studies in Mathematics, 55*, 103–132.

Sfard, A. (1991). On the dual nature of mathematical conceptions: Reflections on processes and objects as different sides of the same coin. *Educational Studies in Mathematics, 22*, 1–36.

Sierpinska, A. (1990). Some remarks on understanding in mathematics. *For the Learning of Mathematics, 10*(3), 24–36.

Star, J. (2000). On the relationship between knowing and doing in procedural learning. In B. Fishman & S. O'Connor-Divelbiss (Eds.), *Fourth International Conference of the Learning Sciences* (pp. 80–86). Mahwah, NJ: Erlbaum.

Strauss, A., & Corbin, J. (1998). *Basics of qualitative research, techniques and procedures for developing grounded theory.* Thousand Oaks, CA: Sage Publications.

Szydlik, J. (2000). Mathematical beliefs and conceptual understanding of the limit of a function. *Journal for Research in Mathematics Education, 31*(3), 258–276.

Tall, D. (1980). Mathematical intuition, with special reference to limiting processes. In R. Karplus (Ed.), *Proceedings of the Fourth International Conference for the Psychology of Mathematics Education* (pp. 170–176). Berkeley, CA: PME.

Tall, D. (1991). The psychology of advanced mathematical thinking. In D. Tall (Ed.), *Advanced mathematical thinking* (pp. 3–21). Dordrecht, The Netherlands: Kluwer Academic Publishers.

Tall, D. (2004). Thinking through three worlds of mathematics. In M. Johnsen Høines & A. B. Fugelstad (Eds.), *Proceedings of the 28th Conference of the International Group for the Psychol-*

ogy of Mathematics Education (pp. 281–288). Bergen, Norway: Bergen University College.

Tall, D., & Vinner, S. (1981). Concept image and concept definition in mathematics with particular reference to limits and continuity. *Educational Studies in Mathematics, 12,* 151–169.

Viholainen, A. (2006). Why is a discontinuous function differentiable? In J. Novotna, H. Moraová, M. Krátká & N. Stehlíková (Eds.), *Proceedings of the 30 th Conference of the International Group for the Psychology of Mathematics Education* (vol. 5, pp. 329–336). Prague: Charles University.

Vinner, S. (1991). The role of definitions in the teaching and learning of mathematics. In D. Tall (Ed.), *Advanced mathematical thinking* (pp. 65–81). Dordrecht: Kluwer Academic Publishers.

Williams, S. (1991). Models of limit held by college calculus students. *Journal for Research in Mathematics Education, 22*(3), 219–236.

Williams, S. (2001). Predications of the limit concept: An application of repertory grids. *Journal for Research in Mathematics Education, 32*(4), 341–367.

30

THE ROLES OF PRIOR KNOWLEDGE WHEN STUDENTS INTERPRET MATHEMATICAL TEXTS

Magnus Österholm

UMEÅ UNIVERSITY

Abstract: In this chapter I examine what roles different types of prior knowledge have in the comprehension process when reading mathematical texts. Through theoretical analyses, three central aspects are highlighted: cognitive structure, cognitive process, and metacognition. For all these three aspects, questions arise regarding relationships between general and content-specific types of prior knowledge. Some empirical studies are described that study these questions.

Keywords: cognitive perspective, mathematical text, prior knowledge, reading comprehension

INTRODUCTION

It may seem quite obvious that more or better prior knowledge is beneficial when a reader tries to understand a mathematical text, and empirical studies generally show a connection between prior knowledge and performance of different kinds (Dochy, Segers, & Buehl, 1999). However, when examining this statement in more detail it becomes clear that much needs to be theoretically refined and empirically examined about the roles of prior knowledge when interpreting a mathematical text. For example, what is meant by "more" or "better," what types of knowledge are relevant, what are the relationships between content-specific types of knowledge and more general types of knowledge, and, from an empirical perspective, how can you examine whether a person has "more" or "better" prior knowledge and how can you examine different types of knowledge? These questions need not be limited to *if* a specific type of prior knowledge affects reading comprehension, but can also include the attempt of *explaining how* this can happen, in order to get some detailed insight into the process of reading and comprehending a mathematical text. Some of these questions will be addressed in this chapter.

In the following, similarly to Dochy, Segers, and Buehl (1999), prior knowledge refers to the whole of a person's knowledge. In particular, the notion of prior knowledge is not limited to declarative or conceptual knowledge, but also includes procedural and metacognitive components. Of course, all types of prior knowledge cannot be covered here, but different theoretical frameworks regarding reading comprehension and perspectives on different types of knowledge will be used in order to cover some of the types of prior knowledge that seem most relevant for the study of reading comprehension of mathematical texts.

Although prior knowledge is a very multifaceted notion, a literature review showed that "very few authors explain what they mean by the term prior knowl-

The First Sourcebook on Nordic Research in Mathematics Education, pages 431–440

edge or give an adequate description of the methods used to assess prior knowledge" (Dochy et al., 1999, p. 149). Some empirical studies use easily accessible variables, such as school grades or number of courses taken, as a measure of prior knowledge. This is a very imprecise method for examining prior knowledge. For example, it is unclear what type of knowledge is measured; grades at the Swedish upper secondary level can be seen as including judgments of at least six different types of competencies (Palm, Bergqvist, Eriksson, Hellström, & Häggström, 2004). Therefore, it can be difficult to interpret conclusions about the roles of prior knowledge from studies that use grades as a measure of prior knowledge and to compare conclusions from different such studies. Also, when using these more general measures of prior knowledge, it becomes difficult to try to *explain* any possible relationships between prior knowledge and other variables.

In order to gain more specific insight into the roles of prior knowledge, in general and in particular when reading mathematical texts, it is necessary to use specific measures of prior knowledge when conducting empirical studies. The notion of different competencies has been mentioned in relation to grades in the Swedish upper secondary school. Competencies have also been studied from a more general perspective by, for example, Niss and Jensen (2002), and such a framework can be one way to specify what type of knowledge is measured, in order to be able to describe the study of prior knowledge in more detail. Besides being able to give a more detailed description of different types of prior knowledge, in order to explain something it is also necessary to have a close connection between this description and the situation/process that is being studied. Schoenfeld's (1985) description of different types of knowledge that are being used in the problem-solving process can be seen as an example of this type of close connection. Another possibility could be to use a theoretical framework that focuses on the situation/process that is being studied in combination with a general framework that focuses on different types of knowledge (such as the framework by Niss and Jensen, 2002).

As with all phenomena, the roles of prior knowledge when reading can be studied from many different perspectives—for example, by using different theoretical frameworks. In this chapter, the reading of mathematical texts will be studied from a cognitive perspective, where a theoretical framework focusing on the comprehension process (Kintsch, 1998) will be used, together with different perspectives on knowledge (more details will be given in the section "Theoretical Accounts").

The supposed beneficial effect of more or better prior knowledge was mentioned above, but there can also be a negative effect of prior knowledge. This has, for example, been studied when focusing on conceptual change, where some type of prior knowledge can be seen as a hindrance to learning (e.g., see Merenluoto, 2005). The work of Schoenfeld (1985) can be seen as an example of a positive perspective, when focusing on the use of different types of prior knowledge when solving problems, such as resources, heuristics, and control. However, his framework also includes a component—belief system—that can act as a hindrance in the problem-solving process, for example by the possibility that "students' beliefs about the nature of mathematical thinking... prevent them from solving a problem that they are capable of solving with ease" (Schoenfeld, 1985, pp. 35–36). Other studies have focused on reading comprehension (or reading ability) as a type of prior knowledge in relation to the solving of some tasks or problems. For example, a conclusion from an analysis of the Nordic results from PISA is that, at least for some types of items, "the influence of reading proficiency on students' ability to solve maths items was obvious" (Roe & Taube, 2006, p. 154). However, this conclusion is solely based on the study of correlations between test results, and a qualitative analysis of the relationship between reading and problem solving resulted in the conclusion that the relationship is more complex, where the two processes (i.e., reading and problem solving) are more intertwined (Österholm, 2007). This highlights the necessity of studying the reading process in itself, and not always reducing reading to a prerequisite for problem solving.

PURPOSE AND METHOD

How can it be explained that different persons create different understandings when reading the same text? This question functions as a background to this chapter, where I will examine, from a cognitive perspective, what roles different types of prior knowledge have in the comprehension process when reading mathematical texts. Relationships between different types of prior knowledge will also be analyzed, particularly relationships between content-specific types of prior knowledge and more general types.

The analyses will be made from theoretical as well as empirical perspectives. In the theoretical part, models for describing and explaining the utilization of prior knowledge will be discussed. In particular, a framework describing the *comprehension process*

(Kintsch, 1998) will be used, together with theories focusing on different types of prior knowledge, such as Schoenfeld's (1985) framework and the theory of *content literacy* (McKenna & Robinson, 1990). The theoretical perspective will function as a foundation for the empirical studies, where, for example, different methods for examining prior knowledge are used in order to test the validity of the theoretical perspectives on the one hand, and to examine relationships between different types of prior knowledge and reading comprehension on the other hand.

THEORETICAL ACCOUNTS

In this section, a general theory of reading comprehension will be described, focusing on the process of creating meaning from reading a text (Kintsch, 1998). From this theory, some relevant types of prior knowledge will be inferred. Thereafter, more types of prior knowledge will be discussed, relating to the different types of prior knowledge described in the framework of Schoenfeld (1985).

The Process of Reading Comprehension

One way to model both the meaning of text content and also mental structure and content (i.e., memory) is to use associative networks of propositions, which can be viewed as "an important theoretical and methodological advance in text comprehension" (Lorch & van den Broek, 1997, p. 222). An associative network is built up by nodes and links between nodes that "are unlabeled and vary in strength" (Kintsch, 1998, p. 74). A node can be a singular word, which gets its meaning from the existing links to other nodes, or a type of meaning unit (a proposition). The activation and utilization of prior knowledge can be seen as the central aspect of reading comprehension, where, for example, most of the individual words need to be recognized in order to create meaning from the sentences in a text. Much of this activation and utilization happens in a quite automatic and unconscious manner (Jefferies, Lambdon Ralph, & Baddely, 2004; Kintsch, Patel, & Ericsson, 1999). This process can be described by the CI model (Kintsch, 1998), which consists of two steps: construction (C) and integration (I). Due to the limits of working memory, only small parts of a text can be processed at one moment. From such a small part of the text, an associative network of propositions is created in working memory, together with possible existing associations in long-term memory (i.e., prior knowledge of the reader). This is the construction phase of the process. The activation of prior knowledge happens through local associations (i.e., only based on single nodes), and many of the activated nodes in long-term memory can therefore be irrelevant for a specific situation. In the integration phase, a so called spreading activation process occurs, in which "clusters of highly interconnected propositions attract most of the activation in the network, thus deactivating sparsely interconnected portions of the network as well as nodes with negative links" (Kintsch, 1992, p. 153).

This integration phase gives the content of working memory after reading the present part of the text. Thereafter, nodes with highest activation are kept in working memory when the next part of the text is processed in the same manner. This cyclic process makes it possible to create a coherent representation of the text (i.e., a meaningful interpretation of the text), which also can be connected to parts of long-term memory if such associations have been possible. The CI model can be described in more detail, and when simulating reading comprehension using this model, the result is very similar to empirical data when actual students recall texts that have been read (Kintsch, 1998).

Based on this model of the comprehension process, a specific type of test of prior knowledge can be created. Since the essential aspect of the process is the activation of associative links, this could be used in a test of prior knowledge by letting readers, prior to reading the text, see singular words or expressions from the text and give their spontaneous associations to each word/expression. A high number of associations could be seen as beneficial since these can create a more interconnected network, which could give better conditions for creating a more coherent mental representation of the text. However, the *type* of associations must also be highly relevant, and perhaps more relevant than the *number* of associations, since a totally random association seems unlikely to contribute to the creation of a coherent mental representation of a text. Thus, associations should be relevant; that is, they ought to deal somehow with an appropriate meaning of the word. Also, in order to be able to create a well-organized and coherent mental representation, it seems reasonable that it is better if associations also are well organized and coherent. Langer (1984) has created a framework that describes organization of knowledge, which can be used in order to categorize associations according to whether prior knowledge is highly, partially, or diffusely organized. Empirical studies by Langer also showed a relevance of organization of knowledge in relation to reading.

This type of test of prior knowledge focuses on the specific content of a text since it uses specific words from the text. However, there are also other types of associations that could influence the comprehension process. For example, the type of text that is being read or the situation in which the text is read could activate certain types of nodes from long-term memory that could be included in the reading process and thereby affect the mental representation of the text. If the reader has some specific beliefs about mathematics or mathematical texts and is reading a mathematical text, these beliefs could be activated and affect the comprehension process, and thus affect the interpretation of the text.

Types of Prior Knowledge

The types of prior knowledge that have been discussed so far are the ones that can be activated through associations made in a specific situation when reading a specific text. These types of prior knowledge lie close to the categories of resources and belief system in the framework of Schoenfeld (1985). However, in order to understand a text one also needs to be able to read, which includes the utilization of fundamental techniques for "decoding" the symbolic codes in texts and perhaps also the utilization of certain reading strategies in order to make sense of a text. This type of prior knowledge lies close to Schoenfeld's notion of heuristics.

There seems to be a general agreement that students need to be taught how to read mathematical texts (Burton & Morgan, 2000; Konior, 1993; Krygowska, 1969); that is, that one needs to be able to use certain specific types of reading strategies for the comprehension of mathematical texts. However, this belief does not seem to be based on research, since there seems not to be much research done about the more detailed use of mathematical texts in learning situations (Fenwick, 2001; Laborde, 1990; Love & Pimm, 1996). Thus, the question whether it is necessary to develop a special kind of reading ability for mathematical texts is worth examining. McKenna and Robinson's (1990) notion of *content literacy* describes the possible necessity of such specific types of reading abilities for different domains (or for different types of texts in general). McKenna and Robinson distinguish between three components of content literacy: general literacy skills, content-specific literacy skills, and prior knowledge of content. Both the general and the content-specific literacy skills can be assumed to refer to some more general type of knowledge that is not dependent on the detailed content of a specific text. This type of knowledge is primarily used to "decode" texts. The third component of content literacy, prior knowledge of content, refers to knowledge that is connected to the content of a specific text. The activation through associations discussed earlier can of course include the component of prior knowledge of content, but can perhaps also include more general aspects: that the organization of knowledge discussed earlier might be a general property of a person's long-term memory, and also that the process of activation might differ between persons—for example, regarding speed and accuracy. These more general aspects are located within the components of general and content-specific literacy skills, together with any other types of techniques or strategies needed to interpret texts in general and texts within a specific domain.

Perhaps there are properties of mathematics that create certain special properties of mathematical texts, which in turn create a need for content-specific literacy skills in mathematics. The symbolic language is perhaps the most obvious property of mathematics that might create a need for such literacy skills. To understand a text written purely with mathematical symbols one might not even need general literacy skills, that is, one might not need to know how to "read" (in the sense perhaps most commonly used, i.e., reading texts written in a natural language). However, the perhaps most common mathematical texts use both the symbolic language and a natural language, where one would seem to need both general and content-specific literacy skills. But what about mathematical texts that do not contain any symbols at all? Are there other special properties of these texts (since they are mathematical texts) that demand other types of content-specific literacy skills, or would it be enough for the reader to use general literacy skills? These questions will be addressed in the empirical part of this chapter.

The fourth component from Schoenfeld's (1985) framework, which has not been discussed yet, is control. This component refers mainly to metacognitive aspects, such as planning and monitoring, which can be regarded as a more general type of prior knowledge and not connected to the content of a specific text. However, whether metacognitive knowledge can be regarded as domain general (and thus included in general literacy skills) or domain specific (and thus included in content-specific literacy skills) has been investigated without coming to any clear-cut result (Veenman, Van Hout-Wolters, & Afflerbach, 2006).

Monitoring can be seen as the crucial part of the control component, in the process of self-regulation, since "failure to adequately evaluate comprehension . . . may halt the execution of strategy use, resulting in less promising learning" (Lin & Zabrucky, 1998, p. 388). In order to evaluate the level of comprehension, one needs some kind of criteria for comprehension. However, these criteria might be quite implicit in a reading situation, activated only at an unconscious level (van den Broek, Risden, & Husebye-Hartmann, 1995)—for example, resulting in a feeling of (not) understanding, without performing any explicit actions to evaluate the comprehension. Epistemological beliefs are suggested to act as a foundation for criteria for comprehension (Hofer & Pintrich, 1997), and a qualitative study of students' use of criteria for comprehension when reading mathematical texts and their beliefs noted several possible connections between these two variables (Österholm, 2006b). Thus, beliefs have been noted to be able to affect reading comprehension in two different ways. On the one hand, beliefs can be activated and utilized in the reading process through associations, and on the other hand, the influence can happen via self-regulation.

The discussion about different types of prior knowledge can be summarized by the following three aspects. For each of these three aspects, there is an open question about the relationships between general and content-specific aspects. Such questions will be examined in the next section.

- Cognitive structure—associations to words and phrases in a text: Is the organization of knowledge a general property of a person's long-term memory, or is it more closely connected to specific content?
- The cognitive process—content literacy skills: Is there a need for some content-specific literacy skills when reading mathematical texts, or is the main component of the comprehension process for these texts the general literacy skills?
- Metacognition: Is metacognitive knowledge a general type of knowledge that can be applied in all domains and all situations, or is this type of knowledge (in particular, the ability to evaluate one's own knowledge) more dependant on the specific situation?

EMPIRICAL ACCOUNTS

In this section, two empirical studies will be described that use a similar procedure for collecting and analyzing data. This procedure will not be described and problemized in detail here due to space limitations, but references will be given to other publications where more details can be found.

Methods and Purposes

In the two empirical studies, hereafter referred to as study A and B, students from the end of the upper secondary level and the beginning of the university level voluntarily participated. The students worked individually with self-instructed material in which they gave all answers in writing. The procedure, which took about one hour, can for the two studies be summarized as follows.

In study A (for more details, see Österholm, 2006a; Österholm, 2006b, chapter 4), in which 95 students participated, the students first gave associations to specific words and phrases. Thereafter they read two texts, one mathematical and one historical. After each text they answered questions about the content of the text, in order to test their reading comprehension. Two versions of the mathematical text were used; both versions presented to the reader some basic concepts in group theory (*mathematical system* and *group*), but one text does so without using any specific mathematical symbols while the other uses such symbols. The two different mathematical texts were randomly distributed among participants. The historical text describes the events in Russia during the Russian Revolution, and it does not use any mathematical symbols.

In study B (for more details, see Österholm, 2006b, chapter 8), in which 91 students participated, the students first gave associations to specific words and phrases, after which they also answered a questionnaire about their epistemological beliefs related to mathematics. Thereafter, they read two texts, one about the concept of absolute value and one about the procedure of partial fraction decomposition. After each text the students first got to evaluate their own comprehension, after which they also answered questions about the content of the text, in order to test their reading comprehension.

For reading comprehension, one can distinguish between three different levels or components of comprehension: the surface component, the textbase, and the situation model (Kintsch, 1998). The surface component includes the memory of specific words and phrases from the text, where the *meaning* of these words/phrases is not included. The textbase includes the meaning of the text, that is, the semantic structure of the text, and it "consists of those elements and relations that are directly derived from the text itself... without adding anything that is not explicitly specified in the text" (Kintsch, 1998, p. 103). The situation model is an elaboration of the textbase using prior knowledge that connects to and extends the specific content of a text. The two studies (A and B) both used different types of questions in the test of reading comprehension—some aiming for testing a created textbase and some for testing a created situation model. However, since this chapter focuses on students' interpretation of the text content, and not on how this is integrated with other types of prior knowledge, the questions that will be used in the analysis are those testing a created textbase. These questions are constructed so that they could be answered with some explicit information given in the text. But in order not to measure pure memorization of the texts (i.e., the surface component), the questions are not formulated in ways that resemble specific formulations given in the texts.

For all mentioned parts of the procedure (the test of associations, the test of beliefs, the test of reading comprehension, and the self-evaluation) the students' answers were analyzed in order to create a quantitative measure of each variable. For the test of associations, the theory of organization of knowledge (Langer, 1984) was used in order to categorize students' answers, where higher level of organization gave higher points. For the test of beliefs, a Likert scale was used, where higher points were given to responses that more fully agreed with certain statements. For the test of reading comprehension, points were given according to how complete and correct the answers were in relation to the content of the text. Finally, for the self-evaluation, the answers themselves were of a quantitative kind, where the students answered on a four-point scale how much of the text they had comprehended, and also how many of five questions and five tasks they thought they could answer correctly.

For all measures, the scores from several questions were used to build up the total score of a specific variable. The detailed statistical analyses then used are studies of correlations between different variables and of comparisons between different subgroups of students regarding the same variable.

Regarding the test of associations, one purpose of using this special kind of test of prior knowledge is to test the validity of the theoretical framework used, by examining whether there is a positive correlation between this test and the test of reading comprehension. The test of associations will also be examined in relation to a more general type of measure of prior knowledge—the students' grades—in order to examine if the theoretically grounded measure has a stronger connection to reading comprehension.

The test of associations was divided into two parts, one part each for the two texts read by the students. Thus, in study A, one part of the test examines associations to mathematical content and the other part to historical content. Relationships between these two parts and the reading comprehension of the two texts will be studied in order to examine whether this type of test seems to measure a more general type of prior knowledge (organization of long-term memory) or a more content-specific type of prior knowledge.

The effect of beliefs on reading comprehension will be studied by examining connections between beliefs and reading comprehension, and also between beliefs and other measures of prior knowledge. Four different types of epistemological beliefs are studied: to what extent one regards mathematics as (1) focused on symbols, (2) authoritarian, (3) focused on facts, and (4) focused on procedures and methods. In a previous study, these types of beliefs were noted as relevant when students read mathematical texts (Österholm, 2006b, chapter 6). Beliefs are not directly connected to the specific content of the texts used in the studies, and are thereby more general types of prior knowledge (but perhaps specific to mathematics).

A fifth type of belief was also used in the questionnaire: belief about one's own ability and level of knowledge in mathematics. This type of belief will primarily be part of the study of metacognitive aspects since other studies have noted the relevance of this type of belief when readers try to evaluate their own reading comprehension (the effect of domain familiarity—Glenberg & Epstein, 1987).

Relationships between the measures of reading comprehension of different texts will be studied in order to examine the generality of reading ability. In study A, texts from different domains are studied in order to examine if there is a need for some content-specific literacy skills for mathematical texts. In particular, the effect of the use of symbols in mathematical texts is studied. In study B, another type of variation for texts within the domain of mathematics is studied by using texts that focus on different types of knowledge—conceptual or procedural.

Finally, regarding the study of metacognitive aspects, there are methodological difficulties when trying to examine students' ability to evaluate their own reading comprehension. Here, within study B, analyses will focus on the characterization of students' self-evaluations. In order to study a more direct connection between reading comprehension and the ability to evaluate one's own reading comprehension, each participant would need to read more than two texts. Due to the length of the data-collecting procedure, this was not done in study B.

Results and Analysis

The analysis will focus on relations between two variables in one group of participants, using the Pearson correlation coefficient (r), and on comparisons between two subgroups of participants regarding the same variable, using the t-test. Unless otherwise noted, all reported correlations and differences are statistically significant at level 0.05.

For all texts, that is, the three in study A and the two in study B, there is a positive correlation between the test of associations and the test of reading comprehension (with *r*-values between 0.23 and 0.36). This shows that this type of test of prior knowledge is relevant in relation to reading comprehension. Also, since the values of the correlation coefficients are of similar size for all texts, the test of associations seems relevant for texts in general, that is, that associations play a significant role in the reading process.

There is also a positive correlation between the students' mathematics grades and the test of reading comprehension for the text about group theory in study A and the text about absolute value in study B (with *r*-values between 0.37 and 0.48), but not for the text about partial fraction decomposition in study B (although the value is close to significance: $r = 0.21$, $p = 0.06$). Whether this difference in significant correlations is due to the fact that the texts focus on different types of knowledge (conceptual or procedural) is unclear, due to the rather small difference in the statistical values. Also, due to the breadth of grades as a measure of prior knowledge, such a possible difference is difficult to interpret.

When partial correlations are studied for the texts focusing on conceptual knowledge, both the partial correlation between the test of associations and the test of reading comprehension when controlling for the grades and also the partial correlation between the grades and the test of reading comprehension when controlling for the test of associations are still significant. The decrease of the *r*-value is around 0.08 when calculating the partial correlation compared to the ordinary correlation. Thus, the two types of measures of prior knowledge (the test of associations and the grades) are both relevant for reading comprehension, and they can be seen as two separate sources of variation in reading comprehension.

In order to study the nature of the test of associations regarding relationships between general and content-specific knowledge, correlations between all different tests of associations and all tests of reading comprehension are examined (separately for study A and B since different students participated in the two studies). In both studies there is a positive correlation between the two different tests of associations, which suggests that this type of test to some degree measures a general type of prior knowledge. This conclusion is strengthened by existing correlations between reading comprehension for a text and a test of associations not created for the same text. In study B this type of correlation exists for both texts, while in study A there is a correlation between the test of associations for the history text and reading comprehension for the mathematics texts, but not between the test of associations for the mathematics texts and reading comprehension for the history text. Since these types of "mixed" correlations have appeared both for different types of mathematical texts (in study B, texts focusing on different types of knowledge) and also partly for different domains (in study A, mathematics and history), the test of associations seems to measure a general type of prior knowledge. However, since not all tests of associations correlated with all tests of reading comprehension, the test of associations cannot solely measure a general type of prior knowledge; it must also have a more content-specific component.

Five different types of beliefs (four epistemological and one about own level of knowledge in mathematics) are used in study B, and these beliefs are examined in relation to the two different measures of prior knowledge (grades and test of associations) and to the reading comprehension of the two different texts (one focusing on conceptual knowledge and the other on procedural).

The students' grades correlate negatively with all four types of epistemological beliefs (with *r*-values between −0.25 and −0.47) and positively with the belief about one's own level of knowledge ($r = 0.64$). The test of associations correlates negatively with two types of epistemological beliefs (with *r*-values −0.23 and −0.27) and positively with the belief about one's own level of knowledge ($r = 0.32$). Thus, there is a stronger connection between grades and beliefs than between the test of associations and beliefs. When calculating partial correlations, this difference becomes even clearer. There are no significant partial correlations between the test of associations and beliefs when controlling for grades, and all partial correlations except one (which lies close to significance, $p = 0.056$) between grades and beliefs are significant when controlling for the test of associations.

Two of the four types of epistemological beliefs have a significant correlation with reading comprehension. The belief that mathematics is focused on facts correlates negatively to the reading comprehension of both texts, while the belief that mathematics is authoritarian correlates negatively only to the reading comprehension of the text about absolute value. In order to test if epistemological beliefs can be seen as a relevant and separate type of prior knowledge in relation to reading comprehension of mathematical texts, partial correlations between beliefs and reading comprehension for the two types of text are calculated when controlling for grades and the test of associations. In this case, one correlation is significant: the correlation between the belief that mathematics is focused on facts and the reading comprehension of the text about partial fraction decomposition. Firstly, these analyses show a difference between the two types of texts regarding what types of prior knowledge are relevant for the comprehension. Secondly, although some correlations existed between beliefs and reading comprehension, in general there are no clear and independent effects of beliefs on reading comprehension of mathematical texts in these data.

In study A, there is a correlation between reading comprehension of the mathematical text without symbols and the historical text, while no correlation exists between reading comprehension of the mathematical text with symbols and the historical text. These results show a similarity in reading comprehension between the mathematical text without symbols and the historical text, and a difference in reading comprehension between the mathematical text with symbols and the historical text. The existing correlation, which exists also when controlling for the tests of associations, can be interpreted as showing general literacy skills in function since it shows a connection between reading comprehension of texts from different domains that lies outside the effect of organization/content of prior knowledge. Since previous analyses of the effect of different measures of prior knowledge have shown a similarity between all different texts, the non-existing correlation regarding reading comprehension must be due to some other type of knowledge or ability. A comparison between the two groups who read the two different mathematical texts, with or without symbols, shows a significant difference regarding their results on the test of reading comprehension, with a better result for the students who read the text without symbols.

In study B, there is a correlation between reading comprehension of the two types of mathematical texts. However, this correlation vanishes when controlling for the tests of associations, which shows that the connection between these two types of texts mainly comes from an effect of organization/content of prior knowledge.

Table 30.1 shows a summary of relationships between students' self-evaluations and their reading comprehension. As previously noted, results from other studies have shown that students sometimes seem to use more general types of insights into their own knowledge (such as their beliefs about their own level of domain knowledge) instead of utilizing the specific experience of reading the text when performing a self-evaluation of reading comprehension. Therefore, partial correlations are also calculated where grades and belief about one's own level of knowledge are controlled for. Compared to the values in Table 30.1, a general decrease of the correlation coefficients then occurs (of magnitude around 0.1), but the significances and proportions are the same.

TABLE 30.1 Correlation Coefficient Between Students' Self-Evaluations and Reading Comprehension, Depending on Type of Text and Whether the Text Was Read First or Second

Text	Read first	Read second
Absolute value	0.58*	0.73*
Part. fract. decomp.	0.24	0.42*

* Significant correlation (level 0.05).

From the values in Table 30.1, at least two relationships can be noted. There is a general difference in the magnitude of the correlation coefficient (i.e., regarding how well students manage to evaluate their own comprehension) when comparing, on the one hand, the two different texts and, on the other hand, whether a text is read first or second. These results suggest that the ability to evaluate one's own reading comprehension is not a general type of ability, but that the evaluation is more dependent on the specific type of text and the specific type of situation.

CONCLUSIONS

By using aspects from theoretical perspectives of the reading process, in this chapter I have shown the possibility of creating a specific test, and a simple test, that is relevant in relation to reading comprehension: the test of associations. A similar result has also been noted by Langer (1984). Indirectly, this type of result strengthens the validity of Kintsch's (1998) theory of comprehension, in particular that associations play a relevant role in the process of reading comprehension. Grades have also shown to be connected to reading comprehension, but this measure of prior knowledge can include many different types of knowledge, which makes it difficult to use grades in analyses when trying to get more in-depth theoretical insight into the process of reading comprehension. The more specific construct of the test of associations is more suitable for such theoretical endeavors. However, there are still some uncertainties about the test of associations regarding relationships between general and content-specific aspects of knowledge.

The test of associations, as it has been constructed in this chapter, includes both general and content-specific aspects of prior knowledge. The organization of knowledge is the general aspect and the use of specific words is the content-specific aspect. In several cases, the general aspect seems to dominate in the connection to reading comprehension since the same test has an equally strong connection to texts with different content. The content-specific aspect was also evident on one occasion since one test was connected to reading comprehension of the text using the words included in the test but was not connected to reading comprehension of another text. From the present study it is not possible to explain why different tests have such different properties in relation to different texts—whether this is due to some type of effect of differences between domains (the content-specific aspects appeared when studying mathematics and history texts) or if some other property of the words is responsible. More studies are needed for this—for example, to study the use of this type of test of prior knowledge for several different types of texts where the words used in the texts are chosen in a well-defined manner. One could also use another type of theory, instead of the theory of organization of knowledge, when judging the quality of associations and examine how this affects the connection to reading comprehension. Such a study could show other relevant properties of prior knowledge, which could be general or content-specific, depending on what type of theory is used.

Beliefs, and epistemological beliefs in particular, have been noted as possibly relevant in relation to reading comprehension in two ways: through activation by associations or through a process of self-regulation. Among four different types of epistemological beliefs, the belief that mathematical knowledge is focused on facts seems most relevant for affecting reading comprehension; the stronger this belief is, the worse the result is on a test of reading comprehension. The results also show a difference between two different types of text regarding the effect of beliefs. Thus, beliefs seem to have quite a weak effect on reading comprehension, and this small effect is not similar for texts in general.

Reading ability does not seem to be a homogenous construct, since the results have shown a lack of correlation between reading comprehension for different types of texts. However, the main difference does not appear to be between texts from different domains, but other properties of texts seem more relevant, in particular the use of symbols in mathematical texts. It is unclear whether the results regarding reading comprehension for the mathematical text with symbols are due to an inability among the participants (i.e., that they need some kind of reading ability they do not possess) or if the participants were using some sort of special reading strategy for this text. Such possible strategy was not beneficial since the reading comprehension was worse for this text compared to the text without symbols. The results from another study about students' reading behavior for mathematical texts show that students tend to focus on symbols when such exist in a text (Österholm, 2006b). This behavior could explain the previous results regarding mathematical texts with or without symbols: If some parts of a text (natural language) do not get the same attention as others (symbols), this could of course cause the reader to miss certain parts of the content of the text, which could explain the worse result on a test of reading comprehension for the text with symbols. Also, if a text does not include any symbols, the students cannot utilize this special type of reading strategy and will probably read the text in a fashion more similar to "ordinary" texts, which would explain the connection regarding reading comprehension between the mathematical text without symbols and the historical text.

Regarding a possible need for some content-specific literacy skills, it seems more relevant to see the variety among mathematical texts than to focus on mathematical texts as something special compared to texts from other domains. This conclusion is strengthened by the result from study B, where the correlation between reading comprehension for two types of mathematical texts that focus on different types of knowledge mainly was due to an effect of existing organization/content of associations.

The result showing a general improvement in self-evaluation for the second text read compared to the first text read highlights a type of ability that is adaptive for different situations. This improvement happened for two types of texts, thus showing a general type of prior knowledge: the ability to adjust one's self-evaluation for the second text based on the experience of reading, evaluating, and answering questions about the first text. A more content-specific aspect of this metacognitive ability also exists, since the results showed a difference between the two types of texts regarding the accuracy of students' self-evaluations.

From the results presented here, it is difficult to give a complete picture of reading comprehension for mathematical texts, such as how all different types of prior knowledge are connected to each other and in conjunction to reading comprehension. This was also not the purpose of this chapter, but it has given results of a more "local" nature, where the different types of prior knowledge have been examined in relation to reading comprehension. However, there is one perspective that has been present for all types of prior knowledge, giving the opportunity to also give a somewhat more comprehensive picture: the relationships between general and content-specific aspects of prior knowledge. A non-general, and thereby more content-specific, aspect has been noted for all types of prior knowledge: associations, beliefs, reading ability, and metacognition. Most of these types of prior knowledge have also shown a more general aspect: associations, reading ability, and metacognition. Thus, in general there is a need to consider both general and content-specific aspects of different types of prior knowledge. However, the content-specificity does not seem to come mainly from the distinction of texts from different domains; other types of differences between texts are more relevant.

REFERENCES

Burton, L. & Morgan, C. (2000). Mathematicians writing. *Journal for Research in Mathematics Education, 31*, 429–453.

Dochy, F., Segers, M., & Buehl, M. M. (1999). The relation between assessment practices and outcomes of studies: The case of research on prior knowledge. *Review of Educational Research, 69*, 145–186.

Fenwick, C. (2001). Students and their learning from reading. *Humanistic Mathematics Network Journal, 24*, 52–58.

Glenberg, A. M. & Epstein, W. (1987). Inexpert calibration of comprehension. *Memory & Cognition, 15*(1), 84–93.

Hofer, B. K. & Pintrich, P. R. (1997). The development of epistemological theories: Beliefs about knowledge and knowing and their relation to learning. *Review of Educational Research, 67*(1), 88–140.

Jefferies, E., Lambdon Ralph, M. A., & Baddely, A. D. (2004). Automatic and controlled processing in sentence recall: The role of long-term and working memory. *Journal of Memory and Language, 51*, 623–643.

Kintsch, W. (1992). A cognitive architecture for comprehension. In H. L. Pick, P. van den Broek & D. C. Knill (Eds.), *Cognition: Conceptual and methodological issues* (pp. 143–163). Washington, DC: American Psychological Association.

Kintsch, W. (1998). *Comprehension: A paradigm for cognition.* Cambridge: Cambridge University Press.

Kintsch, W., Patel, V. L., & Ericsson, K. A. (1999). The role of long-term working memory in text comprehension. *Psychologia, 42*, 186–198.

Konior, J. (1993). Research into the construction of mathematical texts. *Educational Studies in Mathematics, 24*, 251–256.

Krygowska, Z. (1969). Le texte mathématique dans l'enseignement. *Educational Studies in Mathematics, 2*, 360–370.

Laborde, C. (1990). Language and mathematics. In P. Nesher & J. Kilpatrick (Eds.), *Mathematics and cognition: A research synthesis by The International Group for the Psychology of Mathematics Education* (pp. 53–69). Cambridge: Cambridge University Press.

Langer, J. A. (1984). Examining background knowledge and text comprehension. *Reading Research Quarterly, 19*, 468–481.

Lin, L.-M. & Zabrucky, K. M. (1998). Calibration of comprehension: Research and implications for education and instruction. *Contemporary Educational Psychology, 23*, 345–391.

Lorch Jr., R. F. & van den Broek, P. (1997). Understanding reading comprehension: Current and future contributions of cognitive science. *Contemporary Educational Psychology, 22*, 213–246.

Love, E. & Pimm, D. (1996). "This is so": A text on texts. In A. J. Bishop et al. (Eds.), *International Handbook of Mathematics Education* (pp. 371–409). Dordrecht, The Netherlands: Kluwer.

McKenna, M. C. & Robinson, R. D. (1990). Content literacy: A definition and implications. *Journal of Reading, 34*, 184–186.

Merenluoto, K. (2005). Conceptual change in mathematics. *Nordic Studies in Mathematics Education, 10*(2), 17–34.

Niss, M. & Jensen, T. H. (Eds.) (2002). *Kompetencer og matematiklæring: Ideer og inspiration til udvikling af matematikundervisning i Danmark* (Report no 18—2002). Copenhagen, Denmark: Undervisningsministeriets forlag. Retrieved August 5, 2008, from http://pub.uvm.dk/2002/kom/hel.pdf

Österholm, M. (2006a) Characterizing reading comprehension of mathematical texts. *Educational Studies in Mathematics, 63*, 325–346.

Österholm, M. (2006b) *Kognitiva och metakognitiva perspektiv på läsförståelse inom matematik [Cognitive and metacognitive perspectives on reading comprehension in mathematics].* Doctoral dissertation, Department of Mathematics, Linköping University, Sweden. Retrieved August 5, 2008 from http://urn.kb.se/resolve?urn=urn:nbn:se:liu:diva-7674

Österholm, M. (2007). A reading comprehension perspective on problem solving. In C. Bergsten & B. Grevholm (Eds.), *Developing and researching quality in mathematics teaching and learning. Proceedings of MADIF 5, the 5th Swedish Mathematics Education Research Seminar* (pp. 136–145). Linköping, Sweden: SMDF.

Palm, T., Bergqvist, E., Eriksson, I., Hellström, T., & Häggström, C.-M. (2004). *En tolkning av målen med den svenska gymnasiematematiken och tolkningens konsekvenser för uppgiftskonstruktion* (Pm nr 199). Umeå, Sweden: Department of Educational Measurement, Umeå University. Retrieved August 5, 2008, from http://www.umu.se/edmeas/publikationer/pdf/Pm%20nr%20199.pdf

Roe, A. & Taube, K. (2006). How can reading abilities explain differences in maths performances? In J. Mejding & A. Roe (Eds.), *Northern lights on PISA 2003—A reflection from the Nordic countries* (pp. 145–157). Copenhagen: Nordic Council of Ministers. Retrieved August 5, 2008, from: http://www.norden.org/pub/uddannelse/uddannelse/sk/ TN2006523.pdf

Schoenfeld, A. H. (1985). *Mathematical problem solving.* Orlando: Academic Press.

Van den Broek, P., Risden, K., & Husebye-Hartmann, E. (1995). The role of readers' standards for coherence in the generation of inferences during reading. In R. F. Lorch & E. J. O'Brien (Eds.), *Sources of Coherence in Reading* (pp. 353–373). Hillsdale, NJ: Lawrence Erlbaum Associates.

Veenman, M. V. J., Van Hout-Wolters, B. H. A. M., & Afflerbach, P. (2006). Metacognition and learning: Conceptual and methodological considerations. *Metacognition and Learning, 1*, 3–14.

SECTION III

Icelandic Research in Mathematics Education

Guðbjörg Pálsdóttir and Bharath Sriraman, *Section Editors*

31

MATHEMATICS EDUCATION IN ICELAND

Explaining the Non-Homogeneity in a Homogenous System

Guðný Helga Gunnarsdóttir and Guðbjörg Pálsdóttir

UNIVERSITY OF ICELAND, SCHOOL OF EDUCATION

Bharath Sriraman

THE UNIVERSITY OF MONTANA

Abstract: In this opening salvo to the Icelandic section, we synthesize some of the themes present in the different chapters and address aspects that the reader may not realize about Icelandic mathematics education. We briefly outline distinct features and anomalies within mathematics education in Iceland, which in spite of the homogeneity of the system reveals many threads of thought in terms of reform and implementation of new ideas.

Keywords: mathematics education in Iceland

Bjarnadottir, in her opening chapter, outlines the history of mathematics education in Iceland by examining mathematics textbooks and bureaucratic/political decisions from different periods regarding the textbooks that are sanctioned by the state for use in schools. In her chapter, the influence of Danish regulations and universities on Icelandic syllabi is explained in addition to the effects of New Math. Bjarnadottir ends the chapter with the lament that achievement of Icelandic pupils is lacking and there is considerable room for improvement in spite of recent advances in the areas of curriculum development and mathematics education policy.

Mathematics education in Iceland is subject to detailed plans from the central government, and the curriculum guidelines are typically authorless; that is, the documents do not include the names of authors like documents from North America often do. In the compulsory school there is mainly one book series to teach from, which limits options that teachers might seek. As reported by Gunnarsdóttir and Pálsdóttir in their chapter, in the last ten years a new series is being implemented, but teachers can choose from the materials and put an emphasis on what they consider important. The assessment assignments included in the materials have, for example, not been widely used. However, even though a new curriculum is in place, new fads often find a place and even get implemented—for example, Global Education Model of Schooling (GEMS—see http://www.gemsedu.

The First Sourcebook on Nordic Research in Mathematics Education, pages 443–445

org/). This particular model is old news now, but it reveals the tension in the system between individualized teaching and whole-class teaching, which is what teaching materials typically suggest, and guidelines for that style of teaching are provided for teachers.

In his doctoral dissertation and ensuing writings, Savola reports that the Icelandic educational system is in transition and discusses the upheavals in the system from societal and institutional forces. In his video-based study comparing Finnish and Icelandic primary schools, he studied the connections between pedagogical theory and actual classroom practices in these two countries. Among his findings was the trend of learner-based instructional strategies becoming the norm in most Icelandic schools at the expense of content-related public discourse, namely sacrificing whole-class discussions to attend to individual needs. Savola points that Tomlinson's (1999) ideas about *differentiated instruction* became popular in Iceland but not implemented the way the model was intended—that is, teachers attending to dissimilar learning styles, with tiered activities and materials accommodating the diverse learning needs of the students. Instead, he found that students were taught using the same instruction methods and learning materials, and differentiation translated to mean students moving ahead in the same book at a different pace.

In spite of these anomalies, there are no large tensions per se—that is to say, between the teacher education universities, teaching materials, and the curricular guide from the authorities. The development for teaching children up to 16 years and their teachers has mostly been influenced by the teacher educators in Reykjavik and Akureyri. As reported in many of the chapters in this section, practicing teachers are trying to learn about new ideas and implement them. Another salient point in the system is that reform is more or less tried uniformly—everyone goes with the flow. Teachers are willing and interested in trying out new things, but that doesn't necessarily mean big changes in what content has highest priority or which approach and thinking is valued. But teaching methods get attention, hands on activities, and games and calculators are the trend. There is also an emphasis in the importance of children being able to explain their thinking and solutions and problem solving.

Other major influences to mathematics education in Iceland have been the National tests—*Námsmatsstofnun*, psychometricians in charge of large-scale testing and reporting on them, and last but not least the teachers themselves. Námsmatsstofnun is responsible for PISA, which briefly brought Iceland into the spotlight because of the achievement levels of girls in relation to boys (Halldorsson & Olafsson, 2009; Palsdottir & Sriraman, 2010; Steinthorsdottir & Sriraman, 2008). Such reports have supported the need for change and pointed out the need to challenge good students more and that rote learning and exercises won't help. However, the national tests in Iceland place a lot of emphasis on procedural knowledge, the criticism of which is dealt with by claiming that it is difficult to test all areas. In general, though, PISA problems have had some influence on the problems on the standardized tests for the 10th grade.

Another distinct feature of the Icelandic system is that there is no tension between teacher education and curriculum. This can be considered a negative thing for the following reasons.

Teacher educators have been in the leading role in making the teacher education both in service and pre-service, making the curricula. Their professional knowledge and development has had its way. The people at the top or those responsible are few in number and they have similar ideology and views on mathematics learning, though they sometimes disagree. It is negative in the sense that few people with little resources get to decide, and newer people who come into the field are either trained by or encouraged by the very same people. This prevents a heterogeneity of thought or perspectives in mathematics education. In general, choices are restricted—since Icelandic society is small, it is not possible to have many choices, and with the teachers having more education, they should be able to search outside the country for opportunities. Teacher educators try to give teachers access to research and ideas so they can be more independent and make justifiable choices. The system is by and large vulnerable since it is not grounded in research in Iceland itself but based on teachers' practice and trends from the U.S. and other Nordic countries.

The few teacher educators in the country are not in conflict and try to follow what goes on in the outside world and apply what they think is relevant for society in general. Differences do not get discussed to the extent they should, but work gets accomplished together. Unlike other Nordic countries and elsewhere, in Iceland, choices at the policy level are made by teacher educators.

Since mathematics education research in Iceland is relatively new, reporting of existing teaching/research is lacking. University teacher educators' identities are still being shaped from being practitioners and field experiences with other researchers in mathematics education, which explains the lack of citations to prior studies done within Iceland in many of the chapters in this section. Mathematics education research is limited because there are few people at the university level, and the multiple duties heaped on them curtail time for research. Recently the number of researchers has been growing in terms of faculty and graduate students at the master's and doctoral level. Also the university has started to exert more pressure on faculty to publish their research, which is changing the culture and attitude towards research in mathematics education. However, university teacher educators have engaged in plenty of curriculum development, field testing, and self-study without the necessity for reporting to the outside. Also, professional development and in-service training are slowly being reported in lectures and presentations at conferences or teachers' meetings.

In May 2008, there was a Nordic Teacher Education Conference in Iceland in which several authors from this section (Palsdottir, Gunnarsdottir, and Kristinsdottir) presented their self-analysis of professional development as teacher educators. So far in mathematics education there have been three PhD recipients in the history of the country; currently one of the university mathematics educators is enrolled as a PhD student, and there are approximately a dozen master's holders of mathematics education. This explains the lack of research in mathematics education in Iceland. We conclude this section highlighting the common threads in the chapters comprising mathematics education in Iceland. These threads are:

- Focus on teachers and focus on own teaching
- Self-analysis of many years of teaching practices
- Teacher education under development with researchers self-examining their role as teacher educator
- Relating existing research to research in other countries research—this suggests that development is not occurring out of the blue but building on existing traditions elsewhere.

REFERENCES

Halldorsson, A.M., & Olafsson, R. F. (2009). The case of Iceland in PISA: Girls' educational advantage. *European Educational Research Journal, 8*(1), 34–53.

Palsdottir, G., & Sriraman, B. (2010). Commentary on J. Jacob's Feminist Pedagogy and Mathematics. In B. Sriraman & L. English (Eds.), *Theories of mathematics education: Seeking new frontiers* (pp. 467–475). New York: Springer.

Steinthorsdottir, O. & Sriraman. B (2008). Exploring gender factors related to PISA 2003 results in Iceland: a youth interview study. *ZDM—The International Journal on Mathematics Education, 40*(4), 591–600.

Tomlinson, C. A. (1999). *The differentiated classroom: Responding to the needs of all learners.* Alexandria, VA: ASCD.

32

THE HISTORY OF PUBLIC EDUCATION IN MATHEMATICS IN ICELAND AND ITS RELATIONS TO SECONDARY EDUCATION

Kristín Bjarnadóttir

UNIVERSITY OF ICELAND—SCHOOL OF EDUCATION

Abstract: The history of mathematics education in Iceland is traced, mainly by examining mathematics textbooks from different periods and considering official decisions that have exerted influence on the choice of textbooks. While Danish regulations and universities exerted their influence on the high schools, they themselves influenced the syllabus of lower school levels, whose choice was further limited by small means of a state textbook publishing house. The introduction of "New Math" became a spark releasing all school levels from stagnation in mathematics education, but it caused also a temporary turbulence. Recent research reveals that Icelandic adolescent pupils have reached average or above in their mathematics performance. However, excellent performance is missing, and more mathematics teachers with the required education are needed.

Keywords: arithmetic, textbooks, state textbook publishing house, stagnation period, teacher education

INTRODUCTION

Iceland, settled by the Vikings around the year 900, became a part of the Danish realm in the late 14th century, remaining until 1944. From the mid-18th century, public education in reading and Christendom was monitored by parish priests and provided by the families. Adherents of the Enlightenment in the late 18th century made an attempt to modernize the society. A considerable amount of educational material was produced, which included arithmetic textbooks, written in a European style, and aimed at business.

A few primary schools were established in the largest towns starting in the 1860s, and two lower secondary schools were established in the 1880s. However, the main rule was that children were to be educated within the families. Legislation on public education in writing and arithmetic, was passed in 1880. Its implementation was the responsibility of the families under the supervision of parish priests. It soon emerged that the task was beyond their capacities.

The 1880 legislation released the initiative and creativity of the non-mathematical educated class, theologians, and teachers. A number of arithmetic textbooks were composed for primary education and the emerging lower secondary education, and geometry textbooks as well. The European commercial arithmetic was tacitly adopted and adjusted to Icelandic context. Its justification was training young people to cope with the growing trade in Iceland's relatively late industrialization.

The First Sourcebook on Nordic Research in Mathematics Education, pages 447–465

The growth continued after the 1907 Education Act, when education for 10–14-year-old children became free of charge, paid for by the local communities. Schools were built, and itinerant schools were run in the remotest of rural areas. They were first completely superseded by boarding schools and school buses in the 1970s. In 1946 the parliament passed new school legislation, shortening the former six-year college-boung secondary schools to four-year high schools and creating a route from the lower secondary level into them by a national entrance examination. High expectations, which were only later realized, were attached to that legislation.

Several official decisions in the period 1920s to 1940s hindered natural development of public mathematics education. They made it dependent on the upper school level and caused a standstill situation for decades. The impact of these decisions remained into the 1960s and 1970s when people's ideas of admission to further and higher education were rapidly changing. As changes were long overdue by then, Iceland fell prey to the international "New Math" reform wave and had to cope for a decade with ideas alien to teachers and parents before a real domestic reform could take place. The reasons for the consequential official decisions will be considered and analyzed by the historical method; examining legislation, official documents and arithmetic textbooks.

THE PERIOD 1780–1880

Enlightenment

By the mid-18th century it had become clear that Icelanders had lagged behind other European nations in most respects. The Enlightenment movement brought increased interest in education that aimed at raising the standards of Icelanders in all walks of life, in order to enhance their independence vis-à-vis foreign authorities and merchants. From the 1740s, requirements in arithmetic at the two Icelandic cathedral schools were set to the four arithmetic operations in whole numbers and fractions, and in the late 1770s Icelanders in Copenhagen established the Society of the Learned Arts, which laid emphasis on practical matters. Under its influence, handbooks of various kinds were published—among them two substantial arithmetic textbooks.

One of the most prolific writers of the Society of the Learned Arts was Ó. Olavius (c. 1741–1788), who wrote an arithmetic textbook called *Greinilig Vegleidsla til Talnalistarinnar* (*A Clear Guidance to the Number Art*), modeled after a comprehensive German textbook, *Der Demonstrativen Rechnenkunst* by Christlieb von Clausberg (1732, 1748, 1762). Olavius's aims, which were clearly utilitarian (Bjarnadóttir, 2007a), were declared in the forewords:

> About the computing art, this important and for the human being indispensable knowledge one has up to now not had any instruction in the country's language. . . . But who cannot realize the harmful consequences thereof, especially when trade is done with foreign merchants and experienced calculators, who perhaps sometimes could and wanted to earn more than they should? (Olavius, 1780, pp. vi–ix)

Clear Guidance was an excellent textbook, but it did not receive the respect it deserved. The author made some comments on the misery of Icelanders, which offended many a good countryman of his (Bjarnadóttir, 2007a). Also, another no less substantial arithmetic textbook was published only five years later, *Stutt Undirvísun í Reikníngslistinni og Algebra / A Short Teaching in the Art of Computation and Algebra* (1785), which was immediately authorized as a textbook for the two cathedral schools. Its author was Ó. Stefánsson, a lawyer and later governor of Iceland, who had more opportunities to promote his book than Olavius, who lived in Denmark and died there in 1788. However, no one was available to teach arithmetic. The pupils in Reykjavík Latin School were given the governor's book, but it was up to them whether they opened it or not (Helgason, 1907).

Iceland thus had two good arithmetic textbooks by the mid-1780s, at the time when huge volcanic eruptions broke out, killed the livestock, and caused famine, and the misery of Icelanders reached its apex. One of the two cathedral schools was closed down and moved to Reykjavík, the growing capital, and by 1800, the other one was closed down too. In the following decades, arithmetic teaching at the school was minimal. Remarkable as it was to have a choice of two good printed arithmetic textbooks at that time, no proper arithmetic textbook was published again until 1841. Registrations of private collections in the 1820s point to the fact that copies of the two books were preserved far into the 19th century among the general public.

The Latin School

In the period 1800–1930 only one Latin School existed in the country. Mathematician B. Gunnlaugsson came to the school in 1822 to serve there for 40 years. Gunnlaugsson's presence changed mathematics education at the school from next to nil to adapting to the requirements of the Royal Directorate of the University and the Latin Schools in Denmark. His teaching did not affect public mathematics education to any degree, as public education was still within the families with no direct connection to the Latin School. Gunnlaugsson used Danish textbooks, a tradition that remained in college-bound education until the 1970s.

During the decade following Gunnlaugsson's retirement in 1862, the Danish system of upper secondary schools was revised to be divided into a mathematics–science stream and a language–history stream. At that time, the Icelandic school was too small to be divided into two streams. A school affairs board proposed one combined stream as previously; political lobbyism at highest level, however, resulted in the choice of the language–history stream, announced in regulations for the Reykjavík Latin School in 1877. Behind that decision were the headmaster and his colleagues who taught modern and ancient languages, while there is also evidence that the mathematics teaching at the school at that time was not in good shape and did not appeal to its pupils (Bjarnadóttir, 2006a, 2007c). The decision delayed the preparation of students for science and engineering studies, not a good fate for a small nation without durable buildings, roads, and bridges in a huge country compared to its population.

THE PERIOD 1880–1900s

Concurrently with the reduction of mathematics teaching at the Reykjavík School, there was progress in public education. It was in most respects different from that in Denmark and did not adhere to Danish legislation on public schools, passed already in 1814. Publishing of basic arithmetic textbooks flourished after the 1880 legislation giving responsibility for public education in writing and arithmetic to the families, and especially in the early 1900s.

Two lower secondary schools were established, one in the northern Iceland in 1880, and the other, a private institution, in the vicinity of Reykjavík in the south in 1882. The northern school was granted permission in the early 1900s for its graduates to enter the upper division of the Reykjavík Latin School. Teacher training was established in 1892 as an affiliate to the lower secondary school in the south, and the Teacher Training College was established in 1908. In the early 20th century, several private lower secondary schools, called people's schools, were established. Public primary and lower secondary education remained without influence or pressure from the Latin School on upper secondary school level until the 1920s.

Textbooks by E. and H. Briem

Of special interest are textbooks from the late 19th century, written by two brothers, the Reverends E. and H. Briem. E. Briem wrote an arithmetic textbook, *Reikningsbók* (*Arithmetic*), first published in 1869, which was to have great influence into the 1910s. H. Briem wrote two books on geometry: plane geometry and stereometry. Both of them declared in their forewords that they would avoid scientific proofs and concentrate on what they considered important for the "working life" in Iceland.

E. Briem said that parallel to progress in society, increased commerce, and more diverse industry, the people's need for a solid knowledge of arithmetic increased as well; further, they would not have full use of that knowledge unless they computed easily without errors, which demanded considerable practice, so revising more than once or twice was of the utmost necessity—"especially those who have had instruction are apt to forget again, if they do not regularly revise what they have learnt" (E. Briem, 1880, p. iv). This sentence expresses the author's suspicion that instruction by others may be vulnerable, and not as effective as self-instruction. The author said that, as he avoided giving reasons for the rules, he was generally not able to use foreign books as models. The author's avoidance of reasoning seems to be a reaction to what he saw as excessive formalism in the foreign textbooks. The book was used at the lower grades in the Reykjavík Latin School, at the lower secondary schools, and for home studies as well.

Some youngsters of that period went far in their self studies. In an autobiography, A. Arnalds (1949) tells that he went through E. Briem's *Arithmetic* by himself and learned without external instruction about whole number operations, common fractions, decimal fractions, *regula de tri* in various forms, area and volume calculations, percentages, and interests. As the parish priest knew that Arnalds had been working hard on arithmetic, he gave him the problem for his confirmation to compute the sum to be expected if a penny carried 4% interests from the year Christ was born. Arnalds only knew the method to find the length of the period needed to double the sum, which in this case was 17 years and several months. From this he calculated up to the year 1886, coming up with a 30 digit figure, taking him more than an hour to find.

The lack of formalism did not harm E. Briem's *Arithmetic* too much, while H. Briem's *Flatamálsfræði* (*Plane Geometry*) would have gained from more rigidity. It is mainly a collection of definitions and demonstrations. As an example, the sum of angles in a triangle is only demonstrated for a right triangle as one-half of a square, postulated with four right angles.

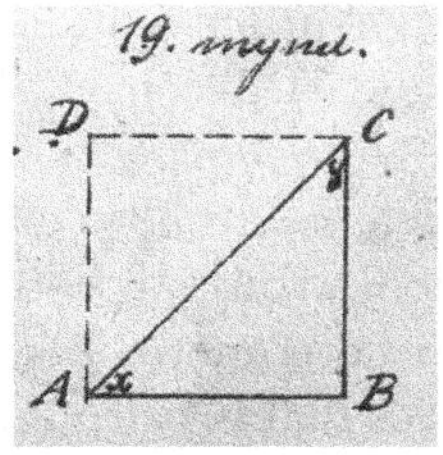

> All the angles in a triangle are 180° in total. In the triangle ABC (diagram 19) CB is perpendicular to AB, therefore the angle B = 1R [R a right angle], furthermore CB is equal to AB; by drawing the triangle ADC equally large and similar to the triangle ABC [congruence had not yet been defined], one may see that *x* and *y* each are the half of a right angle, therefore the sum of the angles in the triangle is 2R. The same concerns all triangles, as the larger or smaller one of the angles is, the others (one of them or both) become smaller or larger. (Briem, 1889, p. 14)

In this text a diagram is referred to, but because of a high printing cost, all diagrams are printed together as an attachment at the back of the book with references in the main text. Clearly the author appeals to the intuition of the reader to see that the angles *x* and CAD are complementary, as well as *y* and ACD. Furthermore, the triangle ABC is a special case of an isosceles right triangle, but the reader is invited to take its attributes as universal. The author had introduced parallel lines and their angles to a transversal line and so was able to present the regular proof of the sum of angles in a triangle but preferred to do it this way.

As this was the first textbook on the subject written in Icelandic, the language was a great obstacle, and a number of new terms, most of them foreign words, were adapted to the Icelandic language system. The author taught the book himself at the lower secondary school in the north. H. Briem's *Þykkvamálsfræði* (*Sterometry*) was also a collection of definitions and formulas, but was probably of more practical use for "working life." It was taught by the author at the northern lower secondary school and by his brother at the Reykjavík Latin School as well.

Danish textbooks were used at the Latin School when the Briem brothers were pupils there. Textbooks by G. F. Ursin were used there for a long time, at least from the 1840s, and E. Briem mentioned Ursin's arithmetic textbook in his foreword as a source for his own problems. However, the brothers seem to have been unanimous that the axiomatic approach they became acquainted with in the Danish textbooks would not suit their readers—the general public who was in urgent need for practical knowledge. Indeed, axiomatic approach to geometry was never presented in public education until in textbooks in the spirit of the "New Math" in the 1970s, and then rather informally.

In 1877, *Lærebog i den elementære Plangeometry* (*A Textbook in Elementary Plane Geometry*), by the Danish Professor Julius Petersen (1943), was taken for teaching in the Reykjavík Latin School. It was translated into Icelandic in 1943 during World War II, as new supplies could not be brought to Iceland from Denmark due to the war. It remained as a basic textbook in plane geometry for nearly a century. More modern Danish textbooks were used from 1967, at the time of the New Math, but the Danish language was an obstacle and Petersen's translated textbook appeared again in the 1970s for a while. It may be considered as the main introduction to axiomatic approach for the few Icelanders who were admitted to the elitist school level.

Textbooks for the Primary Level

After the 1880 legislation, textbooks intended for children and their teachers were published. Some of them were only outlines for teachers, but in 1890 Morten Hansen published his *Reikningsbók handa alþýðuskólum* (*Arithmetic for People's Schools*). In his textbook he introduced the abacus, which he had been distributing, according to his foreword. The guide to the abacus was mainly taken from a guide made by a Mr. J.A. Bonnevie in Trondheim in 1886. The textbook itself was modeled after the textbooks of arithmetic teacher Chr. Hansen in Odense, mainly *Tavleregning-Opgaver II* and *III*. Chr. Hansen's textbooks dominated the Danish market of arithmetic textbooks from around 1860 into the 20th century (Hansen, 2002). The public education was thus under indirect Danish influence as well.

Morten Hansen's textbook was republished many times (6th edition in 1911). The guide to the abacus was omitted after the first edition. Instructions were minimal and mainly directed to the teacher. The computation methods were conventional, as far as can be seen. How were the teaching methods? Children were expected to pay attention, be obedient, and answer when the teacher asked them questions (Halldórsson, 2001), which probably means that they have been expected to simulate what was demonstrated to them in crowded classes, up to 40 in each class.

Dr. Daníelsson

In 1906, Iceland's first university graduate in mathematics, Dr. Ó. Daníelsson, published an arithmetic textbook, *Reikningsbók* (*Arithmetic*). It was quite elementary with straightforward explanations of common algorithms on the four operations in whole numbers and fractions, but it contained worthwhile problems for solving.

Daníelsson became, in 1908, the first mathematics teacher at the new Teacher Training College, where he built up mathematics training of primary teachers. He published a second, more advanced edition of his *Arithmetic* in 1914, to suit the student teachers, of whom most had only short schooling, such as itinerant primary school of several weeks a year, and possibly a year or two in one of the lower secondary schools.

In 1919, Daníelsson became the main mathematics teacher at the six-year Reykjavík School, now Reykjavík High School, when a mathematics stream was established after 42 years of absence of higher mathematics. The main reason for its establishment was to train future engineers that the country needed badly. This was the highest post available for a mathematician, as the University of Iceland, established in 1911, was only a professional school in theology, law, and medicine, with a small department of humanities. In 1920, Daníelsson published a new edition of his book, more suitable for beginners in the lowest grades of his school. By that edition Daníelsson had omitted reasoning to support procedures, such as the Euclidian Algorithm, which had been contained in his first edition. Experience may have contributed to Dr. Daníelsson's opinion that explanations of so theoretic topics were futile, so he decided to lay his main emphasis on problem solving in his textbook.

Dr. Daníelsson also wrote a textbook for his pupils on geometry, which was his specialization. There he was at odds with most of his Icelandic contemporaries, who saw only the utilitarian aspects of mathematics education:

> I have been aware, that the purpose of mathematics education in schools is completely hidden from some intellectuals; they think that the goal of the geometry teaching is something in the direction of teaching people to measure cabbage gardens or grass fields.... No, the purpose... is to train the pupil in precision of his thinking and at the same time his inventiveness, which no other subject trains him in to the same degree. (Daníelsson, 1920b, pp. iii–iv)

In the late 1920s the government, under the leadership of Minister of Education J. Jónsson, a former teacher trainer, established a number of public lower secondary schools, both in rural and the small scattered urban areas. The aims of these schools were utilitarian, to prepare young people for working life, and theoretical considerations were not on their agenda. Arithmetic was, for example, mainly to be concerned with bookkeeping. At the same time, Minister J. Jónsson restricted the admission to Reykjavík High School to 25 new pupils a year in order to lead pupils to his practical schools, which he considered more useful for them and for the country that was endangered by too many officials. Furthermore, he elevated the northern lower secondary school to a high school, not the least to counterbalance the Reykjavík School and its emphasis on ancient studies (Friðriksson, 1992).

By the 1930s Dr. Daníelsson had acquired a superior position on the secondary school level as the main mathematics teacher of the Reykjavík High School and the proctor of mathematics teaching in the new northern high school. No mathematics teacher was his equal in education and status. On the other hand, he had to submit to regulations adapted to regulations for Danish high schools and to requirements of Danish universities.

Dr. Daníelsson's position was still further strengthened by the restricted admission to the Reykjavík School. That action created a fierce competition for the few seats in the school producing future professionals. Gradually Daníelsson's *Arithmetic* was taken up in some of the practical lower secondary schools, presumably in case the pupils would make attempts to enter the upper level of the high school. Ex-Minister Jónsson continued to emphasize in parliamentary discussions that influences by the high schools on other lower secondary schools were to be minimized, but without being able to reverse the development he himself had initiated.

MATHEMATICS TEXTBOOKS IN THE 1900s–1960s

All mathematics textbooks for primary and lower secondary level in this period were arithmetic textbooks, most of them containing no geometry except area and volume computations. The textbooks were written in the European tradition of arithmetic textbooks, with introduction to the number concept, numeration, and basic instruction in the four operations, basically with utilitarian aims, such as commerce (Swetz, 1992). The majority of these textbooks were written by the largest professional class, theologians and priests, while several were written by members of the growing class of primary teachers.

Before the establishment of Iceland Teacher Training College in 1908, the options for future primary teachers were to attend a teacher training college in Denmark or elsewhere abroad. Two teachers and highly regarded school leaders, Ö. Sigurðsson (1900) and S. Jónsson (1906), went that way. Both of them studied at teacher training seminaries in Copenhagen. Sigurðsson had also stayed in Chicago and brought new ideas about teaching from there. Both of them wrote, in 1900 and 1906 respectively, arithmetic textbooks, which were important contributions to public education. However, their books were very small and had the disadvantage not to present, as Dr. Daníelsson did, the decimal measuring system that was adopted in Iceland in 1907 and became the main concern in arithmetic teaching in the following years.

In spite of the classical content of the textbooks of this period and their frugal finish, they offered considerable variety in approaches, and some discussion on didactics among primary teachers took place. The variety concerned algorithms and mental arithmetic, presentation of proportional calculations, divisibility, and effects of the place-value notation on the appearance of the numbers.

Reikningsbók/Arithmetic by Ó. Daníelsson

Daníelsson's *Reikningsbók* (*Arithmetic*) was in many respects more challenging and mathematically rigid than contemporary textbooks. Its 1920 edition became a leading textbook for lower secondary schools and the Teacher Training College through the 1950s. It was not abandoned until the mid-1970s. In his foreword, he said:

> This little booklet is intended, on my part, to be an attempt to compensate for two drawbacks which I think characterize most or all of our arithmetic textbooks; one is that they give no explanations at all, not even of the simplest computation methods, and therefore many learn the methods by heart without understanding their reasons; and more so as many of those who work at teaching may lack sufficient skills to explain the arithmetic down to its roots, without having for that any support from the textbooks. But the other drawback is that their exercises are generally too easy, and each of them is most often aimed at only one computation method. The pupil can therefore guess the method without understanding the problem. This I have wanted to prevent. I have after each paragraph made several exercises in adopting simultaneously the methods that have previously appeared, and generally speaking I have tried to choose the exercises so that they may not be solved unless they are fully understood. (Daníelsson, 1906, pp. iii–iv)

According to the author, the main characteristics of Daníelsson's *Arithmetic* with respect to other arithmetic textbooks of the time were thus the following: Daníelsson considered it important that the pupils should understand the exercises and the algorithms of the four operations and avoid rote learning. Daníelsson explained common algorithms by taking examples from the daily life of the early 20th century. In the textbook's first edition, he offered alternative algorithms for mental calculations, which he omitted in later editions. In teaching ratios and proportions, he may have suspected that his explanations of the ratio concept were difficult to follow. Therefore he may have emphasized a procedural approach, but different from that of E. Briem, which may be traced back to the 18th century. Daníelsson's emphasis on understanding was manifested mainly in cleverly composed exercises, avoiding pupils being able to imitate without understanding.

In spite of Daníelsson's emphasis on avoiding rote learning, examples of unexplained procedures may be found in his book. In the case of adding fractions, the author presented the necessity of finding a common denominator divisible by all the denominators as the main idea to be understood. Often it sufficed to multiply all the denominators together. However, for the lowest common denominator in the case of, for example,

$$\frac{1}{5}+\frac{5}{6}+\frac{2}{3}+\frac{4}{9}+\frac{5}{16}+\frac{3}{12}+\frac{7}{12}+\frac{3}{10}$$

the denominators were to be written in a line: 5 – 6 – 3 – 9 – 16 – 4 – 12 – 10. Denominators that divided another one were then obviously to be ruled out, in this case 5 (dividing 10), and 3, 4, and 6 (dividing 12). Thereafter the remaining denominators were to be divided by the lowest number dividing two or more of them, and the others copied unchanged:

$$\begin{array}{r|l} 2 & 9 - 16 - 12 - 10 \\ \hline 2 & 9 - 8 - 6 - 5 \\ \hline 3 & 9 - 4 - 3 - 5 \\ \hline & 3 - 4 - 1 - 5 \end{array}$$

By now all the remaining numbers and the divisors were to be multiplied together: $2 \cdot 2 \cdot 3 \cdot 3 \cdot 4 \cdot 1 \cdot 5 = 720$ (Daníelsson, 1920a, pp. 28–29). Another example is presented in Figure 32.1.

This procedure is found in a number of 18th- and 19th-century textbooks. As was the custom, Daníelsson left it unexplained, in spite of his declared aim of furthering understanding. In a couple of other contemporary textbooks the procedure was supported by prime factoring, a topic that Daníelsson did not touch until he had proved rigidly that the number of prime factors could not exceed the number itself, a fact that one might expect pupils to find unnecessary to prove. Thus, in spite of Dr. Daníelsson's intention to avoid rote learning, his rigid mathematical training seems to have hindered him from introducing a comprehensible procedure in this particular case.

Dæmi:

	720			
$2\,^{5}/_{6}$	— 120	— 600		
$3\,^{7}/_{12}$	— 60	— 420		
$4\,^{5}/_{16}$	— 45	— 225		
$5\,^{1}/_{9}$	— 80	— 80		
$7\,^{11}/_{15}$	— 48	— 528		
$3\,^{17}/_{20}$	— 36	— 612		

$27\,^{305}/_{720} = 27\,^{61}/_{144}$ 720) 2465 (3
2160
305

2| 6—12—16—9—15—20
2| 6— 8—9—15—10
3| 3— 4—9—15— 5
4—3— 5

$5 \cdot 3 \cdot 4 \cdot 3 \cdot 2 \cdot 2 = 720$

Figure 32.1 Another example of finding the least common denominator (Daníelsson, 1906, p. 59).

In his first edition of *Arithmetic*, Daníelsson (1906) presented Euclid's algorithm, supported by reasoning on one example but without a formal proof, in order to find the greatest common divisor. The divisibility by three was also explained by the base-ten place-value notation. Both explanations had been omitted in the 1920 and later editions, where Euclid's algorithm was only presented as a procedure, as shown in Figure 32.2.

A large part of Daníelsson's later editions of *Arithmetic* concerned ratio and proportions, presented in the form of the Rule of Three, as had been the custom since 17th-century textbooks. Daníelsson took comprehensive examples to illustrate that type of problem. However, the main emphasis was on procedural knowledge, how to arrange the numbers above and below a line segment used to designate a ratio (Daníelsson, 1938). Cases where the Rule of Three did not apply were only mentioned after 43 pages.

1) Þegar ekki er augljóst, með hvaða tölu skuli stytta brotið, má finna hana á þennan hátt:

Fyrst er minni tölunni deilt í þá stærri, þá afganginum í þá minni, þá afganginum, sem þar verður, í þá tölu, sem síðast var deilt með, o. s. frv., þangað til einhver afgangur gengur upp. Þá má stytta brotið með honum. Eigi t. d. að stytta brotið $^{536}/_{871}$, er »styttingartalan« fundin þannig:

536) 871 (1
536
335) 536 (1
335
201) 335 (1
201
134) 201 (1
134
67) 134 (2
134

Styttingartalan er þá 67, og þegar brotið $^{536}/_{871}$ er stytt með henni, kemur $^{8}/_{13}$ út.

Figure 32.2 In a nutshell the text says that this is how you find the greatest common divisor, in this case 67 (Daníelsson, 1938, p. 28)

The last part of Daníelsson's *Arithmetic* dealt with some geometry. Dr. Daníelsson's area of speciality was geometry, but as this was an arithmetic textbook, only six pages were devoted to its basic concepts and the proof of the sum of the angles in a triangle. The remainder concerned area and volume computations, and the Pythagorean Theorem, which was proved by a well-known geometric proof.

The topics of Daníelsson's well-composed exercises were the daily life of early 20th-century Iceland. People traveled on boats and horses, and sold and bought sheep, horses, and hay. Their salaries were paid in boots or frocks. For the post-WWII, generations these were echoes of a distant world.

Other Early 20th-Century Textbooks

Three other early arithmetic textbooks may be compared to that of Daníelsson. They were written the Rev. S. Á. Gíslason (1911), primary teacher Steingrímur Arason (1928), and by the Rev. J. Jónasson (1906), whose book was somewhat more advanced than the other two.

S.Á. Gíslason's *Reikningsbók* (*Arithmetic*) (1911–1914) was published in six small volumes. Its first four volumes were authorized in 1929 as a textbook series for the primary school level. Gíslason's approach to fractions differs from those of Arason and Daníelsson, in that Gíslason emphasized mental arithmetic, and he introduced prime numbers and prime factoring. From that he developed the concepts of a common divisor and a common multiple and illustrated how to use the lowest common multiple as a common denominator. Afterwards he demonstrated also the ancient procedural method given by Daníelsson (Gíslason, 1911), as presented in Figure 32.3. Gíslason also introduced ordering of fractions, which Daníelsson did not.

In the last two volumes of S. Á. Gíslason's *Arithmetic*, intended for the secondary school level, the concept of ratio is explained, and proportions are dealt with as equations. Gíslason emphasized finding the unit, thus solving proportional problems in two steps, without mentioning the concept of the Rule of Three or any procedural method until the problems were composed of several ratios. Area is dealt with before a general introduction to geometric concepts, which have not so small treatment, while the Pythagorean Theorem is only presented by the case of 3 – 4 – 5 as the lengths of the sides of a right-angled triangle.

Gíslason's *Arithmetic* is by modern standards a good textbook series from an educational point of view. It was more comprehensive in its six (small) volumes than Arason's in one larger volume, which may

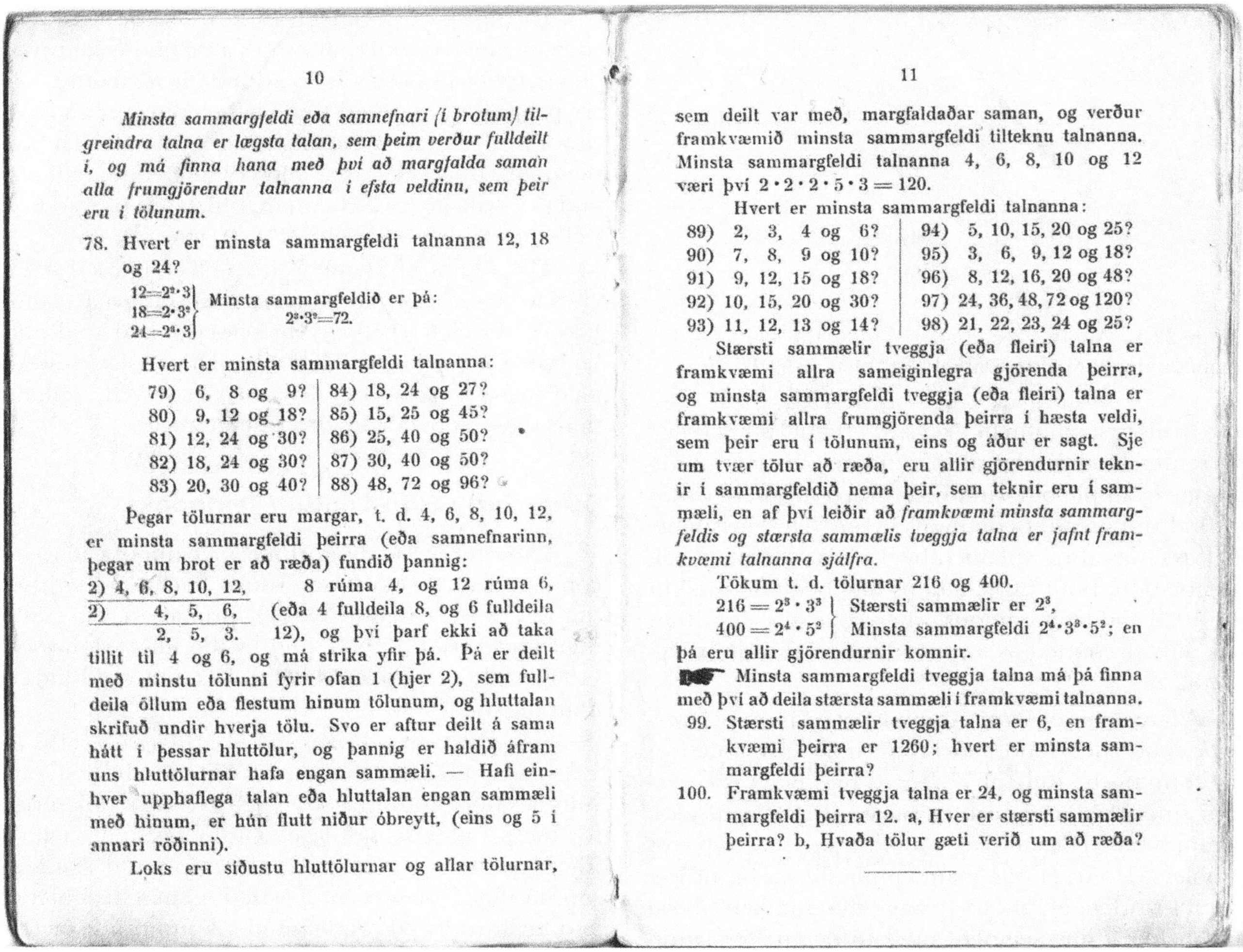

10

Minsta sammargfeldi eða samnefnari (í brotum) tilgreindra talna er lægsta talan, sem þeim verður fulldeilt í, og má finna hana með því að margfalda saman alla frumgjörendur talnanna í efsta veldinu, sem þeir eru í tölunum.

78. Hvert er minsta sammargfeldi talnanna 12, 18 og 24?

$12 = 2^2 \cdot 3$, $18 = 2 \cdot 3^2$, $24 = 2^3 \cdot 3$ } Minsta sammargfeldið er þá: $2^3 \cdot 3^2 = 72$.

Hvert er minsta sammargfeldi talnanna:

79) 6, 8 og 9?	84) 18, 24 og 27?
80) 9, 12 og 18?	85) 15, 25 og 45?
81) 12, 24 og 30?	86) 25, 40 og 50?
82) 18, 24 og 30?	87) 30, 40 og 50?
83) 20, 30 og 40?	88) 48, 72 og 96?

Þegar tölurnar eru margar, t. d. 4, 6, 8, 10, 12, er minsta sammargfeldi þeirra (eða samnefnarinn, þegar um brot er að ræða) fundið þannig:

2) 4, 6, 8, 10, 12,
2) 4, 5, 6,
2, 5, 3.

8 rúma 4, og 12 rúma 6, (eða 4 fulldeila 8, og 6 fulldeila 12), og því þarf ekki að taka tillit til 4 og 6, og má strika yfir þá. Þá er deilt með minstu tölunni fyrir ofan 1 (hjer 2), sem fulldeila öllum eða flestum hinum tölunum, og hluttalan skrifuð undir hverja tölu. Svo er aftur deilt á sama hátt í þessar hluttölur, og þannig er haldið áfram uns hluttölurnar hafa engan sammæli. — Hafi einhver upphaflega talan eða hluttalan engan sammæli með hinum, er hún flutt niður óbreytt, (eins og 5 í annari röðinni).

Loks eru síðustu hluttölurnar og allar tölurnar,

11

sem deilt var með, margfaldaðar saman, og verður framkvæmið minsta sammargfeldi tilteknu talnanna. Minsta sammargfeldi talnanna 4, 6, 8, 10 og 12 væri því $2 \cdot 2 \cdot 2 \cdot 5 \cdot 3 = 120$.

Hvert er minsta sammargfeldi talnanna:

89) 2, 3, 4 og 6?	94) 5, 10, 15, 20 og 25?
90) 7, 8, 9 og 10?	95) 3, 6, 9, 12 og 18?
91) 9, 12, 15 og 18?	96) 8, 12, 16, 20 og 48?
92) 10, 15, 20 og 30?	97) 24, 36, 48, 72 og 120?
93) 11, 12, 13 og 14?	98) 21, 22, 23, 24 og 25?

Stærsti sammælir tveggja (eða fleiri) talna er framkvæmi allra sameiginlegra gjörenda þeirra, og minsta sammargfeldi tveggja (eða fleiri) talna er framkvæmi allra frumgjörenda þeirra í hæsta veldi, sem þeir eru í tölunum, eins og áður er sagt. Sje um tvær tölur að ræða, eru allir gjörendurnir teknir í sammargfeldið nema þeir, sem teknir eru í sammæli, en af því leiðir að *framkvæmi minsta sammargfeldis og stærsta sammælis tveggja talna er jafnt framkvæmi talnanna sjálfra.*

Tökum t. d. tölurnar 216 og 400.

$216 = 2^3 \cdot 3^3$, $400 = 2^4 \cdot 5^2$ } Stærsti sammælir er 2^3, Minsta sammargfeldi $2^4 \cdot 3^3 \cdot 5^2$; en þá eru allir gjörendurnir komnir.

☛ Minsta sammargfeldi tveggja talna má þá finna með því að deila stærsta sammæli í framkvæmi talnanna.

99. Stærsti sammælir tveggja talna er 6, en framkvæmi þeirra er 1260; hvert er minsta sammargfeldi þeirra?

100. Framkvæmi tveggja talna er 24, og minsta sammargfeldi þeirra 12. a, Hver er stærsti sammælir þeirra? b, Hvaða tölur gæti verið um að ræða?

Figure 32.3 In the text it says that finding the least common denominator is done by prime factoring if the factors are few, such as 12, 18, and 24. In the case of a number of factors, such as 4, 6, 8, 10, 12 the procedural method is recommended. The attention is drawn to the fact that the product of the least common denominator and the greatest common divisor of two numbers is equal to the product of the same numbers. Example: 216 and 400 (Gíslason, 1911, pp. 10–11).

have ruled Gíslason's series out in the 1930s for its total price.

Arason's *Reikningsbók* (*Arithmetic*) was worked from an earlier edition in cooperation with another author (Brynjólfsson & Arason, 1914). It was one of three textbook series that were authorized for primary school in 1929, along with those of S. Á. Gíslason and E. Bjarnason. The book is concise, in view of its intended use in primary school. For example, fractions are treated in ten and a half pages without using a single diagram (Arason, 1928, pp. 78–89), except one far earlier (p. 60). Operations, such as cancelling, are demonstrated but not explained, and finding the lowest common denominator is treated similarly to the procedure in Daníelsson's *Arithmetic*. Proportions are presented as the Rule of Three with procedural approach. Teachers were not too fond of Arason's *Arithmetic* (Indriðadóttir, 1995).

J. Jónasson's *Arithmetic*, first published in 1906, was intended for lower secondary education. It begins, however, with a basic explanation of the four operations. This is natural, as it was published before the 1907 school legislation. Jónasson spent eight and a half pages on prime numbers and prime factoring as an introduction to fractions, using it to find the greatest common divisor and least common multiple (see Figure 32.4). Rules on divisibility were explained by base-ten place-value notation (Jónasson, 1906). Jónasson also emphasized the two-step method of finding a unit in proportional equations, but is ultimately closer to procedural method than S. Á. Gíslason.

The textbooks' sphere of distribution proved to coincide with the authors' sphere of authority. J. Jónasson taught in the northern lower secondary school where his book was used during his time, while S. Á. Gíslason taught his books in a commercial school

58. gr. – **Minsti samdeilandi** eða **minsta sammargfeldi** *heitir hin minsta tala, sem tvær eða fleiri tölur ganga upp í afgangslaust,* eða sem tvær eða fleiri tölur má margfalda svo, að báðar eða allar gefi sama pródúkt. Hann finst þannig: Leys allar tölurnar upp í frumfaktora sína; *minsti samdeilandi er pródúktið af frumfaktorum talnanna, svo mörgum af hverjum, sem þeir koma flestir fyrir í hverri tölu.*

a) Hver er minsti samdeilandi talnanna 36 og 48?

36 = 2 . 2 . *3* . *3*.
48 = *2* . *2* . *2* . *2* . 3.

Hjer koma 2 oftast fyrir í 48, enn 3 í 36; minsti samdeilandi er því 2 . 2 . 2 . 2 . 3 . 3 = $2^4 . 3^2$ = 144. (36 og 48 ganga upp í 144; 36 og 48 má margfalda svo (með 4 og 3), að pródúktið sje 144.)

b) Finn minsta samdeilanda 40, 150, 252 og 675.

40 = *2* . *2* . *2* . 5.
150 = 2 . 2 . *5* . *5*.
252 = 2 . 2 . 3 . 3 . *7*.
675 = *3* . *3* . *3* . 5 . 5.

2 koma oftast fyrir í 1. línu, 3 í 4. línu, 5 í 2. línu og 7 í 3. línu (5 koma og fyrir jafnoft í 4. línu, enn því er slept). Minsti samdeilandi er því 2 . 2 . 2 . 3 . 3 . 3 . 5 . 5 . 7 = $2^3 . 3^3 . 5^2 . 7$ = 37800.

59. gr. – Venjulegast er þetta þó reiknað öðruvísi. Tölurnar eru ritaðar í röð frá vinstri til hægri handar, athugað, hvort nokkur þeirra gangi upp í annari, og ef svo er, þá er sú tala strikuð út. Síðan er tekin lægsta frumtala, sem upp gengur í tveim eða fleirum, og deilt með henni þeim tölum, sem hægt er, og kvótinn ritaður neðan undir. Svo er aðgætt enn, hvort nokkur gangi upp, og dregið út ef svo er. Þegar hin fyrsta tala gengur ekki lengur upp í neinni tölu, er byrjað á næstu frumtölu og farið eins að. Svo er haldið áfram, þangað til alt er upp gengið, og allir frumfaktorar eða frumtölur, eins og þær koma flestar fyrir í hverri tölu, eru orðnar að deilum á þennan hátt; skal svo margfalda þær saman; pródúkt þeirra er minsti samdeilandi talnanna. Dæmið b) í 57. gr. er hjer reiknað út á þenna hátt; fyrst er deilt með 2, þangað til allar tölurnar eru orðnar ójafnar; svo er deilt með 3 meðan það er hægt; á því bili ganga bæði 5 og 25 úr; síðan er deilt með 5 og síðast með 7; þá er alt upp gengið; faktorarnir eða deilarnir eru 2 . 2 . 2 . 3 . 3 . 3 . 5 . 5 . 7 = $2^3 . 3^3 . 5^2 . 7$, og minsti samdeilandi = 378000 eins og í 57. gr.

2) 40 – 150 – 252 – 675
2) 20 – 75 – 126 – 675
2) 10 – 75 – 63 – 675
3) [5] – 75 – 63 – 675
3) » – [25] – 21 – 225
3) » – » – 7 – 75
5) » – » – 7 – 25
5) » – » – 7 – 5
7) » – » – 7 – 1
» – » – 1 – »

a) Finn minsta samdeilanda fyrir 12, 21, 35, 60 og 84.

Hjer ganga 12 upp í 60 og 21 í 84, og ganga þær tölur því úr. Síðast ganga bæði 5 og 7 upp í 35, og þegar því er deilt með 5 og kvótanum með 7, er alt uppgengið, og eftir eru $2^2 . 3 . 5 . 7$ = 420.

2) [12] – [21] – 35 – 60 – 84
2) » – » – 35 – 30 – 42
3) » – » – 35 – 15 – 21
5) » – » – 35 – [5] – [7]
7) » – » – 7 – » – »
» – » – 1 – » – »

Figure 32.4 Jónasson illustrated two methods of finding the least common denominator: by prime factoring and by repeated division by a common prime factor.

and a school for girls. And, as mentioned earlier, Dr. Daníelsson adapted his textbooks to his pupils, first at the Teacher Training College and later at the Reykjavík High School.

Jónasson's *Arithmetic* disappeared in favor of Daníelsson's *Arithmetic* at the lower secondary school level when Daníelsson had become the greatest authority in secondary school mathematics after 1920. After that Daníelsson's interpretation of school mathematics dominated Icelandic mathematics education until the late 1960s, when the New Math reform movement reached Iceland.

Reikningsbók (*Arithmetic*) by E. Bjarnason

By school legislation acts in 1907 and 1926, primary schools were gradually established around Iceland. Three arithmetic textbook series were officially authorized for use in schools in 1929. The third of these arithmetic textbook series, *Reikningsbók/Arithmetic* by Elías Bjarnason, was first published in 1927–1929 in two volumes. In 1939 it was chosen for free distribution along with Arason's *Arithmetic* by the newly established State Textbook Publishing House, after which there was no market for other textbooks. By then Bjarnason's *Arithmetic* was expanded to four thin volumes. Eventually it became the only choice of arithmetic books for the 10–13 years age group in all Icelandic primary schools until the 1970s.

The author, E. Bjarnason, was one of the first graduates of the Iceland Teacher Training College, where his teacher was Dr. Daníelsson. Bjarnason claimed in the foreword to his book that he had avoided any discrepancy between his book and that of Daníelsson, who had most graciously read Bjarnason's book and offered good advice. In fact, Bjarnason's textbook may be seen as an abridged and simplified version of Daníelsson's, adapted to younger pupils. Bjarnason also favored emphasizing exactness and speed by training of one algorithm only for each operation without any alternatives. He considered a short and concise demonstration of a certain algorithms desirable, and a choice of algorithms undesirable (Bjarnason, 1927). Arithmetic learning should lead to accuracy and certainty, the desirable attributes of the early 20th century.

Bjarnason followed Daníelsson's example in the unexplained procedures of finding the lowest common denominator and Euclid's algorithm for the greatest common divisor, without any further comments. Bjarnason also presented as a memorizing rule the fact that 3 divides a number if it divides its transversal sum, and in parentheses that it also applied to 9. Discussing reasoning for such facts with children was not on Bjarnason's agenda. Schools in Reykjavík, where the author taught, were heavily packed with children at that time, and the author may have considered too many explanations confusing in that environment.

In the last part of Bjarnason's series, Daníelsson's approach to ratio and proportions is copied. After 1950 that part of the series was not widely used in compulsory education. Looking at the 100 exercises, it is appalling to observe how little they had to do with the world of 13-year-old children, reflecting instead the adult world of the 1920s.

In total, Bjarnason's textbook bears more resemblance to a handbook of how to do things than a tool for furthering mathematical thinking—regrettably, considering how long a life it had.

THE STATE TEXTBOOK PUBLISHING HOUSE

The 1930s were a difficult period in many countries. The Great Depression caused poverty among the working class and the general public. As a result of that situation in Iceland, a State Textbook Publishing House was established in 1937 by decision of parliament, in order to distribute textbooks free of charge to children at compulsory school level up to the age of 14, thus reducing the burden on families of paying educational expenses. The publications were to be funded by a special textbook levy for each family with children, 5–8 *krónur* (crowns) in the first two years, thereafter 5 *krónur*.

The execution of the project was entrusted to a state printing press, and its director became the executive of the project (Alþingistíðindi, 1936). The publishing house composed or developed new textbooks on its own initiative and bought the right to publish textbooks already on market, such as the 1927–1929 *Arithmetic* by E. Bjarnason and Arason's 1928 *Arithmetic*. Of them, Bjarnason's *Arithmetic* soon became the only choice.

Most of the new titles were published in the period 1937–1941 (Sigurgeirsson, 1987), after which increasing inflation caused serious difficulties in funding the project. The parliament was reluctant to increase the textbook levy, so the amount of available money declined in value. In 1956 new legislation on the State Textbook Publishing House was passed (Stjórnartíðindi, 1956), entailing that the state would provide one-third of the funding. The bill was presented by the following speech of Minister of Education, B. Benediktsson, who had children in school at that time:

> It has been a great obstacle to reform in these matters that it has not been possible to pay adequate royalties for [text]books, so therefore people have been much more reluctant to undertake that task. . . . Then one must also admit, that the books have been very plain in their external finish and in fact, exceedingly so . . . of course it is necessary for children, not least for the youngest ones, and becomes the more necessary as more things distract them from the studies, that the books be reasonably well made. . . . (Alþingistíðindi, 1955, pp. 885–886)

The revision of the legislation came 20 years after the 1936 legislation on a state monopoly of textbook publishing, long after the Great Depression and its political aftermath. Mathematics education stagnated in textbooks rooted in the 1920s. The decision to establish a state publishing company was thus to have an impact on the general conception of school mathematics, choice of mathematics topics and their presentation, the relation between the social settings presented in textbooks and the reality of the times of the users, teacher training, and teachers' preparation for new topics and different presentation, when at last the syllabus began to change in the late 1960s.

THE PERIOD 1946–1960s

By education legislation in 1946, compulsory schooling was extended to the age of 15, and the formerly six-year college-bound secondary school became a four-year high school. An entrance examination into it was to be arranged all around the country, as a measure towards providing equal opportunities for education. The syllabus of what had been the second year of Reykjavík six-year school was adopted as a syllabus for the entrance examination, presumably as a concession to the two college-bound secondary schools. The mathematics textbooks to be used were Daníelsson's 1920 *Arithmetic* and 1927 *Algebra*. This syllabus remained in place for 30 years, until 1976, with alternative options of New Math-oriented topics from 1967.

Daníelsson's *Arithmetic* thus continued to dominate the non-compulsory lower secondary level, but it was considered too difficult for its compulsory part. A couple of textbook series on the same topics were published. One of them was *Reikningsbók* (*Arithmetic*) *II A* by J. Á. Gissurarson and S. Guðmundsson (1949–1950), published after 1956 by the State Textbook Publishing House. It was used as an alternative to Daníelsson's *Arithmetic* for preparation to the entrance examination and had therefore to cover the same topics.

Arithmetic by Gissurarson et al. was harshly criticized for poor explanations in the teachers' journal (Gestsson, 1962). Ratio and proportions were introduced within the framework of the Rule of Three. The total introduction and explanation covered one page, as compared with four pages by Daníelsson. In continuation there were exercises with examples in between, and more variations of the rule and interests and percentages on the following 50 pages. The topics of the exercises were similar to Daníelsson's; so far as they had content, they reflected the neo-urban society with agricultural background. The list of procedures was probably intended to be concise and foolproof, but nothing is said about ratio or proportions until implicitly in connection with partition of properties. The limitations of the rule were not mentioned; nor were cases where the quantities are not proportional or cases where large portions might be bargained to a lower price than small ones. Rules for area and perimeter of common plane figures were explained by diagrams and the Pythagorean Theorem by a diagram of a triangle with sides of lengths 3, 4, and 5 and squares on the sides.

Reikningsbók (*Arithmetic*) by K. Gíslason (1962) for the lower secondary level was published by the State Textbook Publishing House in 1962 as another alternative to Daníelsson's *Arithmetic*. Gíslason's exercises were easier than those of Gissurarson and Guðmundsson. K. Gíslason introduced the Rule of Three in five pages, followed by 120 exercises and further sections on interest and percentages with similar procedures. The treatment is clear, emphasizing the two-step method of first finding the unit, and is careful to explain and prevent common mistakes. However, no mention is made of cases when using the Rule of Three might be inappropriate or of ratio or proportions until ratio is treated in connection with proportional partition of properties. Area, perimeter, and volume computations were similar to those of Gissurarson and Guðmundsson.

The arithmetic textbooks by Gissurarson and Guðmundsson and by K. Gíslason did not contain any topics that were not covered in Daníelsson's 1920 edition, but were simplified and slightly modernized versions of it. They covered perimeter and area calculations of common geometric figures, as well as the Pythagorean Theorem, and volume calculations of rectangular prisms, pyramids, cones, and spheres, largely without explanations or proofs. Mathematical proofs of any kind were scarcely seen in textbooks for the lower secondary level, while geometrical formulae were supported by diagrams.

The former diversity in choice of textbooks was reduced by the establishment of the State Textbook Publishing House for the primary level in 1937. Choosing books for free distribution ruled out creation of new primary level arithmetic textbooks. Accepting Daníelsson's textbook in 1946 as the prescribed curriculum of the national entrance examination to the upper secondary level, which remained unaltered for two decades, together with Bjarnason's monopoly at the primary level, resulted in the total authority of Daníelsson's early 20th century ideas of arithmetic teaching. This remained into the late 1960s, when a release was found in the New Math reform wave, which broke rather harshly on Iceland (Bjarnadóttir, 2006a, 2007c).

LANDMARKS IN MATHEMATICS TEACHER EDUCATION

Iceland belonged to the Danish realm until 1944, and Icelandic teacher education has its roots in Denmark. After the implementation of the 1907 school legislation, the Iceland Teacher Training College was established in 1908. Many teacher students in the first few years had no previous schooling, so they had to study elementary arithmetic. Only minimal algebra (equations) and geometry (mainly area, volume, and Pythagorean Theorem) were taught at the Teacher Training College until after 1946 and 1962, respectively. Some mathematical content knowledge and pedagogical content knowledge of primary-school level student teachers was ensured in Dr. Daníelsson's time in 1908–1920. From 1929 until 1962, the college was served by part-time mathematics teachers, some without special training in mathematics. Dr. Daníelsson's 1906 arithmetic textbook, modified in 1914 and 1920, and from 1946 his 1927 algebra textbook, were the basis of teacher training in mathematics until the 1960s. For teachers at lower secondary schools, study of theology or another university subject was the accepted preparation (Bjarnadóttir, 2006a, 2007c).

By the 1946 Education Legislation Act, some university education, preferably in the teaching subject, was required for tenure at the lower secondary level, and in 1950 a B.A. program in mathematics and physics was established as a part of a program for engineering students at the University of Iceland. Only 15 teachers from the program ever taught mathematics at the lower secondary level, but they proved to be a strong force to build up mathematics education for the college-bound stream of the level (Bjarnadóttir, 2006a, 2007c).

The Teacher Training College became Iceland University of Education in 1971. It merged with the University of Iceland and became its School of Education in 2008. After the 1974 Education Legislation Act, the requirements for teaching at lower secondary level became a B.Ed. degree from the University of Education or equivalent, preferably with mathematics and mathematics education as one of either one or two electives, and for teaching at the primary school level the same degree with any field of choice. These requirements have remained the same since that time but will advance to a M.Ed. degree in 2011.

In the academic year 2003–2004, 33% of mathematics teachers in grades 8-10 of compulsory school, the lower secondary level, had a B.Ed. degree with mathematics as an elective, and 2% had a B.Sc. degree in mathematics in addition to fulfilling requirements in pedagogy and didactics. At the upper secondary level, 46% had a B.Sc. degree or higher qualification in mathematics (Menntamálaráðuneytið, 2005).

The majority of teacher students at the University of Iceland, School of Education, are graduates from social science streams at the upper secondary school level, streams without specialization in mathematics. In 2005, only 18% of first-year students had completed 18 mathematics credits or more out of 140 credits earned for a four-year study in an upper secondary school. In 1999 the Ministry of Education published a new national upper-secondary-level curriculum, which lowered the minimum number of mathematics credits required at all streams. The number of students completing fewer than 12 credits in mathematics rose from 12% to 39% among students enrolled at the University of Education in 2005, according to whether they graduated from upper secondary school before 2003 or later, when the new national curriculum had come into effect (Bjarnadóttir, 2006b).

"NEW MATH" IN 1960s–1970s

By 1966 the State Textbook Publishing House had not taken any initiative in updating the primary school arithmetic syllabus, and it only offered a simplified alternative to Daníelsson's *Arithmetic* for the lower secondary level. At that time the "New Math" reform movement reached the Icelandic mathematical community. The four Nordic countries— Denmark, Finland, Norway, and Sweden—established a cooperation to produce mathematics teaching material. A textbook series by the Danish author Agnete Bundgaard, containing extremely theoretical material (Høyrup, 1979), was translated into Icelandic and implemented at the primary level.

By 1972 the new Danish syllabus was applied at the whole primary level for half the population. Algebraic concepts, such as the commutative, associative, and distributive laws; place-value notation to the base five; prime numbers; permutation of three digits; and the transverse sum and its relation to the nine times table were all introduced before the close of the third grade. In the following years basic concepts of set theory were introduced: pairing, subsets, intersection, and union; in addition to various place-value systems; linear functions; and geometry of points, lines, and planes in a set-theoretical framework (Bjarnadóttir, 2006a, 2007c). Set theoretical notion was used to find common divisors, as shown in Figure 32.5.

Most primary-level teachers had little acquaintance with these concepts, not to speak of changes in working methods, as explained in the following quotation from a handbook for teachers to the new syllabus:

> The role of mathematics in developing language sense has increased greatly by the introduction of the new attitudes to mathematics teaching upon which this textbook is based. Teachers' discussions with pupils about the topics will now become a much greater factor in the teaching than it has been in most places in arithmetic teaching up to this time. Secondly, mathematics makes strong demands for clear and logical use of the language, as logical thinking is the prime condition for mathematical thinking.... [T]he value of mathematics teaching in compulsory schools is not only concerned with the pupils learning to solve specific types of exercises that simulate various problems of daily life. The other purpose is much more important: that the effort put into solving mathematical problems may develop the pupils' accurate and logical thinking, to enable them to react correctly to various problems and handle them correctly. (Gíslason, 1967, p. 7)

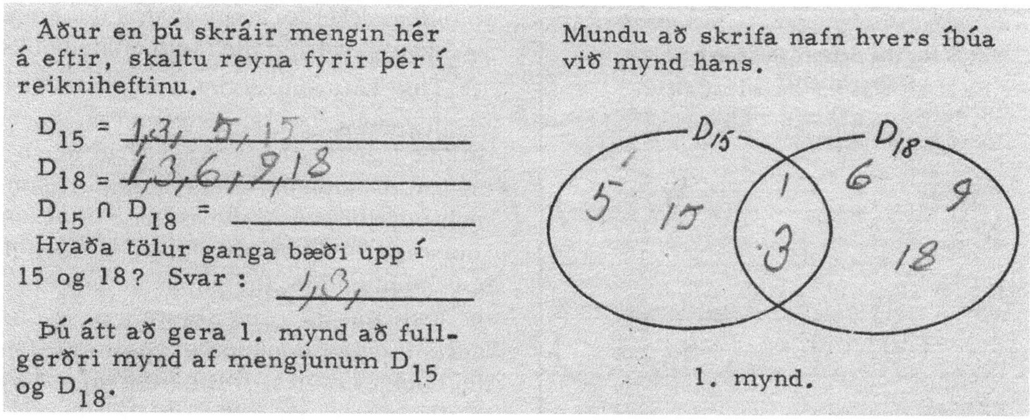

Figure 32.5 Bundgaard's method of finding common divisors of 15 and 18.

The quotation points to a change in attitudes to the desirable way of teaching mathematics—away from emphasizing computational procedures, towards conducting discussions and training in mathematical thinking. Teachers may have had difficulties in adapting to these new methods as may be gathered from the following quotation by G. Arnlaugsson, the main leader of the New Math reform movement:

> Many teachers in the primary and lower secondary schools have never in their studies met mathematical thinking. ... Mathematics... should... be the tool to train the child in logical thinking. If this is clear to the teacher, and he/she has an overview of the coherence of the topics of arithmetic that he/she is teaching, he/she could doubtless achieve a better result than... now, even if there were few actual changes in the syllabus. (Arnlaugsson, 1967, pp. 43–44)

The topics were not new in themselves, only the set-theoretical framework. They had shown up in various 18th-, 19th-, and early 20th-century textbooks, but not in the Daníelsson, Bjarnason, Guðmundsson & Gissurarson, or K. Gíslason series, which had monopolized the Icelandic market of mathematics textbooks for the compulsory level for many decades. In addition, new and different algorithms were introduced, especially for multi-digit multiplication.

As practically all primary level teachers had studied only at the Teacher Training College, where mathematics education was minimal after Dr. Daníelsson left the school in 1920, they had to work hard to cope with this new material. Two-week in-service courses each year were of some help. The initiative for creating new learning material was transferred from the State Textbook Publishing House to the Ministry of Education. A new department of School Research and Development, SRD, was established at the Ministry to conduct the developmental work in mathematics, and other school subjects.

The New Math soon had a reaction. The experience of a completely different material from what they had grown up with and taught inspired a group of teachers to create new teaching material, which was prepared by the SRD department. A new series of mathematics textbooks was written and carefully brought into life by experimental publications and test-teaching. The Danish Bundgaard material dwindled out in the 1970s, and its end marked the end of Danish influence on Icelandic mathematics education.

The SRD enjoyed the favor of the Minister of Education, and a number of young and creative teachers were recruited there. In Figures 32.6 and 32.7 are graphs of the ever-increasing parliamentary allocations to the SRD and the State Textbook Publishing

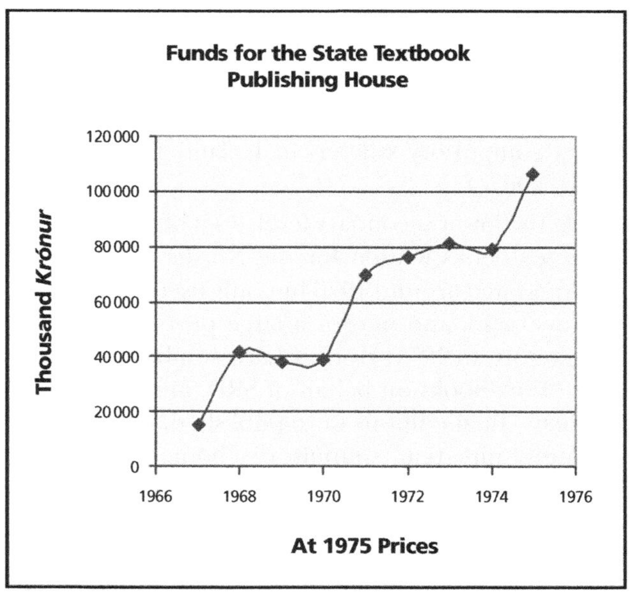

Figure 32.6 The State Textbook Publishing House.

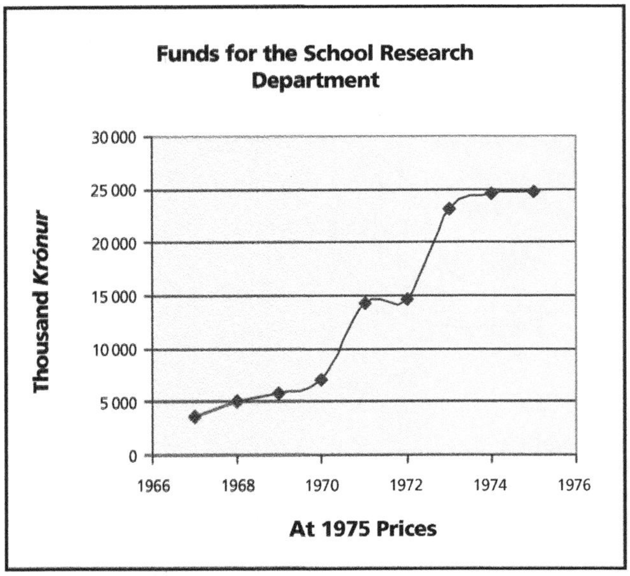

Figure 32.7 The School Research Department.

House during the years 1967–1975, recalculated to 1975 prices (Bjarnadóttir, 2006a, 2007c).

A preliminary national curriculum for primary level mathematics was created in 1970. Geometry appeared in the curriculum proposal without any further definition. The committee assigned to write the proposal said in its introduction that geometry should be taught in the fourth to sixth grade, well prepared by giving the children opportunities to view the tasks on their own, but it did not feel confident enough to make any suggestions about topics in geometry, as no experience was available in Iceland. When this was written, early in 1970, the first groups of pupils using the New Math Bundgaard material had not yet reached the textbook volumes containing geometry. They had even not been published yet. Geometry, with the exception of measurements and computations of length, area, and volume, had never been included in the compulsory syllabus in Iceland (Bjarnadóttir, 2006a, 2007c).

At the lower secondary level, a set of Swedish New Math textbooks written for the Nordic Cooperation was translated around 1970 but only used for a couple of years or so, and not for a large proportion of the year cohort. In 1971, Hörður Lárusson began to write a set of textbooks on behalf of SRD, and in the next few years, final editions were published, intended for the future nine-year compulsory school, under school legislation of 1974. The books were on similar topics to the Swedish ones. In Lárusson's ninth grade book, probability and statistics appeared for the first time in Icelandic lower secondary textbooks. Lárusson's series was used as the main textbook series for the lower secondary level for more than a decade at the lower secondary level and the ninth grade book longer.

The Lárusson series comprised textbooks for reading and books of exercises. Some teachers dropped the textbook, which had much more reading text than the previous conventional ones, and had their pupils use only the exercise book. On in-service courses for teachers the author stressed that this was not the intention: the pupils were primarily to study the text, and the exercises were of secondary importance. If either book was to be dropped, it should be the book of exercises. Different attitudes crystallized in these exchanges of opinions. The teachers may have wanted to interpret the topics on their own as they used to do. By pupils studying the text, the teachers' role was mainly to help the pupils with what they could not find out on their own and motivate them, which sometimes proved harder than to explain, as not all the teachers were enthusiastic about this new syllabus themselves.

AFTER THE NEW MATH: THE PERIOD 1980s–1990s

A new textbook series for the primary level, *Stærðfræði handa grunnskólum* (*Mathematics for Compulsory School*) written as a reaction to Bundgaard's theoretical textbook series, had more variety than the Bjarnason series. It emphasized understanding the place-value number system and introduced variables early, but turned back to the conventional algorithms. Geometry had its proper place, intuitive classical geometry as well as metric, together with the coordinate system. The books were printed in color as far as printing technique allowed to do economically at that time, one color in addition to black on each page. The initial complaints of teachers about the new primary level textbook series were that it was too versatile, that there were too frequent shifts between topics, and too few exercises. The idea was to resist behaviorism, to initiate discussions, and to avoid blind training of procedures. There are indications that the understanding it was intended to promote received less attention as teachers had gained routine in their use (Bjarnadóttir, 2007b).

By the end of 1990s, when new material was being developed, a series of extra material had been published with a number of training exercises by the National Centre for Educational Material, NCEM, the successor of the State Textbook Publishing House from 1979. The NCEM was another monopoly enterprise. The monopoly was released in 2007 by law (Stjórnartíðindi, 2007).

The fact that the population of Iceland is only 300,000 and each cohort counts around 4,000 makes it difficult to keep up competition on the versatile teaching material demanded at present times. In the late 1980s the NCEM offered a choice of two very different textbook series for the lower secondary level, a translated Swedish series, *Almenn stærðfræði* (*General mathematics*) by Björk, Björksten, Brolin, Ernestam, and Ljungström (1987), and its own production, a continuation of the series *Stærðfræði handa grunnskólum* (*Mathematics for Compulsory School*) *7–9*, each with its own subtitle, written by A. Kristjánsdóttir, Magnúsdóttir, Birgisdóttir, Ármannsson, and Bjarnadóttir (1987–1991). The series was never completed; only five books of six were published, and it did not reach great distribution after the full Swedish series was published. In this series, finding the least common multiple was delayed until grade 7 for age 13. Two methods were presented, one of them by using prime factoring. The other is graphical, illustrated on Figure 32.8 below.

In 1999, a new national curriculum was published, influenced by international currents, such as *Standards*, laid by the National Council of Teachers of Mathematics in the United States. In 2008, a new Icelandic series for the compulsory level, adjusted to the new national curriculum, had been completed up to grade 10, named *Átta-tíu* (*Eight-ten*) (Pálsdóttir & Gunnarsdóttir, 2005–2008) for the three uppermost grades. While older material for the primary level was taken out of the market by the NCEM, the translated Swedish books were still kept in choice as an alternative for the lower secondary level.

Finding the least common multiple was still a topic for the 13 year age group, now in eighth grade. Intuitive search, supported by diagrams, is encouraged, while the method of using prime factoring is introduced in *Átta-tíu*, see Figure 32.9.

In a recent master's thesis study, the three textbook series, *Mathematics for Compulsory School 7* (Kristjánsdóttir et al., 1988–1989), *Eight–Ten 8* (Pálsdóttir & Gunnarsdóttir, 2005), and *General Mathematics 8* (Björk et al., 1987), all written for the present grade 8, were compared by R. Halldórsdóttir (2008). Halldórsdóttir's conclusion was that the *General mathematics* was most influenced by behaviorism, while the *Eight–Ten 8* adhered to constructivism and *Mathematics for Compulsory School* was somewhere in-between, adhering to information processing. The *General Mathematics* would be easiest for teachers to conduct without a special training. Considering the relatively few teachers who have mathematics as an elective, this, in addition to general conservatism, may explain the reluctance of teachers to accept the two Icelandic series. In those, the authors' wishes were obviously to share their enthusiasm with the pupils, while parts of the series, especially the *Eight–Ten*, were a little too overloaded: too much material introduced in the text, according to Halldórsdóttir (2008). This is a problem related to the short academic year in Iceland; much is to be presented in a short time to keep up with the international curriculum, knowing that many children in the adolescent age are capable of absorbing many ideas, once the settings are favorable. This makes heavy demands on the teachers, to sort out what is relevant and possible for each and every pupil.

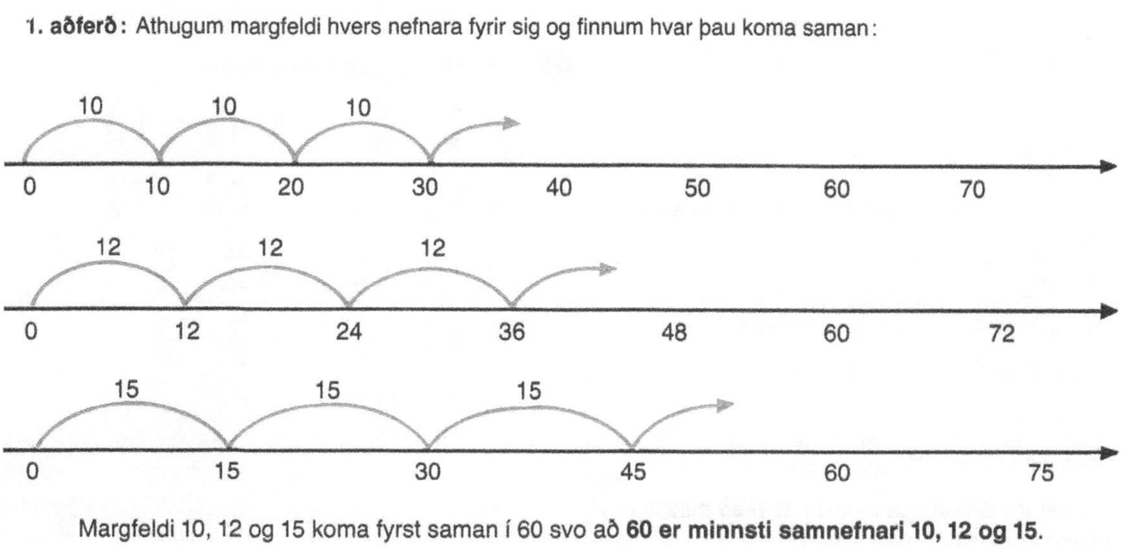

Figure 32.8 Method 1 in finding the least common multiple. Method 2 was using prime factoring (Kristjánsdóttir et al. 1988, p. 39).

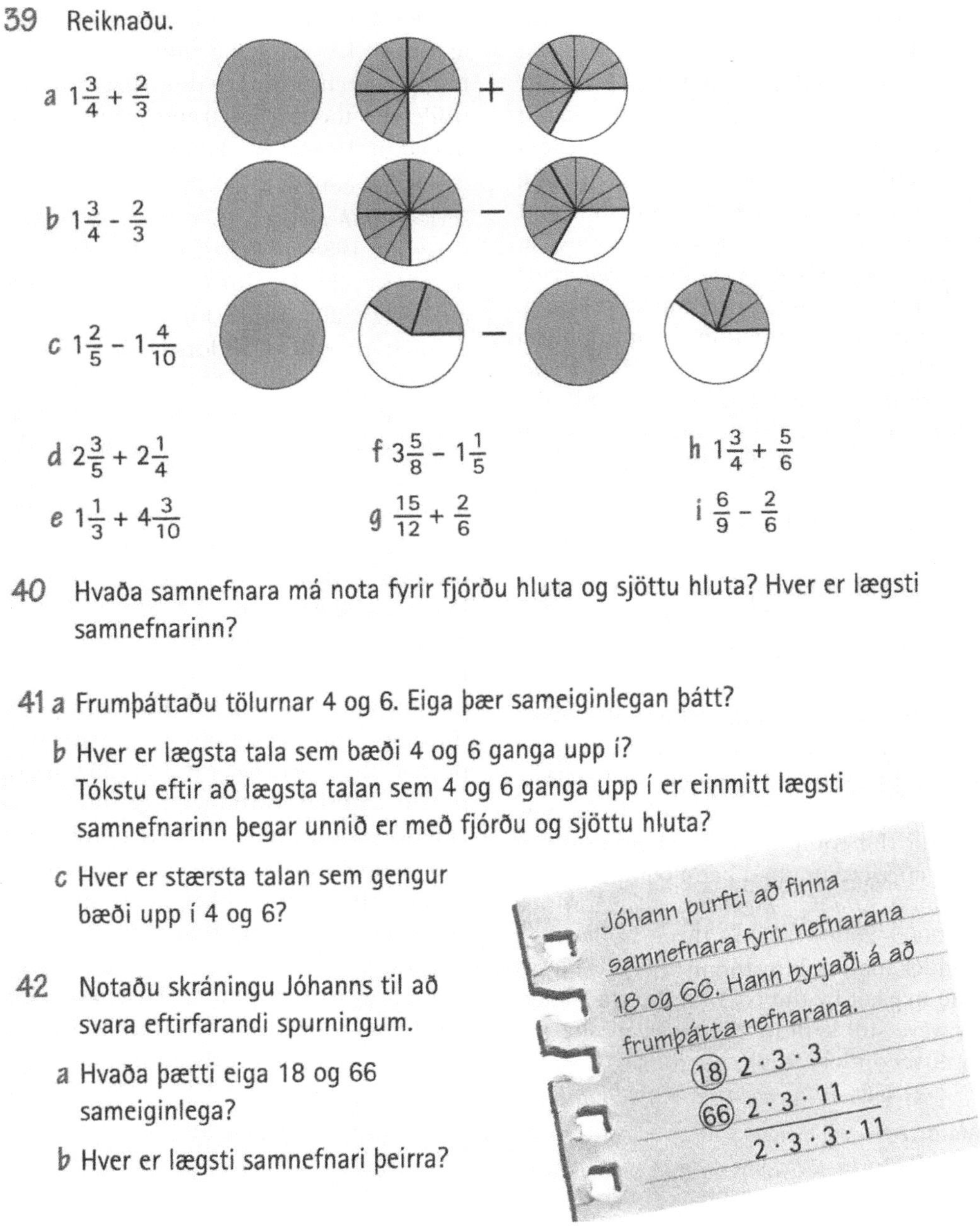

39 Reiknaðu.

a $1\frac{3}{4} + \frac{2}{3}$

b $1\frac{3}{4} - \frac{2}{3}$

c $1\frac{2}{5} - 1\frac{4}{10}$

d $2\frac{3}{5} + 2\frac{1}{4}$ f $3\frac{5}{8} - 1\frac{1}{5}$ h $1\frac{3}{4} + \frac{5}{6}$

e $1\frac{1}{3} + 4\frac{3}{10}$ g $\frac{15}{12} + \frac{2}{6}$ i $\frac{6}{9} - \frac{2}{6}$

40 Hvaða samnefnara má nota fyrir fjórðu hluta og sjöttu hluta? Hver er lægsti samnefnarinn?

41 a Frumþáttaðu tölurnar 4 og 6. Eiga þær sameiginlegan þátt?

b Hver er lægsta tala sem bæði 4 og 6 ganga upp í?
Tókstu eftir að lægsta talan sem 4 og 6 ganga upp í er einmitt lægsti samnefnarinn þegar unnið er með fjórðu og sjöttu hluta?

c Hver er stærsta talan sem gengur bæði upp í 4 og 6?

42 Notaðu skráningu Jóhanns til að svara eftirfarandi spurningum.

a Hvaða þætti eiga 18 og 66 sameiginlega?

b Hver er lægsti samnefnari þeirra?

Jóhann þurfti að finna samnefnara fyrir nefnarana 18 og 66. Hann byrjaði á að frumþátta nefnarana.

(18) $2 \cdot 3 \cdot 3$

(66) $2 \cdot 3 \cdot 11$

$2 \cdot 3 \cdot 3 \cdot 11$

Figure 32.9 Finding the least common multiple by fraction models and by prime factoring (Pálsdóttir & Gunnarsdóttir, 2005, p. 52).

TEACHING METHODS

In this chapter, teaching methods have been hinted at, but no research exists on which to base any conclusion. The traditional method was presumably that the teacher explained new procedures to the degree that the pupils could practice on their own. Some later textbooks are clearly adapted to certain working methods, but the teachers may have been stuck to what they have found most effective in their previous work.

In the New Math wave, teachers were encouraged to conduct discussion with the pupils. In the *Mathematics for Compulsory School* series, the emphasis was on a few exercises at a time, avoiding drill, but gradually extra training booklets were prepared. Pupils inventing their own procedures in order to deepen their understanding was not suggested until in the *Eight–Ten* series and in their complementary series for the primary level. In the same series, the emphasis is on a problem-solving approach by class introduction, group cooperation and discussion, and final summing up in class. On the other hand, the leading method, prescribed by many local educational authorities, has lately been individualized work, for which the *Eight–Ten* is less well suited. If the problem-solving approach is to become an effective teaching method, then conducting fruitful discussion with and between students in class is one of the most urgent tasks to devote attention to in mathematics teacher education.

INTERNATIONAL COMPARISON STUDIES

In TIMSS, the Third International Mathematics and Science Study in 1995, Icelandic eighth-graders' achievement in mathematics did not differ significantly from that of pupils in Denmark, the United States, Scotland, Latvia, Spain, Greece, Romania, Lithuania, and Cyprus. The international average score was 515, while Icelandic pupils scored 487, placing it at 32 of 41 countries (International Association for the Evaluation of Educational Achievement, 1996). As the score was below the average and significantly lower than that of the neighboring countries of Norway, Sweden and England, this achievement was considered unsatisfactory by educational authorities and the public, and it sparked considerable discussion.

Gradually, mathematics education in Iceland has begun to be measurable on the same scale as other nations. One possible explanation is that the unfavorable results in TIMSS were a challenge to the teachers to focus more on mathematics. In spite of the mathematical background of many teachers being weak, the nature of their general education is tailored to respond to new challenges. This may have been proved by satisfying results in the Programme for International Student Assessment (PISA) comparison study in 2003. In its mathematics part, Iceland's placement was no. 13–17 out of 41 countries and no. 10–14 out of 29 OECD countries (Námsmatsstofnun, 2004), on similar terms as Czech Republic, Denmark, France, Sweden, and Britain, and significantly better than Norway, in addition to being well above the average.

A lack of mathematical training of teachers may, however, be reflected in the PISA 2003 results. Iceland's above-average performance was mainly based on a relatively large group at level 4, 23.2% (OECD, 2004). Low percentage of Icelandic pupils on two uppermost performance levels 5 and 6 causes concern. The OECD average for level 6 of highest score was 4.0% (SD = 0.1) and the average sum on levels 5 and 6 was 13.1%. Comparison to pupils in Denmark and Czech Republic, whose total scores were similar to Iceland, and to Finland, is found in Table 32.1.

TABLE 32.1 Percentages of Students at Levels 5 and 6 Compared to Level 1 and Below at the PISA 2003 Comparison Study

Country	Level 6	Levels 5 and 6	Level 1 and below
Iceland	3.7% (SE = 0.4)	15.4%	15.0%
Denmark	4.1% (SE = 0.5)	16.2%	15.4%
Czech Republic	5.3% (SE = 0.5)	18.2%	16.6%
Finland	6.7% (SE = 0.5)	23.4%	6.8%

Looking at the three countries' performances, Icelandic pupils score the lowest on levels 5 and 6, but also at the two lowest levels. In Finland the score is totally different (OECD, 2004). Considering the similar social status of the Nordic countries it raises questions why pupils in Iceland, and Denmark for the same reason, are not on a level similar to Finland in the international PISA survey. Is it related to teacher education, their ways of working or their working conditions or some other educational policy?

SUMMARY AND CONCLUSIONS

This account is a short summary of the history of public education in mathematics in Iceland, mainly done by exploring learning material in arithmetic. Explicit goals of mathematics education were not stated, neither by the 1880 and 1907 legislation nor in the 1929 national curriculum or a draft to such in 1948. The real goals have to be read between the lines of the texts and the authors' statements in forewords of textbooks. They were primarily utilitarian, to aid the general public in coping with life in a context of growing commerce and rising industrialization. The material was true to the European tradition of textbooks, but originally some variety in presentation was offered. In the early 20th century, teachers had some choice of textbooks, but the distribution of textbooks often depended on the position, status, and sphere of authority of the textbook authors.

Gradually the choice reduced. Three political decisions of social-democratic origin, intended to ease the access to education for the general public, had serious consequences. The decision to restrict the admission to the dominating Reykjavík college-bound secondary school ensured the superiority of its employees to govern the syllabus of the primary and lower secondary education. The establishment of the State Textbook Publishing House resulted in a limited choice of learning material, especially of arithmetic textbooks. Finally, the decision to choose the syllabus of the former lower secondary division of the Reykjavík college-bound secondary school as that of a national entrance examination into upper secondary school level kept the syllabus of the lower secondary education rigid for a quarter of a century, resulting in no progress.

Dr. Daníelsson was a unique entrepreneur in textbook writing and restoring upper secondary mathematics education after four decades of its absence. But he was also a gatekeeper to higher professional education. In his position of power, arithmetic textbooks written or approved by him became the permitted and feasible syllabus in Icelandic primary and lower secondary education far into the second half of the 20th century. The textbooks in question did not have many faults in themselves, but by their monopoly position, teachers and pupils were kept from becoming acquainted with any other presentation than approved by Dr. Daníelsson long after his retirement. Emphasis on certain "correct" algorithmic procedures prevailed, and basic ideas, such as prime numbers or prime factoring, were unknown to teachers and pupils, not because Dr. Daníelsson did not know them, but because he did not consider them timely at the primary and lower secondary level.

Dr. Daníelsson's influence remained well into the 1960s and 1970s when people's ideas of admission to further and higher education was rapidly changing. As this influence was long overdue by then, Iceland fell prey to the international New Math reform wave, to drop into a difficult turmoil for one more decade before a real domestic reform could take place.

The view that mathematics education was to serve utilitarian purposes has dominated Icelandic educational policy. Its role in compulsory education was to aid the public in commerce, and higher mathematics served mainly to fulfill the requirements of the University of Copenhagen and later the present Technical University in Copenhagen and its predecessors. When that was no longer necessary in the 1870s, Icelanders saw no reason to practice it at their sole Latin school until the need for training Icelandic engineers became obvious in the early 20th century. The role of mathematics to train the mind was rarely mentioned except by Dr. Daníelsson and his pupil, Arnlaugsson, at the introduction of the New Math. Otherwise such a role may have been considered an intellectual luxury.

Iceland was dependent of Denmark in most respects until the early 1940s. Danish influence on mathematics education prevailed into the 1970s when the primary level Bundgaard material was abandoned and a full program of engineering studies was established at the University of Iceland. In spite of a temporary economic crisis in the early 21st century, Iceland has developed far from its slender means of the early 20th century, when hardly any infrastructure or schools existed, also considering mathematics education. National curricula and textbooks are up to international standards, as is education of mathematics teachers. However, the PISA comparison study witnesses a lack of excellent performance of Icelandic pupils. The lack of teachers with the required mathematics training to teach mathematics in grades 8 to 10 of compulsory school (Menntamálaráðuneytið, 2005) points to the fact that too few teacher students are enthusiastic enough about mathematics to undertake the strenuous work of mastering it sufficiently to teach it to excellence. Improving that situation will be a task to elaborate on in the coming years.

REFERENCES

Alþingistíðindi [Parliamentary reports]. (1936). Frumvarp til laga um ríkisútgáfu námsbóka [A bill on a state textbook publishing house], A no. 247, pp. 599–602.

Alþingistíðindi [Parliamentary reports]. (1955). Ríkisútgáfa námsbóka [State textbook publishing house], B no. 35, pp. 884–900.

Arason, S. (1928). *Reikningsbók handa alþýðuskólum* (*Arithmetic for people's schools*) (4th edition). Reykjavík: Gutenberg.

Arnlaugsson, G. (1967). Ný viðhorf í reikningskennslu [New attitudes in arithmetic teaching]. *Menntamál* [Educational matters], *40*(1), 40–51.

Arnalds, A. (1949). *Minningar* [*Memoirs*]. Reykjavík: Hlaðbúð.

Bjarnadóttir, K. (2006a). *Mathematical education in Iceland in historical context—Socio-economic demands and influences.* Reykjavík: Háskólaútgáfan.

Bjarnadóttir, K. (2006b). Menntun stærðfræðikennara í ljósi sögunnar [Education of mathematics teachers in the light of history]. In *Nám í nýju samhengi* [*Learning in a new context*] (pp. 25–38). Rannsóknarstofnun: Kennaraháskóla Íslands.

Bjarnadóttir, K. (2007a). Greinileg vegleiðsla til talnalistarinnar [Clear guidance to the number art]. *Vefnir,* an electronic journal of the Society for 18th Century Research. Retrieved June 27, 2008 from http://vefnir.bok.hi.is/2007/kb.pdf.

Bjarnadóttir, K. (2007b). The concept of understanding in mathematics textbooks in Iceland. In C. Bergsten, B. Grevholm, H.S. Måsøvall and F. Rønning (Eds.), *Relating practice and research in mathematics education. Proceedings of NORMA 05, fourth Nordic conference on mathematics education* (pp. 163–175). Trondheim: Tapir Academic Press.

Bjarnadóttir, K. (2007c). *Mathematical education in Iceland in historical context—Socio-economic demands and influences.* Retrieved July 29, 2008 from http://rudar.ruc.dk/bitstream/1800/2914/1/Chapter0_IMFUFA.pdf.

Bjarnason, E. (1927–1929). *Reikningsbók* [*Arithmetic*]. Reykjavík: Bókaverslun Guðm. Gamalíelssonar.

Björk, L., Björksten, C, Brolin, H. Ernestam, A., & Ljungström, L. (1987). *Almenn stærðfræði fyrir grunnskóla [General mathematics for compulsory school].* Reykjavík: Námsgagnastofnun.

Briem, E. (1880). *Reikningsbók* [*Arithmetic*]. (2nd part, 2nd ed., rev.). Reykjavík: Einar Þórðarson and Eiríkur Briem.

Briem, H. (1889). *Flatamálsfræði handa alþýðuskólum* [*Plane geometry for people's schools*]. Reykjavík: [s.n.].

Brynjólfsson J. & Arason, S. (1914). *Reikningsbók handa alþýðuskólum* [*Arithmetic for people's schools*]. Reykjavík: [s.n.].

Daníelsson, Ó. (1906). *Reikningsbók* [*Arithmetic*]. Reykjavík: Sigurður Kristjánsson.

Daníelsson, Ó. (1920a). *Reikningsbók* [*Arithmetic*]. Reykjavík: Arinbjörn Sveinbjarnarson. (Republished until 1956).

Daníelsson, Ó. (1920b). *Um flatarmyndir* [*On plane geometric figures*]. Reykjavík: Guðmundur Gamalíelsson.

Daníelsson, Ó. (1927). *Kenslubók í algebru* [A textbook in algebra]. Akureyri: Bókaverslun Þorsteins M. Jónssonar. (Republished until 1971).

Daníelsson, Ó. (1938). *Reikningsbók* [*Arithmetic*] (5th ed.). Reykjavík: Ísafoldarprentsmiðja h.f.

Gestsson, G.O. (1962). Reikningskennsla [Arithmetic teaching]. *Menntamál* [*Educational matters*], *35*(2), 114–137.

Gíslason, K. (1962). *Reikningsbók handa framhaldskólum* [*Arithmetic for secondary schools*] (Vol. 2). Reykjavík: Ríkisútgáfa námsbóka.

Gíslason, K. (1967). Um málnotkun [On language use]. In A. Bundgaard, *Stærðfræði. Reikningur. Handbók kennara* [*Mathematics. Arithmetic. Teachers' handbook*] (vol. 1, p. 7). Reykjavík: Ríkisútgáfa námsbóka.

Gíslason, S.Á. (1911–1914). *Reikningsbók* [*Arithmetic*] I–VI. Reykjavík: Bókaverslun Sigfúsar Eymundssonar.

Gissurarson, J.Á. & Guðmundsson, S. (1949–1950). *Reikningsbók handa framhaldsskólum* [*Arithmetic for secondary schools*] *II A.* Reykjavík: Ísafoldarprentsmiðja.

Friðriksson, G. (1992). *Dómsmálaráðherrann. Saga Jónasar Jónssonar frá Hriflu*, II. Reykjavík: Iðunn.

Halldórsdóttir, R. (2008). *Stærðfræði fyrir 8. bekk, námsbækur—námskrá [Mathematics for 8th grade, textbooks—curriculum].* A master's thesis. Reykjavík: University of Iceland, School of Education.

Halldórsson, Á. (2001). *Saga barnaskóla í Reykjavík til 1930* [*The history of primary schools in Reykjavík until 1930*]. Reykjavík: Kennaraháskóli Íslands.

Hansen, H.C. (2002). *Fra forstandens slibesten til borgerens værktøj. Regning og matematik i folkets skole 1739–1958* [*From the grindstone of the mind to the citizens useful tool. Arithmetic and mathematics in compulsory school 1739-1958*]. Aalborg: Aalborg University.

Hansen, M. (1890). *Reikningsbók handa alþýðuskólum [Arithmetic for people's schools].* Reykjavík: Ísafoldarprentsmiðja.

Helgason, Á. (1907). Frásagnir um skólalíf á Íslandi um aldamót 18. og 19. aldar. In *Safn til sögu Íslands og íslenskra bókmennta að fornu og nýju,* IV (pp. 74–98). Copenhagen and Reykjavík: Hið íslenska bókmenntafélag.

Høyrup, J. (1979). Historien om den nye matematik i Danmark—en skitse [The history of the new mathematics in Denmark—a sketch]. In P. Bollerslev (Ed.), *Den ny matematik i Danmark* [*The new mathematics in Denmark]* (pp. 49–65). Copenhagen: Gyldendal.

Indriðadóttir, K. (1995). Hugmyndir Steingríms Arasonar um kennslu og skólastarf. Uppruni og afdrif [Steingrímur Arason's ideas on teaching and schoolwork. Origin and end]. *Uppeldi og menntun* [*Upbringing and education*] *4*, 9–33.

International Association for the Evaluation of Educational Achievement. (1996, November). *Mathematics Achievement in the Middle School Years: IEA's Third International Mathematics and Science Study (TIMSS).* Chestnut Hill, MA: TIMSS International Study Centre, Boston College.

Jónasson, J. (1906). *Reikningsbók* [*Arithmetic*]. Akureyri: Bókaverzlun Odds Björnssonar.

Jónsson, S. (1906). *Reikningsbók handa unglingum* [*Arithmetic for adolescents*]. Reykjavík: Gutenberg.

Kristjánsdóttir, A., Magnúsdóttir, Á., Birgisdóttir, B., Ármannsson, I., & Bjarnadóttir, K. (1987–1991). *Stærðfræði handa grunnskólum* [*Mathematics for Compulsory School*] *7–9.* Reykjavík: Námsgagnastofnun.

Menntamálaráðuneytið. (2005). *Menntun kennara í stærðfræði- og náttúrufræðigreinum í grunn- og framhaldsskólum 2003–2004* [*Education of teachers in mathematical and nature subjects in compulsory and upper secondary schools 2003–2004*]. Reykjavík: Menntamálaráðuneytið.

Námsmatsstofnun (Educational Testing Institute). (2004). *Stærðfræði við lok grunnskóla* [Mathematics by the end of compulsory school]. Retrieved on June 28, 2009 from http://www.namsmat.is/vefur/rannsoknir/pisa_kynning2003.html.

OECD. (2004). *Learning for tomorrow's world. FirstrResults from PISA 2003.* Paris.

Olavius, Ó. (1780). *Greinilig Vegleidsla til Talnalistarinnar með fjórum höfudgreinum hennar og þriggja lida Reglu skipud eptir Landsvísu og Kauplögum Íslendínga.* Copenhagen: [s.n.].

Pálsdóttir, G. & Gunnarsdóttir, H. G. (2005–2008). *Átta–tíu: Stærðfræði* [*Eight–ten: Mathematics*]. Reykjavík: Námsgagnastofnun.

Petersen, J. (1943). *Kennslubók í rúmfræði handa gagnfræðaskólum* [*A textbook on geometry for secondary schools*]. (Sigurkarl Stefánsson, Trans.). Reykjavík: Ísafoldarprentsmiðja.

Sigurðsson, Ö. (1900). *Reikningsbók handa börnum* [*Arithmetic for children*]. Reykjavík: Ísafoldarprentsmiðja.

Sigurgeirsson, I. (Ed.) (1987). *Námsefni í 50 ár: Ríkisútgáfa námsbóka, Fræðslumyndasafn ríkisins, Námsgagnastofnun, 1937–1987. Afmælisrit* [*Learning material for 50 years: State Textbook Publishing House, State Film Library, Centre for Educational Material, 1937–1987. A celebration publication*]. Reykjavík: Námsgagnastofnun.

Stjórnartíðindi (Governmental reports). (1956). Lög um ríkisútgáfu námsbóka [Legislation on a state textbook publishing house], no. 51, April 7.

Stjórnartíðindi (Governmental reports). (2007). Lög um Námsgagnastofnun [Legislation on a national centre for educational materials], no. 71, March 28.

Swetz, F. (1992). Fifteenth and sixteenth century arithmetic texts: What can we learn from them? *Science & Education, 1,* 365–378.

33

MATHEMATICS TEACHER EDUCATION AT ICELAND UNIVERSITY OF EDUCATION

Recent Development of Mathematics Education Courses

Guðný Helga Gunnarsdóttir

UNIVERSITY OF ICELAND, SCHOOL OF EDUCATION

Guðbjörg Pálsdóttir

UNIVERSITY OF ICELAND, SCHOOL OF EDUCATION

Abstract: Research on learning has put the role of teachers and their influence on students' learning into focus. Consequently, in the last decades, teacher educators have developed lot of theories, and more research has been done on teacher education. This has marked the structure and practice of mathematics teacher education. Teacher educators have been developing courses and trying out new ideas. At the Iceland University of Education, the teacher education for compulsory school has been developing towards more specialization as the research has revealed greater need for teachers' knowledge of the subject and the teaching and learning of it. It is important to analyze how theories and research influence practice in mathematics teacher education. In the course of mathematics teaching and the professional development of mathematics, new ideas and ways of thinking are stressed through greater emphasis on research and students' reflective thinking. Research on mathematics teacher education has shown that educating mathematics teachers is a complex task and that it needs to take into account many factors. Therefore teacher educators need to be working on putting more research into practice.

Keywords: Teacher education, professional development, teacher knowledge, Lesson Study

Teacher education has been changing and developing in the last decades. The authors of this chapter have worked in teacher education both in pre-service and in-service education for more than 20 years. We have been teaching different mathematics education courses and have been developing them in cooperation with other colleagues. In Iceland University of Education[1] (IUE), there have been frequent changes in the structure of teacher education that have challenged us to reorganize mathematics teacher education and the content and structure of the courses. Not much has been written about the change process or the ideas teachers base their decisions on when building up courses. In this chapter we give an overview of the structure of the mathematics teacher education in Iceland and report on our study of one of the courses.

The First Sourcebook on Nordic Research in Mathematics Education, pages 467–477

MATHEMATICS TEACHER EDUCATION AT ICELAND UNIVERSITY OF EDUCATION

Education in Iceland for compulsory school teachers has been at the Bachelor of Education level since 1971. At IUE teacher educators have participated in structuring teacher education, and teachers of mathematics and mathematics education have taken responsibility for the curriculum development in the specialization of the mathematics teachers. The university writes its own curriculum, but accreditation from the Ministry of Education is required.

At IUE teacher education has been under constant development and students have been able to specialize in mathematics teaching and learning to different degrees (see Figure 33.1 for degree structure). The most recent reforms open more possibilities for specialization both within educational studies and certain professional areas or subjects. The new reforms also allow students to commence graduate studies directly after finishing undergraduate degrees.

IUE places great emphasis on teacher education as a research based professional education. Professional depth and comprehensive overview characterize the studies and it is also considered of great importance that students are active and responsible for their studies. Connection to research, as well as to the field, and focus on creativity and communication are the three threads that are required to be intertwined in the content of all courses. In the university policy there is also an emphasis on increased research activities on behalf of the university teachers. There is pressure on them to do educational research in their field and publish their results in respected scientific publications.

The B.Ed. degree consists of 80 ECTS[2] in pedagogy and didactics, 80 ECTS in specialization and 20

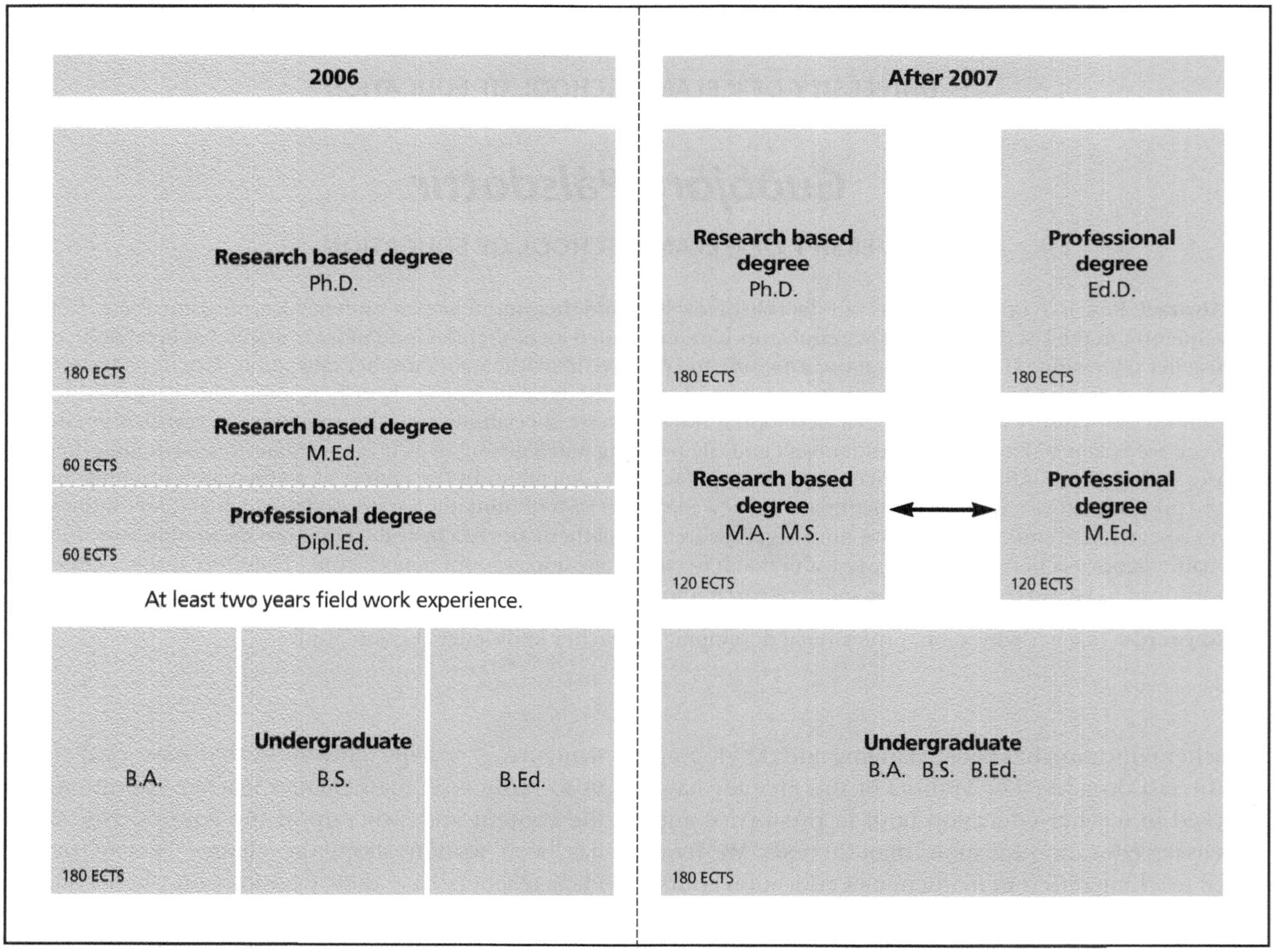

Figure 33.1 IUE degree structure. *Note:* A new law on teacher education was accepted by the pairlament on the 29th of May 2008. According to the law all teachers up to university level have to have a masters degree from an university that has been acknowledgded for the field of study by the minister of education. http://www.althingi.is/altext/135/s/1258.html.

TABLE 33.1 Course Overview

Year	Fall	Spring
1		Numbers and number operations
2	Mathematics teaching and learning in grades 7–10	Numbers and number theory
	Geometry	Mathematics teaching and learning for all
3	Algebra	Calculus
	Research on mathematics learning	B. Ed. Thesis

ECTS for studies of own choice (Stefnumótun Kennaraháskóla Íslands 2005–2010).

Students who specialize in mathematics and mathematics education take eight courses. Each course is of 10 ECTS. Table 33.1 gives an overview of the courses.

Students start their specialization in the spring term of their first year. In the first course the focus is on mathematics learning and teaching of numbers and operations. The students also deepen their content knowledge of numbers, ratio, proportions, probability, and statistics. They are also expected to increase their understanding of children's as well as their own mathematical learning. The students observe, plan, and teach mathematics in inclusive classrooms. In the second year the students concentrate on their specialization. They deepen their knowledge in both geometry and number theory and are introduced to theories and ideas about mathematics teaching and learning in compulsory schools. This is done in close connection to their teaching practice. Students have their own home school where they take most of their teaching practice and become active members of the school community.

In the last year, students study algebra and are introduced to calculus. They also get to know mathematics education as a research field; study some research in the field and try to conduct research on their own practice. The final project in the studies is a 10 ECTS B. Ed. thesis. There the students decide their own area of study, which is very often related to their specialization. The project can be a small research study, development of teaching materials, or an essay based on the scientific literature.

In previous curricula, we have been responsible for a course called *Mathematics teaching and the professional development of the mathematics teachers.* Development of new courses called *Mathematics teaching and learning in grades 7–10* and *Mathematics teaching and learning for all* are to be based on that experience. In this chapter we will describe and analyze the course *Mathematics teaching and the professional development of mathematics teacher.* We will look into the background of the course, the content, and the ways of working. We will also explore the reasons for our choice of structure, content, readings, and assignments. The chapter is meta-reflective in nature.

The teacher education program at IUE has used the concurrent model in the education of compulsory school teachers. Teacher students specialize in their chosen subject along with general pedagogical studies. The subject study is a combination of courses in mathematics and mathematics education. This is similar to the structure of teacher education programs for compulsory education in many of the other Nordic countries.

RESEARCH ON TEACHER EDUCATION

In recent years there has been considerable focus on teachers in mathematics education research (Clarke, Emanuelsson, Jablonka, & Mok, 2006; Grevholm, 2004, 2006, 2007; Hiebert, Morris, & Glass, 2003; Hill, Ball, & Schilling, 2008; Kilpatrick, 2004; Sfard, 2005). There also seems to be a shift in focus from what works to why and how teachers do what they do (Sfard, 2005; Sherin, Sherin, & Madanes, 2000).

Wood (Wood & Berry, 2003), one of the editors of *The Journal of Mathematics Teacher Education*, "argued for the importance of generating and sharing knowledge about the complexity in mathematics teaching" (p. 195). In the 11 years of this journal's existence, it has played an important role in building up a knowledge base for teacher educators who wish to improve the education of mathematics teachers. The collaboration of teachers, teacher educators, and students seems to be a central issue in many of the research studies reported on in the journal. Team work, learning communities, networks, and design research are often suggested as ways to help teachers to tackle and meet the complexity of teaching mathematics (Jaworski, 2005, 2006; Krainer, 2003; Wood, 2002; Wood & Berry, 2003).

Groups of experts in mathematics education who have defined important mathematical competencies or proficiencies for pupils have also defined which competencies or proficiencies teachers need to have to be able to teach pupils so that they achieve these competencies (Kilpatrick, 2004; Niss, 2004). In the Danish KOM project[3] (Niss, 2004), one of the questions asked was "What does it mean to be a good mathematics teacher?" The first answer was: "A good mathematics teacher is one who can effectively foster the development of mathematical competencies with her/his students" (p. 188). In trying to specify what this really meant, six competencies were identified:

- Curriculum competency
- Teaching competency
- Uncovering of learning competency
- Assessment competency
- Collaboration competency
- Professional development competency

The first four competencies focus on the students, their backgrounds, and opportunities, while the last two focus on the teacher and her or his position and opportunities within her or his professional environment (Niss, 2004).

In the report *Adding it up* (Kilpatrick, 2004), proficiency in teaching mathematics is characterized as five intertwined strands similar to the strands of mathematical proficiency for students. One of the aims of the committee that wrote the report was to provide research-based recommendations on teacher education. The five proposed strands for proficiency in teaching mathematics were:

- Conceptual understanding of core knowledge
- Fluency in carrying out basic instructional routines
- Strategic competence in planning instruction and solving problems that arise during instruction
- Adaptive reasoning in justifying and explaining one's practices and reflecting on those practices so as to improve them
- Productive disposition toward mathematics, teaching, learning, and the improvement of teaching practice.

The committee made it clear that professional learning opportunities play a central role when trying to build programs that are supposed to help teachers develop their knowledge to teach. This applies both to in-service and pre-service teacher education.

Even though Kilpatrick uses the word proficiency and Niss uses the word competency, it is clear that there are many common threads and similar views on what it means to be a good mathematics teacher. The choice of words is probably rather a reflection of how the same words can have a different meaning in different contexts and cultures than being different views on mathematics teaching. These views on what it means to be a good mathematics teacher also make it clear that educating mathematics teachers is a complicated task.

Grevholm (2006, 2007) also emphasizes that teaching is a complicated profession and that it is only possible to prepare pre-service students to a certain extent. She points out that it will never be possible to meet all needs, demands, and expectations in pre-service teacher education. Therefore, teacher students have to build up knowledge on different ways to develop professionally and to seek further education. To get insight into research communities and pedagogical content knowledge is another important factor. This can be done by emphasizing subject-based didactics in both pre-service education and professional development of teachers. For mathematics teachers, mathematics education or didactics of mathematics therefore plays a central role.

Grevholm (2006) finds it important that teacher students build up a professional identity as teachers. According to her, the professional identity of mathematics teachers is based on five main components of teacher knowledge:

- A professional language
- A personal view on beliefs about knowledge and learning
- Knowledge about classroom management methods and material
- Competence to judge and diagnose pupils' learning in mathematics
- Knowledge in mathematics related to teaching.

If students are to start to develop this kind of knowledge in their pre-service teacher education, then teacher education needs to be research-based and linked to practice. The students also have to learn to share their experiences with others and create learning communities themselves that will support their learning as teachers (Grevholm, 2006; Hiebert, Morris, & Glass, 2003). Teacher students have to realize that in their pre-service education they are only starting to create a knowledge base and that they are entering a profession where lifelong learning is essential.

An important factor in research based mathematics teacher education is the use of research based literature for students to get an insight into the work and methods used by researchers. They also have to learn to conduct small research projects themselves and get the background for starting researcher education. Students have to learn to present ideas in writing in a scientific way and reason and present their ideas in public in a professional way (Grevholm, 2004).

Shulman's (1986, 1987) ideas of a knowledge base for teaching, particularly his ideas of pedagogical content knowledge, have played a central role in teacher education. Pedagogical content knowledge "represents the blending of content and pedagogy into an understanding of how particular topics, problems, or issues are organized, represented, and adapted to the diverse interests and abilities of learners, and presented for instruction" (Shulman, 1987, p. 8).

Other researchers have built on these ideas (Hill et al., 2007; Shuhua, Kulm, & Wu, 2004). Ball and her colleagues at the University of Michigan have been developing practice-based theory of mathematical knowledge for teaching. They emphasized that mathematics teachers need a particular kind of highly specialized knowledge that requires an intertwining of content and pedagogy. They call this "mathematical knowledge for teaching." It consists of both subject matter knowledge and pedagogical content knowledge. The subject matter knowledge includes both common content knowledge (the mathematical knowledge of the school curriculum) and specialized content knowledge (the knowledge of mathematics needed specifically for the work of teaching). Pedagogical content knowledge includes knowledge of students and content and knowledge of teaching and content.

Shuhua, Kulm, and Wu (2004) put forward ideas about a network of pedagogical content knowledge that includes three main components. They are knowledge of content, knowledge of curriculum, and knowledge of teaching. They see knowledge of teaching as the core component in this network, a component that is based on knowledge of students' thinking, content, and curriculum.

One of the main tasks of teacher education programs is to develop among the students a knowledge of teaching. If students are to obtain a profound knowledge of teaching and to develop a practice that is different from what they themselves experienced as pupils, they need learning opportunities that are more powerful than simply reading and talking about new pedagogical ideas. The best learning environment is when the students are given the opportunity to do, study, and reflect in collaboration with others and by looking closely at the pupils and their work and share what they see with others. This requires a close connection to schools and teaching practices. Students need both opportunities to try out their ideas in practice and knowledge and support to reflect upon and interpret practice (Darling-Hammond, 1998).

Hiebert, Morris, and Glass (2003) propose a teacher preparation program that focuses on helping students acquire the tools needed to learn to teach rather than the finished competencies of effective teaching. The two main goals of the program are, firstly, that the students become mathematically proficient and, secondly, that they learn to teach in a manner that can help their students become mathematically proficient. They suggest that all lessons prospective teachers give should be treated as experiments and that they also need to know how to learn from these classroom teaching experiments. They also describe the learning environments that the prospective teachers must learn to create in order to sustain their own and other teachers' learning. Teacher students need to become responsible for their own professional development when they enter into the profession of teaching and also for the development of their own profession. Teacher preparation programs can only be the starting point of this long-term process, and they have to make the students responsible for creating and sustaining a knowledge base for mathematics teaching.

Geirsdottir (2008) claims that every school has its own local pedagogical discourse and that the teachers are very powerful in the transformation of this local pedagogical discourse to the students. The teachers' conceptions of the discipline acquired during their own time of studying the discipline, their experience of teaching, and the disciplines' organizational saga and culture have great impact on development of the local pedagogical discourse at university level. The teachers are influenced by several factors in their institution, from their discipline, and from the society. Teachers have an interest in changing the curriculum, to offer courses on new topics, and to introduce new approaches in their disciplines. Other factors, like the decision of the IUE to intertwine the three threads (i.e., to maintain a strong connection to research and to the field, and to emphasize creativity and communication) within the content of all courses, also influence new curriculum. Geirsdottir (2008) finds it important that teachers engage in developing new curriculum, and that they analyze and discuss in groups the socio-cultural reality that the curriculum is based on. The curriculum is created in a social context, and the teachers need to be aware of what influences their decisions on courses content and teaching approach. It is important that teachers question their own and their institutions' practice and take up relevant issues for the development of the curriculum. The process of creating a curriculum is complex, and curriculum decisions have strong impact on the students learning. Teachers have different ideas, and conflicts can arise when they discuss what is essential for each student and how to develop their local pedagogic discourse.

In mathematics teacher education there are many factors that need to be taken into consideration in the organizations of the studies. These factors need to be intertwined to form a coherent and flexible teacher education. In the next section we analyze our mathematics education courses and describe what has influenced the design and development of these courses.

TEACHING MATHEMATICS AND THE PROFESSIONAL DEVELOPMENT OF MATHEMATICS TEACHERS—AN EXAMPLE

The course, *Teaching Mathematics and the Professional Development of Mathematics Teachers*, has been in development for some years. Students attend the course in their final year and it is intertwined with teaching practice in grades 7–10. Emphasis is placed on connection to the field, especially to the schools where the teacher students have their teaching practice. The aim of the course is also to give the students opportunity to practice studying and discussing issues related to the teaching and learning of mathematics. This is considered important for the students who are building up their professional identity and going through a learning process in order to be able to take responsibility for their own professional growth.

> **Course description:**
> In this course students learn about teaching and learning mathematics in lower secondary school. The course is closely related to teaching practice in mathematics in secondary school (grades 7–10). Emphasis is put on mathematics for all students and varied ways to plan teaching in cooperation with others.
> *Curriculum IUE 2007–2008*

The course is 6 ECTS and the teaching practice is 10 ECTS. It is organized in three periods. It begins with a seven-week period where the students make observations in classrooms, study teaching materials, get introduced to journals and web-based materials for teachers about mathematics teaching and learning, and read about research in mathematics education. This is done along with preparations for the teaching practice. There the students are encouraged to work on this main question: What ideas do I want to try out in my teaching practice?

Focus is put on the teaching and learning of algebra in order to create a common ground for discussions on different teaching approaches. All the students are supposed to prepare at least a two week period of algebra teaching in one class.

The practice period is five weeks, and the students work in close cooperation with their supervising teachers. Here they get a chance to try out their ideas and work with three different classes for the whole period. They are encouraged to try out ideas that are new to them and ideas they have read about in the course literature. The students are also supposed to include the use of ICT in their teaching plans. Both the supervising teachers and the university teachers support the students when they are developing their ideas and finding ways to implement them in practice. The collaboration with their fellow students, both during the preparations and when it comes to the actual teaching, is also very important because it gives students many opportunities for discussions with their supervising teacher, other students, and the teachers at the university. The students practice teaching and communicating with pupils as well. They experience how to approach different mathematical content and meet pupils at different levels. The last two weeks of the term are used to evaluate and reflect on the students' experience of the teaching practice. The students read about other teachers' experiences and relate them to their own experiences. Focus is put on the teachers' role and how important it is for teachers to be a part of a progressive environment where they can reflect on their practice and develop ideas in cooperation with other teachers. The students discuss the idea that teacher education is a lifelong learning-process and that they have to take responsibility for their own professional development. They are introduced to professional learning opportunities for teachers and post-graduate studies.

The choice of course literature is of great importance. According to our experience, the first texts the students study influence their beliefs about the subject and what issues they find important to take up for discussion. In teacher education students will get familiar with the research field of mathematics education. They are introduced to examples of research and how it can support teachers in practice. Very little has been written in Icelandic on mathematics education, and the students have to learn the vocabulary of the field both in Icelandic and foreign languages. Most of the course literature is in English, but some of the readings come from the other Nordic countries. In a homogeneous language community like Iceland, it is important for professionals to be able to read about their field in foreign languages if they are to develop professionally.

A good starting point to build up and strengthen this competence is to read about the teaching and learning of mathematics in journals and magazines written for mathematics teachers. These texts offer the students opportunities of reading about other teachers' experiences and give them ideas for their teaching practice. Many of the journals are published by national organizations of mathematics teachers who also regularly hold conferences and have websites with resources for teachers. The journals from National Council of Teachers of Mathematics (NCTM) and Association of Teachers of Mathematics (ATM) have given teachers inspiration and ideas to use in their teaching and access to new topics in the mathematics teaching and learning debate. The three Nordic mathematics teachers' journals *Matematik*, *Nämnaren*, and *Tangenten* have also inspired Icelandic teachers. The students are introduced to these journals and are encouraged to read them regularly.

The students also have to learn to read and study scientific papers. Research journals have many interesting articles where different topics are taken up for examination and discussions. The students see examples of how researchers in mathematics education build their research on the theories students have studied in pedagogy, psychology, and sociology. They also experience how researchers develop these theories and new ones and use them in their research in mathematics education. The articles themselves reveal many interesting research results that can guide the students in many ways in their work as teachers. It is of great value for them to have the opportunity to reflect on and discuss research in a learning community.

The focus in the literature of the course has been on different teaching approaches, the teacher's role, and developmental work in mathematics teaching. For the last three years, the students have studied the book *Connecting Mathematical Ideas* (Boaler & Humphreys, 2005). The book and the accompanying classroom CD are the result of cooperation between mathematics teacher Cathy Humphreys and researcher Jo Boaler for a whole school year, and they describe in detail the interaction and ongoing development of teacher and researcher. The benefits of cooperation and mutual support between teachers become evident to the students, and they also get ideas about how teachers and researchers can work together, which can help them bridge the gap between theory and practice. From this book students get ideas of how it is possible to structure and implement change in mathematics education. In their course assignments they are supposed to work further with those ideas. In order to help them understand classroom practice, the first assignment in the course is to visit a mathematics class for one lesson and write down a detailed description of how the lesson proceeds. This gives a good background for their discussion of Cathy's lessons. In the second assignment, the students work in groups and write a journal about their discussions and reflections while reading the book and watching the CD. They are supposed to reflect both upon Cathy's experiences and thoughts and on Jo's comments. These comments add more theoretical concepts to their discussions. Jo's analysis and use of theory in the comments helps the students to make connections and to use Cathy's case as a prototypical example.

The third assignment is a teaching plan and evaluation of the plan. In the teaching plan, the students have to combine theoretical and practical knowledge. They have to show that they have a good understanding of the content and how pupils build up their understanding of important concepts in that area. They also have to argue for their choice of teaching approaches. The series of lessons the students plan are supposed to be holistic, coherent, and adapted to the pupils and their circumstances. The students teach in pairs, and they reflect both on their own teaching and on the teaching of their partners with the practice teacher. In the evaluation of the teaching plan, the students analyze their teaching and reflect on their experiences.

The last assignment is supposed to give students an opportunity to gather their knowledge, ideas, and thinking and argue for their views on mathematics teaching and learning. By attending various courses in teacher education, the students have developed their knowledge of schools and teaching. At this point they should be able to present their own professional theory based on their theoretical knowledge and experiences. In the assignment, the students are asked to select between (1)writing an article about certain areas of mathematics education for teachers journals, a policy for the mathematics teaching in a school, a booklet, or webpage for parents about new ideas on mathematics learning and reform in mathematics teaching; (2) make an introduction and set up a plan for a reading circle for mathematics teachers, or (3) write a book review of a book about mathematics education for teachers. The assignment ideally results in a product that can be used in real settings and that makes it real and worthwhile for the student to work on. It should reflect that the students have developed the competence to discuss mathematics education as professionals and are aware of their opportunities and responsibilities for their own professional development.

Example of Assignment 4

1. You are supposed to write an article in a journal for mathematics teachers about reforms in mathematic education in the last decade. You should choose to use as an example:
   a. One particular content area
   b. Processes and approaches
   c. Ideas about mathematical competencies
2. You recommend to group of teachers that they form a reading group around the book *Connecting Mathematical Ideas.* Argue for your choice of book and give a picture of Jo Boaler as an experienced researcher. You can use her web page as a source for information http://www.sussex.ac.uk/education/profile205572.html
3. You are supposed to make a booklet or a webpage for parents about recent reforms in mathematics teaching and learning in grades 7–10 according to a new curriculum and new teaching materials. What is important to address in such a booklet and why? Give an idea of the structure of the booklet and write a chapter on important changes of emphasis and content.
4. In the book *Developing Thinking in Algebra,* you have been challenged by many algebraic tasks. How has this work influenced you as a teacher of algebra? Has this work affected your beliefs and ideas on the content and ways of working in algebra? Would you recommend this book to other mathematics teachers? Why?

These four course assignments are meant to help students build up their competence and to recognize that teaching is a complicated task. Being a teacher requires researching your own teaching and constantly seeking ways to develop and improve it. One also needs to be aware of new research and development in the field and create and use opportunities for continuous professional development.

The teaching approach used in the course focuses on creating a learning community among the students. The university teachers play an active role in this community as participants in discussions and as group leaders or experts. The students are supposed to be active and bring in their knowledge, views, and thoughts about mathematics education. One way to support the building of a learning community is the use of ideas from Lesson Study (Lewis, 2002). This approach has created a good starting point for fruitful learning communities within the frame of the course. The ideas behind Lesson Study are clear and give all participants in the group opportunity to open up for their considerations about teaching and at the same time the strength in the group is used.

The main idea of Lesson Study is that a group of teachers, with support from an expert, develop a teaching plan for one lesson. The structure of the lesson and the role of the teacher are often in focus. The Lesson Study process can be described as a cycle (see Figure 33.2).

In the first phase, the group has to discuss what the content of the lesson should be. When the content has been defined, participants are meant to explore it, both in regard with what it means to acquire understanding of the content and how it can be approached in teaching. The participants are in control, and they make decisions about the process. One important factor is that the participants deepen their knowledge of the content and possible teaching approaches. They communicate, do research, work together, make decisions, plan teaching, and experience the advantages of participating in a learning community. The research lesson is taught by one of the participants where the other participants are present and take notes. After the lesson, the experience is discussed and an expert joins the group discussions. On that basis the lesson plan is revised and the lesson taught again for a similar group. The cycle can be repeated several times.

The students use the Lesson Study cycle, and the university teachers play the role of the expert. This

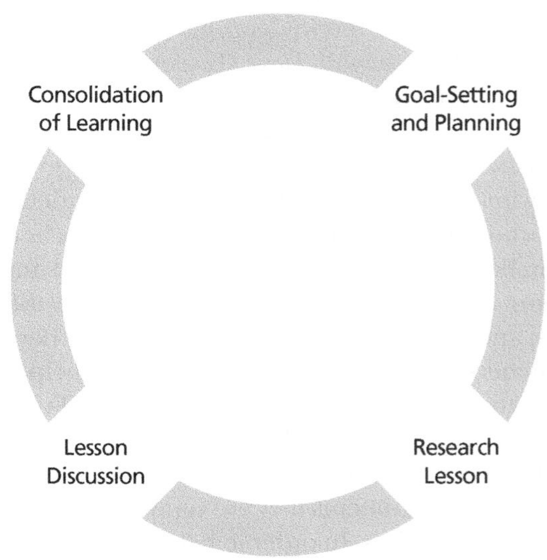

Figure 33.2 Lesson Study Cycle.

has led to discussions of teaching experiences with a different focus. The teachings of the content and the interactions between the pupils and the teacher have become more central. More students are involved in the same lesson practice, which gives them opportunities to discuss the connection between theory and praxis from a different perspective.

Different groups of students have attended the course. Students at the IUE often have very different backgrounds and experiences, which plays a big role when using ideas like the Lesson Study. Last year two different groups attended the course at the same time. The first group was a group of teacher students with valuable experience from different areas but little experience of teaching. The other group was a group of experienced teachers who had been teaching mathematics without having a specialization in mathematics teaching and learning. The experienced teachers were specializing in mathematics and mathematics education as a part-time study along with teaching. It was very interesting to see how differently these two groups approached their studies.

The main purpose of the experienced teachers for their studies was to develop and change their teaching practice and to build up their knowledge in mathematics and mathematics education. They were focused on the teacher's role and their own development as mathematics educators. They were self-confident as teachers but wanted to develop their ideas about mathematics teaching and learning and were aware of that they had much to learn regarding both context and ways to work. They believed that they were capable of and responsible for changing their own teaching and thereby their pupils' learning. The Lesson Study process helped the teachers to find ways to research their own teaching and cooperate with their colleagues in developing their teaching. Furthermore, they were developing their skills in taking notes and gained a great deal from discussing and studying each other's ways of noticing and taking notes. They also claimed that they had learned a great deal about themselves as teachers and that they were more conscious of their ideas and intentions. They found Lesson Study a useful professional development tool that could add to their professionalism and satisfaction in their work. They also claimed that they wanted to continue using Lesson Study and introduce the ideas in their own school or school district and saw it as a possibility to continue to work together.

The teacher students used Lesson Study when preparing for their teaching practice. They were building up their identity as teachers, especially in mathematics, and finding out about their own views and beliefs of teaching and learning. They had gotten a lot of ideas for teaching methods and tasks from discussions and from the literature used in the course but were wondering how to act as teachers in different situations. The students were developing their competence to be analytical and to combine their theoretical and practical knowledge. Often, teacher students do not take the necessary time to analyze and reflect before they make decisions. Therefore it is useful for them to think through one lesson in detail and look at it from many different angles. The interaction with the pupils where the teacher is expected to promote the pupils' learning is often thought to be the most difficult part of the teacher's role. Students in their teaching practice often use most of their energy to carry out their own teaching plan and find it difficult at the same time to interact with the class and respond to individual pupils.

The Lesson Study approach gave the students the opportunity to concentrate more on the pupils and interactions in the classroom during the planning, teaching, and evaluation of the lesson. Moreover, it helped the students to take their own teaching seriously and to look at it as an experiment that one can learn from. In the Lesson Study, the emphasis is that learning to teach is a lifelong process and that teaching can always be improved. The students also learned how they can benefit from developing teaching in cooperation with others.

DISCUSSION

In designing the course *Teaching Mathematics and the Professional Development of Mathematics Teacher,* we have built on literature and research on mathematics teacher education, especially from the U.S. and the other Nordic countries.

The structure of the course reflects the fact that it is possible in mathematics education courses to develop the six competencies identified in the Danish KOM project as important for the education of mathematics teachers (Niss, 2004). An important factor is the close connection to practice, which helps the students to develop their competencies regarding the teaching, learning, and assessing subjects. An emphasis on collaboration in preparing for practice makes it possible for the students to develop their competencies regarding assessment, collaboration, and professional development. The structure should also make it possible for the students to develop a professional language, a personal view about knowledge and learning, and knowledge about classroom management and material, which are important components of any teacher's knowledge, according to Grevholm (2006).

The choice of literature for the course gives a clear indication of the studies being research-based (Grevholm, 2004). It should also give the students some ideas about where they can access literature and materials that can help them to develop professionally and take responsibility for their own professional development (Hiebert et al., 2003; Niss, 2004). Through their readings and discussions of literature, the students develop their professional language and develop understanding of important concepts in mathematics education (Grevholm, 2006).

The course literature choices also are indicative of how much emphasis is placed on building up the students' knowledge of teaching, which is considered a very important competency and the main factor of pedagogical content knowledge (Niss, 2004; Shuhua, Kulm, & Wu, 2004). The study of the literature also gives students an opportunity to build up personal views and beliefs about knowledge and learning and gives them something to relate to when building up and reconsidering their views on teaching and learning mathematics (Grevholm, 2004, 2006, 2007).

Through the course assignments the students get some experience in conducting small research projects and in writing about their experiences and ideas in a scientific manner for different target groups (Grevholm, 2004). Furthermore, through the assignments, the students combine their knowledge of content, curriculum, and teaching (Shuhua, Kulm, & Wu, 2004) and develop their competencies as mathematics teachers (Niss, 2004). The assignments also prepare the students for being responsible about their own professional development and learn to work in collaboration with others, both with respect to professional discussions and when trying to implement ideas in practice (Hiebert et al., 2003; Niss, 2004). The assignments connected to the teaching practice help the students to develop their proficiency in teaching mathematics. There they plan, justify, and gain fluency in carrying out their teaching (Kilpatrick, 2004).

The choice of teaching approach reflects our ideas on how important it is that student teachers develop a productive disposition towards mathematics teaching and learning and how significant it is constantly to seek to improve practice (Kilpatrick, 2004). The chosen teaching approach highlights the emphasis that is put on intertwining the three threads—connection to research, connections to the field, and creativity and communication. This creates powerful learning opportunities for the students where they can, in collaboration with others, do, study, and reflect on their teaching and on their pupils' learning (Darling-Hammond, 1998). In the teaching approach much emphasis is placed on the development of the competency to collaborate with others and to evaluate pupils' learning in mathematics (Grevholm, 2007; Niss, 2004). The use of Lesson Study as a teaching approach also indicates that teaching experiments create good opportunities for teachers and teacher students to learn from their own practice. It is a lifelong process to learn to teach, and therefore it is important that teacher students learn to learn to teach (Hiebert et al., 2003). The students need to be introduced to ways to improve and develop their teaching. By introducing them to the Lesson Study and ideas about teachers researching their own practice, we try to help the teacher students to start working on their own professional theory and development.

We claim that it is a part of the local pedagogic discourse at IUE constantly to seek ways to improve the teacher education. Frequent reforms and encouragement to develop both course content and teaching approaches are a part of the culture of the institution. IUE has, for instance, played a leading role in developing distance teacher education, and it has been possible to finish a B.Ed. degree though a distance education program since 1993. It is probably due to the nature of the teacher education that the teachers at IUE are often engaged in discussing, reflecting on, and developing their courses in cooperation with other colleagues. In our view, it is essential that teacher educators write about their experiences and that they make the effort to analyze and interpret them, thereby making them accessible to others. That can give those of us who are in mathematics teacher education better opportunities to discuss our teaching with others and for others to reflect on their own experiences but also to discuss teacher education in general.

NOTES

1. Iceland University of Education merged with the University of Iceland on first of July 2008. Teacher education will be a faculty within a School of Education.
2. A full academic year of studies is 60 ECTS.
3. The Danish KOM project (Competencies and the learning of mathematics) was launched by the Ministry of Education and the Nationa Council for Science and Mathematics in 2000–2002.

REFERENCES

Boaler, J., & Humphreys, C. (2005). *Connecting Mathematical Ideas.* Portsmouth, NH: Heineman.

Clarke, D., Emanuelsson, J., Jablonka, E., & Mok, I. A. C. (Eds.). (2006). *Making Connections: Comparing Mathematics Classrooms Around the World* (Vol. 2). Rotterdam: Sense Publishers.

Darling-Hammond, L. (1998). Teacher Learning That Supports Student Learning. *Educational Leadership, 55*(5), 6–11.

Geirsdottir, G. (2008). *We are caught up in our own world. Conceptions of curriculum within three different disciplines at the University of Iceland.* Reykjavík: Iceland University of Education.

Grevholm, B. (2004). What does it mean for mathematics teacher education to be research based? In R. Stässer, G. Brandell, B. Grevholm, & O. Helenius (Eds.), *Educating for the Futrue. Proceedings of an International Symposium on Mathematics Teacher Education* (pp. 119–133). Göteborg: The Royal Swedish Academy of Sciences.

Grevholm, B. (2006). Matematikdidaktikens möjligheter i en forskningsbaserat lärarutbildning. In S. Ongstad (Ed.), *Fag og didaktik i lærerutdanning. Kunnskap i grenseland* (pp. 183–203). Oslo: Universitetsforlaget.

Grevholm, B. (2007). Why is mathematics teacher education problematic? In A. Andzans, D. Bonka & G. Lace (Eds.), *Retrospective and perspectives. Proceedings, May 11-12, 2007, Riga.* Riga: Latvijas Universitate.

Hiebert, J., Morris, A. K., & Glass, B. (2003). Learning to learn to teach: An "experiment" model for teaching and teacher preparations in mathematics. *Journal of Mathematics Teacher Education, 6*(3), 201–222.

Hill, H., Ball, D. L., & Schilling, S. G. (2008). Unpacking "pedagogical content knowledge": Conceptualizing and measuring teachers´ topic-specific knowledge of students. *Journal for Research in Mathematics Education, 39*(4), 372–400.

Jaworski, B. (2005). New beginnings and recurring themes. *Journal of Mathematics Teacher Education, 8*, 1–3.

Jaworski, B. (2006). Theory and practice in mathematics teaching development: Critical inquiry as a mode of learning in teaching. *Journal of Mathematics Teacher Education, 9*, 187–211.

Kilpatrick, J. (2004). Promoting the proficiency of U.S. mathematics teachers through centers for learning and teaching. In R. Stässer, G. Brandell, B. Grevholm & O. Helenius (Eds.), *Educating for the future. Proceedings of an International Symposium on Mathematics Teacher Education* (pp. 143–154). Göteborg: The Royal Swedish Academy of Sciences.

Krainer, K. (2003). Teams, communities & networks. *Journal of Mathematics Teacher Education, 6*, 93–105.

Lewis, C. C. (2002). *Lesson Study: A handbook for teacher-led instructional change.* Philadelphia, PA: Research for Better Schools, Inc.

Niss, M. (2004). The Danish "KOM" project and possible consequenses for teacher education. In R. Stässer, G. Brandell, B. Grevholm & O. Helenius (Eds.), *Educating for the Future. Proceedings of an International Symposium on Mathematics Teacher Education* (pp. 179–190). Göteborg: The Royal Swedish Academy of Sciences.

Sfard, A. (2005). What could be more practical than good research? *Educational Studies in Mathematics, 58*, 393–413.

Sherin, M., G., Sherin, B., L., & Madanes, R. (2000). Exploring diverse accounts of teacher knowledge. *Journal of Mathematical Behavior 18*(3), 357–375.

Shuhua, A., Kulm, G., & Wu, Z. (2004). The pedagogical content knowledge of middle school mathematics teachers in China and the U.S. *Journal of Mathematics Teacher Education, 7*, 145–172.

Shulman, L., S. (1986). Those who understand: Knowledge growth in teaching. *Educational Researcher, 15*(2), 4–14.

Shulman, L., S. (1987). Knowledge and teaching: Foundations of the new reform. . *Harvard Educational Review, 57*, 1–22.

Stefnumótun Kennaraháskóla Íslands 2005–2010. (2005). Policy of Iceland University of Education 2005–2010. Reykjavík: Kennaraháskóli Íslands.

Wood, T. (2002). Demand for complexity and sophistication: Generating and sharing knowledge about teaching. *Journal of Mathematics Teacher Education, 5*, 201–203.

Wood, T., & Berry, B. (2003). What does "design research" offer mathematics teacher education? *Journal of Mathematics Teacher Education, 6*, 195–199.

34

DEVELOPING TEAM-TEACHING THROUGH SELF-STUDY

Hafdís Guðjónsdóttir, Jónína Vala Kristinsdóttir, and Edda Óskarsdóttir

UNIVERSITY OF ICELAND—SCHOOL OF EDUCATION

Abstract: This chapter focuses on team-teaching through self-study of teacher education. The purpose of the collaborative self-study inquiry was to articulate how teachers' learning about how to teach all children mathematics influences the manner in which we as teacher educators teach. This is a report from three teacher educators (special and mathematics educators) who have created and developed a graduate course that intends to prepare teachers to teach mathematics to diverse groups of students. Mathematics education and special education have developed from different disciplines, and our conclusion is that the different knowledge we bring with us makes the team stronger than the individuals who make up the team. In the chapter we report the complexities of our teaching and the challenges and dilemmas we have met on the way and how our team teaching and collaboration has helped us see from a broad range of view how our students are learning and how we have responded to them.

Keywords: Teacher education, team-teaching, self-study, diversity

INTRODUCTION

The diversity of students in Icelandic classrooms is greater than ever, as students with disabilities have entered their neighborhood schools and immigration has brought in students for whom Icelandic is not their native tongue. Work and workplaces are changing, and the skills people will need in the future will be different from current and past experience, although we do not yet know exactly what they will be. These changing social conditions as well as the change of student groups have put increasing pressure on teachers to modify their practices. Teachers are challenged by having to differentiate teaching in response to a spectrum of different students and learning needs. They are challenged to adopt flexible approaches for learning and focus on problem solving methods. Responding to these changes in isolation will not give as many opportunities as if done in collaboration with colleagues. Team teaching is one way for teachers and teacher educators to respond to this demand. It is an approach in which two or more teachers are jointly responsible for course content, lesson activities, and assessment as well as collaborating on the teaching, and it should be a part of school cultures that encourage professionals to work together (Barth, 2006).

In this chapter we will explain the development of our collaboration and how we have used self-study methodology to help us create a graduate course at the University of Iceland—School of Education where we team teach in order to not only teach about team teaching but also to team teach in action. The perspective of our teaching is constructivist. The course focus is mathematics for all and the belief that

The First Sourcebook on Nordic Research in Mathematics Education, pages 479–486

all children can learn mathematics if they are allowed and supported to construct their own understanding. Consistent with our belief, we rely on the teachers to find their way of learning to understand children's mathematical learning although our responsibility is to guide them. We provide them with tools to critically reflect on their practice, explore children's learning and respond to their own analyzes.

Throughout this chapter we refer to our students as teachers, teacher learners, or participants; to their students as students or children; and to us as teacher educators.

DEVELOPING COLLABORATION

Two of us, Hafdís and Jónína, began our professional collaboration in June 1995 when we attended a course within the Cognitively Guided Instruction (CGI) project (Carpenter, Fennema, & Franke, 1995; Fennema et al., 1996). The project is based on an integrated program of research focused on the development of students' mathematics thinking; on instruction that influences that development; on teachers' knowledge and beliefs that influence their instructional practices; and on the way that teachers' knowledge, beliefs, and practices are influenced by their understanding of students' mathematical thinking.

We were both primary school teachers and graduate students at that time. Hafdís was working on her doctoral studies at the University of Oregon and Jónína on her self-study for her master's degree at the Iceland University of Education (IUE). The following year we organized a follow-up course for the teacherswho attended the CGI course with the goal of providing opportunity to discuss our learning in the classroom. Two years later we organized a second follow-up course. We soon found that our beliefs on teaching were similar, and the constructive ideology of the CGI project was in harmony with our experience of teaching. In the fall of 1995 we attended a graduate course on theories of mathematics education. As we studied together we realized better and better that our professional working theory is similar (Dalmau & Guðjónsdóttir, 2002).

Working on our goals as primary school teachers and graduate students at different universities, we found it important to have a forum for discussing our work. For two semesters we met every week for two or three hours to discuss our work and ideas about learning and teaching. For the last decade we have been teacher educators and have been encouraging collaboration between programs in general and special education. Edda Óskarsdóttir and Hafdís met at the University of Oregon when we were both working on our master's degrees in special education in 1991. There we had the opportunity to reflect on our studies and learn from each other. In 2001 Hafdís and Edda got the opportunity to work together on developing a graduate course, together with Edda Óskarsdóttir on learning difficulties in mathematics from the year 2001. We asked Jónína to join us to work on changing the focus of the course from learning difficulties in mathematics to children's abilities to learn mathematics.

METHODS

This research is a self-study of our teaching as we team teach a graduate course: *Mathematics for all.* Self-study positions the teacher as an inquirer and a learner (Samaras & Freese, 2006), and the purpose of the self-study inquiry is to better understand one's own practice, the renewal of programs, and personal/professional development (Kosnik, Beck, Freese, & Samaras, 2005). The focus of this study is on ourselves as professionals and our part in the development of the course. In doing so we believe that we cannot leave the participants out of the research or the story, and therefore we present vignettes of examples from the interaction between us the teacher educators and between us and the teachers. We believe that by systematically and critically examining our actions and the context, we will gather a clearer understanding of the complexities and dilemmas we are coping with in our teaching. The purpose of this chapter is to articulate how the teachers' learning about how to teach all children mathematics influences the manner in which we as teacher educators teach. Tom Russell (1997) claims that the way teacher educators teach sends a message about how student teachers can teach. It is complicated to teach about teaching, but the constant reviewing and reconstruction that keeps alive what we do is what makes the job so interesting (Kroll, 2007).

According to Bodone, Guðjónsdóttir, and Dalmau (2004), self-study does not exist without collaboration, although it takes different paths. The process of action–reflection–learning–action that is at the heart of collaborative self-study research provided three significant opportunities during the inquiry and renewal of the course:

- collection of data about teaching and learning
- deeper understanding of how we can teach teachers to analyze their students' learning

- the review and improvement of our team teaching and research collaboration

From the beginning, we developed the course through a self-study approach, and over five years of teaching we gathered data from multiple sources, including all the material from the courses, student projects, discussions from class, and documentation of our teaching, discussions, and reflections. We use the findings to continuously develop the course. The research is as a spiral where data collection, analyses, interpretation, and teaching are intertwined. We reflect on our teaching and students' work, analyze, and respond to it as we plan our actions for next class or next time we teach the course. We use teachers' cases and commentary to understand what and how they learn about their students' learning. From our own analyzes and interpretation we decide what and how to teach next, or the reading we want to bring into the class.

To make the data that we base our study on accessible and clear in this chapter, we offer vignettes from our classrooms that are drawn from our field notes, teachers' work and our collaborative reflections (Loughran, 2007). The vignettes are used to bring to life certain situations or episodes from our teaching and collaboration and are then as such used both as a method and a form of analysis (Loughran, Berry, & Tudball, 2005).

The double collaboration offers opportunities to see our teaching, students' work, and responses from different perspectives. By team teaching, we experience what happens in the classroom, and because of our different roles and theoretical background it gives us different perspectives. Thus we are able to respond in a way that enhances our teaching and deepens our research.

TEAM TEACHING MATHEMATICS FOR ALL

The course we are focusing on is a graduate course and the participants are teachers (preschool, primary, and secondary) and social educators. The course is mainly organized through distance learning, although participants meet on campus for five full days also. Internet-based software is used for teaching and learning between campus meetings. Analysis of our original courses revealed that participants found it important to understand learning disabilities in mathematics. They also recognized the importance of understanding and challenging their beliefs, knowledge, and practices. We learned that some participants felt uncertain about mathematics and were not confident in teaching the subject (Guðjónsdóttir & Kristinsdóttir, 2007). The emphasis in the first years of the course was from the perspective of special education and thus on learning disabilities, the appearances of the difficulties, diagnosings and teaching algorithm. Throughout the course development we have continued to work on reversing the emphasis from diagnosing children's weaknesses, to exploring how children learn mathematics and to build the teaching on students' strengths, not weakness.

As we continue the study and the development of the course, we have learned more about teacher dilemmas related to their beliefs and attitudes, their professional understanding, and the ways they themselves learned mathematics when they were children (Guðjónsdóttir & Kristinsdóttir, 2007). The teachers' conceptions of mathematics, teaching, and learning, learned from their educational experiences, make it difficult for them to act in new ways to foster a different quality of mathematics learning for students (Crawford & Adler, 1996). Teachers need more knowledge and experience of how to explore and experiment with mathematics and engage in discussions about their own mathematical thinking. They also need knowledge of the development of children's mathematical learning (Bredcamp, 2004).

Through the years we have been jointly responsible for the course content, presentations; and grading; interacted in front of the class; and discussed specific problems from different points of view (Schafer, 2000–2001; Wallace, 2007). We naturally add to each other's teaching if we believe that we can add different experience or perspective; trying to build a learning community where differences are respected can be discussed and used as a support with our understanding and learning. By constant reflection on our work we have been able to revise the course with the intention of building a community where our students interact, share their ideas, and draw conclusions from their explorations in mathematics and in their teaching.

Participants begin the course by emphasizing students' diversity and their many different ways of learning to solve mathematical problems. We also emphasize the teachers' own investigations of mathematics and pose a mathematical problem to them to solve. They work on the problem, discuss in small groups and present their solution to the whole group. During this task it is crucial that we are two or three in the classroom working as teacher educators and researchers. The team teaching gives us the opportunity to change roles back and forth from teaching to researching.

Anton baked doughnuts for his 12 grandchildren and put them into bags. Some of the doughnuts were big and others small. He decided that the big ones would go 7 in a bag and the small ones 8 in a bag. He packed 91 doughnuts into the 12 bags.

How many bags contained big doughnuts and how many small ones?

Jónína was in charge of the lesson and Hafdís joined her. We asked the teachers to think about the problem for a while and then discuss their ideas in small groups. We walked around the room and discussed with the groups. A group of secondary teachers was discussing different ways to solve the problem and found that the most efficient way would be to make two equations with two unknown variables. A group of preschool teachers was listening to their discussions and seemed frustrated hearing the mathematical language they used.

Jónína asked the preschool teachers about their ideas for solving the problem, and they admitted that they didn't have a clue how to solve it. She then asked them if it would help them to draw a picture of the situation, but they still seemed confused. She suggested that they should try to draw a picture of the situation and then went to a room next door, where she kept all kinds of mathematical models, to fetch unifix-cubes. Hafdís was aware of the confusion in the group and continued the discussion while Jónína was away. When Jónína came back with the cubes, Hafdís felt that the teachers were so frustrated that they most of all wanted to leave the room. But Jónína was standing in the door with a tray of 100 unifix-cubes, which she put it on their table, so there was no way of escaping. She asked them if they could use the cubes to solve the problem. They hesitated for a while and then gradually began by first taking nine cubes away and then making 12 groups out of the rest.

Jónína asked the preschool teachers if they were willing to share their experience. They hesitated and then started by apologizing for solving the problems like young children. This gave us the opportunity to discuss different ways to solve problems and the diverse solution strategies the teachers used. Both of us participated in the discussion as well as the other teachers. We were happy to find that the secondary teachers supported the preschool teachers by admitting that the preschool teachers' way of solving the problem helped them to realize that their students might gain from having various tools and mathematical models available in mathematics lessons. We discussed the situation that arose in our classroom with the teachers. Do similar situations arise in their classrooms? Can they learn from this experience in dealing with teaching in diverse classrooms? Do grouping methods affect the learning in the classroom? We did not divide the teachers into groups so those who knew each other were working together. Would we have had a different experience if we had grouped them differently? We also discussed the mathematics that they were exploring and how they dealt differently with the problem depending on their mathematical background. Can children learn different things by dealing with a problem like this? Is this a valid problem for their students? What would they learn from solving it? Our interplay during this session was natural, and we supported each other constantly in the discussion where our strengths in different fields supported us in the discussion.

This vignette highlights our way of teaching. The interplay during this session was natural, and our strengths in different fields helped us create an interactive learning environment. By walking around and participating in the small group discussions, we managed to gain a clear picture of the mathematical situation in the whole group, use it for the discussion, and point out that it is natural that students in every group think about and solve problems in different ways.

During the course, teachers write cases both from their real practice and course-related projects (Kruger & Cherednichenko, 2006). The cases are about students' strategies as they solve mathematical problems. The cases are analyzed, evaluated, and related to pedagogical and professional working theories. According to their findings, the teachers design and adapt teaching and learning for diverse groups of students.

To better understand how our teaching is impacting our students, we ask them to discuss and record their reflections on their learning and perspectives. Their solution strategies differ profoundly, and we can discuss how their understanding of the problem reflects their way of learning mathematics. We assess and discuss the way the teacher learners solve mathematical problems, and from there, we relate their personal experience to their experience with the children. We encourage them to connect to children's learning and how they might understand the problem (Guðjónsdóttir & Kristinsdóttir, 2007).

> What I found interesting was when we solved the problem in different ways. Then I thought: If we can solve problems in different ways then the children must be able to do it and even in more different ways than we do. They find a strategy that they understand and can even explain how they did it. Their understanding develops and if they invent something themselves they will remember it. Then they will use this, even without knowing it, in their lives (learning) and get used to investigate and explore "hidden" sides of something. They will be more open, because their understanding was respected. (Birna)

This teacher thus related her experience to her own teaching and the capacities and interests of children she works with. She observed that children are more likely to understand and use strategies they have invented themselves. Another teacher demonstrated her awareness of how necessary it is to give students the opportunity to do things their way:

> ...I have learned so much, and the greatest experience was to discover that problems can be solved in incredibly many ways; and how children use different ways and all kinds of hands on materials to solve them....I'm a swimming trainer for handicapped people and I have always believed that it doesn't really matter how you swim. It is much more important to be able to swim....The goal is that the students can use their capability to swim to escape from drowning and as a tool for training their body and for a very few as athletes. In the same way we want everyone to be able to solve mathematical problems, in their own way. (Linda)

A team of two teachers worked together on a project during the course; they reported the following:

> The problems we made and posed to the students are in coherence with the problems in the course textbook. After we posed the problems many questions have arisen about how we could have phrased them differently to get various and more solutions from the students. We would have liked to explore multiplication and division closer and also more complex problem solving and even composite problems. But it should be taking into account that we are analyzing students in this way for the first time.... (Gunnar and Hanna)

Their reflection on their work indicates that they are aware of that teachers can learn from interacting with students and hence make problems according to their level of understanding; this can be practiced in the context of teaching mathematics.

Hafdís and Edda have a long experience of collaboration; as they are both special educators they prefer to teach together about difficulties in mathematics, interfere as they explain and give examples, and participate in the dialogue. Edda is presenting a lesson on assessing mathematics difficulties, and Hafdís is present in the room.

Edda is discussing how students' mathematical abilities can be mapped using different material and different approaches. How this can be used to inform teaching—to make a lesson plan tailored to students' needs. A teacher asks for an example of how one can teach students with such broad range of abilities in one lesson. Hafdís says that she has an example of how this can be accomplished and tells about a mathematics lesson in 5th grade that she was teaching. The lesson was about measurements and ratio. The students were working in groups with measurements of a building that they were then supposed to make a model of in the ratio of 1:100. One student in that class was working on learning about different forms, how to use the ruler, and counting to twenty so his goal in that lesson was to sort the pieces that his group made and measure them with his ruler and write the outcome down.

This vignette is an example of how our different teaching experiences can give our lessons more depth and practicality. Additionally, the teacher who is in charge of the teaching is often occupied with that and has difficulties calling examples from her memory, but her colleague who is observing has more space.

To highlight our point about team teaching and integrating two different worlds of education, here is another vignette from our course:

One day during our teaching, as Hafdís introduced to the participants a teaching module from Gudrun Malmer (1999), an experienced educator and a researcher from Sweden, an example of team teaching came up unexpectedly. Malmer has divided the teaching into six different steps, as explained in Figure 34.1.

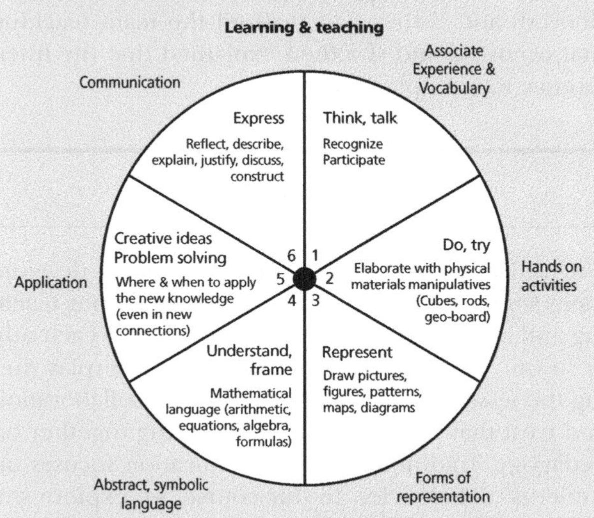

Figure 34.1

Hafdís, who is in charge of the teaching, is standing in front of the blackboard explaining the steps with different examples. The participants add examples from their own experience, and the discussions are lively and full of ideas. We are to step five, about how to transfer how to transfer knowledge learned at school into the daily life. How we use equations, algebra, or different formulas in our everyday activities. The faces become blank; the enthusiasm that we felt before is fading out, and there aren't any suggestions. Then Jónína, who has been a part of the class although not saying anything, steps in and says: "I feel like you do not understand what Hafdís is talking about." She walks towards the blackboard and writes triangles on it, connected together like construction beams. She asks if they have seen such constructions and gets lots of responses and discussions about why triangles are suitable where strength is needed. She then asks them how many rods we need to make a construction of 5 triangles and then probes them to calculate how many are needed to make a beam with 47 triangles. This leads the teachers to find a formula for the calculations and in doing so, connect to Malmer's model.

In this vignette Hafdís was not ready with an example for step 5 thinking so that the teacher learners would come up with them as before. However, Jónína, being the author who knows the mathematics textbooks used in schools well, was able to bring in an example from them and create a learning moment for all of us. In the end this session gave opportunities for discussions and new understanding. As this incident ended, Hafdís decided to use it for reflection in action with the teachers and highlight the team teaching. She asked the participants what they had experienced, and if they had realized the team teaching that occurred and she then explained that the interference was not planned.

DISCUSSION

Our self-study and inquiry into the course development supports us in reflecting together on our teaching and also in reflecting and responding to each other in our teaching (Schön,1987). Our interplay during the sessions is a result of a long-time collaboration and trust that we have built by reflecting together on pedagogy. Traditionally special education focuses on students' deficiencies. In our course, we explore students' abilities in mathematics. Traditions and knowledge from two disciplines are brought together to create a vision on teaching all children mathematics.

Jaworski (2006) discusses the way inquiry can become a tool for critical awareness: "In communities of inquiry, we all engage with inquiry as a tool to develop meta-knowing, a form of critical awareness that manifests itself in inquiry as a way of being" (p. 208). Our analysis and interpretations have influenced ongoing development of the course and to emphasize children's mathematical development and understanding plus teachers' capacity to evaluate and promote students' learning through analysis of their engagement in authentic mathematical problems—to move from special education's traditional focus on diagnosing and working on deficiencies to more authentic assessment, where the emphasis is on observing how students solve their problems and having them explain and discuss their thinking and their work.

All our work, our pedagogy, and the trust we have managed to build enables us to collaborate during our teaching. Writing about our research enhances our understanding of how our collaboration and team teaching has grown to develop a community of inquiry where we reflect on our work together. It has also affected the learning community that we have developed along with the teachers. We have experienced that when the teachers explore in mathematics their confidence in solving problems increases. Additionally, their understanding of how children use diverse ways to solve mathematical problems expands. Problem-solving activities may help teachers to experience and discuss difficulties similar to those met by students in class and understand the importance of evaluating the process in mathematical activities (Boero, Dapueto, & Parenti, 1996; Crespo & Sinclair, 2008).

The teachers' responses and their discussions both in sessions and on the web imply that they gradually become a part of our community of inquiry as we urge them to participate seriously in discussions that help them reflect on their way of learning and teaching.

While discussing our learning and reflecting on our experiences during the course, we often speculate why we find it natural to collaborate, what our philosophy of education is, and how it relates. We are all educated as classroom teachers, but at different programs; Hafdís and Edda later specialized in special education and Jónína in general pedagogy and mathematics education. We have all worked for long time as primary school teachers, though in different schools and within different school cultures. At the same time, we are all interested in mathematics and inclusive education and have studied together at workshops and university courses on the matter.

Our team is made up of instructors from different academic fields. It is therefore important to notice that we bring into our teaching different strengths; however, we have an insight into each other's interests and specialties also. We bring a variety of perspectives to the subject under consideration (Schafer, 2000–2001). Mathematics education and special education have developed from different disciplines and the knowledge we bring with us makes the team stronger than the individuals who make up the team. As we prepare our course, we each bring in different knowledge, but at the same time our pedagogy and way of teaching overlaps. According to Pugach (2005), perspectives and attitudes of teachers involved in inclusive education are more positive towards diverse groups of students than those who have not had the opportunity.

CONCLUSIONS

Our collaboration has helped us see from a broad range of views how our students are learning, and in so doing we believe that we have managed to respond to them in a more professional way. We have discussed our responses to our students and helped each other understand their learning.

We have given our students access to research on children's mathematical development and understanding and enhanced their capacity to evaluate and promote students' learning through analysis of their engagement in authentic mathematical problems. We have encouraged them to reflect on, and criticize, their classroom practice in the light of other's research (Cockburn, 2008).

According to Loughran (2007), self-study can and should make a major contribution to improving the quality of teacher education. The close relationship between teaching and research is a motivation for self-study because knowledge development can then be closely tied to one's own teaching, the results of which can then shape the manner in which subsequent practice is conducted. By writing about our research and our collaboration we believe that we can make the new knowledge meaningful to others.

REFERENCES

Barth, R. S. (2006). Improving the relationships within the schoolhouse. *Educational leadership 63*(6), 8–13.

Bodone, F., Guðjónsdóttir, H., & Dalmau, M. C. (2004). Revisioning and recreating practice: Collaboration in self-study. In J. J. Loughran, M. L. E. Hamilton, V. LaBoskey & T. Russell (Eds.), *International handbook of self-study of teaching and teacher education practices* (pp. 743–784). Boston: Kluwer Academic Publishers.

Boero, P., Dapueto, C., & Parenti, L. (1996). Didactics of mathematics and the professional knowledge of teachers. In A. J. Bishop (Ed.), *International handbook of mathematics education* (pp. 1097–1121). Dordrecht: Kluwer Academic Publishers.

Bredcamp, S. (2004). Standards for preschool and kindergarten mathematics education. In D. H. Clements & J. Samara (Eds.), *Engaging young children in mathematics* (pp. 77–82). Mahwah, NJ: Lawrence Erlbaum Associates.

Carpenter, T., Fennema, E., & Franke, M. L. (1995). *Children's thinking about whole numbers.* Madison, WI: Wisconsin Center for Educational Research, University of Wisconsin

Cockburn, A. D. (2008). How can research be used to inform and improve mathematics teaching practice? *Journal of Mathematics Teacher Education, 11*(5), 343–347.

Crawford, K., & Adler, J. (1996). Teachers as researchers in mathematics education. In A. J. Bishop (Ed.), *International handbook of mathematics education* (pp. 1187–1205). Dordrecht: Kluwer Academic Publishers.

Crespo, S., & Sinclair, N. (2008). What makes a problem mathematically interesting? Inviting prospective teachers to pose better problems. *Journal of Mathematics Teacher Education, 11*(5), 395–415.

Dalmau, M. C., & Guðjónsdóttir, H. (2002). Framing professional discourse with teachers: Professional working theory. In J. Loughran & T. Russell (Eds.), *Improving teacher education practices through self-study* (pp. 102–129). London: Routledge/Falmer.

Fennema, E., Carpenter, T., Franke, M. L., Levi, L., Jacobs, V. R., & Empson, S. B. (1996). A longitudinal study of learning to use children's thinking in mathematics instruction. *Journal for research in mathematics education, 27*, 403–434.

Guðjónsdóttir, H., & Kristinsdóttir, J. V. (2007). Preparing teachers to teach all children mathematics. In P. A. Bartolo, A. M. Lous, & T. Hofsäss (Eds.), *Responding to student diversity: Teacher education and classroom practice. Proceedings of the International Conference on Teacher Education for Responding to Student Diversity* (pp. 44–61). Malta: Faculty of Education, University of Malta.

Jacobs, V. R., Franke, M. L., Carpenter, T., Levi, L., & Battey, D. (2007). Professional development focused on children's algebraic reasoning in elementary school. *Journal for Research in Mathematics Education, 38*(3), 258–288.

Jaworski, B. (2006). Theory and practice in mathematics teaching development: Critical inquiry as a mode of learning in teaching. *Journal of Mathematics Teacher Education, 9*, 187–211.

Kosnik, C., Beck, C., Freese, A. & Samaras, A. (2005) *Making a difference in teacher education through self-study: Studies of personal, professional and program renewal.* Dordrecht: Springer.

Kroll, L. (2007) Central nature of theory in my practice. In T. Russell & J. Loughran (Eds.), *Enacting a pedagogy of teacher education: Values, relationships and practices* (pp. 95–105). London: Routledge Taylor & Francis Group.

Kruger, T., & Cherednichenko, B. (2006). Social justice and teacher education: Re-defining the curriculum. *The International Journal of Learning* (12), 1–8.

Loughran, J. (2007). Researching teacher education practices: Responding to the challenges, demands, and expectations of self-study. *Journal of Teacher Education, 58*(1), 12–20.

Loughran, J., Berry, A., & Tudball, L. (2005) Learning about teaching. In C. Kosnik, C. Beck, A. Freese and A. Samaras (Eds.) *Making a difference in teacher education through self-study: Studies of personal, professional and program renewal* (pp. 203–226). Dordrecht: Springer.

Malmer, G. (1999). *Bra matematik för alla.* Lund, Sweden: Studentlitteratur.

Pugach, M. C. (2005) Research on preparing general education teachers to work with students with disabilities. In M. Cochran-Smith & K. M. Zeichner (Eds.) *Studying teacher education: The report of AERA panel on research and teacher education* (pp. 549–590). Mahwah, NJ: Lawrence Erlbaum Associates.

Russell, T. (1997) Teaching teachers: How I teach IS the message. In J. Loughran & T. Russell (Eds.), *Teaching about teaching: Purpose, passion and pedagogy in teacher education.* (pp. 32–47) London: Falmer Press.

Samaras, A. & Freese, A. (2006). *Self-study of teaching practices.* New York: Peter Lang Primer.

Schafer, I. (2000–2001). *Team teaching: Education for the future.* Retrieved March, 14, 2008 from http://www.usao.edu/~facshaferi/teamteaching.htm

Schön, D. A. (1987). *Educating the reflective practitioner.* Oxford: Jossey-Bass.

Wallace, J. J. (2007). Effects of interdisciplinary teaching team configuration upon the social bonding of middle school students [Electronic Version]. *Research in Middle Level Education,* 5. © 2007 National Middle School Association.

CHAPTER 35

TEACHERS' DEVELOPMENT IN MATHEMATICS TEACHING THROUGH REFLECTIVE DISCUSSIONS

Jónína Vala Kristinsdóttir

UNIVERSITY OF ICELAND—SCHOOL OF EDUCATION

Abstract: The focus of this case study was on developmental work initiated by classroom teachers in a primary school. The developmental process was not intentional to begin with but progressed as the teachers' interest in their students' development of mathematical understanding grew. Their cooperation started when two of them came back from a course on children's mathematical thinking and were inspired to use their knowledge from the course in their teaching. Through constant discussions and reflections on their students' solution strategies and reflection on their teaching, the teachers built a community of inquiry where they developed their understanding of their students' learning and managed to create classroom cultures that promoted learning.

Keywords: Mathematics teaching, teacher development, children's learning, reflective discussions

INTRODUCTION

Four classroom teachers[1] worked closely on developing their mathematics teaching for thirteen years. Their collaboration in developing their teaching was not intentional to begin with but emerged from their enthusiasm in learning to understand the development of their students' mathematical learning. Their cooperation started in 1995 after Björk and Íris attended an in-service professional development course on learning mathematics. They were eager to use their knowledge from the course in their teaching. At weekly meetings with their colleagues (all teaching first grade at the time), they discussed their work and inspired their co-teachers to collaborate.

The course was on Cognitively Guided Instruction (CGI), which is an integrated program focused on the development of students' mathematics thinking; instruction that influences that development; teachers' knowledge and beliefs that influence their instructional practices; and the way that teachers' knowledge, beliefs, and practices are influenced by their understanding of students' mathematical thinking (Carpenter & Moser, 1984; Carpenter, Fennema, & Franke, 1995; Carpenter, Fennema, Franke, Levi, & Empson, 1999). CGI grew out of inquiry on explicit knowledge about the development of children's mathematical thinking (Carpenter & Moser, 1984; Carpenter, 1985), which was used as a context to study teachers' knowledge of students' mathematical thinking (Carpenter, Fennema, Peterson, & Carey, 1988). The rationale was that teachers might use knowledge of students' mathematical thinking in making instructional decisions. In a series of studies (Carpenter, Fennema, Peterson, Chiang & Franke, 1989; Fennema, Franke, Carpenter, & Carey, 1993; Fennema et al., 1996), it was found that learning to

The First Sourcebook on Nordic Research in Mathematics Education, pages 487–494

understand the development of children's mathematical thinking could lead to fundamental changes in teachers' beliefs and practices and that these changes were reflected in students' learning.

There was coherence in the change of the teachers' beliefs in the CGI studies and their way of working with their students. When children began to show increased learning, the teachers continued to implement new methodologies that resulted in improved learning, and so the cycle of developmental change continued. When they found that their students were able to find their own solution strategies when given meaningful problems to work on, they relied on the children's thinking. Developing an understanding of children's thinking provides a basis for change, but change occurs as teachers attempt to apply their knowledge to understand their own students. The CGI study provides strong evidence that knowledge of children's thinking is a powerful tool that enables teachers to transform this knowledge and use it to change instruction. It also appears that this knowledge is not static and acquired outside the classrooms in workshops, but dynamic and ever-growing, and can probably only be acquired in the context of teaching mathematics (Fennema et al., 1993; Fennema et al., 1996).

The most important role for the teacher is creating a classroom in which all students can reflect on mathematics and communicate their thoughts and actions. Building a community of mathematical practice requires teachers to take the lead in establishing appropriate expectations and norms (Hiebert et al., 1997).

In classroom cultures that promote mathematical learning, all students have a voice and are supported to develop their understanding of mathematics through exploring, investigating, discussing, reflecting, and drawing conclusions. Reflective discussions should support teachers in their teaching:

> The teacher's assessment of a student's conceptual structures does not have to be a blind conjecture. If one starts from the assumption that students generally try to make sense of their experience, it is usually possible to get some idea of how they think. The more experience with learners a teacher has gathered, the better the chance to make an educational guess about what a particular student's thinking might be and to hypothesize what Vygotsky aptly called 'the zone of proximal development'. It is only after working with a particular student for a considerable time, that the teacher may gain confidence in his or her conceptual portrait of that individual. (Glasersfeld, 1995, p. 187)

The enthusiasm of Björk and Íris with regard to using their knowledge from the CGI course and their reflection on the children's learning had an impact on the collaboration with their colleagues as other teachers at the school became interested in their work. In 1997, the four teachers applied for funding to work with all the teachers at their school on reviewing their mathematics teaching and asked me, the author, to be their adviser. In 2003 they again applied for funding to be able to engage more in their developmental work under my supervision.

The four teachers worked closely together for 10 years as classroom teachers in grades 1–7. Sara and Tinna then each had a leave of absence from the school for one year. When they came back they joined a new team and got a grade 5 class. They did not participate in the CGI course in 1995 and have not participated in many in-service courses on mathematics education. Their main interest is language and social sciences. Björk and Íris kept on working together and became a part of a new team of grade 5 teachers. They both participated in the CGI course that influenced the development from the beginning. They had taken part in courses on mathematics teaching and other developmental work in mathematics and science. During the fifth year of their co-learning, a researcher carried out a study on children's understanding of fractions in Björk's and Írisi's classrooms. The four teachers collaborated with the researcher while planning the instruction for the fraction classes. The researcher invited Sara and Tinna to be a part of the team, which they accepted, but they were not involved with the research to the same extent as Björk and Íris.

Although this developmental work was not planned to begin with and there was no intention to research the teacher work neither from my side nor the teachers', I followed it closely from the beginning as their adviser. This included watching lessons and interviewing teachers several times from the year 1997. In 2007 the four teachers were interviewed again and observed in their classrooms with the intention of getting a deeper understanding of their work and how it had developed through the years.

The research questions were:

- How is teacher learning about student understanding of mathematics reflected in teaching?
- How do the teachers perceive that their professional discourse has affected their teaching?
- What are the main obstacles perceived by the teachers in their developmental work?

These questions arose from my previous knowledge about the work of the teachers and ongoing work with them, as well as earlier interviews and classroom observations.

METHODS

Before interviewing the teachers, previous transcriptions of interviews from 1997, 2005, and spring 2007 were studied as well as an interview with the principal of the school in 2005. I also read notes from former classroom observations and went through a collection of the children's work. The information gathered by investigating the earlier data provided a scaffold for me in the interviews that were semi-structured, in-depth interviews (Taylor & Bogdan, 1998). The interviews were flexible, and I responded to the teacher answers by probing further or clarifying significant issues that emerged in the course of the interviews (Bryman, 2004; Kvale, 1996).

The purpose of the interviews was to obtain information about the teachers' beliefs and practices with respect to interpreting the meaning of their descriptions of their mathematics teaching and their goals with their work (Pring, 2000). They described their current situation, collaboration with their colleagues, their students' ways of learning mathematics, and their way of teaching. They were urged to describe their own view of the process and understanding of how their collaboration has influenced their work. Their stories are narratives where knowledge is embodied in storytelling (Kvale, 1996)

The interviews were audio taped and transcribed. While transcribing the interviews, initial themes emerged and were generated through coding (Creswell, 1998; Maxwell, 2005). On re-reading the transcribed interviews, new themes emerged, and finally they were condensed into four themes.

During the observations in each classroom that followed the interviews, I was a participant observer (Bogdan & Biklen, 1992) and tried to be aware of the communication and the learning community established in the classroom. Themes drawn from the interviews helped me make structural questions to focus the observation (Spradley, 1980). When the teachers led class discussions I was a passive observer, and while the children were working at their tasks I walked around, observed their way of working and communicating, and discussed with them in order to get an understanding of their way of approaching the tasks. Written artifacts consisted of minimal notes during the lessons, but as soon as possible after each lesson I wrote down detailed notes. Copies of the children's written work and tasks given by the teachers were also procured for analysis. The field notes were analyzed by generating themes in similar ways as the interviews.

RESULTS

The main themes that emerged from the interviews and observations are:

- analysis of children's learning
- teaching method
- parents' views
- professional discourse and collaboration

Analysis of Children's Learning

Relying on children's ability to construct their own knowledge was the core of their discussions. They argued about teaching traditional algorithms or relying on that the children could construct their own algorithms.

> There were times when we were concerned if we were on the right track. It was the subtraction and the division. We then found how important the collegial support was. (Íris)

Björk reported again and again that she would not give up, while the others were more skeptical, especially when the children were struggling with subtraction algorithms and multiplication and division.

> The disagreement was about the children's constructions of mathematical knowledge. I believe that the children can find their own way of solving problems. The others were not always as patient and two of them wanted to teach the children the traditional algorithms. When the children were struggling with the subtraction we had a difficult time. We discussed the children's strategies and managed to support each other in not giving up. (Björk)

In visiting Björk's class an example of how she analyzed student learning was observed:

> When I visited Björk her students were starting a project where they were asked to respond to an order from a pet store that wanted aquariums in three different sizes: 20, 30, and 60 liters, and three different shapes respectively. Björk started the lesson by referring to their earlier experiments with volume and asking them to review their work from last week. The children worked in groups of three, and Björk encouraged them to experiment with cardboard boxes they had collected. Some of the children needed to explore with the cardboard, and one group went to measure the

aquarium in the corridor. When they came back they discussed their findings with Björk, who encouraged them to draw conclusions from the measurements of the aquarium. Some of the children started directly with the calculations and used their knowledge gained from former experiment about the capacity of 1 liter.

During the 60-minute-long lesson she stopped the children's work twice and asked them to report on their work to the whole class. She said that she had noticed that some of them were facing problems and wanted them to discuss them. There was a lively discussion for a few minutes where they discussed things they had discovered. The students wrote their calculations on a sheet of paper and drew pictures of the different aquariums. At the end of the lesson Björk said that they would be working on this problem the whole week and would report on their findings when they had finished.

I only saw the beginning of the work with the aquariums in Björk's class. She indicated that even though the children had experimented a lot with volume and capacity, many of them still had only a vague understanding of the dimensions of solids. She was aware of each student's level of understanding and knew that some of them had not reached the level of abstract thinking needed to link between volume (cm^3) and capacity (liters). She believed that children need to construct their own knowledge and that hands-on activities will help them make the bridge between concrete and abstract thinking. In the interview with her, she stressed this often by stating that children need to understand what they are doing and that rote learning and traditional algorithms will not help them. She also believes that all children gain from collaborating in the classroom and discussing their way of solving problems.

Teaching Methods

In the ten years all four teachers worked together and had developed a close relationship with the special needs teacher who often worked with them in their classrooms, planned together with them and worked with all children in their classes. Sometimes they divided the classes into five groups and the special needs teacher worked with one group. They experimented with different kinds of grouping (gender, ability, random etc.) and drew conclusions about their work from this experiment.

> We sometimes split the classes into groups and mixed children from different classes. We then learned to know all the children in the same age group, and they learned to work with each other. (Sara)

When they were teaching different age groups they missed the collaboration and their reflective discussions:

> What I miss most now is the learning community that we had built together, our discussions. In the new team we have not managed to do this even though we discuss our work. It is more about practical things than the children's learning. (Sara)
>
> We do not develop as teachers in the same way as before. We are always supporting the new teachers in the team; they constantly ask us what to do and why we do teach the way we do. We are always repeating the same again and again. (Íris)

In a visit to Íris I saw an example of how a teacher could respond to an individual student.

Her students had finished the project on the aquariums and were reporting their findings to the class. They came to the blackboard and showed their drawings and wrote their calculations on the blackboard. I noticed that one boy was always asking his classmates for clarifications and probing for the scales of their drawings. When his group was reporting on its solutions, he reported that they had started with the 20 liter aquarium and found all the dividends of 20,000 because that helped them find all possible different sizes of the aquarium. Many children asked him for clarification. Íris asked him to explain his thinking, and he answered that he had factorized 20,000 and found the result was $2^5 \times 5^4$. The children did not understand his explanations. Íris then offered to write down his thinking on the blackboard and asked him to describe carefully every step in his thinking. The other children now could follow his thinking and expressed that none of them had discovered this clever way to solve the problem. Íris told them that later this year they will explore factorizing further, and many students said that they looked forward to that.

When Íris was queried about the boy she said that he had autistic symptoms. His mathematical thinking was far ahead of the other children's, but socially he was not mature. She believed that the other students learned much mathematically by communicating with him, and he was learning to socialize by participating in group work and whole class discussions.

When I visited Sara, the students were working in groups on triangle problems in the textbook.

They drew their constructions on a sheet of A3 paper. Two boys who have severe learning difficulties were with the special needs teacher in another classroom. Two children who usually have a different curriculum were working with the class on the triangle problem. The boy continued to work with his group, but the girl soon gave up and went back to her textbook. I asked her why she didn't finish the work with the others, and she said that she liked better to work on her own. She said she didn't understand the triangle problems. The problems she was working on in her textbook were multiplication problems, and she also found them difficult.

While the children were working, Sara discussed with individual children and groups and sometimes stopped the children's work to ask them to report to the whole class on problems they were facing. The children discussed lively their findings and problems that had aroused. By the end of the lesson she said that they would finish the work the following day and then they would report on their work.

Sara said that the students who were with the special needs teacher would not have been able to participate in this lesson. They had difficulties with concentrating on their work and working in groups. She said that the children who have a different curriculum sometimes joined the others when they worked on problems, but usually they worked alone with their textbook. They had problems with following the same curriculum as the rest of the class.

Parent's Views

Some of the parents were skeptical about this way of teaching. Tinna has always found it difficult to meet the parents' criticism and finds that she lacks confidence in reasoning on her way of teaching mathematics.

> And then parents complain that their children are not challenged enough. Can't they get more tasks, more challenging tasks? (Tinna)

Björk and Íris led the discussions at parental meetings. They described carefully how the children solved problems and showed the children's work to their parents.

> And then at one parental meeting one of the fathers who had been so skeptical with our work raised his hand. And I thought, what is going to happen now! He is an engineer and he had argued that he knew how to teach mathematics. But now he had changed his mind. He told us how he had felt the year before and how skeptical he was. Then he had decided to follow his son's work and then he understood what I had been talking about. He saw how his son developed in his mathematical thinking and that he knew exactly what he was doing and could describe it to others.... Then another father said that we were far ahead of other teachers and that is why people criticize us. He said that it is always difficult to be a pioneer. (Björk)

Parents' views could affect the way teachers worked. When I visited Tinna, her students were working on fractions:

They were all working individually within the same chapter in the textbook, though on different pages. Some students discussed their work in pairs, but most of them worked alone. The lesson went on smoothly and most of the children worked effectively, sometimes consulting with a classmate or the teacher. One boy complained that the boy sitting next to him was copying his work. Tinna discussed the problem with the boys and helped them find a way to work together.

Tinna stopped the children's work twice to ask students to report on problems they were facing. The children reported on different things, and Tinna pointed out what they could learn from each other's explanations although they were not all working on the same problems. There was no whole class discussion by the end of the lesson.

In the ensuing discussion with Tinna after the lesson, she reflected that during the lesson the children were almost all working on similar problems and that she could have planned the lesson differently. If they had all been working on the same problems, they could have a common discussion and learn from each other. That was what she used to do when she had support from the other teachers. She felt so much pressure both from the parents and the children that each of them should work through the textbook on their own page, so they hardly ever had a chance for whole class discussions in mathematics classes. Tinna was not content with this way of working but was responding to the pressure from the parents. She referred to two lessons where she had managed to have a mathematical discussion with the whole class where they discussed things that were relevant to their work and reported that the children enjoyed it. She wished that she had time to do that more often.

Professional Discourse

All four teachers stressed that their *professional discourse* supported them in their teaching.

> Our discussions were constructive. We have different backgrounds and different beliefs but we were always able to work closely together. We all participated in the discussions although we didn't always agree. (Sara)

> We discussed everything, how we worked with the children and their contributions, what they said and what they thought. We were much more aware of their thinking than before. I gained from participating in the discussions about mathematics teaching in this group. We always discussed our opinion. The discussions about the mathematical teaching influenced my language teaching. (Tinna)

> We discussed how things were going and how different children solved the problems in various ways. We were always discussing mathematics at our meetings. (Björk)

Sara and Tinna expressed their need for support from Björk and Íris. They found it difficult to include all children in the mathematical activities in the classroom. Some of their students followed a different curriculum in special needs classes. Both of them said that they wished that they could differentiate their teaching and include the children with learning difficulties as they had done earlier. They found parents' expectations difficult to deal with and felt a growing pressure from the parents.

Íris and Björk were confident in their teaching but missed the creative community with Sara and Tinna. All children were included in the activities in their classrooms, and they still had the enthusiasm to include every student in mathematical explorations. They saw parents as a resource and emphasized how important it was to explain to them their way of teaching and how the children constructed their knowledge.

The teachers' development influenced other teachers in the school, who then also reviewed their mathematics teaching. Björk, Sara, and Íris assisted other teachers in their classes, and Íris and Björk also led courses for teachers at the school about children's mathematical thinking and how to promote mathematical learning. Some of the teachers were skeptical about this way of teaching mathematics, and there were times when the four felt that other teachers envied them for their close collaboration and success. This was clearly evident when results from national tests in grades 4 and 7 arrived and the students in their classes scored far above the average in the whole country.

DISCUSSION

From working with these four teachers for thirteen years, I learned that their collaboration had supported them in developing their mathematical teaching and in their reliance on their students' abilities to construct their own mathematical knowledge in a classroom culture that promotes mathematical learning.

The teachers' collaborative learning helped change their beliefs about mathematics teaching and learning and their classroom practices. When teachers saw that their students were able to find their own solution strategies when given meaningful problems to work on, they were more willing to rely on the children's thinking. This is in coherence with the teachers in the CGI studies (Fennema et al., 1996). Developing an understanding of children's thinking provided a basis for change, and change occurred as the teachers attempted to apply their knowledge to understand their own students.

The four teachers helped each other to gain confidence in constructing a conceptual portrait of all the students in their classes and used their knowledge to guide them in their teaching. Not only were they aware of every student's zone of proximal development, they also used their knowledge to scaffold the children and help them scaffold each other. In Íris's lesson this was clearly reflected when she supported the boy who could not explain his thinking in reporting to the class. The other three teachers all encouraged their students to explore, investigate, draw conclusions, and discuss problems they were facing or things they had discovered both within their groups and with the whole class and in that way support each other in their learning.

One major concern arising from the data indicated that when Sara and Tinna no longer worked with Björk and Íris which contributed to their lack of confidence in including all children in activities with the whole class. In the interviews they reported that their development in mathematics teaching had affected their teaching in other subjects and they still find it easy to include all children in other subjects but in mathematics classes some children are excluded from the classroom culture. The terms integration and inclusion have been used to describe different levels of including children in school. According to Ainscow (1995), integration is about making a limited number of additional arrangements for individual pupils with special educational needs in schools, but the schools do not change much. Inclusion implies more radical changes through which schools restructure themselves to be able to embrace all children.

My interpretation of what I have observed is that the teachers managed to include all students in their mathematics teaching when they all four worked together with one special needs teacher. The school culture can be described by the term *integration*, and the other special needs teachers do not see themselves as a part of the learning community in the overall school environment. Their focus was on learning difficulties and not on building on children's strengths. When

Sara and Tinna didn't have the collegial support anymore, they sent their low-achieving students to special needs classes where they worked with curriculum that was different from their classmates.

Parents' views were important. The teachers were developing a way of teaching mathematics that differed from what most of the parents had experienced in school. Some of the parents were skeptical, and the teachers found it important to explain to them their way of working. In the excerpt from the interview with Björk, she discusses this dilemma and how they managed to convince the parents that their children were learning mathematics in an effective way. Tinna expressed many times in the interview that she felt the parents' expectations were overwhelming and they conflicted with her ideology of teaching and the school culture in general. She felt a pressure from many parents who wanted their children to go through the textbooks in their own pace and should not be held back by having to work with children who were not as quick as their own. This prevented her from working with the whole class on problems and discussing with them important mathematical concepts as she used to do and she believes is important.

CONCLUSIONS

Results from the study indicate that teacher discussions about their own students' way of learning mathematics and reflections on their teaching can influence teaching in diverse classrooms. During weekly meetings, the teachers discussed the children's work and reflected on the way they solved their problems. Such reflection became an important part of community building, providing opportunity for common ways of thinking to emerge and for developmental concepts to grow within a community (Jaworski, 2006).

Through constant discussions of the children's solution strategies and reflection on their teaching, the four teachers developed their understanding of their students' learning. Systematic reflection on mathematical interactions that focus on students' learning and understanding of processes, as well as on one's own interaction behavior, represents an essential professional competence of teachers (Mason, 2002).

It is important to note that teacher beliefs on mathematics teaching and learning and teacher practice can change, as we have seen in this study. The results cohere with the work by Goos (2004), that by participating in a dialogue about their beliefs on teaching and learning mathematics the teachers managed to build a learning community of inquiry where they have developed an understanding of their students' learning and managed to create classroom cultures that promote understanding and help all children make sense of mathematics.

NOTE

1. The teachers in this chapter have been given the names Björk, Íris, Sara, and Tinna.

REFERENCES

Ainscow, M. (1995). Education for all: Making it happen. *Support for Learning, 10*(4), 147–154.

Bogdan, R. C. & Biklen, S. C. (1992). *Qualitative research for education* (2nd ed.). Boston: Allyn and Bacon.

Bryman, A. (2004). *Social research methods* (2nd ed.). Oxford: Oxford University Press.

Carpenter, T. (1985). Learning to add and subtract: An exercise in problem solving. In E. A. Silver (Ed.), *Teaching and learning mathematical problem solving: Multiple research perspectives* (pp. 17–40). Hillsdale, NJ: Lawrence Erlbaum Associates.

Carpenter, T., Fenema, E., Peterson, P. L., & Carey, C. A. (1988). Teachers´pedagogical content knowledge of students´problem solving in elementary arithmetic. *Journal for research in mathematics Education 19*, 385–401.

Carpenter, T., Fennema, E., & Franke, M. L. (1995). *Children's thinking about whole numbers.* Madison: Wisconsin Center for Educational Research, University of Wisconsin.

Carpenter, T., Fennema, E., Franke, M. L., Levi, L., & Empson, S. B. (1999). *Children´s mathematics: Cognitively guided instruction.* Portsmouth, NH: Heineman.

Carpenter, T., Fennema, E., Peterson, P. L., Chiang, C. P., & Franke, M. L. (1989). Using knowledge of children´s mathematics thinking in classroom teaching: An experimental study. *American Educational Research Journal, 26*, 499–531.

Carpenter, T., & Moser, J. M. (1984). The aquisition of addition and subtraction concepts in grades one through three. *Journal for Research in Mathematics Education, 15*(3), 179–202.

Creswell, J. W. (1998). Data analysis and representation. In J. W. Creswell (Ed.), *Qualitative inquiry and research design: Choosing among five traditions* (pp. 139–165). Thousand Oaks, CA: Sage.

Fennema, E., Carpenter, T. P. Franke, M. L., & Carey, D. A. (1993). Using children's mathematical knowledge in instruction. *American Educational Research Journal, 30*, 555–583.

Fennema, E., Carpenter, T., Franke, M. L., Levi, L., Jacobs, V. R. & Empson, S. B. (1996). A longitudinal study of learning to use children's thinking in mathematics instruction. *Journal for Research in Mathematics Education, 27*, 403–434.

Glasersfeld, E. v. (1995). *Radical constructivism. A way of knowing and learning.* London: RoutledgeFalmer.

Goos, M. (2004). Learning mathematics in a classroom community of inquiry. *Journal for Research in Mathematics Education,* 35(4), 258–291.

Hiebert, J., Carpenter, T. P., Fennema, E., Fuson, K. C., Wearne, D., Murray, H., Oliver, A. & Human, P. (1997). *Making*

Sense. Teaching and learning mathematics with understanding. Portsmouth, NH: Heinemann.

Jaworski, B. (2006). Theory and practice in mathematics teaching development: Critical inquiry as a mode of learning in teaching. *Journal of Mathematics Teacher Education, 9,* 187–211.

Kvale, S. (1996). *Inter views. An introduction to qualitative research interviewing.* Thousand Oaks, CA: Sage Publications.

Mason, J. (2002). *Researching your own practice. The dicipline of noticing.* London: Routledge Falmer.

Maxwell, J. A. (2005). *Qualitative research design. An interactive approach.* Thousand Oaks, CA: Sage Publications.

Pring, R. (2000). *Philosophy of educational research.* London: Sage.

Spradley, J. P. (1980). *Participant observation.* Orlando: Harcourt Brace Jovanovich.

Taylor, S. J. & Bogdan, R. (1998). *Introduction to qualitative research methods. A guidebook and resource* (3rd ed.). New York: Wiley.

36

RELATING LIFE STORY AND PEDAGOGICAL VISION

A Teacher's Voice[1]

Katrín Friðriksdóttir
SNÆLANDSSKÓLI (COMPULSORY SCHOOL)

Sigrún Aðalbjarnardóttir
UNIVERSITY OF ICELAND

Abstract: In this chapter we focus on the relationship between a teacher's life story and his or her pedagogical vision. We follow one teacher, Marta, and explore her ideas about teaching, her aims and teaching activities, and how her ideas relate to her work with her students. Further, we consider how her aims and teaching activities relate to her own life story. By emphasizing teachers' life history in relation to their aims and teaching activities, we can gain a deeper understanding of their pedagogical visions, which can be used as a basis for supporting their professional growth and school development.

Keywords: Life story, Pedagogical vision, Professional growth

> At least in working with the younger elementary-school children, whom I have taught for about 30 years, I feel that the social factors—supporting students' positive self-identity, promoting their social competence and skills, and fostering their well-being—are the basis for all other learning, all subject matters such as Icelandic, mathematics, social studies...
>
> —Marta, teacher

PROFESSIONAL GROWTH AND REFLECTION

In recent decades researchers have become increasingly interested in teacher professional development (e.g., Fullan, 2001; Hargreaves, 1998). They have focused on the importance of teacher professional growth as one of the prerequisites for school development. In this discussion about changes in the work of schools and its development, a strong emphasis has been placed on teachers reflecting on their own teaching (Grimmet & Crehan, 1992; Schön, 1987). Based on this reflection, teachers are expected to become more aware of their own pedagogical vision and pursue their professional growth (e.g., Bartlett, 1990; Fullan & Hargreaves, 1992; Zeichner, 1994). It is assumed that understanding how teachers think about their profession, and how they perform and grow in it, can provide new perspectives on school development.

The First Sourcebook on Nordic Research in Mathematics Education, pages 495–503

The underlying theoretical framework for teacher professional development lies in the philosophical tradition of phenomenology with an emphasis on hermeneutics (Heidegger, 1962): that is, the study of ways to interpret human experience and place it into context. The assumption is that each person (e.g., a student, a teacher) is active in constructing her own meaning of the knowledge, experience, and reality she is in. Within social constructivism (Vygotsky, 1978), a special emphasis is placed on the idea that the meaning a person makes of her life experiences relies on her interactions with her social, historical, and cultural environment. In other words, the meaning she makes is socially and culturally bound.

To better understand teachers' reflections on their teaching and the sense of meaning they make regarding their work, researchers have increasingly emphasized the use of interpretive methods (e.g., Richardson, 1994; Schön, 1987; Shulman, 1987). In this spirit, reflective teaching has become a form of inquiry used in in-service and pre-service approaches to teacher development (e.g., Brookfield, 1995; Korthagen, 1993). Accordingly, in studying teacher professional development and simultaneously school development, several theorists have focused their attention on teachers' reflections on their values, aims, and teaching activities (e.g., Adalbjarnardóttir, 1999; Ingvarsdóttir, 2005; Johnston, 1997). For example, teachers can broaden and deepen their own views on their aims, values, and teaching activities by reflecting constructively on their vision and work (Adalbjarnardóttir, 1999, 2007). Thus, it is expected that teachers who have opportunities to reflect on their work will be more likely to acquire a clearer picture of their pedagogical vision and that they will progress in their work with their students, which again will be mirrored in their students' progress and achievement.

Studies indicate that one way to improve the way teachers provide instructions during math lessons is to help teachers understand the mathematical thought processes of their students (Fennema et al., 1996; Strutchens, 2007). This understanding is important if we believe that each student creates his or her own knowledge. Reflecting on mathematics teaching enables teachers to think about the type of questions they ask students in order to help them to better understand mathematical tasks. Teachers who reflect on their own teaching methods tend to look back at their own thoughts and actions while they were students. They are therefore likely to focus on and understand their own students' thinking processes while solving math problems (Kristinsdóttir, 2006; Strutchens, 2007).

Case studies of how teachers reflect on their work may help other teachers and school professionals to mirror themselves in others' pedagogical visions in their process of professional growth. In this chapter, we provide a case study that can be used in such a way.

LIFE STORY

The increased focus on teachers' professionalism, such as their theories of teaching, aims, and methods (Adalbjarnardóttir, 1999; Fullan & Hargreaves, 1992; Gall, Borg, & Gall, 1996; Ingvarsdóttir, 2005), has evoked interest among scholars in teachers' life stories.

Several researchers have pointed out that experiences in childhood such as parental nurture (Butt, McCue, Raymond, & Yamagishi, 1992) and school attendance (Aitken & Mildon, 1992; Fischler, 1999), as well as formal education and various experiences in everyday life (Connelly & Clandinin, 1990; Goodson & Walker, 1991; Kelchtermans, 1994), can have a great impact on teachers' ideas about teaching and education. Many of the ideas that adults have about teachers and teaching originate from their own childhood (Fullan & Hargreaves, 1992; Goodson & Walker, 1991; Knowles, 1992; Mitchell & Webber, 1996). The adults have all experienced being children at school, and as students they have gathered ideas about teaching and education. The home environment, such as parents' attitudes about school and education, has also colored their ideas (Mitchell & Webber, 1996). Interestingly, there is often a gap between what student teachers say they would like to do in the classroom and what they actually do there when they are in charge (Aitken & Mildon, 1992; Fischler, 1994; Huberman, 1995). The main reason for this incongruity is probably that student teachers bring with them ideas about school that they formed earlier in their lives, and then find it hard to let go of those ideas once they start teaching.

In this light, in recent years life stories have become more common in exploring teachers' ideas about teaching. When using a life-story approach, the person (here the teacher) constructs his or her own life/career experiences into a story that is meaningful to him or her. The focus is not on the facts in their lives, but rather on what these facts mean to the teachers (Sikes, 1985). The purpose is to gather information about their personal and professional experiences and to encourage them to reflect on what this has meant to them as a teacher. The researcher tries to understand what the teacher thinks has influenced his or her thoughts, ideas, attitudes, and actions (Butt et al., 1992; Measor & Sikes, 1992). The focus has been on how a teacher views his or her own childhood, school attendance, teacher education, teaching experience, and upbringing experience at home, and how those elements have affected his or her work with the students (e.g., Butt et al., 1992).

Butt and his colleagues (1992) are among those who point out that telling life stories is a useful way to understand how teachers think about their own teaching and career and what makes them think that way. They claim that in order to understand what goes on in the classroom, we need to look at how teachers' thoughts, ideas, acts, and professional knowledge change in light of their experience both in their professional and private lives. This, they claim, provides the basis for the teachers' own speculations about how they can change their teaching and develop it.

By emphasizing teachers' life history in relation to their aims and teaching activities, we can gain a deeper understanding of their pedagogical visions, which can be used as a base for supporting their professional growth and school development (Adalbjarnardóttir, 2007; Aitken & Mildon, 1992; Connelly & Clandinin, 1990; Fischler, 1999).

Many teachers have ideas about what they regard as "good" teaching, which is part of their self-understanding. Reflecting on their teaching may change their understanding of good teaching. And as they better understand the attitudes and values on which they base their work, they are more capable of developing their work (Butt et al., 1992; Huberman, 1995). Because such reflection and professional growth is an ongoing process, it is important to look at teaching as a life-long development. Fennema and her colleagues (1996) pointed out that teachers who took part in math programs focusing on children's learning process often changed their teaching methods and ideas about what they regard as good teaching. For them it became vital to be collaborators in knowledge development. Teachers do not always have to be the possessors of all knowledge but to respect and value students' thinking (Mau, 2007).

Below, Marta's reflection on her life story in relation to her pedagogical vision and work with students is presented.

MARTA'S LIFE STORY AND PEDAGOGICAL VISION

Marta was in her early fifties when we met her and she had been an elementary school teacher for nearly thirty years. During the time we visited her, she was teaching children aged six and nine, the age group she had frequently taught. She teaches them most of their subjects, including mathematics. Marta is cheerful with a good sense of humor; she is open, sincere, and expressive about her teaching.

Interviews and Classroom Observations

Semi-structured and follow-up interviews were used to discern Marta's attitudes towards teaching, her aims, and how she worked towards attaining these aims. We tried to understand why her aims and teaching activities were important to her. During the follow-up interviews we focused on her as person and how she thought her life story had influenced her, both as a person and a teacher. All four interviews took place at the school where she worked.

We also used classroom observations to explore how Marta worked with her students. In addition, we looked for connections between her aims, the teaching methods she said she adhered to, and her daily work with her students. In total, we conducted twenty observations in her classroom.

Analysis

We analyzed the data in multiple ways using qualitative methods (Bogdan & Biklen, 1992). We started to analyze our interviews with Marta to explore how she felt events in her life story had affected her perspectives on teaching and work with her students. The analysis focused on presenting her life story in an honest way (Kvale, 1996). The major issues in her life story were her upbringing, school attendance, teacher education, teaching experience, the upbringing of her own children, and personality. Secondly, we thematically analyzed what she said about her aims and teaching activities (Kvale, 1996). Third, we analyzed the classroom observations both to see how she interacted with her students and how her aims and proposed teaching methods came into practice.

Why Did Marta Become a Teacher?

Marta was very young when she decided to become a teacher. She has never regretted that decision:

> I was seven when I decided to become a teacher. My first teacher had this influence on me this way. I will never forget her; she was gentle and kind in her interactions with the children. My childhood play changed after that in the way that my dolls became the students and I was teaching them from my desk. I spent hours teaching my dolls this way.

Here Marta refers to her very first teacher in school. She still remembers how this teacher interacted with the children and how safe she herself felt in the classroom. Later in our interviews Marta talked about the classroom as a workplace where each child should feel safe.

A year later Marta got a new teacher, who had a different teaching style that Marta disliked. This teacher discriminated against the children, and Marta felt bad about that, even though she herself was not treated unjustly. Thinking back, she now believes most of the children were uneasy about this, not only those who where unjustly treated. Now Marta herself is very determined to be gentle and kind in interactions with her students and objects to methods that let the teachers favor some children at the cost of others.

This second teacher taught Marta for the next six years, and all that time she was unsatisfied with the teaching methods the teacher used. Still, she remained determined to become a teacher: "After I had such a good role model in my first teacher, I did not see any other profession. Other teachers did not change my mind." Although Marta did not like the methods this teacher used she says that overall her school attendance was pleasant.

Now, nearly forty-five years, later Marta has been working as a teacher for about thirty years. When asked why she works as a teacher she says:

> Each child is very special and I find it very important to support each of them well. It gives me great joy and pleasure to work with children, and I hope to manage to share with them some knowledge and also—which to me is just as valuable—to teach them something about social relations.... Getting to share their victories is a privilege.... For me, teaching has always seemed the right thing to do.

Thus, it looks like Marta made the right decision at the age of seven.

How Did Marta Become the Teacher She Is?

When we first interviewed Marta we were intrigued to notice that she immediately connected her ideas about teaching as a profession with her own life story. This is rare among teachers we have interviewed. She looked at some events in her life and told us how she thought they had affected her professional awareness and pedagogical vision, even before we asked her about such connections.

Going back to the five main factors we focus on in individual life history (childhood, school attendance, teacher's philosophical and pedagogical theories, teaching experiences, and various everyday experiences) we see how Marta feels each of them affects her as a teacher. Marta believes that the factors that had the greatest effect on her work with children are her own childhood, school attendance, and teaching experience. We have already seen how she thinks her own experience from primary school affected her as a teacher, as her decision to become a teacher grew out of her early school attendance. Below, we take a closer look at the other issues that relate to her view on teaching.

How Does Marta Think Her Childhood Affected Her as a Teacher?

Marta feels that three main factors in her childhood experience have affected her as a teacher. They are all linked to her immediate family.

Marta lost her father when she was only two years old, and she was raised by close relatives she refers to as parents. She felt different from the other children; she was ashamed that she did not live in a traditional family and that her parents were older than those of the other children: "For me this was difficult. I was ashamed at not being like the rest of the children." Now she feels this experience has helped her to understand and work with children who have lost a parent, or those who, for any other reason, do not live in traditional families.

Marta talks a lot about peoples' attitudes towards the children she grew up with. Both her parents and other relatives showed children a lot of respect and took time to listen to them. She says: "In my family the attitude towards children is very special. Each child is respected as a person and valued no less than adults." When Marta grew older she was surprised to find out that it was not the case in every family. This attitude meant a lot to her and after she became a teacher she has tried to respect and value each child the way he or she is.

When Marta was a child, the society expected children to be polite: "Children were expected to sit, listen and obey. I did not want to be so strict as a mother and a teacher. But I have the tendency to go to the other extreme and to be too soft and permissive." Here is another way she feels her upbringing has affected her as a teacher.

How Does Marta Think That Her Teacher Education and Teaching Experience Has Affected Her as a Teacher?

Marta feels quite sure that her own teacher education has affected her as a teacher. Although many years have passed, she can recall some philosophical and pedagogical theories from her formal teacher education that she believes have influenced her teaching. She admits, however, that she does not often think about them now.

She feels that her experience both as a teacher and as a participant in continuing education has had great effects on her as a teacher and on her work with her students. Also, she feels her teaching experience has had a great influence on her as a teacher: "My teaching has changed a lot through the years. Even though I have similar ideas that guide me, my teaching methods are different, and my experience is a 'treasure' that I constantly use." Then she gives examples of how new curriculum and methods, seminars and workshops have dramatically changed her work with students through the years. In this context she mentions, for example, how the storyline method and a new methodology in social studies brought her closer to her students. And she says a year-long in-service training program for practicing teachers, which focused on promoting students' social and moral growth, improved her communication with her students. In addition, she mentions that curriculum evaluation has improved her understanding of teaching.

Overall, Marta feels she is closer to her students now than when she started her teaching career. She feels more willing to let the students take part in the lessons and influence their direction.

She explains how a group of difficult students taught her a lot about communication and relationships in the classroom. She says she has sometimes been exhausted at the end of the day and often wondered whether this work was worth the effort:

> Sometimes you have to put a lot of energy into working with groups or individuals. Every now and then you wonder whether it is worth all the effort. I have come to the conclusion that it always is. Just a good memory for the child makes it worth it. If a child can look back and feel good about himself or herself, then it is worth all the effort.

When sharing these experiences Marta made it clear that being a teacher has not always been easy for her though she never has regretted the decision.

What Is Marta's Passion as a Teacher?

As we saw from Marta's quote in the beginning of this chapter, she believes that the basis for all learning is to "support students' positive self-identity, promote their social competence and skills, as well as to foster their well-being." If the child does not have an intact self-identity, he or she cannot study the subject matter, including mathematics.

Further, when asked what she thinks is most important for students to learn in elementary school, she clearly focuses on their personal, social, and moral growth. When discussing her aims in teaching, her themes or key words are self-identity, respect, and social competence and skills.

First, regarding the importance of self-identity, she says: "It is important for a child to feel good about him/herself, both for their present growth and their future."

Second, when talking about the importance of respect, she says:

> It is important for children to learn to show respect for themselves and others in their work and play. They have to respect others so that they can interact successfully. It is also important for their self-respect and self-esteem that they show each other respect.

Third, regarding the importance of social competence and skills she says:

> The aim is that students improve in their social relationships, that they can work with each other, can put themselves in each other's shoes.... Social competence and skills are the keys to all relationships. The ability to communicate is essential in our society; the ability to be capable of looking at things from different points of view, and improve social relationships. By having such competence we have the possibility to do better in life; this I think is quite crucial.

In considering Marta's aims we see that she is concerned about her students, both in their present life and their future. Although her main focus is on the child himself, it is clear that she believes that the society will benefit if the individuals are able to interact successfully. She says:

> We do not know the society that awaits these children as they grow up. But whatever it will be like it will be important that they can pick up and create knowledge. People with rich social competence and skills as well as positive self-identity are more likely to become effective and productive individuals.

Later in the same interview, Marta gave an example of the importance of self-identity and social skills:

> Soon these children will have to make decisions that can affect their whole life, for example concerning smoking or drug abuse. Then it is very important that they feel good about themselves and that they are able to listen to others and look at things from different points of view.

In short, Marta's teaching aims refer to important individual life competencies and skills that she contextualizes with a concern for students' well-being, with regard to both their current daily lives and their futures. Moreover, we observe similar themes in her life story and in her aims with a major focus on positive identity, respect, and social competence and skills in relationships.

We asked Marta what she thought would be the best way in teaching to attain her aims. She responded by saying that in order for a student to acquire a more positive identity, self-respect, and respect for others and to develop his or her social competence and skills, the student had to feel cared for and respected by the teacher. She added: "It is very important that the teacher cares for each child in the classroom and makes them all feel that they are unique and deserve the best."

Marta knows, however, that it is not always easy for the teacher to be loving and caring, especially if children show difficult behavior or have found it difficult to bond with other people: "The teacher has to work on his or her own feelings towards that child, respect and care for him or her." Marta points out that each student has his or her own background, needs, and skills. In order to meet their diversity, she emphasizes the use of various teaching activities.

Further, she says, regarding collaboration with parents:

> In my work I have to focus on, and facilitate, students' self-confidence in every subject. Often it is helpful to work with parents on such matters. In everyday life parents have many oppertunities to discuss mathematical problems. At home you run into various mathematical problems. It can make a huge difference if you, as a teacher, provide the parents with ideas on how they can discuss mathematics with their children in their daily routine.

When asked about the main objectives of mathematics learning, she says that it is most important that her students become able to use their mathematical knowledge while learning, playing, and working. Marta reflects back at her own childhood and says:

> When I was learning mathematics we had to repeatedily solve very similar examples by using a certain method. For me it sometimes felt more like cross-stitching or something like that. I did not always understand what I was doing or why, but I knew how to solve these examples, such as division or multiplication, write the answers systematically, and make it look tidy in my workbook. We were meant to write our answers on the right page of the book. My father was into mathematics and once he made me show some calculations on the opposite page to check how or if I understood what I was doing. At school my teacher took this page as an example of messy work and showed it to all my classmates. I think I was around 10 to 12 years of age then and I never showed my calculations after that.

Marta is pleased to say that the teaching methods have changed through the years. She explains that she tries to build on her students' understanding and work with hands-on material to strengthen their understanding. She says:

> In my math teaching I focus on discussions, both when we are faced with something new and also by encouraging the students to explain how they came to a certain conclusion. Often we write down different ways to solve a certain task and discuss them. Some years ago I did most of the talking, but now I make an effort to get them to talk while I try to understand the way they think.

Moreover, she integrates all subject matters such as "Icelandic, mathematics, social studies, natural sciences, music, art, play, manual work, etc." This integration she calls "holistic activities."

When observing Marta's classroom work, we did not focus on any particular subject, and in fact she often taught many subjects at a time in line with the holistic approach she holds. The students often worked on a particular theme for a period of time using methods such as storyline. During the time we were observing her lessons, the theme was *the car* and the students' task was to find out what it was made of, in which countries it was made, and so on. They were expected to write, read, and listen to descriptions of cars and stories about them, reflect on their personal and environmental effects, wonder about different types and functions of cars, and work with concepts like transportation, energy, engine, price, time, distance, and fuel.

The students worked individually, in pairs, and in groups. It seemed to us that they were given a lot of freedom in the classroom and were encouraged to ask for and receive help from the other students. This is one way to strengthen students' social competence and skills as well as their knowledge and understanding of the subject matter. Sometimes they were all working on the same subject, but more often they were working on different things, depending on their interests or abilities. Sometimes some of them were doing math while others were writing stories.

When solving mathematical tasks, they often worked in groups of four. For example, the tasks involved counting cars at nearby car parks and finding out how many wheels those cars had altogether. They investigated different types of car tires, focusing on patterns and texture. She asked them to make a guess on how far different toy cars would drive down a slope, and they tested their hypothesis by measuring the distance. They built cars of Lego cubes, drew them accordingly on paper, and stitched a picture of them. The more competent students got sometimes deep into their tasks, while some of the other students benefitted from the group work and the discussions within the groups.

Each week they had to do some homework and at least once a week it was based on mathematics. In part, everyone in the classroom had the same homework but in part it differed from one student to another mainly based on their capability. This shows us that Marta is aware of her students' diversity and that she tries to take that into account when she plans lessons and homework.

Frequently we observed how Marta motivated the students by using their own experiences and guided them with open questions to evoke their thinking.

One Monday morning Icelandic was on the schedule. Marta started the day by asking whether someone would like to tell the others about his or her weekend. One boy had been to a birthday party at his cousin's home. He talked about a computer game his cousin wanted to buy using his birthday money. The other children were also interested in this computer and took part in the discussion. The boy knew exactly how much money his cousin got for his birthday, the amount of money he had before, and the exact price of the computer. Marta used this information to ask the students how they could find out whether he had enough money to buy the computer, how they could find out how much more money was needed, and what the boy could do to get such an amount of money. Questions like these provoked others, and when the discussion came to an end the students had found a conclusion by using different methods such as addition, subtraction, and multiplication. This lesson was planned to be a lesson in Icelandic, but turned out also to be a lesson in mathematics where the initiative came from the students. This is an example of Marta's willingness to let her students take part in the lessons, building on their interests and supporting their communication skills.

Another thing that intrigued us when we observed Marta at work with her students was her choice of words when talking to and about her students. She uses words like "sweetie," "loved," "beautiful," and "friend" when addressing her students. Her students seemed to be used to this and when we asked, she said this is what she grew up with. When she wrote notes in her students' workbooks she was very positive; an example is "Excellent, Diana. This is absolutely brilliant." This use of words is one of her many ways to make her students feel good about themselves, show them respect, and to foster their positive self-identity.

LIFE STORY AND PEDAGOGICAL VISION

The purpose of our study was to acquire a deeper understanding of the relationship between teachers' life stories and their pedagogical vision. Here, we have followed one teacher and focused on her reflections on how her life story relates to her pedagogical vision and daily work with students. Marta represents a teacher who very clearly connects specific factors in her life story to her pedagogical vision, and she easily expresses and shares these connections with others.

In a way we can say that teaching was a calling for Marta, based on her experiences in childhood. She was only seven years old when she decided to become a teacher, wanting to follow the steps of her own teacher, whom she admired greatly. She still feels that teaching has been the right profession for her. She mentions several values that have followed her from childhood which she brings into her teaching, such as respect and care for children.

Marta has a broad and a deep view on her role as a teacher. In her aims we see that she is concerned both about supporting the students in their daily life as well as preparing them for the future, referring to important individual life competencies and skills.

With the purpose of attaining her aims to promote students' social, moral, and academic growth, she uses various teaching strategies and activities. In this regard she contextualizes her aims and teaching strategies by emphasizing the individual needs of each child, based on their different social backgrounds and competencies.

In short, there is a good harmony between her aims, her ideas about how she suggests to attain those aims, and the way she works and interacts with her students in the classroom. Her case provides essential insights into the pedagogical vision of a highly professional teacher. As such, Marta is among those teachers who may serve as a model for other teachers.

Her case, along with other case studies, can be viewed as a mirror that both pre-service and in-service teachers—whatever subject matter they teach—can use effectively to reflect on their own vision, values, aims, and teaching strategies in relation to their own life story in the eternal search for effective and responsible teaching.

Reflecting on their own profession is of great importance for teachers' personal and professional growth as well as for understanding how teachers think about their profession in general and in supporting them to grow in their profession In this process, teacher' life stories play an important role in how they construct their ideas and practice (e.g., Adalbjarnardóttir, 2007; Aitken & Mildon, 1992; Connelly & Clandinin, 1990; Fischler, 1998).

Certainly, there are differences in the way each teacher reflects on and makes meaning of his or her teaching and change (e.g., Korthagen, 2001). With regard to math teachers, it should be interesting, informative, and instructive to see whether there is something special in the ways they reflect on their teaching and relate their vision to their own life story: Do they share visions similar to those of other teachers, or is there something special about teaching mathematics that gives them a unique position among teachers?

NOTE

1. The work presented in this chapter was supported by grants from the Research Student Fund of the Icelandic Council of Science. The study is a part of the first author's MA thesis under the supervision of the second author. We gratefully thank the teacher Marta (pseudonym), who kindly consented to participate in this case study.

REFERENCES

Adalbjarnardóttir, S. (1999). Tracing the developmental processes of teachers and students: A sociomoral approach in school. *Scandinavian Journal of Educational Research, 43*(1), 57–79.

Adalbjarnardóttir, S. (2007). *Virðing og umhyggja. Ákall 21. aldar [Respect and care: The challenge of the 21st century].* Reykjavík: Heimskringla.

Aitken, J., & Mildon, D. (1992). Teacher education and the developing teacher: The role of personal knowledge. In M. Fullan & A. Hargreaves (Eds.), *Teacher development and educational change* (pp. 10–35). London: Falmer Press.

Bartlett, L. (1990). Teacher development through reflective teaching. In J. C. Richards & D. Nunan (Eds). *Second Language Teacher Education* (pp. 202–214). Cambridge: Cambridge University Press.

Bogdan, R., & Biklen, S. K. (1992). *Qualitative research for education: An introduction to theory and methods.* Boston: Allyn and Bacon.

Brookfield, S. D. (1995). *Becoming a critically reflective teacher.* San Francisco: Jossey-Bass.

Butt, R., McCue, G., Raymond, D., & Yamagishi, L. (1992). Collaborative autobiography and the teachers voice. In I. F. Goodson (Ed.), *Studying teachers lives* (pp. 18–51). London: Routledge.

Connelly, F. M., & Clandinin, D. J. (1990). Stories of experience and narrative inquiry. *Educational Research, 22*(11), 5–12.

Fennema, E., Carpenter, T.P., Franke, M.L., Levi, L., Jacobs, V.R., & Empson, S.B. (1996). A longitudinal study of learning to use children's thinking in mathematics instruction. *Journal for Research in Mathematics Education, 27,* 403–434.

Fischler, H. (1994). Concerning the difference between intention and action: Teachers conceptions and actions in physics teaching. In I. Carlgren, G. Handal & S. Vaage (Eds.), *Teachers´ minds and actions: Research on teachers thinking and practice* (pp. 165–180). London: Falmer Press.

Fischler. H. (1999). The impact of teaching experience on student-teachers and beginning teachers conceptions of teaching and learning science. In J. Loughren (Ed.) *Researching teaching: Methodologies and practices for understanding pedagogy* (pp. 172–197). London: Falmer Press.

Fullan, M. (2001). *The new meaning of educational change* (3rd ed.). New York: Teachers College, Columbia University.

Fullan, M., & Hargreaves, A. (1992). Teacher development and educational change. In M. Fullan & A. Hargreaves (Eds.), *Teacher development and educational change* (pp. 1–9). London: Falmer Press.

Gall, J., Borg, W., & Gall, M. (1996). *Educational research.* New York: Longman.

Goodson, I., & Walker, R. (1991). *Biography, identity and schooling.* London: Falmer Press.

Grimmet, P., & Crehan, P. (1992). The nature of collegiality in teacher development. In M. Fullan & A. Hargreaves (Eds.), *Teacher development and educational change.* (pp. 56–82). London: Falmer Press.

Hargreaves, A. (1998). *Changing teachers, changing times: Teachers' work and culture in the postmodern age* (2nd ed.). London: Cassell

Heidegger, M. (1962). *Being and time* (J. Macguarrie & E. Robinson, Trans.). New York: Harper Collins. (Originally published 1927)

Huberman, M. (1995). Professional careers and personal development. In T. Guskey & M. Huberman (Eds.), *Professional development in education. New paradigm and practice* (pp. 193–224). New York: Teacher Collage Press.

Ingvarsdóttir, H. (2005). You cannot put academic learning in the freezer for nine years: A case study of a woman's life history. In Úlfar Hauksson (Ed.), *Rannsóknir í félagsvísindum* VI [Research in social sciences] (pp. 683–693). Reykjavík: Félagsvísindastofnun Háskóla Íslands

Johnston, M. (1997). *Contradictions in collaboration: New thinking on school/university partnerships.* New York: Teachers College Press.

Kelchtermans, G. (1994). Biographical methods in the study of teachers' knowledge. In I. Carlgren, G. Handal, & S. Vaage (Eds.), *Teachers' minds and actions. Research on teachers' thinking and practice* (pp. 93–106). London: Falmer Press.

Knowles, J. K. (1992). Models for understanding pre-service and beginning teachers´ biographies: Illustration from case studies. In I. Goodson (Ed.), *Studying teachers lives* (pp. 209–234). London: Routledge.

Korthagen, F. A. J. (1993). The role of reflection on teachers´ professional development. In L. Kremer- Hayon, C. H. Vonk, & R. Fessler (Eds.), *Teacher professional development. A multiple perspective approach* (pp. 133–145). Amsterdam/ Lisse: Swets & Zeitlinger.

Korthagen, F. A. J. (2001). *Linking practice and theory: The pedagogy of realistic teacher education.* London: Lawrence Erlbaum.

Kristinsdóttir. J. V. (2006). Að læra af eigin kennslu.[Learning from one's own experience] Rannsókn kennara á eigin stærðfræðikennslu á yngsta stigi grunnskóla. *Uppeldi og menntun, 15*(1), 43–64.

Kvale, S. (1996). *Interviews: An introduction to qualitative research interviewing.* Thousand Oaks, CA: Sage Publications

Mau, S. T. (2007). Under the microscope: Looking closely at one's own teaching and learning. In G.Martin, M. Strutchens, & P. Elliott (Eds.), *The learning of mathematics.* (pp. 345–357). Reston, VA: National Council of Teachers of Mathematics.

Measor, L., & Sikes, P. (1992). Visiting lives. Ethics and methodology in life history. In I. Goodson (Ed.), *Studying teachers' lives* (pp. 209–234). London: Routledge.

Mitchell, C., & Webber, S. (1996). Using drawings to interrogate professional identity and the popular culture of teaching. In I. Goodson & A. Hargreaves (Eds.), *Teachers' professional lives* (pp. 109–126). London: Routledge.

Richardson, V. (1994). Conducting research on practice. *Educational Researcher, 23,* 5–10.

Schön, D. A. (1987). *Educating the reflective practitioner: Toward a new design for teaching and learning in the professions.* San Fransisco: Jossey-Bass.

Shulman, L. S. (1987). Knowledge and teaching: Foundations of the new reforms. *Harvard Educational Review,* 57, 1–22.

Sikes, P. (1985). The life cycle of the teacher. In S. J. Ball, & I. F. Goodson (Eds.), *Teachers' lives and careers* (pp. 27–60). London: Falmer Press.

Strutchens, M. (2007). Reflection on mathematics teaching and learning. In G. Martin, M. Strutchens, & P. Elliott (Eds.), *The learning of mathematics* (pp. 317–319). Reston, VA: National Council of Teachers of Mathematics.

Vygotsky, L (1978). *Mind in society: The development of higher psychological progress.* (Michael Cole, Ed.). Cambridge, MA: Harvard University Press.

Zeichner, K. M. (1994). Research on teacher thinking and different views of reflective practice in teaching and teacher education. In I. Carlgren, G. Handal & S. Vaage (Eds.), *Teachers' minds and actions—Research on teachers' thinking* (pp. 9–27). London: The Falmer Press.

37

PROFESSIONAL DEVELOPMENT

How Did Teachers' Participation in the Professional Development Program Influence Their Teaching Practices?

Guðrún Angantýsdóttir

MENNTASKÓLI KÓPAVOGS – UPPER SECONDARY SCHOOL, ICELAND

Abstract: This chapter addresses issues related to action research. The research was about how four teachers learned to use new course materials in mathematics, and its goal was to discover how the new course materials could contribute to teachers' professional development. In the research, the teachers got temporary support from the researcher, in the way students learn mathematics. We observed together how teachers learn mathematical topics and methods that can be found in the new course materials. They developed various teaching methods and thoughtful evaluation techniques, observations and evaluations on their own work, and the work of each other.

The results indicate that the teachers felt pleased with how they learned the new teaching techniques and felt that they had grown professionally.

Keywords: Professional development, action research, teaching methods

INTRODUCTION

In the last decades of the 20th century, continuing education for Icelandic teachers was mostly done either through summer courses that were a few days in duration or through longer afternoon courses during the academic year. These courses were organized by parties outside of the schools, although in many cases they were organized because of critique from teachers or special interest groups. Usually course offerings were sent to the schools, and the teachers were able to pick the courses they wanted to take. For that reason teachers mostly took courses out of interest or need. When the courses concluded, it depended entirely on the teachers whether they used their newfound knowledge or not. Both Icelandic and foreign researchers have criticized this arrangement, as they believe there should be a link between a teachers' continuing education and the needs of the schools they work in. If such a link existed, researchers believe that continuing education would have a greater impact on teachers' careers as well as the school they belong to (Aðalbjarnardóttir, 2002; Eggertsdóttir, 1999; Fullan, 2001; Hannesson, Eggertsdóttir, & Sigþórsson, 2002; Kemmis, 1988; Liberman, 1995). In addition, the research focus has shifted from understanding teacher behavior in class to understanding their ideas about their own work. Teachers have increasingly worked on understanding themselves by thinking about what they teach, and why and how (Hitchcock & Hughes, 1995).

The First Sourcebook on Nordic Research in Mathematics Education, pages 505–518

In this research, the researcher looked at how four teachers learned to use new course materials in mathematics, and its goal was to discover how the new course materials could contribute to a teachers' professional development. In the years 2002–2003, new course materials in mathematics were published for 5th grade students: *Geisli 1*. The researcher, also in the capacity of one of three co-authors of the new educational materials, was interested in encouraging mid-level teachers to observe these new ideas in mathematical teaching, and particularly interested in the reaction of teachers who taught it. The course materials required the teachers to use many different teaching techniques and carefully design their lessons and students' learning environment (Pálsdóttir, Angantýsdóttir, & Kristinsdóttir, 2002). Preparations began in the fall of 2003, which included drawing up a research plan, conducting a literature survey, and obtaining permission from the study site to carry out the research. None of the teachers had any further mathematical education besides what was taught in the core studies at the Icelandic teachers' academy (Kennarahaskoli Islands). The youngest teacher had been teaching for only two years, but the oldest had been teaching for twelve. The research was done in the spring semester of 2004, from January to June.

METHODS FOR ACTION RESEARCH

With action research, there is emphasis on the creation of knowledge, where teaching, school research, development of schools, and the assessment of school work are intertwined together into one and the same cycle. This is also, at the same time, a way for schools to do some self-assessment to improve the quality of their school work. In this way teachers are made to be a part of developing and forming new ideas with their students and creating a culture of research within the school where the emphasis is on continuous learning. In this way I, as a researcher, have an opportunity to observe all the stages of the research, since it is my own work that is being researched. I am steadily considering, revising, and improving my work as the research progresses. One must bear in mind that action research has limited value of generalization as the research reflects the factors the researcher thinks are important (Elliott, 1991).

According to McNiff (2002), action research is a term that refers to a practical way of looking at one's own work to check that it is as one would like it to be. Because action research is done by the practitioner, it is often referred to as practitioner-based research, and because it involves the practitioner thinking about and reflecting on his or her own work, it can also be called a form of self-reflective practice (McNidd, Lomax & Whitehead, 1996).

When the author began this research, it was important to give teachers all the theoretical consultations and support they would need. In addition the goals of the research were:

- To make teaching in courses so clear that the teachers realize the full potential of these new mathematical course materials and discover how students understand what it is that they learn
- To make teachers use varied teaching methods and teaching aids as needed
- To make teachers use varied assessment tasks
- To give teachers a positive attitude towards mathematics
- To increase peer-to-peer interaction among teachers

Fullan (2001) believes that any improvement in school work directed towards improving quality of teaching depends on the teacher, as it is the teacher who puts these improvements into practice. He emphasizes that the teacher understands the personal evolution that takes place when their own work is studied or examined. He believes that a teacher must realize that conducting a study is not merely to evaluate methodology but also adaptation and independent implementation of the methodology. Done well, the study can increase teacher's self-confidence and independence. He believes that teachers' career development and school development must go hand in hand, and school development should be a part of teachers' career development as soon as teachers begin their educational studies. He says that effective change process and effective teacher development are one and the same (Fullan, 1996). Sigþórsson (1999) points out the importance of teachers' cooperation and believes that a teacher must be an expert in three areas:

- In teaching and communication with students, as each student can study through home-schooling (school without separation)
- In working cooperatively, supporting his colleagues
- In leadership, by spearheading improvements in school governance (Sigþórsson, 1999)

By studying one's own work, the main focus is on creating knowledge, bringing together teaching, research, and school evaluations into the same process. Keeping a diary with our thoughts was something I emphasized both to myself and the teachers. The author made field notes and studied mathematical instruction in the classroom with the teachers who participated in the research, and they were given a check list to be able to evaluate their teaching. In our meetings the teachers decided, after reading and discussing in pairs, what they wanted to look at and investigate. When they had watched each other's teaching, they discussed and considered it in our regular meetings. As there was good cooperation and confidence between the teachers, they decided to discuss and evaluate their results mutually.

All research data were categorized into the following categories: *Experience, knowledge, expectations, attitude, cooperation, security/insecurity, facilities, study materials,* and *time.* The categories were compared to find common themes and how they supported each other, so it was possible to read a story on teachers' thoughts regarding the transformation and the progress of the study. The data were regularly studied during the course of the study through meetings, field observations, and journal review. Every effort was made by the author to be unbiased and honest when gathering and analyzing data (Hitchcock & Hughes, 1995).

The information collected was investigated for:

- the education of the participating teachers
- the changes teachers make when they observe their own and their colleagues teaching when teaching the new syllabus
- the author's role as a researcher in the research

McKernan (1991) says that action research has implications because it can serve to improve problematic social situations. It can enhance the personal understanding of the researcher, and it can serve to illuminate the researcher's social surroundings, or the environment and conditions in which he or she works. The goal is to deepen the researchers' understanding of the work.

The process or the action plan that the researchers have in mind is:

4. Reflect → 1. Plan
↑ ↓
3. Observe ← 2. Act

McKernan says that the action plan shows action reflection as a cycle (see Figure 37.1). This action research cycle can turn into new action research cycles, as new areas of investigation emerge. It is possible to imagine a series of cycles to show the processes of developing practice.

The processes can be shown as a spiral of cycles, where one issue forms the basis of another and, as one question is addressed, the answer to it generates new questions (McKernan, 1991). The results can give teachers a better vision of their own work and increase their ability to determine what can and should be changed. They also increase teachers' understanding of their students and lead to a better attitude towards cooperation and overall school work.

This study had three research cycles. In the first circle, the main emphasis was on ideology of mathematical instruction, course materials, and teaching aids. In the second circle, the emphasis was on teaching methods and the role of the teacher in mathemati-

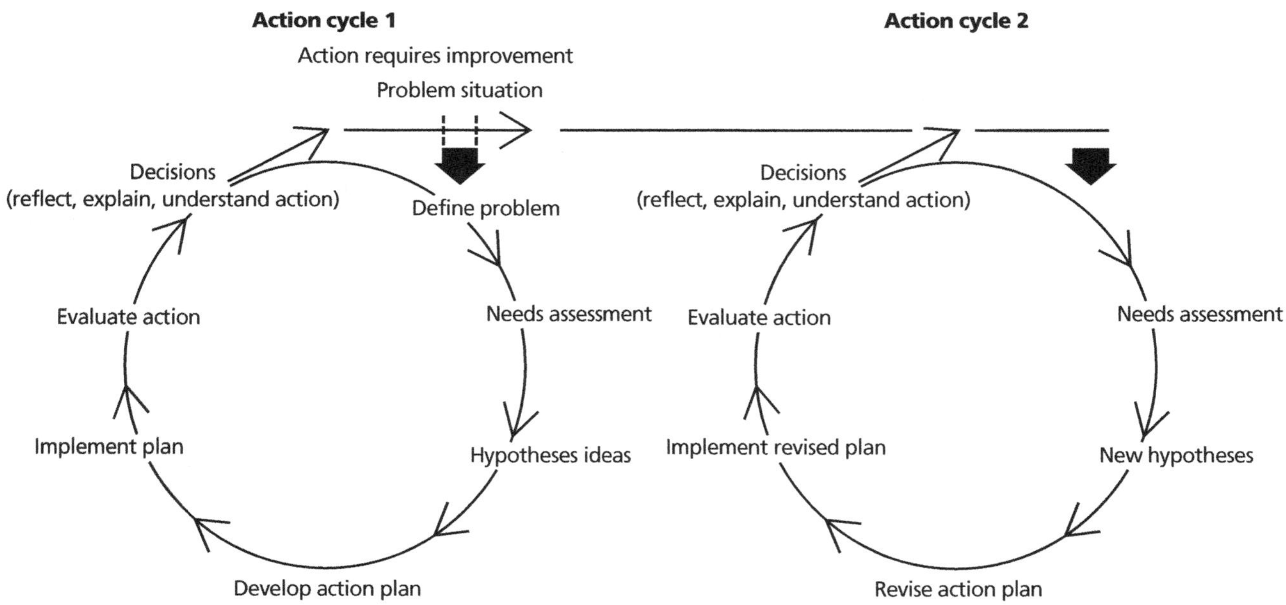

Figure 37.1 Action reflection as a cycle.

cal instruction. Lastly, in the third circle, the emphasis was on course evaluations. The team met 15 times, and the meeting agenda was centered around carrier research and mathematical instruction. Each meeting usually ran for one to two hours. We discussed how students learned mathematics and how well teachers were able to use the new course materials, *Geisli 1* (Pálsdóttir, Angantýsdóttir, Kristinsdóttir, 2002a,b,c,d). During the meetings the author distributed reading materials on mathematical studies and mathematical instruction and carried out action research. The reading materials ensured that teachers familiarized themselves with number theory, mathematical education, diverse teaching methods, and course evaluations. In addition, teachers studied mathematical education goals for mid-level students from the main mathematical study guide. We always discussed the reading materials, usually about a week after the teachers got these reading materials. The teachers themselves had to evaluate their own and others' teaching methods. They were encouraged to study these new methods so that they could continue to meet and discuss their own teaching even after the completion of this research.

THE MAIN CONCLUSIONS FROM THE PRELIMINARY SURVEY

From the conclusions of the quantitative questionnaire about work experience, teachers' attitudes about mathematics and math teaching, the way they use teaching materials in math teaching, and short interviews with the teachers before the research began, it is possible to conclude the information shown in Table 37.1.

TABLE 37.1

Teaching in the autumn semester	In the beginning of a traditional lesson the teacher would introduce new topics which would be followed by the students solving problems in their schoolbooks working either alone or in pairs. The teachers found it very important that the students would learn mathematical rules and traditional methods of doing math and emphasized the importance of understanding them. The teachers did not have access to a computer lab and did not use the counteractive games which can be found on the website for Geisli. If the teacher's book suggested a practical project for the students then they fulfilled that.
Attitude to math teaching	Teachers found it important that students had the opportunity to explain their results to others and found that conversation was an important factor of math learning. They felt that math learning was based on how new knowledge was built upon previously learned facts and that group work was an important part of math learning. The teachers found that when they themselves were in school the emphasis was on math training but not understanding.
Difficulties that teachers had in the autumn semester	The teachers said that they had felt insecure when they started doing the problem solving in the beginning of Geisli 1A. They said that they had helped students along in the search for a solution and solved the problems on the blackboard in cooperation with the students instead of watching the students solve the problems on their own as suggested in the teacher's book. The teachers found it important to teach the students traditional methods of doing multiplication and division rather than encouraging them to find their own methods. They did not realize that it was possible to use various ways in division, for example dividing in stages or using repeated subtraction, but they emphasized the fact that multiplication and division are opposite operations.
The use of teaching materials and teacher's books	When working within the syllabus they did not always use teaching materials. The teachers distributed teaching materials to the students whenever they thought it necessary. The students used calculators when they were working at research projects or playing math-games where the need for calculators was at hand or when they were solving problems which involved a lot of calculations. The teachers usually used the guidelines from the teacher's book and followed suggestions from the student's books if they encouraged the use of various teaching materials for solving problems.
Attitude to teaching materials	In general the teachers found that emphasis should be on the use of various teaching materials. The teachers were positive towards the use of calculators in teaching and thought they had a positive effect. They were also very positive towards the use of computers in teaching.
Attitude to Geisli	The teachers were pleased with the diversity of projects in Geisli 1A. They had not had significant difficulties in adapting to the new syllabus but had felt insecure. Therefore they were thankful to have the opportunity to gain more confidence towards it by participating in an action research.
Attitude to course assessment	The teachers found it important that students regularly undergo written tests (assessments). They found that continuous assessment was important but had not organized a systematic listing for it. They had not looked into the information about assessment on the Geisli website.
The methods teachers use when calculating	They either calculated in their minds or used methods they were taught in school. The mathematical teaching they received in school is what moulded them as teachers in the beginning. They found they were restricted to using the methods they were taught in school when teaching the new syllabus rather than using other methods.

One teacher said about her attitude to the new curriculum:

> At first I was very skeptical. I was thinking about what was going on when I first heard it being discussed in the staff room. But when I got to know the new material myself, my attitude towards it changed. There are so many new aspects and the use of various teaching materials has increased a lot which makes it easier for us to explain things in a manipulative way to the students. The projects are more rational. It is very logical to teach in this way. I find the students understand everything better. For example, before the students were just taught that a circle is 360° but now they are much more aware about the nature of the circle after having learnt about it in a more manipulative way.

TEACHERS' LEARNING AND RESULTS OF RESEARCH CYCLES

Aðalbjarnardóttir (1999) thinks that the development of teachers' professional awareness involves each teachers reflecting on their job with devotion and sincerity and studying their ideas, objectives, and methods of teaching. In meetings during the research there was emphasis on looking at and reflecting on number sense, operations, diverse teaching methods, and different ways of assessment.

Research Cycle 1

In the research cycle shown in Figure 37.2, the emphasis was on working with decimal fractions in *Geisli 1B* (Pálsdóttir et al., 2002b,d). The teachers experienced certain inferiority towards working with the number sense of decimal fractions. It was decided to give this topic special time. The author found some reading material about number sense and how students build it up. The teachers discussed the use of models in teaching and different teaching methods. By discussing these topics the teachers could acquire information, experience, assistance and support from each other.

> *Teacher 1*: When I was in school I never understood decimal fractions. We learned certain rules about how to write them and how to do calculations with them, i.e., that the decimal point should be in the same place when doing addition or subtraction and one should count the digits behind the decimal point in multiplication. I never got any understanding on the subject, but there was a lot of emphasis on the training. I have difficulty in explaining to students topics I never understood myself in school.

Teachers mostly agreed about this and discussed the difficulties in teaching decimal fractions.

> *Teacher 2*: In the older curriculum facts were given and students were trained in writing decimal fractions. Here there is more emphasis on the students understanding what each digit stands *for*.

> *Teacher 3*: It wasn't as difficult to teach fractions earlier this winter as we had fraction cards which students could arrange together to make a whole. By using the fraction cards, the concept of fractions was more visible to the students and they were able to relate to how big each part of a fraction was of a whole.

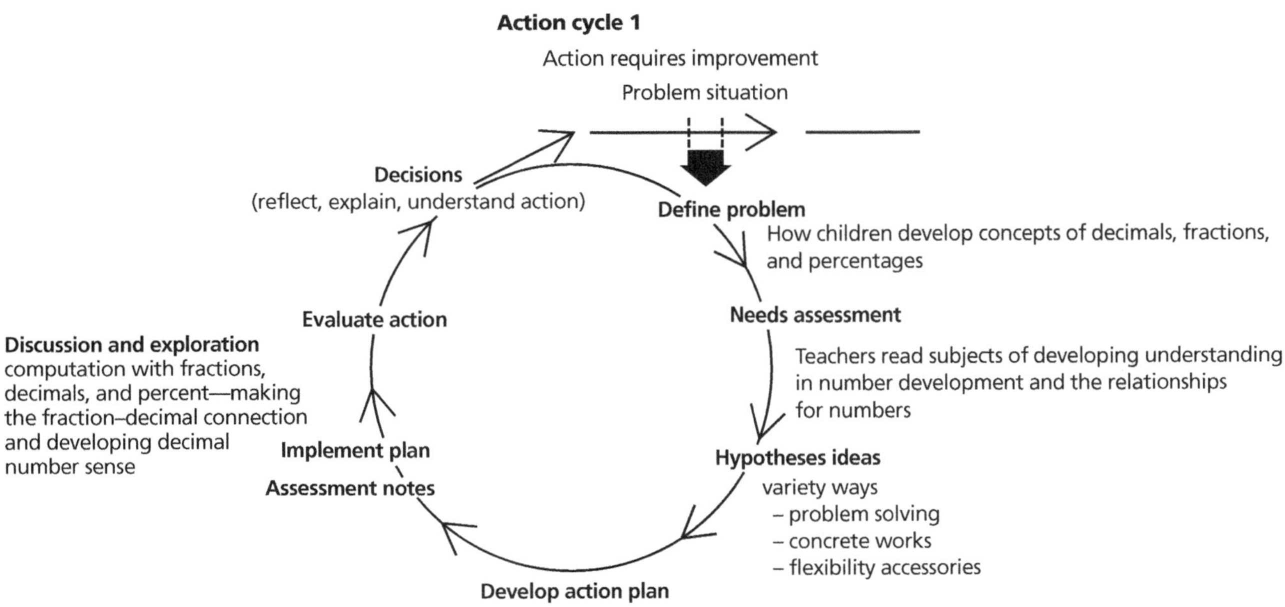

Figure 37.2 Research cycle 1.

The teachers agreed on the importance of going to a manipulative stage to help students obtain understanding of numbers that are between whole numbers.

> *Teacher 4*: I told my students to look at their ruler and wonder about the little lines that were between the numbers on the ruler. Most of the children knew that these lines meant millimeters. Then we started wondering how we could write them as centimeters. A few of them knew how numbers were written with decimal points and could inform the other children. That's how they realized that there were numbers that existed between any other two whole numbers.

Different points of view and the experience of teachers explain the changes that happen with general consideration. Concurrently the students underwent a test to examine their understanding of fractions. The author took notes from the lessons in which the tests were done and spoke to the teachers after the lessons. In our next meeting we discussed the results of our projects. The teachers agreed that the students' answers helped them assess the students' number sense and would assist them at their level. They realized that they would have to give the students more time to solve the problems and emphasized that the students should use models. The teachers realized that the understanding of the meaning of each digit in a number is the foundation for the students to be able to compare number sizes and calculate with them. They therefore found it important to emphasize on building a good understanding of number sense in a diverse way and to try to meet each student in their search for knowledge. At the end of the cycle, which took about three weeks, the teachers presented to the students a practical project from *Geisli 1B* that involved fractions (Pálsdóttir et al., 2002b,d). Notes about the project are presented below:

> The kids took a group test in fractions. There were three together in a group, and they played a game involving fractions. The cooperation went well and they were also very interested. Their understanding was on different levels. I am sure they learned something from this project, and those who had little understanding before have gained more and better understanding. I was very pleased with the game and will use it again.

The teachers found that this work gave them a lot of experience and were fairly pleased with their students' results.

Research Cycle 2

Here is a description of a part of the research (Research cycle 2; see Figure 37.3) when the teachers compared different ways and possibilities of math teaching and discussed the role of the math teacher. The knowledge and experience of a teacher is formed by the experience they acquired when studying mathematics and the surroundings and the experience they had acquired through their work (Conney, 1999). The way the teachers themselves had learned mathematics was studied to realize the teacher's attitudes towards math and whether that had an affect on how they taught math.

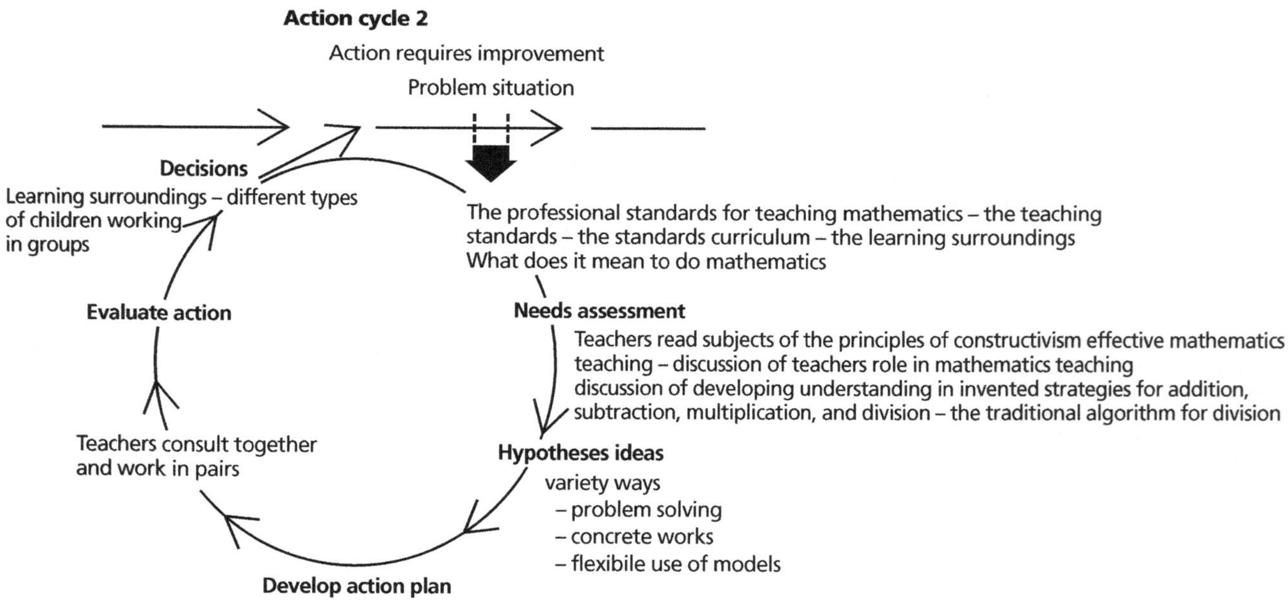

Figure 37.3 Research cycle 2.

In our meetings we discussed the contents of our reading material about constructivist theories, the role of a math teacher, and how problem solving is used in math teaching. The teachers received reading material about constructivism, the role of a math teacher, number sense, and operational understanding. By reading and discussion they gained new angles in operational understanding and realized, for example that it is possible to use various methods in division. By considering and observing in the classroom they realized that it is important to figure out the way students think and the way they develop their understanding of mathematical operations. It was decided to look into and define further the role of a teacher. The teachers received a list of questions they had to answer, and they looked into their own teaching by filling out an assessment checklist prepared by the author. It was emphasized that the teachers would keep notes on how their math teaching progressed over a certain period of time. The teachers made pauses in their work regularly within the classroom, looked around, and made notes about their experience.

They worked in pairs, visiting each other's classrooms and taking notes of what they saw and experienced there. The aim was to bring out common factors in teachers' teaching experiences. By working in pairs, the teachers were especially observing diverse teaching methods. They visited each other's classroom twice. The first time, they took notes about what they observed during a traditional math lesson. The second time, they had organized five mathematical stations and observed what went on in each other's classrooms and took notes. The teachers were pleased with working in pairs. They found that cooperation of this kind strengthened them as teachers, and they enjoyed discussing their experiences. They felt that with this kind of cooperation, they were more aware of the fact that teachers were usually experiencing similar things in their jobs.

In a group interview after their teamwork had taken place, we had discussions about the role of the math teacher. In this way they received an overall view of their job and could compare their experiences of common projects and give each other advice or find common solutions to certain projects.

These observations helped them in trying new ways to increase the students' independence. They emphasized allowing the students to try to solve the problems on their own before explaining to them what to do. Here is a short description from a diary on a math project after this work had taken place:

> *Teacher 3*: The first thing I noticed when I looked at this page before the lesson started was that I would get a lot of questions about the basketball pictures and many students would soon give up. So I decided to allow them to work in pairs and told them to really try to solve the problem on their own and they weren't allowed to ask me any questions. Then they were supposed to come and show me their answers. This went better than I had hoped for, as I had found the problem difficult. Many of them were fairly quick to come up with the solution—but not all of them. They didn't have any difficulties with the other problems on the page.

> *Teacher 2*: Teacher 1 came into my classroom and I started the lesson and explained what they were supposed to do at each station. I had already divided the students into five groups. They started working and everything went very well. The kids were very quick to realize what they were supposed to do and they were hard working and active. I didn't have to assist them much. The problem-solving group surprised me the most. I had read the problems and thought they were rather difficult, but they solved them in a much shorter time than I had expected and they didn't find them so difficult. I enjoy this kind of work very much and the kids find they are playing but not learning although they are learning quite a lot.

> **A visit to Teacher 1**: I then went into a lesson with teacher 1. She had already started her lesson and had divided her students into groups. The kids started working and their work went very well. They started their projects right away and enjoyed themselves. I can't say much else. I sat, observed and took notes. Teacher 1 walked around the classroom and assisted those who needed help. It was the same with these kids as it was with mine; they enjoyed themselves and got carried away.

Research Cycle 3

As a researcher, the author emphasized to the teachers the need to discuss and try diverse ways of assessment. During the research the teachers tried using written tests, manual tests, group tests, home tests, and continuous assessment and had the students assess themselves. Most of these assessment projects can be found in the project folder that comes with *Geisli 1*. When the teachers were asked about their views about course assessment after having tried and discussed different approaches, they said they were pleased with using diverse ways. Lack of time was the main reason we couldn't look into course assessment as much as we would have liked to. In our discussions at our meetings, there was a positive attitude toward diverse course assessment:

> *Teacher 2*: My opinions on assessment have changed. I feel we have to try different ways, not only to use written tests but to try diverse ways. . . . I was surprised

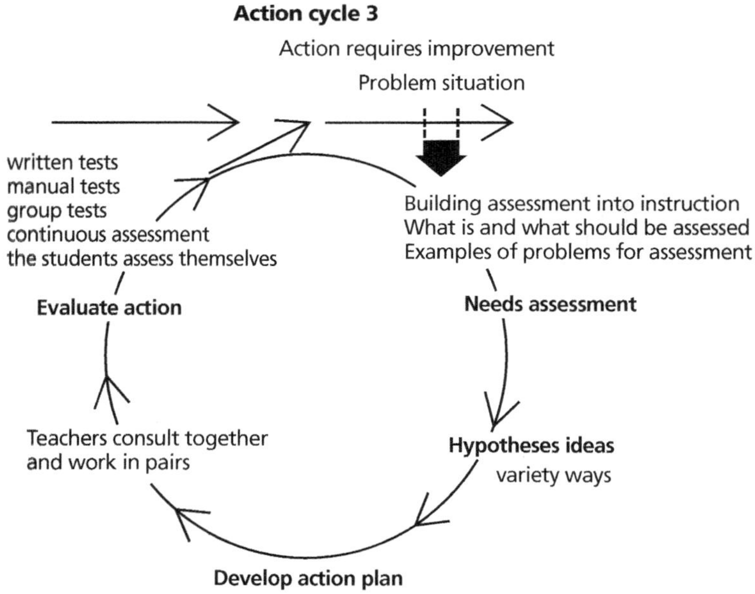

Figure 37.4 Research cycle 3.

by the group test, and the same can be said about the home test. I didn't believe in it at first because the kids could have someone solve their problems for them at home. But now I see the purpose. This is a good way to get parents to be responsible and for them to observe their children, and the kids solved all the problems by themselves. Those who didn't get assistance at home were those who needed it the most.

Teacher 4: I enjoyed the fraction assessment project very much. It makes you stronger as a teacher to try such group projects. One knew more or less how things would go and I knew who could solve the problems. But I was surprised to see that there were more students who could do a lot more than I thought they could.

Teacher 1: We decided to make notes on the back of the home tests on whether the students had received help with them at home. It came as a surprise to me that most of the students had received very little help at home, and I could see from the projects that they used their own methods. They used many different methods. I find it difficult to assess the projects. I think this way of testing is very interesting and I think it helps parents see what their children are doing in school.

At the end of the third research cycle (Figure 37.4) the teachers reviewed how they managed to reach their own goals. Here is an example of Teacher 1's speculations:

- *I aimed at getting my students more independent in mathematics*:
  I tried to activate them into assisting each other. By doing that I think they started trying to solve problems more on their own.
- *To use diverse teaching methods*:
  I think this goal was reached best in "the carousel" and in the book Akrabadabra because then I dared to" take off" with the kids.
- *Innovating cooperation between students*:
  That succeeded in "the carousel" and I had them working a lot in pairs.

TEACHERS' IDEAS ABOUT THE CHANGING PROCESS

Guðjónsdóttir (2004) says that when a teacher develops his own professional work theory, he shows professionalism in his job. She defines professional consideration as a process that builds on certain key factors, which are knowledge of teaching and pedagogy, awareness of social culture and the historical effect of education, the understanding of systematic consideration, and the connection of necessary preconditions and theoretical understanding. A teacher reinforces his professionalism by realizing the constant interaction of professional knowledge, performance, consideration, and ethical values. The success of professional development depends on the depth of professional consideration and the involvement of teachers who share their knowledge by common discussion and consideration (Guðjónsdóttir, 2000).

When the teachers were asked about what part of the research cycle they found most important, they agreed on it being the discussions and thoughts on the subject in our meetings. In our meetings they felt most encouraged and supported. There they realized that teachers are in fact generally facing similar problems. They emphasized the importance of having the opportunity to meet and discuss new teaching materials and teaching methods. The also said that they felt more sure of themselves in their lessons and found they were better teachers than before. They said that they hadn't developed any ideas about this research beforehand, but that it had met their needs completely.

> *Teacher 1*: I find that the course has made me much stronger. It has helped me to hear that we are all engaging in the same things. When one shares one's problems and hears about others from the other teachers, one often hears solutions from others or one comes up with solutions that can help others. This is very important as everything is so new.... I think it is very important that teachers have an opportunity like this to compare notes. It makes you more confident. I feel good about having had the chance to get to know the new syllabus so well.
>
> *Teacher 4*: One is probably teaching more methods in one project than before. Before the students learned about percentages, fractions, and decimal fractions separately, without them being connected in any way. But now we work with all these concepts together and the kids can see how connected they all are. By talking together and wondering one can see this all so clearly. One becomes aware of how their understanding builds up with such discussions. It is so good to take one concept and analyze it together.
>
> *Teacher 2*: One goes on a course and listens and often there are clever things one hears about but there is no support. Here we are talking together in a small group, we share our thoughts, carry out our ideas, and discuss our results.
>
> *Teacher 3*: I thought I was fairly good at math before, but I found I was very insecure this autumn. I somehow felt I wasn't good enough in math to teach this new material, but now I feel surer about myself. I find I have learned a lot, gained more points of view, and think I am a more qualified teacher.
>
> *Teacher 4*: The course developed in a way that helped us. I found it most significant to be able to exchange views. I found I jumped into something in the autumn but on this course I felt more confident. I felt more confident and my views changed. Before I was often stuck in one method, but now I see that I can use many ways. I find I am a better teacher.

DIFFICULTIES IN THE RESEARCH

It is important to discuss the difficulties that came up during the research but were solved successfully. They could have affected the results if we hadn't agreed on a solution. During the research the author was aware of a heavy work load on the teachers. This problem was solved by speaking to their school principal and asking for a reduction in teaching hours for them. They were allowed to reduce their working hours by one hour a week during the research period. One hour a week isn't much compensation for all the extra work that accompanied the research, and it would have been better if their teaching duties could have been made less.

Here are some entries from the author's diary:

> March 1.
>
> I felt that the teachers were irritated and discontent. I spoke to them and asked them to tell me what was bothering them....am unhappy because I find the work connected to the research too much, too much reading material and too many things to register...said that other teachers participating in action researches in their school got a reduced work load. I have asked for an interview with the principal and will be meeting him tomorrow. This must be cleared up. My teachers must also receive a reduced work load...I am going to try to cut down the theoretical text for the time being and go better into discussions and work that will benefit the teachers.
>
> March 2.
>
> I spoke to the principal at ten o'clock. He agreed to pay the teachers one extra hour per week during the research period. He attended our meeting and informed them of our conclusions. Everybody was pleased with the outcome.

The author was often aware of anxiety among the teachers because of certain demands made on them. Most of the anxiety was about keeping a diary, but as soon as they mastered that everything went well. The teachers were a long time getting started with keeping their diaries, and the amount they wrote down differed a lot. They also felt uneasy about the concept of working in pairs, and the author tried to introduce that concept as smoothly as possible by giving them some reading material about it. The teachers were also very uneasy about visiting each other. The felt they didn't have much to offer and were in fact nervous about having someone visit their classroom. Trust and cooperation were therefore important. Their experience after their visits was very positive. They said they had learned a lot and found that collaboration like this would give them more results than the self-evaluation and rating lists they had tried previously.

> *Teacher 3*: I felt insecure when I heard them talk about working in pairs. What did I, a newly graduated teacher, have to offer? However, the experience was very positive. I was very conscious about my teaching while my partner in the research was in my classroom. I gave my lessons great thought and I can still remember those lessons well. I was aware of many things and it surprised me when my partner said that he had the same speculations as I had so we could discuss them and support each other.

> *Teacher 2*: The discussion about working in pairs scared me. I wouldn't have liked just anyone into my classroom. But this group was very likeable so I wouldn't have minded any of them for a visit. I was very aware of what I was doing while my partner was in my classroom and I realized things I hadn't been aware of before. I am very positive towards work of this kind and have learned a great deal by having a fellow teacher visit my classroom. This kind of work gives us a chance to speculate about our work as we become so aware of what we are doing when someone else is observing our work and when we observe the work of other teachers. It was also good to be able to talk about this within our group. We are in fact all dealing with the same things every day and we can help each other in this way.On the whole I think working in pairs is a very good way to look into one's own work and improve it and to see how others do theirs. One sees new ways and methods of teaching which one hadn't thought of by oneself. It is also a good opportunity to get pointers about things one does incorrectly or could do in a different way, a way that one hadn't thought about before.

The worst enemy of the research was lack of time. The teachers were very busy and although the meetings were within their working hours we lacked time for theoretical reading. Sometimes the author was too demanding towards the teachers and could feel their uneasiness when too much was asked of them. The greatest difficulty encountered was to get the teachers to realize the diverse ways of using the operations, especially in division. The teachers were very stuck in using the methods they grew up with and using traditional algorithms.

Notes from the author's diary illustrate this:

> February 11.
>
> This is a very enjoyable meeting. Everybody is very active and has a lot to talk about. The discussions about algorithms were very useful. Teachers are very stuck in using their traditional method. It is the way they were taught to use in division and there is nothing to comment on that. They did, however, find it very interesting to see how many other methods you can use instead. They had never thought about that before. I think it is very important to look well at this aspect because I think most teachers haven't thought about the possibilities of using other methods in division than the one we learned ourselves in school.

The author realized that the teachers had too much reading material, which was taken into consideration in consultations with them.

> *Teacher 3*: Possibly it would have been better to fragment the theoretical part more; it was off-putting at first but one got used to it. It was a bit much at first and I didn't realize the amount of work involved so I wasn't as positive in the beginning. In the beginning I found this all quite difficult, i.e., to keep a diary, think about the theoretical part, and be a participant in interviews and to express my opinions.
>
> To be able to get involved with all the aspects connected to mathematics, i.e., the theoretical aspects, the methodology, the diverse teaching methods, and assessment methods, has opened new dimensions for me. This is very good as it becomes useful in all subjects in school and in life itself.

THE ROLE OF THE RESEARCHER

Fullan emphasizes that the role of an instructor in work supervision must:

- build up trust
- analyze the needs of those he is guiding
- face a certain development process
- provide and use reading material for others
- build up capability and self-confidence to finish the project
- direct the project (Fullan, 1995).

The author's role as the researcher was to organize and take care of the research. In fact, my job was one work cycle in the research cycle of this action research. The job can be divided into three parts:

1. The making of the research plan, the preparation of the course and the planning of the research
2. The execution of the research and the gathering of information
3. Considering and interpreting the research information

Figure 37.5 shows my position as a researcher in the research.

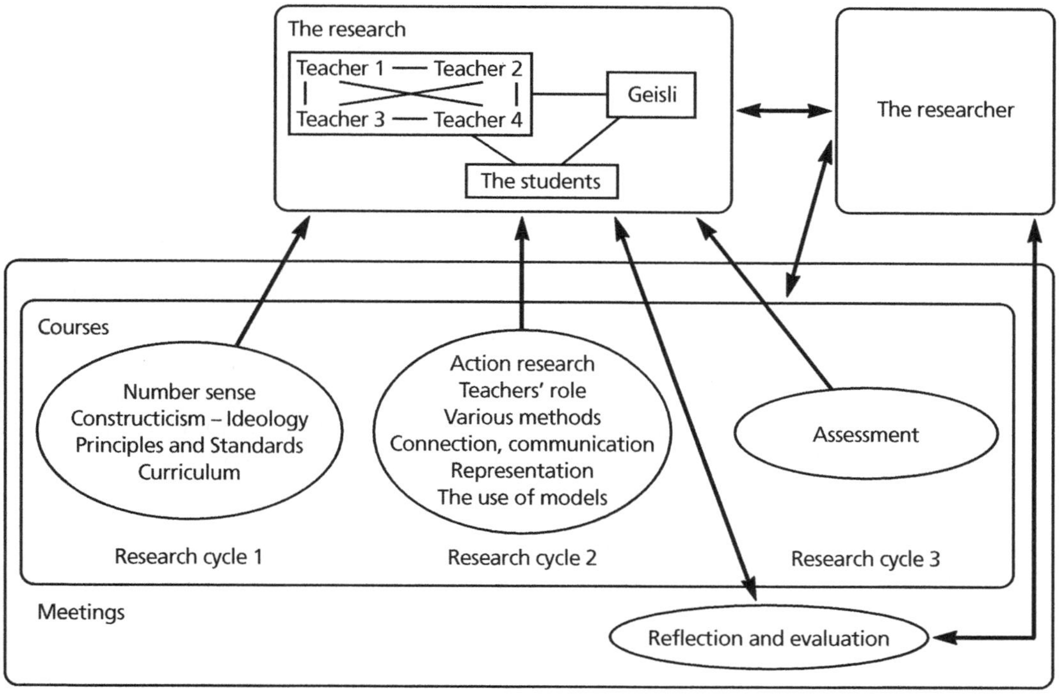

Figure 37.5

The author was interested in the teachers' active participation in the course and hoped that they would share their work with one another and reveal to others how they taught certain topics in *Geisli*. The most important factors were:

- the ideology of educational theories, the curriculum, and a the new syllabus
- teaching methods and the use of models in mathematics
- assessment

Even though the main purpose of the research was for teachers to adapt to the new math syllabus, it was also a way for them to acquire new teaching methods. The teachers felt that the regular meetings satisfied them the most. There the author got the chance to assist the teachers in adapting to new and demanding projects and, in doing so, helped the teachers to become more capable and independent in their math teaching. It was important to make sure that the teachers were made to feel secure and independent with the new curriculum. This was done by considering and discussing common projects, getting to know the new syllabus better, and getting to know the main ideas of scholars on how children learn math. Trust between the author and the teachers built up easily during the research. The author did, however, feel that they were afraid of ruining the research if they didn't write enough in their diaries. They felt guilty when they were asked to write down everything that went on in their lessons in their diaries and when they were encouraged to read the reading material given to them. Often there was little time to tend to this as they had a heavy work load and the research wasn't planned as a part of their job. During our meetings the teachers were not afraid of asking for advice, and we always found mutual solutions to any problems that came up in their work.

> *Teacher 2*: I found it a great advantage to be able to make our own decisions about topics we were not sure about. For example when we were working with decimal fractions. You (the researcher) gave us advice which helped us work at increasing the students understanding. I found that I learnt a great deal by doing that and the way we discussed the matters and realized that we all felt unsure. We didn't stop until we were pleased with the results.

The author gradually tried to withdraw from the research but felt that the teachers needed more support and advice. The teachers were asked if they disliked being manipulated too much, but they replied by saying they wanted the author to lead their work. As the research progressed, though, the author gradually withdrew and encouraged the teachers to have discussions within their group and make suggestions of their own. The author did, however, participate in the discussions and gave advice if needed. One of the teachers said in a group interview, "We find it good to be able to come to you for advice as you can see our difficulties clearly and come up with solutions promptly."

The teachers were asked to evaluate the research cycle at the end of the semester and write it down in their diaries. The all agreed that with this work they had acquired new teaching methods they were pleased with. They all felt that their professional knowledge in mathematics and math teaching had increased and that that knowledge would strengthen them as math teachers in the future.

> *Teacher 3 (diary)*: To meet and discuss matters, about math, and the way we feel helps us to succeed better in our math teaching. One is also more open towards ones co-workers, isn't embarrassed about asking questions and trying to learn from each other. It is also a good feeling if something one says can help a co-worker. I thought I would never be able to tell a teacher who has been working longer than I have about how I find it best to do something.... This course has strengthened me in my job and will continue to do so. I could think of continuing in similar work.

> *Teacher 2 (diary)*: I found the advantages of this course were that we worked together in a small group and got know and trust each other well. We could exchange opinions and talk about the things we found difficult or had problems with. I wouldn't do that in a larger group or with people I didn't trust. It was also good to meet up regularly and discuss matters. I found this different to other courses I have attended as they are usually crowded, superficial and impersonal and usually too theoretic.

When teachers were asked to list the main disadvantages of the research, they mentioned lack of time as the main problem, and they would have liked to discuss more practical ways of teaching.

> *Teacher 3 (diary)*: This was difficult; the lack of time is what comes first to mind. There was a lot of work involved which would have been better to disperse over a whole winter. I find it difficult to write down how I have done but I find it easier as I write these words. To obtain the technique of keeping a diary of one's thoughts is difficult, but I know it helps so I will try to continue doing so. To have meetings at the end of a work day wasn't good; I was tired and not always in the mood. Maybe it would be better to use free times in our working day and make sure that those involved in the research get joint free times.

> *Teacher 2 (diary)*: I feel that the main flaws being that we could have been in closer contact with the new teaching material. Then I mean to have the book itself (Geisli 1B) and take a look at each page, ponder over them and come up with ideas of what could be done with each page. Try out new ideas and meet afterwards and see how things went with everybody else. This is what I missed the most in this course and in other courses I have attended.

It was difficult to intertwine the subjects of the course with the teachers' development in a way that would suit or be utilized completely in their work. The subject had to be approached gradually so it would not trigger the teachers' defense mechanism and possibly prevent any development within the schools. The author believed it important to support the teachers but also to be aware of giving the correct guidance to aim at a set plan and systematic work. The author often wondered about the best way to guide the discussions and how much leading was necessary in the research cycle. The author found the way the discussions were executed a very important factor in the research. Teachers are fairly used to discussing their work, but the author emphasized that the discussions should be well considered and that they should result in systematic school work. The teachers discussed their experiences in the classroom and connected them to the understanding of scholars on education and teaching. With common executions, the teachers were able to adapt to the new syllabus and enhance their mathematical understanding and math learning. It was felt that if the author's support had subsided during the research cycle, it would have ended with unsuccessful results. However, too much support had its dangers as well. The author therefore tried to find the so-called golden middle. The teachers wanted to share their experiences and with their discussions they could share and learn a lot from each other.

CONCLUSION

The goals of this action research were met because the teachers realized the meaning of math teaching and the way the students' understanding develops, as can be seen in the first research cycle when the emphasis was on students' number sense. The teachers used diverse teaching methods and were efficient in using the appropriate teaching materials, as can be seen in the second research cycle. They used diverse ways of assessment and showed interest in continuing that in their teaching in the future. They developed good cooperation and were very positive towards continuing working in pairs and felt they had strengthened their professional maturity. After discussing and reading theoretical articles, the teachers demonstrated changes of attitude and they often discussed, in our meetings, various methods the students used in problem solving.

Teacher 1: One has started to think a lot more about how to approach the new material than before. One is more awake and thinks about the students' understanding more than before. One has started looking for different methods in multiplication. The kids can easily use diverse methods. When we discuss certain problems they keep coming up with different methods to use. I perhaps introduce one method and they bring more and on the blackboard we suddenly have maybe four methods. It also helps the ones that never express themselves when they see the various methods of the others.

What the author considers to be the most important factor in the teachers' work is that they acquire a way to see how their students go about building upon their mathematical knowledge and to trust them to do so. At the beginning, the teachers had felt incompetent towards the subject and the many methods students can use in embracing the subject. Their anxiety diminished gradually and they felt confident in assessing their students' knowledge.

Teacher 4: Now I can see I shouldn't keep telling them how to do things. At first I felt I had to be in control of what the students were doing. I didn't trust them in what they were doing as I was insecure myself.

The author realized that this project was more extensive than was initially planned in the beginning and had difficulty in limiting the research. The author is, however, confident that this kind of arrangement can be successful and believes that it is important for teachers to get the opportunity to work in this way when a new syllabus is introduced. What was most pleasing was the fact that the research and the teaching were intertwined into a process that molded the teachers' teaching methods. In light of this experience, the author thinks the following aspects should be kept in mind when approaching work of a similar kind:

- School masters must show interest and good will when their teachers are adapting to new teaching methods or working with a new syllabus.
- There must be an allowance for extra time within the teachers' working hours.
- Teachers in all classes must have an opportunity to develop their own methods in cooperation with other teachers, within the same department, which would benefit the students.
- Teachers must show initiative and interest in such cooperation.
- It would be preferable to have a teachers' consultant, who might be working at the school, to lead a group of teachers into cooperation building on similar ideas as above and to give them support and encouragement.

If these factors are kept in mind, the author is convinced that cooperation of this kind will lead to a development in math teaching that will benefit our students.

REFERENCES

Aðalbjarnardóttir, S. (1999). Þróun fagvitundar kennara. Að efla félagsþroska og samskiptahæfni nemenda [Teacher Development. Practices and Support Students Public and Diplomacy Development]. In H. S. Kjartansson, H. Ragnarsdóttir, K. Indriðadóttir, & Ó. J. Proppé (Eds.), *Steinar í vörðu til heiðurs Þuríði J. Kristjánsdóttur sjötugri*, (pp. 259–267). Reykjavík: Rannsóknarstofnun Kennaraháskóla Íslands.

Aðalbjarnardóttir, S. (2002). Í eilífri leit. Virðing og fagmennska kennara [Always Searching. Respect and Professionalism], *Netla –Veftímarit um uppeldi og menntun*. Rannsóknarstofnun Kennaraháskóla Íslands. Retrieved June 7, 2008, from http://netla.khi.is/greinar/2002/005/03/index.htmConney, T. J. (1999). Conceptualizing teachers´ ways of knowing. *Educational Studies in Mathematics, 38*, 163–187.

Eggertsdóttir, R. (1999). Frá námskeiði til skólastofu [From a Course to a Schoolroom]. In H. S. Kjartansson, H. Ragnarsdóttir, K. Indriðadóttir, & Ó. J. Proppé (Eds.), *Steinar í vörðu. Til heiðurs Þuríði J. Kristjánsdóttur sjötugri*, (pp. 271–292). Reykjavík: Rannsóknarstofnun Kennaraháskóla Íslands.

Elliott, J. (1991). *Action research for educational change*. Philadelphia: Open University Press.

Fullan, M. G. (1995). *The limits and the potential of professional development*. New York: Teachers College Press.

Fullan, M. G. (1996). *The new meaning of educational change*. London: Cassel.

Fullan, M. G. (2001). *The new meaning of educational change* (3rd ed.). London: RoutledgeFalmer.

Guðjónsdóttir, H. (2000). *Responsive professional practice: Teachers analyze theoretical and ethical dimensions of their work in diverse classrooms*. Ann Arbor, MI: UMI Dissertation Services.

Guðjónsdóttir, H. (2004). Kennarar ígrunda og rannsaka eigið starf [Teachers Analyze Theoretical and Ethical Dimencions of tehir Work Indeverse Classrooms]. *Tímarit um menntarannsóknir, 1*, 27–38.

Hannesson, J. B., Eggertsdóttir, R., & Sigþórsson, R. (2002). *Aukin gæði náms: Skóli sem lærir* [More quality educational matters. School who learn]. Reykjavík: Rannsóknarstofnun Kennaraháskóla Íslands.

Hitchcock, G. & Hughes, D. (1995). *Research and the teacher. A qualitative introduction to school-based research* (2nd ed.). London: Routledge.

Kemmis, S. (1988). Action research. In J. P. Keeves (Ed.), *Educational research, methodology and measurement: An international handbook* (pp. 173–180). Oxford: Pergamont Press.

Lieberman, A. (1995). *Practices that support teacher development. Transforming conceptions of teacher learning*. Retrieved June 7, 2008, from http://www.questia.com/googleScholar

.qst;jsessionid=LxvRcGQpc1QfgXTtQPwVQhY5fk8MpQp GF7s2Fpftnfg3yyy2Yvvg!1454331138?docId=5001646243M

cKernan, J. (1991). *Curriculum action research. A handbook of methods and resources for the reflective practitioner.* London: Kogan Page.

McNiff, J., Lomax, P., & Whitehead, J. (1996). *You and your action research project.* New York: Routledge Falmer.

McNiff, J. (2002). *Action research for professional development. Concise advice for new action researchers.* Retrieved June 7, 2008, from http://www.jeanmcniff.com/booklet1.html.

Pálsdóttir, G., Angantýsdóttir,G., & Kristinsdóttir, J. V. (2002a). *Geisli. 1A.* Reykjavík: Námsgagnastofnun.

Pálsdóttir, G., Angantýsdóttir, G., & Kristinsdóttir, J. V. (2002b). *Geisli. 1B.* Reykjavík: Námsgagnastofnun.

Pálsdóttir, G., Angantýsdóttir, G., & Kristinsdóttir, J. V. (2002c). *Geisli. 1A. Kennsluleiðbeiningar* [Education instruction]. Retrieved June 7, 2008, from http://www.nams.is/stae_midstig/index.htmPálsdóttir, G., Angantýsdóttir, G., & Kristinsdóttir, J. V. (2002d). *Geisli. 1B. Kennsluleiðbeiningar* [Education instruction]. Retrieved June 7, 2008, from http://www.nams.is/stae_midstig/index.htm

Sigþórsson, R. (1999). Sterkur þekkist af sínum verkum [Powerful is known from his work]. Hugleiðing um sjálfsmynd og sérfræði kennarastéttarinnar á nýrri öld. *Ný menntamál, 17*(3), 27–32.

38

STRUCTURES OF FINNISH AND ICELANDIC MATHEMATICS LESSONS

Lasse Savola

TEACHERS COLLEGE, COLUMBIA UNIVERSITY
REYKJAVIK UNIVERSITY

Abstract: Videos from forty Finnish and Icelandic mathematics lessons are analyzed using a two-pass coding method. The method, which uses ideas from two major video-based comparative studies, the 1999 TIMSS Video Study and the Learner's Perspective Study, offers a way to investigate the different forms of classroom interaction by which teachers attempt to achieve their pedagogical goals. Two lessons from ten randomly chosen mathematics teachers of 14- and 15-year-olds were recorded in each country. Despite the modest sample size, some national patterns and cross-national differences can be detected. The Finnish lessons in the sample essentially follow the conventional *Review–Lesson–Practice* script, whereas more than half of the recorded Icelandic lessons exhibit versions of *individualized learning*, a learner-based instructional philosophy. Public content-related discourse can be missing entirely from these lessons; instead, the teacher tutors the students one-on-one. This is in contrast with the Finnish lessons, where teacher-led activities, which often involve student participation, are emphasized.

Keywords: Lesson structure, Comparative studies, Finland and Iceland

This chapter presents a video-based analysis of the structures of forty Finnish and Icelandic mathematics lessons. The first of the six sections contains a rationale for the study. The method of lesson structure analysis is described in the second section. The third section explains the data collection process. Results, including lesson diagrams, make up the fourth section. Some interpretations and implications of the findings as well as limitations of the study are discussed in the fifth section, and the final section suggests directions for future research.

RATIONALE

Finland and Iceland are noteworthy within the domain of mathematics education. These countries—like all Nordic countries—share many characteristics, yet their performances in the PISA assessments differ considerably. While both Finland's and Iceland's PISA scores show very small between-school differences, the Finnish students have fared much better than their Icelandic counterparts in all three rounds of PISA studies. Furthermore, Iceland is the only country where the girls have significantly outperformed the boys in mathematics (OECD, 2004, 2007b).

Finland and Iceland have much in common. Both have subsisted in the fringes of Europe only to flourish in recent decades as they have reaped benefits from globalization and the digital revolution. The populations of these welfare states are some of the most homogeneous in the world. Universal social rights and competitive economies helped boost Finland and Iceland into becoming some of the most "citizen-friendly" and affluent countries in the world. Prior to the global financial crisis, Iceland ranked first and Finland twelfth in the United Nations' Hu-

The First Sourcebook on Nordic Research in Mathematics Education, pages 519–537

man Development Index list that takes into account such quality-of-life variables as life expectancy, access to education, and GDP (United Nations, 2008).

People in the Nordic countries exhibit similar patterns of educational attainment. They tend to study relatively little when they are young, but they often remain as students well into their adult lives. A Finnish child can expect to receive only 5,523 hours of school instruction between the ages of 7 and 14, the lowest hour total in the OECD. In Iceland, children in that age range typically receive 6,277 instructional hours, which is also well below the OECD average of 6,898 hours (OECD, 2007a, Chart D1.1). According to PISA 2003 data, an average Finnish youth studies 4.9 hours weekly outside of school—the least in the OECD—while in Iceland 5.9 hours per week is normal for primary school students. In 2005, 43% of Finnish people aged 20 to 29 were at least part-time students. This is the highest percentage among the OECD countries. Iceland's 37% ranks third behind Denmark, where 38% of people in their twenties are students (OECD, 2007a, Table C2.2). Overall, Finns and Icelanders spend on average about 20 years of their lives in school, placing them—along with the other Nordic nations—at the top of the OECD countries with regard to years of educational attainment (OECD, 2005, Chart C1.2).

Parallels between Finland and Iceland are more difficult to draw when it comes to academic achievement as measured by recent international assessments. Finnish 15-year-olds have excelled in all areas of the PISA studies. In PISA 2006, for instance, Finland had the highest scores in mathematics and science among the OECD countries, while ranking second in reading after South Korea. In the same test, Icelandic students scored significantly below the OECD average of 500 in reading and science and just above it in mathematics (OECD, 2007b). The differing trends for the PISA scores for Finland and Iceland can be seen in Figure 38.1.[1]

Student assessment results are inevitably influenced by a network of interrelated variables. Some of the factors that likely have contributed to Finland's recent success are teacher-related: the good quality of the teacher education programs, the relatively high status of educators—and the respect they command—within the Finnish society, and the amount of creative freedom they enjoy in schools. School-based curricula with active learning pedagogies, the structure of the educational system, a national endeavor (the LUMA project) to improve the teaching and learning of natural sciences and mathematics, effective special education starting with the early grades, students' own interests and leisure activities, the nature of the Finnish language, the Finns' ethos of obedience, and Finland's unique socio-historical and cultural heritage have also been identified as possible contributors (Björkqvist, 2006; Malaty, 2006; Simola, 2005; Välijärvi et al., 2007).

Factors behind Iceland's declining PISA scores and the unusual gender differences are elusive. Despite spending 45% more money per primary school student than is the OECD average, the Icelandic educational system is facing concerns about student achievement (OECD, 2007a, Table B1.1). Reasons for the recent declines are uncertain. On the other hand, although definite answers do not yet exist, some plausible explanations for the gender differences, such as the "Jokkmokk-effect" and the gendered discourse explanation, have at least been ruled out (Halldórsson, Ólafsson, & Björnsson, 2007; Ólafsson, Halldórsson, & Björnsson, 2006; Steinthorsdottir & Sriraman, 2007). Although the gender differences are smaller in

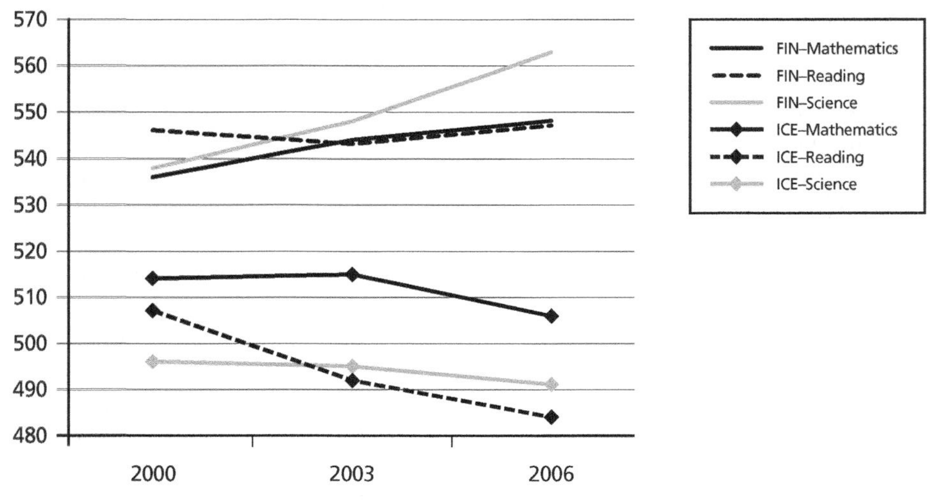

Figure 38.1 PISA 2000, 2003, and 2006 scores for Finland and Iceland (OECD, 2002, 2004, 2007b)

PISA 2006 than they were in PISA 2003, Icelandic girls are still outperforming their male counterparts on all scales, including mathematics. The recent reduction in the gender gap is unfortunately not due to an increase in the boys' scores—Icelandic boys' mathematics scores dropped 4 points from 2003 to 2006—but instead to a steep 15-point decline in the girls' scores (OECD, 2007b, Table 6.3b). It is clear that the current challenges of the Icelandic educational system concern all students, not just the boys.

Given the similarities between the two countries on the one hand and the differences in their students' achievement scores on the other, a comparative look at the educational systems of Finland and Iceland seems justified. Since Finnish and Icelandic students spend so little time studying outside of school, it may be particularly helpful to investigate the similarities and differences between classroom practices in these countries.

It is reasonable to ask whether video analysis is the appropriate tool to study mathematics teaching in Finland and Iceland. Other tools, such as surveys and on-the-spot coding of classroom observations, may also be useful. However, as discussed in recent literature (e.g., Erickson, 2006; Goldman, Pea, Barron, & Derry, 2007), video footage offers superior data-yielding capabilities in classroom research and can be effective in detecting patterns in instructional practices. One approach to carrying out international comparative studies in education would be to compare national curricula. But a study of Finnish and Icelandic mathematics curricula would probably not prove very helpful. Olsen (2006) writes,

> To some extent Finnish students have the same relative strengths and weaknesses as their Nordic peers. This implies that overall, the Finnish students are stronger than their Nordic peers in all aspects of mathematics covered by PISA. Hence, if the data from large scale international comparative assessments are perceived as a resource for learning from others, this finding implies that detailed studies of the subject matter of the curriculum are not necessarily the way ahead. (p. 33)

METHOD OF LESSON STRUCTURE ANALYSIS

Introduction

Video-based classroom studies can effectively complement other pedagogical research (see, e.g., Erickson, 2006). Large video-based classroom studies such as the 1999 TIMSS Video Study (Hiebert et al., 2003) and the Learner's Perspective Study [LPS] (Clarke, Emanuelsson, Jablonka, & Mok, 2006) have answered many questions about how to use video technology to study classroom practices, but they also have raised new issues. For instance, it is not clear which aspects of classroom practice offer the most useful evidence for cross-national comparison. How should quantitative and qualitative video-based research methods be combined in the analysis of classroom practices? What criteria should be used in the evaluation of video-based classroom studies?

Classrooms are complex, dynamic settings. Therefore classroom video footage is always multilayered and rich, filled with nuances and subtleties. Deciding which aspects of the *behavior stream* (Barker, 1963) to investigate is a crucial task. Indeed, it is possible to scrutinize classroom video footage with respect to countless research variables. In this study, the dimensions of the pedagogical *function* and the *form* of classroom interaction are used to model classroom practices and, more specifically, to reveal structures in lessons.

Considering lesson structures is not a new idea. Johann Friedrich Herbart, a follower of Kant, first postulated a cyclical sequence of learning steps in 1835 (Dunkel, 1969). Herbart believed that knowledge acquisition is comprised of formal stages that are linked together by association and that must be completed in order for learning to occur. Many understood his idea to imply that all teachers should follow the same step sequence in their teaching. As a result, lessons were often designed to follow Herbart's formal stages of learning: (1) creating cognitive clarity of previously learned material, (2) stimulating new knowledge elements by associating them to knowledge already possessed, (3) systematizing these associations, and (4) applying the new knowledge (see, e.g., Dunkel, 1969; Oser & Baeriswyl, 2001). Herbart's ideas shaped classroom practices significantly in Europe and the U.S., especially during the first half of the twentieth century. In fact, his ideas about learning are still prevalent in many classrooms today.

Lesson structure analysis is an important part of many video-based pedagogical studies. Examples of different coding schemes for lesson structure can be found in classroom studies such as the 1995 and 1999 TIMSS Video Studies (Hiebert et al., 2003; Stigler, Gonzales, Kawanaka, Knoll, & Serrano, 1999), as well as the LPS (Clarke, Emanuelsson et al., 2006). The approach used here combines aspects of the TIMSS and the LPS studies. For example, the LPS also uses the function and form dimensions; however, the analysis there is based on *lesson elements*—not whole lessons as in the TIMSS and in this study—and is more qualitative in nature. Some of the coding criteria in the current method were adapted from the TIMSS studies and Jablonka's (2004) habilitation thesis.

The two-pass coding method presented in this section can be used to identify patterns in classroom practices in terms of the pedagogical functions and the forms of social participation. The interaction of the two dimensions—function and form—is of particular interest, as the main purpose of the method is to offer a way to investigate the different forms of classroom interaction teachers use in attempting to meet their pedagogical goals.

The categories for the function-variable (first pass) are fixed, but the ones for the form-variable (second pass) are sample-sensitive. One of the strengths of this open-ended method is its ability to capture unique, yet often subtle, classroom practices. After all, a research tool in a comparative study is only effective if it can reveal meaningful differences and similarities across a sample. Whereas a method of lesson structure analysis using only predetermined categories may miss subtle yet critical differences in classroom practices, the adaptable method described here can be used to distinguish even slight variations in patterns of classroom behavior. The output of this coding method is a set of data that can be utilized in a variety of ways, including many types of statistical analyses and visual representations.[2] The method is intended to be generalizable across a variety of applications, including analyses of lessons given in subjects other than mathematics.

Lesson structure analysis has its limitations.[3] Although it can yield valuable information about classroom practices, lesson structure analysis does not shed light on many interesting and relevant facets of teaching and learning. For instance, the success of the pedagogical strategies employed and the types of cognitive structures provoked by the instruction generally cannot be deduced from the coded data. Moreover, researchers should avoid making hasty cause/effect pairings based on lesson structure analysis and other dimensions of interest such as PISA results.

First Pass: Function

The first pass of the coding method is based on Herbart's formal stages of learning. It is nearly identical to the *purpose* variable from the 1999 TIMSS Video Study. The *purpose* variable is a *coverage code*—every moment of the lesson is coded—and has three mutually exclusive categories: *addressing content from a previous lesson(s)*, *introducing new content*, and *practicing/applying/consolidating content introduced in the current lesson* (Hiebert et al., 2003; LessonLab, 2003). In the first pass of the coding scheme discussed in this chapter, these categories are referred to as *review*, *introducing new content*, and *practicing/applying*. An additional category, *other*, is included in the set of categories for the first pass variable because the aforementioned three were deemed non-exhaustive and over-inclusive even for a rough analysis of lesson structures.

The first pass variable, the function variable, lacks sensitivity to nuances in order to give a general idea about the main purpose of lesson segments. Thus the first pass should only be considered a starting point to lesson structure analysis. Although the four categories can be seen as covering each moment of any mathematics lesson, only limited information can be drawn from this kind of simplified analysis.

Many of the coding criteria for the first pass are adapted from the 1999 TIMSS Video Study. For coding purposes, the lesson starts when the teacher begins the first public statement and it stops when the teacher concludes the last such statement. If the lesson begins with a problem-solving phase, and the starting point is not clear, coding starts when half of the students are sitting down and working. If the ending point is not clear, coding ends when the bell rings or when half of the students have stopped working.

A segment less than 30 seconds in length is not coded, unless it is in the very beginning or the end of the lesson. If a segment lasts for less than 30 seconds, and the segments before and after it are of the same type, the short segment is simply ignored. However, if the segments adjacent to the short segment are different, the short segment is merged with the segment following it. A segment is considered to start at the beginning of its announcement. For example, if the teacher says, "Take out your books, notebooks, and pencils because we're going to start practicing the new idea now," the code for *practicing/applying* should begin as the teacher says "Take." The length of the coding interval is one second.

Table 38.1 lists the types of classroom activities that are included under each first pass category. A more detailed discussion of these categories and the coding criteria is presented elsewhere (Savola, 2008).

Second Pass: Form

Classroom interaction can have a significant effect on learning. Following Vygotsky's lead, educational researchers have recognized the importance of the socio-cultural context of learning environments (e.g., Cobb & Bauersfeld, 1995; Lave & Wenger, 1991). Bauersfeld (1980) considers *human interaction* to be constitutive and one of the "hidden dimensions in the so-called reality of a mathematics classroom" (p. 23). He asserts,

> Teaching and learning mathematics is realized through *human interaction*. It is a kind of mutual influ-

TABLE 38.1 First Pass Categories

Review	• Discussing content covered in a previous lesson • An average student is expected to know the content • Going over homework • Going over a quiz/test
Introducing new content	• Students are not expected to know all of the content • Exploring or demonstrating a new idea, procedure, or a type of problem • Reading or writing about new content • Only the first problem counts here. Subsequent examples of the same type count as practice ("Can an average student accomplish the task without having to make leaps?") • Summary is part of the segment only if it directly follows the demonstration
Practicing/applying	• Working on the types of problems that were introduced that day • Students are expected to know how to work with the new idea(s) using the content presented that day and previously • Assignment of practice problems
Other	• If none of the other three categories apply, or if two or three apply • Classroom management (roll calls, getting settled, disciplinary action, technical set-up, etc.) • Mathematics management (announcing tests or homework, goal statements, distributing/collecting materials or technology, etc.) • Interruptions (fire drill, etc.) • Public social talk • Announcements over the intercom or TV

> encing, and interdependence of the actions of both teacher and student on many levels. It is not a unilateral sender-receiver relation. Inevitably the student's initial meeting with mathematics is mediated through parents, playmates, teachers. The student's reconstruction of mathematical meaning is a construction via social negotiation about what is meant and about which performance of meaning gets the teacher's (or the peer's) sanction. How can we expect to find adequate information about teaching and learning when we neglect the interactive constitution of individual meanings? (Bauersfeld, 1980, p. 35)

The second pass of this coding scheme concerns the forms of social interaction and classroom participation. It is intended to shed light on the "hidden dimension" discussed above. This part of lesson structure analysis is closely related to what was done during the first coding pass. Having seen the lessons and classified each moment in them as belonging to one of the four categories of the first pass variable, the researcher is in a position to devise and carry out a finer categorization. The categories of the second pass variable—that is, the form variable—stem from asking "Who is doing what?" and "How are the participants interacting?" They vary from sample to sample according to the range of actions and interactions of the teachers and their students on the recordings. The creation of appropriate categories is not straightforward, and it is almost certain that two coders would conceive different sets of categories. The basic idea is that each significant type of participatory action by the teacher and the students should be classifiable with an appropriate label, and that significantly different types of actions are not "lumped" together.

During the second coding pass the lesson segments classified using the four categories of the first pass variable are further divided by the kind of classroom participation they exhibit. There is one exception: Lesson components classified as *other* are generally divided further using non-form-based categories such as *classroom management*, *mathematics management*, and *interruption*. With the other three categories, important decisions about the level of detail of social interaction to be included in the coding system must be made since not all kinds of interaction can be represented. The smaller the differences in the classroom practices within the sample are, the finer the distinctions between the second pass categories can be. While there are no limits to the number of categories that can be used, a "golden mean" for each set of videos probably exists. This depends not only on the recorded lessons, but also on the agenda of the researcher. During their development, the coding categories should be checked against specific cases to ensure that they are capturing meaningful differences and not creating misleading contrasts (Angelillo, Rogoff, & Chavajay, 2007).

Coding for the form variable can be more challenging than coding for the pedagogical function. The coding categories are formulated specifically for each sample, a task whose difficulty tends to increase with the sample size. In addition, the endpoints of the form segments can be harder to identify than those of the function segments. This is partly because different forms of classroom interaction can have considerable overlap, whereas the teacher's intentions about pedagogical goals more often are clearly expressed and followed. Carefully defined, mutually exclusive second pass categories help with this process.

DATA COLLECTION

About the Teachers

The twenty teachers involved in the study were randomly selected from among the Finnish and Icelandic mathematics teachers who taught 14- and 15-year-olds during the academic year 2006–2007. This age range is the one targeted by the PISA studies, and it corresponds to 8th and 9th graders in Finland and 9th and 10th graders in Iceland. Twelve of the twenty teachers were observed teaching both 14- and 15-year-olds. The remaining teachers were observed in two classes of the same age group (4 cases), or a class with 13-year-olds was included (4 cases).

At the time of the taping, the mean age of the teachers was 46.2. Their ages and their years of teaching experience range from 29 to 66 years and 2 to 35 years, respectively. Although the Finnish teachers in the recordings are younger than their Icelandic counterparts—44.4 years of age as opposed to 47.9—they have more teaching experience: 18.2 years versus 15.1 years. Six of the Icelandic teachers and seven of the Finnish teachers are female.

Sampling: Finland

The ten Finnish teachers who participated in this study were randomly selected. Because comprehensive lists of Finnish mathematics teachers or primary schools were not readily available, the first step of the selection process involved the choosing of ten educational providers. Typically these are municipalities (i.e., cities, towns, or villages) or, less frequently, foundations. The educational providers, which vary considerably in size, were weighted according to the number of students they serve (Finnish Department of Education, 2005). Then ten providers of education were chosen using a random number generator. In effect, ten "dummy" students were randomly selected, and the towns in which they attend school listed.

The schools were chosen next. If the selected educational provider oversees only one primary school with 14- and 15-year-olds, that school was asked to participate in the study. If, on the other hand, the educational provider operates multiple primary schools, the schools were listed alphabetically and one of them randomly chosen. The principals of the selected schools were contacted. They were asked to participate and to provide a list of active mathematics teachers. If the principal agreed to the study and submitted the said list, a teacher was then randomly chosen. If the teacher was not willing to cooperate, another one from the same school was chosen instead.

To obtain the sample of Finnish schools, a total of thirty principals had to be contacted. Only two of the originally selected ten schools ended up in the final sample. Whenever a school declined, another school from the same educational provider would be invited to participate. In two instances, every school in a municipality refused to take part. In those cases a municipality of similar size in the same geographical region was selected. The reasons for the low participation rates of the Finnish schools can only be speculated. Perhaps some of the Finnish educators feel as if they have already contributed enough to research, since many research projects have focused on the Finnish educational system following its success in the PISA studies.

Despite the relatively low rate of cooperation, this sample can be considered to sufficiently represent the Finnish mathematics teachers—at least to a degree afforded by the modest sample size. The teachers were randomly selected by the researcher, who was not familiar with any of them or their professional reputations. Furthermore, in almost all cases, the teacher selected first agreed to participate in the study; the principals were more likely to refuse to co-operate.

Sampling: Iceland

The Icelandic teachers were selected using a two-step process: first the school, then the teacher. A list of Icelandic primary schools together with their attendance figures was available from an online source (Statistics Iceland, 2005). Schools were assigned weights according to the number of students they serve. A random number generator was then used to select the participating schools. Of the original set of ten schools, eight were willing to participate. Two additional schools were chosen with the above method to replace the schools that did not want to take part in the study. As in the Finnish sample, the principals were asked to list his or her school's active mathematics teachers. One teacher was then randomly selected. If the teacher was not willing to cooperate, another one from the same school was chosen as a replacement; however, this happened only with one school.

Recording Procedure

The two-camera recording set-up described below was used to capture all forty lessons in the sample. One videographer can carry out the recordings using this recording design. The video cameras are situated in the back of the classroom as shown in Figure 38.2.

The action camera (A) follows the teacher (T), while the fixed camera (F) captures as much of the classroom as possible. This recording design utilizes three microphones: a wireless lavaliere microphone on the teacher and an external microphone connected to each camera. The two main audio tracks are recorded by the action camera, while the sound captured by the fixed camera serves as backup. See Savola (2008) for more information about recording techniques.

The students' faces are typically not visible in the recordings. In order to protect the identities of the approximately one thousand students who participated in the study, the cameras were placed in the back of the classroom. In addition, the parents of the students were given the option of having their child take part in another educational experience instead of their usual mathematics lesson; however, none of the parents chose to keep their child from attending the lesson.

Typicality of Lessons

Based on the teachers' judgments, the lessons recorded for the purposes of this study were close to typical mathematics lessons. The teachers were asked to follow their normal lesson plans and not to arrange anything special for the day of the recording. The observations were scheduled for a day[4] when there were no tests or only a review for a test.

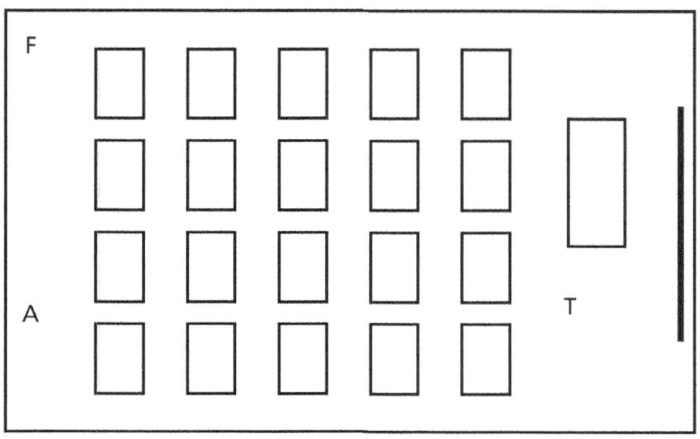

Figure 38.2 The recording set-up.

Whenever it was logistically possible, a "warm-up" lesson was recorded to allow the teachers some time to become accustomed to being filmed. In the case of eight of the twenty schools, three lessons were observed and videoed, but only the last two were coded and used as part of the data set. This was done in order to minimize the effects the video cameras may have on classroom practices. In the remaining twelve schools only two lessons were recorded. Scheduling to observe three mathematics classes in one day was more difficult in Finland since there most mathematics teachers teach other subjects, such as natural sciences or information technology, in addition to mathematics. There seemed to be no major differences between the omitted "warm-up" lessons and the lessons used for the study. Nevertheless, recording an extra lesson is a recommended precautionary measure to lessen the camera effects and thus increase the verisimilitude—or trueness-to-life—of the recordings.

Based on post-observation teacher interviews, the practices of the teachers and the students were in most cases not significantly changed by the presence of the researcher and the cameras. During the interview each teacher commented on the typicality of the lessons in general as well as the typicality of student behavior. All but two of the teachers in the sample said that the lessons were typical for the most part.[5] Teacher perceptions of student behavior during the classroom recordings are summarized in Table 38.2. The Finnish students were slightly more influenced by the presence of the cameras. Students were not interviewed, although that would have given a more complete picture of the typicality of the recorded lessons.

There were above-normal levels of student activity—mainly rowdiness—in two of the Finnish lessons. In both instances the teacher remarked that a small group of students felt the need to "perform" for the cameras. This was not a phenomenon that any of the Icelandic teachers recognized despite the fact that a larger proportion of the Icelandic classrooms appeared to have discipline problems. The reduction

TABLE 38.2 Teacher Perceptions of Student Behavior

	Finland	Iceland	Total
Students behaved normally	14 (70%)	17 (85%)	31 (77.5%)
Students were more active than normal	2 (10%)	0 (0%)	2 (5%)
Students were less active than normal	4 (20%)	3 (15%)	7 (17.5%)
TOTALS	20	20	40

in student activity in the four Finnish lessons was attributed to shyness, being afraid to provide wrong answers while being videotaped, and overall better-than-normal behavior of the students. According to their teachers, students behaved better than normal in three Icelandic lessons. There were no mentions of shyness or being afraid to provide wrong answers within the Icelandic sample. The figures in Table 38.2 are consistent with those found in the TIMSS 1999 Video Study (Hiebert et al., 2003).

RESULTS

Overview

Based on the analysis of the lessons in the sample, there are differences in the ways in which Finnish and Icelandic mathematics teachers conduct their classes. The Finnish mathematics lessons exemplify the *Review–Lesson–Practice* [RLP] script and are fairly uniform in their functional structure. In contrast, Icelandic mathematics teachers seem to be working from two distinct pedagogical philosophies: nine of the lessons in the sample essentially follow the RLP structure, whereas eleven lessons are conducted using versions of the *individualized learning* [IL] instructional strategy, a learner-based pedagogical paradigm.

Many Finnish teachers seem to promote collaborative learning activities. Those teachers often favor class discussions and student presentations over delivering monologues in the front of the classroom. This is especially apparent during lesson segments dedicated to reviewing material from previous lessons. Furthermore, Finnish mathematics teachers spend significantly more time introducing new content than do their Icelandic counterparts.

The following section offers an explanation of the coding categories specific to this sample. Visualizations of the coded lessons are presented in the form of *lesson diagrams*. A summary of the initial results of the coding process is followed by a section on intercoder reliability.

Coding Categories

The first pass-categories are *review, introducing new content, practicing/applying,* and *other*. The categories for the second pass were created specifically for this sample according to the guidelines discussed earlier. This section explains the coding criteria for the second pass in some detail. Tables 38.3 through 38.6 show the various categorizations, and Table 38.7 presents a summary of all the coding categories for this particular sample.

The *review* segments in the lessons in this sample are divided into three categories based on classroom interaction: "Teacher discusses examples or a concept in the front," "Students write solutions on the board," and "Class works together on a problem." The teacher may elicit responses from the students during his or her presentation. For example, the teacher may ask the students to recall a certain formula. In contrast, during an episode when the class is working together, the teacher would be more likely to lead the discussion by asking questions such as "What do we do next?" or "What do you think?" The amount of creative freedom with which the teacher empowers the class can be an important factor while coding the *review* segments. Open-ended questions can reveal higher levels of such empowerment and should trigger the "Class works together" code. However, the occurrence of open-ended questions is not a necessary condition for that code; teachers can accomplish a collaborative learning atmosphere in various ways. "Students write solutions on the board" includes the time during which the teacher assigns problems to be done on the board. Table 38.3 displays the three *review* categories for the lessons in this sample.

The lesson segments during which new material is introduced, that is, the "lessons," are further categorized during the second coding pass. Five categories of classroom participation during lesson segments dealing with new content are distinguished for this sample. The first category, "Teacher presents a lesson, intermittent questions," is used for the segments where the teacher presents new content while the students pay attention and answer the teacher's sporadic questions. Students may also ask questions of the teacher. The teacher may use various types of technology—blackboard, overhead projector, document reader, PowerPoint presentations, etc.—in delivering the lesson. In one Finnish lesson, FIN-7A, the teacher demonstrated a concept using a student volunteer. The "Teacher presents" code was applied for this 45-second segment, although one of the students had a physically active role in the presentation.

TABLE 38.3 ***Review* Categories**

Review	Teacher discusses examples or a concept in the front
	Students write solutions on the board
	Class works together on a problem

The second code, "Teacher elicits responses from the class by asking a series of connected questions," identifies segments where the teacher draws the students into the lesson via the Socratic method or another type of questioning technique involving a series of related questions (Jablonka, 2004). These types of segments are often conversational in nature. One or two *Initiation–Response–Evaluation* sequences are not enough to trigger this code (see, e.g., Hiebert & Grouws, 2007).

"Students work on a new type of problem, teacher helps" code applies for two distinct pedagogical situations: (1) Students are asked to work on a new type of problem—the first example—directly following an instructional segment, and (2) students explore a new concept via the "problem of the day" approach. The latter was observed in only one lesson, FIN-9B. The new material in this lesson was delivered via a problem-solving activity not unlike the *Structured problem-solving* pedagogical strategy used by many Japanese mathematics teachers (Clarke, Mesiti, Jablonka, & Shimizu, 2006).

"Students copy text" code covers two types of activities: (1) There is so much writing on the board that the students need time to catch up with their note taking, and (2) students are asked to copy text from a book. "Students read the book" was used in only one lesson.

Four categories of *practicing/applying* are identified in the lessons in this sample. "Teacher discusses examples in the front" is used for any teacher-centered segments where the new content for the day is practiced. It is used for any teacher-presented examples beyond the first one, which should be coded as *introducing new content* (Table 38.4). If the role of the teacher is closer to a "guide on the side" as opposed to a "sage on the stage," it may be that the class is practicing the new content collaboratively. In this case the "The class, led by the teacher, works together on a problem" code would be appropriate.

TABLE 38.4 ***Introducing New Content* Categories**

Introducing new content	Teacher presents new content, intermittent questions
	Teacher elicits responses from the class by asking a series of connected questions
	Students work on a new type of problem, teacher helps
	Students copy text from the board or the book (no other interaction)
	Students read the book (no other interaction)

TABLE 38.5 ***Practicing/Applying* Categories**

Practicing/ applying	Teacher discusses examples in the front
	Students work on practice problems individually or in small groups, teacher helps
	Students work on practice problems individually or in small groups, teacher does not help
	The class, led by the teacher, works together on a problem

During *practicing/applying* segments where the students work on problems, the teacher may or may not actively offer help to the students. Some teachers simply have the students work while they, for instance, prepare materials for the next class, set up technology, or perform bookkeeping duties. Gauging student progress is considered helping. Table 38.5 summarizes the form categories for *practicing/applying*.

This sample requires an extended range of *other* categories. This is mainly due to the prevalence of the individualized learning [IL] pedagogical strategy employed by several Icelandic teachers. During their lessons there is often no public instructional discourse. Instead, each student works on problems at his or her own pace for the duration of the lesson while the teacher walks around the room and helps. Therefore the pedagogical function of any particular moment is undeterminable; it is impossible to assign a label such as reviewing or working with a new concept to the class as a whole during IL segments.

Three distinct forms within IL are identified: "Kikan-shido" (i.e., between-desks instruction: students work at their desks while the teacher walks around and helps) (O'Keefe, Xu, & Clarke, 2006), "Teacher presents in the front," and "Student presents in the front." The latter two categories are used when the students are individually working on problems, but either the teacher or one of the students presents material on the board.

"Homework/progress check" is used to accommodate for two specific pedagogical devices. Several of the Finnish lessons contained a segment during which the teacher checked each student's notebook to see whether the homework assignments had been completed. The same category was used for the verbal progress check some Icelandic teachers conduct at the beginning of an IL style lesson.

The "Interruption" code can be applied to various situations such as a visit to the classroom by someone

TABLE 38.6 ***Other*** **Categories**

Other	Classroom management
	Mathematics management
	Homework/progress check
	Interruption
	Social talk
	Individualized learning—Kikan-shido
	Individualized learning—Teacher presents in the front
	Individualized learning—Student presents in the front

or an announcement over the intercom. One teacher had to leave the classroom to take a student to another location to take a test. This also was considered an interruption to classroom activities, as were the "morning services" delivered during two of the Finnish lessons; one was broadcast over the intercom, the other via school TV.

The eight *other* categories, together with their visual codes, are summarized in Table 38.6. The checkerboard patterns visually unify the IL segments.

There were three Finnish lessons where one code was not sufficient to describe the key participatory actions. In these cases one or more students were writing solutions to homework problems on the board while the teacher visited each student's desk to check whether they had completed the assignment for the day. Since these are significant events in a lesson, both "Students write solutions on the board" and "Homework/progress check" codes were assigned to these lesson segments.

Lesson Diagrams

This section contains illustrations of the coded lessons. A summary of all twenty codes, a necessary companion to interpreting the lesson diagrams, is presented as Table 38.7. Time is the third dimension in these diagrams. Lessons FIN-1A and FIN-1B were given by the Finnish teacher FIN-1, and so on.

TABLE 38.7 Summary of Codes

Review	Teacher discusses examples or a concept in the front
	Students write solutions on the board
	Class works together on a problem
Introducing new material	Teacher presents new material, intermittent questions
	Teacher elicits responses from the class by asking a series of connected questions
	Students work on a new type of problem, teacher helps
	Students copy text from the board or the book (no other interaction)
	Students read the book (no other interaction)
Practicing/applying	Teacher discusses examples in the front
	Students work on practice problems, teacher helps
	Students work on practice problems, teacher does not help
	Class works together on a problem
Other	Classroom management
	Mathematics management
	Homework/progress check
	Interruption
	Social talk
	Individualized learning—Kikan-shido
	Individualized learning—Teacher presents in the front
	Individualized learning—Student presents in the front

First Results

This section offers a brief summary of the preliminary findings from this video study. The focus is

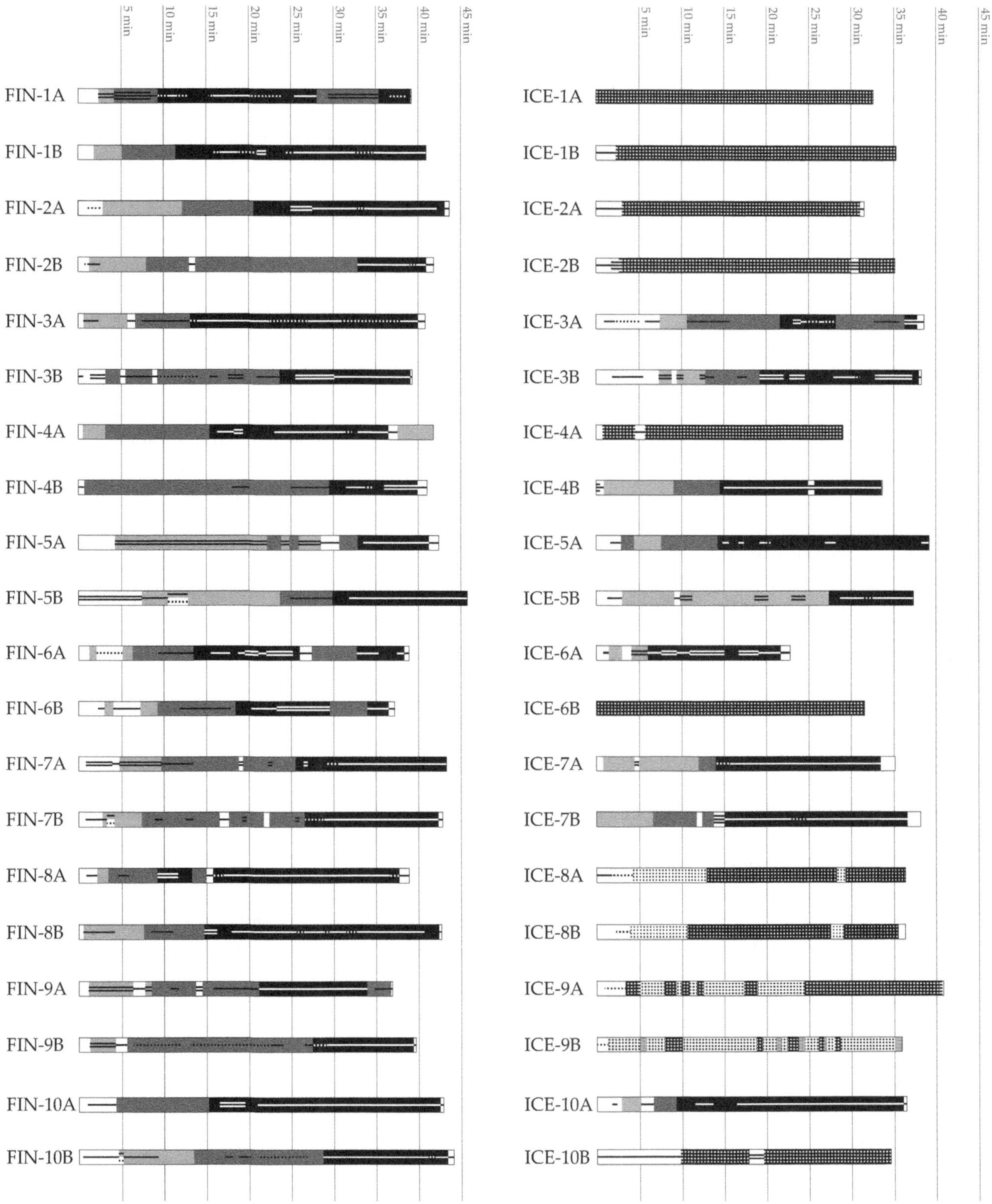
5 min
10 min
15 min
20 min
25 min
30 min
35 min
40 min
45 min
FIN-1A
FIN-1B
FIN-2A
FIN-2B
FIN-3A
FIN-3B
FIN-4A
FIN-4B
FIN-5A
FIN-5B
FIN-6A
FIN-6B
FIN-7A
FIN-7B
FIN-8A
FIN-8B
FIN-9A
FIN-9B
FIN-10A
FIN-10B
ICE-1A
ICE-1B
ICE-2A
ICE-2B
ICE-3A
ICE-3B
ICE-4A
ICE-4B
ICE-5A
ICE-5B
ICE-6A
ICE-6B
ICE-7A
ICE-7B
ICE-8A
ICE-8B
ICE-9A
ICE-9B
ICE-10A
ICE-10B

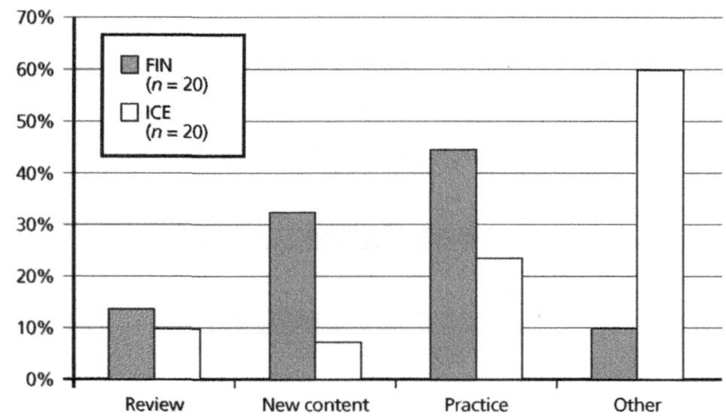

Figure 38.3 First pass categories: Finland and Iceland.

on the relative frequencies of the various codes. The first coding pass sheds light on the lessons with regard to the pedagogical functions. Figure 38.3 shows how class time is distributed over the four first pass categories in the Finnish and Icelandic lessons.

Eleven of the twenty Icelandic lessons exhibit versions of the IL strategy. Since IL segments are coded as *other*, that category is by far the most prevalent for Iceland. Figure 38.4 displays the comparison between the Finnish lessons and the nine non-IL Icelandic lessons, which essentially follow the RLP script. This set of lessons is denoted with an asterisk: Iceland* or ICE*.

Based on this "apples-to-apples" comparison, the Finnish mathematics teachers appear to spend less time reviewing and practicing and more time introducing new material than do the Icelandic teachers not using IL. In all, the Finnish teachers in the sample spent 13.6% of lesson time reviewing, while the percentage for Iceland* is 20.7. New content was being taught 32.2% of the time in the Finnish lessons versus 15.7% in the Iceland* lessons.

The relative frequencies of the first pass categories in the two countries can be compared using the proportion of time spent on each of the categories in each lesson. Applying such analysis, the only statistically significant difference at the .05-level is for the category of *introducing new content*, for which a t-test yields t = 2.78 and p = .010.

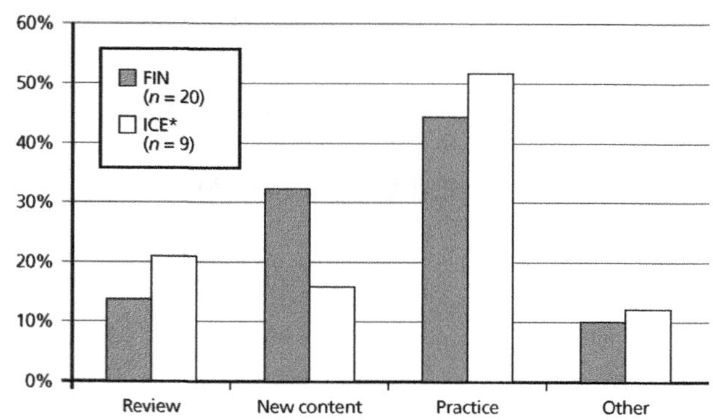

Figure 38.4 First pass categories: Finland and Iceland*.

Figures 38.5 through 38.8 show the prevalence of the various forms of classroom interaction for each of the first pass categories for Finland and Iceland*.

In this sample, 50.1% of the *review* segments within the Finnish lessons can be classified as having the form "Teacher discusses examples or a concept in the front." In contrast, 85.5% of the time spent reviewing during the Iceland* lessons feature that form of interaction. This is statistically significant when considered at the level of the lesson: t = –2.80 and p = .010. Notably, no student presentations took place during the Iceland* lessons.

Figure 38.6 shows the relative frequencies of the various forms of *introducing new content*. Overall, when introducing new material, Finnish teachers spent 62.7% of the time presenting in the front without involving the students for more than an occasional question. For Iceland*, the percentage is 79.6. This difference is nearly significant at the .05-level: t = –1.96 and p = .061.

Approximately 10% of the time when the Finnish teachers were introducing new material, they asked their students to work on the new type of problem rather than first demonstrating how they should do it.

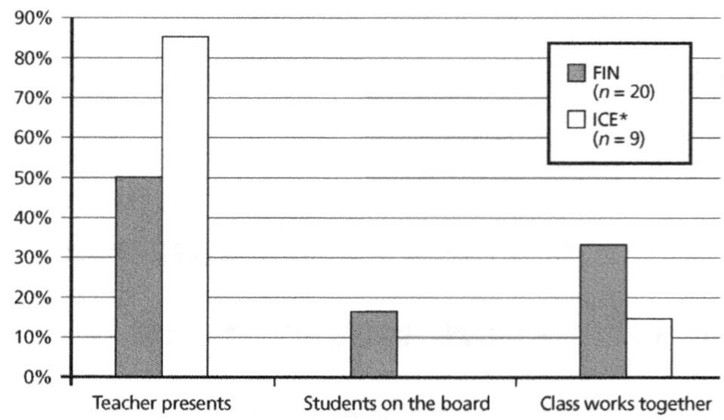

Figure 38.5 Forms of *review*: Finland and Iceland*.

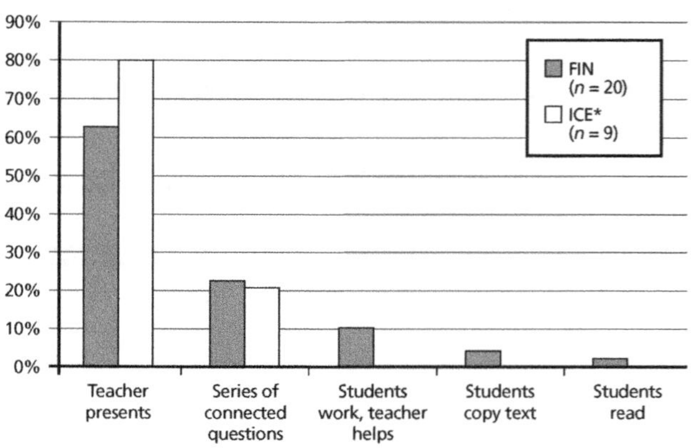

Figure 38.6 Forms of *introducing new content*: Finland and Iceland*.

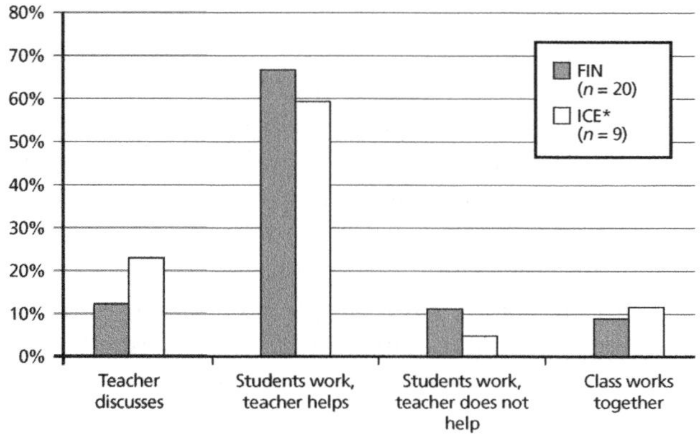

Figure 38.7 Forms of *practicing/applying*: Finland and Iceland*,

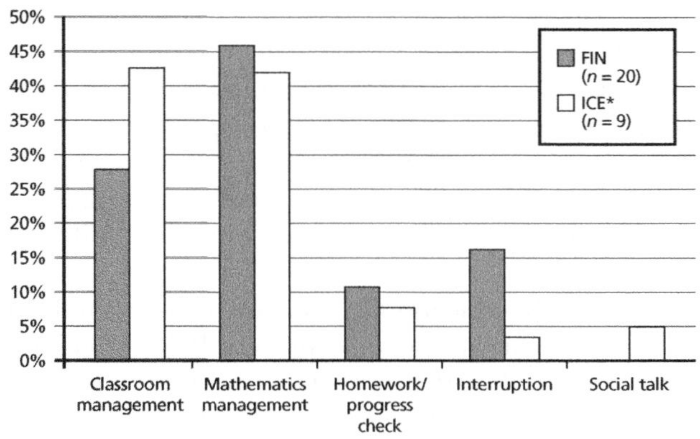

Figure 38.8 Forms of *other*: Finland and Iceland*.

However, in all but one class, FIN-9B, the teacher did show the students how to get started.

In *review* and *introducing new content* segments combined, the Finnish teachers presented 59% of the time, whereas the Iceland* teachers spent 83% of that time presenting in the front. This is significant at the lesson level with t = –2.146 and $p = .044$.

Figure 38.7 shows the relative frequencies of the various forms of *practicing/applying*. The most common form is Kikan-shido, or between-desks instruction. The differences in the proportions are not statistically significant, although Finnish teachers seem to, again, spend less time presenting on the board than the Iceland* teachers.

Though not statistically significant at the lesson level, there may also be differences in the relative frequencies of the teachers' administrative actions (Figure 38.8). Some of the differences can perhaps be explained by cultural factors. For example, there were less discipline problems in the Finnish classrooms, and the classes settled down faster there. This may be one reason why the time spent on managing the classroom is lower in Finland than in Iceland. The two "morning services" that were coded as "Interruption"

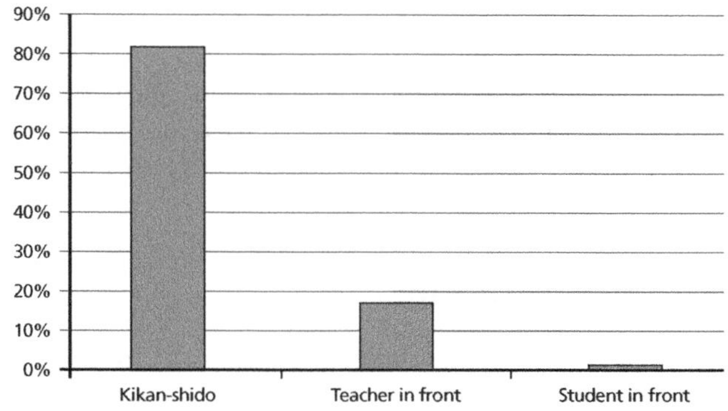

Figure 38.9 Prevalence of the interaction forms during *individualized learning*.

explain much of the difference for that code. There were no episodes of social talk during the recorded Finnish mathematics lessons.

Eleven of the twenty Icelandic lessons were conducted using the IL pedagogical strategy. Figure 38.9 shows the relative frequencies of the forms of classroom interaction the teachers of these lessons chose to employ.

Kikan-shido is clearly the most common form of interaction during IL, as 82.1% of the time the IL teachers essentially tutored the students one-on-one. Seven of the eleven IL lessons contained no public content-related discourse. The lessons from two IL teachers, ICE-8 and ICE-9, stand out. During the IL segments of these lessons, either the teacher or one of the students presented examples 45.2% of the time. Both teachers directed their public instruction to the below-average students. These teachers' methods will be discussed in more detail below.

Figure 38.10 displays the proportions of class time in the Finnish and Icelandic lessons during which a student was presenting in the front. This seems to be a much more common activity in the Finnish classrooms. In fact, only one Icelandic lesson, ICE-9B, featured student presentations. On the other hand, students presented solutions on the board in six of the

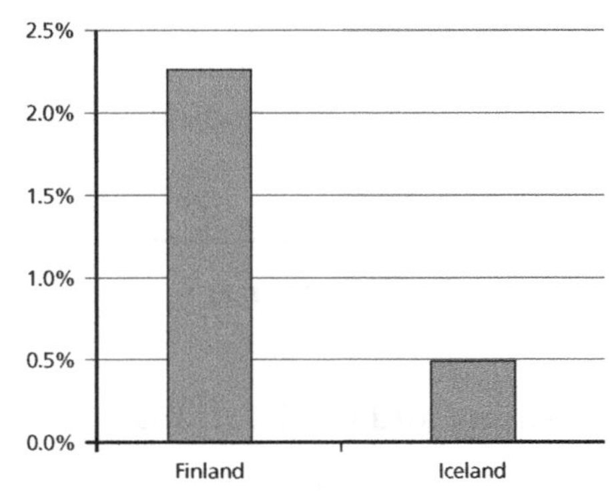

Figure 38.10 Prevalence of student presentations.

Finnish lessons. In two others, the teacher received no volunteers for a call to the board as some of the Finnish students may have felt shy being videotaped. The overall percentage of time devoted to student presentations for the lessons in the sample is 2.24 for Finland and 0.48 for Iceland.

Inter-Coder Reliability

As with any coding system, inter-coder reliability needs to be established for this method of lesson structure analysis. It can be done by having a second person code a portion of the lessons and comparing the results. A mathematics educator studied the coding guidelines created by the author and independently coded two Finnish and two Icelandic lessons in the sample. A total of 9,036 seconds, or just over 2.5 hours, of classroom footage was coded by the second coder.

The two coders agreed on both coding passes at above satisfactory levels. As expected, there were more disagreements in the second coding pass. Table 38.8 displays the Cohen's Kappa index between the coders for each pass.

TABLE 38.8 Cohen's Kappa

	Cohen's Kappa
First Pass	0.968
Second Pass	0.881

Cohen's Kappa is regarded as a more accurate measure of inter-coder reliability than simple percentage agreement, as it takes into account the possibility that the coders agreed by chance. According to criteria of the TIMSS 1999 Video Study, a Cohen's Kappa of 0.7 or higher indicates an acceptable level of agreement (Jacobs et al., 2003).

Though careful translations of the lesson transcripts were made, some of the disagreements may have been due to the fact that the second coder understands neither Finnish nor Icelandic, while the first coder is a native Finnish speaker and also has some understanding of spoken Icelandic. However, most of the disagreements probably stem from different interpretations of lesson activities.

DISCUSSION

Interpretations and Implications

Finnish and Icelandic mathematics teachers seem to be on separate tracks. Based on the video evidence, Finnish teachers are rather traditional and pedagogically conservative in the classroom. This is in contrast to the Icelandic teachers, many of whom use progressive-minded, learner-based instructional strategies. The classroom practices in Finland include a substantial social component, while many students in Iceland are getting used to learning independently, without significant collaboration with others.

Despite being given ample freedom within the national curricular guidelines (Björkqvist, 2006), Finnish mathematics teachers appear to have found a common tune in their classrooms. By and large the Finnish lessons in the sample follow the *Review–Lesson–Practice* [RLP] script. Activities during *review* and *introducing new content* segments are led by the teacher, who regularly engages the class by asking questions, provoking discussions, or by having the students present solutions on the board. The most common form of interaction during the *practicing/applying* segments is Kikan-shido, or between-desks instruction, during which the teacher gauges the students' learning progress and those in need can ask for a hint or additional instruction.

A British educational research team visited 50 Finnish primary schools in the mid-1990s. They reported of highly disciplined and ordered lessons, where catering to the students' individual differences was not a priority:

> Whole classes following line by line what is written in the textbook, at a pace determined by the teacher. Rows and rows of children all doing the same thing in the same way, whether it be art, mathematics or geography. We have moved from school to school and seen almost identical lessons, you could have swapped the teachers over and the children would never have noticed the difference.... [I]n both the lower and upper comprehensive school, we did not see much evidence of, for example, student-centered learning or independent learning. (Norris et al., 1996, as quoted in Simola, 2005, p. 462)

To some extent, data from the current study support these aspects of the British report. Naturally there are differences between teachers and their methods, but it does seem that the Finnish teachers conduct their classes in fairly uniform ways. They are comfortable with their teaching methods, which evidently are effective for the students in Finland under the unique conditions that exist in that country.

The Icelandic educational system is in transition. In recent years the more traditional methods of teaching have begun to give way to versions of "*einstaklingsmiðað nám,*" or *individualized learning* [IL]. IL has become—from the top down—the nationally endorsed pedagogical philosophy (Sigurgeirsson, 2003). While there are many facets to IL, probably the most important is the increased responsibility on the part of the learner; students are expected to assume control of their own learning process. Based on the recorded lessons, Icelandic mathematics teachers are having varying success in implementing IL in their classrooms. Furthermore, there appears to be confusion as to what IL teaching should entail.

The Icelandic educational authorities have promoted Tomlinson's ideas about *differentiated instruction* (Sigurgeirsson, 2003). Tomlinson (e.g., 1999) calls attention to the students' dissimilar learning needs and suggests that teachers use, for instance, differentiated curricular materials and tiered activities to accommodate them. In a *differentiated classroom*, the content, the learning process, as well as the product—what the students know and can do at the end of the learning process—vary for each student. In such a classroom, the teacher responds to the learning style, instructional needs, interests, and readiness of each unique learner (Tomlinson, 1999). Based on the evidence from the video study, Tomlinson's ideas have been only partially implemented in the Icelandic mathematics classrooms. The reality is that the students are usually taught using the same instructional methods and learning materials; they are only moving ahead in the book at somewhat different speeds.[6]

Teachers and administrators should keep in mind that IL teaching is challenging. It is not simply omitting the public lesson and tutoring the students instead. In fact, using IL in the classroom probably requires more planning and effort from the teacher than does teaching using more conventional methods. The IL teacher should find ways to vary the curriculum, curricular materials, motivators, scaffolding, expectations, and ways of assessment to suit each student's needs. The teacher has to actively operate within the *zone of proximal development* (Vygotsky, 1978) of each student. Furthermore, the shift in the responsibility for learning—from the teacher to the student—must be structured and monitored by the teacher. If there are fifteen or more students in a class,[7] the total amount of individual guidance needed can become overwhelming for a teacher, unless his or her lessons are well planned. As a result, in some classes many or all students can be left without enough guidance.

Minimally guided instruction has not been shown to be effective (Kirschner, Sweller, & Clark, 2006; Mayer, 2004). The *cognitive load*-based argument against minimally guided instructional strategies, such as *pure discovery*, essentially states that if novice learners are left to learn how to work problems by themselves from start to finish, much of their cognitive capacity and working memory is spent grasping trivialities before more meaningful learning can begin (see, e.g., the special cognitive load theory issue of *Instructional Science*, 2004, *32*(1–2)). It is thus more effective to guide the learning process by, for instance, scaffolding, or by letting students study worked-out examples before tackling more complex tasks by themselves.

Based on the recorded lessons, the learner-based teaching strategy so prevalent in Iceland may be more aptly termed *independent learning* rather than *individualized learning*. This is because in the IL lessons many students received only minimal attention from their teacher; thus they essentially studied mathematics on their own. Because teaching people to be autonomous students of mathematics is particularly challenging, some experts recommend that classroom instruction of this subject be more thoroughly guided by the teacher than other subjects (Harvey & Chickie-Wolfe, 2007; Hmelo-Silver, Duncan, & Chinn, 2007). Harvey and Chickie-Wolfe (2007) suggest the following steps for students learning mathematics in an IL classroom: (1) observe and recognize success in another person's performance, (2) emulate and adopt patterns and processes, (3) self-monitor while practicing the strategies, and (4) *self-regulate*—a cyclical problem-solving process including *preparation*, *performance*, and *appraisal* (Pintrich, Boekaerts, & Zeidner, 2000)—when adapting to various applications. Direct instruction on a just-in-time basis can also be used as a pedagogical tool in learner-based learning environments (Edelson, 2001).

Social interaction and the use of language in the classroom facilitate the students' constitution of mathematical meanings (e.g., Cobb & Bauersfeld, 1995; Steffe & Gale, 1995). The Finnish lessons in the sample include many segments during which the teacher stimulates whole-class discussions about the topic at hand. Also, Finnish students regularly are asked to present their solutions in front of the class. These lesson elements are not as common in the Icelandic RLP lessons, and they are almost entirely missing from the IL lessons. Overall, Finnish mathematics teachers in the sample provided more opportunities for learning through classroom interaction and the use of language than their Icelandic counterparts.

Some learner-based instructional methods, such as *problem-based learning* and *inquiry learning*, promote social interaction and the use of language through collaborations (e.g., Hmelo-Silver et al., 2007). In the Icelandic mathematics classrooms it is common for the students to sit in clusters. This, however, does not ensure any of the benefits that collaboration can yield. As Slavin (1983) points out in his meta-study, all forms of collaborative learning are not equally effective; group study without group rewards and a strong sense of individual accountability is not linked with increased student achievement. Additionally, students should be guided through the collaborative process to keep the discussion focused on the targeted skills (Palinscar & Brown, 1984).

Although successfully implementing Tomlinson's pedagogical strategies in a classroom of twenty or so students may be practically impossible, they may be implementable in classes specially designed for this purpose. If the teachers are adequately trained in IL teaching and they have a better support system—with clear recommendations for instructional design—and if the student-to-teacher ratio is around 10:1, it is conceivable that a differentiated learning environment could be effective. However, this would be very expensive. Given the current economic climate in Iceland, it is unlikely that there will be enough funds to properly carry out this IL experiment.

So should Iceland and other countries aspiring to do better in international assessments[8] start, or return to, teaching like the Finnish mathematics teachers? While some of the classroom practices from the Finnish schools may prove effective elsewhere, we must keep in mind that there is no "one-size-fits-all" educational system; what works in Finland will not necessarily work anywhere else. Cultures—educational and other—are "situated contextual organisms" (Goldman, 2007, p. 33) that have the ability to adapt and morph only within certain limits. Finland is, after all, a bit different from its Nordic neighbors and other nations. It is a border country that has gone through three bloody wars in the last one hundred years. It fought against Russia, whose influences still permeate the Finnish culture at all levels. Perhaps it is because of this unique socio-historical background that the Finns are rather obedient and allow a sense of authoritarianism, all the while maintaining a democratic, Western society (Simola, 2005). This bodes well for the Finnish schools where traditional instructional methods are still prevalent.

Limitations of the Study

Due to its limited scope and methodology, any findings from this study should above all be considered grounds for more comprehensive research endeavors to increase our understanding of the Finnish and Icelandic educational systems and cultures. In addition to general limitations to lesson structure analysis such as those mentioned in introduction to the section on methods, this study has more specific limitations. Some of them are mentioned here.

Only one lesson per class was included in the study. This approach cannot give a comprehensive picture of typical instructional practice in any one classroom. After all, lessons are parts of a sequence in which each session serves a certain purpose. The LPS study examined such sequences of 10 or more lessons (Clarke, Emanuelsson et al., 2006). By doing so, the LPS team was able to capture a more complete representation of practices in the classrooms they studied. However, it is hoped that the set of 20 lessons recorded in each country can satisfactorily be used to represent classroom practices in Finland and Iceland. Naturally, a larger sample would have been preferred.

Any analysis of classroom video is subject to "tunnel vision" (Erickson, 2006) due to the orientation of the researcher or researchers. Since this study was done by only one person, it lacks the benefits of the multiple perspectives present in many larger studies. Furthermore, a richer analysis of the classroom practices could have been done using a combination of quantitative and qualitative approaches.

The author is a native speaker of Finnish and only has some understanding of Icelandic. This may have played a role in the video analysis. However, the rather high inter-coder reliability figures indicate that the language issue was probably not significant. With the aid of adequate translations, this type of lesson analysis can feasibly be conducted by a researcher who does not speak the language of the classrooms. What is more, this study could probably be replicated in other countries without major methodological adjustments.

SUGGESTIONS FOR FURTHER RESEARCH

This study has raised more questions than it has answered. For instance, the Icelandic gender issue, the role of homework in the Icelandic IL classrooms, and various extensions of the lesson structure analysis can yield many ideas for future studies. This final section puts forth some ideas for further research.

The Icelandic gender enigma keeps puzzling researchers (Ólafsson et al., 2006; Steinthorsdottir & Sriraman, 2007). An interesting phenomenon surfaces when the boys' and girls' PISA scores from the three main areas are summed into a *total score*, and the differences between assessments are determined.[9] Table 38.9 shows the differences in total scores for Iceland between assessments.

While the boys lost ground between each pair of assessments, the girls did so only between 2003 and 2006. Why did these major declines happen first for

TABLE 38.9 Differences in the Icelandic Total Scores Between Assessments

	PISA 2000–PISA 2003	PISA 2003–PISA 2006
Girls	0	–34
Boys	–34	–11

the boys and only then for the girls? The Icelandic students' homework habits may be a factor.

PISA data obtained from Námsmatsstofnun (the Icelandic Educational Testing Institute) suggest that Icelandic students are now spending less time doing mathematics homework than before. The changes between the 2003 levels and the 2006 levels are considerable and more pronounced for the girls. Figure 38.11 shows the changes in the mathematics homework levels from 2003 to 2006. In 2003, 40% of the girls and 50% of the boys did less than two hours of homework each week; in 2006, 64% of girls and 70% of boys reported doing so. In 2003, 34% more boys than girls reported to doing no mathematics homework at all; the difference had decreased to 24% by 2006. The percentage of students who spend six or more hours doing mathematics homework dropped from 16.2% in 2003 to a mere 1.6% in 2006. What has prompted the Icelandic students, and especially the girls, to spend so much less time than before doing homework? What is the role of homework in the Icelandic curriculum, especially for the IL teachers? How do the teachers monitor their students' homework completion?

The video study offers some empirical evidence. In only two of the Icelandic lessons in the sample is homework mentioned at the beginning of the lesson, and in only one of them, an RLP lesson, are homework problems discussed publicly; in the other lesson, the teacher collected a written assignment that was due that day. In comparison, homework was mentioned in fifteen of the twenty Finnish lessons, and problems were discussed in twelve of them. In addition, it is common for the Finnish teacher to verify the level of homework completion directly from the students' notebooks. It seems that the Finnish students are asked to do short yet frequent homework assignments. They also are directly held accountable for their homework responsibilities. Perhaps the Icelandic students are also held accountable, but this is not apparent from the recorded lessons. If the students were held more directly accountable for completing their homework assignments, would the out-of-school effort levels increase?[10]

Since the Icelandic mathematics teachers set aside so much class time for individual seatwork, a look at the teacher–student interactions that take place during these sessions may be in order. The teacher–student interactions can be analyzed using various quantitative and qualitative approaches. A preliminary analysis of fourteen[11] Icelandic mathematics lessons with 622 teacher–student interactions shows that teachers do not treat girls and boys equally during problem-solving sessions. In these lessons almost all teachers, especially the males, paid more individual attention to the girls. On average, these Icelandic teachers assisted each girl for 88 seconds and each boy for 64 seconds per class. These means differ significantly with $t = 3.78$ and $p = 0.002$. Possible explanations for this phenomenon include: (1) By now some Icelandic teachers have higher learning expectations of girls which been shown to increase attention from teachers (Meece, Glienke, & Burg, 2006), and (2) teachers help those who ask for help, not necessarily those who need it

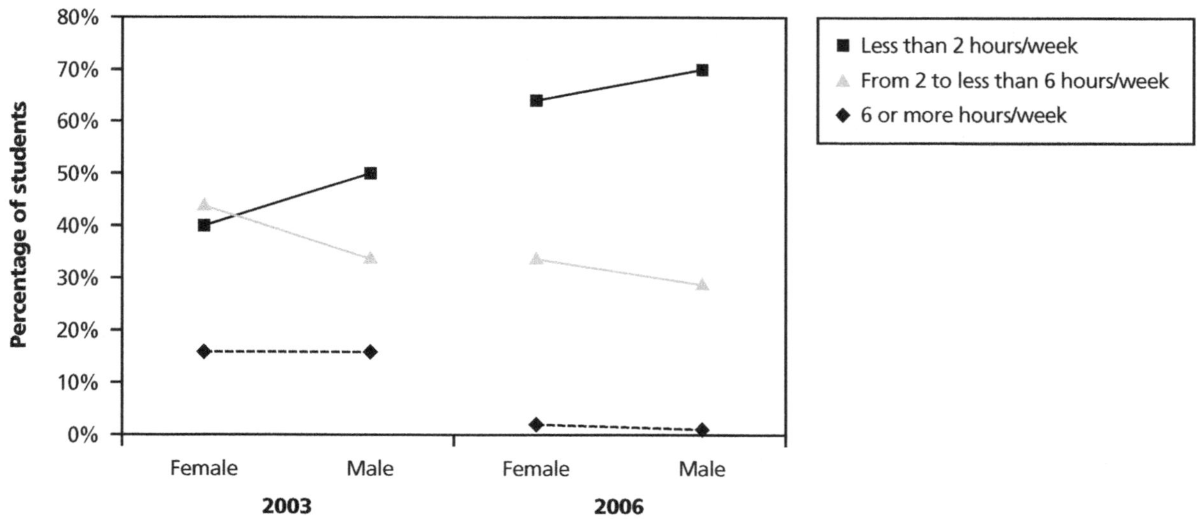

Figure 38.11 Time spent on mathematics homework each week.

the most. Girls tend to ask for help more and they also seem to be more appreciative of it. A qualitative discourse analysis of these interactions could help answer questions such as: Do the teachers pay the same kind of attention to male and female students? Is the individual assistance perhaps more procedure-related to one gender and more conceptual for the other? What kind of scaffolding is evident in the teacher–student interactions? Do these interactions veer off-topic more frequently for one gender? What proportion of students receives no individual attention at all during seatwork, and why? Some Finnish teachers help their students rather systematically by interacting with each student during seatwork, and not just those who asked for help. But are the teacher–student interactions similar in Finland as in Iceland? It would also be helpful to know what some of the Icelandic teachers' and students' attitudes and beliefs about gender roles in the mathematics classroom are.

The instructional methods used by the teacher in ICE-9A and ICE-9B are effective. This IL teacher, ICE-9, can be associated with outstanding PISA 2003 mathematics scores.[12] What makes this teacher's scores even more noteworthy is that the school is located in an isolated fishing village, and rural schools generally fare worse in achievement tests than the urban schools (Ólafsson et al., 2006). In addition, the standard deviation of the PISA scores is by far the smallest within all Icelandic schools. These were true "No child left behind" lessons. Although they were IL lessons, ICE-9A and ICE-9B included public content-related discourse in the form of teacher and student presentations. While letting the faster students work independently, the teacher controlled and scaffolded the learning process for the slower students. Worked-out examples were presented on the board by the teacher as well as the students. All students were able to study the examples; in fact, it was required of the slower students. Others could use them as a "safety net." There were tiered activities (Tomlinson, 1999). Students sat by themselves, but tutoring relationships or "study buddies," a pedagogical device recommended by IL research (Harvey & Chickie-Wolfe, 2007; Rafoth, 1999), were encouraged. ICE-9 was successful in creating "a community of learning." Further research is necessary to identify factors that contribute to this teacher's apparent success. Are the instructional methods used by this teacher transferable to other classrooms in Iceland and elsewhere?

Based on this study, two main instructional philosophies prevail in Iceland: *individualized learning* [IL] and *direct instruction* based on Herbart's formal stages of learning. It is natural to ask what effects the choice of instructional method has on the quality and quantity of learning. Do students who have learned to study independently have different study habits? A longitudinal study can be used to determine whether the IL learners preserve and benefit from their study habits later in life. Also, do the IL learners retain information differently than those who have been taught by direct instruction?

Data from lesson structure analysis can be linked with factors such as curricular materials, content, the level of the teachers' pedagogical and content knowledge, the physical set-up of the classroom, technology, the teachers' intentions about planning and structuring lessons, and student feedback. Combining variables such as the above with lesson structures can yield interesting information about classroom practices.

This study focused on the relative frequencies of lesson segments. An additional approach to analyzing lessons would be to consider the sequencing of lesson segments. The learning experiences supported by lessons with similar relative frequencies for lesson segments can be quite different depending on how those segments are sequenced.

Countries, states, linguistic regions, and so on can be compared to one another using the method of lesson structure analysis described in this chapter. In addition, this method can be used by institutions of teacher education to monitor the effectiveness of their training programs. For example, it may be instructive to compare the IL methods taught at the Icelandic institutions of teacher education with the ones observable in the field.

NOTES

1. The science scores are not fully comparable across assessments because the science framework was only developed for PISA 2006. Since the mathematics framework was developed for PISA 2003, the first round of mathematics scores is not fully comparable with the latter two. The reading scores are comparable across all three assessments.
2. Videograph (Rimmele, 2002), the coding software that was used in this study, is capable of exporting data into SPSS. However, it does not produce useful visual representations. Those can be made with, for instance, Adobe Illustrator.
3. For a discussion about the problems and shortcomings of video analysis in classroom research, see Savola (2008).
4. The lessons were typically recorded during one day. However, there were four instances when this was not possible, and the recordings were conducted over two days.
5. One Finnish teacher remarked that the students were so afraid to raise their hands and participate that this may have changed the way the lesson was run. The other not-so-typical lesson was attended by only 5 of the 12 students due to a flu epidemic.

6. Based on teacher interviews, the individual differences in progress usually are not significant; however, there are exceptions.
7. The mean class size for the Icelandic mathematics classrooms in this sample is 16.6. However, this figure does not include the students who were absent the day of the recording. Probably closer to the truth is 19.7, the OECD's estimate for the mean class size in the Icelandic lower secondary schools (OECD, 2007a, Table D2.1).
8. No single assessment can fully measure the effectiveness of teaching. For example, there are skills, such as the ability to work in groups, that are not measured by the major assessments. However, due to space concerns, this chapter does not focus on the pros and cons of PISA or any other assessment.
9. See Footnote 1 for an issue regarding between-assessment comparisons.
10. Time spent on homework may not be a good predictor of academic success. This is true at the level of the individual—high-achieving students may feel like they don't have to do homework while weaker students do a lot of work at home to keep up with the class—as well as cross-nationally. For example, students in Finland did the least amount of homework according to PISA 2003 data, yet they scored very high in all areas of the test (OECD, 2005, Table D1.3).
11. From the twenty recorded Icelandic lessons, only those that were attended by 10 or more students and that included at least 10 minutes of seatwork were included.
12. PISA 2006 scores are not available for this school.

REFERENCES

Angelillo, C., Rogoff, B., & Chavajay, P. (2007). Examining shared endeavors by abstracting video coding schemes with fidelity to cases. In R. Goldman, R. D. Pea, B. Barron & S. J. Derry (Eds.), *Video research in the learning sciences* (pp. 189–206). Mahwah, NJ: Lawrence Erlbaum Associates.

Barker, R. (1963). *The stream of behavior: Explorations of its structure & content.* New York: Appleton-Century-Crofts.

Bauersfeld, H. (1980). Hidden dimensions in the so-called reality of a mathematics classroom. *Educational Studies in Mathematics, 11*(1), 23–41.

Björkqvist, O. (2006). Mathematics education in Finland: What makes it work? In A. Rogerson (Ed.), *Proceedings of the International Conference of the Mathematics Education into the 21st Century Project on "Reform, Revolution and Paradigm Shifts in Mathematics Education"* (pp. 45–48). Johor Bahru, Malaysia: Universiti Teknologi Malaysia.

Clarke, D., Emanuelsson, J., Jablonka, E., & Mok, I. A. C. (Eds.). (2006). *Making connections: Comparing mathematics classrooms around the world.* Rotterdam, The Netherlands: Sense Publishers.

Clarke, D., Mesiti, C., Jablonka, E., & Shimizu, Y. (2006). Addressing the challenge of legitimate international comparisons: Lesson structures in the USA, Germany and Japan. In D. Clarke, J. Emanuelsson, E. Jablonka & I. A. C. Mok (Eds.), *Making connections: Comparing mathematics classrooms around the world* (pp. 23–45). Rotterdam, The Netherlands: Sense Publishers.

Cobb, P., & Bauersfeld, H. (Eds.). (1995). *The emergence of mathematical meaning: Interaction in classroom cultures.* Hillsdale, NJ: Lawrence Erlbaum Associates.

Dunkel, H. B. (1969). *Herbart and education.* New York: Random House.

Edelson, D. C. (2001). Learning-for-use: A framework for integrating content and process learning in the design of inquiry activities. *Journal of Research in Science Teaching, 38,* 355–385.

Erickson, F. (2006). Definition and analysis of data from videotape: Some research procedures and their rationales. In J. Green, G. Camilli, & P. Ellmore (Eds.), *Handbook of complementary methods in education research* (pp. 177–191). Mahwah, NJ: Lawrence Erlbaum Associates.

Finnish Department of Education. (2005). Peruskoulun oppilasmäärä vuosiluokilla 1–9 sekä esiopetuksessa ja lisäopetuksessa syksyllä (aikasarja). Retrieved August 20, 2006 from https://www.data.oph.fi/wera/wera

Goldman, R. (2007). Video representations and the perspectivity framework: Epistemology, ethnography, evaluation, and ethics. In R. Goldman, R. D. Pea, B. Barron, & S. J. Derry (Eds.), *Video research in the learning sciences* (pp. 3–37). Mahwah, NJ: Lawrence Erlbaum Associates.

Goldman, R., Pea, R. D., Barron, B., & Derry, S. J. (Eds.). (2007). *Video research in the learning sciences.* Mahwah, NJ: Lawrence Erlbaum Associates.

Halldórsson, A. M., Ólafsson, R. F., & Björnsson, J. K. (2007). *Færni og þekking nemenda við lok grunnskóla: Helstu niðurstöður PISA 2006 í náttúrufræði, stærðfræði og lesskilningi.* Reykjavik: Námsmatsstofnun.

Harvey, V. S., & Chickie-Wolfe, L. (2007). *Fostering independent learning: Practical strategies to promote student success.* New York: Guilford.

Hiebert, J., Gallimore, R., Garnier, H., Givvin, K. B., Hollingsworth, H., Jacobs, J., et al. (2003). *Teaching mathematics in seven countries: Results from the TIMSS 1999 video study.* U.S. Department of Education. Washington, DC: National Center for Education Statistics.

Hiebert, J., & Grouws, D. A. (2007). The effects of classroom mathematics teaching on students' learning. In F. Lester (Ed.), *Second handbook of research on mathematics teaching and learning* (pp. 371–404). NCTM. Charlotte, NC: Information Age Publishing.

Hmelo-Silver, C., Duncan, R. G., & Chinn, C. (2007). Scaffolding and achievement in problem-based and inquiry learning: A response to Kirschner, Sweller, and Clark (2006). *Educational Psychologist, 42*(2), 99–107.

Jablonka, E. (2004). *Structure in diversity: Initiation into mathematical practice in classrooms from Germany, Hong Kong and the United States.* (Habilitationsschrift). Berlin: Freie Universität Berlin.

Jacobs, J., Garnier, H., Gallimore, R., Hollingsworth, H., Givvin, K. B., Rust, K., et al. (2003). *TIMSS 1999 Video Study technical report. Volume 1: Mathematics.* U.S. Department of Education. Washington, DC: National Center for Education Statistics.

Kirschner, P., Sweller, J., & Clark, R. (2006). Why minimal guidance during instruction does not work: An analysis of the failure of constructivist, discovery, problem-based, experiential, and inquiry-based teaching. *Educational Psychologist, 41*(2), 75–86.

Lave, J., & Wenger, E. (1991). *Situated learning: Legitimate peripheral participation.* Cambridge, UK: Cambridge University Press.

LessonLab. (2003). TIMSS-R video math coding manual [Electronic Version]. Retrieved July 20, 2006 from http://www.lessonlab.com/TIMMS/codedev.htm

Malaty, G. (2006). Mitkä ovat Suomen PISA-menestyksen taustalla olevat syyt? (In Finnish) [Electronic Version]. *Solmu,* Special issue 2/2005–2006 from http://solmu.math.helsinki.fi/2006/erik2/malaty.pdf.

Mayer, R. (2004). Should there be a three-strikes rule against pure discovery learning? The case for guided methods instruction. *American Psychologist, 59*(1), 14–19.

Meece, J. L., Glienke, B. B., & Burg, S. (2006). Gender and motivation. *Journal of School Psychology, 44,* 351–373.

O'Keefe, C., Xu, L. H., & Clarke, D. (2006). Kikan-shido: Between desks instruction. In D. Clarke, J. Emanuelsson, E. Jablonka, & I. A. C. Mok (Eds.), *Making connections: Comparing mathematics classrooms around the world* (pp. 73–105). Rotterdam, The Netherlands: Sense Publishers.

OECD. (2002). *Education at a glance: OECD indicators 2002.* Paris: Author.

OECD. (2004). *Learning for tomorrow's world: First results from PISA 2003.* Paris: Author.

OECD. (2005). *Education at a glance: OECD indicators 2005.* Paris: Author.

OECD. (2007a). *Education at a glance: OECD indicators 2007.* Paris: Author.

OECD. (2007b). *PISA 2006 science competencies for tomorrow's world.* Paris: Author.

Ólafsson, R. F., Halldórsson, A. M., & Björnsson, J. K. (2006). Gender and the urban-rural differences in mathematics and reading: An overview of PISA 2003 results in Iceland. In J. Mejding & A. Roe (Eds.), *Nordic lights on PISA 2003: A reflection from the Nordic countries* (pp. 185–198). Copenhagen: Nordic Council of Ministers.

Olsen, R. (2006). A Nordic profile of mathematics achievement: Myth or reality? In J. Mejding & A. Roe (Eds.), *Nordic lights on PISA 2003: A reflection from the Nordic countries* (pp. 33–45). Copenhagen: Nordic Council of Ministers.

Oser, F. K., & Baeriswyl, F. J. (2001). Choreographies of teaching: Bridging instruction to learning. In V. Richardson (Ed.), *Handbook of research on teaching* (4th ed., pp. 103–1065). Washington, DC: American Educational Research Association.

Palinscar, A. S., & Brown, A. L. (1984). Reciprocal teaching of comprehension-fostering and monitoring activities. *Cognition and Instruction, 1,* 117–175.

Pintrich, P., Boekaerts, M., & Zeidner, M. (Eds.). (2000). *Self-regulation: Theory, research, and applications.* Mahwah, NJ: Lawrence Erlbaum Associates.

Rafoth, M. A. (1999). *Inspiring independent learning: Successful classroom strategies.* Washington, DC: National Education Association.

Rimmele, R. (2002). *Videograph. Multimedia-Player zur Kodierung von Videos.* Kiel, Germany: IPN.

Savola, L. (2008). *Video-based analysis of mathematics classroom practice: Examples from Finland and Iceland.* Doctoral dissertation. Columbia University. ProQuest/UMI (AAT 3317607).

Sigurgeirsson, I. (2003). Skref í átt til einstaklingsmiðaðs náms. Retrieved June 12, 2008, from http://starfsfolk.khi.is/ingvar/namskeid/Fraedslumidstod/vefur/

Simola, H. (2005). The Finnish miracle of PISA: Historical and sociological remarks on teaching and teacher education. *Comparative Education, 41*(4), 455–470.

Slavin, R. (1983). When does cooperative learning increase student achievement? *Psychological Bulletin, 94,* 429–445.

Statistics Iceland. (2005). Pupils in compulsory schools by grade and school 2001–2005. Retrieved August 10, 2006, from http://www.statice.is

Steffe, L. P., & Gale, J. (Eds.). (1995). *Constructivism in education.* Hillsdale, NJ: Lawrence Erlbaum Associates.

Steinthorsdottir, O. B., & Sriraman, B. (2007). Iceland and rural/urban girls: PISA 2003 examined from an emancipatory viewpoint. In B. Sriraman (Ed.), *International perspectives on social justice in mathematics education* (pp. 169–178). The Montana Mathematics Enthusiast: Monograph Series in Mathematics Education. Monograph 1. Charlotte, NC: Information Age Publishing.

Stigler, J., Gonzales, P., Kawanaka, T., Knoll, S., & Serrano, A. (1999). *The TIMSS 1995 videotape classroom study: Methods and findings from an exploratory research project on eighth-grade mathematics instruction in Germany, Japan, and the United States.* US Dept of Education. Washington, DC: National Center for Education Statistics.

Tomlinson, C. A. (1999). *The differentiated classroom: Responding to the needs of all learners.* Alexandria, VA: ASCD.

United Nations. (2008). *Human development report 2008.* New York: UNDP Report.

Välijärvi, J., Kupari, P., Linnakylä, P., Reinikainen, P., Sulkunen, S., Törnroos, J., et al. (2007). *The Finnish success in PISA and some reasons behind it 2: PISA 2003.* Jyväskylä, Finland: Institute for Educational Research, University of Jyväskylä.

Vygotsky, L. S. (1978). *Mind in society: The development of higher psychological processes.* Cambridge, MA: Harvard University Press.

39

THE IMPLEMENTATION OF THE INTENDED CURRICULUM IN TEACHING MATERIALS

Authors' Perspective

Guðný Helga Gunnarsdóttir

Guðbjörg Pálsdóttir

UNIVERSITY OF ICELAND SCHOOL OF EDUCATION

Abstract: Teaching materials are influential and used by most teachers as guidelines in their teaching. It is a challenge for a small nation like Iceland to provide high-quality teaching materials to teachers and schools. Most materials available are published by a state-run publishing house. There is a limited choice of teaching materials in mathematics for compulsory schools with materials being typically renewed every twenty years. In the last ten years, the authors of this chapter have taken part in the development of new teaching materials on the basis of new national curriculum guidelines in mathematics. Mathematical processes have gotten a special focus in the new guidelines, and here the authors analyze the teaching materials according to those objectives and discuss the challenges in their implementation.

Keywords: Curriculum reform, textbooks development, mathematical processes

Official documents like curriculum guides reflect society's educational policy and influence the implemented curriculum and pupils' achievement. In February 1999, a new reform-oriented National Curriculum Guide in Mathematics was published by the Ministry of Education in Iceland. The main change in this new curriculum was that objectives for mathematics were divided into two categories: objectives for mathematical processes, and objectives for content.

It was clearly stated in the curriculum that both categories were to be given equal attention, and pupils were supposed to view both as an active part of learning mathematics (Aðalnámskrá grunnskóla, stærðfræði, 1999). The curriculum guide was revised in 2007 but only minor changes were made.

In Iceland, Námsgagnastofnun, The National Centre for Educational Materials (NCEM)[1] publishes educational materials in accordance with the National Curriculum Guides. Since 1999 NCEM has published around 100 new titles in mathematics consisting of textbooks, workbooks, teacher guides, compendiums with work cards and assessment materials, interactive web programs, worksheets, and thematic booklets. The authors of this chapter have been involved in the development and writing of 80% of these materials. In this chapter we use the words "teaching materials"

The First Sourcebook on Nordic Research in Mathematics Education, pages 539–549

to emphasize that the materials consist of more than just textbooks for pupils. We find it important to look at all these materials as a whole.

In the chapter we aim to give an insight into how and by whom educational materials in Iceland are developed and published. We think it is of interest to others to find out how a small nation with only 300,000 citizens supplies the schools with good quality educational materials in its own language. We also discuss how we have interpreted the ideology and emphasis in The National Curriculum Guides in Mathematics both regarding to how content is dealt with and how it is presented to pupils. This is illustrated with some examples from the textbooks.

BACKGROUND

Who Are the Authors?

In a study like this where authors reflect on their own work, it is important for the readers to know their background. This self-analysis is made by the authors and therefore it is necessary to give the readers some ideas of their own professional theory. The authors themselves also need to be aware of what affects their ideas, understanding, and what they choose to place emphasis on.

Guðný Helga Gunnarsdóttir (1952–) is an assistant professor in mathematics education at The School of Education, University of Iceland. She started her career as a compulsory school teacher in 1975 and worked as a teacher for over 20 years. Early in her career she became involved in writing teaching materials and consulting mathematics teachers. She has worked in The Ministry of Education, The National Centre for Educational Materials, and since 1980, in pre-service and in-service teacher education at Iceland University of Education—now University of Iceland, School of Education.

Guðbjörg Pálsdóttir (1956–) is an assistant professor in mathematics education at The School of Education, University of Iceland. She finished her teacher education in 1978 and was a teacher in compulsory schools for over 20 years. She has been participating in pre-service and in-service for teachers since 1980 and for the last ten years has been writing the new teaching material in mathematics.

In small countries, the same people are often involved in the same sector at different levels. We, the authors, have both been engaged in teacher education at the biggest teacher training institution for more than 25 years and have also been responsible for a great deal of in-service training of the mathematics teachers. At the same time we have been teaching children, writing teaching materials, and been active in the mathematics teachers society. This creates a unique situation for us to build an overview and to create coherence for the mathematics teachers. We have many threads in our hands and are aware of the huge responsibility that comes with multiple layers of responsibility. Often we have felt this to be demanding but at the same time challenging.

We started working together in 1979 when we developed our first teaching materials. Since then we have been working for a shift in emphasis in the mathematical learning and teaching in compulsory schools in Iceland. In the first in-service courses, we focused on introducing manipulatives to teachers and giving them ideas about how to use them in teaching. Later the focus was on tasks and multiple teaching approaches. We have learned through our experience and from research that it is not enough for teachers to be introduced to new ideas (Gunnarsdóttir, 2002). They need to work with the ideas themselves and with their pupils and filter them into their own knowledge to be able to make them their own and use them creatively. In doing that, teachers need time and support both from their own environment and professionals. Teaching materials are one of the most important resources teachers use, and therefore it is vital that they have access to good quality materials.

We saw the writing of teaching materials as an opportunity to support teachers in their professional development. In the teachers' guide we could give the teachers access to theories, research, and discussions on mathematics education in their own language and thereby they could develop important tools for their professional development. In the textbooks and interactive materials we could give examples of rich tasks and introduce different ways of working with mathematics. The teachers can use the teaching materials as a base, but they have to construct the teaching process themselves. Here their creativity and knowledge comes in, since they are the ones who know the pupils and the actual circumstances. We also wanted teachers to get an opportunity to learn about mathematical learning from observations and interactions with their pupils. Therefore we made many tasks that required the pupils to explain, work in groups, and discuss their understanding with both fellow pupils and the teacher. In the assessment tasks we wanted to give the teachers some tools that gave them opportunities to listen to pupils and analyze their work. We wanted to work towards better mathematics teaching and learning in Iceland, and we think it is worthwhile to try to develop good teaching materials even though we know that teaching materials are only one of the factors that influence pupils' opportunities to learn mathematics.

The National Curriculum Guide

The national curriculum guide for compulsory school in mathematics from 1999 and 2007 is divided into two main sections. The first section presents the background and the ideology that the teaching and learning of mathematics should be based on and summarizes the final objectives for pupils leaving the compulsory school. In the second section, the intermediate objectives for grades 1–4, 5–7, and 8–10 are then presented. Mathematics is a mandatory school subject, and the pupils are supposed to use 1/6 of their school time studying the subject. The curriculum is based on the idea that mathematics is an important part of the culture and that people use mathematics both to understand the world around them better and to challenge their cognitive development.

The objectives are divided in two main categories: mathematical processes and content. The mathematical processes are mathematics as a language, problem solving, logic and reasoning, and connection to daily life. The content is divided in six areas: numbers, calculations, proportions and percent, patterns and algebra, geometry, and statistics and probability. Both categories should be given equal attention, and the learning of the different processes and content should be intertwined.

The curriculum guide gives some ideas for the development of teaching materials. They should include basic material and some supplementary materials where pupils get opportunities to deepen their knowledge in some areas and broaden their view on the subject according to their needs and interest. The pupils are to study the basic material collaboratively, discuss and do investigations. In the teaching materials pupils should be presented to multiple ways to work with tasks and projects. The tasks should be challenging and multistage. Teachers should use multiple teaching methods, including individual, team, and collaborative learning activities. ICT is supposed to be an important part of the teaching and learning of mathematics. Assessment should be built on the demands that are established in the objectives of the national curriculum guide. Assessment has to take into account all the important factors of mathematical power, and various approaches should be used. Different ways that are suitable to evaluate different parts/fields of the subject have been developed, like interviews, self-evaluation, standardized tests, oral tests, written tests, projects, essays, evaluations schemes, and checklists (Aðalnámskrá grunnskóla, stærðfræði, 1999, 2007).

In the national curriculum guide, the intended curriculum is stated. It is an official document that gives guidelines for the school system. The school system is given some time to implement the curriculum guide every time it changes. In the mean time, the NCEM has the role of developing teaching materials that interpret the curriculum guide. The authors of the teaching materials structure and choose tasks and approaches that should help teachers to give the pupils good opportunities to fulfill the objectives. In the teaching materials the teachers should have a valid resource for implementing the curriculum. The authors influence, through their teaching materials, the teachers' interpretation of the curriculum guide, especially where only one group of authors are given the task. Therefore it is important both for the authors and the school system that the authors' interpretation is visible and introduced to the teachers—for example, in the teachers' guide.

The main changes in the national curriculum guide from 1999 and 2007 compared to prior guide from 1989 are the division of the objectives into two main categories, giving the mathematical processes special attention. Focus is placed on the methods used in investigations with emphasis on the approaches, mathematical thinking, and discussions of the solutions and ways used to find them. There is not much change in the content areas, but some content areas get a different view and emphasis. The biggest change within the content areas is in numbers and algebra. The focus is more on numeracy and using algebra to describe relations, not just to use numbers in calculations and to get good at manipulating symbols in algebra. It is also noticeable that proportion and percent is an independent group. In the guide, as has been revealed, there is also a discussion of the ideology for instruction and the teaching material. This discussion is the base for the interpretation used when designing the teaching material according to the national curriculum guide.

Publication of Educational Materials in Iceland

The National Centre for Educational Materials (NCEM) provides the compulsory schools (6–16 years) in Iceland with all kinds of educational materials: books, online materials, DVDs, and CDs. NCEM is a state-run publishing house, under the auspices of the Ministry of Culture and Education. The market for educational materials in Iceland is small, with around 4,300 pupils in each grade. Until now there has been very little supply of educational materials from other publishers for the schools to choose from, and the schools have had limited resources to buy materials from other publishers. Through the years it has often been considered whether to privatize the publication of educational materials. In 1995 the educational system was decentralized, and the local authorities were made responsible for the operation and the cost of compulsory schooling. Even though the state is still legally responsible for policymaking, evaluation, development, and supervision of the educational system, it has been questioned whether it should be publishing educational materials. The National Curriculum Guide set by the Ministry of Education is the framework for schools to work within, but the schools are also to set their own priorities and develop their own school curriculum. Limited choice of educational materials can constrain the schools when setting their own curriculum.

When the work of NCEM is looked into there seems to be a general agreement that the company is providing the schools with good quality educational materials for the money allocated by the Icelandic Parliament in the financial law every year. The budget is around 380 millions ISK or 9,000 ISK (US $100) per pupil, and approximately 800,000 copies are distributed by NCEM to the schools free of charge. But it is difficult for the company to meet all demands from the schools, and they would like to have some more choice. However, there seems to be a political agreement that educational materials for compulsory schools are to be free of charge and financed by the state (Frumvarp til laga um námsgögn ásamt athugasemdum, 2006–2007).

In 2007, a new law on educational materials[2] was set by the Icelandic Parliament. According to the law, NCEM is to get a similar amount of money by state financial law and is still supposed to be the main provider of educational materials for compulsory schools. Two new funds financed by the state were established to secure more variety and supply. Both funds have their own board appointed by the Minister of Education with representatives from teachers' unions and Icelandic Association of Local Authorities. One fund is supposed to give compulsory schools some money to buy educational materials from other publishers. The amount each school gets is dependent on the number of pupils, but it is allowed to give the smallest schools some benefits. This fund got 100 million ISK by state financial law for 2008. The other fund was supposed to encourage development and renewal of educational materials for all school levels except the university level. The fund got 56.6 millions ISK by state financial law for 2008 and money is granted to projects by the fund's board once a year based on applications. In reality this means an increase of approximately 100 millions ISK towards production and development of educational materials.

The population of Iceland is very small so the market for educational material is limited. Though the state has opened for financial support to private companies it is difficult to profit from making materials for the compulsory school. Therefore it is not likely that there will be a big production and it will be important to give NCEM the possibility to develop materials.

THE TEACHING MATERIALS

In 1999, NCEM had started to publish new teaching materials in mathematics according to the new curriculum guide. It was decided to start to publish materials for grade one and then follow that up with new materials for that age group every year. Now ten years later new teaching materials for all the ten years of compulsory school have been published.

For grades 1–4 it was decided to translate and adapt Danish teaching materials[3] to the Icelandic curriculum guide. This proved to be complicated. Many changes had to be made both in the content and also in the illustrations. Even though Icelandic and Danish culture are similar in many ways there are also differences which become very clear when making teaching materials in mathematics that relate to the daily life of 6- to 10-year-old children. One important factor is for instance a difference in currency as 1 DDK was at that time equal to 10 ISK. The curriculum guides in Iceland and Denmark are very different. The Danish guide is not as detailed as the Icelandic guide and the emphasis in content for example in reasoning and probability is not the same. It meant that it was necessary to change the Danish material and write some additional sections.

It was therefore decided by the NCEM to make Icelandic teaching materials for grades 5–10. NCEM appointed a working group in 2000 that provided a general framework for new teaching materials in mathematics for grades 5–10. In this group there were three compulsory school teachers, an assistant professor in mathematics education from Iceland University of Education, one upper secondary school teacher in mathematics and an editor of the curriculum guide, and the editor of mathematics textbooks at NCEM. One of the authors was a member of this group. The working group of course pointed out the curriculum guide as the general framework for the new materials. It also pointed out that because the guide is very detailed the authors have to be careful not to treat all concepts mentioned in the curriculum as isolated pieces of knowledge but to treat them in a wider context. The working group recommended authors to put a lot of emphasis in the mathematical processes. The group also recommended varied tasks that would give pupils the possibility to work at different levels both individually and in groups, tasks that encourage pupils to use investigative approach and that the teachers should be able to teach mathematics through problem solving. The group also found it important that the textbooks included some texts about mathematics and the history of mathematics for pupils to read and discuss and not necessarily with tasks to be worked out (Ástvaldsdóttir et al., 2000).

The Icelandic teaching materials are divided into two series. The materials for grade 5–7 are called *Geisli*, and the materials for grades 8–10 are called *Átta-10*. In designing the materials *Geisli*, the authors worked with the teachers' guides as a focal point. Studying the guides should give the teachers an overview and give them background so they can choose which tasks to use and in what way. The textbooks and the compendium should give examples of varied and rich tasks for the pupils and give possibilities for multiple teaching approaches. The pupils should get tasks from all the categories in the national curriculum guide in a holistic way. There were four authors who participated in writing the materials.

The teaching materials *Átta-10* are built on the prior materials *Eining* and *Geisli* and are written for the last three years of compulsory school.[4] In this chapter we use these materials to demonstrate the authors' interpretation of the intentions in the curriculum guide. These materials, except for the thematic booklets and the interactive programs, are written only by the two of us. The themes for the thematic booklets have been chosen by us in cooperation with the editor and a group of specialists[5] who have read and commented on all the materials we have written for pupils. An overview of the materials for grades 8–10 can be found in Figure 39.1. There is access on the webpage (http://www.nams.is/atta-tiu/atta-tiu.htm) to all web-based materials.

The teacher guides are central in the teaching materials. There the authors explain the philosophy and ideas behind the materials and refer teachers to texts about important aspects of mathematics education. Very little is being written in Icelandic on mathematics education, so the Icelandic teachers usually have to read texts in English or the Nordic languages. The teacher guides also provide the teachers with background information about the topic being dealt with in each chapter and give them ideas about where they can find more support materials because many teachers have weak mathematical backgrounds, and some of the topics are new at this level. In the guides the teachers are also given some ideas about how they can plan a teaching process based on the materials provided, but it is made clear that this is only an example and that every teacher has to plan teaching according to their pupils and situation. They also get some ideas and examples about assessment, and an emphasis is put on multiple ways of assessing mathematical knowledge.

Átta-10, materials for grades 8–10

- Grade 8—two textbooks (112 p. each)
- Grade 9—two textbooks (112 p. each)
- Grade 10—two textbooks (112 p. each)
- Thematic booklets (16–32 p. each)
  Spreadsheet, The golden ratio, Measurement, Problem solving
- Teacher guides, solutions and other materials for teachers and pupils on the webpage.
- Interactive programs, a list of important concepts, a text on assessment.
- Assessment tasks published on a CD.

Figure 39.1

The main goal for the creation of these new teaching materials in mathematics is to implement the curriculum from 1999/2007. The materials are written to support teachers in that process. The new teaching materials are replacing textbooks that have been in use since 1987. These textbooks were translated from Swedish in 1987–1988 and are very traditional textbooks where procedures are explained and lot of exercises given. In the TIMSS 1995 Mathematics Curricula Study these textbooks for 8th grade were analyzed and it revealed that 60% of textbook blocks were devoted to whole number operation and almost none to patterns, relations, and functions. Performance expectations like mathematical reasoning and complex communication get almost no attention in the 8th grade textbooks. The main emphasis is on knowing and using vocabulary and using routine procedures (Schmidt, McKnight, Valverde, Houang, & Wiley, 1997).

Research in the early 1990s showed that little whole class teaching was taking place in many Icelandic mathematics classes in the middle school. The pupils were mostly working individually with the textbooks (Sigurgeirsson, 1994). Recent research in grades 8 and 9 shows similar pattern where pupils are working individually and the teacher is circling in the class helping individuals. The pupils are working in their own pace according to a teaching plan. Sometimes the teacher gives explanations on the board for individuals or small groups or brings up for discussion items many pupils are struggling with. In most schools there are regular tests accompanied by a public review process (Savola, 2008).

For the Icelandic teachers, the new materials make a big change. They have for over 25 years only had one textbook for every school year and some short tests. Many of the teachers know those textbooks by heart and have developed ways to use them in their teaching. It is therefore a big effort to change to more complicated teaching materials, but it is necessary if they are to follow the curriculum. The authors have tried to make it easier for the teachers by placing emphasis on the teaching guide and concentrating on one mathematical topic in most chapters.

The teacher plays an important part in creating a good learning environment for the pupils. Even though the teaching materials are an important factor, they are only one of the resources the teachers have in creating good opportunities for their pupils to learn. The objectives on mathematical processes are a new focus for the teachers, though reasoning and conceptual understanding have always been important for mathematical learning.

MATHEMATICAL PROCESSES

Ways of working and thinking with mathematics have always been considered a part of mathematical learning. Now emphasis is put on mathematical processes in many curriculum guidelines, and they are considered key elements in mathematical learning. The mathematical processes are influenced by the culture and discourse in society. The use of mathematics, the role of discussion in society, the school culture, and the structure of the society influence how the mathematical processes are interpreted and represented (Niss, 2004; NCTM, 2000). In the Icelandic context, it was a big challenge to take the objectives for the mathematical processes as represented in the curriculum guidelines and implement them in the teaching materials in accordance with the culture and society. In the writing of the materials, we decided to put an emphasis on these processes as expected in the curriculum guide and suggested by NCEM's working group in 2000. In this section we present the main ideas from the curriculum guide for each of the mathematical processes: mathematics as a language, problem solving, logic and reasoning, and connection to daily life. We give some examples of tasks from the pupils' textbooks, *Átta-10: Stærðfræði 1–6*,[6] to give an idea of how we have interpreted and implemented the curriculum in the teaching materials.

Mathematics and Language

Pupils are supposed to be able to read and write numerical information and mathematical texts. An emphasis is also put on communication and discussions of mathematical content and ways of working. The pupils should be able to explain their solutions and strategies both orally and in writing.

Prime numbers, Átta-10, Stærðfræði 1

Write a text about all you know about prime numbers and composite numbers.

The pupils are given some hints about what to include in their texts and that they can seek for information on the Internet.

Angles, Átta-10, Stærðfræði 6 Group-work

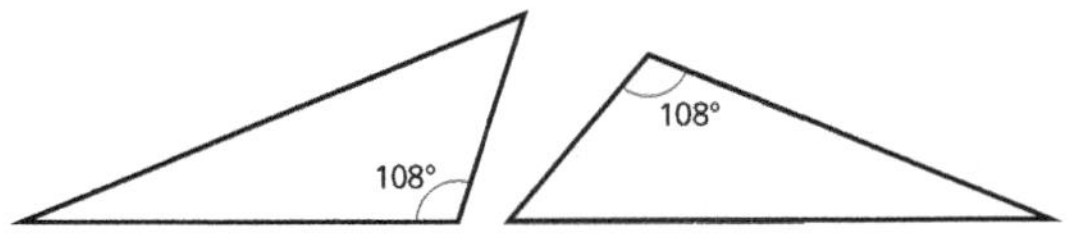

It is given that two triangles have one angle of equal size. Investigate what more information you need to be able to find out whether the two triangles are similar. Make a presentation where you tell about your investigations and solutions.

Numbers, Átta-10, Stærðfræði 3

Pupils read a 3-page text about number sets and the historical development of number systems and answer some reading comprehension questions based on the texts.

Polygons and circles, Átta-10, Stærðfræði 4

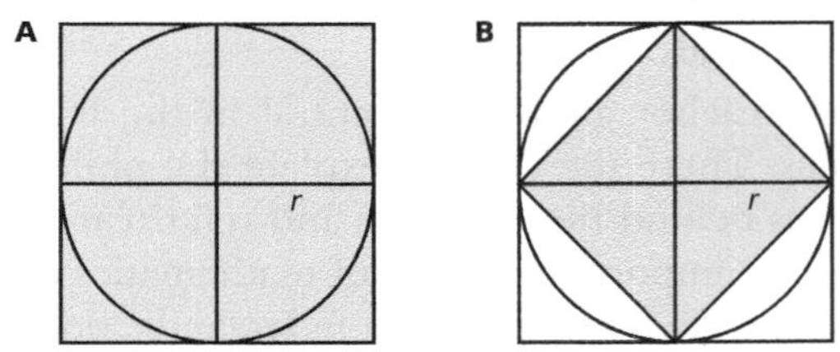

*Explain why picture A shows that the area of a circle must be less than 4 * r^2 and picture B shows that it must be more than 2 * r^2.*

Statistics and probability, Átta-10, Stærðfræði 4

Information about the use of phones, mobile phones, and sms messages during the last seven years is represented graphically in different ways. Pupils are asked to interpret the information and to represent the information differently. They compare the different representations and are asked to argue for how they think it is best to present this kind of information. They are also encouraged to seek up-to-date information.

Problem Solving

Pupils should get to know ways to solve both word problems and more complex problems that take time. They should experience the benefits of working with others when solving complicated problems. The pupils should meet real problems where they have to find ways to solve them and then interpret the solutions. They should also experience that there can be many different ways to solve the same problem and be able to present their own solution strategies and examine and discuss others. The pupils should learn some solution strategies for solving problems that they can use in other areas.

Reasoning Átta-10, Stærðfræði 6

In a chapter about reasoning pupils are asked consider different ways to solve problems and justify solutions. They solve problems and describe different solution processes.

Wooden boards are sold in two lengths, 8 feet and 10 feet. Simbi bought 90 boards and their combined length was 844 feet. How many board of each length did he buy?

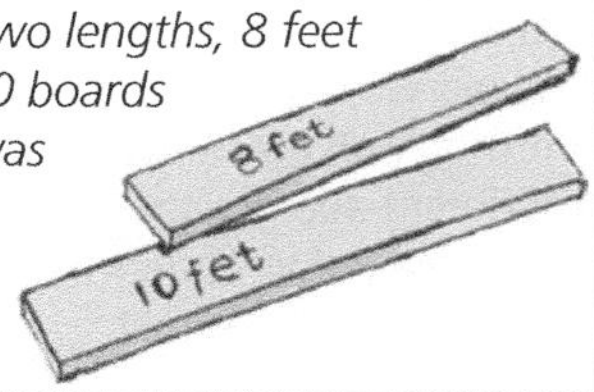

Space, Átta-10, Stærðfræði 3

On their web-pages shipping companies provide their customers with calculators they can use to estimate the volume of shipments. These are the estimations for some common home appliances.

A three seat sofa	1,7 m^3
Television	0,3 m^3
A single bed	1,05 m^3
A chest of drawers	0,4 m^3
A fridge	1,14 m^3
A bicycle	0,35 m^3

Are these estimations realistic in your opinion? Justify your answer.

Teenagers and finances, Átta-10, Stærðfræði 6

In the chapter *Teenagers and Finances* the pupils investigate different aspects of finances like wages, working hours, taxes, savings and expenses. They work in groups on some open problems and are encouraged to use a spreadsheet for calculations.

Numbers and number theory, Átta-10, Stærðfræði 6

Three consecutive whole numbers are multiplied together and the number in the middle is added.

Example $3 * 4 * 5 + 4 = 4^3$

Find some numbers where this relationship is also true.

Does this relationship apply, regardless of which three consecutive numbers you choose?

Show the relationship by letting n symbolize the middle number.

Models, Átta-10, Stærðfræði 5

In a chapter about mathematical modeling, the pupils work with a model that is used to find out how many parking places are needed for 50 square meters of an industrial estate and 35 square meters of a shopping estate. They try out the model and evaluate the results. They then make their own models that can be used to solve various real problems.

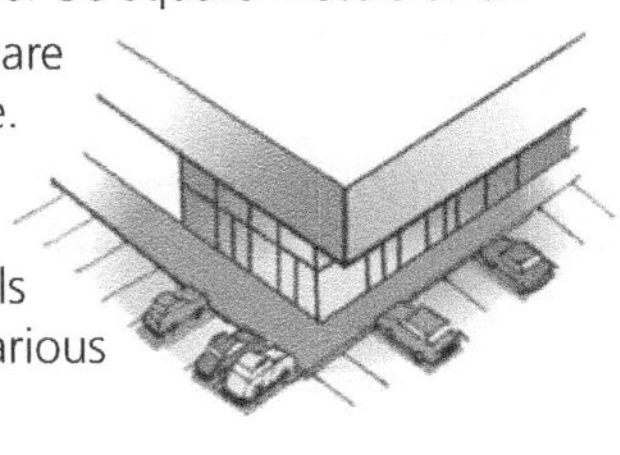

Reasoning

Pupils should recognize the importance of reasoning in mathematics and get to know some simple proofs. They should be able to use simple reasoning, for instance, when justifying and explaining their solution strategies or algorithms. At this stage, pupils should get to know mathematics as a life field of study where research is being done and new discoveries are being made.

Reasoning, Átta-10, Stærðfræði 6

Show with reasoning that this statement is true.

If the sum of two numbers is higher than 18, one of them must be higher than 9.

Polygons and circles, Átta-10, Stærðfræði 4

A formal proof that opposite angles are equal is presented to the pupils, and they are asked to read the proof and explain it to each other.

Fractions, Átta-10, Stærðfræði 3

What common denominator did Fjóla choose?

a) $\frac{2}{3}+\frac{5}{18}=\frac{12}{x}+\frac{5}{x}$ b) $\frac{7}{9}-\frac{5}{6}=\frac{14}{x}-\frac{15}{x}$

Could she have chosen a lower common denominator? Justify your answer.

Fractions, Átta-10, Stærðfræði 1

The pupils in 8th grade are learning to add fractions. They are calculating $\frac{5}{6}+\frac{1}{2}$.

Look at their different solutions. How would you have solved this task?

What do you think is the best solution? Justify your answer.

Polygons and circles, Átta-10, Stærðfræði 4

In a chapter about polygons, pupils investigate polygons and their properties. They have previously worked with tessellations. Here they are investigating certain pentagons that tessellate and are told about how some of them were discovered by Marjorie Rice, a housewife from San Diego in the 1970s.

Connections to Other Fields

Pupils should recognize that mathematics plays an important role in daily life of human beings and in society. It is connected to many other subjects and can be used to describe and explain phenomena. It is an important goal that pupils understand that knowing mathematics is an important life skill. Pupils should get to know the history of mathematics, as it is well suited to explain that discoveries in mathematics often originate from the need of human beings to understand their environment.

In the textbooks the pupils get to know the history of mathematics through short texts about some important aspects in the development of mathematics and about some of the mathematicians who have shaped this history. They read about the development of pi, Eratosthenes, Pythagoras, Descartes, Archimedes, Gauss, Abel, and Sierpinski, just to give a few examples.

Three dimensions, Átta-10, Stærðfræði 1

In 2004 the Taipei tower in Taipei was the highest building in the world. Use the information given to estimate its volume. Compare the volume with the volume of a small apartment block you know. Information is given about higher buildings that are under construction to encourage pupils to investigate this issue further.

Percentages, Átta-10, Stærðfræði 1

Many people have to use decimals and percentages in their work. Work in small groups. You can choose to be a shopkeeper, a nutritionist, a news reporter, a cook in a school kitchen, or manager of a small airline that regularly offers some good travel deals. Give some examples about how you would use decimals and percentages in your work.

Statistics, Átta-10, Stærðfræði 6

In a chapter about statistics, pupils work with data concerning driving and accidents caused by different age groups. They are also encouraged to gather their own data and analyze that.

You are going to investigate people's views of the idea whether to raise the age limit for a driving license from 17 years to 18 years.

a. *What information would it be important to gather?*
b. *Design a form that would be easy to use to gather and register information.*
c. *What is important to have in mind when designing the form?*

How would you choose a sample for this investigation?

Number operations, Átta-10, Stærðfræði 2

Give some examples about situations in daily life where you feel it is sufficient to use rounding and estimation.

> Space, Átta-10, Stærðfræði 3
>
> Páll is going to visit a family in the States. He receives a letter from Jimmy who is his age where he describes his home, his school, and his daily routines using U.S. measurement units. Páll has to interpret this information and then to write a letter to Jimmy with similar information about his daily life using U.S. units.

Here we have given some examples of tasks that can meet some of the objectives in the mathematical processes. Tasks can be characterized in many different ways, and the same task can often be used to meet multiple objectives. The tasks we have chosen also reflect different content areas and age levels. Some of the examples are tasks where the pupils work alone or in pairs, and others are problems or investigations where the pupils are supposed to work in groups. This should give the readers some idea of how we have interpreted and implemented the curriculum in the teaching materials. How the materials have been interpreted and implemented by the teachers is another matter and not a part of our study.

DISCUSSION

The process from writing a curriculum guide (the intended curriculum) to pupils learning (the attained curriculum) is long and influenced by the interpretation of many actors. The intended curriculum is interpreted by the authors of teaching materials and, to some extent, the teachers. More often the teachers use the teaching materials as a guide to their interpretation of the curriculum, combined with their views on what is important in mathematics learning. The pupils rarely have the opportunity to interpret the curriculum directly. They meet it through the interpretation of the teaching materials and their teachers. The authority of the textbooks is well known. Research has shown that textbooks (teaching materials) influence *what* to teach and are also a primary source for *how* to present content. They are an influential provider of educational opportunities for the pupils (Johansson, 2006; Pepin, 2008; Schmidt et al., 1997; Valverde, Bianchi, Wolfe, Schmidt, & Houang, 2002). It is therefore a big responsibility to write teaching materials where the authors represent to a whole generation of pupils their version of what mathematics and mathematics learning is about. This is especially sensitive in a small country like Iceland where the supply of teaching materials is limited.

Being aware of the authority of teaching materials, we have tried in our teachers' guides to support and encourage the teachers to use the materials flexibly and design their lessons according to their situation. We have pointed out other resources than the textbooks that the teachers can use in their planning and instruction. We emphasize that it is not possible for authors of teaching materials to make a teaching plan that suits all teaching situations. Teachers always have to base their teaching upon an analysis of the situation, the pupils, and the objectives. The teaching materials can help them in this process but never replace the teacher. We feel that the curriculum guide supports this interpretation of the role of the teacher. In the teaching guide we discuss ways to meet the different needs of pupils. The main idea is that pupils can work with the same content on different levels and develop their own connections through reflections and discussions.

Mathematics learning should be meaningful to all pupils. The problem of making mathematics meaningful cannot only be solved by putting an emphasis on mathematics in real-life contexts (Alrø & Skovsmose, 2004; Howson, 2005). We give pupils opportunities to construct their mathematical ideas by dealing with them in different contexts and representations. We find it important that pupils find meaning both dealing with real-life context and mathematical context. Tasks that are related to the pupils' daily life often take up issues that can be of interest to them, such as movies, computers, music, sports, and technology, just to mention a few examples. The pupils are asked to seek and use information on the Internet or other media. When real information is used in tasks, they are encouraged to check the information and update it if necessary.

In the teaching materials we have tried to develop challenging tasks that take some time to be solved and encourage pupils to make use of and develop their mathematical thinking. They are encouraged to reason and justify their solutions, both orally and in writing, and to use their previous knowledge to make conjectures, explore relationships, solve problems and create new tasks. Research has shown that the cognitive demand of tasks and the time spent on each task influence the pupils' opportunities to learn by dealing with the tasks (Boaler & Humphreys, 2005; Stiegler & Hiebert, 1999; Smith & Stein, 1998).

The emphasis on mathematical processes in the curriculum guidelines and the fact that such an emphasis is new in Iceland made it easier for us to write materials that correspond with our ideas about what is important in mathematics teaching and learning. Investigative approaches and discussions which we feel are very important factors if mathematical learning is to take place can easily be justified within the frame the curriculum guide provides. This we demonstrate in the section about mathematical processes by giving some examples from the pupils textbooks.

It has been pointed out that the new teaching materials are very different to the materials used in Iceland; for the last 20 years, in grades 8-10 (Halldórsdóttir, 2009). For the pupils the change is not as big because they have been using materials based on the same ideas since first grade. When they start in 8th grade their background and experience in mathematics is often different from their teachers' expectations. The teachers who have not taught in the lower grades are facing a new situation both regarding the pupils and the teaching materials. In the teachers guide the teachers are encouraged to get acquainted to the changes made in the teaching of mathematics in the lower grades and to use a diagnostic teaching approach.

In Iceland teaching materials are produced by a state run publishing house. That gives authors some freedom but at the same time it is restricting. The restriction for the authors in Iceland is the curriculum guide and their working conditions. Authors who do not have to compete on a market have the possibility to set standards and make progressive materials even though they have the responsibility of writing materials that can be used by all teachers and pupils in the country. This becomes even more real when the authors are involved in in-service and pre service teacher education. In most other countries several series of teaching materials in mathematics are competing for the market and authors often have to make some compromises that make the materials more attractive to the market.

It has been rewarding for us as authors to analyze our work and write this chapter. At this moment it is not possible for us to evaluate to what extent we have succeeded in writing teaching materials that support teachers in implementing the curriculum and developing new ways of working in mathematics teaching based on research. The process of writing and publishing this new series of teaching materials in mathematics is now finished and it is up to the teachers to interpret and implement them. The teachers seem to be ready to try out the materials and discuss ways to implement them. This creates opportunities to study the relationship between the intended, implemented and attained curriculum in the future.

NOTES

1. Námsgagnastofnun. http://www.nams.is/enska/
2. Lög um námsgögn nr. 71/2007
3. Faktor – Malling Beck A/S
4. For pupils in grades 8–10 (13–16 years old).
5. Two lower secondary school teachers and two mathematics teacher educatior.
6. In this section we refer to the pupils textbooks Átta-10, Stærðfræði 1-6. Further information about the textbooks is to be found in the references.

REFERENCES

Aðalnámskrá grunnskóla, stærðfræði. [National Curriculum Guide, Mathematics] (1999). Reykjavík: Menntamálaráðuneytið.

Aðalnámskrá grunnskóla, stærðfræði. [National Curriculum Guide, Mathematics] (2007). Reykjavík: Menntamálaráðuneytið.

Alrø, H. & Skovsmose, O. (2004). *Dialogue and learning in mathematics education.* Dordrecht, The Netherlands: Kluwer Academic Publishers.

Ástvaldsdóttir, Þ., Birgisson, G.., Bjarnadóttir, K., Finnbogadóttir, H., Marteinsdóttir, J., & Pálsdóttir, G. (2000). *Tillaga starfshóps í stærðfræði fyrir 5.-10. bekk.* [Proposal of a working group for mathematics in grades 5–10] Reykjavík: Námsgagnastofnun.

Boaler, J. & Humphreys, C. (2005). *Connecting Mathematical Ideas.* Portsmouth, NH: Heinemann.

Frumvarp til laga um námsgögn ásamt athugasemdum [A law proposal on educational materials with comments]. (2006–2007). Lagt fyrir Alþingi á 133. löggjafarþingi 2006–2007. Þskj. 772–511. mál.

Gunnarsdóttir, G. H. (2002). *Stærðfræðiátakið í Hafnarfirði: Áhrif á starfsþróun kennara.* [The Mathematics Project in Hafnarfjörður. Influence on teachers professional development] [Unpublished M.Ed. Thesis]. Reykjavík: Kennaraháskóli Íslands.

Gunnarsdóttir, G. H. & Pálsdóttir, G. (2005). *Átta-10: Stærðfræði 1* [Eight -10: Mathematics 1]. Reykjavík: Námsgagnastofnun.

Gunnarsdóttir, G. H. & Pálsdóttir, G. (2006a). *Átta-10: Stærðfræði 3* [Eight -10: Mathematics 3]. Reykjavík: Námsgagnastofnun.

Gunnarsdóttir, G. H. & Pálsdóttir, G. (2006b). *Átta-10: Stærðfræði 2* [Eight -10: Mathematics 6]. Reykjavík: Námsgagnastofnun.

Gunnarsdóttir, G. H. & Pálsdóttir, G. (2007a). *Átta-10: Stærðfræði 5* [Eight -10: Mathematics 5]. Reykjavík: Námsgagnastofnun.

Gunnarsdóttir, G. H. & Pálsdóttir, G. (2007b). *Átta-10: Stærðfræði 4* [Eight -10: Mathematics 4]. Reykjavík: Námsgagnastofnun.

Gunnarsdóttir, G. H. & Pálsdóttir, G. (2008). *Átta-10: Stærðfræði 6* [Eight -10: Mathematics 6]. Reykjavík: Námsgagnastofnun.

Halldórsdóttir, R. (2009). Breytingar á kennslubókum í stærðfræði fyrir 8. bekk síðasta aldarfjórðung [Changes on teaching material in mathematics for 8th grade the last 25 years]. *Flatarmál, rit Flatar, samtaka stærðfræðikennara. 16*(2), 32–35.

Howson, G. (2005). "Meaning" in school mathematics. In J. Kilpatrick, C. Hoyles, & O. Skovsmose (Eds.), *Meaning in mathematics education* (pp. 17–38). Dordrecht, The Netherlands: Kluwer Academic Publishers.

Johansson, M. (2006). *Teaching mathematics with textbooks: A classroom and curricular perspective.* Luleå: Luleå University of Technology.

Johansson, M. (2003). *Textbooks in mathematics education: A study of textbooks as the potentially implemented curriculum.* Luleå: Luleå University of Technology.

National Council of Teachers of Mathematics. (2000). *Principles and standards for school mathematics.* Reston, VA: Author.

Niss, M. (2004). The Danish "KOM" project and possible consequences for teacher education. In R. Strässer, G. Brandell, B. Grevholm, & O. Helenius (Eds.), *Educating for the Future* (pp. 179–109). Göteborg: The Royal Swedish Academy of Sciences.

Pepin, B. (2008). *Mathematical tasks in textbooks: Developing an analytical tool based on connectivity.* Paper presented at ICME 11 in Monterrey, Mexico. Retrieved on July 1st 2008 from http://dg.icme11.org/document/get/245.

Savola, L. T. (2008). *Video based analysis of mathematics classroom practice: Examples from Finland and Iceland.* Ph.D. Thesis, Columbia University: New York. Retrieved on August 10th, 2008 from http://ru.is/publications/SoHE/LasseSavola2008.pdf.

Schmidt, W. H., McKnight, C. C., Valverde, A. G., Houang, R. T., & Wiley, E. D. (1997). *Many visions, many aims. Volume 1.* Dordrecht, The Netherlands: Kluwer Academic Publishers.

Sigurgeirsson, I. (1994). *Notkun námsefnis og viðhorf til Námsgagnastofnunar í 200 grunnskólum* [The Use of Teaching Materials and Views on The National Centre for Educational Materials]. Reykjavík: Rannsóknarstofnun Kennaraháskóla Íslands. Retrieved on June 25th 2008 from http://starfsfolk.khi.is/ingvar/namsefni/Notkun_namsefnis.doc.

Stein, M. K., & Smith, M. S. (1998). Mathematical task as a framework for reflection: From research to practice. *Mathematics Teaching in the Middle School, 3*(4), 268–275.

Stiegler, J. V., & Hiebert, J. (1999). *The teaching gap: Best ideas from the world teachers for improving of education in the classroom.* New York: The Free Press.

Valverde, G. A., Bianchi, L. J., Wolfe, R. G., Schmidt, W. H., & Houang, R. T. (2002). *According to the book. Using TIMSS to investigate he translation of policy into practice through the world of textbooks.* Dordrecht: Kluwer Academic Publishers.

SECTION IV

Danish Research in Mathematics Education

Bettina Dahl and Bharath Sriraman, *Section Editors*

CHAPTER 40

MATHEMATICS EDUCATION RESEARCH IN DENMARK

A Brief Introduction

Bettina Dahl
AARHUS UNIVERSITY

Bharath Sriraman
THE UNIVERSITY OF MONTANA

This section presents nine chapters describing and discussing recent mathematics education research at primary, secondary, and tertiary levels as well as adult mathematics education in Denmark. The chapters represent a wide range of research methodologies such as qualitative, quantitative, historical, sociological, and philosophical accounts. Most of the twelve authors are university researchers from all corners of Denmark, while others are from the university college system or reside outside of Denmark. All this contributes to a very broad and comprehensive view of contemporary Danish mathematics education research.

In the chapter "Learning Mathematics Through Modeling: The Case of the Integral Concept," Blomhøj and Kjeldsen investigate mathematical modeling as a means for learning mathematics. They analyze and discuss how modeling has the potential of supporting students' learning of mathematical concepts, and how these potentials can be realized in teaching situations. In another chapter entitled "Mathematical Modeling as Goal in Mathematics Education: Developing Modeling Competency Through Project Work," the same two authors argue that problem-oriented, student-guided project work in interdisciplinary science studies is a powerful pedagogical strategy for developing students' mathematical modeling competency. These two chapters thus focus at tertiary level mathematics teaching in modeling at the university. The chapter by Lindenskov and Andresen entitled "Recent Development in School Mathematics' Roles and Relations" also relates to modeling. They however focus on upper secondary teaching and discuss the new construct, multidisciplinarity, which was recently introduced into the curriculum. They do a curriculum and qualitative study and relate it to three key characteristics on educational culture in Denmark—namely, regulations and the teachers' autonomy, educational aims and goals, and lastly the students' attitude towards mathematics. They also discuss the potentials of multidisciplinary mathematics teaching. In the chapter "Preparing In-Service Teachers for an Interdisciplinary Curriculum," Michelsen works within a similar area as he also discusses interdisciplinarity and the modeling competence. He introduces the didactical framework and the outcomes of a recently developed educational program for in-service teachers from pri-

The First Sourcebook on Nordic Research in Mathematics Education, pages 553–554

mary and lower secondary levels and how to prepare them for a more interdisciplinary curriculum in mathematics and science. These chapters use qualitative research methods.

Dahl (who is also the editor of this volume) writes the chapter "Can the Two Different Types of Mathematics Teacher Preparation in Denmark Integrate? A Discussion of Research-Based Teaching, Mathematics Competencies, and Content." This chapter also focuses on teacher preparation and compares the systems for compulsory and lower secondary education with that for upper secondary teaching. The two systems are very different. The chapter does a curriculum and quantitative analysis discussing the competencies and the progression of mathematics competencies in the curricula as well as the idea of research-based teaching.

A chapter by Wedege is entitled "Adults Learning Mathematics: Research and Education in Denmark." She discusses the borderland between mathematics education and adult education as a new research field. The subject area encompasses formal adult mathematics education as well as adults' non-formal mathematics learning in the communities of everyday practice, such as the workplace. The key concept is numeracy, and the chapter presents an overview of Danish research in adults learning mathematics with a focus on the interplay between research and education.

The next two chapters move away from a mere focus on classroom research and curriculum study. The chapter by Alrø, Skovsmose, and Valero, "A Learning Landscape: Building Perspectives on Mathematics Learning in Multicultural Classrooms," discusses the notion that students' mathematics learning cannot be understood by looking solely at mathematical classroom activities. The concept of "learning landscape" serves as a theoretical and methodological tool. They draw on data and results of an empirical study on students' foregrounds with 45 students in two 8th grade multicultural classes in Denmark and show the dialectical relationship between each dimension of the landscape and the whole of it, and how, as a whole, it can help us come closer to better theorizations. Skovsmose, in a separate chapter called "Critical Mathematics Education: In Terms of Concerns," continues some of this discussion. He characterizes critical mathematics education in terms of five corners—for instance, the fact that mathematics education operates in a globalized world and that globalization includes ghettoizing as well as the fact that much research in mathematics education has concentrated on an affluent teaching–learning environment, which has brought about a prototypical bias in theory building. The concluding chapter by Lange and Meaney is a comparative analysis of sense-making activities that children in Denmark and New Zealand were involved in. These authors explore factors that enable the enaction of agency in children. Similar to the comparative study of classrooms in Iceland and Finland included in previous section featuring Iceland, the value of such studies lies in the fact that basic constructs in mathematics education can be comprehensively studied in different national and cultural contexts.

41

MATHEMATICAL MODELING AS GOAL IN MATHEMATICS EDUCATION

Developing Modelling Competency through Project Work

Morten Blomhøj and Tinne Hoff Kjeldsen

IMFUFA, DEPARTMENT OF SCIENCE, SYSTEMS, AND MODELS, ROSKILDE UNIVERSITY

Abstract: Problem-oriented, student-guided project work in interdisciplinary science studies is a powerful pedagogical strategy for developing students' mathematical modeling competency. Based on experiences from supervising modeling projects at a two-year introductory science study program at Roskilde University (Nat-Bas), the qualities of this holistic approach to teaching mathematical modeling are analyzed and discussed. From a definition of mathematical modeling competency and analytic descriptions of a modeling process and of the epistemological structure in applications of mathematical models in extra mathematical contexts, key didactical challenges are identified. Two types of students' modeling projects, emphasizing respectively the societal use of a model and the modeling process, are illustrated and analyzed through concrete examples, and their relations to the key challenges in developing modeling competency are discussed.

Keywords: Mathematical modeling as goal, project work, modeling competency, modeling process, societal applications

INTRODUCTION: THE AIM AND METHOD OF THE CHAPTER

The aim of this chapter is to discuss the development of university students' mathematical modeling competency with special focus on the following three educational challenges:

1. How can we challenge and support students to work with the entire modeling process in a way that supports their autonomy?
2. How can we give sufficient support to students' work with the processes of mathematization and mathematical analysis as integrated elements of teaching mathematical modeling?
3. How can students' internal and external reflections in their modeling activities be developed?

This will be done in two steps: first, based on theories of the modeling process and modeling competency, we explain in detail how these challenges are to be understood, and we argue for their significance and importance in developing students' modeling competency. Second, we discuss how and in what sense the three challenges can be met through problem-oriented, student-directed project work. This discussion is based on a huge set of empirical material, col-

The First Sourcebook on Nordic Research in Mathematics Education, pages 555–567

lected over the past twenty years, of project reports on problems solved through mathematical modeling by students in the interdisciplinary two-year introductory science program (Nat-Bas) at Roskilde University (Niss, 2001).

Hence, we link theories and reflections on developing students' modelling competency in the educational system with real, authentic "classroom" projects, thereby providing empirical evidence for our discussion on how and in what sense these challenges can be met.

Methodologically, we apply in-depth analysis of four project reports that, in different ways, are representative of modeling projects done within the Nat-Bas program. The analyses presented in the present chapter are a result of a comparison and discussion of analyses of the project reports performed independently by both authors. Both authors of this chapter have extensive experiences in supervising such projects. In fact, the ones chosen for the in-depth analysis were supervised by the first author. So, besides the written reports, our reflections are based on these experiences, our experiences from the last decade from many other modeling projects for which we have functioned as opponent supervisor, and from informal discussions with students during their project studies and other supervisors. It should be clear that the empirical evidence we are presenting and analyzing is not the result of controlled teaching experiments especially designed to answer the three educational challenges we formulated above. Rather, we are presenting a form of teaching based research that takes place within the problem-oriented, student-directed project work of the practice of the Nat-Bas program.

In the next section we introduce the theories of modeling competency and the modelling process, and formulate the first and the second educational challenge. Then follows a section where we discuss the role of two kinds of reflections in modelling competency and formulate the third educational challenge. In the following section we introduce the Nat-Bas program with special emphasis on two types of mathematical modeling projects that are common within the program. Four project reports are analyzed in depth with respect to how and in what sense the three educational challenges in developing students' modeling competency are met. We end the chapter with a discussion of how and in what sense the Nat-Bas program meets the challenges of developing mathematical modeling competency in the educational system together with a critical discussion of the educational set-up and the crucial role of the supervisor for the success of the endeavor.

DEVELOPING MATHEMATICAL MODELING COMPETENCY

Mathematical modeling is a complex interdisciplinary activity, which is intricately connected to the learning of mathematics. In the context of the Danish KOM project (KOM stands for KOMpetence in Danish) under the auspices of the Ministry of Education, mathematical modeling competency is seen as one of eight main competencies spanning together with three types of overview and judgement mathematical competency in general (Niss & Jensen, 2005); see Figure 41.1.

Within the framework of mathematical competency, modeling competency can be defined as: "A person's insightful readiness to autonomously carrying through all aspects of a mathematical modelling process in a certain context and to reflect on the modelling process and the use of the model" (Blomhøj & Jensen, 2003, p. 127). In this definition, mathematical modeling competency refers to *a mathematical modeling process* and therefore, in order to unfold the elements involved in modeling competency it is necessary to first analyze a modeling process. Although a mathematical modeling process can be conceptualized in different ways for different purposes (Niss, Blum, & Galbraith, 2007), it is quite uncontroversial to consider a modeling process as composed of six sub-processes; see Figure 41.2.

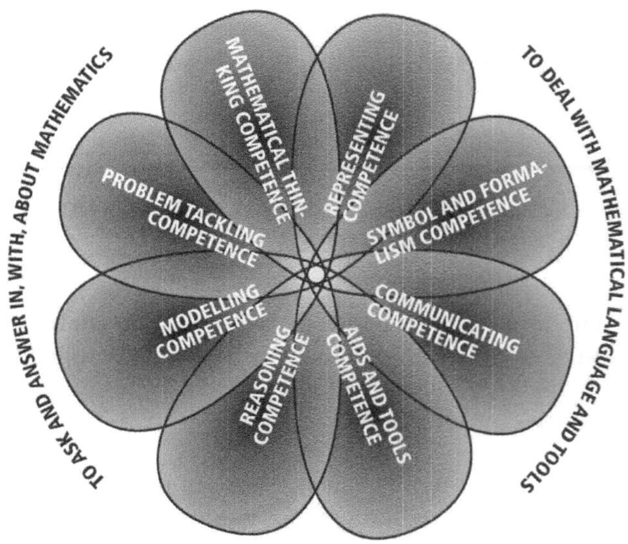

Figure 41.1 A visual representation—the "KOM flower"—of the eight mathematical competencies that are presented, exemplified, and argued to span mathematical competency in the KOM project (Niss & Jensen, 2005).

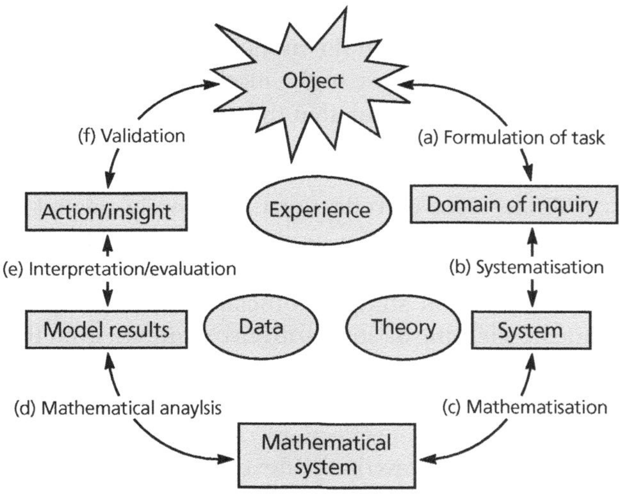

Figure 41.2 A visual representation of a mathematical modelling process (Blomhøj, 2004, p. 148).

a. Formulation of a task (more or less explicit) that is related to a perceived reality and influenced by the modeler's interests. Through this process, the object of the modeling process is constructed. The object can be reconstructed as a result of the modeling process. However, it is the object and the formulated task that guides the identification and construction of a domain of inquiry.
b. Selection and construction of the relevant objects, relations, and so on from the domain of inquiry, and idealization of these, in order to make a mathematical representation possible
c. Transformation and translation of selected objects and relations from their initial mode of appearance to mathematics by further abstraction and idealization
d. Using mathematical methods to achieve mathematical results and conclusions
e. Interpretation of these as results and conclusions regarding the system or the initiating domain of inquiry
f. Evaluating the validity of the model by comparison with data (observed or predicted) and/or with already established knowledge (theoretically based or shared/personal experience based)

The modeling process should not be understood as a linear process with fixed start and ending points. In real-life modeling as well as in modeling taking place in an educational context, a modeling process need not follow the logical sequence of the sub-processes indicated in Figure 41.2. The model depicted in Figure 41.2 is a tool for analyzing the structure of the modeling process. The circular diagram with the double arrows describes a cyclical process where reflections along the process can lead to changes in previous processes, thereby initiating a new loop in the modeling cycle. It should be noticed that the term *mathematical system* is used here to describe the mathematical representation of the model. This is in order to be able to define the model as a relationship created through the modeling process between the object and the mathematical system.

It follows immediately from this unfolding of the modeling process that the different sub-processes involve different forms of reasoning and reflections. Therefore, in order to pursue modeling competency as an educational goal, regardless of educational level, teaching approaches where students are often challenged sufficiently to work with the entire modeling process are needed. Of course, mathematical modeling competency may not be equally important as an educational goal at all levels, but if the goal is to make students able to perform mathematical modeling at whatever technical level, they need to work with mathematical modeling in full flesh. In other words, modeling competency does not follow from knowing the relevant mathematics and having insights into the relevant part of reality.

Moreover, to develop mathematical modeling competency, as defined above, it is not enough that students are challenged to work with the entire modeling process; it also involves developing students' *autonomy* in relation to carrying through an entire modeling process. This aspect is made part of the definition of modeling competency to emphasize mathematical modeling competency as a gestalt, a unifying whole, that cannot be captured through the ability to master each of the involved sub-processes separately. As a consequence, mathematical modeling competency cannot be developed through atomization and separate training of the involved sub-processes. However, as has been extensively discussed in the literature since Niss (1989), there is a delicate balance to maintain between working with the more mathematically demanding inner parts of the modeling process (i.e., process (c) and (d) in Figure 41.2), which of course are more closely connected to the learning of mathematics, and a more holistic approach where students are set free to work with the entire modeling process.[1]

The definition of mathematical modeling competency then poses a dilemma for developing students' modeling competency in an educational setting. On the one hand, students need to work with the entire modeling process as a unifying whole in a way that does not atomize into working with the sub-processes separately. On the other hand, there is a recognized need to pay special attention to the learning of the inner parts of the modeling process. This dilemma gives rise to the first two of the three educational challenges in developing modeling competency that were mentioned in the introduction:

1. How can we challenge and support students to work with the entire modeling process in a way that supports their autonomy?
2. How can we give sufficient support to students' work with the processes of mathematization and mathematical analysis as integrated elements of teaching mathematical modeling?

THE ROLE OF REFLECTIONS IN MODELING COMPETENCY

Our definition of modeling competency explicitly includes reflections on (1) a modeling process and (2) the use, or possible uses, of a model in extra-mathematical contexts. This is essential and calls for some elaboration.

Reflections on a modeling process can take place in parallel with the actual modeling process. On the first level it has to do with noticing what is going on in the process, and here a model of the modeling process (like Figure 41.2) can help students to structure their reflections in accordance with the sub-processes. These reflections can be guided by questions such as:

- Why did we formulate the problem like we did?
- What is the purpose of our modeling process or our model?
- Which important elements did we include in our system and why?
- Which were disregarded and why?
- What were our reasons for mathematizing the processes in the system like we did?
- On what grounds did we estimate the parameters?
- Why did we perform exactly these calculations with our model?
- How sensitive are our model results towards changes in parameters and initial conditions?
- Why did we interpret the results the way we did?
- What type of knowledge or insights have we gained through the modeling process and through the use of the model?
- Why do we think that the model is valid in relation to our problem or why not?

These types of reflections are closely related to the documentation of a modeling process. However, as can be experienced from supervising modeling projects or from studying official modeling reports made in scientific or administrative contexts, even this type of first-level reflection does by no means follow automatically with the construction of a model.

These first-level reflections can lead to critique of the modeling process on the next level: "Could we have decided differently?", "What would be the consequences of this or that change in the modelling process?" and so on. These types of reflections and critique are directed towards particular sub-processes in the modeling process and can therefore be integrated in the modeling process. Accordingly, this type of reflection and critique may be labeled "internal reflections."

Of course, reflections of a modeling process can also be exercised retrospectively based on a written documentation of a modeling process in which the students have not been involved. However, such an endeavor presupposes that the students themselves simulate the entire modeling process in order to make the appropriate reflections and critique of the modeling process.

Another type of reflection is concerned with the use or possible uses of a mathematical model in societal or technical contexts. This type of reflection necessarily involves the context in which the model is being used or could be used and can therefore not be integrated in or structured by means of the modeling process. Accordingly, this type of reflection could be labeled "external reflections."

In general, when a mathematical model is being applied in a societal context, there are always some interests involved. Interests related to power and political issues, economical issues, and ethical issues could be built into the model through the modeling process and influence the actual application of the model and the consequences of the application, intended or unintended. The modeling process creates a complex structural relationship between two entities of different epistemological nature, namely the object that is to be modeled and the mathematical system. The object that is modeled through a mathematical modeling process is not given beforehand and is not simply a part of reality, nor is it a formal system. The object is constructed through the modeling process. Of course, there is always some part of reality in focus in the beginning of a modeling process, but the exact delimitation and perception of the object to be modeled are created during the process under strong influence of the modeler's interests and the theoretical basis of the model. The mathematical system, on the other hand, is a formal system that can be analyzed and manipulated by means of mathematics and is formed through a process of abstraction and idealization. Theory, broadly understood as what is known or taken for granted also in the form of available data, is influencing the modeling process in an intricate interplay with the interests behind the model and its intended applications.

The structural relationship between the object, the mathematical system and this application is depicted in Figure 41.3. When a mathematical model is being applied in a societal or technological context, the model is often disconnected from the modeling process and the mathematical system is taken to be identical with the object.

Except for modeling within the framework of a mathematized theory as found in physics and in certain parts of chemistry, it is not possible to secure the model's validity theoretically. As a consequence, the validity of a concrete application of a mathematical model has to be judged on the basis of the modeling process behind the model and/or empirical testing. Therefore, it is an important part of modeling competency to know about the structural relationship between the object that is actually modeled and the mathematical system, connected to a mathematical model and to be able to identify the model and the modeling process behind an application.

Mathematical models of different kinds and complexity are playing important roles in the functioning and forming of societies based on high technologies. Therefore, the development of an expert's as well as a layman's competency to reflect upon and criticize the use of mathematical models in decision making and technological investigations is becoming imperative for the maintaining and further development of democracy. Accordingly, external reflection and critique are explicitly included in the definition of mathematical modeling competency.

Application of a mathematical model to a real-life problem should be considered as a process of its own that is separate from and more or less independent of the modeling process. The application process tends to cause changes in the context of the real-life problem as well as in the model when it is adjusted to the problem.

When a mathematical model is developed for intended applications, its development is influenced by certain interests of the acting subjects. The modeling process that lies behind the model can be theoretically and/or empirically well or less well founded. However, when the model is applied in a political and/or technological investigation, the model is often separated from the modeling process and the possible critical reflections connected with the process. The model may be used for particular purposes that might be different from what it was constructed for. Moreover, the application process can be influenced by yet other interests that were not present in the modeling process. Therefore, the process of applying a mathematical model in general tends to cause (1) a reformulation of the problem in hand in order to be adequate for investigation by means of mathematical modeling, (2) changes in the discourse about the problem in the direction that is either pro or contra to the model and possible adjustments of the model, (3) a limitation of the possible actions taken into consideration to those

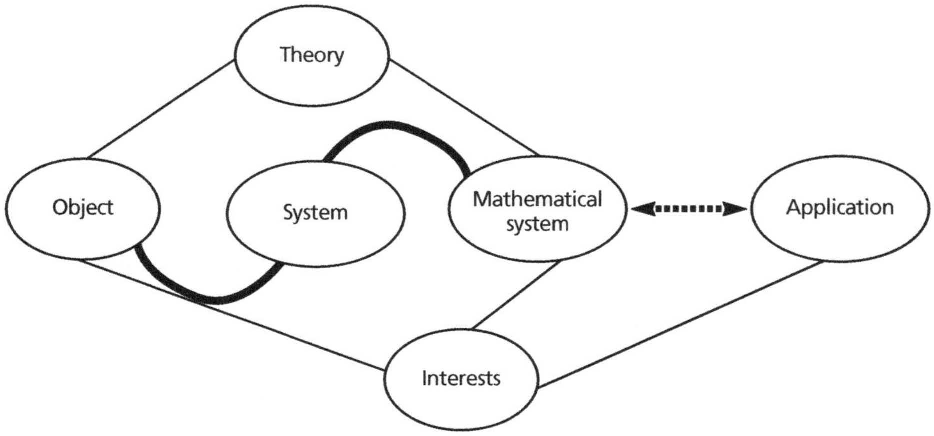

Figure 41.3 Application of a mathematical model to a real-life problem is a process more or less independent of the modeling process. The application process tends to cause changes in the context of the real-life problem as well as in the mathematical system when the model is adjusted to the problem.

that can easily be evaluated in the model, and/or (4) a delimitation of the group of people who can take part in the discussion and act as a basis of critique (Skovsmose, 1990, pp. 129–130). Awareness of and experiences with such general phenomena can provide a strong basis for students' external reflections concerning the use of mathematical models.

Some mathematics educators may find that such reflections are of a non mathematical nature and therefore should not be part of the teaching of mathematical modelling. However, as will be illustrated in the examples to follow, how model results are used in a concrete political decision or in a technical design can often only be understood and criticized on the basis of a thorough analysis of the mathematical modelling process leading to the model and the model result. In other words, mathematical modelling competency is needed in order to understand and criticize decisions based on mathematical models. Therefore, developing mathematical modeling competency includes the obligation to foster among students an ability to reflect critically upon actual or possible applications of a mathematical model. Of course the priority of this obligation depends on the educational level. Hence, a third educational challenge in connection with the development of mathematical modeling competency can be formulated:

3. How can we develop students' internal and external reflections in their modeling activities?

MATHEMATICAL MODELING PROJECTS AT THE NAT-BAS PROGRAM

The problem-oriented project work at the two year introductory and interdisciplinary study program in science (Nat-Bas) at Roskilde University can be seen as an extreme example of how to organize a study at university level, that challenges students to work with all phases of a mathematical modeling process as a unifying whole and hence supports their development of modeling competency and in particular their autonomy. Also, the two types of reflections connected to mathematical modeling are explicitly included in the intentions for the project work at Nat-Bas.

In each of the four semesters of the two-year program the students work on a problem-oriented, student-directed project in groups of three to seven students. Each project runs for the whole semester and covers half the credit (15 etcs, 60 etcs equals one year of full study activity). In addition to the projects, the students study two subject-oriented, traditional first-year university science courses. The project work is organized in houses with typically 60–80 students.

For supervising the projects a group of 10–12 professors from the various science disciplines (molecular biology, environmental science, physics, mathematics, etc.) are allocated to a house. In the beginning of each semester, the project groups are formed by the students based on their subject matter interests and social preferences. Before a deadline (after approximately two weeks), the groups have to hand in a preliminary version of the problem and a brief description of the project that the group has decided to work with. Based on these preliminary project descriptions, the supervisors are distributed among the groups to best match the supervisors' competencies and the subject matter of the projects. Normally, the supervisor will meet with his or her group once a week for one to two hours and, if necessary, can quite easily get help from other supervisors in the house or from experts within the project area at the university or at other universities or industries.

The projects of the first three semesters have to comply with the so-called semester theme, but it is in fact possible for the students to work with projects that involve different aspects of mathematical modeling in all four semesters of the program. However, in the third semester the perspectives are of a didactical, historical, or philosophical nature and not directly on the modeling process or the application of a model in a particular context (see the paper by Kjeldsen & Blomhøj, 2009). Therefore, only the semester themes for the first and the second semester are of interest here:

1st semester: Application of natural sciences in technology and society

2nd semester: Models, theories, and experiments in the natural sciences (Floutrup & Jensen, 2006, p. 28)

In the following we present and discuss two types of common mathematical modeling projects at Nat-Bas fulfilling respectively the first and second semester theme for the project work. Both types of projects can also be carried through in the fourth semester where there is no theme that constrains the projects work. We refer to the paper by Blomhøj and Kjeldsen (2009) for general background information on the organization of the project work.

On average, 25% of the projects in semester 1, 2, and 4 contain a substantial element of mathematical modeling and related reflections—every year 8–10 modeling projects are conducted by groups of four to seven students. Many of the students who continue their studies in mathematics, physics, chemistry, or biology have completed two projects involving mathematical modeling. In addition, many of the students also follow a two-semester course on mathematical modeling as part of their Nat-Bas program. All in all, mathematical modeling plays a prominent role in the interdisciplinary science study program at Nat-Bas.

Based on our supervising experiences, as explained in the introduction, we shall characterize two different types of modeling projects focusing on societal or technological use of a given model and on the modeling process itself. Each type of project will be illustrated with two concrete examples.

PROJECTS EMPHASIZING THE APPLICATION OF A PARTICULAR MODEL IN A SOCIETAL CONTEXT

In post-modern societies where production, administration, and societal planning involve advanced technology, mathematical modelling plays a crucial role. Mathematical models are developed in interplay with sciences and technology with the purpose of describing, predicting or prescribing societal or technological problems or systems (Davis & Hersh, 1986). Mathematical modeling is a forceful and often indispensable approach whenever we wish to make quantitative prognoses or predictions of the effects of certain actions in the natural environment or in societal systems. Questions such as these cannot be dealt with properly without mathematical modeling:

- How will the Danish population develop the next 10 years?
- How large a percentage of our electricity consumption can be produced by windmills?
- Should we vaccinate against influenza in public schools?
- Where and how should we invest in the transportation infrastructure?

However, as already explained, applications of mathematical models in relation to such societal and technological issues are not at all neutral. Therefore, mathematical modeling projects under the theme of the first semester at Nat-Bas should provide the students with exemplary experiences with the application process—that is, external reflections on the relevance and function of applications of mathematical models in relation to societal problems. The following examples serve as illustrations. The first project analyzes an authentic case where an important decision concerning traffic planning in the city of Roskilde was taken on the basis of a mathematical model, a type of modeling project that is quite common at Nat-Bas. In the second project the students are carrying out a modeling process themselves with the purpose of evaluating different vaccination strategies against influenza and comparing them to the one presently used in Denmark where primarily only older generations are vaccinated.

The Use of a Traffic Model in the City of Roskilde—The Case of "Ny Østergade"

This project, which, even though it was a fourth-semester project, fulfilled the theme of the first semester, was conducted by two students. The problem that guided the students' work took its point of departure in a contemporary political conflict between the local administration in the city of Roskilde and a group of citizens living very close to where a new road connection was planned (Jarbøl & Kofoed, 1996). The decision of where exactly this new road connection should be built was based on results that came out of a mathematical modeling process that also incorporated other traffic regulations. The model was developed for the city by a consulting firm, and the students analyzed the role of the model in the decision as well as in the discussion of the decision.

In the modeling process, the modelers from the consulting firm divided the city into 57 zones and, based on interviews and traffic counting, a 57×57 tour matrix was set up to describe the total traffic flow within and in/out of the city per day. The matrix contains the estimated number of tours between each pair of zones where some of the zones represent "ports" leading in and out of the city.

The tours were distributed on a road net that represented all the main roads, both those within the city and those leading in and out of the city. The planned new road connection and some other planned regulations were included in the road net. In the model, the traffic was distributed on the road net according to three principles: (a) for all roads the traffic allocated should be less than the capacity of the road in question, (b) all tours should be placed on the net so that the calculated time of transportation for each tour is minimized, and (c) for each road element in every tour the time of transportation is calculated in minutes according to the formula:

$$T_i = 60 \cdot \frac{L_i}{v_i} + c_i L_i,$$

where L_i is the length of the i-piece of the considered road measured in kilometers, v_i is the average speed at that piece of road in kilometers per hour, and c_i signifies that some extra time is used depending on the type of road. The value of c_i in the particular case of the planned new road connection was 0.3 minutes/kilometer.

In their report, the students described in detail the model and its assumptions. The model's behavior, the empirical basis for the tour matrix, and the estimation of the model parameters were also analyzed.

To structure their analysis, the students used a model of a modeling process similar to the one presented in Figure 41.2. As part of their analysis the students identified and discussed the possible reasons for and effects of the different assumptions of the model, and these internal reflections formed the basis on which the students, later in their report, were able to recognize, as will be explained below, that the model had been used beyond its scope of validity in the decision process. By analyzing the model used by the authorities at Roskilde to decide where to build the road, the students worked with the entire modeling process. The point of departure for the students' work was to examine and understand the role the model played in the decision process and the public discussion of the decision, which is a complex problem that was not atomized beforehand into a list of given tasks related to the modeling process. Hence, the problem orientation and the student-directed way of teaching in the project work under the first semester theme ensured that the students were forced to work with modeling in "full flesh" in a way that supported their autonomy.

Regarding the challenge of how to develop students' external reflections in their modeling activities, the students of this project necessarily had to make external reflections in order to understand the role of the model in the application process. The students discovered that the model was used at two different occasions to predict the traffic on the new road to be in the one case clearly above 10,000 cars per day and in the other case clearly below 10,000 cars per day. They found out that the high estimate was used in an argumentation for the effect the new road would have in order to reduce the traffic in the center of the city and thereby for the construction project to get support from national governmental sources, whereas the low estimate was used to prevent the road project from being subject to an EU procedure that evaluates the effects on the environment. This EU procedure, which the group of citizens had wished for, only applies for road projects with a predicted traffic flow above 10,000 cars per day. From their analysis of the modeling process, the students realized that the two (contradicting) results were obtained simply by using two different values for the parameter v_i for the new road connection—a change for which there was no valid foundation since the parameter v_i for a non-exiting road cannot be made subject to empirical control; see Figure 41.4.

Based on interviews with one of the modellers from the consulting bureau, local politicians and the spokesperson for the group of citizens, the students analysed the discourse pro and contra the new road connection. They concluded that the model played a central role in the discourse and that the results from the model were used beyond its scope of validity.[2]

Figure 41.4 The road map of Roskilde illustrating the situation in the model where two different tours are connecting the same two points A and B. The connection south of the railway includes the new road (the blue part). Since the two connections are very close in length the model's distribution of traffic is very sensitive towards changes in the parameter v for the new road.

Influenza Epidemic—A Mathematical Modeling Project

In Denmark, individuals over the age of 65 are offered free vaccination against influenza each year before the winter season, and this is the only public program against influenza. Investigations indicate that vaccinating children would decrease the mortality of older people. Because of these findings, a group of four students decided to investigate whether influenza epidemics could be avoided by applying another vaccination strategy than the one presently used by the Danish authorities. The group decided to evaluate the effect of different vaccination strategies by means of mathematical modeling (Jørgensen et al., 2002).

The group realized that in order to model the dynamic of influenza epidemics over many years and to investigate different vaccination strategies they needed to develop an age-structured model allowing for age-depending parameters for the spread of the infection. Assuming that the population was in equilibrium they developed a cohort model with the following five age groups: 1–5, 6–15, 16–25, 26–64, and 65–75. The main argument for the age groups was that the population within each group should be as homogeneous as possible with respect to their pattern of social contacts. The students modeled the dynamic of the possible yearly influenza epidemic by means of a SIR model according to the compartment diagram in Figure 41.5.

Based on research literature, the students estimated rates for effective contacts within and in between all five age groups. Using the SIR model structure the students developed a model that could simulate influenza epidemic on a yearly basis. For each year in the model it was calculated whether or not the threshold for an epidemic to occur was exceeded. If so, an influenza epidemic was simulated and the number of S, I and R in each age group was updated. A new cohort of susceptible one-year-old children was born into the model population and the other groups were adjusted according to their growing age. To balance the model population, the cohort of 76-year-old persons was leaving the model. To simulate the genetic changes in influenza virus from year to year and the related partial immunity, the students introduced a flow from the group of immune back to the group of susceptible.

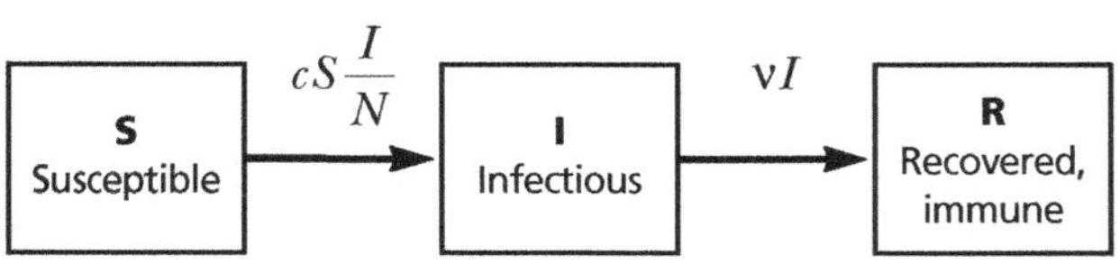

Figure 41.5 Shows the dynamic for a SIR-model for an infectious disease with immunity like influenza. The flow goes from susceptible to infectious to immune. c is the rate of effective contacts, ν is the rate of recovery and N is the total population. An epidemic can only occur if the flow into I is larger than the flow out of I.

The students tested the model and they were able to reproduce some observed phenomena such as the period of 2–3 years between influenza epidemics and that the 6–15-year-old group has the largest prevalence during an epidemic. They used the model to evaluate different vaccination strategies assuming different degrees of immunity after vaccination. Vaccination of all people over 65 years only reduced the overall prevalence in the population during an epidemic by 4%, while the same number of vaccinations given to the 6–15-year-olds reduced the prevalence by 31%. Still, people over 65 years would be slightly better protected if they were all vaccinated compared to only vaccination of the 6–15-year-olds. Moreover, the model predicts that influenza epidemics could virtually be avoided by vaccination of the 6–15-year-olds every year. However, such a strategy presupposes that it is possible to develop and administrate an effective vaccine against the expected type of influenza each year.

In contrast to the first case, these students constructed a model on their own—not totally from scratch, but by remodeling the standard SIR model for spread of diseases. In this project, the students had to study some epidemiology of infectious diseases. Based on this, the group was actually able to complete and argue for the remodeling, the introduction of the age groups, and the significance of these adjustments of the SIR model, through internal reflections on the modeling process as they carried it out. To be more specific, the students' internal reflections are apparent in their project report where they discussed in detail the assumptions made in their modeling process and argued that the model should not be expected to produce valid predictions of the level of prevalence of influenza under different vaccination strategies.

However, they argued that the model gives strong arguments for a change of strategy towards vaccination of population groups with high contact rates. The students also discussed the ethical questions related to the vaccination of healthy young people who are expected to recover easily from influenza, and leaving older people with a risk of dying from influenza protected only by the bulk immunity in the population. The purpose of the model was to see if it could suggest a "better" vaccination program, and the students' evaluation and discussion of the use of the model as an argument to implement a new vaccination program brought out external reflections as described above.

As was the case in the previously discussed project, through the problem-oriented and student-directed way of teaching in the Nat-Bas program, students who choose to work with mathematical modeling almost automatically get to work with the entire modeling processes. In both these projects the students worked in detail and quite autonomously with the entire mathematical modeling process guided by their problem and not the least by their own interest in the projects; the first group of students through their analysis and simulation of a modeling process behind an existing model and the second group through their own modeling. Therefore, it is fair to say that the projects have contributed to the development of the students' modeling competency. However, in relation to the theme for the first semester, the important thing is that in both projects the students got first-hand experiences with the use of mathematical models in a societal problem situation. And in both cases the groups performed relevant external reflections concerning the role of the model as basis for decisions of societal importance.

The point in this type of project is that students get an insight into an often very complex mathematical model and the actual or possible applications of the model in a social or technological context. Moreover, in the best cases students get first-hand experiences with reflecting critically on the model's validity in relation to concrete applications. The point is not to criticize some concrete political decisions but to understand mathematical modeling as an often indispensable but also not unproblematic tool for investigating complex societal and technological problems.

PROJECTS EMPHASIZING THE MODELING PROCESS

Mathematical modeling is playing a very important role in many other scientific disciplines. Not only in connection with applications of theories from other disciplines, but even more importantly, mathematical modeling is often woven together with the formulation and definition of theoretical key concepts and with standard experimental practices. Quantitative experiments or interpretation of experimental measurements always involve some kind of mathematical model explicitly or implicitly. Of course, this is especially evident in the natural sciences—particularly physics and parts of chemistry, where mathematical modeling is integrated in the theoretical framework and the scientific practice to such an extent that it is hard to identify where the mathematical modeling starts and ends.

The general idea with mathematical modeling projects under the theme of the second semester at Nat-Bas is that students should get first-hand experiences with the interplay between theoretical concepts, mathematical modeling, and experiments. To achieve this, as was the case also for modeling projects done under the first semester theme, students need to either set up a mathematical model themselves or analyze in detail the modeling process behind an existing model. The difference lies in the use of the models. In first semester modeling projects, the application process is directed towards societal or technological decisions, whereas in second semester modeling projects, the model is used to get insights into the behavior of a complex phenomenon in another scientific field. This is exemplified below by two projects in which the students followed each of the two approaches of either analyzing and existing model or building one themselves.

Conjugative Plasmid Transfer

This project was a second semester project with a substantial experimental element, which is compulsory in one of the four projects (Barascuk et al., 2001). Inspired by a scientific paper (Andrup, Smidt, Andersen, & Boe, 1998), a group of eight students investigated the rate of transfer of plasmids with an antibiotic resistance gene in a bacterial system. The paper presented the hypothesis that the rate of transfer of plasmids between bacteria by conjugation would follow the well-known Michaelis–Menten model for the kinetic of first-order enzymatic catalysis. In this model the rate of catalysis V varies with the concentration of substrate $[S]$ according to the formula

$$V = V_{\max} \cdot \frac{[S]}{[S]+k_m},$$

where k_m [mol/l] is the substrate concentration that corresponds with a rate V equals to the half of the maximum rate, $V_{\max}$ [$\text{time}^{-1}\text{mol}^{-1}$], depending of course on the amount of enzyme.

The students analyzed carefully all the assumptions behind the theoretical deduction of this model for enzymatic catalysis, leading them to reflect on the modeling process. For example, for each particular assumption they discussed in detail the degree to which the conjugative system could be expected to comply with the assumption. They argued that, from what is known about the conjugation, the model should be valid with some approximation also for plasmid transfer by conjugation. In this modeling process the donor cell (the bacteria with the plasmid) and the recipient cell were considered to be analogous to the enzyme and to the substrate, respectively. Hence, the number of plasmid transfer per time per donor cell, *V*, as function of the concentration of recipient cells should follow the Michaelis–Menten model.

The students tested this hypothesis experimentally in the lab, and they were able to confirm the model at least for some of the experiments. In addition, they were able to estimate the model parameters by means of linear regression on transformed data.

In this project, the students worked intensively with theory (in microbiology), mathematical modeling, and experimental investigations. The use of the model in this context, to gain new knowledge about conjugative systems, led the students to external reflections. Their experience was that the modeling process and the model were necessary for forming the hypothesis and for designing the experiment, and that the experiment on the other hand was necessary for testing the model and for estimating the parameters. They also found that the model with its theoretical foundation and the experimental results gave rise to new research questions about the characterization of different conjugative systems in terms of model parameters.

Why Is There Always a Queue in the Morning Traffic at the Highway Passing by Roskilde University? And What Can Be Done to Overcome the Problem?

This question was the point of departure for a second semester project that was made by a group of four students (Andersen et al., 1995). To gain insights into the queuing phenomenon, the group, in which three of the students had some programming skills, decided that they would try to model the highway traffic mathematically by modeling each individual car and its behavior in relation to the surrounding traffic. The students designed the model according to the following eight points, where 1–4 have to do with the generation of vehicles in the model and 5–8 with the dynamical simulation of the traffic in small discrete time intervals:

1. The type of vehicle was decided—that is, the length, according to three classes: standard private cars, large vehicles, and trucks.
2. The speed, at which the vehicle would like to travel if possible, was taken randomly from an empirical estimated normal distribution depending of the type of vehicle.
3. The destination of the vehicle was determined—one of three possible junctions or the end of the simulated highway segment. The frequencies of the four possible destinations were estimated from available traffic data and traffic counting made by the group.
4. According to the frequencies estimated in (3), the vehicles were placed either at the beginning of the segment or at one of the junctions.
5. The vehicles were moved according to their speed and their actual position on the road (with two lanes) taking in to account the distance to the nearest vehicle in front.
6. The speed for each vehicle was updated according to the laws of mechanics with some particular security distance as a parameter and with realistic values for acceleration and de-acceleration.
7. It was checked just before each junction if it is time for the vehicle to leave the highway.
8. For each vehicle it was decided whether a change of lane should occur in the next time interval.

The group succeeded in implementing the model in C++ with a graphing facility that showed the movements of each individual vehicle as lighting dots on the screen. The model was tested against traffic data from the morning peak hour and the model simulated the traffic outflow from the highway segment within a 3% tolerance. The students experimented with the model to investigate the effect of changes in the system, such as changes in the frequencies of incoming vehicles, and concluded that the capacity of the highway segment often would be exceeded in the morning peak hour.

In relation to the theme of the second semester, the main point in both projects is that the students became deeply involved in the modeling process and related internal reflections. In the plasmid transfer project, the students experienced how a research question in microbiology can give rise to a mathematical modeling process. Through their detailed discussions of the biological assumptions that should apply to the conjugative system for it to be modeled by the Michaelis–Menten model the students gained first-hand experiences with the interplay between theoretical knowledge and empirical data in the construction of a mathematical model. The students even realized that a model can be a prerequisite for analyzing experimental data or for designing relevant experiments. In the traffic modeling project, the students actually carried through an entire modeling process from scratch. In their modeling work, the students used their competencies in programming in a very effective way that enabled them to maintain a great deal of autonomy throughout the entire process. They engaged in detailed considerations on how to delimit and model the characteristics of the different types of vehicles and the variation among the individual vehicles. In this work, they experienced the interplay between the delimitation of the system and available data for parameter estimation. The students also developed their own way to test the model against traffic data. In both projects, the students experienced the complex interplay between theoretical knowledge, empirical data or experiment, and a mathematical model.

DISCUSSION

The first challenge in developing mathematical modeling competency as defined in this chapter is to ensure that students are actually working with all the elements of a modeling process. As illustrated with the four projects, the problem-oriented, student-guided project work at Nat-Bas where the students are free to pursue problems that are not constrained by a certain curriculum, does in fact provide excellent opportunities for students to work with the entire modeling process, which is an important precondition for developing the students' autonomy in relation to mathematical modeling. In the project work, students have time and support from their supervisor to really go into depth with a problem involving a mathematical modeling process. Nat-Bas is an interdisciplinary study program in science, and therefore it is completely adequate for the students to study the subject areas in medicine, molecular biology, or in other disciplines, which plays an important role in the modelling process.

As for the second challenge we have to realize that the project work cannot do the job alone. Students also need to work more systematically with the inner parts of the modeling process as addresses in the second challenge. This is a difficult task to handle within the project work, and often it does not happen. The project work is not an optimal teaching strategy to support students to work more systematically with those parts of the modeling process. In the Nat-Bas program this is instead taken care of in an elementary modeling course covering two semesters,[3] and in more advanced modeling courses included in the mathematics bachelor program.

The third challenge concerning students' internal and external reflections in relation to their modeling activities is addressed in the thematic organization of the project work at Nat-Bas. Mathematical modeling projects in the first and second semester have to be exemplary in relation to external and internal reflections, respectively. But, as we have seen, both types of reflection occur, though with different emphasis, in both types of projects.

A fourth challenge can also be identified in relation to modeling projects in the context of the interdisciplinary Nat-Bas program. The students' free choice of problem for their projects has to be balanced with the possibility for them to get hands-on experience with the subject areas involved, so they can carry out a modeling process or analyze a given model on a solid ground of knowledge. This is of course enhanced, if the subject area is within the supervisor's field of research.

The problem-oriented, student-directed way of organizing half of the studies in the Nat-Bas program does not in itself guarantee that the third challenge is met in modeling projects. That needs to be specifically addressed by the supervisor, who throughout the entire process of the project work has to keep challenging the students to reflect upon the modeling process they are either construction themselves or simulating based on an existent model. In fact, in order to integrate internal reflections and even critique into students' modeling projects, this needs to be given special attention not only in the supervising process, but also in evaluations of the projects. Many students have difficulties with handling the uncertainties in the modeling process, and it takes some experience for them to recognize that their reflections and critique of their own modeling process add quality to their work. Concerning the external reflections, the challenge is to develop students' ability to make a balanced analysis and judgement about the (possible) use or misuse of a mathematical model in a societal or technologic context, or as a tool to gain insights into the behavior of (complex) systems in other sciences.

The project work and the clear focus on mathematical modelling competency come with a price—less time for teaching important and fundamental science disciplines. However, we see mathematical modelling together with experimentation as the two main constituents in our interdisciplinary introductory science study program at Roskilde University and—as we hope this paper demonstrates—for good reasons.

NOTES

1. In Blomhøj & Jensen (2007) the dilemma of teaching autonomy directed specifically towards modeling competency

is discussed with the focus on upper secondary mathematics teaching.

2. This project is also discussed by Skovsmose (2005, p. 93).
3. This course is described and discussed in Blomhøj & Jensen (2003) and in Ottesen (2001).

REFERENCES

Andersen, M., et al. (1995). *Trafik modellering.* Unpublished student project report, 2nd semester, Nat-Bas, Roskilde University, Denmark.

Andrup, L., Smidt, L., Andersen, K., & Boe, L. (1998). Kinetics of conjugative transfer: A study of the plasmid pXO16 from Bacillus Thuringiensis subsp. Israelensis. *Plasmid, 40,* 30–43.

Barascuk, N., et al. (2001). *Plasmidoverførsel.* Student project report, 2nd semester, Nat-Bas, Roskilde University, Denmark.

Blomhøj, M., & Jensen, T. H. (2003). Developing mathematical modelling competence: Conceptual clarification and educational planning, *Teaching Mathematics and its applications, 22*(3), 123–139.

Blomhøj, M. (2004). Mathematical modelling—A theory for practice. In B. Clarke et al. (Eds.), *International perspectives on learning and teaching mathematics* (pp. 145–160). Gothenburg, Sweden: National Center for Mathematics Education, University of Gothenburg.

Blomhøj, M., & Jensen, T. H. (2007). What's all the fuss about competences? Experiences with using a competence perspective on mathematics education to develop the teaching of mathematical modelling. In W. Blum (Ed.), *Modelling and applications in mathematics education. The 14th ICMI-study* (pp. 45–56). New York: Springer-Verlag.

Blomhøj, M., & Kjeldsen, T. H. (2009). Project organised science studies at university level: Exemplarity and interdisciplinarity. *ZDM—The International Journal on Mathematics Education, 41*(1–2), 183–198.

Davis, P. J., & Hersh, R. (1986). *Descartes' dream.* London: Harvester Press.

Floutrup, S., & Jensen, J. H. (Eds.). (2006). *Studievejledningen 2006–2007* [The study guide 2006–2007]. Study Board for the Basic Studies Programme in Natural Sciences, House 13.1, Roskilde University, Denmark.

Jarbøl, M., & Kofoed, M. (1996). *En trafikmodel i Roskilde kommune—en analyse af modellen bag beslutningsprocessen omkring omfartsvejen Ny Østergade.* Student project report, 4th semester, Nat-Bas, Roskilde University, Denmark.

Jørgensen, C. et al. (2002). *Influenzaepidemier—et matematisk modelleringsprojekt* (IMFUFA-text no. 410). Roskilde: Roskilde University.

Kjeldsen, T. H., & Blomhøj, M. (2009). Integrating history and philosophy in mathematics education at university level through problem oriented project work. *ZDM— The International Journal on Mathematics Education, 41,* 87–104.

Niss, M. (1989). Aims and scope of applications and modelling in mathematics curricula. In W. Blum et al. (Eds.), *Applications and modelling in learning and teaching mathematics* (pp. 22–31). Chicester, UK: Horwood.

Niss, M. (2001). University mathematics based on problem-oriented student projects: 25 years of experience with the Roskilde model. In D. Holton (Ed.), *The teaching and learning of mathematics at university level: an ICMI study* (pp. 405–422). Dordrecht: Kluwer Academic Publishers.

Niss, M., & Jensen, T. H. (Eds.) (2005). Competencies and Mathematical Learning—Ideas and inspiration for the development of mathematics teaching and learning in Denmark. English translation of part I-VI of Niss & Jensen (2002): *Kompetencer og matematiklæring – Idéer og inspiration til udvikling af matematikundervisning i Danmark, Uddannelsesstyrelsens temahæfteserie nr.* 18 – 2002. Copenhagen: Danish Ministry of Education.

Niss, M., Blum, W., & Galbraith, P. (2007). Introduction. In W. Blum, P. Galbraith, H.-W. Henn, & M. Niss (Eds.), *Modelling and applications in mathematics education. The 14th ICMI-study* (pp. 3–32). New York: Springer-Verlag.

Ottesen, J. (2001). Do not ask what mathematics can do for modelling. Ask what modelling can do for mathematics! In D. Holton (Ed.), *The teaching and learning of mathematics at university level* (pp. 335–346). Dordrecht: Kluwer Academic Publishers.

Skovsmose, O. (1990). *Ud over matematikken.* København: Systime.

Skovsmose, O. (2005). *Travelling through education. Uncertainty, mathematics, responsibility.* Rotterdam: Sense Publishers.

42

LEARNING MATHEMATICS THROUGH MODELING

The Case of the Integral Concept

Morten Blomhøj and Tinne Hoff Kjeldsen

IMFUFA, DEPARTMENT OF SCIENCE, SYSTEMS AND MODELS, ROSKILDE UNIVERSITY

Abstract: In this chapter we investigate mathematical modeling as a means for learning mathematics. We analyze and discuss how modeling has the potential of supporting students' learning of mathematical concepts, and how these potentials can be realized in teaching situations. The theoretical framework is our understanding of the modeling process and different theories of difficulties of learning mathematical concepts. The institutional context is an elementary modeling course for first-year university students in the natural sciences. Based on our experiences in developing and teaching this course as well as on student reports and our observations of students at work, we argue that the modeling approach has the potential to challenge and enhance students' learning of mathematical concepts involved in their modeling activities. It seems that students' contextualized reflections in the modeling situations are especially important for the realization of the learning potentials. We have identified reflections where students use the context in the modeling situation as a means for making sense of their mathematical conceptions. To illustrate this type of reflection, we present and analyze dialogues from a modeling project that is designed especially to support students' learning of the integral concept.

Keywords: Learning mathematics through modeling, course development, students' dialogues, learning, reflections

INTRODUCTION

One of the main reasons given for teaching mathematical modeling is to support students' learning of mathematics (Blomhøj, 2004; Niss, 1989). In this line of argument, mathematical modeling is seen as a didactic means for learning mathematics, and not as an objective in its own right. This is exactly the focus of this chapter.[1] We are interested in researching modeling as a tool for learning mathematics both for a better understanding of general aspects of mathematical learning and for developing our own teaching in mathematical modeling.

From a research perspective, this approach to learning mathematics is interesting because it has the ability to connect both to fundamental questions about the epistemological relations between mathematical concepts and reality, and to specific difficulties identified in the process of learning mathematics. Through modeling, mathematics is used to describe, understand, predict, and prescribe the reality we are living in. Hence, modeling can create connections in the learning process between the students' extra-mathematical experiences and the mathematics involved in their modeling activities. Moreover, observing students' modeling activities opens a window to their mathematical conceptions, because in the

The First Sourcebook on Nordic Research in Mathematics Education, pages 569–581

modeling process students are challenged to use different aspects of their mathematical conceptions and to bring them into play from new perspectives and in new situations. Therefore, from a research perspective it is of interest to design and analyze modeling activities with respect to their potentials to challenge and investigate students' mathematical conceptions.

From a developmental perspective we are interested in improving our own teaching in mathematical modeling. In our institutional context, the modeling course, which is described below, serves as a possible (and probably last) gateway for students with a relatively weak mathematical background into having mathematics as a future tool in their further science studies in university. Therefore, we are especially interested in investigating how the modeling approach can be used to support and challenge the students' conceptual learning in our context.

Accordingly, the objective for the study presented in this chapter is to answer the two following general research questions:

1. In what sense can mathematical modeling activities support and strengthen students' learning of the integral concept?
2. How can possible learning potentials be realized through interactions with students during their modeling activities?

Of course, these questions could be asked with equal relevance for other mathematical concepts as well. However, in the endeavor for students to develop a coherent understanding of calculus as a tool for mathematical modeling, the integral concept, together of course with the derivative, is a key concept.

In addition, the integral concept seems to be more readily admissible through modeling in our sense than, for example, a concept such as a K-group. Philosophers of mathematics are discussing whether mathematical objects can be said to be of different natures. Panza (2003) distinguishes between objects that can be both defined and exhibited and objects that can only be defined. Mathematicians have different ways of exhibiting objects. Some objects can be exhibited directly—for example, by drawings—whereas other objects are exhibited by showing that a definition makes sense in a given context (Carter, 2004). This discussion raises the question whether objects that can be directly exhibited are more readily admissible through modeling than objects that cannot be exhibited directly. Another way to talk about different natures of mathematical objects is to look into how an object is introduced. If we do that historically, it can be argued that the early history of the integral concept shows that the integral emerged in connection with solving quadrature problems; it was introduced as an abstraction of a sum. Carter (personal communication) suggests that objects introduced by abstraction are more readily admissible through modeling than objects introduced as consequences of inner mathematical relationships. In the present chapter, we will not go farther into this discussion of the nature of mathematical objects than observing that these philosophical considerations have consequences for modeling as a didactic means for learning mathematics.

However, before going into details with the integral concept, we first briefly describe the mathematical modeling course that constitutes the empirical basis for our research and introduce our understanding of the modeling process. Second, we link our research to present studies on the learning of mathematics in a modeling context as well as to general theories on the learning of mathematical concepts. We discuss how modeling activities can create situations in which students can engage in reflections about their mathematical conceptions in the context of the modeling process and how students through such reflections can develop their understanding of mathematical concepts. Third, we introduce the term "pedagogical observations," and explain how our research has progressed and our teaching enhanced through interplay between theory, pedagogical observations, and development of practice. Fourth, we illustrate our approach by a detailed analysis of a learning situation where groups of students modelled the CO_2-balance in a lake. We demonstrate how this approach, together with the teaching strategies discussed, can support and challenge students' understanding of the integral concept. Finally, we provide some conclusions about our general research questions.

THE EMPIRICAL BASIS—AN UNDERGRADUATE MODELING COURSE

The empirical basis for our research has emerged during eight years of cooperation between the authors (and other colleagues) in developing and teaching an elementary mathematical modeling course, called BASE,[2] for first-year university students. The course is part of a two-year introductory science study program at Roskilde University, Denmark.

The BASE course is designed in accordance with the following theoretical understanding of the concepts of a mathematical model and of the modelling process. By a mathematical model we understand a relationship between a non-mathematical object on the one hand and on the other hand a collection of mathematical objects and relations which can be interpreted as representing certain features of the non-mathematical object. This means that a model is always a model of something, and therefore, behind every mathematical model there is a modelling process, which can be more or less explicit.

Analytically, the modeling process can be described as consisting of six sub-processes as illustrated in Figure 42.1. As described by Blomhøj and Kjeldsen (this volume), the six sub-processes consist of:

a. Formulation of a task (more or less explicit) that is related to a perceived reality and influenced by the modeler's interests. Through this process, the object of the modeling process is constructed. The object can be reconstructed as a result of the modeling process. However, it is the object and the formulated task that guide the identification and construction of a domain of inquiry.
b. Selection and construction of the relevant objects, relations, etc. from the domain of inquiry, and idealization of these, in order to make a mathematical representation possible
c. Transformation and translation of selected objects and relations from their initial mode of appearance to mathematics by further abstraction and idealization
d. Using mathematical methods to achieve mathematical results and conclusions.
e. Interpretation of these as results and conclusions regarding the system or the initiating domain of inquiry
f. Evaluating the validity of the model by comparison with data (observed or predicted) and/or with already established knowledge (theoretically based or shared/personal experience-based)

The core target group for the BASE course is the fraction of students in the science program (approximately 1/3) who feel a need for strengthening their understanding of calculus from their upper secondary education, more mathematically advanced students taking the course out of interest in mathematical modeling, and those who need a mathematical modeling course for their further studies. In the Danish high school system, especially under the regulations that came into force in 2005, students are introduced to mathematical models and modeling (UVM, 2009). However, we cannot expect the students to have experience with doing mathematical modelling themselves.

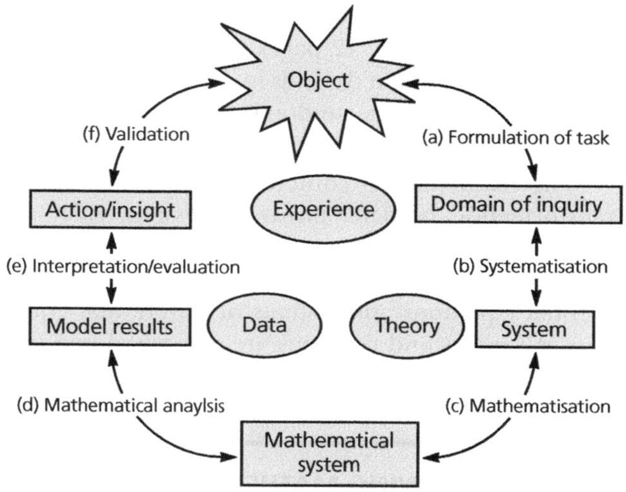

Figure 42.1 A model of a mathematical modeling process. See Blomhøj & Kjeldsen (2006) for a detailed description of the process.

The course is designed to serve the following three purposes:

1. To show that mathematical modeling is relevant—often indispensable—in studying and applying theory in the natural sciences and hereby improve students' appreciation of mathematics and motivate them to learn mathematics and modeling
2. To support students' construction of important and fundamental mathematical concepts and methods connected to the fields of calculus, differential equations, and systems of differential equations
3. To contribute to the development of the students' mathematical modeling competency

In this chapter we will focus on the second purpose of the course and only in relation to the integral concept.[3]

The course is taught in classes of roughly 40 students and includes two sessions of two and a half hours every week for 25 weeks over two semesters. As a special feature, the course is structured around project work with modeling problems. Each student work with six such projects, called mini-projects, which are generally intended to make the student work on the inner parts of the mathematical modeling process that are the sub-processes (c), (d), and (e) as described above (see Figure 42.1), not as isolated parts, but imbedded in the entire modeling process. The mini-projects are made in groups of three to four students over a period of three weeks, during which there will also be normal lectures with exercises and homework. Typically, students can choose among three different problem areas for a mini-project. For each project, a group produces a written report (approximately 12 pages). At the end of each semester every student defends in an oral exam one of the three mini-projects, randomly chosen, in which he or she has been engaged during that semester.

The mathematical content of the course covers transformation of data, linear regression, calculus, and, as the main components, numerical, analytical, and qualitative methods for solving and analyzing differential equations and systems of differential equations. All topics and concepts are presented and discussed as tools for mathematical modeling.

LEARNING MATHEMATICS IN A MODELING COURSE

The Realistic Mathematics Education programme (RME) has developed a practice of developmental research where different real life or experimental contexts are used as points of departure for carefully planned courses that place the process of mathematization as the key process in the learning of mathematics (Treffers, 1987). In RME, mathematization is seen as a central human activity both in the process of developing mathematics and in the process of learning mathematics. The key concepts of RME are horizontal and vertical mathematization. Horizontal mathematization refers to the process of using mathematics describe an everyday, "realistic" situation. The transfer of meaning from the perceived real situation expressed in everyday language to the mathematical description occurs horizontally between the two domains. Vertical mathematization, on the other hand, refers to the process, for example, of proving theorems and defining mathematical concepts, that is, activities that take place solely inside the domain of mathematics. Its purpose is to introduce students to the architecture of mathematical knowledge. Horizontal and vertical mathematization highlight the similarities between the process of mathematization as one of the crucial steps in a mathematical modeling process (process (c) in Figure 42.1), and the learning process in which the learner through social interplay constructs mathematical knowledge. The RME courses are then designed to support students' learning through interplay between horizontal and vertical mathematization (Gravemeijer, 1998; Gravemeijer & Doorman, 1999).

We find the approach of RME interesting and inspiring, and indeed it has been and still is a very successful program, for both research and teaching. However, our course differs from the RME program in at least two essential ways: Ours is a university course, whereas the RME program focuses primarily on lower and upper secondary levels, and the main focus in our course is on the development of students' modeling competency, whereas the RME focus is on the learning of mathematics.

In this chapter we argue that there are specific potentials for learning mathematics in modeling situations where students are focusing on a modeling problem and thereby are using the modeling context to make sense of their mathematical conceptions. The recent work by Zbiek & Conner (2006) has a similar focus but at a different educational level and context. In general, the effects of modeling activities on students' conceptual learning are not that well researched.

In our course we use modeling situations to help students overcome specific learning problems. Through developing and analyzing modeling problems and by observing the students' activities, we try to create modeling situations in which the students' mathematical conceptions are challenged. This is illustrated and discussed in relation to the integral concept in the final section. Such modeling situations contain potential for learning. However, this potential is not realized automatically. The students' reflections about their mathematical conceptions of the relevant concept(s) in the modeling process seem to be crucial for this realization to happen. These types of reflections are situated in students' actual modeling activities, but directed towards their understanding of the mathematical concepts they encounter. In order to facilitate such reflections, the teacher needs to be aware of the potentials and be prepared to support and challenge the students' reflections through dialogues. These reflections cannot be "forced" onto the students' modeling activities by formulating explicit requirements in the text that frames the students' modeling task. They are generated by shifts in the objectives for students' activity from solving the modeling problem towards their own understanding of the mathematics involved in their activity, and this is by nature a spontaneous process.

The reflections we want the students' to engage in are not part of the modeling process; they are cognitive processes related to the modeling process where the modelers reflect on the mathematics used in the modeling process. During these reflections the modelers interpret new meaning into the mathematics used in the modeling process. The reflection process is a social process, where the interplay between the students and their interactions with the teacher play an important role; however, for each student the reflections are directed towards the student's mathematical understanding of his or her activity. The results of such reflections occur in the mind of the individual learner, but will be evident only in dialogues taking place during the modeling activities or eventually in the learner's subsequent activities.

When the students engage in these types of reflections about their understanding of the involved mathematical concepts in the context of the modeling process, their conceptual images (Tall & Vinner, 1981) are generally broadened. This results in a strengthening of their conceptual understandings. Students are not necessarily introduced to mathematical concepts or methods that they have not encountered before (e.g., in their upper secondary curriculum), but through these reflections, they build stronger and more elaborated networks as mental representations of their mathematical concepts and experiences. In terms of the network metaphor, students may not necessarily create new nodes in their mental networks, but they do indeed make new connections and strengthen already existing connections. The design of our modeling situations and our analysis of students' modeling activities build on a constructivist view of learning that takes the contextual character of cognition into account (Ryve, 2006). Hence, we recognize that students' reflections by nature are contextualized in a broad sense.

The modeling projects, the mini-projects, of the BASE course are developed or chosen deliberately to create the best possible opportunities for challenging and enlarging students' understanding of what we consider to be the central mathematical concepts and methods in our modeling course in such a way that students are forced to reflect upon these in their modeling activities. Through their work with the mini-projects, the students' reflections in and about the modeling process and its context are united with their learning of mathematical concepts. This "unitedness" makes it possible to create situations of learning where students "see" the involved mathematical concepts from new sides. The potentials for learning mathematics in connection with the mini-projects do in general not realize themselves automatically. In our experience as teachers, we need to pay very careful attention to the learning potentials related to each of the mini-projects in order to be able to notice when the potentials can be pursued in the contemplations of a particular group of students and to find ways to challenge students to engage in the relevant reflections. A consequence of this is, of course, that the development of both the course material and our teaching practice is an ongoing process.

INTERPLAY BETWEEN PRACTICE, THEORY, AND OBSERVATIONS

The relation between the modelers' mathematical conceptions and the mathematical system, as well as the reflections described above, relate to our first research question: *In what sense can mathematical modeling activities support and strengthen students' learning of the integral concept?*

The second question, *How can possible learning potentials be realized through interactions with students during their modeling activities?*, relates to the implementation of such possible learning potentials in our teaching practice. We find this last question very interesting and challenging both from our point of view as teachers and from a research perspective. Three important issues related to the development of practice, theory, and pedagogical observations are addressed.

First, the course structure has benefited both from our close collaboration in its development and from our having team-taught the course several times. This collaborative effort has given us the opportunity to discuss in great detail the potentials and pitfalls related to each of the mini-projects based on our common experiences.

Secondly, we have drawn on theories about the teaching and learning of mathematical modeling (Blum & Leiß, 2007; Blum & Niss, 1991; Niss, 1989), and on theories about concept formation (Sfard, 1991); concept images (Tall & Vinner, 1981; Vinner & Dreyfus, 1989); and the importance of the epistemological relations between symbol, concept, and object in the learning of mathematics (Steinbring, 1989). These theoretical approaches have served as bases for recognizing the needs and potentials for the various kinds of reflections related to each mini-project and for analyzing the learning difficulties students face when reflecting mathematically on their modeling activities.

Thirdly, from time to time, we have made what we call pedagogical observations to closely analyze students' learning processes, while interacting with them as they worked on the mini-projects. Pedagogical observations are an integrated part of our planning and implementation of teaching situations. They are developed in a kind of dialectic relationship between observations of students' dialogues when they engage in the modeling activities of the different mini-projects and our analyses of these dialogues with respect to the students' mathematical conceptions. Based on such observations and analyses, we construct dialogues as part of our preparation for a specific teaching situation that revolves around a particular mini-project. The dialogues are constructed such that we, through them, deliberately try to bring the students into situations that reveal their misconception or limited conception some of the mathematical concepts involved. In this way we create situations where students can engage in the kind of reflections discussed above.

The first time we taught the course together we made observations during our dialogues with the groups, and we recorded our observations, reconstructing together typical parts of dialogues that we had taken part in or overheard in relation to the different projects. Our observations helped us critically evaluate some of the projects in terms of their mathematical challenges in relation to students' mathematical knowledge, or in terms of the way textual presentation set the scene for students' modelling activities and subsequent reflections. This type of critical evaluation gave rise to improvements in the mini-projects and grounds for developing new mini-projects.

Some of our observations pinpointed specific difficulties in students' modeling activities for the various mini-projects. We discussed and analyzed these observations theoretically, and in subsequent interactions with groups working with the same mini-projects we have deliberately, through our pedagogical observations as explained above, in dialogues with the students, tried to confront them with these particular difficulties and, of course, also tried to find ways to support students in overcoming these difficulties. To a large extent it has been possible for us to continually rediscover the difficulties that we have identified in each mini-project when teaching the course with new students, which suggests that these difficulties in some sense are general.

Particular types of difficulties emerge in relation to the processes of mathematization (c), of analyzing the mathematical system (d), and interpreting the mathematical system and the mathematical results (e) (Figure 42.1). These processes are characterized by the fact that students need to draw on parts of their understanding and images of mathematical concepts that are not well-developed by the students. In these situations the students' focus changes from the task of modeling a particular problem towards their own understanding of the involved mathematical concepts. Students experience what Tall and Vinner (1989) refer to as a cognitive conflict in their evoked concept images. Such situations create frustrations in students, but at the same time these situations contain opportunities for the students to engage in reflections about their mathematical conception in the context of the modeling problem. In these moments of reflection, students are highly motivated to solve the conflict, and therefore they are very receptive to support offered by the teacher or their fellow students.

Accordingly, we see it as a great benefit that the mini-projects are able to create such situations during the group work. Of course this should be balanced against the course goal of enabling students to work fairly independently in setting up and analyzing mathematical models in each mini-project.

In the following section we illustrate how students' modeling activities can create incidences of cognitive conflict while supporting the students' engagement in the above-described reflections and eventually supporting their conceptual learning of the integral concept. Moreover, we show how these opportunities can be realized through dialogues between a group of students and a teacher when the teacher is well aware of the possible cognitive conflicts in particular modeling situations.

SUPPORTING THE LEARNING OF THE INTEGRAL CONCEPT THROUGH MODELING ACTIVITIES

In the mini-project "CO_2 Balance of a Lake," students get to work with central features of the integral concept in ways that challenge their images of this concept and address important aspects of the concept definition. The project has been designed especially with the purpose of challenging students' understanding of the concept of the definite integral to handle some problems we have identified over the years—problems that pose difficulties for students to use the concept as a modeling tool.

Students' Perception of the Integral Concept Prior to the Project

In ordinary classroom teaching where we introduce the definite integral, the focuses are on the interpretation of the integral concept in different modeling situations and on methods of numerical integration. Students complete exercises in analytical integration, but this is not trained extensively. Nevertheless, even after these few exercises students arrive at a clear and firm conviction that "real" integration should follow this schema: Find an anti-derivative, insert the endpoints of the integration interval in the anti-derivative and subtract the two magnitudes. This is typically how they understand the integral concept—as the (analytical) process of integration—and it results in a definite number. For many students, their concept image is limited to this operation and to the interpretation of the integral as "the area under the graph." Surprisingly, this is true both for students who never encountered the integral concept in secondary school as well as others who have been taught the subject before. Apparently, strong forces lead students to this very instrumental and limited understanding of the concept of integration.[4]

Employing this limited concept image has the consequence that students will not be able to evaluate the definite integral of a function not given by an analytical expression or a function to which an analytical anti-derivative cannot be determined. Usually students have not experienced this consequence as a conflict, presumably due to their limited concept image of a function. It is our experience that in the beginning of the course students cannot imagine functions not given by an analytical expression. Even though they have often seen lab-prints of graphs, they do not consider these as representations of functions. Nor can they imagine the existence of analytically given integrable functions for which an analytical expression for the anti-derivatives cannot be found.

Most students are confused about the arbitrary constant C appearing in the general formula for the anti-derivative. When should it be taken into account and when can it be left out? Apparently, it is not included in the formula:

$$\int_{t_0}^{T} f(t)\,dt = [F(t)]_{t_0}^{T} = F(T) - F(t_0)$$

Based on such experiences we have postponed the introduction of the integral concept until after we have introduced single ordinary differential equations as a way to model dynamical phenomena. We use a compartment notation to represent such dynamical systems (Blomhøj & Jensen, 2003); see Figure 42.2.

In situations where the net rate of change for a given compartment is known for a particular period of time, a system as illustrated in Figure 42.2 can be modeled by a differential equation according to the "masse balance" or "in–out" principle. In the case of Figure 42.2, one gets:

$$F'(t) = f(t)$$

If $f(t)$ is known and integrable in the considered interval, $F(t)$ can be found by integration:

$$F(t) = \int f(t)\,dt + C$$

In this formula, the arbitrary constant C is somewhat mystical for many students. Therefore we spend some time at the course setting up and interpreting initial

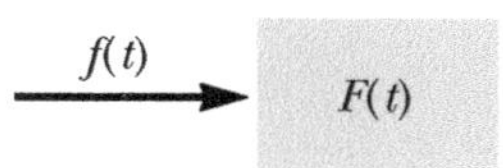

Figure 42.2 This compartment diagram represents a situation where the rate of change of some quantity $F(t)$ is given by a function $f(t)$ of time.

value problems in relation to the definite integral. In the general situation we should find $F(t)$ at some particular time T knowing the value $F_0 = F(t_0)$. By reading the compartment diagram, in different situations, it is easy for most students to set up equations such as:

$$F(T) = F(t_0) + \int_{t_0}^{T} f(t)\,dt$$

What we have at time T is what we found at time t_0 plus the definite integral of the net rate of change in the interval from t_0 to T. Moreover, this compartment-based interpretation provides the students with an intuitive basis for understanding the formula for the definite integral.

However, in our experience, for students to understand fully the concept of the definite integral, the concept of the anti-derivative, the significance of the constant, and the interpretations of these concepts in different problem situations, students need to work with modeling activities that challenge exactly these aspects of the integral concept. As will be seen below, the mini-project of modeling the CO_2 balance of a lake is designed to provide such challenges.

Before students are set free to work on the project, we try to support their deficient or fragile understanding of the definite integral as an accumulated sum by down-playing analytical integration. Instead we spend a considerable amount of time on numerical integration—by hand as well as by means of the mathematical software MatLab.

In the exercises where they integrate by hand, we typically use examples of velocity–time graphs, where interpretations of the definite integral as area and distance traveled play dominating roles. Students integrate by sectioning in smaller rectangles the area between the velocity graph, the ordinate axe, and the lines determined by the endpoints of the time interval in question. The integral can then be estimated by "counting rectangles." Students' reactions towards these exercises clearly show that their concept image is challenged. They are very uncomfortable with this rough and "childish" method of integration. It goes fundamentally against their conception of what calculus and mathematics are about.

The advantage of these kinds of exercises is that they directly illustrate the idea of the concept of the integral as an accumulated sum. They also support the idea of the integral as an area, which has a physical interpretation when we integrate velocity functions over a time interval, but in many situations the idea of the integral as an area cannot be given a meaningful physical interpretation. The disadvantage is that if students are not introduced to examples other than velocity–time examples, the area understanding can be all dominating, which will pose problems when students are placed in situations where this understanding does not make physical sense.

Modeling of the CO_2 Balance of a Lake

The object is to analyze the CO_2 balance in a lake. The rate at which plants add or remove CO_2 from the lake measures the lake's biological activity. Plants use CO_2 in photosynthesis during the day, while they produce CO_2 at night. The animal life in the lake produces CO_2 continuously due to respiration. Biologists are interested in how the net rate of change of CO_2 changes over a 24-hour period. The rate of change is measured in mmol CO_2 per liter of water per hour.

On the basis of a set of data, partly displayed in Table 42.1, consisting of three measures of the rate of change every second hour for 24 hours, a measure of 2.600 mmol CO_2 in the lake at dawn, and a measure of the rate of change to 0 mmol/l/hour also at dawn, students are asked to analyze the CO_2 balance in the lake.

Besides these introductory remarks and the data table, we establish a framework for students' work through a number of helping questions and by means of challenges given to the groups in dialogues with the teacher during their modeling work. The students have already been prepared to produce a plot of the data to get a first impression of the situation (see Figure 42.3).

In the first part of the mini-project the help questions are designed to make the students interpret the plot and realize what information they can read off directly. Here we only comment on the students' work on the following question: What can be concluded from the data material about the sign of the rate of change of CO_2 over the 24-hour period, and what information does that give about the life in the lake?

That the sign of the rate of change is negative during the day and positive during the night is straight-

TABLE 42.1 Indicates the Data Material Given to the Students

Time (Hours after dawn)	**CO_2-rate of change** (mmol/l/hour)	**Time** (Hours after dawn)	**CO_2-rate of change** (mmol/l/hour)
0.666	–0.027	12.666	0.028
1.333	–0.048	13.333	0.058
.	.	.	.
.	.	.	.
12.000	0.000	24.000	0.000

Note: Data express the rate of change in CO_2 concentration (mmol/l/hour) as a function of time after dawn in 40 minutes intervals

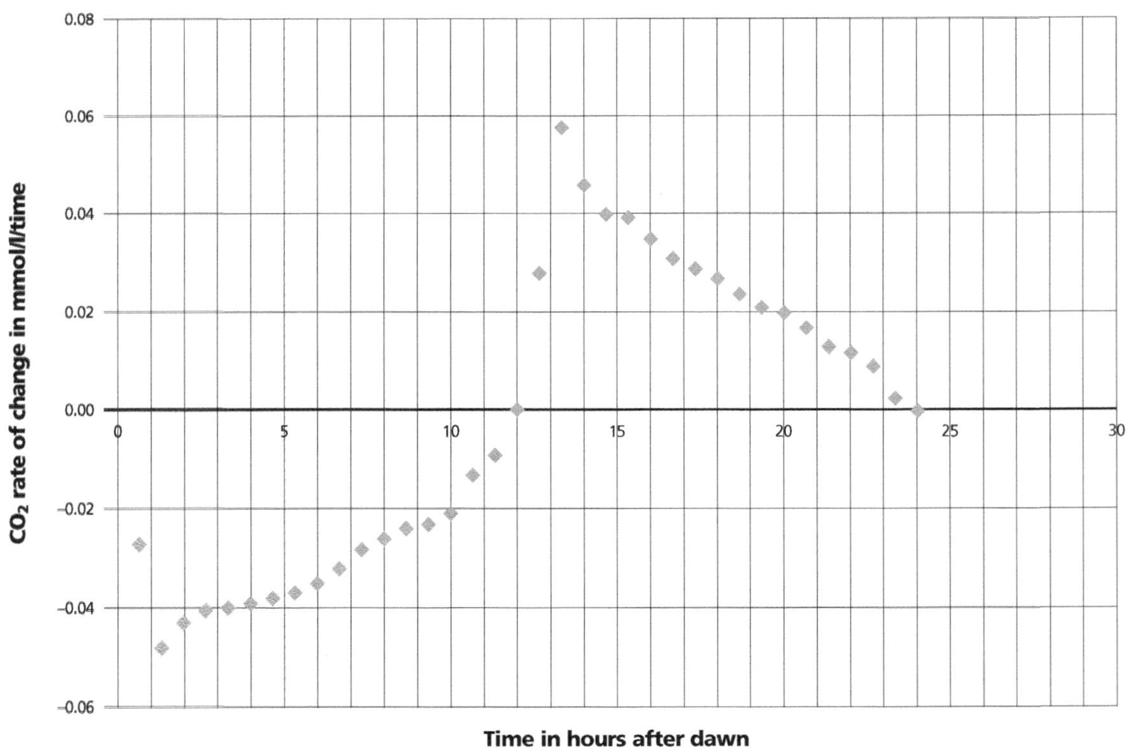

Figure 42.3 Shows a plot of the data indicated in Table 42.1: The rate of change in CO_2 concentration (mmol/l/hour) as a function of time after dawn in 40 minutes intervals.

forward for all students, but what that says about the CO_2 content in the lake is not as clear. Students then become aware that they do not totally understand what the CO_2 rate of change actually tells them about the system, and they are not completely sure about the mathematical relationship between the CO_2 rate of change and the CO_2 content in mmol/l.

Our experience is that it takes some time before they agree among each other that the CO_2 rate of change is the derivative of the CO_2 content as a function of time, and in discussions students very rarely use the given units in their argumentation.

Next they need to understand which information about the CO_2 content can be derived from the fact that the CO_2 rate of change is negative during the day. This is not as straightforward as one might think even though they have already determined that the CO_2 rate of change is the derivative of the CO_2 content. In spite of that we often see, students interpret the plot of data as showing that the CO_2 content as a function of time is decreasing the first one hour and 20 minutes after dawn, after which it will increase until a little over 13 hours, and then it decrease again.[5]

What happens here is that the students are mixing up the graphical picture of the CO_2 rate of change displayed in their plot of data with the CO_2 content. To deduce facts about a function from a graphical representation of its derivative is not very strongly represented in the students' conceptual framework.

At this stage in the project, students do not think of the CO_2 content as an anti-derivative to the CO_2 rate of change. Instead they think of the CO_2 rate of change as the derivative of the CO_2 content. Their concept images of these two different ways of looking at the relation between the CO_2 rate of change and the CO_2 content is generally not integrated at this point. One of the purposes of this modelling project is precisely to put the students in situations where they get opportunities to create a cognitive connection between their concept images of these two phenomena.

In the second part of the project students are asked additional questions. For example, when will the CO_2 content in the water be at its lowest, and how much CO_2 will be in the water when that happens? Is the lake in equilibrium with regard to the CO_2 content? How can this question be decided graphically? How much CO_2 was released to the water during the 12 hours of night and how much was removed during the 12 hours of daytime? These questions all require that students are able to regard the CO_2 content as an anti-derivative of the CO_2 rate of change. They are forced to shift focus and view their plot of data as a plot of a function about whose anti-derivative they are searching for information. At this point, as mentioned above, only a few students have a concept image so well developed that it does not matter to them whether they use one way of considering the data plot or the other. For most students their concept images of derivative functions and of anti-derivatives are not connected. There is a considerable difference in the two points of view.

During their modeling activities in the two first parts of the project, students establish connections between their concept images of derivative functions and anti-derivatives; connections that further develop and support their understanding of the integral concept.

In the second part of the project, the students also experience the numerical integration method of "counting rectangles" as a valuable and sensible tool, which in fact is capable of answering relevant and significant questions. This supports an addition to their understanding of the integral concept beyond analytical integration and extends their area interpretation of the integral further than "a distance travelled." Students are here experiencing a situation where they have only a discrete set of data and not an analytical given integrable function, so with an integral concept limited to analytical integration they cannot in practice get very far.

The Use of Pedagogical Observations

We have made extensive use of pedagogical observations as a tool in the course to get closer to students' learning processes and to get a deeper insight into the problems of learning various mathematical concepts. The modeling project is an excellent context for this tool because it effectively creates situations where the observer can get insights into the challenges facing students as they try to understand and learn. The modeling context stimulates students to talk about the mathematics, formulate problems and questions of a mathematical character, and test their understanding in group discussions with each other.

Below we will show in detail how, in the third part of the mini-project, we have used pedagogical observations to get insights into and challenge students' understanding of the integral concept.

The object in the third (and last) part of the project is to address both students' perceptions of the definite integral as an accumulated sum and the above-mentioned problem they have with understanding the role of the arbitrary constant *C*. Students are asked to estimate the CO_2 content in the water every 20 minutes over the 24-hour period and produce a graph of the CO_2 content as a function of time. For most students this either creates a conflict with their concept image or reveals a huge gap in their understanding. They realize that they need to integrate the function (the CO_2 rate of change) represented by their plot of data, but that is followed by their confusion caused by vague or missing connections between the concept definition and their concept image of the definite integral, as well as their problems with the distinction between the definite integral and anti-derivative functions.

The following dialogues are reconstructed from our experiences and observations over time with groups of students working on the project:[6]

The teacher has been summoned by the group. The students are stumped.

T: What are you being asked to calculate?
S: The CO_2 content of the lake.
T: Correct. How can you determine that?
S: Well, we know the rate of change ...
S: That means we know f'...
S: Is it f we have to find? (Everyone looks at the teacher)
T: I don't know. It is the CO_2 content you are asked to find.
S: We know the velocity so it *is* f we have to find.
S: We must integrate.
T: Yes, what is it you need to integrate?
S: Our function.
T: What function?
S: This one (points at the plot of data)
T: Yes, that's correct.

Then the teacher leaves. After a while the teacher is called back in:

T: How is it going? Have you integrated your function?
S: No.
T: Why not?
S: We do not have an expression for the function.
T: OK—but aren't there other ways to integrate?
S: Yes, numerically, and we did that, but we got –0.025 which cannot be true. The CO_2 content cannot be negative.
T: How did you arrive at this number?
S: We integrated the function by counting rectangles. The ones below the ordinate axe are negative.

Here the students clearly have problems understanding how the integral concept is connected with the problem of determining the function of the anti-derivative.

T: Let's try to focus on what it is you have to find.
S: We have to find the integral.
T: Yes, but why is it that you want to integrate?
S: ... ??? ...
T: What does it say in the text?
S: Oh yes, we have to find the CO_2 content.
T: Yes, when?
S: ...???...What do you mean?
T: Is it at midnight...or...?
S: Yes...no...it is at dawn.
T: Where does it say that?
S: Ehh...it doesn't.
T: No, what does it say?

Rereading the text they realize that it is the concentration for every 20 minutes they need to find.

S: But then we have to calculate many numbers!
T: Yes.
S: But the integral is just one number...right? How can it then become a whole function?
T: How will you estimate the CO_2 content at time 0.666?
S: Integrate.
T: Yes, but what?
S: The function—the rate of change.
T: Yes, but where from and where to?
S: Ohhh—from 0 to 0.666.
T: Yes.

The teacher leaves again—and is called back in after a while.

S: We still don't get it. Do we need to count rectangles all over?
T: What do you mean by "all over"?
S: We have done it for the first data point from 0 to 0.666. And we also did it for the second data point from 0.666 to 1.333. Is that then the CO_2 content at time 1.333?
T: That is a good question.

After some discussion back and forth the group realises that they must add the first integral from 0 to 0.666 to the second to get the CO_2 content at time 1.333. It is only at this point that the students really understand that they can tabulate the anti-derivative (the CO_2 content) by successively adding "the next column" of rectangles. They realize why there is no conflict between the definite integral being *one* number and the possibility of tabulating the entire anti-derivative function by integration even though that requires many numbers.

> **S:** That is a huge calculation . . .
>
> **T:** Yes, but the function to be integrated is tabulated in the data set, so maybe you can use one of your MatLab programs.
>
> **S:** Oh yes, numerical integration.
>
> **T:** Yes, and then you can also consider what to do about the 20th minutes.[7]
>
> **S:** The 20th minutes?

The teacher leaves.

After such discussions, at some point, most of the groups are able to take over the control of their modeling work to produce the requested graph, to use it, and to reflect on it in relation to biological questions about the lake.

One particular group decided to use Excel to integrate instead of one of their MatLab programs. They did that because they felt that in Excel they were able to keep track of what was added when and to what. It is noteworthy that this group eventually realized by themselves that they had to add the initial amount of CO_2 in the lake, and that this amount was the constant *C*. Some of the other groups at first hand overlooked that, and found negative amounts of CO_2 in the water, which gave rise to comments about the feasibility of these results in relation to the context.

DOCUMENTATION: WHAT DID THEY LEARN?

This mini-project does not in itself introduce new mathematics to the course. Instead, students' conceptual learning of mathematics is supported through the connections the students are challenged to explore regarding concepts such as the derivative function, anti-derivative function, the definite integral, and their mutual relations. The modeling problem challenges students' perception of these concepts because it forces them to reflect on the concepts and perspectives they need to employ to answer the different questions. During this project, students' abilities to apply the concepts in particular contexts go far beyond how they had applied these concepts in ordinary mathematical exercises. The project enables them to juggle the concepts and to interpret them in relation to each other and in relation to a perceived reality.

On a more concrete level, the project demystifies matters regarding the arbitrary constant *C* in the formula for the anti-derivative. Through area observations, students are able to answer many questions regarding amounts of CO_2 being removed or added to the lake. In such questions, the constant *C*, or the initial concentration of CO_2, plays no role whatsoever. But when it comes to determine the actual content of CO_2 in the lake at a given time, the constant and its interpretation play a crucial role. The constant determines the actual level of CO_2, and that idea is encountered in the last part of the mini-project, where students realize that they cannot solve the problem without taking the initial level of CO_2 in the lake at time $t = 0$ into consideration. On the basis of this experience, they are able to read new meaning into the general formula for anti-derivatives interpreting the constant *C* as the value of the function $F(t)$ at some chosen time $t = t_0$. It is within the modeling context that this becomes clear for students. They found that without the constant they calculated only increasing and decreasing amounts of CO_2, and because they calculated with signs they ended up finding negative amount of CO_2 in the lake. This contra-intuitive result was what triggered their reflections.

Students' understanding of the concept of the integral got connected to the definition of the concept, and they experienced the usefulness of integrating by counting unit rectangles. This is most clearly demonstrated by the group of students who decided to use Excel to tabulate the function of the CO_2 content. They chose Excel because they felt that they could see and control what the computer actually calculated. This group of students understood the integral concept at a much deeper level than before the mini-project, so that at the end they were able to transfer the definition of the definite integral to new modeling situations.

As stated earlier, these reflections are cognitive processes, and in our constructivist view of learning, we recognize the contextual character of cognition, and therefore we do not presuppose transfer of students' conceptual understanding of the integral concept developed in relation to a particular mini-project to other modeling situations or to other types of contexts. The question of transfer has to be investigated separately. However, we have some soft evidence supporting the notion that students who have worked with the CO_2 mini-project are able later in the course to use the integral concept in other modeling projects and in written test problems. Indeed, students by themselves often refer to the CO_2 mini-project when discussing issues concerning the integral concept later in the course.

CONCLUSION

In What Sense Can Mathematical Modeling Activities Support and Strengthen Students' Learning of the Integral Concept?

Mathematical modeling provides opportunities for students to learn mathematics in different ways. In this chapter we have focused on situations where the students reflect on their own mathematical understanding of the integral concept in relation to modeling activities related to the processes of mathematization, analysis of the mathematical system, and the interpretation of the results that are processes (c), (d), and (e) in Figure 42.1. The learning potentials of such reflections are triggered by students activating different aspects of their conceptual knowledge, experiencing possible conflicts in their understandings, and enlarging their concept images by connecting the concepts to concrete experiences in their modeling activities or by connecting different mathematical aspects of the concepts involved.

How Can Such Possible Learning Potentials Be Realized Through Interactions with Students During Their Modeling Activities?

In course planning and in developing of modeling projects for students to work with, one can definitely look for such learning potentials. However, enhancing these potentials of course has to be balanced against the goal of developing modeling competency. It is our experience that the projects have to be genuine modeling projects in order for the students to engage in reflections about their mathematical understandings. For us, pedagogical observations and analyses have played important roles in the process of supporting and challenging students in the text for each of the mini-projects. However, we recognise that the text can never ensure that the learning potentials are realised. The student-teacher interaction during the modelling activity is absolutely crucial. Here again, the circling process of observing, analyzing, and interacting with students is a fruitful practice for developing a modeling course in which students' learning of mathematical concepts is supported. The weakness is that this form of developmental practice is very difficult to hand over to new teachers of the course.

NOTES

1. It goes without saying that there are other equally (or maybe even more) important arguments for teaching mathematical modeling (Niss, 1989).
2. BASE stands for Basic Analysis, Simulation, and Experiment. The course material consists of 450 pages of notes and a collection of modeling problems for the mini-projects (Blomhøj, Kjeldsen, & Ottesen, 2004).
3. The purposes of the course should be seen in connection with the general structure and idea of the introductory study program in science at Roskilde University, where half of the study time is project-organized and problem-oriented as described and analyzed by Blomhøj and Kjeldsen (2009).
4. According to the model for formation of mathematical concepts by Anna Sfard (1991), it may not be that surprising since the process aspect of mathematical concepts seems to have precedence over the object aspect in the early phase of the learning process.
5. The fact that the CO_2 rate of change is negative during the day means that CO_2 is removed from the lake during those hours from which it can be concluded that the lake properly contains plants (it is implicitly given that the water content of the lake is constant).
6. T stands for the teacher. S stands for an arbitrary student.
7. The students only have data for every 40th minute, but are here required to estimate the CO_2 content at every 20 minutes.

REFERENCES

Blomhøj, M. (2004). Mathematical modelling—A theory for practice. In B. Clarke et al. (Eds.), *International perspectives on learning and teaching mathematics* (pp. 145–160). Gothenburg, Sweden: National Center for Mathematics Education, University of Gothenburg.

Blomhøj, M., & Jensen, T. H. (2003). Developing mathematical modelling competence: Conceptual clarification and educational planning, *Teaching Mathematics and its Applications, 22*(3), 123–139.

Blomhøj, M., Kjeldsen, T. H., & Ottesen. J. (2004). *BASE—Et grundkursus i matematisk modellering* (5th ed.). (Notes for the BASE course—A course on mathematical modelling at the Bachelor Science Study Program at Roskilde University). Roskilde: Roskilde University, IMFUFA.

Blomhøj, M., & Kjeldsen, T. H. (2006). Learning mathematical modelling through project work–Experiences from an in-service course for upper secondary teachers. *Zentralblatt für Didaktik der Mathematik, 38*(2), 163–177.

Blomhøj, M., & Kjeldsen, T. H. (2009). Project organised science studies at university level—Exemplarity and interdisciplinarity. *Zentralblatt für Didaktik der Mathematik, 41*(1–2), 183–198.

Blum, W., & Leiß, D. (2007). How do students and teachers deal with mathematical modelling problems? In C. Haines, P., Galbraith, W., Blum, & S. Khan (Eds.), *Mathematical Modelling (ICTMA12): Education, Engineering and Economics.* Chichester, UK: Horwood Press.

Blum, W., & Niss, M. (1991). Applied mathematical problem solving, modelling, applications and links to other subjects: state, trends and issues in mathematics instruction. *Educational Studies in Mathematics, 22*(1), 37–69.

Carter, J. (2004). Ontology and mathematical practice. *Philosophia Mathematica, 12,* 244–267.

Gravemeijer, K. (1998). Developmental research as a research method. In: A. Sierpinska, A., & J. Kilpatrick (Eds.), *Mathematics education as a research domain: A search for identity* (pp. 277–296). Dordrecht: Kluwer.

Gravemeijer, K., & Doorman, M. (1999). Context problems in realistic mathematics education: A calculus course as an example. *Educational Studies in Mathematics, 39,* 111–129.

Niss, M. (1989). Aims and scope of applications and modelling in mathematics curricula. In W. Blum et al. (Eds.), *Applications and modelling in learning and teaching mathematics* (pp. 22–31). Chicester, UK: Horwood.

Panza, M. (2003). Mathematical proofs. *Synthese, 134,* 119–158.

Ryve, A. (2006). Making explicit the analysis of students' mathematical discourses—Revisiting a newly developed methodological framework. *Educational Studies in Mathematics, 62,* 191–210.

Sfard, A. (1991). On the dual nature of mathematical conceptions: Reflections of processes and objects as different sides of the same coin. *Educational Studies in Mathematics, 22,* 1–36.

Steinbring, H. (1989). Routine and meaning in the mathematics classroom. *For the Learning of Mathematics, 9*(1), 24–33.

Tall, D., & Vinner, S. (1981). Concept image and concept definition in mathematics, with particular reference to limits and continuity. *Educational Studies in Mathematics, 12,* 151–169.

Treffers, A. (1987). *Three dimensions. A model of goal and theory description in mathematics education.* Dordrecht: Kluwer.

UVM. (2009). *Undervisningsvejledning matematik A—Stx.* Copenhagen: The Danish Ministry of Education. Retrieved August 31, 2009, from http://www.uvm.dk/~/media/Files/Udd/Gym/PDF08/Vejledninger/stx/080701_matematik_A_stx_vejledning.ashx /.

Vinner, S., & Dreyfus, T. (1989). Images and definitions of the concept of function. *Journal for Research in Mathematics Education, 20,* 356–366.

Zbiek, R. M., & Conner, A. (2006). Beyond motivation: Exploring mathematical modelling as a context for deepening students' understandings of curricular mathematics. *Educational Studies in Mathematics, 63,* 89–112.

43

RECENT DEVELOPMENT IN SCHOOL MATHEMATICS' ROLES AND RELATIONS

Lena Lindenskov

DANISH SCHOOL OF EDUCATION, AARHUS UNIVERSITY

Mette Andresen

HEAD OF NAVIMAT, UNIVERSITY COLLEGE COPENHAGEN

Abstract: This article presents the enrollment of mathematics in a new construct, multi disciplinarity, which was recently introduced into Danish upper secondary schools with academically oriented programs. We relate collaboration between school subjects to key characteristics of three perspectives on educational culture in Denmark: regulations and the teachers' autonomy, educational aims and goals, and students' attitudes towards mathematics. We choose these three perspectives to sketch a national profile of Danish educational policy and school practice. Next, the potentials of multi-disciplinary mathematics teaching at all levels are discussed. We see students' formation of a profound understanding of mathematical activities and theory as the main point and discuss this issue from a theoretical perspective combining the ideas of Realistic Mathematics Education (RME) with a philosophical approach to mathematical reflections. Modeling perspectives are pivotal in this discussion. Finally, we present a prospective research and development project, planned to run in upper secondary school as a pilot (6–8 teachers) in 2008–2009 and will be extended in 2009–2010. The project is based on discussions similar to the above mentioned, and it deals with mathematics in multidisciplinary teaching projects with three other faculties: human sciences, societal sciences, and natural sciences, and its aim is to inquire how learning potential in mathematics is realized in a number of cases of good practice. We expect teaching materials for such projects to be one of the outcomes of the project, which is a collaborative project between Danish Science Gymnasiums (DASG) and the National Center of Knowledge about Mathematics Education (NAVIMAT). We conclude that multidisciplinary teaching on the one hand may fit well with the professional culture of upper secondary schools with their academic, subject-focused learning environment; as well, it may serve to inspire collaboration between subjects in primary and lower secondary school.

Keywords: Mathematics' role in multi-disciplinary projects, Potentials of multi-disciplinarity, Students' attitudes towards mathematics, Mathematical reflections and modeling, the Vioxx case as a pilot

NATIONAL PROFILE OF DANISH EDUCATION

In the following section, we relate collaboration between school subjects to key characteristics of three perspectives on educational culture in Denmark: regulations and the teachers' autonomy, educational aims and goals, and students' attitudes towards mathematics. We choose these three perspectives to sketch a national profile of Danish educational policy and school practice.

The First Sourcebook on Nordic Research in Mathematics Education, pages 583–596

Regulations and the Teachers' Autonomy

The overall aims, content, and evaluation of primary and secondary school teaching are strictly framed by governmental regulations. In contrast, the detailed planning within this framework is handled autonomously by the teachers.

Primary and lower secondary schools[1] are undivided (comprehensive) schools, where the formation of classes is mainly based on the age of the pupil and not on his or her subject-specific proficiency, which is possible because the Danish primary and lower secondary school is not an examination-oriented school.[2] *The Folkeskole (Consolidation) Act* gives the Minister the authority to establish the regulations and aims regarding education and end objectives, grade level objectives, and descriptions in subjects and compulsory topics. The binding national objectives are supplemented by binding local curricula and descriptions approved by municipal boards. The Ministry of Education prepares guidance on curricula and the descriptions of the educational development designed for achieving the grade level and the end objectives to support teachers and ensure quality teaching and learning guidelines; however, the guidance provided by the ministry is not binding.[3]

As a new phenomenon, national testing is being developed in these years, including ten obligatory national tests, of which there are two mathematics tests in grade level 3 and 6. Politicians and the Ministry declare that the aim is to gain an understanding of the individual students' acquisition of knowledge and skills so that the education process can be planned to a greater extent according to the individual students' strengths, weaknesses, and potential.[4]

At the conclusion of grade levels 9 and 10, Folkeskole students sit for school-leaving examinations. This is compulsory after grade level 9 but voluntary after the voluntary grade level 10. School leaving written examination in mathematics traditionally is compulsory for all. In addition, from 1997 until 2006 school leaving oral examinations in mathematics after grade level 9 were compulsory for all students. After 2006 mathematics is less prioritised, as no students sits for oral examination in mathematics. Instead other subjects are now given higher priority, such as natural science and subjects from the humanities group. From a mathematics educators perspective this priority change is deplorable and many teachers, teacher trainers and researchers in mathematics education have criticized the decision.

Part of the school-leaving examinations is interdisciplinary. At the 9th and 10th grade levels, a mandatory project assignment gives students the opportunity to complete and present an interdisciplinary project. The project assignment is assessed in a written statement on the content, working process, and presentation of the final result. The written statement affords a broader and more detailed assessment of the student's ability. At the student's request, a grade can also be given. The assessment of the project assignment can be indicated in the leaving certificate.

Teachers' autonomy in this system is made possible by the Danish concept of curriculum. The term curriculum covers national and municipal aims, briefly stated in the regulations, supplied by identification of central knowledge and skill areas, and descriptions of the educational development. The regulations are designed in accordance with the grade level and end objectives. The choice of instructional materials, of teaching methods, of lesson plans, or specific goals for individual students is not prescribed by any formal regulations: It is up to the individual teacher and teacher teams to decide about this within the limits of the individual school's economical resources. It is not uncommon in Denmark that teachers are involved with preparation of instructional materials for their own teaching. Another aspect of this structure is the possibility for the teachers to realize the prescribed intention of including the students in planning, as described in the Folkeskole (Consolidation) Act §18 (4) for primary and lower secondary school:

> §18 (4) At each form level and in each subject, the teacher and the pupil shall cooperate continuously on determining the targets they are trying to meet. The work of the pupil shall be organised under due consideration of these targets. The establishment of working methods and the selection of subject-matter shall as far as possible take place in cooperation between the teachers and the students. (Official English version, 2003)

This prescription of cooperation between the teachers and the students is implemented in Danish classroom practices in forms of informal, trustful, and cooperative teacher–student relations. The prescription of cooperation reflects the issues of participation in democracy, taking responsibility and being heard, which are also stressed in the regulations.

In accordance with the described teacher autonomy within national regulations, the introduction of the new construct, multidisciplinarity, was prescribed in the national upper secondary school curriculum by governmental regulations in 2005. Before the reform, each teacher planned and taught lessons in the two subjects he or she was specialized in, without them being obliged to collaborate on the content with colleagues specialized in other subjects. After the reform every single teacher now has an obligation to participate in such collaboration. This introduction was not a revolution, as such collaboration was frequently carried out before, on a voluntary basis, either in smaller, self-chosen groups of teachers or as a school's whole team of teachers in certain subjects. Many experiments where made around the country with different models for the collaboration.

Educational Aims and Goals

The regulations prescribe, among other things, a double aim of the students' outcome: preparation for further education as well as commitment to learn more. Further, expected outcomes are to enhance students' self-confidence and their ability to participate in, and demonstrate mutual responsibility towards, a free and democratic society. The final sentence in The Act's §1 for primary and lower secondary school points at values underpinning national political regulations and local school practices: "The daily activities of the school must, therefore, be conducted in a spirit of intellectual freedom, equality and democracy."

School failure in primary and lower secondary school is an almost non-existent phenomenon, although the Act allows for the possibility—with the consent of the child's parents—to repeat a school year, for example, if the student has been away for a longer period of time or other reasons make this relevant. Recently, school-leaving examinations have become compulsory after form level 9 but are still voluntary after form level 10. Standard rules exist for all examinations, including the compulsory written mathematics examination, to ensure uniformity throughout the country. The tasks for the written examination are the same for every student, the Ministry of Education organizes the formulation and marking of all tasks. Each examination subject is assessed on its own merit; results cannot be summed up to give an average mark, and in the same way, there are no pass or fail criteria. However, in all education following lower secondary school pass and fail criteria exist.

The national regulations, along with school practices, place a very high priority on the use of mathematics and mathematical modeling as well as the relationships between mathematics and other spheres of life. This is also evident in upper secondary academically oriented programs, even at the highest level of mathematics in the general section, which is termed Danish Mathematics A–stx. We cite from the departmental order (Danish Ministry of Education, 2008), which shows that the identity of the subject is described with focus on abstractness and the broad use of mathematics:

> Mathematics builds upon abstraction and logical thinking and holds a wide range of methods for modelling and problem solving. Mathematics is indispensable in many trades, in science and technology, in medicine and ecology, in economy and social science, and as basis for political decisions. Mathematics is also substantial in everyday life. The broad applications of mathematics are rooted in the subject's abstract nature and reflect the experiences that many widely different phenomena behave homogeneously. Often when hypotheses and theories are formulated in the language of mathematics new insight is gained. Mathematics has accompanied the cultural development since the earliest civilisations and mankind's first considerations of number and form. Mathematical science has developed in a steadily interaction between applications and building up theory. (1.1 Identity/our translation)

The aim is described with a focus on interactions, problem coping, reasoning, democracy, and further education:

> It is the aim that the students gain knowledge of important aspects of mathematics' interaction with culture, science and technology. They shall acquire insight in how mathematics can contribute to understanding, formulating and treating problems in different subjects, as well as insight in mathematical reasoning. Hereby the students shall be able to better conduct themselves to others' use of mathematics and acquire sufficient competences to accomplish further education, where mathematics forms part. (1.2 Aims/our translation)

Based on the regulations' description of identity and aim, we argue that teaching that addresses mathematics in interaction with other subjects will not, necessarily, contradict teaching that is aligned with the aforementioned identity of mathematics.

Mathematics in interaction with other subjects is an issue also seen from an international perspective. As part of the Danish governmental globalization efforts, a commission was formed and asked to evaluate school mathematics regulations and practices for grade level 4–6. One of the recommendations coming from the commission concerned relations between mathematics and other school subjects (our translation):

> School principals shall organise initiatives to make teachers in other subjects able "to see the mathematics" in their own subjects, in order to contribute to the efforts of improve the students' mathematical competences. It is common that school mathematics lives its own life without much relation to other subjects. It is, however, important that both mathematics teachers and other teachers become aware of how mathematics and other subjects can relate (interact/interplay) [Danish: spille sammen]. Although mathematics

> teachers are responsible for mathematics teaching, teachers in other subjects can importantly contribute to students' understanding and use of mathematics in many different settings. This requires that mathematics teachers are familiar with applications of their subject and that the schools initiate that teachers in other subjects become involved in the efforts to raise the students' mathematical competences. (Danmarks Evalueringsinstitut, 2006, p. 11)

In our view the construct of multidisciplinarity can serve as part of the school initiatives recommended by the Commission.

Students' Attitudes Towards Mathematics

The international comparisons in the PISA study include a student questionnaire on attitudinal aspects. The measurement reveals that Danish students' interest and enjoyment in mathematics and their mathematics anxiety is unusually placed in an OECD context. Figures 43.1 and 43.2 are constructed with PISA 2003 database as source.

Looking at Figure 43.1 it is remarkable that Finnish, Norwegian, and Icelandic students show averages of interest and enjoyment in mathematics below the OECD average; conversely, the Swedish and Danish show averages above the OECD average. Especially Danish students show a high average score of interest and enjoyment only surpassed by Turkish and Mexican students. Except for Mexico, scores for interest and enjoyment covariate positively with performance scores for every country. In all Nordic countries the covariations are relatively strong. We conclude that Danish students' interest and enjoyment in mathematics is exceptionally high.

Figure 43.2 underscores the fact that students' anxiety in mathematics is relatively low in all Nordic countries, although the average for Norwegian students corresponds to the OECD average. Although Mexican and Turkish students show high interest and enjoyment, they also show high anxiety. Swedish students show the lowest anxiety among all participating countries and Danish students the second lowest. The student anxiety data given in the OECD report covariates negatively with performance in all countries. We conclude that Danish students' mathematics anxiety is exceptionally low.

Attitude measurements in international comparisons such as those from the OECD report call for critical attention. Do students across nations and cultures understand the questions alike, do they define mathematics the same way, and do they value the same aspects of mathematics? Very little is known about these crucial issues. We have been involved in several investigations into Danish students' view of mathematics. In a research project, Lindenskov and Gronemann followed some form level 10 classes where teachers were involved in developing new mathematics instructional materials in a scenario setting.

The students' activities differed quite a bit from their former school mathematics experience. In a scenario of students' private economy and decision making, the students made the choice among financial websites on the World Wide Web, they formulated their own questions about their own personal situation in a scenario ten years ahead, and they reflected upon results. The necessary calculations behind the results were carried out by the website, not by the students themselves. The students in the follow-up interviews all stressed that the scenario approach was very motivating. The observations we made in the classroom indicated that students' perception of and performance in mathematics will change when the focus of the lessons changes to scenarios of possible authentic settings. At least for most of the students, a change was observed away from carrying out calculations towards formulating problems and reflecting on results given on the webpage.

Also, the teachers explained in interviews that they found the students highly motivated by the group-work organization of exploring scenario, as illustrated by one teacher' s quote:

> I think, what they liked the most, was the discussions in groups. They liked to be given a task, which is not ready made to analyse and answer, so that student group discussions became necessary. I think they found it really exciting (...) Many students told me themselves that they found the dynamics among the students really good. (Lindenskov & Gronemann, 2004)

All teachers emphasized it as an important aim for their teaching that the students develop self-confidence and independence, although some teachers said that for a proportion of the class—maybe as many as one third of the students in some form level 10 class—it is extraordinarily demanding and challenging to engage independently in open tasks.

One teacher explained:

> Students who so to speak "played" with the real world inherited in the scenarios, learnt a lot. But more hesitating students did not develop, you know, this ownership to their findings. Although they produced themselves their questions and findings, so potentially they could see them as meaningful and even having impact of their own actual financial decisions solution, they still kept narrowly focused on figuring out what the teacher want them to do, and where they can find

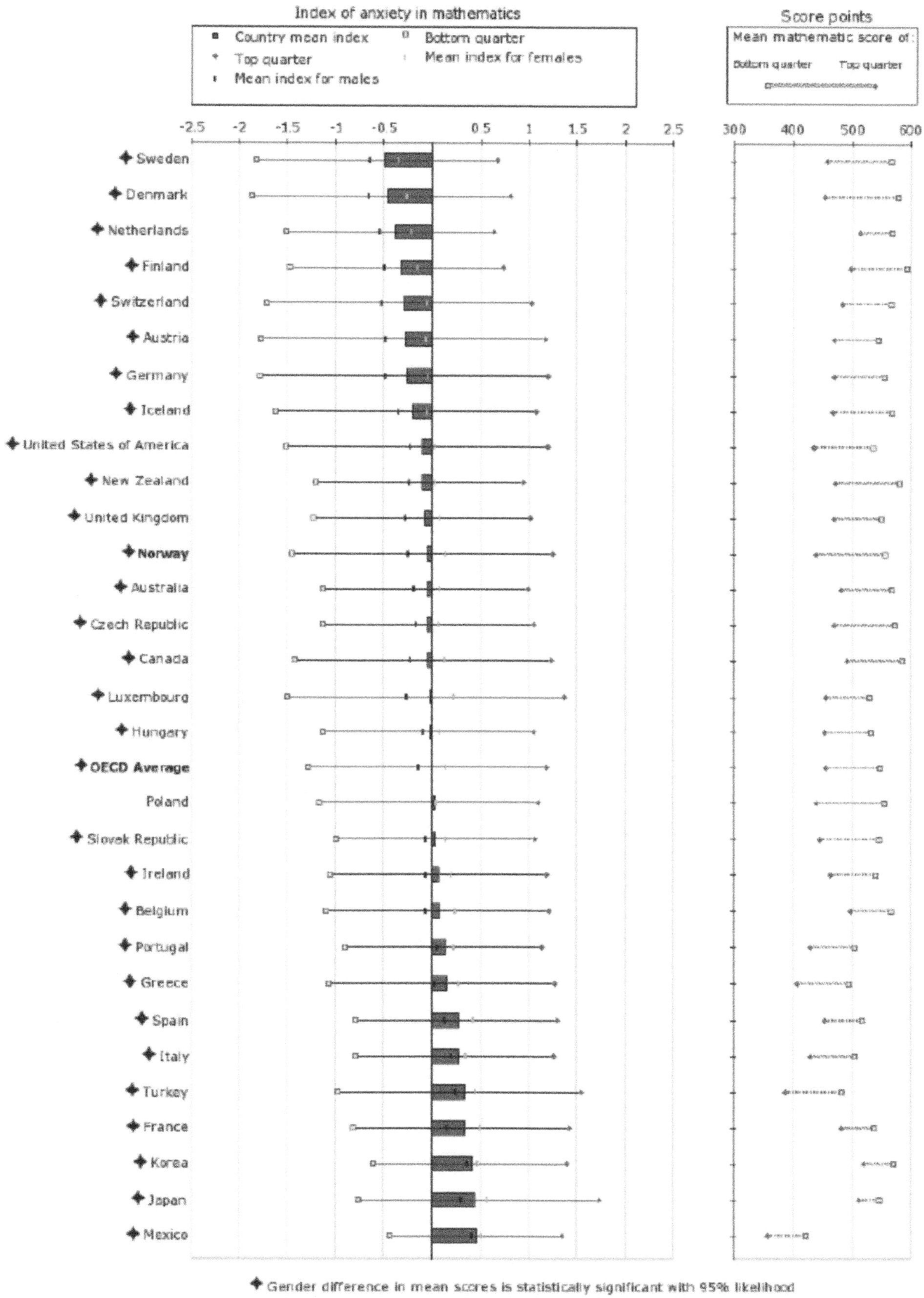

Figure 43.1 Students' interest in and enjoyment of mathematics in 28 countries (Source: PISA 2003 Database).

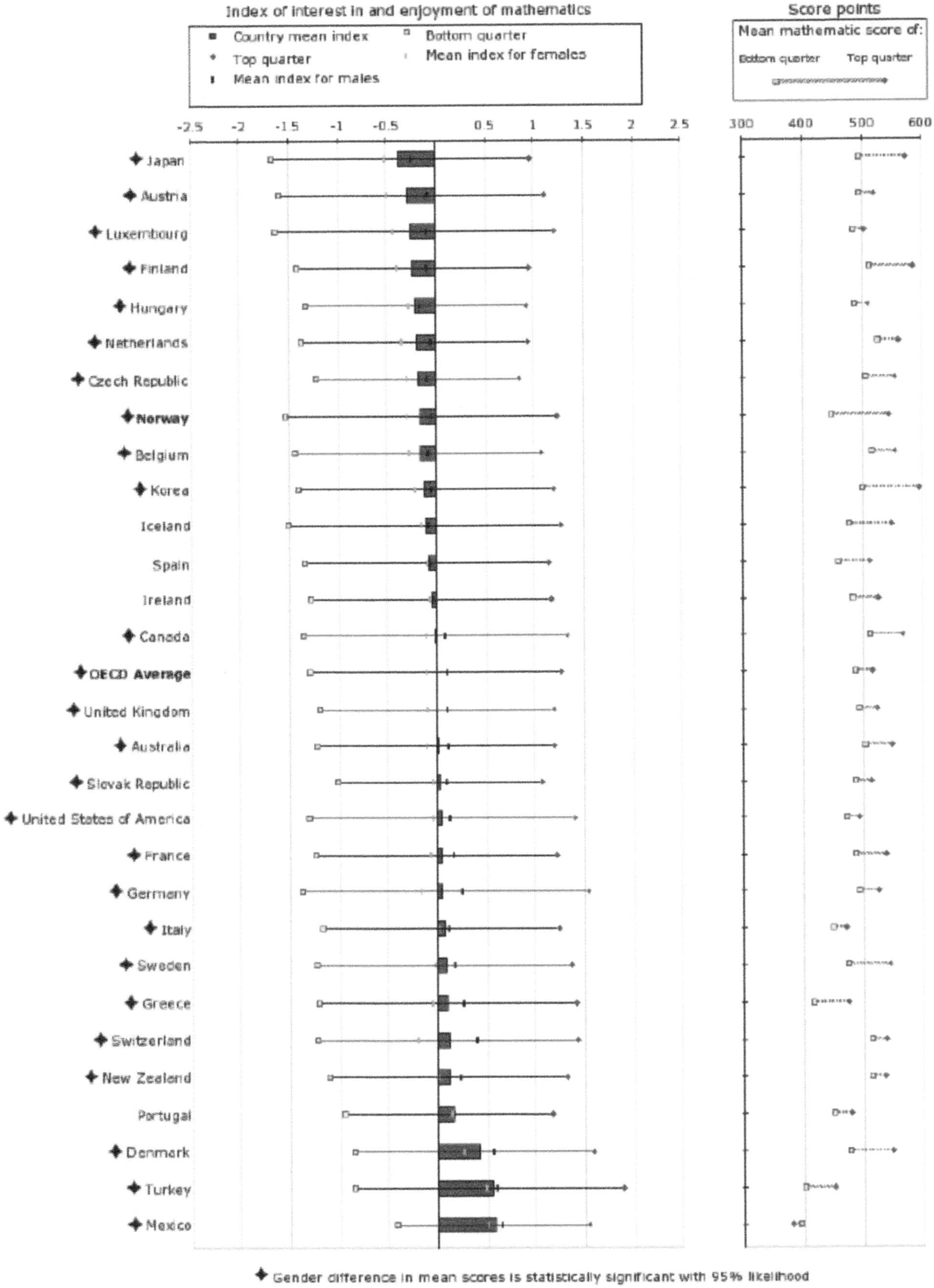

Figure 43.2 Students' anxiety in mathematics in 28 countries (Source: PISA 2003 Database).

> help for how to solve the problems. You see, they don't skip their former attitudes, they don't dare to.

Another teacher said:

> Some students cannot accept that many of the tasks have several solutions. We would hope that in form level 10 the students are sufficiently mature to recognize that among the many possible solutions, they should present what they themselves found appropriate. In this kind of assignments, you should not ask the teacher "is this correct?", but the fact is that many students address a question like this to the teacher. (Lindenskov & Gronemann, 2004)

When asked what the students found most challenging and difficult, most teachers answered like the following quote:

> Oh you see—the most challenging and difficult, I think, was for the students to recognize that they themselves should make so many choices concerning conditions and methods to use in coping with the scenario problems. (Lindenskov & Gronemann, 2004)

In another research project, Andresen and Lindenskov investigated perceptions of mathematics among students starting in upper secondary school. Our investigations revealed that mathematics was perceived as a school subject with collaborative explorations of the world around us. The dominant perception among the students was that problem-solving activities are more relevant and interesting than the training of skills. The students found that problem solving is meant to cope with and get insight into real-world phenomena. They saw mathematics as a means to explore real-world phenomena as they manifest themselves in other school subjects and in settings outside school. This does not mean that skills training activities do not have any advantage compared to problem-solving activities. Half of all students found that training mathematical skills raise your confidence in performing specific procedures, which is not attained by solving problems. We also investigated possible differences in the individual students' perceptions of mathematical activities across different branches of mathematics, and we found that individual students' perceptions of relevance and accessibility are stable across mathematical branches (Andresen & Lindenskov, 2008).

Teachers' Perceptions

In another research project, Lindenskov focused on mathematics teachers in basic adult education and their perceptions of relations between everyday life and mathematics education. These perceptions form a significant frame of what becomes discursively possible and how everyday-like assignments could be interpreted. The research showed four distinctive types of perceptions among teachers (Lindenskov, 2003). Although the project addressed teachers in adult education, the results can illuminate the possible roles for mathematics in multidisciplinary teaching. In the first type, it is stressed that mathematics is a specific discourse concerning generalized qualities of abstract entities, and everyday life as well as other subjects may *motivate students,* but should not be part of curricular aims and goals. In the second type, only advanced mathematics is perceived to be applicable in other academic subjects, while simple mathematics—for instance, mathematics in upper secondary school—is justified as a necessary precondition to learn more complicated mathematics, which shows up at the tertiary level. In the third type, mathematics is to be used everywhere in other school subjects and outside school. This means that all mathematical assignments dressed in everyday clothing or dressed in other subjects' clothing potentially develop students' mathematical conceptions and processes. In the fourth type, mathematics is utilized with quite different intentions and perspectives highly influenced and close-knitted with specific situations, problems, and school subjects. In our view these perceptions frame what becomes discursively possible and how mathematics assignments in a multidisciplinary setting could be possibly interpreted by teachers and students.

Cross-Disciplinarity and Multidisciplinarity

As part of the national teacher guidance developed by the Ministry of Education for upper secondary school, 124 so-called "paradigmatic examples" of teaching sequences (paradigmatiske undervisningsforløb) are provided as inspiration for teachers. It is remarkable that almost half of all the teaching sequences (51) involve cooperation with other subjects.[5] Many different models have been suggested to guide collaboration between school subjects. Danish discussions such as Nielsen (1992) have frequently used two concepts, which have been seen as contradictory:

- One is formal cross-disciplinarity, where the set of subjects is chosen first, and then afterwards the themes are decided and developed. The disadvantage of the formal cross-disciplinarity is that the subjects chosen do not add to the theme.
- The other is functional cross-disciplinarity, where the themes for the collaboration to be properly studied need more than one

subject, and where the theme is chosen first and predicts which subjects are being involved The disadvantages of the functional cross-disciplinarity is the risk of no subjective knowledge being developed, besides only common sense, and that some subjects are never involved in collaborations (Jensen, 1995; Nielsen, 1992).

The reform in 2005 was intended to capitalize on both formal and functional cross-disciplinarity concepts' advantages and escape their individual disadvantages. This attempt is demonstrated by the introduction of a new concept in the Danish educational discourse: multidisciplinarity. The reform reference group repeats and underlines in a report that the aim behind the compulsory multidisciplinarity is to strengthen the individual subject through the cooperation with other subjects (Foelgegruppen, 2008).

Hence, the intention is that multidisciplinary teaching provides opportunity to let the subjects mutually fertilize and, in parallel, shape a clear picture in the students' minds of the characteristics and identity of the individual subjects.

The Potential of Multidisciplinarity

Some of the potentials of multidisciplinary mathematics teaching were discussed in Andresen and Lindenskov (2008). We see potentials achieving a number of different goals.

i. Students' motivation and interest. Multidisciplinary projects can stimulate the students' interest and engagement in mathematics because the usefulness of the mathematics taught and its links with the students' own experienced world are in constant request in Danish school. These issues were dealt with in the first part of this chapter. Multidisciplinary projects set the stage for the teaching of useful applications of mathematics in authentic, daily life settings. Hence, such projects can serve to meet the students' requests and to improve their desired understanding of connections between subjects and the world outside school.

ii. Transfer. The authors report on their observations of the classroom that indicated

   a. Students' awareness of the possibilities to transfer concepts and results between subjects
   b. Students' consciousness about benefits, traps, and misunderstandings caused by such transfer
   c. Students' reflections upon the relations between the project's subjects

The students seemed to build relations between the subjects in parallel with their formation of concepts and new skills belonging to the single subject.

iii. Implementation. Multidisciplinarity can be seen as a means to revise the role of school mathematics and, thereby, to embed students' mathematical competence into a broad and reflected view of math and science. Compared to cross-disciplinarity (formal or functional), multidisciplinarity has better odds for successful implementation because it resonates with the following four main elements of Fullan and Hargreaves's (1992) framework for understanding teacher development: (1) the teachers' purpose, (2) the teachers as a person, (3) the real-world context for the teacher's work and (4) the culture of teaching

So, we argue that multidisciplinary teaching has important potentials for improving students' motivation and interest and for an enhanced transfer between subjects. We expect multidisciplinarity to be successfully implemented, and we expect it to serve potentially as a means in the future to support the embedding of the students' competencies into broad and reflected view on mathematics.

Modeling for Concept Formation

In addition to this, the didactical potentials of a multidisciplinary project rest on the role of mathematical modeling and reflections for concept formation. Mathematical models in multidisciplinary projects play a double role: on the one hand, the model can serve as the link between subjects and daily life, authentic problems, and so on, dealt with above. On the other hand, modeling plays an important role for concept formation. The role of modeling for concept formation in learning mathematics is apparent in the domain-specific instruction theory for realistic mathematics education (RME). RME is rooted in Hans Freudenthal's idea of "mathematics as a human activity" (Freudenthal, 1991). According to this theory, students should be given the opportunity to reinvent mathematics by mathematizing—mathematizing subject matter from reality (horizontal mathematizing) and mathematizing math-

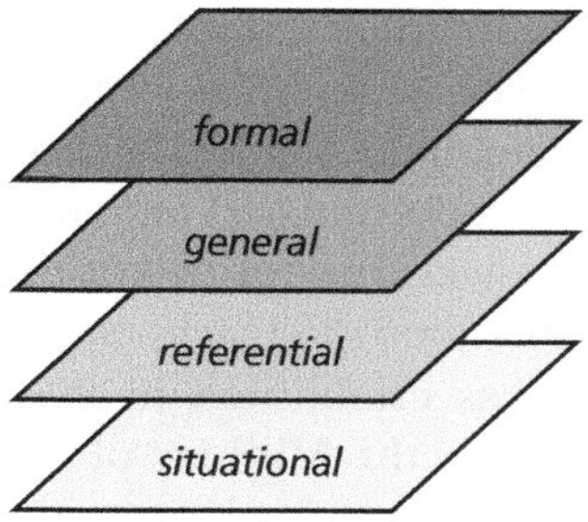

Figure 43.3 Levels of activity (Gravemeijer & Stephan, 2002, p. 159).

ematical matter (vertical mathematizing). This implies that the students develop a high level of intellectual autonomy. Hence, the core principle is that mathematics can and should be learned on one's own authority, through one's own mental activities. Horizontal and vertical mathematizing may be modeled by the passing of four levels of activity (see Figure 43.3).

A new mathematical reality is created at each level. From this point of view, all mathematical activity concerns modeling, and it means little to try to discern theoretically between learning, applying, or developing new mathematics. Strict borderlines between the three are not to be drawn. In general, the use of the term "modeling," therefore, has to be specified, since it depends on the context. In this chapter, though, we still use the terms in the "common way" unless stated otherwise.

Modeling and Mathematical Reflections

Reflection is the driving force for the process of mathematical modeling in the sense of progressive mathematizing (Gravemeijer & Stephan, 2002). In line with this, Andresen and Froelund (2008) discuss how to make the students' philosophical reflections explicit, as a tool for mathematical reasoning, and thereby strengthen the students' consciousness of the art of reflection and of the relationship between reflection and learning. In line with Schoenfeld's results that "performance on many tasks is positively correlated with the degree of one's metaknowledge" (1985, p. 138), Andresen and Froelund suggest the explication of mathematical reflections as a tool for learning. The use of philosophical reflections as a tool for mathematical reasoning was recently discussed (Prediger, 2007). Prediger's discussion was based on the stratification (Neubrand, 2000) of reflective practice in mathematics into four levels:

1. The level of the mathematician
2. The level of the deliberately working mathematician
3. The level of the philosopher of mathematics
4. The level of the epistemologist.

Andresen and Froelund (2008) argue for the teaching of mathematics based on the use of a reflection guide, which combines Neubrand's four levels of reflection with the four levels of activity in Gravemeijer's model. Thought-provoking questions at all four levels should give rise to reflections and initiate the passing from one level to the next in Gravemeijer's model.

Teaching in a multidisciplinary setting, such as in the project involving statistics and social sciences described in the last section of this chapter, provides a design that particularly favors explication of mathematical reflections. We shall return to this point in the discussion of the example.

Potential Drawbacks and How to Avoid Them

Teaching multidisciplinary projects, hence, is a promising prospect. We also see some potential drawbacks. In some aspects, the impact of multidisciplinarity on the students' view of mathematics is comparable to the impact of the use of computers. The 2006 reform also imposed the introduction of the compulsory use of computer algebra systems (CAS) in mathematics. This introduction will, obviously, cause comprehensive changes now and in the future; such as the teaching of models and modeling where the use of CAS opens up a wider range of topics and the numerical treatments of a variety of models, as in the case of differential equations models. CAS has the potential for a huge extension and development of the teaching of models and technical modeling in the sense of comparing a number of models and fitting them with a set of data (Andresen, 2007a). It also has potential to support students' model recognition and capability to understand and criticize authentic use of ready-made models in different contexts.

Results from our previous research, however, show that in general, the use of CAS tends to change focus of attention to the technical and practical aspects of upper secondary school mathematics. This tendency results from the individual teachers' choices based on preferences, habits, and CAS competencies. In general, teaching with a computer is centered upon solving tasks, whereas the reading of proofs and theoretical treatments in general are carried out without use of computer (Andresen, 2006).

There is a potential danger, therefore, that the same trend might direct the multidisciplinary teaching into a skills-based view of mathematics by the students, at the expense of giving the students a more profound insight into mathematical activities, theory, and knowledge. To avoid this, the students' more technical and practical view on models and modeling, should be balanced by explicit reflections upon the use of models and upon the modeling process, that is, upon horizontal and vertical mathematizing.

The didactical potential of a multidisciplinary project, therefore, depends on its design: The design has to ensure that the project's modeling processes are visible to the students as well as providing the opportunity to make students' mathematical reflections explicit during classroom discussion, and so on.

AN ONGOING PROJECT

The rest of this chapter presents a (prospective) research- and development project on mathematics in multidisciplinary teaching, to be conducted between 2008–2010 collaboration between Danish Science Gymnasiums (DASG)[6] and National Center of Knowledge about Mathematics Education (NAVIMAT).[7]

Partners in Collaboration

DASG is a network, incorporating about 36 Danish Upper Secondary Schools (out of 200[8]). The network aims to increase the public's interest in mathematics and science, particularly focusing on literacy in and knowledge about mathematics and science and, in general, to encourage a larger number of students to engage with mathematics, science, and technical education. There are five pivotal themes for the network's activities, which are:

- To support the innovation of teaching and learning methods and to develop teaching materials based on educational research outcomes and on recent education theory and ideas
- To arrange courses, seminars, and conferences to support the teachers' professional development
- To frame collaboration between educational researchers and practitioners in school
- Strengthen partnerships between schools, universities, and companies
- To draw attention to examples of good teaching practice in mathematics and science

The Lundbeck Foundation in 2006 provided a five-year sponsorship for the DASG network's activities,[9] and each gymnasium in DASG has applied for its membership on a voluntary basis. The members were chosen with consideration for geographical distribution and size. Membership implies an obligation by the school to spend resources, in the form of teachers' working hours, on participation in at least one of the 5–8 sub-projects, which are formulated and announced every year. The sub-projects run for two or three years, and each one involves about 25 teachers.

NAVIMAT was established in 2008 on a grant of three years from the Ministry of Education, with the mission to empower and strengthen the teaching and learning of mathematics at all levels from kindergarten to teacher education. NAVIMAT has its headquarters at University College Copenhagen and organizes, among other activities, a dozen collaboration and development projects at university colleges in Denmark. The NAVIMAT development projects are based on research and involve researchers in mathematics education.

The DASG–NAVIMAT Project

The collaboration between DASG and NAVIMAT encompasses a two-stage project. During the first year, three different types of teaching materials will be produced and tried out in a pilot; each of these materials represents the interplay between mathematics and one of the three participating faculties—human sciences, social sciences, and natural sciences. During the next stage, the following year, trials and evaluations of revised teaching materials from the pilot will be offered to the DASG schools as sub-projects. The revised versions of these materials hopefully will be published by NAVIMAT to provide inspiration for teachers at the conclusion of the trials.

NAVIMAT will form teams to prepare the materials by putting two or three teachers, one or two researchers in mathematics education, and one professional specialist in contact. The teams will then, autonomously, plan and make arrangements for their work during the first year of the project. DASG will organize joint seminars for all the teams during this stage, for the exchange of ideas and experiences so far.

The professional specialists are selected depending on the mathematics teachers' choice of subject—for example, an engineer if mathematics and physics collaborate, a statistician or an economist for multidisciplinary teaching that involves social science, and so on. The professional specialist's role in the team is to provide inspiration and expertise with regard to the content of the teaching materials. The mathematics education researcher is will provide inspiration and expertise with regard to the design of the materials and with the evaluation of the teaching experiment. The researcher is responsible for development and formulation of guidelines for good practice in multidisciplinary teaching. The team's teachers design and produce the teaching materials and carry out the teaching sequences. The teachers participate in the evaluation and discuss the results with the researchers.

New Challenges to the School Subject Mathematics

The aim of the project is to produce examples of good teaching and, thereby, to throw light on and demonstrate what works, as examples of good teaching practice, for the learning of mathematics in multidisciplinary contexts. The current interest is a consequence of the recent reform in upper secondary school, mentioned previously: In mathematics, the reform's tendency towards multidisciplinarity gave rise to focused attention on the role of mathematics in different types of collaborations. In line with our discussion in the previous paragraph, groups of mathematics teachers share their experiences of the subject in a service role where it mainly serves as a tool for calculation in those new, multidisciplinary environments. Alternatively, the multidisciplinary projects involve cultural, historical, or philosophical aspects that are important but not at the heart of mathematics taught in schools. To balance this tendency, there is a need for advice and ideas about how to empower the learning of what one might call "core mathematics" within a multidisciplinary teaching context.

Content of the Pilot: The Vioxx Case

We will now describe a preparatory design and trial of teaching materials leading up to the pilot, as a means of demonstrating the tenets of multidisciplinary teaching in the project. The materials for this pilot used the Vioxx case, described below, as its starting point and concentrated on probability theory and statistics in mathematics. The envisioned collaborator was social scientist, although philosophical ethics or chemistry may have also been appropriate.

Vioxx was a pain-reducing drug produced by Merck, and the case was about the statistical estimation of its long-term effects. In such cases it is impossible to carry out large-scale trials to determine the serious or long-term effects of drugs such as Vioxx. Therefore, when the drug is approved, such trials may be substituted by statistical inquiry of the population of users. For such inquiries, though, statistical models suitable for large-scale trials have to be modified, and, in particular, the criteria for the acceptance or rejection of hypotheses must be changed. Hence, the Vioxx case served as a context for the students in mathematics to study probability value (*p*-value), statistical significance and confidence intervals.

Vioxx, which was withdrawn from the U.S. market in 2004, is part of the class of drugs known as nonsteroidal anti-inflammatory drugs (NSAIDs). Vioxx was used to reduce pain, inflammation, and stiffness caused by osteoarthritis; to manage acute pain in adults; to treat migraines; and to treat menstrual pain. Merck, the manufacturer of Vioxx, announced a voluntary withdrawal of the drug from the U.S. and worldwide market, due to safety concerns of an increased risk of cardiovascular events (including heart attack and stroke) in patients taking Vioxx.

According to the U.S. Food and Drug Administration (FDA)'s website,[10] FDA originally approved Vioxx in May 1999. The original safety database included approximately 5,000 patients on Vioxx and did not show an increased risk of heart attack or stroke. A second study was primarily designed to look at the side effects of Vioxx, such as stomach ulcers and bleeding, and was submitted to the FDA in June 2000. The second study showed that patients taking Vioxx had fewer stomach ulcers and bleeding than patients taking naproxen, another NSAID; however, the study also showed a greater number of heart attacks in patients taking Vioxx. This second study was discussed at a February 2001 Arthritis Advisory Committee, and the new safety information from this study was added to the labeling for Vioxx in April 2002. Merck then began to conduct longer-term trials to obtain more data on the risk of heart attack and stroke with long-term users of Vioxx.

Merck's decision to withdraw Vioxx from the market was based on new data from this later trial, in which Vioxx was compared to a placebo (sugar-pill). The purpose of the trial was to see if Vioxx 25 mg was effective in preventing the recurrence of colon polyps. This trial was stopped early because there was an increased risk of serious cardiovascular events, such as heart attacks and strokes, first observed after 18 months of continuous treatment with Vioxx compared with placebo.

The Vioxx case attracted public attention since a large number of people had been taking Vioxx, and among them, some had heart attacks. Heart attack victims and surviving relatives took legal action and were, in a number of cases, rewarded. For example, John McDarby, 77, and his wife were rewarded a $4.5 million dollar verdict and $9 million in punitive damages from a New Jersey jury in one of the first Vioxx trial cases against Merck.[11] The controversial question for the judgement about Merck's responsibility was to determine whether data were sufficient to validate any hypothesis about correlation between Vioxx and the heart attacks.

From the viewpoint of mathematics, binomial distribution, Poisson distribution, and normal distribution are sufficiently strong tools to deal with these issues. Data from the original and from the later trials are available on Merck's website, and the determination rests on decisions about level of significance and the confidence intervals.

First Experiences from Teaching the Vioxx Case

For the preparatory teaching experiment, the teacher designed a sequence of about twenty lessons. This teacher has a master's degree including economics, so he agreed to spend some time and efforts on the inclusion of societal economic aspects in his teaching. The design was based on preceding discussions at a two-day seminar on authentic mathematics in upper secondary school and, subsequently, in a team with another mathematics teacher, a bio statistician, and a researcher in mathematics education. This small group gathered twice during the semester where the experiment took place, for inspiration, the exchange of ideas, and evaluation.

The class spent about twenty lessons on this part of the course. The students had no prior experiences with probability or statistics. Consequently, the teacher decided to spend the major part of the time on introducing and teaching basic terms and relations within these branches. The teacher decided to base the introduction and the training on the textbook with additional tasks collected from the web. In addition, the teacher and the bio statistician prepared a spreadsheet for the students to experiment with distribution, confidence intervals, and correlation coefficients.

In parallel, the students learned about the Vioxx case. Different aspects of the case were discussed in the class; in particular, the weighting between ethical and economical aspects and the role of mathematizing in such cases were examined and debated.

The challenge for the teacher, thus, was to combine the following three elements:

1. The mathematical content: introduction and basic training of terms and relations in probability and statistics. The content was taught in line with common practice in this class, based on the same textbook.
2. The role of mathematics in the Vioxx case. In the Vioxx case the process of mathematizing, obviously, was an issue of debate because of its implications for clients, the Merck Company, etc. Thus, the case did not serve as a bare illustration of a "neutral application" of mathematics. On the contrary, the case intended to draw attention to the modeling process itself.
3. A look from outside at the societal role of mathematics. Development, testing, and application of medical treatments are based on the use of bio statistics and play an important role in healthcare at individual and societal levels, although it implies ethical and economical perspectives. Besides, public discussion of these issues may be seen as one element of democracy.

Resulting Outcomes

We now discuss some of the resulting outcomes related to the above-mentioned points.

Mathematical Content

During the teaching experiment, the students showed a great deal of interest in the subject and in the Vioxx case. According to the teacher, the students were so eager to understand and to feel comfortable with the mathematical terms and relations and as a consequence, the class had to spend more time than expected on the technical-mathematical part of the course. For example, they spent six lessons just on working with level of significance. The teacher noted that this part of the sequence worked very well for the students.

Mathematical Modeling

The teacher indicated that the discussions stayed at the level of ready-made models. No attempts were made to modify the binomial distribution or to critically sort out the website's data. Modeling as such appeared not to be self-explanatory; on the contrary, every step had to be pointed out explicitly if the students were expected to be aware of it. For example, it was complicated for the students to make mathematical decisions, such as stating the level of significance. The teaching experiment, evidently, intended to demonstrate exactly that point to the students, so the stage was set to go deeper into the complex questions. The class spent time on changing the levels of significance and studied the consequences and effects. But they did not have time to follow these studies up.

Societal Role of Mathematics

The teacher had the impression that the envisioned "look from outside" on mathematics and its role would give input to very interesting and fruitful lessons on societal economy, law, and issues about democracy, public opinion, and politics. It could be great, according to the teacher, to arrange a replay of one of the big hearings as a game with students arguing for and against. To complete this, teachers from both subjects would have to collaborate. The instructional materials for such a replay could be found on the various websites but it should, preferably, be prepared in the enlarged team, including a teacher of social sciences.

The DASG–NAVIMAT Pilot Project Conclusion

The big challenge for the pilot project was to make the three elements listed above as a coherent and convincing whole. Part (1), the mathematics content, was marked by its status at the introduction. The teacher who carried out the preparatory experiment recommended starting the pilot project when the students had already had an introduction to basic probability and statistics. They would then be able to concentrate on the role of mathematics and to work deliberately with models and modeling. Connections, then, would be established more easily between the mathematical activities and the other elements of the Vioxx case. Such connections could be established, based on the teacher's guiding questions and sub-tasks, focusing on specific aspects of the distribution, the level of significance or the parameters influencing the basic probabilities, and so on, combined with special points of interest from a social science point of view.

To sum up, the teacher recommended that part (1) precede the proper multidisciplinary part, which should then combine parts (2) and (3). The complete project should build up to a student role play of one of the big hearings, with arguments and a final decision in the form of a verdict.

The pilot, scheduled to run in 2008–2009, was designed to encompass at least two more experiments; one is most likely to combine physics with mathematics and the other is most likely to combine philosophy with mathematics. From the perspective of this article, part (2) in the scheme above is the most interesting. In the case of physics, mathematics' role for building a concrete model of how nature is experienced, aiming to explain and predict observations and measurements, will be the issue of this part of the teaching experiment. In the case of philosophy, part (2) of the scheme will consider mathematical reflections as an example of a method for how knowledge is gained through philosophical reflections, while mathematical reflections will serve as a means to sharpen the students' consciousness about the very characteristics of mathematical activity.

FINAL COMMENTS

Mathematics in multidisciplinary settings is requested by formal regulations. Seen from mathematical educational perspectives, this approach is desirable.

We conclude that multidisciplinary teaching on the one hand may fit well with the professional culture of upper secondary schools with their academic, subject-focused learning environment; as well, it may serve to inspire collaboration between subjects in primary and lower secondary school.

Consequently, we expect multidisciplinarity to be successfully implemented and further developed in the near future and that the Danish mathematics education can serve as a valuable frontrunner in this area.

NOTES

1. http://www.eng.uvm.dk/~/media/Files/English/Fact%20 sheets/080101_fact_sheet_the_folkeskole.ashx
2. This means that school failure is almost non-existing. With the consent of the child's parents it is possible to repeat a school year—that is, if the student has been away for a longer period of time or other reasons make this relevant.
3. A description (in Danish) with binding national objectives and national guidance of each subject in primary and lower secondary school is provided in the subject's folder under the title "Common Objectives" (Fælles Mål) at http://www.faellesmaal.uvm.dk/.
4. Students must complete the following tests: Danish, with a focus on reading in form levels 2, 4, 6 and 8; English in form level 7; Mathematics in form level 3 and 6; Geography in form level 8; Biology in form level 8; Physics/chemistry in form level 8.
5. http://www.emu.dk/gym/fag/ma/undervisningsforloeb/paradigm.html
6. http://www.emu.dk/gym/tvaers/sciencegym/english.html (retrieved 22 May 2008)
7. www.navimat.dk (in Danish, retrieved. 22 May 2008)
8. 155 general and 42 technical upper secondary schools (gymnasiums)
9. Description of the Lundbeck Foundation in Danish at http://www.lundbeckfonden.dk/ (retrieved 27 May 2008)
10. http://www.fda.gov/cder/drug/infopage/vioxx/vioxxQA.htm (retrieved June 2008)
11. http://www.weitzlux.com/vioxx/trial_402790.html (retrieved June 2008)

REFERENCES

Andresen, M. (2006). *Taking advantage of computer use for increased flexibility of mathematical conceptions.* Unpublished doctoral dissertation, Danish University of Education.

Andresen, M. (2007a). Modelling with the software 'Derive' to support a constructivist approach to teaching. *International Electronic Journal of Mathematics Education, 2*(1).

Andresen, M. (2007b). Understandings of 'modelling.' In D. Pitta-Pantazi & G. Philippou (Eds.), *Proceedings of the Fifth Congress of the European Society for Research in Mathematics Education. 2–26 February 2007, Larnaca* (pp. 2042–2049). ISBN 978-9963-671-25-0.

Andresen, M., & Froelund, S. (2008). Philosophical reflections made explicit as a tool for mathematical reasoning. *Proceedings of 5th International Colloquium on the Didactics of Mathematics, University of Crete, 17–19 April 2008, Crete.* 11 pages. Department of Education, University of Crete ISBN 978-960-87898-3-8.

Andresen, M., & Lindenskov, L. (2008). New roles for mathematics in multi-disciplinary, upper secondary school projects. *ZDM—The International Journal on Mathematics Education, 41*(1), 213–222.

Danish Ministry of Education. (2008). *Departmental Order for Mathematics at academic upper secondary school.* Copenhagen: Author. [BEK nr 741 af 30/06/2008, Appendix 35]. Retrieved May 26, 2010 at https://www.retsinformation.dk/Forms/R0710.aspx?id=120566#Bil35

Danmarks Evalueringsinstitut. (2006). *Matematik på grundskolens mellemtrin—Skolernes arbejde med at udvikle elevernes Matematikkompetencer.* Copenhagen: Danmarks Evalueringsinstitut.

The Folkeskole (Consolidation) Act. (2003). Retrieved 21/8 2009 from http://pub.uvm.dk/2003/consolidation.html (English version) and from https://www.retsinformation.dk/Forms/R0710.aspx?id=125580 (newest Danish version)

Freudenthal, H. (1991). *Revisiting mathematics education: China lectures.* Dordrecht: Kluwer.

Fullan, M., & Hargreaves, A. (Eds.). (1992). *Teacher development and educational change.* London: RoutledgeFalmer.

Foelgegruppen for reformen af de gymnasiale uddannelser (Reform reference group). (2008). *Rapport nr. 7 til Undervisningsministeren (Report to the Minister of Education).* Retrieved May 26, 2010 from http://www.uvm.dk/~/media/Files/Udd/Gym/PDF08/Politiske%20oplaeg%20og%20aftaler/081029_Foelgegruppen_Rapport_7.ashx

Gravemeijer, K., & Stephan, M. (2002). Emergent models as an instructional design heuristic. In K. Gravemeijer, R. Lehrer, R., B. van Oers, B., L. Verschaffel, L. Gravemeijer et al. (Eds.), *Symbolizing, modelling and tool use in mathematics education* (pp. 145–169). Dordrecht: Kluwer Academic Publishers.

Jensen, B. E. (1995). Tværfaglighed. *Undervisningsministeriets tidsskrift Uddannelse,* 9. Copenhagen: Ministry of Education.

Lindenskov, L. (2003). Tutors' approaches to mathematics to live and work: A model of four types. In J. Maasz & W. Schloeglman (Eds.), *Mathematics to Live and Work in our World* (pp. 146–153). Linz, Austria: Universitätsverlag Rudolf Trauner.

Lindenskov, L., & Gronemann, K. (2004). *Matematikportal til 10. klasse: ITMF-projekt 140.* Copenhagen: Danmarks Pædagogiske Universitet. Retrieved May 26, 2010 from http://www.dpu.dk/everest/Publications/Forskning/Medier%20og%20IT/ITMFx2Drapporter/20041215111732/CurrentVersion/ITMF440.pdf

Neubrand, M. (2000). Reflecting as a didaktik construction. In I. Westbury, S. Hopmann, & K. Riquarts (Eds.), *Teaching as a reflective practice. The German didaktik tradition* (pp. 251–267). New Jersey.

Nielsen, V. O. (1992). Om tværfaglighed og faglighed i Folkeskolens undervisning. In S. Breiting, et al. (Eds.), *Undervisning kunne være et alternativ. Om tværfaglig undervisning i folkeskolen.* Vejie, Denmark: Kroghs Forlag.

Prediger, S. (2007). Philosophical reflections in mathematics classrooms. In: François, K. & J.-P. van Bendegem (Eds.), *Philosophical dimensions in mathematics education* (pp. 43–58). New York: Springer.

Schoenfeld, A. H. (1985). *Mathematical problem Solving.* Orlando, FL: Academic Press.

44

PREPARING IN-SERVICE TEACHERS FOR AN INTERDISCIPLINARY CURRICULUM

Claus Michelsen

CENTRE FOR SCIENCE AND MATHEMATICS EDUCATION, UNIVERSITY OF SOUTHERN DENMARK

Abstract: A wave of mathematics and science[1] reforms aiming at a more interdisciplinary curriculum is going on in Denmark. To implement the objectives of this reform, educational cooperation across the traditional boundaries between the subjects of mathematics and science both at the level of subject matter as well as at the level of pedagogy is required. This is far from a trivial task, partly due to a lack of framework for integrating productive ideas from a variety of theoretical and practical perspectives on the relations between the subjects. To prepare the teachers for a more interdisciplinary curriculum in mathematics and science, the University of Southern Denmark has developed educational programs for in-service teachers from the Danish lower and upper secondary school. The programs are designed around a didactical framework for coordination and mutual interaction of the subjects of mathematics and science developed at the University of Southern Denmark. During the courses the teachers get involved in design, implementation, and evaluation of innovative interdisciplinary instructional sequences that deal with a wide range of aspects of mathematics and science, such as modern science and the importance of science in society. In the programs, contemporary science and mathematics education research serve as a basis for the design and development of warranted practices with which the teachers experiment in their classrooms. The chapter introduces the didactical framework and the outcomes of offering a program that is intimately tied to (1) contemporary science and mathematics education research, (2) modern science and mathematics, and 3) the teacher's practices in the classroom.

Keywords: Interdisciplinarity, educational reconstruction, in-service teacher training, professional development, teacher–researcher cooperation

INTRODUCTION

It is an impossible situation that the mathematician teaches a mathematics that cannot be applied and the physicist applies a mathematics that has not been taught by the mathematician.

—Freudenthal, 1973, p. 553

Primary and lower secondary education are in Denmark structured as a 9-year comprehensive basic school (Grundskolen) followed by a voluntary 10th school year. About 50% percent of each class continues after completion of basic school to the academically oriented 3-year general or technical upper secondary school (Gymnasium). While the teachers of basic school are educated at teacher colleges, the teachers of general upper secondary school are educated at universities.

A wave of mathematics and science reforms aiming at a more interdisciplinary curriculum is going on in Denmark. Since the start of the new millennium, there has been a growing interest in mathematics educational relations to the subjects of biology, chemistry, and physics. This interest is mainly due to initiatives from the government aiming at a

The First Sourcebook on Nordic Research in Mathematics Education, pages 597–612

more interdisciplinary curriculum in both lower and upper secondary education. In August 2005, a structural reform was introduced in general and technical upper secondary education. The reform implies that after a common introductory half-year period, all students must choose a so-called subject package, where the participating subjects form a coherent program. To prepare students for studies in science, mathematics, and technology in tertiary education, some of the packages have as their core the subjects of physics, chemistry, biology, and mathematics. In February 2008, a group of mathematics and science educators appointed by the government launched an *action plan for mathematics and science education* (Ministry of Education, 2008). Among other things, it recommended initiating actions with aim of development of

- New and innovative content and new forms of instruction at primary and secondary education
- Common learning goals for mathematics, science, and technology
- Innovation-focused training for teacher students and in-service teachers

One of the challenges of the reform wave is thus to replace the current monodisciplinary approach with an interdisciplinary approach, where mathematics, biology, chemistry, physics, and technology are woven continuous together.

Reform movements are enacted on various levels, and a large variety of agents are involved: teachers, teacher trainers, researchers, administrators, and so on. In the end, however, it will be the will be the teachers who will have to make the innovation come through. It will be the teachers who have to establish the reformed practice in their classrooms.

To implement the objectives of the reforms educational cooperation across the traditional boundaries between the subjects of mathematics and natural sciences both at the level of subject matter as well as at the level of pedagogy is required. This is far from a trivial task, partly due to a lack of framework for integrating productive ideas from a variety of theoretical and practical perspectives on the relations between the subjects of natural sciences. To the grasp the challenges from the reform wave there is a need for:

- A didactical framework for interdisciplinary activities
- In-service teacher training with focus on interdisciplinary activities
- Prototypes of meaningful interdisciplinary teaching sequences

To prepare the teachers for a more interdisciplinary curriculum in mathematics and science, The University of Southern Denmark have developed course programs for in-service teachers from the Danish lower and upper secondary school grounded in a didactical framework for interdisciplinary activities. In the following sections I present and describe the didactical framework and two programs for in-service teachers.

EXPANSION OF DOMAIN: MATHEMATICS IN AN INTERDISCIPLINARY CONTEXT

In the middle of the twentieth century an attempt was made to separate mathematics and physics. The consequences turned out to be catastrophic. Whole generations of mathematicians grew up without knowing half of their science and, of course, in total ignorance to any other sciences. They first began teaching their ugly scholastic pseudo-mathematics to their students, then to the schoolchildren

—Arnold, 1998, p. 229

Mathematics plays a crucial role in science. This role is brought about predominantly through the building, employment, and assessment of mathematical models. To celebrate the World Mathematical Year 2000, the American Mathematical Society, under the auspices of the International Mathematical Union, published the book *Mathematics: Frontiers and Perspectives.* In the preface of the book, the world-famous mathematician Michael Atiyah (1999) reflects over mathematics in the 21s century. The close symbiotic relationship between mathematics and physics is underscored with a forecast for the 21st century pointing towards a rise to new heights. Atiyah also expects that mathematics will have a useful part to play in some of the major challenges of biology, like understanding the brain and in handling of the vast data bank produced by the Human Genome Project. Although biology has historically been the least mathematical of the natural sciences, it has never been divorced from mathematics.

Mathematics is a gatekeeper to many mainstream studies and careers, and when students move through education and on into adult life they encounter issues in politics, finances, environment, and health that rely on the mathematical techniques that were learned in secondary school. A certain transparency of the transition to mathematics beyond upper secondary education is necessary if we want the students to realize the importance of mathematics. This role should also be reflected in educational settings. The instruction should promote the students' motivation for and interest in technology, science, and scientific ways of thinking by including authentic interdisciplinary complexes in the instruction. But this is not the rule in secondary education, where the focus in mathematics instruction is on relatively specialized algebraic techniques that students rarely remember beyond the final exam.

One might ask whether contemporary mathematics education prepares the students well for the application of mathematics in other subjects, their civic obligations, their future employment, and personal needs. Steen (2005) highlights this by pointing out that the absence of strong curricular ties between the biological and quantitative sciences misinterprets contemporary biological research; deskills many biology, computer science, and mathematics students; and fails to prepare them to collaborate on significant problems. Lesh and Sriraman (2005) point at the dramatic change in the nature of problem-solving activities during the past twenty years, and at the difficulties in recruiting students capable of graduate level interdisciplinary studies such as mathematical biology and bio-informatics. They suggest a bottom-up solution based on modeling of complex systems that occur in real-life situations from the early grades. This opens up the possibility for more up-to-date mathematics content by a shift in perspective: Rather than approaching mathematics by first teaching what is to be learned and then applying these concepts in realistic situations, the mathematization approach first puts students in sense-making situations where the concepts that they develop on their own are later decontextualized and formalized.

Interdisciplinary Competences

The interdisciplinary reform wave in Denmark calls for an innovation of the content provided on the basis of target management rather than content management. This draws attention to new ways of describing curricula. In the so-called *KOM report* (Competences and Learning in Mathematics) the notion of competence is introduced as a basis for describing and analyzing mathematics instruction from kindergarten to tertiary level. The KOM report defines mathematical competence as an insightful preparedness to act properly in situations that contain a particular kind of mathematical challenge. Eight mathematical competences are identified, aiming at describing what it means to master mathematics: (1) mathematical thinking competence, (2) problem-solving competence, (3) modeling competence, (4) reasoning competence, (5) representational competence, (6) symbolizing and formalizing competence, (7) communication competence, and (8) auxiliary tools competence (Niss & Jensen, 2002). In a similar report, *Fremtidens Naturfaglige Uddannelser* (Science Education for the Future), four science competences are identified and described: (1) empirical competence, (2) representational competence, (3) modeling competence, and (4) reflective competence. It is recommended that all secondary level students should be offered a specific science program including the subjects of physics, chemistry, and biology on an equal footing. The report doesn't deal with the role of mathematics in science education—actually mathematics is hardly mentioned in the report (Andersen, Bush, Horst, & Troelsen, 2003). The KOM report shortly touches the relations between mathematics and science by pointing at the fact that although a growing number of subjects include ingredients from mathematics, it is still difficult for both teachers of mathematics and teachers of other subjects to see the use of mathematics in other subjects—partly due to the use of concepts and language. This paradox of relevance leads to the *problem of isolation*, which works to the disadvantage of both mathematics and other subjects that could profit from a conscious inclusion of mathematical competences (Niss & Jensen, 2002). But although the reports only to a modest degree deal with the relations between mathematics and science, two competences are present in both reports—the modeling and the representational competence. This indicates that the modeling and representational competences are the links between mathematics and science, and we can identify them as *interdisciplinary competences.*

Expansion of Domain

The transfer of mathematical ideas, concepts, and procedures is one of the biggest challenges in education. It is well known that it is difficult for students to apply concepts, ideas, and procedures learned in mathematics in a new and unanticipated situation, either in or out of school. The lack of coordination between the curricula of mathematics and science is one of the primary causes of the students' difficulty in applying mathematics in other subjects like physics, chemistry, and biology. Further, students get a distorted picture of how mathematics concepts are built, how they grow and how they are used in other sciences, and so they tend to adopt a negative attitude towards mathematics.

Many mathematics and science educators are in favor of a more realistic education, where modeling activities are used to treat concepts in realistic, everyday life contexts (e.g., de Lange, 1996; Gilbert & Boulter, 1998). But little attention has been paid to the potential of such activities to bridge the gap between mathematics and science. A cause of this might be the very simplistic way problem solving often has been conceptualized. There are two levels: (1) the realistic, everyday life contexts and applications of mathematics, and (2) the abstract world of mathematical concepts and operations. According to this perspective problem solving involves a translation from the physical reality into the abstract mathematical world. A consequence is that mathematics may be and often is learned separately from its applications. As an alternative to this approach Lester and Kehle (2003) suggest to subsume problem solving within the much broader category of mathematical activity.

There are a number of challenges for the transfer of mathematical skills or tasks to an applied context. These include different methods of calculations, types of representations, and ways to represent data. Attempts to develop skills in transferring from mathematics to other subjects have not been successful, mainly because of the lack of adequate integration of mathematical skills or tasks within the particular context. According to Niss (1999), a significant example of the major findings of research in mathematics education is the key role of domain specificity. The student's conception of a mathematical concept is determined by the set of specific domains in which that concept has been introduced for the student. When a concept is introduced in a narrow mathematical domain, the student may see it as a formal object with arbitrary rules. This results in the recognized difficulty of applying the concept in new settings. Michelsen (2001, 2005a) suggests the idea of *expanding the domain* as a way to transcend the problem of domain specificity. The idea is to include contexts from science to broaden the domain where a concept is introduced. Interdisciplinary activities between mathematics and science offer a great variety of domain relations and context settings that can serve as a basis for developing a more practical and coherent structure of a mathematical concept. Lobato (2003) addresses the central educational issue of transfer learning and criticizes the traditional models of transfer for relying too heavily on the determination of transfer from an expert's point of view. She argues for an actor-oriented framework, where one assumes that the students are making connections between situations nearly all the time, guided by the aspects they find personally salient. As a consequence the critical issue is to design an instructional environment that supports the students' construction of personal relations of similarities across situations.

THE DIDACTICAL FRAMEWORK: HORIZONTAL LINKING AND VERTICAL STRUCTURING

Mathematics has arisen and arises through mathematising.

—Freudenthal, 1991, p. 66

Issues related to the integration of topics from mathematics and science are complex, because they comprise two apparently different components, an extra-mathematical and a mathematical context. The integration must take into account that didactical[2] problems with the various topics have more consequences than the respective curricula might suggest. There exist numerous general models of integration that offer a theoretical grounding for the diverse perspectives associated with integration of mathematics and science. These models most often take the form of a continuum reaching from the discipline of mathematics over balanced mathematics and science to the discipline of science. Generally speaking, the models are based on two assumptions: (1) teaching mathematics in relation to science supports students' learning by providing meaningful contexts in which the students can see the application of abstract mathematical concepts—and as a side effect this helps students instill positive attitudes towards mathematics, and (2) mathematics offers science education the tools for quantifying, representing, and analyzing phenomena from science—and as a side effect fosters a sense of objectivity in science. This often results in connections between mathematics and science, which is imbedded within the context of real-world applications, and as such the connections tend to be contextual rather than conceptual (Berlin, 2003). *The Principles and Standards for School Mathematics* recommends a restoring of the proportions between the mastery of algorithms and conceptual understanding, in favor of the second aspect (NCTM, 2000). It is of great importance for the students of mathematics to have opportunities to apply their mathematical skills to the solution of scientific problems. To avoid a separation between application of mathematics and conceptual aspects of mathematics, contextual problems should also be exploited as meaningful starting points from which conceptual structures in mathematics and science can emerge. For that purpose a coherent didactical framework including both a contextual and a conceptual approach is needed.

Contextual and Conceptual Approaches

There exist several frameworks that include both contextual and conceptual approaches. The concept of emergent model suggested by Gravemeijer (1997) has as the departing point situation specific problems, which are subsequently modelled. Context problems first offer the students the opportunity to develop situation-specific methods and symbolizations. Then the methods and symbols are modelled from a mathematical perspective and in this sense the models emerge from the students' activity. The models first come to the fore as a model of the situation, and then the model gradually becomes an entity on it own right and starts to serve as a model for mathematical reasoning. The shift presented from "model of" to "model for" should concur with a shift in the way students perceive and think about the models, from models that derive their meaning from the context situation modeled to thinking about the mathematical content.

According to Freudenthal (1991), mathematizing is the key process of mathematical activity. There are two types of mathematization in an educational context—horizontal and vertical mathematization. In horizontal mathematization, the students come up with mathematical tools that can help to organize and solve a problem located in a real-life situation. Vertical mathematization is the process of reorganization within the mathematical system itself, like finding shortcuts and discovering connections between concepts and strategies, and the application of these discoveries. This approach exposes mathematics as an activity without losing sight of mathematics as a product. The strong emphasis on mathematization makes the approach a promising theoretical framework to design instructional sequences that strengthen the relation between mathematics and science. This does not mean that that all models are or should be mathematical, but mathematization makes it possible to transfer the modeling process from one place to another.

Horizontal Linking and Vertical Structuring

Michelsen (2005a) proposes an extension of the distinction between horizontal and vertical mathematization to a didactical framework for integrated mathematics and science courses consisting of two phases: *horizontal linking* and *vertical structuring*. The idea of expansion of domain and the description of the modeling competence as an interdisciplinary competence entails that situations from science are embedded in the contexts to be mathematized; this is a horizontal linking of mathematics and science. The vertical mathematization must include a vertical structuring—that is, the conceptual anchoring of the general model in the systematic and framework of mathematics and science, respectively. The didactical framework emphasizes mathematizing and modeling as ideas that cut across mathematics and science and encompasses both contextual and conceptual aspects. In the horizontal phase, thematic integration is used to connect concept and process skills of mathematic science by modeling activities. Also in this phase, explicit connections are established between the process skills of mathematics and science. For example, the facility with graphical, tabular, algebraic, and written representations of functions is explicitly linked to data analysis. The vertical phase is characterized by a conceptual anchoring of the concepts and process skills from the horizontal phase by creating languages and symbol systems that allow the students to move about logically and analytically within mathematics and science, without reference back into the contextual phase. The shift from the horizontal to the vertical phase thus might concur with a shift from integrated instruction to subject-oriented instruction. It should be stressed that the didactical framework is iterative. Once the concepts and skills are conceptually anchored in the respective subjects, they can evolve in a new interdisciplinary context as a part of a horizontal linking.

Radioactivity and Exponential Growth in an Interdisciplinary Context

To investigate the applicability of the didactical framework, a team of science and mathematics teachers from the region of southern Denmark and educational researchers from the University of Southern Denmark work on developing integrated tasks for mathematics and science, which enables students to construct mathematical models for problem solving in an interdisciplinary context between mathematics and science. As a result of this work, there exist examples of integrated mathematics–science courses that encourage the notion that mathematical functions can be used to model relationships in which two quantities vary. The first course was designed and implemented in 2000 as a six-week integrated mathematics–physics course for a group of 26 grade 10 students. The course was centered on exponential growth and radioactivity. The exponential behavior of the accumulation of the variable number of nuclei is a very common phenomenon in our physical, biological, and social environments and therefore the same model, structure, and behavior can be found in the study of other parts of the real world. The overall curricular unit was designed to integrate three components: the gathering of data from a physical experiment, development and exploration of mathematical models (verbal, symbolic, graphical, and tabular), and a reflective discourse about the mathematical content of the models. To make the structure and content of mathematics and physics as explicit as possible, course material was organized around a small number of basic models embedded in context problems. Instruction was designed to make the students familiar with the structure and use of the basic models. Beginning with an experimental situation, the students had to make explicit their own representations of a physical event, choose variables, and pose relationships among them. In this manner (e.g., acting with the tabular representation of data), the tabular form changes from functioning as a model of acting with the collected data to functioning as a model for reasoning about the relations between two variables. By using this approach, the students developed a concept of function, which included both the traditional and static correspondence notion, that treats functions as a mapping from x to a corresponding y, and the more dynamical covariational notion that relates the change in x to the change in y. This facilitated the students' functional thinking and demonstrated how by concentrating on the concept of functions, students can lay a foundation for two essential mathematical understandings: the ability to model authentic phenomena mathematically and the capacity to create graphic, linguistic, and numerical representations of mathematical information.

It was the basic idea of the integrated course to use experimental contexts from physics, to expand the domain for the acquisition of the concept of exponential functions. This expansion offered the students an environment where a reasoned analysis of authentic situations required moving between the different representations of the quantitative relations in the situations, and where passages between different settings are experienced as natural moves. An important aspect of modeling is that it allows students to visualize abstract concepts by creating structures through which they can explore and experiment. For instance, the students' investigation of the simultaneous decay of two different radioactive sources naturally led to the idea of summation of two functions. The contexts from physics thus provided the students with an experience of the actions that created the need for the application of mathematizing, and the study of change offered by this approach gave the students a deeper understanding of the ways in which changes in quantities can be represented and of the concept of rate of change.

To illuminate how the students' engagement in modeling activities contributed to experience and manifest their interdisciplinary competences, we point out that the students made reference to both mathematics and physics and used representations transcending the boundaries between the two subjects. The students' combination of perspectives from both subjects often helped them to overcome problems in the modeling process. When modeling the absorption of light in water, a group of students was blocked by the problem of assigning the correct value to a^0. According to the students, the value was 0, but then reference to the intensity of light at the surface of the water made them change the value to 1, which was then checked on their calculator. The following excerpt from a student's written report shows that the models constructed by the students had both a formal and concrete status (translation from Danish by the author):

> We saw that cooling of the hot liquid resulted in a decreasing exponential function. Therefore the graph has the form of $y = ba^x + c$. In the formula the value of c must be the room temperature; else we had to cool the liquid with something colder than room temperature. The graph of heating the ice water was a graph that was opposite to the cooling graph. If you place the temperatures from both graphs at the same time you get a horizontal graph.

In this description the function represented by a formal expression; it is not considered as a static entity but as a manipulative object that can be transformed. And through the transition to the more formal and generalized model, the extended model does not become detached from the original model (Michelsen, 2001).

Right now the team of teachers and researchers work within the PARSEL Project[3] (Holbrook, 2008) and the ScienceMath Project[4] (Beckmann, 2007), with their efforts centered on the identification of interdisciplinary contexts from the subjects of mathematics, physics, chemistry, and biology suitable for a modeling approach based on the didactical framework. The websites of the two projects[5] offer interdisciplinary teaching modules ready for teacher use in total or in part. This naturally leads us to consider the challenges teachers face when interdisciplinary activities are put on the agenda.

INTERDISCIPLINARY INSTRUCTION: A NEW CHALLENGE FOR IN-SERVICE TEACHERS

> *Science and mathematics are related naturally and logically in the real world. The integration of science and mathematics has the potential to help students perceive the relevance and power of this union.*
>
> —Berlin & White, 1998, p. 499

In the last decade there has been an increased focus on science education in general and science education in the secondary school in particular. Not only do the students lack interest in learning science, but science teachers find it challenging to connect the science in school to the everyday life of the students. From the Danish government there has consequently been a call for an improved professional development of science and mathematics teachers (Andersen et al., 2003; Niss & Jensen, 2002). In 2006 the Ministry of Education's committee for preparing a strategy for science education in elementary school proposed to educate municipal science education consultants (Ministry of Education, 2006). To meet this challenge, Michelsen and Sahl-Madsen (2006) suggest the development of a research-based master's degree program[6] in mathematics and science education for in-service teachers.

In the communities of mathematics and science education researchers, it is generally believed that teachers do not use educational research to improve their teaching. The problem here is that someone outside the classroom decides what is wrong and what changes teachers have to make. Improvement of the teacher learning process requires that teachers' experiences are acknowledged and built upon. Descriptions of practice by researchers are often decontextualized, and therefore make little sense to the teachers. Their considerations are much broader and more contextual than the researchers' theoretical orientation can count for. Taking the perspective of change in teaching practice and the use of research in the process, Richardson (1990) argues that research should provide teachers not just with findings in the form of activities that work, but also with ways of thinking and empirical premises related to thinking and learning. In this way research becomes a basis for the development of warranted practices with which the teachers may experiment in their classrooms. Teachers exercise considerable control over the decision of whether and how to implement a change in teaching practice, and any intervention should acknowledge this control and help teachers to understand and be held accountable for the intervention.

Cooney and Krainer (1996) think of in-service teacher education as a crucible for theory and practice. This places a twofold demand on the educators. It requires attention to the state-of-art developments in the field of educational research, and these developments must be seen as relevant to the teachers' practical concerns. McIntyre (2005) argues that we must recognize that there is a very large gap between the kind of knowledge that good scholarly educational research can provide and the kind of that teachers most use in classroom teaching. In the conventional mechanistic paradigm educational change is viewed as a linear process. This approach is criticized by teacher education researchers (Hoban 2002; Korthagen, Kessels, Koster, Lagerwerf & Wubbels, 2001) for often producing disappointing results. According to Hoban (2002), in an educative environment that empowers teachers to generate their own knowledge and products, the teachers can manage their own learning and set their own directions for change. Is this view educational change is essentially about teachers learning to do something differently.

To these ends the Center for Science and Mathematics Education at the University of Southern Denmark has conducted the projects *The Modeling Course*[7] and *Science and Mathematics Teachers of the Future*[8] with the aim to develop and implement training programs in mathematics and science education for in-service upper secondary and lower secondary teachers. It is the core idea of the programs to involve teachers in design, implementation, and evaluation of innovative instructional sequences that deal with a wide range of aspects of mathematics and science. Both programs are based on a theoretical framework developed by a group of researchers from the Center for Science and Mathematics Education at University of Southern Denmark.

Theoretical Framework

As a rule the majority of the teachers are interested in improving and enriching their teaching methods, but as Tyack and Cuban (1995) point on: teachers need help in adapting or developing new instructional practices. McLaughlin, Black-Hawkins, Brindley, McIntyre, and Taber (2006) emphasize the importance of inclusiveness. Experiences from the past century, both in schools and in research, show that innovations in schools can more readily survive if the recipients, mainly the teachers, have been involved early in the decision making. This calls for a strategy for teachers' professional development aimed at motivating teachers to use more effective practices. According to Mamlok-Naaman, Navon, Carmeli, and Hofstein (2005), action research is an effective means of helping teachers to reflect on their practice, if they are provided with an environment of support, collegiality, and a chance to collaborate with researchers and other teachers. Teachers experience a new dimension of professional development trough action research in three main areas: (1) implementation of change through action research, (2) having a sense of being a part of professional community, and (3) having contact with academic experts. Michelsen (2005b) points out that in design-based educational research, teachers and researchers collaborate to produce meaningful change in the classroom practice. This means that goals and design constraints are drawn from the local context, which leads to the suggestion of a design strategy that deliberately creates opportunities for the stakeholders to influence the design process and focus on adaptation to already existing practices. The design process thus calls for the cultivation of the ongoing relationships between teachers and researchers. In this context, pre-service as well as in-service teachers play a crucial role. With the rationale of supporting teachers in participating in and contributing to the design process, there is a clear-cut need for including instructional design in teacher education.

The model of *educational reconstruction*, developed by Kattmann, Duit, Gropengießer, and Komorek (1996) offers a promising frame for involving teachers in action research with a focus on designing, implementing and evaluating innovative instructional sequences and was adopted as the underlying educational approach. This model proposes a cyclic process of analysing scientific contents, studying student's perspectives and developing sequences of instruction. Two reasons for choosing this approach can be spelled out. First, the triadic model of educational reconstruction allows for change-of-practice processes to be vehicles for connecting knowledge of scientific and pedagogical content and classroom practice. Thus the model would, in theory, facilitate the development of warranted teaching practices. Second, the model was originally developed as a research model for science education research, and as such it fits naturally with the aim of equipping the participating teachers so as to implement action research-type projects. According to the model of educational reconstruction, what the science educator does in this process is to reconstruct scientific knowledge "in order to make the science point of view understandable and meaningful to learners" (Kattmann et al., 1996, p. 3). And it fleshes out a beneficial process of reconstruction involving (1) analyzing content structures so as to identify salient concepts and their relations; (2) investigating students' pre-scientific conceptions—both cognitive and affective; and (3) developing instructional sequences on the basis of the first two steps (Kattmann et al., 1996).

In the context of the programs, an approach was adopted in which the participating teachers are guided through multiple cycles of developing, implementing, and evaluating instructional sequences. The model of educational reconstruction comes to the fore in the process of developing such sequences, and in this respect, the model of reconstruction has shaped the content and curriculum of the programs. In order to properly analyze content structures, one must have a firm background of content knowledge

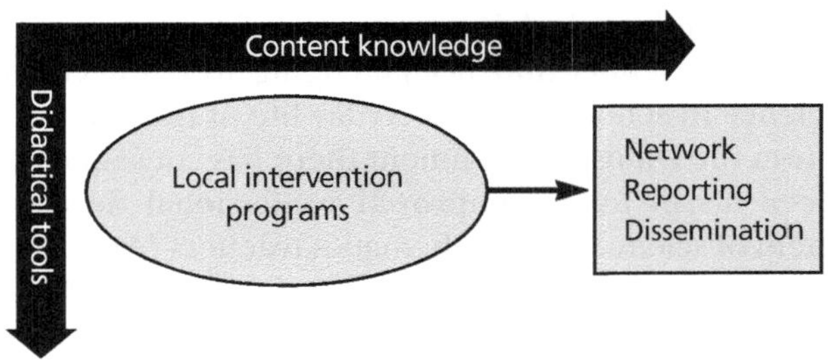

Figure 44.1 A diagram of the educational outlook behind the project *Science Teachers of the Future* in Denmark.

across mathematics and the sciences. Being able to reconstruct a specific content means in the first instance to be able to navigate the field to which that content belongs. Further, a proper investigation of students' pre-scientific conceptions requires both analytical investigatory tools as well as background knowledge in the dialectics of the psychology of learning. To this end, the programs offer to participating teachers an analytical toolbox of scientific and mathematical content knowledge and a background in contemporary educational research. In addition, the participating teachers are introduced to multifarious tools to render successful their reconstruction—such as insights into the usage of a variety of teaching formats and the usage of alternative learning environments. It is thought that the parallel expansion in, on the one hand, the dimension of knowledge of subject matter content and, on the other hand, in the dimension of educational tools will equip teachers properly for the development, implementation, and evaluation of instructional sequences (see Figure 44.1).

This process is thought to result in dissemination of the sequences; increased networking of the participating teachers; and teachers' reporting to academic experts, each other, and other teachers, thus facilitating a lasting way of sharing their ideas and experiences with their colleagues and having contact with academic experts (Michelsen & Nielsen, 2008).

THE MODELING COURSE

With all the institutional demands on their time, energies, and attention, teachers need help in adapting or developing new instructional practices. Some changes are very hard to do alone.

—Tyack & Cuban, 1995, p. 138

The development of *The Modeling Course* for upper secondary teachers was initiated in the autumn of 2004, where 12 teachers participated in a pilot study. Based on the experiences from the pilot study, the first version of the course was developed in the autumn of 2005. The course's objectives were as follows:

- To develop prototypes of interdisciplinary instructional units centered on modeling activities and including at least two subjects
- To encourage the creation of communication between the subjects of biology, chemistry, mathematics, and physics
- To introduce the teachers to the System Dynamics approach to modeling in an interdisciplinary context

In the spring of 2006, 14 teachers with several years of teaching experience participated in the course. The course was structured as three sessions: a three-day workshop, a two-day workshop, and an open seminar. During the two workshops the teachers were accommodated in a hostel. All the sessions included inspirational lessons by modeling experts from the Faculty of Science at University of Southern Denmark.

The Modeling Workshops

The first workshop began with a presentation of the didactical model of horizontal linking and vertical structuring and the model of educational reconstruction. Then the software Powersim was introduced. Powersim is a tool for System Dynamics modeling. During the workshop, the teachers were shown examples of modeling with Powersim and worked in groups with the program.

To effectively create interdisciplinary teaching units, teachers from different subjects need to collaborate. Therefore, the first workshop ended with creating seven interdisciplinary teams of two to three teachers representing different subjects. Once the teams were established, the members faced the challenging task of designing a scenario for an interdisciplinary instructional unit centered on modeling activities and including at least two of the subjects of physics, chemistry, biology, and mathematics.

The second workshop was held one month after the first. Each of the teams gave a preliminary report about their instructional unit and got comments from the other teams. The participants were also introduced to a guide for reporting on their project. Then the teams had a month and a half for writing a report with a description of their interdisciplinary instructional unit, preparing a presentation of the unit at a seminar, and implementation of some of their ideas in the classroom.

The Modeling Seminar for Teachers

A key factor in successful innovations is the degree to which the innovation is perceived to be better than the existing program it hopes to supersede. Lesh and Sriraman (2005) introduce the useful survival law that determines the continuing existence of innovative programs and curriculum materials. Usefulness involves going beyond being powerful in a specific situation and for a specific purpose to also be sharable with other people and reusable in other situations. What is at stake is making the improvements available to a larger community. To meet this challenge, the teams presented at the final session their interdisciplinary projects for discussion at an open seminar at the University of Southern Denmark attended by 32 upper secondary school teachers. Table 44.1 shows the projects presented.

The table shows that there were different approaches to projects ranging from a focus on the software to a focus on the modeling process. Looking at Table 44.1, it is not unfair to say that the topics of the projects are ones that belong in the traditional content of the subjects. Therefore it is obvious to include a more up-to-date interdisciplinary content including technological and socio-scientific issues in the next version of the modeling course. After the seminar reports, presentations, and lecture notes from the teams were made accessible on a website.

Experiences from the Modeling Course

Modeling provides a generic methodology that can serve as a common denominator for learning subjects such as physics, biology, chemistry, and mathematics. The teachers welcomed the opportunity to explore new ways to engage students in interdisciplinary modeling activities. There was agreement among the teachers that the course supported active innovation

TABLE 44.1 The Interdisciplinary Projects Presented at the Seminar

Project	Subjects involved
Mathematical modeling with Powersim	Maths and biology
Models—interplay between reality and simulation	Maths and physics
Data sampling and modelling	Maths, chemistry and biology
Modelling an inclined throw	Maths and physics
Traffic and kinematics	Maths and physics
Mathematical modeling and enzyme kinematics	Maths, biology and chemistry

and intervention in classrooms. The System Dynamics approach was, according to the teachers, a new way to enhance students' understanding of complex systems and formal thinking. But the integration of System Dynamics into teaching is also challenging and time-consuming. From our experience, the course structure with workshops and an open seminar made it possible for the teachers to share their ideas and experiences with their colleagues and to have contact with academic experts in the fields of modeling and educational research. However, to get full profit from interdisciplinary modeling activities, further research on the constraints and possibilities of the cooperation between the subjects of physics, chemistry, biology, and mathematics is needed.

SCIENCE AND MATHEMATICS TEACHERS OF THE FUTURE

Probably nothing within a school has more impact on students in terms of skills development, self-confidence, or classroom behaviour than the personal and professional growth of their teachers.

—Barth, 1996, p. 49

In January 2006, the two-year project *Science and Mathematics Teachers of the Future* was initiated with the aim to develop and implement a master's degree program for in-service teachers from lower secondary school. On the basis of the results of this pilot project, the Danish Accreditation Counsel in January 2008 approved the program as the first Danish master's degree program in mathematics and science education for in-service teachers from lower secondary school.

The strategic aim of the program is to provide a sufficient number of teachers with skills and competences in helping students to learn mathematics and science in a rational way that reflects the need of improving the proficiency and competence level in the Danish lower secondary school. The *Science and Mathematics Teachers of the Future* project involved teachers as key partners in the development of the master's program by involving teachers in trying out and taking ownership of the sequences developed in the program. When teachers work together in ways that provide professional support for one another, practice is improved (Loughran, 2006). Reflecting about practice through collaboration with trusted colleagues makes the tacit explicit and develops knowledge, skills, and expertise in practice. As a consequence, the master's program is organized as workshops and open seminars with the purpose of making it possible for the teachers to share their ideas and experiences with their colleagues and to have contact with academic experts in the fields of science, mathematics, and educational research. During the development of the master's program, it was prominently emphasized that the aim of the collaboration of educational researchers and teachers is to produce meaningful change in the teachers' classroom practice. The pilot project was initiated with participation of 24 science and mathematics teachers.

The Structure and Content of the Program Offered in the Pilot Project

The program offered in the pilot project is a one-year, full-time equivalent study program on the graduate level and conducted over four semesters. Each semester constitutes an individual unit with an overall theme. In each semester the teachers attend workshops, discussion groups, seminars, and lectures. Each semester is concluded with an examination, and the program as a whole is concluded with a thesis. The pilot project began with the participation of 24 teachers from four municipalities in the region of southern Denmark. Prior to launching the project, a developmental support group was formed with members who represented the four municipalities, two of the regional university colleges, and a regional science center.

In order to facilitate processes of educational reconstruction the program offers a wide range of input (see Figure 44.2). First, by introducing the teachers to the newest research in different mathematics and science fields, the teachers not only familiarize themselves with the state-of-the-art research, they also learn how the scientist works to produce new results. This aspect was provided by lectures, workshops, and seminars by and with researchers in science and in science and mathematics education from the University of Southern Denmark. Two key themes have been continuously stressed in this respect: (1) the historicity of science and mathematics in general and the histories behind the concurrent research products; and (2) the everyday work of a science and/or mathematics researcher.

Second, the teachers were introduced to educational theories specifically aimed at science and mathematics education. This might also mark a novel opportunity for teachers since Danish compulsory teachers are educated in general educational theory at the university colleges and only to a very small degree do they get acquainted with research-based, course-specific educational theory. Third, the teachers were introduced to different aspects of, and ways

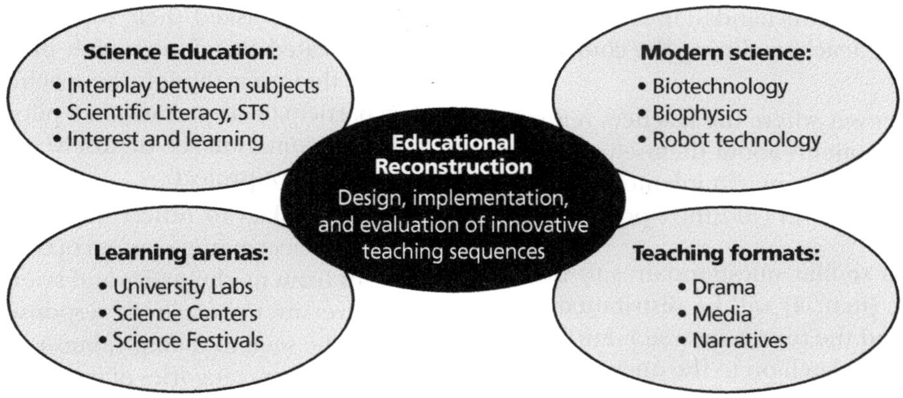

Figure 44.2 A diagram of the different tools offered to teachers for their educational reconstruction (STS is an acronym for Science Technology and Society).

of thinking about, the relation between science and society. This topic is challenging for students to work with and equally challenging for the teacher to convey (Sjøberg, 2005). In the program, the teachers experience the forefront of scientific research and how it relates to and impacts on society. Fourth, the program offers an overview of, and work with, multiple approaches and teaching strategies, such as multifarious learning environments. The teachers were introduced to what Sjøberg (2005) calls the three dimensions of science: namely, (1) the products of mathematics and science—in the recent scientific results, (2) the processes of mathematics and science—in the presentation from the scientists on how they reached their results, and (3) the role of mathematics and science in society—in seeing that the recent result of science is applied in society. These three dimensions are introduced explicitly with the aim that the teachers will let these dimensions play a part in the designed sequences. Fifth, the teachers were given courses in participatory action research and multiple forms of data collection. Thus they acquired the tools for implementing and evaluating new reconstruction sequences in their own classroom. This was the examination assignment for the second semester: a written report on how they individually had developed a sequence, and how they implemented and evaluated it on the basis of their own data collection.

In order to facilitate that the teachers report and disseminate their work in the program, the teachers will have the opportunity to publish their third semester written report in a special issue of the *Paper Series* of the Centre for Science and Mathematics Education, at the University of Southern Denmark. Further it is expected that the teacher's work on their final thesis will be presented to other teachers at a seminar.

Case Study—The Teachers' Reflections on the Pilot Project

It was decided that the in-house evaluation of the structure and curriculum of each semester should be merged with a more focused qualitative research study into the salient themes of the degree program from the perspective of the teachers. The present study adheres to the basic tenets of the naturalistic inquiry paradigm (Lincoln & Guba, 2000), allowing for a study process in which categories were stipulated as embracing clusters of emergent themes. Given the number of participating teachers, and because of the nature of this project as being a pilot project, a general inductive approach (Bryman & Burgess, 1994; Dey, 1993) was adopted through which possible key themes could emerge and guide further subsequent studies as well as the continuous development of the degree program.

The data consisted of:

1. Transcribed videotaped semi-structured interviews conducted prior to the first semester of the degree program. A total of eight teachers were interviewed on their reasons for commencing the program and on their expectations.
2. Two rounds of 20-item questionnaires with open-ended questions. These questionnaires were distributed to the participating teachers at the end of the first semester and again at the end of the second semester. The questions focused on the teachers' experiences with and attitudes towards the structure and content of the completed semester as well as their expectations for the following semester.

3. Written assignments handed in by the participating teachers during the course of study.
4. A two-page essay where the teachers reflected on five components about themselves: (1) self-image, (2) self-esteem, (3) job motivation, (4) task perception, and (5) future perspective.

It is expected that similar questionnaires to the one mentioned above (item 2) will be distributed at the end of the third and the fourth semester. Further, follow-up interviews that latch on to the ones mentioned above (item 1) are planned to be conducted at the end of the entire project.

A qualitative analysis of the data was performed using the general inductive approach. After systematic reading of the collected data, codes and categories were assigned in order to identify possible key themes emerging from the data (Byrne, 2002; Creswell, 1998, 2002). The identified codes and categories were then renegotiated and a second coding process was undertaken. The frequencies of the occurrence of the individual themes were subsequently recorded.

The Teachers' Experiences

The teachers' general evaluations of both the first and second semester of the program are overly positive about both their gains and the close contact to researchers of science and science education. A clear progression is indicated, according to which teachers after two semesters feel:

- More inspired
- More motivated
- More secure in planning their teaching
- More secure in giving reasons for their pedagogical decisions

Further, the teachers generally report that they have become more sensitive to the intricate and deep relation:

- Between their subject(s) and other subjects
- Mathematics and science in general to society

A few points that might be improved, however, are mentioned by a number of teachers. In the teachers' evaluation of the first semester, around half of the teachers express that information about the aim, structure and content of the program could have been made more explicit prior to the program. However, they seem to ascribe this lack of information to the fact that the program is indeed a pilot. Taken in conjunction with the fact that the majority of the teachers have asked their supervisors to participate, it is indicated that future such pilot projects should beware that even though the teachers are aware that they participate as partners in a pilot project, it is possible that some will see themselves as participating in not just a pilot project.

When asked to reflect on the outcomes with respect to subject matter content of the sessions involving lectures from mathematics and science researchers at the university, the teachers' responses for both semesters can be seen as falling (almost evenly distributed) within the four categories of more content knowledge, inspiration for their teaching, more knowledge about the processes of science, and a deeper understanding of the connections between subjects. Statements of the following type are recurring (translation from Danish by the author):

> I have gained a greater insight into the research work at the university. (Teacher 14, 2nd Semester)

> I have gained a wider foundation in the other subjects, e.g., biology, to which the visits to SDU [University of Southern Denmark] were very inspiring and important as motive force. (Teacher 15, 1st Semester)

> I have been inspired on [different] scientific areas through the guest lecturers. It has been really exiting. (Teacher 20, 1st Semester)

In general, however, it seems that the teachers place even greater value on what they have gained with respect to theories of learning and educational tools. This interpretation is supported by a look at the teachers' explication of their expectations. Roughly there seems to be a pattern of different expectations towards gains in subject matter content and towards pedagogical content: While the subject matter content part of the program is expected by the teachers to be inspirational for their teaching, the pedagogical content part is expected to provide a solid background in theories of learning that affords reflections of their practice leading to change.

Returning to the teachers' reflections on what they have gained with respect to theories of learning and educational tools, after the first semester their responses fall within the categories of more solid background knowledge, provoking reflections on their own practice, acquisition of new teaching strategies, and brushing up on already acquired knowledge. Statements of the following type are recurring (translation from Danish by the author):

> After nine years of practice it is really good to both read new and reread educational literature and be

> able to latch it on to a self-perceived reality. (Teacher 12, 1st Semester)

> It is almost a new world which has opened itself up.... Today I have far more course-specific educational competencies when I have to plan my teaching in the science subjects. (Teacher 22, 1st Semester)

> I have gained more overview and connection in what I have "looked at" in other contexts. And I have learned something new. I have, among other things, read about scaffolding (theoretical), and it has had direct influence on my practice. (Teacher 21, 1st Semester)

> It has been [a] cause for considerations of my own practice (Teacher 20, 1st Semester)

There is a clear progression from the first semester to the second in this respect. Here two categories describe the majority of the teachers' responses: solid knowledge of multiple theories of learning and their individual tenets, and affordance of tools for practice. In general, the teachers feel much more able to distinguish among theories of learning on the basis of basic assumptions and contexts. Not surprisingly, the teachers seem to have chosen one or two concrete theories as favorite objects for further study, and to some extend also as favorite worldviews of their own. There is also an increased tendency of teachers to be more concrete in terms of how what they have learned has changed their practice (translation from Danish by the author):

> [I have gained] a greater insight into the necessity of student involvement. Story line sequences or variations of such have received great respect in my mind. (Teacher 5, 2nd Semester)

> I have been presented a "bouquet" of theories from which I can pick. I have gained a foundation for comparison to when I consider or discuss theories of learning. The homework assignment caused me to work in depth with a single theory (flow). (Teacher 9, 2nd Semester)

> I have become more aware of using some of the "knowledge" about learning which at the moment is derived from modern neuroscience. (Teacher 3, 2nd Semester)

After the first two semesters the majority not only feel they been moved to reflect on their own practice, but actually express that they have changed their everyday practice. Only a few teachers who feel that the participation in the program is very time consuming explicitly mention that they have not yet had the time available for changing their everyday practice. An important insight might be found in this progression. It might well be that genuine and well reasoned change of practice is more profitably facilitated by a long term study of the theories of learning and educational tools in which teachers' studies includes work with both the historical/dialectical field of learning theories as well as in-depth work with a self-chosen learning theory.

The teachers to a wide extent stress a discovered importance of discussing/disseminating what they have learned with/to other teachers. In general there seems to be generally agreed upon focus on the importance of professional networks and cooperation between teachers, and some teachers already have taken initiative to establish subject networks on their school. Further, around half of the teachers have concrete examples of types of future networks and projects they would like to participate in. Such examples include networks/projects between schools and universities; municipal and regional science teaching networks/projects; and projects such as PARSEL in which the educational researchers at the University of Southern Denmark are participating partners.

The last points can be seen in conjunction to a progression from the end of the first semester, where the teachers, in general, wanted to learn more about how to perform graduate level studies, to the end of the second semester where the teachers, in general, look forward to learn more about doing participatory action-research in their own practice. This not only indicates a progression in self-efficacy with respect to what could be called study-specific issues, it also indicates that the participating are very serious about their desire of understanding and improving the practice of their own practice as well as the regional science teaching in general.

Some further notable progressions deserve to be mentioned. First, a recurring theme seems to be the progression from the teachers having their focus on and interest in one or maybe two subjects, to teachers expressing a discovered insight of the connectedness of their subject(s) to (1) other science subjects, (2) subjects from the humanities and social science subjects, and (3) society as a whole. And there is a perceived importance of using and highlighting these connections in classroom teaching. Second, in conjunction with the will to discuss educational approaches with their colleagues in professional networks, there is a tendency that the teachers feel more secure about issues concerning such discussions—for example, a number of teachers explicitly state that they feel more secure in arguing for their decisions about planning their teaching.

To summarize, after two full semesters into the degree program, the key themes emerging from the teachers' evaluations of the program involve categories such as development of and reflection on practice, connectedness of their science subject(s) to others and to society, professional networking, inspiration, and research into their own practice; furthermore, there has been a perceivable cognitive and affective progression on the side of the teachers within the fields denoted by these themes (Michelsen, Nielsen & Petersen, 2008).

CONCLUSIONS AND IMPLICATIONS

Issues related to interdisciplinary activities involving mathematics are complex because they comprise two apparently different components, an extra-mathematical and a mathematical context. But if we as mathematics educators take the stance that mathematics has value for solving meaningful problems or even improving society, then we have to design learning environments that are meaningful to and valuable for the students. Putting a question to the content of mathematics education opens new frontiers for researchers in mathematics education to explore. In this chapter, educational programs for in-service teachers are used to describe a didactical framework with the aim of helping and supporting teachers to improve instruction. The programs represent an integrated approach that combines content and instructional presentation with actual scientific specimens presented by practicing scientists.

In view of the growth of research in mathematics education over the last decades, it is remarkable that only little attention has been paid to research on the educational relations between mathematics and other subjects. In the approach presented in this chapter, the choice is a model for the curriculum that integrates across the disciplines of mathematics and science. Integrated means really interdisciplinary, and both applied and conceptual mathematics are part of the curriculum. This does not mean that all mathematics concepts should be developed in an interdisciplinary context. There are topics in mathematics that are the result of structuring within the mathematical system. But the more connections that exist among facts, concepts, and procedures, the better the students' understanding, and the more the curriculum is presented as a cohesive program with a range of tentacles, the more likely the students will have a rounded, effective, and meaningful education. Interdisciplinary competences could be the generic methodology that acts as a common denominator for disciplines, such as mathematics and science. And there is room for considerable improvement. However, an identification and clarification of the interdisciplinary competences is needed as well as further research on the constraints and possibilities for the integration of mathematics and science.

Interdisciplinary activities entail new challenges for teachers. It is essential for teachers to be provided with experiences that allow them to deepen, extend, and share their own knowledge and understanding of their teaching practice. Korthagen and colleagues (2001) distinguish between two types of knowledge in teacher education: episteme and phronesis. Episteme is characterized as abstract, objective, a propositional knowledge. Phronesis is perceptual knowledge, consisting of assertions of a general nature that apply to many different situations in the practice of teaching. Phronesis is being developed through experience, and its value is related to the particular person using this knowledge for action. According to Korthagen et al. (ibid), the development of phronesis is most important for teacher education. This aim is achieved with the modeling course and the master's program through involving the teachers in the design, implementation, and evaluation of innovative instructional sequences that deal with a wide range of aspects of mathematics and science. It is indicated that long-term professional development programs have a positive effect on how teachers think about, and are actually able to perform cogently, practice change. A future topic that deserves to be explored is the modulation of teacher identity through such programs. According to Bjuland, Cestari, and Borgersen (2008), positive change of identity involves (1) *integration* of old practice within practice changes, (2) a *transformation* from an asymmetrical relation of teachers to educational researchers to a symmetrical relation in which the teacher conceives of herself as being on par with the researcher, and (3) increased *reflection* on the aspects linked to learning processes. All of these elements in some way emerge as themes in our data from the teachers. In future studies it would be interesting to enquire into identity modulation in the context of the long-term professional development program provided in our programs for in-service teachers.

NOTES

1. The term *science* is in this chapter used as a common denominator for the subjects of physics, chemistry and biology. These subjects are in Denmark taught separately, not as a common science course
2. In Europe, there seems to be a preference for using the label the didactics of mathematics, inspired by names such as "Didaktik der Mathematik" (German), "didactique des mathématiques" (French), "didáctica de las matemáticas"

(Spanish), "matematikdidaktik" (Scandinavian languages) (Niss, 1999).

3. PARSEL is an acronym for Popularity And Relevance (of) Science Education (for scientific) Literacy. The project is supported by a European Commission grant SAS6-CT-2006-042922-PARSEL.
4. The project ScienceMath is an interdisciplinary European co-operation project for the promotion of mathematical and scientific literacy. It is supported by the European Commission within programme Comenius 2.1.
5. www.parsel.nu and www.sciencemath.ph-gmuend.de
6. In the Danish context of professional development, the term "master's degree" refers to a one-year full-time equivalent professional development study on the graduate level. A master degree is studied at the university whereas the university colleges are responsible for the education of teachers for the elementary and lower secondary school.
7. The project is financially supported by the Danish Ministry of Education.
8. The project is financially supported by the European social fund.

REFERENCES

Andersen, N. O., Busch, H., Horst, S., & Troelsen, R. (2003). *Fremtidens naturfaglige uddannelser. Uddannelsestyrelsens temahæfteserie 7.* Copenhagen: The Ministry of Education.

Arnold, V. I. (1998). On teaching mathematics. *Russian Math. Surveys, 53,* 229–236.

Atiyah, M. (1999). Preface. In V. Arnold, M. Atiyah, P. Lax, & N. Mazur (Eds.), *Mathematics: Frontiers and perspectives* (pp. vii–xi). Providence, RI: American Mathematical Society.

Barth, R. S. (1996). *Building a community of learners. Conversation 96.* CA: California School Leadership Center—South Bay School Leadership Team Development Seminar Series; Seminar 10.

Beckmann, A. (2007). ScienceMath—ein Fächerübergreifendes europäisches Projekt. In *Gesellschaft für Didaktik der Mathematik Beiträge zum Mathematikunterricht 2007* (pp. 209–212). Berlin: Verlag Franzbecker.

Berlin, D. F. (2003). Integrated mathematics: From models to practice. In S. A. McGraw (Ed.), *Integrated mathematics. Choices and challenges* (pp. 43–58). Reston, VA: National Council of Teachers of Mathematics.

Berlin, D. F., & White, A. L. (1998). Integrated science and mathematics education: Evolution and implications of a theoretical model. In B. J. Fraser & K. Tobin (Eds.), *International handbook of science education* (pp. 499–512). Dordrecht: Kluwer Academic Publishers.

Bjuland, R., Cestari, M. L., & Borgersen, H. E. (2008, April). *The constitution of mathematics teacher identity.* Paper presented at the 5th Nordic Conference on Research in Mathematics Education, Copenhagen, Denmark. Retrieved April 29, 2008, from http://www.dpu.dk/Everest/Publications/Medarbejdere/mmi/norma/regular%20papers/20080211115412/CurrentVersion/Bjuland(B).rtf

Bryman, A., & Burgess, R. G. (Eds.). (1994). *Analyzing qualitative data.* London: Routledge.

Byrne, D. (2002). *Interpreting quantitative data.* Thousand Oaks, CA: Sage.

Cooney, T., & Krainer, K. (1996). Inservice mathematics teacher education: The importance of listening. In A. J. Bishop et al. (Eds.), *International Handbook of Mathematics* Education (pp. 1155–1185). Dordrecht: Kluwer Academic Publishers.

Creswell, J. W. (1998). *Qualitative inquiry and research design: Choosing among five traditions.* Thousand Oaks, CA: Sage.

Creswell, J. W. (2002). *Educational research: Planning, conducting, and evaluating quantitative and qualitative research.* Upper Saddle River, NJ: Pearson Education.

de Lange, Jan (1996). Using and applying mathematics in education. In A. J. Bishop et al. (Eds.), *International handbook of mathematics education* (pp. 49–97). Dordrecht: Kluwer Academic Publishers.

Dey, I. (1993). *Qualitative data analysis: A user-friendly guide for social scientists.* London: Routledge.

Freudenthal, H. (1973). *Mathematics as an educational task.* Dordrecht: Reidel.

Freudenthal, H. (1991). *Revisiting mathematics education. China lectures.* Dordrecht: Kluwer Academic Publishers.

Gilbert, J. K., & Boulter, C. J. (1998). Learning science through models and modelling. In B. J. Fraser & K. Tobin (Eds.), *International handbook of science education* (pp. 53–66). Dordrecht: Kluwer Academic Publishers.

Gravemeijer, K. (1997). Mediating between concrete and abstract. In Nunes & Bryant (Eds.), *Learning and teaching mathematics: An international perspective* (pp. 315–345). Hove, UK: Psychology Press.

Hoban, G. F. (2002). *Teacher learning for educational change. A systems thinking approach.* Philadelphia: Open University Press.

Holbrook, J. (2008). Introduction to the special issue of *Science Education International* devoted to PARSEL. *Science Education International, 19*(3), 257–266.

Kattman, U., Duit, R., Gropengießer, H., & Komorek, M. (1996, April). *Educational reconstruction—Bringing together issues of scientific clarification and students' conceptions.* Paper presented at the Annual Meeting of the National Association of Research in Science Teaching (NARST), St. Louis, MO.

Korthagen, F. A. J., Kessels, J., Koster, B., Lagerwerf, B., & Wubbels, T. (2001). *Linking practice and theory. The pedagogy of realistic teacher education.* Mahwah, NJ: Lawrence Erlbaum Associates.

Lesh, R., & Sriraman, B. (2005). Mathematics education as a design science. *Zentralblatt für Didaktik der Mathematik, 37*(6), 490–505.

Lincoln, Y. S., & Guba, E. G. (2000). Paradigmatic controversies, contradictions, and emerging confluences. In N. K. Denzin, & Y. S. Lincoln (Eds.), *Handbook of qualitative research* (2nd ed.) (pp. 163–188). Thousand Oaks, CA: Sage.

Lester, F. K., & Kehle, P. E. (2003). From problem solving to modeling: The evolution of thinking about research on complex mathematical activity. In R. Lesh & H. Doerr (Eds.), *Beyond constructivism. Models and modeling perspectives on mathematical problem solving, learning, and teaching* (pp. 501–517). Mahwah, NJ: Lawrence Erlbaum Associates.

Lobato, J. (2003). How design experiments can inform a rethinking of transfer and vice fersa. *Educational Researcher, 31*(1), 17–20.

Loughran, J. (2006). *Understanding teaching and learning about teaching.* London: Routledge.

Mamlok-Naaman, R., Navon, O., Carmeli, M., & Hofstein, A. (2005). Chemistry teachers research their own work: Two case studies. In K. Boersma et al. (Eds.), *Research and the quality of science education* (pp. 141–155). New York: Springer.

McIntyre, D. (2005). Bridging the gap between research and practice. *Cambridge Journal of Education, 35*(3), 357–383.

McLaughlin, C., Black-Hawkins, K., Brindley, S., McIntyre, D., & Taber, K. S. (2006). *Researching schools. Stories from a schools–university partnership for educational research.* London: Routledge.

Michelsen, C. (2001). *Begrebsdannelse ved Domæneudvidelse.* Odense: Syddansk Universitet.

Michelsen, C. (2005a). Expanding the domain: Variables and functions in an interdisciplinary context between mathematics and physics. In A. Beckmann, C. Michelsen, & B. Sriraman (Eds.), *Proceedings of the First International Symposium of Mathematics and its Connection to the Arts and Sciences* (pp. 201–214). Berlin: Verlag Franzbecker.

Michelsen, C. (2005b). Commentary on Lesh and Sriraman: Mathematics education as a design science. *Zentralblatt für Didaktik der Mathematik, 38*(1), 73–76.

Michelsen, C., & Sahl-Madsen, C. (2006). Fremtidens naturfaglige lærere. *Uddannelse, 7,* 45–48.

Michelsen, C., & Nielsen, J. A. (2008). Between teaching and researching: Envisaging ownership benefits of involving teachers from an in-service teacher training program in the PARSEL project. *Science Education International, 19*(3), 315–324.

Michelsen, C., Nielsen, J. A., & Petersen, M. R. (2008). Science and mathematics teachers of the future. *Interaccoes, 9,* 97–112

Ministry of Education. (2006). *Fremtidens naturfag i folkeskolen.* Ministry of Education. Retrieved April 29, 2008, from http://www.uvm.dk/06/documents/nat.pdf.

Ministry of Education. (2008). *Et fælles loft—National strategi for natur, teknik og sundhed.* Copenhagen: Undervisningsministeriet.

NCTM. (2000). *Principles and standards for school mathematics.* Reston, VA: National Council of Teachers of Mathematics.

Niss, M. (1999). Aspects of the nature and state of research in mathematics education. *Educational Studies in Mathematics, 40,* 1–24.

Niss, M., & Jensen, T. H. (Eds.). (2002). *Kompetencer og matematiklæring* (Uddannelsesstyrelsens temahæfteserie, 18). Copenhagen: Ministry of Education.

Richardson, V. (1990). Significant and worthwhile change in teaching practice. *Educational Researcher, 1*(7), 10–15.

Sjøberg, S. (2005). *Naturfag som almendannelse – en kritisk fagdidaktik.* Århus: Klim.

Steen, L. A. (2005). (Ed.). *Math & bio 2010. Linking undergraduate disciplines.* Washington, DC: The Mathematical Association of America.

Tyack, D., & Cuban, L. (1995). *Tinkering toward Utopia. A century of public school reform.* Harvard, MA: Harvard University Press.

45

CAN TWO DIFFERENT TYPES OF MATHEMATICS TEACHER PREPARATION IN DENMARK INTEGRATE?

A Discussion of Research-Based Teaching and Mathematics Competencies

Bettina Dahl

AARHUS UNIVERSITY, FACULTY OF SCIENCE, CENTRE FOR SCIENCE EDUCATION

Abstract: This chapter begins with a description of the structure of tertiary Danish mathematics teacher training for compulsory (grades 1–9) and high school (grades 10–12) levels. Tertiary level refers to the universities and the university colleges. The chapter discusses two quite different Bachelor degrees. One is the research-based university bachelor from which students may enter a graduate program, which is required for the prospective high school teachers. The other is the university college profession bachelor that prepares teachers for the compulsory level. The difference between mathematics as a university subject and mathematics as a *teaching* subject being taught at the university colleges is also described. Then the chapter moves to a discussion of attempts to merge the structure from 2007 and discusses the mathematics competencies and the progression of mathematics competencies in the curricula as well as the idea of research-based teaching.

Keywords: research-based teaching, tertiary mathematics, SOLO Taxonomy, profession Bachelor, teacher training

INTRODUCTION

This chapter focuses on the two very different teacher training programs in Denmark. One program prepares teachers for high school, grades 10–12, in the universities' research-based mathematics courses, and the other program prepares teachers for compulsory school, grades 1–9 (with an optional 10th grade), in the non-university teacher training at the "university colleges" (sometimes also referred to as teacher training colleges). I[1] will also discuss various suggestions from 2007 of "crossing over"—that is, having compulsory teacher training at the universities. The chapter will finally discuss both the mathematics content being taught, the mathematics competencies, and the progression of such. This includes discussing the background of the teachers at the universities and the university colleges.

THE STRUCTURE OF DANISH HIGHER EDUCATION LEADING TO TEACHER TRAINING

Before entering a description and discussion of the above-mentioned things, I will give a description of

The First Sourcebook on Nordic Research in Mathematics Education, pages 613–625

the Danish higher education system. The focus is on the universities' mathematics programs and how they can lead to high school teacher training and on the university colleges' teacher training programs for the compulsory level.

Universities and High School Teacher Training

The Danish universities in general follow the 3–5–8 European model of higher education (Murdock, 2003). This means that a bachelor's degree is obtained after three years of full-time undergraduate study. The years are measured by ECTS (European Credit and Transfer System), which is a quantitative measure for student work load. Sixty ECTS is thus equivalent to one year of full-time study. Following the bachelor's degree, a candidatus/candidata[2] degree can be obtained after two years of graduate study. After completion of these five years of study, a PhD can be pursued through a three-year (180 ECTS) program during which the PhD fellow is considered a (junior) part of the academic staff[3] at the university (Ministry of Science, Technology, and Innovation, 2008). This is also sometimes called the 5+3 way of achieving a PhD, and it is the typical route to take.[4] The university teachers would have a PhD in the subject they are teaching but usually not experience in high school teaching or educational research. Also, the mathematics programs are the same regardless of whether or not the students are going to be a high school teacher or another kind of professional mathematician. There may, however, sometimes be some education courses or theses that the students can choose.

Students who aim to become high school teachers should usually study two subjects. The concepts of "minor subject" and "major subject" are used to denote which of the two was weighted the most. A Danish minor subject generally amounts to 1.5 years of full-time study (90 ECTS) at undergraduate level while a Danish major subject generally amounts to 3.5 years of study (210 ECTS)—that is, 1.5 years of undergraduate study (half of the bachelor) plus two years at graduate level namely the two years it takes to obtain the candidatus degree. The Danish major and minor subjects are thus quite different from the U.S. where the same concepts—major and minor—denote what *undergraduate* topics have been studied how much (Dahl, 2005).[5]

In Denmark, studies do not have a "university core curriculum" as seen in the U.S., for instance. A core curriculum is a body of mandatory courses that are considered central for all students in order for them to become educated citizens.[6] U.S. universities have different core curricula, but they are all to be taken during the undergraduate study. The Danish university students, however, usually focus on their subjects of study right from the beginning of the first year, although some universities have a common science and mathematics entry program of up to two years after which students specialize.

High School Teacher Training in Mathematics

The Danish high school (upper secondary) lasts three year following nine (or an optional 10th) years of compulsory education beginning at the age of seven. The teacher education for high school teachers consisted (until summer 2009) of a two-year in-service program after completing the candidatus program. These two years took place at a high school and consisted of three parts: (1) student teaching in the co-operating teacher's (mentor's) classes, (2) courses in theoretical pedagogy offered by the university with the ministerial permission to offer such, and (3) teaching his or her own classes. Therefore, the two years are not as full-time student, but a mix between "real" teaching, being under supervision, and coursework. Usually the high school teacher teaches two topics and is required to have studied at least 120 ECTS (Ministry of Science, Technology, and Innovation, 2006) in order to achieve the official Subject Competence to teach high school, but national rules and regulations also require that the study include certain topics. Since the Danish minor is shorter (90 ECTS), the prospective teacher should study a supplementation of 30 ECTS as part of the two years. Of the 120 ECTS, the distribution of topics within mathematics is as follows:

- Core topics, 60 ECTS: Calculus, analysis, geometry, linear algebra, algebra, probability and statistics.
- Depth topics, 30 ECTS: Selected areas of significance for research, development, or dissemination. The topics should be put into perspective and be a progression of the core topics. The topics should be chosen by the university but should include mathematical modeling in science, one or more further topics within the cope topics supplemented with discrete mathematics, algorithms, or dynamic systems.
- Breadth topics, 20 ECTS: The history of mathematics, introductory programming IT, mathematical modeling from one or more topic areas
- Education and theory of science, 10 ECTS

However, starting in summer 2009, the two-year program was reduced to one year. The supplementation has been abolished, and now the one year (60 ECTS) is equally divided between student teaching in the cooperating teacher's classes, courses in theoretical pedagogy, and teaching one's own classes (Ministry of Education, 2008b).

Compulsory School Teacher Training at the University Colleges

The system for compulsory school teacher training is very different from the one for high school. This education takes place at four-year university colleges where the students get a profession bachelor. The Ministry of Education (2004) defines a profession bachelor as a profession-aimed tertiary education program usually lasting three to four years, including a minimum half a year of practice. A profession bachelor combines theory and practice and is aimed at a specific vocation or job. University colleges offer a wide range of profession bachelor degrees, one of which is the compulsory teacher training program. This bachelor degree does not give access to candidatus programs at the universities since the programs at the universities are academic and the teaching there is research-based, which I will return to below. However, the profession bachelor does give access to the pedagogical candidatus programs in either education or a subject like mathematics only offered at the School of Education (DPU) at Aarhus University (AU), which I will return to below. A new teacher education program went into force January 2007 (Ministry of Education, 2006). The education lasts four years (240 ECTS) and consists of the following elements:

- One Big Main Subject (*Liniefag*): This has to be either Danish (age specific: grades 1–6 or 4–10), mathematics (age specific: grades 1–6 or 4–10), nature/technology or physics/chemistry (72 ECTS).
- One or two extra Main Subjects: Either one more Big Main Subject (of 72 ECTS) such as English, history, or physical education or two Small Main Subjects (of 36 ECTS) such as art, biology, Danish as a second language, French, geography, housecraft, Christianity/religion, material design, music, social science, special teaching, or German.
- Bachelor report in connection with a Main Subject (10 ECTS).
- General education (*Didaktik*), psychology, and pedagogy (33 ECTS)
- Christianity, life information, and citizenship (17 ECTS)
- Internship/student teaching: 24 weeks (36 ECTS) spread out on all four years and in all Main Subjects.

The teachers at the university colleges would usually either have a candidatus degree in the humanities, social sciences, natural sciences, or a combination of these, or they would have a profession bachelor in teaching followed by a candidatus in education from DPU. It is not required that the teacher has a PhD, and in fact most of them would not have one. The teachers are also not expected to do research, but they are expected to participate in development projects—for instance, in collaboration with university researchers. They are not allowed to do research without having a university researcher attached to the project.

The mathematics content is focused around the areas of numbers, algebra, functions, geometry, probability, and statistics. The subject being taught, however, is the *teaching subject*, and the focus is therefore on the mathematics that the prospective teachers are actually going to teach. It is not "university mathematics" such as calculus, linear algebra, partial differential equations, statistical analysis of the multivariate normal distribution, and so on. Such topics are not part of the curriculum. Rather, the focus is on children's learning; the importance of the learning environment, including teaching and learning forms, materials, and physical contexts; planning of teaching and evaluation thereof; and the mathematics content that is being taught during grades 1–10. This means that the focus is twofold, divided equally between mathematics content and mathematics education. Besides the content areas, the curricula also state a number of learning objectives—competencies—that the students should acquire. The objectives for grades 1–6 and 4–10 are identical, whereas the objectives for grades 4–6, which are the common grades for all who have any of the mathematics main subjects, are a little more elaborate (Ministry of Education, 2007). I return to these below. The mathematics course therefore integrates mathematics with subject "Didaktik" (*Fach Didaktik*). The concept of *didaktik* needs some further comments:

> In Europe, there seems to be a preference for using the label *the didactics of mathematics*, inspired by names such as 'Didaktik der Mathematik' (German), 'didactique des mathématiques' (French), 'didáctica de las matemáticas' (Spanish), 'matematikdidaktik' (Scandinavian languages), and their analogues in most European languages, in spite of the slightly oblique connotations attached to the term 'didactical' in English. (Niss, 1999, pp. 1–2)

Niss hence uses the *didaktik* concept interchangeably with "mathematics education." However, there is more to it than just words: "Didaktik has been the most important tool for planning, enacting, and thinking about teaching in most of northern and central Europe.... [I]n the English-speaking world... the issues Didaktik address are presented in the different framework of 'curriculum and method' or 'curriculum and instruction'" (Hopmann & Riquarts, 2000, p. 3).

Didaktik is usually presented as the didaktik triangle (Figure 45.1), which is a tool to structure the whole field of didaktik research and theory (Hopmann & Riquarts, 2000). The didaktik triangle is widely used in the Danish compulsory teacher education programs and also used in the theoretical pedagogy courses for high school teachers. Another characteristic of the Danish compulsory teacher education is that it consists of three parts: practice, general didaktik, and the teaching subject (Figure 45.2).

These three components are supposed to integrate and create the prospective teachers' *teaching competence.* Therefore the Main Subjects are being taught simultaneously with the subject didaktik. The concept of subject didaktik has similarity with Shulman's (1986) concept of pedagogical content knowledge (PCK), but it is not the same (Nielsen & Pontoppidan, 2009). Together with general didaktik, which includes psychology, pedagogy as well as the internships and student teaching at schools form the central elements of compulsory teacher education.

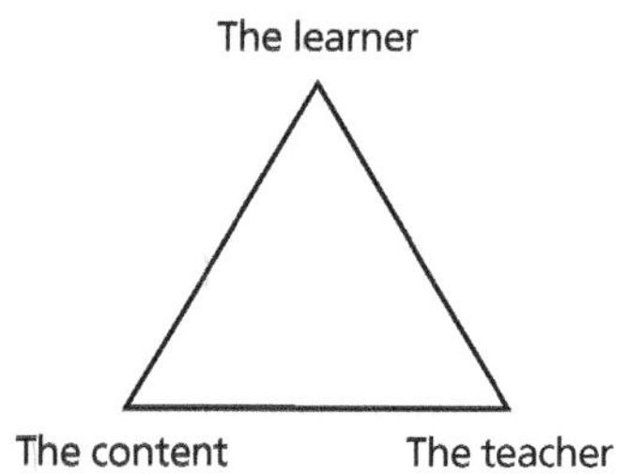

Figure 45.1 The *Didaktik* triangle. Drawn by author, inspired from Hopmann & Riquarts (2000).

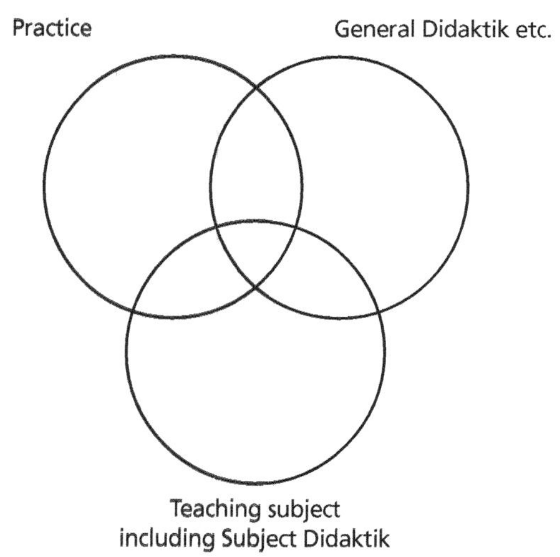

Figure 45.2 The three parts on which Danish compulsory teacher education rest and what creates Teacher Competence (Nielsen & Pontoppidan, 2009, p. 5). Drawn by author, inspired from Nielsen & Pontoppidan (2009).

In terms of high school teacher preparation, the candidatus program in mathematics does not focus on the teaching subject but on the subject itself, and although there is a requirement of 10 ECTS of education and theory of science courses, these courses vary from university to university. There is no national standard as to what these education courses should include, and hence they may or may not focus on the teaching subject. The focus on the teaching subject happens to a larger extent during the two-year in-service program. Furthermore, the high school teacher only has the right to teach in his two topics, whereas the compulsory teacher in principle can teach all the school's topics.

Overall Structure

In Figure 45.3, we see a sketch of the general routes of the Danish tertiary education system, particularly in relation to the teacher training systems described above. The figure illustrates that if, for instance, a student first takes a university bachelor degree, then he can move on to a candidatus degree. Most students would enroll in an ordinary candidatus program, but they can also choose the candidatus pædagogiæ degree in education, which is only offered at DPU. Both types of candidatus degrees can later give access to a PhD program. Only the ordinary candidatus degree gives access to the high school teacher program. A person with a profession bachelor does not have access to ordinary candidatus programs, only to the candidatus pædagogiæ. The reason is that the profession bachelor is not research-based and academic and not focused on the subject in itself, but on the teaching subject. The box mentioning high school teacher training has "2 yrs" in quotation marks. This is to denote that it is not a program that consists of two full years of coursework or student teaching; in fact, it also includes a large proportion of teaching own classes as a normal employee.[7]

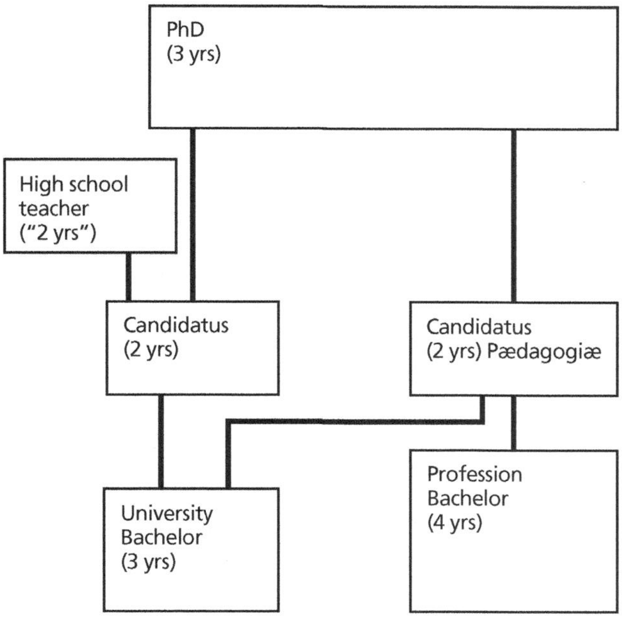

Figure 45.3 Sketch of the Danish tertiary education system focusing on the education subjects and teacher training. *Note:* Drawn by the author but based on a discussion at a Research Group Meeting with colleagues at the Steno Department for Studies of Science and Science Education at the Faculty of Science and the Research Unit for Mathematics, ICT, and Science Education at the School of Education (DPU), both at Aarhus University (AU); 5–6 February 2008.

THE TWO-TRACK TEACHER TRAINING SYSTEMS—CAN THEY INTEGRATE?

As stated above, one of the key factors that separate universities from other tertiary education in Denmark is that university teaching is research-based. Another difference is that the universities teach the subject while the university colleges' teacher training programmes teach the *teaching* subject. This section will first discuss what research-based teaching actually means, then it will move to a discussion of the mathematical content and competencies being taught both places, and finally it will discuss possible ways of integrating the systems.

Research-Based Teaching versus a Profession Bachelor?

The academic staff at the Danish universities have both the right and duty to do research and research-based teaching is a central service from the universities to society. But what is research-based teaching? Usually one mentions three characteristics (Laursen, 1998):

1. The teachers are active researchers within the area.
2. The teachers teach the newest knowledge and research results.
3. The teaching method during the education program is increasingly research-like.

The Ministry of Education (2008a) states that the "knowledge field" for the university bachelor "should have research-based knowledge about theory, method and practise within one or more knowledge areas." A person who holds the candidatus degree is supposed to be able to relate to and apply his subject's research and to some extent to be able to do such research. The final thesis is where he demonstrates this (Jensen, 2001). He must therefore be able to handle knowledge, maintain his own qualifications, change, and contribute to change. University education is furthermore supposed to have a progression where the student increasingly becomes independent, and the goal is that the student at some point will/can be a researcher himself. The ideal of research-based teaching has its roots in Humboldt's[8] ideal that teaching should take place in close connection with research. Humboldt saw the act of studying as a search for knowledge and truth, and this search should take place in collaboration between teachers and students. Doing research was most of all a *Bildung*, where the individual became a better person by searching for truth. *Bildung* is a qualification for a reasonable degree of self-determination, freedom, and responsibility for individual thought and decisions. However it is not about individualism since the notion of *Bildung* also includes concepts such as mankind and the world in general. These two parts of the concept of *Bildung* are linked together (Klafki, 2000).

Contrary to this is the university colleges, where a person holding a profession bachelor degree "should have development-based knowledge about the profession and the area's practise and use of theory and method" (Ministry of Education, 2008a). This definition is in line with Dale (1998), for instance, who states that in a profession, you have formal training, well-defined limits of knowledge within theory and practice, motivation, formalized state authorization, and vocation ethics, and a meta-critique should also develop. However, the concept of "development-based" knowledge is not as well-defined as that of research-based knowledge.

Hence, prospective high school teachers have a large amount of research-based pure mathematics knowledge, and they have research competencies within mathematics. Particularly the high school teachers with a Danish major would be highly qualified mathematicians since they would have spent around 3.5 years (210 ECTS) working full-time with university mathematics, but also the teachers with a university bachelor would have a considerable mathematics experience of 1.5 years (90 ECTS) full time—particularly the ones with supplementation (altogether 120 ECTS). However, the education/pedagogical part of the preparation is quite small in comparison since it consists of 10 ECTS (divided with theory of science) during the university degree and 20 ECTS theoretical pedagogy during the in-service program following a candidatus degree. Contrary to this are the compulsory school teachers who have acquired a profession bachelor (240 ECTS) and therefore have a strong background in the teaching subjects for grades 1–10. If the teacher has mathematics for grades 1–10, he will have 144 ECTS in mathematics, but if he chooses either grades 1–6 or 4–10, he would have 72 ECTS. However, this mathematics is not like the mathematics that university students take; instead it is the teaching subject mathematics, which includes subject didaktik. He furthermore has 33 ECTS of general didaktik.

Thus, the preparations for these two types of teacher are very different both in content of mathematics, amount of education/pedagogy, and the overall purpose of the education being research-based (universities) or seen as a profession education (university colleges). Structurally the systems are also so different that it is in fact two separate routes with usually no possibility of changing path without starting all over.

Mathematics Contents and Competencies and Who Best to Teach Them?

As stated above, the mathematical *content* of the two different teacher preparations are quite different. However, the curricula also address mathematical *competencies*. Such competencies denote what the students are supposed to be able to do with the mathematical content. In terms of mathematical competencies taught at university level, I will refer to some research that I did in collaboration with Brabrand.[9] In Denmark, there is not one single mathematics curriculum for all universities. We, however, studied *inter alia* mathematics competencies and competence progression at the Faculties of Science at AU and the University of Southern Denmark (SDU) for all the science departments (Brabrand & Dahl, 2008, 2009). The object of study was the new course objectives—intended learning outcomes (ILO)—formulated in the language of competencies. The SOLO taxonomy (e.g., Biggs, 2003) was the analytical tool to measure the level of competence and the university teachers had all been instructed in formulating ILOs using the SOLO taxonomy. This taxonomy operates with five numbered increasing levels of competencies, 1–5. SOLO is short for Structure of the Observed Learning Outcome and was developed aimed at research-based higher education teaching (see Appendix A and B for more about the SOLO taxonomy and the analysis). What we learned was that, first of all, there appeared to be hardly any SOLO progression in the mathematics courses from bachelor to candidatus level and furthermore, the SOLO level at all levels is moderate—around "SOLO 3." Two other calculation methods used by Dahl gave a rather similar result. See Tables 45.1 and 45.2.

In terms of the SOLO progression, one might argue (see also Brabrand & Dahl, 2008, 2009) that the SOLO taxonomy might not rightfully describe mathematical progression and that it does not in itself take the content into account—that is, it is naturally more "difficult" to interpret or evaluate some very abstract,

TABLE 45.1 SOLO-Averages at Undergraduate and Graduate Mathematics Course at AU and SDU

	AU				SDU			
	All AU	**U-grad**	**Grad**	**Progression**	**All SDU**	**U-grad**	**Grad**	**Progression**
SOLO average	3.1	3.2	2.9	−0.3	2.8	2.8	2.9	0.1

Source: Based on Brabrand & Dahl (2008, 2009) (See Appendix A)

TABLE 45.2 SOLO-Averages at Undergraduate and Graduate Mathematics Courses at AU and SDU Taken Together

	Method 1			Method 2		
	Undergrad	**Graduate**	**Progression**	**Undergrad**	**Graduate**	**Progression**
SOLO average	3.3/3.4	3.1	−0.2/−0.3	3.3	2.9/3.3	0/−0.4

Note: This uses two other weighing methods than in Table 45.1 (see Appendix B)

high-level graduate mathematics than more simple mathematics. However, taking Laursen's (1998) points into consideration, his last point refers to the teaching methods being increasingly research-like. Research abilities include the ability to see new structures, analyze, make hypotheses, and so on, to an ever larger extent. These are all competencies that the SOLO model puts on a higher level (SOLO 4–5) than being able to perform procedures, reconstruct proofs by others, and so on (SOLO 2–3). Hence, one can expect that to some extent the mathematical competencies would show evidence of progressing towards higher SOLO levels, or at least overall be higher that SOLO 3. There may be several reasons for these low levels. One is that research-based teaching to a large extent follows the two first points of Laursen (1998), namely that it is active researchers who teach and that they teach the newest knowledge. The competencies, or ILOs, are exactly where we see evidence of, or lack of, Laursen's (1998) last point. Another reason for the low SOLO level might be that the investigation did not include the PhD level, which in all likelihood would have shown a much higher SOLO level since this is a *research* degree. However, the nature of mathematics might require that it is necessary to have all such new knowledge internalized before being able to do research with it, and it might therefore take more time before the university student is able to do mathematical research, at least with the very abstract university mathematics content even though primary and secondary pupils might be able to do research-line explorations with the less abstract mathematics they meet.[10] This might be the case, particularly since mathematics is a vertical subject with knowledge and methods building upon each other. Alternatively, it could also reflect the teaching tradition at these two universities.

Such an analysis of official curricula thus focus at the "matter meant," rather than the "matter taught" or "matter learned" (Bauersfeld, 1979, pp. 204–206). The former is not necessarily the same as the two latter. However owing to a new Danish grade scale which was implemented in the whole education sector during 2006–2007, grades are now to be given based on how well the students meet the curricula formulated intended learning outcomes which are stated in the terminology of competencies. This means that the "matter meant" strongly effects the examination and grading wherefore it is relevant to study the curricula and make predictions for teaching and learning. All this being said, it seems that research-abilities like the ones mentioned as the last point by Laursen (1998) and also granted a high SOLO number, do not play a large role at the mathematics curricula at the Danish universities AU and SDU.

I did a similar SOLO analysis of the objectives to the mathematics curricula of the teacher training at the university colleges (Ministry of Education, 2007). Appendix C gives the details of the analysis. The SOLO double weigh average for the university colleges' mathematics compulsory teacher training is 3.8. This is much larger than for university level—between 2.8 and 3.4—around 3. One needs to be very careful comparing across systems since another style of curriculum writing might obscure the comparison. Hence, we cannot conclude that the difference in SOLO level is something like "3.8 – 3 = 0.8." But seeing the university college teacher training profession bachelor from within itself, it *is* remarkable that the mathematics competencies seem so "research-based" (i.e., much over 3) in SOLO terms given that the profession bachelor is not a research-based education.

What makes the picture even more interesting is that the teachers at the universities are researchers in mathematics, while the teachers at the university colleges are not researchers, neither in mathematics nor mathematics education/didaktik. The university college teachers would most likely not have a PhD, but instead they would usually have a candidatus in mathematics from a university, and hence be very familiar with the subject of mathematics, but not have learned about the teaching subject of mathematics while at university, and they would also be familiar with research-based mathematics (at least to the moderate SOLO level). The teachers from the university colleges without a candidatus in mathematics would usually have a profession bachelor followed by a canidatus pædagogiæ and would have rather small knowledge about the subject mathematics after high school level but instead be very familiar with the teaching subject, also through a teaching that is research-based during their candidatus pædagogiæ degree. Hence, just looking at the university and university college systems separately, each of them seems to have some inconsistencies in terms of the teaching qualifications and the curricula for future school teachers.

Can the Systems Integrate?

For instance, during 2007 and 2008 there was a huge debate in Denmark about having compulsory teacher training at the university. It was argued by some that if a student starts a four-year profession bachelor in order to become a compulsory school teacher, he cannot later transfer to a university without starting all over. It is, in fact, a one-track system. Furthermore, Denmark experienced a shortage of teachers, so new discussions about the role of the universities in educating teachers and new more flexible structures of the tertiary system arose. There were also some who raised concerns about the lack of subject, in contrast to the "teaching subject," in the profession bachelor program. In the autumn of 2007, AU applied to the Ministry of Education for permission to educate compulsory school teachers. Other Danish universities announced that in case AU got the permission, they too wanted to train teachers (Theisen, 2007). AU's program was titled "Academic Bachelor of Teaching" and was structured as a three-year university bachelor degree for one-subject compulsory teachers, followed by a two-year candidatus program divided into two tracks—one for multiple-subject compulsory teachers and the other for high school teaching. The purpose was, among other things, to ensure greater subject knowledge through the teaching being research-based as well as a higher amount of subject courses—105 ECTS. Teachers with a profession bachelor qualification would also be able to enter this candidatus program. This program was, however, not approved. But the idea did not end there, and in 2009 the Ministry of Education officially welcomed applications for trial programs with the purpose of developing compulsory teacher training in collaboration with universities and university colleges (Ministry of Education, 2009). Time will tell if some programs will overcome the difficulties discussed above. However, what seems not to be discussed is taking the Humboldt ideal and applying it to university colleges. If the university colleges are given the right to do research and the teachers are given the possibility/duty to acquire a PhD, the teaching at the university colleges might benefit from having research-trained teachers teach the competencies that appear to require a research-based teaching.

CONCLUSION

It seems that the profession bachelor at the university colleges aim at the higher SOLO competencies, i.e. more research-based teaching particularly in relation to the last point of Laursen (1998) which states that the teaching method during the education is increasingly research-like, at least on paper. However in relation to the first point, namely that active researchers within the area teaches, the teachers at the university colleges are not required to do research, and most would not have research training. In terms of the second point, that the newest research-based knowledge is being taught, there is no reason to doubt, that this should not be the case since many university mathematics education researchers take part in writing the text books and other material used at the university colleges and various collaboration take place. However, in terms of points one and three, it raises the question of a discrepancy between the "research-abilities" of the people teaching at the university colleges and the high research-based nature of what they are actually supposed to teach. In other words, non-research trained teachers are expected to teach a curricula, which seems to be quite strongly aimed at research-based teaching, including making the students' learn to work research-wise. Even without the use of the SOLO Taxonomy, Appendix C shows a curriculum which is definitely focused on making the students reflective etc. and not just know the basic methods. The teachers with a Candidatus degree do have a research-based background from when they studied their subject; however the mathematics subject being taught at the university colleges is not the same as at the university. Instead it is, see Figure 2, the teaching subject that is being taught at the university colleges' teacher training program. Various attempts are discussed, particularly since 2007, to integrate the two systems, and the Ministry of Education is open to trial programs where universities and university colleges collaborate and, for instance, some of the training of compulsory teachers takes place at a university. However, I also argue that it might also be an option to enhance the research abilities of the teachers of the university colleges.

ACKNOWLEDGEMENT

The author would like to thank the following people for kindly having provided feedback on various versions of this chapter: Hanne Møller Andersen, Michael E. Caspersen, Lars Brian Krogh, Birgitte Lund Nielsen, Birgitte Pontoppidan, and Henning Westphael. The faults remain my own.

NOTES

1. The full name of author is Bettina Dahl Søndergaard (Soendergaard). Dahl is, however, the author name.
2. The two different names in Latin indicate feminine (candidata) and masculine (candidatus). For matters of convenience, I will for the remaining part of the chapter just use the masculine inflection.
3. The term "academic staff" denotes all assistant, associate, and full professors. It is equivalent to the term "faculty" used in a U.S. context.
4. At Aarhus University (AU) Faculty of Science, a 4+4 route is widely used and, on occasion, also a 3+5 route. Here, students enter a PhD program after their 4th or 3rd year at university. The candidatus degree is achieved at some point during the PhD program.

5. Some Danish universities also have programs where the student can study two subjects equally, and, for instance, a candidatus in Biology traditionally consists of five years of only biology.
6. For instance, at Virginia Tech (2005), the purpose of the core curriculum is stated as follows: "While the major may be expected to provide in-depth study in one discipline, the core curriculum is designed to introduce the student to a range of traditions, modes of thinking and inquiry, and issues of central human importance now and for the future." The student here needs to study a certain number of courses within the following areas, with the credit hours in brackets: Writing and Discourse (6), Ideas, Cultural Traditions, and Values (6), Society and Human Behavior (6), Scientific Reasoning and Discovery (8), Quantitative and Symbolic Reasoning (6), Creativity and Aesthetic Experience (1), Critical Issues in a Global Context (3). Hence, the Virginia Tech Core Curriculum consists of 36 U.S. credit hours, or approximately 72 ECTS (Dahl, 2005).
7. Figure 45.3 has left out a number of things such as the variation of doctoral programs, since this is not relevant when giving an overview of the Danish tertiary education system in relation to education programs and teacher training. It has also left out various master's courses. When the term "master" is used in a Danish context, it denotes a professional qualification based on either a bachelor or candidatus degree plus some years of work experience. It is at the level of a candidatus level, but it is not the same. It does not mean quite the same as the international use of the word "Master." See the chapter by Michelsen (this volume) for such an example.
8. Wilhelm von Humboldt (1767–1835) was a Prussian education minister that established a new university in Berlin in 1810. The university carries his name: Humboldt–Universität zu Berlin.
9. Claus Brabrand is an Associate Professor at the IT University of Copenhagen.
10. An investigation of the curricula statements of mathematics competencies at grade 9 in the Danish compulsory school showed that it had a SOLO average of 3.5 (Dahl, 2009).

REFERENCES

Bauersfeld, H. (1979). Research related to the mathematical learning process. In International Commission on Mathematics Instruction, ICMI (Eds.), *New trends in mathematics teaching* (Vol. IV, pp. 199–213). Paris: UNESCO.

Biggs, J. B. (2003). *Teaching for quality learning at university*. Philadelphia: Open University Press.

Brabrand, C., & Dahl, B. (2008). Constructive alignment and the SOLO taxonomy: A comparative study of university competences in computer science vs. mathematics. In R. Lister & Simon (Eds.), *Conferences in research and practice in information technology, 88* (pp. 3–17). Sidney: Australian Computer Society, Inc

Brabrand, C., & Dahl, B. (2009). Using the SOLO taxonomy to analyze competence progression of university science curricula. *Higher Education, 58*(4), 531–549.

Dahl, B. (2008). Improving the Danish university education system: A comparison of policy borrowing from an outcome-based framework in 1973 and 2003. In *Proceedings, 15th Improving Student Learning Symposium (ISL)* (pp. 255–266). Oxford: The Oxford Centre for Staff and Learning Developments (OCSLD).

Dahl, B. (2005). A comparison of the Danish and the Virginia secondary teacher education system: Their values and emphasis on mathematics content knowledge. *The Montana Mathematics Enthusiast, 2*(2), 93–106.

Dahl, B. (2009). Transition problems in mathematics that face students moving from compulsory through to tertiary level education in Denmark: Mismatch of competencies and progression. In M. Tzekaki, M. Kaldrimidou, & C. Sakonidis (Eds.), *Proceedings, the 33th Conference of the International Group for the Psychology of Mathematics Education (PME33)* (pp. 369–376). Thessaloniki, Greece: Aristotle University of Thessaloniki & University of Macedonia.

Dale, E. L. (1998). *Pædagogik og professionalitet*. Århus: Klim.

Hopmann, S., & Riquarts, K. (2000). Starting a dialogue: A beginning conversation between didaktik and the curriculum traditions. In I. Westbury, S. Hopmann, & K. Riquarts (Eds.), *Teaching as a reflective practice: The German didaktik tradition* (pp. 3–11). London: Lawrence Erlbaum.

Jensen, H. S. (2001). Forskningsbaseret undervisning—Undervisningsbaseret forskning. *Uddannelse 3/2001*. Copenhagen: Ministry of Education.

Klafki, W. (2000). The significance of classical theories of bildung for a contemporary concept of allgemeinbildung. In I. Westbury, S. Hopmann, & K. Riquarts (Eds.), *Teaching as a reflective practice: The German didaktik tradition* (pp. 85–107). London: Lawrence Erlbaum.

Laursen, P. F. (1998). Forskningsbaseret undervisning—og læring. In T. S. Gabrielsen & P. F. Laursen (Eds.), *At undervise i Humaniora* (pp. 93–109). Frederiksberg: Samfundslitteratur.

Ministry of Education. (2004). *Professionsbachelor. Uddannelse 3/2004*. Copenhagen: Ministry of Education.

Ministry of Education. (2006). *Lov om uddannelsen til professionsbachelor som lærer i folkeskolen* (LOV 579 of 09/06/2006. Law. AT002600).

Ministry of Education. (2007). *Læreruddannelsesbekendtgørelsen* (BEK nr 219 af 12/03/2007. Ministerial order. AT002851).

Ministry of Education. (2008a). *Bekendtgørelse om akkreditering og godkendelse af erhvervsakademiuddannelser og professionsbacheloruddannelser* (BEK nr 684 af 27/06/2008. Ministerial order. AT003438).

Ministry of Education. (2008b). *Pædagogikumbekendtgørelsen* (BEK nr 18 af 14/01/2009 Ministerial order. AT003607).

Ministry of Education. (2009). *Nyheder 20 February 2009, 6/2009.* Retrieved March 28, 2009 from http://www.uvm.dk/~/media/Files/Aktuelt/Ugenyt09/06%2020%20februar%20 2009.ashx

Ministry of Science, Technology, and Innovation. (2006). *Retningslinjer for universitetsuddannelser rettet mod undervisning i de gymnasiale uddannelser (Faglige mindstekrav)* (VEJ 5 of 18/01/2006. Ministerial Guidance. BD000256).

Ministry of Science, Technology, and Innovation. (2008). *PhD-bekendtgørelsen* (BEK 18 of 14/01/2008. Ministerial order. BD000389).

Murdock, J. (2003). Standardisation and differentiation in the levels of diplomas in higher education systems in Europe. In D. Phillips, & H. Ertl (Eds.), *Implementing European Union education and training policy—A comparative study of issues in four member states* (pp. 265–276). Dordrecht: Kluwer Academic.

Nielsen, B. L., & Pontoppidan, B. (2009). Veje til styrkelse af de lærerstuderendes undervisningskompetence i naturfag. *MONA (Særnummer)*, 4–13.

Niss, M. (1999). Aspects of the nature and state of research in mathematics education. *Educational Studies in Mathematics, 40*, 1–24.

Shulman, L. S. (1986). Those who understand: Knowledge growth in teaching. *Educational Researcher, 15*(5), 4–14.

Theisen, T. (2007, August 9). Flere universiteter vil uddanne lærere. *Folkeskolen.dk* (9 August 2007).

Virginia Tech. (2005). *Virginia Tech 2005–06 Undergraduate Course Catalog and Academic Policies.* Retrieved September 8, 2009 from http://www.undergradcatalog.registrar.vt.edu/0506/acapolicies/corrcur.html

APPENDIX A: THE SOLO TAXONOMY AND THE ANALYSIS OF PROGRESSION

The SOLO taxonomy was developed by studying the learning outcomes of higher education research-based teaching. The learning outcomes are progressively more complex; first there are quantitative improvements, as when students become able to deal with more aspects, then there are qualitative improvements when the details integrate to form a structure. This gives a five-step process (Biggs, 2003, pp. 34–53):

SOLO 1—Pre-Structural Level: The student has not any kind of understanding or uses irrelevant information. He may have acquired scattered pieces of information.

SOLO 2—Uni-Structural Level: The student can only make obvious single connections such as use terminology, recite, or perform simple instructions/algorithms.

SOLO 3—Multi-Structural Level: The student can deal with several aspects but sees these independently. He can describe, classify, combine, structure, execute procedures, and so on.

SOLO 4—Relational Level: The student understands relations between several aspects and how they fit together to create a whole. He can compare, relate, analyze, explain things in terms of cause and effect, and so on.

SOLO 5—Extended Abstract Level: The student can generalize structure beyond what was given and may transfer ideas to new areas. He may have the competence to hypothesize, criticize, theorize, and so on.

Brabrand and Dahl made a list of verbs representing various competencies. The list was developed after advice from other SOLO-experienced colleagues at AU (e.g., Biggs), and after scrutiny of the course descriptions seen in context. Part of the list is seen in Table A.1. The ILOs that were added to the usual course descriptions are shown in Figure A.1.

Brabrand and Dahl therefore defined SOLO progression as moving still further up in SOLO levels and calculated SOLO averages. This method rests on several assumptions, such as if the SOLO classification in Table A.1 is credible, if the SOLO taxonomy is an appropriate measure for progression, if each statement of ILO is weighted the same, and if the equal numeric steps between adjacent SOLO levels reflect equal steps in progression. These assumptions are addressed in detail in Brabrand and Dahl (2009). For the course in Figure A.1, the SOLO average (double weighted) is: $(3 + 2.5 + 3 + 4 + 2.5 + 3) / 6 = 3.0$.

TABLE A.1 Examples of Competencies (Verbs) with SOLO 2–5

SOLO 2	SOLO 3	SOLO 4	SOLO 5
Arrange	Account for	Analyze	Assess
Calculate	Characterize	Argue	Criticize
Choose	Combine	Compare	Delimit
Connect	Conduct	Conclude	Discuss
Define	Describe	Construct	Estimate
Find	Formulate	Decode	Evaluate
Identify	Handle	Develop	Interpret
Operate	Illustrate	Explain	Judge
Recognize	Prove	Model	Predict
Reproduce	Solve	Plan	Put into Perspective
Seek	Uncover	Structure	Reflect
Translate	Use method	Use theory	Reason

Source: Based on Biggs (2003, p. 48) and Brabrand & Dahl (2008, 2009).

Introduction to Mathematical Analysis. 5 ECTS. Undergraduate level

By the end of the course, the student is expected to be able to

- *formulate* (3) mathematical statements precisely using mathematical terminology and language
- *reproduce* (2) and *illustrate* (3) definitions of mathematical concepts
- *use* (3) basic techniques, results, and concepts in order to *solve* (3) given exercises
- *argue* (4) for the steps in problem solving
- *reproduce* (2) central results and *prove* (3) them in a logical and detailed way
- *use* (3) basic techniques, results, and concepts on concrete examples.

Figure A.1 Example of a mathematics course with SOLO-level added to the competencies.

APPENDIX B

Tables B.1 and B.2 show two calculation methods done by Dahl that do not use the double weighted SOLO average in Appendix A. However, the results in Appendix B have not been published before. Table B.1 shows the ten most used SOLO competencies in all the mathematics courses at the two universities, and how many times they were used at undergraduate and graduate courses. Table B.2 gives an overview of which competencies are used more at what level.

Table B.2 seems to confirm the picture from Table 45.1 since the numbers appear to be averaging at 3, and the overall SOLO level decreases slightly when moving from undergraduate to graduate level. However, to finally decide on this, we need to do some calculations since each of the numbers in the third column (SOLO levels) should not be weighted the same. There are two ways of weighting. One way, Method 1, is to take into consideration how often a competence is used, for instance, at graduate level compared to undergraduate level. Table B.3 illustrates this.

Another way to weigh, Method 2, is to take into consideration how often a particular competence occurs at graduate level, for instance, compared to the other competencies mainly used at graduate level. Table B.4 illustrates this method.

One might argue that the calculations in Table B.3 and B.4 complement each other, each taking the weakness of the other into consideration. The results also confirm each other, and they are quite similar to the results in Table 45.1. The calculation methods behind Table 45.1 and Tables B.3 and B.4 hence validate each other. The numbers in Tables B.3–B.4 are higher than the numbers in Table 45.1, which is due to the different techniques and assumptions. However, the overall pattern, regression in SOLO competences from undergraduate to graduate level, remains the same. In general, the conclusion is that there does not appear to be SOLO progression from undergraduate to graduate level of mathematics courses at the two universities and that the overall SOLO level is moderate—around 3.

TABLE B.1 Number of Times Each of the Ten Most Used Competencies Are Used in Graduate and Undergraduate Courses (AU and SDU are taken together)

Competence	SOLO	Total	Undergraduate (U)	Undergraduate of total	Graduate (G)	Graduate of total	Mainly G or U
Reproduce	2	39	16	41%	23	59%	G
Use method	3	67	29	43%	38	57%	G
Solve	3	26	18	69%	8	31%	U
Formulate	3	26	16	62%	10	38%	U
Prove	3	25	6	24%	19	76%	G
Account for	3	16	3	19%	13	81%	G
Illustrate	3	8	7	88%	1	12%	U
Combine	3	8	0	0%	8	100%	G
Argue	4	18	15	83%	3	17%	U
Compare	4	17	1	6%	16	94%	G

TABLE B.2 The Ten Most Used SOLO Competencies and at What Level They Are Mostly Used

Level	Competencies	SOLO levels
Undergraduate	Solve, formulate, argue, *illustrate*	3 3 4 *3*
Graduate	Prove, account for, compare, *combine*, use method, reproduce	3 3 4 *3* 3 2

Note: Underline indicates that the competencies had a distribution that was less distinct than 40–60%. Italic means that the competence was used less than ten times altogether which makes the percentage distribution very sensitive to small changes. AU and SDU are taken together

TABLE B.3 Method 1: Weighted by How Often a Competence Is Used at One Level Compared to the Other

Undergraduate:

(3 · 0.69 + 3 · 0.62 + 4 · 0.83 + *3 · 0.88*) / 3.02 = 9.89 / 3.02 = **3.3**

3.02 is the sum of all the weights (0.69 + 0.62 + 0.83 + 0.88), see Table B.2

Graduate:

(3 · 0.76 + 3 · 0.81 + 4 · 0.94 + *3 · 1.0* + 3 · 0.57 + 2 · 0.59) / 4.67 = 14.36 / 4.67 = **3.1**

4.67 is the sum of all the weights (0.76 + 0.81 + 0.94 + 1 + 0.57 + 0.59), see Table B.2

One might argue, that the competences 'illustrate' and 'combine' (in italic) weight too heavily (88%-100%) compared to the number of times it actually occurs (8 times each).

Undergraduate, without "illustrate":

(3 · 0.69 + 3 · 0.62 + 4 · 0.83) / 2.14 = 7.25 / 2.14 = **3.4**

2.14 is the sum of all the weights (0.69 + 0.62 + 0.83), see Table B.2

Graduate, without "combine":

(3 · 0.76 + 3 · 0.81 + 4 · 0.94 + 3 · 0.57 + 2 · 0.59) / 3.67 = 11.36 / 3.67 = **3.1**

3.67 is the sum of all the weights (0.76 + 0.81 + 0.94 + 0.57 + 0.59), see Table B.2

Hence, we see that the averages remain rather stable even taking that into consideration

TABLE B.4 Method 2: Weighted by How Often a Competence Occurs Compared to Others at the Same Level

Undergraduate:

3 · 18/56 + 3 · 16/56 + 4 · 15/56 + 3 · 7/56 = **3.3**

The number '56' is the sum of 18, 16, 15, 7, which is the total number of times these four competencies occur in the undergraduate curricula, see Table B.2

Graduate:

3 · 19/117 + 3 · 13/117 + 4 · 16/117 + 3 · 8/117 + 3 · 38/117 + 2 · 23/117 = **2.9**

The number '117' is the sum of 19, 13, 16, 8, 38, 23, which is the total number of times these six competencies occurred in graduate curricula, see Table B.2

Using the same calculation method with just the four first competencies gives 3 · 19/56 + 3 · 13/56 + 4 · 16/56 + 3 · 8/56 = **3.3**.
"Use method" and "reproduce" are not as clearly graduate as the first four competencies—but they occur so many times and may weigh disproportional.

APPENDIX C

In Table C.1, the learning objectives for grades 4–6 are seen, which are mandatory for all prospective mathematics teachers. The ones for grades 1–6 and 4–10 are identical and also quite similar to the ones for grades 4–6. I used the method of double-weighted average behind Table 45.1. The results in Appendix C have not been published before. The competencies in the learning objectives have been assigned a SOLO level based on Table A.1. The spaces with NA refer to the original SOLO verb classification in Brabrand and Dahl (2009), where NA indicates either non-operational verbs such as communicational activities (to present, to read), or professional activities (to work with, to participate in).

TABLE C.1 Calculation of SOLO Average for Mathematics Teacher Preparation (grades 4–6), Profession Bachelor (translation by the author)

	The students are supposed to have acquired	Averages, A–I
AA	Competence to *account for* (3) depth and connection between the school subjects content areas at grades 4–6 and the parts of the research subject that support the teacher profession in this area.	3
BB	Competence to answer and ask in, with and about mathematics; this means to *follow* (NA) and *evaluate* (5) various types of mathematical reasoning; *develop* (4) and *argue* (4) mathematically, including proofs, *have experience with* (NA); *choose* (2) strategies and tools to formulation and solution of mathematical problems, and *delimit* (5), *structure* (4), *mathematise* (NA), *interpret* (5), and *criticize* (5) mathematical models.	(5 + 4 + 4 + 2 + 5 + 4 + 5 + 5)/8 = 34/8 = 4.25
CC	Competence to *understand* (NA) and *use* (3) adequate representations; *connect* (2) between daily language and mathematical language; *decode* (4), *translate* (2), and *handle* (3) symbolic statement knowing about the particular efficient symbolism play in mathematics; *use* (3) and *evaluate* (5) it in a mathematical and pedagogical connection.	(3 + 2 + 4 + 2 + 3 + 3 + 5)/7 = 22/7 = 3.142857
DD	Competence to *understand* (NA), *analyse* (4), and *assess* (5) the frames and rules for the subject mathematics nationally and locally, and *construct* (4) and *explain* (4) learning and teaching objectives.	(4 + 5 + 4 + 4) / 4 = 17/4 = 4.25
EE	Competence to *explain* (4), *plan* (4), and *conduct* (3) teaching in mathematics in interaction with the pupils; *find* (2), *assess* (5), and *develop* (4) teaching materials to mathematics and *motivate* (NA) and *inspire* (NA) the pupils to be actively engaged in mathematical activity.	(4 + 4 + 3 + 2 + 5 + 4)/6 = 22/6 = 3.67
FF	Competence to *uncover* (3), *assess* (5) and *characterize* (3) the pupils' mathematical achievements and competencies; *use* (4) a large selection of tools to evaluation and have knowledge of their validity and reliability and *uncover* (3) pupils' learning strategies and attitudes to mathematics in relation to progression and differentiation in the teaching.	(3 + 5 + 3 + 4 + 3)/5 = 18/5 = 3.6
GG	Competence to *communicate* (NA) and *collaborate* (NA) with colleagues and people outside the school, for instance parents, administrators and authorities about issues relating to the teaching of mathematics.	NA
HH	Competence to *develop* (4) own mathematical insight and subject Didaktik competence through *reflection* (5) over own teaching and identification of areas needed to be developed.	(4 + 5)/2 = 4.5
II	Competence to *use* (4) subject Didaktik knowledge and skills in interaction with general didactical, psychological and pedagogical theories with the purpose of *handling* (3) teaching mathematics and other tasks for a teacher in a concrete practice, and *analyse* (4) and *reflect* (5) on practice.	(4 + 3 + 4 + 5)/4 = 16/4 = 4.00
Double weighted average		(3 + 4.25 + 3.14 + 4.25 + 3.67 +3.6 + 4.5 + 4)/8 = 30.4/8 = **3.8**

46

ADULTS LEARNING MATHEMATICS

Research and Education in Denmark

Tine Wedege

MALMÖ UNIVERSITY

Abstract: In the borderland between mathematics education and adult education, a new research field has been cultivated, and an international research forum is formed: Adults Learning Mathematics. The subject area encompasses formal adult mathematics education as well as adults' informal mathematics learning in the communities of everyday practice—for example, the workplace. The key concept is *numeracy* and the problem field is related to mathematics and lifelong education. Danish researchers have played a role with their theoretical and empirical studies and in their engagement in the international community of research and in international networks and projects. This chapter presents an overview of Danish research in adults learning mathematics with a focus on the interplay between research and education. These are the key issues: What adults know; why adults do (or do not) learn mathematics; why we teach mathematics to adults; and how to teach or what is "best practice" in adult mathematics education. The epistemological concept of *problematique* is used as an analytical tool in this overview.

Keywords: adult numeracy, justification, lifelong education, resistance, workplace mathematics.

Why do you always talk about adults' problems with mathematics and never about the problems of mathematics education with adults?

—Poul Hansen, Federation of Workers in Denmark, 1995

In the 1990s, a new research field was cultivated in the borderland between research in mathematics education and in adult education. The subject area investigated encompasses formal adult mathematics education as well as adults' non-formal mathematics learning in the communities of everyday practice, such as the workplace. In this context, the term *adults* refers to people who start, resume, or continue their education in formal, informal, or non-formal settings, beyond the normal age of schooling in their societies. In 1994, an international research forum was formed with the name *Adults Learning Mathematics* (ALM), and Danish researchers have played a role in this work with theoretical and empirical studies, engagement in building an international community of research and co-operation in international networks and projects. As a starting point in 1995, the problem field of research in Denmark was related to basic adult education: Adult Vocational Training and Formal Adult Education. In 2000, the Ministry of Education invited two of the researchers, Lena Lindenskov and Tine Wedege, to develop a new educational program, *Preparatory Adult Education in Mathematics* (PAE), in accordance with the trend of "lifelong learning" in education policy, where "lifelong" indicates education that takes place in all stages and spheres of life. In the 21th century, the right and the obligation concerning education does not stop with childhood and youth but also includes adult life. The aim of the program PAE is that students develop numeracy, which is defined as the functional mathematical skills and understanding that in principle all

The First Sourcebook on Nordic Research in Mathematics Education, pages 627–650

people in society need to have. The development of education and teacher education was research-based, and an operational model of numeracy was the pivotal point in this work. Another education researcher, Lene Østergaard Johansen, followed the developmental process with a focus on the justification problem with a discourse analytical approach. As a statistician, Inge Henningsen has had a critical look at concept constructions and quantitative surveys; she has also taken a gender perspective.

The development of the field of practice is generally seen as an important criterion of relevance in mathematics education research. However, this is not just the ability of research to answer the problems in the field of practice, but also to criticise and reformulate these problems. The article presents an overview of Danish research in adults learning mathematics with a focus on the interplay between research and education. The organization of the article follows the structure of the epistemological frame of "problematique" as it is defined in Wedege (1997b; 2000a; 2001b; 2006) and presented below with subject area and problem field in its historical and theoretical contexts. One of the basic principles of this framework is that research is always an answer to questions—whether these are explicit or implicit. The French philosopher Gaston Bachelard (1927) puts it like this: The sense of the problem is the nerve of scientific progress. This chapter is presented in five major sections: (1) the international scientific and ideological context is given for the Danish research in adult mathematics education around the turn of the 21st century; (2) the subject area of the research is presented—that is, mathematics teaching and learning related to the Danish adult education and training system; (3) the problem field of the research with issues coming from adults learning and knowing mathematics in school and everyday life is presented; (4) the problematique of the Danish research for reformulating and answering the following questions is presented: What mathematics do adults know; why do adults (not) learn mathematics; and why teach mathematics to adults. Finally (5), in relation to the question what does "best practice" mean in adult mathematics education, the article concludes with a discussion on if—and how—the Danish problematique of numeracy is reflected in educational practices.

SCIENTIFIC AND IDEOLOGICAL CONTEXTS

Around the turn of the 21st century, international developments in the scientific and ideological contexts prepared the ground for the Danish research in adult mathematics education, which focus on adults with short formal educational stories. The international "Adults Learning Mathematics" community of practice and research grew with the same main focus and was accepted as a domain within mathematics education research—as will be discussed below. Danish research in adult education with a focus on vocational qualifications and competences was growing in the 1990s, and— internationally—a third generation of "lifelong learning" as a guiding principle for restructuring education has taken over with consequences for adult mathematics education. At the crossroad of science, politics, and ideology, international surveys on adult skills in literacy and numeracy have been carried out, with consequences for education and research.

Research in Adult Education and in Mathematics Education

Adult education has been a growth area, and the same is true of research into adult education. However, adult mathematics education is a relatively uncultivated area of research, but it has also been an area of increasing activity over the last 15 years. In 1994, at the initiative of British Diana Coben, among others, the international forum for research Adults Learning Mathematics (ALM) was formed (Coben, 1995). During the following years a community of research was created within ALM, which has been a continuous frame of reference and cooperation in the Danish research. On a personal level, ALM members like Australian Gail FitzSimons, British Jeff Evans, and Dutch Mieke van Groenestijn have been research partners. At the 8th International Congress on Mathematical Education (ICME 8), in 1996, for the first time a working group was organized around the theme of "adults returning to mathematics education" with FitzSimons as chief organizer. In the international handbook of mathematics education, also from 1996, there was an entry for "adults" for the first time ever in a reference work on mathematics education research. And there was a whole chapter devoted to the theme of "adults and mathematics." The authors characterized the research field as one showing "great heterogeneity" (FitzSimons, Jungwirth, Maasz, & Schlöglmann, 1996). In the second international handbook from 2003, "lifelong mathematics education" was the theme of a chapter with focus on adults (FitzSimons, O'Donoghue, & Coben, 2003). Meanwhile, a study group at the ICME congresses has become a tradition. At ICME10 in Copenhagen 2004, the name was "Adult Lifelong Education in Mathematics," and Evans and

Wedege were the chief organizers (Wedege, Evans, FitzSimons, Civil, & Schlöglmann, 2008).

In adult education research, the subject area encompasses formal adult mathematics education as well as adults' informal mathematics learning in the communities of everyday practice, such as the workplace. The development of adult education research to an independent academic field was closely associated with the institutionalization of adult learning. But, although the development of the field of practice is an important criterion for relevance, research does not only have the ability to answer the problems in the field of practice, but also to criticize and reformulate these problems (Olesen & Rasmussen, 1996). Within the field of mathematics education research, relevance to the practice of teaching or learning mathematics is also a criterion of quality. The subject field is constituted by the problem field of mathematics education "in all its complexity." However, a critical, sociomathematical approach might be opened up when studies concern the functions of mathematics education in society and in people's lives (Wedege, 2003). One may find Nordic examples in constructions such as "folk mathematics" (Mellin-Olsen, 1987) and "critical mathematics education" (Skovsmose, 1994).

It is an important part of the self-conception in the research field of adult education that it cannot be subordinated within a disciplinary context (such as a sub-discipline in psychology or sociology), but that inter-disciplinarity is a significant feature (Olesen & Rasmussen, 1996). The field of mathematics education research also makes use of concepts, methods, and results from other disciplines (psychology, sociology, mathematics, linguistics, anthropology, philosophy). In adult vocational and further education, the reasons for teaching and learning mathematics are to be found outside mathematics. That is another reason why interdisciplinarity is essential, both in education and research, and reconstruction of conceptual and theoretical frameworks from other disciplines is a central task (Wedege, 2000a; Wedege & Evans, 2006).

In addition, mathematics education research has a specific relationship to mathematics as a scientific discipline, as a social phenomenon, and as a school subject (Niss, 1994). What is recognized as mathematics, and what is not, is important to research, and it is also a political question—a question about mathematics and power (Mellin-Olsen, 1987; FitzSimons, 2002). Wedege (2003) has argued that conceptions like ethnomathematics have expanded the problem field of mathematics education and paved the way for studies of adults' mathematics in everyday, working, and societal life. Of importance for Danish research is also Lindenskov's (1993) study of students' everyday knowledge and mathematical knowledge—especially her demonstration and exploration of the student's personal "mathematics curriculum."

In Danish adult education research and development, in the late 1980s and the 1990s, the theoretical construction of a general (vocational) qualification concept was a driving force, as adult education is closely connected with work as an individual and a social phenomenon (Olesen, 1994). In the classical German concept, *qualification* is purely determined by labor market requirements, but the Danish researchers Olesen and Illeris and colleagues questioned this construction: "Qualifications must be understood in their duality between the objective demands that determine them and the subjective embedding that constitutes their conditions of existence" (Illeris et al., 1994, p. 28).

In importing "qualification" from Danish adult education research and redefining the concept in the context of adult mathematics education (Lindenskov, 1996; Wedege, 1995; 2000a; 2002b), Lindenskov and Wedege had a perspective grounded on a research interest in this conflict. In the period considered in this chapter (from 1995 to 2008), the international education discourse has changed from "qualification" to "competence," and today the term "competence" is almost hegemonic in educational discourses with "mathematical literacy" and "numeracy" as prominent examples of constructions (Wedege, 2001a).

A general framework for understanding adults' learning in the Danish research was found in Illeris's (2003) comprehensive theory, which combined a variety of learning theories and was based on two fundamental assumptions: (1) All learning includes two basic processes: an external interaction process between the learner and his or her social, cultural and material environment, and an internal psychological process of acquisition and elaboration; and (2) all learning includes three dimensions embedded in a societally situated context: the cognitive dimension of knowledge and skills, the emotional dimension of feelings and motivation, and the social dimension of communication and co-operation.

The Danish researchers were importing and reconstructing theory from adult education research, but they found their identity as members of the community of practice and research of Adults Learning Mathematics.

Six Statements on ALM Research

As mentioned above, the field of adults and mathematics was characterized as having "great heterogeneity" (FitzSimons et al., 1996). Wedege has argued that this is due to lack of a "grand narrative" concerning adults and mathematics and due to the great complexity of the subject area (Wedege, Benn & Maasz, 1998). However, her preliminary reconnaissance in the scientific landscape to identify "Adults Learning Mathematics" as a subject area, as a research domain, and as a field of practice led to the conclusion that it makes sense to speak about ALM as "a community of practice and research," in spite of this heterogeneity. The reconnaissance resulted in five statements (1–5) on the international research forum of ALM (Wedege, 2000a, 2001b). In her reading of ALM literature, Johansen (2006) has added a sixth statement on the discourse within the research community of ALM.

(1) Preliminary place in the scientific landscape: The ALM community of practice and research is accepted as a domain within mathematics education research.

Since 1997, the identity of the research domain in ALM has been debated at the annual conferences (Wedege, 1997b). An important question in the debate was this: Where is the research domain situated? (Wedege, Benn, & Maasz, 1998). This is not just an academic question without any consequences for practice and research. When we know where we are—or want to be—in the scientific landscape, we know something about scientific legitimacy and about criteria of quality and relevance. Wedege (2000a, 2001b) has argued and claimed that "adults learning mathematics" is situated in the borderland between research in mathematics education and in adult education from where we import and reconstruct concepts, theories, methods and findings. FitzSimons (2002) sees her work on the borderlands of three fields: mathematics education, adult education and vocational education—connecting all three. And she continues: "Perhaps it is due to these borderland crossings that my research is often located at the margins of each of these research communities" (p. 2). However, as documented above, ALM is accepted as a research domain of some of the central academic institutions within mathematics education research.

(2) Subject area: The learner is the focus of the ALM studies, and her or his "numeracy" is understood as mathematical knowledge.

In the aim of the research forum it is stated: "Within ALM we understand the term 'mathematics' to include 'numeracy'" (Coben, 1995). Adult numeracy is the main construct in the subject field. In English-speaking countries, the term "numeracy" is used for certain basic skills and understandings in mathematics that people need in various situations in their daily lives. Numeracy is a key word in basic adult mathematics education. As a concept, however, numeracy is deeply contested in politics, education, and research. Nevertheless, as an analytical concept, adult numeracy may be considered as mathematical activity in its cultural and historical context. (For a review of international research and related literature on adult numeracy, see Coben et al., 2003.) Numeracy versus mathematics has been debated in a series of ALM papers as well as in the European Network for Motivational Mathematics for Adults (EMMA, 2005–2007). Using the analytical framework of Bernstein, FitzSimons (2002) has argued that numeracy is a *horizontal discourse* that draws upon foundations of mathematical knowledge developed by individuals over a lifetime of personal experience and enculturation but which, unlike the *vertical discourse* of the discipline of mathematics, relies on common sense and is specific to and dependent upon contexts.

(3) Problem field: Research questions in mathematics education are integrated with general adult education questions in ALM and the studies are interdisciplinary.

The focus on the adult learner with capacities developed throughout a lived life is argued and developed with reference to the principles of andragogy—that is, that adults need to be involved in the planning and evaluation of their instruction, and that experience (including mistakes) provides the basis for learning activities (Benn, 1997; Lindenskov, 2006a). Research questions related to this might concern the adult's personal relationship with mathematics, such as anxiety or blocks. Other central phenomena to be studied are adult numeracy and activities in working and everyday life. This calls for theories and conceptual frameworks outside mathematics education and is another reason for interdisciplinarity. The construction or reconstruction of conceptual frameworks is an important task in research. Lave's anthropological theory of situated learning and Engeström's psychological theory of expansive learning are two examples of general theories that have been used and re-interpreted in studies of adults' mathematics in work (Wedege, 1999; FitzSimons, 2002).

(4) Two perspectives: The duality between the general and subjective perspective is either implicit or explicit in all ALM problematiques.

Two different lines of approach are possible and intertwined in the research: *the general line of approach,* starting either with societal and labor market requirements with regard to adults' mathematics-containing competences or with demands from the academic discipline (transformed into "school mathematics"), versus *the subjective line of approach* starting with adults' need for mathematics-containing competences and their beliefs and attitudes towards mathematics (Wedege, 2000a). The general approach is obviously to be found in international surveys on adult literacy and numeracy like OECD (2000, 2005). This approach is also represented by a large-scale British investigation of the mathematical needs of adult life initiated in order to make recommendations concerning the curriculum in primary and secondary schools. The title of the report was "Mathematics Counts" (Cockcroft, 1982), and it is to be found among the references in nearly all research papers on adults and mathematics in the 1990s. Fifteen years later, *Adults Count Too* was the title of a book written by a prominent British ALM member, Roseanne Benn (1997). The two books examine the low level of numeracy in society, but the approaches are quite different. The approach of Benn was the subjective, starting with the adults. She argued that mathematics is not a value-free construct, but is imbued with elitist notions that exclude and mystify. She recognizes but rejects the discourse of mathematics for purposes of social control, where mathematical literacy is seen as a way of maintaining the status quo and producing conformist and economically productive citizens. Similarly, she rejects the approach where any problem with mathematics is located within the learner rather than the system (Benn, 1997). However, to understand the affective and social conditions for people's learning processes in mathematics, one has to take both dimensions into account, and five years after *Adults Count Too,* another ALM member, FitzSimons (2002), published her doctoral thesis entitled "What counts as mathematics?" In her discussion of technologies of power in adult and vocational education one may find the dialectic between the two approaches—the general and the subjective.

(5) Justification problem: The overall purpose of ALM practice and research is the "empowerment" of adults learning mathematics.

The fifth statement was a conclusion of the debate of Adults Learning Mathematics as a community of practice and research at ALM5 (Wedege et al., 1998). This is also reflected in some of the titles of the yearly conferences—for example, ALM8 in Denmark: "Numeracy for empowerment and democracy?" (Johansen & Wedege, 2002) and ALM15 in 2008 in the U.S.: "A declaration of numeracy: Empowering adults through mathematics education." The former opened a debate. The main issue addressed was numeracy as a possible answer to questions of empowerment and democracy in the broadest sense of these terms. No single answer was found, and new questions were raised. On the basis of a discussion of two conceptions of democracy, Lindenskov and Valero (2002) presented examples of educational practices that could easily end up generating disempowerment of the students, although the opposite was intended. On the basis of her studies of the justification problem in adult mathematics education, Johansen (2002b) presented and discussed the consequences of different conceptions of active citizenship.

(6) Discourse: Two grand stories, "the school subject curriculum model" and "the ethno curriculum model," are located in the researcher–practitioner discourse.

The two "stories" were constructed by Johansen (2006) on the basis of her reading and interpretation of the discourses found in the ALM literature. According to her, the first story, based on "the school subject curriculum model," is often saying what "we" in ALM do not want, or else it is presented as a model for "adult independent learning" where adults study school mathematics at home. The focus and starting point in this model for teaching adults numeracy is their lack of formal and basic mathematical skills. The second story, based on "the ethno curriculum model," talks about "an ideal model" of adult mathematics education, which we in ALM really want, according to Johansen. In this model, the focus and starting point is that adults are already in possession of mathematics-containing knowledge before they enter a numeracy course, even though they may not be aware of their own competences and even though they may think that "math is something I cannot do." Johansen concludes that—despite the lack of a "grand narrative"—it is possible to talk about an ALM Discourse. In the European project *Adult Life Mathematics Across Borders* (ALMAB), which aims at exchanging good teaching materials and good practices between countries, a paradox of two similar trends in adult mathematics education and research has been faced. The project confronts the trend towards ethnomathematics, which acknowledges mathematics of different cultural groups, with the trend towards internationalization—or globalization—which tends to neglect cultural characteristics (Lindenskov & Hermeler, 2003).

Lifelong Adult Education

The rhetoric of lifelong learning is supposed to structure education in the Nordic countries, and the rights and the obligations concerning education do not stop with childhood and youth but include adult life. According to Rubenson (2001), in the late 1960s, UNESCO introduced lifelong learning as a utopian-*humanistic* guiding principle for restructuring education. The concept disappeared from the policy debate but reappeared in the late 1980s in a different context and in a different form. The debate was now driven by an interest based on an *economistic* worldview, emphasizing the importance of highly developed human capital, and science and technology. From the first to the second generation, lifelong learning had changed from a utopian idea to an economic imperative. This restricted, economistic view has been severely criticized and, from the late 1990s, it seems that a third generation (*economistic-social cohesion*) with active citizenship and employability as two equally important aims for lifelong learning—at least on the rhetorical level—is taking over. I find that the change in educational discourse from qualification to competence mentioned above is a possible consequence of change from the second to the third generation of the notion of lifelong learning. It is also possible to see the theme formulated by the international program committee for the debate of Discussion Group 3, at ICME-10 in Copenhagen 2004 (Lindenskov, Villavicencio, Garfunkel, Polla, & Rampal, 2008), as an illustration of one of the dilemmas built into the third generation: Mathematics for whom and why?—the balance between "mathematics for all" and "mathematics for high level mathematical activity."

As a discourse of education, lifelong learning assumes that learning takes place in all spheres of life, not only in schools and institutions: as formal and non-formal learning in education and training programs, and as informal learning in the workplace and the adults' everyday lives. According to the terminology of UNESCO (2000), *informal education* means the truly lifelong process whereby people acquire attitudes, values, skills, and knowledge from daily experience and the educative influences and resources in his or her environment—from family and neighbors, from work and play. For the most part, this process is relatively unorganized and unsystematic. *Formal education* refers to the highly hierarchically structured, chronologically graded educational system, from primary school through to university and includes a variety of specialized programs and institutions for technical and professional training. *Non-formal education* is defined as any educational activity organized outside the established formal system that is intended to serve identifiable learning objectives.

The first generation of lifelong learning saw a strong role for civil society, while the second generation privileged the market, downplayed the role of the state, and almost totally neglected civil society. In the third generation, the market still has a central role, but the responsibilities of the individual and the state are also visible. According to Olesen (2002), in adult and continuing education there seem to be two parallel and combined processes going on: an institutionalizing process adding schools for adults to the schools for children and adolescents, and a de-institutionalizing process with a focus on adults' learning processes outside schools.

As we have seen, the concept of lifelong learning is not formed on a national level but on an international level (e.g., UNESCO, OECD, EU). It is a globalized idea being reinterpreted on the national level. In the official Danish documents concerning lifelong learning from the late 1990s, one rediscovers the discourses of the third generation as well as in the documents at the beginning of the 21st century:

> Excellent development and learning opportunities for children, young people and adults shall strengthen personal development, employment and active participation in society of the individual.... This is crucial for raising the level of education and strengthening competitiveness and cohesion in Danish society. (Danish Ministry of Education, 2007, p. 1)

Lindenskov (2006b) has stressed how numeracy and mathematical literacy have a central place side by side with literacy in political documents from UNESCO and from the European Commission on the key competences required in the society of tomorrow. In Denmark, this is illustrated by the new program of Preparatory Adult Education (PAE) with literacy and basic mathematics offered to adults since 2001. The new tendencies in the third generation of "lifelong learning," with a combination of economics and social cohesion, are also visible in this programme. During the political debate and the educational planning process of PAE Mathematics, "active citizenship," "employability," and "personal needs" were used as equivalent arguments (Johansen, 2002b). An obvious danger of lifelong learning as a political project is that learning for active citizenship and democracy is reduced to an individual project. From this point of view, it is important to notice that the following statement was formulated by the Danish government in the Bill of Preparatory Adult Education (Forberedende Voksen-Undervisning), in 2000:

> Further development and maintenance of the individual's skills are not *only* an individual and private affair and responsibility. It is *also* a common societal responsibility. PAE encompasses both a democratic aspect to maintain and promote the development of active citizenship and an economic perspective linked to the demands and needs of the labour market. (Danish Ministry of Education, 2000, p. 7)

International Surveys on Adults' Basic Skills.

At the crossroad between science, policy, and ideology are the OECD surveys on adult literacy and numeracy, which in this context are named "basic skills" or "key competences." The survey results play a major role in the public debate on education and labor market policies, and they have consequences for political decisions regarding the education system and resources for research in adults' skills. Jablonka (2003) investigated different perspectives on mathematical literacy and numeracy. Her message is that any given conception of mathematical literacy promotes a particular social practice. For example, she argues that the attempts in OECD surveys (e.g., PISA) of developing a cross-cultural definition of mathematical literacy for the purpose of generating measurable standards promote mathematical literacy for developing human capital; that is, these constructs belong to the second—economistic—generation of lifelong learning. This also holds true for the construction of literacy in the International Adult Literacy Survey (IALS), where the aim was to obtain uniform and thus comparable results in the 1990s from 20 countries about the populations' skills, which were categorized into five levels within each of the three proficiency scales—prose, document, and quantitative literacy (OECD, 2000). Compared with adults in other countries, the populations in the Nordic countries had high test scores. Almost three quarters of the Danish population scored at the three highest levels in quantitative literacy, while approximately two thirds were at those levels with respect to document literacy. Just over half of the population had prose literacy skills at the three highest levels.

However, despite the generally good levels for Denmark, the literacy skills of a fairly large number of Danes were deemed inadequate when judged on the basis of the test used in the OECD survey. Depending on the scale in question, between 28% and 46% of the population between 16 and 66 years were thus claimed to have inadequate literacy skills for the demands made on them at work and in everyday life (Jensen & Holm, 2000). In Denmark these results, published in 2000, were used by the social-democratic government of the time as an argument in the political debate prior to their adult educational reform in which a very important element was to strengthen basic skills such as literacy and numeracy. The new program, Preparatory Adult Education, with mathematics as one of the lines, was offered starting in 2001.

The subsequent OECD survey, the Adult Literacy and Life Skills Survey (ALL), might be seen as a corrective to the IALS; and a framework for adult numeracy was constructed by a group of ALM members (see Groenestijn, 2002) on the basis of the following definition: *Numeracy* is the knowledge and skills required to effectively manage the mathematical demands of diverse situations. The title of the first concluding report from ALL, "Learning a Living," reflects the age of lifelong learning, and "change" is the theme which is struck in the preface:

> Change is a defining feature of modern life. Technologies change, the organization of work changes, terms of trade change, communities evolve and social roles change as individuals negotiate the life course. Hence change is unavoidable. It obliges individuals, families, schools, firms and nations to adapt.... The ability to adapt to change depends, to a large extent, on the pool of skills upon which individuals, institutions and nations can call. (OECD, 2005, p. 3)

Here, the ability to adapt is emphasized as in the qualification analysis of the labor market in 1980s. Even though people's societal life is described as one of the contexts for numerate behavior in the theoretical framework of numeracy, I find more second-generation development of human capital than active citizenship in this discourse. The validity of the OECD surveys has been questioned from various perspectives inside and outside the ALM community. Some of this criticism is presented below.

These were—and still are—the scientific, political, and ideological contexts of Danish research and developmental work in adult mathematics education.

SUBJECT AREA: ADULT MATHEMATICS EDUCATION

By *subject area,* I mean the area to be investigated in research. Following Niss (1999), I describe the subject area in *classical mathematics education research* as stretching between and combining the following three main areas: (1) *the teaching of mathematics* with focus on matters pertaining to organized attempts to transmit or bring about mathematical knowledge, skills, insights, competencies, and so forth, to well-defined categories of students; (2) *the learning of mathematics* with focus on what happens around, in and with students who engage in acquiring such knowledge, skills, and so on, with particular regard to the processes and products of learning; and (3) *the outcomes* (results and consequences) of the teaching and the learning of mathematics, respectively. Besides these three interrelated areas, the field of mathematics education research itself is made the subject of investigation in theoretical and empirical studies. In classical research, the interest is mainly children and adolescents in formal education.

At any given time, the subject area is "always-already" structured and delimited by the concrete forms of practice and knowledge that are currently regarded as mathematics teaching, learning and knowing. When mathematical knowledge is not only seen as academic mathematics and school mathematics but also as everyday mathematics, then mathematics teaching and learning encompasses informal activities outside schools and education institutions. If people's attitudes are seen as important for the learning and the knowing of mathematics, then the affective dimension is a possible focus when learning and outcomes are studied.

The subject area is both delimited and widened in *adult mathematics education research* (see Wedege, 2000a). The three main areas are: (1) *adult mathematics teaching*, which encompasses mathematics-containing instruction in vocational education (e.g., apprenticeship); (2) *adults learning mathematics*, where the context for learning is widened to everyday and working life—from formal and non-formal to informal learning; and (3) *adult mathematical knowledge and adults' attitudes towards mathematics*, where mathematics-containing competences are seen as a prerequisite for learning, not only as outcomes, and where the affective dimension of people's relationship with mathematics is included.

In the Danish research, the subject areas are found in and around the adult education and training system where mathematics teaching, learning, and knowledge/attitudes are investigated, and in workplaces where knowledge/attitudes are studied. In the following, I give a short presentation of the adult education system with a special focus on basic adult mathematics education.

Formal and Non-Formal Education

In Denmark, adult education is extensive and based on a long tradition, as in the other Nordic countries. It is provided in many different forms and under many different auspices, ranging from national or municipal adult education to labor market and staff training and competence development at work, and also ranging from liberal adult education activities (non-formal education) to qualifying general as well as vocationally oriented adult education (formal education). The adult education system is built up in parallel with the mainstream education system, and mathematics is offered to adults at lower and higher secondary levels (see Figure 46.1). In the area of Vocational Education and Training (VET), there is a long-standing tradition of involving the trade unions and the employers' associations, who have been attributed significant influence. Publicly financed adult education can be roughly divided into three main categories (see Figure 46.1):

- Adult liberal education (e.g., basic mathematics courses at evening schools)
- General adult education (e.g., numeracy and mathematics at primary and secondary levels)
- Vocationally oriented adult education and training from VET level (e.g., a mathematics-containing truck-driving course) to the highest academic level (e.g., a master's degree in mathematics)

Only the education in the last two sectors offers formal qualifications.

One intention within the adult education system is that the work and life experiences of the adult play an important role in connection with the organization of the education programs as well as in their content, profile, and duration.

Mathematics instruction covers a wide variety of activities. *Mathematics instruction* (organized communication of a mathematical subject area, either as a single-subject or as an independent subject or module within an educational programme) can be found in all three sectors, as well as *mathematics-containing instruction* (communication of a single or interdisciplinary subject area where mathematics is an integrated but identifiable part) (Wedege, 2000a). In adult and continuing training programmes, the objectives range from a purely job-related or concrete supplementary training view, over a broad vocational view to a societal view. It is possible to locate two different objectives in mathematics and mathematics-containing instruction: knowledge and abilities concerning mathematics as a subject in itself, and mathematics in relation to other subjects. Furthermore, the instruction can be organised in two different ways: mathematics taught as a separate subject or as a part of and integrated into other subjects (see Figure 46.2).

In Danish research the subject area within formal education is found in *Adult Vocational Training* (courses giving formal qualifications to semi-skilled workers, skilled workers, and middle management) and in *Formal Adult Education* (general education at lower secondary level organized as single subject). Since 2000, the program *Preparatory Adult Education* (PAE, Da: Forberedende Voksenundervisning) and the teachers

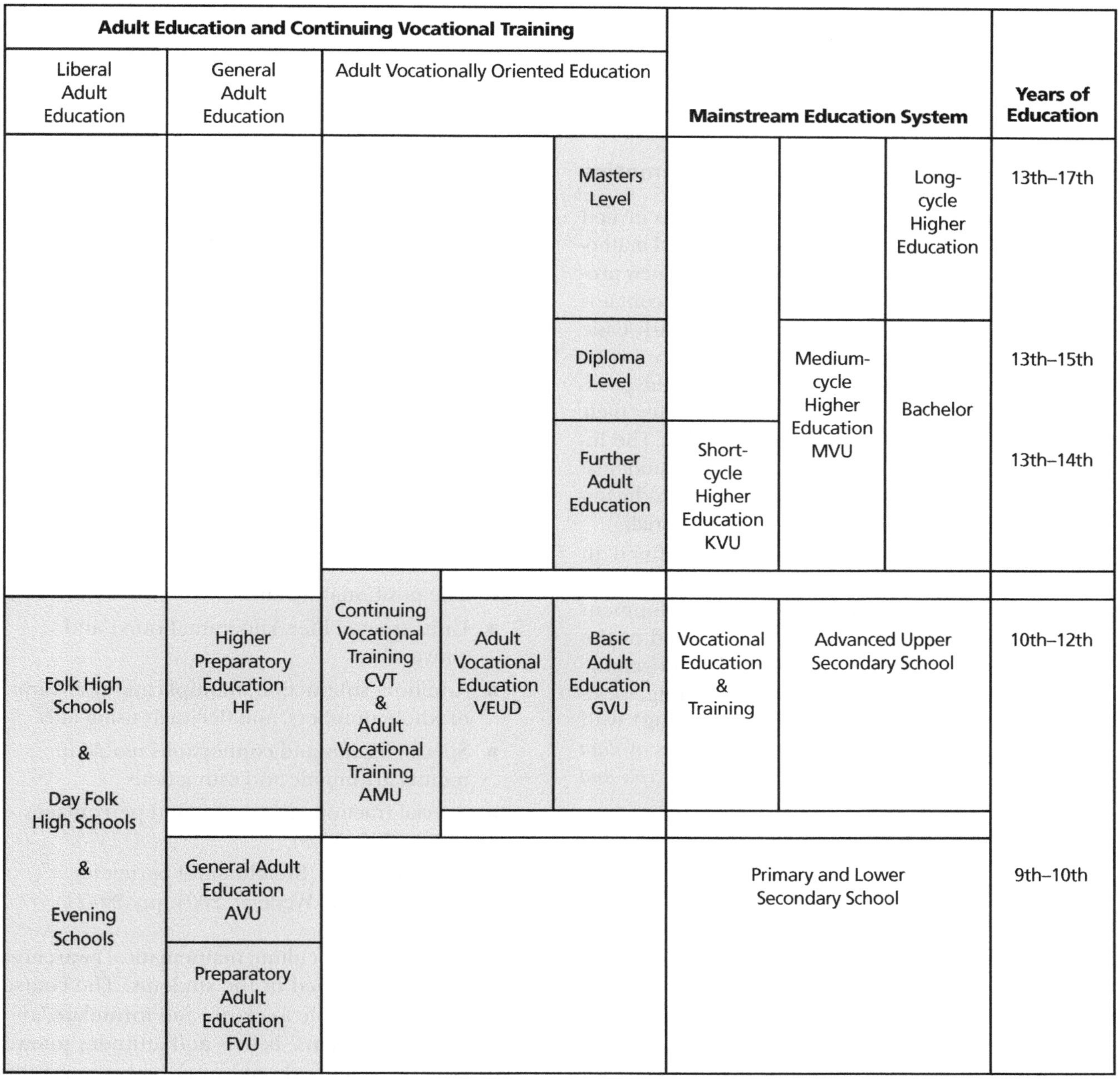

Figure 46.1 The Danish adult education and training system at the beginning of the 21st century (Source: www.uvm.dk).

Objective Organisation	**Focus on mathematics**	**Focus on other subjects**
Mathematics as a separate subject	General Adult Education Higher Preparatory Education	Adult Vocational Training & Education
Mathematics integrated in other subjects	General & Preparatory Adult Education Adult Vocational Education	Adult Vocational Training & Education

Figure 46.2 Two dimensions of formal mathematics and mathematics-containing instruction in Danish Adult Education (Wedege, 1998a, p. 182).

involved play a special role as it has both been subject of developmental work (Lindenskov & Wedege, 2001; Wedege, 2007) and of research (Henningsen & Wedege, 2003; Johansen, 2006; Lindenskov, 2006a).

Preparatory Adult Education in Mathematics

In January 2000, the Ministry of Education invited Lindenskov and Wedege to develop the national mathematics curriculum and teacher training for the new program, Preparatory Adult Education, which also contains a literacy curriculum. Since the autumn of 2001, basic preparatory mathematics has been offered to adults. The purpose is formulated to ensure students the possibility of clarifying, improving, and supplementing their functional arithmetic and mathematical skills. The intention of the education is to increase the students' possibilities of coping with, processing and producing mathematics containing information and materials.

A specific terminology is used and defined in the curriculum of PAE mathematics. The aim is formulated as the adult students' further development of their *numeracy* (a term that will be defined in the next section). The content is described as a dynamic interplay between a series of *activities* (counting, locating, measuring, designing, playing, explaining) with inspiration from Bishop (1988), various types of *data and media*, as well as selected *mathematical concepts and operations* (see Figure 46.3).

PAE mathematics has two levels: *figures and quantity* (level 1) and *patterns and relations* (level 2), which in addition include the area of form and dimension, as well as data and chance. At level 1, the content is described like this:

Activities:

- Counting (reading numbers, sorting, comparing, calculating)
- Measuring (surveying, weighing, comparing, calculating, converting)
- Locating (denoting place, time, and direction)
- Playing (following rules and strategies)

Data and media:

- Amount, time, price, discount, loss, weight, temperature, volume, length, distance, numbers, and dates
- Written information and communication (informative and instructive texts, reference and fill-in texts), oral information and communication, concrete materials

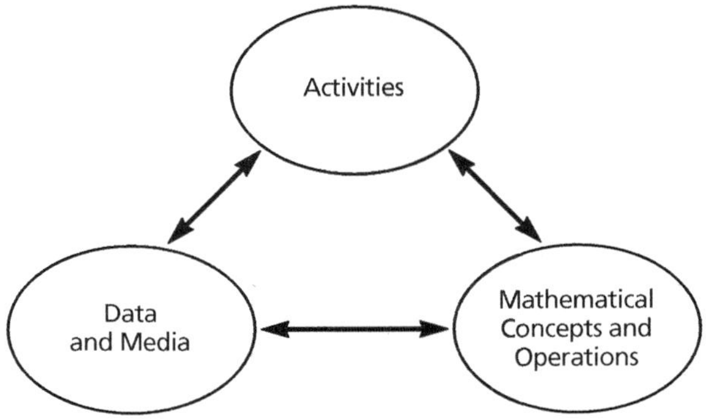

Figure 46.3 The content of the curriculum presented as a dynamic interplay (Lindenskov & Wedege, 2001, p. 19).

Mathematical operations and concepts:

- Cardinal numbers, ordinal numbers and codes and numbers
- Less than, more than, equal to, the same as, and different from
- The positional system
- Units (meter, liter, kilogram, hours) and conversion
- Addition, subtraction, multiplication, division of whole numbers, and decimals using aids
- Special figures and connections usable in mental arithmetic and estimations
- Special fractions (¼, ½, ¾,) and percentages (10%, 25%, 50%)
- Length, height, breadth, and perimeter (Lindenskov & Wedege, 2001, pp. 20–21)

According to the curriculum, mathematical awareness is cultivated and trained in the students. The course aims to make it possible to clarify and formulate, and maybe change, students' beliefs and attitudes toward mathematics. Students should work with several different kinds of contexts. In addition to the mathematical context, they should work with everyday and societal contexts. The class decides upon the choice of context for whole-group activities. With regard to individual activities, the individual students choose their contexts on the basis of what they need to learn.

The organization, concrete aims, and content should be arranged so that the background and foreground of the students take centre place. Dialogue is used to clarify and make use of the students' background and foreground. The relevance of the content is made clear by concrete connections to activities outside education. The way the problems are posed and formulated as well as the problem-solving methods should be authentic in relation to the chosen context.

PROBLEM FIELD OF ADULTS LEARNING AND KNOWING MATHEMATICS

The subject area of "adult mathematics education" is being cultivated, research problems are formulated in a dialectic relationship with practice, and the subject area is growing into a subject field. When the researchers explicitly start to locate and formulate problems concerning phenomena within the subject area, then it is further structured, and the *subject field*, that is, the field that is actually investigated, is created. Taking their point of departure in a specific position, the researchers adopt a certain view of the subject and identify a *problem field* concerning the subject area by locating and formulating *problem complexes*. In ALM research the learner is in focus and the problems might be, for example, "how can adults' everyday experiences constitute the basis for learning mathematics?" (Lindenskov, 1998b); "how can adults' mathematical understandings become identified with common sense?" (Wedege, 2002a); "how can the discrepancy between the adults' self-assessed proficiency and the results of the international surveys be explained?" (Henningsen, 2008a); and "who are the potential participants in PAE mathematics?" (Johansen, 2002c). The main subject fields in mathematics education research (teaching, learning, and knowing mathematics) are structured from the perspective of the three classical didactical questions (what, how, and why). In "Adults Learning Mathematics" they are supplemented by questions like "Who are the (potential) students in adult mathematics education?" and "Where do adult learn and use mathematics and under what circumstances?"

Problem Complexes

From the beginning, two main problem complexes have given fuel to the Danish research: (1) adults' cognitive, social, and affective relationships with mathematics, and (2) adult mathematics in vocational training and in the workplace. *Profiles in the Mathematics of Adults* was the name of the first Danish developmental project combined with research in adult mathematics education. The project crossed the two programs Adult Vocational Training and General Adult Education, and the purpose was to create a basis for advising the individual student returning to mathematics education on a point of departure for his or her mathematical skills development. The aim was to develop guidance materials for students and teachers, which include adults' capacities structured in working and everyday life (Lindenskov, 1998b; Wedege, 1997a). As a starting point in this project, a series of potential problems on the students' relations to numbers and mathematics was formulated as items in a survey conducted at the adult and vocational training/education centers. This survey confirmed that the teachers agreed upon the following: many students are blocked in relation to numbers and mathematics; for the students theory is one thing, practice is quite another; and many students lack fundamental skills in arithmetic and mathematics. In the next project, *Vocational Mathematics* (Wedege, 1998b), these three issues were reformulated with a subjective approach and investigated in qualitative interviews with 45 students at the Adult Vocational Training Centers. A general problem for understanding mathematics located in the teaching was then formulated on the basis of the adults' stories: "It is because I do miss pictures..." (Lindenskov, 1996). The students' blockage and resistance was another issue investigated. In adult education, resistance to learning is a well-known phenomenon. Often adults' resistance in the learning situation has to do with the fact that they have experienced themselves as competent persons without mathematics, and that mathematics has not been perceived as relevant to their life project (Wedege & Evans, 2006). This problem—among others—is reformulated and studied in the international project *Adults Learning Mathematics in School and Everyday Life*, which aims at establishing an interdisciplinary theoretical framework to describe, analyze, and understand social and affective conditions of adults' learning processes.

Another problem is mathematics as a gate-keeper. To many adults, the attitude is, "Mathematics—that's what I can't do," and the door named mathematics to further education is closed (Wedege, 2002a). The issue of exclusion and inclusion in and by adult mathematics education is thematized by Johansen (2002a) and also by Lindenskov (2006b), who—in an article on mathematics difficulties in general—clarifies a distinction between inclusion and integration of students with learning difficulties. According to her, both notions imply that everybody—students with special needs as well as others—are admitted and participate in the school. However, inclusion is more extensive than integration because the process of inclusion is two-sided. It involves more than adaptation, and it challenges the way mathematics is taught "normally" in the school. Another issue related to equality is gender mainstreaming, discussed by Henningsen (2008b). Mainstreaming has been widely adopted by the international community, and a growing body of research explores how pluralism and multiculturalism in mathematics education affect people's mathematical understanding, attitudes, and participation. Gender mainstreaming investigates how cultural and economic factors influence the formation of female and male differences in these areas. However, as Henningsen (2002, 2008b) has shown, gender mainstreaming of adult mathematics education is not on the agenda anywhere today—neither in curriculum planning nor in international surveys.

As a consequence of the involvement in reforms and in teacher education programs in adult mathematics education, it was also found relevant to formulate problem complexes on the teachers' attitudes towards mathematics (Henningsen & Wedege, 2003) and on their approaches to teaching mathematics to adults (Lindenskov, 2003b). Defining dilemma as a "right–right" choice situation, Lindenskov and Hansen (2001) present two dilemmas that they consider as general in teaching adult basic mathematics. The first dilemma is how to meet the adult learners' already established conceptions and procedures. It is often recommended to help adults become aware of their methods of calculating, approaching, and solving problems in mathematics. But their observations and interviews in a classroom challenged this idea. The second dilemma concerns the influence of the learners. According to the Danish curricula, the students participate in the ongoing planning of the course (for example, they decide upon the themes—art, food, transport, etc.), but how are the students provided the adequate information upon which to base their decisions? Lindenskov and Hansen do not see one and only one right answer, but they expect teachers to be able to reflect on several good answers.

One of the fundamental reasons given for providing mathematics education refers to the labor market. Mathematics education should provide students with qualifications for their working lives and contribute to the technological and socio-economic development of society (Johansen, 2002a). However, in international policy reports concerning the requirements of technological change in society, the general categories of qualifications and competences are described in isolation from the technological contexts of workplaces. As a consequence, complexity of workplace competences is reduced, and the use of mathematical knowledge in workplace situation is seen as a simple question of knowledge transfer. This exemplifies two problems that the researchers agree upon (see Wedege, 2004):

- Mathematics is integrated in the workplace activities and often hidden in technology (e.g., FitzSimons, 2002; Strässer & Zevenbergen, 1996; Wedege, 2000a).
- The so-called "transfer" of mathematics between school and workplace—and vice versa—is not a straightforward affair (e.g., Evans, 2000; FitzSimons & Wedege, 2007; Lindenskov, 1996, 1998a; Wedege, 1999, 2000b, 2002b).

In the Danish AVT programs, it has been a tradition since the late 1980s to cooperate with researchers in adult education (e.g., Olesen and Illeris) in educational planning and teacher training. However, as late as the mid 1990s, there had not really been any research in adult mathematics education, either in Denmark or internationally. No help was to be found to answer questions like "What mathematics should be taught in adult vocational training, and how?" Hence, the abovementioned analytical project, *Vocational Mathematics,* was initiated to address numbers and mathematics in Adult Vocational Training (Wedege, 1998b). One central issue was in line with the general problem complex on transfer and concerned differences between mathematics in school and in the workplace. Another issue was in line with the problem complex on mathematics being integrated within the workplace activity and concerned the workers' mathematics-containing competences.

Terminology

The subject field is also structured, opened, and delimited by the researchers' concepts and terminology. In English-speaking countries, "numeracy" is a key word in basic adult education, but in the mid 1990s, the Danish language did not have a single expression corresponding to the term "numeracy." Nevertheless, Lindenskov and Wedege (1997) chose to use the term *numeralitet,* which was later adopted by the Danish Ministry of Education. In their definition:

- Numeracy consists of functional mathematical skills and understanding that in principle all people need to have.
- Numeracy changes in time and space along with social change and technological development (Lindenskov & Wedege, 2001, p. 5).

It is the expression "in principle" that makes possible a general assessment of adult numeracy and the development of general courses in numeracy. All adults who participate in a numeracy course will, in fact, have their own foregrounds, their own backgrounds and needs, and their own strategies for learning. In the above definition, numeracy is historically and culturally determined, and it changes along with social change and technological development: Numeracy in Denmark 2008 might be different from numeracy in Togo 1978. It is this definition of numeracy that is adopted as the aim in the national curriculum of PAE-mathematics.

The term *everyday mathematics* is another example of a phenomenon in the subject field to be investigated. As a matter of fact, it was debated during the developmental process of PAE about whether to name the course "everyday mathematics" instead of "mathematics" (see Johansen, 2006, p. 195). The subject field is elucidated when Lindenskov (2004) inquires: "What do we mean by everyday mathematics in adult education?" In this article she discusses three categories of subjects/phenomena named "everyday mathematics": *titles* for educational programs, national curricula, and mathematics textbooks; *pictures* provided by course mathematics when phenomena, things, and activities from outside school are represented in learning materials and communication within the classroom; and *activities* involving "real-world" mathematics as in folk mathematics, street mathematics, and ethnomathematics.

Another term often used in literacy and numeracy studies is *functional*. In PAE mathematics the aim was formulated as intending the students to develop functional mathematical skills and understandings. When Johansen (2004a) asks what "functional skills and understanding" mean in adult mathematics, she distinguishes between four different analytical domains in adults' lives (democratic, school, workplace, and everyday life), and she studies functional skills from four different discursive perspectives (society/politicians', researchers', mathematics teachers', and individuals'). Her analysis emphasizes that skills and understanding can be functional in one domain from one perspective and not functional in another domain or from another perspective.

Context is a term often used in mathematics education and is relevant, for example, to problems of transfer/transition and to these clarifications of what is "functional." Wedege (1999) asks, what does "context" mean in mathematics education? She locates two meanings and proposes a terminological distinction. One is where context represents reality in tasks, word problems, examples, textbooks, teaching materials, and so on (e.g., everyday life situations evoked in a problem-solving task). Wedege calls this type *task context*. The other fundamental meaning has to do with contexts for learning, using and knowing mathematics (e.g., school, everyday life, workplace), or contexts of mathematics education (e.g., education system, education policies), and this type is named *situation context*.

In order to investigate the relationship between mathematics education and technology in the workplace, it is necessary to have broad conceptions of mathematical knowledge and of technology in the workplace. Realizing that there is more to technological development than new machinery, Wedege (1995, 2000c) defines *technology* in the labor market as consisting of three elements: technique, human competences and qualifications, and work organization—and of their dynamic interrelations. *Technique* is used in the broader sense to include not only tools, machines, and technical equipment, but also cultural techniques (such as language and time management), and techniques for deliberate structuring of the working process (for instance, in Taylor's "scientific management" and ISO 9000 industrial quality certification). *Work organization* is used to designate the way in which tasks, functions, responsibility, and competence are structured in the workplace.

Mathematics-containing technology is technology where mathematics is an integrated but potentially identifiable part. This term "mathematics-containing", which is used above, is imported from German, translated and defined (Wedege, 2000a). With the term "mathematikhaltigen Weiterbildung," Jungwirth, Maasz, and Schlöglmann (1995) opened a subject field for research on vocationally oriented adult education in which mathematics is an integral part.

PROBLEMATIQUE FOR RESEARCH IN ADULTS LEARNING MATHEMATICS

"Could there be a specific problematique for research in adult mathematics education?" This question was debated at the Fourth International Conference on Adults Learning Mathematics in Limerick (Wedege, 1997b). The research area in adult mathematics education research stretches between the phenomena of adult mathematics teaching, learning, and knowing and adults' attitudes and emotions towards mathematics. Subject fields are structured by problems and research questions formulated on the basis of a specific theoretical and/or methodological approach, and a systematically linked problem field, a *problematique* for research in mathematics education, is construed. Any problematique is, for example, based on a specific conception of mathematical knowledge, of how to learn mathematics and of global reasons for mathematics education (Wedege, 2006). As demonstrated through the six statements on ALM research presented above, her answer to the initial question was that there are a series of common and specific characteristics in the research within the community of ALM (Wedege, 2001b). But in adult mathematics education there are also incompatible problematiques. As an example, results from an international survey like Adult Literacy and Life Skills (OECD, 2005)—presuming that it is meaningful to measure people's numeracy with the same tool across different countries and cultures—cannot complement results from comparative investigations from an ethnomathematical perspective, which emphasizes the local and the culturally specific (Bishop, 1988). In both cases, the subject area is people's mathematical everyday competences, but the subject field (what is actually investigated) is construed differently because of different research interests and questions, theories and methodologies.

In the light of this short presentation of a concept and a context, it is relevant to ask if there exists a problematique for Danish research in adults learning mathematics. In her study of the justification given explicitly and implicitly through the discourses of the politicians and the researchers involved in PAE, Johansen (2005b, 2006) presents arguments which might be understood as a positive answer to this question. Her analytical framework includes the four didactical questions: Who is going to learn? What is going to be learned? How is it going to be taught? Why should it be taught? The analysis of the discourses, from the period 1996–2001, revealed that Lindenskov and Wedege gave the same or compatible answers concerning who, what, how, and why. Thus, a preliminary conclusion might be that they work within the same problematique. In what follows we shall see that the background for this—and what is actually constituting "our" problematique—is the construction of numeracy as a key concept and an analytical tool based on socio-cultural theories and research.

Numeracy as a Key Concept and an Analytical Tool

Our general definition of numeracy as an everyday competence needed in principle by all people in society, presented above, was formulated within a general approach (societal needs). But in order to capture the complexity of numeracy in people's everyday lives within a subjective approach (individual needs), Lindenskov and Wedege have given it an operational definition and constructed a working model with four analytical dimensions: context, media, skills and understanding, and personal intention. This model has been used and further developed as an analytical tool in Danish adult mathematics education and research (e.g., Johansen, 2004b, 2006; Lindenskov, 1998a,b; Lindenskov & Wedege, 2001; Wedege, 1998b, 2000a, 2004).

The genesis of the working model began in the project *Profile in Mathematics of Adults*, juxtaposing concepts of literacy in international studies with central concepts in mathematics education research. However, working with the problems of adults' profile in mathematics, Lindenskov realized that there is more to adult numeracy than mathematical skills and understandings (Lindenskov & Wedege, 1997). Thus, during the research and development work in the Adult Vocational Training system, they (Lindenskov & Wedege, 2001; Wedege, 1998b) developed an operative model for numeracy with four interrelated dimensions for the study of adult numeracy:

- *Media:* (a) written information and communication, (b) oral information and communication, (c) concrete materials, (d) time, and (e) processes
- *Context*—in the meaning of situation context: (a) working life, (b) family life, (c) educational context, (d) social life, and (e) leisure
- *Personal intention:* (a) to inform/be informed, (b) to construe, (c) to evaluate, (d) to understand, and (e) to practice, etc.
- *Skills & Understanding*: Dealing with and having a sense of (a) quantity and numbers, (b) dimension and form, (c) patterns and relations, (d) data and chance, (e) change, and (f) models.

Our construction of the operative model of Numeracy was based on the paradigmatic socio-cultural studies such as Scribner (1984), Lave (1988), and Nunes, Schliemann, and Carraher (1993); on conceptions of functional literacy such as OECD (2000); and on the six "big mathematical ideas" presented and discussed by Steen (1990). As mentioned, this model has been used and further developed as an analytical tool. During the development of PAE mathematics, inspiration was found in Bishop's (1988) cross-cultural studies of mathematical components in everyday activity and mathematical activities such as counting, measuring, locating were added to the fourth dimension (skills and understanding) of the Numeracy model (see Figure 46.4).

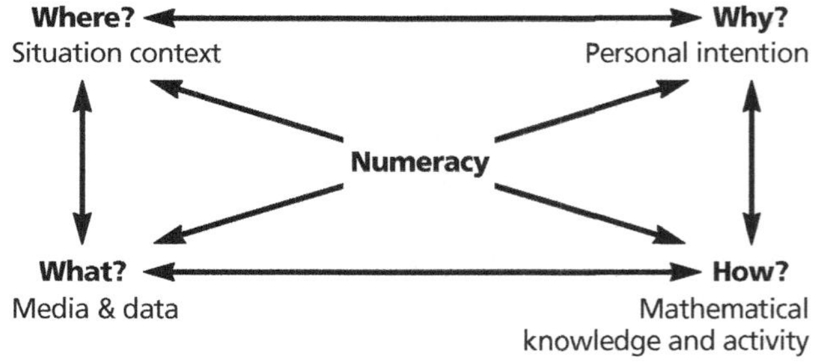

Figure 46.4 Four analytical dimensions of Numeracy (Wedege, 2004 p. 113; revised from Figure 3 in Lindenskov and Wedege, 2001 p. 6).

In what follows, I use the term "Numeracy" (with a capital N) to refer to the concept of numeracy as defined in this operational model and "numeracy" (with a small n) to refer to the underlying conception of numeracy as defined above.

What Mathematics Do Adults Know?

As shown by Johansen (2006), it is a basic conception in Lindenskov and Wedege's research and developmental work that adults returning to basic mathematics education are competent people managing their jobs and daily lives and taking part in society. This is unlike those Danish politicians to whom the potential students in basic adult education are mainly defined as people lacking basic skills. In the political discourse the target group of PAE mathematics was construed as a group of people being excluded from society (Johansen, 2005b). This construction fits in with the results of the international surveys. In the Norwegian report on Adult Literacy and Life skills (ALL), for example, it is stated that a total of one third of the adult population's "literacy skills are defined by international experts as inadequate when compared with what is required in today's labour market and everyday life" (Henningsen, 2008a, p. 75). However, when asked if they have the necessary proficiency in numeracy to meet the demands of their main job, more than 95% of the adults in the survey actually agree that they do. This discrepancy is not treated seriously in the report, and Henningsen questions whether it is ethically defensible—on the basis of what she calls "dubious statistics"—to "narrate" big groups of adults in the labor market as excluded from society and lacking basic skills.

Johansen (2002c) adapted the results from Second International Adult Literacy Survey (SIALS) in Denmark in a report designed for the Ministry of Education to be an analysis of the target group for PAE. The idea was to draw the profile of the potential target group, which was estimated to be people on proficiency levels 1 and 2 in document and quantitative literacy. Johansen concluded that, according to this, not only were 418,000 of the unskilled and semi-skilled workers potential students, but also 325,000 skilled workers. However, Johansen's analysis also reveals some limitations in the validity of the survey as a basis for describing the target group of PAE. Firstly, she shows that the kinds of numeracy skills being measured in SIALS are not consistent with the functional skills and understandings defined as numeracy in PAE. Secondly, her analysis of a series of the test items shows that it is not the calculation skills that are tested but rather the competences of interpreting and understanding different types of documents with different task contexts. Johansen (2004b) asks: How is it possible to assess adults' functional numeracy skills? She compares the methods used in the International Adult Literacy Survey and in the national test of PAE mathematics using the four dimensions of Numeracy (context, media, intentions, skills and understandings). Johansen finds, among other things, that it is the adults' knowledge of the task context rather than their mathematical skills that is tested in the international survey.

This also exemplifies another problem in the international surveys. As demonstrated by Lave (1988), there is a difference between the ability to calculate in school and in everyday life. The mathematical competence required to solve formal mathematics exercises differs from the mathematics-containing competence required for solving practical tasks even though they might involve similar types of mathematical ideas and techniques. In a student survey at the AVT centers, Lindenskov (1998a) investigated possible differences and came up with the result that the 108 semi-skilled students in the survey were, generally speaking, stronger in practical calculation with "artifacts" on the table than in the corresponding formal tasks. Findings from an Australian study of numeracy in the context of chemical spraying and handling demonstrate that mathematically straightforward skills become transformed into workplace numeracy competence, when the complexities associated with successful task completion as well as the supportive role of mediating artifacts and the workplace community of practice are taken into account (FitzSimons & Wedege, 2007).

In education research, the overall interest in studying adults' mathematics *in* the workplace is mathematics education *for* the workplace—whether a general or a subjective approach is employed (Wedege, 2004). With a general approach, adults' use of mathematics in the workplace has been investigated in international surveys like the International Adult Literacy Survey, where respondents were asked to report how often they used mathematics defined as (1) measuring or estimating the size or weight of objects and (2) calculating costs, prices, or budgets in their job (OECD, 2000). In a local survey at the Danish Adult Vocational Training Centers with 160 respondents (semi-skilled and skilled workers), 75% reported frequently counting (blanks or money), 73% calculating (+, –, ×, ÷), and 61% measuring (length or thickness) (Lindenskov, 1998a). However, the design of these kinds of questionnaires with more specific questions about adults' use of mathematics presumes some knowledge of mathematics-containing activities in the workplace and of the specific language—for example, a question about reading a "diagram" or a "graph," which in the workers' discourse could be known as a "chart" (Evans, 2000).

Traditional survey and interview studies do not reveal mathematics at work because mathematics is hidden from the perception of the worker by artifacts (material tools, workplace procedures, or organizational features) or hidden from the consciousness of the adults due to their self-perception and/or conception of mathematics (Evans, 2000; FitzSimons, 2002; Lindenskov, 1996; Wedege, 1999, 2002a,b). When you ask the general question: "Do you use mathematics in your work?" The answer is often "no." But the answer might be "yes" if you ask more specific questions based on knowledge about what is actually happening in the workplace—such as mixing liquids or concrete in the right proportion (Lindenskov, 1998).

It is an assumption in the research that numeracy as a cultural technique, like literacy, is a basic general qualification for the labor market (Wedege, 2000b, 2002b). Thus, it makes sense to talk about "adult numeracy in the labor market," and the four questions and perspectives of Numeracy in Figure 46.4 (where, what, why, and how) might help in investigating the complexity. In the analytical concept of Numeracy, there is an inevitable tension between the subjective embeddedness (numeracy as a personal competence) and the general determination (numeracy as a general qualification).

In order to analyze and describe numeracy in the labor market, Wedege (2000) has investigated selected firms within four lines of industry: building and construction, the commercial and clerical area, the metal industry, and transport. The object was to identify and describe mathematics in semi-skilled job functions and to analyze how mathematics knowledge at work is interwoven with the workers' competences. In processing the data, an observation form was used as an operational tool constructed on the basis of Numeracy (for methodology, see Wedege, 2004). One of the working hypotheses in this study was that there are systematic differences between mathematics (or numeracy) at the workplace and in mathematical instruction structured by task completion. This statement was developed and documented on the basis of Wedege's own and others' research (Wedege, 2000a, 2002b). At the workplace, the "tasks" result from satisfactorily completing a working task where the numbers are to be found or constructed with the relevant units of measurement (e.g., hours, kg, mm) given. It is the working tasks and functions in a given technological context which control and structure the process, not the particular task itself. Some of these tasks look like school tasks (the procedure is given in the work instruction), but the experienced worker has his or her own routines and methods of measurement, calculation and judgment. Circumstances in the production might cause deviations from the instruction or, for example, that the number of random samples in the quality control be raised or reduced. It is characteristic that tasks are solved in different ways and that different procedures and solutions might be satisfactory. In the workplace solving tasks is a joint matter: you have to collaborate, not compete. Solving tasks has always practical consequences: a product, a working plan, distribution of products, a price, and so on.

Why Do Adults (Not) Learn Mathematics?

The viewpoint in the idea of lifelong learning demands a rupture with the limited conception of learning and knowledge. Individual and collective learning processes do not only take place as schooling within formal education; the focus has to been moved to learning in the workplace and in everyday life. In workplace studies, we have seen that adults develop to a great extent their mathematics-containing competences and qualifications through participation in communities of everyday practice (FitzSimons & Wedege, 2007). Nevertheless, it appears that their beliefs about mathematics are primarily related to their school experiences, and mathematics is experienced by many adults as something that others can do, but that they themselves neither can do nor need (Wedege, 2002a). There is an apparent contradiction between many adults' problematical relationships with mathematics in formal settings and their "mathematics-containing" competences in everyday life. Resistance is a well-known phenomenon in adult education today, one which is experienced by adults as a field of tension between felt needs concerning what one wants to learn—or has to learn—and constraints. However, there is very little research done on the subject, and resistance is often explained purely as a lack of motivation and the symptom described as non-learning. In order to investigate adults' resistance to learning, Evans and Wedege have taken into account this set of conflicts between the needs and constraints in adults' lives. They see people's resistance as interrelated with their motivation and their competence, and thus as containing the potential to be a crucial factor in all types of learning (Wedege & Evans, 2006). Lindenskov relates the concept of resistance to Skovsmose's notion of foreground for learning mathematics. She places resistance in the context of a problematique on inclusion, where the individual is a subject with intentions and with beliefs about the actual and future situation. Here the actual relationship between individual and instruction is crucial. However, blocks to learning are placed in a problematique "on segregation and integration where the individual as an object has a learning past, which the teacher has to take into account while organizing the instruction" (Lindenskov, 2006a, p. 87, my translation).

It is obvious that two approaches are needed in research on adults learning mathematics: the general (society's requirements with regard to mathematical competences), and the subjective (adults' need for mathematical competences). Wedege (1999) discusses the conditions under which mathematics has classically "operated" in the realm of adult education and lifelong learning. Bourdieu's (1980) theory of habitus and Lave's (1993) theory of situated learning both break with sociological and psychological problematiques about socialization and learning, respectively. They both aim at challenging the dichotomies of subject–object and actor–structure. Both are critical of phenomenology and structuralism while simultaneously having social relations as the focus of their subject areas. Bourdieu set himself the task of constructing a theory of action as social practice and Lave a theory of learning as an integral part of social practice. By adapting and analyzing an interview with a 75-year-old woman about mathematics in her life, Wedege illustrates and discusses, by using the two analytical concepts of "legitimate peripheral participation" and "habitus," their suitability for analyzing adults' mathematics knowledge in different situation contexts. She claims that the concept of habitus, which was developed and belongs in a sociological problematique, can be imported into a didactic problematique about adults' learning mathematics together with the concept of situated learning.

Complexity is a characteristic of the problem field in mathematics education and gender is one of the dimensions that researchers have to decide upon. As stressed by Henningsen (2002, 2008b), very few studies on adults and mathematics are designed with gender in the foreground. Gender is an important analytical dimension in the ongoing work on people's motivation and resistance to learn mathematics (Wedege & Evans, 2006). But so far Evans and Wedege have not designed an empirical study with gender in the foreground. They have access to rich empirical data from 81 semi-structured interviews with the students (2/3 female and 1/3 male) from an English research project on adult students' reasons for learning mathematics. In this project, gender is in the background: none of the research questions are about gender. However, in a pilot case study on one of these students, Wedege (2008) has tried to bring gender viewpoints into a small part of the data using an analytical framework of structural, symbolic, personal, and interactional gender.

Why Teach Mathematics to Adults?

Could one imagine that mathematics teaching and learning provide semi-skilled workers with the possibility to develop technological competences that may place them as subjects in relation to technology in their workplace? That is, to have a personal competence that changes inability to initiative and power. And, if the answer is in the affirmative, how should mathematics teaching and learning be organized? This was the basic research question in Wedege's doctoral work, which indirectly produced an answer to the so-called justification problem: why teach mathematics to adults in vocational training and education (Wedege, 1995; 2000a). In the doctoral work planned by Johansen, "Bildung" (Danish: dannelse) was the key concept, and the purpose was "looking for the concept of Bildung in mathematics education of adults." This is reflected in some of her publications (Johansen, 2002a, 2002b, 2005b), where aspects of the concept of Bildung, which is central in Nordic education and educational research, are pinned down: Bildung can be associated with "forming" and "shaping" on the one hand, and with an "ideal" on the other hand. The process, which can be individual and active or social and more passive, refers to people developing throughout a lifetime. The ideal for human development refers to the goal of forming or shaping; a theory of Bildung is founded on a conception of humanity and society, and it includes a concept of Bildung, and the means for reaching this Bildung. Johansen (2002b) confronts two ideals of active citizenship in democratic countries: one definition presented by Benn (2001) at the ALM7 conference in which "the key components of citizenship are civic engagement and public discourse" (p. 155), and another definition given by the European Council in a political context in which "employability—the capacity to secure and keep employment" is seen as the core dimension of active citizenship (Commission of the European Communities, 2000 p. 5). The first definition focuses on communication and participation in society, locally as well as nationally. In the second definition, the council links active citizenship with employability and, hence, only includes the labor force.

As mentioned above, Johansen has used the concept of Bildung as a tool for exploring goals of teaching numeracy to adults who lack basic mathematical skills. For example, the concept of Bildung captures the importance of not separating goals from their ways of formation—that is, product from process, or "what" from "how"—in education. Her working hypothesis was that there exist in society different and conflicting reasons as to why adults who lack basic mathematical skills need to learn mathematics. These conflicting reasons are bound to different discourses about what mathematics education is, and to various assumptions about the benefits that adults can get out of it (Johansen, 2003, 2006). Her specific research interest was in the reasons for providing mathematics education to adults with a lack of basic skills. This issue is an aspect of the justification problem: Johansen investigates the general reasons given by the system or society—global as well as local (Denmark and PAE). It is especially in reform periods that reasons for mathematics education are formulated explicitly, and the adult education reform in Denmark, which included PAE, was her case study.

In her analysis, Johansen (2005b, 2006) distinguishes between a political discourse and a researcher/curriculum planner discourse, and she demonstrates that the reasons constructed in the two discourses conflict in the constructions of the target group (who), of the output of education (what) and of "best practice" (how).

As argued by Johansen "why" and "what" are closely related in mathematics education. In a 2001 survey, the potential PAE teachers' views of mathematics were studied (Henningsen & Wedege, 2003). The material was analyzed using a combination of quantitative and qualitative methods. An interesting result in relation to the debate on citizenship and democracy was this: A factor analysis and an analysis of the raw scores both seemed to indicate that teachers with no specialized education in mathematics and teachers with students on the lowest level tend to see mathematics as closed and undemocratic as compared to teachers with specialized mathematics education and teachers with students on a higher level in mathematics. In the qualitative part of the survey, the descriptions of mathematics in the teachers' essays centered on three different types of answers, which were tentatively denoted: *everyday mathematics, curriculum mathematics,* and *mathematics in the world.* In accordance with the rhetoric of Danish basic adult education in mathematics, the majority of the teachers' essays were in the first category of everyday mathematics.

HOW TO TEACH IN ADULT MATHEMATICS EDUCATION

Research and developmental work in Adults Learning Mathematics is ultimately about how to teach mathematics to—or with—adults in a way that promotes their learning of mathematics. But, as we have seen, there are many rough workings before the result: "What mathematics" and "Why mathematics" are two of the value-laden questions that we have to answer. Johansen (2005a, 2006) has argued that it is possible to locate three models for "best practice" in the Discourse of ALM and that these models are grounded in conceptions of "who" and "why." In the European developmental project *Adults Learning Mathematics Across Borders* (ALMAB), the participants realized that they had to go for good practices in the plural:

> Analysing and conceptualising country differences of developed ideas and methods is a two-edged sword. It is a risk that our engagement in co-operation and our common research for "best practice" (singular) . . . , will narrow our view so we cannot perceive that there might be no best practice which is best for all countries and cultures. Even not for European countries, and even not for the four ALMAB countries [Belgium, Denmark, the Netherlands, Norway]. (Lindenskov & Hermeler, 2003 p. 217)

In ALMAB, teachers and researchers were working together across borders, and it is a hypothesis that adult mathematics education also has the potential to improve intercultural understanding, partly by understanding learning mathematics as cultural, and partly by understanding culture with mathematics as a means (Lindenskov, 2003a). Teachers and researchers also cooperated in the European follow up project *Mathematics in Action* (MIA), where they investigated commonalities across differences in adult mathematics education. MIA has partners from seven countries, and the aim is to support teachers in adult education in order to improve the quality of learning and teaching of mathematics in adult education and to raise success rates of adult learners, by widening learning opportunities (Groenestijn & Lindenskov, 2007).

The Danish studies in the project *Vocational Mathematics* provided pieces of information for an answer to the question of how to arrange mathematics-containing teaching in AVT so that it supports the students and gives them the opportunity to exploit their potential. These pieces, which have been subject to further development in PAE, can be characterized under the two main titles of "relevance" and "visibility" of mathematics:

- The mathematics teaching in AVT has to be *relevant* to the participants' vocationally oriented needs for learning. Thus, the content has to be prioritized and arranged carefully in order to achieve the action-oriented aim of the education.
- Use of mathematical ideas and techniques has to be *visible* in AVT teaching in a way that the

students' everyday competences for employing these ideas and techniques in the work (their numeracy) are made visible (Wedege, 1998b, 2007).

Four years later, the principles for organization and content of PAE mathematics were presented and discussed in the Teaching Guidelines written by Lindenskov together with a group of teachers and published by the Danish Ministry of Education (2002). In this publication, one may find the four basic assumptions concerning adults' relationship with mathematics on which the course is built:

- Adults' numeracy has great influence on their participation in education, working life and societal life, and their personal organization in everyday life. However, many adults are not aware of this.
- Adults learn better when the mathematics education is meaningful—that is, the content and the methods used are authentic and relevant.
- Many adults had bad/negative school experiences with mathematics. This might cause blocks when they return to learn mathematics. Adults' resistance towards learning is also a well-known phenomenon.
- Adults learn in different ways. Thus they profit from different learning activities and materials.

In the Teaching Guidelines, the operational model of Numeracy is presented and discussed in relation to the teaching practice, and the term numeracy is used through the whole publication.

As stated in the Act of Preparatory Adult Education (PAE), the educational programme was evaluated in 2004 by the Danish Evaluation Institute (EVA). Numeracy is the aim of the mathematics course and the pivotal point of the teacher education program, which was compulsory. Nevertheless, neither teachers nor school leaders mentioned the word "numeracy" during the whole evaluation process. As someone involved in the development of the curriculum in 2000–2001 and in the evaluation of 2004–2005 as one of the five members of the evaluation group,[1] Wedege (2007) questions whether the model of Numeracy as described above was actually implemented in the teaching and learning practices of PAE mathematics—she formulated this as the "success" or "failure" of Numeracy as a tool in adult mathematics education.

According to Mertens (2005, p. 49), "Evaluations are conducted on the merit and worth of programs in the public domain, which are themselves responses to prioritized individual and community needs that resulted from political decisions." Thus politics and science are inherently intertwined. Here, *worth* refers to the value of an object in relation to a purpose: Is what the program does important? *Merit* refers to the excellence of an object as assessed by its intrinsic qualities or performance: How well does the program perform?

The aim of EVA's (2005) evaluation was to evaluate the strong and weak points of PAE and to assess whether the implementation of the Danish act on preparatory adult education is living up to its purpose; that is, it was mainly to evaluate the worth of the program. In relation to her reflections on success or failure, Wedege found the following results relevant: It appears that many students have benefited greatly from the course, both personally and socially, and a number of them have used PAE as a springboard for further education. A general example was of adults now being able to help their children with their homework, and happy people saying that, after all, they were really able to learn mathematics. Although the word numeracy was not mentioned by the teachers during the evaluation process, Wedege was sure that many students further developed their numeracy as a result of PAE mathematics. The first statement of the famous *Math Anxiety Bill of Rights* is this: "I have the right to learn at my own pace and not feel put down or stupid if I'm slower than someone else" (Tobias, 1993, p. 226). A common remark from the students was that they felt on a level with the teachers; they were actually treated as competent adults. Hence, it seems that participation in PAE can open the door marked "mathematics" to people. On the one hand, the adults get at chance of changing their beliefs about mathematics and their self-perception in relation to mathematics and, on the other, they have the possibility of formally qualifying to continue their education.

The main purpose of the evaluation was to assess the worth of PAE; however, the report also contains results concerning the merit. For example, it is observed that, in general, PAE is characterized by dedicated teachers and managers capable of creating a successful environment for adult education in which the students feel safe and where teaching is based on the students' needs and qualifications (EVA, 2005). According to the evaluation group's opinion, the teachers in general are well qualified and capable of carrying out their teaching on the basis of the qualifications and needs of the individual students. However, among the mathematics teachers, there are many examples of this not being the case. Here are a few examples: In the curriculum it is required that concrete material (e.g., juice, rice, wood, fabric) should be one of the media used in combination with activities and mathematical operations or concepts. In a locker marked with the words "concrete material" in a well-equipped classrooms, Wedege found only plastic gadgets normally used in children's mathematics classrooms. When students spoke in general about mathematics they often used the terms "equation," "x," and "y," which were not in the curriculum. In the following statement, Wedege found another example of how the students' views of mathematics and their perception of mathematics had not changed: A woman who was fired after many years in the same job said, "For the last 32 years I have only worked in LEGO's design department. I cannot do any mathematics." It seems that to her mathematics is still "what I cannot do" (see Wedege, 2002a).

The purpose of the PAE evaluation was not to investigate whether the educational planners' ideas, concepts, and design as manifested in the curriculum were being implemented in the teaching and learning practices of PAE mathematics. However, being an educational planner and researcher, not a politician, one of Wedege's personal interests was to assess the intrinsic qualities or performance of the education (merit), particularly the implementation of the operational model, Numeracy.

In her paper "Balancing the Unbalanceable," Sfard (2003) sees the U.S. NCTM (2000) Standards as a result of a serious and comprehensive attempt to teach "mathematics with a human face." However, she states: "Success of educational ideas, however, is never a simple function of the ideas themselves. There is no direct route from general curricular principles to successful instruction" (p. 354). The conclusion of Wedege's reflections is in keeping with this statement. In the case of PAE mathematics, we do not deal with reform of an education program already institutionalized. This was a new program that replaced another in adult basic mathematics. However, the schools and the mathematics teachers were the same. If we look at the teachers' views of mathematics before the start of PAE mathematics, we may find a reason why nobody talked about numeracy during the whole evaluation procedure (both self evaluation and in interviews). In most teachers' conceptions, mathematics is associated with everyday mathematics (Henningsen & Wedege, 2003). Thus, one might expect that they do not find it necessary to use the term "numeracy" instead of "mathematics." At one of the Ministry's information meetings on PAE mathematics, a teacher whispered to a colleague: "We go to this meeting; we listen, and we go home doing what we are used to doing." The new rhetoric was interpreted in ways that fit with the current practices in adult mathematics education and resulted in little change.

With the focus on the worth of the new adult mathematics education, "success" may be the response to the initial question whether the working model of Numeracy was actually implemented in the teaching and learning practices. What PAE mathematics does is important: Many students have benefited greatly from this course. But, looking at the merit of the new program, the response may be "failure." PAE mathematics does not perform according to the curriculum with a dynamic interplay between the activities, the data and media, and the mathematical concepts and operations.

CONCLUSION

The starting point of ALM research and education is the subjective approach with a focus on the individual adult learner and her or his competences and needs. But we do not have to neglect the general approach on numeracy with requirements from society (democracy and labor market) and from mathematics in the adult and further education system. It is common among teachers and researchers in mathematics education to claim that mathematics is everywhere and thus important for all to learn. However, there has not been much research done to legitimize and to qualify this claim, which is almost like a creed. Nevertheless, according to the surveys, 25% of the Danish adult population is defined as having inadequate basic skills by "international experts." We need a much grander scale of empirical research into people's everyday lives,combining the two viewpoints—social requirements and individual needs—in order to respect and to provide possibilities for adults to further develop their important but unrecognized mathematical skills and knowledge.

ACKNOWLEDGEMENTS

Many thanks to Jeff Evans, Gail E. FitzSimons, Inge Henningsen, and the editor who gave valuable comments to an earlier version of this chapter.

NOTE

1. EVA needed a Danish researcher with expertise within the area of adult mathematics education. Wedege's involvement in the development of PAE mathematics and non-involvement in the implementation of the curriculum were assessed and they concluded that she was impartial.

REFERENCES

Bachelard, G. (1927). *Essai sur la connaissance approchée.* Paris: Vrin.

Benn, R. (1997). *Adults count too. Mathematics for empowerment.* Leicester: NIACE.

Benn, R. (2001). All for one and one for all: Citizenship and math education. In K. Safford & M. Schmitt (Eds.), *Conversation between researchers and practitioners, The 7th International Conference on Adults Learning Mathematics (ALM7)* (pp. 155–159). Medford, MA: Tufts University.

Bishop, A. J. (1988). *Mathematical enculturation. A cultural perspective on mathematics education.* Dordrecht: Kluwer Academic Publishers.

Bourdieu, P. (1980). *Le sens pratique.* Paris : Les éditions de minuit.

Coben, D. (Ed.). (1995). *ALM-1: Proceedings of the inaugural conference of Adults Learning Maths—A Research Forum, 22–24 July 1994 at Fircroft College Birmingham, UK.* London: Goldsmiths University of London.

Coben, D. with contributions by Colwell, D., Macrae, S., Boaler, J., Brown, M., & Rhodes, V. (2003). *Adult numeracy: A review of research and related literature.* London: National Research and Development Centre for Adult Literacy and Numeracy.

Cockroft, W. H. (Chairman of the Committee of Inquiry into the Teaching of Mathematics in Schools). (1982). *Mathematics counts.* London: Her Majesty's Stationery Office.

Commission of the European Communities. (2000). *A memorandum of lifelong learning.* Brussels: Commission of the European Communities.

Danish Ministry of Education (2000). *Forslag til Lov om forberedende voksenundervisning (FVU-loven), Lovforslag nr. L 252.* Copenhagen: J. H. Schultz Grafisk.

Danish Ministry of Education. (2002). *Undervisningsvejledning. FVU-matematik.* Retrieved September 9, 2009 from www.uvm.dk (Forberende voksenundervisning i matematik)

Danish Ministry of Education. (2007). *Denmark's strategy for lifelong learning: Education and lifelong skills upgrading for all.* Copenhagen: The Danish Ministry of Education, National Education Authority, Department of Adult Vocational Training.

EMMA. (2005–2007). *European Network for Motivational Mathematics for Adults.* Retrieved September 9, 2009 from www.statvoks.no/emma/

EVA. (2005). *Forberedende voksenundervisning. Evalueringsrapport.* Copenhagen: The Danish Evaluation Institute.

Evans, J. (2000). *Adults' mathematical thinking and emotions. A study of numerate practices.* London: RoutledgeFalmer.

FitzSimons, G. E. (2002). *What counts as mathematics? Technologies of power in adult and vocational education.* Dordrecht: Kluwer Academic Publishers.

FitzSimons, G. E., Jungwirth, H., Maasz, J., & Schlöglmann, W. (1996). Adults and mathematics (adult numeracy). In A. J. Bishop et al. (Eds.), *International handbook of mathematics education* (pp. 755–784). Dordrecht: Kluwer Academic Publishers.

FitzSimons, G. E., O'Donoghue, J., & Coben, D. (2003). Lifelong mathematics education. In A. J. Bishop, et al. (Eds.), *Second international handbook of mathematics education* (pp. 103–142). Dordrecht: Kluwer Academic Publishers.

FitzSimons, G. E., & Wedege, T. (2007). Developing numeracy in the workplace. *Nordic Studies in Mathematics Education, 12*(1), 49–66.

Groenestijn, M. (2002). *A gateway to numeracy: A study of numeracy in adult basic education.* Utrecht: Utrecht CDβ Press, Utrecht University.

Groenestijn, M. v., & Lindenskov, L. (Eds.). (2007). *Mathematics in action: Commonalities across differences. A handbook for teachers in adult education.* Utrecht, NI: ALL Foundation.

Henningsen, I. (2002). Gender in ALM: Women and men learning mathematics. In L. Ø. Johansen & T. Wedege (Eds.), *Numeracy for empowerment and democracy? Proceedings of the 8th International Conference on Adults Learning Mathematics (ALM8)* (pp. 223–232). Roskilde: Centre for Research in Learning Mathematics, Roskilde University.

Henningsen, I. (2008a). Adults just don't know how stupid they are. Dubious statistics in studies of adult literacy and numeracy. In V. Seabright & I. Seabright (Eds.), *Crossing borders: Research, reflection and practice in adults learning mathematics. Proceedings of the 13th International Conference on Adults Learning Mathematics (ALM13), July 2006, Queens University, Belfast* (pp. 66–77). Limerick: CAMET & Institute of Technology Tallaght Dublin.

Henningsen, I. (2008b). Gender mainstreaming of adult mathematics education: Opportunities and challenges. *Adults Learning Mathematics: An International Journal, 3*(1), 32–40.

Henningsen, I., & Wedege, T. (2003). Values and mathematics: Attitudes among teachers in adult education. In J. Maasz & W. Schlöglmann (Eds.), *Mathematics to live and work in our world. Proceedings of the 10th International Conference on Adults Learning Mathematics (ALM10)* (pp. 110–118). Linz, Austria: Johannes Kepler Universität.

Illeris, K., Andersen, V., Kjærdsgaard, C., Larsen, K., Olesen, H. S., & Ulriksen, L. (1994). *Qualifications and living people. 4th report from the General Qualification Project.* Roskilde: Adult Education Research Group. Roskilde University.

Illeris, K. (2003). Towards a contemporary and comprehensive theory of learning. *International Journal of Lifelong Education, 22*(4), 396–406.

Jablonka, E. (2003). Mathematical literacy. In Bishop, A. J. et al. (Eds.), *Second international handbook of mathematics education* (pp. 75–102). Dordrecht: Kluwer Academic Publishers.

Jensen, T. P., & Holm, A. (2000). *Danskernes læse-regne-færdigheder: i et internationalt lys.* Copenhagen: AKF-forlaget.

Johansen, L. Ø. (2002a). Why teach math to the excluded? In P. Valero & O. Skovsmose (Eds.), *Proceedings of the Third International Mathematics Education and Society Conference: MES 3* (pp. 330–340). Copenhagen: Centre for Research in Learning Mathematics, the Danish University of Education.

Johansen, L. Ø. (2002b). The concept of Bildung and goals of numeracy teaching. *Literacy and Numeracy Studies: An International Journal in the Education and Training of Adults, 11*(2), 37–46.

Johansen, L. Ø. (2002c). *Målgruppeanalyse: En undersøgelse af resultaterne fra SIALS.* Roskilde: Centre for Research in Learning Mathematics, Roskilde University.

Johansen, L. Ø. (2004a). Functional skills and understanding: What does it mean in adult mathematics? *Topic Study Group 6: Adult and lifelong mathematics education, The 10th International Congress on Mathematical Education.* Retrieved September 9, 2009 from www.icme-organisers.dk/tsg06

Johansen, L. Ø. (2004b). Voksnes regnefærdigheder/numeralitet: Hvordan testes det? In A. Engström (Ed.), *Democ-*

racy and participation: A challenge for special needs education in mathematics. Proceedings of the 2th Nordic Research Conference on Special Needs Education in Mathematics (pp. 301–316). Örebro: Department of education, Örebro University.

Johansen, L. Ø. (2005a). "Best Practice" i matematikundervisning for voksne. In T. Wedege (Ed.), *Grunnleggende voksenundervisning i matematikk: Til glede og styrke?* In *Vurdering i matematikk – Hvorfor og hvordan? Fra småskole til voksenopplæring. Konferansserapport no. 3.* (pp. 237–245). Trondheim: Nasjonalt Senter for Matematikk i Opplæringen.

Johansen, L. Ø. (2005b). Bildung: A pseudojustification or a hidden reason for adult mathematics? In L. Lindberg (Ed.), *'Bildning' and/or training. Proceedings of the 11th International Conference on adults learning Mathematics in Kungälv* (pp. 50–60). Göteborg: Göteborg University, Department of Education.

Johansen, L. Ø. (2006). *Hvorfor skal voksne tilbydes undervisning i matematik? En diskursanalytisk tilgang til begrundelsesproblemet.* Unpublished doctoral thesis. Aalborg University, Denmark.

Jungwirth, H., Maasz, J., & Schlöglmann, W. (1995). *Mathematik in der Weiterbildung. Abschlussbericht zum Forschungsprojekt.* Linz, Austria: Johannes Kepler Universität.

Johansen, L. Ø., & Wedege, T. (Eds.). (2002). *Numeracy for empowerment and democracy? Proceedings of the 8th International Conference on Adults Learning Mathematics (ALM8), 28–30 June 2001.* Roskilde: Centre for Research in Learning Mathematics, Roskilde University.

Lave, J. (1988). *Cognition in practice. Mind, mathematics and culture in everyday life.* Cambridge: Cambridge University Press.

Lave, J. (1993). The practice of learning. In S. Chaiklin & J. Lave (eds.), *Understanding practice. Perspectives on activity and context* (pp. 3–34). New York: Cambridge University Press.

Lindenskov, L. (1993). *Hverdagsviden og matematik: læreprocesser i skolen.* (Doctoral dissertation). Roskilde: IMFUFA text no. 349, Roskilde University.

Lindenskov, L. (1996). *"Det er fordi jeg mangler billeder... ": AMU-kursisters oplevelser og potentialer i faglig regning og matematik.* Copenhagen: The Danish Labour Market Authority.

Lindenskov, L. (1998a). Kursistundersøgelsen. Delrapport 2. In L. Lindenskov & T. Wedege (Eds.), *Tre rapporter fra FAGMAT: et analyse- og udviklingsprojekt om tal og faglig matematik i arbejdsmarkedsuddannelserne.* (IMFUFA-text no. 349). Roskilde: Roskilde University.

Lindenskov, L. (1998b). Developing guidance material to uncover a mathematical profile of adult participants on a crane course. In D. Coben & J. O'Donoghue (Eds.), *Adults learning mathematics. Proceedings of ALM-4 the fourth conference of adults learning maths: A research forum, July 97, Limerick* (pp. 129–133). London: Goldsmiths University of London.

Lindenskov, L. (2003a). Cultural understanding. In M.v. Groenestijn. (Ed.), *Adults Learning Mathematics Across Borders. A Grundtvig project of Belgium, Denmark, the Netherlands and Norway* (pp. 23–30). S'Hertogenbosch, Holland: Cinop.

Lindenskov, L. (2003b). Tutors' approaches to mathematics to live and work: A model of four types. In J. Maasz & W. Schloeglman (Eds.), *Mathematics to live and work in our world. The 10th international conference on adults learning mathematics (ALM10)* (pp. 146–153). Linz: Universitätsverlag Rudolf Trauner.

Lindenskov, L. (2004). What do we mean by everyday mathematics in adult education? In B. Clarke et al. (Eds.), *International perspectives on learning and teaching mathematics* (pp. 431–446). Göteborg: National Center for Mathematics Education.

Lindenskov, L. (2006a). Deltagernes egne metoder og intentioner? In O. Skovsmose & M. Blomhøj (Eds.), *Kunne det tænkes? Om matematiklæring* (pp. 63–79). Copenhagen: Malling Beck.

Lindenskov, L. (2006b). Matematikvanskeligheder i inkluderende undervisning for børn, unge og voksne. *Nordic Studies in Mathematics Education, 11*(4), 65–95.

Lindenskov, L., & Hansen, E. P. (2001). Two dilemmas in communicating mathematics in adult basic courses: How to meet pre-knowledge of adult learners and how to support democratic classroom decisions. In M. J. Schmidt & K. Safford-Ramus (Eds.), *Conversation between researchers and practitioners. The 7th International Conference on Adults Learning Mathematics (ALM7)* (pp. 29–33). Medford , MA: Tufts University.

Lindenskov, L., & Hermeler, B. (2003). The ALMAB-project. Opportunities and risks in exchange across borders. In J. Evans et al. (Eds.), *Policies and practices for adults learning mathematics: Opportunities and risks. The 9th International Conference on Adults Learning Mathematics (ALM9)* (pp. 214–220). London: King's College London.

Lindenskov, L., & Valero, P. (2002). (Dis)empowering forces in everyday mathematics: Challenges to democracy. In L. Ø. Johansen & T. Wedege (Eds.), *Numeracy for empowerment and democracy? Proceedings of the 8th international conference on adults learning mathematics* (pp. 33–41). Roskilde: Centre for Research in Learning Mathematics, Roskilde University.

Lindenskov, L., & Wedege, T. (1997). Numeralitet: En kompetence i hverdag og undervisning. *Voksenuddannelse, 31,* 18–19.

Lindenskov, L., & Wedege, T. (2001). Numeracy as an analytical tool in adult education and research. *Centre for Research in Learning Mathematics, Publication no. 31.* Roskilde: Roskilde University.

Lindenskov, L., Villavicencio, M. U., Garfunkel, S., Polla, G., & Rampal, A. (2008). Mathematics for whom and why? The balance between 'mathematics for all' and 'for high level mathematical activity' (Discussion Group 3). In M. Niss (Ed.), *Proceedings of the 10th International Congress on Mathematical Education, 4–11 July, 2004* (pp. 433–437). Roskilde: IMFUFA, Department of Science, Systems and Models, Roskilde University.

Mellin-Olsen, S. (1987). *The politics of mathematics education.* Dordrecht: Kluwer Academic Publisher.

Mertens, D. M. (2005). *Research and evaluation in education and psychology. Integrating, diversity with quantitative, qualitative, and mixed methods.* London: Sage Publications.

NCTM National Council of Teachers of Mathematics. (2000). *Principles and standards for school mathematics.* Reston, VA: NCTM.

Niss, M. (1994). Mathematics in society. In R. Biehler et al. (Eds.), *Didactics of Mathematics as a Scientific Discipline* (pp. 367–378). Dordrecht: Kluwer Academic Publishers.

Niss, M. (1999). Aspects of the nature and state of research in mathematics education. *Educational Studies in Mathematics, 40,* 1–24.

Nunes, T., Schliemann, A. D., & Carraher, D. W. (1993). *Street mathematics and school mathematics.* Cambridge: Cambridge University Press.

OECD. (2000). *Literacy in the information age: Final report of the Adults Literacy Survey.* Paris: Statistics Canada.

OECD. (2005). *Learning a living: First results of the adult literacy and life skills survey.* (Chap. Measuring numeracy in ALL, pp. 291–300). Paris: Statistics Canada.

Olesen, H. S. (1994). Qualification Research. Basic Concepts and Danish Research. In S. Tøsse et al. (Eds.), *Social change and adult education research. Adult education research in Nordic countries 1992/93* (pp. 104–113). Trondheim: Tapir.

Olesen, H. S. (2002). Lifelong learning—A political agenda! Also a research agenda? In Johansen, L. Ø., & Wedege, T. (Eds.), *Numeracy for empowerment and democracy? Proceedings of the 8th International Conference on Adults Learning Mathematics (ALM8)* (pp. 10–20). Roskilde: Centre for Research in Learning Mathematics, Roskilde University.

Olesen, H. S., & Rasmussen, P. (Eds.). (1996). *Theoretical issues in adult education—Danish research and experiences.* Frederiksberg: Roskilde University Press.

Rubenson, K. (2001). Lifelong learning for all: Challenges and limitations of public policies. *The Swedish Ministry of Education and Science European Conference: Adult lifelong learning in a Europe of knowledge; Eskildstuna March 23–25, 2001* (pp. 29–39). Stockholm: Ministry of Education and Science in Sweden.

Scribner, S. (1984). Studying working intelligence. In B. Rogoff & J. Lave (Eds.), *Everyday cognition: Its development in social context* (pp. 9–40). Cambridge: Harvard University Press.

Sfard, A. (2003). Balancing the unbalancable: The NCTM standards in light of theories of learning. In J. Kilpatrick, W. G. Martin & D. Schifter (Eds.), *A research companionto principles and standards for school mathematics* (pp. 353–392). Reston, VA: NCTM.

Skovsmose, O. (1994). *Towards a philosophy of critical mathematics education.* Dordrecht: Kluwer Academics Publishers.

Steen, L. A. (Ed.). (1990). *On the shoulders of giants. New approaches to numeracy.* Washington, DC: National Academy Press.

Strässer, R., & Zevenbergen, R. (1996). Further mathematics education. In A. J. Bishop et al. (Eds.), *International handbook of mathematics education* (pp. 647–674). Dortrecht: Kluwer Academic Publishers.

Tobias, S. (1993). *Overcoming math anxiety. (Revised and expanded).* New York: Norton.

UNESCO. (2000). *World education report 2000: The right to education: Towards education for all throughout life.* Paris: UNESCO Publishing.

Wedege, T. (1995). Teknologi, kvalifikationer og matematik. *Nordic Studies in Mathematics Education, 3*(2), 29–51.

Wedege, T. (1997a). Profile in mathematics of adults returning to education. In G. FitzSimons (Ed.), *Adults returning to study mathematics. Papers from Working Group 18, ICME 8* (pp. 152–160). Adelaide: The Australian Association of Mathematics Teachers.

Wedege, T. (1997b). Could there be a specific problematique for research in adults' mathematics education? In D. Coben & J. O'Donoghue (Eds.), *Adults learning mathematics. Proceedings of ALM-4 the Fourth Conference of Adults Learning Maths—A Research Forum, July 97, Limerick* (pp. 210–217). London: Goldsmiths University of London.

Wedege, T. (1998a). Adults knowing and learning mathematics. Introduction to a new field of research between adult education and mathematics education. In S. Tøsse et al. (Eds.), *Corporate and nonformal learning. Adult education research in Nordic countries* (pp. 177–197). Trondheim: Tapir Forlag.

Wedege, T. (1998b). *Fagmat. Et analyseprojekt om tal og faglig matematik i arbejdsmarkedsuddannelserne.* Copenhagen: The Danish Labour Market Authority. (Translated, updated and generalized to a broader European context in: Wedege, T. (2007). *Vocational Mathematics. Numbers and mathematics in AVT-teaching.* Malmö: Malmö University. Available 09.09.09 at www.vox.no/emma)

Wedege, T. (1999). To know—or not to know—mathematics, that is a question of context. *Educational Studies in Mathematics. 39*(1–3), 205–227.

Wedege, T. (2000a). *Matematikviden og teknologiske kompetencer hos kortuddannede voksne: Rekognosceringer og konstruktioner i grænselandet mellem matematikkens didaktik og forskning i voksenuddannelse.* (Doctoral dissertation). Roskilde: Roskilde University, IMFUFA Text no. 381.

Wedege, T. (2000b). Mathematics knowledge as a vocational qualification. In A. Bessot & J. Ridgway (Eds.), *Education for mathematics in the workplace* (pp. 127–136). Dordrecht: Kluwer Academic Publishers.

Wedege, T. (2000c). Technology, competences and mathematics. In D. Coben, G. FitzSimons & J. O'Donoghue (Eds.), *Perspectives on adults learning mathematics: Research and practice* (pp. 192–209). Dordrecht: Kluwer Academic Publishers.

Wedege, T. (2001a). 'Competence' as a construction in adult and mathematics education. In G. FitzSimons, J. O'Donoghue & D. Coben (Eds.), *Adult and lifelong education in mathematics. Papers from WGA6, 9th International Congress on Mathematical Education (ICME9)* (pp. 21–30). Sydney: Language Australia.

Wedege, T. (2001b). Epistemological questions about research and practice in ALM. In K. Safford & M. Schmitt (Eds.), *Conversation between researchers and practitioners. The 7th International Conference on Adults Learning Mathematics (ALM7)* (pp. 47–56). Medford, Massachusetts, USA: Tufts University.

Wedege, T. (2002a). "Mathematics—that's what I can't do"—People's affective and social relationship with mathematics. *Literacy and Numeracy Studies: An International Journal of Education and Training of Adults, 11*(2), 63–78.

Wedege, T. (2002b). Numeracy as a basic qualification in semi-skilled jobs. *For the Learning of Mathematics—An International Journal of Mathematics Education, 22*(3), 23–28.

Wedege, T. (2003). Sociomathematics: People and mathematics in society. *Adults Learning Maths—Newsletter, 20,* 1–4.

Wedege, T. (2004). Mathematics at work: Researching adults' mathematics-containing competences. *Nordic Studies in Mathematics Education, 9*(2), 101–122.

Wedege, T. (2006). Hvorfor stave problematik med q?—hvad, hvordan og hvorfor i matematikkens didaktik. In O. Skovsmose & M. Blomhøj (Eds.), *Kunne det tænkes?—om matematiklæring* (pp. 312–332). Copenhagen : Malling Beck.

Wedege, T. (2007). Numeracy as a tool in adult mathematics education: Success or failure? In C. Bergsten & B. Grevholm (Eds.), *Developing and researching quality in mathematics teaching and learning. Proceedings of MADIF5* (pp. 123–135). Linköping: SMDF.

Wedege, T. (2008). A gender perspective on adults' motivation (and resistance) to learn mathematics. In T. Maguire, N. Colleran, N., O Gill & J. O'Donoghue (Eds.), *The changing face of adults mathematics education: learning from the past, planning for the future. Proceedings of the 14th International conference of Adults Learning Mathematics (ALM)* (pp. 289–298). Limerick: CAMET, University of Limerick.

Wedege, T., & Evans, J. (2006). Adults' resistance to learn in school versus adults' competences in work: The case of mathematics. *Adults Learning Mathematics: an International Journal 1*(2), 28–43.

Wedege, T., Benn, R., & Maazs, J. (1998). 'Adults Learning Mathematics' as community of practice and research. In M. Groenestijn & D. Coben (Eds.), *Mathematics as part of lifelong learning. Adults Learning Maths—A Research Forum. Proceedings of the Fifth Conference of ALM, ALM-5, Utrecht* (pp. 54–63). London: Goldsmiths University of London.

Wedege, T., Evans, J., FitzSimons, G. E., Civil, M., & Schlöglmann, W. (2008). Adult and lifelong mathematics education (Topic Study Group 8). In M. Niss (Eds.), *Proceedings of the 10th International Congress on Mathematical Education, 4–11 July, 2004* (pp. 315–318). Roskilde: IMFUFA, Department of Science, Systems and Models, Roskilde University.

47

A LEARNING LANDSCAPE

Building Perspectives on Mathematics Learning in Multicultural Classrooms

Helle Alrø, Ole Skovsmose, and Paola Valero

AALBORG UNIVERSITY

Abstract: Students' mathematics learning cannot be understood by looking solely at mathematical classroom activities. We discuss this claim in a multicultural context using the notion of "learning landscape." This notion serves as a theoretical and methodological tool that both defines a research perspective and sketches a field of empirical research. In this chapter we focus on the notion and illustrate its usefulness when researching mathematical learning in multicultural contexts. We draw on data and results of an empirical study on students' foregrounds with 45 teenage students in two 8th grade multicultural classes in Denmark. We show the dialectical relationship between each dimension of the landscape and the whole of it, and how, as a whole, it can help us come closer to better theorizations.

Keywords: multicultural context, learning landscape, student foreground, 8th grade class, learning from diversity

"Whether it is reading, mathematics or science, the picture is the same: The academic qualifications of bilingual immigrants are especially worse than those of their Danish classmates" (Rockwool fondens forskningsenhed, 2007a, p. 2, our translation). This is the opening sentence of the executive summary of a study that, using the Danish PISA results from 2005, focused on the competencies in reading, mathematics, and science of bilingual immigrants from non-Western countries in relation to their Danish counterparts (Egelund & Tranæs, 2007). The results of this study are condensed in the following way:

> The conditions in the ethnic families are very significant for the bad school results: "The parents' educational level, their income and their status in the labor market generally influence the children's learning. And the weaker situation of immigrant families in these areas is the main explanation for the great difference between the school performance of students with Danish and with other ethnic backgrounds." (Rockwool fondens forskningsenhed, 2007b, p. 1, our translation)

The results of this investigation, called *PISA Ethnic*, in Denmark resonate with what a similar study for the whole of PISA has concluded:

> Despite these strong learning dispositions immigrant students often perform at significantly lower levels than their native peers in key school subjects, such as mathematics, reading and science, as well as in general problem-solving skills. The differences are most pronounced in Austria, Belgium, Denmark, France, Germany, the Netherlands and Switzerland. (OECD, 2006, p. 3)

In this OECD report, the low performance of immigrant students is explained in terms of students' fam-

The First Sourcebook on Nordic Research in Mathematics Education, pages 651–669

ily background and the language spoken at home, as well as in relation to the countries' approaches to immigration and the integration of immigrants.

These two reports and their findings exemplify one of the pressing and current political, social, and educational discussions in Denmark and many other European countries: the perceived conflict between native national cultures and the expansion of ethnic minorities that, due to a myriad of reasons, have migrated to Europe in recent decades. In particular relation to mathematics education, there has been little attention paid to the challenges that the increase in the ethnic and linguistic diversity of students pose to both teachers and researchers (Abreu, César, Gorgorió, & Valero, 2007). While multiculturality has become an expanding characteristic of many European classrooms, the attention given to the implications of diversity on the processes of mathematics education have been limited in terms of the amount of research dedicated to this topic, as well as in terms of the kinds of explanations and interpretations provided.

The reports also illustrate some of the main explanatory factors given by existing research to the challenges of multiculturality. First of all, deficiencies in the performance of students belonging to ethnic minorities are seen to be anchored in students' own personal characteristics, most of which have to do with their lack of mastery of the language of instruction (e.g., Kaiser, 2003), or with culturally poor family backgrounds. Immigrant students' low performances are being attributed to the students', families', and, all in all, the minorities' lacks and deficiencies in relation to the dominant culture; and as a consequence, a discourse of deficiency characterizes many of these types of explanations.

In contrast to this, the results of the OECD study mentioned above, however, also point to other factors far outside the (mathematics) classrooms that contribute to the situation, namely the national policies on integration of immigrants—that is, the very same policies of the dominant culture are implicated in the generation of practices that do not favor immigrants' possibilities for achieving highly in school subjects.

Our experience in the project "Learning From Diversity" (Alrø, Skovsmose, & Valero, 2003, 2005) has led us to realize that mathematics education in multicultural settings is more complex than what many of these reports and most of the literature on the issue actually assume it to be. We faced the challenge of devising ways of understanding classrooms, their participants, and processes, and grasping the multiple relationships of factors and actors constituting mathematics education in multicultural settings. In this chapter we concentrate on exploring the notion of *learning landscape*, a theoretical and methodological tool that we have constructed to research mathematics education in multicultural settings. We illustrate how this notion can be put in operation when analysing and interpreting empirical material from our two-year investigations in Danish classrooms. We conclude with some general reflections about how the learning landscape can be seen in relation to a theory of (mathematics) learning.

CONCEPTUAL DIMENSIONS OF A LEARNING LANDSCAPE

What should be looked at when one researches an educational phenomenon? We talk about a learning landscape, referring to the complex situation we want to study. We use the word "landscape" in order to emphasize the undefined nature of what we are looking at. A landscape can be seen in many different ways. We can look at its many hills, for where to put up a tent, for possibilities for walking, or for where to sit in order to enjoy the many nuances of the color green. When looking, we look in a certain way, and in this way we are structuring the landscape. We can choose a lot of different perspectives, but still we cannot see whatever we want.

What we construct as a research object depends on how we construct it. The fact that what we see depends on how we see it is a simple truth that applies to research. Research data, whether recorded, collected, produced, or constructed, are heavily formed by the perspective adopted by the researchers. "Perspectivism" is a well-established philosophic idea, which also applies to research in mathematics education.

The concept of a learning landscape has a double meaning. We devised it as a theoretical and methodological construct that takes hold of the dialectic constitutive relationship between a research perspective and a research object. As a research perspective, the notion responds to the view that (mathematics) education is a set of complex social practices constituted in a multiplicity of sites of action.[1] This perspective implies the construction of a research field as a network of dynamically interrelated, interdependent sites of mathematics education. As a research object, the learning landscape allows pointing to the research foci of particular theoretical and empirical studies. Thus, the notion of learning landscape combines a research perspective and a research field. Our conceptual dimensions shape not only the way we look at a learning landscape, but also shape the very landscape itself. What we see depends on these dimensions.

Our choice of these dimensions is theoretically as well as empirically informed. We look at practice from a theoretical reflection of specific dimensions, and the other way around, our observations inform the theoretical development of these and potential new dimensions. The choice of dimensions reflects the following dilemma. One could choose too few dimensions and in this way construct a scope of research that is possible to manage. One could also choose too many dimensions that make it impossible to formulate any specific insight about the educational situation. A too-narrow perspective could turn the researcher blind, for instance, to political and cultural complexities in which the processes of learning are situated. A too-wide perspective, however, which opens for such complexities, could establish confusion instead of insight. We try to handle this dilemma through the choice of nine dimensions that occur in our studies of learning mathematics in a multicultural setting. Let us shortly present the nine dimensions.

Students' Foregrounds

Students' foregrounds are the students' interpretations of their learning possibilities and life opportunities, in relation to what the socio-political context seems to make acceptable for and available to them (Alrø & Skovsmose, 2002; Skovsmose, 1994, 2005a, 2005b; Valero, 2004). Thus, the foreground is not a simple factual given to the person. Rather it is a personally interpreted experience of future possibilities within the social and political frame of a person's life situation. "Students' foreground" is, on the one hand, a theoretical notion, which we have elaborated through notions such as "learning in action" and "intentionality," and on the other hand, students' foregrounds are an experienced socio-political complexity (Alrø, Skovsmose & Valero, 2009).

Students' Construction of Identity

The stories that students tell about themselves as participants in educational and learning spaces constitute their construction of identity (Sfard & Prusak, 2006), in particular in relation to the diversity of cultural elements available in those settings. In a similar way, the notion of identity is a theoretical construct, which has received much attention in the recent research studies, as well as an empirical action-structuring entity (Abreu, 2005; Chronaki, 2005; Gorgorió & Planas, 2005). The featuring of identity in relation to the study of learning is associated with a conception of learning as change in who people are and how they define themselves, rather than simply in terms of cognitive change (Wenger, 1998).

Teachers' Perspectives, Opinions, and Priorities of Teaching

Individual perspectives of teachers format not only their instructional activity but also their general role as educators of a group of students (Andrews, Yee, Greenhough, Hughes & Winter, 2005; César & Favilli, 2005). Teachers' perspectives, opinions, and priorities have been explored theoretically, for instance in terms of teachers' systems of beliefs. And also in this case the notion of belief establishes ways of looking at teacher opinions and priorities as acted out in the classroom or in teacher descriptions of students and of classroom activities.

The Content of Learning

In our case, the content is the *mathematical topics* for classroom interaction, including forms and priorities for how mathematics should be taught and learned (Chassapis & Chatzivasileiou, 2008; Powell, 2002). Discussing learning means discussing learning something, in this case mathematics. However, as soon we try to grasp this field of learning, the notion of mathematics gets slippery. One can think of mathematics as a school subject, as presented in textbooks and by teachers. However, one can also consider students' mathematics, referring to the actual mathematical notions, with which the students operate. (Some might claim that students' mathematics includes many misconceptions, but this does not exclude it from being a form of mathematics.) One can operate with the notion of street mathematics, which has been explored through ethnomathematical studies, and immediately we become aware of the very many different practices that include mathematics: the mathematics of the supermarket, of the family economy, of different working practices, of engineering, not to forget the (pure and applied) advanced mathematics. There are very many different practices that include mathematics in different formats, and there are possible transitions between these practices. Some of these transitions might be well known, but some might be completely unknown to many, not least to students who occupy themselves with a school mathematics practice that they cannot relate to any other practice.

Tools or Resources for Learning

Students and teachers have tools available that might be considered relevant for the learning of mathematics (Borba & Villarreal, 2005). It is a general epistemological point that any development of knowledge depends on the tools that might be available. This assumption was, for instance, emphasized by the Danish physicist Niels Bohr. Our knowledge of physical phenomena, not least within atomic theory, depends on our approach to these phenomena. Physical theories reflect strategies and conditions of experimentation, the available technology, and so on. Our tools of cognition make part of the processes through which we construct knowledge. This also applies to mathematics. Our mathematical knowledge reflects the tools and technology available. For a long period of time, paper, pencil, ruler, and compass have been the "defining" technologies of geometry. The computer has changed this, and therefore it has also changed the geometry itself, for instance, with respect to procedures for identifying geometrical guesses, strategies for exploration, ways of visualizing and documenting, and also forms of proving. The computer makes up part of the educational practice in some mathematics classrooms, together with other tools for learning: pocket calculator, schoolbooks, access to library, and so on. Processes of getting to know (i.e. learning processes) cannot be interpreted without considering the tools that are available. The worldwide distribution of tools is also a distribution of learning opportunities.

Classroom Interaction

Classroom cultures are constituted and expressed among students and between students and teacher through classroom interaction (Alrø & Skovsmose, 2002; Gorgorió, Planas & Vilella, 2002). Classroom interaction makes up a part of any processes of learning inside school. This statement is in no way remarkable, as much educational research emphasizes the importance of communication between teacher and students and among students. In previous research we have had this focus in theoretical and empirical studies of dialogue and learning in mathematics education.

Parents and Family

Parents and family serve as role models or authorities for students' priorities in relation to learning and schooling (Civil, Planas, & Quintos, 2005; Gutstein, 2003; Hawighorst, 2005). Parents and family refer to those in closest relationship to a person, who affect both the background and the foreground of this person. Parents and family not only refer to a certain group of persons, but also to the values and cultural priorities of which the family might be a carrier.

Friends

Friends and peers, though, can be experienced as even closer than parents, especially among teenagers, and are the reference groups for the construction of students' identities (Bishop, 2002; Baber, 2007).

Public Discourses

There is a great deal of public discourse that deals with immigrants, schooling, and multiculturalism and their repercussion in school practices (Baber, 2007; Martín-Rojo, 2003). The term "public discourse" refers to ways of talking in a society about certain issues. In the public discourse in Denmark, for instance, immigrants are often talked about as "ethnic people," "new Danes," or "Danes with other ethnic background" as opposed to being "ethnic Danish," which denotes the "original" Danish. In the public school discourse, immigrant children are named "bilingual children." This could connote an understanding of immigrant children as more skilled—capable of more than one language—but in fact it refers to a problematic diversity that has to be overcome in order for these children to learn in school.

Dynamic Learning Landscape

Considering these nine dimensions of a learning landscape, it is possible to pursue different kinds of studies. And maybe we could even discover new dimensions. A learning landscape is dynamic. The dimensions are interconnected, and this means that when we concentrate on one particular dimension, light will be shed on other dimensions as well.[2] We will consider how these dimensions might help to provide an understanding of a learning situation. In this sense the dimensions turn into dimensions for interpreting learning. We will consider to what extent the dimensions can be considered dimensions of practice and of a more general theory of learning.

The dimensions of a learning landscape reflect the way we see things. They represent dimensions of our perspective. According to radical perspectivism, nothing exists but perspectives. Our perspectivism does not turn into this radical form. We think of perspectives as being perspectives *on* something, although we cannot find ways of talking about this "something" without applying a *particular* perspective. (Thus, we do not assume that one can establish a privileged perspective that helps to reveal deeper truths about the "something.") This also means that we can talk about learning and the complexities of learning from different perspectives. We could have chosen other dimensions, like language, culture, conflict, or ethnicity. We will certainly assume that concentrating on such dimensions would have helped us to see other things. We do not choose to use such almost self-evident dimensions for two reasons. First, we are in need of talking about language, culture, conflict and ethnicity, with respect to all the dimensions. Thus, conflict concerns all the dimensions of our learning landscape. We consider notions like language, culture, conflict and ethnicity as belonging to a more widespread conceptual network. (This is illustrated in Figure 47.1.) Second, we hope that the dimensions we have chosen to apply could help us in *crystallizing an insight*, which the self-evident dimensions might not when taken in isolation.

We can visualize the learning landscape symbolically with a daisy flower that has many petals. The daisy symbolises the learning landscape. The petals of the flower may be different in size and they are not all visible at the same time. And when the flower is in bloom other petals may pop up.

We have put the notions of language, culture, conflict, and ethnicity into the center of the flower, while we have symbolized our nine dimensions of the learning landscape as petals of the flower. As previously indicated, we need to use some general and connecting concepts that move across the particular dimensions. These notions we have put in the center of the flower. They are all contested concepts; they can be elaborated in many ways. These concepts we apply

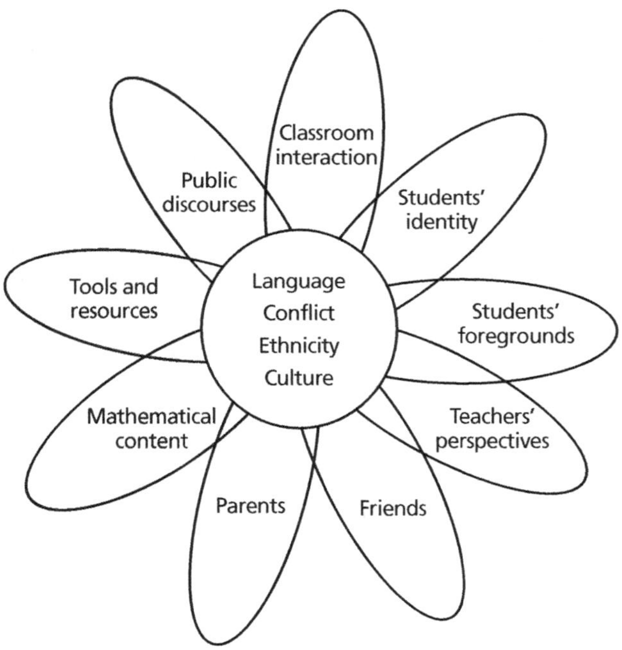

Figure 47.1 The flower-representation of the learning landscape, including its nine dimensions and its cluster of connecting concepts.

whenever it appears relevant with respect to the nine dimensions of the learning landscape.

In what follows we refer to an empirical study among teenage students in a Danish public school. As the study is part of the project "Learning from Diversity," we have chosen to visit a multicultural school.

THE SCHOOL AND THE CLASSES

Mælkevejen Skole is a primary and lower-secondary school situated in a suburb of a bigger city. The suburb, as in many other Danish suburban areas, has a high concentration of immigrant and low-income population. The school buildings are one-story houses made of concrete, built in the 1970s. The school area is rather big consisting of several houses. In one we find the 8th and 9th grade. In this building there is also a computer lab, which at the time of the research was temporarily closed because of hooligan acts. In the middle of the building there is a hall with tables and benches that are used for group work during the lessons and for relaxation during the breaks. There are also two tables for table tennis available.

Mælkevejen Skole hosts students from 29 countries from all over the world. We collaborated with two 8th grade classes that became 9th grade before we finished our collaboration. The classes have 24 and 21 students, respectively. Half of them originate from other countries than Denmark. They come from Iceland, Ireland, Iraq, Greenland, Lebanon, Malaysia, Somalia, Turkey, and Vietnam. The classrooms, however, look like typical Danish classrooms with high desks and chairs, blackboard, and a cabinet for books and materials. There are no artifacts that indicate the manifold of a multicultural classroom, except for the names on the students' pigeon holes. The mathematics teachers are two men. One is a very experienced teacher who has been working at the school almost from its very beginning. He is close to going on pension. The other one is a young novice teacher, graduated recently from the teacher training college. He is in his early thirties. Both teachers are responsible for the upper level of lower secondary school (from grades 7th to 9th), and they sometimes collaborate in giving tips and sharing activities with both classes.

EMPIRICAL STUDIES

The empirical material for this study includes three different but interconnected contacts with the teachers and students in the two classes in Mælkevejen

Skole. The first empirical study was conducted by Helle Alrø and focused on the students' foregrounds (Alrø, Skovsmose & Valero, 2009). The second study, conducted by Paola Valero, was inspired by critical ethnography (Carspecken, 1996) and included classroom observations for two months as well as assistance with respect to specific teaching tasks. A third study, conducted by Ole Skovmose, included an element of experimental teaching (Alrø & Skovsmose, 2002) where Ole, in consultation with the teachers, was allowed to set a teaching sequence with the students. Ole acted as a teacher and the teachers assisted him. In the following, the empirical material collected through Helle's foregrounds investigations serves as the principal resource, while the material emerging from Paola's and Ole's field work is used to provide supplementary information to support the findings of the foregrounds investigation.

The foregrounds investigation as conducted by Helle contained five phases. In the first phase, Helle was present in the mathematics lessons during a couple of weeks. The purpose of this phase was to do classroom observations and to get in contact with students and teachers. In the second phase of looking into the future, the students were asked to imagine themselves in ten years. What would they look like? Where would they live and with whom? Who would their friends be and what would their interests be? Would they go to school or to work, and if so which kind of school or work? The students were asked to write a story about themselves and their imagined future lives. The third phase was a questionnaire related to their present school experiences in general and to mathematics in specific. These first three phases were used to produce the topics for an inter-view[3] with each student about his or her foregrounds. The fourth phase, the inter-views, were organized as an exploration of the students' foregrounds with respect to mathematics. The principal issues raised through the inter-views were related to the students' answers of the questionnaire and to the storytelling part. We wanted to compare students' school experiences, especially in mathematics, with their interpretation of their future life. How do they consider their possibilities in life, their foregrounds, and how could this interpretation influence and be influenced by their school experiences? Thus, we wanted to compare learning motives inside school and possible activities and practices outside school. The inter-views were organized in such a way that they could shed more light on the first dimension of the learning landscape, the students' foregrounds. However, the inter-views contained remarks and comments with respect to several dimensions of the learning landscape, and it is not possible to isolate each dimension from the others. Finally, in the fifth phase, Helle, Ole, and Paola inter-viewed the two teachers and asked them to give a guess of what the students would like to do after finishing 9th grade. The two teachers shared with us their views about each student, his or her strengths and weaknesses, and a guess of what they could be good at and would like to do in the future, based on their appreciation of students' capacities. The intention of this conversation with the teachers was to have background material about how the teachers see the students' foregrounds.

Paola's critical-ethnography-inspired study consisted of a two-month period of field work at Mælkevejen Skole, where Paola was present in the school in a variety of locations. Besides attending the mathematics classes of the two groups of students, she was also present in the staffroom. Concerning the classroom observations, the teachers allowed Paola to be present in the room, to talk to students during classes about the tasks they were engaging in, and ask questions about the content of the lessons and also raise questions of a more general type—Why do you think it is important to do exercises that train you to the exam at the end of grade 9? The teachers and Paola agreed that, in the same generous way that the teachers gave time to talk to Paola, Paola was also willing to assist the teachers in whichever way they found it convenient. The result was Paola engaging in correcting some exercises that the students were given, leading a class session with the students, and covering for one of the teachers for two weeks while he was absent for paternity leave. Besides observations and many informal conversations with teachers and students, Paola also conducted conversations with students and teachers based on her observations, with the purpose of deepening her understanding and interpretation of some episodes. She also talked to other teachers in the school and to the headmaster in order to have a better sense of the institutional priorities of the school concerning education, in general, and the education of students from ethnic minorities, in particular.

The idea of the experimental teaching conducted by Ole was to establish a situation different from the one the students were accustomed to. The students were presented with the idea of "Math Mornings":[4] "The alarm clock is ringing at 7 o'clock. You wake up. It is Monday morning. You have to be at school at 8 o'clock, and the first lesson is mathematics. During this hour, from when you wake up and until you get to school, you look for mathematics. It could be mathematics in any form. It is not always easy to see mathematics, and to see it clearly, not on a Monday morning. Mathematics could easily hide behind many things, and in many things. In the alarm clock, maybe?

In the bathroom? You have to look for it everywhere. Maybe you bike to school. Could there be mathematics in biking? Anyway, here is a one-time camera. Each of you gets one. And as soon as you get glimpse of mathematics Monday morning, you take a picture. We will then use these photos during the week. We are going to investigate the mathematics you have seen the first hour Monday morning."

The introduction to the project had this form. The students were puzzled. They made some clarifying questions. The introduction was given by Ole during the last math lesson before the weekend. The one-time cameras were handed out, and the students left for the weekend. During the project, which lasted for the following week, the students established a list of issues where they had seen mathematics. This list was based on the pictures taken by each member of the group (traffic sign, cross road, a pile of oranges, the hand wash in the bathroom, etc.) Next they selected a few of the issues for further investigation. They produced a poster that showed some of their work. Finally, the students wrote a statement with the title "In my opinion..." where they could make whatever comments about the project they wanted. So the empirical material from Math Mornings was the material produced by the students including their opinions and Ole´s experiences in organizing the one-week project. The teacher was present some of the time during the project.

The different empirical material produced through the three studies has been discussed by all of us in order to identify salient elements in practices of the teachers and students in the school, and in order to have an indication of their relevance for the learning landscape. More particularly, our approach to analyzing the inter-views refers to pragmatics in the philosophy of language (Austin, 1962; Searle, 1969; Wittgenstein, 1953) and its interpretation in linguistic pragmatics and socio-pragmatics (Levinson, 1983; Wunderlich, 1975). Thus, the issue of analysis is the use and function of verbal and non-verbal language in context. This qualitative analysis of data includes descriptions and quotations as a basis for interpreting the meaning production in context. The data of analysis is language in use, and our interpretations can be validated in the quotations and perhaps be challenged or analyzed differently from other research perspectives.

THE LEARNING LANDSCAPE IN OPERATION

We will try to illustrate the dimensions of our learning landscape by providing some glimpses of the empirical material. Naturally, the point is not that a particular glimpse can be related to one and only one dimension. The point is instead to present the extension of the landscape and to show that the whole landscape is relevant for interpreting the students' learning activities. As mentioned above, we concentrate on the material emerging form the foregrounds investigations, but complement it with information from the ethnographic study and the teaching experiment study.

During the inter-viewing it became clear that there are a lot of dimensions that seem to be related to the students' foregrounds and to their motives for learning. These dimensions all seem to have influenced and have been influenced by the students' conceptions of their future lives. These are dimensions that are included in our learning landscape. Some dimensions are touched upon in the inter-viewing more than others. This is due to the fact that the learning landscape is both deductively developed from theorizing and inductively developed from the inter-viewing. This means that when we talk with students about their school experiences and foregrounds, other dimensions become important.

Students' Foregrounds

From Helle´s inter-views, it is possible to get an insight in some of the students' hopes and expectations for the future. Minh likes to go to school.[5] He considers school and mathematics to be very important for his future life, as he wants to become a banker or real estate agent: "Something with mathematics—I love numbers."

Natascha is very fond of moving things around in her room and making things better and more beautiful. She estimates the size of the furniture and she calculates and makes drawings of the room. She has just got herself a new room of 25 m^2, so there is a lot of work for her to do. Natascha wants to be an architect or estate agent. She knows that for her dreams to come true she is supposed to be very good at mathematics. And so she is. In the case of Natascha there seems to be a close connection of her interests, skills, and hopes for the future—that is, her foregrounds. And mathematics certainly plays an important role in these foregrounds.

Peter dreams about being a famous guitarist, maybe an architect or an engineer if this is not too boring. Boredom is the worst thing Peter could think of, and he would certainly not become a banker or somebody who is supposed to take care of other people's economy. He wants to become something "unusual." But he would also like to have wife and kids and a huge villa—if possible in the United States.

Ryan and Kim both want to become computer scientists. They are well aware that it takes a lot of mathematics, though, to become a computer scientist. In the first place it is a precondition to get good marks in mathematics. "But you also need mathematics for programming and for estimating codes and things like that," Ryan says. They think that being a computer scientist is a well-paid job, but the important thing is having a job that you like.

Thus, many students have a clear imagination of their future education and professional life that is connected with certain ambitions and that includes mathematics. However, a large group of other students do not know if they are going to use mathematics for the profession that they imagine to go for, but most of them mention that mathematics is important "for almost everything." Laura, for instance, wants to become a dentist, and she thinks that she is going to use mathematics, but how and why she cannot imagine at all. The same counts for Monica, who wants to be a psychologist.

Most of the students see no hindrance at all for realizing their ambitions, although Laura and Monica point to the importance of getting good marks. "The pressure of too much homework if you are not capable enough could certainly give you some problems of reaching what you want. The only hindrance is yourself."

These examples illustrate that the notion of students' foregrounds is important in order to understand their motives for learning.

Students' Identities and Cultural Diversities

Mælkevejen Skole hosts students from 29 countries and many different cultures. In the inter-views, however, none of the students touches upon that subject by themselves. And when Helle asks what difference it makes, most of the students as their first answer claim that: "It makes no difference!" and "It is no big deal!" They have obviously no focus on cultural diversity. But when Helle asks about advantages and disadvantages of being a multicultural class, some students express an opinion. Natascha likes the presence of children from all over the world: "I think everybody should have the right to go to school and to learn, no matter where they come from, what color they are or language they speak. Everybody should have the right to go to school and get on with their lives. Cultural differences do not mean anything to me." When asked of advantages she says that it is nice to hear about different cultures, for instance how they celebrate Christmas in Vietnam. Tanja thinks that it is exciting with different cultures in school. "You can learn from each others' different ways of doing things."

Although the students seem to be very including and emphasize that cultural diversity "doesn't matter," it is interesting to observe how their statements often correspond with an "us-and-them" discourse:

Gro: It is not really important. We just talk to them.

Hilda: Maybe their opinions are different, but that doesn't matter. People are just not the same.

Helle: Do you see any advantages by different cultures being represented in your class?

Gro: You could probably change your mind about things.

Actually, Hilda herself is from Iceland and strictly taken not Danish, but obviously she does not include herself in the "them" category.

Ida says: "It is not important, I think. They can do the same things as we can do, they are going to learn the same things that we are going to learn, and they more or less use their spare time the way we do." She does not think that cultural differences are used in school, but they have touched upon different issues, "where they told us about their country and we tasted some of their food and things like that."

Lisa says: "I also count the foreign children for Danish, which means, I don't think you should discriminate against them. You can learn about their countries." But according to the students, multiculturality is not much in focus either during the lessons or in the breaks. Minh puts it this way: "It is not important to me. We are all human beings, aren't we?"

The students are well aware of cultural differences, but they seem to agree that to take specific notice is not important. Directly addressed on this subject, however, they can only come to think of a very few examples of "learning from diversity." This indicates that they represent the "sameness" approach in the Danish public discourse about integration, which implies making "them" just like "us" (Alrø, Skovsmose, & Valero, 2005, p. 1146). This is an important observation in order to interpret what is happening in the classroom: in particular, the phenomenon that diversity is not used as a resource for teaching and learning. Students' identities and cultural diversities are important as the students' approaches to learning reflect how they identify themselves in a cultural setting. And as we have seen, one possibility is to try to neglect cultural differences.

TEACHERS' PERSPECTIVES AND PRIORITIES

When asked about their guesses for what it would be likely for students to get engaged with in the future, the teachers had very clear ideas about what could be possible for all students. Having been their mathematics teachers for three years in a row, the teachers thought they knew well about the students' strengths and weaknesses, as well as how they could match their own personal wishes. Few of the students seem to have clear outstanding capacities.

> The boy is very smart and he could do almost whatever he likes. He should certainly continue with high school.

> She has a very high moral of labor; she performs over the mean, bright, and dares to open her mouth and express her opinion. She is nice and it is good to have her in my class. She comes form a solid, well-rooted family, and she will follow high school.

Many others the teachers considered to be capable, but it was less clear to foresee in which direction they could go. Some students were said to have good capacities, but had other obstacles against them:

> She is a middle student; very quiet. It's a pity she is so quiet. She is very creative and writes very well. But she is not so much engaged with mathematics as she could.

And some other students are simply seen to have difficulties going through school:

> She is a bit silly, but she is always happy. We have agreed with her mother that when she leaves 9th grade the girl should master the four operations. Probably we won't get there. Still she works persistently.

Given the organization of the Danish primary and lower secondary school, the mathematics teachers' views and priorities are not the only important factor in influencing students' views about themselves and their futures. While the mathematics teacher appreciated Minh's ability in mathematics and thought he could almost study whatever he wanted, Minh's Danish teacher saw obstacles for him. Having been the class teacher for the boy all the way from 1st grade to 9th grade, she felt she knew well the boy's strengths and weaknesses. He was clever at maths; that was true. But it was difficult to know whether Minh's expressed wish of studying medicine was the boy's real wish or his parents' desire for him. (At the time Helle met Minh, he wanted to become a banker or real state agent; but when Paola met Minh he expressed hesitation. Probably he should study medicine.) Advising him—and his parents—about how to continue with his studies after compulsory school was a complicated issue. The Danish teacher said:

> So when he shows hesitation, I think: If he takes an education in commerce, a three-year study, he will be in school three years and one in practical training in a bank, that is something he can complete. Then he can build on top of that if he wishes. But if we push him to go to high school and get a certification in mathematics, and later on six or seven years at the university, then I become more hesitant about whether he will be able to cope with that and succeed in finishing. And then he would drop out of school and be left without any education at all. I would worry about him not having any certificate.

But why would he drop out? In reality, Minh's weakness was his lack of mastery of the Danish language. "He is not very articulated, he . . . hmm . . . he is not very good at formulating himself in written form. He is not a catastrophe, but he is not good. We have given him special support in the past but it has not really helped much," the Danish teacher says. In general, the Danish teacher thinks that it is difficult for students who do not speak Danish at home to get acquainted with many words and with the right syntactic structure of Danish, particularly if they are not fluent in oral Danish or if they do not communicate in writing at home. And even if they did, their parents cannot really correct them. Minh is not an exception. "Students can get better; Minh has in fact improved. But he will never have a fully developed language like many others in the classroom," the Danish teacher adds. In a school such as Mælkevejen Skole, such a problem is common to many "bilingual students," as well as to many students coming from "weak homes" or "homes without many resources," according to the Danish teacher.

The Mathematical Content

Many students refer to shopping when asked about their use of mathematics in their daily lives. Some of them also refer to weighing and measuring while cooking and baking. Quite a few students, though, state that they only use mathematics when doing their homework.

> **Helle:** Do you use mathematics outside school?
> **Thao:** No, what am I going to use that for?

Many students mention that they find mathematics interesting when it is easy (i.e., when they can figure it out), but boring when it is too difficult. Some students find mathematics rather difficult. To Gro and Hilda, a good mathematics lesson is when they understand what is going on. But unfortunately these girls do not like mathematics—they do not understand what is going on! "Pythagoras wasn't that difficult," Gro says. Addition, subtraction, multiplication, and division are also no problem. But calculating percentage "is just not my cup of tea," Hilda says. She thinks, though, that it is very important to learn it. Calculating 25% or 50% of something goes quite all right, "but 65%!!"

The two of them agree that mathematics is very important in your everyday life. For instance, you will have to know "if you have enough money to go to the movies." And when you grow up you will have to think about "going to work, earning money, paying rent, and all that stuff." Both girls want to become agricultural technicians, so they have to become good at biology, mathematics, and physics, they say.

Tanja does not like mathematics, but actually she does not know why. "Maybe it is because I like to talk a lot, and mathematics is not a talk subject." She thinks she is doing OK in mathematics, but she spends a lot more time on it than the other students, she says. She uses mathematics in her everyday life, especially if she is going to buy things—"for instance if there is a sale." Some professions might need a lot of mathematical knowledge, she says, but she is not interested in such professions.

Peter is fond of mathematics; especially he likes big exercises with lots of calculations that require his thinking. He does not like doing exercises from the textbook. He likes to work outside the classroom, for instance, when they measure things and go back and calculate. He is very good at mathematics, he says. "Sometimes I just feel like making an equation," he says with at smile. He is obviously fascinated by the aesthetics of mathematics.

Lisa thinks that mathematics is very useful in your everyday life. "Financial things at home etc., it has to go around." So she is well aware that is takes mathematical skills to run a home.

Mathematics is important in order to get admission to a higher education. Many students stress this point when asked about the importance of learning mathematics. For those students, however, it is not mathematics per se, but mathematics and good marks as a ticket to the education they want. Razia is not sure what it would take to become a nurse, but she suggests biology, "and then you use mathematics, I think. You use mathematics almost for everything." Razia is not able to point to any mathematical content of a nurse's job. Thus, mathematics per se plays an inferior role in Razia's foregrounds. But being good at mathematics and doing well in school are certainly important parts of her educational plans.

Annie agrees. She wants to be good at mathematics because "almost everything has to do with mathematics, no matter if you want to be a cook or a designer."

Michael wants to become an architect. He likes to draw straight lines, and he is fascinated by the thought that the durability of a house depends on the drawing:

Helle: How do you manage to find out how to draw such drawings?
Michael: It is . . . it is degrees and scales and things like that.
Helle: So you will have to calculate a lot of things?
Michael: Yes, you need some mathematics.

This is Michael's motivation for learning mathematics. He can manage almost everything in mathematics, but sometimes he loses energy—for example, when doing equations. He is not sure if an architect uses equations, but "it is good to manage in mathematics," he states. Michael's favorite exercises are scales and drawings—for instance, in a system of coordinates. "I think it is fun to calculate while drawing," he says.

Ahmed claims that he wants to be a programmer: "In the future everything will have to do with computers," he says. He thinks that mathematics is easy and that he is good at it, but he is not so fond of it, although he knows that he is going to use a lot of mathematics to become a programmer:

Helle: But you said you don't like mathematics.
Ahmed: No, but . . . that's life.

It seems obvious that ideas about mathematics make up part of students' motives for learning mathematics. However, the students' ideas about how mathematics might relate to their aspirations are rather confusing. The value of mathematics operates in an instrumental form, not as a motive based on an insight about mathematics gained as part of the mathematics lessons.

Tools

Only a few students touch upon the computer as a tool in mathematics education, and those who do are not really interested.

Ryan and Kim use a lot of their spare time sitting at the computer. They have already made their own web pages. In school, however, they do not use computers a lot. "In mathematics we have just used the computer once, I think," Kim says. They do not really know if they use mathematics when using the computer—well, perhaps in strategy games they have to estimate a few things, but that is nothing much, they say. But they agree that they learn a lot of English by playing computer games.

Ida and Hannah agree that computers are not important in mathematics education. "They can do nothing more than a calculator," they claim. "You do not make any progress in mathematics by using a computer . . . you don't gain more knowledge . . . it is just a tool." Johnny agrees: "The calculator does the job. It is more important to use your head." Lisa adds to that: "You can only use Excel. You cannot hand in your homework on a computer for instance."

That the students do not see computers as being of any significance reflects at least two phenomena. First, that the students have not previously experienced what it could mean to use a mathematical software. The only use they have made of the computer is the use of spread sheet, not for making any mathematical experimentation, but just as a means of solving some given exercises.

During the project Math Mornings, Ole brings a computer into the classroom. The computer has a program by means of which one can make sketches of mathematical graphs. The students observe (certainly helped by the teacher) that one could represent the use of water for a morning shower as a linear graph. The function:

$$f(x) = ax + b$$

is then interpreted in the following way. The value x signifies the time (measured in minutes) one stays in the shower. The $f(x)$ measures the amount of water measured in liters. The parameter b refers to the amount of water that is used while you are waiting for the water to get warm enough before you enter. And the parameter a reflects how much you open for the water. To Ole's surprise, the students are not particularly interested in working with the computer representation of their morning shower function. Maybe they are just not used to that kind of investigations in mathematics.

A second phenomenon is that the sufficiency of tools seems to be taken as a given. Just as the adequacy of colors, papers, scissors, and whatever material needed for producing the posters is taken as given. Naturally, the one-time camera is something extra, but not an extra that brings about any particular surprise, since most of the students already have a camera function in their mobile phones. In a Danish classroom, materials do usually not seem to be a limitation. The students also have access to a lab where they could explore the internet and write and print their comments.

Classroom Interaction

The two classes have very different patterns of interaction and communication between teacher and students. The more experienced teacher managed classroom communication following a pattern of teacher commands or questions addressed to the whole class or to particular students that were followed by student(s)' answers to the teachers' requests. He selected topics from the textbook and made short presentations of key concepts, and gave students different sets of tasks—either from the textbook, from other textbooks, or from samples of 9th grade exam sets. The students had to work mainly individually on the tasks, although they were allowed to consult to each other, normally to the students sitting nearby. Even if the classroom was organized in rows of student desks, students tended to cluster desks together to be close to those mates of their predilection. The class was disciplined, organized, and quiet.

It seemed as if the class was clustered around different groupings of students. On the right side of the room mainly girls were clustered together, while boys sat to the left flank of the room. Within gendered groups there also seem to be different categories: the popular girls, the two girls from Greenland, the cool girls, the not-so-engaged girls, the Danish boys, the two Turkish boys, and some lonely rangers without a clear grouping who were sitting in the vicinity of different groups but within the same gender area.

The class run by the novice teacher followed a quite different interaction pattern, where students worked mainly in groups based on the tasks that the teacher had selected for them from newer textbooks, and open activities that he himself had constructed. He introduced shortly the tasks and explained shortly what they were expected to do. Students organized how they wanted: alone, or together with their favorite mates. The teacher went around helping students who signaled a need for help by raising their hands or calling for the teacher. The room seemed more chaotic and noisy, and students did not always keep on task.

Similarly to the other class, the students' groupings were also visible, though gender clustering was not as marked as in the other class. The most engaged girls sat together at the right back of the class. The quiet girls tended to sit at the front middle of the room. A group of middle engaged boys and girls sat in the middle of the room. The not-so-engaged boys sat at the left middle of the room. They moved around and generated conversations among groupings.

In the interactions between teachers and students and among students, it was not possible to observe patterns of communications where one immediately became aware of cultural differences playing a role. Maybe a careful investigation of communication inside groupings could reveal some other aspects, but at the communicative surface, cultural differences did not manifest themselves so clearly as other studies have documented (e.g., Gorgorió, Planas & Vilella, 2002). However, teachers were aware that sometimes their choice of task for the students might engage students differently and, therefore, provide more opportunities for some students than others to participate. Talking about that, the more experienced teacher mentions an episode when he proposed a task that built on Greenland. He observed clearly that the Greenland girls, who were normally quiet and a bit retracted from participating in the class, were more active and actually could provide information to the rest of the class. Some other times, tasks building around football may interest boys more than girls; but tasks building on horse caring and racing favors girls. "I can't always engage them all; therefore I try to select a variety of topics," he says.

When observing and talking with the groups during the Math Mornings, Ole did not become aware of any situation where cultural diversities surfaced. One seems tempted to conclude that cultural diversities are considered to have no significance for the classroom communication—and as a consequence, no significance for the learning of mathematics. One reason for this could be found in the procedures at Mælkevejens skole for integrating students into the regular classes. New students who do not have Danish as mother tongue are integrated through special classes, where they get the opportunity to learn Danish. Only after a period in these special classes do they enter the regular classes.

Parents and Family

Minh does not think that his parents think he is good at mathematics. They do not tell him so, but he does not feel good enough and he seems to be very sorry about that. Minh's parents are very ambitious on behalf of Minh. The father owns a grill-bar, and he wants his son to get a well-paid job, so he wants him to become a doctor. Minh himself would rather get a job that he is interested in and happy about, but he knows that his father will get disappointed if he chooses "only" to become a banker.

Paul wants to become a car mechanic or an engineer—or, rather, his mother wants him to become an engineer. He himself has no ambitions of becoming an engineer. He likes cars a lot and would rather like to become a mechanic.

Hannah's grandmother always tells her to become an inventor. And as the predictions of the grandmother have more or less come true as for the rest of the family, Hannah has some confidence in her grandmother's words. Ann thinks about becoming an engineer, something about research and professorship: "Something big so that you can travel and meet with many adventures." Both of their parents happen to be engineers and "mathematics nerds."

Razia is well aware of the importance for a Muslim woman of being educated. Her parents have a big say in this question. Her father has directly told her to get an education. Her father's opinion must be of big importance, since Razia claims that in her culture it is not possible to say no to one's parents. She wants to become a nurse: "My father is a nurse, and therefore he can tell me many things." More important, maybe, is her perception of her mother's situation. In some sense, she does really not want to end up like her mother:

> My mother, she has no education. So she says, "I do everything for you. You are having an education. You should not become like me."...Because in Iraq, if you don't get an education, then you marry early. My mother got married when she was 14 years old. And therefore she is afraid.

In Iraq it is rather difficult for a woman to get educated, especially in the part of the country where she lived. So she is quite clear on the fact that she wants to get educated in Denmark before she goes back to Iraq in the future.

In different ways, parents seem to have a major influence on the students' learning motives. The students' ways of engaging in their schoolwork might be in harmony somewhat with their parents' priorities, or it might represent a reaction. But it is difficult to ignore the parents' perspectives as making part of an interpretation of the students' attitudes towards learning.

Friends and Peers

During the inter-views, the students were asked if they like to go to school and what they like or dislike about it. "I like to go to school because of my friends and because it is important to learn in order to get on in life," Natascha claims. Her point coins the most frequent answer among these 15-year-olds: The most important thing about going to school is to get friends and to meet with your friends. Therefore, the breaks are considered to be most important. Like Natascha, some other students also refer to the importance of learning, but the role of friends seems to occur immediately to the students' minds when asked about school. On the other hand, friends can also be an important explanation for students who do not like school. Fatima's answer to the question why she does not like school is:

Fatima: I am not sure, but I haven't found the right friends, yet.
Helle: Do you have some friends in your class?
Fatima: Yes, Sofie and Janneth—I sit next to them. But right now, we are on bad terms.

Some claims about going to school may relate to the fact that we have to do with teenagers. "I don't like going to school, especially not in the mornings. I am all too tired. But I like being together with my friends," Peter says. Johnny confirms this attitude. He thinks the worst thing about going to school is that it takes so long. "You have to be there all day. There are lots of other things you could spend your time on."

Rune: It is not that it is too boring, but it IS boring. But it does not have to be fun, though. It is OK. I do not get teased. There are no problems in our class, so . . .
Helle. No problems? This sounds good. But you think it is boring. How?
Rune: Well, ehh, I will have to admit that I am one of the skilled students. And sometimes we are going to listen to teacher explanations that we have already heard before. Then some of us get bored when we wait for the teacher to get finished.
Helle: I see. Well, so there is too little challenge, is that it?
Rune: Yes!

Many students claim that they find school boring. As for Rune this seems to be due to too little educational challenge compared to his expectations and skills. But there are also other reasons for students to get bored in school. Tanja could certainly find a more interesting place to spend her time. School is just boring, especially homework, but also when the teacher "talks and talks and talks." On the other hand this condition is seen as a necessity for their future life: "otherwise you would not know anything at all about what is going on in the world."

Paul has been into bad company, and he works consciously on getting rid of this. He is interested in playing football and he plays at the picked team level in a biggish club where he practices four times a week.

Alexander has a special story to tell as he was bullied by a clique of four boys during three years from the 3rd to the 6th grade. "Well, at first, at the beginning, I didn't do anything about it. And then I told my parents and my teacher." By now the situation has changed. Alexander does not get bullied anymore, but actually, he has no friends in class. Not to make easy conclusions, but the inter-viewer got rather shaken to hear about his future plans and wishes. He wants to become either a commando or a fighter pilot.

Friends can have good as well as bad influence on students' preferences. But the solid stress of the importance of being with friends when going to school also indicates an important influence of friends for students' learning motives.

Public Discourses

A couple of the students touch upon elements of the dominant public discourse on immigrants and cultural diversity in Denmark.

> Susie: We learn about other cultures, but we do not look at them differently. For instance, we have heard about the Ramadan, and we have had food from foreign countries. This is very good, so that you not believe it is like in the news . . . that they are just somebody who kills other people or things like that. It is good to know how they really are. The picture in the news has nothing to do with the children we know.

Susie argues against the media discourse of immigrants as mostly portraying minority youth as being criminals. The children in her class are not like that.

Another strong issue in the public discourse is the "headscarf-girls." Razia, who wears a headscarf for religious reasons, comments on that: "There are many, very many who don't like headscarves . . . it is about racism and so on." Razia transfers her claim of Danes not liking her headscarf into a more generalized interpretation of racism.

These examples indicate that prejudice as expressed in the public discourse might have serious consequences, especially for immigrant students' learning motives.

Some of the teachers in the school also express opinions, which correspond with public ways of talking about immigrants in the media. The example of the Danish teacher when commenting on Minh's dilemma about his future studies clearly shows the point also raised by the PISA ethnic report cited at the beginning of this paper (Rockwool fondens forskningsenhed, 2007b): Some immigrant families do not have the cultural capital needed to succeed in the Danish school and that is expressed in the low type of linguistic competence these students are able to gain. The teacher also mentions the very high expectations that immigrant parents often have for their children. They are sometimes even unrealistic, and certainly higher than the expectations of Danish parents. This resonates with what Baber (2007) found to be the case about how some Pakistani families perceive their belonging to Danish society: If an immigrant wants to be accepted and be successful in Denmark, it is not enough to be good; one has to be much better than Danes.

THE IMPORTANCE OF MULTI-DIMENSIONAL ANALYSIS

In the previous section, we addressed each of the dimensions of the learning landscape. In this section we will address a couple of issues moving across the dimensions. In fact, our point is that the dimensions should not be addressed in isolation, but as dimensions integrated in a learning landscape. Thus, the flower in Figure 47.1 is not only composed of the individual petals but of their interconnections. In order to illustrate the connectedness of the petals, we comment on two examples from our study. Through these examples we want to emphasize that our understanding of learning moves across the dimensions of the learning landscape.

The Importance of Mathematics in Students' Foregrounds

During the Math Morning experiment, some students did not show much enthusiasm. They were present and somehow active, but in a rather reserved way. One way of trying to motivate students is to relate what they are doing to their foregrounds and to their background. After the students had identified a range of issues using their camera, they had to decide what issues to explore further. One difficulty for the students was to conceptualize what it could mean to explore a topic. And what topic to explore? What could be experienced as relevant? Here Ole found it obvious to ask what the students intended to do later. What were they interested in studying? Several answers, like becoming a designer or an architect, could then become related to some of the issues that could be explored. In these conversations, however, Ole did not find it proper to ask about the students' cultural background. A conversation about what could be the meaning of an activity never turned in the direction of background in order to be substantiated.

It has been emphasized in some research that the students' background is essential for the students' construction of meaning with respect to their activities in the classroom. It might well be so. However, our study also points in a different direction—namely, that meaning production can be related to the foregrounds of the students. It was not background but rather foregrounds that were used as motivating references for the students. Foregrounds were used in providing meaning to the classroom activities. It appears that the students' foregrounds and the meaning they may associate to the different activities in the classroom can be closely related.

Much analysis of mathematics education has pointed out that mathematics makes up part of many different practices: from shopping and a variety of work situations (accounting, banking, money transfer) to more complex tasks (architecture, design, management, decision making). In this way identifying mathematics everywhere presupposes that one develops a broader interpretation of what to refer to as mathematics. It comes to include formal reasoning in many situations. According to our investigations, such a broad interpretation of mathematics is not part of the school mathematics tradition, as this was acted out in the educational practice we observed.

In the inter-views only a few students pointed to the relevance of specific mathematical content for their future life. Many students, though, reproduced a common construction in the political discourse, that mathematics per se is an immanent and absolutely necessary good. Only a few expressed a direct connection between school subjects and relevant professional skills. This was true for Minh, who wants to become a banker or work in real estate, as he "loves numbers." Actually, Minh's father wants him to become a doctor, and the mathematics teacher is convinced that he would succeed. But the Danish teacher claims that Minh will never come to master the Danish language well enough to go through the study of medicine. This judgment seems to become an obstacle for Minh to fulfill his father's wishes and perhaps his own as well.

Michael, who wants to become an architect, finds mathematical drawings very meaningful; Ryan and Kim, who are going to be computer scientists, know that they are going to use mathematics, but they attach no specific meaning to learning mathematics. The view is more general—"You use mathematics for almost everything"—and external—"It is important to get good marks in mathematics."

This means that the students' motives for learning mathematics easily become *instrumental*. The importance of getting good marks is the only plausible value to assign to the school subject, if no everyday practice including mathematics is explored. As mentioned, Laura wants to become a dentist, and Olivia wants to be a psychologist, but none of them can imagine anything about how mathematics could be part of this, even though they assume that mathematics could be relevant. The possible relevance of mathematics is well-hidden in the school mathematics tradition.

As a consequence of this, it might become difficult for some students to see that mathematics could play any significant role in their future. If meaning for students somehow is related to their foregrounds, then instrumentalism could become a strict implication of the fact that mathematics appears as a clean and isolated subject.[6] When asked about it, they might acknowledge such relevance, but apparently they have not experienced transitions between classroom practices including mathematics and out-of-school practices including mathematics. Still, when asked, they are aware of such possible *transitions*. The school mathematics tradition seems to obstruct the transition between the in-school practices of mathematics (solving exercises, finding the right answer, etc.) and out-of-school practices including mathematics.

Diversities Are Ignored

When the students have left the classroom, there is nothing left that reflects that we are working with a multicultural setting. No pictures on the notice board reveal that there are students with different cultural backgrounds. We have already observed that the patterns of communication do not reveal cultural diversities, and taking a look at the physical aspects of the classroom, there is no sign of cultural diversity either. One could think of diversities as a resource for learning and understanding. One could also consider diversities as a cause of possible troubles and difficulties. We did not see diversities made explicit in any way. Our overall impression is that diversities were both hidden and ignored.

The students in our study claim that cultural differences are not important, although when asked, they reproduce the "them-and-us" discourse, which is a common construction in the Danish public discourse. One implication of this discourse is that integration of newcomers is understood as an assimilation process that implies making "them" just like "us." In this discourse, successful integration means no visible cultural diversity. This can be done by ignoring diversities and focusing on sameness, and of course from a Danish perspective. The potentials of diversity are not considered relevant, neither for society in general nor for learning processes in school.

Why are diversities ignored? One may find one set of explanations in the public discourse in Denmark. Here differences are not always presented in positive terms, or as openings for educational possibilities. The public discourses where cultural differences are presented in unquestioned positive terms are found in advertisements for tourism and in the picturesque description of travel experiences. However, when it comes to the discussion of immigrants in Denmark, most often the public discourse includes the assumption that differences have to be eliminated; in particular, students of immigrant families have to be made Danish as quickly as possible, as only when similarities have been established can an efficient educational program be carried out. This also implies that the success of the integration program of Mælkevejen Skole could be judged by the fact that the teachers, when planning and conducting classroom activities, can in fact ignore diversities. When the students did emphasize that they do not think of differences, they refer to the same assumption: When differences can be ignored, the integration program has been successful.

However, even if there are no differences to be observed directly in the classroom, one can find differences when one considers other aspects of the situation. One of the principal observations made by Sikunder Ali Baber (2007) in his study of students in Denmark from immigrant families from Pakistan is that they express what can be referred to as a polarization of their foregrounds. They find it important to perform better than Danish students in order to be able to obtain average possibilities in Danish society. If an immigrant student's school performances are average or below average, he or she will have only few future possibilities in Denmark. One can only expect to get the poorest jobs. Baber's point is not whether or not these observations reflect any actual features of the Danish labor market (this could be clarified only through an elaborate investigation of this market). The point is that this polarization is expected by the students as well as by their parents. These foregrounds frame the students' school activities. This could result in different strategies: It could mean that students put an extra effort in school work; or they might have an attitude of resignation; they might also pretend that they have high ambitions, and so on. The point is that when one tries to understand the students' motives for learning, then one has to address the complexities of the learning situation. One has to establish an analysis that moves across different dimensions of the learning landscape.

DIMENSIONS OF A THEORY OF LEARNING?

Is the landscape of investigation too extensive? Does it include too many dimensions? Or too few? When theorizing, one must try to manage complexities, but a theorizing could drown in complexities. We find that the number of dimensions we have suggested expands considerably the scope for investigation, which has characterized much theory building within mathematics education research. Our point is that learning mathematics in a multicultural setting cannot be understood by looking solely at classroom activities. At the same time, we hope that the landscape is not more complex than we, ourselves, somehow are able to find our way around.

The landscape is developed with a particular reference to learning mathematics in a multicultural setting. However, it is our hope that it could serve as a way of structuring an understanding of learning in many other situations than the one we have addressed. We find that the dimensions we have presented could turn into *dimensions of a theory of learning.* They represent a way of seeing learning, and seeing learning in contexts (and *contexts* is plural).

We can see the dimensions as representing different contexts of learning:

1. Learning is located in a context of possibilities (or lack of possibilities) and the learner's way of seeing these possibilities. Therefore we included *foregrounds* as a dimension of the learning landscape.
2. Learners are persons and not, say, empty vessels; therefore we include the dimension of *identity.*
3. The processes of learning are framed by the *teachers' discourses and priorities.*
4. Learning means the learning of something, and in this case we concentrate on *mathematics.*
5. Learning is a tool-driven phenomenon. One can use paper, pencil, compass. One can use any kind of information and communication technologies. Some contexts of learning are affluent; in some cases poverty sets the condition. Therefore we need the dimension of *tool.*
6. Learning takes place in a context of communication, and we operate with *classroom communication* as a dimension.
7. Learning relates not only to foreground but also to background, and we have the *family* as a dimension.
8. Certainly, *friends* make part of the learning context.
9. And so does the *public discourse.*

So, one can see the learning landscape as representing the variety of contexts in which a process of learning takes place.

We certainly have to be careful not to claim too much about the adequacy of a learning landscape as providing an adequate framing for a theory of learning. However, any such adequacy depends on what one sees as the task of theorizing.

According to positivism, and to logical positivism and neo-positivism in particular, the task of theory is to make *explanations*, and making explanations is identified by making predictions. Thus, a theory of learning could make predictions about possible learning gains, obtained through a particular educational approach. One can also pursue the hermeneutic trend according to which the task of a theory is to make *interpretations.* A theory of learning, therefore, has to facilitate interpretations of learning processes. If one wants to follow a critical trend, one can claim that the task of a theory is to make *changes*, or at least to make changes possible. In our conception, theorizing with respect to learning, and also with respect to learning mathematics, is to make interpretations that facilitate the identification of new educational possibilities.

In the *Pisa Ethnic* and the OECD report, we find conclusions that seem to correlate the weak performances of students from the so-called ethnic families with features of their family background. Such a line of explanation has strong implications. In particular, this means that reasons for the performances are not to be found in school or in society. This approach of "personalizing" the causes for certain forms of performances reminds us of patterns of explanations of black students' weak school performances, as observed in what can be referred to as "white research in black education" during the apartheid era in South Africa.[7] There were many weak performances that needed "explanation"—an explanation adjusted to the apartheid framework. Thus, one line of explanation draws attention to the dominant father of the black family, who was assumed to obstruct the creativity of the child. In this way the explanation, pointing to features of the background of the students, turned into an explaining away of the devastating impact the apartheid regime had on the black students' opportunities for learning and for their motives for learning. Thus, a huge range of further studies was simply not accessible to black students—any kind of engineering, for instance. "White research in black education" exercised a particular perspective on the situation, and it placed the causes for the weak performances of black students as having nothing to do with the apartheid society. In this way, "white research in black education" provides an absolution of the apartheid regime with respect to any impact on the educational system. It is important to be careful with such attempts to provide explanations of educational failures. We find that it is important to provide a comprehensive interpretation of the situation, and this is what we suggest through the introduction of the notion of learning landscape. We find that it would provide a different way of looking at education during the apartheid era. It also provides comprehensive ways of interpreting students' performances in general.

In the first section, "Conceptual Dimensions of a Learning Landscape," we emphasized that the conceptual framework of a study also helps to construct what is being studied. Thus, the students' foregrounds are not an entity that exists independently of the conception of students' foregrounds. We find it important to develop a conceptual landscape that makes research sensitive to the complexity of the learning situation, including complexities caused by cultural, social, political, and economic structures within which learning is situated.

The learning landscape is something we are researching, but it is also constituted through the way we are researching. Thus the dimensions of the learning landscape represent our way of looking at the landscape, and at the same time the dimensions constitute the landscape. Therefore, it is important to return to the question of whether we could have provided a different dimensioning of the landscape. Certainly, we could have done so. The argument for choosing one and not a different dimensioning cannot be justified in a direct way. However, there might be an indirect justification that has to do with the observations, which a particular dimensioning of a learning landscape makes possible. It has to do with the insight that might be crystallized. So our "justification" of dimension is to be found in the sections where we present some of our observations. Does a different dimensioning provide other observations? We certainly think so. Even though one perspective might reveal *something*, it will never reveal *everything*. Other perspectives are most welcome.

ACKNOWLEDGEMENTS

This chapter makes up part of the research project "Learning from Diversity," funded by the Danish Research Council for Humanities and Aalborg University.

NOTES

1. We are inspired in the concept of network of mathematics education practices (Valero, 2007) referring to the whole series of practices, in multiple sites and at multiple levels of action, where the social meaning of the learning and teaching of mathematics is constituted.
2. It must be emphasized that the dimensions are not proper dimensions in a mathematical sense of the word as they are interacting and mutually influencing each other. Thus, the public discourse has a structuring effect on, for instance, students' foregrounds, students' identities, teachers' perspectives, etc.
3. We prefer the notion of "inter-view" instead of "interview" in order to emphasize the character of "looking at something together" in the conversation (Kvale, 1996).
4. The idea of "Math Mornings" has been invented by Michael Skånstrøm. It is presented in Skovsmose and Skånstrøm (2002) and elaborated in greater details in Blomhøj and Skånstrøm (2006).
5. All names in this article except for the author names are pseudonyms.
6. The notion of instrumentalism has been discussed by Mellin-Olsen (1981).
7. For a critical discussion of educational research in Apartheid South Africa, see Khuzwayo (2000).

REFERENCES

Abreu, G. (2005). Cultural identities in the multiethnic mathematical classroom. In M. Bosch (Ed.), *Proceedings of the Fourth Congress of the European Society for Research in Mathematics Education* (pp. 1131–1140). Barcelona: ERME—Universitat Ramon Llull. Retrieved on April 9th, 2008, from http://ermeweb.free.fr/CERME4/

Abreu, G. d., César, M., Gorgorió, N., & Valero, P. (2007). Issues and challenges in researching mathematics education in multicultural settings. In M. Bosch (Ed.), *Proceedings of the IV Congress of the European Society for Research in Mathematics Education* (pp. 1125–1130). Barcelona: Universitat Ramon Llull—ERME.

Alrø, H., & Skovsmose, O. (2002). *Dialogue and learning in mathematics education: Intention, reflection, critique.* Dordrecht: Kluwer.

Alrø, H., Skovsmose, O., & Valero, P. (2003). Communication, conflict and mathematics education in the multicultural classroom. In M. A. Mariotti (Ed.), *Proceedings of the Third Conference of the European Society for Research in Mathematics Education.* Bellaria, Italy: Pisa University—ERME. Retrieved April 9, 2008, from http://ermeweb.free.fr/CERME3/tableofcontents_cerme3.html

Alrø, H., Skovsmose, O., & Valero, P. (2005). Culture, diversity and conflict in landscapes of mathematics learning. In M. Bosch (Ed.), *Proceedings of the Fourth Congress of the European Society for Research in Mathematics Education* (pp. 1141–1152). Barcelona: ERME—Universitat Ramon Llull. Retrieved April 9, 2008, from http://ermeweb.free.fr/CERME4/.

Alrø, H., Skovsmose, O., & Valero, P. (2009). Inter-viewing foregrounds. In M. César, & K. Kumpulainen (Eds.), *Social interactions in multicultural settings* (pp. 13–37). Leuven, Belgium: EARLI Books. Retrieved June 2010 from http://ermeweb.free.fr/CERME3/tableofcontents_cerme3.html

Andrews, J., Yee, W. C., Greenhough, P., Hughes, M., & Winter, J. (2005). Teachers' funds of knowledge and the teaching and learning of mathematics in multi-ethnic primary schools: Two teachers' views of linking home and school. *Zentralblatt für Didaktik der Mathematik (ZDM) International Reviews on Mathematical Education, 37*(2), 71–80.

Austin, J. L. (1962). *How to do things with words.* Oxford: Oxford University Press.

Baber, S. (2007). *Interplay of citizenship, education and mathematics: Formation of foregrounds of Pakistani immigrants in Denmark.* Unpublished PhD thesis, Aalborg University, Aalborg.

Bishop, A. J. (2002). The transition experience of immigrant secondary school students: Dilemmas and decisions. In G. de Abreu, A. J. Bishop, & N. C. Presmeg (Eds.), *Transitions between contexts of mathematical practices* (pp. 23–52). Dordrecht: Kluwer.

Blomhøj, M., & Skånstrøm, M. (2006). Matematik Morgener—Matematisk modellering i praksis. In O. Skovsmose & M. Blomhøj (Eds.), *Kunne det tænkes?* (pp. 7–23). Copenhagen: Malling Beck.

Borba, M. C., & Villarreal, M. E. (2005). *Humans-with-media and the reorganization of mathematical thinking: Information and communication technologies, modeling, visualization, and experimentation.* New York: Springer.

Carspecken, P. F. (1996). *Critical ethnography in eductaional research.* London: Routledge.

César, M., & Favilli, F. (2005). Diversity seen through teachers' eyes: Discourses about multicultural classes. In M. Bosch (Ed.), *Proceedings of the Fourth Congress of the European Society for Research in Mathematics Education* (pp. 1153–1164). Barcelona: ERME—Universitat Ramon Llull. Retrieved on April 9th, 2008, from http://ermeweb.free.fr/CERME4/

Chassapis, D., & Chatzivasileiou, E. (2008). Socio-cultural influences on children's conceptions of chance and probability. In J. F. Matos, P. Valero, & K. Yasukawa (Eds.), *Proceedings of the Fifth International Mathematics Education and Society Conference* (pp. 197–206). Lisbon: Centro de Investigaçao em Educaçao, Universidade de Lisboa—Department of Education, Learning and Philosophy, Aalborg University. Retrieved on April 9th, 2008, from http://www.mes5.learning.aau.dk/

Chronaki, A. (2005). Learning about "learning identities" in the school arithmetic practice: The experience of two young minority Gypsy girls in the Greek context of education. *European Journal of Psychology of Education, XX*(1), 61–74.

Civil, M., Planas, N., & Quintos, B. (2005). Immigrant parents' perspectives on their children's mathematics education. *Zentralblatt für Didaktik der Mathematik (ZDM) International Reviews on Mathematical Education, 37*(2), 81–89.

Egelund, N., & Tranæs, T. (Eds.). (2007). *PISA Etnisk 2005. Kompetencer hos danske og etniske elever i 9. klasser i Danmark 2005.* Odense: Rockwool Fondens Forskningsenhed og Syd dansk Universitetsforlag.

Gorgorió, N., & Planas, N. (2005). Cultural distance and identities-in-construction within the multicultural mathematics classroom. *Zentralblatt für Didaktik der Mathematik (ZDM) International Reviews on Mathematical Education, 37*(2), 64–71.

Gorgorió, N., Planas, N., & Vilella, X. (2002). Immigrant children learning mathematics in mainstream schools. In G. de Abreu, A. J. Bishop, & N. C. Presmeg (Eds.), *Transitions between contexts of Mathematical Practices* (pp. 23–52). Dordrecht: Kluwer.

Gutstein, E. (2003). Teaching and learning mathematics for social justice in an urban, Latino school. *Journal for Research in Mathematics Education, 34*(1), 37–73.

Hawighorst, B. (2005). Parents' views on mathematics and the learning of mathematics—An intercultural comparative study. *Zentralblatt für Didaktik der Mathematik (ZDM) International Reviews on Mathematical Education, 37*(2), 90–100.

Kaiser, G. (2003). Learning mathematics within the context of linguistic and cultural diversity—An empirical study. In M. A. Mariotti (Ed.), *Proceedings of the Third Conference of the* European Society for Research in Mathematics Education (CERME 3). Bellaria, Italy: Pisa University—ERME.

Khuzwayo, H. (2000). *Selected views and critical perspectives: An account of mathematics education in South Africa from 1948 to 1994.* Unpublished doctoral dissertation, Aalborg University.

Kvale, S. (1996). *Inter-views: An Introduction to Qualitative Research Interviewing.* Thousand Oaks, CA: Sage Publishers.

Levinson, S. C. (1983). *Pragmatics.* Cambridge: Cambridge University Press.

Martín-Rojo, L. (2003). *¿Asimilar o integrar? Dilemas ante el miltilingüalismo en las aulas.* Madrid: Ministerio de Educación, Cultura y Deporte de Espana.

Mellin-Olsen, S. (1981). Instrumentalism as an educational concept. *Educational Studies in Mathematics, 12,* 351–367.

OECD. (2006). *Where immigrant students succeed—A comparative review of performance and engagement in PISA 2003.* Paris: Author.

Powell, A. (2002). Ethnomathematics and the challenges of racism in mathematics education. In P. Valero, & O. Skovsmose (Eds.), *Proceedings of the Third International Mathematics Education and Society Conference* (pp. 15–28). Copenhagen: Centre for Research in Learning Mathematics, Danish University of Education, Roskilde University and Aalborg University. Retrieved in April 9th 2008, from http://www.mes3.learning.aau.dk/.

Rockwool fondens forskningsenhed. (2007a). Etniske elever gør ikke danske børn dårligere [Electronic Version]. *Nyt fra Rockwool fondens forskningsenhed, Maj 2007,* 20. Retrieved May 2008 from http://www.rff.dk/publikationer/nyhedsbreve/2007.

Rockwool fondens forskningsenhed. (2007b). Rettelser til PISA Etnisk: Indvandrerunge har mere forældrekontakt end antaget [Electronic Version]. *Nyt fra Rockwool fondens*forskningsenhed, September 2007, 4. Retrieved May 2008 from http://www.rff.dk/publikationer/nyhedsbreve/2007.

Searle, J. (1969). *Speech Acts.* Cambridge: Cambridge University Press.

Sfard, A., & Prusak, A. (2006). Telling identities: In search of an analytical tool for investigating learning as a culturally shaped activity, *Educational Researcher, 34*(4), 14–22.

Skovsmose, O. (1994). *Towards a philosophy of critical mathematics education.* Dordrecht: Kluwer.

Skovsmose, O. (2005a). Foregrounds and the politics of learning obstacles. *For the Learning of Mathematics, 25*(1), 4–10.

Skovsmose, O. (2005b). Meaning in mathematics education. In J. Kilpatrick, C. Hoyles, & O. Skovsmose, in collaboration with P. Valero (Eds.), *Meaning in mathematics education* (pp. 85–102). New York: Springer.

Skovsmose, O., & Skånstrøm, M. (2002). Matematik-morgener, *Matematik, 4,* 23–28.

Valero, P. (2004). Postmodernism as an attitude of critique to dominant mathematics education research. In M. Walshaw (Ed.), *Mathematics education within the postmodern* (pp. 35–54). Greenwich, CT: Information Age.

Valero, P. (2007). A socio-political look at equity in the school organization of mathematics education. *Zentralblatt für Didaktik der Mathematik. The Intentional Journal on Mathematics Education, 39*(3), 225–233.

Wenger, E. (1998). *Communities of practices. Learning, meaning, and identity.* Cambridge: Cambridge University.

Wittgenstein, L. (1953). *Philosophical investigations.* Oxford: Basil Blackwell.

Wunderlich, D. (1975). *Linguistische Pragmatik.* Wiesbaden: Athenaion.

48

CRITICAL MATHEMATICS EDUCATION

In Terms of Concerns

Ole Skovsmose

AALBORG UNIVERSITY

Abstract: In this chapter I try to characterize critical mathematics education in terms of five concerns: (1) To recognize the socially critical position of mathematics education. Mathematics education can play very different social roles: It can ensure an adaptation of students to the demand of a neo-liberal knowledge market; it may also provide political empowerment and critical citizenship. (2) To recognize that mathematics education operates in a globalized world and that globalization includes ghettoizing. Much research in mathematics education has concentrated on affluent teaching–learning environments, which has brought about a prototypical bias in theory building. Critical mathematics education tries to counteract this bias and address mathematics education in all its possible variations around the world. (3) To recognize the critical position of mathematics itself. The conception of mathematics as an uncontroversial rationality can be questioned through investigations of mathematics in action: in advanced technological enterprises, in work processes, and in daily life situations. As with any form of action, so also mathematics-based actions are in need of critical reflections. (4) To explore educational possibilities. Such an exploration can be done in cooperative forms and address: cooperation, contexts and resources, landscapes of investigation, and mathemacy. (5) To recognize that critique is an expression of uncertainty. No critical activity can be founded on a solid epistemic foundation.

Keywords: Critical mathematics education, prototypical research, mathematics in action, researching possibilities, critique as uncertainty.

At the beginning of the 1970s I became interested in critical education. At that time, strongly inspired by the students' movement, critical education included a preoccupation with university education. It reacted to the so-called professor-driven curriculum and questioned the claimed neutrality and objectivity of science. Instead, university study was suggested to serve not only academic but also political and social interests. Problem-based and project-organized studies were pointed out as one way of realizing this. The problem should emerge from real-life situations, while the project organization should ensure that students became owners of their learning processes. Almost immediately the idea of critical education spread to all parts of the educational system. Not only university studies, but any form of education could be reorganized, and project work, group work, and thematic approaches were tried out in primary, secondary, and tertiary education.

Initially the critical approach to education concentrated on subjects from the humanities and the social sciences, but soon also mathematics and science education were addressed. In Scandinavia, Germany, and other countries, critical mathematics education developed with reference to issues such as women and mathematics, the role of the teacher, ideology, social experience, project work, and so on (see Volk, 1979, who brought together many of the initial

The First Sourcebook on Nordic Research in Mathematics Education, pages 671–682

contributions to critical mathematics education; see also Damerow, Elwitz, Keitel, & Zimmer, 1974; Mellin-Olsen, 1977).

The theoretical inspiration for critical education comes from different sources. One is found in Critical Theory as elaborated by the Frankfurt School.[1] On this basis, Jürgen Habermas (1971) suggested that sciences were constituted through certain knowledge-guided interests, and that social sciences could be guided by an interest of emancipation. With reference to this suggestion it was straightforward to formulate critical education as being guided by an emancipatory interest (see, for instance, Mollenhauer, 1973). This interest was, still according to Habermas, different from the technical interest that was assumed to guide natural sciences. This observation implied that Critical Theory incorporated a suspicious look at natural sciences and technical rationality as represented by mathematics. How could sciences guided by a technical interest serve any emancipation? From the perspective of Critical Theory, the notion of critical mathematics education came to appear dubious, if not contradictory.

During the second half of 1970s and during the 1980s, I struggled to formulate a conception of what critical mathematics education could mean.[2] A result of this effort was encapsulated in *Towards a Philosophy of Critical Mathematics Education* (Skovsmose, 1994), which was written during the period 1991–1992 and published in 1994. It was a great satisfaction for me to provide this more elaborated presentation of ideas that pointed towards a critical mathematics education.[3]

However, my perspective on critical mathematics education continued to change. In 1994, I became involved in a mathematics education project in South Africa that lasted for about six years. What could a mathematics education for democracy and social justice mean in a situation where the apartheid regime had just been abandoned? It was all too obvious that the notion of critical mathematics education, as developed with primarily European and North American references, had no simple educational validity in this context (see also the discussion in Vithal, 2007). Critical mathematics education had to be reformulated within a broader outlook. In 1994, I visited for the first time the post-graduate program in Mathematics Education in Rio Claro in Brazil. Here I came to understand better what an educational preoccupation with cultural conflicts and diversities could mean for mathematics education.[4] Critical mathematics education was in need of radical reformulations. But what form of reformulation was it in need of?

I will address this question through a discussion of concerns having to do with: (1) the critical position of mathematics education; (2) globalization and ghettoizing; (3) the critical position of mathematics; (4) exploration of educational possibilities; and (5) critique as an expression of uncertainty.[5] In terms of such concerns I will try to characterize critical mathematics education.

FIRST CONCERN: THE CRITICAL POSITION OF MATHEMATICS EDUCATION

Mathematics education can mean many different things to many different people and in many different circumstances. In literature we find descriptions of the mathematics teacher as being a tyrant, whose despotism appears established through the very fact that we have to do with mathematics. There seem to be some elements in being a mathematics teacher that can bring out the most undesirable characteristics of a teacher's personality. Students who are not able to follow the correct mathematical reasoning can be exposed, through a strict questioning by the teacher, as lacking even the most simple thinking capacity.

This negative image of mathematics, so common in literature, can be generalized, and there is no lack of suggestions for how to interpret mathematics education as serving questionable socio-political roles. It has been emphasized that mathematics education serves not only as a gate keeper but also as a gate provider ensuring only a selected few get access to further education. Mathematics education may serve to cultivate a particular form of reasoning, mastery of which may imply that one becomes accommodated to certain forms of social order (see, for instance, Bourdieu, 1996; Neyland, 2007).

Knowledge is now part of the market economy, and in the so-called knowledge society, submerged in a world-wide neoliberal informational economy, particular competencies are of particular value. Mathematics is integrated in many different work practices, and a labor force with mathematics-flavored competencies is very much in demand. As a consequence, mathematics education can help to meet the different demands of today's labor market, and it is possible to identify mathematics education as a functional device that operates according to standards established through dominant economic priorities. Furthermore, mathematics education seems able to provide a unique labeling of individual competencies (see Skovsmose, 2008).

There are, however, a great many indications that provide a quite different picture of the possible functions of mathematics education. Many studies point out that it has the potential to empower students; naturally, such a formulation needs to specify what "empowerment" might mean. One can think of it in terms of how individuals become positioned in the labor market; not least through measured mathematical competencies, students might be able to improve their individual career opportunities. One can also see empowerment as referring to critical citizenship, a notion that, like empowerment, is in urgent need of being clarified. Renuka Vithal outlines what mathematics education for empowerment could mean in a South African context, while Eric Gutstein discusses empowerment with reference to a U.S. context, referring to Paulo Freire's notion of reading the world as being open for changes (see Vithal, 2003; Gutstein, 2006).[6] The notion of mathematics education for social justice has been elaborated on lately.[7] This notion appears straightforward, as hardly anybody would dream of arguing in favor of a mathematics education for social injustice. However, the notion of "justice" includes complexities as profound as "empowerment" and "critical citizenship."

Thus it appears that mathematics education can serve very many and very different functions. It can mean disempowerment and suppression; it can cultivate an almost blind adaptation to priorities of the labor market; it can serve personal priorities; it may establish steps towards social justice. Mathematics education does not contain any firmly established quintessence, but can be developed in all possible directions. I do not see any particular qualities of mathematics education deriving from the fact that we have to intimately connect with *mathematics*. I do not relate mathematics education to any optimistic position claiming the existence of an intrinsic connection between mathematics education and, say, democratic values. Nor do I claim that mathematics education per se will serve anti-democratic interests. The socio-political roles of mathematics education are neither fixed nor determined. In this sense I talk about mathematics education as being *critical*.[8]

The critical nature of mathematics education is a challenge to any form of mathematics education. It is a challenge, however, that is acknowledged by critical mathematics education. As a consequence, it is a concern of critical mathematics education to try to be explicit with respect to the diversity of functions mathematics education might exercise, depending on the socio-political, cultural, and economic context. For this reason, we struggle with notions like "empowerment," "equality," and "justice," which, as I will return to later, are highly explosive concepts.

SECOND CONCERN: GLOBALIZATION INCLUDES GHETTOIZING

The book *O berço da desigualdade* (*The cradle of inequality*) shows photos of children in schools all over the world. The photos are taken by Sebastião Salgado (2005), while the accompanying text is written by Cristovam Buarque. From the front cover of the book, a girl is looking at us. She looked at me from the window in a book shop in Rio Claro in 2006, and I went to buy the book. The girl is sitting at an old school desk. Her clothes have been white, but they are now dirty. So are her bare feet. The girl's hand has a firm grip on the pencil. She seems ready to take a note of what I might have to say.

This and many other photographs from the book remind us that learning takes place in very different environments. Some photographs show a "classroom" located in the dusty shadow of a big tree. A blackboard is balancing on an easel; a teacher is standing next to it; children and young people are sitting around in the sand; a camel is passing through the scenery. Other pictures show crowded, dark classrooms without any of the educational equipment that so often is taken for granted. In one picture a teacher instructs the children, who are really paying attention to what he is saying, about different types of land mines that might be scattered around in their neighborhood. A different picture shows children with firewood in their arms as they have to make sure that their cold classroom can be heated during winter time.

When we consider the vast majority of research literature in mathematics education, we meet a quite different learning environment. I have referred to the research-created classroom paradigm as the prototype mathematics classroom (see Skovsmose, 2006, 2008). I could also have referred to it as a research stereotype: it becomes established through the implicit paradigmatic criteria for selecting data and for conducting research in general. The prototype mathematics classroom emerges from the many transcriptions that have been selected to enter research literature and to document how children struggle with learning mathematics. Some have difficulties, but they try to come to grasp the mathematical content. Transcriptions show discussions between teachers and students and among students. The conversation concentrates on mathematical issues; there are only a few side-remarks (which might provide the transcriptions with a nice humorous flavor). The prototypical classroom has the required equipment: if there is need for pocket calculators, there are some; an overhead projector is available, if needed; and so is the relevant number of computers, if the possible learning gains of information and communication technologies are going to be addressed in the research. The events reported on in research are not located in any violent arena. No dramatic poverty interrupts the healthy atmosphere of the classroom. The students engaged in the mathematical investigations are not influenced by drug problems. Research in mathematics education seems to presuppose not only a certain affluence but also that good educational order predominates.[9]

The research-created classroom standards do not look like most pictures from *O berço da desigualdade*. However, if we take a look the world around, the most common sites for learning mathematics are not those portrayed through prototypical research literature. Instead, there are numbers and figures that indicate that the impression one might get from *O berço da desigualdade* reflects the global diversity in a much more profound way than prototypical research literature (see, for instance UNESCO, 2000; World Bank, 2006). The prototype classroom seems to signify a heavy bias in mathematics education research. Naturally, this does not mean that research addressing affluent mathematics learning environments is not relevant. However, it has to be realized that many of the theoretical constructs that emerge in theories of learning and teaching mathematics might be biased. I find this bias to be a principal concern to critical mathematics education.

Let me try to formulate this concern from the perspective of globalization. Manuel Castells (1998) has presented the notion of Fourth World, referring to the poorest parts of the world. It includes Brazilian favelas, African townships as well as slums in any metropolis of the world: New York, London, Tokyo. The Fourth World is not located in any particular geographic area. It is a closure, where people get isolated and partly immobilized. Townships in the Apartheid era in South African can illustrate this point. They were (and still are) scattered around big cities, as black labor force was necessary for white industry and business to operate and succeed. However, the townships were also isolated from each other, as connectedness could turn into a threat for the apartheid regime. The "geography of apartheid" includes a disconnection and isolation of slums; this becomes repeated, however, in today's world-wide "geography of poverty," including the segregation of the fast-growing super-ghettoes.

Within the globalized world one could also consider the situation in the most affluent neighborhoods. Like poverty, affluence is distributed worldwide. For simplicity, we can talk about the 0.1 World (leaving the First, Second and Third World to a different enumeration). This 0.1 World stretches around the globe like a network: There are wealthy neighborhoods in the U.S., but certainly also in Brazil, Tanzania, India, and many other countries. The 0.1 World is characterized through affluence of material goods as well as of services, information, and security. People from the 0.1 World are traveling, in contrast to people from the Fourth World. While the Fourth World is a closure, the 0.1 World is an opening; but it is an opening within limits. The 0.1 World needs a protection as accumulation of wealth needs to be protected, and this can be understood in the most direct way. The wealthy *condominos* in Brazil, for example, are surrounded by walls, and many new walls are erected in order to separate the Fourth from the 0.1 World.

Across the world, the Fourth and the 0.1 World can exist almost side by side. This is the case in most metropolises. Certainly there are also many borderlands from whence one can look into both worlds. One can know what affluence is available in the 0.1 World, as well as one can experience next door the devastating life conditions of the Fourth World. I find a borderland position to be the most common life condition in our globalized world.[10]

Often one talks about globalization from a 0.1-World perspective. But this is a narrow one. Processes of globalization, including processes of ghettoizing, concern everybody: the Fourth World, the 0.1 World, as well as any borderland. Anybody who can operate as devoted consumers becomes integrated in the neoliberal informational economy, while others become excluded. Sites for production become relocated around the globe according to cost–benefit calculations: Where is the salary lowest? Where do we find a labor force with the required qualifications? Where can we locate the polluting component of a production process? How can we avoid becoming subjected to costly insurances charges? And so it goes. Processes of globalization include a dynamic of exploitation. The processes that establish the 0.1 World also produce the Fourth World as a by-product: globalization includes ghettoizing.

The dramatic interaction between the Fourth and the 0.1 Worlds as well as all kind of borderlands concerns systems of education, not least mathematics education (see Skovsmose, 2007). This education can function in very many different ways, depending on the context. What may count as mathematics education for social justice to some students in some contexts may count quite differently to other students in other contexts. The processes of inclusion and exclusion can mean one thing to Indian students in Brazil, and quite different things to favela students in Brazil or to immigrant students in Denmark, just to mention a few examples (see, for instance, Baber, 2007; Skovsmose, Alrø, & Valero, 2008; Skovsmose, Scandiuzzi, Valero, & Alrø, 2008). Thus Sikunder Ali Baber has investigated immigrant students' experiences of possibilities in a Danish context. He points out that they often see their future as polarized: They have to perform well above the schools' average in order to gain just average opportunities; if not, they have slim chances to pursue any ambitions. They experience that a form of general presupposition obstructs their future opportunities in the Danish society, and mathematics education plays a crucial role in the formation of both opportunities and obstructions.

The critical position of mathematics education can be acted out in very different ways. While prototypical thinking in mathematics education research tends to assume a 0.1-World perspective, it is important to critical mathematics education to acknowledge a diversity of perspectives.

THIRD CONCERN: THE CRITICAL POSITION OF MATHEMATICS

Descartes has been called the first modern philosopher, and we can associate modernity with a profound development of a mathematics-based insight into nature. Modernity includes a celebration of the natural sciences, including a thrust in the direction of mathematical rationality. This thrust is defining what I refer to as the *modern conception of mathematics.* This conception includes a variety of elements, of which I want to highlight three: one concerning nature, one concerning technology, and one concerning purity.

The distinction between appearance and reality was crucial for both Galilei (1957) and Descartes (1998). We become familiar with the world of appearance through our senses; but according to Galilei and Descartes, this familiarity does not provide any foundation for knowledge, as appearance is no adequate source for insight. Behind the world of appearance there is a world of realities, which, however, is not accessible through our senses. Our senses cannot grasp this world, but we can grasp it through mathematics. Mathematics provides us with proper means for describing the real world, being composed of units of matter characterized through position, shape, and movement. As part of the modern outlook, mathematics becomes recognized as the true language of science, appropriate for revealing the mechanical features of reality. The industrial revolution was soon to follow the scientific revolution, and immediately mathematics demonstrated massive powers within all technological disciplines. Mathematics became an indispensible tool in all forms of engineering enterprises. It turned into not only the principal language of science, but also the principle language of technology. Mathematics became recognized as the rationality of an all-embracing, technology-driven progress (or assumed progress). This observation, however, did not prevent mathematics from being ascribed a beauty of its own, which could be cultivated for its own sake. This claim of beauty became part of the paradigmatic assumptions of mathematical institutions concentrating on "pure" mathematics. These observations about the power of mathematics with respect to obtaining insight in nature and realizing technological enterprises together with its almost magic preservation of purity (inconsistent as they might be) define what I refer to as the modern conception of mathematics.

This conception provides a strong framing of mathematics education. It implies that mathematics education deals with a superior rationality, and, as a consequence, the mathematics educator can take up the position as a zealous ambassador of mathematics. The modern conception of mathematics constituted an integral part of Felix Klein's concern for broadening mathematics education; it also constituted part of the Bourbaki-inspired Modern Mathematics Movement; and it underpinned Hans Freudenthal's concern for establishing mathematical processes as central to mathematics education (and not the mathematical structures as cultivated by the Modern Mathematics Movement). According to the modern conception of mathematics, the overall aim of mathematics education is to make students appreciate mathematics.

According to critical mathematics education, the *modern conception* of mathematics cannot be upheld. Instead one may talk about a *critical conception of mathematics.* This conception can be illuminated through a discussion of mathematics in action. Mathematics makes part of technological inventions, economic procedures, automatizations of work processes, management, decision making, and daily life routines. Mathematics in action can serve very many different purposes. It can operate powerfully, but in very different ways. There is no intrinsic goodness to be associated with such actions, nor any intrinsic badness. Mathematics can operate as part of actions with the most sinister implications as well as of actions that appear marvelous. In this sense I see mathematics as critical. It is a rationality without any spine, without any essence. In order to address the spectrum of mathematics in action, one can and must address a variety of issues. Here I restrict myself to indicating how mathematics in action can bring about possibilities, strategies, facts, as well as contingencies.[11]

Through mathematics, new technology becomes *possible*. Mathematics operates as part of what we can refer to as the technological imagination. Establishing of the computer technology was based on mathematics. Through mathematics, it was possible to conceptualize the computer, even before the first one was constructed. One can see the whole emergence of information and communication technologies as resourced by a mathematics-based investigation of possibilities—in particular, possibilities for formalization and for turning operations into algorithms. In almost any discipline, technological creatively is deeply rooted in mathematics.

Through mathematics one can establish new *strategies* within a great many different domains. One can think of any kind of business enterprises. What promotion strategy should one choose? What price policy could be adopted? What form of network of distribution of goods would be feasible? Different business and marketing strategies are thus investigated through mathematical modeling. Thus mathematical techniques facilitate the broad application of cost–benefit analysis. And not only business issues can be addressed this way, but also, for instance, medical and environmental policy-making have been subjected to such calculations.

When implemented, strategies turn into *facts*. Facts can take the form of schemes for payment, which any costumer will have to face. It can take the form of schemes for production that the workers have to comply with. In this way, mathematics comes to operate as a fact-producing device.

The emerging of the risk society is a permanent reminder that not only facts but also *contingencies* are fabricated, and mathematics is positioned in the nucleolus of this fabrication. Risks are brought into existence as something that could go wrong when procedures have been automatized: it could be safety procedures for the running of an atomic power plant; it could be procedures operating when the automatic pilot takes over part of the landing procedures. Processes are automatized by being formalized and brought into an algorithmic format, but any formalization includes blind spots. In this way formalizations come to encompass a risk fabrication.

A careful investigation of mathematics in action does not bring about uniform celebration of any mathematical rationality; nor does it make room for a uniform negative critique of this rationality. A careful investigation of mathematics in action helps to reveal the critical position of mathematics in a variety of technological, economic, and social enterprises. I find it important to get beyond the modern conception of mathematics, which suggests a uniform trust in mathematical rationality. I propose instead to recognize the critical position of mathematics, as mathematics makes part of very different forms of actions, which, like any other form of actions might turn out in all possible ways. It can even be that mathematics in action brings about possibilities with respect to which we might be missing any pre-existing set of interpretation about what to count as "good" or "bad." Mathematics in action might bring us deep into uncertainties.

To me, the critical conception of mathematics has profound implications for mathematics education. One cannot assume a blind trust in mathematical rationality assuming that the aim of mathematics education is (simply) to make students appreciate mathematics. The role of mathematics educators as zealous ambassadors of mathematics has come to an end. Instead it becomes important to establish possibilities in mathematics education to address mathematics in action in all its diversities. One such suggestion is presented by Morten Blomhøj (2003), who illustrates how issues of mathematics-based traffic planning can be addressed in mathematics education. Traffic planning includes many tensions: Traffic efficiency might serve the economic growth of a neighborhood, but such efficiency might contradict a concern for environmental priorities; noise and pollution might also be brought to the neighborhood. Such confrontation might be addressed through the mathematical modeling on which the traffic planning is based. The modeling will bring about numbers, figures, and "facts" that enter the specialized as well as the public discourses, and certainly they will be referred to in the political decision-making processes. In traffic planning, the mathematical rationality operates in a complex space of different interests, but not as a neutral and pure arbiter. Instead it may be the conveyer of particular perspectives and interests dressed up in neutral and formal looking numbers and figures. All such issues related to mathematics in action forms part of mathematics education, in such a case one needs to deal with the critical position of mathematics.[12]

FOURTH CONCERN: EXPLORE EDUCATIONAL POSSIBILITIES

The notion of "education for social justice" might serve as inspiration for an exploration of educational possibilities, as can notions such as "education for empowerment" and "education for critical citizenship." However, such notions are what I have called explosive in the sense that any attempt to clarify them brings us to use many others just as uncontrollable.[13] Thus, in order to specify what social justice might mean, one could come to refer to notions such as "empowerment," "equality," "justice," and so on. This does not mean, however, that one should try to abstain from operating with explosive concepts. Critical mathematics education acknowledges both the importance of formulating educational visions in terms of such notions, but also that this brings us far beyond the possibility of establishing visions on any firm ground.

Educational possibilities refer not only to what exists and can be observed, but also to what does not exist, but may be possible.[14] A formulation of educational possibilities may draw on a pedagogical imagination, which can be related to the notion of sociological imagination as described by Wright Mills (1959). While a sociological imagination refers to a capacity to formulate alternatives to a social given, a pedagogical imagination represents a capacity to formulate alternatives to an educational given. Such an imagination can be guided by, say, an idea about social justice, but it might be in need of several other guiding ideas. I will restrict myself to making a few comments about *cooperation*; *context*, and *resource sensitivity*; *landscapes of investigation*; and *mathemacy*. In this way I might illustrate what a search for educational possibilities, related to critical mathematics education, might include.[15]

I see *cooperation* between researcher, student teachers, teachers, and students as a principle element in an exploration of educational possibilities. Naturally, the cooperation can take very different forms. In connection with the South African project, we discussed the organization of such an exploration, and we found that pedagogical imagination plays an important role. Naturally, several groups of people can be involved in the formulation of such imagination. We found that not only researchers and teachers, but also student teachers and students played an essential role.[16] An educational imagination can certainly draw on theoretical insight, but there are many other sources for such an imagination. This is carefully demonstrated through the examples of project work which were elaborated as part of the first steps in mathematics education leaving the apartheid era (Bopape, 2002; Vithal, 2003). Any search for educational possibilities has to acknowledge that very different sets of conditions exist. In this case, establishing a multicultural classroom became a key issue. To get out of the apartheid segregation of people, including the separation of school children according to a racist classification, was a basic step out of the gloomy shadow of apartheid. Taking such steps is part of a collective process, presupposing theory–practice cooperation.

Whenever one provides a suggestion for educational changes, one has to be receptive to the educational context, and critical mathematics education operates with a high degree of *context sensitivity* when exploring educational possibilities (see, e.g., Valero, 2007). One has to be aware of the fact that ghettoizing is part of the set of processes within globalization and that mathematics education operates within this tension. I was once told that in a poor neighborhood in Barcelona a critical mathematics education program was implemented.[17] It was developed with particular reference to the contexts of the students. Examples were developed with reference to situations they knew about. It was ensured that different real-life examples were carefully addressed. It all sounded promising. However, the implication of this (so-called) critical mathematics education program was that no children from this neighborhood subsequently had any possibility of entering any form of further education. The mathematics they knew about was not meeting what was demanded as the entrance level of any further education course. In this case, the well-intended and carefully situated curriculum turned out to be an obstacle for students who might hope to move on to further education. The high degree of situatedness came to operate as part of a continued process of ghettoizing. A generalized form of this problem might apply to many ethnomathematical-inspired educational programs. Through their engravings of traditions and cultural specificities, such programs might come to produce an actual function of exclusion. The implication of this is naturally not that one should give up trying to situate the educational content, nor that one should abandon the ethnomathematical program. Alternatively, one should always keep in mind that the actual functioning of mathematics education in a certain context has to be addressed in its specificity. There is no hope of formulating any universal critical mathematics education curriculum.

When one tries to formulate educational possibilities one has to exercise a *resource sensitivity*. Mathematics education presupposes investments, and in many cases, computers enter the classroom. Often they are celebrated as ensuring a new powerful learning environment. Students can be engaged in mathematical activities; they can perform experiments and construct more meaning for their developing mathematical notions. Computers can ensure motivation as well as "learning efficiency," or so goes the celebration. What is seldom addressed, however, is the implication of this observation for the majority of the world's children, who learn mathematics in non-prototypical classrooms without any computer in sight. Are they left behind? Do we witness a new form of "efficient" social exclusion? For critical mathematical education it is important to address what specific resources as well as lack of such resources might mean for the teaching and learning of mathematics. In particular, it is important to analyse how computers might enter the classroom in poor neighbourhoods (see Penteado & Skovsmose, 2009). For me it is important to formulate educational possibilities without assuming the affluence that dominates prototypical thinking.

Establishing *landscapes of investigation* is a way of providing opportunities for new classroom activities (Alrø & Skovsmose, 2002; Skovsmose, 2001). There are many forms of landscapes of investigation, and I see them as learning milieus that move beyond the framework set by the exercise paradigm. Instead of organizing teaching–learning processes around tasks set by exercises, one can organize them around processes of inquiry. Such processes could take the form of project work and thematic approaches, but many other forms of classroom organization are possible. Landscapes of investigations provide a general invitation for students to enter processes of inquiry as part of their learning processes. This also makes dialogue an important feature of classroom interaction, both in the interaction between teachers and students and among students. In order to be more specific about what dialogue could mean, Helle Alrø and I have identified different dialogic acts that help to characterize a dialogue.[18] Our point is that certain qualities of learning become cultivated when one operates with inquiries and dialogues. If we should explore possibilities for mathematics education to support the development of critical citizenship, it would be relevant to explore how inquiry dialogical processes can come to constitute classroom activities.

Much concern among critical mathematics educators has been expressed in terms of empowerment. As mentioned previously, we have here another troublesome explosive concept. I prefer to raise the issue of empowerment with reference to a different, although also explosive, concept, namely *mathemacy*. This notion can be related to the Freire's notion of literacy, which refers not only to a capacity in reading and writing, but also to a capacity of interpreting a complex social-political context. One can provide a similar rich interpretation of mathemacy.[19] This includes a capacity for dealing with mathematical ideas, a capacity for applying mathematics, and a capacity for reflecting on such applications. I see mathemacy as an integrated set of competencies through which one can read the world (in terms of numbers and figures), and read it as being open to change. When mathematics is brought in action, one acts with reference to numbers and figures. Actions mean making choices, decisions, designs, and predictions, and one can address what it could mean to act responsibly with reference to such numbers and figures. To me, mathemacy includes competencies in both acting and reflecting on what is done, and in this sense one can relate mathemacy to the notion of critical citizenship.

FIFTH CONCERN: CRITIQUE AS UNCERTAINTY AND POSSIBILITY

One can see critique as a way of dealing with a critical situation. Both "crisis" and "critique" are derived from the Greek word *krinein*, which refers to "separating," "judging," and "deciding." A "critical situation" or a "crisis" brings about a need for action and involvement—that is, a need for critique. To me it is important that critical mathematics education addresses both the critical position of mathematics education and that of mathematics.

Throughout history, one can find many different attempts to formulate what a critical approach could mean (in philosophy, economy, and social sciences). Many forms of critique have been elaborated from certain well-defined positions assuming that it is possible to establish knowledge-with-certainty based on a solid foundation. René Descartes wanted to establish epistemic certainty, and through his philosophy critique was expressed through a universal doubt that swept away anything that could be doubted; in this way critique was supposed to provide a foundation for knowledge-with-certainty. Immanuel Kant (1933) tried to establish epistemic certainty through a *Critique of Pure Reason*, which should specify general conditions for obtaining knowledge. Karl Marx presented a critique that was broader in scope. He did not restrict himself to criticizing epistemic structures, as for instance conceptions of economy. He also wanted to address the very economic structures themselves by a critique and to establish this critique as a solid foundation. When, later on, Marxists criticized the political economy, their critique assumed the existence of a foundation, namely the one provided by Marx. Thus one can see a Marxist orthodoxy as assuming the existence of a basis for addressing political and economic issues. All of these notions of critique related to the works of Descartes, Kant, and Marx, different as they are, share an element of epistemic foundationalism. One tries to establish a critique assuming the existence of a solid foundation for the critical enterprise. Such a conception associates to a critical activity a solid dose of a know-all attitude.

The notion of critique as established through Critical Theory is partly inspired by Marx, but it also initiated a movement beyond Marxist orthodoxy. Thus, the critique of modernity formulated by Theodor W. Adorno and Max Horkheimer (2002) in *Dialectic of Enlightenment* is far from expressing any epistemic certainty. Instead, it initiates a critique of the modern outlook and the idea that modernity can be associated with a universal form of progress. Michel Foucault demonstrated how the modern outlook has included presumptions, ideas, and approaches that hardly can signify progress, but appear to be of the most dubious kind. I find that his archaeology of knowledge and analysis of the "order of things" inspires a breaking away from a search for establishing a critical position on a certain foundation (see, e.g., Foucault, 1989, 1994). A further step beyond any foundationalism with respect to critique was taken by Jacques Derrida (1974, 1998), who brought processes of deconstruction to apply to any kind of assumptions, and by Slavoj Žižek (2009), who introduced a virtual conceptual anarchism in his interpretation of Marxist ideas.

The idea of establishing certainty on a solid foundation represents a modernist myth. Instead I see critique as related to *uncertainty*. This observation applies to any form of critical education, and also to critical mathematics education. It is important to acknowledge that the social roles of both mathematics and mathematics education are open-ended. Mathematics in action is without "essence," and so is mathematics education. These enterprises are not destined to play any "good" or "bad" role in social development. They can function in very many different ways. They are both operating from critical positions. One can be concerned about this, but there are no sources from which one can formulate a solid strategy for dealing with critical positions. One cannot formulate guidelines for educational design and in this way realize a critical mathematics education. However, this does not mean that one needs to abstain from searching for educational alternatives—but it does mean this is an uncertain search. "Critique" might have lost contact with any essence.

ACKNOWLEDGEMENTS

I want to thank my colleagues in the research group at Aalborg University—Martin Krabbe Sillasen, Cristina Carulla, Maziar Etemadi, Troels Lange, Diana Stentoft Rees, Sanne Almeborg, Anita Lindkvist Pedersen, Ole Ravn Christensen, Paola Valero, Lene Johansen, and Steffen Elmose—and Peter Gates and Miriam Godoy Penteado for inspiring discussions and critical comments and suggestions for improving this paper.

NOTES

1. Freire (1972, 1974) was also among important sources of inspiration.
2. A first result of this work in found in Skovsmose (1980, 1981a, 1981b).
3. Many have contributed to the formulation of a critical mathematics education; see, in particular, Frankenstein (1983, 1989, 1995); and Powel and Frankenstein (Eds.) (1997). See also Mellin-Olsen (1987), who explored the political dimension of mathematics education.
4. The ethnomathematical approach was highly inspiring, although some of its theoretical limits became accentuated when confronted with the South African post-apartheid situation. See Vithal and Skovsmose (1997).
5. For a new orientation of critical mathematics education, see also Skovsmose (2005b, 2006, 2007).
6. Frankenstein (1998) and Powell (2002) have discussed what empowerment could mean for marginalized adults in a U.S. metropolis. Alrø and I have discussed empowerment in terms of mathemacy, with reference to students in a Danish context (Alrø & Skovsmose, 2002). For a critical discussion of the notion of power with respect to mathematics education, see Christensen, Stentoft, and Valero (2007).
7. For a discussion for mathematics education for social justice see, for instance, Cotton (2004); Atweh and Keitel (2007); and Sriraman (Ed.) (2008). See also Gates (2006) for a careful and critical discussion of how equity and social justice have been addressed through the history of Psychology of Mathematics Education, and Gates and Jorgensen (2009).
8. For a discussion of the "critical" position of mathematics education see also Skovsmose and Valero (2001, 2002), where the relationship between mathematics education and democracy is presented as being critical.
9. There is certainly research which challenges the paradigmatic assumptions forming the prototypical research. See in particular Vithal and Valero (2003), Valero and Zevenbergen (Eds.) (2004), and Valero (2004).
10. For a discussion of borderland position, see Skovsmose, Scandiuzzi, Valero, and Alrø (2008). For a discussion for the related notion of students' foreground, see Skovsmose (1994, 2005a); and Alrø, Skovsmose and Valero (2009).
11. For a discussion of mathematics in action see Skovsmose (2005b, 2009b); and Christensen, Skovsmose, and Yasukawa (2008). For a discussion of mathematics power and materializations, see also Keitel (2006) and Fischer (2006, 2007).
12. See also Greer (2008), who opens up a critical discussion of how mathematics comes to enter the public discourse. It is important that such issues become part of mathematics education if one should challenge the modern assumption of mathematics.
13. For a discussion of explosive concepts, see Skovsmose (2005b, pp. 195–197).
14. For a discussion for critical research as addressing not only what exists but also what could come to exists, see Vithal (2003), Skovsmose and Borba (2004), and Skovsmose (2009a).

15. See, for instance, Araújo (Ed.) (2007) and Appelbaum (2007) for the exploration of educational possibilities within the scope of critical mathematics education.
16. The involvement of students in the exploration of educational possibilities is illustrated in the project "City Planning" organized by Denival Biotto Filho and Miriam Godoy Penteado from São Paulo State University; see Biotto Filho (2007). The importance of a researcher–teacher cooperation is also illustrated by the projects described in Alrø and Skovsmose (2002). Such cooperation makes an essential part of the so-called Agder project, as described in Jaworski (2006). See also Gutstein (2008).
17. In communication with Núria Gorgorió and Núria Planas.
18. These acts are: getting in contact, locating, identifying, advocating, thinking aloud, reformulating, challenging, and evaluation. The dialogic acts are put together in the Inquiry Co-operation Model. See Alrø and Skovsmose (2002).
19. For a discussion of mathemacy with particular reference to literacy, see Skovsmose (1994). See also Jablonka (2003).

REFERENCES

Alrø, H., & Skovsmose, O. (2002). *Dialogue and learning in mathematics education: Intention, reflection, critique.* Dordrecht: Kluwer.

Alrø, H., Skovsmose, O., & Valero, P. (2009). Inter-viewing foregrounds: Students' motives for learning in a multicultural setting. In M. César & K. Kumpulainen (Eds.), *Social interactions in multicultural settings* (pp. 13–37). Rotterdam: Sense Publishers.

Appelbaum, P. (2007). *Embracing mathematics: On becoming a teacher and changing with mathematics.* New York: Routledge.

Araújo, J. de L. (Ed.). (2007). *Educação matemática crítica: Reflexões e diálogos.* Belo Horizonte, Brazil: Argumentum.

Atweh, B., & Keitel, C. (2007). Social (in)justice and international collaboration in mathematics education. In B. Atweh et al. (Eds.), *Internationalisation and globalisation in mathematics and science education* (pp. 95–111). New York: Springer.

Baber, S. (2007). *Interplay of citizenship, education and mathematics: Formation of foregrounds of Pakistani immigrants in Denmark.* Unpublished doctoral dissertation, Aalborg University, Denmark.

Biotto Filho, D. (2007). Trabalho com projetos e o papel sócio-político da matemática. *Anais IX Encontro Nacional de Educação Matemática: Diálogos entre a pesquisa e a prática educativa.* Belo Horizonte, Brazil: Uni-BH.

Blomhøj, M. (2003). Modellering som undervisningsform. In O. Skovsmose, & M. Blomhøj (Eds.), *Kan det virkelig passe? Om matematiklæring* (pp. 51–71). Copenhagen: L & R Uddannelse.

Bopape, M. (2002). *Mathematics school based in-service training (SBINSET): A study of factors contributing towards success or failure of SBINSET in the South African school context.* Unpublished doctoral dissertation, Aalborg University, Denmark.

Bourdieu, P. (1996). *The state nobility: Elite schools in the field of power* (with the collaboration of Monique de Saint Martin). Cambridge: Polity Press.

Castells, M. (1998). *The information age: Economy, society and culture. Volume III, End of Millennium.* Oxford: Blackwell.

Christensen, O. R., Skovsmose, O., & Yasukawa, K. (2008). The mathematical state of the world: Explorations into the characteristics of mathematical descriptions. *Alexandria: Journal of Science and Technology Education 1*(1), 77–90.

Christensen, O. R., Stentoft, D., & Valero, P. (2007). Power distribution in the network of mathematics education practices. In E. de Freitas, & K. Nolan (Eds.), *Opening the research text: Critical insights and in(ter)ventions into mathematics education* (pp. 131–146). New York: Springer.

Cotton, T. (2004). What can I say, and what can I do? Archaeology, narrative, and assessment. In M. Walshaw (Ed.), *Mathematics education within the postmodern* (pp. 35–54). Greenwich, CT: Information Age Publishing.

Damerow, P., Elwitz, U., Keitel, C., & Zimmer, J. (1974). *Elementarmathematik: Lernen für die praksis? Eine exemplarische fersuch zur bestimmung fachüberschreitender curriculumziele.* Stuttgart: Ernst Klet Verlag.

Derrida, J. (1974). *Of grammatology.* Corrected edition. Translated by Gayatri Chakravorty Spivak. Baltimore: Johns Hopkins University Press. (Original French edition 1967.)

Derrida, J. (1998). *Writing and difference.* Translated with an Introduction and Additional Notes by Alan Bass. Chicago: The University of Chicago Press. (Original French edition 1967.)

Descartes, R. (1998). *The World and Other Writings.* Translated and Edited by Stephen Gaukroger. Cambridge: Cambridge University Press. (Original work published 1664.)

Fischer, R. (2006). Materialization and Organisation: Towards a Cultural Anthropology of Mathematics. In J. Maasz, & W. Schloegelmann (Eds.), *New Mathematics Education Research and Practice* (pp. 23–32). Rotterdam: Sense Publishers.

Fischer, R. (2007). Technology, mathematics and consciousness of society. In U. Gellert & E. Jablonka (Eds.), *Mathematisation–Demathematisation: Social, philosophical, sociological and educational ramifications* (pp. 67–80). Rotterdam: Sense Publishers.

Foucault, M. (1989). *The archaeology of knowledge.* London: Routledge. (First French edition 1969.)

Foucault, M. (1994). *The order of things: An archaeology of the human sciences.* New York: Vintage Books. (First French edition 1966.)

Frankenstein, M. (1983). Critical mathematics education: An application of Paola Freire's epistemology. *Journal of Education, 164*(3), 315–339. (Reprinted in I. Shor (Ed.) (1987). *Freire for the classroom* (pp. 180–210). Portsmouth, NH: Boyton and Cook Publishers).

Frankenstein, M. (1989). Relearning mathematics: A different third R—Radical maths. London: Free Association Books.

Frankenstein, M. (1995). Equity in mathematics education: Class in the world outside the class. In W. Secada, E. Fennema, & L. Adajian (Eds.), New directions for equity in mathematics education (pp. 165–190). Cambridge: Cambridge University Press.

Frankenstein, M. (1998). Reading the world with maths: Goals for a critical mathematical literacy curriculum. In P. Gates (Ed.), *Proceedings of the First International Mathematics Education and Society Conference* (pp. 180–189). Nottingham: Centre for the Study of Mathematics Education, Nottingham University.

Freire, P. (1972). *Pedagogy of the oppressed.* Harmondsworth: Penguin Books.

Freire, P. (1974). *Cultural action for freedom.* Harmondsworth: Penguin Books.

Gallilei, G. (1957). *Discoveries and opinions of Galileo.* Translated with an Introduction and Notes by Stillman Drake. New York: Anchor Books.

Gates, P. (2006). The place of equity and social justice in the history of PME. In A. Gutérrez, & P. Boero (Eds.), *Handbook of research on the psychology of mathematics education: Past, present and future* (pp. 367–402). Rotterdam: Sense Publishers.

Gates, P., & Jorgensen (Zevenbergen), R. (2009). Foregrounding social justice in mathematics teacher education. *Journal of Mathematics Teacher Education, 12*(3), 161–170.

Greer, B. (2008). Discounting Iraqi deaths: A societal and educational disgrace. In J. F. Matos, P. Valero, & K. Yasukawa (Eds.), *Proceedings of the Fifth International Mathematics Education and Society Conference I–II* (pp. 293–301). Centro de Investigação em Educação, Universidade de Lisboa, and Department of Education, Learning and Philosophy, Aalborg University.

Gutstein, E. (2006). *Reading and writing the world with mathematics: Toward a pedagogy for social justice.* New York and London: Routledge.

Gutstein, E. (2008). Building political relationships with students. In E. de Freitas, & K. Nolan (Eds.), *Opening the research text: Critical insights and in(ter)ventions into mathematics education* (pp. 189–204). New York: Springer.

Habermas, J. (1971). *Knowledge and human interests.* Boston: Beacon Press. (Original German edition 1968.)

Horkheimer, M. & Adorno. T. W. (2002). *Dialectic of enlightenment.* Translated by John Cumming. New York: Continuum. (First published 1947).

Jablonka, E. (2003). Mathematical literacy. In A. J. Bishop, M.A. Clements, C. Keitel, J. Kilpatrick, & F. K. S. Leung (Eds.), *Second international handbook of mathematics education* (pp. 75–102). Dordrecht: Kluwer.

Jaworski B. (2006). Theory and practice in mathematics teaching development: Critical inquiry as a mode of learning in teaching. *Journal of Mathematics Teacher Education, 9*(2), 187–211.

Kant, I. (1933). *Critique of pure reason.* Translated by Norman Kemp Smith. London: MacMillan. (Original German edition 1791.)

Keitel, C. (2006). Mathematics, knowledge and political power. In J. Maasz, & W. Schloegelmann (Eds.), *New mathematics education research and practice* (pp. 11–22). Rotterdam: Sense Publishers.

Mellin-Olsen, S. (1977). *Indlæring som social proces.* Copenhagen: Rhodos.

Mellin-Olsen, S. (1987). *The politics of mathematics education.* Dordrecht: Reidel Publishing Company.

Mollenhauer, K. (1973). *Erziehung und emanzipation.* Mündchen: Juventa Verlag.

Neyland, J. (2007). Globalisations, ethics and mathematics education. In B. Atweh et al. (Eds.), *Internationalisation and globalisation in mathematics and science education* (pp. 113–128). New York: Springer.

Penteado, M. G., & Skovsmose, O. (2009). How to drag with a worn-out mouse? Searching for social justice through collaboration. *Journal for Mathematics Teacher Education, 12*(3), 217–230.

Powell, A. (2002). Ethnomathematics and the challenges of racism in mathematics education. In P. Valero, & O. Skovsmose (Eds.), *Proceedings of the Third International Mathematics Education and Society Conference* (pp. 15–28). Copenhagen, Roskilde and Aalborg: Centre for Research in Learning Mathematics, Danish University of Education, Roskilde University and Aalborg University.

Powell, A., & Frankenstein, M. (Eds.) (1997). *Ethnomathematics: Challenging Eurocentrism in mathematics education.* Albany: State University of New York Press.

Salgado, S. & Buarque, C. (2005). *O berço da desigualdade.* São Paulo: Editora UNESCO no Brasil.

Skovsmose, O. (1980). *Forandringer i matematikundervisningen.* Copenhagen: Gyldendal.

Skovsmose, O. (1981a). *Matematikundervisning og kritisk pædagogik.* Copenhagen: Gyldendal.

Skovsmose, O. (1981b). *Alternativer og matematikundervisning.* Copenhagen: Gyldendal.

Skovsmose, O. (1994). *Towards a philosophy of critical mathematics education.* Dordrecht: Kluwer Academic Publishers. (Translated into Spanish by Paola Valero as *Hacia una Filosofía de la Educación Matemática Crítica.* Bogotá: Una Empresa Docente, Universidad de los Andes.)

Skovsmose, O. (2001). Landscapes of investigation. *ZDM The International Journal on Mathematics Education, 33*(4), 123–132.

Skovsmose, O. (2005a). Foregrounds and politics of learning obstacles. *For the Learning of Mathematics, 25*(1), 4–10. Reprinted in U. Gellert, & E. Jablonka (Eds.), *Mathematisation–demathematisation: Social, philosophical, sociological and educational ramifications* (pp. 81–94). Rotterdam: Sense Publishers.

Skovsmose, O. (2005b). *Travelling through education: Uncertainty, mathematics, responsibility.* Rotterdam: Sense Publishers. Translated into Portuguese by Maria Aparecida Viggiani Bicudo as *Educação Crítica: Incerteza, Matemática, Responsabilidade.* São Paulo: Cortez Editora.

Skovsmose, O. (2006). Research, practice, uncertainty and responsibility. *Journal of Mathematical Behaviour, 25*(4), 267–284.

Skovsmose, O. (2007). Mathematical literacy and globalisation. In B. Atweh et al. (Eds.), *Internationalisation and globalisation in mathematics and science education* (pp. 3–18). New York: Springer.

Skovsmose, O. (2008). Mathematics education in a knowledge market. In E. de Freitas, & K. Nolan (Eds.), *Opening the research text: Critical insights and in(ter)ventions into mathematics education* (pp. 159–174). New York: Springer.

Skovsmose, O. (2009a). Researching possibilities. In M. Setati, R. Vithal, C. Malcolm, & R. Dhunpath (Eds.), *Researching possibilities in mathematics, science and technology education* (pp. 105–119). New York: Nova Science Publishers.

Skovsmose, O. (2009b). *In doubt: About language, mathematics, knowledge and life-worlds.* Rotterdam: Sense Publishers.

Skovsmose, O., Alrø, H., & Valero, P. in collaboration with Silvério, A. P. & Scandiuzzi, P. P. (2008). "Before you divide you have to add": Inter-viewing Indian students' foregrounds. In B. Sriraman (Ed.), *International perspectives on social justice in mathematics education* (pp. 209–230) The Montana Mathematics Enthusiast, Monograph 1. Charlotte, NC: Information Age Publishing.

Skovsmose, O., & Borba, M. (2004). Research methodology and critical mathematics education. In P. Valero & R. Zevenbergen (Eds.), *Researching the socio-political dimensions of mathematics education: Issues of power in theory and methodology* (pp. 207–226). Dordrecht: Kluwer Academic Publishers.

Skovsmose, O., Scandiuzzi, P. P., Valero, P., & Alrø, H. (2008). Learning mathematics in a borderland position: Students' foregrounds and intentionality in a Brazilian favela. *Journal of Urban Mathematics Education, 1*(1), 35–59.

Skovsmose, O., & Valero, P. (2001). Breaking political neutrality: The critical engagement of mathematics education with democracy. In B. Atweh, H. Forgasz, & B. Nebres (Eds.), Sociocultural Research on Mathematics Education (pp. 37–55). Mahwah, NJ: Lawrence Erlbaum.

Skovsmose, O., & Valero, P. (2002). Democratic access to powerful mathematical ideas. In L. English (Ed.), *Handbook of international research in mathematics education* (pp. 383–407). Mahwah, NJ: Lawrence Erlbaum Associates.

Sriraman, B. (Ed.) (2008). *International perspectives on social justice in mathematics education.* The Montana Mathematics Enthusiast, Monograph 1. Charlotte, NC: Information Age Publishing.

UNESCO. (2000). *Education for all: Statistical assessment 2000.* Paris: UNESCO. Retrieved May 2, 2007 from http://unesdoc.unesco.org/images/0012/001204/120472e.pdf

Valero, P. (2004). Postmodernism as an attitude of critique to dominant mathematics education research. In M. Walshaw (Ed.), *Mathematics education within the postmodern* (pp. 35–54). Greenwich, CT: Information Age Publishing.

Valero, P. (2007). In between the global and the local: The politics of mathematics education reform in a globalized society. In B. Atweh et al. (Eds.), *Internationalisation and globalisation in mathematics and science education* (pp. 421–439). New York: Springer.

Valero, P., & Zevenbergen, R. (Eds.) (2004). *Researching the socio-political dimensions of mathematics education: Issues of power in theory and methodology.* Dordrecht: Kluwer.

Vithal, R. (2003). *In search of a pedagogy of conflict and dialogue for mathematics education.* Dordrecht: Kluwer.

Vithal, R. (2007). The 'uncivilised' scientist. In U. Gellert, & E. Jablonka (Eds.), *Mathematisation–demathematisation: Social, philosophical, sociological and educational ramifications* (pp. 95–105). Rotterdam: Sense Publishers.

Vithal, R., & Skovsmose, O. (1997). The end of innocence: A critique of 'ethnomathematics'? *Educational Studies in Mathematics, 34*(2), 131–157.

Vithal, R., & Valero, P. (2003). Researching mathematics education in situations of social and political conflict. In A. J. Bishop, M.A. Clements, C. Keitel, J. Kilpatrick, & F. K. S. Leung (Eds.), *Second international handbook of mathematics education* (pp. 545–591). Dordrecht: Kluwer.

Volk, D. (Ed.) (1979). *Kritische stichwörter zum mathematikunterricht.* München: Wilhelm Fink Verlag.

World Bank. (2006). *World Development Report 2006: Equity and development.* Washington and Oxford: The World Bank/Oxford University Press.

Wright Mills, C. (1959). *The sociological imagination.* Oxford: Oxford University Press.

Žižek, S. (2009). *The parallax view.* Cambridge, MA: The MIT Press.

49

IF A QUARTER CRASHES, SO IT DIES

Children's Meaning Making in Mathematics Lessons

Troels Lange
AALBORG UNIVERSITY

Tamsin Meaney
CHARLES STURT UNIVERSITY

Abstract: In this chapter, we present extracts from two classrooms in Denmark and New Zealand and discuss how the children made sense of mathematics activities. From a premise that children were constantly involved in meaning making, we explore how the ways in which mathematics activities were constituted influenced the meaning that children ascribed to them. We widen the discussion to look at whether information about children's backgrounds obscures or enlightens our understanding, as researchers, of this meaning making process. In understanding the children's ascription of meaning to mathematical activities, we conclude that what is important is how children enact their agency, as a consequence of interpreting past, present, and future activities.

Keywords: children's meaning making, didactical contract, foregrounds and backgrounds, instrumental and relational understanding

INTRODUCTION

Over the last century, research has highlighted a variety of contributions to the process of gaining mathematically valuable learning. In this chapter, we use classroom excerpts from two countries to consider how children's interpretations of mathematical activities affect their meaning ascription. We speculate on how knowing more about the backgrounds of the children can alter our impressions as researchers about the meaning-making process. Our suggestion is that, like classroom norms, children's background is one of a myriad of conflicting influences on whether they gain mathematically useful learning from participating in classroom activity. In order to understand how meaning is ascribed to mathematical activity and whether it is likely to lead to the gaining of valuable mathematical understanding, we instead need to look at how children enact their agency.

Efforts to improve students' mathematical learning in classrooms have been of concern since mathematics education research began. For example, Hamilton (1924), who reviewed Thorndike's books *The Psychology of Arithmetic* (1922) and *Psychology of Algebra* (1923), noted that British teachers reformed their teaching based on common sense approaches and a general feeling for the aims of school mathematics, whereas their American peers based their reforms on the results of psychological experiments. Although he believed that both approaches were problematic, he felt that the American approach resulted in too much emphasis being placed on the "relatively mechanical aspects of learning" (Hamilton, 1924,

The First Sourcebook on Nordic Research in Mathematics Education, pages 683–697

p. 174). This did not lead to what he considered to be valuable mathematics.

Throughout the twentieth century, researchers continued to try to identify features associated with teachers or teaching that contributed to students learning valuable mathematics. Skemp (1976) proposed that mathematics teachers could be divided into two groups. One group taught for *instrumental understanding* and concentrated on students possessing rules and being able to apply them. The other group taught for *relational understanding*—"knowing what to do and why" (Skemp, 1976, p. 21). This required students to understand conceptually and to link this understanding to other mathematical concepts. He considered this type of understanding to be the most valuable for students' ongoing learning of mathematics.

Didactical Contract—Widening the Perspective to Interactions

Another aspect that received attention was the mutual and mostly tacit understanding between teachers and students about what constitutes an appropriate approach to mathematics teaching and learning. This was conceptualized as a *didactical contract* by Brousseau (Brousseau, Balacheff, Cooper, Sutherland, & Warfield, 1997) and as *classroom social norms* by Cobb, Stephan, McClain, and Gravemeijer (2001). Blomhøj (1995) stated that for teaching and learning to occur, a didactical contract needs to be in place between the teacher and students. This type of contract frames the activity and interactions in the classroom. It develops over time from a balancing of the parties' different positions, obligations, and perspectives. Consequently, the teacher organizes the teaching in ways that are compatible with these regulations and the students' mathematical and socio-psychological background. The teacher evaluates the students' learning, and the students on their part do their best to fulfill the requirements made of them. Blomhøj (1995) saw the relationship between teacher and students and an avoidance of failure as the forces driving the development and maintenance of a didactical contract:

> The mutual dependence between teacher and student... lead to both the teacher and student having a strong incentive to succeed with their shared project: *the student's learning.* The interplay between teacher and students may be restricted in a range of ways and degrees, for example, physical conditions and time limitations, the difficulty and importance of topics, the teacher's and the students' mathematical basis, curriculum and exam provisions, their own expectations and those of the environment. The fewer degrees of freedom that are at their disposal in this project, the more both parties will focus on avoiding failure—that is, recognition that the learning project has failed. In the course of time, with this charged atmosphere as a driving force, a didactical contract is built between teacher and students through the teaching. The observance of this contract in a sense may be said to constitute an insurance against the learning project ending in failure—at least for the great majority of the students. (Blomhøj, 1995, p. 17; translated by TL, italics in original)

The notion of the didactical contract originated within the French school of "Didactique des Mathématiques" and is linked to the theory of didactical situations of Brousseau (Brousseau et al., 1997; Wedege & Skott, 2007). Balacheff (1990) loosened this linkage to more generally include the social interactions:

> The rules of social interaction in the mathematics classroom include such issues as the legitimacy of the problem, its connection with the current classroom activity, and the responsibilities of both teacher and pupils with respect to what constitutes a solution or to what is true. We call this set of rules a *didactical contract.* A rule belongs to the set, if it plays a role in the pupils' understanding of the related problem and thus in the constitution of the knowledge they construct. (p. 260; italics in original)

In contrast to Balacheff (1990), Blomhøj (1995) emphasized the socio-political framing and the social-psychological dynamics of the social interaction. According to Blomhøj (1995), the didactical contract constructed in traditional mathematics teaching implied that:

- the teacher carefully goes through the methods and algorithms that are presented in the textbook
- the teacher only sets problems for which the students have tools to solve beforehand
- a problem is solved when its component questions are answered
- the required answers can be brief—for example, a number, a figure or at most a short sentence
- the students have a right to get the teacher's judgment when a problem is solved
- the students' learning is judged only from whether they can solve the posed problems
- the students on their part do their best to solve the posed problems (p. 17–18. Lange's translation)

However, the establishment of a set of rules of the game does not eliminate all misunderstandings.

All students in a class may not hold the same perspectives, and "rules" that work for most students may not work for all. In regard to instrumental and relational understanding, Skemp (1976) mentioned two types of mismatches. One is a student who aims for instrumental understanding while the teacher teaches for relational understanding. The other is the reverse. For a child who wants one kind of understanding but is facilitated at school into developing the other, the consequences can be frustration and possible damage to the student's learning (Skemp, 1976).

Didactical contracts are rarely discussed explicitly, so mismatches can occur without any awareness by participants. Especially when the teacher and students are new to each other, they may not know how much they share perspectives and even that there may be different perspectives to consider. When participants hold their own perspective as the "natural" one, then it is not negotiable. Herbel-Eisenmann (2003) described how misunderstandings endure for a long time without participants ever being aware that differences existed.

School learning of whatever type requires a meeting of two agents—teachers and students—and thus the coordination of the different processes of teaching and learning. The didactical contract facilitates this co-ordination. If an undiscussed didactical contract is in place, then unexpected outcomes may occur. For example, Cobb, Wood, Yackel, and McNeal (1992) suggested that in one classroom that they filmed, the students came "to view mathematics in school as an activity that involves manipulating symbols that do not signify anything when their teachers would honestly like them to learn with understanding" (p. 581). Nuthall (2007) stated that students do not learn content as such, but rather learn about what constitutes classroom activity—"students learn what they do" (p. 36). Doing exercises from a textbook, using concrete equipment, or fighting to gain the teacher's attention can be the understandings that children learn, rather than the actual knowledge of how subtraction or addition works. For Nuthall (2007), successful teaching was only likely to occur if the teacher made use of knowledge of students' own culture, such as their friendship groupings as well as fashion and music interests.

As agents in their own lives, students continually make sense of the activities in which they are engaged. Seeing education as a meeting of agents highlights the "spaces" that learning environments provide for students' agency to unfold in relation to learning activities. The question then is, how does the existing didactical contract impede or support the active involvement of the learner?

Discussions about the didactical contract have concentrated on what happens in classrooms. However, Herbel-Eisenmann (2003) was critical of approaches that only focussed on the socio-cultural environment of the mathematics classrooms:

> [A]t least two components that seem to be important to understanding mathematics classrooms are not addressed: 1) students are part of other communities of practice that influence how they come to participate and be part of the schooling system (or not) . . . and 2) teachers typically know a lot about the context in which they teach and this knowledge effects what they do/say in their classrooms. (Herbel-Eisenmann, 2003, p. 9)

Valero (2004) went further by criticizing the discursive constructions of students as "schizomathematicslearner." She felt that in most mathematics education research reports, students were seen as mainly cognitive subjects, neither fully human nor social beings situated "in a particular historical time, geographical location, and social position" (Valero, 2004 p. 44). She proposed "a 'realized' view of students as whole learners, who have multiple motives for learning, and who live in a broad context which influences their intentions to participate in school mathematics practices," which would help "recognize the *agency* that students have in the whole educational enterprise" (p. 48; italics in original).

For students to learn valuable mathematics such as relational understanding, it is not just a matter of the teacher providing "good teaching." Teaching and learning is complicated and requires teachers and students to share a set of perspectives about the learning process and about mathematics. Although understanding the contribution of classroom practices to this process provides valuable information, there is also a need to explore other contributing factors outside the immediate classroom situation. Figure 49.1 shows how Skovsmose (2005a) related dispositions to the socio-political situation.

ASCRIBING MEANING IN CLASSROOM MATHEMATICAL ACTIVITIES

In recent times, research into mathematics education has shown that classroom learning cannot be extrapolated from its surroundings. Figure 49.1 illustrates the ideas of Ole Skovsmose, who, in writing with Helle Alrø, provided a comprehensive description of how the ascribing of meaning to different activities is influenced by a number of different components. Skovs-

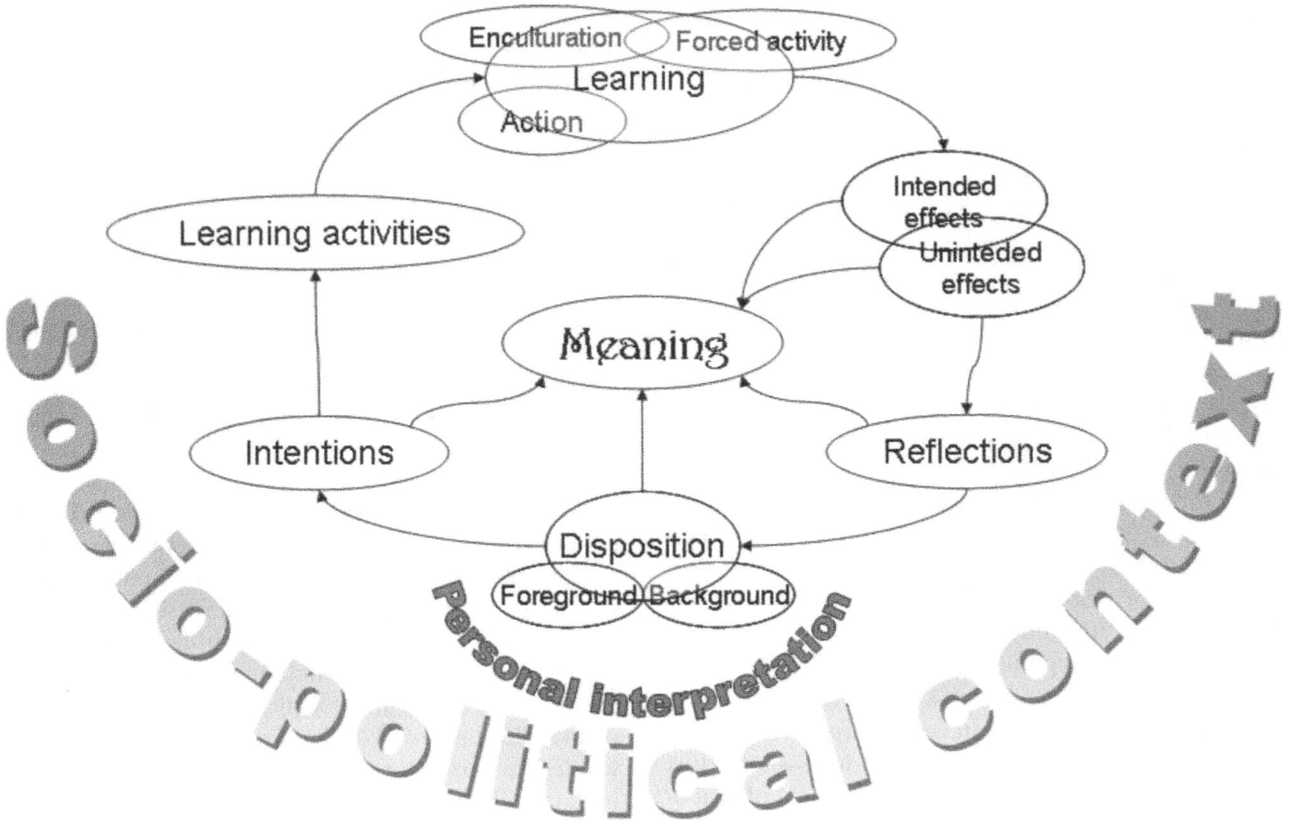

Figure 49.1 An illustration of the complexity of meaning in learning.

mose (2005b) wrote, "Meaning in learning comes to refer to a relationship between the dispositions of the learner, the intentions of the learner, the intended and unintended effects of learning activities, and the learner's reflections on these effects" (p. 93).

> By the *foreground* of a person I understand the opportunities, which the social, political and cultural situation provides for this person. However, not the opportunities as they might exist in any socially well-defined or 'objective' form, but the opportunities as perceived by the person. Nor does the background of a person exist in any 'objective' way. Although the background refers to what a person has done and experienced (such as situations the person has been involved in, the cultural context, the socio-political context and the family traditions), then background is still interpreted by the person. Taken together, I refer to the foreground and the background of a person as the person's *dispositions.* (p. 6–7)

The dispositions of a person "embody propensities that become manifest in actions, choices, priorities, perspectives, and practices" (Skovsmose, 2005a, p. 7). These propensities may be contradictory because the person may conceptualize different foregrounds and backgrounds at different times and situations.

In the center of Figure 49.1, dispositions are interconnected with intentions, learning environments, and reflections on effects of learning activities to contribute to students' development of meaning. Alrø and Skovsmose (2002) were interested in learning as action. They took intentionality as a defining element of action, thereby separating action from mere activity. If the learning situation allowed the active involvement of students, the resulting learning process could be one of action. Students identify with the intentions of the learning activity, and thus joint ownership and shared perspectives between teacher and students could develop. Intentional learning acts constitute forms of learning that are described as action. The meaning ascriptions resulting from learning-as-action would be different from those where students do not engage willingly.

Alrø and Skovsmose (2002) did not believe that all learning came from the active involvement of the learner. Much learning, including what occurs in schools, happens by enculturation or assimilation, where children adopt knowledge or skills without much awareness that they have done so. In everyday life, the acquisition of a first language would be a prime example of this. In schools, Nuthall's (2007) finding that children first of all learn what they do is evidence of such forms or learning. In situations where the learner has no say in what or how they learn, then this learning is forced activity (Alrø & Skovsmose, 2002). An example would be the rote learning of times tables.

In considering students' agency in different learning environments, Alrø and Skovsmose (2002) identified two paradigms of school mathematics education. One is the Exercise Paradigm and the other is Landscapes of Investigation. In the exercise paradigm, goals and activities are related to the mastering of different procedures for tackling sets of exercises. This would lead to Skemp's (1976) instrumental understanding and a "traditional" didactical contract (Blomhøj, 1995). Within the exercise paradigm, students have limited options to express their agency in relation to the mathematical subject matter (cf. Boaler & Greeno, 2000).

On the other hand, landscapes of learning enable learning as action (Alrø & Skovsmose, 2002). Instead of exercises, a "scene" is set so students can ask questions and become part of an inquiry process. The key component is that a cooperative inquiry is undertaken so that a "landscape" is explored with results and conclusions not being determined in advance. Processes of inquiry can only start with students' previous understandings.

Investigating landscapes cannot be a forced activity, as students must be owners and active participants. As ownership, participation, and intentions cannot be forced upon students, the teacher's role is to invite students to cooperate, and students can accept or reject the invitation. Their reasons for doing so could be very complex. Acceptance could come from interest in the topic of inquiry or the "scene," concerns about relationships to classmates or teacher, or fear of exclusion. Students need good reasons to enter a landscape of investigation, so that their intentions become connected to the learning process. The teacher cannot force the students to accept the invitation, nor control the reasons why students accept. Teachers, of course, can influence the students' choices and reasons by making use of the students' experiences and previously acquired understandings when designing landscapes of investigation. However, the "working conditions" provided by the school setting as perceived by the teacher need to encourage, sustain, and valorize such creativity in design.

If students are not invited to engage in meaningful learning acts, the field is left open to all sorts of other meaning productions, such as *underground intentions* (Alrø & Skovsmose, 2002), which "refer to the students' zooming-out of the official classroom activity... partly setting an alternative scene for what is going on in the classroom" (p. 158). These intentions could result in students learning how to manoeuvre themselves so that they sit next to the right person or to finish tasks quickly in order to prevent homework from making inroads on their leisure time.

The forms of communication that go with the exercise paradigm and landscapes of investigation are different. Within the exercise paradigm, the learning activities focus on getting correct answers to exercises. Communication involves the teacher checking the students' answers, pointing out mistakes, and guiding them to the right answer. The teacher is the authority, along with the textbook and its answer key. The students' contributions are often minimal as they concentrate on guessing what the teacher would like to hear (Alrø & Skovsmose, 2002; Blomhøj, 1995). This communication pattern is one reason for the exercise paradigm being in some sense a comfort zone. Each participant has well-defined obligations, goals, and ways of assessing whether the obligations are fulfilled and to what extent the goals are met. On the other hand, landscapes of investigation are risk zones with ill-defined rules. Not only could invitations be accepted or rejected, the investigations themselves do not involve finding one specific result: knowledge develops differently for each participant. In cooperative activity, the communication between teacher and students must be explorative and dialogic, recognizing that they are trying to understand each other's perspective.

For students to gain valuable mathematical understandings, they need to be part of a classroom where they have opportunities to engage in activities that require genuine investigation. The relationship between the teacher and the students is influenced both from outside and inside the classroom. These influences contribute to students accepting invitations to participate, as well as the way that they actually participate. In the next sections, we analyze extracts from mathematics lessons in different countries, suggesting how the students' perspectives contribute to their ascription of meaning to the activities. In the first case, we illustrate how the student's perception of the didactical contract is very difficult to change. In the second case, we discuss how underground intentions arise when students do not know what to do in a more inquiry-based process.

METHODOLOGY

Our focus is on students' perspectives, and so the descriptions of the learning activities do not draw inferences about the teaching practices. Teachers are restrained and facilitated in a number of ways, and it would be unfair to portray the teachers' actions or non-actions as entirely matters of individual and free choice.

The following sections describe the situation for two mathematics learners in different countries. From conversations about our respective research projects, in Denmark and New Zealand, it became clear that although the classrooms were situated differently, the mathematics learning was more likely to be instrumental understanding in both cases. It was the differences in the situations, but with the similarity of outcomes, that fascinated us.

The following sections use extracts from larger projects. As part of an investigation of children's perceptions of being in difficulties with mathematics, Troels Lange interviewed 10–11-year-old children in a Danish 4th grade classroom (see Lange, 2007). He observed the class for nine months on a weekly basis, including keeping field notes and audio recording of some of his interactions with the children during the lessons. He was seen by the students as another adult who could provide help with doing mathematics.

The mathematics lessons typically began with a whole-class activity addressing the students' homework, practicing number skills (e.g., multiplication tables), or introducing the next mathematical topic or type of exercises. In this part of the lesson, the teacher tried to engage with all of the students. In the remaining time, the students did exercises individually, but with the opportunity to talk to their neighbors. These exercises were usually done from the textbook, though sometimes, the teacher brought in other activities such as worksheets. Work that was not completed in class was set for homework.

Tamsin Meaney investigated the acquisition of mathematics language by a 6–7-year-old child in New Zealand (see Meaney, 2007). For one day a week for twenty weeks, both at home and during math lessons, the child was audio recorded. The class discussion was captured on another voice recorder. The child's mother was the research assistant for the project and so attended the mathematics lessons to ensure that they were recorded. The mother was a trained secondary English teacher and also provided assistance to children who asked for help.

These mathematics lessons began with the children sitting on a mat on the floor, in front of a whiteboard. After the introduction, the children did activities in groups. One group worked with the teacher, while the other groups completed worksheets or activities in pairs. After a period of time, the teacher swapped the groups and spent time checking on the pair work.

Neither situation would necessarily be typical of Danish or New Zealand classrooms. However, the extracts illustrate typical situations from our larger data sets, and they allow us to talk generally about these children's learning of mathematics. The extracts are quite long because we wanted children's own voices rather than the researcher's interpretations of them to be presented first. When focusing on children's meaning making, their perspectives and ways of expressing themselves should be the starting points. The Danish transcript is translated into English fairly literally and sometimes at the expense of "good" English or what would resemble native English children's speech.

Each extract is analyzed individually to consider how students ascribe meaning to the activities. We describe how we see the didactical contract as being evident in the extracts. Following the individual analyses of the extracts, we re-analyze the data after introducing more information about the children's dispositions. From this, we consider how students' agency is manifested in the extracts and the impact that this has on their learning of valuable mathematics.

CORRECTING ERRORS AND ADDING DECIMALS IN A DANISH CLASSROOM

In March 2007, seven months into grade 4, the initial activity in this lesson involved the teacher correcting the homework by reading the answer key. The homework was textbook exercises that the students had begun in the previous lesson. Some exercises referred to drawings of sticks of different lengths labelled alphabetically. The exercises were stated as $a + b$, $c - d$, $c + b + a$ etc.

> The teacher instructed the students to put a tick if an answer was correct and a minus if it was wrong. A student asked if it was okay if there was "one centimeter between." The teacher responded by checking the student's measurement of stick *a* in the first exercise. The student had measured it as 6.6. Most students agreed to this but one student said seven. The teacher said that it should not be that far away, before realizing that the student meant 6.7 and not seven centimeters. The teacher would accept 6.5 and 6.7, and it was OK if they were "one on either side." The answers to the exercises could vary and be a little different from what she said. For instance, her answer to the first exercise, $a + b$, was 9.4 centimetres. It was fine and right if they had 9.3. Then there would be a difference of one millimeter. It would even be okay if they had 9.5 or 9.6. It could vary two millimetres. The answer was allowed to vary 0.2 both ways because it could be difficult to measure down to the millime-

> ter. She could accept that. Hence, because it was two lines, both of which they maybe had measured a little short she would accept 0.2 on either side: that is, anything between 9.6 and 9.2 was OK. Asked by a student if 9.1 were OK, she said it was a bit on the high side but not altogether wrong. Then the teacher started reading the correct answers and the students marked their own answers as right or wrong. [Field notes and audio recording 3:10–7:00 from 7/3-2007]

The extract illustrates the exercise paradigm, the operation of a traditional didactical contract, and an emerging instrumental understanding of decimals. The textbook activity focused on obtaining correct answers. Although the drawings of the sticks provided a semi-real context, their actual purpose was for students to produce numbers to substitute for the letters. There was no suggestion that the sticks, their lengths, or the sums and differences of their lengths, were of intrinsic interest. These closed exercises allowed students to practice various measurement and number skills. The students almost never, and the teacher only sometimes, mentioned the units (cm, mm) of the lengths. Without a need to consider seriously the context, the students—and possibly the teacher—conceived the exercise as a mere exercise in numbers. The questions that students asked were limited to gaining the correct answers. Hence, the activity was within the exercise paradigm.

The didactical contract had many features in common with the rules of the traditional didactical contract listed above. In previous lessons, the teacher had introduced the students to methods and procedures needed to solve the exercises. The exercises were solved when their component parts were completed—in this case, when the sticks were measured and the numbers added/subtracted. The students did not need to make sense of the exercise as a whole, for instance interpreting an answer as the combined length of two sticks. Answers were brief, just a number, possibly without a unit. All participants positioned the teacher with the authority and the responsibility to judge the correct answers. Finally, the students did their best to solve the posed problems.

The operation of the exercise paradigm and a traditional didactical contract meant that the students asked about accuracy in terms of correctness—that is, from the perspective of students' "rights" to know the rules. They did not display interest in mathematical features of calculations with the measured quantities. Although the teacher's response indicated awareness of these issues, she eventually presented acceptable divergences for sums and differences between numbers. Uncertainty became a matter of teacher authority instead of a topic for discussion. However, this happened in response to the students call for "legal" advice on acceptable accuracy.

Mathematics education research has shown that a common understanding of decimals is that they are composed of two numbers, one before and one after the decimal point (e.g., Brekke, 1996). This understanding works well in common contexts like money (e.g., kroner and øre, dollars and cents) and measurements (e.g., meter and centimeter, kilogram and gram). The question "is it okay if there is one centimeter between" and the following discussion seem to reflect an emerging understanding of this type. Despite the word *centimeter*, the student's question was about a divergence of one from the "second" number, the digit after the decimal point. The contribution "seven" from the other student suggested a similar understanding. Although the teacher certainly held a mathematical concept of decimals, she understood the students, and when expressing that 6.5 and 6.7 was close enough to 6.6, she phrased it in two-number terms as "one on either side." The examples suggest that a two-number conception of decimals was developing.

Following this interaction, the students corrected errors in their homework. Troels talked with Research Child 1 (RC1) while she did this and another exercise on adding decimal numbers.

1	**RC1:**	You shall just correct these. That one
2	**Troels:**	Which ones are you in doubt about?
3	**RC1:**	It is this one. There is a mistake because I plussed them *[instead of 'minused' in 'a – b' exercise]*
4	**Troels:**	Oh. And now you have corrected them?
5	**RC1:**	Yes
6	**Troels:**	But does it look right?
7	**RC1:**	You must tell if it is right or wrong
8	**Troels:**	Must I tell if it is right. (yes) Okay–It seems, it seems quite right to me
9		*[Pause for 59 seconds while RC1 calculated]*
10	**RC1:**	You said that it was right (mmm). Then it is just that *['a + b + c' type; RC1 has 20.9 cm instead of 21.1 cm. The addition of the three measurement numbers is correct.]*
11	**Troels:**	But isn't it right what you have added? Haven't you added the numbers correctly?
12	**RC1:**	No because, I did not...
13	**Troels:**	You had 20.9. Doesn't it give 20.9?
14	**RC1:**	I don't know
15	**Troels:**	Have a look at the numbers

16		*[Pause for 17 seconds while RC1 adds the numbers again and gets the same result as before]*
17	**Troels:**	So you have added the numbers okay
18	**RC1:**	What have I done wrong then?
19	**Troels:**	What could you have done wrong?
20	**RC1:**	I don't know
21	**Troels:**	What was it you should do? What was the problem about?
22	**RC1:**	That you should just plus them
23	**Troels:**	Which ones should you plus
24	**RC1:**	Those that are there *[points to 'a + b + c']*
25	**Troels:**	Well but what is it that are there?
26	**RC1:**	It is those and that and that (okay) nothing special
27	**Troels:**	Could it be that you measured a little inexactly?
28	**RC1:**	I don't think so
29	**Troels:**	You don't think so?
30	**RC1:**	No (Okay)
31		. . .
32		*[A little later RC1 adds 39.7 and 33.6 and gets 733. The numbers are vertically aligned.]*
33	**RC1:**	Is it right?
34	**Troels:**	Let's try and have look at it. That is about forty and that is about thirty (yes). And forty and thirty is about?
35	**RC1:**	Seventy__
36	**Troels:**	Seventy yes. And there it says 733
37	**RC1:**	Is it then right?
38	**Troels:**	If it were money it would be about forty kroner plus about thirty kroner. That would make about seventy kroner, wouldn't it?
39	**RC1:**	But is it right?
40	**Troels:**	But can it be true that thirty nine and thirty three is seven hundred and something?
41	**RC1:**	No but this is thirteen [the tenths 6 + 7]
42	**Troels:**	That is right
43	**RC1:**	And then I put a one up there (yes that is also right) and then these two are right and then I put a one on top and then is also gets right
44	**Troels:**	That is right. That is right but isn't it right that it is seven hundred and thirty three there?
45	**RC1:**	Yes
46	**Troels:**	But it cannot be right that thirty and thirty [that] that gives seven hundred
47	**RC1:**	Okay (about to erase)
48	**Troels:**	Don't erase because what else could be wrong? It is right what you have done
49	**RC1:**	I don't know
50	**Troels:**	If you put a point between the threes would it then make a difference?
51	**RC1:**	Yes (Okay)

Given that the teacher was positioned as an authority, it is not surprising that RC1 wanted Troels to correct the answers (lines 1, 7, 33, 39). RC1 clearly perceived mathematics to be about getting right answers, and judgment on their accuracy was the responsibility of an external authority. Consequently, it was RC1's right to have the answers judged by this authority. In the first part, RC1 requested Troels to "correct these" (1) and tell if it was "right or wrong" (7), and in the second part he was asked twice, "is it right?" (33, 39). Troels' attempt to direct RC1's attention to the relation between the numbers to add and the answer was in vain (36, 38, 40, 44, 46). RC1 did all the component calculations correctly, but the unreasonableness that was so obvious to him was not significant for RC1 (34–41, 44–47).

The didactical contract construed the teacher as being responsible for evaluating the students' answers, not the students themselves. Troels, as a teacher-like person, was expected to exert that authority. His persistent attempts to evade this expectation by inviting RC1 to reason was not sufficient to upset the didactical contract (2, 8, 34, 38). Eventually, he gave up and complied (27, 50). The child's insistence on maintaining the existing rules of the game resulted in Troels changing his behavior. However, by refusing to conform to the rules of the game, RC1's perception of these rules was exposed, at least to Troels.

RC1 began by realizing that the numbers had been *plussed* instead of *minused* (3). The Danish equivalents to these non-standard verbs are linguistic inventions of mathematics classes and appear frequently in mathematics classroom discourses. These verbs indicate the actions that should be undertaken when plus and minus symbols occur in exercises. However, their common usage suggests a separation of the mathematical operations from an understanding of why textbook questions require their application.

The use of these verbs and their interchange (3) suggested that RC1 interpreted the task as one to produce numbers, which both *plussing* and *minusing* do, and not to solve authentic problems. When the answer diverged from that of the answer key, the drawing of the sticks and their links to a semi-reality did not contribute to understanding what caused the divergence. As well, the teacher's rules for acceptability of the answer ("0.2 both ways") was not applied, maybe because RC1 did not realize that the answer, 20.9 (cm), was within acceptable distance from the expected 21.1 (cm) (10–20). It could be that with a two-number conception of decimals, the application of "between-ness" as the measure of divergence is not transparent. Between the numbers to the left and right of the decimal point, there is one (20 vs. 21) or eight (9 vs. 1), both of which are quite different from zero point two or even two.

Being a clever student working in a didactical contract similar to the traditional one described by Blomhøj (1995), RC1 had learned to filter out the context of the problem, and look for the numbers involved and clues for identifying the expected operation. RC1 did the assigned homework, wrote neatly in the notebook, and strived to solve the posed task, all as prescribed by the didactical contract. No zooming-out occurred and no visible underground intentions developed. Over the course of time, RC1 is likely to learn where to put the decimal point when adding decimal numbers. Consequently, RC1 is likely to continue developing instrumental mathematical understanding.

How did this series of events evolve? Under a didactical contract such as this, students blinkered themselves into conceiving of learning mathematics as a question of figuring out what the teacher/textbook/worksheet wanted them to do. Gaining relational understanding of mathematical ideas was not an option that they saw as belonging to their foregrounds. Their conceptions of the didactical contract meant that there was little likelihood of constructing intentions for learning of mathematics and thus getting involved in learning-as-action. Deprived of the possibility of intentional learning acts and meaning construction of valuable mathematics, learning took place as forced activity.

DOING FRACTIONS IN A NEW ZEALAND CLASSROOM

The learning environment in the New Zealand classroom was different. However, Research Child 2 (RC2) did not appear to learn any more conventional mathematics than RC1. The students had to fold a paper circle twice so that four equal quarters were produced. They then had to cut and glue the newly found quarters onto posters, which already showed two halves from a previous lesson. The students had individual posters, but worked in pairs. Although only one answer was anticipated, students were expected to manipulate the shape so that they discovered how to find four quarters. The teacher seemed to expect students to evaluate whether they achieved the required outcome. Her role was to facilitate the students into doing this. Thus, the learning environment was more a landscape of investigation than belonging to the exercise paradigm (Alrø & Skovsmose, 2002). However, this extract illustrates that although the teacher may have wanted to operate in a landscape of investigation, students' reasons for accepting the invitation to participate could be co-opted by other influences. These affected students' learning intentions so that underground intentions developed.

The extract comes from the second day of recording at the end of July 2005. The teacher began the lesson by discussing fractions and emphasized that the quarters had to be equal in area. Then the children were separated into groups.

52	**Teacher:**	Okay, and freeze. I just need to change a couple of partners over, Partner, could you work with RC2 please, . . . Okay. Quarters. . . .
53	**Partner:**	I like the fraction. *[Quite noisy as class getting into pairs]*
54		. . .
55	**RC2:**	Look I can cut it in half. Like that, and that, done. *[Pause]*
56	**Partner:**	RC2?
57	**RC2:**	Mm?
58	**Partner:**	If the quarter, get it, the quarter crashes, so it dies, so we have it next
59	**RC2:**	What?
60	**Partner:**	Instead of quarters.
61	**RC2:**	I'm going to find the half one, here's the half thing.
62		. . .
63	**RC2**	I'm going to have the whole pizza.
64	**Partner:**	Where's the whole pizza?
65	**RC2:**	This, mine.
66	**Partner:**	I can't see, I can't see. I get this.
67	**RC2:**	I get this and that.
68	**Partner:**	Oh, and these two whole, whole, whole *[makes gulping noises]*
69	**RC2:**	I got the whole pizza.
70	**Partner:**	Who ordered a whole pizza? I ordered three whole pizzas.
71	**RC2:**	Yep
72	**Teacher:**	Remember the paper you have to show two ways of cutting that into quarters.
73	**RC2:**	Oh, I was, two ways.
74		. . .
75	**RC2:**	Where on earth is the pieces?
76		. . .

77 **RC2:** What? Two ways of cutting in quarters.
78 **Partner:** Cutting in quarters
79 **RC2:** Mum, what are two ways of cutting this into quarters?
80 **Partner:** Impossible.
81 **Mum:** Think about the way you did it on the poster. Maybe make this whole piece into quarters.
82 **RC2:** I've got one. This is, I'll show you, like this, and that.
83 **Partner:** Where is it? Where is it? Where are you naughty thing?
84 **Teacher:** Ok, where are you up to Partner?
85 **Partner:** I'm going like before.
86 **Teacher:** I beg your pardon, you've been doing that for a long time, where's the pen gone?
87 **Partner:** I don't know....
88 **Teacher:** You write a quarter.
89 **RC2:** I've got one to do, I just need to find my scissors.
90 **Teacher:** Remember we did the folding and cutting?
91 **RC2:** Yeah, I'm doing that, and then.
92 **Teacher:** Have you folded it RC2, to check?
93 **RC2:** Oh.
94 **Teacher:** Fold, fold it and check if you're going to do it that way. Oh, what's happened? Are you going to have quarters? I'll give you a new piece and you can try it again.
95 ...
96 **Teacher:** Ok, well, have you've done your quarters in your head? *[to RC2 and Partner]*
97 **RC2:** Squares. Look what I did.
98 **Teacher:** Ok, how many pieces have you got?
99 **RC2:** Four.
100 **Teacher:** So what's the, hold up one quarter. Ok, put it back down. Show me one quarter in the pizza, which one would be a quarter?
101 **RC2:** Oh no, not that one. That's a half.
102 **Teacher:** That's a half.
103 **RC2:** Ummm.
104 **Teacher:** Have you got one quarter? You haven't have you, could you cut that pizza into quarters.
105 **RC2:** Yep.
106 **Teacher:** That's it, well done.

The activity revolved around active, hands-on problem solving. However, as the activity progressed, gaining relational mathematics understanding become a secondary focus for both the students and the teacher. The teacher wanted the students to produce four quarters by folding the paper and thus gain a relational understanding that (1) each of the four parts was equal in size, (2) each one was called a quarter, (3) and four quarters made a whole. When they realized that they could not easily do the task, the children engaged in imaginative play that contributed to them interacting socially, while at the same time not being seen by the teacher as being off-task.

The teacher's contributions suggested that she believed that students would learn that quarters were four equal parts of a whole, if they folded the paper appropriately and then cut them out. She had done a lot of preparatory work at the start of the lesson with physically cutting objects such as broccoli that could not produce equal quarters and discussing how fraction parts needed to be equal in size. As the teacher expected the activity would lead to understanding, giving direct instructions on how to produce the quarters was only to be used as a last resort. However, by the end of the extract, the teacher was channelled into doing this by the students' inability to complete the task. Consequently, the aim changed to producing four equal parts. Blomhøj (1995) suggested that in the didactical contract that operates in a traditional mathematics classroom, the teacher "develops a whole arsenal of small instructions and corrections to problem solving" (p. 18) that are used to ensure that students gain success. The students come to rely on these being implemented. The teacher in this class may have wanted to operate in a landscape of investigation. However, her reversion to funnelling questions and commands (72, 88, 90, 92, 94, 98, 100, 104) showed she adopted a more traditional approach in order for the children to be successful. Rather than the teacher driving what happened in the class, the children's lack of success resulted in the teacher changing her practices in the same way that RC1 had altered how Troels responded.

Although RC2 began by focusing on what had to be done (55), Partner quickly turned the pair work into imaginative play by talking about crashing quarters (58). Then RC2 described the shapes as pizzas (63), probably because the teacher had described the paper shape in this way earlier. RC2 accepted that the teacher's intervention meant that they had to return their focus to the task (73). RC2, like RC1 in the previous extract, tried hard to fulfill the requirements of the task. However, she had little understanding of what was required of her (77, 79) and she only produced the four quarters after specific instruction from the teacher (94).

Alrø and Skovsmose (2002) described a resistant group in a 9th grade mathematics class who, instead of engaging in the task, chose to zoom out. The students thought that they were doing what the teacher required as well as doing some of what they wanted to do. Nonetheless, very little conventional mathematics was learned by that group. The situation for RC2 was different. The zooming-out to imaginative play with Partner was not done in tandem with completing the activity or from deliberate resistance to participating. RC2's confusion over what quarters were and the two ways of folding contributed to the move into imaginative play and zooming-out.

Nuthall's (2007) comment that students learn what they do suggests that RC2 and her partner did learn to produce four equal parts. However, it was uncertain that either of them ascribed to this activity a relational understanding about what a quarter was and how it related to a whole. They also learned that they could reduce the opportunities for being in trouble for being off-task by making statements such as "I've got one to do, I just need to find my scissors" (89). These comments suggested to the teacher that the children were engaged, even if they were doing the task much more slowly than other children. The remarks, however, restricted the type of support that the teacher offered because she did not realize that they had little idea of what to do. Even when the teacher directly challenged Partner about the amount of work done (86), it was unlikely that she recognized that they had not produced the quarters because they did not know how. Her later interventions became more directive, but never really made connections with the relational understanding about fractions. Although at the end of the extract, RC2 realized that two halves had been produced rather than four quarters, no connection was made between halves and quarters.

The lack of strategies for gaining clarification about the task contributed to the move from a landscape of investigation into a form of the exercise paradigm. RC2 asked the mother for help but not the teacher, nor Partner. In none of the recorded mathematics lessons did RC2 request help, including when the mother was not in the class, and at times insisted that pair work did not involve working together (Meaney, 2007). The child's reticence to ask the teacher for help, combined with the teacher's belief that the children would learn more if they did it themselves, resulted in a more instrumental understanding about quarters being gained. Although an invitation to a landscape of investigation was accepted (53), RC2 and Partner did not have skills to zoom in on the task by asking questions about it and how it related to their learning.

RC2 and Partner accepted the teacher's invitation to participate in a landscape of investigation. However, the students' incomprehension about what was required meant that they quickly moved into activity that they could do—imaginative play. The teacher's belief that the activity would lead to relational understanding was distorted by the invitation's lack of correspondence with the students' prior understanding. Consequently, the activity was reduced to cutting a shape into four equal parts and thus became forced activity. RC2's underground intention of playing while staying out of trouble was from a very different base than Alrø and Skovsmose's (2002) students. Previous classroom experiences contributed to RC2's beliefs about the importance of being seen to be doing the task and not asking clarifying questions. However, these were not the only contributions to what influenced the change of the activity from one of being a landscape of investigation to one of forced activity.

DISPOSITIONS TO LEARN

Although the contexts for the lessons were quite different, the outcomes for the students were similar—that is, instrumental rather than relational understandings. Both students ended up being engaged in "forced activities" rather than ones in which their own learning intentions came into play. However, it was both students' reactions, RC1's ignoring of Troels' requests to "think again about what could be wrong" and RC2's inability to produce the four quarters, that resulted in the adults changing their aims. This contributed to the development of instrumental understanding.

RC1's aim was to produce numbers when responding to problems. Skemp (1976) described how mismatches between students' and teachers' expectations of the need for relational and instrumental mathematical understanding could result in frustration. RC1's frustration occurred because the teacher-like person did not realize that unreasonableness in answers was something that could be ignored. This person also did not fulfill his obligations to the didactical contract because he refused to confirm or deny the correctness of the answers. Although RC1 knew the answer did not match the key, what the child felt was needed was not provided by Troels. The problem solving context, and how RC1 perceives the appropriate way to operate, formed the didactical contract. This did not lead to relational understandings of mathematics.

The type of problem solving that RC2 was asked to engage in was different. It conformed to Blomhøj's (1995) suggested rules for the didactical contract in traditional mathematics classrooms only in some ways. Therefore, RC2 had more opportunities for gaining relational understandings about mathematics. However, RC2's interpretation of the didactical contract was that if the task was too difficult to do, then it was necessary to be seen to do something. This was more appropriate than asking for help from the teacher or Partner. Consequently, RC2 participated in zooming-out behavior. This example suggests that it is not just resistant groups, such as those described by Alrø and Skovsmose (2002), that zoom out and thus lessen their learning opportunities. It would be interesting to know whether students such as RC2, if not given opportunities in future classes to engage with their teacher to change the didactical contract, become members of resistant groups when they enter high school.

Previous classroom experiences and expectations about future behavior required of them will have contributed to both students' interpretations of how to interact in the learning environment. However, this is not the complete story. In the following sections, we hypothesize about how knowing more about the children's backgrounds will affect our understanding of their learning intentions and thus the meaning that they ascribe to what they are doing. We ponder whether it adds to or modifies our interpretation if some socio-political features are taken into account or whether it could also just become an excuse for attributing the lack of learning to the students. Although this consideration of background as affecting dispositions should be balanced by also considering foregrounds, we do not have information from the children themselves about these.

Backgrounds and Their Influence on Understanding How Meaning is Ascribed

In discussing the classroom episodes, we deliberately restricted the information about our students to the bare minimum of their age and country. We did not disclose their gender by naming them Research Child 1 and 2. We said nothing about their family background, apart from a little about RC2's mother in her capacity of research assistant. Now we ask if it makes a difference to us and to others if we describe both children as girls whose families are immigrants in their respective countries.

Being Girls

There is evidence to suggest that girls in some parts of the world approach how they operate in a classroom differently than boys do (Kenney-Benson, Pomerantz, Ryan, & Patrick, 2006). Girls were more likely to "refrain from disruptive classroom behaviour" and this contributed to "greater effortful learning over time" (p. 11) Females also generally rate their mathematical abilities lower than males (Alton-Lee & Praat, 2000). If RC1 and RC2 are representative of the girls in these other studies, it may be that this influenced their ascribing of instrumental mathematical understanding to the activities in which they participated.

RC1 concentrated hard on solving all the questions, which is what she perceived as being required of her in a mathematics classroom. This seems to match with the suggestion that girls make more effort at their learning. She also wanted Troels to check the correctness of her answer and would not be encouraged to evaluate her own answers. This could be interpreted as RC1 showing little confidence in her own mathematical ability.

RC2 did not participate in disruptive behavior, even though she was unsure about what she should do. Her imaginative play stayed under the radar of the teacher. It is difficult to know whether, if she had understood what to do, she would have made a more concerted effort to learn. Over the course of the twenty weeks, RC2 rarely got in trouble even though she was often confused about what she had to do. It was unlikely that she learned any relational understandings in mathematics in this period.

Of course, with only one interaction analyzed for both girls, it is not possible to understand fully how gender affects these students. It is possible that the typical behaviors associated with being female are visible in the extracts. However, it may be that other aspects of the students' backgrounds and foregrounds also contributed to the meaning ascriptions attached to the activities.

Being Immigrants

As well as being girls, both RC1 and RC2 were from immigrant backgrounds. RC1's family was part of a minority of Muslim immigrants from the Middle East in Denmark. It was likely that her parents were refugees from the civil war in Lebanon. RC1 had been born in Denmark. RC2's father had come to New Zealand from Samoa to study, while RC2's mother had been born in New Zealand, but of Samoan parents. In both families, the children spoke the dominant language, Danish and English respectively, but also the language of their parents. RC1 had more fluency in Arabic that RC2 had in Samoan.

The culture of Middle Eastern minority families in Denmark is in some respects different from that of the majority of ethnic Danish families. According to Khader (2006), a Danish politician who migrated from Syria when he was a child, the family is the basic, central, almost holy, social unit in the Middle East Islamic societies. The father is responsible for bringing up the children according to Islamic prescriptions. Maintaining the honor of the family is essential for all family members. Migrants from the Middle East bring this type of family culture with them, and even if adjustments are made over the course of time, central features persist. Hence, children in Muslim families of Middle East origin are positioned differently in the family compared to mainstream Danish majority families.

These attitudes are reflected in the small amount of mathematics education research that has been done in this area in Denmark. Alrø, Skovsmose, and Valero (2008) interviewed a Muslim Kurdish girl who had recently migrated to Denmark with her family. The girl explained that, in her culture, children could not say no to their parents. The same type of relation between children and adults was reflected in Troels' extensive interviews with students in RC1's class (Lange, submitted). He asked them if there were differences as to what children could decide in the different school subjects. Some children, all of Middle East minority background, including RC1, said that there were no differences, because children cannot say no to what the teacher says. The teacher tells them what to do and then the children do it.

A didactical contract focusing on getting correct answers to exercises that had little inherent interest and fostering instrumental understanding could be seen as coherent with RC1's family background. Both required "obedience" to an external and, in principle, unquestionable authority and a corresponding "demand" from the child to know what the rules were. If this speculation was reasonable, it would be an example of knowledge of students' background adding to our interpretation.

There are similar features in the Samoan background of RC2's family. Jones' (1988) research into Pacific Island girls in high school classes suggested that they "learned from their parents that the teacher, like the priest or pastor, holds valuable knowledge and as such is to be respected, not questioned by mere students" (p. 149). RC2, as a Samoan girl, although much younger than the girls that Jones studied, may also have perceived the teacher in this light. This may have contributed to her unwillingness to seek clarification about the task. It may also have affected her desire to be seen as doing what was required, and thus not get into trouble with the teacher.

However, Skovsmose (2005b) stated that the background of a person is a *personal interpretation* of the person's socio-political context and not the context as such in some "objective" form. It is possible that RC1 interpreted her minority Middle East family context in a way that made her particularly susceptible and conducive to the construction of a traditional didactical contract, and once established, reinforced her adherence to it. It may also be that RC2's Samoan background made her reluctant to attract the teacher's attention either by asking a question or by being seen as off-task. This contributed to the teacher reforming her interactions so that they more closely aligned with the exercise paradigm. Other students could interpret similar "objective" backgrounds in different ways, depending on the other experiences and foregrounds that they bring with them. Therefore, our analysis is only relevant if the girls themselves saw it as being about them.

Consequently, we do not want to suggest a deterministic general relation between being of Middle East descent and adhering to a traditional didactical contract. Nor do we want to suggest that girls of Samoan background were all likely to react in mathematics classrooms in similar ways. To do so would be to reduce people to categories and unreasonably simplify a very complex matter.

CONCLUSION

In this chapter, we considered how the didactical contract operating in classrooms affected the meaning ascriptions made by two students to the learning activities in which they engage. Rather than just concentrating on the snapshot of what happened in the two classroom extracts, we added more information about the students to see how this affected our interpretation. Information about the students' backgrounds did provide supplementary explanations for the children's interpretations of how they should act. However, like knowledge of the didactical contract, this information is only a part of what students use when ascribing meaning to the activities in which they engage. Teaching and learning are complex processes. Simplifying them in order to determine uncomplicated explanations as to why so many children do not gain valuable mathematical understanding is unlikely to help either student or teacher.

The didactical contract provides information about how the classroom culture affects the teaching and learning processes. This metaphor highlights the existence of a set of rules framing interactions, even when these rules are fluid, as is the case in landscapes of investigation, and their nature a "tacit negotiated compromise." The didactical contract provides a frame for teaching and learning to take place. However, its relative stability does not mean that it is non-negotiable. Over the course of time, it simultaneously shapes the participants' conceptions of school mathematics teaching and learning, and these conceptions shape the frame. From our perspective of the students' meaning-ascribing processes, we see how the students are framed and how they try to affect the framing. In the case of RC1, the frame is clearly a traditional mathematical contract. In the case of RC2, the contract is not traditional in the beginning but easily reverts when difficulties are encountered.

Schools are ambiguous places for children. On one hand, they are important social and societal institutions aiming at *the good* for society and children. On the other hand, they are created and controlled by adults and do not necessarily nor always accord with children's wants, needs, and wishes. For us, both children have agency; they are not powerless. In both cases, they play off the teacher against the school setting.

RC1 brushed off Troels' attempts to engage her in conversations that she did not see as relevant to doing mathematical activities. Although she listened respectfully, she refused to be drawn into irrelevant side issues. Instead, she insisted on her right to have her answers evaluated and to be provided with direct instruction so she could get the "correct" answer. This was, after all, the purpose of the activity. Knowing the inherent appropriateness of her response, she could exert her agency in a respectful but forceful manner. Troels' alternatives were not to be taken seriously as they were not part of the way that mathematics teaching and learning was framed by the important mathematical authority, her own teacher, and her previous experiences.

In the case of RC2, the teacher intended that the children discover something about quarters. However, she was squeezed between wanting the children to be successful and the fact they had not finished the task with little time remaining. Achieving the product of the activity, the shape cut in four equal pieces, became the goal, and the intended (relational) understanding of mathematical concepts was sacrificed. It was the children's reactions that resulted in the teacher making changes to her approach. RC2 may have been invited into the activity, but her conceptual background, her lack of understanding of wholes and parts, meant she could not engage. Maybe her background contributed to RC2's being cut off from zooming-in to the activity. Faced with an unintelligible demand, and conscious of her non-comprehension ("Mum, what is cutting in quarters?"), her response was to appear busy to avoid teacher attention. Her busyness was around the mere manipulative aspects of the exercise. The children were reinforcing the experience that eventually the teacher would give in and tell them what to do. They had not learned what was intended—it was incomprehensible to them anyway—but they had produced the outer signs of learning. They might sense that the teacher was not quite happy about it, but they were safe and could not be blamed.

Since research into mathematics education began, researchers have been keen to inform teaching practices about how to increase the likelihood of students gaining valuable mathematics learning. However, Alrø and Skovsmose's (2002) ideas about what contributes to students' meaning ascriptions illustrate the complexity of the learning process. Knowledge of how the didactical contract operates provides some information. However, as the extracts of our two girls on either side of the world show, the didactical contract provides only some information about how meaning is ascribed. Concentrating on the girls' backgrounds also only provides limited information, with the significant risk that taken in isolation this could stereotype the lack of relational learning as being the responsibility of the girls and their home cultures or gender.

Our analysis instead turned to the students' enactment of their agency. Alrø and Skovsmose (2002) stated that students should be equal partners in investigations and this would lead to students' agency, resulting in valuable mathematics learning being achieved. However, other demands such as trying to cater to the needs of a diverse group and time restraints can result in students' enactment of their agency in ways that resulted in a lack of relational mathematical understanding being learned. Instead, what needs to be done is for teachers to find ways to discuss mathematics and mathematics learning so that students have more options for using their agency to form valuable learning intentions. In Figure 49.1, we left undiscussed the impact of reflection on meaning ascriptions. As a consequence of our analysis, it seems clear that if these two children were to gain valuable mathematics understandings, then they needed to engage with their teachers in reflecting about the supports and impediments for this to occur. Teaching is not something done by teachers that is completely independent of learning done by students. Yet without joint reflection, this is what classroom participants can end up doing with neither aware of that the other's perception of the situation is different. Joint reflection may well contribute to the teacher being able to harness the student's agency for a common goal.

ACKNOWLEDGEMENTS

We wish to thank Naomi Ingram, Tracey Smith, Paola Valero, and Arthur Winzenried for helpful comments during the writing of the chapter.

REFERENCES

Alrø, H. & Skovsmose, O. (2002). *Dialogue and learning in mathematics education: Intention, reflection, critique.* Boston: Kluwer Academic Publishers.

Alrø, H., Skovsmose, O., & Valero, P. (2009). Inter-viewing foregrounds: Students' motives for learning in a multicultural setting. In M.César & K. Kumpalainen (Eds.), *Social interactions in multicultural settings* (pp. 13–37). Rotterdam: Sense Publishers.

Alton-Lee, A. & Praat, A. (2000). *Explaining and addressing gender differences in the New Zealand Compulsory School Sector.* Welllington, New Zealand: Ministry of Education.

Balacheff, N. (1990). Towards a problématique for research in mathematics teaching. *Journal for Research in Mathematics Education, 21*(4), 258–272.

Blomhøj, M. (1995). Den didaktiske kontrakt i matematikundervisningen. *Kognition og pædagogik, 4*(3), 16–25.

Boaler, J. & Greeno, J. G. (2000). Identity, agency, and knowing in mathematical worlds. In J. Boaler (Ed.), *Multiple perspectives on mathematics teaching and learning* (pp. 171–200). Westport, CT: Ablex.

Brekke, G. (1996). A decimal number is a pair of whole numbers. In L. Puig & A. Gutierrez (Eds.), *Proceedings of the 20th Conference of the International Group for the Psychology of Mathematics Education,* (Vol 2, pp. 137–144). Valencia, Spain: International Group for the Psychology of Mathematics Education.

Brousseau, G., Balacheff, N., Cooper, M., Sutherland, R., & Warfield, V. (1997). *Theory of didactical situations in mathematics: Didactique des mathématiques, 1970–1990.* Dordrecht: Kluwer.

Cobb, P., Stephan, M., McClain, K., & Gravemeijer, K. (2001). Participating in classroom mathematical practices. *The Journal of the Learning Sciences, 10*(1/2), 113–163.

Cobb, P., Wood, T., Yackel, E., & McNeal, B. (1992). Characteristics of classroom mathematics traditions: An interactional analysis. *American Educational Research Journal, 29*(3), 573–604.

Hamilton, E. H. (1924). Reviewed works. *The Mathematical Gazette, 12*(171), 174–176.

Herbel-Eisenmann, B. (2003). *Examining "norms" in mathematics education literature: refining the lens.* Retrieved June 21, 2007, from http://www.msu.edu/~jansenam/NCTM2003.html

Jones, A. (1988). Which girls are 'learning to lose'? Gender, class, race and talking in the classroom. In S.Middleton (Ed.), *Women and Education in Aotearoa* (pp. 143–152). Wellington, New Zealand: Allen & Unwin.

Kenney-Benson, G. A., Pomerantz, E. M., Ryan, A. M., & Patrick, H. (2006). Sex differences in math performance: The role of children's approach to schoolwork. *Developmental Psychology, 42*(1), 11–26.

Khader, N. (2006). *Ære og skam : det islamiske familie- og livsmønster i Danmark og Mellemøsten.* (3rd ed.). København: Borgen.

Lange, T. (2007). Students' perspectives on learning difficulties in mathematics. In L. Ø. Johansen (Ed.), *Proceedings of the 3rd Nordic Research Conference on Special Needs Education in Mathematics* (pp. 171–182). Aalborg: Adgangskursus, Aalborg Universitet.

Lange, T. (2009 forthcoming). "Tell them that we like to decide for ourselves"—Children's agency in mathematics education. In *Proceedings of the Sixth Congress of the European Society for Research in Mathematics Education.*

Meaney, T. (2007). Authority relations in the acquisition of the mathematics register at home and at school. In D. Pitta-Pantazi & G. Philippou (Eds.), *Proceedings of the Fifth Congress of the European Society for Research in Mathematics Education* (pp. 1230–1239). Department of Education, University of Cyprus: European Society for Research in Mathematics Education. Retrieved May 26, 2010 from http://ermeweb.free.fr/Cerme5.pdf

Nuthall, G. (2007). *The hidden lives of learners.* Wellington: New Zealand Council for Educational Research.

Skemp, R. (1976). Relational and instrumental understanding. *Mathematics Teaching, 77,* 20–26.

Skovsmose, O. (2005a). Foreground and politics of learning obstacles. *For the Learning of Mathematics, 25*(1), 4–10.

Skovsmose, O. (2005b). Meaning in mathematics education. In J.Kilpatrick, C. Hoyles, & O. Skovsmose (Eds.), *Meaning in mathematics education* (pp. 83–100). New York: Springer Science.

Thorndike, E. L. (1922). *Psychology of arithmetic.* New York: Macmillan.

Thorndike, E. L., Cobb, M. V., Symonds, J. S., Wald, E., & Woodyard, E. (1923). *Psychology of algebra.* New York: Macmillan.

Valero, P. (2004). Postmodernism as an attitude of critique to dominant mathematics education research. In M.Walshaw (Ed.), *Mathematics education within the postmodern* (pp. 35–54). Greenwich, CT: Information Age Publishers.

Wedege, T. & Skott, J. (2007). Potential for change of views in the mathematics classroom. In D. Pitta-Pantazi & G. Philippou (Eds.), *Proceedings of the Fifth Congress of the European Society for Research in Mathematics Education,* (pp. 389–398). Department of Education, University of Cyprus: European Research in Mathematics Education. Retrieved May 26, 2010 from http://ermeweb.free.fr/Cerme5.pdf

SECTION V

Contributions from Finland

Lenni Haapasalo, *Section Editor*

50

MAKING MATHEMATICS THROUGH PROGRESSIVE TECHNOLOGY

Lasse Eronen and Lenni Haapasalo

UNIVERSITY OF EASTERN FINLAND

Abstract: The chapter illustrates results and experiences of the ClassPad project emphasizing the making of informal mathematics within "Minimalist Instruction" approach and within the framework of eight main activities and motives, which have been proved to be sustainable in the history of human thinking processes and for the generation of new mathematics. The results suggest that mathematical activities with a progressive pocket computer, which allow the change of representations and simultaneous activation of conceptual and procedural knowledge, can shift students' mathematical identity even during a short period of time within these motives and activities into creative direction, opening new progressive ways to arrange teaching and learning of mathematics. As regards the instructional design in classroom, the chapter gives examples how to combine quasi-systematic model and minimalist instruction.

Keywords: conceptual, heuristics, minimalist instruction, procedural, technology-based

INTRODUCTION

Fixing the defects in educational research mentioned by Kadijevich (2004) will require the use of modern technology in all its forms. However, today even small pocket computers, such as ClassPad made by Casio (see http://www.classpad.org), can be utilized for those demands (cf. Haapasalo, 2004; Haapasalo & Kadijevich, 2003; Kadijevich et al., 2005). ClassPad allows, namely, the use of the drag-and drop technology, where the student can easily manipulate mathematical objects between two windows, illustrating two different forms of mathematical representation. In many cases this means forming links between conceptual and procedural knowledge, which is a relevant perspective from which to evaluate the sustainability of educational technology (cf. Haapasalo & Siekkinen, 2005).

When considering technology-based learning to implement and reinforce creative thinking, the focus has been shifted from a technology-oriented viewpoint to a humanistic view, stressing cognitive, affective, and social variables involved in the learning process. By using examples from mathematics and early childhood education, Haapasalo and Siekkinen (2005) presented support for their five hypotheses: (1) Technology can enhance learning skills (metacognitions) among teachers and students; (2) it is reasonable to utilize minimalist instruction, especially when technology is involved; (3) technology can shift learning from the classroom to free time; (4) technology-based learning can benefit from the "learning by design" principle; and (5) the most appropriate way to implement technology in teacher training is to use it as a solid part of knowledge structure and of student pedagogical thinking. With regard to the impact of pedagogical philosophy and knowledge structures, learning may be based on a developmental approach that assumes that procedural[1] knowledge is based on conceptual knowledge or on an educational approach that assumes the opposite.

The First Sourcebook on Nordic Research in Mathematics Education, pages 701–710

When planning and creating learning environments, the teacher usually encounters a conflict between these two knowledge types: how much students should understand before they are able to do, and vice versa (see Haapasalo, 2003).

Concerning technology-based learning, the first challenge arises from the structure of the topic to be learned, whereas the other is caused by the instructional variables required for technology use. The fact that students seem to learn mathematical as well as technical skills effectively outside school forces us to ask if there is something wrong inside school as far as the question of "how to learn" is concerned. A detailed analysis of the Finnish TIMSS and PISA results, for example, reveals that it is not necessarily school teaching that impacts student performance (see Kupari, 2003; Törnroos, 2003). This makes us ask which factors in our education are important for the development of thinking abilities. If we accept the assumption that the main task of education is to promote a skillful "drive" along knowledge networks so as to support pupils in making use of their rich activities outside school, it seems appropriate to look for a suitable balance between conceptual and procedural knowledge.

The logical relation between the two knowledge types in the *developmental approach* is based on a *genetic view* (G—i.e., procedural knowledge is necessary for the conceptual) or a *simultaneous activation view* (SA—i.e., procedural knowledge is necessary and sufficient for conceptual knowledge). It seems appropriate to claim that the goal of any education should be to invest in conceptual knowledge from the very beginning. If so, the logical basis of this *educational approach* is the *dynamic interaction view* (DI—i.e., conceptual knowledge is necessary for the procedural), or, again, the simultaneous activation view. Such a view assumes that the learner has opportunities to activate simultaneously conceptual and procedural features of the current topic. By "activating," we mean certain mental or concrete manipulations of the representations of each knowledge type. Being at the intersection of two complementary approaches, the simultaneous activation view is loaded with special challenges to utilize modern progressive technology.

BACKGROUND

Getting the Idea

The starting point of our ClassPad project was that Casio Europe donated 25 ClassPad calculators for making empirical research within the guidelines highlighted by the second author during his consulting work with Casio (cf. Haapasalo & Kadijevich, 2003; Haapasalo, 2004; Haapasalo et al., 2004). On the basis of these papers and the second author's experiences with his own children it was from the first beginning clear that the focus of the project would be not only on the school teaching but also on students' free time. The first author accepted this progressive challenge in May 2005, when his students were in 8th class. There were only few days time to plan what could be done during the summer holiday as a preliminary pilot study. The only possibility was to accept very Minimalist Instruction when utilizing ClassPad features. Our idea was to profit from the experiences gained from this spontaneous effort when planning ClassPad environments in the 9th grade classroom after the summer holiday.

Dilemma Between Systematization and Minimalism

We considered it appropriate to take the Zimmermann (2003) network of sustainable mathematical activities (see Figure 51.1 in the chapter by Haapasalo & Eronen, this volume, p. 712) as an element in our theoretical framework for the structuring of learning environments and for analyzing students' cognitive and affective variables—especially the "find" and "play" corners, which represent heuristic activities in the sense of Pólya. Within these activities, our ClassPad study focuses on "changing representation," which is not only a powerful thinking tool to enhance problem-solving processes, but it might also promote links between procedural and conceptual knowledge. So, conducting research with ClassPad might be a solid response to the neglected areas mentioned in Kadijevich (2004).

It is the right-hand half of the octagon that emphasizes creative human activities, which very often run optimally without any external instruction or demand. Students frequently neglect teachers' tutoring, or they feel they do not have time to learn how to use technical tools. Teachers similarly feel they do not have time to teach how these tools should be used. This problem becomes even more severe when the versatility of advanced technology cannot be accessed without first reading heavy manuals. The term *minimalist instruction*, introduced by Carroll (1990), is crucial not only for teachers but also for those who write manuals and help menus for the software. Carroll observed that learners often tend to "jump the gun." They avoid careful planning, resist detailed systems of instructional steps, tend to be subject to learning interference from similar tasks, and have difficulty recognizing, diagnosing, and recovering from their errors. We pick up the following characteristics of minimalist instruction (cf. Lambrecht, 1999; van der Meij & Carroll, 1998) keeping in mind the fostering of problem-solving abilities. These features of *minimalism* include several varieties of the constructivist view, offering also instructional assumptions (cf. Carroll & Van der Meij, 1998; Duffy & Jonassen, 1992; Duffy, Lowyck, Jonassen, & Welsh, 1993; Mirel, 1998; Redish, 1998):[2]

- Learning is modeled and coached for students with unscripted teacher responses.
- Learning goals are determined from real tasks, stressing doing and exploring.
- Errors cannot be avoided and should be used for instruction.
- Learners construct multiple perspectives or solutions through discussion and collaboration.
- Learning focuses on the process of knowledge construction and development of reflexive awareness of that process.
- A criterion for success is the transfer of learning and a change in students' action potential.
- The assessment is ongoing and based on learners' needs.

For too many students, one of the basic difficulties for the learning of mathematics is that very often entities appear as objects as well as processes.[3] By using concrete examples from upper secondary mathematics education, Kadijevich, Haapasalo, and Hvorecky (2005) point out that students should, among other things, be educated to respect the following two requirements: (1) when utilizing mathematics, don't forget available tool(s); when making use of tool(s), don't forget the underlying mathematics; and (2) to solve the assigned task, use, whenever possible, a process approach as well as an object approach, working with different representations (algebraic and graphical, for example). These demands can be realized if the teacher has fundamental know-how of the relation between conceptual and procedural knowledge. According to Rittle-Johnson and Koedinger (2004), the two knowledge types seem to develop iteratively, where a change of problem representation influences their relation (cf. Rittle-Johnson, Siegler, & Alibali, 2001). Such development was assumed in the pedagogical model developed by the first author. This model of the interplay between the two knowledge types makes use of spontaneous procedural knowledge as well as the simultaneous activation of conceptual and procedural knowledge. Figure 50.1 represents a sophisticated interplay between conceptual and procedural knowledge within the framework of the *MODEM* project.[4]

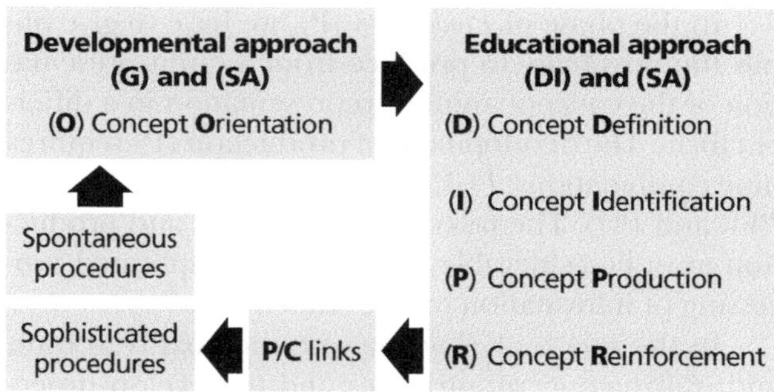

Figure 50.1 A sophisticated interplay between developmental and educational approach.

Orientation (O) forms the first phase of the systematic concept building. It basically utilizes a developmental approach: The interpretations of the situation can be based on mental models of the pupils, coming, more or less, from their naïve procedural ideas. These act like a wake-up voltage in an electric circuit that triggers another, much more powerful current, which is then amplified again. The procedural and conceptual knowledge types start to support each other, offering a nice opportunity to use the principle of simultaneous activation, for example. This principle, being at the intersection of the logical definitions of the two approaches, connects the developmental approach and the educational approach in a most natural way.

The role of the concept *Definition* (D) is to offer students opportunities to make their own investigations, to express the results of their investigation—especially in verbal forms in each case—and to argue about these results within the collaborative teams and between the teams. As a result of social construction, a definition for the concept is born, meaning that students try to fix the relevant determiners of the concept in verbal, symbolic, and graphic forms. Especially in the phases of orientation and definition, creative thinking and productive work are needed. The next phases of concept building utilize the principle of dynamic interaction. The idea is to give students a sufficient number of opportunities to construct concept attributes and procedural knowledge based on them.

In the phase of *Identification* (I), we have to give students opportunities to train themselves in identifying concept attributes in verbal (V), symbolic (S), and graphic (G) forms. For this we need six kinds of tasks (I): *IVV, IVG, IVS, IGG, ISS,* and *ISG*. During the learning process, the teacher must be ready, if necessary, to begin with tasks that require distinguishing between only two elements before going on to the identification of several elements.

In the phase of *Production* (P), we have to give pupils the possibility to produce from a given presentation of the concept another representation in a different form. The development of production (P) requires nine combinations: *PGV, PGS, PGG, PSG, PSV, PSS, PVS, PVV,* and *PVG.* The tasks of identification and production must be achievable without any complicated processing of information on the student's part.

In the phase of *Reinforcement,* the goal is to train and utilize concept attributes and to develop procedural knowledge to be used in problem solving and applications.

AIMS AND METHODS

The aim of the first stage of the *ClassPad project*[5] was to get experiences of what kind of impact minimalist instruction causes among 8th grade students, how they could use ClassPad voluntarily in their free time during their summer holiday, and how these activities would shift each of the eight Zimmermann components concerning the following questions:

1. How strong would each component appear in the student's view of mathematics (i.e., mathematical profile)?
2. How well does the student think he or she is in doing mathematics on each component (i.e., mathematical identity profile)?
3. How strong is the impact of computers in making of mathematics on each of the components (i.e., ICT-profile)?

Data for each type of *profiles* (in Figures 50.4–50.6) were collected by a web-based questionnaire.[6] Students had the opportunity to use—if they wanted to do so—ClassPad during their summer holiday. But there was not any kind of duty to use it. The only duty was to write a portfolio of their working according to certain guidelines. Immediately four students volunteered for this study and after a few days the number increased to fifteen. Those students were given a leaflet including some problems connected to the basic features of linear function, which is one of the most important concepts in the curriculum of grade 9. Before giving ClassPads to those students at the end of May, a cognitive test was administered to the students concerning the topic, and profiles were measured by using a web-based questionnaire. After the summer holiday these measurements were repeated in August. Furthermore, students were interviewed to get more detailed information by their answers and experience. The cognitive tests consisted of the sub-phases of the mathematical concept building in the right-hand box of Figure 50.2, utilizing multiple representations between verbal, symbolic, and graphic forms on the levels of identification and production.

In *Stage 2* the whole class, including the above-mentioned voluntary students, studied the basic features of linear function as a part of their normal school curriculum of 9th grade only using ClassPad. All mathematics student teachers were involved in the project to make their educational studies (see Haapasalo & Eronen, this volume). Besides the affective variables above, students' cognitive results were measured at the end of the working period by using the same test as above. The aim was to find out (1) how ClassPad can be used in classroom when studying topics of normal school curriculum; (2) how does working with ClassPad change students' profiles; (3) how does working with ClassPad affect students' cognitive scores; and (4) what could be the optimal way to combine minimalist instruction and a systematic pedagogical model.

RESULTS

Fostering Metacognitive Skills

Haapasalo (2003, 2007) gives examples of very pure metacognitive abilities among students and teachers. However, already during the first stage of ClassPad project, we noticed that students seem to be capable of solving much more complicated situations outside the classroom than in school by using the simultaneous activation principle. The sample below, taken from a student portfolio, reveals that the method is really used by students and that they have quite sophisticated metacognitive abilities when utilizing ClassPad in their free time. We invite the reader to focus on the basic hints mentioned in Pólya's (1973) checklist, and the sustainability of students' work.

6th session 15.07.2005. Time 00.27.
(duration 1h 15 min)

- The equation of a line is $y = 1.613x–0.5992$
- By changing the equation to $y = 2x–0.5992$ the angle between the line and y-axis becomes smaller
- By changing the equation to $y = 1x–0.5992$, the angle between the line and y-axis becomes bigger.
- The equation is $y = 1.613x–0.5992$.
- I change the equation to $y = 1.613x–0.4$.
- I don't see any changes.

- I change the equation to $y = 1.613x–4$, the line moves to the same direction away from origin.
- When changing the equation to $y = 1.613x+4$, the line moves in the same way, but to another direction on x-axis with equal distance from the origin.
- I will continue in the morning. Time is now 01.42.

7th session 15.07.2005. Time 11.46.
(duration 2 h 37 min)

- I try again, and now I do it with a little bit easier equation than yesterday. I think that now I probably can do it.
- The equation is $y = 2x–3$
- I move the line to the origin.
- By changing the equation to $y = 2x–5$ the line moves "forward to the positive side of the x-axis."
- By changing the equation to $y = 2x–1$ the line has gone to the same direction as before, but the distance is smaller.
- It was something like this -> Amazing . . .

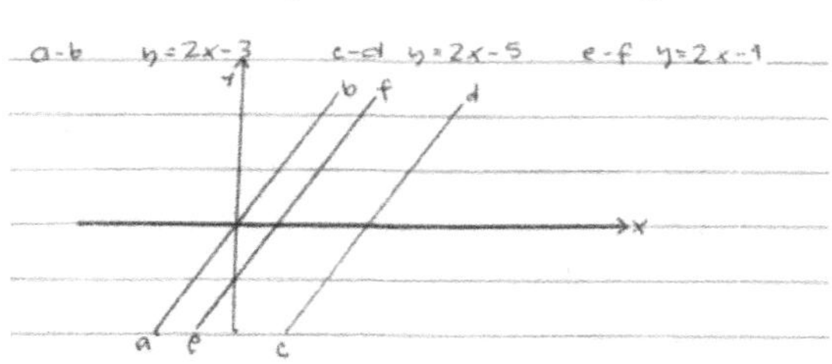

- Here is probably the right answer, but the negative numbers at the end of the equation depend on the y-axis.
- Maybe it's a better idea to think the meaning of that other number, at first . . . or I mean the slope! Because yesterday I already knew that it interacts to the angle, but I didn't realize, that it is the slope.

Shifting Students' Profiles

As an example, we give results concerning affective behavior of the very same student who produced the portfolio examples above. She can neither be considered to be especially motivated in learning mathematics during her studies in grade 7 nor in grade 8. Figure 50.2a represents her view of mathematics before and after the ClassPad work, showing some shift from "apply" to "find" and "argue" corners. The figure in the middle shows that there is the same kind of shifting in her self-confidence. The interview revealed expressions like: *"Now I know better and see things in different light."* However, students' conceptions of the computer's role in doing mathematics seems to be more interesting. Working with ClassPad has shifted the profile even more to a creative direction, the biggest change being in playing. In the interview the student expressed, for example: *"In May I could not even think to play ClassPad in summer holiday. However, I noticed that it was very capable for playing with mathematics."* Maybe the most surprising shift occurred in the "calculate" dimension. The student became aware of the versatility of the technical tool, which was connected with a decrease of the relative intensity of the counting belief: *"ClassPad is suitable for calculating, but if you want to learn how to calculate, you have to do something by hand."*

The study of students' Zimmermann profiles is an interesting and challenging task. When analyzing all students' profiles at stage 2 of the project, we noticed that there is some kind of average shift in the whole class. Furthermore, after ClassPad work, students were

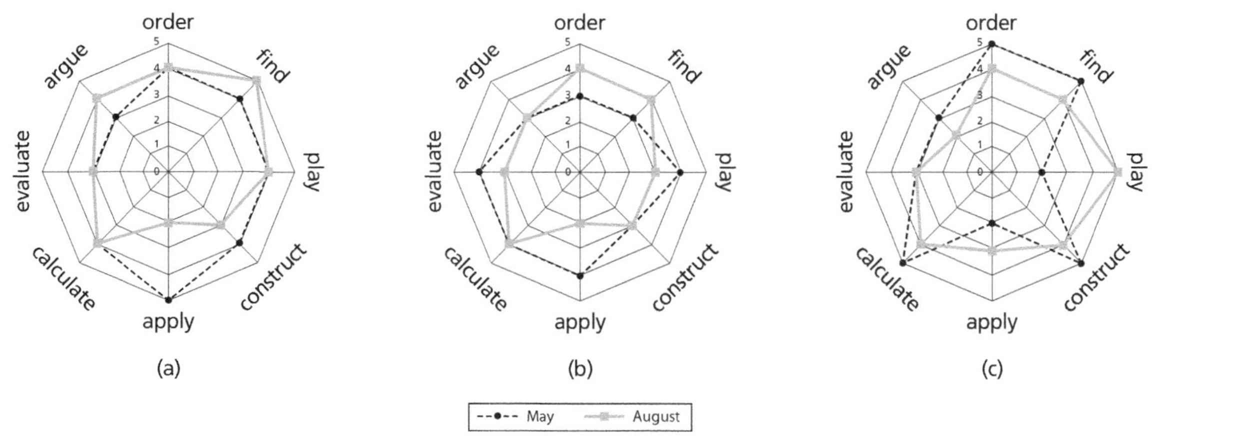

Figure 50.2a–c Zimmermann- profiles before and after her ClassPad work. Mathematical profile on the left, identity profile in the middle, and ICT-profile on the right.

sure about their answers. However, when analyzing their responses individually, there is a huge differentiation. There are many interesting methodological challenges to be solved. It might be clear that quantitative analysis needs support from qualitative methods. On the other hand, a quantitative finding can trigger interesting hypotheses—for example, when a good, conceptually oriented peer teacher was teaching her procedurally bounded classmate (cf. Järvelä, 2003). Figure 50.3 shows that this kind of peer teaching period enriched the profile of the peer teacher but eroded her mathematical self-confidence. Interestingly, the profiles of those two students seem to run in opposite directions. The most interesting—even alarming—finding is that at the same time when the classmate seems to think (perhaps wrongly) she could find, apply, and argue better than at the beginning, her peer teacher's own self-confidence in making mathematics seems to degenerate (see Figure 50.4). This kind of finding might be extremely interesting not only when thinking about the fostering of problem-solving abilities, but also when considering the mental problems among teachers.[7] It is—again—a symptom that behaviourist ways of teaching can damage both student and teacher.

Interplay of Systematization and Minimalism

Emphasizing the genesis of heuristic processes and the ability of students to develop intuition and mathematical ideas within a constructivist or minimalist approach can hardly be reached without a thorough planning of the learning environment from the teacher's side. For this planning, empirically tested, more or less systematic pedagogical models such as MODEM, can be helpful. When planning a constructivist approach to the mathematical concepts under consideration, the focus is on the left-hand side of Figure 50.1. On the other hand, when offering students opportunities to construct links between representation forms of a specific concept, the focus is on the right-hand box, in which the stages of mathematical concept building are illustrated. In learning situations, however, students must have freedom to "jump the gun." They must have freedom to choose the problems that they want to learn to solve within continuous self-evaluation instead of relying on express guidance from the teachers.

We tried to solve the above-mentioned dilemma by organizing different task types into a "problem buffet."[8]

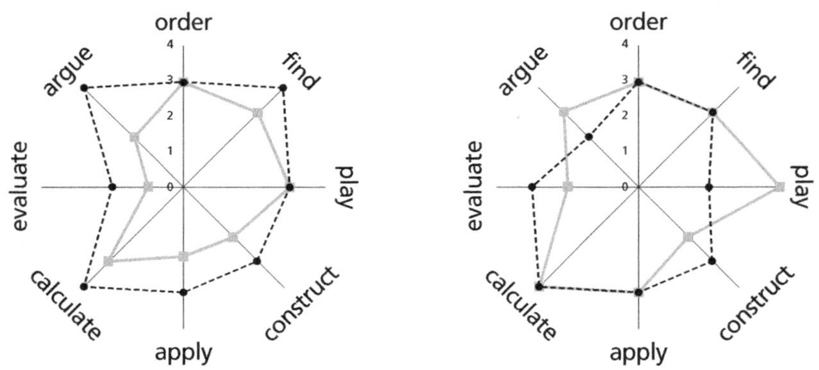

Figure 50.3 Mathematical profiles of the peer-teacher (on the right) and her classmate (on the left) at the beginning (dashed) and at the end of the working period.

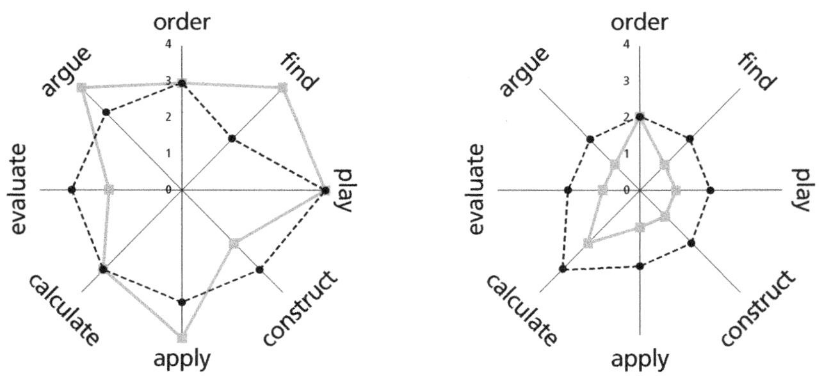

Figure 50.4 Identity profiles of the peer-teacher (on the right) and her classmate (on the left) at the beginning (dashed) and at the end of the working period.

To go for linear function, one student team, for example, initially selected the following quite complicated problem series on optimizing mobile phone costs:

> GSM operator A offers free calls for total fixed price 17,90€ pro month. Operator B does not charge any basic fee whereas cost for every call is 0,10€/min. Can you make a graphic representation to compare which of these offers might be the most suitable one for you?

After realizing that the (partly linear) cost models appeared too difficult for them, they then chose a new, much easier, problem set, which happened to consist of identification tasks—the first and lowest level of the concept building within the systematic MODEM framework, which was on the planning of the learning environments. This example shows that a sophisticated interplay between a systematic approach and a minimalist approach can be achieved even by simple pedagogical solutions. Note this important feature of minimalism: We did not want to regulate students' work by recommending to them an easier sub-problem, for example. Rather, it was students' internal motivation that regulated their task choice. Figure 50.5 represents the path of another student team, how students selected different kinds of problem types from the buffet. Recalling the abbreviations referred to in Figure 50.2, we notice that this team did not utilize the MODEM framework in an optimal way. The team went directly to PSG tasks and also selected from that list those eleven tasks (#1–# 11) more or less randomly (as it was the case within "orientation tasks"). Students evidently liked the amazing ClassPad drag-and-drop function, which automatically performed the PSG action when in selected twelve PSG tasks in a seemingly random order.

Cognitive Skills

Students' scores in all test items were significantly higher after the working period than in the pre-test (see Figure 50.6). The most remarkable changes were in the tasks types (PVV, PSV, PVS, PSS, and IVV). Furthermore, in most test items, students' scores were significantly higher than those of students in conventional school teaching, revealed by the MODEM 1 study in 1987 by using the very same test. However, it is appropriate to mention that after their working period with ClassPad, students also worked two hours with the CAL software mentioned in footnote 2. Even though this probably caused a positive effect on learning results, we felt that it was our responsibility to give students opportunity to use this software, which was available at the school.

One of the main results of the MODEM studies was that the production of a verbal form (i.e., PGV, PSV, and PVV) and identification between verbal forms (IVV) have the highest reliabilities when measuring the conceptual understanding (see Haapasalo, 2003). The fact that the lowest scores appeared in the production from graphic to symbolic form (PGS), for example, might indicate that drag-and-drop activities need a careful reconsideration when learning with ClassPad. Work without any textbooks within interaction between minimalism and systematization was also successful concerning students' cognitive development. Students scored significantly better in all test items after the ClassPad working than in the pre-test. They also showed remarkable procedural skills not only connected to the linear function, but to other function types. These results are presented in more detail by Eronen and Haapasalo (2006). From the viewpoint of problem solving, it is appropriate to mention that students clearly liked the feeling that they had reached action potential, which was described to be one of the main aspects in assessment within minimalism. The students seemed to like a free architecture of learning without any pre-set goals or tutoring from teacher's side.

LOOKING BACK

Recalling the introduction, the findings presented in this chapter give support to the many modern features and trends that are available using educational technology. The most remarkable result is that our students showed quite sophisticated metacogni-

→	**P**SG (#1)	→	**P**SG(#25)	→	**P**SG(#3)	→	**P**SG(#47)	→	**O** (#13)
→	**O** (#21)	→	**O** (#32)	→	**P**SG(#52)	→	**I** VG(all)	→	**P**SG(#6)
→	**I** VS(all)	→	**I** SS(all)	→	**R** (#1)	→	**R** (#1)	→	**R**(#1)
→	**P**SG(#74)	→	**P**SG(#89)	→	**P**SG(#98)	→	**P**SG(#1011)	→	**P**SG(#1110)
→	**P** VV(all)	→	**P** GV(all)	→	**P**SG(#12)	→	**P** GG(all)	→	**P** VV(all)

Figure 50.5 An example of a "classpath" when picking up selecting tasks.

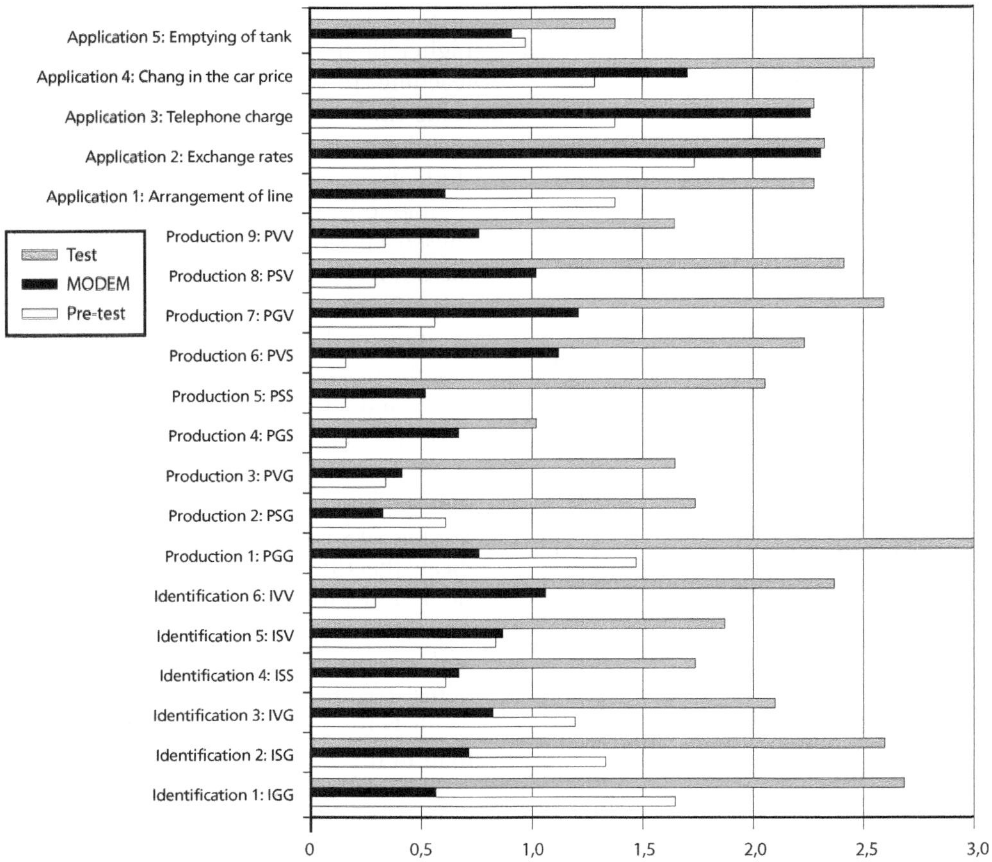

Figure 50.6 Students' cognitive results compared with those of MODEM 1 study.

tive skills when utilizing ClassPad. They used many of the basic heuristics emphasized in Pólya's (1973) checklist more or less spontaneously, especially in their free time when constructing their mathematical knowledge. Therefore, the role of school, not only concerning the fostering of problem-solving abilities, but mathematics education as a whole, might need a thorough reconsideration. Sophisticated hardware and software offer us real opportunities for changing our educational paradigm in very radical ways. Therefore teachers and students should be made aware of the technological developments occurring outside the classroom. If students' mathematical profiles can be shifted into more creative directions, this might increase their self-confidence and confirm their decision to use modern technology.[9] Even though the imagination of the user of a progressive technical tool might limit invention of environments within simultaneous activation, for example, most operations can be complicated to carry out without first learning the basic routines for using the equipment. The ClassPad manual consists of more than 650 pages, which is a non-optimal user interface in the sense of Carroll (1990) and Norman (1986). The versatility of the tool can hinder finding the essential elements that should be altered and tested because each of the intelligent menus actually contains a huge amount of conceptual (mathematical) knowledge. The fact that students in the ClassPad project expressed no particular difficulties with mathematical issues (but rather with more or less technical ones) points to many interesting pedagogical problems connected to minimalism. Moving from an old study culture towards a modern technology-based one is full of cognitive, emotional, and social problems, which are linked to strong paradigmatic tensions emphasized by Haapasalo (2008): (1) Objectivism vs. Radical Constructivism, (2) Developmental Approach emphasizing procedural knowledge vs. Educational Approach stressing conceptual knowledge, (3) Gagne's Systematization emphasizing guided learning vs. Minimalist Instruction emphasizing student's volition to learn, (4) Instrumentation where technology is shaping the actions of doing mathematics vs. Instrumentalization where technology is shaping also the mathematical objects, (5) Learning by Instructional Materials vs. Learning by Design, (6) Teaching mathematical contents vs. Emhasizing sustainable heuristics from the history of mathematics, and (7) Looking internal problems of mathematics education vs. Applying business principles to shift the bad reputation of mathematics.

NOTES

1. For being able to consider the learning from dynamic point of view, we adopted knowledge type characterization of Haapasalo and Kadijevich (2000):
   - Procedural knowledge denotes dynamic and successful utilization of particular rules, algorithms or procedures within relevant representation forms. This usually requires not only the knowledge of the objects being utilized, but also the knowledge of format and syntax for the representational system(s) expressing them.
   - Conceptual knowledge denotes knowledge of and a skilful "drive" along particular networks, the elements of which can be concepts, rules (algorithms, procedures, etc.), and even problems (a solved problem may introduce a new concept or rule) given in various representation forms.
2. The assumptions, characteristics and methods of minimalism are opposite to those of Gagne (1985).
3. Grey and Tall (1993) speak about "procepts" and "proceptual thinking."
4. See Haapasalo (2003) and http://www.joensuu.fi/lenni/modemempe.html. To see the MODEM-framework, the CAL-software is freely downloadable at http://www.joensuu.fi/lenni/programs.html
5. The ClassPad project has been described in detail in Eronen and Haapasalo (2006).
6. To measure each of the so called Zimmermann-profiles, the questionnaire used Likert-type scaling; see http://www.joensuu.fi/lenni/zprofile/Z-Questionnaire.html
7. When typing "teacher's mental problems," Google suggested more than 3 million sites on January 10th, 2007.
8. Just to note that it perhaps would be even more optimal if students could be actively involved in planning and organizing the buffet.
9. The profiles of the student teachers involved in the project showed the same trend.

REFERENCES

Carroll, J. M. (1990). *The Nurnberg funnel: Designing minimalist instruction for practical computer skill.* Cambridge, MA: The MIT Press.

Carroll, J. M. (Ed.). (1998). *Minimalism beyond the Nurnberg funnel.* Cambridge, MA: The MIT Press.

Carroll, J.M. & Van der Meij, H. (1998). Ten misconceptions about minimalism. In J. M. Carroll (Ed.), *Minimalism beyond the Nurnberg Funnel* (pp. 55–90). Cambridge, MA: The MIT Press.

Duffy, T. M. & Jonassen, D. H. (Eds.). (1992). *Constructivism and the technology of instruction: A conversation.* Hillsdale, NJ: Erlbaum.

Duffy, T. M., Lowyck, J., Jonassen, D. H., & Welsh, T. M. (Eds.). (1993). *Designing environments for constructive learning.* New York: Springer.

Eronen, L. & Haapasalo, L. (2006). Shifting from textbook tasks to mathematics making. In T. Asunta & J. Viiri (Eds.), *Pathways into research-based teaching and learning in mathematics and science education* (pp. 75-92). University of Jyväskylä. Department of Teacher Education. Research Report 84.

Gagne, R. M. (1985). *The conditions of learning and theory of instruction* (4th ed.). New York: Holt, Rinehart & Winston.

Gray, E. & Tall, D. (1993). Success and failure in mathematics: The flexible meaning of symbols as process and concept. *Mathematics Teaching, 142,* 6–10.

Haapasalo, L. (2003). The conflict between conceptual and procedural knowledge: Should we need to understand in order to be able to do, or vice versa? In L. Haapasalo & K. Sormunen (Eds.), *Towards meaningful mathematics and science education* (pp. 1–20). Proceedings on the 19th Symposium of the Finnish Mathematics and Science Education Research Association. University of Joensuu. Bulletins of the Faculty of Education, 86.

Haapasalo, L. (2007). Adapting Mathematics Education to the Needs of ICT. *The Electronic Journal of Mathematics and Technology, 1*(1), 1–10. Retrieved May 28, 2010 from http://www.radford.edu/~scorwin/eJMT/Content/Papers/eJMT_v1n1p1.pdf.

Haapasalo, L. (2008). Perspectives on Instrumental Orchestration and Assessment—from Challenges to Opportunities. Plenary Speech in the 13th Asian Technology Conference in Mathematics (ATCM 2008), December 15-19, 2008, Suan Sunandha Rajabhat University, Bangkok, Thailand. Retrieved May 28, 2010 from http://atcm.mathandtech.org/EP2008/papers_invited/2412008_15968.pdf.

Haapasalo, L. & Kadijevich, Dj. (2000). Two types of mathematical knowledge and their relation. *Journal für Mathematik-Didaktik, 21*(2), 139–157.

Haapasalo, L. & Kadijevich, Dj. (2003). Using innovative technology for revitalizing formal and informal mathematics: A special view on the interplay between procedural and conceptual knowledge. *The Teaching of Mathematics, 6*(2), pp. 81–89.

Haapasalo, L. & Siekkinen, M. (2005). Looking Technology Supported Environments from Conceptual and Procedural Perspectives. In H. Ruokamo, P. Hyvönen, M. Lehtonen & S. Tella (Eds.), *Teaching-Studying-Learning (TSL) Processes and Mobile Technologies – Multi-, Inter-, and Transdisciplinary (MIT) Research Approaches* (pp. 57–66). Proceedings of the 12th International NBE Conference. University of Lappland. Publications in Education *11.*

Haapasalo, L., Zimmermann, B., & Rehlich, H. (2004). A Vversatile Ttool to Ppromote Llink between Ccreative Pproduction and Cconceptual Uunderstanding. *The Teaching of Mathematics*, 7(2), 61–70.

Järvelä, J. (2003). *Conceptual and procedural knowledge by learning of basic skills in information technology.* Unpublished master's thesis. University of Joensuu, Faculty of Education.

Kadijevich, Dj. (2004). *Improving mathematics education: Neglected topics and further research directions.* University of Joensuu. Publications in Education 101. Retrieved May 28, 2010 from www.joensuu.fi/research/index.html

Kadijevich, Dj., Haapasalo, L., & Hvorecky, J. (2005). Using technology in applications and modelling. *Teaching Mathematics and Its Applications, 24*(2–3), 114–122.

Kupari, P. (2003). Finnish students' mathematical literacy in the PISA 2000 study. In L. Haapasalo & K. Sormunen (Eds.), *Towards meaningful mathematics and science education* (pp. 83–90). Proceedings on the 19th Symposium of the

Finnish Mathematics and Science Education Research Association. University of Joensuu. Bulletins of the Faculty of Education 86.

Lambrecht, J. J. (1999). *Developing employment-related office technology skills (MDS-1199).* Berkeley: National Center for Research in Vocational Education, University of California.

Mirel, B. (1998). Minimalism for complex tasks. In J. M. Carroll, (Ed.), *Minimalism beyond the Nurnberg funnel* (pp. 179–218). Cambridge, MA: The MIT Press.

Norman, D. (1986). Cognitive engineering. In D. Norman & S. Draper (Eds.), *User cantered system design* (pp. 36–61). Hillsdale, NJ: Erlbaum.

Pólya, G. (1973). *How to solve it* (2nd ed.). New York: Doubleday.

Redish, J. (1998). Minimalism in technical communication: Some issues to consider. In J. M. Carroll (Ed.), *Minimalism beyond the Nurnberg funnel* (pp. 219–245). Cambridge, MA: The MIT Press. pp.

Rittle-Johnson, B. & Koedinger, K. (2004). Comparing instructional strategies for integrating conceptual and procedural knowledge. In D. Mewborn, P. Sztajn, D. White, H. Hiegel, R. Bryant, & K. Nooney (Eds.), *Proceedings of the Twenty-fourth Annual Meeting of the North American Chapters of the International Group for the Psychology of Mathematics Education* (pp. 969–978). Columbus, OH: ERIC Clearinghouse for Science, Mathematics, and Environmental Education.

Rittle-Johnson, B., Siegler, R., & Alibali, M. (2001). Developing conceptual understanding and procedural skill in mathematics: an iterative process. *Journal of Educational Psychology, 93*(2), pp. 346–362.

Törnroos, J. (2003). Finnish mathematics results in TIMMS 1999—A more detailed analysis. In L. Haapasalo & K. Sormunen (Eds.), *Towards meaningful mathematics and science education* (pp. 64–73). Proceedings on the 19th Symposium of the Finnish Mathematics and Science Education Research Association. University of Joensuu. Bulletins of the Faculty of Education 86. pp.

van der Meij, H., & Carroll, J. M. (1998). Principles and heuristics for designing minimalist instruction. In J. M. Carroll, (Ed.), *Minimalism beyond the Nurnberg funnel* (pp. 19–53). Cambridge, MA: The MIT Press.

Zimmermann, B. (2003). On the genesis of mathematics and mathematical thinking—A network of motives and activities drawn from the history of mathematics. In L. Haapasalo & K. Sormunen (Eds.), *Towards meaningful mathematics and science education* (pp. 29–47). Proceedings on the 19th Symposium of the Finnish Mathematics and Science Education Research Association 86. University of Joensuu. Bulletins of the Faculty of Education 86.

51

DESIGN OF PEDAGOGICAL STUDIES TO SHIFT MATHEMATICAL PROFILES AMONG STUDENT TEACHERS

Lenni Haapasalo and Lasse Eronen

UNIVERSITY OF EASTERN FINLAND

Abstract: The first part of the chapter discusses three kinds of mathematical profiles typical of students when using as framework the eight main activities that have been proven to be sustainable throughout the history of human thinking processes and the generation of new mathematics. The study demonstrates that these profiles of student teachers of elementary education and mathematics students are quite deteriorated. The second part of the chapter represents examples of empirical studies that student teachers of mathematics conducted as a part of their pedagogical studies during their involvement with the so-called ClassPad project. Those studies suggest that integrating progressive technology as a part of students' pedagogical thinking can shift student profiles.

Keywords: assessment, constructivism, mathematics, problem solving, socio-constructivism, teacher education

INTRODUCTION

Kadijevich (2004) highlights four areas that have been neglected in the research of mathematics education: (1) promoting the human face of mathematics; (2) relating procedural and conceptual mathematical knowledge; (3) utilizing mathematical modelling in a humanistic, technologically supported way; and (4) promoting technology-based learning through applications and modeling, multimedia design, and online collaboration.

As regards the human face of mathematics, it is relevant to consider which kinds of heuristics have been sustainable for mankind. Figure 51.1 shows eight main motives and activities drawn from the history of mathematics, having been proved to lead very often to new mathematical results at different times and in different cultures for more than 5,000 years. We took this network of activities (called the *Zimmermann profile* hereafter) as an element in our theoretical framework for the structuring of learning environments and for analysing student's cognitive and affective variables. We quantified each activity by looking at the distance of the activity from the center, which tells us how strong a student thinks this activity is represented by each of the following three profiles:

- *Math profile:* How strong each activity appears to be to the student when using the term "mathematics"
- *Identity profile:* How well the student thinks he or she is performing each of the activities
- *Techno profile:* How suitable a computer is for performing each of the activities.

The First Sourcebook on Nordic Research in Mathematics Education, pages 711–717

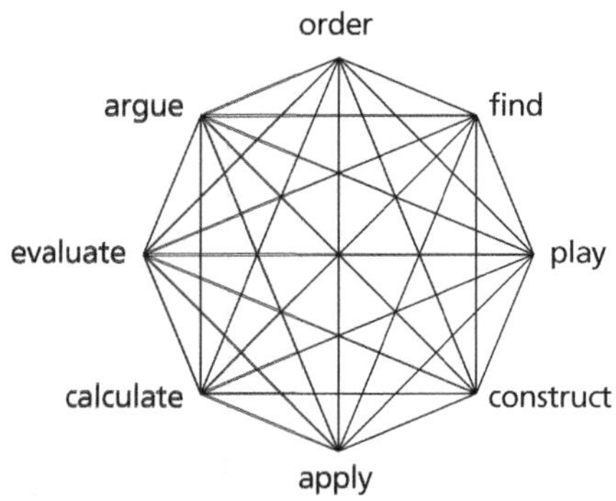

Figure 51.1 Activities and thinking tools which proved to be especially successful in producing new mathematics (Zimmermann 2003, 42)

BACKGROUND

Motivated by our findings (see Eronen & Haapasalo, this volume) that voluntarily working with a modern ClassPad calculator during the summer holiday shifted 8th grade students' profiles in a creative direction (i.e., increased the role of the first four activities above), we decided to use the same instrument for two student groups who had started their university studies. The first one consisted of student teachers of elementary education conducting their educational studies in the Faculty of Education. The other group of students had just started to study mathematics in the Mathematics Faculty and were not connected with educational studies. None of the students belonged to both of these categories. The study was designed and administrated by the authors during the study year 2005–2006 at the Faculty of Education.

AIMS AND METHODS

We aimed to discover (1) how the profiles of the two groups differ from each other, and (2) how the profiles change during the students' first year of study. To measure their profiles, an identical web-based questionnaire was used in September 2005 and in April 2006. The first questionnaire was completed by mathematics students (N = 66) and students of elementary education (N = 116). The numbers of the samples at the second stage were 22 and 50, respectively. For each of the three profiles and each of the eight activities, there were questions with a 4-step Likert-type scaling (altogether 3 × 8 questions; see Eronen & Haapasalo, 2005). For every question, the student also had to express (within the same Likert scale) how sure he or she was about his or her opinion. From the student's answer and the certainty of this answer, a new weighted variable was defined that was used to span student profiles. Because of asymmetric and discontinuous data, the Mann–Whitney Test and Sign test (2-tailed) was used for statistical analysis. To discover how the students understood the questions, sub-samples from both of the samples were also interviewed.

STUDENT PROFILES IN 2005

The math profiles of both of the target groups were very similar, emphasizing calculation and argumentation (see Figure 51.2). There was a significant difference only in application, which appears to be stronger among student elementary teachers than among mathematics students. The identity profiles followed the shape of the math profiles, the components being a little bit smaller throughout, especially in finding and argumentation. A plausible explanation might be that first-year students do not think they are good in mathematics, especially concerning these two activities. Evaluation appears weakly within all of the three profiles, which might be caused by the rather general definition of this term. Interviews support this explanation. The techno profiles did not differ significantly (Figure 51.2). Students believed that a computer mainly supports calculation, but argumentation very weakly. This might be interpreted as an implication of school teaching that over-emphasizes paper-and-pencil work and denies the use of modern technology in matriculation examination, for example. In relation to playing, perhaps this term should have been explained in more detail in the test questions.

Profile Shifts Among the Student Teachers of Elementary Education

Figure 51.3 shows shifts for the math profiles among student teachers of elementary education. A significant shift appeared for playing (an increase) and for calculating (a decrease), the latter being the only decreasing dimension. In 2006, the students emphasized application, argumentation, and playing, whereas evaluation appeared only weakly. Students were more sure of their answers than in 2005. The figure in the middle also shows a shift in the identity profiles. Significant growth occurred in finding, construction, and evaluation. The most remarkable shifts took place in the techno profiles, there were especially significant changes in finding, playing, construction, and argumentation.

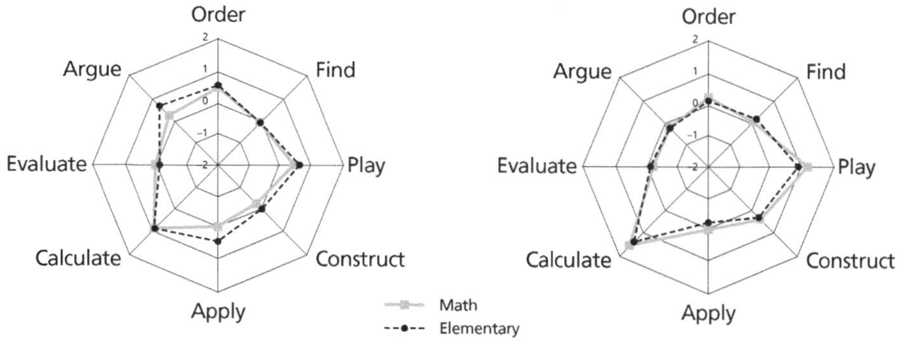

Figure 51.2 Students' math profiles (left) and techno profiles (right) in 2005.

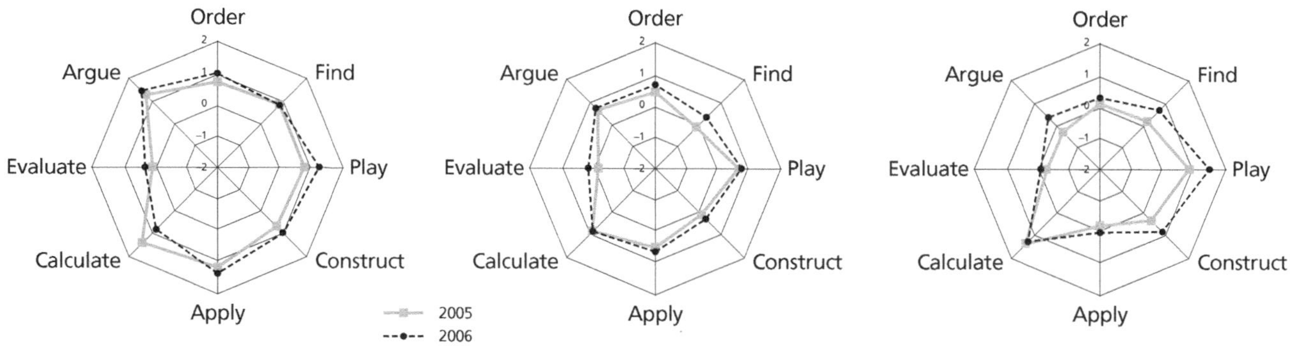

Figure 51.3 Profile shifts among the student teachers of elementary education: math profile on the left, identity profile in the middle, and techno profile on the right.

Profile Shifts Among the Mathematics Students

There were not any significant shifts in any of the components for either the math profiles or the techno profiles among the mathematics students. In addition, in the identity profile, the only significant change appeared in calculation, which surprisingly showed an increasing trend. These findings can be interpreted as suggestive of a very old-fashioned "paper media culture" in the department of mathematics, stressing pure mathematics without any great interest in placing emphasis on applications or the making of one's own hypotheses, for example. This even manifests itself through the students' expressions in everyday language, when they speak about "calculating exercises" even though the question would be about a course of abstract topology.

Comparing the 2006 Profiles

If we take a closer look at the students' profiles for 2006, the student teachers of elementary education show more rounded math profiles and identity profiles than the mathematics students, especially in relation to the creative components on the right-hand side of the octagon (see Figure 51.4). Significant differences appear in playing, construction, application, and calculation (the latter appearing to be stronger, as mentioned above). In the techno profile, the above trend appears in finding, construction, argumentation, playing, and ordering. For the first three, the difference is significant.

PEDAGOGICAL STUDIES WITHIN SOCIO-CONSTRUCTIVIST TECHNOLOGY-BASED ENVIRONMENTS

In Eronen & Haapasalo (this volume), we represented a portfolio sample that shows how students utilized ClassPad properties in a sophisticated way without any need for tutoring from the teacher. We also described the results of the first two stages of the ClassPad project, which focuses on learning mathematics in the 8th and 9th grades through this tool with minimal instruction. Our study suggests that performing mathematics with ClassPad shifted the students' mathematical and identity profiles according to the Zimmermann activities in a positive direction, opening-up new progressive ways of arranging teaching and learning. Encouraged by these kinds of findings, we decided to involve all student teachers of mathematics in this project

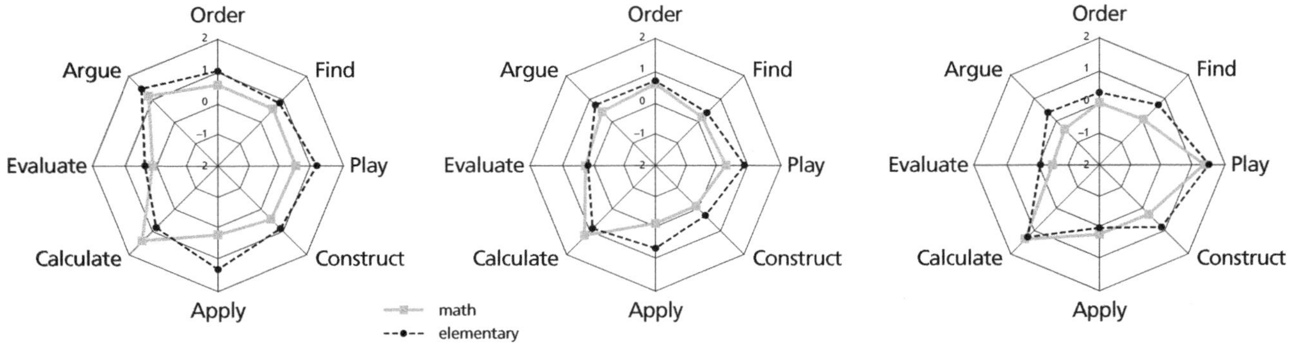

Figure 51.4 Profiles of both groups in 2006. Math profile on the left, identity profile in the middle, and techno profile on the right.

during the study year 2005–2006. In this chapter, we present some of the empirical studies that the students planned and conducted in order to link theoretical studies and practical teaching. Our starting point was to emphasize the following categories related to what modern technology can maintain and promote (see Haapasalo, 2007): links between conceptual and procedural knowledge, interplay between systematic approaches and minimalist instruction, sustainable components of mathematics making, metacognition and problem-solving skills, and learning by design. We will describe why these categories are relevant to teacher education. At the end of each section we briefly represent the sub-studies, including some interesting cognitive and affective results (gained by using quantitative vs. qualitative methods, respectively). To avoid misunderstanding, when speaking of "research teams," we are referring to the student teachers of mathematics, and the term "pupil team," refers to 9th-grade students.

PROMOTING CONCEPTUAL AND PROCEDURAL KNOWLEDGE

There is a basic conflict between conceptual and procedural knowledge (see Note 1 on page 709 in Eronen and Haapasalo, in this volume): how much students should understand before they are able to do, and vice versa. In relation to ICT-based learning, the first challenge arises from the structure of the topic to be learned, whereas the second is caused by the instructional variables required for technology use. These two knowledge types seem to develop iteratively, where a change of problem representation influences their relation. Relating different representations not only supports the development of conceptual knowledge, but also relates procedural and conceptual knowledge (cf. Haapasalo, 2003). To coordinate the process and object features of mathematical knowledge, multiple forms of representation need to be utilized and connected, especially with the aid of modern technological tools. Such a development was assumed in the pedagogical model of the MODEM project, as outlined, in Haapasalo and Silfverberg (this volume). This quasi-systematic model of the interplay between the two knowledge types makes use of spontaneous procedural knowledge and the simultaneous activation of conceptual and procedural knowledge.

One research team (Ruokolainen & Kukkonen, 2006) used the MODEM framework to plan learning and assessment material for basic concepts of statistics (e.g., organizing and representing data, mean, standard deviation, mode, median). Another research team (Hurskainen et al., 2006) did the same concerning the basic features of linear function. A third research team (Tilli & Vauhkonen 2006) measured pupils' understanding of this concept by using the same test as in the MODEM 1 study (cf. Haapasalo, 2003). Some of the main results are represented in Eronen and Haapasalo (2006). However, to summarize here: students' scores for the ClassPad test group were significantly higher than for those students undergoing conventional school teaching. Furthermore, these students also showed high procedural skills. A postponed test, after 5 months, revealed that this scoring level remained consistent, and for many students it even improved.

INTERPLAY BETWEEN SYSTEMATIC APPROACHES AND MINIMALIST INSTRUCTION

The above-mentioned research teams who designed the learning environments for the learning of linear function and statistics combined the features of the systematic approach and minimalism. Team Karttunen, Nevalainen, Pirkonen, and Tillonen (2006) observed and videotaped pupils' learning processes and analyzed the thinking processes, which appeared among pupils when studying linear functions. Team Ruokolainen and Kukkonen (2006) did the same concerning the learning of basic statistical concepts. Both of these studies offer interesting findings concerning not only the interplay between the minimalist instruction and systematic approaches but also affective variables and group dynamics within student pairs.

SUSTAINABLE COMPONENTS OF MATHEMATICS MAKING

Two research teams were involved in studying the students' Zimmermann profiles. The first one (Sormunen, Turkulainen, & Ylinen 2006) measured the profiles among student teachers of elementary education and first-year mathematics students. Those results were used in the above-mentioned chapter in this volume. The second team (Hakkarainen Sivonen, & Pellikka, 2006) analyzed profiles among 9th-grade students. We would like to use the profiles of two students who worked as peers with ClassPad but radically differed in their learning styles (see Järvelä & Haapasalo, 2005). The first one, who classified herself as a conceptually oriented person (i.e., prefers understanding), acted as a peer teacher, whereas the second one, who mainly listened, was classified as a procedurally bounded learner (i.e., stayed with concrete working procedures). At the beginning, the mathematical identity of the learner was even larger than that of the teacher, especially in relation to finding and argumentation. The more the peer teacher explained, it was only her own mathematical profile that showed some tendency to expand but only towards the direction of play. This, however, is not as interesting as what happened to both the students' individual self-confidence to be able to do mathematics. After the work, the learner seemed to believe he could do more: find, apply and argue. However, surprisingly, the self-confidence of the peer teacher seemed to collapse in almost all dimensions.

PROMOTING METACOGNITION AND PROBLEM-SOLVING SKILLS

This category is the primary idea in the work of many of our research teams. Findings support the view that technology-based learning promotes problem-solving skills (cf. Haapasalo, 2007; Eronen & Haapasalo, this volume). The acceptance of constructivist views on teaching and learning mathematics entails emphasizing the genesis of heuristic processes and the ability of students to develop intuition and mathematical ideas. The ClassPad project emphasizes informal rather than formal mathematics within the "minimalist instruction" approach and an eight-fold framework of activities. Within those activities it focuses on "changing representation," which is a powerful thinking tool to promote links between procedural and conceptual knowledge and to enhance problem-solving abilities.

LOOKING BACK

For those who want to obtain a solid view of mathematics, it is important to know how and from where mathematical knowledge and mathematical thinking might appear and come into life and action. Based on a "human laboratory within 5,000 years," Zimmermann's activities can be considered as a relevant framework not only for the teaching and learning of mathematics but also for what could be called "mathematical education" in a more general sense. Besides the fact that our results reveal a very stereotypical view of mathematics among both sample groups, the most disturbing finding is that the answer to the provocative question, "Does mathematics education contaminate mathematical education?" might be affirmative. Conventional mathematics education (in the sense of mathematics teaching) seems to act in a counter-productive way in relation to the extension of students' views of mathematics. However, a progressive shift towards the promotion of creative activities within a constructivist framework using modern technology seems not only to change students' views of mathematics but also to increase their self-confidence in using mathematics and the utilization of technology.

As Silfverberg and Haapasalo (this volume) emphasize, teachers usually have difficulties to shift their stereotypic views on teaching and learning, even more difficulties to go for progressive technology. We did not choose a progressive way to integrate students' pedagogical studies because this question is easy, but rather because it is hard. After having been used to ready-planned materials and objectivist–behaviorist teaching during their own learning history, it was the demand of continuous improvisation that seemed to cause uncertainty, especially at the beginning of the project. However, at the end students expressed mainly positive opinions in relation to this effort, which they described as progressive, challenging, and worth the risk. With respect to the students' own research concerns, the most important aspect was that they gained experience in conducting relevant, valid, original, and rigorous educational research, where mathematics is not just a place holder (cf. Sierpinska, 1993). Even though there were gaps in their data analysis and interpretation of the results, we also learned a lot as supervisors. The project gave us a large amount of new research ideas and, owing to the students' kind permission, we can also use their data and pre-analysis to improve our methods.

Our studies are efforts to develop an instrument based on Zimmermann's activity components in the guidelines of the criteria for relevant and viable educational research, as highlighted by Sierpinska (1993). One of our many challenges is to develop the test so as to measure not only the eight Zimmermann dimensions but also links between them. The analysis revealed that the questionnaire needs improvements, especially in relation to the definition of the terms "evaluate" and "play." Furthermore, the Likert scaling and combination of the students' answers with their certainty needs a thorough re-evaluation. After this, our next plan is to measure the test's reliability by using bigger samples. In this study, less than 50% of the original samples answered the second questionnaire. In addition, the large amount of extra, so-called control questions in our test could have decreased the test's reliability, if the students became bored while answering.

ACKNOWLEDGEMENTS

We would like to express our gratitude to the following students for their fruitful cooperation and their ability to take conscious risks when developing progressive methods in teacher education: Markus Hakkarainen, Olli Hurskainen, Markku Hyttinen, Janne Juntunen, Kimmo Karttunen, Timo Kukkonen, Elina Lampela, Maarit Nevalainen, Marja Pellikka, Kaisa Pirkonen, Sampo Pylkkö, Antti Ruokolainen, Susanna Sivonen, Marko Sormunen, Henri Tanskanen, Jannamari Tillonen, Jenni Turkulainen, Sari Vauhkonen, and Teemu Ylinen.

REFERENCES

Eronen, L., & Haapasalo, L. (2005). *WWW-based questionnaire for measuring of student profiles*. Retrieved May 28, 2010 from http://www.joensuu.fi/lenni/zprofile/Z-Questionnaire.html

Eronen, L., & Haapasalo, L. (2006). Shifting from textbook tasks to mathematics making. In T. Asunta & J. Viiri (Eds.), *Pathways into research-based teaching and learning in mathematics and science education* (pp. 75–92). University of Jyväskylä. Department of Teacher Education. Research Report 84.

Eronen, L., & Haapasalo. L. (2007). A shift in mathematical profiles among student teachers of elementary education and students of mathematics. To appear in the Proceedings of XXIII Annual Symposium of the Finnish Mathematics and Science Education Research Association. Åbo Akademi, Vaasa, Finland.

Haapasalo, L. (2003). The conflict between conceptual and procedural knowledge: Should we need to understand in order to be able to do, or vice versa? In L. Haapasalo & K. Sormunen (Eds.), *Towards meaningful mathematics and science education* (pp. 1–20). University of Joensuu. Bulletins of the Faculty of Education 86.

Haapasalo, L. (2007). Adapting Mathematics Education to the Needs of ICT. *The Electronic Journal of Mathematics and Technology, 1*(1), 1–10. Retrieved May 28, 2010 from http://www.radford.edu/~scorwin/eJMT/Content/Papers/eJMT_v1n1p1.pdf.

Järvelä, J. & Haapasalo, L. (2005). Three Types of Orientations by Learning Basic Routines in ICT. In M. Julkunen (Ed.) *Learning and Instruction on Multiple Context and Settings III* (pp. 205–217). Bulletins of the Faculty of Education 96.

Kadijevich, Dj. (2004). *Improving Mathematics Education: Neglected Topics and Further Research Directions*. University of Joensuu. Publications in Education 101. Retrieved May 28, 2010 from http://www.joensuu.fi/research/index.html.

Sierpinska, A. (1993). Criteria for Scientific Quality and Relevance in the Didactics of Mathematics. In G.Nissen & M. Blomhof (Eds.), *Criteria for scientific quality and relevance in the didactics of mathematics.* (pp. 35–74). Roskilde: IMFUFA. pp.

Silfverberg, H., & Haapasalo, L. (2009). Painful Paradigm Shift in the Teaching and Learning of Mathematics. In this volume.

Zimmermann, B. (2003). On the genesis of mathematics and mathematical thinking—A network of motives and activities drawn from the history of mathematics. In L. Haapasalo & K. Sormunen (Eds.), *Towards Meaningful Mathematics and Science Education* (pp. 29–47). Proceedings on the 19th Symposium of the Finnish Mathematics and Science Education Research Association, University of Joensuu. Bulletins of the Faculty of Education 86.

STUDENTS' FINNISH STUDY REPORTS IN THE DEPARTMENT OF APPLIED EDUCATION OF THE UNIVERSITY OF JOENSUU

Hakkarainen, M., Sivonen, S., & Pellikka, M. (2006). *Oppilaiden affektiivisen ajattelun ja matematiikkakuvan kehittyminen.* [Shifts in affections and mathematical profiles among students]

Hurskainen, O., Tanskanen, H., Lampela, E., & Pylkkö, S. (2006). *Empiirinen tutkimus käsitteenmuodostusprosessin mukaisen materiaalin kehittämisestä lineaarisen funktion peruskäsitteiden opiskeluun.* [Design of learning material for linear function within quasi-systematic model of concept building]

Hyttinen, M., & Juntunen, J. (2006). *Oppilaiden kognitiivisten taitojen kehittyminen käsitteessä taulukkolaskennan ja tilastomuuttujien perusteet.* [Cognitive development among students when learning basic concepts of statistics]

Karttunen, K., Nevalainen, M., Pirkonen, K., & Tillonen, J. (2006). *Empiirinen tutkimus dynaamisesta opiskeluseurannasta opiskeltaessa lineaarisen funktion peruskäsitteitä.* [Empirical study of dynamical assessment when learning linear function]

Ruokolainen, A., & Kukkonen, T. (2006). *Empiirinen tutkimus proressiivisen opiskelumateriaalin toimivuudesta eräiden tilastollisten peruskäsitteiden opiskelussa.* [Empirical testing of quasisystematic learning material when learning basic concepts of statistics]

Sormunen, M., Turkulainen, J., & Ylinen, T. (2006). *Luokanopettajaopiskelijoiden ja matematiikan opiskelijoiden matematiikkakäsityksistä.* [Views of mathematics among student elementary teachers and student mathematics teachers]

Tilli, K., & Vauhkonen, S. (2006). *Oppilaiden kognitiivisten taitojen kehittyminen lineaarisen funktion käsitteessä.* [Cognitive development among students when learning linear function]

52

THE MATHEMATICAL DEVELOPMENT OF CHILDREN AGED BETWEEN 6 AND 15

Lisen Häggblom

ÅBO AKADEMI, VAASA

Abstract: The overall aim of this longitudinal investigation is to chart and analyze how children solve mathematical problems at different ages and to describe how knowledge and skills change during the period when they attend comprehensive school. The target group consists of Finland–Swedish children aged between 6 and 15. Most children's knowledge develops during the first school years in accordance with the demands of the curriculum. When the mathematical content becomes more complex, a kind of tie-breaking mechanism comes into effect, for instance due to an increased demand for complex components of thought.

Keywords: arithmetic operations, assessment, curriculum, number sense, problem solving

INTRODUCTION

The present investigation started in connection with the ELMA project (Projektet om ELevanpassad MAtematikundervisning [the project of teaching mathematics adjusted to the pupils]), at the Faculty of Education, Åbo Akademi University, and the primary school level of Vasa övningsskola. The aim was to develop and assess a form of teaching mathematics that considers the different abilities of pupils to adapt and develop mathematical knowledge (Häggblom, 1994). While participating in the project, my interest in a follow-up investigation into the development of pupils' knowledge of mathematics was aroused and deepened. My thoughts centred around what there is in the nature of mathematics that in some cases leads to success and in other cases to failure. In my investigation I have taken this complex of problems as my starting point.

The overall aim of the present investigation was to chart and analyze how children solve mathematical problems at different ages and to describe how their knowledge and skills change during the period when they attend comprehensive school. Between the years 1988 and 1998, I followed a group of Finland-Swedish children from two regions in the Swedish-speaking part of Finland (Häggblom, 2000).[1] By following up the same group through the higher forms of the lower secondary school, the investigation took on a longitudinal perspective (Bergman & Magnusson, 1990).

THEORETICAL BACKGROUND

A wider view of what mathematics is and what school mathematics should comprise has led to a development of new forms of assessment and fresh ideas as regards bringing out pupils' different modes of learn-

The First Sourcebook on Nordic Research in Mathematics Education, pages 719–729

ing (Björkqvist, 1997; Kulm, 1991; Morgan, 2000; Niss, 1993; Webb, 1992). It is no longer enough to find out whether a pupil is right or wrong in his or her counting. Nor is it enough to point to individual variations in the results, but underlying strategies of thinking should be focused on in the teaching of mathematics (Ginsburg, Jacobs, & Lopez, 1993; Häggblom, 2006; Mulhern 1989; Romberg, 2004) and *uncovering the underlying strategies for thinking* in the learning of mathematics (Alrø & Skovsmose, 1993; Hughes, 1984; Morgan, 2000). The investigation therefore came to land in the field of tension between a behaviorist and a constructivist view of learning (von Glasersfeld, 1995; Kilpatrick, 1993). In this study, two forms of assessment were used: one competence-oriented and one communication-oriented (Imsen, 1992). The choice was determined by the view of learning and assessment that is to be found in the school today.

In the communication-oriented assessment, the children's written accounts are used as a basis for describing the variations in the children's ways of calculating. Mathematics is a mental activity and an intellectually demanding subject. The fact that an individual's thoughts are invisible makes it all the more important to uncover the thinking process in order to assess the *quality in the mathematical proficiency* of the pupil. An active use of language thus becomes a support in the learning process of the pupils (Bauersfeld, 1998; Häggblom, 2006). When the pupils use their language to describe a phenomenon, a concept or a strategy for counting, their self-reflection develops (Ernest, 1998) and their knowledge becomes visible in the form of language. When the pupils get to formulate their thoughts they stop and reflect on their knowledge. At the same time, they develop their proficiency in using the correct terminology and the language of mathematics becomes the pupil's knowledge.

Within a learning culture founded on the abovementioned view on learning, assessment forms an integral part of teaching. Romberg (2004) characterizes an assessment with principles where the purpose is to increase learning by giving pupils the opportunity to show what they know rather than what they do not know. The assessment includes multifaceted and varying forms in which the pupils explain and document their achievements. Assessment becomes a kind of tool for helping the pupils to develop their knowledge in a varied way. The assessment is constructed to become a support for the pupils to reflect on their learning as well as becoming more conscious of their progression, their possibilities and their needs. The method involves limitations in the material for analysis, since all children do not account for their solutions. I nevertheless chose this method, not only because of the extensive data material but above all in order to study variations in the pupils' solutions. The model chosen for the analysis brought out the variations clearly and a kind of pattern of calculations emerged.

On the basis of the contents of the curriculum (Utbildningsstyrelsen, 1994) the investigation was focused on three fields of knowledge: *number sense, arithmetical operations and text-based mathematical problems*. These three variables are included in Magne and Thörn's (1987) so-called cognitive taxonomy for the teaching of elementary mathematics. The contents comprise a limited part of what is contained in the teaching of mathematics at school. In order to obtain as consistent tools of assessment as possible, it was necessary to delimit the contents. In my opinion the three selected variables constitute a unit where the knowledge of numbers and number relations forms a basis for arithmetical operations which can then be used for the solution of text-based problems.

On the basis of the three variables, the following research problems were specified:

I. Number sense

1. What knowledge do children have of numbers and what difficulties are connected with children's number sense?
2. How does children's number sense change?

II. Arithmetical operations

1. How do children perform numerical operations and what difficulties occur in performing such operations?
2. How does children's performance of arithmetical operations change?

III. Text-based problems

1. How do children solve text-based tasks and what difficulties are associated with children's solutions of such problems?
2. How does children's ability to solve text-based problems change?

Children constituted a group of pupils who attended Finland-Swedish kindergartens, continued at a neighboring primary school, and later attended a secondary school. Starting from a Finland-Swedish special problem, the different language affinity of the pupils, in this investigation I paid attention to differences in the achievements of Swedish-speaking and bilingual children. The result was also analyzed from a gender point of view. Results in differences between boys and girls vary widely from country to country and over time (Björkqvist, 1997; Fennema & Peterson, 1985; Grevholm, 1996; Hyde, Fennema, & Lamon,

1990; Skaalvik & Rankin, 1994). The longitudinal perspective made studies of changes in the children's standard of attainment possible. The aim of this investigation was not to describe the causal pattern of the results, although this would have been interesting.

METHODS

When the investigation was initiated, the group consisted of 139 children. The children attended Swedish-speaking kindergartens and were to start their compulsory education by attending Swedish-speaking schools. Having finished primary school, the number of pupils in the group amounted to 119, and at the end of the secondary school stage 113 pupils took part in the final assessment. The number of boys and girls was quite evenly distributed. When the investigation started, 59% of the pupils were categorized as Swedish-speaking and 41% as bilingual. At the end of the investigation, the corresponding number was 58% and 42%, respectively. The language of the children was determined on the basis of the following criteria: If the child spoke Swedish with both parents, he or she was categorized as a member of the Swedish-speaking group. The child was categorized as bilingual if he or she spoke Swedish with either parent and Finnish or some other language with the other parent. If the child spoke Finnish with both parents, the child was also classified as belonging to the bilinguals.

For the data collection, math tests were created for measurement at 6 years, 9 years, 12 years, and 15 years old. The tests were conducted as standardized tests; however, the tests offer the possibility of presenting strategies for different solutions (Steffe & Cobb, 1988). The tests were created according to principles of the current curriculum (Skolstyrelsen, 1985; Utbildningsstyrelsen, 1994) and according to the three variables that had been chosen for the study (Anghileri & Johnson, 1988; Behr, Harel, Post, & Lesh, 1992; Brown, 1981; Kieren, 1993; Magne, 1990; Moser, 1988). The Cronbachs alfa-coefficient, which has been used as an index for the reliability calculations, varied between 0.77 and 0.80 for number sense, between 0.74 and 0.86 for arithmetical operations, and between 0.68 and 0.81 for word problems.

On the basis of quantitative methods, the degree of difficulty of the problems and the differences in means between the genders and between the language groups were analysed. Changes in standards of achievement were described by means of regression analysis and calculation of residuals. For the qualitative analyses, variations in the pupils' calculating strategies have been analysed on the basis of their written responses to the question and the solutions were categorized depending on strategies and error analyzing.

RESULTS

Number Sense

Already when the children start attending school, they display great variations in their sense and use of numbers. At the age of six, the total score achieved by the children varied between 7 and 30 points, the mean being 22.1 points. The frequency of solutions was 74%. Most children used numbers ranging from 0 to 6—that is, they made comparisons and combined the right figure with the right number of objects. The abstract sequence of numbers differentiates the children. A great number of the children showed great confidence in counting numbers up to 32 and bigger numbers and in writing two-digit numbers. Variations could be noted both in the writing and the order of the digits of two- and three-digit numbers. On the whole, children would find it easier to mention a number that follows than a number that precedes a given number. In some children, the informal knowledge had not yet developed to the point of distinction and exactness that characterizes mathematical knowledge.

Between the ages of 6 and 7, numbers were counted with increasing confidence. The knowledge of the number sequence also developed within ranges of numbers that do not belong to the curriculum for the first school year. The development of knowledge during the first school year was so marked and the transfer of learning was so great that certain pupils develop knowledge of such mathematical matters as have not yet been dealt with in the school course. The frequency of solutions within the range of items that had been taught reaches among the 7-year-olds a standard exceeding 90%. The problems included transposition of figures and reversal of the order of figures as well as lack of confidence in the counting of numbers, difficulties in writing out numbers, and linguistic confusion connected with the names of numbers. In connection with number sequences, some mix-up and misinterpretation could be noted concerning the expressions "the number preceding" and "the number following."

At the age of 9, pupils had attended school for three years. According to the curriculum, their conception of numbers at this age should comprise natural numbers within the range of 0–10,000. At least 90% of the pupils could write numbers over one thousand in figures and write three- and four-digit numbers in order of size, continue number sequences forwards, and indicate what number comes before and what number follows another number. The scores varied between 3 and 17 points, the mean score being 13.9 points. The frequency of solutions was 81%. Among the wrong answers, linguistic difficulties and difficulties concerned with number sequences could be noted. Linguistic problems were exemplified by interpreting expressions like "biggest number" and "continue" as well as by confusion connected with the names of numbers. Numbers involving the transition between hundreds and between thousands were unclear to some children. The names of numerical units were mixed up. Problems involving combinations of numerical units and calculating operations represent both the lowest number of solutions and the greatest proportion of unanswered questions.

For the 12-year-olds, the range of numbers was expanded to six-digit whole numbers and decimal numbers and fractions. Measuring shows a mean of 13.9 points and the scores varied between 0 and 15 points. The frequency of solutions was 66%. About 70% of the pupils wrote six-digit whole numbers (and bigger numbers). The proportion of pupils who can interpret fractional models varied depending on the complexity of the model. Fractional problems involving quantities with the same denominator were solved by 90% of the pupils while models which had different denominators were solved by no more than 40%. The proportion of pupils who could handle decimal numbers varied between 60% and 70%. Deficient understanding of the positional system leads to difficulties in the handling of big numbers, the names of numerical units and grasping decimal structures. The handling of fractional models showed both deficient understanding of fractional structures and difficulties in the handling of symbols. Linguistic difficulties were noticeable in connection with the names of both numbers and numerical units. There were great variations in pupils' abilities to interpret and give answers in written form to the request to "state how you thought" or "show how you thought." At the age of 12 half the number of pupils solved at least 70% of the problems and one third solved at most half of the problems. About 30% of the whole group under study displayed obvious difficulties, which was revealed by unanswered questions and wrong answers.

At the age of 15, the mean score is 10.4 points, and the individual scores varied between 1 and 17. The frequency of solutions was 61%. At the age of 15, between 50% and 80% of the pupils could handle the positional system and fractional structures, and half the students could do percentage calculations and numbers expressed as powers of 10. The pupils in the group of 15-year-olds reached a better result in most tasks originally given to the 12-year-old group. The percentage of solutions of these tasks shows an increase of between 10 and 20 percentage units, but the frequency of solved primary-level tasks seldom exceeds 80%. This suggests that the understanding of a topic is developed later in the case of certain pupils. There were, however, a number of secondary-level pupils who still had difficulties with tasks typical of the primary level. With regard to tasks typical of the secondary level, the frequency of solved problems was about 50%. These problems involve, for instance, the converting of a fraction into decimal form, writing and interpreting a number as a power, and the calculation of percentages. The fact that only about half the number of pupils reached half the total score indicates an uncertain number sense.

Between the primary and the secondary levels, the number field becomes more complex as new structures are added and knowledge of more and more complex components of thought is required. The decreased frequency of solutions can be seen as a sign of gradual inability to cope. The number of pupils in the inquiry concerning 15-year-olds who manage to solve the problems typical of the curriculum of that school year decreases, as does the number of those who attain maximal points. The complex thought functions that according to Magne (1990) cause this failure are, in the present investigation, with respect to the number sense variable, identified with fractional and decimal structures for the 12-year-olds and with percentages and powers structures for the 15-year-olds. Solutions within these structures varied depending on the way the task had been formulated.

The change in number sense between the ages of 6 and 15 can be described as a development towards more complex and demanding contents. The percentage of solutions of a problem typical of the curriculum in question grows when the test is given once more at a higher age level, but the percentage of solutions seldom exceeded 80%. When comparing identical problems between two age levels, a marked improvement is noted in the writing of big numbers, continuing number sequences, and handling decimal and fractional structures. As regards the ability to handle the names of the numerical units, very little improvement takes place, and with regard to some tasks, the result actually becomes poorer. The investigation shows that not all pupils can adopt and process the mathematical substance intended to be learned during a certain year, but that the development takes place only later.

Arithmetical Operations

The arithmetical operations (the ASMD variable) assessed involve a progression in substance stretching from natural numbers in the number field from 0 to 10,000 in the study concerned with 9-year-olds, to decimal and fractional structures in connection with the 12- and 15-year-olds and powers and real numbers in connection with the 15-year-olds.

At the age of 9, the scores varied from 10 to 48 points, with a mean of 37.0 points. The frequency of solutions was 77%. When the pupils were free to choose a method of calculation, they used both mental calculation and written resolutions. Problems that are traditionally solved in written form were solved by means of mental calculation by 30–50% of the pupils.

Variations in the pupils' calculating strategies have been analysed on the basis of their written responses to the question, "How do you think when you calculate . . . ?" For each problem there was a strategy that at least half the number of pupils chooses. The strategies for addition and subtraction followed certain patterns. The pupils who got their answers wrong used similar strategies as those who calculated correctly, but calculating errors, omitted numbers, difficulties in the handling of symbols and transposition errors or copying errors result in wrong answers.

In the study concerned with the 12-year-olds, the scores varied between 0 and 16 points, the mean being 10.2 points. The frequency of solutions was 64%. The proportion of pupils who did not arrive at a solution varied from 2% (a problem for 9-year-olds) to 34%. On the whole, a low percentage of solutions could be noted for problems typical of the current school year—that is, problems involving fractions and decimals. The proportion of strategies accounted for varies depending on the numerical structure of the problem and the choice of calculating method. Certain numerical combinations lead to written strategies, others to mental calculations. Every task was solved by a particular strategy chosen by at least half the number of pupils. Problems involving division lead on the whole to low frequences of solution, and the number of unsolved tasks varies from 25% to 40%. The difficulties were mainly due to the handling of decimals.

At the age of 15, the mean total score was 12.1 points and the individual scores varied from 1 to 19 points. The frequency of solutions was 64%. About 60% of the pupils could solve multiplication problems with both factors in decimal form. Division was the most difficult arithmetical operation and the percentage of solutions varied from 38% to 59%. The difficulties lie in the handling of powers, fractions and decimals. In addition, simple calculating errors occur, as well as errors in reduction and confusion in the handling of power items. Increased complexity of numerical values combined with faulty use of symbols implies that the difficulties persist or increase.

The mathematical problems typical of various school classes tend to lead to lower percentages of solutions as pupils move up. The result is connected with the pupils' number sense, where more and more complex mathematical structures lead to new difficulties. Fractions and decimal numbers differentiate the pupils because they require more and more complex mental components for their solution. Engström (1997) sees the causes of pupils' difficulties with rational numbers as an N-distraction, which means that overgeneralization of arithmetical rules from the field of natural numbers continues to influence pupils through the years.

Text-Based Problems

Children's confident handling of numbers when they start school is revealed in their resolution of problems presented orally. At the age of 6, the mean score was 11.3 points and the individual scores varied between 2 and 14 points. The frequency of solutions was 81%. The number field imposes limitations. Problems belonging to the numerical range of 0–10 were solved by most children without the support of concrete material. When the range exceeded 10, differences occurred. *The children generally understand the linguistic content of the counting events presented and choose the correct arithmetical operation with discernment.* The word "more" in connection with comparisons was misunderstood by some children.

At the age of 9, the mean score of the variable text-based problems was 15.1 points and the scores varied between 1 and 22 points, the maximum being 23 points. The frequency of solutions was 65%. The proportion of pupils leaving text-based problems unanswered varied between 1% and 25%. The arithmetical operations were done on paper or were accounted for by being marked and/or done by mental calculation. The difficulties were of three kinds and were concerned with understanding the content of the text, performing numerical operations, and responding to the correct question. Only a minor proportion of the wrong answers are due to the choice of wrong arithmetical operation.

In the study concerned with the 12-year-olds, the text-based problems gave a total score of 12 points. The pupils' individual scores varied between 1 and 12 points, the mean being 8.1 points. The frequency of solutions was 68%. The proportion of pupils leaving the text-based problems unanswered varied between

0% and 18%. The solutions were presented in written form, they were mentally calculated or the arithmetical operation to be used is just indicated. Wrong answers were due to arithmetical errors, choice of wrong arithmetical operation, errors in copying, some omitted part operation or failure to answer the right question.

The maximum score of the variable text-based problems for the 15-year-olds was 14 points and the children's scores varied between 0 and 13 points. The frequency of the solutions was 54%. The proportion of pupils who left problems unanswered varied between 0% and 55%. The solutions varied depending on the semantic structure of the text and the numerical values. The solutions indicate variations in numerical handling, the use of equations, testing methods and the use of pictorial models. Difficulties in numerical calculations (arithmetical errors), the choice of wrong method or an omitted partial operation led to wrong answers.

In the variable text-based problems, the gradual inability to cope was not equally clear as it was with the other variables. The choice of text-based problems was carried out in such a way that the results at different age levels would be comparable with respect to arithmetical operation and the structure of the text. When the same task recurs a later year, the frequency of solutions increases, but when the structure of the text is retained and the numerical values are adjusted to the requirements of the current year, the frequency of solutions generally decreases. Between the ages of 9 and 12, the frequency of solutions usually increased by up to 40 percentage points, while the increase between ages 12 and 15 was seldom higher than 10 percentage units. This implies that even if a development takes place during the lower secondary school stage, it is very marginal. One of the reasons may be that numerical values adapted to the curricula for the different years involve more difficult numerical structures. According to Cobb (1986), pupils experience problems with higher numerical values so abstract that they cannot grasp the connection between numbers and solutions.

TRENDS OF DEVELOPMENT

Differences in the Development of Girls and Boys

With regard to the variable number sense, girls on the whole show greater tendency to change than boys. The differences in achievement between the genders were leveled out during the primary school stage to be once more increased during the lower secondary school stage. The statistically significant differences that were found between the genders at the age of 6 in favor of the boys ($p = 0.006$) no longer exist at the age of 12 ($p = 0.493$). At the lower secondary school stage, the girls become inferior to the boys, so that there were again significant differences between the genders in favor of the boys at the age of 15 ($p = 0.045$). Both boys and girls presented almost the same means at the age of 15 as they did at the age of 6.

The mean of the ASMD variable was higher for the boys than for the girls at the age of 9, and the difference was statistically almost significant ($p = 0.051$). During the rest of their school attendance period, the differences between the genders were leveled out more and more. At the ages of 12 and 15 the boys had a somewhat higher mean than the girls, but the difference is not statistically significant ($p = 0.333$ resp. $p = 0.253$). According to Marshall and Smith (1987), girls' greater confidence in carrying out arithmetical operations is due to the fact that they often develop procedural knowledge by sitting still.

With regard to the variable text-based problems, the results fluctuate. At the ages of 6, 9, and 15, boys obtained a higher mean than the girls, and the differences were statistically significant at ages 9 ($p = 0.026$) and 15 ($p = 0.023$). At the age of 12, the differences had been leveled out to such an extent that the girls even achieved somewhat better results than the boys.

Since research on gender differences and mathematical achievements shows different results depending on country and content, the result cannot be compared with those of previous studies, but is to be seen as a supplement to other investigations. With this investigation, I contribute two interesting results to the debate on differences in girls' and boys' development in mathematics. At the early age of 6, boys presented better mean achievements than girls, and the differences were statistically significant. Kimball's (1989) conclusion that boys have greater experience of mathematics than girls even before they start attending school is thus obviously supported. The girls progressed more quickly than the boys during the primary school stage, while the boys made better progress at the upper levels. One of the reasons may be that the teaching methods at the lower levels favor the girls' development (cf. Ahlberg, 1996), while the contrary is true at the higher levels. It may also be a result of differences between the teaching methods of female and male teachers, since the number of female teachers is higher in the lower grades. The means of the variables number sense and text-based problems were almost the same for the respective groups at the ages of 6 and 15, even if the values have fluctuated during the period of school attendance.

Differences in Development Between Language Groups

For the variable number sense there were, at the age of 6, statistically significant differences ($p = 0.041$) in means between the language groups in favor of the Swedish-speaking children. The Swedish-speaking children thus had an advantage when starting school. During the primary school stage, the bilinguals make such great progress in relation to the Swedish-speaking children that the differences have been leveled out by the time they reach the age of 12. During the lower secondary school stage, the differences again increased so much that there were once more statistically significant differences in means at the age of 15 in favor of Swedish-speaking pupils. When the result is analyzed in relation to the data within the variable number sense, it can be noted that a certain degree of language proficiency is needed in order to be able to solve the problems. The interpretation of the variations in the result can, as far as the bilingual children are concerned, be thought to be due to a connection between unclear terminological expressions and insufficiently developed language ability.

The Swedish-speaking pupils had a higher mean for the variable arithmetical operations at all age levels, but the differences were not statistically significant. The fact that the differences were leveled out more and more towards the end of the period of school attendance can be interpreted to mean that the language background is not decisive regarding pupils' ability to perform arithmetical operations since the mathematical symbols are not tied to any particular language.

The variable text-based problems showed a favorable trend with regard to the bilingual pupils during their time at the comprehensive school. At the age of 6, the differences were statistically significant in favor of the Swedish-speaking pupils ($p = 0.003$). While attending the primary school, the bilingual pupils made such great progress that the difference at the ages of 9 and 12 were no longer statistically significant. At the end of the lower secondary school period, the bilinguals even reached a slightly higher mean level than the Swedish-speaking pupils. The result can on the one hand be interpreted to imply that even if reading comprehension is an important prerequisite for the solution of text-based problems, it is not of equally great importance as other factors seem to be for achieving correct results. On the other hand, the outcome of the study indicates that the pupils' reading comprehension during their later school years is sufficient for them to understand the text of the problems that formed part of the investigation.

In testing reading comprehension at the ages of 9, 12, and 15, there were statistically significant differences in the means in favor of the Swedish-speaking pupils. However, there were no statistically significant differences between the language groups with regard to the ASMD variable. The correlation between the ASMD variable and the variable of text-based problems varied between 0.62 and 0.79, while the correlation between reading comprehension and text-based problems varied between 0.38 and 0.47. The error analyses reveal that most failures to solve text-based problems are about 80% due to arithmetical errors or copying errors. Even if the bilingual pupils in general perform less well in reading comprehension than their Swedish-speaking peers, reading comprehension cannot be considered more of a reason for the failures of the bilingual pupils than for the Swedish-speaking pupils.

General Trends of Development

The trends of development discussed above point to the difficulties of unambiguously describing how the knowledge of the pupils develops and changes during their time at school. I outline a pattern of change in order to answer the question I have been asked repeatedly in the course of my research, namely whether it is possible at an early stage to decide how a pupil will manage at school.

The children's results with regard to the variable number sense at the age of 6 have very little value as predictors of future development. When the whole sample of pupils under study was divided into achievement groups of roughly the same size, classified as over-, under-, and medial-achievers, it can be noted that the pupils change groups at different age levels. For example, with regard to number sense, it can be established that 37% ($n = 13$) of the children who are under-achievers at the age of 6 belong to the same under-achievement group at both 12 and 15 years of age. The group consists of 8 girls and 5 boys. Of 40 over-achievers aged 6, 23% ($n = 9$) retained their position as over-achievers at age 12 as well as 15. The group consists of 6 boys and 3 girls. The result suggests that the transfer from within the group of under-achievers was less than from within the group of over-achievers. The corresponding change takes place with regard to the ASMD variable and the text-based problems variable. The proportion of pupils who belong to the same achievement group during the whole of their time at school was less than 20%.

DISCUSSION

The present investigation was initiated by my reflections on the great differences that exist between the mathematical experiences of different individuals. In the course of my investigation, my awareness of what it is in the nature of mathematics that leads to failure has been increased, but I have not been able to verify what it is that leads to success. I have followed the development of the pupils in a sample group and have seen a dynamic pattern of change that in some respects is obvious and in other respects unpredictable. The varying landscape that according to Jonassen and Grabowski (1993) characterizes the differences between individuals has in certain respects assumed sharper outlines.

Studying a course of development and a change which in many respects is self-evident is complicated by the fact that the pupils' activity is related to so many different factors. The school as a learning environment affects the activity of the pupil and his or her learning, knowledge, and attitudes. Choosing to study the knowledge of pupils in isolation from their learning context was partly due to the longitudinal character of the investigation, partly to an ethical attitude to the research task. The pupils attend schools that have their own profiles and varying environments. In the course of the long period of attending school, there will be many changes in the teaching staff, which will result in changes in the teaching methods, the modes of working, and the teaching aids as well. Even if I had chosen to compare the development of the pupils' knowledge in relation to their school, it would not have been possible to interpret the results unambiguously because of many contributing factors in the pupils' environment.

Limiting mathematical knowledge to three variables, number sense, arithmetical operations, and text-based problems can be seen as a weakness in the investigation. In addition to the bases for the choice of variables that I have illustrated in the introductory part of the thesis, I wish to mention the fact that the extensive subject matter of mathematics makes a selection justified. By reducing the number of variables, the number of data could be increased, which resulted in the reliability coefficient of the individual variables being relatively good.

The structure of the thesis can be discussed at a level of principle. The constructivist approach that dominates in mathematical-didactic research implies an assessment that makes the children's learning tangible. This investigation therefore came to occupy a position in the field of tension between a behaviorist attitude with positivistic traditions of assessment and a constructivist attitude where thinking and the development of knowledge are assessed. In qualitative analyses, I found certain patterns characteristic of the pupils' wrong answers even if I could not penetrate more deeply to find the reasons for these specific erroneous answers. The interviewing method might have given a better basis for an illustration of the pupils' arithmetical activities. Because of my choice of research methods I have, however, achieved informative results in the product and process assessment that Imsen (1992) calls for. When both these methods of assessment are used for different purposes, they supplement each other.

With reference to the results of the investigation and the choice of methods of analysis, I want to draw attention to both teacher research on "the child's world of mathematics" and school-related development research.

The didactic implications of the investigation are summed up from a school perspective, an individual perspective, and a learning–theoretical perspective. From a school perspective, children show good knowledge when starting school, in combination with great individual differences concerning the need for richly varied teaching. By richly varied teaching I refer to a form of teaching where mathematics is seen from a wider perspective. At a macro level I wish to emphasize teaching in accordance with the model Coburn (1989) uses when he speaks of a global perspective of what counting implies. When the model is broken down at micro level with the focus on concept formation, the model of different ways of representing concepts (Lesh, Post, & Behr, 1987) appears as a clue to the network that determines the choice of mode of working. The interaction that exists between various forms of language usage, model formation, and connections with reality is of importance for the way in which children learn. The gradual inability to cope indicates that there is a discrepancy between the mathematics curriculum and the pupils' possibilities of learning. This is one of the problems within the subject of mathematics that schools ought to pay attention to. A wider view of what mathematics is and what should be included in the teaching of mathematics is a challenge for the whole basic stage (years 1–9). The National Core Curriculum for Basic Education of 2004 (Utbildningsstyrelsen, 2004) offers guidelines to look at assessment from a process-orientated perspective. As assessment seeks new forms, it is important to elucidate the purpose of assessment. The kind of assessment that is carried out continuously throughout the learning process needs to support the pupil's learning and help him or her to develop clear aims. The pupils can only assess themselves if they have a clear view of what the goals are. This means that pupils gradually should develop their awareness and ability to assess their own learning. For this, the pupils need the help of a teacher who can analyze their forms of thinking. In social interaction the thoughts of pupils are uncovered and thus become objects of assessment. Such a model of assessment coincides with both the view of pupils' learning in social interaction and the specific characteristics of mathematics. At the same time, it becomes a quality assurance and helps the teachers to assess their own work.

Of the two forms of assessment used in the inquiry, I would, from an individual perspective, like to focus on the communication-oriented model as a means of bringing pupils' counting into focus. The model implies both written and oral communication. The use of language is not only emphasized in connection with concept formation (Lesh, Post, & Behr, 1987; Vergnaud, 1998) but also serves a diagnostic purpose (Keranto, 1998). The different ability of pupils to verbalize a solution points to the need for a systematic use of language. Conscious linguistic training not only increases the pupils' communicative skills but also makes their thoughts tangible and available for assessment. At every stage there is some reason for finding the optimal situations for linguistic training, so that inner forms of representation are translated into a linguistic and symbolical form—so-called external representations. Oral and written communication makes it possible to assess wrong answers. Teaching is often directed at bringing out correct solutions, and in most cases there is no room for analyses of wrong answers in teaching. Above all, when solving text-based problems, it often happens that the pupil focuses his or her erratic calculation on an assumed inability to solve text-based problems even if the error lies in the performance of the arithmetical operation itself. All pupils should be given the possibility to develop mathematical reasoning, and in this process the role of the teacher becomes very central. The task of the teacher is to interpret and to uncover the pupil's language in written form both to the pupil himself or herself and to other pupils in class. Simultaneously there is an assessment of the pupil's perceptions and of the quality of the thinking process. Many pupils need help with expressing their own thoughts in written form.

Even from the theoretical perspective of learning, language has a prominent place in the teaching of mathematics (Kilpatrick, Swafford, & Findell, 2001). Why is it that some children so obviously even at the age of 9 can describe how they think when performing certain arithmetical operations while others just state that they "count" or "that it comes automatically"? Perhaps these children lack the experience of verbalizing their knowledge. Answers that are given quickly are termed direct-answers by Magne (1998), and he sees a danger in such answers if they are wrong. Magne terms the phenomenon direct-answer rigidity and considers this to prevent children from thinking mathematically. When assessing such direct answers one can only say that the answer is right or wrong. If a teacher, know how a pupil reasons, one can help him/her to develop further. The teacher can even use one pupil's thinking mistakes to understand another pupil's way of thinking or to make another pupil aware of his or her thinking and solution. In this process the pupils' competencies of reasoning, thinking and communication become extremely central. By interpreting the pupil's handling of symbols, one can assess the quality in the pupil's learning and in some cases notice a logically and conceptually correct way of expressing a phenomenon, even if the formal exactness is lacking. The pupil is then about to develop an understanding of the concept, but has not yet been able to develop all the representational forms of the concept. From a longitudinal perspective, it is possible to find patterns that recur in the pupils' answers and show individual misconceptions that stem from a long time back.

When pupils communicate their thinking, they develop their awareness of their own thought processes, their metacognition. Even if the purpose of this investigation was not to assess pupils' metacognitive ability, it came to the fore in studying the children's descriptions of their counting strategies. By allowing the pupils to reflect on their solutions, they are given some experience of verbalizing their knowledge and become aware of their thinking. Since the viability of the pupils' knowledge is the starting point for assessing the learning of mathematics according to a social–constructivist view of learning, the metacognitive development of knowledge will be of great importance for an assessment. It facilitates the analysis of pupils' wrong answers.

Under the motto "I can help a child to develop his or her thinking if I know how the child thinks," I see both possibilities and limitations in each individual classroom. When, in the theoretical frame of reference, I illustrated the characteristic features of mathematics, the essential property of its language and conceptual system emerged. There are children who can verbalize their mathematical thinking and there are children who cannot. What are the background factors that cause these variations, and how do these entities agree with the children's results and development? Are there differences between boys' and girls' language manipulation in solving mathematical problems? Since linguistic communication is a natural part of the teaching of mathematics at school, many possibilities of research within a single class will open from the perspective of the teacher.

NOTE

This chapter is a short version of my PhD thesis (Räknespår, 2000): Barns matematiska utveckling från 6 till 15 års ålder. Åbo: Åbo Akademi University press.

REFERENCES

Ahlberg, A. (1996). Undervisningsprocessens betydelse för pojkars och flickors lärande. *Nomad, 2/3*(4), 7–30.

Alrø, H. & Skovsmose, O. (1993). Det var icke meningen!—om kommunikation i matematikundervisningen. *NOMAD, 1*(2), 6–29.

Anghileri, J. & Johnson, D. C. (1988). Arithmetic operations on whole numbers: Multiplication and division. In T. R. Post (Ed.), *Teaching mathematics in grades K–8* (pp. 146–187). Newton, MA: Allyn & Bacon.

Bauersfeld, H. (1998). Radikalkonstruktivism, interaktionism och matematikundervisning. In A. Engström (Ed.), *Matematik och reflektion* (pp. 54–81). En introduktion till konstruktivismen inom matematikdidaktiken. Lund: Studentlitteratur.

Behr, M., Harel, G., Post, T., & Lesh, R. (1992). Rational number, ratio, and proportion. In D. Grouws (Ed.), *Handbook of research on mathematics teaching and learning* (pp. 296–333). New York: Macmillan.

Bergman, L. R. & Magnusson, D. (1990). *Data quality in longitudinal research.* Cambridge: Cambridge University Press.

Björkqvist, O. (1997). *Utvärdering av matematikkunskaperna i årskurs 7 i grundskolan.* Publikationer från Pedagogiska fakulteten vid Åbo Akademi, nr 23. Vasa: Institutionen för lärarutbildning.

Brown, M. (1981). Number operations. In K. M. Hart (Ed.), *Children´s understanding of mathematics* (pp. 11–16). London: John Murray.

Cobb, P. (1986). Context, goals, beliefs and learning mathematics. *For the Learning of Mathematics, 6*(2), 2–9.

Coburn, T. G. (1989). The role of computation in the changing mathematics curriculum. In P. R. Trafton & A. P. Shulte (Eds.), *New directions for elementary school mathematics* (pp. 43–56). Reston, VA: NCTM.

Engström, A. (1997). *Reflektivt tänkande i matematik.* Om elevers konstruktioner av bråk. Stockholm: Almqvist & Wiksell International.

Ernest, P. (1998). Vad är konstruktivism? In A. Engström (Ed.), *Matematik och reflektion. En introduktion till konstruktivismen inom matematikdidaktiken.* (pp. 54–81). Lund: Studentlitteratur.

Fennema, E. & Peterson, P. (1985). Autonomous learning behaviour: A possible explanation of gender-related differences. In L. C. Wilkinson & C. B. Marrett (Eds.), *Gender influences in classroom interaction* (pp. 17–35). New York: Academic Press.

Ginsburg, H. P., Jacobs, S. F., & Lopez, L. S. (1993). Assessing mathematical thinking and learning potential in primary grade children. In M. Niss (Ed.), *Investigations into assessment in mathematics education. An ICMI study* (pp. 157–167). Dordrecht, The Netherlands: Kluwer.

Grevholm, B. (1996). Vad vet fröken om flickor och pojkar i matematiken? In L. Lindberg & B. Grevholm (Eds.), *Kvinnor och matematik. Konferensrapport* (pp. 77–92). Göteborgs universitet. Institutionen för ämnesdidaktik. Avdelningen för matematik.

Hughes, M. (1984). What is difficult about learning arithmetic? In M. Donaldson, R. Grieve, & C. Pratt (Eds.), *Early childhood development and education* (pp. 204–222). Oxford: Blackwell.

Hyde, J. S., Fennema, E., & Lamon, S. J. (1990). Gender differences in mathematics performance: A meta-analysis. *Psychological Bulletin, 107,* 139–155.

Häggblom, L. (1994). Mathematics on the child`s condition—A constructivist approach at school. In M. Ahtee & E. Pehkonen (Eds.), *Constructivist viewpoints for school teaching and learning in mathematics and science* (pp. 93–98). Helsinki: Department of Teacher Education University of Helsinki.

Häggblom, L. (2000). *Räknespår. Barns matematiska utveckling från 6 till 15 års ålder.* Åbo: Åbo Akademis förlag.

Häggblom, L. (2006). Utvärdering I lärandeprocessen. In L. Häggblom, L. Burman, & A-S. Röj-Lindberg (Eds.), *Perspektiv på Kunskapens och lärandets villkor. Festskrift tillägnad professor Ole Björkqvist* (pp. 125–132). Specialutgåva från Pedagogiska fakulteten vid Åbo Akademi, nr 1/2006. Vasa: Pedagogiska fakulteten.

Imsen, G. (1992). *Elevens värld.* Lund: Studentlitteratur.

Jonassen, D. H. & Grabowski, B. L. (1993). *Handbook of individual differences, learning, and instruction.* London: Erlbaum.

Keranto, T. (1998). Kriittinen ajattelu ja tieteentuntemus matematiikan opetuksessa. In P. Räsänen, P. Kupari, T. Ahonen, & P. Malinen (Eds.), *Matematiikka näkökulmia opettamiseen ja oppimiseen* (pp. 18–38). Jyväskylä: Niilo Mäki Instituutti & Koulutuksen tutkimuslaitos.

Kieren, T. E. (1993). Rational and fractional numbers: From quotient fields to recursive understanding. In T. P. Carpenter, E. Fennema, & T. A. Romberg (Eds.), *Rational numbers. An integration of research* (pp. 49–84).Hillsdale, NJ: Erlbaum.

Kilpatrick, J. (1993). Beyond face value: Assessing research in mathematics education. In G. Nissen & M. Blomhøj (Eds.), *Criteria for scientific quality and relevance in the didactics of mathematics* (pp. 15–34). Roskilde: Roskilde University, IMFUFA.

Kilpatrick, J., Swafford, J., & Findell, B. (2001). *Adding it up: Helping children learn mathematics.* Washington, DC: National Academy Press.

Kimball, M. (1989). A new Perspective on Women´s Maths Achievement. *Psychological Bulletin, 105*(2), 198–214.

Kulm, G. (1991). New directions for mathematics assessment. In G. Kulm (Ed.), *Assessing higher order thinking in mathematics* (pp. 71–78). Washington, DC: American Association for the Advancement of Science.

Lesh, R., Post, T., & Behr, M. (1987). Representation and translations among representations in mathematics learning and problem solving. In C. Janvier (Ed.), *Problems of representation in the teaching and learning of mathematics* (pp. 33–40). London: Erlbaum.

Magne, O. & Thörn, K. (1987). *En kognitiv taxonomi för matematikundervisningen*: Del 1. Pedagogisk-psykologiska problem. Nr. 471. Malmö: Lärarhögskolan.

Magne, O. (1990). *Medelsta-matematik. Hur väl behärskar grundskolans elever lärostoffet enligt lgr 69 och lgr 80?* Pedagogisk-psykologiska problem, nr 539. Institutionen för pedagogik och specialmetodik. Lärarhögskolan. Lunds Universitet.

Magne, O. (1998). *Att lyckas med matematik i grundskolan*. Lund: Studentlitteratur.

Marshall, S. P. & Smith, J. D. (1987). Sex differences in learning mathematics: A longitudinal study with item and error analyses. *Journal of Educational Psychology, 79*, 372–383.

Morgan, C. (2000). Better assessment in mathematics education? A social perspective. In J. Boaler (Ed.), *Multiple perspectives on mathematics teaching and learning* (pp. 225–242). Norwood, NJ: Ablex.

Moser, J. M. (1988). Arithmetic operations on whole numbers: Addition and subtraction. In T. Post (Ed.), *Teaching mathematics in grades K–8* (pp. 111–145). Newton, MA: Allyn & Bacon.

Mulhern, G. (1989). Between the Ears: Making Inferences about Internal Processes. In B. Greer & G. Mulhern (Ed.), *New Directions in Mathematical Education*. London: Routledge, pp. 27-33.

Niss, M. (1993). Assessment in mathematics education and its effects: An introduction. In M. Niss (Ed.), *Investigations into assessment in mathematics education. An ICMI study* (pp. 1–30).Dordrecht, The Netherlands: Kluwer.

Romberg, T. A. (2004). Classroom assessment studies. In B.Clarke, D.M. Clarke, G. Emanuelsson, B. Johansson, D.V. Lamdin, F.K. Lester, A. Wallby, & K. Wallby (Eds.), *International perspectives on learning and teaching mathematics* (pp. 585–600). Göteborg: Göteborg University, National Center for Mathematics Education.

Skaalvik, E.M. & Rankin, R.J. (1994). Gender differences in mathematics and verbal achievement, self-perception and motivation. *British Journal of Educational Psychology, 64*, 419–428.

Skolstyrelsen. (1985). *Grunderna för grundskolans läroplan 1985*. Helsingfors: Statens Tryckericentral.

Steffe, L. P. & Cobb, P. (1988). *Construction of arithmetical meanings and strategies*. New York: Springer.

Utbildningsstyrelsen. (1994). *Grunderna för grundskolans läroplan.1994*. Helsingfors: Tryckericentralen.

Utbildningsstyrelsen. (2004). *Grunderna för läroplanen för den grundläggande utbildningen 2004*. Helsingfors: Tryckericentralen.

Webb, N. L. (1992). Assessment of students' knowledge of mathematics: Steps toward a theory. In D. A. Grouws (Ed.), *Handbook of research on mathematics learning and teaching*. (pp. 127–146). New York: Macmillan.

Vergnaud, G. (1998). A comprehensive theory of representation for mathematics education. *Journal of Mathematical Behavior, 17*(2), 167–181.

von Glasersfeld, E. (1995). A constructivist approach to teaching. In L. P. Steffe & J. Gale (Eds.), *Constructivism in education* (pp. 3–16). Hillsdale, NJ: Lawrence Erlbaum.

53

PAINFUL PARADIGM SHIFTS IN THE TEACHING AND LEARNING OF MATHEMATICS

Harry Silfverberg

UNIVERSITY OF TAMPERE

Lenni Haapasalo

UNIVERSITY OF EASTERN FINLAND

Abstract: The chapter discusses trends in the paradigm shifts of mathematics teaching and how those trends seem to be changing the Finnish teaching culture in schools. The considerations are based on a study to examine the most typical conceptions of what the teaching of the advanced course of mathematics at the upper secondary school nowadays in Finland usually is like, what it should be like, and how the students believe that it is developing during the next 15 years.

Keywords: constructivism, mathematics learning, mathematics teaching, teaching paradigm, technology-based

INTRODUCTION

Many features that can be seen to have a direct or at least an indirect impact on the contents and forms of mathematics teaching at schools have changed significantly during the last two decades. The three paradigm shifts that we think have changed our basic conceptions about the culture of mathematics teaching are: (1) paradigms of mathematics teaching and learning in general, (2) paradigms of the educational use of ICT, and (3) paradigms of the role of subject-specific technology in the mathematics teaching process at schools.

In this chapter, we will briefly present some of the development trends that correspond to the above-mentioned three paradigm shifts. Both the educational policy makers' and ordinary teachers' positive or negative attitudes to the possibilities opening with these paradigm shifts determines in many respects how the culture of mathematics teaching is presented in the curricula and in what form it comes true in practice. Scenarios of mathematics education can be studied by cross-tabulating in a matrix form the possible shifts in the subject-specific and general ICT developments against the possible shifts in the conceptions of teaching and learning mathematics. The matrix can be used to concretize the present teaching culture but also the scenarios of the wanted or possible trends of mathematics education as well. We will present some results of a small study where the first author applied this matrix as a research instru-

The First Sourcebook on Nordic Research in Mathematics Education, pages 731–739

ment examining student teachers' conceptions about the way the advanced course of mathematics at the upper secondary school level seems to be taught in Finland nowadays. Students were also asked how they think it should be taught and what kind of changes they foresee happening by the year 2020.

PARADIGM SHIFTS AFFECTING THE CULTURE OF MATHEMATICS TEACHING

Views of the Learning of Mathematics

During the last two decades the general paradigm of teaching and learning has fundamentally changed in mathematics as in many other fields of education: first from the behaviorist knowledge transmission to the constructivist paradigm emphasizing cognitive processes, and then to the socio-constructivist (socio-cultural) paradigm emphasizing social factors in constructing shared knowledge (cf. Ravitz, Becker, & Wong, 2000). However, the real development in schools is not that straightforward. According to Ernest (2005), for example, at different times many views such as cognitivism, enactivism/bodily cognition, radical and social constructivism, and socio-cultural theories have with different emphasis directed both the research of mathematics education and its practice. Recently, some researchers have also speculated about the extensions and reinterpretations of the above-mentioned constructivist views such as communal constructivism (Holmes, Tangney, Fitzgibbon, Savage, & Mehan, 2001; Leask & Younie, 2001) and generative learning (Bannan-Ritland, Dabbagh, & Murphy, 2000). Some guesses at what will come after socio-constructivism, when we have built the infrastructures of the information society, have also been proposed. Brown (2006), for example, calls the new emerging paradigm *navigationism.*

From the point of view of this chapter, the most central paradigm shifts relating to the conceptions of learning are the shifts from behaviorism to individual-constructivism and then from individualistic constructivism to the socio-constructivism. Because behaviorism and the two kinds of constructivism mentioned here can be considered as generally known, they are not discussed in this context in more detail (cf. Björkqvist, 1996; Chen, 2003; Ernest, 2005; Lerman, 2000; Núñez, Edwards, & Matos, 1999; Palicsar, 1998; Ravitz et al., 2000).

It is important to note that the changes in the paradigms concerning learning do not necessarily lead to similar changes in the practice of mathematics teaching. The report of Ravitz and colleagues (2000), for example, shows that in the U.S. the culture of the traditional direct teaching is still living strongly, especially in the teaching of mathematics, although the paradigms of teaching and learning mathematics to which the mathematics education community is officially committed are based on the constructivist ideas. It also seems that the cultures of teaching shown by the same teacher in different contexts can express a commitment to different paradigms of learning.

Changes in the Conceptions of the Possibilities of the Educational Use of Information and Communication Technology

The development of general ICT technology has crucially changed the conceptions of computer-based teaching and learning in general (cf. Aviram & Talmi, 2005) and raised a number of new challenges for instance to pre-service and in-service teacher training (Cheng, 2004). In his doctoral thesis, Alajääski (2000) divided the development of computer assisted teaching so far in Finland into three stages. At the first stage in the 1980s, the teaching of the computer science in schools was considered important as such. Even young children were taught programming languages like Logo and others. From the end of the 1980s, the focus moved to the use of more general tools like word processing, drawing software, and spreadsheets. At that time, the use of information technology also began to spread as a part of all school subjects, and from the early 1990s in the third stage of computer-assisted teaching, the focus on teaching and learning began to shift to using computers more and more as a tool. As the schools got networked, people started to use more and more so-called open learning environments in order to enrich normal teaching. When broadband networks became common, the possibilities for the offering of distance education also essentially improved. According to Alajääski, the adapting of the education sector as a part of the structure of the information society also began in this third phase of development. The view that Alajääski offers of the development seen from the Finnish perspective corresponds well to the three of the four stages presented by Koschmann (1996, 2001) from a more general context.

The Information Society Program for Education, Training and Research for the years 2004–2006 is, in Finland, geared to developing all citizens' information society knowledge and skills to enable educational institutions to use information and communications technology (ICT) in a versatile way in their activities, and to promote social innovation through the use of ICT. According to Vainionpää (2006), the dominating paradigm of the current educational use of an information and communication technology (ICT) is a so-called v-paradigm, where the use of virtual learning environments offered by the Internet is central. Later, the m-paradigm based on mobile interfaces and making net-resources reachable nearly everywhere will probably modify the v-paradigm further. The still more distant future will perhaps bring a reality that is in accordance with the so-called u-paradigm, where ICT integrates into the body, clothes, and buildings, so even people themselves will become a part of the data network.

Goos, Galbraith, Renshaw, and Geiger (2003) distinguish four fundamental attitudes to the use of technology in general and into the educational use of technology in particular. According to Goos and colleagues, technology can be seen as (1) a master determining the operation, (2) a servant assisting the operation, (3) an equal partner, or (4) an extension of the self that helps the individual to exceed limits that would be unreachable without the technology. Aviram and Talmi (2005) distinguish technocratic, reformist, and holistic paradigms of the educational use of ICT. According to the technocratic view, the use of ICT does not essentially bring anything new to the teaching. The structures and functions of the present school organization will not significantly change with the technological development. According to a reformist view, the new technology can be used to reform teaching, primarily by developing more active didactic approaches to teaching. The ones representing a holistic paradigm are unanimous about the fact that the development of ICT will cause major pressures for change of the whole school system and that the large-scale transition to the information society will radically change not only our conceptions of the information, learning, work, communication, and co-operation, but also our conception of the identity of an individual.

In our opinion, the transition from the local educational use of information technology into the use of ICT that utilizes data networks can be considered as one of the major shifts in the paradigm of the educational use of ICT. The information sources that are quite easily available for teachers and for pupils have multiplied, the number of the applications that are freely usable for everybody has grown, the possibilities of distance learning and of independent studying have improved significantly, and communality has gained new forms in learning. Despite the fact that all this has happened, we consider that the shifts in the educational use of ICT have not yet essentially changed the mainstream culture of mathematics teaching in Finland. Likewise, many foreign evaluations (see, for example, Barron, Kemker, Harmes, & Kalaydjian, 2003, p. 496) have shown that the use of ICT seems clearly to be more infrequent in mathematics than in most other school subjects.

Conceptions of the Educational Use of Subject-Specific Technology

The rapid development of subject-specific ICT applications and of the so-called handheld technology has brought a number of new possibilities to the field of mathematics education and made possible first the transition from paper-and-pencil mathematics to numeric calculation carried out by machines and then the transition from the world of numeric mathematics and production of static graphics to the world of the symbolic algebra and the dynamic geometry implemented within interactive learning environments. In the following, we will call the area of ICT that includes the tools developed to be used especially in mathematics teaching as *Mathematics Teaching Technology* (MTT). Referring both to Nilsson & Naeve (2003) and Haapasalo & Silfverberg (2007), ICT-based mathematics education is becoming nowadays a very broad concept. So far it has expanded to include at least the following solutions:

- computer algebra systems (CAS), dynamical geometry (DGS), and dynamical statistics (DSS)
- spreadsheets, drawing programs, and other versatile tools for mathematical modeling
- online databases of available software, instruction, research, statistics, history, etc.
- online communication in all of its synchronous and asynchronous forms
- new kinds of environments to read, write and publish, including tools for support
- tools for the utilizing of the world-wide web: search engines, etc.
- online experiments and simulations in diverse forms of digital educational content
- online libraries containing books, learning objects, other teaching materials, etc.
- learning management systems (*LMS*), which are used to manage students and course materials
- assessment instruments identifying the learner's progress: digital portfolios and online tests with automatic checking
- virtual worlds in the form of three-dimensional immersive environments offering, for example, shared exhibitions or other forms of collaborative functionality
- automatic proofing, etc.

These applications can be used online via networks or be locally run by computers, modern calculators and communicators. As a consequence of the paradigm shifts described in the introduction, the character of teaching and learning mathematics is also changing to be more distributive (independent of time, place, and formal modes), more constructivist (learner-centered or even learning community-centered), more dependent on cultural factors, more technologically enhanced, and based more and more on hypermedia instead of "papermedia" (cf. Merilampi, 2002). Even though MTT offers a large number of different applications to the teaching of mathematics, according to our view, it still has not had a very significant effect on the change in the culture of the teaching of mathematics in Finland. The real paradigmatic change in the culture of the mathematics teaching probably is waiting for the time when the use of symbolic algebra (CAS) will be allowed in school mathematics and the skills to use CAS will also be tested in the national matriculation tests.

COMBINING THE SIMULTANEOUS PARADIGM SHIFTS

Silfverberg's (2006) aim has been to develop a simple conceptual framework to examine simultaneous paradigm shifts, which modify our conceptions of the culture of mathematics teaching. Figure 53.1 shows a classification of five different views of how mathematics is learned best and how it should therefore be taught. With this classification, answers were looked for to two basic questions concerning the learning and teaching process: (1) Who are the real actors in this process? and (2) who are the primary decision makers in this process? The first question addresses the level of learner-centricity and the social character of the learning process, and the second, the autonomy and empowering of the learner in this process.

The implementation of ICT in schools can proceed mainly in two basic ways. Schools either develop mainly local learning environments emphasizing mostly individual learning in traditional settings or they put more effort in developing open, net-based learning environments primarily emphasizing team learning and learning independent of time, place and formal modes. In applying ICT to the practice of mathematics education we distinguish three different paradigmatic choices. Firstly, we may want to restrict the use of technology only to the numeric and graphic (handheld) technology with strict rules as to when this is allowed (in fact, this is the present situation in Finland regarding the advanced course of math on the upper secondary school level). Secondly, we may allow all kinds of use of technology, provided that it is based totally on numeric processing. Thirdly, we may allow the use of any technology including CAS, DGS, and so on and base our math curriculum on this presupposition. Figure 53.2 shows a kind of decision tree leading to six mutually different cultures of educational use of ICT. By cross-tabulating the categorizations shown in Figures 53.1 and 53.2, we get a matrix presented in Figure 53.3.

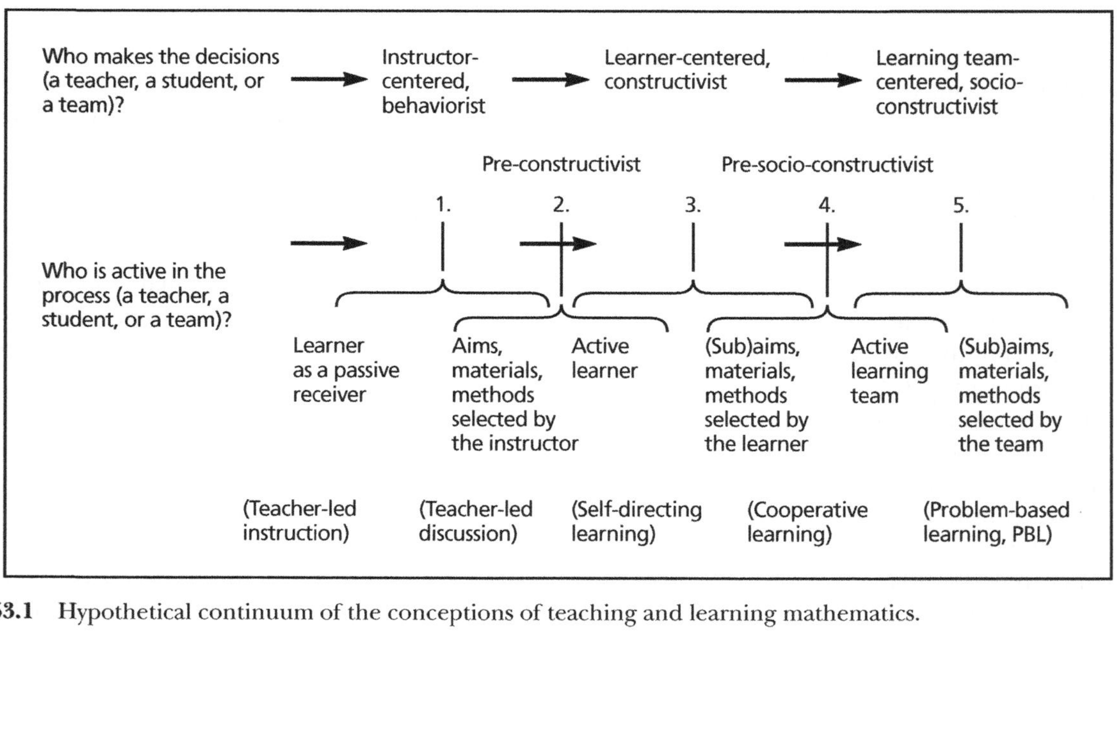

Figure 53.1 Hypothetical continuum of the conceptions of teaching and learning mathematics.

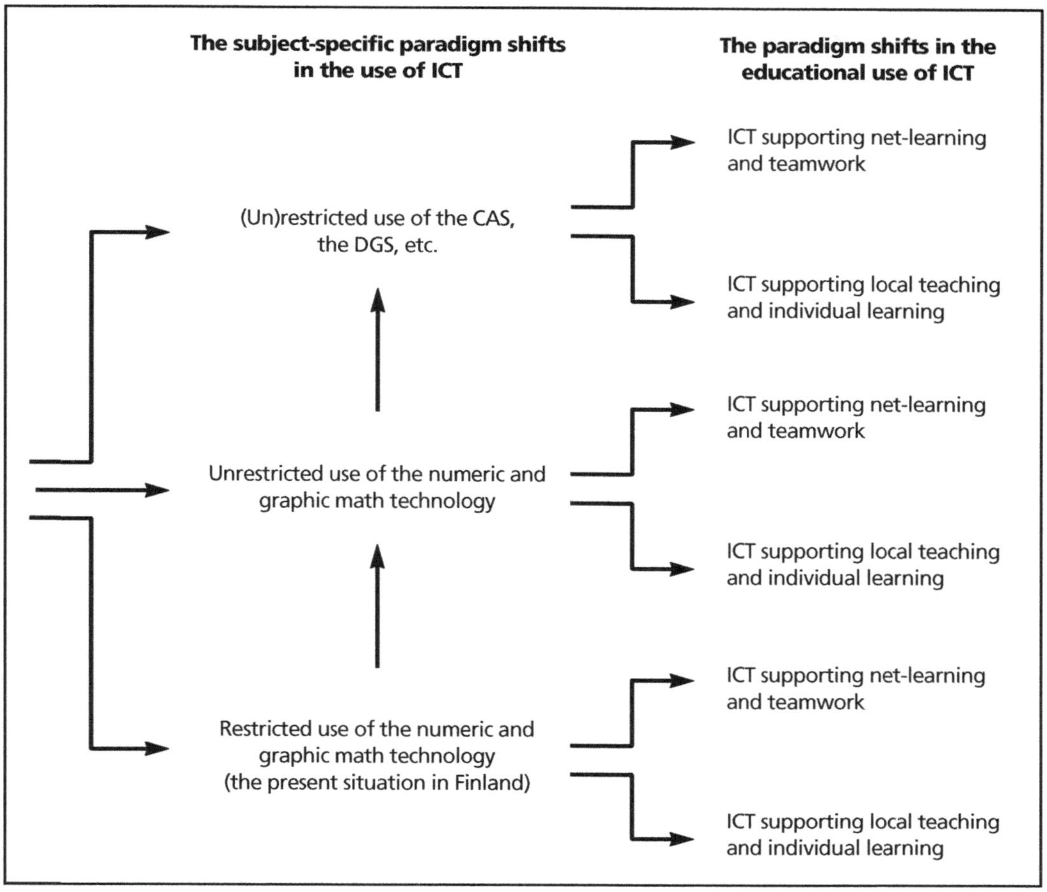

Figure 53.2 The choices leading to different cultures of the educational use of the *ICT.*

STUDENT TEACHERS' CONCEPTIONS OF TEACHING

Aims and Methods

The study aimed to find out student teachers' conceptions of teaching the advanced course of mathematics at the upper secondary school. The empirical data were collected from 43 mathematics and science teacher trainees with two questionnaires—A and B. The first one contained 20 items that addressed, from different perspectives, the features of the teaching culture of the upper secondary school level advanced course of mathematics nowadays and its possible development trends from now to the year 2020. As in this chapter we will restrict our discussion to the results obtained from questionnaire B, we do not present the details of questionnaire A.

Questionnaire B contained simply the matrix in Figure 53.3, and verbal instructions were given on how it should be filled in. Before students answered these questionnaires, three lessons were given to them about the paradigm shifts that were presented in Figure 53.3 and their (potential) effects on the birth of different cultures of mathematics teaching. An attempt was made to present the different cultures of mathematics teaching as neutrally as possible during these lessons. After the lecture ended, the data were collected so that each student first filled out questionnaire A and then questionnaire B by circling those cells that in his or her opinion corresponded to the teaching culture of mathematics in such a way that

1. reflects the prevailing practice of teaching the advanced course of mathematics in most upper secondary schools in Finland during the academic year 2005–2006
2. reflects the change in practice that should have taken place in most upper secondary schools in Finland in the academic year 2005–2006
3. reflects the practice that he or she believes will be prevailing in the year 2020 in most upper secondary schools.

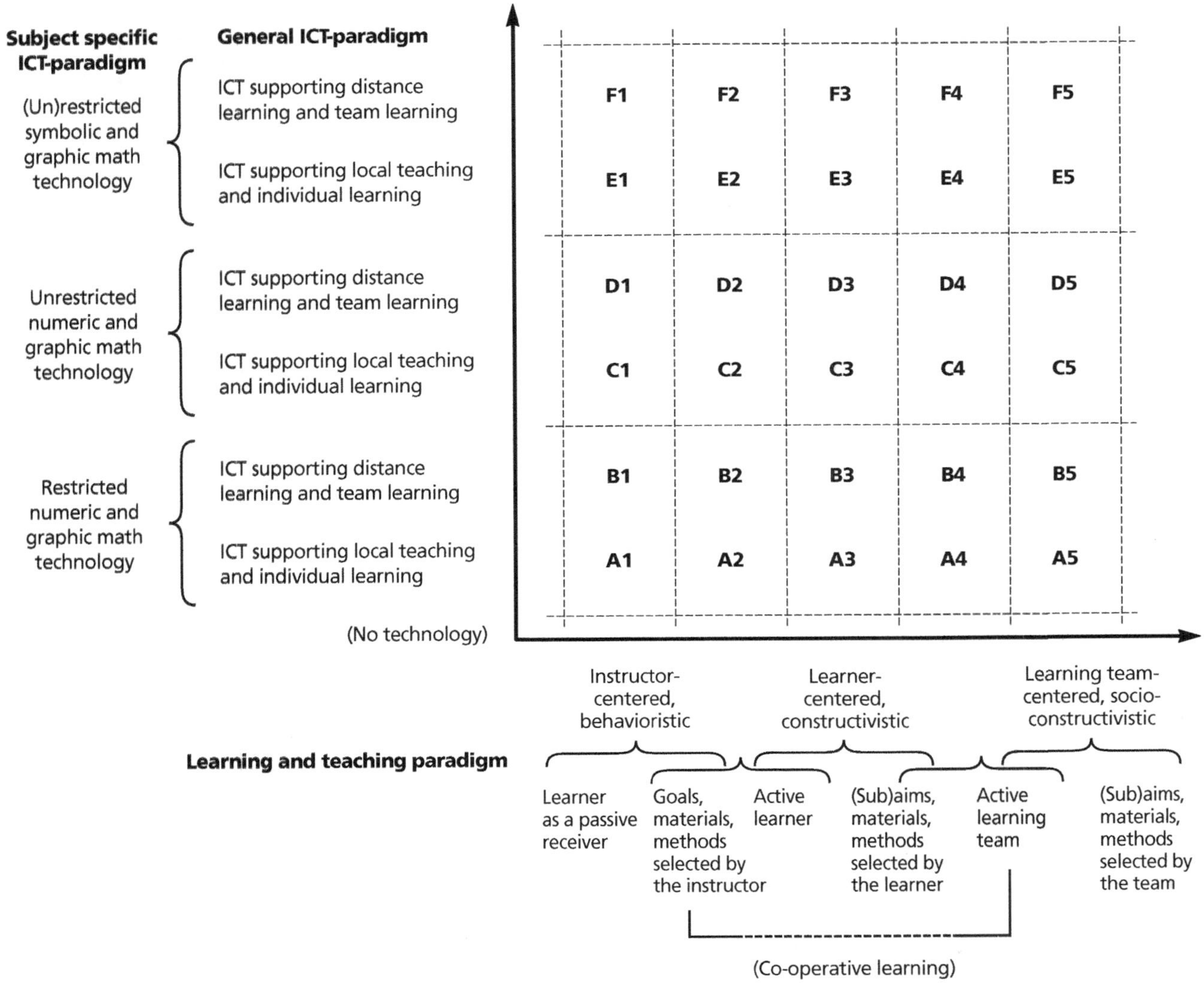

Figure 53.3 The matrix of paradigmatic choices in mathematics education.

In the testing situation it was especially emphasized to the students that the cells chosen for the areas *A*, *B* and *C* did not need to form a uniform area and that the different cultures of mathematics teaching can come true simultaneously.

Results

From the answers to questionnaire B, the relative frequencies for each cell were calculated for each alternative A, B and C. Figures 53.4–53.6 present the distributions of the students' choices in the different parts of the matrix. The darker the cell is the higher its relative frequency is.

The students considered the present teaching culture of the advanced course of mathematics in upper secondary school mostly quite traditional. However, plenty of dispersion was found in the conceptions. On the basis of the classifications of Figure 53.4 the majority of the students seem to think that the teaching of the advanced course of mathematics should be more student-centered and more constructivist than it actually is at present. Only a few students presented hopes for an addition of the use of technology. According to the same students, the way of teaching should be more constructive, and ICT should support net-based teaching and team learning (cf. Figure 53.3).

The answers of most students to alternative *C* deviated quite distinctly from the answers that were given to alternatives *A* and *B*. Many students believed that in the teaching and studying of the advanced course of mathematics, the use of technical tools by the year 2020 will become essentially more common than today, leading, for example, to free use of symbolic calculation. Many also believed that teamwork will become more common in the studying of mathematics. Likewise, it was com-

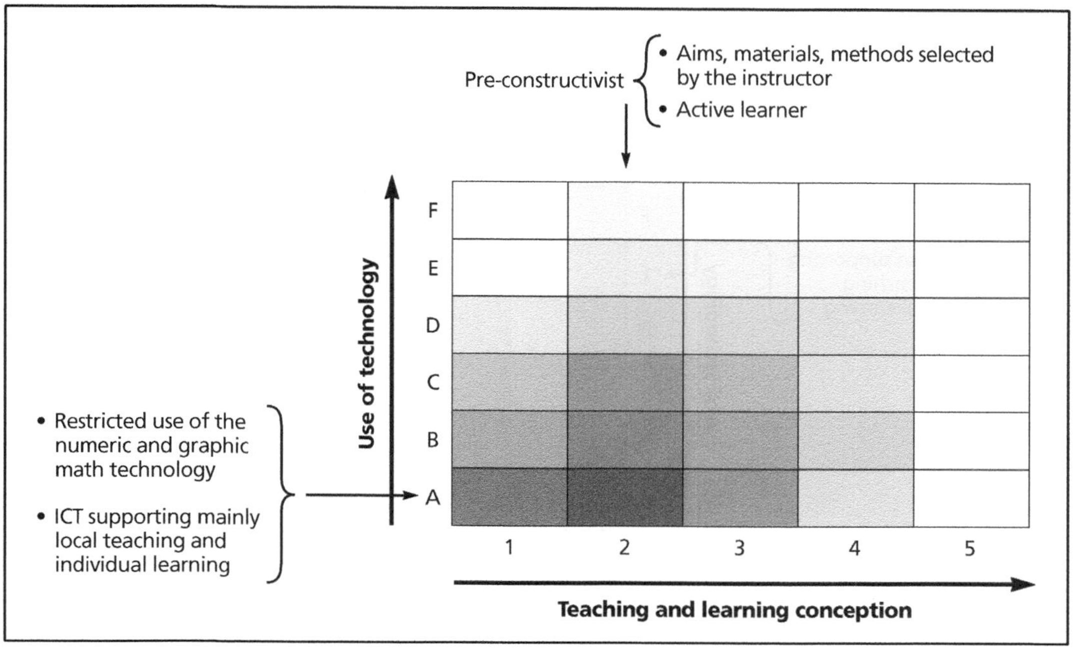

Figure 53.4 The distribution of the students' choices in A.

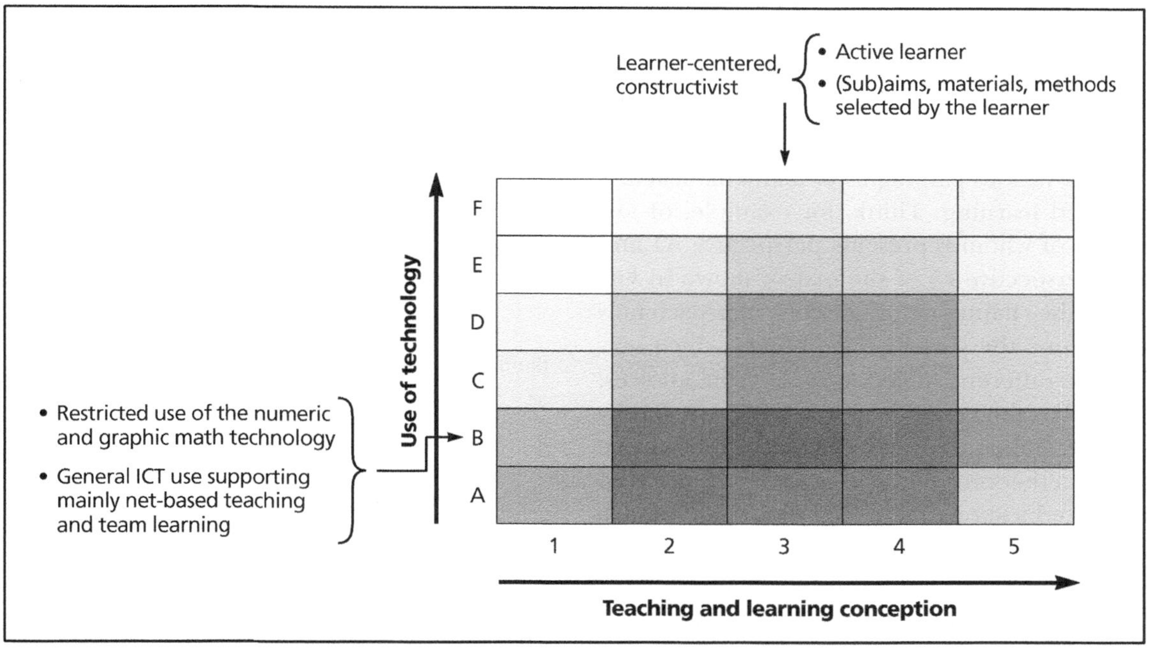

Figure 53.5 The distribution of the students' choices in B.

monly believed that the responsibility for the learning will shift more and more to the learner himself or to the learner community to which the learner belongs.

DISCUSSION

It is clear that good teaching of mathematics is related to many things. During the last decades our conceptions of how mathematics is learned and how it should be taught have changed so significantly that these changes can be called paradigm shifts. Likewise, because of the rapid technical development, we have been forced to re-evaluate our conceptions of the possibilities of the educational use of ICT. In the crossswell of the paradigm shifts, we are forced to answer, for instance, to the following big question: How do the new technology and the changing paradigms of

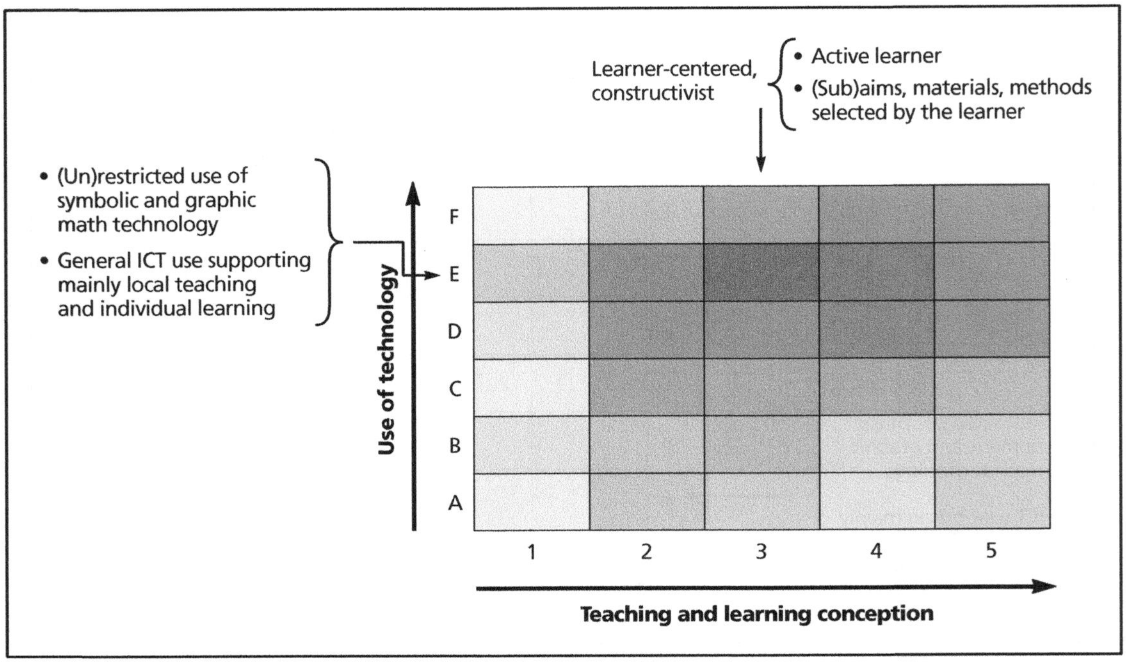

Figure 53.6 The distribution of the students' choices in C.

learning and teaching transform the way mathematics is taught and learned and the conception of competence in mathematics? Obviously we will get rather different answers to this question from the people committed to the various paradigms of mathematical and ICT-enhanced learning. Think, for example, of two people, one of whom represents perspective A1 and the other perspective F5 of the matrix shown in Figure 53.1. In the chapter we offered one approach how we can examine the paradigmatic bonds, which may lead to quite different subject-specific value choices. We believe that the conceptual space constructed by cross-tabulating the various paradigmatic shifts can be used both as a theoretical tool in the political decision making and educational planning and as an empirical tool by which we can study different conceptions of the present and the desired future realizations of the culture of teaching math in schools. Many of the prospective teachers of mathematics believed that the paradigm shifts actually change the culture of teaching mathematics in the long run. However, much must happen before the changes are possible. The curricula must be renewed, the paradigm shifts must be taken into account in textbooks, the teachers' professional competence has to be raised, and teachers ought to be inspired to take responsibility for carrying out the renewing process. As regards teacher education, perhaps the most optimal way is to include the usage of technology in the students' pedagogical studies (cf. Eskelinen & Haapasalo, 2006a, 2006b).

REFERENCES

Alajääski, J. (2000). *Tietokoneopetukseen liittyvän osaamisen ja vaikuttavuuden kehittyminen informaatioteknologiaan perustuvassa opetusympäristössä—Opettajien itsearviointiin perustuva tutkimus Rauman normaalikoulussa.* Tampereen teknillinen korkeakoulu. Julkaisuja 287.

Aviram, A. & Talmi, D. (2005). The impact of ICT on education: The missing discourse between three different paradigms. *E-Learning, 2*(2), 169–191.

Bannan-Ritland, B., Dabbagh, N., & Murphy, K. (2000). Learning object systems as constructivist learning environments: Related assumptions, theories, and applications. In D. A. Wiley (Ed.), *The Instructional Use of Learning Objects: Online Version.* Retrieved June 2, 2010 from http://reusability.org/read/chapters/bannan-ritland.doc.

Barron, A.E., Kemker, K., Harmes, C., & Kalaydjian, K. (2003). Large-scale research study on technology in K–12 schools: Technology integration as it relates to the national technology standards. *Journal of Research on Technology in Education, 35*(4), 489–507.

Björkqvist, O. (Ed.). (1996, July). *Mathematics teaching from a constructivist point of view.* Paper presented at the International Congress on Mathematical Education, Seville, Spain. Faculty of Education Report No. 3.

Brown, T. H. (2006). *Beyond constructivism: Navigationism in the knowledge era.*

Chen, C. (2003). A constructivist approach to teaching: Implications in teaching computer networking. *Information Technology, Learning, and Performance Journal, 2*(2), 17–27.

Cheng, Y. C. 2004. *Three Waves of Teacher Education & Development: Paradigm Shift in Applying ICT.* Keynote Speech presented at The Scandinavian/Asian Pacific Conference on "The Challenge of Integrating ICT in Teacher Education" organized by School of Education and Communication,

Jönköping University June 2–4, 2004, Jönköping Sweden. Retrieved June 2, 2010 from home.ied.edu.hk/~yccheng/doc/speeches/2-4jun04.pdf.

Ernest, P. (2005). Paradigm shifts in mathematics education. In A. Rogerson (Ed.), *Proceedings of the Eighth International Conference. Reform, Revolution and Paradigm Shifts in Mathematics Education.* Johor Bahru: Universiti Teknologi Malaysia.

Eskelinen, P. & Haapasalo, L. (2006a). Teachers and students as designers of their learning environments. In T. Asunta & J. Viiri (Eds.), *Pathways into research-based teaching and learning in mathematics and science education* (pp. 107–115). University of Jyväskylä, Department of Teacher Education. Research Report 84.

Eskelinen, P. & Haapasalo, L. (2006b). Socioconstructivist collaborative group processes in teacher education. Paper presented at the XXIII Annual Symposium of the Finnish Mathematics and Science Education Research Association. Åbo Akademi, Vaasa, Finland, October 27–28, 2006.

Goos., M., Galbraith, P., Renshaw, P., & Geiger, V. (2003). Perspectives on technology mediated learning in secondary school mathematics classrooms. *Journal of Mathematical Behavior, 22*(1), 73–89.

Haapasalo, L. & Silfverberg, H. (2007). Technology enriched mathematics education. In E. Pehkonen, M. Ahtee, & J. Lavonen (Eds.), *How Finns learn mathematics and science* (pp. 163–180). Rotterdam, The Netherlands: Sense Publishing.

Holmes, B., Tangney, B., Fitzgibbon, A., Savage, T., & Mehan, S. (2001). Communal constructivism: Students constructing learning for as well as with others. In J. Price, D. Willis, N.E. Davis, & J. Willis (Eds.), *Proceedings of the 12th International Conference of the Society for Information Technology and Teacher Education* (SITE 2001) (pp. 3114–3119). Charlottesville: Association for the Advancement of Computing in Education. Retrieved June 2, 2010 from http://www.cs.tcd.ie/publications/tech-reports/reports.01/TCD-CS-2001-04.pdf

Koschmann, T. (Ed.). (1996). *CSCL: Theory and practice of an emerging paradigm.* Mahwah, NJ: Lawrence Erlbaum Associates.

Koschmann, T. (2001). Revisiting the paradigms of instructional technology. In G. Kennedy, M. Keppell, C. McNaught, & T. Petrovic (Eds.), *Meeting at the crossroads. Proceedings of the 18th Annual Conference of the Australian Society for Computers in Learning in Tertiary Education* (pp. 15–22). The University of Melbourne: Biomedical Multimedia Unit. (http://www.ascilite.org.au/conferences/melbourne01/pdf/papers/koschmannt.pdf)

Lerman, S. (2000). The social turn in mathematics education research. In J. Boaler (Ed.), *Multiple perspectives on mathematics teaching and learning* (pp. 19–44). Westport, CT: Ablex.

Leask, M. & Younie, S. (2001). Communal constructivist theory: Information and communications technology pedagogy and internationalisation of the curriculum. *Journal of Information Technology for Teacher Education, 10*(1–2), 117–134.

Merilampi, R.-S. (2002). School in the grips of change—Media education in Finland. *eFinland*, September13, 2002. Retrieved June 2, 2010 from http://www.e.finland.fi/netcomm/news/showarticleed70.html?intNWSAID=8534

Nilsson, M. & Naeve, A. (2003). *On designing a global infrastructure for ICT-enhanced mathematics education.* Paper presented at the Nordic Pre-conference to the International Congress on Mathematics. Education (PICME 10). Växjö, Sweden, May 9–11, 2003. Retrieved June 2, 2010 from http://cid.nada.kth.se/pdf/CID-237.pdf.

Núñez, R.E., Edwards, L.D. & Matos, F. (1999). Embodied cognition as grounding for situatedness and context in mathematics education. *Educational Studies in Mathematics, 39*(1–3), 45–65.

Palicsar, A.S. (1998). Socio-constructivist perspectives on teaching and learning. *Annual Review of Psychology, 49*, 345–375.

Ravitz, J., Becker, H.J., & Wong, Y.-T. (2000). *Constructivist-compatible beliefs and practices among U.S. teachers. teaching, learning and computing: 1998 Survey. Report #4.* Retrieved June 2, 2010 from http://www.crito.uci.edu/TLC/FINDINGS/REPORT4/

Silfverberg, H. (2006). Pohdintaa paradigmavaihdosten vaikutuksista matematiikan opettamisen kulttuuriin. In L. Häggblom, L. Burman, & A.-S. Röj-Lindberg (Eds.), *Perspektiv på Kunskapens och lärandets villkor* (pp. 89–100). Festskrift tillägnad professor Ole Björkqvist, Pedagogiska Fakulteten vid Åbo Akademi.

Vainionpää, J. (2006). *Erilaiset oppijat ja oppimateriaalit verkko-opiskelussa.* Tampere: Tampere University Press.

Printed and bound by CPI Group (UK) Ltd, Croydon, CR0 4YY

15/09/2026

14962373-0005